ANNUAL REVIEW OF
BIOCHEMISTRY

EDITORIAL COMMITTEE (1975)

ANNUAL REVIEW OF BIOCHEMISTRY

ESMOND E. SNELL, *Editor*
University of California, Berkeley

PAUL D. BOYER, *Associate Editor*
University of California, Los Angeles

ALTON MEISTER, *Associate Editor*
Cornell University Medical College

CHARLES C. RICHARDSON, *Associate Editor*
Harvard Medical School

VOLUME 44

1975

ANNUAL REVIEWS INC.   4139 EL CAMINO WAY   PALO ALTO, CALIFORNIA 94306

ANNUAL REVIEWS INC.
Palo Alto, California, USA

International Standard Book Number 0-8243-0844-1
Library of Congress Catalog Card Number 32-25093

FILMSET BY TYPESETTING SERVICES LTD, GLASGOW, SCOTLAND
PRINTED AND BOUND IN THE UNITED STATES OF AMERICA

ANNUAL REVIEWS INC. is a nonprofit corporation established to promote the advancement of the sciences. Beginning in 1932 with the *Annual Review of Biochemistry,* the Company has pursued as its principal function the publication of high quality, reasonably priced Annual Review volumes. The volumes are organized by Editors and Editorial Committees who invite qualified authors to contribute critical articles reviewing significant developments within each major discipline.

Annual Reviews Inc. is administered by a Board of Directors whose members serve without compensation.

Annual Reviews are published in the following sciences: Anthropology, Astronomy and Astrophysics, Biochemistry, Biophysics and Bioengineering, Earth and Planetary Sciences, Ecology and Systematics, Entomology, Fluid Mechanics, Genetics, Materials Science, Medicine, Microbiology, Nuclear Science, Pharmacology, Physical Chemistry, Physiology, Phytopathology, Plant Physiology, Psychology, and Sociology. The *Annual Review of Energy* will begin publication in 1976. In addition, two special volumes have been published by Annual Reviews Inc.: *History of Entomology* (1973) and *The Excitement and Fascination of Science* (1965).

CONTENTS

Annual Reviews Inc. and the Editors of its publications assume no responsibility for the statements expressed by the contributors to this Review.

REPRINTS

The conspicuous number aligned in the margin with the title of each article in this volume is a key for use in ordering reprints. Available reprints are priced at the uniform rate of $1 each postpaid. Effective January 1, 1975, the minimum acceptable reprint order is 10 reprints and/or $10.00, prepaid. A quantity discount is available.

A FEVER OF REASON ×872

The Early Way

Erwin Chargaff

Department of Biochemistry, College of Physicians and Surgeons,
Columbia University, New York, New York 10032

La jeunesse est une ivresse continuelle : c'est la fièvre de la raison.

La Rochefoucauld

CONTENTS

WHITE BLOOD, RED SNOW

When these pages appear it will be just about thirty years since the atomic bombs fell on Hiroshima and Nagasaki. I was then forty years old, poorly paid, and still an assistant professor at Columbia University; I had already published nearly ninety papers; I had a good laboratory and a few gifted young collaborators; and I was getting ready to begin the study of the nucleic acids. A yearly grant of $6000 from the Markle Foundation was the seal of my earthly success.

It is difficult to describe the effect that the triumph of nuclear physics had on me. (I have recently seen a film made by the Japanese at that time, and all the horror was revived, if "revive" is the correct word in front of megadeath.) It was an early evening in August 1945—was it the sixth? We were spending the summer in South Brooksville, Maine and had gone on an after dinner walk where Penobscot Bay could be seen in all its sunset loveliness. We met a man who told us that he had heard something on the radio about a new kind of bomb which had been dropped in Japan. Next day, the *New York Times* had all the details. But the details have never stopped coming in since that day.

1

The double horror of two Japanese city names grew for me into another kind of double horror: an estranging awareness of what America was capable of, the country that five years before had given me its citizenship; a nauseating terror at the direction in which the natural sciences were going. Never far from an apocalyptic vision of the world, I saw the end of the essence of mankind; an end brought nearer, or even made possible, by the profession to which I belonged. In my view, all natural sciences were as one; and if one science could no longer plead innocence, none could. The time had long gone when you could say that you had become a scientist because you wanted to learn more about nature. You would immediately be asked: "Why do you want to know more about nature? Do we not know enough?" And you would be lured into the expected answer: "No, we don't know enough; but when we do, we shall improve, we shall exploit nature. We shall be the masters of the universe." And even if you did not give this silly answer, you felt inwardly that the evil do-gooders may get away with it, were it not for death, the great eraser of stupidities. For had not Bacon assured me that knowledge was power, and Nietzsche that this was what I wanted all my life? Of course, both are completely wrong, as far as I am concerned; and there is more wisdom in one of Tolstoy's folk tales than in the entire *Novum Organum* (with *Zarathustra* added without regret).

In 1945, hence, I proved a sentimental fool; and Mr. Truman could safely have classified me among the whimpering idiots he did not wish admitted to the presidential office; for I felt that no man has the right to decree so much suffering, and that science, in providing and sharpening his knife and in upholding his arm, had incurred a guilt of which it will never again rid itself. It was at that time that the nexus between science and murder became clear to me.

That this was not the first, and not the greatest, slaughter of the innocent in our times dawned on me only later and very gradually. The governments of the world, both friend and enemy, had very successfully, and for multiple reasons of their own, concealed all knowledge of the German extermination factories. Such names as Auschwitz, Belsen, Buchenwald, and the rest of this infernal ABC of suffocation and incineration, down to Westerbork and Yanov, fell only slowly on my consciousness, like blood drops from hell.

In the first years of this century the great Léon Bloy looked at science—and what a tiny giant it then was!—and this is what he wrote (1):

La science pour aller vite, la science pour jouir, la science pour tuer!

In the meantime we have gone faster, we have enjoyed it less, and we have killed more. The Nazi experiment in eugenics—"the elimination of racially inferior elements"—was the outgrowth of the same kind of mechanistic thinking that, in an outwardly very different form, contributed to what most people would consider the glories of modern science. The diabolical dialectics of progress change causes into symptoms, symptoms into causes; the distinction between torturer and victim becomes merely a function of the plane of vision. Humanity has not learned— if I were a true scientist, i.e. an optimist, I should insert here the adverb "yet"— how to call a halt to this vertiginous tumble into the geometrical progression of disasters which we call progress.

This was not the kind of science envisioned by me when I made my choice; we shall come to that later. At that time, I certainly did not know that the ideal was to eat your cake and sell it, while making it also an object of worship. The year 1945 changed my entire attitude toward science or, at any rate, the kind of science that surrounded me. Even when I was young, my inclinations always were in favor of critique and scepticism, as shown by some of the comprehensive reviews I wrote very early: one on the chemistry of the tubercle bacillus (2) and the other on lipoproteins (3); but even I was not prepared for the orgy of maximalist superficialities that was soon to engulf biology. (When the so-called think tanks began to replace the thought processes of human beings I used to call them the asceptic tanks.)

THE ADVANTAGE OF BEING UNCOMFORTABLE

When I was younger and people sometimes still told me the truth, I was often called a misfit; and all I could do was to nod sadly and affirmatively. For it is a fact that, with only a few glorious exceptions, I have not fitted well into the country and the society in which I had to live; into the language in which I had to converse; yes, even into the century in which I was born. This has certainly been the fate of many people throughout history; and our inhuman century, so full of enormous world wars, unprecedented devastations, heartrending dislocations, has added more than its share to the sum of human misery. But not everybody is born with a stone in the shoe.

There also accrue, however, great benefits to the outsider; there is some comfort in being uncomfortable. If one is left alone in the sense of solitude, one is also left alone in the sense of bother. Having never in my life received a call from another university—and this more than my sedentary habits or the rightly undefinable charm of Columbia University probably explains why I have remained here for nearly forty years—has preserved me from the upheaval of frequent moving. Never having filled a post at any of the professional societies to which I belong has protected me from having to make these horribly inane and vapid speeches with which our statesmen, scientific or otherwise, are expected to hypnotize the populace. If I have never belonged to the house of peers that goes by the name of "study section," I cannot complain, for "peer review" has been decent to me, and I have not lacked scientific support.

Nevertheless, if at one time or another I have brushed a few colleagues the wrong way, I must apologize: I had not realized that they were covered with fur.

THE OUTSIDER AT THE INSIDE

I really do not know why I was asked to write a prefatory chapter. This perplexity should not be considered as an instance of arrogant humility; I cannot serve as an example for younger scientists to follow. What I could teach cannot be learned. I have never been a "100% scientist." My reading has always been shamefully non-professional. I do not own an attaché case, nor do I carry it home at night, full of journals and papers to read. I like long vacations, and a catalog of my activities in general would be a scandal in the ears of the apostles of cost effectiveness. I do not play the recorder, nor do I like to attend NATO workshops on a Greek island or a Sicilian mountain top; this shows that I am not even a molecular biologist. In fact, the list of what I have not got makes up the American Dream. Readers, if any, will conclude rightly that the *Gradus ad Parnassum* will have to be learned at somebody else's feet.

To sum up, as I wrote recently, I have always tried to maintain my amateur status (4). I am not even sure that I comply with my own definition of a good teacher: he learned much, he taught more. Of one thing I am certain, namely, that a good teacher can only have dissident pupils; and in this respect I may have done some good.

I have often referred to myself as an outsider at the inside of science. Our certified scientific accountants may say correctly that they have no use for such outsiders. Well, they don't, but science does. Every activity of the human mind has, throughout history, given rise to criticism within its own ranks; and some, for instance philosophy, consist to a large extent of criticism of what went on before and of the conceptual basis on which they rest. Only science has become complacent in our times; it slumbers beatifically in euphoric orthodoxy, disregarding with disdain the few timid voices of apprehension. These may, however, be the forerunners of horrible storms to come.

Our scientific mass society regards the outsider with little tenderness. Nowhere, however, is the penalty on even the mildest case of nonconformity higher than in the United States. I have lived in that country for 42 years or so, and most of my scientific work and nearly all my teaching have taken place there. Whether some of my scientific observations are worth anything remains to be seen. But whatever the future decides—and I am afraid it will have other worries—I cannot help but find it remarkable that almost all the recognition my work received has come from Europe.

A BAD NIGHT FOR A CHILD TO BE OUT

I started this article with the middle of my life, and it is now time for me to go back. I was born on August 11, 1905 in Czernowitz, at that time a provincial capital of the Austrian monarchy. I have always thought with great pity of my wonderful parents; they had a harder life than I have had. My father, Hermann Chargaff (1870–1934), had inherited modest wealth and a small private bank from his father. He had studied medicine at the University of Vienna but had to give it up because of my grandfather's early death. My mother's name was Rosa Silberstein. She was born in 1878 and died, only God knows where and when, having been deported into nothingness from Vienna in 1943.[1] She lives as a gentle and merciful figure in the memories of my childhood, embodying for me, more than anyone else I ever met, what the Latin language calls, out of its very heart, *misericordia.*

[1] An Austrian scoundrel-physician and a heartless American consul combined forces to prevent her from joining me in New York before the war broke out.

When I was born, my parents were well off and I grew up in what the present abominable sociologico-economic jargon would call an upper middle class family. In subsequent years the capital of my father's small bank vanished, mostly owing to the misplaced trust my father had put in employees and customers; he liquidated the firm and had to seek employment. According to one of the numerous highly untrustworthy family legends, some of the embezzled or otherwise defalcated money ended up in America, contributing to the early glory of Hollywood. I could have wished it a better use.

These were the last peaceful years of a century that will surely be remembered in history (if there will still be history) as the century of mass slaughter. It is true, I missed the Boer War and the Russian-Japanese War; but from the time I was seven years old my life has been accompanied, an incessant bourdon as it were, by reports from battlefields, by daily body counts, by tales of slaughter. The first film I saw, a newsreel in 1912, showed a troop train in the Balkan War, and the engine came at me with frightening speed, accelerated by the hammering of the pianist. Later, when I was older, science seemed a refuge from the horrors, but these have caught up with me.

My memories of the city of my birth are dim. Colors keep coming back to me: black and rose; the bright costumes of the Ruthenian peasants who came to the market; the park of the episcopal palace; never again was anything so green in my life. And then the garden behind our house: there was a tiny grotto in it, and all the dangers of medieval chivalry were relived shiveringly in the dead-serious world of a dreamy child. Rather dull to reality, I lived in a world of my own making; and if it was not as well-furnished as Mörike's Orplid or the dream world of the Brontë children, I had to build it all by myself, for I had few friends.

And then came 1914. We were spending the summer in Zoppot on the Baltic Sea. One afternoon at the end of June, we were sitting near a tennis court, watching the younger sons of Emperor Wilhelm II playing tennis. An adjutant came and whispered something into the imperial ears. They threw down their rackets and went away: the Austrian Archduke Franz Ferdinand had been assassinated. The lamps that went out all over Europe during that summer have never been lit again.

When the summer was over and we were due to return, there was no home anymore: Czernowitz was about to be occupied by the Russian army. We went to Vienna, a city that in many respects I have always considered as my home town; at any rate, it is in Vienna that my father is buried, it was from Vienna that my mother was taken away.

EXPERIMENTAL STATION FOR THE END OF THE WORLD

The Austro-Hungarian monarchy, whose evening glow I barely experienced, was a truly unique institution. The nuptial skills of the Habsburgs, immortalized in a celebrated hexameter, really had as little to do with it as the well-known Viennese *Gemütlichkeit* which is often a thin crust over a truly bestial ferocity. Prince Metternich—the Kissinger of the 19th Century, only better looking—was no more responsible for it than was Haydn or Mozart or Schubert, Stifter or Nestroy or

Trakl. The empire, much more humanized by its subjugated Slavic components than by its Germanic, let alone its Hungarian masters, was actually held together by the patina it had acquired, more or less accidentally, through many centuries. When I opened my eyes first and saw it, the monarchy was in an extremely unstable equilibrium. A passage from one of Kleist's letters (Nov. 16, 1800) comes to mind. He had been passing through an arched gateway: "Why, I thought, does the vault not collapse, though entirely without support? It stands, I replied, because all the stones want to fall down at the same time." The Antonine repose of the late monarchy was fictitious; but like all genuine fiction, it lived a life of its own. I suppose it had to break up; its disappearance did not make for a better world.

A description of what it meant to live during the dying years of the Austrian monarchy, and especially in Vienna, has often been attempted, seldom successfully. The odor exhaled by all official buildings, a mixture of wilted roses and fermenting urine, cannot be duplicated except in dreams; the combination of easy-going *Schlamperei*, sycophantic good-naturedness, and ferocious brutality was probably as unique as was the instinctive search for the middle way, the willingness to accept a compromise as long as it was advantageous to the party proposing it. I suspect, however, that every *bas-empire* will develop similar channels of blissful decrepitude. Child though I was, I soon became a not unobserving spectator, for my eyes had been opened early.

Rummaging through my uncle's books one day in 1915 or 1916, I came across a recent issue of *Die Fackel,* a periodical edited and at that time entirely written by Karl Kraus. An avid extracurricular reader even then, I tried to understand, though it was not easy. Besides, the text was full of white patches: the censor had done his work. For Karl Kraus, the greatest satirical and polemical writer of our times, was a fearless critic of the war and of the society that had given rise to it. His was the deepest and, next to Kierkegaard, the only influence on my formative years; and his ethical teachings and view of mankind, of language, of poetry, have never left my heart. He made me sensitive to platitudes, he taught me to take care of words as if they were little children, to weigh the consequences of what I said as if I were testifying under the oath of eternity. For my growing years he became a sort of portable Last Judgment. This apocalyptic writer—the title of this chapter comes from one of his descriptions of Austria—was truly my only teacher; and when many years later I dedicated a collection of essays (5) to his memory I acquitted myself of a small share of a grateful debt. Several people in this country who had noticed the dedication asked me whether it was to a former high school teacher of mine. I said yes.

Having learned from Karl Kraus how heavy words can be, I have always lamented my enforced separation from the language in which my mother spoke to me when I was a child. I have never let myself be torn away from the German language, nor have I ever declared war on it; but there is an unavoidable estrangement. This is not compensated by my having learned, in the meantime, many languages; one of which, French, I spoke better when I was four than I do now (*Fräuleins* from Fribourg or Neuchâtel saw to it). There exist mysterious links between language and the human brain; and the heartless and brutal way in which language is used in our times, as if it were only a power tool in public relations, a shortcut from sly producer to gullible consumer, has always seemed to me the most threatening portent of incipient bestialization. It is frightening to observe that a progressive aphasia, not organically determined, appears to overtake large numbers of people, especially in this country, who seem to be unable to express themselves except by hoarse barks and (undeleted) expletives. The gift of tongues, not explainable on the basis of natural selection, is the true attribute of *Menschwerdung* (hominization); and it is only fitting that it is revoked shortly before the tails are beginning to grow.

Once in my early life I experienced Austria in its expiring glory, and that was in 1916 when the 86-year-old Emperor Franz Joseph died and was laid to rest with all the pomp of the Spanish baroque. The spectacle impressed me deeply, though it may have been only a Makart copy of a Greco original. The riderless black horses tripped their way through my dreams for quite some time.

I received most of my education in one of the excellent *Gymnasiums* which Vienna at that time possessed, in the Maximiliansgymnasium in the ninth district. The instruction was limited in scope, but on a very high level. In particular, I loved the classical languages and was very good at them. I had excellent teachers and I have not forgotten their names: Latin, Lackenbacher; Greek, Natansky; German, Zellweker; History, Valentin Pollak; Mathematics, Manlik. These were the principal subjects in addition to some philosophy, a little physics, and a ridiculous quantity of "natural history." Of chemistry and the rest of the natural sciences there was nothing. I was one of those horrible types who enjoy school; I had a good memory and learned easily.

The Vienna theater, and especially the *Burgtheater,* had been great in the 19th century, but I saw no more than the rapidly disappearing rear lights. Still, I remember my first *Iphigenie* with Hedwig Bleibtreu. Music was still great: unforgettable evenings in the *Hofoper,* later called *Staatsoper,* with Jeritza as Tosca, Mayr as Leporello, with Richard Strauss conducting Mozart or his own works, with Franz Schalk conducting *Fidelio*; and later the terrible battles with the "Stieglitz gang," a semilicensed claque which tyrannized the queues for the standing room; unforgettable afternoons with the Rosé Quartet or with the Philharmonic Symphony conducted by Nikisch, Weingartner, or Bruno Walter. Such names as Schönberg, Webern, Berg, I did not hear mentioned in Vienna. The audience went, reluctantly, up to Gustav Mahler and stopped there.

Altogether the stratification of cultural life was truly remarkable; except perhaps for literature, we lived much more in the past than in the present. Every day on my way to school I passed the house in the Berggasse where at the entrance door a plaque announced the office of "Dr. S. Freud." But this meant nothing to me; I had not heard the name of the man who had discovered entire continents of the soul which, arguably, may better have been left undiscovered. That great work was done around me in many disciplines, in philosophy and linguistics, in the history of art and economics, in mathematics, etc, escaped me entirely. The name of Wittgenstein, for instance, became known to me only when I lived in New York.

The flavor of life in Vienna at that time can be gathered from a few novels, such as Musil's *Der Mann ohne Eigenschaften* or Roth's *Radetzkymarsch*. The intellectual history of Austria has been summarized in an excellent book by my friend Albert Fuchs, written shortly before he died at an early age (6).

NO HERCULES, NO CROSSROADS

My generation in central Europe will always be marked for me as the children of the Great Inflation. The extent to which the value of money was wiped out in Austria and Germany can hardly be imagined, although as I write this the beginnings of a similar process become noticeable, at any rate in all capitalist countries. Savings and pensions disappeared into the darkening sky that was to unload itself finally in the horrible thunderstorm that the Hitler regime represented for central and western Europe. When an insurance policy that my father had acquired in 1902 was redeemed twenty years later it amounted to the price of one trolley ticket. When in the summer of 1923, before entering the university, I made my "Maturareise" through Germany, one had to eat with the utmost celerity because prices often were augmented in the middle of the meal. My parents were not exceptional in being completely impoverished.

I was eighteen and the world was before me, as the silly saying goes. Actually, the world never is before anyone, nor does it ever look darker than when one is eighteen. The future scientist should at this moment be able to tell stories out of his brief past, how he always knew that he wanted to be a chemist or a lepidopterist; how he could be nothing else, having blown himself up at six years of age in his basement laboratory or having captured, in tender years, a butterfly of such splendor and rarity as to make Mr. Nabokov blanch with envy. I can offer nothing of the sort. Being gifted for many things, I was gifted for nothing. Indolent, shy, and sensitive, I had built my ambush where no game would ever pass.

It was quite clear to everybody that I should have to enter the university and acquire a doctor's degree. This had the advantage of postponing the unpleasant decision about my future by four years or so and also of equipping me with the indispensable prefix without which a middle class Austrian of my generation would have felt naked. Quite different from more advanced civilizations where this appellative is reserved for medical businessmen, in Vienna the doctor's title became a fundamental part of one's persona, and it has stuck to me even down to the current New York telephone directory, this particular form of amputation being much too painful.

There remained the decision in which faculty to register. Decisions usually are not made as a consequence of profound deliberations, but by much more casual routes which then are subjected to a post-factum rationalization. This was certainly my case. There were four, or later five, faculties at the university: philosophy, law, medicine, theology, and later also political science. In addition, there was the Technische Hochschule with its several branches of engineering, but here, unless you worked long years for the Dr.-Ing., you got only the degree of "Ingenieur," much less useful for impressing hotel concierges, barbers, etc. I rejected medicine, since I felt it to be incompatible with my temperament, and law, partly for the same reason and partly because I did not want to become a businessman. Teaching in any form also appeared repulsive to me. I was not irresistibly attracted by anything else, so I chose chemistry for essentially frivolous reasons which winners at high school science fairs are asked to contemplate with disgust. The reasons were: 1. chemistry was one of the subjects I knew least about, never having studied any of it before; 2. in the Vienna of 1923 the only natural science offering some hope of employment was chemistry; and 3. like almost all Viennese I had a rich uncle, but unlike most other uncles he owned alcohol refineries and similar things in Poland, and there were vague promises of future splendor. But even before I had started on my dissertation, the uncle was dead and the alcoholic hopes had evaporated in the hot summer of 1926.

I had conceived a harebrained scheme: I would register simultaneously at two universities—my good grades exempted me from practically all tuition—and study chemistry at the Institute of Technology and at the same time follow courses at the University in the history of literature and English philology. In this way, I hoped, I could acquire in parallel both a chemical engineer's degree and a Dr. phil. The arrangement worked for one year but then began to give signs of breaking down owing to difficult logistics; therefore, I transferred the study of chemistry to the university from which I received the Dr. phil. in chemistry in 1928.

I do not believe that the University of Vienna in my time, 1923 to 1928, could still have been called outstanding. The collapse of the Austro-Hungarian Empire; the turmoil of the 1918 revolution, although it was not much of a revolution; the frightful economic disorganization of the postwar years; the sudden restriction of the pool of talents to a few small Alpine provinces; and the complete politicization of all appointments and promotions tended to produce a brotherhood of chummy incompetents. The faculty of medicine still formed an exception, and there were a few very bright lights here and there. But on the whole the aspect was dismal. It must be admitted, even under the best of circumstances, there is something frightening and bizarre about the aspect of a university, that caravanserai of disconnected and mutually incompatible specialties, in which the patrimony of the West is being dispensed in innumerable tiny vials of many different colors to hordes of mostly reluctant recipients. This grotesque aspect is reinforced in America where the concentration of the "campus" emphasizes even more the character of a spiritual hotel. The European universities functioned, in my time at any rate, more as offices for the issuance of various licenses.

As I had begun the study of chemistry almost without any knowledge of what I was getting into, I could not help falling under the spell of the novelty and the coherence of a mature and fully developed exact science. It is true, it may have been the sort of attraction exerted by a football game; but there it was, and I disliked it much less than I had expected. The shock of getting into dimensions of which I could never have dreamed was probably lessened by the old-fashioned type of instruction we were getting, especially in the introductory lectures. The revolution in chemical theory, which marked the 1920s, passed me by nearly unnoticed, and I have never been good at "electron pushing." The only intrusion of modernism took place in the infrequent colloquia; and I listened to lectures by many of the great in physics and chemistry. But not a single American journal was kept in the Chemistry Library; and when I once inquired about the *Journal of the American Chemical Society,* I was informed that nothing worthwhile was being published there.

Looking back—and when you get old this is all you can do—I must say that I have not learned much from my teachers. In the strictest sense of the word I have had none. During almost my entire life I have myself been much more of a teacher than a pupil; and even this, in the complete moral and intellectual collapse of our time, may not amount to much. The sciences are extremely pedigree conscious; and the road to the top of Mount Olympus is paved with letters of recommendation, friendly whispers at meetings, telephone calls at night. From all this I have never been able to benefit. I am, to an unusual extent, my own product. In contrast, I remember having been at a scientific meeting together with four prominent colleagues, each of whom could rightly claim to have been the favorite pupil of Meyerhof.

Thus, I have not been the pupil, favorite or otherwise, of any of the great establishment figures of the past, unable to exploit this glory from my own cradle to the master's grave and beyond. This I have never regretted. If there is such a thing as a great scientist—I have met in my life perhaps one or two to whom I should have granted this attribute—this greatness can certainly not be transferred by what is commonly called teaching. What the disciples learn are mannerisms, tricks of the trade, ways to make a career, or perhaps, in the rarest cases, a critical view of the meaning of scientific evidence and its interpretation. A real teacher can teach through his example—this is what the ducklings get from their mothers— or, most infrequently, through the intensity and the originality of his view, or his vision, of nature.

Who, then, were my professors? The Institute of Physical Chemistry was directed by old Wegscheider (he certainly was then much younger than I am as I write this), a very typical Austrian *Hofrat,* courteous and grouchy-benign, unemphatic, but not uninsidious. I could not say that he succeeded in making physical chemistry sound as interesting and important as it deserved. Only a few years later, when I lived in Berlin, I realized what could have been made of it. The Professor of Organic Chemistry was E. Späth. He was a good organic chemist and a great authority on alkaloids, but not exactly inspiring as a spiritual example. The narrow slit through which the scientist, if he wants to be successful, must view nature constricts, if this goes on for a long time, his entire character; and, more often than not, he ends by becoming what the German language so appositely calls a *Fachidiot.* It was not easy to be accepted by Späth as a doctor's student, it also cost a lot of money (graduate students had to pay for all chemicals and apparatus required in the course of their work); and so I did not even try. I must, however, say that Späth treated me decently throughout my studies; and in the final examinations, the *Rigorosum,* which came after the completion of the dissertation, when he examined me for two hours in organic chemistry, he gave me a summa cum laude.

I was very anxious to be able to support myself soon, and it was clear to me that I had to choose a thesis sponsor whose problems were known to require neither much time nor money. My choice fell on Fritz Feigl who at that time had just been promoted to "Extraordinarius" in Späth's institute. He looked much more like an Italian tenor than a scientist and was a very nice and decent man. His interests were divided between politics—he was a social-democratic councilor in the Vienna City Council—and the chemistry of metal-organic complex compounds. The first contributed to his economic well-being and the second to the development of the technique of spot tests on which he wrote a well-known treatise. Our heartless centrifugal century propelled this typical Viennese all the way to Rio de Janeiro where he lived since 1939 and where he died after a long, active, and, I hope, reasonably happy life.

My dissertation, completed at the end of 1927, dealt with organic silver complexes and with the action of iodine on azides. My first two scientific publications described part of my work (7, 8). The most interesting portion, namely the discovery that organic sulfhydryl derivatives catalyzed the oxidation of sodium azide by iodine, was not published at that time. Many years later I fell back on this reaction as a device for the demonstration, by paper chromatography, of sulfur-containing amino acids (9).

In the early summer of 1928 I received my Dr. phil. degree from the University of Vienna. The great decision was about to be made, as usual, on insufficient grounds and in an aleatory fashion. Actually, this decision was never made; I floated from one thing to the next.

IL GRAN RIFIUTO

The decision, of course, was to determine what I was going to do now. There were almost no suitable positions to be found in Austria. Truncated by the, in many respects well-deserved, loss of the war, this megalocephalic dwarf had inherited most of the German-speaking system of advanced education erected in the course of centuries by the large monarchy. The production of academically trained people continued at a high rate, but there was no place for them to go; they had to be exported. Most went, partly for linguistic reasons, to Germany where the outlook for employment in the industry, let alone the universities, was far from good. A few went to the successor countries of the monarchy: Czechoslovakia, Hungary, Poland.

The year in which it fell to me to decide on my future, 1928, was an ominous year. Black clouds had begun to gather everywhere. America was getting ready to elect "The Great Engineer" as its next President. The postwar boom, in which even central Europe had participated after the stabilization of its currencies, had dissipated itself. The beasts of the abyss, held both on leash and in reserve by the German industry, were beginning to dream the noble dream of the night of the long knives. They were soon to be let loose to begin collecting blood. The workers were confused and poorly led. One year earlier, in 1927, I could witness the first huge street riots in Vienna; they were suppressed in a most cruel and bloody manner by the icy monsignor who led the Austrian cabinet, truly an exponent of the *ecclesia militans*. Thus, I have been sensitized early against such slogans as "law and order." All that they produce in the end is a *Chile con sangre*. But to be entirely just I should mention that, listening to the parliamentary twaddle and verbal tricks with which the social-democrats of the world claimed to fight the growth of fascism, I jotted down, at that time, one of my first aphorisms: "Austrian social democracy: in case of rain, the revolution will take place in the hall."

Somehow I wanted to get away from it all, at least for some time, into another country, into another language. But the whole was governed by a sort of fairy-tale logic: I would take the first thing that offered itself, whether industry, research, or teaching. As if, in the tale, the boy were told to go out in the street and follow the first animal he met. The first animal that came along in that Brothers Grimm world of mine was called "research"; and so I have stayed with it all my life. It has always been my habit to float passively where the currents would take me. Whenever the currents stopped, I got stuck. That it was research that was first pulled out of the pack of cards probably suited my unacknowledged preferences; I have always been longing for a remote ivory tower (with air conditioning and running cold and hot water). But, all jokes aside, in at least one respect getting into research in 1928 was quite different from what it has been during, say, the last twenty years. I have recently tried to describe this change of atmosphere (10) and I do not want to repeat myself here. Perhaps the most important difference is that, when I got in, the selection of apprentices still operated through a sort of pledge of eternal poverty. (That at the same time some of the sorcerers who administered the pledge were quite well-off failed to strike our young and inexperienced eyes.)

What I did not realize for a long time was the enormous pull of the vortex into which I let myself be carried. By the time I was 23, I was wont to distinguish strictly between what one did with one's head and what one considered as one's profession. Chemistry was my profession and I hoped it would feed and sustain me; and not only me, for I was getting ready to get married. But at the same time I thought of myself as a writer. I had written a great deal—if not mountains, then respectable hills of paper; a little of it had been published; more would have been but for my timidity and lack of contacts. Had I not left Vienna, tearing myself away from the German language, and, even more (what a tremendous "more" this is!), had not the entire world collapsed into the most bloody barbarism under the leadership of that very same German language, there might have been one more mediocre German writer. The spiritual economy of the world being obscure to me, I cannot gauge the loss or gain. In any event, the pull that science exerts even on a critical and questioning mind proved immensely stronger than I had expected; and this is what, in greater words, the title of this section means to express.

BLUEBIRD OF HAPPINESS

This is how it all came about. I was in the middle of learning Danish—there was a rumor that Sörensen had an opening at the Carlsberg Laboratories in Copenhagen —and had just begun to master its most disagreeable phonetic specialty, the glottal stop, that timid death rattle of an expiring introvert, when a more solid rumor reached me. One of the Physiological Chemistry professors at the Medical School, S. Fränkel, had just then returned from a lecture tour in the United States and brought back the news that T. B. Johnson of Yale University had a research fellowship available for a young man willing to assist R. J. Anderson in his work on the lipids of tubercle bacilli. I knew English quite well at that time, having acquired a stilted form of upper class English with the help of two Cambridge spinsters who ran a small school. But I knew nearly nothing of America, and what I knew was not conducive to learning more. As a child I had read Cooper, Poe, and Mark Twain in mostly very bad translations, and also Whitman's poems with little enthusiasm. Some of Dreiser's and Sinclair Lewis' books I had read in the original. The maudlin films out of Hollywood made me sick, although I made an exception for Greta Garbo. But I loved Charlie Chaplin, Buster Keaton, and Harold Lloyd; out of that threatening continent, somber and dehumanized, there seemed to arise a wind of the freedom of the absurd.

In any event, I applied for the job and, to my great discomfiture, was accepted. The "Milton Campbell Research Fellowship in Organic Chemistry" paid $2,000 per year, in ten monthly installments.[2] I was to start in autumn. As I knew nothing about lipids, a short stay in Fränkel's laboratory was supposed to make me know and love them, but it accomplished neither. As the time of my departure grew nearer, so grew my fears. I was afraid of going to a country that was younger than most of Vienna's toilets. Others would try to console me, telling me that I should be surprised and that America would turn out much better than I expected. But I remained doubtful, applying to that Promised Land the immortal saying of one of Vienna's wits, Anton Kuh: "Wie der kleine Moritz sich Amerika vorstellt, so ist es."[3] I did not prove wrong.

The giant liner "Leviathan" brought me to New York. No sooner had I approached the Land of the Free than I found myself in jail. A remarkably gruff immigration officer took a look at my passport in which my name was embellished with a doctor's title, as for reasons outlined before it had to be; then he looked at the "student's visa," which a far from charming American consul in Vienna had handed me as if it were the holy grail; the officer's face reflected painful and somber cerebration; and out of the side of his thin mouth came the words "Ellis Island."

Imprisoned in that noteworthy American concentration camp, I had an excellent view of the Statue of Liberty. I thought that this conjunction of jail and monument was not accidental; it had the purpose of teaching some of the detained immigrants the advantages of dialectical thinking. But the view, early in the morning, of the fog-shrouded seascape was enchanting, and the plaintive sounds of the foghorns

[2] This was about one sixth of what a Sterling Professor, holder of an elite professorship at Yale, made. Despite the lapse of 45 years, the span between the income of a beginning "post-doc" and that of a very full professor has remained about the same.

[3] "Little Maurice," an important figure in the Austrian jocular universe, is an awfully obtuse boy, the typical "terrible simplifier"; and therefore is often right when the sages falter in their complicated constructions.

and the cries of the sea gulls were a melancholy accompaniment to an America that would never be.

One or two days later I was brought before a tribunal presided over by a very big black lady who was assisted by two somnolent uniformed elderly gentlemen wearing what looked to me like Salvation Army fatigues. The verdict— immediate deportation—was prompt, for the case was clear: I was revealed as a double impostor. If I was a doctor, I could not be a student; if I was a student, how could I be a doctor? I stammered something about Faust, despite multiple doctorates, having been an eternal student. I might as well have tried playing pinochle with a Martian. The whole thing could have been a scene from Jarry's *Ubu Roi*. I was taken back to my quarters, which were really sixteenths, sent a telegram to Yale University whose counsel intervened in Washington, and was set free after a few days. Whether sternness was overcome by reasonable argument or by something more potent, I have never been able to ascertain. I may, in fact, have been some sort of a test case, for I belonged to the early crop of "post-docs" who at that time began to flock to the United States in ever increasing numbers.

In New Haven, Treat B. Johnson, Sterling Professor of Chemistry, and hence six times as powerful as my insignificant ego, met me at the railway station. He was a decent and kind man, very much a remnant of an older and better America whose last traces were then still visible; and he tried to lighten my first painful days on this excessively new continent. Later, when I became interested in the nucleic acids, I realized how important his work on the chemistry of purines and pyrimidines had been. Johnson took me to his own house as a guest for a few days. In my room, there was a sort of embroidered panel on the wall. It showed a bluebird and underneath it said: "May the bluebird of happiness find an eternal resting place in your home." I was moved by the trust that America sets in bluebirds. Where I came from, birds were of the utmost grayness.

THE END OF THE BEGINNING

Rudolph J. Anderson looked very much like a British army officer in reluctantly civilian clothes. Born in Sweden, but brought up in New Orleans, he represented a peculiar mixture of national and cultural characteristics. He was an excellent experimental chemist, and it was from him that I learned the respect for matter, the care for quantity in essentially qualitative investigations, and the reverence for accuracy in observation and description. If everybody needs a teacher, he was that in my case; and yet I hesitate to call him so, for I do not believe that my future course was influenced by him. A teacher is one who can show you the way to yourself; and this no one has done for me.

I worked two years with Anderson, remaining in the Yale chemistry department from 1928 to 1930. He had not long before come to Yale to set up a program of research on the chemical composition of tubercle bacilli and other acid-fast micro-organisms. My stay was quite productive: I published seven papers with Anderson; the most interesting dealt with the discovery of the peculiar branched chain fatty acids, tuberculostearic and phthioic acids (11, 12), and with the quite complex lipopolysaccharides of the tubercle bacillus (13). In addition, I found the time to pursue entirely independent studies on iodine cyanide (14) and on organic iodine compounds (15) and also, rediscovering Tswett's chromatographic separation methods some time before R. Kuhn and E. Lederer, on the carotenoid pigments of the timothy bacillus (16).

Despite the lure of an assistant professorship of chemistry at Duke University, at that time excessively devoted to tobacco research, we wanted to return to Europe and left the United States in the summer of 1930. Going to Berlin, I found, miraculously, a good position almost without delay. I was appointed *Assistent* at the Bacteriology Department of the University whose chief, Martin Hahn (of Buchner and Hahn fame), treated me during our entire association with an incredible benevolence. I was given an apartment in the Institute—a few steps away from the *Reichstag* whose flames were soon to illuminate the onset of the Third Reich—I was entirely independent in my research and even began to have collaborators.

My stay at the University of Berlin—October 1930 until April 1933—was perhaps the happiest time of my life; and if Germany had not got into the hands of the cannibals, I might have stayed much longer. Work went on in many directions. Two of the most substantial pieces, a study of the lipids of the Bacillus Calmette-Guérin (BCG) (17) and a detailed investigation of the fat and phosphatide fractions of diphtheria bacteria (18), were designed for my *Habilitationsschrift,* the treatise whose submission precedes the appointment as *Privatdozent.* At the Medical School to which my Institute belonged, this title, i.e. the right to give lectures, was limited to holders of the Dr. med. degree. Therefore, Martin Hahn arranged for me to become *Privatdozent* at the Berlin Institute of Technology. At the end of January 1933 the Black Plague has assumed the government of Germany; and one week later I could absurdly have been seen trotting to Charlottenburg with a careful parcel, in order to deposit my *magnum opus* in the Technische Hochschule. By the time the appointment was taken up I was, however, far from Berlin, having left for Paris long ago. One look at the style and the physiognomy of the new power was enough. I wish I could have remained so light-footed in later predicaments.

That I could so easily transfer my activities to Paris was due to another piece of work I had done. Geheimrat Hahn was one of the court experts in the well-known "Lübeck case," in which several physicians were prosecuted for being responsible for the death of a large number of babies who had been fed cultures of virulent tubercle bacilli instead of BCG vaccine. Hahn asked me to undertake the chemical portion of his report; and I believe that my studies contributed materially to an understanding of what had really happened. This work was published (19), and Calmette, who was naturally happy about the proof that his BCG preparations were not responsible for the catastrophe, read my paper. In March 1933 I received an entirely unsolicited letter from him, inviting me to come to the Institut Pasteur. In the middle of April we were in Paris.

Calmette, who was deputy director of the Institut Pasteur, was a charming, good hearted, and very intelligent man in his early seventies. He was extremely hard of hearing, a deplorable fact that he wished to be ignored, which made conversation difficult. The tuberculosis section which he headed occupied a separate, at that time modern building and was the only part of the Pasteur Institute in which up-to-date-work could be done. The main building of the Institut Pasteur, on the other side of rue Dutot, was beyond description.[4] Its director was an extremely frugal mummified octagenarian, Émile Roux, who, I was told, had done some distinguished work forty years ago. Salaries were very low, and without the help of the Rockefeller Foundation I should myself soon have looked like the director. I was informed by my colleagues that requests for more money were useless, but that after the third intervention with Dr. Roux he saw to it that one received the Légion d'honneur as a consolation prize. Unfortunately, at the end of 1933, after only my second visit to him, Roux died, preceded by Calmette; and so I have remained entirely undecorated with the "petit ruban." Although the principal building in my time lacked toilets, it contained a rather tasteless crypt—in a peculiar second empire byzantine style—devoted to Louis Pasteur; and there I participated in the death vigil, first for Calmette, then for Roux, being assigned, as a junior member, the period from 3:00 to 4:00 in the morning.

I did some work at the Institut Pasteur (not very much) on bacterial pigments and polysaccharides. The wonderful city of Paris lived at that time, perhaps, its last genuine moments before losing its French tears and its French laughter, becoming teutonized, americanized, pompidized, etc. But the shadows began to fall, and the Institut Pasteur got an insignificant director who wore a skull cap. I knew I had to go; and with the help of Harry Sobotka of the Mount Sinai Hospital of New York I left Paris at the end of 1934 and sailed again, to my own amazement, to America.

But this is another story; and if the surrogate god of modern science, the Great Idol Chanceandnecessity, permits it, I may tell it another time. In any event, after much searching, it turned out in 1935 that Hans Clarke had a little job for me at Columbia.

THE SILENCE OF THE HEAVENS

I came to biochemistry through chemistry; I came to chemistry, partly by the labyrinthine routes that I have related, and partly through the youthfully romantic notion that the natural sciences had something to do with nature. What I liked about chemistry was its clarity surrounded by darkness; what attracted me, slowly and hesitatingly, to biology was its darkness surrounded by the brightness of the givenness of nature, the holiness of life. And so I have always oscillated between the brightness of reality and the darkness of the unknowable. When Pascal speaks of God in hiding, *Deus absconditus,* we hear not only the profound existential thinker, but also the great searcher for the reality of the world. I consider this unquenchable resonance as the greatest gift that can be bestowed on a naturalist.

---

[4] It is not even necessary for me to attempt an account of this labyrinth of torture chambers of rabbits, guinea pigs, and mice. This has been done, with masterful maliciousness, in one of the greatest French novels of this century, in Céline's *Voyage au bout de la nuit* (pp. 275–79 of the Pléiade edition).

When I look back on my early way in science, on the problems I studied, on the papers I published—and even more perhaps on those things that never got into print—I notice a freedom of movement, a lack of guild-imposed narrowness, whose existence in my youth I myself, as I write this, had almost forgotten. The world of science was open before us to a degree that has become incredible now, when pages and pages of application papers must justify the plan of investigating ("in depth") the thirty-fifth foot of the centipede; and one is judged by a jury of one's peers who are all centipedists. I would say that most of the great scientists of the past could not have arisen, that in fact, most sciences could not have been founded, if the present utility-drunk and goal-directed attitude had prevailed.

It is clear that to meditate on the whole of nature, or even on the whole of living nature, is not a road that the natural sciences could long have traveled. This is the way of the poet, the philosopher, the seer. A division of labor had to take place. But the overfragmentation of the vision of nature—or actually its complete disappearance among the majority of scientists—has created a Humpty-Dumpty world that must become ever more unmanageable as more and tinier pieces are broken off, "for closer inspection," from the continuum of nature. The consequence of the excessive specialization, which often brings us news that nobody cares to hear, has been that in revisiting a field with which one has been very familiar, say, ten or twenty years ago, one feels like an intruder in one's own bathroom, with 24 grim experts sharing the tub.

Profounder men than I have failed to diagnose, let alone cure, the disease that has infected us all. If I may be orphic for a moment, I should say that the ostensible goals have obliterated the real origins of our search. Without a firm center we flounder. The wonderful, inconceivably intricate tapestry is being taken apart strand by strand; each thread is being pulled out, torn up, and analyzed; and at the end even the memory of the design is lost and can no longer be recalled. What has become of an enterprise that started by being an exploration of the *gesta Dei per naturam*?

To follow the acts of God by way of nature is itself an act that can never be completed. Kepler knew this and so did many others, but it is now being forgotten. In general, it is hoped that our road will lead to understanding; mostly it only leads to explanations. The difference between these two terms is also being forgotten: a sleight of hand that I have considered in a recent essay (20, 21). Einstein is somewhere quoted as having said: "The ununderstandable about nature is that it is understandable." I think, he should have said: "that it is explainable." These are two very different things, for we understand very little about nature. Even the most exact of our exact sciences floats above axiomatic abysses that cannot be explored. It is true, when one's reason runs a fever, one believes, as in a dream, to grasp this understanding; but when one wakes up and the fever is gone, all one is left with are litanies of shallowness.

In our time, so-called laws of nature are being fabricated on the assembly line. But how often is the regularity of these "laws of nature" only the reflection of the regularity of the method employed in their formulation? Lately, many tricks have been found out about nature, but these tricks seem to have been specially produced by nature for the imbeciles to find out; and there is no Maimonides to guide them out of their confusion. In other words, science is still faced with the age-old predicament, the lack of ultimate verification. It is written in the *Analects* of Confucius (XII, 19): "The Master said, Heaven does not speak."

Literature Cited

1. Bloy, L. 1902. *Exégèse des lieux communs,* 1ère série, CXXIV. Paris: Mercure de France
2. Chargaff, E. 1931. *Naturwissenschaften* 19:202–6
3. Chargaff, E. 1944. *Advan. Protein. Chem.* 1:1–24
4. Chargaff, E. 1973. *Perspect. Biol. Med.* 16:486–502
5. Chargaff, E. 1963. *Essays on Nucleic Acids.* Amsterdam, London, New York: Elsevier. 211 pp.
6. Fuchs, A. 1949. *Geistige Strömungen in Österreich, 1867–1918.* Wien: Globus. 320 pp.
7. Feigl, F., Chargaff, E. 1928. *Monatsh. Chem.* 49:417–28
8. Feigl, F., Chargaff, E. 1928. *Z. Anal. Chem.* 74:376–80
9. Chargaff, E., Levine, C., Green, C. 1948 *J. Biol. Chem.* 175:67–71
10. Chargaff, E. 1974. *Nature* 248:776–79
11. Anderson, R. J., Chargaff, E. 1929. *J. Biol. Chem.* 84:703–17
12. Anderson, R. J., Chargaff, E. 1929. *J. Biol. Chem.* 85:77–88
13. Chargaff, E., Anderson, R. J. 1930. *Z. Physiol. Chem.* 191:172–78
14. Chargaff, E. 1928. *J. Am. Chem. Soc.* 51:1999–2002
15. Chargaff, E. 1929. *Biochem. Z.* 215:69–78
16. Chargaff, E. 1930. *Zentralbl. Bakteriol.* 119:121–23
17. Chargaff, E. 1933. *Z. Physiol. Chem.* 217:115–37
18. Chargaff, E. 1933. *Z. Physiol. Chem.* 218:223–40
19. Chargaff, E., Dieryck, J. 1932. *Biochem. Z.* 255:319–29
20. Chargaff, E. 1970. *Experientia* 26:810–16
21. Chargaff, E. 1971. *Science* 172:637–42

ENZYMATIC REPAIR OF DNA   ×873

Lawrence Grossman, Andrew Braun, Ross Feldberg, and Inga Mahler
Graduate Department of Biochemistry, Brandeis University,
Waltham, Massachusetts 02154

CONTENTS

INTRODUCTION

The ability of a cell to survive in an environment specifically damaging to its DNA can be attributed to a variety of inherent repair mechanisms.

19

There is a form of repair in which alterations are directly *reversed* to their original form. This reversibility is exemplified by the photoreactivation of ultraviolet (UV)-induced pyrimidine dimers. Kelner (1) and Dulbecco (2) respectively discovered that when inactivated by UV light, cells and bacteriophages can be reactivated upon exposure to visible light during a postirradiation period. This phenomenon is attributable to the action of an enzyme, photolyase (photoreactivating enzyme), which is able to monomerize the UV-induced pyrimidine dimers in the presence of 320–370 nm light. Photolyases have been isolated and purified from a variety of organisms and are virtually ubiquitous (3, 4).

Dilution of damage can be effected through a series of sister chromatid exchanges, controlled by recombinational mechanisms as a postreplication event (5). In this form of repair, replication proceeds to the point of damage, stops, and resumes at the point of the next initiation site, resulting in a gap in the newly synthesized daughter strand. It is presumed that those strands containing damaged regions exchange with undamaged regions of other DNA strands, resulting in the eventual dilution of such damage.

The specific *removal* of damage from modified DNA was discovered by Boyce & Howard-Flanders (6), Setlow & Carrier (7), and Riklis (8). The UV photoproducts, pyrimidine-pyrimidine dimers, are excised from a DNA molecule through the sequential action of a variety of nucleases. At present this latter mechanism of repair is the better understood at the enzymatic level and is the subject of this review.

Classification of Damage

The varieties of damage imposed on DNA are classified into two general categories in order to facilitate enzyme classification related to repair specificities.

Monoadducts will be generally defined as single base modifications in which an addition reaction, adduct formation, or a chemical transformation leads to the alteration of a single nitrogenous base. Examples of this general type of single base damage are: monoadducts formed as a result of the mutagenic action of hydroxylamine, in which N^4OH cytidylate residues are formed in DNA (9, 10); deamination reactions mediated by nitrous acid with either purine or pyrimidine exocyclic amino groups (11); bisulfite effecting the specific deamination of cytosine (12); the incorporation of base analogs into DNA, whose tautomeric changes cause mispairing; and the formation of 5,6-dihydroxydihydrothymine monoadducts (thymine glycol) from OsO_4 treatment or γ radiation (13).

Another class of damage found in DNA exposed to either radiation, intercalating agents, or certain carcinogens is the formation of *diadducts,* in which more than one nitrogenous base is involved in the final chemical product. Pyrimidine-pyrimidine dimers arising from UV light (14), the formation of crosslinks between thymine residues in the presence of psoralen and light (15), or the treatment of DNA with mitomycin C (16) are examples of this form of damage. Crosslinks arising from such treatment can result in inter- or intrastrand diadduct formation. It is the latter form of damage that this review addresses.

Classification of Nucleases

The excision repair mechanisms described in this review involve a variety of nucleases that show specificity for the primary or secondary chemical event. The structural changes arising from the action of repair also affect the activity of exonucleases involved in repair. For these reasons, a redefinition of the nucleases and how they may act in repair is germane to this review.

Exonucleases are phosphodiesterases that require a terminus for hydrolysis. Further subclassifications can be made according to specificities for their initiation at 3′ or 5′ termini, or whether hydrolysis occurs at a terminal or internal phosphodiester bond. The ability of such nucleases to participate in the removal of damaged nucleotides is therefore dependent on such specificities.

An endonuclease is a phosphodiesterase that does not require a terminus for hydrolytic activity. Its specificity may be confined to substrate conformation, nucleotide sequence, nucleic acid species, and the presence of modified nucleotides. To indicate whether such nucleases are able to hydrolyze on the 3′ (*a*) or the 5′ (*b*) side of a phosphodiester bond would necessitate additional classification.

An endonuclease that specifically acts on damaged DNA resulting in correctional pathways in vivo is defined in this review as a correctional endonuclease or correndonuclease. *A correndonuclease I-type endonuclease is defined as one that is specific for damaged DNAs possessing monoadduct derivatives. Those correctional endonucleases that are specific for diadduct modified regions of DNA are referred to as correndonuclease II.*

Much of the current interest and knowledge of repair mechanisms has been confined to UV-induced photoproducts, their effects, and subsequent removal. The progress already made is a reflection of a good understanding of the fundamentals of UV photochemistry which allows for control of photoproduct formation in DNA and ease of analysis (17, 18). Consequently, in the ensuing discussions, removal of the primary UV photoproducts, pyrimidine-pyrimidine dimers, is emphasized.

EXCISION REPAIR CYCLE

uvr Genes Controlling Repair in Escherichia coli

In *E. coli*, alterations in five genes result in mutants that share the phenotypic properties of being UV sensitive (UV⁻), mitomycin sensitive (Mˢ), and host cell reactivation negative (HCR⁻). These mutants, which map at discrete sites on the chromosome (Table 1), are classified as *uvrA, uvrB, uvrC, uvrD,* and *uvrE. uvrA, uvrB,* and *uvrC* are unable to excise pyrimidine dimers from their DNA or from UV-irradiated phage DNA. *uvrD* and *uvrE* have been included in this group of excision repair defective mutants chiefly on the basis of their reduced hcr activity.

uvrA, uvrB, and *uvrC* appear very similar in their sensitivity to UV, their resistance to γ irradiation, and sensitivity to host cell reactivation, as well as induction of

Table 1   *Escherichia coli* genetic loci affecting UV repair

Excision loci	Map position (minutes)[a]
uvrA	81
uvrB	18
uvrC	36
Other hcr⁻ loci	
uvrD	75
uvrE = mutU	75
Recombination loci	
recA	51
recB	54
recC	54
recF = uvrF	73
sbcA	unmapped
sbcB	38
Polymerase loci	
polA = polI	76
polB = polII	2
dnaE = polIII	4

[a] Map positions from Taylor & Trotter (31).

phage λ. In addition, efforts to demonstrate a sequential action of these gene products have yielded negative results (19, 20). In spite of these phenotypic similarities, recent experiments have shown that whereas *uvrA* and *uvrB* mutants lack an endonuclease specific for UV-irradiated DNA, *uvrC* mutants contain wild-type levels of this particular enzyme activity (21). Work reported with DNA extracted from *uvrC* mutants shows the presence of single strand breaks after treatment with mitomycin C (22) and UV irradiation (23). This evidence suggests that in mutants of *uvrC* the incision of damaged DNA does occur. The *uvrC* defect thus might possibly affect a step following incision.

The mutants classified as *uvrD* differ from *uvrA, uvrB,* and *uvrC* according to their physiological properties. They are somewhat less sensitive to UV irradiation, sensitive to irradiation with γ rays, show rapid and extensive degradation of DNA following exposure to UV irradiation, and *uvrD⁻* is dominant over *uvrD⁺* (24).

Unlike *uvrA, uvrB,* and *uvrC* mutants, which show no enhanced UV sensitivity as double mutants, the double mutant *uvrD uvrB* is approximately three times as sensitive as *uvrB* alone. Since this double mutant shows a significant reduction in DNA degradation, Ogawa et al (24) suggested that the *uvrD* function occurs after the *uvrB* step.

A rather interesting collection of UVˢ strains are those described by Mattern (25) as *uvrE. uvrE* are mutator strains and have been studied in detail by Smirnov & Skavronskaya (26) and Siegel (27). *uvrE* mutants are UV sensitive and exhibit an enhanced mutation rate. A strain containing the double mutation *uvrE uvrA* is just slightly more UV sensitive than the *uvrA* strain alone. The *uvrE* mutants possess normal recombination activity, are slightly more sensitive to γ irradiation than wild-type *E. coli,* and show a slight reduction in hcr ability for phage λ. Host cell reactivation ability for phage T1, on the other hand, appears to be completely normal. Unlike the other classes of *uvr* mutants, *uvrE* is inviable with *E. coli polA* (28).

A strain isolated by Storm & Zaunbrecher (29) and designated *uvrF* probably does not belong among the excision defective mutants discussed in this section, since the *uvrF* strain has normal hcr ability for phage λ, shows no UV induction of λ prophage, and is defective in recombination. The UV sensitivity of *uvrF* arises from some malfunction of the recombination repair pathway, rather than in a defect in excision repair, since *uvrF* and *recF* appear to be identical (30).

From transfection experiments, Taketo et al (32) suggested that *uvrA, uvrB, uvrC,* and *uvrD* mutants might share a common enzymatic defect, since with all of these hosts, in vitro treatment of UV-irradiated ϕA RF transfecting DNA with the phage T4 (V^+) correndonuclease II led to an enhancement in transfection. These results are paradoxical in light of normal correndonuclease II levels in *uvrC* and *uvrD* mutants and may be explicable in terms of mass action effects. The exposure of *uvrC* spheroplasts to excess exogenous levels of fully incised ϕA DNA may simply overwhelm those cellular processes that lead to abortive repair.

In the pathway illustrated in Figure 1, *uvrA* and *uvrB* genes are involved in the incision step, *uvrC* gene activity is localized at a postincision, pre-excision locus, whereas the roles of other *uvr* genes are unknown.

Recombination (rec) Genes Involved in Repair

Another well-studied group of UV-sensitive, mitomycin-sensitive strains of *E. coli* are those designated as *rec.* Unlike the *uvr* mutants, the *rec* strains are defective in genetic recombination, in addition to being sensitive to UV and ionizing radiation. The UV sensitivity of a *uvr rec* strain is extreme where one lethal event per dimer is seen. The lack of pyrimidine dimer excision in a *uvr rec$^+$* strain suggests that the *uvr* and *rec* gene products may mediate separate pathways of DNA repair (6, 33–36).

Studies of photoproduct excision in vivo in isogenic mutants of *recA, recB,* and *recC* clearly show that these strains are not excision defective (37). Mutants of *E. coli* K12, defective in recombinational repair, fall into several groups of which *recA, recB,* and *recC* have been studied in detail (34, 38, 39). Cells lacking the *recA* function are characterized by spontaneous DNA degradation, which is so acutely enhanced by cellular exposure to UV light that a "reckless" event of DNA degradation is observed (40). This state can be cured by introducing *recB* mutants into such strains.

recA cells are not UV inducible for phage λ and are altered in cell division processes (41, 42). *recB* and *recC* mutants show reduced recombination, intermediate UV sensitivity, and little DNA breakdown after irradiation. Both strains are inducible for phage λ (43). Double mutants of either *recA* and a temperature-sensitive *polA* mutation (*polA12*) or a *recB polA12* show conditional lethality (44).

While the product of the *recA* gene has not yet been purified, there is some information concerning its regulatory effects on *recBC* functioning (45). Definitive

DNA Intrastrand Diadduct Repair Cycle

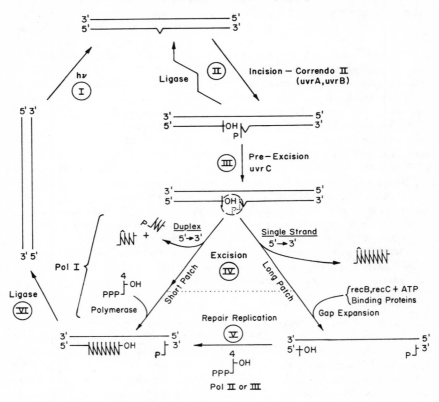

Figure 1   A proposed DNA intrastrand diadduct repair cycle

The four deoxynucleoside triphosphates: deoxyadenosine-5′-triphosphate, deoxyguanosine-5′-triphosphate, deoxycytidine-5′-triphosphate, and thymidine-5′-triphosphate, ATP-adenosine-5′-triphosphate, Pol I; DNA polymerase I of either *E. coli* or *M. luteus*; 5′ → 3′-exonucleolytic activity initiating at a 5′ terminus when hydrolysis proceeds towards a 3′ terminus and ⎯〈⎯ pyrimidine photodimer region of UV-irradiated DNA.

progress has been made, however, on the *recB recC* gene product. These two genes determine a complex nuclease (46–48) consisting of two subunits (49). The enzyme possesses four activities: an ATP-dependent double stranded and single stranded exonuclease, an ATP-stimulated single stranded DNA endonuclease, and a DNA-dependent ATPase (49, 50).

Identification of *recB* and *recC* as the structural genes for this enzyme was carried out by Oishi et al (51). The purification of this multifunctional enzyme activity from the respective temperature-sensitive mutants led Kushner (52) to suggest that of the numerous activities associated with the *recBC* nuclease, only the ATP-dependent, exonuclease hydrolyzing double stranded DNA is abnormally thermolabile.

In 1971, Clark (53) and Kushner et al (54) suggested that the products of the *recB recC* genes were involved in a recombinational pathway (*recBC* pathway) which accounted for approximately 99% of wild-type recombination. The existence of other, minor pathways was suggested at that time. Isolation and genetic analysis of new groups of recombination-defective mutants of *E. coli* K12 by Horii & Clark (55, 56) substantiated the existence of at least two other recombinational pathways: *recF* and *recE*.

All three pathways are blocked in the absence of wild-type *recA* gene product. A detailed discussion of the function of the *rec* genes as well as two controlling genes, *sbcA* and *sbcB* (56, 57), and their function in the control of recombination in *E. coli* can be found in recent reviews by Clark (56, 57).

Gene Products

The enzyme systems available to cells for repair of UV damage through photo-reactivation and recombinational mechanisms have been amply reviewed elsewhere and are not the subject of this paper (58, 59). The general mechanism shown in Figure 1 represents a current view of the excision repair cycle present in *E. coli* and *M. luteus* in which intrastrand diadduct damaged DNA is sequentially removed; the individual steps are detailed in each section of this review.

The excision event represents a series of enzymatic steps including incision by rather specific correndonucleases under potential control by a subsequent post-incision step in which abortive repair is prevented. The concluding step in the excision of damage can occur via two interdependent pathways, which may involve either polymerase-associated exonucleases or unassociated exonucleases involved in the removal of photoproducts. The reinsertion mechanisms may occur either con-comitantly in polymerase I-associated excision pathways (short patch repair), or it can proceed as a result of a stepwise mechanism (long patch repair) when single stranded 5′ → 3′ exonucleases catalyze the removal of photoproduct regions of DNA. Long patch pathways of excision-reinsertion may be controlled by polymerases II and/or III. The restoration of the strand continuity is under the control of a single polynucleotide ligase step.

INCISION STEP: *General*

Endonucleolytic activities that incise DNA containing damaged bases have been found in extracts derived from bacterial, bacteriophage, and mammalian sources. Although many correndonuclease activities have not been fully characterized, a similarity is apparent among most of those activities studied. Correndonucleases generally are small proteins, having molecular weights of less than about 30,000.

All act in the absence of divalent cation, although some enzymes are stimulated by Mg^{2+} or NaCl. All appear to incise close to the damage on the DNA strand.

In the following two sections, the characteristics of specific correndonuclease activities are discussed.

Correndonuclease I

ESCHERICHIA COLI CORRENDONUCLEASE I (*E. coli endonuclease II*)   Extracts of *E. coli* incise and degrade DNA that has been treated with a monofunctional alkylating agent such as methyl methanesulfonate (MMS) (60). Friedberg et al (61) isolated the endonuclease involved, endonuclease II, and found that the enzyme incises alkylated DNA but not UV-irradiated DNA. Subsequent investigations indicated that the enzyme acts on DNA containing apurinic sites (62) and that some alkylated purines are removed from alkylated DNA to yield apurinic sites (63). That the enzyme acts on heavily X-irradiated DNA is probably due to the formation of apurinic sites resulting from the primary damage to the DNA (64).

The enzyme has a sedimentation coefficient of 3.2—indicative of its low molecular weight (61). While endonuclease II functions in the absence of added divalent cations, it is also markedly stimulated by Mg^{2+} and inhibited by EDTA (61).

A similar endonuclease, isolated from *E. coli* by Verly et al (65, 66), appears to act exclusively at apurinic sites. Its general properties, as well as its purification scheme, are very similar to endonuclease II. Recently, Verly et al (67) demonstrated the excision of apurinic sites and the restoration of the DNA integrity by treatment of apurinic DNA with this endonuclease, polymerase I, and polynucleotide ligase.

The structure of the incision produced by endonuclease II is unknown. Since the closely related apurinic site endonuclease appears to yield a priming site for the polymerase I, the incision event may result in the formation of a 3'OH terminus. Hadi et al (68) suggested that endonuclease II incises 5' to the apurinic site, leaving a 3' hydroxyl terminus. However, a similar mammalian enzyme from calf thymus appears to incise 3' to the apurinic site (79).

Mutants defective in endonuclease II have been isolated by Yajko & Weiss (69) and found to be MMS sensitive, but not sensitive to X rays or ultraviolet light. The spontaneous mutation rate in these mutants was reported to be lower than in the wild type.

MICROCOCCUS LUTEUS CORRENDONUCLEASE I (*M. luteus γ-endonuclease*)   Patterson & Setlow showed in 1972 that extracts of *M. luteus* contain an endonucleolytic activity that specifically incises γ- and X-irradiated DNA (70). This γ damage-specific activity could be separated chromatographically from a UV-specific activity also present in these extracts.

Some years earlier, Kushner purified a similar activity from *M. luteus* (71). This enzyme was found to act both on X-irradiated DNA and at UV-produced pyrimidine dimers, and hence appears to typify correndonuclease I as well as correndonuclease II. Kushner found that the *M. luteus* enzyme incised UV-irradiated DNA 5' to pyrimidine dimers, producing a 3'-phosphoryl terminus. The enzyme had a molecular weight of about 17,000, incised in the presence of EDTA, but was stimulated threefold by 10 mM Mg^{2+} (72). The enzyme has been described as "the *M. luteus* UV endonuclease" in several publications (72, 73), but its involvement in repair of UV-irradiated *M. luteus* is not certain. Mutants of *M. luteus,* which lack all endonucleolytic activity on UV-irradiated DNA, are both UV and X-ray sensitive (74). Thus, it is possible that this γ-specific endonuclease is involved in repair of X-ray damage in *M. luteus.*

It should be emphasized that another UV-specific endonucleolytic activity has been purified from *M. luteus* and is substantially different from the "γ-endonuclease." This UV-specific enzyme is discussed in a subsequent section.

ESCHERICHIA COLI CORRENDONUCLEASE I (*γ-endonuclease*)   Extracts of *E. coli* have recently been found to incise and degrade γ-irradiated polyd(A-T) (75). The incision process occurs at normal rates in extracts derived from cells deficient in correndonuclease II and is thus evidence for a new damage-specific endonucleolytic activity. One of the principal products of γ irradiation of DNA is 5,6-dihydroxydihydrothymine (76), a product that can also be produced chemically by treating DNA with osmium tetroxide (13). Extracts of *E. coli* have also been reported to act upon DNA treated with osmium tetroxide (13). Strniste & Wallace demonstrated an endonucleolytic activity that acts on X-irradiated φX174 RFI DNA in extracts of *E. coli* (77).

MICROCOCCUS LUTEUS URACIL-SPECIFIC ENDONUCLEASE   Recently, Carrier & Setlow reported the presence of an unusual correndonuclease in extracts of *M. luteus* (78). This enzyme, which can be chromatographically separated from UV- and γ-specific endonucleases, acts on DNA modified to contain uracil, as well as the uracil containing DNA from the *Bacillus subtilis* phage PB S2. While little is known about this activity at present, it is possible that the enzyme repairs deaminated cytosine and thus could play a role in reducing the rate of spontaneous mutations.

MAMMALIAN APURINIC SITE ENDONUCLEASE   An endonuclease specific for apurinic sites in DNA has been isolated from calf thymus by Ljungquist & Lindahl (79). This enzyme introduces nicks at the 3' side of apurinic sites in DNA and is presumably the first step in hypothetical excision of such sites in vivo. The protein has a molecular weight of 32,000 and is active in the presence of EDTA, although it is strongly stimulated by Mg^{2+} or Mn^{2+}. Ultraviolet- or γ-irradiated DNA is initially resistant to the action of the enzyme, but after the DNA is heated to 70°C for 30 min, apurinic sites appear as secondary lesions. A similar activity has been described in extracts of rat liver (80). The possible biological significance of the apurinic lesion has been discussed by Lindahl & Nyberg (81) and by Verly et al (82).

ESCHERICHIA COLI "MISMATCH" ENDONUCLEASE   Transfection of *E. coli* with heteroduplex λ DNA containing different mutations in both strands yields some progeny without mutations (83, 84). This has been interpreted as evidence for a mismatch repair system in *E. coli.* A possible mechanism for this repair involves excision of the mismatched region and resynthesis. This scheme demands an initial incision in the vicinity of the mismatch. The detection of such a mismatch enzyme has not yet been reported.

MAMMALIAN γ-ENDONUCLEASE   Brent described an activity in HeLa cell extracts that specifically incises PM2 DNA exposed to 3300 rad (85). Bacchetti et al (86) have described a UV-specific correndonuclease that acts upon photoreactivated UV-irradiated DNA. Both that activity and that described by Brent did not require divalent cations and were observed in cell lines derived from individuals with *xeroderma pigmentosum*. Lindahl and his co-workers have pointed out (87), however, that both UV and γ irradiation can introduce a significant number of apurinic sites into DNA. It is possible that the endonuclease observed by Brent and by Bacchetti et al is actually an apurinic site-specific endonuclease.

Correndonuclease II

All known correndonuclease II-type enzymes incise UV-irradiated DNA. In general, some of these enzymes also can incise DNA containing damage involving only a single nitrogenous base. The enzymes described below are classified as correndonuclease II-type enzymes, since it is believed that they are involved in the repair of pyrimidine dimer damage (intrastrand diadduct), as well as crosslink damage (interstrand diadduct).

ESCHERICHIA COLI CORRENDONUCLEASE II (*uvrA, uvrB endonuclease*)   The endonuclease involved in the first step of excision repair has recently been isolated and partially characterized (21). The involvement of this enzyme in UV repair is indicated by its absence in *uvrA* and *uvrB* excision defective mutants. The enzyme is small, appearing to have a molecular weight of less than 14,000. It acts in the presence of 10^{-3} M EDTA and requires an ionic strength of about 50 mM. In an irradiated : unirradiated DNA heteroduplex, the enzyme acts only on the damaged strand (88). The incision event occurs 5' to a pyrimidine dimer and leaves a 3'-hydroxyl terminus, which renders the break sealable by polynucleotide ligase (89, 90).

Binding of the enzyme to its substrate can be measured by using the membrane filter technique of Riggs et al (91). The specific binding of the enzyme to UV irradiated DNA can be prevented by prior treatment of the irradiated DNA with yeast photolyase and light (21). The purified enzyme, which has a K_m of 1.5×10^{-8} M (pyrimidine dimers) (92), is competitively inhibited by caffeine with a K_i of about 10^{-2} M; this is comparable to the inhibitory concentration of this drug in vivo (93). Caffeine inhibits the binding stage of the enzyme reaction. Preliminary data indicate that inhibition of the endonuclease by acriflavin also occurs at concentrations comparable to those found to be inhibitory in vivo (94).

Since *uvrA* and *uvrB* mutants of *E. coli* are more sensitive than wild-type to crosslinking agents such as mitomycin C (95) and psoralen plus light (96), there is reason to believe that the *E. coli* correndonuclease II acts on many diadduct forms of damage. However, there is in vivo evidence that the enzyme also acts on monoadduct damage. Otsuji & Murayama reported that *uvrA* and *uvrB* mutants of *E. coli* are sensitive to monofunctional analogs of mitomycin (97).

Another endonuclease acting on UV-irradiated DNA from *E. coli* has recently been described by Radman (98). This enzyme, referred to as endonuclease III by the author, does not appear to be involved in pyrimidine dimer repair, since it was found to be present in excision defective cells. Many of the properties of the enzyme were similar to the *uvrA uvrB* gene product.

The recent observation by Waldstein et al (99) that UV-specific incision rates in toluene treated *E. coli* cells can be significantly increased by the addition of ATP suggests that another step in the excision program may exist. Since ATP does not appear to enhance the reaction rate of the *E. coli* UV correndonuclease in vitro (100), the ATP enhanced incision rate may be due to either a preparative incision step or to some unknown phosphorylated activator of the system in vivo.

MICROCOCCUS LUTEUS CORRENDONUCLEASE II (*UV endonuclease*)   At least two UV-specific endonuclease activities exist in *M. luteus*. One, described earlier in this review, is a general radiation endonuclease, which acts on both UV- and X-irradiated DNA (a UV-X-correndonuclease). There is also an activity that is similar in its properties to the *E. coli uvrA, uvrB* correndonuclease and does not incise X-irradiated DNA (101). This correndonuclease II incises 5' to the pyrimidine dimer, leaves a 3'-hydroxyl terminus (102), requires 50 mM salt, and acts in the presence of 10^{-3} M EDTA (21). The general radiation correndonuclease also incises UV irradiated DNA 5' to a pyrimidine dimer; however, it incises by a *b* mechanism, leaving a 3'-phosphoryl terminus. This is not a trivial distinction, since a 3'-hydroxyl terminus provides a substrate site for both the DNA polymerase and polynucleotide ligase, whereas a 3'-phosphoryl terminus is refractory to the action of both enzymes.

"UV endonucleases," active on UV-irradiated DNA, have been isolated from extracts of *M. luteus* by several laboratories (103–105). Since it has recently become clear that at least two such activities are present in extracts of *M. luteus,* it is difficult to ascribe substrate specificities to enzymes described by a given laboratory. For example, Patrick & Harm (106) have shown that the *M. luteus* UV endonuclease and yeast photolyase act on all pyrimidine dimers (T̂T, ĈT, ĈC). Unfortunately, it is not possible to determine with certainty which of the two correndonucleases was used in these experiments.

BACTERIOPHAGE T4 CORRENDONUCLEASE II (V^+ *gene UV endonuclease*)   Correndonuclease II, coded for by the phage T4 V gene, is present in large amounts in T4-infected *E. coli* (107–111). This enzyme is similar in many of its properties to the *E. coli uvrA, uvrB* correndonuclease II. It is a small protein of molecular weight about 15,000, which acts in the presence of 10^{-3} M EDTA. Incision is 5' to the pyrimidine dimer and a 3'-hydroxyl terminus is generated (112). The enzyme acts both on single and double stranded UV irradiated DNA (112).

An interesting property of the T4 correndonuclease II may be an ability to incise gross distortions of double stranded DNA. Berger & Benz reported that heteroduplex T4 DNA with one strand containing a deletion is repaired by extracts of T4-infected *E. coli* (113, 114). Extracts of T4 V⁻ infected *E. coli* did not repair these heteroduplexes, implying that the V gene product, the T4 correndonuclease II, incises at the large noncomplementary loops formed in these heteroduplex DNA structures.

Kozinski presented biochemical evidence indicating that the T4 virion contains the V gene product, correndonuclease II (115, 116). In vivo evidence indicates that the T4 virion injects the V gene product into the infected cell (115).

MAMMALIAN CORRENDONUCLEASE II (*UV-specific endonuclease*)   Van Lanker & Tomura recently reported the isolation of a correndonuclease II from rat liver (117). The enzyme, purified to homogeneity, has a molecular weight of 20,000 and is stimulated by Mg^{2+}. The nuclease specifically nicks UV-irradiated and acetyl-aminofluorene treated double stranded DNA to yield 3'-phosphoryl termini. The presence of an endonuclease from cultures of human cells which nick at or near thymine dimers has been detected by Duncan et al (118) using dimer excision as an assay. It is noteworthy that this endonuclease activity is sensitive to freezing. Lytle et al (119), using a bioassay, reported an endonucleolytic activity in human fibroblasts which is absent from a fibroblast line derived from a patient with the repair defective condition, *xeroderma pigmentosum.*

CORRENDONUCLEASE II AS AN ANALYTICAL TOOL   Phage T4 and *M. luteus* correndo-nuclease II activity has been useful in measuring repair in vivo (120, 121). After a postirradiation period of repair, the DNA is extracted from the cells, treated with purified correndonuclease II, and the DNA sedimented through alkaline sucrose. Incisions produced by the in vitro enzymatic treatment represent unrepaired pyrimi-dine dimers. The reduction with time of these correndonuclease II susceptible sites is an indication of the extent of repair in vivo. This method allows for the con-clusion that the early stages of repair are absent in the *uvrB* mutants of *E. coli* and, furthermore, is defective in certain *xeroderma pigmentosum* cell lines.

PRE-EXCISION STEP

Incision Mechanism a   (*5'-phosphoryl group*)

Although the incision produced by the *E. coli* correndonuclease II provides an initiation site for excision and nucleotide incorporation by DNA polymerase I (102), indirect evidence suggests that an intermediate step between incision and excision exists. This evidence is based on the properties of the excision defective *uvrC* mutant. Such mutants are unable to remove pyrimidine dimers from DNA in vivo, although there are indications that single strand incision events are operative in only a transient manner. Since exonuclease VII, correndonuclease II, and poly-merase levels appear to be normal in such mutants, it can be assumed that the capacity for incision and excision is unaffected. The progression of single strand molecular weight changes in *uvrC* mutants is peculiar in that the extent of such breakage is low, whereas the initial rate of single strand break formation is indistin-guishable from wild-type cells (122). What is seen is a rapid, partial decrease in single strand molecular weight, which is quickly restored to that of control chain lengths. DNA isolated from wild-type cells under similar conditions, however, exhibits a rapid and more extensive loss in its single stranded molecular weight, which is fully restored during extended postirradiation time periods.

Seeberg & Rupp (122) prepared double *uvrC* and temperature-sensitive poly-nucleotide ligase mutants. At temperatures restrictive for ligase, the early rate and extent of postirradiation molecular weight losses are similar to those of wild-type cells, followed, however, by the expected accumulation of low molecular weight DNA. At permissive temperatures, however, the molecular weight is rapidly and fully restored. These data, in conjunction with the observations that polynucleotide ligase is capable of resealing the phosphodiester bonds of incised DNA, places polynucleotide ligase in a controlling position during early repair steps. The juxtaposition of a 3'-hydroxyl group and the 5'-phosphoryl group, in association with a pyrimidine dimer containing nucleotide, appears to be sensitive to *E. coli* polynucleotide ligase.

From these two observations, it can be inferred that the *uvrC* gene product prevents resealing of the correndonuclease II (incision) before excision of the pyrimidine dimer has occurred. A number of mechanisms, such as a 5'-polynucleo-tidase, nuclease, or perhaps an unwinding protein affecting the conformation of the damaged strand, can be suggested for its molecular mechanism in restricting ligase activity at this step of repair.

Incision Mechanism b (3'-phosphoryl group)

The incisions produced by the *M. luteus* general radiation correndonuclease II enzyme result in a 3'-phosphoryl terminus, 5' to a pyrimidine dimer (72). This terminus is not suitable as a nucleophilic site necessary for priming by poly-merases. Exonuclease III, because it can act as a phosphomonoesterase at such a 3'-phosphoryl double stranded terminus prior to phosphodiester bond hydrolysis, has the potential for repair capacity (123).

The 3'-polynucleotidase reported by Becker & Hurwitz (124) could conceivably participate at this step in the repair cycle in providing such a nucleophilic site. Mutants defective in exonuclease III, isolated by Milcarek & Weiss (125), show normal UV sensitivity, implying that the correndonuclease II enzymes repairing UV-irradiation damage by endonucleolytic activity either do not act by a *b*-type mechanism or, if 3'-phosphoryl groups were formed, other exonucleases may participate in this step in the cycle.

EXCISION MECHANISMS

It was implicit in the early observations of Nakayama et al (126) that excision was at least a two-step process in *M. luteus*. They found that the purification of a UV-dependent nuclease activity by TEAE-cellulose chromatography resulted in the loss of activity, which could be recovered by the reconstitution of two different fractions eluted from such a column. The first fraction, fraction *A*, did not exhibit any nuclease activity; however, acid-soluble nucleotide release specifically from irradiated DNA required the presence of a second TEAE fraction which contained a nonspecific nuclease activity, fraction *B*. Fraction *B* was active on unirradiated or irradiated denatured DNA, and it was correctly assumed by these investigators that fraction *A* was an endonuclease.

An analogous multistep excision repair system was demonstrated in phage T4-infected *E. coli* (111). Correndonuclease II, isolated and purified to homogeneity, is the V gene product of phage T4. To demonstrate the role of such an endonuclease in excision, crude extracts of cells infected with phage T4 V⁻ were used as the source of exonuclease. The exonucleolytic properties associated with such crude extracts were later identified by Onshima & Sekiguchi (127). Similar types of excision systems seem to exist in *M. luteus, E. coli,* and phage T4-infected *E. coli.*

Two types of exonucleases can execute excision: unassociated exonucleases may act on incised DNA, or DNA polymerase-associated exonucleases may react with such incised irradiated DNA intermediates.

DNA Polymerase-Associated Exonucleases

$3' \rightarrow 5'$ EXONUCLEOLYTIC ACTIVITY   DNA polymerase I of *E. coli* is a multifunctional enzyme contained on a single polypeptide chain of molecular weight 109,000 (128). The *M. luteus* DNA polymerase for all intents and purposes is identical to the *E. coli* enzyme, having a similar molecular weight, nuclease properties, and N-ethylmaleimide insensitivity. Preliminary immunological experiments have demonstrated crossreactivity with antibodies specifically directed against *E. coli* DNA polymerase (129).

In addition to their polymerizing properties, these enzymes have two associated exonucleolytic activities, one of which is a $3' \rightarrow 5'$ nuclease, which at 37°C prefers single stranded DNA possessing a 3'-hydroxyl terminus in which 5' nucleotides are the exclusive product of digestion. An important role for this nuclease activity, in conjunction with the enzyme's polymerizing properties, is its editing functions, in which the exonuclease activity is capable of digesting in a direction opposite to that of polymerization up to the point of hydrogen bond stability. Brutlag & Kornberg (130) demonstrated that synthetic duplex model polymers containing noncomplementary nucleotides at the 3' end of the primer template are removed by the polymerase's $3' \rightarrow 5'$ exonucleolytic activity.

A series of mutator and antimutator mutations of bacteriophage T4 has been isolated in gene 43 (131), the structural gene of the T4 DNA polymerase. Muzyczka et al (132) have shown a close correlation between the mutator properties of the polymerase and reduced $3' \rightarrow 5'$ exonuclease activity. Conversely, antimutator polymerase has an increased $3' \rightarrow 5'$ activity. Hall & Lehman (133) have shown in vitro that the rate of misincorporation of nucleotides by mutator T4 DNA polymerase is statistically greater than that by wild-type polymerases.

$5' \rightarrow 3'$ EXONUCLEOLYTIC ACTIVITY   The $5' \rightarrow 3'$ nuclease activity associated with polymerase I of *E. coli* and *M. luteus* is specific for DNA duplexes. This exonuclease activity is stimulated by neighboring 3'-hydroxyl groups provided during polymerization of the four deoxynucleoside triphosphates (102, 134–136) included in reaction mixtures in which nicked DNA provides priming and potential templating sites. When synthesis and hydrolytic activity proceed at comparable rates, synthesis leads to an extension of the 3'-hydroxyl terminus of a nick and simultaneous removal of nucleotides at the 5′ terminus, which is referred to as nick translation (136). The translation continues in a 5′ → 3′ direction until the nick is located at the extreme 3′ end of the template strand, at which time both activities cease. This process is probably the dominant reaction in vitro at temperatures below 22°C (137). At higher temperatures, however, another process dominates which is expressed by the synthesis of branched, nondenaturable structures. Such aberrant synthetic mechanisms are reflected in an extent of synthesis in excess of one template equivalent (138).

The 5′ → 3′ nuclease initiates hydrolysis of DNA duplexes at internal rather than terminal phosphodiester bonds, regardless of whether such termini are 5′ esterified (139). About 20–25% of the products of 5′-terminated DNA are dinucleotides and longer oligonucleotides with the majority represented by 5′ nucleotides. Low pH favors the production of dinucleotides; at pH 6, approximately 50% of the acid-soluble products are dinucleotides (140).

Exposure of the polymerase to primer templates with a short 5′ noncomplementary terminus results in release of such noncomplementary nucleotides as a fragment (141).

Irradiated poly(dA : dT) incised nonspecifically with pancreatic DNase was hydrolyzed by polymerase I, liberating dimer-containing fragments. The products of exonucleolytic hydrolysis by the polymerase were oligothymidylates of chain lengths ranging from dinucleotides to heptanucleotides containing photoproducts. The spectrum of products released with the labeled irradiated poly(dA : dT) was somewhat different than that with the unirradiated control in which all the radioactive label was isolated as mononucleotides. Comparative results were obtained with irradiated DNA in which dimer-containing fragments larger than trinucleotides were located chromatographically.

Any assignment of cellular involvement in polymerase-mediated repair is dependent on experiments obtained with those mutants lacking the respective nuclease activities. Mutant strains of E. coli with negligible levels of DNA polymerase I (polA), although increasingly sensitive to UV, are considerably more resistant than uvr or rec mutants (142, 143). polA Strains of E. coli, such as P3478, excise thymine-containing dimers to the same rate and extent as do wild-type cells (144). The inference derived from such results is that the UV sensitivity of these mutants may be related to the reduced polymerizing capabilities during reinsertion, and that other mechanisms of excision must be operative in these strains of E. coli. Polymerase I has been purified from a variety of polA mutants. Polymerase I activity in these E. coli mutants was between 0.5–3% of wild-type activity, whereas the associated 5′ → 3′ exonuclease activities remained unaffected (145). These findings account for the normal levels of pyrimidine dimer excision in such mutants.

The size of reinserted regions of DNA from experiments in vivo ranges from 10 nucleotides (short patch repair) to about 3000 nucleotides in length (long patch repair). PolA1 mutations result in a preponderance of long patch repair synthesis (146). These data may be explicable in terms of the elevated ratio of 5′ → 3′ hydrolysis to polymerizing activity in such mutants. However, the resolved 5′ → 3′ exonuclease fragment (3.3S) excises slowly in the absence of concomitant polymerization. The addition of the large (7.7S) polymerizing fragment markedly stimulates pyrimidine dimer removal (147). These findings are in accord with the known properties of the enzyme in nick translation in vitro (136).

It is difficult to conclusively assign an in vivo excision role to the $5' \rightarrow 3'$ exonuclease of polymerase I, since temperature-sensitive mutants in this exonucleolytic activity have been found to be conditionally lethal by a number of laboratories (148, 149). However, a mutant of E. coli with a substantial defect in the $5' \rightarrow 3'$ exonuclease, but with normal polymerizing properties, has been described by Glickman et al (150, 151). This mutant is moderately UV sensitive and shows a somewhat reduced pyrimidine dimer excision capability.

Polymerase III Excision Capabilities

E. coli polymerase III has associated $3' \rightarrow 5'$ and $5' \rightarrow 3'$ hydrolytic activities (152). The $3' \rightarrow 5'$ activity is specific for single stranded DNA and releases 5' mononucleotides, and its activity is inhibited by a 3'-phosphoryl group. The $5' \rightarrow 3'$ activity can initiate hydrolysis on single strands and proceed into a duplex region once initiation has occurred. The $5' \rightarrow 3'$ activity can also initiate hydrolysis of double strand DNA possessing a single stranded 5' terminus. The limit products of hydrolysis of the $5' \rightarrow 3'$ activity are dinucleotides and mononucleotides which may arise from the combined $3' \rightarrow 5'$ and $5' \rightarrow 3'$ activities of polymerase III. The structure of UV-irradiated DNA incised by the M. luteus correndonuclease II is ideally suited for $5' \rightarrow 3'$ exonucleolytic removal of pyrimidine dimers.

Unassociated Exonucleases in Excision

The hydrolysis of the phosphodiester bond 5' to the photoproduct by the correndonuclease II requires that exonucleases involved in excision initiate hydrolysis in a $5' \rightarrow 3'$ directional manner. Moreover, an additional requirement for such exonucleases is that they must be able to hydrolyze internal phosphodiester bonds in order to catalyze the removal of pyrimidine dimers. It would appear that those exonucleases involved in the excision of pyrimidine dimers do not hydrolyze the phosphodiester bond linking the pyrimidine nucleotide dimer residues (153).

The catalytic properties of two generally distinguishable but functionally related exonucleases capable of excising pyrimidine dimers have been identified in M. luteus (72), E. coli (154), T4-infected E. coli (127, 155), rabbit liver nuclei (156, 157), human placenta (158), and mammalian cell lines (118). Both types of exonucleases, a UV exonuclease and exonuclease VII, have specificities limited to denatured DNA and are capable of hydrolyzing UV damaged and undamaged substrates at comparable rates and to the same extent (159). This characteristic property can be used to distinguish those single stranded exonucleases with potential pyrimidine dimer excision capabilities in vitro.

The UV exonuclease from M. luteus can be distinguished from functionally unrelated exonucleases, such as E. coli exonuclease I, venom phosphodiesterase, and bovine spleen phosphodiesterase, whose hydrolytic activities are restrained by photoproducts containing DNA (159). The prokaryotic repair exonucleases already characterized have the unusual capacity of initiating hydrolysis from either 3' or

5' termini. In all cases, the products, whether mononucleotides or oligonucleotides, contain 5' phosphorylated termini. The *M. luteus* UV exonuclease is unique in that it requires divalent cation and has no limit hydrolysis product (72). However, *E. coli* exonuclease VII (154) and a similar type of enzyme purified from *M. luteus* (160) do not require magnesium, thereby allowing for their ability to function optimally in the presence of 10^{-3} M EDTA, permitting its easy detection in crude extracts.

Exonuclease VII can act on denatured DNA, single stranded regions extending from duplex DNA treated with exonuclease III (154), or λ DNA. An exonuclease VII type of enzyme from *M. luteus* cleaves part of the 12 nucleotide long single stranded cohesive ends of λ DNA, leaving three or four nucleotides on each end (161). None of these exonucleases is able to use RNA as a substrate, nor are polyribonucleotide : polydeoxyribonucleotide hybrid polymers hydrolyzed by exonuclease VII.

The eukaryotic exonuclease activity that presumably constitutes the second step in the excision of pyrimidine dimers has been purified from rabbit tissue by Lindahl (156, 157). The enzyme, termed DNase IV, hydrolyzes DNA in a $5' \rightarrow 3'$ direction, releasing oligonucleotides containing five to eight residues. A similar activity has been detected in crude extracts of human cell cultures (118). The importance of such exonucleolytic activities in mammalian cells is emphasized by the fact that, unlike the bacterial polymerases, none of the mammalian DNA polymerases purified to homogeneity have been found to contain any nuclease activity.

The *M. luteus* UV exonuclease (72) and DNase IV (156) of rabbit liver nuclei produce 5' mononucleotides during digestion with no oligonucleotide intermediates. The limit products of exonuclease VII action are oligonucleotides bearing 5'-phosphoryl and 3'-hydroxyl termini in which approximately one third are in the range of dinucleotides to trinucleotides and the majority are in the range of tetranucleotides to dodecamers, in which no mononucleotides have been observed as intermediates during hydrolysis.

The exonuclease VII-type enzymes and polymerase-associated $5' \rightarrow 3'$ exonucleases yield oligonucleotides as an ultimate product, and the UV exonuclease catalyzes the release of dimer-containing oligonucleotides from correndonuclease II incised irradiated DNA (73). Since the *M. luteus* UV exonuclease does not hydrolyze the phosphodiester bond located between thymidylate dimer residues, the initial hydrolytic event must have occurred approximately six nucleotides 3' to the photochemical damage (153). The ratio of nucleotides released per phosphodiester bond originally hydrolyzed by the correndonuclease II provides an estimation of the approximate size of the region distorted by the formation of photoproducts. Although such ratios are somewhat dependent on the source of irradiated DNA substrates, they do correlate moderately well with the distortion sizes predicted spectroscopically (162) and from model building (141). From data obtained with the correndonuclease II of *M. luteus,* the ratio of the number of nucleotides released to the number of phosphodiester bonds broken indicates that the size of the excised region is approximately 6–10 nucleotides in length (73).

The conformational specificities of the polymerase associated and unassociated $5' \rightarrow 3'$ exonucleases are different, which may influence the ability of these enzymes to excise pyrimidine dimers in regions of differing hydrogen bond stabilities. For example, the combined reaction of correndonuclease II and the single stranded specific *M. luteus* UV exonuclease seems to function preferentially in AT-rich regions, judging from the distribution of nucleotides 3' to dimers in excised fragments (73). It would be of interest, therefore, to determine not only the nucleotide compositions of fragments excised by polymerase-associated and -unassociated exonucleases, but environmental effects as well on the course of action of such nucleases. Parenthetically, neither exonuclease VII mutants nor polymerase mutants affected in their $5' \rightarrow 3'$ nuclease activity show marked UV sensitivity or impaired dimer excision capabilities (150, 163). The construction of double mutants will, therefore, be of considerable value in assessing the involvement in vivo of these various enzymatic activities. That significant increases in UV sensitivity are not observed in single mutants may reflect either the interdependency or overlapping specificities of the polymerase-associated exonucleases and the unassociated exonucleases.

REINSERTION MECHANISMS

Following excision of the damaged region, reinsertion of nucleotides is catalyzed by DNA polymerases using the complementary strand as a template. The reinsertion of nucleotides into UV-irradiated *E. coli* was first studied by Pettijohn & Hanawalt (164) and shown to be distinct from semiconservative DNA replication.

Reinsertion events in their experiments are distinguishable from semiconservative replication through the uptake of [^{3}H]-BrdU into DNA having light or intermediate buoyant densities. Since 5-BrdU is more dense than its homolog thymidine, its incorporation into DNA in a semiconservative manner (i.e. normal replication) results in the newly synthesized strand of DNA having a greater density than the template strand. Measurement after shearing to small fragments (0.5×10^6 daltons) of the density of the double stranded DNA by isopycnic centrifugation in CsCl or NaI reveals radioactivity banding as a "hybrid" area of the gradients. After UV irradiation, incorporation of nucleotides takes place in relatively short gaps exposed during excision of the damaged bases. Radioactive 5-BrdU incorporated into small gaps does not significantly increase the density of the DNA and thus appears in the normal density regions of the gradient (164). This operational definition of short patch repair is restricted to regions of approximately 10–30 nucleotides in length (7, 164). Under conditions in which short gap repair is inhibited, as in *polA* mutants, the labeled 5-BrdU is associated with intermediate regions of the gradient which are less dense than the hybrid semiconservatively replicated DNA and more dense than the normal density regions of the gradient. Such regions are approximately 1000–3000 nucleotides long and represent long patch repair (146). Both short and long patch repair occur in wild-type cells, but are absent in excision defective cells. Therefore, two branches of reinsertion are available to *E. coli* cells.

One route of reinsertion is considerably more complicated, more extensive in extent and perhaps, as a consequence, less efficient than the other. This long patch repair (146), or growth medium-dependent (165) pathway of repair, requires *uvrA* (166) and *rec* genes (146, 165), polymerase II (167), and/or polymerase III (168); is ATP dependent in toluene treated cells (169); and may require the further support of specific unwinding proteins [see the review in this volume by Gefter (165a)]. An additional pathway involving *recF* genes (55, 56), which is an exonuclease I-sensitive locus (54), can supplement both the *polA* and *recA, recBC* branch of repair, and is perhaps similar to *recA, recBC* postreplicative repair mechanisms (33–36).

It is difficult to assess the specific molecular determinants that influence the direction of excision and reinsertion into any one of the enzymatic branches of repair. Such determinants may be architectural; for example, enzymatic juxtapositions to damaged DNA might govern repair directions. They might also be influenced by the structural conformations of incised DNAs.

Short Patch Pathway of Repair (uvr, polA, lig)

There is sufficient evidence that the sequence of events controlled by the *uvr* genes is necessary for excision and reinsertion of nucleotides via this pathway of repair (19, 166). The size of the patches observed in vivo from 5-BrdU density labeling is of the same order of magnitude expected from studies in vitro (75). Although it is difficult to determine under normal conditions what proportion of the DNA is repaired by polymerase I, it can be deduced from the data of Cooper & Hanawalt (166) that the majority of repaired DNA is in short patches synthesized via the polymerase I pathway of reinsertion.

Short patch repair has been mimicked under controlled in vitro conditions. Heijneker et al (170) partially restored the transforming activity of UV-irradiated *B. subtilis* DNA by incising the DNA with the *M. luteus* correndonuclease II and then incubating the incised DNA with a mixture of *E. coli* polymerase I and polynucleotide ligase. Roughly 20 nucleotides per pyrimidine dimer were excised at 34°C.

Hamilton et al (102) examined the control in vitro of both excision of pyrimidine dimers and the reinsertion of nucleotides catalyzed by DNA polymerase I of *M. luteus*. Polymerase I-type enzymes, unlike polymerases II and III and phage-infected DNA polymerases, specifically bind at nicks, satisfying an important and specific requirement for the short patch excision process (173). It would appear, therefore, that the nick binding and translating properties of these enzymes provide for the necessary concerted mechanisms of reinsertion and excision. Nick translational conditions result in a stoichiometry of equivalence between removal of photoproducts with associated nucleotides and the reinsertion of nucleotides. Furthermore, polymerase is capable of repairing gaps (171). Incubation of this polymerase with excised DNA, for example, at 10°C in optimal Mg^{2+} leads to a stoichiometric reinsertion-excision reaction which may be controlled by polynucleotide ligase. Under conditions of strand displacement at 37°C, the presence of ligase appears to restrict the stoichiometric ratio to one nucleotide excised per nucleotide polymerized. Therefore, in vitro, the polymerase:correndonuclease II combination in conjunction with ligase is sufficient for the complete repair of single strand breaks associated with incision and restoration of biological activities of UV-irradiated transforming DNA.

When transforming DNA is 50% or less inactivated by UV, there is a quantitative restoration of biological activity by these three enzymes. At higher doses, the maximum repair capabilities by this enzyme system decline considerably. This is attributed to the formation of double strand breaks, arising not from irradiation, but as a consequence of the repair process. The unidirectionality ($5' \rightarrow 3'$) of repair on strands of opposite polarity leads to double strand break formation during the excision process. Hamilton et al (102) have suggested that polyamines, such as spermine, may stabilize the DNA repair intermediates and limit their degradation to prevent double strand breaks and thus increase the reactivation of *B. subtilis* DNA at high UV doses.

Long Patch Pathway of Repair (*uvr, recBC, unwinding protein, polB, dnaE, lig*)

Delineation of repair pathways is operational and defined according to the density position of [^{3}H]-BrdU in gradients where long patches are at the heavy end and short patches are located with the light end of such a gradient among nonreplicated DNAs. There are clearly no well-separated peaks of label, but rather a spectrum of densities which must be visualized, in molecular terms, as path average rather than path specific. Cooper & Hanawalt (146) found, for example, that the UV-induced nonconservative DNA synthesis, rather than becoming limited in *polA* mutants, was in fact considerably stimulated. This stimulated synthesis was dependent on the *uvrA* gene product (169), indicating a divergent, rather than two separate paths of reinsertion. The size of the reinserted patch was at least 100 times greater than that observed in a *polA*-dependent short patch pathway of reinsertion and, furthermore, was dependent on the presence of the *recA, recBC* gene products.

Experiments in toluene treated *E. coli* suggest that both polymerases II and III are involved in UV-induced repair. Since polymerase II lacks a $5' \rightarrow 3'$ exonuclease function (172), its role in repair must be confined to reinsertion reactions. Repair in toluene treated cells seems to be of the long patch type, which is missing if both polymerases II and III are inactivated (169).

A controlling feature of long patch repair may be attributable to the size of the fragment excised by the unassociated exonuclease VII, UV exonucleases or polymerase III initiation of hydrolysis of the single stranded regions of incised DNA. The fragments released by the unassociated exonucleases vary between 6–8 nucleotides in length (73, 152, 154). Once the single strand fragment is removed, the duplex nature of the excised DNA limits further hydrolysis by the exonucleases. The binding properties, however, of polymerases II and III are limited to much larger regions, such that in order for these polymerases to carry out reinsertion, a gap expansion step is necessary (173). This gap expansion is probably catalyzable by the ATP-stimulated double stranded exonuclease activity associated with the multifunctional *recBC* enzyme (exonuclease V) (45, 174–176). It is the ATP-dependent double stranded exonuclease activity that cannot hydrolyze nicked DNAs, but is able to degrade duplex circles containing gaps as short as five nucleotides in length (176). This activity is bidirectional such that hydrolysis occurs in a 5' and 3' direction (177). This gap expansion activity is thermolabile in *recBC* temperature-sensitive mutants. Furthermore, at restrictive temperatures, it is this *recBC* function that confers both UV sensitivity and reduced viability to such mutant *E. coli* cells (52).

The *E. coli* unwinding protein required for the processive polymerization by polymerase II on single stranded DNA (178) is potentially able to assume a protective role in the gap expansion process. The complementary strand of the excised DNA should be susceptible to endonucleolytic attack by the ATP-dependent single stranded endonuclease of the *recBC* enzyme. The *E. coli* unwinding protein, however, is able to inhibit this activity specifically in vitro without affecting the ATP-dependent duplex exonuclease activity, thereby preserving the integrity of the strand opposite to that which is being repaired (179). It may be suggested that the *recA* gene product serves similarly in protecting the complementary strand exposed at this locus in repair or during postreplication repair in an analogous manner to the T4 gene 32 protein, which functions in recombination and replication. Parenthetically, temperature-sensitive gene 32 proteins are extremely UV sensitive under restrictive conditions (183).

ATP is required in addition to the four deoxynucleoside triphosphates to stimulate long patch repair in toluene treated *polA* mutants (167). Such cofactor requirements may partially represent the need for a number of different processes in toluene treated cells. For example, ATP is needed for the enlargement of gaps, required in toluene treated cells for preincision steps, and may also be needed for recombinational mechanisms yet to be elucidated.

COMMENTS ON EUKARYOTIC REPAIR MECHANISMS

Progress in elucidating prokaryotic repair mechanisms has been enhanced by the available mutants affecting various loci of the excision or reinsertion pathways of repair. In the absence of equivalent isogenic strains of cloned mammalian cell lines, those interested in eukaryotic repair pathways are exploiting related mutant skin fibroblasts lines obtained from patients with the photosensitive disease *xeroderma pigmentosum* (XP).

At least five complementation groups of XP and related cell lines have been reported (180, 181), indicating that the numbers of genes controlling repair in mammalian systems must be at least as numerous as the genes controlling UV sensitivity in *E. coli*. In spite of the fact that cell lines from the various complementation groups show some physiological defect in a UV repair system, the relationship between clinical symptoms and biochemical abnormalities in repair has not been clearly established. This lack of correlation is the result of a paucity of information dealing with the enzymatic repair of DNA in normal mammalian cells. The photobiology of mammalian cells, however, has been effectively reviewed recently and for those readers interested in mammalian repair, the review of Cleaver is recommended (182).

ACKNOWLEDGMENTS
The authors are grateful to Mrs. R. Geffen for her devoted and patient secretarial efforts in preparing this review. The work of this laboratory has been supported by grants from the American Cancer Society (NP-8D), the Atomic Energy Commission [AT(11-1)3232-2], the National Institutes of Health (GM-15881-14), and the National Science Foundation (GB-29172X). This is contribution No. 977 of the Graduate Department of Biochemistry.

Literature Cited

1. Kelner, A. 1949. *Proc. Nat. Acad. Sci. USA* 35:73–79
2. Dulbecco, R. 1950. *J. Bacteriol.* 59:329–47
3. Sutherland, B. 1974. *Nature* 248:109–12
4. Regan, J. D. 1969. *Ann. 1st. Super. Sanita* 5:355–59
5. Rupp, W. D., Wilde, C. E., Reno, D. L., Howard-Flanders, P. 1971. *J. Mol. Biol.* 61:25–44
6. Boyce. R. P., Howard-Flanders, P. 1964. *Proc. Nat. Acad. Sci. USA* 51:293–300
7. Setlow, R., Carrier, W. 1964. *Proc. Nat. Acad. Sci. USA* 51:293–300
8. Riklis, E. 1965. *Can. J. Biochem.* 43:1207–19
9. Phillips, J. H., Brown, D. M. 1967. *Progr. Nucl. Acid Res.* 7:349–68
10. Banks, G. R., Brown, D. M., Streeter, D., Grossman, L. 1971. *J. Mol. Biol.* 60:425–39
11. Schuster, H., Wilhelm, R. C. 1963. *Biochim. Biophys. Acta* 68:554–60
12. Shapiro, R., Servis, R. E., Welcher, M. 1970. *J. Am. Chem. Soc.* 92:422–24
13. Hariharan, P. V., Cerutti, P. A. 1974. *Proc. Nat. Acad. Sci. USA* 71:3532–36
14. Beukers, B., Berends, W. 1961. *Biochim. Biophys. Acta* 49:181–89
15. Cole, R. S. 1971. *Biochim. Biophys. Acta* 254:30–39
16. Iyer, V. N., Szybalski, W. 1963. *Proc. Nat. Acad. Sci. USA* 50:355–62
17. Carrier, W. L., Setlow, R. B. 1971. *Methods Enzymol.* 21:230–37
18. Seaman, E., Van Vunakis, H., Levine, L. L. 1972. *J. Biol. Chem.* 247:5709–17
19. Howard-Flanders, P., Boyce, R. P., Theriot, L. 1966. *Genetics* 53:1119–36
20. Mattern, I. E., van Winden, M. P., Rörsch, A. 1965. *Mutat. Res.* 2:111–31
21. Braun, A., Grossman, L. 1974. *Proc. Nat. Acad. Sci. USA* 71:1838–42

22. Kato, T. 1972. *J. Bacteriol.* 112:1237–46
23. Seeberg, E., Johansen, I. 1973. *Mol. Gen. Genet.* 123:173–84
24. Ogawa, H., Shimada, K., Tomizawa, J. 1968. *Mol. Gen. Genet.* 101:227–44
25. Mattern, I. E. 1971. *First European Biophysics Congress*, ed. E. Broda, A. Locker, H. Springer Lederer, 237–40. Verlag der Wiener Medizinischen Akademie
26. Smirnov, G. B., Skavronskaya, A. G. 1971. *Mol. Gen. Genet.* 113:217–21
27. Siegel, E. 1973. *J. Bacteriol.* 113:145–60
28. Siegel, E. 1973. *J. Bacteriol.* 113:161–66
29. Storm, P. K., Zaunbrecher, W. M. 1972. *Mol. Gen. Genet.* 115:89–92
30. Horii, Z., Clark, A. J. Quoted in Ref. 31
31. Taylor, A. L., Trotter, C. D. 1972. *Bacteriol. Rev.* 36:504–24
32. Taketo, A., Yasuda, S., Sekiguchi, M. 1972. *J. Mol. Biol.* 70:1–14
33. Hertman, I. 1969. *Genet. Res.* 14:291–307
34. Howard-Flanders, P., Theriot, L. 1966. *Genetics* 53:1137–50
35. Howard-Flanders, P., Theriot, L., Stedeford, J. B. 1969. *J. Bacteriol.* 97:1134–41
36. Clark, A. J., Margulies, A. D. 1965. *Proc. Nat. Acad. Sci. USA* 53:451–59
37. Shlaes, D. M., Anderson, J. A., Barbour, S. D. 1972. *J. Bacteriol.* 111:723–30
38. Willetts, N. S., Clark, A. J., Low, B. 1969. *J. Bacteriol.* 97:244–49
39. Capaldo-Kimball, F., Barbour, S. D. 1971. *J. Bacteriol.* 106:204–12
40. Clark, A. J., Chamberlin, M., Boyce, R. P., Howard-Flanders, P. 1966. *J. Mol. Biol.* 19:442–54
41. Brooks, K., Clark, A. J. 1967. *J. Virol.* 1:283–93
42. Inouye, M. 1971. *J. Bacteriol.* 106:539–42
43. Willetts, N. S., Clark, A. J. 1969. *J.*

Bacteriol. 100 : 231–39
44. Monk, M., Kinross, J. 1972. *J. Bacteriol.* 109 : 971–78
45. Goldmark, P. J., Linn, S. 1970. *Proc. Nat. Acad. Sci. USA* 67 : 434–41
46. Barbour, S. D., Clark, A. J. 1970. *Proc. Nat. Acad. Sci. USA* 65 : 955–61
47. Buttin, G., Wright, M. 1968. *Cold Spring Harbor Symp. Quant. Biol.* 33 : 259–69
48. Oishi, M. 1969. *Proc. Nat. Acad. Sci. USA* 64 : 1292–99
49. Goldmark, P. J., Linn, S. 1972. *J. Biol. Chem.* 247 : 1849–60
50. Nobrega, F. G., Rola, F. H., Pasetto-Nobrega, M., Oishi, M. 1972. *Proc. Nat. Acad. Sci. USA* 69 : 15–19
51. Lieberman, R. P., Oishi, M. 1973. *Nature New Biol.* 243 : 75–77
52. Kushner, S. R. 1974. *J. Bacteriol.* 120 : 1219–22
53. Clark, A. J. 1971. *Ann. Rev. Microbiol.* 25 : 438–64
54. Kushner, S. R., Nagaishi, H., Templin, A., Clark, A. J. 1971. *Proc. Nat. Acad. Sci. USA* 68 : 824–27
55. Horii, Z., Clark, A. J. 1973. *J. Mol. Biol.* 80 : 327–44
56. Clark, A. J. 1973. *Genet. Suppl. XIII Int. Congr. Genet.* In press
57. Clark, A. J. 1973. *Ann. Rev. Genet.* 7 : 67–86
58. Howard-Flanders, P. 1968. *Ann. Rev. Biochem.* 37 : 175–200
59. Setlow, R. B. 1968. *Progr. Nucl. Acid Res.* 8 : 247–95
60. Friedberg, E. C., Goldthwait, D. A. 1969. *Proc. Nat. Acad. Sci. USA* 62 : 934–40
61. Friedberg, E. C., Hadi, S. M., Goldthwait, D. A. 1969. *J. Biol. Chem.* 244 : 5879–89
62. Hadi, S., Goldthwait, D. A. 1971. *Biochemistry* 10 : 4986–94
63. Kirtikar, D. M., Goldthwait, D. A. 1974. *Proc. Nat. Acad. Sci. USA* 71 : 2022–26
64. Goldthwait, D. A., Kirtikar, D., Hadi, S. M., Friedberg, E. C. 1975. *Molecular Mechanisms for the Repair of DNA,* ed. P. C. Hanawalt, R. B. Setlow. New York : Plenum. In press
65. Verly, W. G., Paquette, Y. 1972. *Can. J. Biochem.* 50 : 217–24
66. Paquette, Y., Crine, P., Verly, W. G. 1972. *Can. J. Biochem.* 50 : 1199–1209
67. Verly, W. G., Gossard, F., Crine, P. 1974. *Proc. Nat. Acad. Sci. USA* 71 : 2273–75
68. Hadi, S. M., Kirtikar, D. M., Goldthwait, D. A. 1973. *Biochemistry* 12 : 2747–54
69. Yajko, D. M., Weiss, B. 1974. *Fed. Proc.* 33 : 1599 (Abstr.)
70. Patterson, M. C., Setlow, R. B. 1972. *Proc. Nat. Acad. Sci. USA* 69 : 2927–31

71. Kushner, S. R. 1970. *UV-Endonuclease : purification, properties and specificity of the enzyme from Micrococcus luteus. The photoproduct excision system of Micrococcus luteus.* PhD thesis. Brandeis Univ., Waltham, Mass. 205 pp.
72. Kaplan, J. C., Kushner, S. R., Grossman, L. 1971. *Biochemistry* 10 : 3315–24
73. Kushner, S. R., Kaplan, J. C., Ono, H., Grossman, L. 1971. *Biochemistry* 10 : 3325–34
74. Mahler, I., Kushner, S. R., Grossman, L. 1971. *Nature New Biol.* 234 : 47–50
75. Cerutti, P. A. 1974. *Photochemistry & Photobiology of Nucleic Acids,* ed. S. Y. Wang, M. Patrick, Chap. 6. New York : Gordon & Breach
76. Hariharan, P. V., Cerutti, P. A. 1972. *J. Mol. Biol.* 66 : 65–81
77. Strniste, G. F., Wallace, S. S. 1975. *Molecular Mechanisms for the Repair of DNA,* ed. P. C. Hanawalt, R. B. Setlow, Basic Life Science Series, Vol. 5. General Ed., A. Hollander. New York : Plenum. In press
78. Carrier, W. L., Setlow, R. B. 1974. *Fed. Proc.* 33 : 1599 (Abstr.)
79. Ljungquist, S., Lindahl, T. 1974. *J. Biol Chem.* 249 : 1530–35
80. Verly, W. G., Paquette, Y. 1973. *Can. J. Biochem.* 51 : 1003–9
81. Lindahl, T., Nyberg, B. 1972. *Biochemistry* 19 : 3610–18
82. Verly, W. G., Paquette, Y., Thibodeau, L. 1973. *Nature New Biol.* 244 : 67–69
83. Wildenberg, J., Meselson, M. 1973. *Symp. Genet. Recomb.,* 48 (Abstr.). Nutley, NJ: Roche Inst. of Mol. Biol.
84. Kaiser, A. D., Hogness, D. S. 1960. *J. Mol. Biol.* 2 : 392–415
85. Brent, T. P. 1973. *Biophys. J.* 13 : 399–401
86. Bacchetti, S., van der Plas, A., Veldhuisen, G. 1972. *Biochem. Biophys. Res. Commun.* 48 : 662–69
87. Ljungquist, S., Anderson, A., Lindahl, T. 1974. *J. Biol. Chem.* 249 : 1536–40
88. Radman, M., Braun, A. Unpublished data
89. Braun, A., Grossman, L. 1974. *Fed. Proc.* 33 : 1599 (Abstr.)
90. Rupp, W D., Seeberg, E., Braun, A. In preparation
91. Riggs, A. D., Suzuki, H., Bourgeois, S. 1970. *J. Mol. Biol.* 48 : 67–83
92. Braun, A. Hopper, P., Grossman, L. See Ref. 77
93. Sideropoulous, A. S., Shankel, D. M. 1968. *J. Bacteriol.* 96 : 198–204
94. Smith, B., Braun, A., Grossman, L. Unpublished data
95. Boyce, R. P., Howard-Flanders, P.

1964. *Z. Verebungslehre* 94:345–50
96. Kohn,K.,Steigbigel, N.,Spears,C. 1965. *Proc. Nat. Acad. Sci. USA* 53:1154–61
97. Otsuji, N., Murayama, I. 1972. *J. Bacteriol.* 109:475–83
98. Radman, M. See Ref. 77
99. Waldstein, E. A., Sharon, R., Ben-Ishai, R. 1974. *Proc. Nat. Acad. Sci. USA* 71:2651–54
100. Braun, A. Unpublished data
101. Riazuddin, S., Grossman, L. Unpublished data
102. Hamilton, L. D. G., Mahler, I., Grossman, L. 1974. *Biochemistry* 13:1886–96
103. Carrier, W. L., Setlow, R. B. 1970. *J. Bacteriol.* 102:178–86
104. Nakayama, H., Okubo, S., Takagi, Y. 1971. *Biochim. Biophys. Acta* 228:67–82
105. Kaplan, J. C., Kushner, S. R., Grossman, L. 1969. *Proc. Nat. Acad. Sci. USA* 63:144–51
106. Patrick, M. H., Harm, H. 1973. *Photochem. Photobiol.* 18:371–86
107. Friedberg, E. C., King, J. J. 1969. *Biochem. Biophys. Res. Commun.* 37:646–51
108. Sekiguchi, M., Yasuda, S., Okubo, S., Nakayama, H., Shimada, K., Takagi, Y. 1970. *J. Mol. Biol.* 47:231–42
109. Yasuda, S., Sekiguchi, Y. 1970. *J. Mol. Biol.* 47:243–55
110. Yasuda, S., Sekiguchi, Y. 1970. *Proc. Nat. Acad. Sci. USA* 67:1839–45
111. Friedberg, E. C., King, J. J. 1971. *J. Bacteriol.* 106:500–7
112. Friedberg, E. C., Minton, K., Durphy, M., Clayton, D. A. See Ref. 77
113. Berger, H., Benz, W. See Ref. 77
114. Benz, W., Berger, H. 1973. *Genetics* 73:1–11
115. Shames, R., Lorkiewicz, Z., Kozinski, A. 1973. *J. Virol.* 12:1–8
116. Shames, R., Kozinski, A. 1974. *Fed. Proc.* 33:1493 (Abstr.)
117. Van Lanker, J., Tomura, T. 1974. *Biochim. Biophys. Acta* 353:99–114
118. Duncan, J., Slor, H., H., Cook, K., Friedberg, E. C. See Ref. 77
119. Lytle, C. D., Luis, S. P., Bockstahler, L. E., Benane, S. G. 1973. *Biophys. Soc.* FPM-G8 (Abstr.)
120. Ganesan, A. K. 1973. *Proc. Nat. Acad. Sci. USA* 70:2753–56
121. Patterson, M. C., Hohman, P. H. M., Slayter, M. L. 1973. *Mutat. Res.* 19:245–56
122. Seeberg, E., Rupp, W. D. See Ref. 77
123. Richardson, C. C., Kornberg, A. 1964. *J. Biol. Chem.* 239:242–50
124. Becker, A., Hurwitz, J. 1967. *J. Biol. Chem.* 242:936–50
125. Milcarek, C., Weiss, B. 1972. *J. Mol. Biol.* 68:303–18
126. Nakayama, H., Okubo, S., Takagi, Y. 1971. *Biochim.Biophys. Acta* 228:67–82
127. Onshima, S., Sekiguchi, M. 1972. *Biochem. Biophys. Res. Commun.* 47:1126–32
128. Englund, P. T. et al 1968. *Cold Spring Harbor Symp. Quant. Biol.* 33:1–9
129. Hamilton, L. 1974. Unpublished data
130. Brutlag, D., Kornberg, A. 1972. *J. Biol. Chem.* 247:241–48
131. Drake, J. W., Abbey, E. F. 1968. *Cold Spring Harbor Symp. Quant. Biol.* 33:339–44
132. Muzyczka, N., Poland, R. L., Bessman, M. J. 1972. *J. Biol. Chem.* 247:7116–22
133. Hall, Z. W., Lehman, I. R. 1968. *J. Mol. Biol.* 36:321–33
134. Zimmerman, B. K. 1966. *J. Biol. Chem.* 241:2035–41
135. Litman, R. M. 1970. *Biochem. Biophys. Res. Commun.* 41:91–98
136. Kelly, R. B., Cozzarelli, N. R., Deutscher, M. P., Lehman, I. R., Kornberg, A. 1970. *J. Biol. Chem.* 245:39–45
137. Dumas, L. B., Darby, G., Sinsheimer, R. L. 1971. *Biochim. Biophys. Acta* 228:407–22
138. Wu, R. 1970. *J. Mol. Biol.* 51:501–21
139. Cozzarelli, N. R., Kelly, R. B., Kornberg, A. 1969. *J. Mol. Biol.* 45:513–31
140. Deutscher, M. P., Kornberg, A. 1969. *J. Biol. Chem.* 244:3029–37
141. Kelly, R. B., Atkinson, M. R., Huberman, J. A., Kornberg, A. 1969. *Nature* 224:495–501
142. deLucia, P., Cairns, J. 1969. *Nature* 224:1164–66
143. Gross, J. D., Gross, M. 1969. *Nature* 224:1166–68
144. Boyle, J. M., Patterson, M. C., Setlow, R. B. 1970. *Nature* 226:708–10
145. Lehman, I. R., Chien, J. R. 1973. *J. Biol. Chem.* 248:7717–23
146. Cooper, P. K., Hanawalt, P. C. 1972. *Proc. Nat. Acad. Sci. USA* 69:1156–60
147. Friedberg, E., Lehman, I. R. See Ref. 77
148. Konrad, E. B., Lehman, I. R. 1974. *Proc. Nat. Acad. Sci. USA* 71:2048–51
149. Olivera, B.,Bonhoeffer, F. 1974. *Nature* 250:513–14
150. Glickman, B. W., van Sluis, C. A., Heijneker, H. L., Rörsch, A. 1973. *Mol. Gen. Genet.* 124:69–82

151. Glickman, B. W. See Ref. 77
152. Livingston, D. M., Richardson, C. C. 1975. *J. Biol. Chem.* In press
153. Kaplan, J. C. 1971. *UV-Exonuclease: purification, properties and specificity of the enzyme from M. luteus.* PhD thesis. Brandeis Univ., Waltham, Mass. 159 pp.
154. Chase, J., Richardson, C. C. 1974. *J. Biol. Chem.* 249:4553–61
155. Sekiguchi, M., Shimizu, K., Sato, K., Yasuda, S., Onshima, S. See Ref. 77
156. Lindahl, T. 1971. *Eur. J. Biochem.* 18:407–14
157. Lindahl, T. 1971. *Eur. J. Biochem.* 18:415–21
158. Doniger, J., Grossman, L. 1974. Unpublished data
159. Grossman, L., Kaplan, J. C., Kushner, J. C., Mahler, I. 1968. *Cold Spring Harbor Symp. Quant. Biol.* 33:229–34
160. Garvik, B., Grossman, L. 1974. Unpublished data
161. Ghangas, G. S., Wu, R. 1975. *J. Biol. Chem.* In press
162. Pearson, M., Johns, H. E. 1966. *J. Mol. Biol.* 20:215–29
163. Chase, J., Richardson, C. C. 1974. Personal communication
164. Pettijohn, D. F., Hanawalt, P. C. 1964. *J. Mol. Biol.* 9:395–410
165. Youngs, D. A., van der Schuren, E., Smith, K. C. 1974. *J. Bacteriol.* 117:717–25
165a. Gefter, M. L. 1975. *Ann. Rev. Biochem.* 44:45–78
166. Cooper, P. K., Hanawalt, P. C. 1972. *J. Mol. Biol.* 67:1–10
167. Masker, W. E., Hanawalt, P. C., Shizuya, H. 1973. *Nature New Biol.* 244:242–43
168. Youngs, D. A., Smith, K. C. 1973. *Nature New Biol.* 244:240–41
169. Masker, W. E., Hanawalt, P. C. 1973. *Proc. Nat. Acad. Sci. USA* 70:129–33
170. Heijneker, H. L., Pannekoek, H., Oosterbaan, R. A., Pouwels, P. H., Bron, S., Arwert, F., Venema, G. 1971. *Proc. Nat. Acad. Sci. USA* 68;2967–71
171. Wu, R., Kaiser, A. D. 1968. *J. Mol. Biol.* 35:523–37
172. Wickner, R. B., Ginsberg, B., Berkower, I., Hurwitz, J. 1972. *J. Biol. Chem.* 247:489–97
173. Gefter, M. L. 1974. *Progr. Nucl. Acid Res.* 14:101–15
174. Goldmark, P. J., Linn, S. 1972. *J. Biol. Chem.* 247:1849–60
175. Karu, A. E., Mackay, V., Goldmark, P. J., Linn, S. 1973. *J. Biol. Chem.* 248:4874–84
176. Linn, S., Mackay, V. See Ref. 77
177. Mackay, V., Linn, S. 1974. *J. Biol. Chem.* 249:4286–94
178. Mackay, V., Linn, S. 1974. Personal communication
179. Molineux, I. J., Gefter, M. L. 1974. *Proc. Nat. Acad. Sci. USA.* 71:3858–62
180. Robbins, J. H., Kraemer, K. H., Lutzner, M. A., Festoff, B. W., Coon, H. G. 1974. *Annals Intern. Med.* 80:221–48
181. Kraemer, K. H., Coon, H. G., Robbins, J. H. 1973. *J. Cell Biol.* 59:176a (Abstr.)
182. Cleaver, J. E. 1973. *Advances in Radiation Biology,* ed. J. T. Lett, H. Adler, M. Zelle, 1–75. New York: Academic
183. Baldy, M. W. 1970. *Virology* 40:272–87

DNA REPLICATION

×874

Malcolm L. Gefter

Department of Biology, Massachusetts Institute of Technology, Cambridge,
Massachusetts 02139

CONTENTS

INTRODUCTION

It is the intention of this article to present the current status of the problem of
DNA replication. The scope of the problem is so great that a comprehensive
review of all literature in this field would fill several volumes. For this reason,
most of the literature cited deals with bacterial systems. An attempt is made to

45

compare and contrast the bacterial systems with those of higher organisms and to develop unifying concepts where possible.

The reader is referred to past reviews (1, 2) for most of the background material. Literature prior to 1971 is included only occasionally for the sake of continuity.

The problem of DNA replication has been attacked by many different approaches, using genetic, physiological, physical, and biochemical methods. The problem of control of the process aside, genetic analysis of DNA replication in *Escherichia coli* has shown that the overall process requires more proteins than biochemical analysis in vitro has produced.

In summary, analysis of temperature-sensitive mutants of *E. coli* has shown that the products of the *dnaA, dnaB, dnaC-D, dnaE,* and *dnaG* are required for the overall process (3). The *dnaA* and *dnaC-D* gene products appear to be required for initiation of new rounds of replication, whereas the products of the *dnaB, dnaE,* and *dnaG* genes are continuously required for chromosome duplication (or replication fork movement). Several different selective and screening techniques have been used to isolate these mutants, and of several hundred such mutants identified, all seem to fall into the above classes (4).[1] Biochemical analysis of DNA duplication in vitro has revealed several proteins (to be described below) that have not yet been identified by genetic analysis. In addition, several mutants of *E. coli* have been isolated that interfere with plasmid or bacteriophage replication but do not appear to affect chromosome replication (5). As will become apparent from the discussion to follow, different self-replicating systems even within *E. coli* may use alternate pathways to achieve duplication of their DNA. Various bacteriophages growing in *E. coli* may use the same mechanism for DNA duplication as that of the bacterial chromosome but may substitute phage-directed proteins for bacterial proteins. To discuss these events, the overall process will be divided into distinct stages: initiation of replication (chromosome initiation), initiation of single DNA chains (initiation required for DNA chain synthesis), elongation of DNA chains, propagation of the replication fork (events required for the simultaneous elongation of both daughter strands), and termination (events required for the separation of daughter chromosomes).

INITIATION OF REPLICATION AND DIRECTION OF REPLICATION FORK MOVEMENT

The replicon model (6) predicted that an autonomously replicating element would initiate replication at a unique site. From a wide variety of experiments using several independent approaches in different systems, it seems certain that this prediction has been borne out. Many such experiments yielded not only the origin of replication but also the direction of the replication fork movement.

In *E. coli,* the origin of replication has been placed at about 70 min on the genetic map (7, 8). In one approach (7), bacteriophage Mu was integrated into

---

[1] *dnaH* = initiation type recently described.

several different locations on the chromosome. Bacteriophage λ was integrated at its normal attachment site. The amount of Mu DNA relative to λ DNA (measured by DNA : DNA hybridization) in exponentially growing cultures was found to be highest when Mu was integrated close to, but clockwise or counterclockwise from, the *ilv* marker at 74 min. A decrease in the amount of Mu DNA was found as the integration site of Mu was moved towards a point 180° away from *ilv*. From this experiment, it was concluded that the origin of replication is near *ilv* and that there are two replication forks moving in opposite directions at equal speed, converging at about 25 min on the genetic map. The same result was obtained when replication was monitored in cells synchronized by amino acid starvation for DNA replication. Further evidence for bidirectional replication in *E. coli* has come from experiments in which mutagenesis, selective for genes at the replication fork, was employed (9). The use of electron microscopic autoradiography has independently shown replication to be bidirectional (10). Replication has been shown to be bidirectional in *Bacillus subtilis* by both electron microscopy (11) and marker frequency measurements (12). The chromosome of *Salmonella typhimurium* also replicates bidirectionally (13).

Several *E. coli* phages have unique origins of replication. Electron microscopy of partially denatured molecules in the act of replication has shown that bacteriophage λ has a unique site for initiation and that the replication fork moves in both directions from the origin. Most molecules have two forks moving in opposite directions (14). The origin of replication is located close to those genes whose products are required for initiation, and its location is the same as that determined for the origin by genetic analysis (15). Genetic experiments with phage T4 suggest a unique origin of replication close to those genes required for phage DNA synthesis (1, 16).

Electron microscopic evidence obtained from replicating bacteriophage T7 molecules has shown that it too has a unique origin of replication and that two replication forks move in opposite directions (17).

In mammalian cells, electron microscopic autoradiography has shown that replication of individual segments of chromosomes occurs in a bidirectional fashion (18). In addition, it appears that those sites that initiate duplication in one round of replication are the same sites used again in a second round (19). Although there are many sites of initiation in a mammalian chromosome, those sites appear to be uniquely determined, and fork movement can occur in both directions from that site. Similarly, the mammalian virus, Simian Virus 40, has a unique origin of replication, and synthesis proceeds bidirectionally from the origin. This result was obtained by two independent methods. Cells containing replicating SV40 were pulse-labeled for increasing amounts of time. Completely replicated molecules were then isolated and digested with a restriction endonuclease. Of the fragments produced, a gradient of radioactive content was found such that it could be deduced that relative to the physical map of SV40, termination occurred at a unique site and that this site was approached from two directions simultaneously. The minimum of radioactive content was presumed to be at the origin (20). Electron microscopic analysis of replicating SV40 molecules that had been digested with a restriction endonuclease producing only a single break per molecule yielded structures that were consistent with the duplication process beginning at a unique site and proceeding bidirectionally (21). Although the use of a unique origin appears to be universal, replication is not always bidirectional.

By utilizing techniques similar to those used to establish the bidirectionality of phage λ DNA replication, it was found that *E. coli* bacteriophages P2 and 186 replicate unidirectionally (22). Another exception has been found in the mitochondrial DNA of mouse LD cells. By observing replicating molecules that are head-to-tail dimers of normal mitochondrial DNA such that they contain two origins 180° apart, it was shown that each origin gives rise to only one replication fork (23).

As discussed below, at the level of deoxynucleotide addition, bidirectional replication does not pose any particular problems not also encountered when considering a unidirectional mode of replication. The fact that most replicating systems appear to duplicate by a bidirectional mode may reflect a symmetry in structure at the origin. Bidirectional replication has the virtue that the origin is completely duplicated before replication is completed. This mechanism allows for control of the process by elements that may interact specifically with the site of origin. Furthermore, chromosomes can begin to segregate without the need for completion of the entire process.

Control of Initiation of Replication

Control over the initiation process in DNA replication is likely to be the major or perhaps the only mechanism for determination of the DNA content of a bacterial cell (24). The products of two genes, *dnaA* and *dnaC-D*, are apparently involved exclusively in the process of initiation. Mutants temperature sensitive in either of these two genes, when transferred to the nonpermissive temperature, will continue ongoing DNA replication but will not allow new rounds to initiate (1) (this is in contrast to temperature-sensitive mutants in genes *dnaB*, *dnaE*, or *dnaG* that result in the immediate cessation of replication on transfer to the nonpermissive temperature). Mutations in the *dnaC-D* gene are reversible in that synthesis halted by high temperature resumes on return of the culture to the permissive temperature (25).

Two reports have suggested that initiation of replication is under negative control and that the presumptive repressor of initiation must be inactivated (26). The replication of the Col E1 plasmid in *E. coli* will cease if it resides in a *dnaA* temperature-sensitive mutant at the nonpermissive temperature. This requirement for the *dnaA* gene product for Col E1 replication can be obviated if the cells are treated with chloramphenicol at the time of the shift to high temperature. Thus, it has been concluded that chloramphenicol blocks the synthesis of the presumptive repressor and that the *dnaA* gene product (the anti-repressor) is no longer essential (27). The process of controlling initiation is clearly complex, and one must be cautious about interpretations based on temperature-sensitive mutants alone. For example, the assembly of a "replicative complex" (a complex of those proteins and other elements analogous to a ribosome that may catalyze replication) may be needed at the beginning of each round of synthesis such that mutations leading to temperature sensitivity of initiation may be temperature sensitive in the assembly process and not in their functional process.

The complexity of the initiation process is shown by several observations. When DNA synthesis is arrested by raising a *dnaB* mutant to the nonpermissive temperature and then allowed to continue by lowering the temperature, the replication fork resumes replication again, but in addition, a new round of synthesis initiates at the origin of replication (28). Similarly, interruption for ongoing replication by thymine deprivation or by conjugation results in the premature reinitiation of replication (29). A comprehensive theoretical treatment of the control of initiation has recently been presented (30).

The Role of Membranes

Aside from the role of proteins in the initiation process, much attention has been paid to the apparent role of the membrane in the initiation process. For example, in *B. subtilis* markers mapping close to the origin of replication can be recovered from that portion of the DNA that sediments rapidly on gentle lysis of the cells (31). Similarly, active DNA synthesis in vitro of bacteriophage T7-infected cells appears to take place only in rapidly sedimenting fractions of the cell lysates (32, 33). Throughout most of the literature, rapid sedimentation of DNA (i.e. with the membrane fraction) has been interpreted as a physical or chemical association of that DNA with the membrane; it is not clear if such conclusions are warranted. To date, no reports have described the chemistry of a DNA-membrane interaction.

It has been suggested by a more direct approach that phospholipids do play a role in the initiation process. *E. coli* treated with a fatty acid analog accumulate saturated fatty acids in their membranes. In addition to other effects on DNA synthesis, cells harboring temperature-sensitive *dnaA* or *dnaC* mutations thus treated become more temperature sensitive. This enhancement in temperature sensitivity can be reversed by adding unsaturated fatty acids to the cells (34). Most genetic or physiological experiments involving alteration of the cell membrane are subject to criticism based on possible alterations of other cellular processes that could secondarily affect DNA synthesis. Thus, it is likely that the role of the membrane in DNA synthesis will remain open until evidence from in vitro studies is obtained that shows either a chemical linkage between DNA and a membrane or that a membrane or lipid-like substance is required for a biochemical reaction involving DNA synthesis.

Transcription Requirement

At the biochemical level, two basic observations have suggested that transcription of DNA into RNA is needed at or close to the origin of replication in order to initiate DNA replication. A repressed λ prophage will not commence DNA replication even though all the gene products required for λ replication are supplied. Mutants of λ that can replicate under these conditions are those that permit transcription to occur through the replication origin site. Transcription per se (not translation) is both necessary and sufficient (35). A similar conclusion was reached by studying the requirements for chromosome initiation in *E. coli*. Just prior to the onset of a new round of replication, there is a time interval in which an active RNA polymerase (as judged by sensitivity to rifampicin and not chloramphenicol) is required but protein synthesis is not. These results suggest that transcription per se is necessary for the initiation of a new round of replication (36), but the specific requirement in the initiation process is not clear. Must a RNA molecule be synthesized to participate in the initiation event either as a primer for DNA chains (to be discussed in detail below) or as a structural element for the replication machinery? Is the presence of the active enzyme at the origin site sufficient (perhaps for unwinding of template strands)? Although the former seems the most plausible, the answer must await demonstration of the initiation process on double stranded DNA in vitro such that the role of the transcription event can be ascertained.

Propagation of DNA Replication

REQUIREMENT FOR REPEATED INITIATION OF DNA CHAINS   We have defined above that movement of the replication fork leads to the duplication of DNA; i.e. that both parental strands are duplicated simultaneously as the replication fork moves. Since the two parental strands are antiparallel, duplication of both strands would require laying down new DNA strands in both the $5' \to 3'$ direction and the $3' \to 5'$ direction. Every polynucleotide polymerase discovered so far will catalyze synthesis in only the $5' \to 3'$ direction. Thus, only one strand can be continuously elongated (the continuous strand) as the replication fork progresses. Experiments initiated by Okazaki and collaborators have shown that the opposite daughter strand is obligatorily layed down in a discontinuous fashion, i.e. that short fragments, ("Okazaki fragments") are continuously synthesized in the $5' \to 3'$ direction, a direction opposite to the movement of the replication fork, and then subsequently joined together to reconstitute the newly made chromosome (37). The overall scheme is depicted in Figure 1; strand I is the continuous strand and strand II is the discontinuous one. This mechanism has been deduced from experiments in which growing *E. coli* cells are pulse-labeled for very short times, on the order of 0.5% of a generation time, and then the size of the newly incorporated radioactivity is determined by sedimentation analysis. This type of analysis has shown that at least 50% of the incorporated radioactivity sediments slowly (about 10S), and that as the pulse time is increased, a significant fraction of the radioactivity sediments as high molecular weight DNA. The conclusion is drawn that at least one and perhaps both strands are layed down by a discontinuous mechanism. Similar results have been obtained with several bacterial strains, *E. coli* bacteriophages, mammalian cells, and mammalian viruses.

In order to prove that the short fragments are intermediates in the synthesis of high molecular weight DNA, a precursor-product relationship must be shown between the two species of DNA. Both in bacterial and mammalian cells, this proof has not been easy to obtain. The main problem is that the precursor pools do not equilibrate during the labeling period, even a 1000-fold dilution of the specific activity of the precursor in the medium is followed by significantly more incorporation into the DNA (38). In fact, it has been shown that if *E. coli* cells are pulse-labeled with different precursors, not all of the radioactivity incorporated into Okazaki fragments flows directly into high molecular weight DNA (39). Therefore, some fraction of the short fragments may turn over and may not be immediate precursors to high molecular weight DNA. This problem has confounded many studies on the influence of mutations or physiological changes on the discontinuous nature of DNA replication. In *B. subtilis,* it appears that if cells are pulse-labeled with thymine and not thymidine, fragments much shorter than 10S are observed as the immediate precursors to higher molecular weight DNA (40). These results taken together suggest a requirement for additional information on a particular replicating system before one can conclude that DNA synthesis is discontinuous and that the short fragments are derived from one or both daughter DNA strands.

If both strands are layed down in a discontinuous fashion, then one would expect that, on the average, those pieces being layed down in the same direction

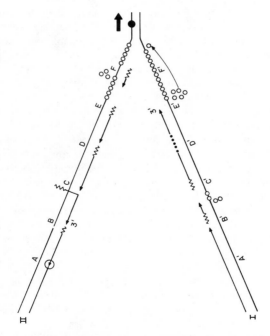

Figure 1 Summary of events occurring at the replication fork. The bold arrow indicates the direction of fork movement. Strand I is the continuous strand and strand II is the discontinuous strand. Sites A–F and A'–F' represent sites for initiation of Okazaki fragments. Solid lines are DNA, jagged and dotted lines are RNA. The open circles are DNA unwinding protein molecules and the solid circle is a superhelix relaxation protein (ω-protein).

as the fork movement (the potential continuous strand) would be larger than those derived from the opposite strand, since in the former case these pieces can be joined to each other before synthesis of the youngest one is completed; this is not possible on the strand being layed down in a retrograde fashion (see Figure 1). A fragment layed down in the interval $B'–C'$ is joined to A' before it has completed synthesis. The fragment layed down in $E–F$ must be completed before it can join to higher molecular weight DNA.

In an in vitro system derived from E. coli, two classes of short fragments have been seen, one class of 10S and the other of about 35S (41). It was further shown that these two classes of molecules indeed represent DNA products synthesized from opposite parental strands. Both classes will not anneal with themselves but will anneal with each other (42). In an experiment performed on growing cells in which bacteriophage λ was integrated, it was shown that the 10S fragments isolated from a logarithmically growing culture would anneal to only one strand of isolated λ DNA, and a class of about 35S annealed exclusively to the opposite strand. Knowing the orientation of the integrated DNA and the direction of movement of the replication fork, it could be concluded, as expected from the above considerations, that the short fragments were derived exclusively from the discontinuous strand and the larger fragments from the continuous strand (43). These results suggest that both strands of the E. coli chromosome are synthesized in a discontinuous fashion. The failure to obtain 10S fragments from both strands probably reflects both the high rate of joining in vivo and a competition on the continuous strand for elongation relative to initiation of fragments.

Discontinuous replication has been observed in many replicating systems, and it has been inferred from observing the percentage of rapidly labeled DNA that both strands are synthesized in a discontinuous fashion (a conclusion not necessarily justified from these data alone, considering possible turnover of fragments and an efficiency of labeling both strands). Similarly, hybridization of short fragments to each of the separated strands of a replicating DNA does not prove that both strands are synthesized discontinuously since bidirectional replication would yield these results regardless. It has been proved both in the case of replicating SV40 DNA in vivo (44, 45) and in replicating polyoma DNA in vitro (46) that both strands are synthesized discontinuously. This was shown by the ability of the fragments to self anneal.

RNA SYNTHESIS CAN INITIATE DNA CHAINS   Given the discontinuous nature of DNA chain growth, new DNA chains must initiate as the replication fork progresses. This type of chain initiation may or may not be related to the requirements for chromosome initiation, a point to which we will return later.

Of all the template-directed DNA polymerizing enzymes studied to date, none has been found capable of initiating DNA chains de novo. All these enzymes require a 3′ OH end of a pre-existing polynucleotide chain (a primer) to which deoxynucleotides are added. This requirement is in contrast to RNA polymerizing enzymes that are able to initiate chains de novo. One solution to the problem of DNA chain initiation resulted from studies on the requirements for conversion of single stranded circular DNA of bacteriophage M13 to a double stranded or replicative form (RF) in vivo (47). These studies showed that infecting M13 DNA could be converted to its replicative form if the cells were treated with chloramphenicol to block protein synthesis. However, inhibition of the cellular RNA polymerase by rifampicin abolished the appearance of RF molecules. Using mutants in which the cellular RNA polymerase was resistant to rifampicin, the cells efficiently converted infecting DNA to its double stranded form.

It was further shown in in vitro studies (48) that RNA synthesis is essential for RF synthesis and that the newly made DNA chain is covalently attached to RNA. The in vitro system capable of synthesizing RF molecules from M13 DNA will be discussed in more detail later. Extending these studies to another bacteriophage ϕX174, which is of the same size as M13 and is also single stranded and circular, showed that the initiation of new DNA chains is not necessarily dependent on RNA polymerase. The conversion of infecting ϕX174 DNA to its replicative form is not abolished by the addition of rifampicin. However, in both the M13 and ϕX174 systems in vitro, evidence was obtained for covalent attachment of DNA chains to RNA chains (49). Since continued replication of the E. coli chromosome is also not affected by rifampicin, it is clear that although RNA may serve as a primer for DNA chain synthesis, the RNA need not be transcribed from DNA by RNA polymerase. This raises the possibility that chromosome initiation which is sensitive to rifampicin may reflect the need for transcription of a primer and that Okazaki fragment synthesis may be primed by a different mechanism.

Evidence has been obtained that Okazaki fragments are initiated with RNA (50). The basic observation is that fragments pulse-labeled for very short times have a characteristic buoyant density in Cs_2SO_4 gradients slightly heavier than single stranded DNA. On treatment with alkali, their density is reduced to that of single stranded DNA, and in vivo there appears to be a rapid removal of the presumptive RNA prior to the completion of their elongation. The density of the fragments is greatest when they are about 4S but becomes characteristic of single stranded DNA when they grow to 10S (the size of the RNA fragment was estimated as 50–100 nucleotides). Further analysis of the RNA-containing fragments has shown that the RNA is on the 5′ end of the fragment and not at the 3′ end or at an internal position. Alkaline hydrolysis does not alter their size, and the fragment is digestible by E. coli exonuclease I, a DNase specific for 3′ hydroxyl termini. The linkage between the polyribonucleotide and the polydeoxyribonucleotide occurs between a ribopyrimidine and a deoxycytidylate residue specifically (51).

Evidence for RNA priming of DNA chains has also been obtained in eukaryotes. In the slime mold Physarum polycephalum, pulse-labeled DNA appears to be attached to RNA (52). In polyoma virus-infected cells, short, polyoma DNA-specific fragments isolated from replicating molecules have the characteristics of DNA-RNA polymers as judged by their buoyant density in Cs_2SO_4 (53). The most definitive evidence obtained so far for RNA priming of nascent DNA chains has come from the analysis of polyoma virus replication in isolated nuclei of infected cells. In short, such nuclear preparations continue replicating polyoma virus DNA in vitro, resulting in the accumulation of synthetic double stranded closed circular molecules (form I DNA). RNA attached to nascent DNA chains has been shown and, following isolation of these molecules from replicating intermediates, the RNA has been analyzed in detail. The RNA linked to DNA itself initiates specifically with a purine ribonucleoside triphosphate. The linkage between ribo- and deoxynucleotide is random with respect to the ribonucleotide and relatively specific for the deoxyribonucleotide in that all four ribonucleotides are found adjacent to a deoxycytidylate residue. The occurrence of deoxycytidylate at the RNA-DNA link was also found in *E. coli* as previously mentioned. In addition, the sequence of this RNA (approximately 10 nucleotides) is not a unique one (54), a result consistent with the assumption that primer synthesis is transcribed from DNA at sites of different sequence and that perhaps the entire process is random. An RNA synthetic event might occur which is terminated spontaneously after limited synthesis or by the addition of deoxycytidylate to its 3′ terminus. The availability of single stranded template as the replication fork progresses may dictate the frequency of RNA initiations. The validity of this assumption remains to be elucidated. It is of interest to note that the transcription event is not sensitive to α-amanitin, a compound that inhibits the RNA polymerase apparently responsible for messenger RNA synthesis in mammalian cells (55).

INITIATION OF REPLICATION IN VITRO   The description of initiation in vitro has so far been limited to the DNA chain initiation. As mentioned above, it is not clear whether this process is analogous to chromosome initiation. Clearly, at least the requirement for transcription is different. The establishment in vitro of a system capable of initiating replication on double stranded DNA has proven to be difficult. Most systems that reflect "true" replication in vitro do not initiate but rather propagate pre-existing replication forks. A possible requirement for a transcriptional event may in part be responsible for this difficulty. Circular molecules isolated from cells or viruses may not have the appropriate super helical density that would promote such a transcription event (56).

A system developed by Bonhoeffer and collaborators (57) (described in detail later) supports bacteriophage λ DNA synthesis. Proteins that are products of genes O and P of λ that are necessary for λ replication are required in this system, and thus their purification is possible using complementation of extracts (58). Another system developed by Kornberg and collaborators (48) will also support λ DNA synthesis dependent on the gene O and gene P products (59). In contrast to the former system, the latter is dependent on exogenous λ DNA for synthesis. This system too has the potential for isolation of those proteins required for chromosome initiation (59). Bacteriophage T7 DNA can also promote DNA synthesis in similar extracts prepared from phage T7-infected cells (60, 61). Even though DNA synthesis is dependent on the gene products of genes 4 and 5 (those required for DNA synthesis in vivo), initiation has not been shown to take place at the origin seen in vivo. The system does, however, clearly show initiation of double stranded DNA synthesis; the product is not covalently attached to its double stranded template (61).

Initiation and complete replication of double stranded circular DNA have recently been obtained in vitro (62). Extracts prepared from cells harboring the colicinogenic factor, Col E1 (a double stranded circular DNA), though not dependent on exogenous DNA, do initiate Col E1 DNA replication and complete replication of this DNA, giving rise to closed circular double stranded product. RNA polymerase appears to be involved in the initiation process in that rifampicin will inhibit initiation but not propagation. Synthesis is discontinuous and the short fragments produced appear to be linked to RNA by density analysis (63). To date, this replicating system seems to be the most promising in terms of understanding the biochemistry of chromosome initiation.

The initial discoveries, described in the previous section, on the role of RNA in DNA synthesis promoted investigations in several different systems for RNA linked to DNA. RNA has been found incorporated covalently into closed circular DNA of mitochondria in HeLa cells (64) and mouse L cells (65). There appears to be at least one segment of about 10 ribonucleotides per DNA molecule. A similar finding was obtained with Col E1 DNA replicated in the presence of chloramphenicol (66). In the latter case, the RNA segment was found to be about 25 nucleotides in length. The RNA could be found in either of the two strands of the DNA with equal probability. Although not as well characterized, RNA has also been found in the DNA of bacteriophage T4 (67, 68).

In many investigations, the presence of RNA in DNA chains was based exclusively on the buoyant density of DNA in Cs_2SO_4 density gradients, in that DNA was found at a slightly heavier density than added marker DNA. The density of the experimental DNA was then shown to shift to that of the marker following RNase or alkali treatment (for example, see 69). It has been shown in our laboratory (L. Sherman) and in the laboratory of Goulian (J. Mendelsohn and M. Goulian, personal communication) that aggregation of RNA to DNA can occur in Cs_2SO_4 gradients and cause the DNA to which the RNA is bound to alter its density. Thus, all such experiments based solely upon density shifts and concluding that RNA is covalently linked to DNA should be viewed with a high degree of caution.

In those cases where it has been proven that RNA is integrated into circular DNA, a primer role for that RNA still must be questioned. If the RNA served as a primer for a circular single stranded DNA, then it must be located on the 5' end of the synthetic DNA. When synthesis proceeds around the circle, the 3' end of the DNA chain approaches the 5' end RNA. Joining of the DNA to the RNA in that configuration cannot be catalyzed by the polynucleotide ligase of E. coli (70). Since this reaction is catalyzed by the T4-induced DNA ligase (70, 71), perhaps the appropriate conditions can be found that will enable the E. coli enzyme to catalyze this reaction. It should be considered that if the RNA is transcribed by RNA polymerase, its 5' terminus bears a triphosphate moiety, and perhaps this activated terminus can serve as a substrate in the E. coli ligase reaction or in a joining reaction catalyzed by an enzyme yet to be described.

An alternative explanation for the occurrence of RNA in a DNA polymer is that ribonucleotides are added by RNA polymerase to a growing DNA chain, a reaction demonstrable in vitro (72). A DNA polymerase present in the reaction or in vivo can then add deoxynucleotides onto the ribonucleotide chain (48, 73–76). This reaction is possible with all known *E. coli* DNA polymerases. Further synthesis by a DNA polymerase can thus lead to a DNA chain containing a covalently joined RNA segment (77).

ELONGATION OF DNA CHAINS

The DNA Polymerizing Enzymes

The basic catalytic properties of the three *E. coli* DNA polymerases have been reviewed (78, 79); therefore, only a summary is presented here. Particular attention is paid to their apparent biological functions.

The basic reaction catalyzed by all known DNA polymerases is the DNA template-directed condensation of deoxyribonucleoside triphosphates. Synthesis proceeds from the 3' OH end of a primer and the chain grows in the 5' → 3' direction with release of one pyrophosphate moiety for each deoxynucleotide incorporated. **The primary differences between these enzymes reside in the speed of catalysis and in nuclease activities associated with the polymerases.**

DNA Polymerase I of Escherichia coli

There are approximately 400 molecules of this enzyme per *E. coli* cell; the enzyme is a single polypeptide chain of 109,000 daltons and appears to contain one atom of zinc per chain (80). The protein will bind to single stranded DNA as well as phosphodiester bond interruptions ("nicks") in double stranded DNA. In addition to its basic DNA synthetic activity, the enzyme will catalyze exonucleolytic degradation of single stranded DNA in the 3' → 5' direction (78).

It seems certain that the 3' → 5' exonuclease activity serves a "proofreading" function in that incorrectly incorporated deoxynucleotides will be removed from the growing chain by this activity prior to continuation of chain elongation (81). This "editing" activity has also been demonstrated in the phage T4-induced DNA polymerase (82) and *E. coli* DNA polymerases II and III (79). In the case of the phage T4-induced enzyme, it has been shown in vivo (83) and in vitro (84) that alteration of the 3' → 5' exonuclease activity affects the rate of misincorporation of deoxynucleotides. Enzymes having low activities of nuclease relative to polymerase have a mutator phenotype in vivo, and those having a high nuclease-to-polymerase ratio have an antimutator phenotype.

The 5' → 3' exonuclease activity of DNA polymerase I has the ability to excise thymine dimers in DNA (see Grossman 84a) as well as ribonucleotides present on the 5' end of a DNA chain (85). This latter activity might be extremely important when considering the possibility of RNA being on the 5' ends of DNA chains at the origin of replication and perhaps at the ends of each Okazaki fragment. The 5' → 3' exonuclease activity can digest the RNA fragment of a previously synthesized Okazaki fragment while simultaneously filling in the created gap with deoxynucleotides (see Figure 1).

It appears from the study of various mutants that DNA polymerase I performs this function in vivo. However, it is not completely clear that this enzyme is indispensable for DNA replication. Mutations in the structural gene for DNA polymerase I (*polA*) (86) can alter the polymerase activity without affecting the $5' \rightarrow 3'$ exonuclease activity (87). Similarly, there is a mutant with an altered $5' \rightarrow 3'$ exonuclease activity that retains the polymerization activity (88). Neither of these mutations is lethal; however, there are physiological consequences to the cell harboring these mutations. Such cells are sensitive to ultraviolet light and appear to join Okazaki fragments at a slower rate than wild-type cells do (1). One must raise the possibility, however, of leakiness of these mutations in vivo (87). One "assay" for leakiness could be an analysis of a system that appears to require, absolutely, the activity of DNA polymerase I in vivo.

The replication of the plasmid Col E1 does not occur in cells harboring a *polA1* (polymerase deficient only) mutation (89, 90) but does replicate in a mutant deficient in the $5' \rightarrow 3'$ activity (27). Therefore, we might conclude that for plasmid replication the polymerase is essential but the nuclease is not. From the studies of these mutants alone, we might conclude that neither function is essential for chromosome replication. Since there are other DNA polymerizing activities in the cell that can replace the activity of the polymerase and there are RNase-H activities (ribonuclease activity specific for a DNA-RNA hybrid structure) (92) that can replace the $5' \rightarrow 3'$ exonuclease function in RNA removal (93, 94), it is possible that DNA polymerase I is dispensable for DNA replication. The existence of alternate pathways for the synthesis of cellular constituents has precedence in biology. The question of the essential nature of this enzyme in replication may not be particularly relevant since the enzyme certainly appears to function in the capacity of fragment maturation when it is present in the cell (91). More direct evidence has also been obtained that suggests that DNA polymerase I functions in vivo to remove ribonucleotides from DNA. Col E1 DNA will, as mentioned previously, accumulate RNA into its closed circular structure when replication proceeds in the presence of chloramphenicol. RNA accumulates in Col E1 DNA when it replicates in a temperature-sensitive polymerase I mutant, *polA107*, at high temperature (95), indicating that DNA polymerase I is normally responsible for excision of RNA from that DNA.

DNA polymerase I is essential for cell viability. A conditionally lethal mutant has been isolated that is defective in the $5' \rightarrow 3'$ exonuclease function of DNA polymerase I. It also accumulates Okazaki fragments at high temperature (96). What cannot be ascertained is whether the failure to join these fragments is in itself the lethal event or if there is another essential function not being catalyzed. This mutant is not "tight" in the sense that 1% of the cells survive at high temperature in broth and the percentage of survival increases in minimal medium. Recently F. Bonhoeffer and B. M. Olivera (personal communication) have isolated a conditionally lethal mutant mapping in the *polA* gene. Their mutant shows less than $10^{-2}\%$ survival in broth and is defective in both polymerase and nuclease functions in vitro. Thus, this enzyme (regardless of the in vitro activities that survive isolation) is essential for cell viability. It would appear that if the lethal event resides in the failure to join Okazaki fragments (which seems most plausible), then the essential activity of this enzyme is its $5' \rightarrow 3'$ exonuclease function. Since replication of the chromosome (accumulation of daughter DNA strands) takes place in such mutants, the polymerase activity may not be important in this process (91).

Other polymerase I-deficient mutants have been obtained both in *B. subtilis* (97) and in *S. typhimurium* (98). They behave essentially the same as the nonlethal *polA1* type of mutation of *E. coli*.

DNA Polymerase II of Escherichia coli

DNA polymerase II is present at a concentration of about 40 molecules per cell. It is composed of a single polypeptide chain of 90,000 daltons (this is the molecular weight in sodium dodecylsulfate obtained in the author's laboratory). The enzyme does not bind tightly to single stranded DNA (98), nor does it bind to nicks in double stranded DNA. The specific activity (units of polymerization catalysis/mg of protein) is about 10% that of polymerase I. Aside from its DNA polymerizing activity, it contains an associated nuclease, digesting single stranded DNA in the $3' \rightarrow 5'$ direction (79).

Both *E. coli* DNA polymerases II and III display a characteristic not observed with DNA polymerase I. Although the basic synthetic activities are the same as polymerase I, polymerases II and III show a decrease in the initial rate of deoxynucleotide addition to a primer as the template strand becomes larger than about 50 nucleotides (79, 99). It is assumed that secondary structure inherent in a single stranded DNA template inhibits propagation by the polymerase. This is consistent with the finding that these enzymes will not displace a strand annealed to the template during synthesis. In the case of DNA polymerase II, this inhibition of propagation can be overcome by the addition of the *E. coli* DNA "unwinding protein" to the DNA (to be discussed in detail later). Beyond this property, the enzyme, DNA polymerase II, does not appear to have any novel properties.

Analysis of this enzyme activity in vivo suggests that it does not necessarily function in DNA replication but may have some role in the repair of ultraviolet light-induced DNA damage. Two mutants were independently isolated that are deficient in DNA polymerase II activity (101, 102), but there has been no detectable phenotype for either one of these mutants other than polymerase II deficiency in vitro. The parameters examined include the ability of various bacteriophages and plasmids to replicate, ultraviolet light sensitivity recombination frequency, and mutation frequency.

Analysis of the repair capability of cells mutant in DNA polymerase I has shown that there is an ATP-dependent ultraviolet light-stimulated incorporation of deoxynucleotides into DNA (103). Part of this incorporation appears to be specific to the activity of DNA polymerase II (104). Thus, the activity of this enzyme in vivo may be restricted to repair of DNA damage. It has also been shown that in the absence of DNA polymerase I activity, DNA polymerase II can elongate Okazaki fragments, and so this enzyme may be part of an alternate pathway responsible for joining fragments into high molecular weight DNA during the replication process (91).

DNA Polymerase III of Escherichia coli

DNA polymerase III is present in *E. coli* at a concentration of about 10 molecules per cell. A recent report indicates that the enzyme is composed of two polypeptide chains, one of 140,000 daltons and one of 40,000 daltons (100). In addition to its DNA synthetic activity, the enzyme has an associated $3' \rightarrow 5'$ exonuclease activity as well as a $5' \rightarrow 3'$ exonuclease activity (105). One distinguishing characteristic of the DNA polymerase activity is its ability to be stimulated by ethanol (106). Perhaps the most important characteristic of this enzyme is its apparent ability to complex with and interact with other cellular proteins involved in DNA replication systems in vitro. Because of this feature, it is difficult to describe the so-called native form of this enzyme since different methods of isolation or assay result in the procurement of different forms of the same enzyme. As described in detail later, all forms isolated so far show enhanced thermolability compared to the wild-type enzyme when they are isolated from temperature-sensitive *dnaE* mutants. Thus, their basic catalytic protein is the product of the *dnaE* gene of *E. coli*. By this definition, the enzyme activity originally isolated (106, 107) and subsequently purified to near homogeneity (100) will be referred to as DNA polymerase III.

As mentioned previously, DNA polymerase III will not efficiently catalyze addition of deoxynucleotides to a primer (DNA or RNA) if the template strand is long. In contrast to DNA polymerase II, the *E. coli* DNA unwinding protein does not release this inhibition of propagation (77). Two protein factors I and II have been isolated which, when added to DNA polymerase III in the presence of ATP, promote synthesis on long templates (100, 108). A complex form of DNA polymerase, DNA polymerase III*, has also been isolated which will also catalyze synthesis on long templates provided that an additional factor, copolymerase III*, is added (109). It has been suggested (108) that polymerase III* is a complex of DNA polymerase III and factor II and that copolymerase III* and factor I are one and the same. There are several discrepancies in the literature concerning this assumption. It is claimed that copolymerase III* will stimulate only polymerase III* and not polymerases I, II, or III, whereas factor I can stimulate all the *E. coli* polymerases. Both proteins are of the same molecular weight (70,000), and it is reasonable to assume that they are the same.

A more serious discrepancy is that the molecular weight of polymerase III* in sodium dodecylsulfate has been reported as 90,000. If this report is correct (109), polymerase III* is not composed of polymerase III (140,000 and 40,000 daltons) and cannot contain factor II since only a single molecular weight species was observed. That polymerase III* contains factor II is only an assumption based upon the catalytic activity of the combined system. It should be considered that perhaps factor II converts polymerase III into polymerase III* and that the two forms of the enzyme differ in physical state only.

Since the activities and characteristics of polymerase III plus factors I and II are so similar to that of polymerase III* plus copolymerase III*, the assignment of the mass of polymerase III* should be re-evaluated.

In the polymerase III*–copolymerase III* system, the role of ATP in promoting synthesis has been determined. Polymerase III* and copolymerase III* bind to the template primer complex in the presence of ATP and spermidine. The complex can then be isolated and shown to contain ATP and spermidine in addition to the proteins. Addition of deoxynucleoside triphosphates results in the release of the bound ATP as ADP and phosphate concomitant with DNA synthesis. Thus, the ATP is only needed to form the protein-template-primer complex and is not needed for the propagation of the DNA chain. It is also suggested that copolymerase III* is not needed for propagation in that antibody directed against copolymerase III* when added during the propagation step does not inhibit the system (110). A curious feature of this system is that ATP and copolymerase III* are not required at all if the template primer is double stranded DNA containing short gaps. If we assume that the only difference in the template primers is the length of the template, such that when it is long both ATP and copolymerase III* are required but when it is short they are not, then the function of the copolymerase III* and ATP hydrolysis is the induction of a conformational change in polymerase III* that allows it to propagate in spite of secondary structure in the template. Another possibility is that in contrast to polymerase III, polymerase III* will bind to single stranded DNA in an ineffective way. Then the difference between the long and short templates systems is the ratio of single stranded DNA to primer termini. Thus, it is possible that copolymerase III* and ATP alter the affinities for binding to single stranded DNA and primer termini such that the kinetic effects observed are obtained. Clearly, more information is necessary to resolve this question.

In addition to the above properties of polymerase III* (or polymerase III plus factor II), it is also capable of propagation in the presence of the *E. coli* DNA unwinding protein, a property not observed with polymerase III. This point is discussed in a subsequent section.

An additional form of DNA polymerase III that requires ATP for synthesis has been independently isolated (111). There have not been sufficient data reported yet on this form to allow a detailed comparison with those discussed above.

The function of DNA polymerase III in vivo is to catalyze the synthesis of DNA in the replication process. DNA polymerase III is the product of the *dnaE* gene of *E. coli* (112, 113). Cells harboring a temperature-sensitive mutation in the *dnaE* gene halt replication immediately on transfer of the cells to the nonpermissive temperature (91). DNA polymerase III itself is thermolabile when isolated from such mutants. Furthermore, mutations in the DNA polymerase III gene can lead to the enhancement of the spontaneous mutation in vivo (114, 115). As mentioned previously, in addition to the *dnaE* gene product, the propagation of replication in vivo requires the products of the *dnaB* and *dnaG* genes as well. The factors described above which interact with DNA polymerase III are not the products of these genes. It remains to be proven by genetic analysis that these factors function in vivo in the replication process. There are, for example, mutator genes (116) in *E. coli,* and their protein products might be assumed to participate in replication.

To summarize the discussion of the three DNA polymerases of *E. coli*, it can be concluded that their basic DNA synthetic activities at the biochemical level are quite similar. The essential requirements, substrates, products, and the direction of synthesis are the same. DNA polymerase III catalyzes the essential elongation step in replication, and DNA polymerase I is involved in the "maturation" of the duplicated DNA into a high molecular weight structure. DNA polymerase II is probably an auxiliary enzyme that can participate in the maturation process or the repair of induced damage in DNA. The essential features of DNA polymerase III that make it indispensable for replication probably reside in its high rate of catalysis (79, 100) as well as its ability to complex with other proteins required for the overall process of replication. The use of mutants alone to describe the biological role of an enzyme is not adequate since the activities in vivo may differ from those observed in vitro (117).

B. *subtilis* also contains three distinctly different DNA polymerases. The DNA polymerase III of *B. subtilis* is sensitive in vitro to a nucleotide analog that inhibits DNA replication in vivo (118, 119). Mutations to analog resistance result in the appearance of analog resistant enzyme in vitro (120). Furthermore, mutations in the DNA polymerase III gene can result in a "mutator" phenotype (121). In vitro studies (122, 123) have shown that the peculiarities of the three *E. coli* polymerases are shared by the three enzymes in *B. subtilis*. Thus, there appears as in *E. coli* an indispensable replication enzyme (DNA polymerase III) and a major in vitro activity (DNA polymerase I) involved in maturation and repair. Further work is necessary to establish the role of DNA polymerase II of *B. subtilis* and to examine the interactions of polymerase III with other proteins.

Mammalian DNA Polymerases

In all mammalian systems studied so far, at least two separable DNA polymerases have been seen (124): a high molecular weight enzyme found predominantly in cytoplasm and a low molecular weight enzyme confined to the nucleus. The high molecular weight enzyme appears to exist in several physical states in that its sedimentation properties are those of a nonhomologous species (6–8S). The low molecular weight enzyme (40,000–50,000 daltons) has been purified to homogeneity from both calf thymus (126) and human KB cells (127). Both the high and low molecular weight enzymes differ from the prokaryote enzymes in that they do not contain associated exonuclease activities and do not exhibit a deoxynucleoside triphosphate–pyrophosphate exchange reaction (128). The low molecular weight enzyme from KB cells displays a template-dependent nucleoside triphosphate to monophosphate conversion activity, characteristic of an associated $3' \to 5'$ exonuclease activity, but none has been found. The mechanism of this reaction remains to be elucidated.

As mentioned earlier, the $3' \to 5'$ exonuclease activity associated with DNA polymerases acts as an editing function in removing mismatched nucleotides from the primer terminus before commencing elongation. The fact that the low molecular weight DNA polymerase will not remove such a mismatch is consistent with the inability to detect a $3' \to 5'$ exonuclease activity in that enzyme (129).

In view of the unavailability of genetic analysis, it has not yet been possible to assign unequivocally specific biological functions to the mammalian enzymes. Various physiological studies have, however, shed some light on this question. Advantage has been taken of the restriction of DNA replication to the S phase of the cell cycle to monitor the activities of the DNA polymerases. Concomitant with thymidine incorporation into DNA of growing mouse L cells, there is an increase in the activity of the cytoplasmic high molecular weight DNA polymerase, while the activity of the low molecular weight nuclear enzyme remains unchanged (125). A similar result was obtained using synchronized HeLa cells. In that report, the activity of a second nuclear enzyme recently discovered, the r-DNA polymerase (an enzyme that prefers a polyribonucleotide template to a polydeoxynucleotide template), appeared to rise just prior to the onset of DNA replication (130). These results taken together would show that if the in vitro activity of a particular polymerase is an accurate reflection of DNA synthesis in vivo, then the cytoplasmic enzyme is responsible for the replication process.

It should be kept in mind that cytoplasmic means that the enzyme activity can be isolated from that subcellular fraction and not that it functions exclusively in that compartment. It has been reported in the sea urchin system that DNA polymerase activity migrates to the nucleus during S phase (131). Several reports (132, 133) suggest that the low and high molecular weight enzymes may be interconvertible. Thus, given the possibilities of enzyme conversion and migration as well as the inability to localize these enzymes in vivo, the question of which enzyme activity is essential for replication will remain an open one. As described later, in vitro systems capable of replication DNA have been obtained with mammalian systems and perhaps the analysis of such systems will shed light on this problem.

PROTEINS OTHER THAN POLYMERASES INVOLVED IN DNA METABOLISM

Polynucleotide Ligase

Given the discontinuous nature of chromosome duplication, the joining of the intermediate fragments to high molecular weight DNA requires the activity of a DNA ligase. In mutants defective in DNA ligase, it can be shown that the joining of Okazaki fragments to high molecular weight DNA is severely impaired (134). This is consistent with the assumption that the ligase isolated in vitro is responsible for the joining reaction observed in vivo. A conditionally lethal (temperature-sensitive) DNA ligase mutant has also been isolated, thereby suggesting that the failure of this enzyme to act in vivo is a lethal event (134, 135). When such a mutant is observed at the nonpermissive temperature, essentially no joining of Okazaki fragments is seen; DNA synthesis, on the other hand, continues for several generation times. It remains to be seen whether or not such synthesis is a result of replication fork movement or the continued synthesis and release of Okazaki fragments from the same sites. It is also possible that the replication fork moves, but that in the absence of ligase activity, elongation of fragments takes place concomitantly with strand displacement such that "older" fragments are released as single stranded DNA and accumulate (see Figure 1). Analysis of this situation by alkaline sucrose gradient sedimentation (the method generally used) would not discriminate this possibility from the former ones. Perhaps such a mechanism is operating to some extent, since the conditionally lethal mutants do show an early increase in viability on transfer to high temperature (134, 135).

The dnaG Protein

As mentioned earlier, the product of the *dnaG* gene is required continuously for the propagation step in DNA replication. A plausible role for this protein has been suggested based on studies of the behavior of *dnaG* temperature-sensitive mutants in vivo and in vitro. The properties of two *dnaG* mutants were that on transfer to the nonpermissive condition, DNA synthesis ceased almost immediately. Okazaki fragments that had been synthesized prior to the transfer were converted into high molecular weight DNA, but no new pieces were initiated. These studies suggested that the *dnaG* protein is necessary for the initiation of synthesis of DNA fragments (136).

In an attempt to understand the biochemistry of this process, the *dnaG* protein was purified and studied. It was isolated by in vitro complementation of extracts requiring the *dnaG* function for complete activity. In one system (cellophane system), maximal activity of *E. coli* replication was restored to an extract prepared from a temperature-sensitive *dnaG* mutant by the addition of wild-type extract. Purification of the activity from wild-type cells resulted in the isolation of the *dnaG* protein (137). In a soluble system able to convert ϕX174 DNA to its replicative form, it was shown that the *dnaG* protein was required. By a similar complementation procedure, the *dnaG* protein was independently isolated. In this study, it was rigorously shown that the protein isolated was in fact the *dnaG* protein because the activity of the protein in vitro was thermolabile when the protein was isolated from a temperature-sensitive *dnaG* mutant (138). Since the complementation assays are based solely on the ability of a protein fraction to stimulate DNA synthesis, it is essential that rigorous proof of the type mentioned above be obtained. The complementations are done against temperature-sensitive mutants, and therefore any factor able to increase the stability of a labile protein will appear to complement.

The *dnaG* protein is composed of a single polypeptide chain of 60,000 daltons. It does not bind to an appreciable extent to DNA. So far, it has not been possible to ascribe any enzymatic function to this protein (138). If the notion of a "replication complex" is correct, it may not be possible to assay a "partial reaction" for each of the compounds. Undoubtedly, this problem will be solved when more such essential proteins are isolated.

The dnaB Protein

By methods similar to those described for the isolation of the *dnaG* protein (requirement for ϕX174 RF formation), the *dnaB* protein has also been isolated.

The protein isolated from a *dnaB* mutant was also shown to be thermolabile compared to the protein isolated from wild-type cells (139). An enzymatic activity has been ascribed to this protein. The protein catalyzes the hydrolysis of ribonucleoside triphosphates to ribonucleoside diphosphates and phosphate. The activity is stimulated twentyfold in the presence of DNA (140).

Details of the action of the *dnaB* protein in vivo have not yet been determined, although it is clear that its activity is continuously needed for the propagation step in DNA replication. It remains to be seen where this DNA-dependent ribonucleoside triphosphatase fits into the overall process.

Superhelix Relaxation Proteins

Movement of the replication fork requires the unwinding of parental DNA strands such that new strands can be laid down. The parental strands must be untwisted one turn for each ten nucleotides laid down. In a closed circular structure, the total number of twists must always remain constant. The strain imposed in a circular structure by untwisting parental strands can be relieved to an extent by supertwisting the unreplicated portion of the molecule. This process has a limit in that the propagation of the replication fork would cease unless the supertwists could be unwound, thereby relieving the strain. Such a relaxation of supertwists can only take place by breaking the circularity and allowing one strand to rotate about the other strand (i.e. the introduction of nicks in a non-replicating portion of the molecule). Circularity can then be restored by ligation and the fork can move on.

It has been elegantly shown that such a process takes place during replication of circular mitochondrial DNA (141) and SV40 DNA (142). In both cases, it was shown that in isolated replicating intermediates, the parental strands are intact, by removing the newly made DNA by denaturation, the parental strands partially re-anneal to generate a highly negatively supertwisted structure with portions of the parental strands remaining denatured. This is the condition expected if during replication the parental strands were broken, untwisted with respect to each other, and then rejoined (i.e. a deficiency in helical turns exists after the newly made DNA is removed).

As mentioned above, this situation could come about by nucleolytic action combined with rapid ligation. Proteins have been isolated that appear to accomplish this reaction. It cannot be shown that they have either nucleolytic or ligation activity. An "untwistase" has been isolated from mammalian cells that will relax both positive and negative supertwists in DNA (143). A protein (the ω protein) has been isolated from *E. coli* that has the ability to relax negatively supertwisted DNA only (144). In neither case has it been possible to demonstrate an intermediate structure (a nicked circle) expected during their action. Thus, these proteins seem ideally suited for superhelix relaxation during replication in that there is apparently no free nicked structure. If the replication fork were to progress into a nick, then circularity would be immediately broken and the biological activity of the structure destroyed.

DNA Binding Proteins

A group of proteins from a wide variety of sources have been isolated by affinity chromatography on single stranded DNA cellulose. These proteins have been called "unwinding" or "DNA binding" proteins. The first such protein isolated, which has the general properties of the group, is the gene 32 protein isolated from bacteriophage T4-infected cells (145). The protein binds strongly and co-operatively to single stranded DNA and thus promotes denaturation (unwinding) of double helical DNA (146). When bound to the single stranded DNA, it prevents intrastrand annealing and maintains the DNA in an extended conformation, thereby promoting renaturation of DNA (145, 147). Genetic, physiological, and biochemical experiments suggest that the gene 32 protein is involved in the recombination and replication of DNA (145). Proteins with similar characteristics to the 32 protein have been isolated from *E. coli* (77, 148), bacteriophage T7-infected *E. coli* (149, 150), mammalian cells (151), and adenovirus-infected cells (152).

In addition to its interaction with DNA alone, the 32 protein also stimulates the phage T4-induced DNA polymerase (153). Similar stimulations of DNA binding proteins of various DNA polymerases have been observed with the *E. coli* protein and DNA polymerase II and the phage T7 protein and the phage T7-induced DNA polymerase (77, 148, 150, 154). A novel feature of the observed stimulations is that a given protein will stimulate only the DNA polymerase isolated from the same source.

As mentioned previously, DNA polymerase II will not propagate when the length of the template strand is greater than 50 nucleotides. If sufficient *E. coli* DNA binding protein is added to complex with all of the single stranded template, maximal stimulation of rate is observed (77). The specificity of this protein-DNA complex is such that DNA polymerases I and III are inhibited in propagation while DNA polymerase II is stimulated by 10–50-fold. In addition, *E. coli* RNA polymerase will not transcribe the DNA-protein complex (77, 155), and various deoxyribonucleases will not digest the complex, but the $3' \to 5'$ exonuclease of DNA polymerase II is active on this complex under the appropriate conditions (98).

A detailed analysis of the interaction of the *E. coli* DNA binding protein with DNA and DNA polymerase II has suggested a model to account for these various observations. DNA polymerase II (90,000 daltons) and the *E. coli* DNA binding protein (22,000 dalton monomer) will form a protein-protein complex in the absence of DNA. DNA binding protein, when bound to DNA, retains its ability to bind to DNA polymerase II in that a ternary complex can be formed between DNA, DNA binding protein, and DNA polymerase II. *E. coli* DNA polymerases I and III and the phage T4-induced DNA polymerase make neither the protein-protein complex nor the ternary complex (98). Thus, it appears that the specific stimulation is due to complex formation. The stimulation of rate and extent of synthesis seen when the protein is added can be explained by removal of secondary structure in the template strand as well as by binding to the polymerase during synthesis such that the enzyme does not dissociate from the template during synthesis.

Although specific stimulation cannot be observed with polymerase III alone, it has been observed with polymerase III plus factors I and II (108) as well as with polymerase III* plus copolymerase III* (155). In the latter case, it has also been observed that in the presence of DNA binding protein, the DNA products obtained are longer than those made in the absence of the protein, a property expected if the DNA binding protein prevents the dissociation of polymerase from template during synthesis. It remains to be seen if complex formation can be demonstrated between the E. coli DNA binding protein and the more complex form of polymerase III, as would be predicted from the model.

In contrast to the behavior of the group of proteins discussed above, which extend the DNA to which it is bound, the gene 5 protein isolated from cells infected with the filamentous bacteriophages appears to collapse the DNA to which it is bound (156, 157). When covered with this protein, DNA templates appear to be inactive (155, 156). This protein functions in vivo to prevent the conversion of single stranded DNA to double stranded DNA, but allows synthesis of single stranded DNA from a duplex template (158). The properties observed in vitro are in keeping with the function of this protein in vivo.

All proteins discussed in this section were isolated either by complementation, in which case their existence was predicted from genetic analysis, or by search for activity in vitro based upon predictions from biological experiments. There have been reports suggesting that some proteins required for DNA replication might be quite difficult to isolate and characterize. These proteins are the so-called cis-acting proteins. Such proteins do not freely diffuse from the DNA which encoded them. Evidence for the existence of such proteins has come from analysis of noncomplementable functions required for DNA replication of bacteriophages P2 (159) and φX174 (154). Recent experiments (N. Kleckner, personal communication) have suggested that under specialized conditions, the gene O product of phage λ may also be cis-acting for early λ replication. If such proteins indeed turn out to be nondiffusible, then their analysis will have to come from the study of replication on endogenous DNAs.

IN VITRO SYSTEMS

Much of our knowledge concerning DNA replication has been deduced from in vitro studies of the process. For many years, it was not possible to observe DNA replication in vitro and studies were limited to analysis of DNA synthesis, mostly of the repair type (i.e. non-semiconservative synthesis). In recent years, several systems have been developed that do carry out semiconservative synthesis in vitro. Although no system is ideal for studying all steps in replication, each one has particular advantages that make it suitable for analysis of a part of replication. They all have one feature in common, and that is the maintenance of a high concentration of cellular constituents.

Nucleotide Permeable Systems

The basic nucleotide permeable system is made by treating cells with toluene (160) or ether (161). The toluene system has been exploited the most. The features of this system are that replication occurring in vivo at the time of the treatment continues in vitro. The essential requirements for replication in vitro are the addition of deoxynucleoside triphosphates and ATP. Synthesis is semiconservative, discontinuous, and sensitive to agents that inhibit replication in vivo. In addition, the temperature sensitivity of mutants in DNA replication is maintained in vitro, with the exception of those required for initiation, since this process apparently does not occur in toluene- or ether-treated cells.

The main advantage of this system is the ability to study the effects of small molecules on replication (macromolecules do not diffuse into such cell preparations). The requirement for ATP in replication was clearly established in this system (1). It appears that ATP is required specifically (i.e. not simply for the regeneration of deoxynucleoside triphosphates). Short fragments of DNA are not converted into high molecular weight DNA in the absence of ATP (162) (at least one role for ATP in replication has been observed with purified components: the "initiation complex" of DNA polymerase III*).

The ability to use phosphorylated substrates for replication has allowed detailed analysis of the linkage of DNA to RNA in nascent DNA fragments (163). It was shown that the specificity of the deoxyribonucleotide function in nascent DNA is maintained in toluene-treated cells. An additional linkage, that of deoxyguanylate to a ribopyrimidine, was observed in vitro but was not seen in vivo. In addition, a more detailed analysis of the RNA joined to DNA was possible in that a partial sequence was obtained (163).

The toluene system has been extended to *B. subtilis* cells (164), and the general features observed are in keeping with those seen in *E. coli*. Recently, it has been shown that if toluene-treated *E. coli* are treated with a non-ionic detergent, they become permeable to macromolecules as well as small molecules (165). This sytem is yet to be exploited.

The Cellophane Disc System

This in vitro system is prepared by lysing cells in concentrated suspension on cellophane membranes. Small molecules can be added by allowing them to diffuse through the cellophane and large molecules can be added to the top of the disc. This system shows a limited capacity to initiate new rounds of replication, but allows for continuation of the replication fork in vitro. As with the toluene system, DNA synthesis is discontinuous and semiconservative. In vitro synthesis is sensitive to a wide variety of agents affecting replication in vivo as well as requiring the various *dna* gene products (57). The system can replicate bacterial DNA, ϕX174 DNA (166), and, as mentioned previously, bacteriophage λ DNA. It has been useful in the isolation of the *dnaE* and *dnaG* proteins, as well as in the determination of the *dnaG* function, and the establishment of the existence of two size classes of Okazaki fragments. With the possible exception of chromosome initiation, there is no obvious reason why this system cannot be exploited yet further for elucidating the biochemistry of DNA replication.

The Single Strand Phage DNA System

This system is prepared by gently lysing cells in a highly concentrated suspension. Almost all the cellular DNA and membranes can be removed from the lysate by centrifugation at high speed. Single stranded bacteriophage DNAs can be added to the lysate and, under the appropriate conditions, be converted into their replicative forms (48). The advantage of this system is that essentially all DNA synthesis is dependent on exogenous DNA. This allowed for the initial demonstration in vitro of the involvement of RNA polymerase in priming DNA chains in the conversion of M13 DNA to RF and in the discovery of an alternate system probably involving RNA in the conversion of ϕX174 DNA to its replicative form (48).

In this system, synthesis of ϕX174 RF requires the functioning of the dnaB, dnaC-D, dnaE, and dnaG proteins (167) and perhaps the dnaA protein (48). Using this system, the dnaB, dnaD, and dnaG proteins were partially purified. In spite of the availability of all the dna gene products, synthesis of ϕX174 RF has not yet been possible using these and other purified components. Although it has been clearly shown that initiation of DNA chains occurs at a unique site on the molecule (155), the primer for this synthesis has not yet been demonstrated. Thus, it remains to be elucidated how these gene products interact with each other and what is the nature of the priming event in ϕX174 RF formation.

In the case of M13 RF formation, all the components required have been purified from the crude system (155). Synthesis of the primer by RNA polymerase occurs at a unique site; DNA synthesis is mediated by polymerase III* and copolymerase III* in the presence of ATP and spermidine. The RNA can be removed by DNA polymerase I and the RFI (closed circular molecule) formed by the action of DNA ligase. The E. coli DNA unwinding protein is required probably to serve two functions: to restrict transcription and to facilitate elongation. The basic steps in primer formation and DNA synthesis were also independently derived by purification from the crude system (108). The results differ only with respect to the nature of the polymerase III (requirement for factor I and II) as mentioned previously.

It has also been found that DNA polymerase II can substitute for DNA polymerase III in the formation of M13 RF (77). From the description of the activity of this enzyme in vitro and its interaction with the unwinding protein, this alternate pathway for the synthesis of M13 RF seems plausible. So far, it has not been possible to obtain a definitive answer as to the requirement for polymerase III or II in RF formation in vivo since mutants apparently lacking all known polymerase activites (by mutation) support the synthesis of M13 FR (168). This result is probably due to leakiness in vivo, and other mutants will have to be isolated in order to resolve this question.

The Bacteriophage T4 System

This system is prepared from bacteriophage T4-infected cells (169). It is also a highly concentrated lysate and T4 DNA synthesis is dependent upon the activities of those genes that are essential for T4 replication in vivo. By complementation in vitro, the products of genes 44 and 62 were isolated as a complex, and the products of genes 41 and 45 purified separately. The system also requires the gene 32 protein (unwinding protein) and gene 43 protein (the polymerase) (169). It should be possible to exploit this system further in order to understand the biochemistry of phage T4 DNA replication.

Isolated Nuclei Systems

A system has been developed from HeLa cells in which isolated nuclei carry on DNA replication at sites active in vivo. This system might be considered analogous to the toluene system in *E. coli* in that the nucleus is permeable to small molecules and phosphorylated substrates. For maximal activity, the system requires ATP and the cytoplasmic fraction, in addition to deoxynucleoside triphosphates. Synthesis occurs at close to the in vivo rate and about 1–2% of the DNA can be replicated (170). This system should prove useful in the identification of factors and substrates required for mammalian DNA replication.

An analogous system has been developed from mouse 3T6 cells infected with polyoma virus (171). As mentioned previously, polyoma virus molecules in the act of replication in vivo continue replication in vitro and the product can be shown to be form I polyoma virus DNA. This system has been exploited to show discontinuous synthesis on both strands of polyoma DNA and the involvement of RNA in the initiation of DNA fragments. The ability to use radioactive substrates for RNA and DNA synthesis allowed for the characterization of the RNA linked to DNA. A similar system from Balb 3T3 cells has been independently derived, and the results on polyoma virus DNA replication are similar to those described above (38, 172).

What remains to be achieved in the mammalian systems is to develop a system that will initiate as well as show a requirement for exogenous DNA. Undoubtedly, progress will be made in that direction in the future.

A SUMMARY OF EVENTS AT THE REPLICATION FORK

This summary is based on the information contained in this article and is meant to express the events that, based on biochemical, genetic, and physiological experiments, are likely to occur at the replication fork (see Figure 1).

Unwinding of the parental strands is aided by the DNA unwinding protein (originally suggested by B. Alberts), and the unwound single strands are covered by this protein. The strain imposed by unwinding parental strands is relieved by the action of a superhelix relaxation protein. The unwound DNA covered by unwinding protein is protected against endonucleolytic cleavage and is not a substrate for transcription. The DNA-protein complex does, however, facilitate elongation of DNA strands.

Sites A through F are sites for initiation of Okazaki fragments. Similarly, sites A′ through F′ may serve for sites of initiation. As a site (F) becomes available, primer synthesis or attachment takes place. The *dnaG* proteins are required in this step. Although in situ synthesis of a primer is likely, it has not been ruled out that a preformed RNA can serve a primer function. For example, it has been shown that the reverse transcriptase of oncornaviruses can utilize a transfer RNA molecule as a primer (J. Dahlberg, personal communication). The *dnaG* protein might facilitate binding of this RNA to a specific site on the template strand, this binding being perhaps independent of sequence but dependent on structure. Alternatively, the *dnaG* protein might function alone or in conjunction with another protein to synthesize a ribonucleotide primer.

The elongation of the primer (interval F–E) as deoxynucleotide is catalyzed by DNA polymerase III (altered by or in conjunction with a protein factor) and an additional factor that promotes binding to the template primer complex. As elongation proceeds, DNA unwinding protein is released and is free to act for unwinding of parental strains again. The number of nucleotides unwound per unit time (speed of the replication fork) is matched by the number of nucleotides layed down per enzyme molecule per unit time. The role of the *dnaB* protein probably functions at some stage up to this point in aiding in either unwinding, primer synthesis, or elongation. In vitro experiments conducted so far would argue for the role of the protein being related to primer formation.

The DNA polymerase III and associated proteins dissociate from the template at or close to the primer for the Okazaki piece synthesized previously (site E). Further elongation can take place by DNA polymerase I or II while DNA polymerase III reassociates with a primer at a new site created. DNA polymerase I digests the RNA primer by action of its associated $5' \rightarrow 3'$ exonuclease activity and simultaneously adds deoxynucleotides in the gap created (8). When all the ribonucleotides are removed such that only a single phosphodiester bond remains to be formed, DNA ligase joins the newly made DNA to form high molecular weight DNA (site A). Alternatively, in the absence of DNA ligase, DNA polymerase I might displace a completed Okazaki fragment while elongation continues (site C) to join its chain to another fragment or to high molecular weight DNA directly. Alternatively, the RNA could be removed by action of a ribonuclease H such that fragment elongation can take place by the action of any of the DNA polymerases, and subsequent ligation occurs as before. This situation would also prevail if the primer is not complementary to the template strand.

The events on the continuous strand (strand I) may be identical to those described for strand II, the discontinuous strand. Elongation of DNA strands is occurring in the same direction as fork movement and may continue without interruption (C′–F′). Elongation may stop while fork movement continues, and a new initiation may take place, B′. The conversion of this fragment into high molecular weight requires the same steps as described above. If elongation were to stop, RNA polymerase may add ribonucleotides onto the DNA chain; this

RNA chain could then be elongated by any of the DNA polymerases, resulting in the incorporation of RNA into the DNA chain (site D'). This RNA would have to be removed by a ribonuclease H activity.

It remains to be determined which of the various alternative schemes is operable. Since the cell has all the enzyme activities required for any of the processes described, all would be possible and perhaps all of them do occur. It is likely that those steps involving inefficient replication or wastage of energy have been selected against.

TERMINATION OF REPLICATION—A MODEL[2]

The model was derived to explain the events that would allow for segregation of two circular products from the replication of a double stranded circular DNA (Figure 2).

Initiation of replication takes place at a fixed site, O, on the molecule. If the molecule replicates bidirectionally, then the events occurring on either side of O are identical and two replication forks progress (Figure 1) in opposite directions around the circle. If replication is unidirectional, then one fork is established moving clockwise or counterclockwise. In any case, the events to follow are independent of the number of replication forks in that termination (segregation) takes place at the origin or at its antipode.

The replication fork(s) progresses around the circle while relaxation of supertwists takes place by nicking and sealing or the action of a relaxation protein. In Figure 2, structure III, we have assumed that replication occurred unidirectionally and that the product, circle A, was duplicated by continuous synthesis and circle B by discontinuous synthesis. In keeping with the requirement for the maintenance of circularity of both parental strands in the limit, the parental strands are interlocked by one twist, and in the limit, the last several nucleotides of each parental strand have not been duplicated (IV).

At this point, in order for the daughter strands to separate, a break in one or both of the parental strands must take place. If this were to happen, then the circularity of one or both products would be broken. Unless a mechanism existed that could reform the circles without loss of DNA, the biological activity of one or both products of replication would be lost. We consider the existence of a specific end-to-end joining activity unlikely. Thus, in order to obtain segregation, a terminal redundancy on one or both strands must be synthesized (this must occur on both strands if breaking of either of the parental strands can take place).

We propose that the terminal redundancy can come about in the following way. In a concerted reaction, the continuous strand (circle A) elongates while breaking the bonds holding the parental strands together. New bonds holding the parental strand together reform at sites O', while both the 5' terminated synthesized

---

[2] This model was conceived by the author and Dr. D. Botstein. A similar model was independently derived by Dr. A. Skalka. The one presented here is a composite of both. The author wishes to thank Drs. M. Fox and A. Levine for critical discussions.

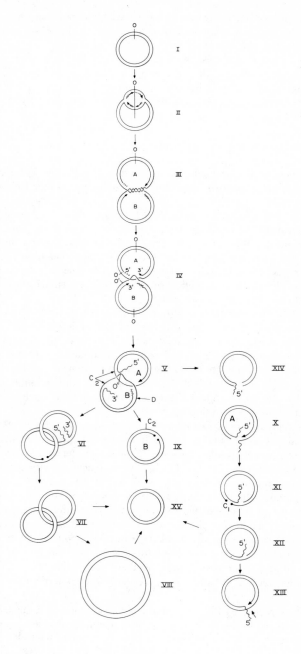

strand (circle A) and the 3′ terminated synthesized strand (circle B) are displaced from their respective complements (this concerted reaction assumes that the energy required to disrupt two helices and reform one is contained in the replication machinery elongating the continuous strand) (V). In circle B, a new initiation site might be opened by the migration of the site of interlocking, aiding in making the migration irreversible.

Migration and elongation continue until a nick is reached in one of the parental strands (site C′ 1 or 2) or a nick is introduced. At this point, the circles separate, generating structures IX and X. Structure IX is circular, containing a gap in the synthesized strand which can be repaired to yield a closed double stranded circle XV. Structure X is linear, but its synthesized strand has a terminal repetition such that an open circular structure XI can be formed. Sealing of the nick in the parental strand gives structure XII, which contains one closed strand and one open strand with a single stranded tail (the 5′ end is loose). This tail can be digested away and subsequent sealing of the synthesized strand yields a closed double stranded circular product XV. (If the nick were introduced in the other parental strand, structure IX with a nick at c_2 and structure XII would result.) Regardless of which parental strand contained the nick, the daughters of replication are not identical. One daughter contains a gap and the other contains a 5′ terminated tail. These products are obtained regardless of uni- or bidirectionality or in which circle 5′ → 3′ elongation took place.

There are other possibilities for the resolution of structure V. If migration and elongation cease prior to encountering a nick in the parental strands, the displaced strands can reform on their respective complements, generating VI which, when repaired as in IX and XII, generate a catenated dimer VII. Such structures have been observed in T4 DNA (173), mitochondrial DNA (23), and SV40 DNA (174). Structure VII could also arise by displacement of parental strands from each other without migration of the interlock, followed by repair and ligation. Structure VII could by recombination be converted to XV or a circular dimer VIII.

If a nick were to be introduced prior to migration or behind the interlock region (site D, structure V), then linear molecules would result. The ends of the linear monomer would depend on the amount of displacement that took place. Linear unit length monomers having ends other than the ends of mature phage DNA have been observed in phage λ-infected cells (175).

In structure V, if continuous elongation and displacement took place such that the complementary displaced strands annealed to each other and no nicks were reached in the parental strands, linear concatemers would result. This "double rolling circle" could generate linear double stranded DNA of lengths that would

←

Figure 2   A model for the segregation of duplicated circular molecules. Structure I is a covalently closed double stranded circular molecule. O is the origin and termination site for unidirectional replication. O′ is the termination site for bidirectional replication. A dotted line represents the primer initiating chromosome replication; the jagged line represents primers for Okazaki fragments. See text for the description of structures II–XV.

be determined by the amount of DNA synthesized on the parental strands. The end of the double stranded linear molecule is the termination site of replication.

If structure XII is not "repaired" to a circle, it can serve as a rolling circle for concameter generation. As mentioned above, one product of segregation always has the structure of a potential rolling circle. Which parental strand participates in this rolling circle is dictated by which synthesized strand circle, A or B, is continuously elongated in the $5' \rightarrow 3'$ direction and not by which parental strand is broken.

If the repair of the segregated circles is defective, then the products would be one circle with a gap at a unique site and one with a tail at a unique site. The latter is not repairable by DNA polymerase and ligase, but the former is. Such structures have been obtained as products of replication in λ, SV40, and ϕX174.

REPLICATING SYSTEMS

All replicating systems studied so far have in common a unique site for the initiation of replication. Replication may proceed in a uni- or bidirectional fashion. Although in most cases replication occurs by a discontinuous mode, mitochondrial DNA is apparently an exception in that replication appears to be largely continuous. Synthesis proceeds from the origin on one strand only, until about half the chromosome is partially duplicated. Retrograde synthesis then begins while the first strand continues. This displacement replication has been described in detail elsewhere (176, 177).

In considering *E. coli* replication as the basic system, it requires transcription for initiation as well as the products of the *dnaA* and *dnaC-D* genes. Fork migration requires the *dnaG* gene product and probably the *dnaB* gene product for fragment initiation and the *dnaE* gene product for DNA synthesis. Systems replicating within *E. coli* can use part or all of these functions. The F factor, for example, encodes its own genes for initiation in that it can replicate in *dnaA* or *dnaC-D* mutants. In fact, when integrated into the chromosome, it can obviate the requirement for these functions in chromosome replication (178). Bacteriophage λ is similar in this since its gene products O and P are necessary and sufficient for its initiation and it too can replicate in *dnaA* and *dnaC-D* mutants (1). Both the F factor and phage λ utilize the elongation functions (*dnaB, E,* and *G*) of the host.

Other replicating systems in *E. coli*—for example, Col E1 and the mini-circular DNA of strain 15—require the *dnaA* and *dnaC-D* functions for initiation but do not require some of the elongation functions. Neither of these plasmids requires *dnaB* (179) or *dnaE* for replication, but both are dependent on DNA polymerase I (180). Col E1 DNA replication does require the *dnaG* function and perhaps to some extent the *dnaE* function (D. Helinski, personal communication). The replication of F does not appear to require transcription (181), whereas Col E1 and minicircular DNA do (182, 183). In contrast, Col E1 replication does not appear to require continued protein synthesis for continued synthesis, whereas the F factor and the chromosome do (184).

The rather diverse ways in which DNA can be replicated are quite evident from analysis of M13 RF formation and φX174 RF formation. Both DNAs are of the same size and structure. The M13 phage DNA is converted to RF using transcription for primer generation. This may be analogous to chromosome initiation for plasmid DNAs. Phage φX174 DNA is converted to RF using apparently all the *dna* gene products for primer generation and elongation, and this may be analogous to chromosome initiation for the bacterial chromosome. Okazaki fragment initiation requires only the *dnaG* function (and perhaps the *dnaB* function).

Thus, it appears that there are many overlapping mechanisms that can lead to duplication of a particular DNA. Whether or not the detailed chemistry of DNA replication is universal remains to be established. We can look forward to the resolution of this question in the coming years as the biochemistry of DNA replication is completely understood. There is reason to be optimistic in that one of the last remaining problems in molecular biology will be solved.

ACKNOWLEDGMENTS

The author is indebted to many members of the Biology Department at the Massachusetts Institute of Technology for their interest and constructive criticisms, in particular Drs. I. J. Molineux and M. S. Fox. The author would also like to thank the many investigators who made the results of their experiments known prior to publication. A special thanks to Mrs. W. Fischer for preparation of the manuscript.

Literature Cited

1. Klein, A., Bonhoeffer, F. 1972. *Ann. Rev. Biochem.* 41:301–32
2. Wells, R. D., Inman, R. B., Eds. 1973. *DNA Synthesis In Vitro*. Baltimore: Univ. Park Press
3. Wechsler, J. A., Gross, J. D. 1971. *Mol. Gen. Genet.* 113:273–84
4. Bonhoeffer, F., Hermann, R., Gloger, L., Schaller, H. 1973. *J. Bacteriol.* 113:1381–88
5. Kingsbury, D. T., Helinski, D. R. 1973. *Genetics* 74:17–31
6. Jacob, F., Brenner, S., Cuzin, F. 1963. *Cold Spring Harbor Symp. Quant. Biol.* 28:329–47
7. Bird, R. E., Louarn, J., Martuscelli, J., Caro, L. 1972. *J. Mol. Biol.* 70:549–66
8. McKenna, W. G., Masters, M. 1972. *Nature* 240:536–39
9. Hohlfeld, R., Vielmetter, W. 1973. *Nature New Biol.* 242:130–32
10. Prescott, D. M., Kuempel, P. L. 1972. *Proc. Nat. Acad. Sci. USA* 69:2842–45
11. Gyurasits, E. B., Wake, R. G. 1973. *J. Mol. Biol.* 73:55–63
12. Hara, H., Yoshikawa, H. 1973. *Nature New Biol.* 244:200–3
13. Fujisawa, T., Eisenstark, A. 1973. *J. Bacteriol.* 115:168–76
14. Schnös, M., Inman, R. B. 1971. *J. Mol. Biol.* 55:31–38
15. Dove, W. F., Inokuchi, H., Stevens, W. F. 1971. In *The Bacteriophage Lambda*, ed. A. D. Hershey, 747–71. Cold Spring Harbor, NY: Cold Spring Harbor Lab.
16. Mosig, G. 1970. *J. Mol. Biol.* 53:503–14
17. Wolfson, J., Dressler, D. 1972. *Proc. Nat. Acad. Sci. USA* 69:2682–86
18. Huberman, J. A., Tsai, A. 1973. *J. Mol. Biol.* 75:5–12
19. Amaldi, F. et al 1973. *Exp. Cell Res.* 80:79–87
20. Danna, K. J., Nathans, D. 1972. *Proc. Nat. Acad. Sci. USA* 69:2097–2100
21. Fareed, G. C., Garon, C. F., Salzman, N. P. 1972. *J. Virol.* 10:484–91
22. Chattoraj, D. K., Inman, R. B. 1973. *Proc. Nat. Acad. Sci. USA* 70:1768–71
23. Kasamatsu, H., Vinograd, J. 1973.

Nature New Biol. 241:103–5
24. Helmstetter, C., Cooper, S., Pierucci, D., Revelas, E. 1968. *Cold Spring Harbor Symp. Quant. Biol.* 33:809–22
25. Schubach, W. H., Whitmer, J. D., Davern, C. I. 1973. *J. Mol. Biol.* 74: 205–21
26. Blau, S., Mordoh, J. 1972. *Proc. Nat. Acad. Sci. USA* 69:2895–98
27. Goebel, W., Schrempf, H. 1972. *Biochim. Biophys. Acta* 262:32–41
28. Worcel, A. 1970. *J. Mol. Biol.* 52: 371–86
29. Haumik, G., Cummings, D. J., Taylor, A. L. 1973. *Biochim. Biophys. Acta* 312:793–99
30. Sompayrac, L., Maaløe, O. 1973. *Nature New Biol.* 241:133–35
31. O'Sullivan, M. A., Sueoka, N. 1972. *J. Mol. Biol.* 69:237–42
32. Strätling, W., Knippers, R. 1971. *Cell J. B.* 20:330–39
33. Center, M. S. 1973. *J. Virol.* 12:847–54
34. Fralick, J. A., Lark, K. G. 1973. *J. Mol. Biol.* 80:459–75
35. Inokuchi, H., Dove, W. F. 1973. *J. Mol. Biol.* 74:721–27
36. Lark, K. G. 1972. *J. Mol. Biol.* 64: 47–60
37. Okazaki, R. et al 1973. In *DNA Synthesis In Vitro,* ed. R. D. Wells, R. B. Inman. Baltimore: Univ Park Press
38. Francke, B., Hunter, T. 1974. *J. Mol. Biol.* 83:99–121
39. Jacobson, M. K., Lark, K. G. 1973. *J. Mol. Biol.* 73:371–96
40. Wang, H. F., Sternglanz, R. 1974. *Nature* 248:147–50
41. Olivera, B. M., Bonhoeffer, F. 1972. *Nature New Biol.* 240:233–35
42. Hermann, R., Huf, J., Bonhoeffer, F. 1972. *Nature New Biol.* 240:235–37
43. Louarn, J. M., Bird, R. E. 1974. *Proc. Nat. Acad. Sci. USA* 71:329–33
44. Fareed, G. C., Salzman, N. P. 1972. *Nature New Biol.* 238:274–77
45. Fareed, G. C., Khoury, G., Salzman, N. P. 1973. *J. Mol. Biol.* 77:457–62
46. Pigiet, V., Winnacker, E. L., Eliasson, R., Reichard, P. 1973. *Nature New Biol.* 245:203–5
47. Brutlag, D., Schekman, R., Kornberg, A. 1971. *Proc. Nat. Acad. Sci. USA* 68: 2826–29
48. Wickner, W., Brutlag, D., Schekman, R., Kornberg, A. 1972. *Proc. Nat. Acad. Sci. USA* 69:965–69
49. Schekman, R. et al 1972. *Proc. Nat. Acad. Sci. USA* 69:2691–95
50. Sugino, A., Hirose, S., Okazaki, R.

1972. *Proc. Nat. Acad. Sci. USA* 69: 1863–67
51. Hirose, S., Okazaki, R., Tamanoi, F. 1973. *J. Mol. Biol.* 77:501–17
52. Waqar, M. A., Huberman, J. A. 1973. *Biochem. Biophys. Res. Commun.* 51: 174–80
53. Sadoff, R. B., Cheevers, W. P. 1973. *Biochem. Biophys. Res. Commun.* 53: 818–23
54. Eliasson, R., Martin, R., Reichard, P. 1974. *Biochem. Biophys. Res. Commun.* 59:307–12
55. Magnusson, G., Pigiet, V., Winnacker, E. L., Abrams, R., Reichard, P. 1973. *Proc. Nat. Acad. Sci. USA* 70:412–15
56. Botchan, P., Wang, J. C., Echols, H. 1973. *Proc. Nat. Acad. Sci. USA* 70: 3077–81
57. Schaller, H. et al 1972. *J. Mol. Biol.* 63:183–200
58. Klein, A., Powling, A. 1972. *Nature New Biol.* 239:71–73
59. Shizuya, H., Richardson, C. C. 1974. *Proc. Nat. Acad. Sci. USA* 71:1758–62
60. Strätling, W., Ferdinand, F. J., Krause, E., Knippers, R. 1973. *Eur. J. Biochem.* 38:160–69
61. Hinkle, D. C., Richardson, C. C. 1974. *J. Biol. Chem.* 249:2974–84
62. Sakakibara, Y., Tomizawa, J. I. 1974. *Proc. Nat. Acad. Sci. USA* 71:802–6
63. Sakakibara, Y., Tomizawa, J. I. 1974. *Proc. Nat. Acad. Sci. USA* 71:1403–7
64. Wong-Staal, F., Mendelsohn, J., Goulian, M. 1973. *Biochem. Biophys. Res. Commun.* 53:140–48
65. Grossman, L., Watson, R., Vinograd, J. 1973. *Proc. Nat. Acad. Sci. USA* 70: 3339–43
66. Williams, P. H., Boyer, H. W., Helinski, D. R. 1973. *Proc. Nat. Acad. Sci. USA* 70:3744–48
67. Buckley, P. J., Kosturko, L. D., Kozinski, A. W. 1972. *Proc. Nat. Acad. Sci. USA* 69:3165–69
68. Speyer, J. F., Chao, J., Chao, L. 1972. *J. Virol.* 10:902–8
69. Fox, R. M., Mendelsohn, J., Barbosa, E., Goulian, M. 1973. *Nature New Biol.* 245:234–37
70. Nath, K., Hurwitz, J. 1974. *J. Biol. Chem.* 249:3680–88
71. Westergaard, O., Brutlag, D., Kornberg, A. 1973. *J. Biol. Chem.* 248: 1361–64
72. Nath, K., Hurwitz, J. 1974. *J. Biol. Chem.* 249:2605–15
73. Cavalieri, L. F., Carroll, E. 1970. *Biochem. Biophys. Res. Commun.* 41:1055–60
74. Keller, W. 1972. *Proc. Nat. Acad. Sci.*

USA 69:1560–64
75. Gefter, M. L. et al. See Ref. 37
76. Roychoudhury, R. 1973. *J. Biol. Chem.* 248:8465–73
77. Molineux, I. J., Friedman, S., Gefter, M. L. 1974. *J. Biol. Chem.* 249:6090–98
78. Kornberg, A. 1969. *Science* 163:1410–14
79. Gefter, M. L. 1974. In *Progress in Nucleic Acid Research and Molecular Biology,* 101–15. New York: Academic
80. Williams, R. O., Loeb, L. A. 1973. *J. Cell Biol.* 58:594–601
81. Brutlag, D., Kornberg, A. 1972. *J. Biol. Chem.* 247:241–48
82. Hershfield, M. S., Nossal, N. G. 1972. *J. Biol. Chem.* 247:3393–3404
83. Vries, F. A. de, Swart-Idenburg, C. J., Waard, A. de. 1972. *Mol. Gen. Genet.* 117:60–71
84. Muzyczka, N., Poland, R. L., Bessman, M. J. 1972. *J. Biol. Chem.* 247:7116–22
84a. Grossman, L., Braun, A., Feldberg, R., Mahler, I. 1975. *Ann. Rev. Biochem.* 44:19–43
85. Westergaard, O., Brutlag, D., Kornberg, A. 1973. *J. Biol. Chem.* 248:1361–64
86. DeLucia, P., Cairns, J. 1969. *Nature* 224:1164–67
87. Lehman, I. R., Chien, J. R. See Ref. 37
88. Glickman, B. W., Shio, C. A., von Heignecker, H. L., Pärsek, A. 1971. *Proc. Nat. Acad. Sci. USA* 68:3150–58
89. Kingsbury, D. T., Helinski, D. R. 1973. *J. Bacteriol.* 114:1116–24
90. Veltkamp, E., Nijkamp, H. J. 1973. *Mol. Gen. Genet.* 125:329–40
91. Tait, R. C., Smith, D. W. 1974. *Nature* 249:116–19
92. Stein, H., Housen, O. 1969. *Science* 116:393–401
93. Miller, H. I., Riggs, A. D., Gill, G. N. 1973. *J. Biol. Chem.* 248:2621–24
94. Berkower, I., Lies, J., Hurwitz, J. 1973. *J. Biol. Chem.* 248:5914–21
95. Goebel, W., Schrempf, H. 1973. *Nature New Biol.* 245:39–41
96. Konrad, E. B., Lehman, I. R. 1974. *Proc. Nat. Acad. Sci. USA* 71:2048–51
97. Gass, K. B., Hill, T. C., Goulian, M., Strauss, B. S., Cozzarelli, N. R. 1971. *J. Bacteriol.* 108:364–74
98. Laipis, P. J., Ganesan, A. T. 1972. *J. Biol. Chem.* 247:5867–71
99. Whitfield, H. J., Levine, G. 1973. *J. Bacteriol.* 116:54–58
100. Livingston, D. M., Hinkle, D. C., Richardson, C. C. 1974. *J. Biol. Chem.* In press
101. Campbell, J. L., Soll, L., Richardson, C. C. 1972. *Proc. Nat. Acad. Sci. USA* 69:2090–94
102. Hirota, Y., Gefter, M., Mindich, L. 1972. *Proc. Nat. Acad. Sci. USA* 69:3238–42
103. Masker, W. E., Hanawalt, P. C. 1973. *Proc. Nat. Acad. Sci. USA* 70:129–33
104. Masker, W., Hanawalt, P., Shizuya, H. 1973. *Nature New Biol.* 244:242–43
105. Livingston, D. M., Richardson, C. C. 1974. *J. Biol. Chem.* In press.
106. Kornberg, T., Gefter, M. L. 1972. *J. Biol. Chem.* 247:5369–75
107. Otto, B. Bonhoeffer, F., Schaller, H. 1973. *Eur. J. Biochem.* 34:440–47
108. Hurwitz, J., Wickner, S. 1974. *Proc. Nat. Acad. Sci. USA* 71:6–10
109. Wickner, W., Schekman, R., Geider, K., Kornberg, A. 1973. *Proc. Nat. Acad. Sci. USA* 70:1764–67
110. Wickner, W., Kornberg, A. 1973. *Proc. Nat. Acad. Sci USA* 70:3679–83
111. Milewski, E., Kohiyama, M. 1973. *J. Mol. Biol.* 78:229–35
112. Nusslein, V., Otto, B., Bonhoeffer, F., Schaller, H. 1971. *Nature New Biol.* 234:285–87
113. Gefter, M. L., Hirota, Y., Kornberg, T., Wechsler, J. A., Barnoux, C. 1971. *Proc. Nat. Acad. Sci. USA* 68:3150–53
114. Hall, R. M., Brammar, W. J. 1973. *Mol. Gen. Genet.* 121:271–76
115. Thompson, R., Broda, P. 1973. *Mol. Gen. Genet.* 127:255–58
116. *Genetics* 1973. (Suppl. 1973) 73:67–80
117. Denhardt, D. T., Iwaya, M., McFadden, G., Schochetman, G. 1973. *Can. J. Biochem.* 51:1588–97
118. Brown, N. C., Wisseman, C. L. III, Matsushita, T. 1972. *Nature New Biol.* 237:72–74
119. Bazill, G. W., Gross, J. D. 1972. *Nature New Biol.* 240:82–83
120. Cozzarelli, N. R., Low, R. L. 1973. *Biochem. Biophys. Res. Commun.* 51:151–57
121. Bazill, G. W., Gross, J. D. 1973. *Nature New Biol.* 243:241–43
122. Ganesan, A. T., Yehle, C. O., Yu, C. C. 1973. *Biochem. Biophys. Res. Commun.* 50:155–63
123. Gass, K. B., Cozzarelli, N. R. 1973. *J. Biol. Chem.* 248:7688–7700
124. Chang, L. M., Bollum, F. J. 1972. *Science* 175:116–19
125. Chang, L. M., Brown, M., Bollum, F. J. 1973. *J. Mol. Biol.* 74:1–8
126. Chang, L. M. 1973. *J. Biol. Chem.* 248:3789–95
127. Wang, T. S., Sedwick, W. D., Korn, D. 1974. *J. Biol. Chem.* 249:841–50
128. Chang, L. M., Bollum, F. J. 1973. *J. Biol. Chem.* 248:3398–3404
129. Chang, L. M. 1973. *J. Biol. Chem.*

78   GEFTER

248 : 6983–92

130. Spodori, S., Weissbach, A. 1974. *J. Mol. Biol.* 86 : 11–20
131. Fansler, B., Loeb, L. A. 1972. *Exp. Cell Res.* 75 : 433–41
132. Lazarus, L. H., Kitron, N. 1973. *J. Mol. Biol.* 81 : 529–34
133. Hecht, N. B. 1973. *Nature New Biol.* 245 : 199–201
134. Gottesman, M. M., Hicks, M. L., Gellert, M. 1973. *J. Mol. Biol.* 77 : 531–47
135. Konrad, E. B., Modrich, P., Lehman, I. R. 1973. *J. Mol. Biol.* 77 : 519–29
136. Lark, K. G., 1972. *Nature New Biol.* 240 : 237–40
137. Klein, A., Nüsslein, V., Otto, B., Powling, A. See Ref. 37
138. Wright, M., Wickner, S., Hurwitz, J. 1973. *Proc. Nat. Acad. Sci. USA* 70 : 3120–24
139. Wright, M., Wickner, S., Hurwitz, J. 1973. *Proc. Nat. Acad. Sci. USA* 70 : 3120–24
140. Wickner, S., Wright, M., Hurwitz, J. 1974. *Proc. Nat. Acad. Sci. USA* 71 : 783–87
141. Robberson, D. L., Clayton, D. A. 1972. *Proc. Nat. Acad. Sci. USA* 69 : 3810–14
142. Salzman, N. P., Sebring, E. D., Radono-vich, M. 1973. *J. Virol.* 12 : 669–76
143. Champoux, J. J., Dulbecco, R. 1972. *Proc. Nat. Acad. Sci. USA* 69 : 143–46
144. Wang, J. C. 1971. *J. Mol. Biol.* 55 : 523–33
145. Alberts, B., Frey, L. 1970. *Nature* 227 : 1313–18
146. Delius, H., Mantell, N. J., Alberts, B. 1972. *J. Mol. Biol.* 67 : 341–50
147. Wackernagel, W., Radding, C. M. 1974. *Proc. Nat. Acad. Sci. USA* 71 : 431–35
148. Sigal, N., Delius, H., Kornberg, T., Gefter, M. L., Alberts, B. 1972. *Proc. Nat. Acad. Sci. USA* 69 : 3537–41
149. Reuben, R. C., Gefter, M. L. 1973. *Proc. Nat. Acad. Sci. USA* 70 : 1846–50
150. Scherzinger, E., Litfin, F., Jost, E. 1973. *Mol. Gen. Genet.* 123 : 247–62
151. Tsai, R. L., Green, H. 1973. *J. Mol. Biol.* 73 : 307
152. Vliet, P. C. Van der, Levine, A. J. 1973. *Nature New Biol.* 246 : 170–74
153. Huberman, J. A., Kornberg, A., Alberts, B. N. 1971. *J. Mol. Biol.* 62 : 39–52
154. Reuben, R. C., Gefter, M. L. 1974. *J. Biol. Chem.* 249 : 3843–50
155. Geider, K., Kornberg, A. 1974. *J. Biol. Chem.* 249 : 3999–4005
156. Oey, J. L., Knippers, R. 1972. *J. Mol. Biol.* 68 : 125–38
157. Alberts, B., Frey, L., Delius, H. 1972. *J. Mol. Biol.* 68 : 139–52

158. Mazur, B. J., Model, P. 1973. *J. Mol. Biol.* 78 : 285–300
159. Lindahl, G. 1970. *Virology* 42 : 522–28
160. Moses, R. E., Richardson, C. C. 1970. *Proc. Nat. Acad. Sci. USA* 67 : 674–81
161. Durwalt, H., Hoffman-Berling, A. 1971. *J. Mol. Biol.* 58 : 755–67
162. Pisetsky, D., Berkower, I., Wickner, R., Hurwitz, J. 1972. *J. Mol. Biol.* 71 : 557–71
163. Sugino, A., Okazaki, R. 1973. *Proc. Nat. Acad. Sci. USA* 70 : 88–92
164. Matsushita, T., White, K. P., Sueoka, N. 1971. *Nature New Biol.* 232 : 111–14
165. Moses, R. E. 1972. *J. Biol. Chem.* 247 : 6031–38
166. Olivera, B., Bonhoeffer, F. 1972. *Proc. Nat. Acad. Sci. USA* 69 : 25–29
167. Wickner, R. B., Wright, M., Wickner, S., Hurwitz, J. 1972. *Proc. Nat. Acad. Sci. USA* 69 : 3233–37
168. Staudenbauer, W. L., Olsen, W. L., Hofschneider, P. H. 1973. *Eur. J. Biochem.* 32 : 247–53
169. Barry, J., Alberts, B. 1972. *Proc. Nat. Acad. Sci. USA* 69 : 2717–21
170. Hershey, H. V., Stieber, J. F., Mueller, G. C. 1973. *Eur. J. Biochem.* 34 : 383–94
171. Winnacker, E. L., Magnusson, G., Reichard, P. 1972. *J. Mol. Biol.* 72 : 523–37
172. Hunter, T., Francke, B. 1974. *J. Mol. Biol.* 83 : 123–30
173. Rhoades, M., Thomas, C. A. Jr. 1968. *J. Mol. Biol.* 37 : 41–61
174. Jaenisch, R., Levine, A. 1973. *J. Mol. Biol.* 73 : 199–212
175. Skalka, A., Poonian, M., Barth, P. 1972. *J. Mol. Biol.* 64 : 541–50
176. Robberson, D. L., Kasamatsu, H., Vinograd, J. 1972. *Proc. Nat. Acad. Sci. USA* 69 : 737–41
177. Berk, A. J., Clayton, D. A. 1974. *J. Mol. Biol.* 86 : 801–24
178. Nishimura, Y., Caro, L., Berg, C. M., Hirota, Y. 1971. *J. Mol. Biol.* 55 : 441–56
179. Goebel, W., Schrempf, H. 1972. *Biochim. Biophys. Acta* 262 : 32–41
180. Goebel, W., Schrempf, H. 1972. *Biochem. Biophys. Res. Commun.* 49 : 591–600
181. Hobom, B., Hobom, G. 1973. *Nature New Biol.* 244 : 265–67
182. Clewell, D. B., Evenchik, B., Cranston, J. W. 1972. *Nature New Biol.* 244 : 265–67
183. Messing, J., Staudenbauer, W. L., Hofschneider, P. H. 1972. *Nature New Biol.* 238 : 202–3
184. Clewell, D. B. 1972. *J. Bacteriol.* 110 : 667–76

CHEMICAL CARCINOGENESIS ✴875

Charles Heidelberger[1]

McArdle Laboratory for Cancer Research,
University of Wisconsin, Madison, Wisconsin 53706

CONTENTS

INTRODUCTION AND SCOPE

It has been 15 years since chemical carcinogenesis was reviewed by Miller &
Miller (1) in this series. Since then enormous strides have been made in the
empirical and theoretical development of this field. By contrast, viral carcinogenesis
has been reviewed twice recently, in 1970 by Green (2) and in 1972 by Eckhart (3).
Because of this long hiatus, it is impossible within the limitations of this chapter
to provide a thorough review or even a summary of all the major developments
in chemical carcinogenesis since 1959. Hence, attention is restricted to certain topics
that I consider vitally important in the development of the field. I have reviewed
elsewhere aspects of my own research and its relevance to this domain (4–8). There

---

[1] American Cancer Society Professor of Oncology.

79

are textbooks available on chemical carcinogenesis by Clayson (9), Hueper & Conway (10), Daudel & Daudel (11), Arcos, Argus & Wolf (12), and Süss, Kinzel & Scribner (13). There is also a series of tomes surveying all compounds that have been tested for carcinogenic activity. It is complete through 1971 (14) and represents a valuable source material.

In this chapter. I will consider the following general topics: 1. chemicals as a major cause of human cancer; 2. metabolic activation of chemical carcinogens; 3. cell culture systems for the study of chemical carcinogenesis; and 4. proposed mechanisms of chemical carcinogenesis.

CHEMICALS AS A MAJOR CAUSE OF HUMAN CANCER[2]

It is tragic, and perhaps appropriate, that it was in humans where chemical carcinogenesis was first documented by Pott in 1775, who called attention to the high incidence of scrotal cancer in the chimney sweeps of London and correctly attributed this to their constant contact with coal tar and soot (15). Since that time, a depressingly large list of compounds that are proven carcinogens in man has emerged and is shown in Table 1. Most of these compounds were incriminated as human carcinogens as the tragic result of industrial exposures. One cannot help being impressed by the variety of chemicals that are human carcinogens; the same is found in carcinogenesis in laboratory animals, where a bewildering number of vastly different chemical structures produce the constellation of biological effects known as cancer. No single molecular feature can be pinpointed as cancer producing but, as will be documented later, some common features of chemical reactivity can be identified. Not all of these chemicals are manmade, and the Millers have reviewed the naturally occurring substances that can induce tumors (30).

This list of human carcinogens is ever increasing. In the last four years three compounds have been added: stilbestrol in 1971 (31), methyl chloromethyl ether in 1973 (21), and vinyl chloride in 1974 (27). Stilbestrol and vinyl chloride were recognized as human carcinogens on the basis of relatively few cases because they induce tumors that are very rarely observed, vaginal adenocarcinomas and angiosarcomas of the liver, respectively. The induction period for chemical carcinogenesis in humans is usually very long; many decades of exposure are often necessary. In order to slow the accession of compounds to this list, it will be

[2] Abbreviations used: BP, benzo[a]pyrene; MCA, 3-methylcholanthrene; BA, benz[a]anthracene; DMBA, 7,12-dimethylbenz[a]anthracene; DBA, dibenz[a,h]anthracene; 7-BrMe-BA, 7-bromomethylbenz[a]anthracene; DAB, p-dimethylaminoazobenzene; AAF, 2-acetylaminofluorene; N-OAc-AAF, N-acetoxy-2-acetylaminofluorene; N-OH-AAF, N-hydroxy-2-acetylaminofluorene; DMN, dimethylnitrosamine; DEN, diethylnitrosamine; NMU, N-nitrosomethylurea; NEU, N-nitrosoethylurea; MNNG, N-methyl-N′-nitro-N-nitrosoguanidine; NQO, 4-nitroquinoline-1-oxide; 7-Me-G, N-7-methylguanine; O-6-Me-G, O^6-methylguanine; 1-Me-A, 1-methyladenine; 3-Me-C, 3-methylcytosine; O-4-Me-T, O^4-methylthymine; PAH, polycyclic aromatic hydrocarbons; NADPH, reduced nicotine adenine dinucleotide; AHH, aryl hydrocarbon hydroxylase; 7,8-BF, 7,8-benzoflavone; MMS, methyl methanesulfonate.

necessary to mount a vastly increased surveillance of the environment to identify and remove potential carcinogens. Such increased testing is included in the US national plan for the conquest of cancer. Since such tests, especially for weak carcinogens, require the use of hundreds of animals kept for their entire life span, it is clear that more economical and less time consuming testing methods must be developed. As discussed below, considerable efforts are now being devoted to developing new screening methodologies.

In all screening systems for chemical carcinogenic activity, it is vitally important to make serious attempts to extrapolate the animal data to man. This is most

Table 1   Chemicals recognized as carcinogenic in man

Chemical Mixtures	Site of Cancers	References
Soots, tars, oils	Skin, lungs	15–17
Cigarette smoke	Lungs	18
Industrial chemicals		
2-Naphthylamine	Urinary bladder	19
Benzidine	Urinary bladder	19
4-Aminobiphenyl	Urinary bladder	20
Chloromethyl methyl ether	Lungs	21
Nickel compounds	Lungs, nasal sinuses	22
Chromium compounds	Lungs	23
Asbestos	Lungs	24, 25
Arsenic compounds	Skin, lungs	26
Vinyl chloride	Liver	27
Drugs		
N,N-bis(2-chloroethyl)-2-naphthylamine	Urinary bladder	28
Bis(2-chloroethyl)sulfide (mustard gas)	Lungs	29
Diethylstilbestrol	Vagina	31
Phenacetin	Renal pelvis	32
Naturally occurring compounds		
Betel nuts	Buccal mucosa	33
Aflatoxins	Liver	34
Potent carcinogens in animals to which human populations are exposed		
Cyclamates	Bladder	35
Sterigmatocystin	Liver	36, 37
Cycasin	Liver	38
Safrole	Liver	39
Pyrrolizidine alkaloids	Liver	40
Nitroso compounds	Esophagus, liver, kidney, stomach	41

difficult to do, but the basis of the evaluation system rests with those compounds that have demonstrated unequivocal activity in several species, including man. The International Agency for Research on Cancer (IARC) in Lyon, France is primarily devoted to this activity and has issued a number of excellent *Monographs on the Evaluation of Carcinogenic Risks of Chemicals to Man.* These have been prepared by the IARC staff and committees of experts and include: Volume 1, Overall Evaluation (42); Volume 2, Some Inorganic and Organometallic Compounds (43); Volume 3, Certain Polycyclic Aromatic Hydrocarbons and Heterocyclic Compounds (44); Volume 4, Some Aromatic Amines, Hydrazines, and Related Substances, N-Nitroso Compounds and Miscellaneous Alkylating Agents (45); and Volume 5, Some Organochlorine Pesticides (46). In these volumes data on the carcinogenicity of individual compounds, their occurrence, and their physical and chemical properties are critically evaluated relative to their possible risk to man. Another valuable source book has been issued by the US National Academy of Sciences, *Biological Effects of Atmospheric Pollutants: Particulate Polycyclic Organic Matter* (47). This volume deals with the chemistry and physics of polycyclics in the atmosphere, analytical methods for their detection, biological properties, screening methods, and human epidemiology.

The areas of epidemiology and geographic pathology have established the most impressive incrimination of chemicals in the environment as a major cause of cancer in man. Space limitations preclude discussion of this fascinating field here. However, the IARC has recently published a definitive textbook on the subject, *Host Environmental Interactions in the Etiology of Cancer in Man* (48). There are also two excellent recent reviews by Higginson (49, 50). The overall conclusion from modern epidemiology is that 80–90% of all human cancers are environmentally caused. Since oncogenic viruses are not highly contagious and since cosmic rays and ultraviolet light are fairly uniformly distributed, chemicals must be the predominant environmental carcinogens. A prime illustration of the importance of the environment in causing human cancer stems from the research of Haenszel & Kurihara (51) on migrant populations. It is well known that in Japan there is a very high incidence of stomach cancer and a low incidence of intestinal cancer, whereas in the United States there is a low incidence of stomach cancer and a high incidence of intestinal cancer. It was found (51) that in the offspring of Japanese immigrants to the US the incidences of these types of cancer were intermediate between the incidences in the two countries. In the second generation immigrants the incidences were essentially the same as in the US. This clearly demonstrates that for stomach and intestinal cancers the major causative factor is environmental rather than genetic. There is now increasing evidence that nitrosamines are a major cause of stomach cancer in man. This is discussed below.

METABOLIC ACTIVATION

General Considerations

It has been known for a long time that many classes of chemical carcinogens become covalently bound to DNA, RNA, and proteins of the cells in target tissues

(1, 4, 52–54). It has become axiomatic that the induction of cancer by chemical carcinogens results from such covalent binding to one or more of these cellular macromolecules. However, despite much intensive research, the macromolecular target(s) has not yet unequivocally been identified. This subject is discussed below. Among the important classes of chemical carcinogens that become covalently bound to tissue macromolecules are polycyclic aromatic hydrocarbons (PAH), aromatic amines, nitrosamines, and aflatoxins; however, none of the individual compounds in these classes become covalently bound when incubated in a test tube with DNA, RNA, or proteins. Therefore, these carcinogens must be metabolized to a chemically reactive form, which then reacts to form covalent macromolecular complexes. Of all the really powerful carcinogens, only the nitrosamides and some alkylating agents are chemically reactive and do not require metabolic activation to exert their noxious effects.

The initial recognition of the importance of metabolic activation; the elucidation of the metabolic activation pathways of aromatic amines, aflatoxins, and safrole; and the generalization of these findings have been achieved by my colleagues, Elizabeth and James Miller. They have reviewed their work in this field (52–54). Their conclusions, now generally accepted, are that 1. all chemical carcinogens that are not themselves chemically reactive must be converted metabolically into a chemically reactive form; 2. the activated metabolite is an electrophilic reagent; and 3. this activated metabolite reacts with nucleophilic groups in cellular macromolecules to initiate carcinogenesis. The fact that all reactive forms thus far characterized are electrophilic represents a generalization that is probably the least common denominator among the structurally diverse chemical carcinogens. Furthermore, the Millers and others have discovered that these metabolic activations are carried out primarily by the microsomal mixed function oxidases. It is an irony of nature that these enzyme systems, which are usually considered detoxifying and drug metabolizing, are the same ones that activate chemicals to carcinogenic and mutagenic forms.

Polycyclic Aromatic Hydrocarbons (PAH)

ARYL HYDROCARBON HYDROXYLASE   It has been known since 1951 that carcinogenic PAH are covalently bound to mouse skin proteins (55–57), mouse skin DNA (58–60), and to DNA, RNA, and proteins of transformable rodent cells in culture (61). In 1969 it was simultaneously and independently discovered by Grover & Sims (62) and by Gelboin (63) that PAH, which did not bind covalently to DNA in the test tube, were bound covalently to DNA added after incubation of the PAH with the liver microsomal mixed function oxidase system. These experiments show the requirement for microsomal activation in order for PAH to become covalently bound to DNA.

The microsomal mixed function oxidases require NADPH and oxygen, they contain cytochromes P-450 and P-448, they have not yet been purified, and their mechanism of action is not completely understood (64, 65). They are very nonspecific and carry out aromatic ring hydroxylations, N-hydroxylations, and oxidative demethylations of a great variety of organic compounds ordinarily not encountered in nature (65). One of the overall results of this enzyme system's action is to convert lipophilic compounds into hydrophilic ones; ordinarily this results in detoxication (64). The mixed function oxidases are found in greatest quantities in the liver, but are also detectable in most other tissues (65). Another property of this enzyme system is that it is inducible; pretreatment of experimental animals with various compounds. of which the most widely studied are PAH and phenobarbital, results in an increase of enzyme levels and activities in the livers (65).

Another of the many names for this enzyme system is aryl hydrocarbon hydroxylase (AHH), and it is assayed by the conversion of benzo[a]pyrene (BP) into 3-hydroxy-BP, which is determined fluorimetrically (66). As mentioned, AHH is induced in the liver by PAH and by phenobarbital, which have different binding sites on the cytochrome P-450 (67). AHH induction has also been demonstrated in epithelial cell cultures from rat liver and requires both RNA and protein synthesis (68). However, in such cultures AHH can be induced in the absence of PAH by inhibiting protein synthesis with cycloheximide or puromycin; these results suggest that there are two events in the induction: the PAH inducer acts primarily at the level of transcription, and a labile protein regulates translation of induction-specific RNA (69).

In addition to the studies done in hepatic systems, AHH is also present and inducible in cultured fibroblasts (66), mouse lungs (70), and several other extra-hepatic tissues.

Extensive work by Nebert and his colleagues has demonstrated that there is a wide difference in the degree of inducibility of AHH among different strains of inbred mice. They found that in some strains the induction of AHH by PAH was an all-or-nothing phenomenon, that this induction was accompanied by the conversion of hepatic low spin to high spin cytochrome P-450, and that induction could be demonstrated in bowel and kidney (71). From these studies they concluded that the induction of AHH in mice is expressed as a simple autosomal dominant trait (71). Moreover, it was found that the induction of the following microsomal cytochrome P-450 oxygenases is controlled at or near the same genetic locus: p-nitroanisole-O-demethylase, 7-ethoxycoumarin-O-deethylase, and 3-methyl-4-methylaminoazo-benzene-N-demethylase (72). However, the situation appears to be more complicated, since Gelboin's group has found that in strains of mice in which hepatic AHH is not induced, it is markedly induced in lung, small intestine, kidney, and skin (73). These latter tissues may be more relevant to hydrocarbon carcinogenesis than liver. Bresnick and his colleagues have studied AHH induction in fetal liver explants from strains of mice, some of which were different from those used by Nebert, and concluded that the genetic control of the induction is multifactorial (70). This view was further supported by in vivo studies of AHH induction in leukemia-susceptible and -resistant strains and in their F_1 hybrids, in which the results could not be explained on the basis of the control being exerted by a single autosomal locus (74). Obviously, further research is required to resolve these controversies.

It is of interest to inquire about the relationship between the inducibility of AHH and the carcinogenic activities of PAH. Franke (75) has carried out an extensive correlation between the inducibility of AHH produced by a series of PAH of known carcinogenic activities and several theoretical parameters related to molecular structure. From these calculations he concluded that the AHH-inducing activity of PAH increases with increasing strength of hydrophobic interactions and that the induction of AHH is the first step in hydrocarbon carcinogenesis (75).

Nebert and co-workers have studied the relation between the induction of AHH and initiation of skin carcinogenesis by 7,12-dimethylbenz[a]anthracene (DMBA) in inbred and hybrid strains of mice. They observed no correlation between the two factors and concluded that if activation of DMBA by skin AHH is required for tumorigenesis, the constitutive levels sufficed (76). Other possible interpretations were that DMBA itself may be the active carcinogen or that DMBA activation may occur through mechanisms not mediated by AHH. The same group also found that there was no correlation between the inducibility of AHH in mouse skins of various strains and the initiation of tumorigenesis by BP (77). Kouri et al (78), on the other hand, studied the hepatic inducibility by MCA of AHH in 14 strains of mice and oncogenesis produced by subcutaneous injection of MCA and found a good correlation between the two processes. However, the results obtained using BP and DMBA as carcinogens showed no obvious correlation. In the noninducible strains that were resistant to MCA oncogenesis, they also found an enhanced expression of the group-specific antigens of C-type RNA oncogenic viruses. The interactions of chemical carcinogens and oncogenic viruses are discussed below. Kouri et al (79) studied MCA subcutaneous tumorigenesis in two strains of mice in which the inducibility of AHH appeared to segregate as a single autosomal dominant trait, and they found a highly significant relationship in which the highly inducible strains were 5–10 times more susceptible to MCA tumorigenesis; this relationship held up in specific backcrosses.

In fetal rat hepatocytes in culture Benedict et al (80) found that the cytotoxicity produced by BP was prevented by phenobarbital, and oncogenic transformation of fetal hamster cells by BP was abolished following induction by BA. These seemingly contradictory facts reflect the complexity of AHH-induced metabolic activation of PAH.

A pioneering series of studies on AHH induction in mitogen-stimulated human lymphocytes has been carried out by Kellermann and co-workers. The induction produced by MCA in these cultures was determined in specimens from 103 healthy donors and induction ratios of 1.4–5.6 were obtained; the values from several samples from individuals were highly reproducible. The population segregated into at least two groups, implying that the inducibility of AHH in humans is also genetically determined (81). The metabolism of BP to water-soluble products in the cultured lymphocytes of the same subjects was also found to be highly correlated with their inducibility of AHH, suggesting that PAH metabolism in humans may be under the same genetic control as AHH inducibility (82). Similar observations on metabolism to water-soluble compounds of human embryo fibroblasts were made by Huberman & Sachs (83), who found that their populations segregated into three groups of low, medium, and high activities. In an important preliminary study, Kellermann et al (84) measured the inducibility of AHH in cultured lymphocytes in normal subjects and found that they segregated into three groups of low, medium, and high inducibility. In the lymphocytes of 50 patients suffering from broncogenic carcinoma, the data indicate that susceptibility to this disease is associated with the higher levels of inducibility. If this finding is substantiated with larger numbers of patients and normal subjects, it may provide the means to identify individuals at risk in whom preventative measures could be instituted.

In addition to studies of the relation between the inducibility of AHH and tumorigenesis, the effects of AHH inhibitors have been investigated. Gelboin et al (85) found that 7,8-benzoflavone (7,8-BF) added to skin homogenates inhibited AHH activity. Kinoshita & Gelboin (86) found that 7,8-BF inhibited mouse skin tumorigenesis produced by repeated applications of DMBA. Furthermore, 7,8-BF also inhibited the binding of labeled DMBA to mouse skin DNA, RNA, and proteins; however, the flavone had little effect on either the binding of BP to mouse skin macromolecules or on skin tumorigenesis by BP (87). Alexandrov & Frayssinet (88) studied the effect of intraperitoneal injections of a number of naphthalene derivatives on the induction of hepatic AHH by MCA. Naphthalene inhibited the induction. 1- and 2-Naphthylamine did not produce hepatic induction of AHH but did cause its induction in lung and kidney preparations, again pointing out the complexity of the system and its genetic control. Unfortunately, 7,8-BF is a two edged sword; not only does it inhibit AHH, but it also induces it, making interpretation of results quite difficult. For example, Wattenberg & Leong (89) observed that 7,8-BF inhibited skin tumorigenesis by BP, which might be due to inhibition of activation. However, they considered that the inhibition of tumorigenesis was the result of the induction of AHH and consequent detoxification of BP. Wattenberg (90) also found that the carcinogenic and toxic effects of DMBA were inhibited by phenolic antioxidants, but this was not related to effects on AHH.

Manipulation of AHH also produces marked effects on malignant transformation of cells in culture. As mentioned above, induction with BA inhibited transformation of fetal hamster cells by BP (80). On the other hand, Marquardt & Heidelberger (91) found that induction of AHH in mouse prostate fibroblasts increased malignant transformation produced by MCA, and that treatment with 7,8-BF abolished the transformation. These results are consistent with the hypothesis that MCA requires metabolic activation by AHH. By contrast, DiPaolo et al (92) found that pretreatment of hamster embryo cells with 7,8-BF increased the transformation produced by BP and MCA; under this schedule the flavone probably induced AHH. Some of this confusion may be resolved when we consider the chemical aspects of the metabolic activation of PAH, where it will be shown that epoxides (arene oxides) are one of the activated metabolites.

The enzyme that converts epoxides into *trans*-dihydrodiols is called epoxide hydrase, and represents an important means for detoxification of epoxides. This subject has been comprehensively and critically reviewed by Oesch (93). Nebert et al (76) showed that levels of epoxide hydrase in various mouse strains were similar and that the hepatic enzyme was not induced by MCA. They could not detect the enzyme in the skin of these mice. Oesch et al (94) found that hepatic epoxide hydrase could be induced by phenobarbital. Although epoxide hydrase is functionally and structurally coupled with AHH, genetic analysis revealed that they are under different genetic control, and hence are not analogous to a bacterial operon; the genetics of epoxide hydrase induction by phenobarbital is too complex for any simple analysis.

METABOLISM OF POLYCYCLIC AROMATIC HYDROCARBONS   From the extensive research of Boyland and Sims has come most of our present understanding of the metabolism of polycyclic aromatic hydrocarbons, as studied most intensively in liver in vivo, liver slices, homogenates, and microsomal preparations. The primary metabolic steps are illustrated in Figure 1 for a typical case, benz[a]anthracene (BA), where the K-region is indicated as a double bond. The products of metabolism are the phenol, the *trans*-dihydrodiol, and the glutathione conjugate; the phenol can then undergo further conjugation with glucuronic acid or sulfate. In 1950 Boyland (95) proposed that an epoxide (arene oxide) intermediate could account for the observed products. The epoxide would be chemically reactive and would not have been isolated as an intermediate, as such experiments were carried out. The evidence for Figure 1 was provided by Boyland & Sims (96), who showed that the metabolism of BA in intact rat livers occurs primarily at the K-region; similar metabolism at the K-region also occurs with dibenz[a,h]anthracene (DBA) (97). The major metabolites of DMBA in rat liver homogenates were the two isomeric hydroxy-methyl derivatives, showing that metabolism occurred primarily on the methyl groups (98). However, Sims subsequently showed that both 7- and 12-methyl-BAs were converted by rat liver homogenates into the K-region phenols, *trans*-dihydrodiols, and glutathione conjugates; the phenols were produced non-enzymatically. The same products plus other ring hydroxylated compounds were

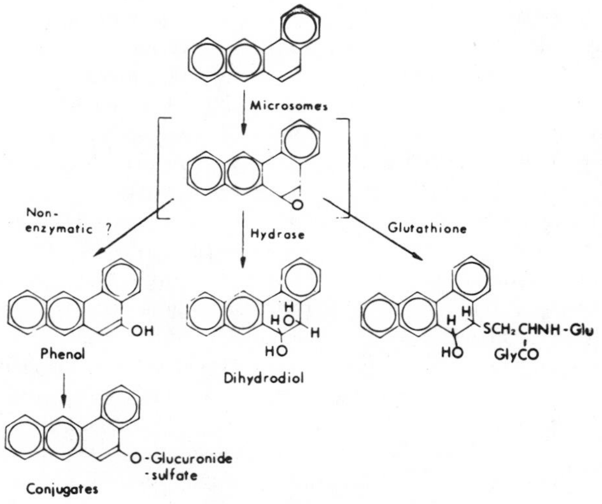

Figure 1   Activation of polycyclic aromatic hydrocarbons (6).

formed in adrenal homogenates (99). Since all the above studies were only qualitative, the use of thin layer chromatography and labeled isotopes was instituted by Sims, who found in the case of DBA that dihydrodiols were produced at the 1,2 and 3,4 positions in addition to the K-region (100); quantitatively, however, the major metabolites were in the form of unidentified water-soluble compounds. The metabolism of DBA, MCA, BP, and DMBA in mouse embryo cells was found to be essentially the same as in rat liver homogenates, again with the major radioactivity being present in water-soluble products (101).

Diamond et al (102) found a direct correlation between the conversion of tritiated DMBA into water-soluble metabolites and the cytotoxicity it exerted against a number of cultured cell lines. Similarly, BP and DMBA were converted into water-soluble products only by those cells that were subject to cytotoxicity, and it appeared that the hydrocarbons were converted into phenols before the water-soluble derivatives (103). Brookes and his colleagues have also done rather similar studies. They found that all compounds, whether oncogenic or not, disappeared at equal rates in low concentrations (104). The disappearance of unchanged BP in primary mouse embryo cultures was first order at low concentrations (105), the oncogenic DBA had a lower rate of metabolism than the non-oncogenic isomer DB[a,c]A (106), and the rate of disappearance of BP from human fibroblasts was constant over many passages in culture. In our laboratory we studied the conversion of several PAH into water-soluble metabolites in cultures of trans-formable and chemically transformed mouse prostate fibroblasts; the former cells carried out more extensive metabolism than the latter. In hamster embryo fibroblasts there was no correlation between the oncogenic activity of a series of 5 PAH and their rate of conversion to water-soluble products (107).

As mentioned above, Boyland in 1950 postulated that epoxides could serve as intermediates in the metabolic conversions of PAH into their known metabolites (95). Once the concept of metabolic activation was recognized, it seemed likely that epoxides were also logical candidates for being a metabolically activated form of PAH, since they are quite reactive and electrophilic in character. Although there had been some evidence that epoxides applied to the skin were less carcinogenic than the parent hydrocarbons, it seemed worthwhile to reinvestigate this possibility. The impetus came from the research of Grover & Sims, who in 1970 demonstrated that K-region epoxides of phenanthrene and DBA reacted to give covalent complexes with DNA and histones, which the corresponding hydrocarbons and dihydrodiols did not do (108). They also showed that labeled K-region epoxides of several PAH were bound covalently to DNA, RNA, and proteins of cultured BHK21 cells (109), and we made similar observations in transformable mouse prostate fibroblasts; the epoxides were always bound to the cellular macro-molecules to a much greater extent than the hydrocarbons, phenols, or dihydrodiols (110).

In 1970 Jerina et al (111) demonstrated that naphthalene oxide is an obligatory intermediate in the microsomal hydroxylation of naphthalene, and suggested that an oxide intermediate would be produced in all microsomal aromatic ring hydroxylations. The role of arene oxides in the "NIH shift" has been reviewed (112).

If epoxides were to be implicated as activated metabolites, it was necessary to isolate them as intermediates in the microsomal metabolism of carcinogenic hydrocarbons. This was accomplished in our laboratory (113) and by Sims and his colleagues (114, 115).

A summary of the research that has been done on isolation in microsomal systems of epoxides of various PAH, the dihydrodiols produced from PAH, the epoxides that have been synthesized, and the dihydrodiols obtained by the microsomal metabolism of the epoxides is presented in Table 2. A number of conclusions can be drawn: 1. all epoxides isolated and characterized have been located at the K-region; 2. epoxides are converted by microsomes into the corresponding dihydrodiols (usually *trans*), hence, the presence of a dihydrodiol can be taken as strong evidence that an epoxide was the intermediate; 3. several non-K-region epoxides have been synthesized, and thus far have been less stable than K-region epoxides (116, 130); 4. metabolism of PAH to dihydrodiols can take place at positions other than the K-region also with presumed epoxide intermediates (116, 130); and 5. in the case of methylated benz[a]anthracene, metabolism to hydroxymethyl compounds can precede epoxide formation. Most of the separations and characterizations described above have been carried out by thin layer chromatography. A significant improvement has been made recently by Selkirk et al (124), who developed quantitative separations of BP and its metabolites by high pressure liquid chromatography.

Swaisland et al (125) have studied the rates of nonenzymatic rearrangements to phenols, alkylation of 4-(p-nitrobenzyl)pyridine, and microsomal conversion to dihydrodiols by a series of K-region epoxides. Although there was considerable variation in the rates of all three processes among the six epoxides studied, none of these rates correlated with the carcinogenicity of the parent hydrocarbon. Such studies should be more rigidly quantitated and extended to non-K-region epoxides as well.

The evidence cited above clearly demonstrates that epoxides are produced during the metabolism of PAH and have the requisite chemical reactivity to qualify them for the role as an ultimate carcinogenic form of PAH. What about their biological activity? As documented below, in many cases epoxides are considerably more active than the parent hydrocarbons at producing malignant transformation of mouse prostate fibroblasts (132, 133) and hamster embryo cells (134). They are also more active than the hydrocarbons in producing mutagenesis in bacteriophage T$_2$ (135), in *Salmonella* (136), and in Chinese hamster cells (137). Are epoxides the only activated forms of PAH? There are other possibilities for which there is suggestive evidence at present.

In 1968 Dipple et al (138) made some theoretical calculations that supported the idea that powerfully carcinogenic methylated hydrocarbons such as DMBA may be metabolically activated on the methyl groups, especially since it is well established that hydroxymethyl compounds are formed by microsomal metabolism (98). Dipple & Slade (139) proposed that bromomethyl derivatives might serve as models for an alkylating group capable of forming a carbonium ion, and they showed that 7-bromomethyl-12-methyl-BA had moderate carcinogenic activity and was capable

Table 2 Metabolism of polycyclic aromatic hydrocarbons and epoxides

Hydrocarbon	Epoxide isolated (Ref.)	Dihydrodiols isolated from hydrocarbons (Ref.)	Epoxides synthesized (Ref.)	Dihydrodiols isolated from epoxides (Ref.)
Benz[a]anthracene	5,6- (114, 120, 123)	5,6- (118)	5,6- (126, 127) 8,9- (116)	5,6- (114, 118) 8,9- (116)
Dibenz[a,h]anthracene	5,6- (113)	5,6- (100) 1,2- (100) 3,4- (100)	5,6- (126)	5,6- (118)
Dibenz[a,c]anthracene		10,11- (117)	10,11- (117)	10,11- (117)
Benzo[a]pyrene	4,5- (115, 123)	4,5- (119, 123) 9,10- (119, 130) 7,8- (119, 130)	4,5- (127–129) 9,10- (130) 7,8- (130)	9,10- (130) 7,8- (130)
Dibenzo[a,h]pyrene		uncharacterized (122)		
Dibenzo[a,i]pyrene		uncharacterized (122)		
7,12-Dimethylbenz[a]-anthracene	5,6- (121)	5,6- (121)	5,6- (127–129, 131)	5,6- (131)
7-Hydroxymethyl-12-methylbenz[a]anthracene	5,6- (121)	5,6- (121)	5,6- (131)	5,6- (131)
7-Methylbenz[a]-anthracene	5,6- (121, 123)	5,6- (123)	5,6- (125)	5,6- (121)

BA

DBA

BP

DB[a,c]A

of alkylating 4-(p-nitrobenzyl)-pyridine. The reactions with DNA are discussed below. Flesher & Sydnor (140) provided some data to suggest that there was formation of 6-hydroxymethyl-BP in liver homogenates incubated with BP; the former compound had carcinogenic activity equal to that of BP and 6-methyl-BP. It was impossible to determine whether BP was first methylated and then hydroxylated or whether there was a direct hydroxymethylation of BP.

There have been model chemical studies to indicate that PAH may be converted into radical cations (141, 142). Nagata et al have carried out experiments in which an electron spin resonance (ESR) free radical signal was demonstrated when BP was incubated with skin homogenates (143), and this signal was subsequently attributed to the 6-phenoxy radical, which was also produced on stirring BP with albumin (144). Lesko et al (145) reported that treatment of BP with I_2 or H_2O_2/Fe^{3+} (which generates free radicals) led to covalent binding to DNA, and that incubation of 6-hydroxy-BP in the presence of light with DNA also resulted in covalent binding. In incubations of BP and I_2, a free radical was also detected by ESR, and a correlation was found between carcinogenic activity in a series of PAH and the production of free radicals on treatment with I_2 (146). Thus, at the present time it appears that metabolic activation of PAH can involve the formation of epoxides, free radicals, and carbonium ions on methyl groups.

COVALENT BINDING OF POLYCYCLIC HYDROCARBONS TO NUCLEIC ACIDS AND PROTEINS   It has already been mentioned that PAH become covalently bound to DNA, RNA, and proteins of mouse skin following topical application, as a result of prior metabolic activation (55–61). Irving has reviewed the binding of many types of chemical carcinogens to DNA (147).

In our laboratory we observed that PAH were bound covalently to a soluble protein fraction obtained on electrophoresis in direct proportion to their carcinogenic activity, and that this protein, called the h protein, was undetectable in tumors induced by PAH (148). This h protein has been extensively purified from mouse skin and characterized (149), and we found that PAH bind to this protein fraction in transformable mouse prostate fibroblasts and are undetectable in chemically transformed cells (150); epoxides also bind specifically to this protein in the former cells. The properties of this slightly basic protein fraction obtained from mouse skin (149) resemble those of "ligandin," which has been characterized from rat liver and binds polycyclic hydrocarbons covalently and other anionic compounds, such as corticosteroid metabolites, noncovalently (151). Ligandin's physiological function has not yet been elucidated. We are about to undertake a direct comparison between the mouse skin h protein and rat liver ligandin. Corbett & Nettesheim (152) treated with Raney nickel the protein fraction to which DMBA and MCA had been bound, and a radioactive hydrocarbon derivative was released, suggesting that the covalent bond to the protein involved a sulfur atom, probably cysteine.

There are also other proteins in skin, liver, and transformable cells to which PAH are bound noncovalently. Some of these proteins resemble in properties, but are not identical to, the corticosteroid binding proteins of mouse liver (A. Sarrif and C. Heidelberger, unpublished). Thus, there may be a specific cellular receptor protein that transports bound PAH to the nucleus of cells, in analogy to the function of steroid receptors. Slaga et al (153), on incubation of MCA with an epidermal mouse skin cytosol followed by polyacrylamide gel electrophoresis, found a peak of radioactivity noncovalently bound to a single protein fraction, and that PAH competed for this site in proportion to their carcinogenic activities. This protein is different from the h protein because it is acidic and closely resembles serum albumin in its properties. Toft et al (154) found that MCA is bound in a rat uterine cytosol preparation in a different fashion from estradiol. They also showed by sucrose gradient sedimentation that MCA was bound to a unique protein fraction in rat lung that was not present in liver; this may be significant, since lung and not liver is a target organ for PAH. This lung protein conjugate had a sedimentation coefficient of 7S at low ionic strength which changed to 5S in the presence of salt—behavior reminiscent of steroid receptors. MCA was also bound to the same fraction in vivo; however, no binding specificity or relation to carcinogenic activity was demonstrated (155).

When DNA and PAH were incubated for long periods of time in the dark, a small amount of noncovalent physical binding was known to occur; when the physical complex was treated with ultraviolet light or I_2 the binding became covalent (145). In cell cultures, Yuspa et al have observed that the binding of DMBA was slightly higher to nonreplicating than to replicating DNA (156), and the same was also found for BA (157). A similar situation exists for in vivo binding of MCA to DNA in mouse mammary hyperplastic nodules (158).

The microsomal-induced binding of BP to calf thymus satellite DNA was found to be about the same as to the main band DNA (159), and the same result was obtained with the binding of DMBA to mouse epidermal satellite DNA in vivo (160).

Bürki et al (161) found that 1,1,1-trichloro-2-propene oxide, which is an inhibitor of epoxide hydrase, inhibited the conversion of MCA to the dihydrodiol and increased the binding of BP and MCA to DNA in microsomal preparations. When the propene oxide was applied simultaneously with MCA to mouse skin, there was an increased formation of tumors. These experiments provide additional evidence that epoxides are metabolically activated forms of PAH. It has been reported by Regan & Cavalieri (162) that when purified nuclei obtained from the livers of rats that had been pretreated with MCA were incubated with several labeled PAH, covalent binding was obtained to the nuclear DNA, suggesting that there may also be AHH activity in nuclei.

Dipple and his colleagues have investigated the binding to DNA of bromomethyl-benz[a]anthracenes as possible models for methyl group activation. In aqueous solution, reaction of 7-bromomethyl-BA with DNA occurred predominantly with the amino groups of adenine and guanine, in contrast to the reactions with the N-7 of guanine and N-1 of adenine which occur with most alkylating agents (163). They also compared the binding to DNA in the test tube of the carcinogenic 7-BrMe-12-Me-BA with that of the noncarcinogenic 7-BrMe-BA; the latter was bound more extensively and gave similar reaction products involving the amino group of the purine bases (164). Almost identical results were obtained by an in vivo comparison of the binding of the same two compounds to mouse skin DNA (165). The main conclusion from this work was that there was no simple correlation between carcinogenic potency of these model compounds with either their total reaction with DNA or an attack on any specific residue.

Grover & Sims (166) studied the binding of four PAH K-region epoxides with various polynucleotides and found much less reaction with apurinic acid than with DNA and RNA. There was also considerably more binding to polyribonucleotides of purines than of pyrimidines; however, there was no correlation between the extent of reaction and carcinogenic activity. In somewhat related work, Jones et al (167) degraded the DNA from hamster embryo fibroblasts that had been incubated with labeled MCA. On acid depurination the radioactivity was released into dialyzable form at the same rate as were the purines. However, the apurinic acids still contained about half of the total radioactivity, implying that there was considerable binding to pyrimidine nucleotides.

Baird & Brookes (168) treated mouse embryo cells with ^{3}H-7-Me-BA; isolated the DNA; degraded the DNA with DNase, phosphodiesterase, and phosphomono-esterase; and chromatographed the products on Sephadex LH-20 columns. Characteristic elution patterns were obtained, although the products corresponding to the labeled degradation products could not be identified. Similar peaks were observed on analysis of the DNA obtained from mouse embryo cells, mouse skin, and human embryo lung cells. These peaks most likely contain hydrocarbon derivatives attached to deoxyribonucleosides. Baird et al (169), using the above techniques, compared the products obtained from 7-Me-BA-treated mouse embryo cells with the DNA obtained from chemical reactions in the test tube with 7-BrMe-BA and with 7-Me-BA-5,6-epoxide. All the elution patterns were different. This finding does not provide support for a K-region epoxide as an intermediate in the binding. However, it also does not exclude a methyl carbonium ion or an epoxide as an intermediate in the binding of the hydrocarbon in the cells. Hydrocarbon derivatives applied externally might be metabolized differently from those present intracellularly, the epoxide might be metabolized in an additional position, or an epoxide produced from the hydrocarbon at other than the K-region may be the reactive intermediate.

A powerful new approach to elucidating the structures of DNA-bound hydro-carbon derivatives has been provided by Daudel and her colleagues, who have constructed an exceptionally sensitive spectrofluorimeter capable of determining the fluorescence spectra of hydrocarbon derivatives bound in minute quantities to DNA. They showed that when 7-Me-BA-5,6-epoxide was reacted directly with DNA, the fluorescence spectrum resembled that of the 5,6-dihydrodiol; thus, the K-region ring of the bound derivative was hydroaromatic (170). By contrast, the DNA obtained from hamster embryo cells incubated with 7-Me-BA had a fluorescence spectrum indicating that the K-region ring was aromatic (171). Although the structure of the hydrocarbon-bound derivative is different from that of the product of the direct reaction of the epoxide with DNA, this does not exclude the possibility that the epoxide is a metabolically activated intermediate; for example, an intracellularly produced epoxide might become dehydrated after reaction with the DNA. It seems likely that considerable additional progress will be made on the structures of the DNA and protein complexes of PAH, which may further elucidate the nature of the activated intermediates.

Aromatic Amines

Our knowledge of the metabolic activation, binding to macromolecules, and mechanism of action of aromatic amines stems largely from the research of the Millers, who have reviewed their work and that in the field (1, 52–54, 172–174), much of which has been carried out with 2-acetylaminofluorene (AAF). This potent carcinogen is activated by microsomal N-hydroxylation to a proximal carcinogenic form, which in rat liver is converted by a sulfotransferase into a highly reactive, electrophilic sulfate ester, which is considered to be the ultimate carcinogenic form; this ester reacts covalently with DNA, RNA, and proteins, and thereby initiates the carcinogenic process. Because of the very high reactivity of the sulfate ester, it is not possible to prepare it chemically and study its reactions critically; however, the synthetic N-acetoxy-AAF (N-OAc-AAF) is more stable, and has been used as a model compound to study a variety of problems relating to the mode of AAF action and its metabolites.

The scheme of reactions involved in these processes is shown in Figure 2 (cf 54). The main reaction of the AAF-N-sulfate with proteins is via methionine to give the pictured sulfonium compound. On alkaline hydrolysis of the proteins, the side chain of methionine is eliminated and the methyl-thio-AAF is produced. Studies

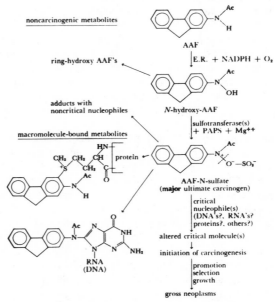

Figure 2   Activation of 2-acetylaminofluorene (54).

with N-OAc-AAF demonstrated that its chemical reaction with guanosine occurred at carbon-8 rather than at nitrogen-7, the usual site of alkylation; this structural assignment was confirmed by chemical synthesis (175). The microsomal N-hydroxylation of AAF in mouse and hamster livers involves cytochrome P-450, and an immune serum against NADPH cytochrome c reductase inhibited the microsomal N-hydroxylation reaction (176).

Other mechanisms of metabolic activation of AAF have also been elucidated. One of these involves the formation of a nitroxide free radical in the presence of peroxidase that can undergo dismutation to give N-OAc-AAF (177). In model studies on the reactions of the sulfate ester of N-hydroxy-2-acetylaminophenanthrene with methionine, adenosine, and guanosine, free radical intermediates were detected (178). Another mechanism of activation involves the participation of an acetyltransferase, which reacts with N-OH-AAF and AAF to give N-OAc-AAF (179, 180). Furthermore, the glucuronide of N-OH-AAF has been suggested by Irving to be a metabolically activated form of AAF (147). Activation by N-hydroxylation of 2-naphthylamine and 4-aminobiphenyl also occurs in dogs and monkeys that are susceptible to bladder tumor induction by these compounds; the N-hydroxy metabolities are excreted in the urine (181).

The scheme for the metabolic activation of the 4-dimethylaminoazobenzene (DAB) hepatocarcinogens has also been elucidated by the Millers, although the activated intermediate has not been isolated, due to its reactivity. Microsomal demethylation of DAB to 4-methylaminoazobenzene (MAB) occurs. The synthetic ester, N-benzoyloxy-MAB, serves as a model for the reactions undergone by the metabolically activated intermediate; these are shown in Figure 3 (53). The

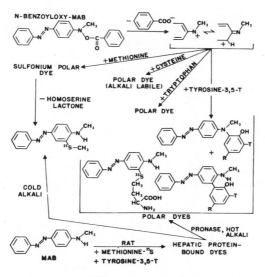

Figure 3   Reactions of N-benzoyloxymethylaminoazobenzene (53).

structures of its reaction products with proteins are shown. Very recently it has been found that N-benzoyloxy-MAB reacts chemically with guanosine and deoxyguanosine at carbon-8, in analogy with the AAF derivative (J. A. Miller, private communication). Lin & Fok (182) have shown that MAB in the presence of potassium persulfate or iodine reacts covalently with nucleic acids, presumably via a free radical reaction.

Chauveau et al (183) carried out an extensive study of the binding of DAB and its noncarcinogenic isomer, 2-Me-DAB, to macromolecules in rat liver following a single oral administration. In general, there was greater binding of the carcinogen to DNA than of the noncarcinogen, and the time course of binding was different. In the case of various proteins, the binding of the noncarcinogen was the highest. They concluded that binding to DNA rather than proteins was correlated with carcinogenicity. A similar comparison was made by Albert & Warwick (184), who found radioactivity associated with all fractions of liver and spleen cells; more label from the carcinogen was associated with the nuclear fraction, whereas the opposite occurred in the mitochondrial fraction. They also found that DAB was extensively bound to acidic nuclear proteins with a higher specific activity than to total histones (185). Kriek has isolated and partially characterized a derivative of AAF which remains bound to rat liver DNA for a long period of time. The bound derivative is associated with deoxyguanosine, but is not identical to the product obtained chemically from N-OAc-AAF (186). He suggested that the structure involves a carbon-carbon bond between C-8 of guanine and the 3-carbon of AAF. What role this derivative plays in the carcinogenic process is not known.

Sorof and his group have worked for many years on a rat liver protein to which DAB and its derivatives are covalently bound, and which is not identical to ligandin (151). They isolated the principal liver azoprotein in rat liver cytosol following feeding repeated or single doses of 3'-Me-DAB; it is relatively basic and is termed slow h_2-5S azoprotein (187). A specific antiserum prepared against this protein gives only a single diffusion band with the purified conjugate. This antiserum reacted with two proteins in the rat liver cytosol, one of which was identical to the purified conjugate, which has been immunologically detected primarily in rat and mouse livers (188). In a quantitative immunological investigation, this protein target was found in azo dye-induced hepatomas in considerably less quantity than in differentiated hepatomas induced by other chemicals or in normal liver (189). The purified azoprotein consists of two identical subunits of molecular weight 44,000 with an average of two azo dye residues bound per subunit; its amino acid composition was given (190). Although several reports have attributed various enzymatic activities to this protein, none have been substantiated, and the role of this protein in azo dye carcinogenesis remains to be elucidated.

A classical series of investigations on the physical, chemical, and biochemical consequences of the covalent reaction of N-OAc-AAF with various nucleic acids and polynucleotides has been carried out by Weinstein, Grunberger, and their collaborators. In their initial study they reacted N-OAc-AAF with mixed *Escherichia coli* tRNAs. As the result of the covalent attachment of fairly small numbers of AAF residues, there was inhibition of acceptor activity for several activated amino acids, inhibition of the codon recognition ability of lysine-tRNA, inhibition of binding of the charged tRNA to ribosomes, and altered chromatographic behavior (191). It was suggested that the AAF substitution at C-8 produced a change of conformation of that nucleoside from *anti* to *syn*, which could account for the observed alteration of properties of the tRNA. They also studied the binding of AAF and N-OH-AAF to rat liver RNAs in vivo. At 12 hr after intraperitoneal injection the tRNA had a higher specific activity than did the 5, 18, and 28S RNAs; at 24 hr the specific activities of all RNAs were comparable (192). They also developed a benzoylated DEAE-cellulose column that completely separated the substituted from the unsubstituted tRNAs, and showed that AAF substitution occurred on several species of tRNA (192). N-OAc-AAF was reacted with oligonucleotides and poly(U_3G), and degradation indicated that only guanine residues were substituted. Such substitutions markedly impaired the function of the oligonucleotides at directing the binding of specific aminoacyl-tRNAs to ribosomes, and also inhibited the template function of the polynucleotide in protein synthesis (193). Major conformational changes involving rotation around the glycosidic bond and alterations in stacking interactions were inferred from circular dichroism spectra of the substituted oligo and polynucleotides (194). Further experiments with circular dichroism and nuclear magnetic resonance (NMR) strengthened the conclusions that major conformational changes resulted from AAF substitution, and demonstrated that there is strong base stacking between the AAF moiety and the base adjacent to the modified guanosine. Computer-generated models of altered conformations were presented (195) to account for the biological effects observed. Several oligonucleotides were prepared and purified in which an AAF-substituted guanosine is adjacent to various specific codons, and these were compared with unsubstituted analogs for their codon recognition properties as measured by the binding to ribosomes of specific charged tRNAs. It was found that if the modified G is part of a codon, that codon is inactivated. If a codon involving an A is adjacent to the modified G, codon recognition is impaired, which did not occur when the modified G is adjacent to a U (196). Their studies on the specific modification of RNAs reached a culmination in the demonstration that limited substitution of *E. coli* fmet-tRNA with N-OAc-AAF occurred only on the guanosine residue at position 20 in the nonhydrogen bonded dihydrouracil loop. This specific substitution resulted in an inhibition of methionine acceptor and transformylase activities, but codon recognition, which occurs at a site in the molecule distant from the substitution, was unaffected (197).

The effects of AAF substitution of DNA have also been studied. Troll et ál (198) isolated DNA from the livers of rats that had been fed AAF and used it as a template for partially purified *Micrococcus lysodeikticus* RNA polymerase and a regenerating liver DNA polymerase system. The DNA from AAF treated livers had a diminished template activity for RNA polymerase, and such activity was absent in DNA treated chemically with N-OAc-AAF. In the in vivo substituted DNA, template activity for the DNA polymerase was unaffected. Zieve (199) found that DNA treated with N-OAc-AAF had markedly decreased template activity for purified *E. coli* RNA polymerase, even at a level of substitution so low that there was no effect on the T_m of the DNA. The inactivation of the template was rapid, and the treated DNA did not affect the transcription of control DNA, indicating that the substituted DNA did not inactivate the enzyme. However, the treated DNA bound more polymerase than normal, and poly(A) synthesis was increased over the control, presumably because the substitution resulted in the production of local regions of denaturation in the DNA template.

Weinstein and his group studied the effect of AAF modifications on the conformation of DNA and double stranded RNAs. They found (200) that N-OAc-AAF reacted more extensively with denatured than with native DNA, and a high ionic environment which stabilizes DNA secondary structure decreased the reaction. The substituted native DNA had a lower T_m and intrinsic viscosity than normal. The susceptibility of double stranded reovirus RNA to ribonuclease was increased by AAF substitution, and the hybridization of E. coli ribosomal RNA to homologous DNA was also inhibited. These observations suggest that local regions of denaturation are produced in the DNA molecule by AAF substitution. On the basis of this and previous physical evidence, they proposed a "base displacement" model, in which substitution produces a rotation around the glycosidic bond to the syn conformation; as a consequence the modified guanine is postulated to be displaced from its normal stacking within the double helix, while the AAF residue is inserted and stacks coplanar to the adjacent base (200). Such a model could lead to base substitution, frameshift, and small deletion mutants that we had observed in T_4 bacteriophage treated with N-OAc-AAF (201).

Fuchs & Daune (202) carried out extensive physical studies on DNA that had been modified with AAF, and they measured the extent of substitution by the effect on the T_m of the DNA. Analysis of T_m curves and circular dichroism led them to conclude that the modified bases are shifted outside the helix, in agreement with Weinstein's base displacement hypothesis. Light scattering studies showed that substitution reduced the radius of gyration, indicating that hinge points were introduced into the DNA molecule. They also demonstrated (202) that a few crosslinks were introduced into the DNA, which were destroyed by alkaline conditions. On the other hand, Chang et al (203) treated denatured T_4 bacteriophage DNA with N-OAc-AAF and measured its renaturation kinetics; from experiments with electrical dichroism, they concluded that the AAF residues were located outside the DNA helix, in contradiction to the previous two studies (201, 202). Hopefully, this discrepancy can soon be resolved, so that the structure of AAF-substituted double helical DNA can be firmly established to serve as a molecular basis for understanding the biological effects of such substitution.

Finally, Maher et al (204) examined the effects of AAF substitution on the biological activity of Bacillus subtilis transforming DNA and found that substitution led to a major inhibition of transforming activity, up to a 100-fold increase in the frequency of mutations in the transforming DNA and a decreasing buoyant density of the DNA in CsCl. The strains of the bacteria that could repair ultraviolet damage could restore 50% of the transforming activity that was lost, and the induced mutations were reversible, suggesting that they were caused by single base pair changes. Similar results were produced by treatment of the transforming DNA with N-benzoyloxy-AAF, AAF-N-sulfate, and N-benzoyloxy-DAB.

Nitrosamines

In 1950 Magee & Barnes found that dimethylnitrosamine fed to rats produced a very high incidence of hepatomas (205). This discovery opened up a very important field of research with many practical as well as theoretical consequences. Simple aliphatic nitrosamines produce a variety of malignant tumors, most commonly in liver, kidney, lung, and esophagus (41, 206), and they are effective carcinogens by several different routes of administration. The whole field of nitrosamines has been reviewed extensively by Magee & Barnes (41) and by others (206–214). The importance of nitrosamines as environmental carcinogens was first suggested by Druckrey et al in 1963 (215), who postulated that secondary aliphatic amines could react in the body with sodium nitrite to give nitrosamines. This in situ formation of powerful carcinogenic compounds could be quite important, since both components of the reaction are ubiquitous in foods. This suggestion has been amply corroborated (41, 45, 48–50, 207, 208, 210–212, 216, 217). The nitrosamines also require microsomal metabolic activation before their carcinogenic and mutagenic activities are exerted (41, 208, 218). As a result of this activation, the nitrosamines alkylate nucleic acids and proteins (41, 209, 219).

Following Druckrey's suggestion (215), it was found that secondary aliphatic amines react with sodium nitrite at acidic pH to give nitrosamines. Mirvish, who has carried out extensive studies of such chemical reactions, has shown that the active nitrosating agent is N_2O_3, that there is a fairly satisfactory correlation between the rate of nitrosation and induction of lung adenomas by feeding amines and nitrite, and that there are a number of natural products that are easily nitrosated. He has also reviewed the field (210). Sander et al were the first to produce stomach tumors when secondary amines and nitrite were fed, and they have reviewed this field (211). However, the production of nitrosamines is not limited to secondary amines, since Lijinsky et al (220) found that some drugs with tertiary amino groups gave rise to dimethylnitrosamine on nitrosation; oxytetracycline and antipyrine gave particularly large yields, and other drugs produced small, but significant amounts (221). Lijinsky et al (217) have studied the conditions under which a series of naturally occurring tertiary amines could be nitrosated and cleaved to give nitrosamines, and in some cases considerable yields were obtained under conditions of physiological pH and temperatures. Although these nitrosation reactions generally proceed more rapidly at acidic pH, it has recently been found that formaldehyde, which is present in the environment, catalyzes the formation of nitrosamines from secondary amines at neutral or basic pH (222).

It is clear from the above considerations that nitrosamines have a very high potential for being human carcinogens. This has been emphasized by Lijinsky & Epstein (216) who pointed out that nitrates occur widely in nature, can be enzymatically reduced to nitrites that are found in green vegetables, and are extensively used as preservatives for fish and meat. Secondary amines are present in fish products and can be produced by pyrolytic reactions of diamines during cooking. Secondary amines can occur in wine, food flavoring, tobacco, toothpaste, etc, and represent potential hazards. Consequently, major efforts are now being devoted to analyses of foods, other materials, and human excreta for nitrosamines. The analytical methods have been critically reviewed by Preussmann & Eisenbrand (213). Since it is known that the Japanese have a very high incidence of stomach cancer (49–51) and eat much fish, nitrosamines become suspect as causative agents. The Japanese foodstuffs are being analyzed for secondary amines, nitrite, and nitrosamines, and are high in these substances (212).

One of the first indications of metabolism in vivo of dimethylnitrosamine (DMN) stems from the experiments of Dutton & Heath (223), who administered ^{14}C-DMN to rats and obtained a high yield of respiratory $^{14}CO_2$. Subsequent work showed that there was considerable incorporation into tissue components, including DNA, RNA, and proteins. The metabolism of nitrosamines has been thoroughly reviewed by Krüger (209). DMN gives rise to formaldehyde in the first step of its metabolism, but in nitrosamines with alkyl groups larger than methyl some α- and β-oxidation is required. However, characterization of the urinary metabolites of a series of dialkyl nitrosamines has revealed that ω-oxidation to the corresponding alcohols and carboxylic acids occurs (224). The primary metabolism of nitrosamines is carried out by the microsomal, cytochrome P-450 mixed function oxidases (41, 208, 225).

It is known that dimethylnitrosamine-^{14}C covalently labels tissue nucleic acids and proteins (41). In 1962 Magee & Farber (219) injected rats with ^{14}C-DMN, isolated the RNA from various tissues, and demonstrated unequivocally the presence of labeled N-7-methylguanine. This proves that DMN acts as a methylating agent for nucleic acids. The mechanism of the metabolic activation and alkylation by DMN was established in an elegant experiment by Lijinsky et al (218). They injected completely deuterated DMN into rats, isolated the liver DNA and RNA, hydrolyzed them, and isolated the 7-methylguanine. The 7-Me-G was analyzed by mass spectrometry and three deuterium atoms were present (Figure 4). It had widely been thought that diazomethane was the metabolically active form; however, if this were true the 7-Me-G would contain only two rather than three deuterium atoms. Thus (Figure 4), the methyl carbonium ion is the active methylating agent and is produced without the intermediate formation of diazomethane.

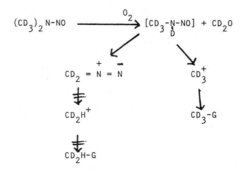

Figure 4   Activation of dimethylnitrosamine (218).

Swann & Magee (226) injected four methylating agents, including DMN, into rats, isolated the total nucleic acids, and analyzed for 7-Me-G. All the compounds produced similar amounts of 7-Me-G, although their carcinogenic activities and tissue specificities were very different. They also found that diethylnitrosamine in vivo and another ethylating agent led to 7-ethyl-G, but again there was no correlation with carcinogenic activity (227). Lijinsky et al (228) observed that some carcinogenic cyclic nitrosamines did not significantly alkylate nucleic acids on the N-7 of guanine. Krüger (229) found a fiftyfold greater formation of 7-Me-G from the odd numbered side chain of di-β-hydroxypropylnitrosamine as compared with di-n-propylnitrosamine, which supports the β-oxidation hypothesis of metabolic activation. Wünderlich et al (230) found that DMN labeled mitochondrial DNA from liver and kidney 4.5 times more than the nuclear DNA of the same tissues; on this basis they suggest that carcinogenesis might result from cytoplasmic mitochondrial mutations.

Hennig et al (231) fractionated by density gradient centrifugation the livers of rats 12 hr after the injection of DMN. They found the highest specific activity in the high molecular weight heterogeneous RNA, which is considered to be the precursor of mRNA; rRNA had the next highest specific activity. The presence of 7-Me-G in mRNA led to a disaggregation of ribosomes and they postulated that this was the cause of the cytotoxic effects of DMN.

Since a single injection of DMN causes hepatomas when given to rats after partial hepatectomy, Craddock (232) studied the binding of labeled DMN under such conditions. For the first time, O-6-Me-G was detected in the DNA, whereas none was detected after injection of methyl methanesulfonate (MMS), a methylating agent that is not carcinogenic to liver. Similarly, O'Connor et al (233) detected O-6-Me-G following DMN and not MMS administration to rats. The O-6-Me-G was lost from the DNA with a half-life of 13 hr, whereas N-7-Me-G had a half-life of 3 days. An additional unstable and uncharacterized product of alkylation of the DNA was also observed. The latter two studies suggest that if alkylation of DNA is important in the carcinogenic process, O-6-Me-G is most likely to be the critical site of alkylation, as had originally been proposed by Loveless (235) for mutagenesis. Schoental (234) has suggested, however, that nitrosamines may be converted metabolically to bifunctional reagents.

Nitrosamides

The aliphatic nitrosamides are powerfully carcinogenic and mutagenic compounds that do not require metabolic activation. Much of the pioneer work in this field, particularly with respect to mechanism of action, has been done by Lawley, who has reviewed the field (236, 237), as have Magee & Barnes (41). These compounds, which are exemplified by nitrosomethylurea (NMU), N-methyl-N'-nitro-N-nitrosoguanidine (MNNG), and methyl methanesulfonate (MMS), have very interesting organ specificity for tumor induction. Members of the series of methyl nitrosoureas induce squamous carcinomas of the forestomach, gastric carcinomas and adenocarcinomas, and various carcinomas of the large and small intestines (238). When ethyl nitrosourea (ENU) was given to pregnant rats, malignant tumors of the nerves and brains were induced in the offspring; Ivankovic called attention to these findings as possible models for transplacental carcinogenesis in humans (239). MNNG has induced cancer of the stomach in rats, hamsters, and dogs; the latter species is being used for model studies for the high incidence of stomach cancer in Japan (240). MNNG produces stomach cancer in rats by a single dose (241), and N-nitroso-N-butylurea is highly leukemogenic in rats (242).

Lawley has produced evidence that the more powerfully carcinogenic NMU and MNNG react with nucleophiles by SN_1 (unimolecular kinetics) mechanisms, and the less carcinogenic alkylating agents, dimethyl sulfate and MMS, react by SN_2 (bimolecular kinetics) mechanisms (237). These compounds produce rather different minor products on reactions with DNA and RNA. Lawley in 1968 (243) demonstrated that NMU and MNNG reacted with DNA in neutral solution to give primarily 7-Me-G and 3-Me-A. He and his colleagues found that the RNA isolated from livers of rats treated with MMS did not contain O-6-Me-G, which was detected following DMN administration (244); these compounds also produced different proportions of the minor bases, 3-Me-C, 1-Me-A, and 7-Me-A. Lawley & Thatcher, in a remarkably thorough study of MNNG in cultured mammalian cells (245), found that the compound is activated by thiols and that the amount of alkylation of the nucleic acids was influenced by the intracellular thiol concentration; the proteins were methylated to a lesser extent than the nucleic acids. Both chemically and in cells, MNNG produced appreciable amounts of O-6-Me-G, whereas dimethyl sulfate did not. Rather similar findings were made by Swann & Magee (226). Lawley & Shah (246), who critically examined the methods for the analysis of small amounts of methylated bases in DNA, identified 3- and 7-Me-A, 1-Me-A, 3-Me-C, and O-6-Me-G as products of nitrosamides, which also reacted with the ribose and phosphate moieties of the nucleic acids to give unidentified products, possibly phosphotriesters, which led to the degradation of RNA. When Chinese hamster ovary cells were treated with labeled O-6-Me-G, it was de-methylated to G, and no O-6-Me-G was found in the nucleic acids (247). Lawley & Shah (248) reacted ^3H and ^{14}C NMU with DNA at various pHs and identified the following products: 3-, 7-, and O-6-Me-G, 1-, 3-, and 7-MeA, 3-Me-C, and 3- and O-4-Me-T. Moreover, the ratio of ^3H to ^{14}C showed that the methyl group was transferred intact without the intermediacy of diazomethane, as had been found for DMN by Lijinsky et al (218). They also found that NMU and NEU given by injection into the portal vein of rats were localized in many tissues and that the methyl and ethyl groups were transferred intact; at the specific activities used, no methylation or ethylation of DNA was detected (249).

Pegg (250) treated several tRNAs with labeled NMU and NEU and found the usual methylated bases. However, when fmet-tRNA was treated, the distribution of methyl groups throughout the molecule was random, indicating that its tertiary structure did not affect the methylation reaction, in contrast to the very specific reaction of N-OAc-AAF with the same tRNA (197). Saito & Sugimura (251) showed that MNNG reacted with the DNA and histones of many rat tissues in vivo following oral administration and that there was no appreciable degradation of the methylated DNA.

Goth & Rajewsky administered labeled NEU to pregnant rats and found 7-Et-G in various tissues of the fetus, particularly in brain, but there was no correlation between the amount of N-7 alkylation and transplacental carcinogenic activity (252). However, they found (253) that O-6-Et-G persisted in the DNA of the fetal brains much longer than it did in liver, and suggested that this may be responsible for the transplacental neurocarcinogenicity of NEU.

Aflatoxins

The aflatoxins are a series of mold metabolites. Wogan & Newberne (254) discovered that aflatoxin B_1 is the most powerful liver carcinogen known for the rat and is hepatocarcinogenic in a wide variety of species (255). Aflatoxins are suspect as causes of hepatomas that are in very high incidence among the Bantus of Africa (49, 50). Wogan and his colleagues have studied the structure-activity relationships among a number of aflatoxins and derivatives (256), as well as their binding to DNA and other nuclear effects (257). Although they could not measure any binding to DNA in vivo, they studied noncovalent binding in the test tube by equilibrium dialysis and found no correlation with carcinogenic activity. DNA so treated inhibited RNA polymerase activity and produced a number of ultrastructural changes in hepatocyte nuclei. Lijinsky et al (258) found covalent binding to liver DNA, RNA, and proteins following administration of aflatoxin B_1 to rats, whereas no covalent binding was observed in chemical reactions, suggesting that the aflatoxins require metabolic activation.

Garner et al (259) found that incubation of aflatoxin B_1 with rat liver microsomes under the conditions for mixed function oxidases, produced a reactive species that was highly toxic to *Salmonella typhimurium*. The product bound covalently to tRNA, and since the addition of tRNA decreased the lethal effect of the metabolite against the bacterium, it was concluded that the same metabolite that was bound was also lethal (260). Gurtoo found spectroscopically that aflatoxin B_1 binds covalently to rat liver microsomes (261). Similarly, using labeled material, Garner found that following action of a microsomal mixed function oxidase system, aflatoxin B_1 is bound covalently to added DNA, tRNA, some polyribo-

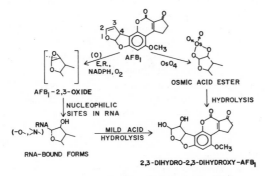

Figure 5   Activation of aflatoxin B_1 (264).

nucleotides, and proteins (262). Garner (263) obtained chemical evidence for a reactive product in hamster liver microsomes that gave rise to two water-soluble products. Although he could find no evidence for a dihydrodiol intermediate, he concluded that an epoxide was the intermediate. However, Swenson and the Millers (264) also incubated aflatoxin B_1 with rat and hamster liver microsomes, isolated the product-bound ribosomal RNA, degraded it with mild acid, and isolated the 2,3-dihydro-2,3-dihydroxy derivative, whose structure was confirmed by synthesis. This provides evidence that the activated metabolite is an epoxide, as shown in Figure 5.

Campbell et al (265) extracted the urine from Philippine children who were known to have ingested aflatoxin B_1 in contaminated peanut butter, and fed the extracts to rainbow trout. The small amount of aflatoxin M_1 present in the urine accounted for all the hepatocarcinogenicity, which suggests that aflatoxin B_1 is excreted mainly as detoxified metabolites or conjugates thereof.

Although several more types of chemical carcinogens require metabolic activation, space limitations preclude their inclusion here.

CHEMICAL ONCOGENESIS IN CELL CULTURES

Because of the obvious advantages of studying the events of carcinogenesis in well-controlled systems free from the influences of the host, it became necessary to develop cell culture model systems in which malignant transformation could be quantitatively investigated. Much has been accomplished, and the subject has recently been reviewed by Casto & DiPaolo (266) and myself (8). Space limitations preclude consideration of many important topics, such as membrane changes, biochemical effects of transformation, somatic cell fusion, etc. Consequently, attention is devoted to a description of the systems themselves, with emphasis on the topics considered above. The vast majority of the research in this field has been done with fibroblasts, which produce sarcomas on inoculation into suitable hosts. Studies on the malignant transformation of epithelial cells that give rise to the clinically much more important carcinomas are in their infancy. Thus, at present it is impossible to determine how many of the conclusions that have been drawn from studies with fibroblasts will also apply to epithelial cells. The semantics and criteria of "transformation" have been discussed by Freeman & Huebner (267).

In 1961 Gey reported on the spontaneous malignant transformation of cultured fibroblasts (268), and Earle & Nettleship found that the transformation they had previously thought to have been produced by PAH in mouse embryo cell lines was actually spontaneous (269), a subject that Sanford has reviewed (270). The next sections briefly summarize important topics covered in my review (8).

Hamster Embryo Fibroblasts

The first quantitative system for oncogenesis in cultures was developed by Berwald & Sachs (271). They showed that various PAH, after addition to primary or secondary cultures of hamster embryo cells, caused the production of transformed colonies which contained cells that piled up in a cross-cross array, in contrast to the flat morphology of nontransformed colonies. They also showed that mass cultures containing transformed cells gave rise to sarcomas on inoculation into hamsters. Huberman & Sachs (272) concluded from the dose-response curve that BP transforms the cells by a single-hit mechanism, and Gelboin et al (273) found that these cells metabolically produced 3-OH-BP, which was considerably more toxic than BP but did not produce transformation. Kuroki & Sato (274) reported that mass cultures of hamster embryo cells were transformed by 4-nitroquinoline-1-oxide (NQO) and that morphological transformation preceded the acquisition of malignancy.

Much research with these cells has been done by DiPaolo and his group, who have developed the quantitative aspects of the system (275, 276) and showed that individual piled-up colonies produced sarcomas in hamsters (277). They studied the effects of transformation on chromosomal morphology (278), as discussed below, and found that a variety of carcinogenic compounds other than PAH also transformed these cells (279).

Mouse Fibroblast Cell Lines

Permanent lines of mouse fibroblasts, used extensively for research on oncogenic viruses, have also been very useful for studies of chemical oncogenesis. In our laboratory we developed a quantitative system using mouse prostate fibroblasts (280), in which epoxides showed greater activity than the parent hydrocarbons (91, 132, 133). The Balb/3T3 mouse embryo cell line developed by Aaronson & Todaro (281) has been used for research on chemical oncogenesis by DiPaolo et al (282) and by Kakunaga (283), who found that in addition to forming piled-up colonies, the transformed cells also gave rise to colonies in soft agar. In our laboratory we have developed cell lines derived from C3H mouse embryos, termed C3H/10T1/2 (284), which are transformed to malignancy by PAH and MNNG (285). Sanders and his colleagues have produced transformation in several cell lines with NMU (286, 287), and he has proposed a mathematical treatment of dose-response curves for transformation (288).

In all these cell lines, scoring of transformation is done by measuring piled-up colonies, which are produced as a result of loss of density-dependent inhibition of cell division, sometimes referred to as a loss of contact inhibition.

Virus-Infected Fibroblastic Cell Lines

Enhanced transformation by chemicals of virus-infected cell lines has been observed by Freeman, Rhim, and their colleagues. They found that diethylnitrosamine transformed rat embryo cells that had been productively infected with Rauscher leukemia virus, whereas no morphological transformation occurred with the chemical or virus alone (289). Similar results were obtained with DMBA (290), and tumors obtained in rats by inoculation of transformed cells contained the group-specific antigen of oncornaviruses (C-type RNA tumor viruses). By contrast, in late passage hamster embryo cells, chemicals such as PAH did produce transformation in the absence of deliberate infection of the cells with oncornaviruses (291). In long passaged rat embryo cells there was a spontaneous "switch on" of group-specific antigens of oncornaviruses (292). In studies of the combined effects of MCA and Rauscher leukemia virus in rat embryo cells, it was found that the virus had to be added prior to MCA in order for transformation to occur (293). Mouse embryo cells infected with the AKR leukemia virus were transformed by extracts of smog (294).

Epithelial Cell Transformation

Most attempts to transform epithelial cells have been made with rat liver cell cultures that are notoriously difficult to maintain in a differentiated state. The pioneer in this field has been Katsuta, who has described many types of liver cell cultures and finally achieved malignant transformation following treatment with NQO (295). He has discussed the likelihood that criteria for transformation that have been used so widely for fibroblasts, such as piling-up and loss of density-dependent inhibition of cell division, will not apply to epithelial cells (296). Spontaneous transformation, which plagues the fibroblast field, also occurs in rat liver cell cultures (297, 298). Iype has carefully developed techniques for culturing differentiated rat liver cells as monolayers and is characterizing them enzymatically and karyotypically as a prelude for chemical transformation investigations (299). The presence of liver-specific cell surface antigens is maintained for at least 6 mo in these cultures (300). When they were treated with NMU, new cell surface antigens were produced that were common to those of MCA and DAB, but were different from those on aflatoxin- and N-OAc-AAF-treated cells (301).

Williams et al (302) described the preparation and culturing of epithelial liver cells from 10-day-old rats, which on prolonged treatment with aflatoxin B_1, N-OH-AAF, and DMBA gave rise to cultures that exhibited no distinct characteristics of transformed cells, but nevertheless gave rise to carcinomas on inoculation into isologous rats. It is interesting that these carcinomas did not histologically resemble hepatomas (303). Montesano et al (304) treated epithelial-like cells obtained from livers of young rats with DMN and MNNG, and while no morphological changes were seen, the treated cells grew in soft agar and produced tumors on inoculation into rats; the DMN-treated cells gave differentiated adenocarcinomas, whereas the MNNG-treated cells produced carcinosarcomas, suggesting that the original cultures contained a mixed population of cells. Yamaguchi & Weinstein (305) reported transformation of rat liver epithelial cells exposed to N-OAc-AAF; these cells grew in soft agar and produced tumors of unspecified histology on inoculation. They were able to select for a clone that was temperature sensitive for the property of growing in soft agar, which the untreated cells did not.

Elias et al (306) have prepared epithelial-like cells from full term mouse fetal skin which, after treatment with DMBA or DMBA plus Tween 80, accelerated their growth rate and remained differentiated in appearance, but gave rise to un-differentiated tumors on inoculation; many ultrastructural features of the cells were described. Vitamin A produced ultrastructural changes in cultures of mouse epidermal cells (307).

Brown has established an epithelial cell culture from rat salivary gland, which was susceptible to the toxicity of DMBA and MCA, and formed piled-up colonies after such treatment (308). When these cells were injected, both carcinomas and sarcomas were obtained, showing that the cells were present as a mixed population. Leuchtenberger & Leuchtenberger (309) found that cigarette smoke condensates produced cytotoxicity, abnormal proliferation, and excess DNA content per cell in cultures of hamster lung cells.

Recent Developments with Hamster Embryo Cells

The following papers have come out since my recent review (8). Markovits et al (310) have succeeded in obtaining morphological transformation with PAH of hamster embryo cells without using feeder layers of irradiated cells as the Sachs and DiPaolo groups do. Lubet et al (311), by a microspectrophotometric method, showed that 3-OH-BP, produced intracellularly from BP, exerted considerable cytotoxicity to these cells and transformed them very poorly, thus confirming earlier results of Gelboin et al (273). Popescu & DiPaolo (312) found autoradiographically that the incorporation of tritiated DMBA into hamster embryo cells was inhibited by cycloheximide.

In hamster embryo cells good correlations have been observed between the carcinogenic activities of various compounds and their ability to produce transformation: PAH (271, 272, 276, 278); epoxides of PAH (134); 4-nitroquinoline-1-oxide (274); arabinosylcytosine, 5-fluoro-2'-deoxyuridine, and hydroxyurea (313); sodium nitrite (314); aflatoxin B_1, N-OAc-AAF, MNNG, and methylazoxymethanol (279). Evidently, hamster embryo cells are capable of activating those of the above compounds that require it. However, DiPaolo et al (279) found that urethane and DMN did not transform these cells, presumably because they were not metabolically activated. Therefore, they devised a system for producing metabolic activation of these and other compounds by giving them to pregnant hamsters and preparing cell cultures from their embryos. They observed transformation in the cultures from the embryos treated transplacentally with urethane and DMN, and not in the controls (317); this host-mediated in vivo, in vitro system was proposed as a general means to obtain metabolic activation and transformation in cells that do not metabolize potential carcinogens; transformation in this system was produced by 12 carcinogens that did not directly transform the cells.

Umeda & Iype (315) have proposed that an improved expression of absolute transformation rates in hamster embryo cells can be obtained by plotting the transformation frequency against relative plating efficiency, rather than against concentration as is commonly done. Finally, Fujimoto (316) has transformed primary C3H mouse embryo cultures with NQO.

Recent Developments with Transformable Cell Lines

A good correlation between carcinogenic activities of compounds and the extent of transformation they produce in culture has been achieved in cell lines: PAH in mouse prostate fibroblasts (280); epoxides of PAH in mouse prostate fibroblasts (91, 132, 133, 318); PAH in Balb/3T3 cells (282, 283); PAH and MNNG in C3H/10T1/2 cells (285); NQO and MNNG in Balb/3T3 cells (283); and NMU in Chinese hamster lung cell lines (286). Further studies must be carried out with some additional means to provide metabolic activation before such cell lines can be validated as primary screens for oncogenic activity.

Marquardt (319) has demonstrated that poly(I): poly(C) treatment of prostate fibroblasts completely prevents transformation by DMBA and MNNG, which could not be ascribed to the interferon-inducing properties of the polymer; the explanation for this effect is presently unknown. Parodi et al (320) have obtained good cloning efficiencies of C57B1/6 mouse embryo cells in microtiter plates with conditioned medium in the presence of MCA, but no transformation was reported. Lasne et al (321) found enhanced transformation in rat embryo cell lines when BP and phorbol esters were applied sequentially, in analogy with initiation and promotion in mouse skin.

It is well known that PAH-induced subcutaneous sarcomas have cell surface tumor-specific transplantation antigens (TSTA) that are individual and noncross-reacting (322). We have found that an analogous situation occurs in mouse prostate fibroblasts (323), where multiple TSTAs were found in individual transformants from a highly cloned population, indicating that these TSTAs are produced during the transformation process (324). Di Mayorca et al (325) have been able to transform BHK21 cells, frequently used in studies of oncogenic viral transformation, with DMN and NMU. They observed that the transformed state was apparently conditional, since the transformed clones grew more slowly and retained a flat appearance at 32°C in contrast to their transformed colonial morphology at 38°C, and concluded that this indicates that malignant transformation is a single gene mutation. However, since the cells grew much more slowly at the lower temperature and since they only carried their cultures for 10 days, it is not clear whether or not piled-up colonies would have been found if the experiment had been continued for longer times.

We have observed a striking cell cycle dependency for transformation in C3H/10T1/2 cells treated with the short acting MNNG. Using four different methods of producing synchrony, we found the sensitive phase to occur about 4 hr prior to the onset of DNA synthesis (326). In critical studies, we were unable to find any direct relationships between oncogenic transformation and the production or repair of single strand breaks in the DNA of asynchronous (327) or synchronized cells (328). By contrast, Marquardt (329), using mouse prostate cells and the mitotic detachment method of synchronization in which the G_1 phase cannot be measured, reported that transformation occurs primarily during the S phase. Whether this disagreement results from the use of different cells or from the different method of synchronization remains to be determined.

Oshiro et al (330) found an increased uptake of 2-deoxy-D-glucose in chemically transformed Balb/3T3 cells as compared to the nontransformed ones. Kuroki & Yamakawa (331) confirmed this result and found an increase in the V_{max} of uptake without a change in K_m. Laug et al (332) found that chemically transformed cells of various kinds, including our C3H/10T1/2 cells, had a much greater activity in inducing fibrinolysis of sera than their nontransformed counterparts. This extends the work of Reich and his colleagues, who had discovered this phenomenon in virally transformed cells (333).

Oncogenesis in culture has also been observed in rat embryo cell lines. Sekely et al (334) obtained malignant transformation of such cells by treatment with various N-hydroxy- and N-acetoxy-AAF derivatives, but not by the parent AAF, and Rhim & Huebner (335) transformed similar cells with DMBA.

Interactions of Chemical Carcinogens and Oncogenic Viruses in Transformation

It has been known for some time that vaccinia virus infections enhance in vivo carcinogenesis (cf 336). On the basis of this and other considerations, Huebner & Todaro have proposed their "oncogene" theory, which among other things postulates that all cells carry the genetic information for malignancy and in most cases also for oncogenic viruses (337, 338). I have referred previously to the systems used by Freeman and Rhim to study chemical transformation in cells infected with oncogenic viruses. Considerable work has been done with such cells since I reviewed the field (8). Oncogenic virus-infected mouse and rat embryo cells were transformed by smog extracts (339). A very extensive study in high passage rat embryo cells expressing oncornaviruses has been carried out by Freeman et al (340) with over 30 compounds of several classes, and they found a good correlation between the production of cell transformation and the in vivo carcinogenic activity of the compounds, with a few exceptions mainly of compounds that the cells did not metabolically activate. Thus, this system seems partially suited for the qualitative identification of potentially oncogenic compounds and might find use as part of a primary screen for environmental carcinogens. Although I have criticized this system for being wholly qualitative (8), a rebuttal has been made (340) in which the authors point out that they express their qualitative conclusion of + or oncogenic activity on the basis of serial dilutions that are a quantitative expression. Similarly, Rhim et al (341) found a good correlation between in vivo carcinogenic activity of several types of chemical and their ability to transform mouse embryo cells infected with a nontransforming AKR leukemia virus; again a lack of metabolic activation of certain types of compounds, such as aminoazo dyes and some nitrosamines, represents a limitation of the usefulness of this system as a primary screen.

Casto and his colleagues have demonstrated that the transformation of hamster embryo cells by adenovirus SA7 is enhanced by PAH (342, 343) and other types of chemical carcinogens (344), again providing a possible primary screening system with the same limitations of lack of metabolic activation of some compounds. Ledinko & Evans (345), using the same system, have found enhancement of adenovirus transformation by MNNG, caffeine, and hydroxylamine.

Some effects of infection of mice with oncornaviruses on chemical oncogenesis have been observed. Salerno et al (346) found that the incidence of MCA-induced sarcomas was diminished in mice infected by wild-type oncornaviruses, but that the modifying effect is influenced by the strains of virus and mice, and by the age and sex of the mice. Whitmire (347) showed that radiation-induced leukemia virus infection reduced the incidence of sarcomas induced by MCA, and that infection with the Graffi leukemia virus also inhibited MCA sarcomagenesis at the proper dosages (348).

Do Chemical Carcinogens "Switch on" Oncogenic Viruses?

One of the tenets of the "oncogene" hypothesis (337, 338) is that chemical carcinogens may exert their effects solely by "switching on" oncogenic viruses that cause the malignant transformation. There is considerable experimental evidence for this phenomenon in certain circumstances. In 1963, Irino et al (349) discovered that cell free filtrates of certain chemically induced lymphomas produced lymphomas on inoculation. Basombrio (350) has found that cell free supernatant fractions from MCA-induced sarcomas injected into newborn Balb/c mice greatly increased the incidence of lymphoid neoplasms. He also found that the tumors used as the source material contained infectious oncornaviruses by the XC test, but he was unable to determine whether the presence of the viruses was necessary or accidental in the induction of the sarcomas.

Igel et al (351) detected the presence of the group-specific antigens of mouse leukemia virus in PAH-induced sarcomas of certain mice strains, and these observations were extended by Whitmire et al (352). As mentioned previously, in strains of mice that were resistant to MCA sarcomagenesis and noninducible for AHH there was an increased expression of oncornavirus group-specific antigens (78). The presence of these antigens and C-type particles in hamster tumors induced by MCA and by polyoma and SV40 viruses was demonstrated by Freeman et al (353). Weinstein et al (354) detected the presence of oncornaviruses in cell cultures obtained from primary rat hepatomas induced by aromatic amines; however, these virus particles had abnormal RNA-directed DNA polymerases.

In summary, much evidence has been obtained that oncornaviruses are present in a number of chemically induced tumors of different types. However, it is not known whether these viruses play a role in causing these tumors or whether they are innocuous passengers.

Turning now to studies carried out in cell cultures, it was found by Freeman et al (355) that hamster embryo cell lines, which were negative for infectious oncornaviruses after transformation by chemicals, on inoculation into hamsters produced tumors that contained oncornaviruses; they concluded that the chemical treatment and activation of the virus were related events. Rhim et al (356) showed that a cell line derived from rat kidney underwent morphological transformation after treatment with DMBA and acquired C-type particles with reverse transcriptase activity. Altanerova & Altaner (357) reported that avian sarcoma viruses were induced by MCA and NQO in hamster cells infected with the Schmidt-Ruppin strain of the avian sarcoma virus, and that these recovered viruses were oncogenic in young hamsters. Freeman et al (358) demonstrated a potentiation of MCA transformation of rat embryo cells on treatment with 5-bromo-2'-deoxyuridine, with concomitant appearance of the group-specific antigens of the rat leukemia virus. They suggested that the induction of the endogenous oncornavirus renders the cells more susceptible to transformation by MCA.

On the other hand, in collaboration with Drs. Ulf Rapp and R. C. Nowinski, we have investigated this phenomenon in transformable cell clones derived from lines of cells obtained from embryos of inbred mice. In a large number of clones of C3H/10T1/2 cells transformed to malignancy with several chemicals, there was no expression of oncornavirus group-specific antigens, nor of infectious virus. With cells derived from AKR mice, nontransformed and transformed clones alike produced infectious mouse leukemia virus (359). From this it can be concluded that the genotype of the mouse determines the expression of oncornaviruses and that the transformed phenotype does not. Moreover, in the C3H/10T1/2 clones, treatment with 5-iodo-2'-deoxyuridine induced the transient formation of a nontransforming oncornavirus (359). These studies suggest that the "switch on" of complete transforming viruses is not involved causally in chemical oncogenic transformation. Nucleic acid hybridization studies are now underway to determine whether the chemical oncogen "switches on" viral nucleic acids.

THEORIES OF THE CELLULAR MECHANISMS OF CHEMICAL ONCOGENESIS

At this stage of knowledge it is possible to formulate four questions about the cellular mechanisms of chemical oncogenesis that are in the process of being answered.

1. *Does the chemical oncogen transform cells to malignancy, or does it select somehow for pre-existing malignant cells, as proposed by Prehn (360)?* The original dose-response curves for the transformation of hamster embryo cells by PAH suggested that the transformation was direct (271, 272, 278). With mouse prostate fibroblasts we demonstrated in single individual cells that MCA produced a very high rate of transformation without toxicity, and concluded that there must have been a direct transformation and not a selection of pre-existing transformed cells (361). Thus, it appears that this question has been answered: the chemical oncogens directly transform cells to malignancy.

2. *Does the chemical oncogen transform cells by itself or by "switching on" an oncogenic virus that carries out the transformation?* This question has been explored above, and it can be concluded that at least in some circumstances the chemical can produce transformation without the intermediacy of complete oncogenic viruses.

3. *If the chemical transforms cells by itself, is the mechanism a mutagenic one that involves an alteration of the primary sequence of DNA, or is the mechanism a perpetuated effect on gene expression?* As was pointed out previously, it is now axiomatic that oncogenesis is initiated as a result of the covalent interaction of the chemical oncogen with a cellular macromolecule. The target molecule has not yet been identified, but the majority of workers in the field believe that it is DNA.

However, as emphasized primarily by Pierce (cf 362), oncogenesis can be considered to result from abnormal differentiation. Space limitations prevent an examination of this theory; it suffices to say that while we do not now understand the mechanisms of normal differentiation, we do know that it is a perpetuated process that does *not* involve a mutation of the DNA, but rather an alteration of gene expression. Pitot and I (363) proposed models based on Jacob and Monod's theory to show that perpetuated changes can result from a single interaction of a carcinogen with a repressor protein.

In 1955 Burdette surveyed the literature on the relationship between carcinogenesis and mutagenesis and concluded that there was no correlation; most potent carcinogens were not mutagenic and many powerful mutagens were not carcinogenic (364). However, with an understanding that most chemical carcinogens require metabolic activation, which is not carried out by the bacteria and bacteriophages in which mutagenesis is most commonly tested, it has been found that in all cases where the metabolically active form has been identified and tested, it is mutagenic. This has been found to be true in bacteriophage T4 (201), bacteria (136), cultured mammalian cells (137), and *Neurospora* (365). The relationship between mutagenesis and carcinogenesis has been reviewed by Miller & Miller (366). Ames et al (367) have found that certain strains of *S. typhimurium* are very sensitive to mutagenesis with chemical carcinogens and undergo frameshift mutations. They have coupled this system with rat liver microsomes to obtain metabolic activation and have proposed this as a general screening test for mutagenic and carcinogenic activities; they state flatly that carcinogenesis results from mutagenesis (368). Recently proposed systems for measuring chemical mutagenesis have been described and evaluated by Stoltz et al (369).

However, correlations (no matter how good they are) and assertions (no matter how strongly put) do not constitute proofs of mechanism. It is still a fact that certain powerful mutagens are not carcinogenic, and careful and critical studies remain to be performed before the mutagenesis theory of carcinogenesis will be proven.

4. *Do chromosomal alterations produce malignancy?* There is much literature in this field that is difficult to evaluate. Since I am not a cytogeneticist, I will restrict consideration without critical evaluation to recent work in cell cultures.

Sachs and his group have postulated that there are chromosomes that cause malignancy (effector, E) and chromosomes that suppress (S) malignancy and that the balance in the number of E and S chromosomes determines whether a cell is or is not malignant. Using hamster embryo cells transformed by SV40 and less malignant revertants that retain the viral genome, they have presented evidence that there are constant chromosomal changes reflecting the malignancy of the clones (370). Similar conclusions were reached from transformed and revertant clones isolated from DMN treated hamster embryo cells (371). Using more sophisticated chromosome banding techniques, they have suggested, based on the examination of a considerable number of transformed and revertant clones, that in the hamster karyotype the E chromosome is the 5_7, that the S chromosome for transformation is the 7_3, and that the S chromosome for malignancy is the 7_2 (372). Benedict (373) has reported that chromosomal changes appear in hamster embryo cells shortly after treatment with PAH, and suggested that these changes may be important in oncogenic transformation.

On the other hand, DiPaolo and his colleagues have done very careful studies of the chromosomes of cells, and while they have found a number of changes, they consider them to be random and not directly concerned in the process of transformation. This research has been done in hamster embryo cells (276, 278), in hamster embryo cells using chromosome banding techniques (374, 375), and in transformed and revertant clones of Balb/3T3 cells, in which certain marker chromosomes were found (376). Obviously, more work must be done to resolve these discrepancies before the role of chromosomal alterations in malignant transformation can be established or ruled out.

Other logical theories for viral and chemical carcinogenesis have been put forth recently. Comings (377) has suggested that cells contain structural genes that are capable of coding for transforming factors and are suppressed by specific regulator genes; he has discussed factors that may regulate the interplay of these genes. He has also suggested that the transforming information can be introduced exogenously, which is different from the "oncogene" theory. Holley (378) has proposed that the crucial change produced early in the process of malignant transformation involves an alteration in the properties of the cell surface membrane such that there would be an increased concentration inside the cells of nutrients that regulate cell growth and division. The studies on increased uptake of deoxyglucose in transformed cells (330, 331) are in accord with this concept. Finally, Temin (379), in an ingenious fashion based primarily on his "protovirus" hypothesis, has theorized on ways in which genes for neoplasia may have originated in evolution, in oncogenic viruses, and in chemically transformed cells.

In conclusion, it is again necessary to point out that space limitations forced me to restrict consideration of chemical carcinogenesis to certain topics that I consider to be of great importance, and that even within those topics the literature citations could not be complete. This is not to downgrade the importance of many other aspects of chemical carcinogenesis and transformation, such as membrane alterations, effects of cyclic AMP on cell properties, immunological effects on tumor induction, the possible role of DNA repair in chemical carcinogenesis, or mechanisms of chemical mutagenesis. All these topics have been dealt with elsewhere. What I have hoped to convey is that there are exciting recent developments in chemical carcinogenesis, and that while very few final and definitive answers to questions about mechanisms have yet emerged, there are now model systems available in cell cultures that give promise to provide such answers. While recognizing the validity of the criticism that carcinogenesis is strictly an in vivo phenomenon, I nevertheless submit that the model systems have considerable practical and theoretical importance, provided that the models are continually and critically scrutinized for their validity and relevance to carcinogenesis in animals and in man. It is interesting in this latter connection that there are currently no convincing reports on transformation of human cells by chemical oncogens. Such a finding would furnish new impetus to this field.

Acknowledgment

I am deeply grateful to Patricia F. Boshell for her invaluable assistance in the preparation of this manuscript.

Literature Cited

1. Miller, E. C., Miller, J. A. 1959. *Ann. Rev. Biochem.* 28:291–320
2. Green, M. 1970. *Ann. Rev. Biochem.* 39:701–56
3. Eckhart, W. 1972. *Ann. Rev. Biochem.* 41:503–16
4. Heidelberger, C. 1970. *Eur. J. Cancer* 6:161–72
5. Heidelberger, C. 1970. *Cancer Res.* 30:1549–69
6. Heidelberger, C. 1972. In *Topics in Chemical Carcinogenesis,* ed. W. Nakahara, S. Takayama, T. Sugimura, S. Odashima, 371–88. Tokyo: Univ. Tokyo Press
7. Heidelberger, C. 1973. *Fed. Proc.* 32:2154–61
8. Heidelberger, C. 1973. *Advan. Cancer Res.* 18:317–66
9. Clayson, D. B. 1962. *Chemical Carcinogenesis.* Boston: Little, Brown. 467 pp.
10. Hueper, W. C., Conway, W. D. 1964. *Chemical Carcinogenesis and Cancers.* Springfield, Ill: Thomas. 744 pp.
11. Daudel, P., Daudel, R. 1966. *Chemical Carcinogenesis and Molecular Biology.* New York: Interscience. 158 pp.
12. Arcos, J. C., Argus, M. F., Wolf, G. 1968. *Chemical Induction of Cancer,* Vol. 1. New York: Academic. 491 pp.
13. Süss, R., Kinzel, V., Scribner, J. D. 1973. *Cancer-Experiments and Concepts.* Heidelberg: Springer. 285 pp.
14. Hartwell, J. L. 1951. *Survey of Compounds which have been Tested for Carcinogenic Activity,* USPHS Publication No. 149; Washington, DC: US GPO; Shubik, P., Hartwell, J. L. 1957. Ibid, No. 149: Suppl. 1; Shubik, P., Hartwell, J. L. 1969. Ibid, No. 149: Suppl. 2; Carcinogenesis Program National Cancer Institute. 1973. Ibid, No. 149:1961–67; Thompson, J. I. and Co. 1971. Ibid, No. 149: 1968–69; Carcinogenesis Program National Cancer Institute. 1974. Ibid, No. 149: 1970–71
15. Pott, P. 1775. In *Chirurgical Observations,* 63. London: Hawkes, Clarke & Collins
16. See historical review in *IARC Monographs on the Evaluation of Carcinogenic Risk of the Chemical to Man.* 1973. 3:25–30. Lyon: International Agency for Research on Cancer
17. Volkmann, R. 1875. In *Beitrag zur Chirurgie.* Leipzig: Breitkopf und Härtel
18. Smoking and Health Report of the Advisory Committee to the Surgeon General of the Public Health Service. 1964. USPHS Publication No. 1103; Doll, R., Hill, A. B. 1956. *Brit. Med. J.* 2:1071–81
19. Case, R. A. M., Hosker, M. E., McDonald, D. B., Pearson, J. T. 1954. *Brit. J. Ind. Med.* 11:75–104
20. Koss, L. G., Melamed, M. R., Kelly, R. E. 1969. *J. Nat. Cancer Inst.* 43:233–43
21. Figueroa, W. G., Raszkowski, R., Weiss, W. 1973. *N. Engl. J. Med.* 288:1096–97
22. Doll, R., Morgan, L. G., Speizer, F. E. 1970. *Brit. J. Cancer* 24:623–34
23. Baetjer, A. M. 1950. *Arch. Ind. Hyg. Occup. Med.* 2:487–516
24. Doll, R. 1955. *Brit. J. Ind. Med.* 12:81–86
25. IARC. 1973. *Biological Effects of Asbestos,* ed. P. Bogovski, J. C. Gilson, V. Timbrell, J. C. Wagner. IARC Scientific Publications No. 8, Lyon
26. Neubauer, O. 1947. *Brit. J. Cancer* 1:192–251
27. *Brit. Med. J.* 1974. Editorial 1:590–91
28. Thiede, T., Chievitz, E., Christensen, B. C. 1964. *Acta Med. Scand.* 175:721–25; Thiede, T., Christensen, B. C. 1969. *Acta Med. Scand.* 185:133–37
29. Wada, S., Miyanishi, M., Nishimoto, Y., Kambe, S., Miller, R. W. 1968. *Lancet* 1:1161–63
30. Miller, E. C., Miller, J. A. 1973. In *Toxicants Occurring Naturally in Foods,* 508–49. Washington DC: Nat. Acad. Sci.
31. Herbst, A. L., Ulfelder, H., Poskanzer, D. C. 1971. *N. Engl. J. Med.* 284:878–81
32. Angervall, L., Bengtsson, U., Zetterluno, C. G., Zsigmond, M. 1969. *Brit. J. Urol.* 41:401–5
33. Muir, C. S., Kirk, R. 1960. *Brit. J. Cancer* 14:597–608
34. Wogan, G. N. 1973. In *Host Environment Interactions in the Etiology of Cancer in Man,* ed. R. Doll, I. Vodopija, 237–41. IARC Scientific Publications No. 7, Lyon
35. Bryan, G. T., Yoshida, O. 1971. *Arch. Environ. Health* 23:6–12
36. Dickens, F., Jones, H. E. H., Waynforth, H. B. 1966. *Brit. J. Cancer* 20:134–44
37. Purchase, I. F. H., Van der Watt, J. J. 1970. *Food Cosmet. Toxicol.* 8:289–95
38. IARC. 1972. *IARC Monographs on the Evaluation of Carcinogenic Risk of Chemicals to Man* 1:157–68
39. IARC. Ibid, 169–74

40. Schoental, R. 1970. *Nature* 227:401–2
41. Magee, P. N., Barnes, J. M. 1967. *Advan. Cancer Res.* 10:163–246
42. IARC. See Ref. 38, Vol. 1. 184 pp.
43. Ibid 1973. Vol. 2. 181 pp.
44. Ibid 1973. Vol. 3. 271 pp.
45. Ibid 1974. Vol. 4. 286 pp.
46. Ibid 1974. Vol. 5. 240 pp.
47. *Biological Effects of Atmospheric Pollutants, Particulate Polycyclic Organic Matter* 1972. Washington DC: Nat. Acad. Sci. 361 pp.
48. Doll, R., Vodopija, I., Eds. 1973. *Host Environment Interactions in the Etiology of Cancer in Man,* IARC Scientific Publication No. 7, Lyon. 464 pp.
49. Higginson, J. 1972. In *Environment and Cancer,* 24th Symp. Fundam. Cancer Res., 69–92. Baltimore: Williams & Wilkins
50. Higginson, J., Muir, C. S. 1973. In *Cancer Medicine,* ed. J. F. Holland, E. Frei III, 241–306. Philadelphia: Lea and Febiger
51. Haenszel, W., Kurihara, M. 1968. *J. Nat. Cancer Inst.* 40:43–68
52. Miller, E. C., Miller, J. A. 1966. *Pharmacol. Rev.* 18:806–38
53. Miller, J. A. 1970. *Cancer Res.* 30:559–76
54. Miller, E. C., Miller, J. A. 1974. In *Molecular Biology of Cancer,* ed. H. Busch, 377–402. New York: Academic
55. Miller, E. C. 1951. *Cancer Res.* 11:100–8
56. Wiest, W. G., Heidelberger, C. 1953. *Cancer Res.* 13:250–54
57. Heidelberger, C., Moldenhauer, M. G. 1956. *Cancer Res.* 16:442–49
58. Brookes, P., Lawley, P. D. 1964. *Nature* 202:781–84
59. Brookes, P. 1966. *Cancer Res.* 26:1994–2003
60. Goshman, L. M., Heidelberger, C. 1967. *Cancer Res.* 27:1678–88
61. Kuroki, T., Heidelberger, C. 1971. *Cancer Res.* 31:2168–76
62. Grover, P. L., Sims, P. 1969. *Biochem. J.* 110:159–60
63. Gelboin, H. V. 1969. *Cancer Res.* 29:1272–76
64. Conney, A. H. 1972. *Science* 178:576–86
65. Gillette, J. R. et al, Eds. 1969. *Microsomes and Drug Oxidations.* New York: Academic. 547 pp.
66. Nebert, D. W., Gelboin, H. V. 1968. *J. Biol. Chem.* 243:6242–49; 6250–61
67. Fujita, T., Shoeman, D. W., Mannering, G. J. 1973. *J. Biol. Chem.* 248:2192–2201
68. Whitlock, J. P., Gelboin, H. V. 1974. *J. Biol. Chem.* 249:2616–23
69. Whitlock, J. P., Gelboin, H. V. 1973. *J. Biol. Chem.* 248:6114–21
70. Bürki, K., Liebelt, A. G., Bresnick, E. 1973. *Arch. Biochem. Biophys.* 158:641–49
71. Nebert, D. W., Gielen, J. E. 1972. *Fed. Proc.* 31:1315–25
72. Nebert, D. W., Considine, N., Owens, I. S. 1973. *Arch. Biochem. Biophys.* 157:148–59
73. Wiebel, F. J., Leutz, J. C., Gelboin, H. V. 1973. *Arch. Biochem. Biophys.* 154:292–94
74. Bürki, K., Liebelt, A. G., Bresnick, E. 1973. *J. Nat. Cancer Inst.* 50:369–80
75. Franke, R. 1973. *Chem. Biol. Interactions* 6:1–17
76. Nebert, D. W., Benedict, W. F., Gielen, J. E. 1972. *Mol. Pharmacol.* 8:374–79
77. Benedict, W. F., Considine, N., Nebert, D. W. 1973. *Mol. Pharmacol.* 9:266–77
78. Kouri, R. E., Salerno, R. A., Whitmire, C. E. 1973. *J. Nat. Cancer Inst.* 50:363–68
79. Kouri, R. E., Ratrie, H., Whitmire, C. E. 1973. *J. Nat. Cancer Inst.* 51:197–200
80. Benedict, W. F., Gielen, J. E., Nebert, D. W. 1972. *Int. J. Cancer* 9:435–51
81. Kellermann, G., Cantrell, E., Shaw, C. R. 1973. *Cancer Res.* 33:1654–56
82. Kellermann, G., Luyten-Kellermann, M., Shaw, C. R. 1973. *Humangenetik* 20:257–63
83. Huberman, E., Sachs, L. 1973. *Int. J. Cancer* 11:412–18
84. Kellermann, G., Shaw, C. R., Luyten-Kellermann, M. 1973. *N. Engl. J. Med.* 289:934–37
85. Gelboin, H. V., Wiebel, F., Diamond, L. 1970. *Science* 170:169–71
86. Kinoshita, N., Gelboin, H. V. 1972. *Proc. Nat. Acad. Sci. USA* 69:824–28
87. Kinoshita, N., Gelboin, H. V. 1972. *Cancer Res.* 32:1329–39
88. Alexandrov, K., Frayssinet, C. 1973. *J. Nat. Cancer Inst.* 51:1967–69
89. Wattenberg, L. W., Leong, J. L. 1970. *Cancer Res.* 30:1922–25
90. Wattenberg, L. W. 1972. *J. Nat. Cancer Inst.* 48:1425–30
91. Marquardt, H., Heidelberger, C. 1972. *Cancer Res.* 32:721–25
92. DiPaolo, J. A., Donovan, P. J., Nelson, R. L. 1971. *Proc. Nat. Acad. Sci. USA* 68:2958–61
93. Oesch, F. 1973. *Xenobiotica* 3:305–40
94. Oesch, F., Morris, N., Daly, J. W., Gielen, J. E., Nebert, D. W. 1973. *Mol. Pharmacol.* 9:692–96
95. Boyland, E. 1950. *Symp. Biochem. Soc.* 5:40–54
96. Boyland, E., Sims, P. 1964. *Biochem. J.*

91 : 493–506
97. Boyland, E., Sims, P. 1965. *Biochem. J.* 97 : 7–16
98. Boyland, E., Sims, P. 1965. *Biochem. J.* 95 : 780–87
99. Sims, P. 1970. *Biochem. Pharmacol.* 19 : 2261–75
100. Sims, P. 1970. *Biochem. Pharmacol.* 19 : 795–818
101. Sims, P. 1970. *Biochem. Pharmacol.* 19 : 285–97
102. Diamond, L., Sardet, C., Rothblatt, G. M. 1968. *Int. J. Cancer* 3 : 838–49
103. Diamond, L. 1971. *Int. J. Cancer* 8 : 451–62
104. Duncan, M., Brookes, P., Dipple, A. 1969. *Int. J. Cancer* 4 : 813–19
105. Duncan, M. E., Brookes, P. 1970. *Int. J. Cancer* 6 : 496–505
106. Duncan, M. E., Brookes, P. 1972. *Int. J. Cancer* 9 : 349–52
107. Huberman, E., Selkirk, J. K., Heidelberger, C. 1971. *Cancer Res.* 31 : 2161–67
108. Grover, P. L., Sims, P. 1970. *Biochem. Pharmacol.* 19 : 2251–59
109. Grover, P. L., Forrester, J. A., Sims, P. 1971. *Biochem. Pharmacol.* 20 : 1297–1302
110. Kuroki, T., Huberman, E., Marquardt, H., Selkirk, J. K., Heidelberger, C., Grover, P. L., Sims, P. 1971–1972. *Chem. Biol. Interactions* 4 : 389–97
111. Jerina, D. M., Daly, J. W., Witkop, B., Zaltzman-Nirenberg, P., Udenfriend, S. 1970. *Biochemistry* 9 : 147–55
112. Daly, J. W., Jerina, D. M., Witkop, B. 1972. *Experientia* 28 : 1129–49
113. Selkirk, J. K., Huberman, E., Heidelberger, C. 1971. *Biochem. Biophys. Res. Commun.* 43 : 1010–16
114. Grover, P. L., Hewer, A., Sims, P. 1971. *FEBS Lett.* 18 : 76–80
115. Grover, P. L., Hewer, A., Sims, P. 1972. *Biochem. Pharmacol.* 21 : 2713–26
116. Sims, P. 1971. *Biochem. J.* 125 : 159–68
117. Sims, P. 1972. *Biochem. J.* 130 : 27–35
118. Sims, P., Grover, P. L., Kuroki, T., Huberman, E., Marquardt, H., Selkirk, J. K., Heidelberger, C. 1973. *Biochem. Pharmacol.* 22 : 1–8
119. Kinoshita, N., Shears, B., Gelboin, H. V. 1973. *Cancer Res.* 33 : 1937–44
120. Grover, P. L., Hewer, A., Sims, P. 1973. *FEBS Lett.* 34 : 63–68
121. Keysell, G. R., Booth, J., Grover, P. L., Hewer, A., Sims, P. 1973. *Biochem. Pharmacol.* 22 : 2853–67
122. Waterfall, J. F., Sims, P. 1973. *Biochem. Pharmacol.* 22 : 2469–83
123. Grover, P. L., Hewer, A., Sims, P.

1974. *Biochem. Pharmacol.* 23 : 323–32 ; Grover, P. L. 1974. *Biochem. Pharmacol.* 23 : 333–43
124. Selkirk, J. K., Croy, R. G., Gelboin, H. V. 1974. *Science* 184 : 169–71
125. Swaisland, A. J., Grover, P. L., Sims, P. 1973. *Biochem. Pharmacol.* 22 : 1547–56
126. Newman, M. S., Blum, S. 1964. *J. Am. Chem. Soc.* 86 : 5598–5600
127. Dansette, P., Jerina, D. M. 1974. *J. Am. Chem. Soc.* 96 : 1224–25
128. Goh, S. H., Harvey, R. G. 1973. *J. Am. Chem. Soc.* 95 : 242–43
129. Cho, H., Harvey, R. G. 1974. *Tetrahedron Lett.* 16 : 1491–94
130. Waterfall, J. F., Sims, P. 1972. *Biochem. J.* 128 : 265–77
131. Sims, P. 1973. *Biochem. J.* 131 : 405–13
132. Grover, P. L., Sims, P., Huberman, E., Marquardt, H., Kuroki, T., Heidelberger, C. 1971. *Proc. Nat. Acad. Sci. USA* 68 : 1098–1101
133. Marquardt, H., Kuroki, T., Huberman, E., Selkirk, J. K., Heidelberger, C., Grover, P. L., Sims, P. 1972. *Cancer Res.* 32 : 716–20
134. Huberman, E., Kuroki, T., Marquardt, H., Selkirk, J. K., Heidelberger, C., Grover, P. L., Sims, P. 1972. *Cancer Res.* 32 : 1391–96
135. Cookson, M. J., Sims, P., Grover, P. L. 1971. *Nature New Biol.* 234 : 186–87
136. Ames, B. N., Sims, P., Grover, P. L. 1972. *Science* 176 : 47–49
137. Huberman, E., Aspiras, L., Heidelberger, C., Grover, P. L., Sims, P. 1971. *Proc. Nat. Acad. Sci. USA* 68 : 3195–99
138. Dipple, A., Lawley, P. D., Brookes, P. 1968. *Eur. J. Cancer* 4 : 493–506
139. Dipple, A., Slade, T. A. 1970. *Eur. J. Cancer* 6 : 417–23
140. Flesher, J. W., Sydnor, K. L. 1973. *Int. J. Cancer* 11 : 433–37
141. Fried, J., Schumm, D. E. 1967. *J. Am. Chem. Soc.* 89 : 5508–9
142. Wilk, M., Girke, W. 1969. In *Physico-Chemical Mechanisms of Carcinogenesis,* Jerusalem Symposium on Quantitative Chemistry and Biology, ed. B. Pullman, E. Bergmann, 1 : 91. Jerusalem : Israel Academy of Sciences and Humanities
143. Nagata, C., Kodama, M., Tageshira, Y. 1967. *Gann* 58 : 493–504
144. Nagata, C., Inomata, M., Kodama, M., Tageshira, Y. 1968. *Gann* 59 : 289–93
145. Lesko, S. A., Hoffman, H. D., Ts'o, P. O. P., Maher, V. M. 1971. *Progr. Mol. Subcell. Biol.* 2 : 347–70

146. Caspary, W., Cohen, B., Lesko, S., Ts'o, P. O. P. 1973. *Biochemistry* 12: 2649–56
147. Irving, C. 1973. *Methods Cancer Res.* 7: 190–244
148. Abell, C. W., Heidelberger, C. 1962. *Cancer Res.* 22: 931–46
149. Tasseron, J. G., Diringer, H., Frohwirth, N., Mirvish, S. S., Heidelberger, C. 1970. *Biochemistry* 9: 1636–44
150. Kuroki, T., Heidelberger, C. 1972. *Biochemistry* 11: 2116–24
151. Litwack, G., Ketterer, B., Arias, I. M. 1971. *Nature* 234: 466–67
152. Corbett, T. H., Nettesheim, P. 1974. *Chem. Biol. Interactions* 8: 285–96
153. Slaga, T. J., Scribner, J. D., Rice, J. M. 1973. *Chem. Biol. Interactions* 7: 51–62
154. Toft, D. O., Spelsberg, T. C. 1972. *Cancer Res.* 32: 2743–46
155. Toft, D. O., Spelsberg, T. C. 1974. *J. Nat. Cancer Inst.* 52: 1351–54
156. Yuspa, S. H., Eaton, S., Morgan, D. L., Bates, R. R. 1969–1970. *Chem. Biol. Interactions* 1: 223–33
157. Yuspa, S. H., Bates, R. R. 1970. *Proc. Soc. Exp. Biol. Med.* 135: 732–34
158. Banerjee, M. R., Kinder, D. L., Wagner, J. E. 1973. *Cancer Res.* 33: 862–66
159. Meunier, M., Chauveau, J. 1973. *FEBS Lett.* 31: 327–31
160. Zeiger, R. S., Salomon, R., Kinoshita, N., Peacock, A. C. 1972. *Cancer Res.* 32: 643–47
161. Bürki, K., Stoming, T. A., Bresnick, E. 1974. *J. Nat. Cancer Inst.* 52: 785–88
162. Regan, E. G., Cavalieri, E. 1974. *Biochem. Biophys. Res. Commun.* 58: 1119–26
163. Dipple, A., Brookes, P., Mackintosh, D. S., Rayman, M. P. 1971. *Biochemistry* 10: 4323–30
164. Rayman, M. P., Dipple, A. 1973. *Biochemistry* 12: 1202–7
165. Rayman, M. P., Dipple, A. 1973. *Biochemistry* 12: 1538–42
166. Grover, P. L., Sims, P. 1973. *Biochem. Pharmacol.* 22: 661–66
167. Jones, P. A., Gevens, W., Hawtrey, A. O. 1973. *Biochem. J.* 135: 375–78
168. Baird, W. M., Brookes, P. 1973. *Cancer Res.* 33: 2378–85
169. Baird, W. M., Dipple, A., Grover, P. L., Sims, P., Brookes, P. 1973. *Cancer Res.* 33: 2386–92
170. Daudel, P., Croisy-Delcey, M., Jacquignon, P., Vigny, P. 1973. *C.R. Acad. Sci. Paris* 277: 2437–39
171. Daudel, P., Croisy-Delcey, M., Alonso-Verduras, C., Duquesne, M., Jacquignon, P., Markovits, P., Vigny, P. 1974. *C.R. Acad. Sci. Paris* 278: 2249–52
172. Miller, J. A., Miller, E. C. 1966. *Lab. Invest.* 15: 217–41
173. Miller, J. A., Miller, E. C. 1967. *Progr. Exp. Tumor Res.* 11: 273–301
174. Miller, J. A., Miller, E. C. 1971. *J. Nat. Cancer Inst.* 47: V–XIV
175. Kriek, E., Miller, J. A., Juhl, U., Miller, E. C. 1967. *Biochemistry* 6: 177–82
176. Thorgeirsson, S. S., Jollow, D. J., Sasame, H. A., Green, I., Mitchell, J. R. 1973. *Mol. Pharmacol.* 9: 398–404
177. King, C. M., Bednar, T. W., Linsmaier-Bednar, E. M. 1973. *Chem. Biol. Interactions* 7: 185–88
178. Scribner, J. D., Naimy, N. K. 1973. *Cancer Res.* 33: 1159–64
179. Bartsch, H., Dworkin, M., Miller, J. A., Miller, E. C. 1972. *Biochim. Biophys. Acta* 286: 272–98
180. Bartsch, H., Dworkin, C., Miller, E. C., Miller, J. A. 1973. *Biochim. Biophys. Acta* 304: 42–55
181. Radomski, J. L., Conzelman, G. M., Rey, A. A., Brill, E. 1973. *J. Nat. Cancer Inst.* 50: 989–95
182. Lin, J.-K., Fok, K.-F. 1973. *Cancer Res.* 33: 529–35
183. Chauveau, J., Meunier, M., Benoit, A. 1974. *Int. J. Cancer* 13: 1–8
184. Albert, A. E., Warwick, G. P. 1972. *Chem. Biol. Interactions* 5: 65–68
185. Albert, A. E., Warwick, G. P. 1972. *Chem. Biol. Interactions* 5: 61–64
186. Kriek, E. 1972. *Cancer Res.* 32: 2(?2–48
187. Sorof, S., Young, E. M. 1973. *Cancer Res.* 33: 2010–13
188. Sani, B. P., Mott, D. M., Szajman, S. M., Sorof, S. 1972. *Biochem. Biophys. Res. Commun.* 49: 1598–1604
189. Mott, D. M., Sani, B. P., Sorof, S. 1973. *Cancer Res.* 33: 2721–25
190. Sorof, S., Sani, B. P., Kish, V. M., Meloche, H. P. 1974. *Biochemistry* 13: 2612–20
191. Fink, L. M., Nishimura, S., Weinstein, I. B. 1970. *Biochemistry* 9: 496–502
192. Agarwal, M. K., Weinstein, I. B. 1970. *Biochemistry* 9: 503–8
193. Grunberger, D., Weinstein, I. B. 1971. *J. Biol. Chem.* 246: 1123–28
194. Grunberger, D., Nelson, J. H., Cantor, C. R., Weinstein, I. B. 1970. *Proc. Nat. Acad. Sci. USA* 66: 488–94
195. Nelson, J. H., Grunberger, D., Cantor, C. R., Weinstein, I. B. 1971. *J. Mol. Biol.* 62: 331–46
196. Grunberger, D., Bobstein, S. H., Weinstein, I. B. 1974. *J. Mol. Biol.* 82:

459–68

197. Fujimura, S., Grunberger, D., Carvajal, G. Weinstein, I. B. 1972. *Biochemistry* 11:3629–35

198. Troll, W., Belman, S., Berkowitz, E., Chmielewicz, Z. F., Ambrus, J. L., Bardos, T. J. 1968. *Biochim. Biophys. Acta* 157:16–24

199. Zieve, F. J. 1973. *Mol. Pharmacol.* 9:658–69

200. Levine, A. F., Fink, L. M., Weinstein, I. B., Grunberger, D. 1974. *Cancer Res.* 34:319–27

201. Corbett, T. H., Heidelberger, C., Dove, W. F. 1970. *Biochem. Pharmacol.* 6:667–79

202. Fuchs, R., Daune, M. 1972. *Biochemistry* 11:2659–66

203. Chang, C.-T., Miller, S. J., Wetmur, J. G. 1974. *Biochemistry* 13:2142–48

204. Maher, V. M., Miller, E. C., Miller, J. A., Szybalski, W. 1968. *Mol. Pharmacol.* 4:411–26

205. Magee, P. N., Barnes, J. M. 1956. *Brit. J. Cancer* 10:114–22

206. Druckrey, H., Preussmann, R., Ivankovic, S., Schmähl, D. 1967. *Z. Krebsforsch.* 69:103–201

207. Magee, P. N., Swann, P. F. 1969. *Brit. Med. Bull.* 25:240–44

208. Magee, P. N. See Ref. 6, 259–75

209. Krüger, F. W. See Ref. 6, 213–32

210. Mirvish, S. S. See Ref. 6, 279–94

211. Sander, J., Bürkle, G., Schweinsberg, F. See Ref. 6, 297–310

212. Ishidate, M. et al. See Ref. 6, 313–21

213. Preussmann, R., Eisenbrand, G. See Ref. 6, 323–40

214. Weisburger, J. H. See Ref. 50, 54–60

215. Druckrey, H., Steinhoff, D., Beuthner, H., Schneider, H., Klärner, P. 1963. *Arzneim. Forsch.* 13:320–23

216. Lijinsky, W., Epstein, S. S. 1970. *Nature* 225:21–23

217. Lijinsky, W., Keefer, L., Conrad, E., Van de Bogart, R. 1972. *J. Nat. Cancer Inst.* 49:1239–49

218. Lijinsky, W., Loo, J., Ross, A. E. 1968. *Nature* 218:1174–75

219. Magee, P. N., Farber, E. 1962. *Biochem. J.* 83:114–24

220. Lijinsky, W., Conrad, E., Van de Bogart, R. 1972. *Nature* 239:165–67

221. Lijinsky, W. 1974. *Cancer Res.* 34:255–58

222. Keefer, L. K., Roller, P. P. 1973. *Science* 181:1245–46

223. Dutton, A. H., Heath, D. F. 1965. *Nature* 178:644

224. Blattmann, L., Preussmann, R. 1973. *Z. Krebsforsch.* 79:3–5

225. Czygan, P., Greim, H., Garro, A. J., Hutterer, F., Schaffner, F., Popper, H., Rosenthal, O., Cooper, D. Y. 1973. *Cancer Res.* 33:2983–86

226. Swann, P. F., Magee, P. N. 1968. *Biochem. J.* 110:39–47

227. Swann, P. F., Magee, P. N. 1971. *Biochem. J.* 125:841–47

228. Lijinsky, W., Keefer, L., Loo, Y., Ross, A. E. 1973. *Cancer Res.* 33:1634–41

229. Krüger, F. W. 1973. *Z. Krebsforsch.* 79:90–97

230. Wünderlich, V., Tetzlaff, I., Graffi, A. 1971–1972. *Chem. Biol. Interactions* 4:81–89

231. Hennig, W., Kunz, W., Petersen, K., Schnieders, B., Krüger, F. W. 1971. *Z. Krebsforsch.* 76:167–80

232. Craddock, V. M. 1973. *Biochim. Biophys. Acta* 312:202–10

233. O'Connor, P. J., Capps, M. J., Craig, A. W. 1973. *Brit. J. Cancer* 27:153–66

234. Schoental, R. 1973. *Brit. J. Cancer* 28:436–39

235. Loveless, A. 1969. *Nature* 223:206–7

236. Lawley, P. D. 1966. *Progr. Nucl. Acid Res. Mol. Biol.* 5:89–131

237. Lawley, P. D. See Ref. 6, 237–56

238. Druckrey, H. See Ref. 6, 73–101

239. Ivankovic, S. See Ref. 6, 463–72

240. Sugimura, T. et al. See Ref. 6, 105–17

241. Hirono, I., Shibuya, C. See Ref. 6, 121–31

242. Odashima, S. See Ref. 6, 477–89

243. Lawley, P. D. 1968. *Nature* 218:580–81

244. O'Connor, P. J., Capps, M. J., Craig, A. W., Lawley, P. D., Shah, S. A. 1972. *Biochem. J.* 129:519–28

245. Lawley, P. D., Thatcher, C. J. 1970. *Biochem. J.* 116:693–707

246. Lawley, P. D., Shah, S. 1972. *Biochem. J.* 128:117–32

247. Miller, C. T., Lawley, P. D., Shah, S. A. 1973. *Biochem. J.* 136:387–93

248. Lawley, P. D., Shah, S. A. 1973. *Chem. Biol. Interactions* 7:115–20

249. Lijinsky, W., Garcia, H., Keefer, L., Loo, J., Ross, A. E. 1972. *Cancer Res.* 32:893–97

250. Pegg, A. E. 1973. *Chem. Biol. Interactions* 6:393–406

251. Saito, T., Sugimura, T. 1973. *Gann* 64:537–43

252. Goth, R., Rajewsky, M. F. 1972. *Cancer Res.* 32:1501–5

253. Goth, R., Rajewsky, M. F. 1974. *Proc. Nat. Acad. Sci. USA* 71:639–43

254. Wogan, G. N., Newberne, P. M. 1967. *Cancer Res.* 27:2370–76

255. Newberne, P. M., Butler, W. H. 1969. *Cancer Res.* 29:236–50

256. Wogan, G. N., Edwards, G. S., Newberne, P. M. 1971. *Cancer Res.* 31: 1936–42
257. Edwards, G. S., Wogan, G. N., Sporn, M. B., Pong, R. S. 1971. *Cancer Res.* 31:1943–50
258. Lijinsky, W., Lee, K. Y., Gallagher, C. H. 1970. *Cancer Res.* 30:2280–83
259. Garner, R. C., Miller, E. C., Miller, J. A. 1971. *Biochem. Biophys. Res. Commun.* 45:774–80
260. Garner, R. C., Miller, E. C., Miller, J. A. 1972. *Cancer Res.* 32:2058–66
261. Gurtoo, H. L. 1973. *Biochem. Biophys. Res. Commun.* 50:649–55
262. Garner, R. C. 1973. *Chem. Biol. Interactions* 6:125–29
263. Garner, R. C. 1973. *FEBS Lett.* 36:261–64
264. Swenson, D. H., Miller, J. A., Miller, E. C. 1973. *Biochem. Biophys. Res. Commun.* 53:1260–67
265. Campbell, T. C., Sinnhuber, R. O., Lee, B. J., Wales, J. H., Salamat, L. 1974. *J. Nat. Cancer Inst.* 52:1647–49
266. Casto, B. C., DiPaolo, J. A. 1973. *Progr. Med. Virol.* 16:1–47
267. Freeman, A. E., Huebner, R. J. 1973. *J. Nat. Cancer Inst.* 50:303–6
268. Gey, G. O. 1941. *Cancer Res.* 1:737
269. Earle, W. R., Nettleship, A. 1969. *J. Nat. Cancer Inst.* 4:213–27
270. Sanford, K. K. 1968. *Nat. Cancer Inst. Monogr.* 26:387–408
271. Berwald, Y., Sachs, L. 1965. *J. Nat. Cancer Inst.* 35:641–61
272. Huberman, E., Sachs, L. 1966. *Proc. Nat. Acad. Sci. USA* 56:1123–29
273. Gelboin, H. V., Huberman, E., Sachs, L. 1969. *Proc. Nat. Acad. Sci. USA* 64:1188–94
274. Kuroki, T., Sato, H. 1968. *J. Nat. Cancer Inst.* 41:53–71
275. DiPaolo, J. A., Donovan, P. J. 1967. *Exp. Cell Res.* 48:361–77
276. DiPaolo, J. A., Donovan, P. J., Nelson, R. L. 1969. *J. Nat. Cancer Inst.* 42:867–74
277. DiPaolo, J. A., Nelson, R. L., Donovan, P. J. 1969. *Science* 165:917–18
278. DiPaolo, J. A., Nelson, R. L., Donovan, P. J. 1971. *Cancer Res.* 31:1118–27
279. DiPaolo, J. A., Nelson, R. L., Donovan, P. J. 1972. *Nature* 235:278–80
280. Chen, T. T., Heidelberger, C. 1969. *Int. J. Cancer* 4:166–78
281. Aaronson, S. A., Todaro, G. J. 1968. *J. Cell. Physiol.* 72:141–48
282. DiPaolo, J. A., Takano, K., Popescu, N. C. 1972. *Cancer Res.* 32:2686–95
283. Kakunaga, T. 1973. *Int. J. Cancer* 12:464–73
284. Reznikoff, C. A., Brankow, D. W., Heidelberger, C. 1973. *Cancer Res.* 33:3231–38
285. Reznikoff, C. A., Bertram, J. S., Brankow, D. W., Heidelberger, C. 1973. *Cancer Res.* 33:3239–49
286. Sanders, F. K., Burford, B. O. 1967. *Nature* 213:1171–73
287. Sanders, F. K., Burford, B. O. 1968. *Nature* 220:448–53
288. Sanders, F. K. See Ref. 6, 429–44
289. Freeman, A. E., Price, P. J., Igel, H. J., Young, J. C., Maryak, J. M., Huebner, R. J. 1970. *J. Nat. Cancer Inst.* 44:65–78
290. Rhim, J. S., Creasy, B., Huebner, R. J. 1971. *Proc. Nat. Acad. Sci. USA* 68:2212–16
291. Rhim, J. S., Cho, H. Y., Joglekar, M. H., Huebner, R. J. 1972. *J. Nat. Cancer Inst.* 48:949–57
292. Rhim, J. S., Vernon, M. L., Huebner, R. J., Turner, H. C., Lane, W. T., Gilden, R. V. 1972. *Proc. Soc. Exp. Biol. Med.* 140:414–19
293. Price, P. J., Suk, W. A., Freeman, A. E. 1972. *Science* 177:1003–4
294. Rhim, J. S., Gordon, R. J., Bryan, R. J., Huebner, R. J. 1973. *Int. J. Cancer* 12:485–92
295. Yamada, T., Takaoka, T., Katsuta, H., Namba, M., Sato, J. 1972. *Jap. J. Exp. Med.* 42:377–88
296. Katsuta, H., Takaoka, T. See Ref. 6, 389–99
297. Oshiro, Y., DiPaolo, J. A. 1974. *J. Cell. Physiol.* 83:193–201
298. Borek, C. 1972. *Proc. Nat. Acad. Sci. USA* 69:956–59
299. Iype, P. T. 1971. *J. Cell. Physiol.* 78:281–88
300. Iype, P. T., Baldwin, R. W., Glaves, D. 1972. *Brit. J. Cancer* 26:6–9
301. Iype, P. T., Baldwin, R. W., Glaves, D. 1973. *Brit. J. Cancer* 27:128–33
302. Williams, G. M., Weisburger, E. K., Weisburger, J. H. 1971. *Exp. Cell Res.* 69:106–12
303. Williams, G. M., Elliott, J. M., Weisburger, J. H. 1973. *Cancer Res.* 33:606–12
304. Montesano, R., Saint-Vincent, L., Tomatis, L. 1973. *Brit. J. Cancer* 28:215–20
305. Yamaguchi, N., Weinstein, I. B. 1974. *Proc. Am. Assoc. Cancer Res.* 15:94
306. Elias, P. M., Yuspa, S. H., Gullino, M., Morgan, D. L., Bates, R. R., Lutzner, M. A. 1974. *J. Invest. Dermatol.* 62:569–81

307. Yuspa, S. H., Harris, C. C. 1974. *Exp. Cell. Res.* 86:95–105
308. Brown, A. M. 1973. *Cancer Res.* 33: 2779–89
309. Leuchtenberger, C., Leuchtenberger, R., Zbinden, I. 1974. *Nature* 247:565–67
310. Markovits, P., Coppey, J., Mazabraud, A., Hubert-Habart, M. 1973. *C.R. Acad. Sci. Paris* 277:1265–68
311. Lubet, R. A., Brown, D. Q., Kouri, R. E. 1973. *Res. Commun. Chem. Pathol. Pharmacol.* 6:929–42
312. Popescu, N. C., DiPaolo, J. A. 1973. *J. Nat. Cancer Inst.* 50:1463–69
313. Jones, P. A., Taderera, J. V., Hawtrey, A. O. 1972. *Eur. J. Cancer* 8:595–99
314. Tsuda, H., Inui, N., Takayama, S. 1973. *Biochem. Biophys. Res. Commun.* 55: 1117–24
315. Umeda, M., Iype, P. T. 1973. *Brit. J. Cancer* 28:71–74
316. Fujimoto, J. 1973. *J. Nat. Cancer Inst.* 50:79–85
317. DiPaolo, J. A., Nelson, R. L., Donovan, P. J., Evans, C. H. 1973. *Arch. Pathol.* 95:380–85
318. Marquardt, H., Sodergren, J. E., Sims, P., Grover, P. L. 1974. *Int. J. Cancer* 13:304–10
319. Marquardt, H. 1973. *Nature New Biol.* 246:228–29
320. Parodi, S., Furlani, A., Scarcia, V., Brambilla, G., Cavanna, M. 1973. *Pharmacol. Res. Commun.* 5:101–9
321. Lasne, C., Gentil, A., Chouroulinkov, I. 1974. *Nature* 247:490–91
322. Prehn, R. T. 1968. *Cancer Res.* 28: 1326–30
323. Mondal, S., Iype, P. T., Griesbach, L. M., Heidelberger, C. 1970. *Cancer Res.* 30:1593–97
324. Embleton, M. J., Heidelberger, C. 1972. *Int. J. Cancer* 9:8–18
325. Di Mayorca, G., Greenblatt, M., Trauthen, T., Soller, A., Giordano, R. 1973. *Proc. Nat. Acad. Sci. USA* 70: 46–49
326. Bertram, J. S., Heidelberger, C. 1974. *Cancer Res.* 34:526–37
327. Peterson, A. R., Bertram, J. S., Heidelberger, C. 1974. *Cancer Res.* 34:1592–99
328. Peterson, A. R., Bertram, J. S., Heidelberger, C. 1974. *Cancer Res.* 34:1600–7
329. Marquardt, H. 1974. *Cancer Res.* 34: 1612–15
330. Oshiro, Y., Gerschenson, L. E., DiPaolo, J. A. 1972. *Cancer Res.* 32: 877–79
331. Kuroki, T., Yamakawa, S. 1974. *Int. J. Cancer* 13:240–45
332. Laug, W. E., Jones, P. A., Benedict, W. F. 1975. *J. Nat. Cancer Inst.* In press
333. Ossowski, L., Unkeless, J. C., Tobia, A., Quigley, J. P., Rifkin, D. B., Reich, E. 1973. *J. Exp. Med.* 137:112–26
334. Sekely, L. I., Malejka-Giganti, D., Gutmann, H. R., Rydell, R. E. 1973. *J. Nat. Cancer Inst.* 50:1337–45
335. Rhim, J. S., Huebner, R. J. 1973. *Cancer Res.* 33:695–700
336. Duran-Reynals, M. L. 1963. *Progr. Exp. Tumor Res.* 3:148–85
337. Huebner, R. J., Todaro, G. J. 1969. *Proc. Nat. Acad. Sci. USA* 64:1087–94
338. Todaro, G. J., Huebner, R. J. 1972. *Proc. Nat. Acad. Sci. USA* 69:1009–15
339. Gordon, R. J., Bryan, R. J., Rhim, J. S., Demoise, C., Wolford, R. G., Freeman, A. E., Huebner, R. J. 1973. *Int. J. Cancer* 12:223–32
340. Freeman, A. E., Weisburger, E. K., Weisburger, J. H., Wolford, R. G., Maryak, J. M., Huebner, R. J. 1973. *J. Nat. Cancer Inst.* 51:799–807
341. Rhim, J. S., Park, D. K., Weisburger, E. K., Weisburger, J. H. 1974. *J. Nat. Cancer Inst.* 52:1167–73
342. Casto, B. C. 1973. *Cancer Res.* 33: 402–7
343. Casto, B. C., Pieczynski, W. J., DiPaolo, J. A. 1973. *Cancer Res.* 33:819–24
344. Casto, B. C., Pieczynski, W. J., DiPaolo, J. A. 1974. *Cancer Res.* 34:72–78
345. Ledinko, N., Evans, M. 1973. *Cancer Res.* 33:2936–38
346. Salerno, R. A., Ramm, G. M., Whitmire, C. E. 1973. *Cancer Res.* 33:69–77
347. Whitmire, C. E. 1973. *J. Nat. Cancer Inst.* 51:473–78
348. Whitmire, C. E., Salerno, R. A. 1973. *Proc. Soc. Exp. Biol. Med.* 144:674–79
349. Irino, S., Ota, Z., Sezaki, T., Suzaki, K. 1963. *Gann* 54:225–37
350. Basombrio, M. A. 1973. *J. Nat. Cancer Inst.* 51:1157–62
351. Igel, H. J., Huebner, R. J., Turner, H. C., Kotin, P., Falk, H. L. 1969. *Science* 166:1636–39
352. Whitmire, C. E., Salerno, R. A., Rabstein, L. S., Huebner, R. J., Turner, H. C. 1971. *J. Nat. Cancer Inst.* 47: 1255–65
353. Freeman, A. E., Kelloff, G. J., Vernon, M. L., Lane, W. T., Capps, W. I., Bumgarner, S. D., Turner, H. C., Huebner, R. J. 1974. *J. Nat. Cancer Inst.* 52:1469–76
354. Weinstein, I. B., Gebert, R., Stadler, U. C., Orenstein, J. M., Axel, R. 1972. *Science* 178:1098–1100

355. Freeman, A. E., Kelloff, G. J., Gilden, R. V., Lane, W. T., Swain, A. P., Huebner, R. J. 1971. *Proc. Nat. Acad. Sci. USA* 68:2386–90

356. Rhim, J. S., Duh, F. G., Cho, H. Y., Elder, E., Vernon, M. L. 1973. *J. Nat. Cancer Inst.* 50:255–61

357. Altanerova, V., Altaner, C. 1972. *Neoplasma* 19:405–12

358. Freeman, A. E., Gilden, R. V., Vernon, M. L., Wolford, R. G., Hugunin, P. E., Huebner, R. J. 1973. *Proc. Nat. Acad. Sci. USA* 70:2415–19

359. Rapp, U., Nowinski, R. C., Reznikoff, C. A., Heidelberger, C. 1975. *Virology.* In press

360. Prehn, R. T. 1964. *J. Nat. Cancer Inst.* 32:1–17

361. Mondal, S., Heidelberger, C. 1970. *Proc. Nat. Acad. Sci. USA* 65:219–25

362. Pierce, G. B., Wallace, C. 1971. *Cancer Res.* 31:127–34

363. Pitot, H. C., Heidelberger, C. 1963. *Cancer Res.* 23:1694–1700

364. Burdette, W. J. 1955. *Cancer Res.* 15:201–26

365. Ong, T-M., de Serres, F. J. 1972. *Cancer Res.* 32:1890–93

366. Miller, E. C., Miller, J. A. 1971. In *Chemical Mutagens—Principles and Methods for their Detection,* ed. A. Hollaender, 1:83–119. New York: Plenum

367. Ames, B. N., Gurney, E. G., Miller, J. A., Bartsch, H. 1972. *Proc. Nat. Acad. Sci. USA* 69:3128–32

368. Ames, B. N., Durston, W. E., Yamasaki, E., Lee, F. D. 1973. *Proc. Nat. Acad. Sci. USA* 70:2281–85

369. Stoltz, D. R., Poirier, L. A., Irving, C. C., Stich, H. F., Weisburger, J. H., Grice, H. C. 1974. *Toxicol. Appl. Pharmacol.* 29:157–80

370. Hitotsumachi, S., Rabinowitz, Z., Sachs, L. 1971. *Nature* 231:511–14

371. Hitotsumachi, S., Rabinowitz, Z., Sachs, L. 1972. *Int. J. Cancer* 9:305–15

372. Yamamoto, T., Rabinowitz, Z., Sachs, L. 1973. *Nature New Biol.* 243:247–50

373. Benedict, W. F. 1972. *J. Nat. Cancer Inst.* 49:585–90

374. DiPaolo, J. A., Popescu, N. C., Nelson, R. L. 1973. *Cancer Res.* 33:3250–58

375. Olinici, C. D., DiPaolo, J. A. 1974. *J. Nat. Cancer Inst.* 52:1627–34

376. Olinici, C. D., Evans, C. H., DiPaolo, J. A. 1974. *J. Cell. Physiol.* 83:401–8

377. Comings, D. E. 1973. *Proc. Nat. Acad. Sci. USA* 70:3324–28

378. Holley, R. W. 1969. *Proc. Nat. Acad. Sci. USA* 69:2840–41

379. Temin, H. M. 1974. *Cancer Res.* 34:2835

PATHWAYS OF CARBON FIXATION IN GREEN PLANTS   ✶876

Israel Zelitch

Department of Biochemistry, The Connecticut Agricultural Experiment Station, New Haven, Connecticut 06504

CONTENTS

INTRODUCTION

When photosynthesis was last reviewed in this series in 1970 by Walker & Crofts (1), photochemistry and photosynthetic electron transport as well as pathways of carbon metabolism were discussed. Since then the great interest in the importance of carbon pathways in controlling net CO_2 fixation justifies a separate perspective of the more recent developments in carbon aspects of photosynthesis.

Rapid progress has been made in a number of subjects, including understanding the enzymology of photosynthetic carboxylation reactions, finding new mechanisms

123

of control of CO_2 assimilation by isolated chloroplasts, and learning more about the metabolism of the so-called C_4 species, which have much faster rates of net photosynthesis at high light intensities in normal air and warm temperatures than do most other plant species. Considerable work has been done on the process of photorespiration, which is rapid in the C_3 species, and on the close relationship between photorespiration and the metabolism of glycolic acid. This subject is important because the rapid photorespiration diminishes net CO_2 fixation greatly in these species, and the possibility of regulating photorespiration therefore offers a rational means of increasing their plant productivity.

Progress in these subjects is evaluated here, but at least one of the more interesting problems has been ignored because of present lack of knowledge. Leaves of several species, for example sunflower and cattail (*Typha latifolia*), have rapid rates of net photosynthesis in spite of their C_3-type metabolism and a rapid photorespiration (2, p. 244). The manner by which such species overcome the handicap of a rapid photorespiration remains an important problem for future research.

RIBULOSE-1,5-DIPHOSPHATE CARBOXYLASE

The reaction catalyzed by ribulose-1,5-diphosphate carboxylase is the primary but not the exclusive reaction by which CO_2 is fixed during photosynthesis. In most species 3-phosphoglyceric acid, the product of this reaction, is the first labeled compound detected after $^{14}CO_2$ fixation. The enzyme has a molecular weight of about 560,000 and accounts for a high proportion of the soluble protein in chloroplasts. Much attention has been drawn to the disconcerting fact that the purified enzyme has a much higher K_m for bicarbonate (CO_2), about 20 mM bicarbonate, than do isolated chloroplasts (K_m about 0.5 mM bicarbonate). In spite of additional diffusion barriers, intact leaves have a still lower overall K_m of about 0.4 mM bicarbonate (9 mM CO_2) during photosynthesis (2), although leaf extracts contain a low K_m form of the carboxylase, as discussed later.

Calvin and his colleagues had postulated that the 3-phosphoglyceric acid produced by the carboxylation of ribulose-1,5-diphosphate was derived from a β-ketoacid intermediate, 2-carboxy-3-ketoribitol-1,5-diphosphate. Siegel & Lane (3), in a highly original paper, described the chemical synthesis of this intermediate and showed that it was cleaved nonenzymatically (at pH 9) or by the enzyme to produce two molecules of 3-phosphoglycerate. The existence of β-ketoacid intermediates was also inferred by the isolation of a substance with the properties expected of the intermediate from reaction mixtures of ribulose-1,5-diphosphate with the enzyme (4).

The enzyme is allosteric with respect to bicarbonate (CO_2) concentration and shows a typical sigmoidal rate curve with increasing amounts of bicarbonate (CO_2) (5). It has eight large subunits (molecular weight 58,200) and eight small subunits (molecular weight about 15,300), each with different C-terminal amino acids (6). The immunological properties of large and small subunits were the same in the enzyme isolated from five species of *Nicotiana* (7). It is believed that the large subunit contains the catalytic site and the small subunit has regulatory functions (6).

Regulation by Metabolites

Fructose-6-phosphate activated ribulose-1,5-diphosphate carboxylase three- to four-fold at bicarbonate concentrations of 10 mM or less and diminished the K_m for bicarbonate by 80%, to 4 mM (8). Unfortunately the absolute reaction rates in these experiments cannot be estimated from available data, but the rates are probably slow compared to rates of carboxylation during photosynthesis. These workers also found that fructose-1,6-diphosphate inhibited the enzymatic carboxylation somewhat and interfered with activation by fructose-6-phosphate. Because reduced ferredoxin and a protein factor obtained from chloroplast extracts stimulated fructose-1,6-diphosphatase activity and hence fructose-6-phosphate formation (9), they suggested that the activation of ribulose-1,5-diphosphate carboxylase activity by light in chloroplasts results when ferredoxin is reduced during photosynthetic electron transport.

The strong inhibition of the purified enzyme by 6-phosphogluconate at a concentration similar to that of ribulose-1,5-diphosphate has been demonstrated (10), but apparently this is true only at saturating bicarbonate levels. At low bicarbonate concentrations 6-phosphogluconate is a better activator of the carboxylase than fructose-6-phosphate (11), indicating how subtle the regulation by intermediates of the Calvin cycle might be.

CO_2 FIXATION BY ISOLATED CHLOROPLASTS

Control of CO_2 fixation by isolated chloroplasts has received considerable attention because it provides a system amenable to simpler analysis than intact tissues. When isolated whole chloroplasts are first illuminated, there is usually a lag of several minutes before maximal rates of CO_2 uptake are attained. This may be partly explained by the light activation of ribulose-1,5-diphosphate carboxylase which could involve conformational changes in the enzyme. Walker and colleagues (12, 13) have shown that the major factors causing the lag are the time required for the concentration of intermediates in the Calvin cycle to increase and the autocatalytic nature of the cycle. This view is supported by the observation that chloroplasts that fix CO_2 at high rates in light cease to fix CO_2 as soon as the light is turned off even though there is still an adequate supply of ribulose-1,5-diphosphate available (14).

Lin & Nobel (15) estimated that the Mg^{2+} concentration of chloroplasts in illuminated pea plants was 10 mM and only about 1 mM in plants kept in darkness, and suggested that this change in concentration could control photosynthetic activity. Jensen (16) was able to stimulate CO_2 fixation by isolated spinach chloroplasts at least 25-fold in darkness by supplying 20 mM Mg^{2+} in the presence of ATP and ribose-5-phosphate or ribulose-5-phosphate, again suggesting that an increase in the Mg^{2+} in the stroma of chloroplasts activated CO_2 fixation during photosynthesis. Preparations of freshly lysed spinach chloroplasts have recently been shown by Bahr & Jensen (17) to contain a low K_m form of ribulose-1,5-disphosphate carboxylase (K_m bicarbonate of 0.5 to 0.8 mM). These preparations can be stabilized for up to 2 hr in the presence of ribose-5-phosphate, ATP, and Mg^{2+}. The carboxylation assay was carried out in a hypotonic medium that contained 25 mM Mg^{2+}. The low K_m form of their enzyme preparation had insufficient activity at normal concentrations of CO_2 in air to account for normal photosynthetic rates. At 9 mM CO_2 (0.4 mM bicarbonate) their reaction rate was 24 μmol of CO_2 per mg chlorophyll·hr, less than 20% of the rate of intact spinach leaves in normal air (2). A low K_m form of ribulose-1,5-diphosphate carboxylase has also been found in extracts of young maize leaves (18).

Photosynthesis by isolated chloroplasts is strongly affected by the relative Mg^{2+} and orthophosphate concentrations in the medium, both being required at about 1 mM, although severe inhibitions are obtained by increasing the amounts of these elements to 10 mM (19, 20). Pyrophosphate does not penetrate the chloroplast envelope membrane, but a Mg^{2+}-dependent pyrophosphatase in chloroplast preparations (21) permits the gradual formation of orthophosphate and allows rapid rates of photosynthesis to occur when pyrophosphate is added; inhibitions caused by high concentrations of orthophosphate are thereby avoided.

The inhibitory action of high concentrations of orthophosphate on photosynthesis by isolated chloroplasts is reversed by micromolar quantities of antimycin A (22), an inhibitor of photophosphorylation. Antimycin also stimulates CO_2 assimilation by isolated chloroplasts 33 to 50% and increases the proportion of 3-phosphoglyceric acid synthesized (23). The control mechanism responsible for the stimulation of carboxylation while photophosphorylation is inhibited is not understood (24). The uptake of CO_2 by chloroplasts is completely inhibited by 10 mM DL-glyceraldehyde, and this results from a block in the conversion of triose phosphate to ribulose-1,5-diphosphate (25).

Rates of CO_2 Uptake

Spinach leaves are capable of photosynthesizing at rates of 114 μmol of CO_2 per mg chlorophyll·hr in normal air (300 ppm CO_2) and 265 μmol per mg chlorophyll·hr at saturating CO_2 concentrations (1000 ppm CO_2) (2). Rates about one half the expected maximum in vivo (120 μmol per mg chlorophyll·hr) have been observed at saturating CO_2 concentrations with isolated pea leaf chloroplasts (26). It might be expected that greater net photosynthetic rates would be achieved with isolated chloroplasts than with leaves, since leaves have additional resistances to CO_2 diffusion. These include the diffusive resistances of the leaf boundary layer, the stomatal resistance, and the diffusive resistance caused by rapid rates of photorespiration in some leaf species (2). Perhaps photosynthetic electron transport limits the rates of CO_2 fixation in isolated chloroplasts.

Leaves of efficient photosynthetic species, such as maize, assimilate CO_2 at rates of 300 μmol per mg chlorophyll·hr or more in normal air and 650 μmol per mg chlorophyll·hr at saturating CO_2 (2). Great difficulty has been experienced in isolating maize chloroplasts with rapid rates of CO_2 uptake. The most active preparations, described by O'Neal and co-workers (27), were obtained from 4- to

Table 1   Rate of transport of various substances across the chloroplast envelope membrane[a]

Fast Transport	Slow Transport
HCO_3^-	Ribulose diphosphate
3-Phosphoglycerate	Sedoheptulose diphosphate
Pentose monophosphates	Phosphoenolpyruvate
Triose phosphates	Fructose-6-phosphate
Orthophosphate	Glucose-6-phosphate
Malate	6-Phosphogluconate
Succinate	Sucrose[b]
Glycolate	Glucose
Glycine	Fructose
Serine	ATP, ADP
Alanine	NADP, NAD
Aspartate	Pyrophosphate
Glutamate	H^+

[a] Compiled from (10, 12, 13, 21, 28–33).
[b] Totally impermeable in intact chloroplasts.

6-day-old maize seedlings and had maximal rates of 45 μmol of CO_2 per mg chlorophyll $\cdot$ hr. The products of CO_2 fixation by these chloroplasts were similar to those obtained with spinach or pea chloroplasts, and there was no evidence of a C-4 carboxylation pathway, although leaves of the same age produced ^{14}C-malate and ^{14}C-aspartate after short time labeling with $^{14}CO_2$.

As indicated, the concentration of intermediates and cations can greatly affect the rates of CO_2 uptake by chloroplasts. Substances differ greatly in their ability to cross the envelope membrane and this may regulate the rate of photosynthesis. Differences in transport across the chloroplast envelope membrane have been established from 1. the chemical nature of the products released from isolated chloroplasts, 2. studies on the effects of various metabolites on CO_2 uptake by chloroplasts, 3. measurement of the quantity of an added substance sedimented with chloroplasts through an inert layer of silicone oil, and 4. measurement of metabolite diffusion into isolated envelope membrane vesicles. Some of these differences in transport are summarized in Table 1. Factors controlling the movement of substances into and out of chloroplasts and how they may be regulated will undoubtedly provide an important area for future research, especially now that such studies are possible with isolated chloroplast envelope membranes (31, 34, 35).

PHOSPHOENOLPYRUVATE CARBOXYLASE AND THE C4 PATHWAY

Earlier experiments demonstrating that leaves of certain efficient photosynthetic species, including sugar cane and maize, have additional carboxylation reactions besides that catalyzed by ribulose-1,5-diphosphate carboxylase have been discussed by Hatch & Slack (36) and Black (37). These efficient species are capable of fixing CO_2 in normal air at rates two to three times that of most species on a leaf area basis, and the first detectable product when $^{14}CO_2$ is supplied is usually malate or aspartate produced from oxaloacetate generated by the phosphoenolpyruvate carboxylase reaction. Many investigations support the view that the carboxylation reaction occurs in the mesophyll cells of such leaves and that malate or aspartate is then transported to bundle sheath cells surrounding the vascular tissue. The C_4 compound is then decarboxylated, and the CO_2 released is fixed once more by reactions associated with the Calvin cycle. As indicated below, there are still a number of difficulties preventing full acceptance of this view of the metabolic events occurring in the efficient C_4 photosynthetic species.

There are few published experiments on the control of phosphoenolpyruvate carboxylase activity or the C_4 pathway. Aspartate, glutamate, and α-ketoglutarate inhibited the carboxylase in extracts of *Chlamydomonas* (38). The activity of the enzyme from leaves of efficient photosynthetic species was increased about 50% by the presence of glycine (39), whereas the maize leaf enzyme was inhibited by the product of the carboxylase reaction, oxaloacetate (40). Pyruvate phosphate dikinase, the enzyme responsible for the synthesis of phosphoenolpyruvate from pyruvate, has been purified to homogeneity from extracts of maize leaf (41). The activity of this enzyme was stimulated by ammonium ions, and the greater activity in extracts of previously illuminated leaves, as found by other workers, was confirmed.

Uncertainties About the C_4 Pathway

The C_4 pathway hypothesis requires the rapid transport of a C_4 compound from the mesophyll cells to the bundle sheath cells and an activity of ribulose-1,5-diphosphate carboxylase in the bundle sheath cells at least as great as the rate of CO_2 uptake by the leaf. By microautoradiography, Osmond (42) showed that after 2 sec of $^{14}CO_2$ fixation a significant quantity of ^{14}C was transported from mesophyll cells to bundle sheath cells in maize and sugar cane leaves. By carrying out experiments involving $^{14}CO_2$ fixation in maize leaf segments followed by a chase with $^{12}CO_2$, Hatch (43) calculated the size of the CO_2 and C_4 pools in these two types of cells on the assumption that the pool was restricted to one or the other of these compartments. Using his assumption about pool restrictions, Hatch estimated that the CO_2 concentration in bundle sheath cells was five times greater than in the other photosynthetic cells. It should be pointed out that Goldsworthy & Day (44) provided $NaH^{14}CO_3$ solutions through the vascular system of maize and an inefficient species (*Phragmites*) in light and could not show that the $H^{14}CO_3^-$ was retained any more effectively within the leaf by maize. These results suggest that CO_2 is not so greatly compartmented within bundle sheath cells.

In pulse-chase experiments with $^{14}CO_2$ and $^{12}CO_2$ in various efficient species, Downton (45) was able to classify plants into two groups. "Malate formers," including maize and sorghum, consistently contained higher activities of NADP-malate dehydrogenase and decreased activity of aspartate aminotransferase than did "aspartate formers" such as *Amaranthus* and *Panicum* species. The chase experiments indicated a large movement of carbon from C_4 acids to phosphorylated sugars. Kennedy & Laetsch (46) showed that such a classification was not valid in *Portulaca oleracea* because the relative amounts of radioactive malate and aspartate changed greatly depending on the length of exposure to $^{14}CO_2$ and the age of the leaf. In one experiment of 2 sec duration, 75% of the total ^{14}C was in aspartate and 25% in malate, whereas after 4 sec 14% was in aspartate and 55% in malate. After 2 sec in $^{14}CO_2$, young leaves would therefore be classified as aspartate formers, while mature leaves produced equal proportions of both C_4 compounds.

Even the naming of a plant as a C_3 or C_4 species has uncertainties. For example, after 1 min exposure to $^{14}CO_2$, sorghum leaves (a C_4 species) usually contained three times as much ^{14}C in malic acid as in 3-phosphoglyceric acid, but after flowering the leaves produced primarily labeled 3-phosphoglyceric acid (47). Kennedy & Laetsch (48) recently found that leaves of carpetweed, *Mollugo verticillata,* had features intermediate between C_3 and C_4 species according to several criteria. After 5 sec of photosynthesis in $^{14}CO_2$ they found 14% of the ^{14}C in aspartate, 6% in malate, 54% in alanine, and 18% in phosphorylated compounds. Other experiments difficult to reconcile with the current hypothesis of the C_4 pathway were described by Laber et al (49), who measured transient changes in the pool sizes of photosynthetic intermediates in maize leaves on addition of CO_2 to leaves previously illuminated in CO_2-free air. During this preliminary period in light, one would expect phosphoenolpyruvate to accumulate, and according to the hypothesis a rapid formation of C_4 acids should result on supplying CO_2. Instead, these investigators found an increase in the pool size of ribulose-1,5-diphosphate in the absence of CO_2 in the preliminary period (as in spinach leaves), and on addition of CO_2 there was a decrease in ribulose-1,5-diphosphate concentration and an increase in 3-phosphoglyceric and phosphoenolpyruvic acids. These results indicate that ribulose-1,5-diphosphate carboxylase served as the primary CO_2 acceptor in both spinach and maize leaves.

The C_4 pathway of photosynthesis as outlined above indicates that the conversion of the C-4 of malate into products of the Calvin cycle occurs in the bundle sheath cells of efficient photosynthetic species. Dittrich et al (50) demonstrated such a conversion using malate-4-^{14}C supplied to isolated bundle sheath strands of crabgrass (*Digitaria sanguinalis*). Their fastest rate was only 3.6 μmol per mg chlorophyll·hr, a value about 1% of that required to account for the hypothesis. The authors believe that the rate of malate uptake by the cells was limiting in the isolated system. Isolated mesophyll cells of such species fix CO_2 by the phosphoenolpyruvate carboxylase reaction, and the oxaloacetate produced can serve as a Hill oxidant in the photochemical production of O_2 (51). In this system, rates of CO_2 fixation in the presence of added phosphoenolpyruvate were about 10% of that demanded by the hypothesis. Slow rates of CO_2 fixation by isolated maize bundle sheath cells (52) and crabgrass mesophyll cells (53) have been described, and 21% O_2 inhibited photosynthesis by the bundle sheath cells but not by the mesophyll cells.

Enzyme Distribution Studies in C_4 Species

Huang & Beevers (54) used seven sequential grinding steps to obtain extracts first from mesophyll and then from bundle sheath cells (these cells are broken with greater difficulty) in several species and assayed these extracts for a number of enzyme activities. Phosphoenolpyruvate carboxylase activity was present largely in the mesophyll cells, and ribulose-1,5-diphosphate carboxylase activity was mostly in the bundle sheath cells of sorghum and *Atriplex rosea* leaves. Similar conclusions were reached by Liu & Black (55) about the localization of these enzymes in isolated cells of crabgrass, based on their specific activity in various fractions. Sequential grinding of millet leaves (*Panicum miliaceum*) also showed a greater specific activity of phosphoenolpyruvate carboxylase in mesophyll cells on a chlorophyll basis, and ribulose-1,5-diphosphate carboxylase was mostly in the bundle sheath cells (56). In the abovementioned investigations the total activity recovered in various fractions was generally not even considered, but Kanai & Edwards (57), in an investigation of enzyme activities in the isolated cell types of maize leaves, could account for only 51% of the total ribulose-1,5-diphosphate carboxylase, and all of their activity was located in the bundle sheath cells. The possible inactivation of this enzyme during isolation of mesophyll cells cannot therefore be ignored in interpreting their findings. Hatch & Kagawa (58) reached similar conclusions about the localization of these two enzymes in leaves of *Atriplex spongiosa* and maize by using sequential grinding techniques, and their results are given as a percentage of total activity in various extracts. About two thirds of the total ribulose-1,5-diphosphate carboxylase activity was found in the bundle sheath cells.

Baldry et al (59) showed that quinones resulting from the oxidation of phenols were released during progressive grinding of sugar cane leaves and that the quinones inactivated carboxylating enzymes in the mesophyll cells unless thioglycolate or β-mercaptoethanol was added to the grinding medium. Bucke & Long (60) used these sulfhydryl reagents during extraction and concluded that the Calvin cycle enzymes were located in the mesophyll cell chloroplasts of maize and sugar cane, whereas phosphoenolpyruvate carboxylase was present in the cytoplasm of these same cells. Coombs & Baldry (61) found that leaves of *Pennisetum purpureum* normally had low levels of phenols and phenol oxidase activity and that mesophyll cells again contained Calvin cycle activity, whereas phosphoenolpyruvate carboxylase was present in the cytoplasm of these cells. In the above studies, however, the activities of Calvin cycle enzymes were generally low.

Poincelot (62) used two different grinding techniques to fractionate the enzyme activities of maize leaves and added suitable sulfhydryl reagents to the extraction medium. He confirmed the source of the enzyme activities by determination of the ratio of chlorophyll a to b, and obtained excellent recoveries in his fractions. On a chlorophyll basis at least 42% of the total ribulose-1,5-diphosphate activity was present in the mesophyll cells, and about 72% of the carbonic anhydrase activity was also found in the mesophyll cells. Other workers (63) found ribulose-1,5-diphosphate carboxylase activity in the mesophyll cells of young (5 to 6 days) maize seedlings but not in older plants.

Table 2 Minimal rates of photorespiration[a]

Species	Method of assay	Temperature (°C.)	Net photosynthesis in normal air (mg CO_2/dm²·hr)	Photorespiration, percent of net photosynthesis in normal air	Ratio of photorespiration to dark respiration	Reference
Soybean[b]	CO_2 release, CO_2-free air	26	35.2	46		77
Soybean	Postillumination CO_2 burst	25	11	75		78
Soybean	$^{14}CO_2$ release, CO_2-free air	24			2.3	79
Soybean	CO_2 release, CO_2-free air	30	18	42	1.9	80
Sunflower	Short time uptake, $^{14}CO_2$ minus $^{12}CO_2$	25	25	60		81
Sunflower	$^{14}CO_2$ release, CO_2-free air	25	28	27	3.5	82
Sugar beet	CO_2 release, CO_2-free air	25	25.2	47	8.5	83
Sugar beet	CO_2 release, CO_2-free air	25	26	40	1.4	80
Tobacco	$^{14}CO_2$ release, CO_2-free air	30	17–25		1.5–6.0	71
Tobacco	CO_2 release, CO_2-free air	25	11	55		70
Tobacco	Extrapolation of net photosynthesis to "zero" CO_2	25	13.7	25	3.5	84
Tobacco	Postillumination CO_2 burst	25.5	16.9	45		85
Tobacco	Postillumination CO_2 burst	33.5	14.8	66		85
Maize	CO_2 release; air passed through leaf	30		0		86
Maize	Uptake of $^{18}O_2$	28–34	11.5	5.7		87
Maize	$^{14}CO_2$ release, CO_2-free air	30			0.10	88
Maize	$^{14}CO_2$ release, CO_2-free air	24			0.15	79
Maize	CO_2 release, CO_2-free air	35		0		89
Maize	CO_2 release, CO_2-free air	35	50	0		80

[a] These values are minimal and underestimates because photorespiration is assayed under conditions of high light intensity where the main flux of the gas (CO_2 or O_2) is in the opposite direction. Dark respiration contributes somewhat to the photorespiration measured. Descriptions of these assays and their limitations are discussed in (2).

[b] Results recalculated by the authors considering internal diffusive resistances; results are the mean values of 20 varieties.

PHOTORESPIRATION

Characteristics

It has been understood for over a decade that illuminated leaves of many species have a rapid rate of respiration (usually considered as CO_2 evolution) that occurs by biochemical reactions differing from the better understood "dark" respiration that takes place mostly in the mitochondria. This photorespiration diminishes net CO_2 uptake greatly because some recently fixed products of photosynthesis are oxidized to CO_2. Photorespiration requires light and occurs best at high levels of O_2 in the atmosphere, is inhibited by high concentrations of CO_2, is very temperature dependent, and has a close connection with the synthesis and further oxidation of glycolic acid (2, 37, 64–68). The efficient C_4 photosynthetic species, such as maize and sugar cane, have much slower rates of photorespiration than the less efficient ones.

Very young tobacco leaves (less than 1 g fresh weight), unlike older leaves, presumably have low rates of photorespiration because net photosynthesis was not increased in 1% O_2 compared with 21% O_2 (69). Differences in photorespiration in tobacco have been described depending on the leaf position on the stalk based on rates of CO_2 released in CO_2-free air, with the fastest photorespiration being obtained with younger leaves near the top of the plant (70). We have not observed any great differences in photorespiration with tobacco leaves taken from different stalk positions of mature plants (71).

The presence of a postillumination burst of CO_2, characteristic of rapid rates of photorespiration, was examined in a number of grasses and was found lacking in all species of the tribe *Andropogeneae* and present only in the genus *Panicum* of the tribe Paniceae (72). There was no correlation between the presence of well-developed grana in the bundle sheath chloroplasts and the existence of a rapid photorespiration. Heichel (73) made the important observation that tobacco leaves starved by being placed in darkness for 14 hr or more lacked the initial CO_2 burst typical of photorespiration, and there was no detectable burst even in 100% O_2. After 20 min of photosynthesis the postillumination burst was restored, demonstrating that starving the leaf decreased the concentration of photorespiratory substrates for a short time.

The efficient C_4 photosynthetic species discriminate less against ^{13}C than do the species inefficient in assimilating CO_2, and the efficient species have a higher ratio of $^{13}C/^{12}C$ on the average in their carbon compounds. This subject has been reviewed by Smith (74), and the differences between species are usually ascribed to enrichment of ^{13}C during the phosphoenolpyruvate carboxylase step (75). The possible role of the recycling of CO_2 with different isotope ratios in different species has not been fully considered in interpreting these results. One attempt to determine the $^{13}C/^{12}C$ ratio of the photorespiratory CO_2 released from tobacco leaves appears, unfortunately, to have been carried out with leaves that were not in contact with water and that were desiccating during the experiment (76).

Rates of Photorespiration

There is overwhelming evidence by various assays that photorespiration in many species occurs at rates at least 50% of CO_2 uptake (Table 2). In addition, blocking photorespiration with biochemical inhibitors of glycolic acid synthesis (90) or oxidation (91) increases net photosynthesis in leaf tissue by 50% or more. Thus, the magnitude of photorespiration must be at least 76 μmol CO_2 per mg chlorophyll $\cdot$ hr, or 114 μmol per g fresh weight $\cdot$ hr, or about 10 mg CO_2 per $dm^2 \cdot$ hr. Any biochemical mechanism of photorespiration must at least be able to account for such rates of photorespiratory CO_2 production.

All assays of photorespiration underestimate it to some extent because the measurements are made on illuminated tissues where the main flux of CO_2 is from the atmosphere to the leaf; yet the measurement must be made in the ambient atmosphere. A number of different assays of photorespiration (Table 2) have confirmed that in several inefficient photosynthetic species photorespiration is 50% of net photosynthesis and that photorespiration is usually three or more times faster than dark respiration. Photorespiration is slow in efficient species like maize, and at high light intensity the respiration is probably no faster than rates expected of dark respiration. There is evidence, discussed more fully later, that the slow rate of photorespiration in maize results mainly from the slow rate of glycolic acid synthesis in this species (92), whereas glycolic acid is produced rapidly enough in species with fast photorespiration to account for its participation in this process. Many investigators suggest that maize may have significant photorespiration but that the CO_2 released is trapped and refixed by the bundle sheath chloroplasts and thus is not measured (65, 87, 93, 94), but this view is entirely speculative. On the other hand, Goldsworthy & Day (44) could find no evidence that internally supplied $^{14}CO_2$ was refixed any more effectively by maize than by a species without bundle sheath chloroplasts, *Phragmites*. Even attempts to demonstrate fast photorespiration in maize by passing air rapidly through the leaf to remove CO_2 (86) or by placing the leaf in 100% O_2 (79) failed to indicate the presence of rapid photorespiration.

Assays of algae and higher aquatic plants underestimate photorespiration even more because the large aqueous barrier to the diffusion of CO_2 permits even greater recycling of released CO_2 to occur than in land plants. Even so, Cheng & Colman (95) have confirmed that the green algae *Chlorella* and especially *Chlamydomonas* show fast rates of photorespiration, as do *Euglena* and certain blue-green algae such as *Anabena flos-aquae* and *Anacystis nidulans*. Other *Anabena* species also have a fast photorespiration (96, 97), but *Coccochloris peniocystis* and *Oscillatoria* had slow rates (95). Photorespiration was also demonstrated in the aquatic angiosperms *Najas flexilis* and *Myriophyllum spicutum* (98).

Many desert species were found to have a low ratio of CO_2 release in light vs dark and presumably a low rate of photorespiration (99). Desert plants, like aquatic plants, may have additional barriers to CO_2 diffusion and a relatively greater recycling of CO_2 because the leaf stomata are often closed during illumination; these species also frequently have fewer stomata. Thus, photorespiration might be underestimated even more in assays of such species.

The CO_2 Compensation Point

When photosynthetic tissues are placed in a closed system at constant temperature and at light intensities above the light compensation point, the CO_2 concentration in the air is decreased to a constant level known as the CO_2 compensation point. At this steady-state level of CO_2 the rates of photosynthetic CO_2 uptake and release of respiratory CO_2 are equal. The efficient C_4 photosynthetic species have a CO_2 compensation point of less than 10 ppm of CO_2, whereas for the inefficient species the value is 40 ppm or greater. The CO_2 compensation point can be defined algebraically as the product of the rate of CO_2 release times the carboxylation resistance (100); thus a high CO_2 compensation point might arise from rapid photorespiration, a less effective photochemistry or carboxylation system, or a combination of both of these. The CO_2 compensation point can be measured quickly, easily, and accurately (101), and this has encouraged its use in spite of the difficulties associated with interpreting the values obtained. Work before 1970 generally indicated that the CO_2 compensation point was fairly constant for a species at a given temperature and O_2 concentration. More recent research has emphasized that a number of environmental factors can influence the value obtained and that the CO_2 compensation point is a less reliable indicator of photorespiration within a species than other assays.

Heichel (73) found that tobacco leaves placed in darkness for 14 hr or longer failed to show a postillumination burst typical of leaves with a fast photorespiration, but the CO_2 compensation point was the same in starved and normal leaves. In another example, the increased rate of CO_2 efflux in CO_2-free air confirmed the faster photorespiration previously described for the tobacco variety JWB mutant compared with JWB wild (102). The mutant seedlings also succumbed more rapidly in 60% O_2, yet the CO_2 compensation point failed to distinguish between these varieties. Other vexing observations about the CO_2 compensation point were made by Herath & Ormrod (103), who showed that when barley and pea plants were grown at 29/21°C day/night temperatures compared to 18/10°, the CO_2 compensation points of leaves measured at 21° were much higher in plants grown at the higher temperatures. In *Amaranthus edulis* plants grown in a greenhouse, the CO_2 compensation point was generally not greater than 8 ppm, but without any reasonable explanation 24 leaves out of 358 examined gave values above 20 ppm and some had values as great as 78 ppm (104).

Discrepancies in the CO_2 compensation point are often found in the literature. Heichel & Musgrave (105) showed that certain maize genotypes grown in the field in a tropical environment and measured outdoors had CO_2 compensation points as great as 20 ppm, whereas Moss et al (106) found that these varieties grown in a temperate environment and measured in the laboratory at 30°C had values that never exceeded 5 ppm. Previous values published for the CO_2 compensation points of woody species by Moss (107) showed high values, 145 ppm for *Acer plantanoides* for example, but Dickmann & Gjerstad (108) found a value of 55 ppm for this species. These workers also obtained a CO_2 compensation point for *Populus deltoides* leaves of 60 ppm, but when the leaves were under water stress or recovering they found values of about 300 ppm. Thus, if increases are observed in the CO_2 compensation point, it is important to determine whether net photosynthesis has been inhibited. A rhythmic change in the CO_2 compensation point with values from 0–5 ppm or as great as 80 ppm were observed in *Bryophyllum* leaves depending on the phase of the rhythm (109).

In barley leaves at the same position on the plant the CO_2 compensation point fluctuated from 55 ppm to 65 ppm between 11 and 26 days after planting (110), but no data are given about possible differences in net photosynthesis in such leaves. There are considerable data showing the CO_2 compensation point to be constant at all light intensities above the light compensation point (2). Fair et al (111) showed that the value rose linearly from 50 to 75 ppm in barley leaves when the illumination was increased from 30,000 to 100,000 lux. Even more surprising is the observation by Grossman & Cresswell (112) that while leaves of maize and sugar cane gave the usual CO_2 compensation point close to zero when the plants were grown on nitrate as the N source, the value increased to 9–14 ppm after 1 to 2 days when plants were transferred to nutrient medium containing ammonia-N before it returned to 0 again after the fourth day. In these experiments, net photosynthesis in normal air increased at the same time the CO_2 compensation point increased so that the ammonia treatment must have increased both respiration and gross photosynthesis. All the results presented above indicate that the CO_2 compensation is a less reliable indicator of photorespiration than other methods and that it is subject to change by a number of environmental factors, most of which are not at all well understood.

Differences in Photorespiration within a Species

Zelitch & Day (113) showed that a slow growing variety of tobacco (JWB mutant) had lower rates of net photosynthesis in normal air and a faster photorespiration compared with its fast growing sibling (JWB wild), thus demonstrating that genetic control, a pleiotropic effect in this instance, was capable of regulating photorespiration within a species. JWB mutant tobacco also has an altered chloroplast structure and decreased chlorophyll content, hence changes in photorespiration in normal appearing plants were sought. Zelitch & Day (71) have more recently described the results of pedigree selections on siblings of several generations of Havanna Seed tobacco plants with slower photorespiration and faster net photosynthesis than is common for this species. Superior plants on selfing produced about 25% of their progeny with slow photorespiration and fast net CO_2 uptake. The percentage was not increased in successive generations, but it was clearly established that plants growing side by side could show about one half the normal rate of photorespiration accompanied by about a 40% faster rate of net photosynthesis. Attempts are being made to increase the proportion of superior plants in a population by producing doubled haploid plants from selections first made for low rates of photorespiration and high rates of CO_2 assimilation in haploid plants obtained by anther culture (114).

It seems clear that decreasing photorespiration need not have an adverse effect upon the biochemical mechanism of carboxylation, but on the basis of unsupported theory, Lorimer & Andrews (115) stated that attempts to reduce or eliminate photorespiration in crop plants by genetic manipulation are unlikely to succeed. In addition to the contrary indications cited above, Wilson (116) has described variations in photorespiration within populations of the grass species *Lolium* that were independent of the growth conditions. A difference in photorespiration between two wheat varieties was also claimed (117). Björkman (94) has carried out the interesting experiment of producing interspecific hybrids between C_3 and C_4 species of *Atriplex*, and his results demonstrate that neither the superior C_4 anatomy nor a high phosphoenolpyruvate carboxylase activity is in itself sufficient to assure the rapid CO_2 uptake characteristic of the efficient species. I believe that the control of glycolic acid metabolism, especially the regulation of its synthesis, may have a greater effect in increasing net photosynthesis in a given species and that this might be accomplished in inefficient photosynthetic species without invoking the complex mechanism evolved by the C_4 photosynthetically efficient plants.

GLYCOLATE METABOLISM

Glycolate Biosynthesis

The rate of glycolate biosynthesis is probably the most important factor controlling photorespiration, but the biochemical reactions responsible for synthesis in any given photosynthetic tissue have not been worked out. A number of reactions are known that produce glycolate, and considerable evidence indicates that at least several of these may occur simultaneously. For a reaction to be considered as an exclusive pathway responsible for a rapid photorespiration, it should be able to account for glycolate synthesis at rates at least 50% of net CO_2 uptake.

RIBULOSE-1,5-DIPHOSPHATE OXYGENASE   The recent discovery that ribulose-1,5-diphosphate carboxylase can catalyze the reaction between the substrate and O_2 to produce phosphoglycolate and phosphoglycerate has received considerable attention (118–120). Properties of the oxygenase reaction have much in common with the well-known inhibition of photosynthetic CO_2 uptake by O_2 (Warburg effect). Plant tissues contain an active phosphoglycolate phosphatase in chloroplasts (121, 122) that could rapidly produce glycolate. These observations have led some investigators to conclude prematurely that the oxygenase reaction can account entirely for the Warburg effect, the synthesis of glycolic acid, and photorespiration.

There are, however, a number of troublesome aspects preventing the acceptance of the ribulose-1,5-diphosphate oxygenase reaction as the major pathway of glycolate synthesis in vivo. Andrews et al (123) showed that there was a rapid incorporation of $^{18}O_2$ into the carboxyl groups of glycine and serine in leaves in light, which is consistent with the occurrence of the oxygenase reaction. However, even when the reaction was carried out in 100% O_2, which would favor the oxygenase, the specific activity of incorporated $^{18}O_2$ was only about one third of that supplied. This indicates that not more than one third of the glycolate was produced by the oxygenase reaction even in 100% O_2.

There are still uncertainties about the stoichiometry of the oxygenase reaction.

One mole of O_2 was taken up for each mole of ribulose-1,5-diphosphate consumed with the purified enzyme (119), and phosphoglycolate and phosphoglycerate were synthesized. The stoichiometry between O_2 uptake and production of phosphoglycolate has not yet been published.

The pH optimum for the oxygenase reaction with purified enzyme is 9.3 and for the carboxylation reaction it is 7.8 (124). There is virtually no carboxylation at pH 9.3 or oxygenation at pH 7.8. Glycolate synthesis and photosynthetic CO_2 uptake occur rapidly in vivo and one does not occur exclusive of the other. Lorimer & Andrews (115) claim that there is a fixed ratio of oxygenase to carboxylase activity, but examination of the literature shows wide variations of this ratio under optimal conditions of assay for each activity and indicates that sufficient glycolate to account for minimal rates of photorespiration could not be produced (Table 2). Thus, Andrews et al (119) found variations in the oxygenase-to-carboxylase ratio between 0.25 and 0.59; Lorimer et al (125) had a ratio of 0.15; Ryan et al (126) found a ratio of 1.0 in crude fractions of tobacco leaves and 0.5 in spinach and bean leaves and that the ratio changed during enzyme purification to about 0.25; plants grown in a medium high in nitrogen had relatively more oxygenase than carboxylase.

Phosphoglycolate, the product of the oxygenase reaction, has not yet been demonstrated to function as an important intermediate in glycolate biosynthesis in kinetic experiments with $^{14}CO_2$ in intact tissues or chloroplasts. Bassham & Kirk (127) allowed illuminated *Chlorella* cells to reach a steady state in air containing $^{14}CO_2$. Upon changing the atmosphere to 100% O_2 only about one third to one half of the glycolate produced could be accounted for by phosphoglycolate formation. Presumably even less phosphoglycolate would be synthesized in normal air than in 100% O_2.

RATE OF SYNTHESIS AND SPECIFIC RADIOACTIVITY FROM $^{14}CO_2$ When illuminated leaf tissue is placed in a solution of α-hydroxysulfonate, glycolate oxidation is blocked and glycolate accumulates at an initial rate of about 70 to 80 μmol per g fresh weight·hr in tobacco and sunflower (92). This is a sufficient rate of glycolate synthesis to account for the minimal rates of photorespiration shown in Table 2. Illuminated maize leaf, however, synthesizes glycolate about 10% as rapidly (90, 92), thus providing a reasonable explanation for the slow rate of photorespiration in this species. When the $^{14}CO_2$ and the sulfonate are added simultaneously to previously illuminated tobacco leaf disks, the specific radioactivity of the carbon atoms of glycolate-^{14}C is about 50% that of the $^{14}CO_2$ supplied (128), showing that glycolate is synthesized rather directly from fixed CO_2. In the presence of glycidate, an inhibitor of glycolate synthesis and photorespiration, the specific radioactivity of glycolate-^{14}C is even higher (90), indicating that some of the dilution results from a contribution to glycolate synthesis by nonradioactive carbon sources. Robinson & Gibbs (129) have recently studied glycolate synthesis from $^{14}CO_2$ by isolated chloroplasts without addition of any biochemical inhibitors. They found the specific radioactivity of the carbon atoms of glycolate-^{14}C to be between 53 and 71% that of the $H^{14}CO_3^-$ supplied. These experiments further confirm that a rather direct synthesis of glycolate from CO_2 must be possible. Keerberg et al (130) found that the specific radioactivity of glycolate and glycine in bean leaves was greater under red than under blue light, and it is difficult to see how the oxygenase reaction could be affected by light of different wavelengths.

EXCRETION BY ALGAE AND EFFECT OF RED AND BLUE LIGHT   Many species of algae synthesize and excrete glycolate rapidly when illuminated in the presence of high concentrations of O_2 and CO_2 (131). Lord et al (132) showed that glycolate excretion occurred in *Chlorella* and *Euglena* in red light but not in blue light. A red light-induced release of glycolate was also found in *Anacystis nidulans* (133), but *Chlorogonium* apparently excreted more in blue than in red light (134). In bean leaves the accumulation of glycolate in the presence of a sulfonate was greater in plants grown in blue than in red light (135), and blue light increased the activity of the glycolate pathway and decreased the $^{14}CO_2$ incorporation into starch (136). Merrett & Lord (137) concluded that red light is needed by algae for glycolate biosynthesis and that the failure of algae to synthesize glycolate in blue light is not easily explained by the ribulose-1,5-diphosphate oxygenase reaction.

OTHER PATHWAYS OF GLYCOLATE SYNTHESIS   Recent experiments support the view that multiple pathways of glycolate synthesis occur in the same tissue. Other mechanisms of synthesis must exist besides the ribulose-1,5-diphosphate oxygenase reaction and the possible direct carboxylation reaction implied by the synthesis of glycolate from $^{14}CO_2$ with high specific radioactivity. For example, I found that organic acids such as acetate-2-^{14}C and especially pyruvate-3-^{14}C were incorporated into glycolate-2-^{14}C with a higher specific radioactivity in maize than tobacco, whereas $^{14}CO_2$ was incorporated much more readily into glycolate-^{14}C in tobacco than in maize (92). The addition of phosphoenolpyruvate stimulated glycolate synthesis in maize but not in tobacco, and glyoxylate-2-^{14}C was easily converted to glycolate-2-^{14}C in both species. These results demonstrate the existence of alternate pathways of glycolate biosynthesis from CO_2 and from organic acids. Glyoxylate may be available for glycolate formation in leaves from several pathways including the isocitrate lyase reaction (138). Eickenbusch & Beck (139) have also presented evidence for the presence of at least two kinds of reactions involved in glycolate formation from $H^{14}CO_3^-$ in isolated spinach chloroplasts; one pathway was unchanged by O_2 concentration up to at least the levels present in air, while the rate of the second reaction increased linearly when the O_2 concentration was raised above the normal air level.

Shain & Gibbs (140) showed that glycolate could be formed in a reconstructed chloroplast system in light in the presence of a transketolase substrate, such as fructose-6-phosphate, together with transketolase, NADP, and ferredoxin. An oxidant, presumably H_2O_2 from the oxidation of reduced ferredoxin or NADPH, was generated in the light that synthesized glycolate from "active glycolaldehyde" at maximal rates of 10 μmol per mg chlorophyll·hr. The addition of ferricyanide to isolated spinach chloroplasts increased the formation of glycolate-^{14}C at low concentrations of $H^{14}CO_3^-$, suggesting the ferricyanide was oxidizing "active glycolaldehyde" produced by the transketolase reaction (141). Osmond & Harris

(142) observed synthesis of glycolate-^{14}C from ribose-5-phosphate-^{14}C or fructose-6-phosphate-^{14}C in maize, sorghum, and *Atriplex spongiosa* as well as the less efficient photosynthetic species *A. hastata*. The rates of glycolate synthesis cannot be calculated from the data presented, but they were undoubtedly very slow under the experimental conditions.

Additional support for the view that intermediates of the Calvin cycle may serve as precursors of glycolate synthesis comes from the interesting work of Robinson & Gibbs (129), who showed that in isolated spinach chloroplasts unlabeled two carbon fragments from ribose-5-phosphate and fructose-1,6-diphosphate are incorporated into glycolate, and the specific radioactivity of glycolate-^{14}C synthesized from $H^{14}CO_3^-$ is thereby decreased.

The Oxidation of Glycolate

In order to be metabolized glycolate must first be oxidized to glyoxylate, since no reaction is known whereby glycolate reacts directly. In higher plants glycolate oxidase is a flavoprotein that reacts with O_2. The reaction rate is greatly dependent on the concentration of O_2, and this enzyme is found mainly in the peroxisomes of green leaves (65). Some investigators find that most of the glycolate oxidase is in the soluble fraction in homogenates of pea leaves (143) and spinach leaves (144). In plant species with well-developed bundle sheaths, glycolate oxidase activity is greater in these cells than in the mesophyll cells (54, 55). Maize leaves oxidized glycolate supplied in darkness, but the rate of CO_2 production was not rapid in relation to rates of photorespiration in inefficient photosynthetic species (145).

Glycolate oxidase of higher plants also oxidizes L-lactate, but not D-lactate, and is insensitive to inhibition by cyanide. Gruber et al (146) have shown that some green algae, such as *Chlorella,* contain a glycolate dehydrogenase (which couples to suitable dyes but not O_2) that oxidizes D-lactate and not L-lactate and is sensitive to cyanide. Many, but not all, green algae contained such a dehydrogenase, whereas green tissues of all the land plants examined and several fresh water angiosperms contained a glycolate oxidase that reacted with L-lactate (147). Certain phenolics such as chlorogenic acid might function as an alternate acceptor to O_2 for glycolate oxidase in tobacco (148), and small stimulations in glycolate oxidase activity have been found in the presence of chlorogenic acid (149). The serological properties of the enzymes oxidizing glycolate were fairly similar from tobacco leaves, a yellow *Chlorella* mutant, and *Euglena* (150). No great differences in protein structure were observed on disc electrophoresis of the comparable enzymes from tobacco, spinach, barley, and maize (151).

Some environmental factors affect glycolate oxidase activity. Thus, plants growing in blue light produced greater activities than in red light (152). There are some indications that the activity is greater when plants are grown at low CO_2 concentrations, but Cooksey (153) found similar glycolate dehydrogenase activity in *Chlamydomonas* cells grown in 5% CO_2 and in normal air. Growing barley seedlings in 1–5% CO_2 for about 5 days decreased the glycolate oxidase activity about 30% compared with plants grown in air (154).

Biochemical Inhibition of Glycolate Oxidase

α-Hydroxysulfonates are aldehyde-bisulfite addition compounds and effective competitive inhibitors of glycolate oxidase (155). When a suitable sulfonate is supplied to illuminated leaf tissue, glycolate oxidase is blocked and glycolate then accumulates at initial rates sufficient to account for photorespiration in tobacco and sunflower (92). Inhibition of glycolate oxidation in tobacco leaf under suitable conditions also blocks photorespiration and brings about large increases in photosynthetic CO_2 uptake (91). Corbett & Wright (156) showed that glyoxylate-bisulfite probably becomes the active inhibitor no matter which sulfonate is supplied.

α-Hydroxysulfonates usually become toxic to leaf tissues and inhibit photosynthesis after about 15 min of exposure, while efficient C_4 photosynthetic species, including *A. spongiosa,* are even more sensitive to the toxicity of the inhibitor (157). A sulfonate also inhibited photosynthetic CO_2 fixation in spinach chloroplasts, but no biochemical explanation for the toxicity was found (158). Barley leaves exposed to an atmosphere containing 5 ppm of SO_2 showed increases in glycolate concentration, presumably because sulfonates were produced in the tissue (159), and Tanaka et al (160) demonstrated that glyoxylate-bisulfite was formed in leaves exposed to $^{35}SO_2$.

Inhibition of Glycolate Synthesis

Leaves of some efficient photosynthetic species, such as maize, normally synthesize glycolate in the light about 10% as rapidly as the less efficient species (90, 92). It is possible that maize synthesizes glycolate slowly because synthesis occurs in the bundle sheath cells where higher than normal CO_2 concentrations may exist (43). It is well known that high concentrations of CO_2 inhibit glycolate synthesis in all species. It would therefore seem appropriate to attempt to diminish the rate of glycolate production by biochemical or genetic means in tissues with rapid rates of photorespiration and thus reproduce what occurs normally in maize.

Goldsworthy (161) showed that isonicotinic acid hydrazide inhibits photorespiration in tobacco leaves and that this resulted because the inhibitor blocked glycolate synthesis (92).

Recently I found that glycidic acid, 2,3-epoxypropionic acid, an epoxide similar in structure to glycolic acid, inhibited glycolate synthesis in tobacco leaf tissue (90). The inhibitor also decreased photorespiration and appeared to be a specific inhibitor of glycolate synthesis because net CO_2 fixation was increased 50% in the presence of the epoxide. The specific radioactivity of glycolate synthesized from $^{14}CO_2$ by tobacco leaf treated with the inhibitor was also increased. Preliminary experiments did not reveal any effect of glycidic acid on the ribulose-1,5-diphosphate oxygenase reaction, but the biochemical mechanism responsible for the action of the inhibitor on glycolate synthesis has not yet been determined. Nevertheless, the use of this inhibitor has confirmed in an independent manner that slowing glycolate synthesis might block photorespiration and increase net photosynthesis in many inefficient photosynthetic species.

The Glycolate Pathway of Carbohydrate Synthesis

The well-known sequence of the glycolate pathway occurs in photosynthetic tissues, and as usually depicted, four molecules of glycolic acid yield one of glucose and two of CO_2 (2). The CO_2 is assumed to arise during the condensation of two glycine molecules to produce serine. The production of CO_2 during photorespiration is undoubtedly associated with reactions of this pathway, but the stoichiometry indicated above permits the loss of only 25% of the glycolate metabolized, and as documented earlier (Table 2), photorespiratory CO_2 often accounts for at least 50% of net CO_2 fixation during photosynthesis. Thus, CO_2 cannot be produced during photorespiration from only this reaction.

The oxidative decarboxylation of glyoxylate in the presence of hydrogen peroxide yields formate and CO_2; this reaction might occur rapidly in chloroplasts (162, 163) and to a lesser extent in leaf peroxisomes (164) and accounts for most of the photorespiratory CO_2. Glycolate (or glyoxylate) and glycine are decarboxylated equally well when added to leaf tissue in light (165, 166), and the oxidation of glycolate occurred rapidly even when the conversion of glycine to serine was inhibited by isonicotinic acid hydrazide (165). Thus the decarboxylation of glycine did not appear essential for fast photorespiration.

The further oxidation of formate to CO_2 probably also contributes to the CO_2 produced during photorespiration (91). Leaf peroxisomes oxidize formate slowly (172) but an active formate dehydrogenase is found in leaf mitochondria (173). Kent et al (174) found an unequal distribution of ^{14}C in serine in *Vicia faba* leaves given $^{14}CO_2$ and suggested that formate might be synthesized directly from CO_2, but in the absence of direct evidence supporting this hypothesis, it might still be assumed that under the slow photosynthetic conditions used in these experiments the formate was derived from the C-2 of glyoxylate.

The decarboxylation of glycine was observed in tobacco leaf mitochondria (167, 168), but the rates were less than 1% of those needed to account for the role of this reaction in photorespiration. The oxidation rate was increased with increasing O_2 concentration (169), but the absolute rates were not given. Bird et al (170, 171) showed that one molecule of ATP was synthesized during the conversion of glycine to serine by tobacco leaf mitochondria and suggested that photorespiration might therefore not be an entirely wasteful process from an energy standpoint. The rates of serine synthesis in their system were about 10% that required for photorespiration, and unlike the photorespiratory process the rates of this reaction were saturated with 2–4% O_2 in the gas phase.

Other pathways of glyoxylate metabolism in addition to those mentioned above might also account for some photorespiratory CO_2. The reaction first found in bacteria whereby two molecules of glyoxylate condense to yield tartronic semialdehyde and CO_2 has been found in the green algae *Euglena* (175) and *Gloromonas* (176) and in the blue-green alga *Anabena* (97). This reaction provides an alternate sequence by which glyoxylate could be decarboxylated to yield glycerate without glycine or serine as intermediates, as in the glycolate pathway. This alternate

142   ZELITCH

pathway has not yet been demonstrated in higher plants, but its presence would not be inconsistent with current knowledge about glycolate metabolism.

ACKNOWLEDGMENT

The helpful assistance of Pamela Beaudette in compiling material for this article is gratefully acknowledged.

Literature Cited

1. Walker, D. A., Crofts, A. R. 1970. *Ann. Rev. Biochem.* 39:389–428
2. Zelitch, I. 1971. *Photosynthesis, Photorespiration, and Plant Productivity.* New York: Academic. 347 pp.
3. Siegel, M. I., Lane, M. D. 1973. *J. Biol. Chem.* 248:5486–98
4. Sjödin, B., Vestermark, A. 1973. *Biochim. Biophys. Acta* 297:165–73
5. Murai, T., Akazawa, T. 1972. *Biochem. Biophys. Res. Commun.* 46:2121–26
6. Sugiyama, T., Ito, T., Akazawa, T. 1971. *Biochemistry* 10:3406–11
7. Kawashima, N., Kwok, S.-Y., Wildman, S. G. 1971. *Biochim. Biophys. Acta* 236:578–86
8. Buchanan, B. B., Schürmann, P. 1972. *FEBS Lett.* 23:157–59
9. Buchanan, B. B., Schürmann, P., Kalberer, P. P. 1971. *J. Biol. Chem.* 246:5952–59
10. Chu, D. K., Bassham, J. A. 1972. *Plant Physiol.* 50:224–27
11. Buchanan, B. B., Schürmann, P. 1973. *J. Biol. Chem.* 248:4956–64
12. Walker, D. A. 1973. *New Phytol.* 72:209–35
13. Walker, D. A., Kosciukiewicz, K., Case, C. 1973. *New Phytol.* 72:237–47
14. Avron, M., Gibbs, M. 1974. *Plant Physiol.* 53:136–39
15. Lin, D. C., Nobel, P. S. 1971. *Arch. Biochem. Biophys.* 145:622–32
16. Jensen, R. G. 1971. *Biochim. Biophys. Acta* 234:360–70
17. Bahr, J. T., Jensen, R. G. 1974. *Plant Physiol.* 53:39–44
18. Bahr, J. T., Jensen, R. G. 1974. *Biochem. Biophys. Res. Commun.* 57:1180–85
19. Avron, M., Gibbs, M. 1974. *Plant Physiol.* 53:140–43
20. Levine, G., Bassham, J. A. 1974. *Biochim. Biophys. Acta* 333:136–40
21. Schwenn, J. D., Lilley, R. M., Walker, D. A. 1973. *Biochim. Biophys. Acta* 325:586–95
22. Champigny, M. L., Miginiac-Maslow, M. 1971. *Biochim. Biophys. Acta* 234:335–43
23. Schacter, B. Z., Gibbs, M., Champigny, M. L. 1971. *Plant Physiol.* 48:443–46
24. Miginiac-Maslow, M., Champigny, M. L. 1971. *Biochim. Biophys. Acta* 234:344–52
25. Stokes, D. M., Walker, D. A. 1972. *Biochem. J.* 128:1147–57
26. Krainova, N. N., Chernov, J. A. 1971. *Fiziol. Rast.* 18:887–92
27. O'Neal, D., Hew, C. S., Latzko, E., Gibbs, M. 1972. *Plant Physiol.* 49:607–14
28. Heber, U. 1974. *Ann. Rev. Plant Physiol.* 25:393–421
29. Heldt, H. W., Rapley, L. 1970. *FEBS Lett.* 10:143–48
30. Kelly, G. J., Gibbs, M. 1973. *Plant Physiol.* 52:674–76
31. Poincelot, R. P. 1974. *Plant Physiol.* 54:520–26
32. Werdan, K., Heldt, H. W. 1972. *Biochim. Biophys. Acta* 283:430–41
33. Heldt, H. W., Sauer, F. 1971. *Biochim. Biophys. Acta* 234:83–91
34. Douce, R., Holtz, R. B., Benson, A. A. 1973. *J. Biol. Chem.* 248:7215–22
35. Poincelot, R. P., Day, P. R. 1974. *Plant Physiol.* 54:780–83
36. Hatch, M. D., Slack, C. R. 1970. *Ann. Rev. Plant Physiol.* 21:141–62
37. Black, C. C. Jr. 1973. *Ann. Rev. Plant Physiol.* 24:253–86
38. Davies, D. D., Patil, K. D. 1973. *Plant Physiol.* 51:1142–44
39. Nishikido, T., Takanashi, H. 1973. *Biochem. Biophys. Res. Commun.* 53:126–33
40. Lowe, J., Slack, C. R. 1971. *Biochim. Biophys. Acta* 235:207–9
41. Sugiyama, T. 1973. *Biochemistry* 12:2862–67
42. Osmond, C. B. 1971. *Aust. J. Biol. Sci.* 24:159–63
43. Hatch, M. D. 1971. *Biochem. J.* 125:425–32
44. Goldsworthy, A., Day, P. R. 1970. *Nature* 228:687–88
45. Downton, W. J. S. 1971. *Can. J. Bot.* 49:1439–42

46. Kennedy, R. A., Laetsch, W. M. 1973. *Planta* 115:113–24
47. Khanna, R., Sinha, S. K. 1973. *Biochem. Biophys. Res. Commun.* 52:121–24
48. Kennedy, R. A., Laetsch, W. M. 1974. *Science* 184:1087–89
49. Laber, L. J., Latzko, E., Gibbs, M. 1974. *J. Biol. Chem.* 249:3436–41
50. Dittrich, P., Salin, M., Black, C. C. 1973. *Biochem. Biophys. Res. Commun.* 55:104–10
51. Salin, M. L., Campbell, W. H., Black, C. C. Jr. 1973. *Proc. Nat. Acad. Sci. USA* 70:3730–34
52. Chollet, R., Ogren, W. L. 1972. *Biochem. Biophys. Res. Commun.* 46:2062–66
53. Chollet, R. 1973. *Biochem. Biophys. Res. Commun.* 55:850–56
54. Huang, A. H. C., Beevers, H. 1972. *Plant Physiol.* 50:242–48
55. Liu, A. Y., Black, C. C. Jr. 1972. *Arch. Biochem. Biophys.* 149:269–80
56. Edwards, G., Gutierrez, M. 1972. *Plant Physiol.* 50:728–32
57. Kanai, R., Edwards, G. E. 1973. *Plant Physiol.* 51:1133–37
58. Hatch, M. D., Kagawa, T. 1973. *Arch. Biochem. Biophys.* 159:842–53
59. Baldry, C. W., Bucke, C., Coombs, J. 1971. *Planta* 97:310–19
60. Bucke, C., Long, S. P. 1971. *Planta* 99:199–210
61. Coombs, J., Baldry, C. W. 1972. *Nature New Biol.* 238:268–70
62. Poincelot, R. P. 1972. *Plant Physiol.* 50:336–40
63. Magomedov, I. M., Chernyadev, I. I., Kovaleva, L. B., Doman, N. G. 1973. *Dokl. Akad. Nauk SSSR* 213:737–38
64. Goldsworthy, A. 1970. *Bot. Rev.* 36:321–40
65. Tolbert, N. E. 1971. *Ann. Rev. Plant Physiol.* 22:45–74
66. Jackson, W. A., Volk, R. J. 1970. *Ann. Rev. Plant Physiol.* 21:385–432
67. Hatch, M. D., Osmond, C. B., Slatyer, R. O., Eds. 1971. *Photosynthesis and Photorespiration.* New York: Wiley-Interscience. 565 pp.
68. Zelitch, I. 1973. *Proc. Nat. Acad. Sci. USA* 70:579–84
69. Salin, M. L., Homann, P. H. 1971. *Plant Physiol.* 48:193–96
70. Kisaki, T. 1973. *Plant Cell Physiol.* 14:505–14
71. Zelitch, I., Day, P. R. 1973. *Plant Physiol.* 52:33–37
72. Brown, R. H., Gracen, V. E. 1972. *Crop Sci.* 12:30–33
73. Heichel, G. 1971. *Plant Physiol.* 48:178–82
74. Smith, B. N. 1972. *BioScience* 22:226–31
75. Whelan, T., Sackett, W. M., Benedict, C. R. 1973. *Plant Physiol.* 51:1051–54
76. Hsu, J. C., Smith, B. N. 1972. *Plant Cell Physiol.* 13:689–94
77. Samish, Y. B., Pallas, J. E. Jr., Dornhoff, G. M., Shibles, R. M. 1972. *Plant Physiol.* 50:28–30
78. Bulley, N. R., Tregunna, E. B. 1971. *Can. J. Bot.* 49:1277–84
79. Laing, W. A., Forde, B. J. 1971. *Planta* 98:221–31
80. Hofstra, G., Hesketh, J. D. 1969. *Planta* 85:228–37
81. Bravdo, B., Canvin, D. T. 1973. *Plant Physiol.* 51 (Suppl.):42 (Abstr.)
82. Ludwig, L. J., Canvin, D. T. 1971. *Plant Physiol.* 48:712–19
83. Terry, N., Ulrich, A. 1973. *Plant Physiol.* 51:783–86
84. Decker, J. P. 1957. *J. Sol. Energy Sci. Eng.* 1:30–33
85. Decker, J. P. 1959. *Plant Physiol.* 34:100–2
86. Troughton, J. H. 1971. *Planta* 100:87–92
87. Volk, R. J., Jackson, W. A. 1972. *Plant Physiol.* 49:218–23
88. Zelitch, I. 1968. *Plant Physiol.* 43:1829–37
89. Moss, D. N. 1966. *Crop Sci.* 6:351–54
90. Zelitch, I. 1974. *Arch. Biochem. Biophys.* 163:367–77
91. Zelitch, I. 1966. *Plant Physiol.* 41:1623–31
92. Zelitch, I. 1973. *Plant Physiol.* 51:299–305
93. Coombs, J. 1973. *Curr. Advan. Plant Sci.* March:1–10
94. Björkman, O. 1973. *Photophysiology* 8:1–63
95. Cheng, K. H., Colman, B. 1974. *Planta* 115:207–12
96. Lex, M., Silvester, W. B., Stewart, W. D. P. 1972. *Proc. Roy. Soc. B* 180:87–102
97. Codd, G. A., Stewart, W. D. P. 1973. *Arch. Mikrobiol.* 94:11–28
98. Hough, R. A., Wetzel, R. G. 1972. *Plant Physiol.* 49:987–90
99. Glagoleva, T. A., Reinus, R. M., Gedemov, T. G., Mokronosov, A. T., Zalenskii, O. V. 1972. *Bot. Zh. Leningrad* 57:1097–1107
100. Bravdo, B.-A. 1968. *Plant Physiol.* 43:479–83
101. Goldsworthy, A., Day, P. R. 1970. *Plant Physiol.* 46:850–51
102. Heichel, G. H. 1973. *Plant Physiol.* 51 (Suppl.):42 (Abstr.)
103. Herath, H. M. W., Ormrod, D. P. 1972.

Plant Physiol. 49:443–44
104. Lester, J. N., Goldsworthy, A. 1973. J. Exp. Bot. 24:1031–34
105. Heichel, G. H., Musgrave, R. B. 1969. Crop Sci. 9:483–86
106. Moss, D. N., Willmer, C. M., Crookston, R. K. 1971. Plant Physiol. 47:847–48
107. Moss, D. N. 1962. Nature 193:587
108. Dickmann, D. I., Gjerstad, D. H. 1973. Can. J. Forest Res. 3:237–42
109. Jones, M. B., Mansfield, T. A. 1972. Planta 103:134–46
110. Fair, P., Tew, J., Cresswell, C. F. 1973. Ann. Bot. 37:831–44
111. Fair, P., Tew, J., Cresswell, C. F. 1974. Ann. Bot. 38:45–52
112. Grossman, D., Cresswell, C. F. 1973. S. Afr. J. Sci. 69:244–46
113. Zelitch, I., Day, P. R. 1968. Plant Physiol. 43:1838–44
114. Kasperbauer, M. J., Collins, G. B. 1971. Crop Sci. 12:98–101
115. Lorimer, G. H., Andrews, T. J. 1973. Nature 243:359
116. Wilson, D. 1972. J. Exp. Bot. 23:517–24
117. Ghildiyal, M. C., Sinha, S. K. 1973. Indian J. Exp. Bot. 11:207–9
118. Bowes, G., Ogren, W. L., Hageman, R. H. 1971. Biochem. Biophys. Res. Commun. 45:716–22
119. Andrews, T. J., Lorimer, G. H., Tolbert, N. E. 1973. Biochemistry 12:11–18
120. Lorimer, G. H., Andrews, T. J., Tolbert, N. E. 1973. Biochemistry 12:18–23
121. Randall, D. D., Tolbert, N. E., Gremel, D. 1971. Plant Physiol. 48:480–87
122. Kerr, M. W., Gear, C. F. 1974. Biochem. Soc. Trans 2:338–40
123. Andrews, T. J., Lorimer, G. H., Tolbert, N. E. 1971. Biochemistry 10:4777–82
124. Tolbert, N. E. 1973. Curr. Top. Cell. Regul. 7:21–50
125. Lorimer, G. H., Andrews, T. J., Tolbert, N. E. 1972. Fed. Proc. 31:1383 (Abstr.)
126. Ryan, F. J., Omata, S., Ku, H. S., Tolbert, N. E. 1973. Plant Physiol. 51 (Suppl.):40 (Abstr.)
127. Bassham, J. A., Kirk, M. 1973. Plant Physiol. 52:407–11
128. Zelitch, I. 1965. J. Biol. Chem. 240:1869–76
129. Robinson, J. M., Gibbs, M. 1974. Plant Physiol. 53:790–97
130. Keerberg, H., Vark, E., Keerberg, O., Parnik, T. 1971. Eesti NSV Tead. Akad. Toim. Biol. 20:350–53
131. Colman, B., Miller, A. G., Grodzinski, B. 1974. Plant Physiol. 53:395–97

132. Lord, J. M., Codd, G. A., Merrett, M. J. 1970. Plant Physiol. 46:855–56
133. Döhler, G., Koch, R. 1972. Planta 105:352–59
134. Stabenau, H. 1972. Biochem. Physiol. Pflanz. 163:42–51
135. Voskresenskaya, N. P., Khodzhiev, A. Kh. 1973. Fiziol. Rast. 20:309–16
136. Voskresenskaya, N. P., Viil, J., Grishina, G. S., Parnik, T. 1971. Fiziol. Rast. 18:488–93
137. Merrett, M. J., Lord, J. M. 1973. New Phytol. 72:751–67
138. Godavari, H. R., Badour, S. S., Waygood, E. R. 1973. Plant Physiol. 51:863–67
139. Eickenbusch, J. D., Beck, E. 1973. FEBS Lett. 31:225–28
140. Shain, Y., Gibbs, M. 1971. Plant Physiol. 48:325–30
141. Mathieu, Y., Tzenova, M. 1973. Photosynthetica 7:395–401
142. Osmond, C. B., Harris, B. 1971. Biochim. Biophys. Acta 234:270–82
143. Kolesnikov, P. A. et al 1973. Fiziol. Rast. 20:521–24
144. Halliwell, B. 1973. Biochem. Soc. Trans. 1:1147–50
145. Heichel, G. H. 1972. Plant Physiol. 49:490–96
146. Gruber, P. J., Frederick, S. E., Tolbert, N. E. 1974. Plant Physiol. 53:167–70
147. Frederick, S. E., Gruber, P. J., Tolbert, N. E. 1973. Plant Physiol. 52:318–23
148. Codd, G. A., Schmid, G. H. 1971. Planta 99:230–39
149. DeJong, D. W. 1974. Can. J. Bot. 1974. 52:209–15
150. Codd, G. A., Schmid, G. H. 1972. Plant Physiol. 50:769–73
151. Grodzinski, B., Colman, B. 1972. Phytochemistry 11:1281–85
152. Voskresenskaya, N. P., Khodzhiev, A. 1972. Dokl. Akad. Nauk Tadzh. SSR 15:60–63
153. Cooksey, K. E. 1971. Plant Physiol. 48:267–69
154. Fair, P., Tew, J., Cresswell, C. F. 1973. Ann. Bot. 37:1035–39
155. Zelitch, I. 1957. J. Biol. Chem. 224:251–60
156. Corbett, J. R., Wright, B. J. 1971. Phytochemistry 10:2015–24
157. Lüttge, V., Osmond, C. B., Ball, E., Brinckmann, E., Kinze, G. 1972. Plant Cell Physiol. 13:505–14
158. Murray, D. R., Bradbeer, J. W. 1971. Phytochemistry 10:1999–2003
159. Spedding, D. J., Thomas, W. J. 1973. Aust. J. Biol. Sci. 26:281–86

160. Tanaka, H., Takanashi, T., Yatazawa, M. 1972. *Water Air Soil Pollut.* 1:205–11

161. Goldsworthy, A. 1966. *Phytochemistry* 5:1013–19

162. Zelitch, I. 1972. *Arch. Biochem. Biophys.* 150:698–707

163. Elstner, E. F., Heupel, A. 1973. *Biochim. Biophys. Acta* 325:182–88

164. Halliwell, B., Butt, V. S. 1974. *Biochem. J.* 138:217–24

165. Zelitch, I. 1972. *Plant Physiol.* 50:109–13

166. Marker, A. F. H., Whittingham, C. P. 1967. *J. Exp. Bot.* 18:732–39

167. Kisaki, T., Imai, A., Tolbert, N. E. 1971. *Plant Cell Physiol.* 12:267–73

168. Kisaki, T., Yoshida, N., Imai, A. 1971. *Plant Cell Physiol.* 12:275–88

169. Kisaki, T., Yano, N. Hirabayashi, S. 1972. *Plant Cell Physiol.* 13·581–84

170. Bird, I. F., Cornelius, M. J., Keys, A. J., Whittingham, C. P. 1972. *Biochem. J.* 128:191–92

171. Bird, I. F., Cornelius, M. J., Keys, A. J., Whittingham, C. P. 1972. *Phytochemistry* 11:1587–94

172. Leek, A. E., Halliwell, B., Butt, V. S. 1972. *Biochim. Biophys. Acta* 286:299–311

173. Halliwell, B. 1974. *Biochem. J.* 138:77–85

174. Kent, S. S., Pinkerton, F. D., Strobel, G. A. 1974. *Plant Physiol.* 53:491–95

175. Murray, D. R., Giovanelli, J., Smillie, R. M. 1971. *Aust. J. Biol. Sci.* 24:23–33

176. Badour, S. S., Waygood, E. R. 1971. *Biochim. Biophys. Acta* 242:493–99

SUPEROXIDE DISMUTASES

×877

Irwin Fridovich

Department of Biochemistry, Duke University Medical Center, Durham,
North Carolina 27710

CONTENTS

INTRODUCTION

There is a bizarre enzymatic activity universally present in respiring cells. The substrate is an unstable free radical that can be present only in minuscule amounts at any instant, and the reaction catalyzed proceeds at a rapid rate even in the absence of the enzyme. Yet the enzyme is essential for the survival of aerobic cells. It catalytically scavenges the superoxide radical, which appears to be an important agent of the toxicity of oxygen, and thus provides a defense against this aspect of oxygen toxicity. The reaction whose rate it enhances is a disproportionation or dismutation of these radicals and may be written

$$O_2^- + O_2^- + 2H^+ \rightarrow H_2O_2 + O_2$$

This activity was discovered only recently (1, 2) and, given the instability of O_2^-, it is not surprising that this finding was long delayed and finally achieved by following chance observations rather than by design. During the past few years there has been a surge of interest in this enzymatic activity and a voluminous literature has accumulated. This is an intellectually exciting situation but it puts the reviewer in an unenviable position. Thus it is clear that the continued rapid appearance of new reports on this subject will inevitably render this review out of date by the time it reaches its readers. Furthermore, speculations which seemed both bright and reasonable at the time of writing may appear trite or quaint at the time of reading.

147

Nevertheless, the attempt must be made. Hopefully this review will carry the story a bit further than did its predecessors (3–7).

SUPEROXIDE RADICAL AND THE DISMUTATION REACTIONS

O_2^- is a common intermediate of oxygen reduction. This is a consequence of the fact that molecular oxygen in its ground state prefers univalent pathways of reduction. The electronic basis for this preference rests upon a spin restriction, which has been discussed by Taube (8). In any case, a number of reactions of interest to biochemists have been shown to generate O_2^-. Among these are the autoxidations of hydroquinones, leucoflavins, and catechol amines (9–16), thiols (17), reduced dyes (9, 18, 19), tetrahydropteridines (20), ferredoxins (21–24), rubredoxin (25), and hemoproteins (26–28). Furthermore, the catalytic actions of several enzymes have been shown to evolve O_2^-. This category includes xanthine oxidase (2, 29, 30), aldehyde oxidase (31, 32), dihydro-orotic dehydrogenase (33), and a group of flavo-protein dehydrogenases (14). In addition, there are several oxidases and hydroxylases inhibited by superoxide dismutase, which suggests that O_2^- is an intermediate in their catalytic cycles. These include tryptophan dioxygenase (34), the reconstituted liver microsomal hydroxylase (35), soluble hydroxylases from *Aspergillus niger* (36), and galactose oxidase (37). Finally, O_2^- is produced by intact granulocytes during the act of phagocytosis (38–41), by illuminated chloroplasts (42–46), and by lyophilized rat liver microsomes (47, 48). Thus, although we remain ignorant of the identity of the quantitatively most significant sources of O_2^- within any given type of cell and of the absolute rates of O_2^- production inside cells, we can feel secure in concluding that respiring cells will produce significant amounts of O_2^-.

The superoxide radical cannot accumulate in aqueous media because it readily undergoes a disproportionation or dismutation reaction. O_2^- is the conjugate base of a weak acid called the hydroperoxyl radical whose pK_a is 4.8. The pH dependence of the spontaneous dismutation can be explained on the basis of this pK_a and on the following reactions (49)

$$HO_2^. + HO_2^. \rightarrow H_2O_2 + O_2 \qquad = 7.6 \times 10^5 \ M^{-1} sec^{-1}$$

$$HO_2^. + O_2^- + H^+ \rightarrow H_2O_2 + O_2 = 8.5 \times 10^7 \ M^{-1} sec^{-1}$$

$$O_2^- + O_2^- + 2H^+ \rightarrow H_2O_2 + O_2 \ < 100 \ M^{-1} sec^{-1}$$

At pH 7.4 the rate constant for the spontaneous dismutation reaction is approximately $2 \times 10^5 \ M^{-1} sec^{-1}$. Can one hope to significantly accelerate the rate of O_2^- decay at this pH through the action of an enzyme such as superoxide dismutase? An affirmative answer to this question depends upon two factors. The first is that the rate constant for the reaction of O_2^- with superoxide dismutase is close to $2 \times 10^9 \ M^{-1} sec^{-1}$. At pH 7.4 this gives the catalyzed reaction an advantage of 10^4 over the spontaneous dismutation. The second point in favor of the enzymatic dismutation is that the concentration of the enzyme inside cells vastly exceeds the steady state concentration of O_2^-. Therefore the collision rate of a superoxide radical with enzyme will be much greater than its rate of collision with another O_2^-. In whole human liver we estimate the concentration of superoxide dismutase to be at least 3×10^{-5} M; whereas the steady state concentration of O_2^- must be at least five orders of magnitude lower. This gives the enzymatic decay of O_2^- another advantage of at least 10^5. Taken together these two factors would make the enzyme-catalyzed decay of O_2^- at pH 7.4 in liver at least 10^9 times faster than the spontaneous decay.

ASSAYS FOR SUPEROXIDE DISMUTASE ACTIVITY

The first requirement for productive study of an enzyme is a convenient and precise assay. In the case of superoxide dismutase, the instability of its substrate has forced circuitous approaches to this goal. One successful strategy combines a reaction that generates O_2^- with an indicating scavenger for this radical. In such a case, the superoxide dismutase competes with the indicating scavenger for the flux of O_2^- and thus inhibits modification of the scavenger. Xanthine oxidase acting aerobically upon xanthine generates O_2^-, which can be detected by its ability to reduce cytochrome c. The reduction of cytochrome c, which can be followed at 550 nm, will then be inhibited by superoxide dismutase. One unit of activity can be defined as the amount that causes 50% inhibition of the reduction of cytochrome c under specified conditions. In the first isolation of superoxide dismutase, this assay provided a sensitivity such that one unit was 0.1 μg/ml of the enzyme (2). The sensitivity of this assay can be increased greatly by raising the pH and by diminishing the concentration of cytochrome c. O_2^- can be introduced into solutions by means other than the xanthine oxidase reaction, and indicating scavengers of O_2^- other than cytochrome c can be very useful. Much of this has been reviewed previously (7). Very simple assays for superoxide dismutase can be achieved on the basis of its ability to inhibit free radical chain oxidations in which O_2^- is either an initiator or a chain-propagating radical. Reactions that have been useful in this way include the autoxidations of sulfite (49a), epinephrine (11), and pyrogallol (13).

It is unusual to assay an enzyme on the basis of its ability to inhibit some observable process. A direct assay in which superoxide dismutase accelerated the rate of substrate consumption or product production would certainly be conceptually preferable to the indirect assays described above. Direct assessment of the catalytic action of superoxide dismutase has been achieved by following the disappearance of O_2^- by means of electron paramagnetic resonance (EPR) (15) or by direct ultraviolet spectrophotometry (50, 51). In the latter cases high concentrations of O_2^- were rapidly generated in oxygenated aqueous solutions by pulse radiolysis, such that direct observations of its ultraviolet absorbance were feasible. These and similar studies have been very important in probing the mechanism of action of superoxide dismutases.

BIOLOGICAL IMPORTANCE OF SUPEROXIDE DISMUTASE

There are good reasons for concluding that respiring cells generate O_2^-. If we knew enough about the reactivities of O_2^- we could probably deduce the essentiality of superoxide dismutase. In fact, the chemistry and biochemistry of O_2^- are still in a primitive state, so it has been necessary to obtain empirical support for the importance of the dismutase. This has been done and there are several lines of evidence indicating that this enzyme is essential for the survival of respiring cells: 1. Only respiring cells can possibly produce O_2^-, hence only they should need a defense against it. Surveys of a variety of microorganisms demonstrated that the aerobic and aerotolerant species contained superoxide dismutase, whereas the obligate anaerobes did not (52). The only aerotolerant species that lacked this enzyme was found unable to utilize oxygen (53). 2. Superoxide dismutase is inducible by oxygen in *Streptococcus faecalis* (54), *Escherichia coli* B (55), and *Saccharomyces cerevisiae* (56). This induction allows manipulation of the intracellular level of the enzyme by variation of the oxygen tension under which the cells are grown. If O_2^- is an important agent of oxygen toxicity and if superoxide dismutase is the defense, then cells with the induced, high levels of the enzyme should thereby be rendered resistant to hyperbaric oxygen. This was shown to be the case (54–56). The strength of this argument is augmented by controls demonstrating that *Bacillus subtilis,* in which oxygen induced catalase but not superoxide dismutase, did not gain resistance towards hyperbaric oxygen from growth under oxygen (55). Yet another control involved *E. coli* K12, which responded to oxygen by increasing its content of catalase and peroxidase but not of superoxide dismutase. Growth under oxygen did not make this strain of *E. coli* resistant towards hyperbaric oxygen (56). 3. Mutants of *E. coli* K12 C-600 with a temperature-sensitive defect in superoxide dismutase exhibited a parallel defect in their tolerance for oxygen. Revertants were found to have regained both oxygen tolerance at the restrictive temperature and the ability to maintain normal intracellular levels of superoxide dismutase at this temperature (4). 4. Streptonigrin, whose lethality towards *E. coli* is augmented by oxygen, has been supposed to act by serving as an intracellular source of O_2^- (57–59). Induction of superoxide dismutase by oxygen might therefore be expected to provide some tolerance towards this antibiotic. This has been observed (55).

The results obtained with microorganisms establish that superoxide dismutase is an essential defense against oxygen and superoxide toxicity. There are data indicating that this is true also of higher organisms. 5. Cultures of fetal calf myoblasts were damaged when exposed to light in the presence of FMN plus EDTA, and superoxide dismutase added to the medium protected these cells against the lethality of this photochemical flux of O_2^- (59a). Anthracene, methylcholanthrene, or benzpyrene, at levels innocuous when applied alone, markedly enhanced the cytotoxic effects of the photochemical flux of radicals, and once again superoxide dismutase protected. Superoxide dismutase was also reported to provide some protection against the deleterious actions of γ irradiation (59a). 6. Rats exposed to 85% oxygen become adapted to it and will then tolerate 100% oxygen, which would prove lethal in the absence of the adaptation. A modest but reproducible increase in lung superoxide dismutase has been shown to parallel adaptation (60). 7. Glutathione induces a swelling of rat liver mitochondria which is prevented by superoxide dismutase (61). 8. Paraquat introduced into rats causes lung damage whatever the route of administration. Since lung is the most aerobic of the tissues, this leads to the speculation that paraquat, like streptonigrin, may enhance oxygen toxicity by acting as an efficient source of O_2^-. Superoxide dismutase injected at frequent intervals has been reported to diminish the toxicity of paraquat (62).

VARIETIES OF SUPEROXIDE DISMUTASES

A common stress progressively applied to a varied biota is likely to force parallel and independent adaptations among the surviving species. The oxygenation of the earth's atmosphere, due to the photosynthetic activity of blue-green algae, certainly must have constituted such a stress to which superoxide dismutase would be one appropriate adaptation. Since a considerable diversification of life forms must have preceded the appearance of blue-green algae, we need not be surprised to find more than one type of superoxide dismutase among present organisms. There are, in fact, several of these enzymes.

Copper- and Zinc-Containing Superoxide Dismutases

The cytosols of eukaryotic cells contain a superoxide dismutase that has a molecular weight of 32,000, is made up of two identical subunits, and contains one Cu^{2+} and one Zn^{2+} per subunit. The properties of this superoxide dismutase have been remarkably resistant to evolutionary modification, and the enzymes obtained from fungi, plants, birds, and mammals (45, 63, 68, 69, 71, 72) are hardly distinguishable except by relatively minor differences in amino acid composition and in the super-hyperfine details of the electron spin resonance (ESR) spectra (72). This superoxide dismutase has been reversibly resolved. Activity is partially restored by replacement of Cu^{2+} alone and is fully restored by replacement of both metals (72). Attempts to substitute other metals for the Cu^{2+} have been uniformly unsuccessful, however the Zn^{2+} may be replaced by Co^{2+}, Hg^{2+}, or Cd^{2+} (72–75) without loss of activity.

Structural analysis of the superoxide dismutase from bovine erythrocytes is now in a very satisfactory state. The complete amino acid sequence is known (76–78) and the X-ray diffraction analysis has progressed to a resolution of 3.0 Å (79). It is thus possible to make models of the structure which specify the positions of each amino acid residue and of the Cu^{2+} and Zn^{2+}. This has been done. The most prominent structural feature of this enzyme is a cylinder whose walls are composed of eight strands of the sequence in an antiparallel β structure. The segments of the sequence involved in this β barrel are residues 2–11, 13–23, 26–35, 38–47, 80–88, 91–100, 112–118, and 142–149. The metals are in close proximity. Indeed, the Cu^{2+} and Zn^{2+} are joined by a common ligand; the imidazole ring of His 61. The Cu^{2+} is relatively exposed to solvent whereas the Zn^{2+} is more nearly buried inside the structure. Aside from the shared His 61, the other groups liganded to Cu^{2+} are His 44, 46, and 118, while the other groups liganded to Zn^{2+} are His 78 and 69 and Asp 81 (79). Detailed knowledge of the superoxide dismutase structure will now facilitate further studies of its mechanism. It is impressive that the proximity of the Cu^{2+} and the Zn^{2+} was predicted (73, 80, 81), as was a ligand field for Cu^{2+} composed of histidines (80, 82–84). The mechanism of this enzyme has been probed primarily by means of pulse radiolysis (50, 51, 85–88), and there is general agreement that the Cu^{2+} is alternately reduced and reoxidized during successive encounters with O_2^-. If direct electron transfer is to occur between O_2^- and Cu^{2+} then the Cu^{2+} must be accessible to water. The structure deduced from sequence and X-ray diffraction analysis indicates that this is the case (79). Results obtained with NMR also indicate that this Cu^{2+} is exposed to solvent (89, 90). Cyanide binds to this Cu^{2+}, and it is the carbon of cyanide which is then immediately liganded to the metal (91). Cyanide is a reversible inhibitor of the Cu^{2+}- and Zn^{2+}-containing superoxide dismutases (50). The oxidation-reduction properties of the Cu^{2+} of superoxide dismutase have been probed. H_2O_2 can reduce this Cu^{2+} but can also irreversibly inactivate the enzyme (92–95) if present at concentrations exceeding 10 μM. The redox interaction between H_2O_2 and the Cu^{2+} must be univalent since the reversal of the superoxide dismutase reaction has been demonstrated (96).

Manganese-Containing Superoxide Dismutases

E. coli B (97) and *Streptococcus mutans* (98) contain a superoxide dismutase which, except for its catalytic activity, appears totally unrelated to the Cu^{2+}- and Zn^{2+}-containing enzyme from eukaryotic cytosols. Thus this bacterial enzyme contains one atom of manganese per subunit, has a molecular weight of 40,000, and is a dimer made up of two subunits of equal size. A manganese-containing superoxide dismutase has been isolated from the luminous mushroom *Pleurotus olearius* and is composed of subunits which, although of identical size, were separable by ion exchange chromatography in the presence of 7.0 M urea (98a). The color and the NMR properties of the *E. coli* enzyme indicate that the manganese it contains is trivalent (99). It appears likely that the mechanism of this bacterial superoxide dismutase will prove similar to that of the Cu^{2+}- and Zn^{2+}-containing enzyme in that it probably depends upon alternate reduction and reoxidation of the active metal. The catalytic activity of the mangani-superoxide dismutase, as a function of pH, has been investigated using the method of kinetic competition (100). At pH 7.0 it is as active as the Cu^{2+}-Zn^{2+} enzyme, but as the pH is raised it becomes progressively less active, whereas the Cu^{2+}-Zn^{2+} enzyme is unaffected by pH in the range 5.5–10.0.

Mitochondria contain a superoxide dismutase strikingly similar to the mangani-enzyme of prokaryotes, with the notable difference that the mitochondrial enzyme contains four subunits instead of two and consequently has a molecular weight of close to 80,000 (71, 101). The similarities in gross properties between the mangani-superoxide dismutases from *E. coli* and chicken liver mitochondria suggested that their relationship might be very close. Partial amino acid sequences have, in fact, been obtained, and they demonstrate that the mitochondrial and the bacterial superoxide dismutases are closely related whereas the mitochondrial and the cytosol enzymes are unrelated (102). This result constitutes powerful support for the theory that mitochondria developed from a prokaryote that entered into an endocellular symbiosis with a protoeukaryote (103). The mangani-superoxide dismutase has recently been isolated from human liver mitochondria and has been found to be very similar to the corresponding enzyme from chicken liver mitochondria (104).

The superoxide dismutase of high molecular weight reported from bovine liver (105) and the isoenzyme B found in a variety of human tissues (106) are undoubtedly the mitochondrial mangani-superoxide dismutase described above. It seems likely that the mangani-superoxide dismutase isolated from *Pleurotus olearius* (98a) will ultimately be found localized in the mitochondria of this eukaryote, since it shares so many properties with the other mitochondrial superoxide dismutases thus far described.

Iron-Containing Superoxide Dismutases

E. coli B actually contains two superoxide dismutases. One of these is the mangani enzyme already discussed while the other is a ferri enzyme localized in the periplasmic space which can be selectively removed from these cells by osmotic shock (107). The gross properties of this enzyme (100, 107) and its amino acid sequence (102) indicate a close relationship to the mangani enzyme from the matrix of *E. coli* and from chicken liver mitochondria. The increase in the superoxide dismutase content of *E. coli* in response to oxygenation is actually due to an increase in the mangani enzyme. Since this was correlated with increased tolerance for hyperbaric oxygen (55), we may conclude that the matrix enzyme serves to scavenge endogenous O_2^-. The levels of the periplasmic ferri enzyme could be modified by nutritional means, and this permitted the demonstration that its role appears to be that of providing a defense against exogenous O_2^- (108). O_2^- has recently been implicated as one of the agents responsible for the bactericidal action of polymorphonuclear leukocytes (38–41). In agreement with this suggestion, *E. coli* B whose periplasmic superoxide dismutase had been elevated by growth in iron-rich media were more resistant towards phagocytic kill than were comparable cells whose ferri-superoxide dismutase levels had been depressed by growth in iron-deficient media (109).

The valence of the iron in this superoxide dismutase has been established as being Fe(III) by EPR (107) and by NMR (99). Very similar enzymes have been found in two species of blue-green algae, *Plectonema boryanum* (110) and *Spirulina platensis* (70), and also in two species of marine bacteria, *Photobacterium sepia* (111) and *Photobacterium leiognathi* (111).

POLEMICS

Every rapidly expanding area of investigation generates polemics. These are usually based upon unrecognized differences in experimental conditions. It is both a beauty and a strength of the scientific method that such disagreements can ultimately be completely resolved by careful experiments, often performed by third parties. A newcomer to the field or a casual reader is likely to be confused by such conflicts while they are yet in progress. To the extent possible, a review should therefore identify and clarify these current polemics.

The subunits of the bovine erythrocyte superoxide dismutase were reported to be covalently linked (112). This conclusion was based upon the observation that the subunits did not dissociate in sodium dodecylsulfate (SDS) unless mercaptoethanol was present. It was subsequently found, however, that this enzyme does not dissociate in SDS, not because its subunits are disulfide bridged, but rather because it is so stable that it is not unfolded by SDS. Indeed, it remains fully active in 1% SDS. It could however be dissociated by heating in SDS or by treatment with SDS plus urea. Mercaptoethanol facilitates dissociation in SDS because an intrachain disulfide bridge contributes to the stability of the holoenzyme (67, 76).

It has been reported that the true biological function of superoxide dismutase is that of scavenging singlet oxygen (113–117). Indeed, it has been suggested that the acronym SOD, used for superoxide dismutase, should stand for singlet oxygen decontaminase (118). This proposal is based upon several instances of the ability of superoxide dismutase to inhibit chemiluminescences, which supposedly originated from singlet oxygen. It is, in fact, difficult to identify the source of a weak chemi-luminescence, so it is not safe to equate chemiluminescence with singlet oxygen. Moreover, these measurements were made with photometers capable of responding only in the visible range, whereas singlet oxygen ($^1\Delta g$) should emit in the near infrared. To explain this, the possibility has been entertained that dimers of singlet oxygen form and then emit a single photon in the visible which combines the excitation energy of both members of the pair. There is, however, no evidence for oxygen dimers, either in the ground state or in the excited state (119). Furthermore, singlet oxygen is rapidly quenched by hydroxylic solvents. In water its lifetime is 1×10^{-6} sec (120). Given this short lifetime and the low concentration of superoxide dismutase, which is effective in inhibiting the reported chemiluminescence, one can calculate that the enzyme would have to quench singlet oxygen ($^1\Delta g$) at a rate at least five orders of magnitude greater than that set by the diffusion limit (121). Reports of quenching singlet oxygen by superoxide dismutase may therefore be questioned on theoretical grounds. There is, in addition, direct evidence that super-oxide dismutase does not quench singlet oxygen (83, 122–123a). The report that supposedly definitely demonstrated the enzymatic quenching of singlet oxygen by superoxide dismutase (117) was actually flawed in several respects. Thus, it depended upon the decomposition of potassium peroxochromate (V) in water as an exclusive source of singlet oxygen, whereas this decomposition also liberates O_2^- (122). Furthermore, it measured the chemiluminescence of luminol, which in buffered aqueous solutions is always dependent upon O_2^- whatever the oxidant and is there-fore always inhibited by superoxide dismutase (124). We may conclude that super-oxide dismutase does not act to catalytically scavenge singlet oxygen. Whether or not singlet oxygen is generated within biological systems is an entirely different question and is at present not settled one way or the other.

The copper- and zinc-containing superoxide dismutases found in the cytosols of such diverse eukaryotes as *Neurospora crassa* (63), *Saccharomyces cerevisiae* (65), spinach (69), cow (2), and man (87) are at once remarkably similar to each other and totally different from the superoxide dismutases containing iron or manganese in prokaryotes (70, 97, 98, 100, 107, 110, 111) or in mitochondria (71, 101). Although these similarities and differences can be noted even at the level of gross composition and properties, they have been made dramatically plain by comparison of amino acid sequences (102) which show that the bacterial and the mitochondrial enzymes are closely related to each other, while being unrelated to the enzyme found in eukaryotic cytosols. These results have clear evolutionary implications which have been discussed (102, 103). It has, however, recently been reported that *Photobacterium leiognathi* contains a superoxide dismutase whose metal components are copper and zinc (124a), and it has been concluded that this argues against the independent evolutionary development of the prokaryotic and the eukaryotic enzymes. This is certainly a fascinating finding but the conclusion it has prompted may be premature. The diagnostic difference between the eukaryotic and prokaryotic types of superoxide dismutase lies in their amino acid sequences, not in the nature of their prosthetic metals. It is entirely possible that the copper- and zinc-containing enzyme reported from *P. leiognathi* could prove to be related to the iron- or manganese-containing enzymes heretofore isolated from prokaryotes or from mitochondria, rather than to the copper- and zinc-containing enzyme heretofore isolated from the cytosols of eukaryotes. This will not be resolved until the sequence data are available.

There have been controversies concerning other aspects of the superoxide dismutase field, such as the presence or absence of tryptophan in the bovine erythrocyte enzyme, the intracellular distribution of superoxide dismutase activity in liver cells, the relative roles of the metal prosthetic groups in the Cu^{2+}- and Zn^{2+}-containing enzyme, and the valence of the manganese in the *E. coli* matrix enzyme. All of these have, in their time, added spice to the pleasures of active research in this area, and all have been resolved by the self-purifying properties of the cooperative-competitive scientific endeavor. These topics have been discussed in this and previous reviews and there is no need to further elaborate upon them here.

SUMMARY AND SPECULATIONS

O_2^- is readily generated by so many spontaneous and enzymatic oxidations that we may assume its production within all respiring organisms. Furthermore superoxide dismutases, which catalytically scavenge O_2^-, are so widespread among respiring organisms that we may assume that O_2^- is a deleterious species whose cytotoxicity has called forth the evolution of defenses, among which are certainly the superoxide dismutases. All of this is supported by the induction of superoxide dismutases by oxygen, by the ability of elevated levels of superoxide dismutase to provide resistance towards the lethality of hyperbaric oxygen and towards the oxygen-dependent lethality of streptonigrin, and finally by the temperature-dependent oxygen sensitivity of temperature-dependent superoxide dismutase mutants. None of this, however, tells us why O_2^- should be cytotoxic. There is the general feeling, based more upon chemical intuition than upon hard facts, that this oxygen-free radical should be reactive towards cellular components; but there is only a little data. Thus O_2^- can certainly act as an oxidant. It causes the oxidation of epinephrine (2), catechols (125, 126), and dehydrogenase-bound NADH (127), and it acts as a chain-carrying species in the autoxidation of epinephrine (11) and pyrogallol (13). O_2^- can also act as a reductant. Its reduction of cytochrome *c* (15, 128) was the basis of the assay for superoxide dismutase activity (2). It also readily reduces the Fe(III) in ferritin (129).

There are other indications of the reactivities of O_2^- in more complex systems. Thus, glutathione causes the swelling of mitochondria, and superoxide dismutase prevents this effect (61). Irradiation of membranes with X rays causes lipid peroxidation, and once again superoxide dismutase protects (130). Isolated inner membranes of mitochondria were peroxidized when incubated with aerobic solutions of glutathione, and superoxide dismutase prevented this lipid peroxidation (116). There have been other reports of the prevention of lipid peroxidation by superoxide dismutase (131, 132). Dialuric acid, which autoxidizes with the production of O_2^-, caused the hemolysis of vitamin E-deficient rat erythrocytes, and superoxide dismutase protected against this cell disruption (133). Bacteria exposed to fluxes of O_2^- were killed, and again superoxide dismutase provided protection (108, 134). Are we to conclude that O_2^- can itself attack the lipid components of membranes?

There is another possibility which seems reasonable and must be further explored. Any reaction generating O_2^- will, by virtue of the spontaneous dismutase reaction, also be generating H_2O_2. Haber & Weiss (135), in studies of the catalytic decomposition of H_2O_2 by iron salts, proposed that O_2^- and H_2O_2 could react as follows: $O_2^- + H_2O_2 \rightarrow OH^- + OH\cdot + O_2$. Since the hydroxyl radical (OH·) is an extraordinarily powerful oxidant, this reaction, if it really occurs in dilute aqueous solutions, could vastly amplify the potential dangers of O_2^-. There are indications that the Haber & Weiss reaction is a reality. Thus the action of xanthine oxidase on xanthine liberated ethylene from methional and both O_2^- and H_2O_2 were necessary for ethylene production (136). Furthermore, compounds that scavenge OH· but not O_2^-, such as ethanol or benzoate, prevented this ethylene production. The hydroxyl radical was also detected in this reaction mixture as an oxidant of ferrocytochrome c which could be intercepted by ethanol, and the depletion of H_2O_2 by O_2^- was demonstrated as an increased recovery of H_2O_2 in the presence of superoxide dismutase (136). Since then, other instances of the apparent production of OH· from O_2^- and H_2O_2 have been reported (137–140). It seems possible that many of the previously reported effects of O_2^- are in reality effects of OH·, generated from the O_2^- by the Haber & Weiss reaction.

It is a sobering thought that the hydroxyl radical, earlier thought of only in connection with the effects of ionizing radiation, may in fact be produced in respiring biological systems. This generates renewed appreciation for the superoxide dismutases, catalases, and peroxidases which, by their combined actions, keep the steady-state concentrations of O_2^- and H_2O_2 vanishingly small and thus minimize the Haber & Weiss reaction and make aerobic life possible.

Literature Cited

1. McCord, J. M., Fridovich, I. 1969. *Fed. Proc.* 28:346
2. McCord, J. M., Fridovich, I. 1969. *J. Biol. Chem.* 244:6049–55
3. Fridovich, I. 1972. *Accounts Chem. Res.* 5:321–25
4. McCord, J. M., Beauchamp, C. O., Goscin, S., Misra, H. P., Fridovich, I.

1973. *Oxidases and Related Redox Systems*, 51–76. Baltimore: Univ. Park Press
5. Fridovich, I. 1974. *Molecular Mechanisms of Oxygen Activation*, 453–77. New York: Academic
6. Fridovich, I. 1973. *Biochem. Soc. Trans.* 1:48–50
7. Fridovich, I. 1974. *Advan. Enzymol.* 41:

35–97
8. Taube, H. 1965. *Oxygen: Chemistry, Structure and Excited States.* Boston: Little, Brown
9. McCord, J. M., Fridovich, I. 1970. *J. Biol. Chem.* 245:1374–77
10. Misra, H. P., Fridovich, I. 1972. *J. Biol. Chem.* 247:188–92
11. Misra, H. P., Fridovich, I. 1972. *J. Biol. Chem.* 247:3170–75
12. Heikkila, R. E., Cohen, G. 1973. *Science* 181:456–57
13. Marklund, S., Marklund, G. 1974. *Eur. J. Biochem.* 47:469–74
14. Massey, V., Strickland, S., Mayhew, S. G., Howell, L. G., Engel, P. C., Matthews, R. G., Schuman, M., Sullivan, P. A. 1969. *Biochem. Biophys. Res. Commun.* 36:891–97
15. Ballou, D., Palmer, G., Massey, V. 1969. *Biochem. Biophys. Res. Commun.* 36:898–904
16. Massey, V., Palmer, G., Ballou, D. See Ref. 4, 25–43
17. Misra, H. P. 1974. *J. Biol. Chem.* 249:2151–55
18. Nishikimi, M., Rao, N. A., Yagi, K. 1972. *Biochem. Biophys. Res. Commun.* 46:849–54
19. Balny, C., Douzou, P. 1974. *Biochem. Biophys. Res. Commun.* 56:386–91
20. Fisher, D. B., Kaufman, S. 1973. *J. Biol. Chem.* 248:4300–4
21. Orme-Johnson, W. H., Beinert, H. 1969. *Biochem. Biophys. Res. Commun.* 36:905–11
22. Nilsson, R., Pick, F. M., Bray, R. C. 1969. *Biochim. Biophys. Acta.* 192:145–48
23. Misra, H. P., Fridovich, I. 1971. *J. Biol. Chem.* 246:6886–90
24. Nakamura, S. 1970. *Biochem. Biophys. Res. Commun.* 41:177–83
25. May, S. W., Abbott, B. J., Felix, A. 1973. *Biochem. Biophys. Res. Commun.* 54:1540–45
26. Misra, H. P., Fridovich, I. 1972. *J. Biol. Chem.* 247:6960–62
27. Wever, R., Oudega, B., VanGelder, B. F. 1973. *Biochim. Biophys. Acta* 302:475–78
28. Wallace, W. J., Maxwell, J. C., Caughey, W. S. 1974. *Biochem. Biophys. Res. Commun.* 57:1104–11
29. McCord, J. M., Fridovich, I. 1968. *J. Biol. Chem.* 243:5753–60
30. Fridovich, I. 1970. *J. Biol. Chem.* 245:4053–57
31. Rajagopalan, K. V., Fridovich, I., Handler, P. 1962. *J. Biol. Chem.* 237:922–28
32. Rajagopalan, K. V., Handler, P. 1964. *J. Biol. Chem.* 239:2022–26
33. Aleman, V., Handler, P. 1967. *J. Biol. Chem.* 242:4087–96
34. Hirata, F., Hayaishi, O. 1971. *J. Biol. Chem.* 246:7825–26
35. Strobel, H. W., Coon, M. J. 1971. *J. Biol. Chem.* 246:7826–29
36. Kumar, R. Prema, Ravindranath, S. D., Vaidyanathan, C. S., Rao, N. Appaji. 1972. *Biochem. Biophys. Res. Commun.* 49:1422–26
37. Hamilton, G. A., Libby, R. D. 1973. *Biochem. Biophys. Res. Commun.* 55:333–40
38. Babior, B. M., Kipnes, R. S., Curnutte, J. T. 1973. *J. Clin. Invest.* 52:741–44
39. Curnutte, J. T., Whitten, D. M., Babior, B. M. 1974. *N. Engl. J. Med.* 290:593–97
40. Johnston, R. B. Jr., Keele, B. B., Webb, L., Kessler, D., Rajagopalan, K. V. 1973. *J. Clin. Invest.* 52:44a (Abstr.)
41. Salin, M. L., McCord, J. M. 1974. *J. Clin. Invest.* 54:1005–9
42. Asada, K., Kiso, K., Yoshikawa, K. 1974. *J. Biol. Chem.* 249:2175–81
43. Allen, J. F., Hall, D. O. 1973. *Biochem. Biophys. Res. Commun.* 52:856–62
44. Asada, K., Kiso, K. 1973. *Eur. J. Biochem.* 33:253–57
45. Asada, K., Kiso, K. 1973. *Agr. Biol. Chem.* 37:453–54
46. Epel, B. L., Neuman, J. 1973. *Biochim. Biophys. Acta* 325:520–29
47. Debey, P., Balny, C. 1973. *Biochemie* 55:329–32
48. Aust, S. D., Roerig, D. L., Pederson, T. C. 1972. *Biochem. Biophys. Res. Commun.* 47:1133–37
49. Behar, D., Czapski, G., Rabani, J., Dorfman, L. M., Schwarz, H. A. 1970. *J. Phys. Chem.* 74:3209–13
49a. McCord, J. M., Fridovich, I. 1969. *J. Biol. Chem.* 244:6056–63
50. Rotilio, G., Bray, R. C., Fielden, E. M. 1972. *Biochim. Biophys. Acta* 268:605–9
51. Klug, D., Rabani, J., Fridovich, I. 1972. *J. Biol. Chem.* 247:4839–42
52. McCord, J. M., Keele, B. B. Jr., Fridovich, I. 1971. *Proc. Nat. Acad. Sci. USA* 68:1024–27
53. Gregory, E. M., Fridovich, I. 1974. *J. Bacteriol.* 117:166–69
54. Gregory, E. M., Fridovich, I. 1973. *J. Bacteriol.* 114:543–48
55. Gregory, E. M., Fridovich, I. 1973. *J. Bacteriol.* 114:1193–97
56. Gregory, E. M., Goscin, S. A., Fridovich, I. 1974. *J. Bacteriol.* 117:456–60
57. White, J. R., Dearman, H. H. 1965. *Proc.*

Nat. Acad. Sci. USA 54 : 887–91
58. White, J. R., White, H. L. 1966. *Biochim. Biophys. Acta* 123 : 648–51
59. White, J. R., Vaughan, T. O., Yeh, W.-S. 1971. *Fed. Proc.* 30 : 1145 (Abstr.)
59a. Michelson, A. M., Buckingham, M. E. 1974. *Biochem. Biophys. Res. Commun.* 58 : 1079–86
60. Crapo, J. D., Tierney, D. L. 1974. *Am. J. Physiol.* 226 : 1401–7
61. Levander, O. A., Morris, V. C., Higgs, D. J. 1974. *Fed. Proc.* 33 : 693 (Abstr.)
62. Autor, A. P. 1974. *Life Sci.* 14 : 1309–19
63. Misra, H. P., Fridovich, I. 1972. *J. Biol. Chem.* 247 : 3410–14
64. Rapp, U., Adams, W. C., Miller, R. W. 1973. *Can. J. Biochem.* 51 : 158–71
65. Goscin, S. A., Fridovich, I. 1972. *Biochim. Biophys. Acta* 289 : 276–83
66. Weser, U., Fretzdorff, A., Prinz, R. 1972. *FEBS Lett.* 27 : 267–69
67. Beauchamp, C. O., Fridovich, I. 1973. *Biochim. Biophys. Acta* 317 : 50–64
68. Sawada, Y., Ohyama, T., Yamazaki, I. 1972. *Biochim. Biophys. Acta* 268 : 305–12
69. Asada, K., Urano, M., Takehashi, M. 1973. *Eur. J. Biochem.* 36 : 257–66
70. Lumsden, J., Hall, D. O. 1974. *Biochem. Biophys. Res. Commun.* 58 : 35–41
71. Weisiger, R. A., Fridovich, I. 1973. *J. Biol. Chem.* 248 : 3582–92
72. Beem, K. M., Rich, W. E., Rajagopalan, K. V. 1974. *J. Biol. Chem.* 249 : 7298–7305
73. Fee, J. A. 1973. *J. Biol. Chem.* 248 : 4229–34
74. Rotilio, G., Calabrese, L., Coleman, J. E. 1973. *J. Biol. Chem.* 248 : 3855–59
75. Forman, H. J., Fridovich, I. 1973. *J. Biol. Chem.* 248 : 2645–49
76. Steinman, H. M., Abernathy, J. L., Hill, R. L. 1974. *J. Biol. Chem.* 249 : 7339–47
77. Evans, H. J., Steinman, H. M., Hill, R. L. 1974. *J. Biol. Chem.* 249 : 7315–25
78. Steinman, H. M., Naik, V. R., Abernathy, J. L., Hill, R. L. 1974. *J. Biol. Chem.* 249 : 7326–38
79. Richardson, J. S., Thomas, K. A., Rubin, B. H., Richardson, D. L. 1974. *Proc. Nat. Acad. Sci. USA* Submitted
80. Fee, J. A., Gaber, B. P. 1972. *J. Biol. Chem.* 247 : 60–65
81. Fee, J. A. 1973. *Biochim. Biophys. Acta* 295 : 107–16
82. Rotilio, G., Morpurgo, L., Giovagnoli, C., Calabrese, L., Mondovi, B. 1972. *Biochemistry* 11 : 2187–92
83. Forman, H. J., Evans, H. J., Hill, R. L., Fridovich, I. 1973. *Biochemistry* 12 : 823–27
84. Stokes, A. M., Hill, H. A. O., Bannister, W. H., Bannister, J. V. 1973. *FEBS*

Lett. 32 : 119–23
85. Klug-Roth, D., Fridovich, I., Rabani, J. 1973. *J. Am. Chem. Soc.* 95 : 2786–90
86. Fielden, E. M., Roberts, P. B., Bray, R. C., Rotilio, G. 1973. *Biochem. Soc. Trans.* 1 : 52–53
87. Bannister, J. V., Bannister, W. H., Bray, R. C., Fielden, E. M., Roberts, P. B., Rotilio, G. 1973. *FEBS Lett.* 32 : 303–6
88. Fielden, E. M., Roberts, P. B., Bray, R. C., Lowe, D. J., Mautner, G. N., Rotilio, G., Calabrese, L. 1974. *Biochem. J.* 139 : 49–60
89. Gaber, B. P., Brown, R. D., Koenig, S. H., Fee, J. A. 1972. *Biochim. Biophys. Acta* 271 : 1–5
90. Boden, N., Holmes, M. C., Knowles, P. F. 1974. *Biochem. Biophys. Res. Commun.* 57 : 845–48
91. Haffner, P. H., Coleman, J. E. 1973. *J. Biol. Chem.* 248 : 6626–29
92. Simonyan, M. A., Nalbandyan, R. M. 1972. *FEBS Lett.* 28 : 22–24
93. Rotilio, G., Morpurgo, L., Calabrese, L., Mondovi, B. 1973. *Biochim. Biophys. Acta* 302 : 229–35
94. Fee, J. A., DiCorleto, P. E. 1973. *Biochemistry* 12 : 4893–99
95. Bray, R. C., Cockle, S. A., Fielden, E. M., Roberts, P. B., Rotilio, G., Calabrese, L. 1974. *Biochem. J.* 139 : 43–48
96. Hodgson, E. K., Fridovich, I. 1973. *Biochem. Biophys. Res. Commun.* 54 : 270–74
97. Keele, B. B. Jr., McCord, J. M., Fridovich, I. 1970. *J. Biol. Chem.* 245 : 6176–81
98. Vance, P. G., Keele, B. B. Jr., Rajagopalan, K. V. 1972. *J. Biol. Chem.* 247 : 4782–86
98a. Lavelle, F., Durosay, P., Michelson, A. M. 1974. *Biochimie* 56 : 451–58
99. Villafranca, J. J., Yost, F. J. Jr., Fridovich, I. 1974. *J. Biol. Chem.* 249 : 3532–36
100. Forman, H. J., Fridovich, I. 1973. *Arch. Biochem. Biophys.* 158 : 396–400
101. Weisiger, R. A., Fridovich, I. 1973. *J. Biol. Chem.* 248 : 4793–96
102. Steinman, H. M., Hill, R. L. 1973. *Proc. Nat. Acad. Sci. USA* 70 : 3725–29
103. Fridovich, I. 1974. *Life Sci.* 14 : 819–26
104. McCord, J. M. Personal communication
105. Marklund, S. 1973. *Acta Chem. Scand.* 27 : 1458–60
106. Beckman, G., Lundgren, E., Tarnvik, A. 1973. *Hum. Hered.* 23 : 338–45
107. Yost, F. J. Jr., Fridovich, I. 1973. *J. Biol. Chem.* 248 : 4905–8
108. Gregory, E. M., Yost, F. J. Jr., Fridovich, I. 1973. *J. Bacteriol.* 115 : 987–91
109. Yost, F. J. Jr., Fridovich, I. 1974.

Arch. Biochem. Biophys. 161:395–401
110. Misra, H. P. 1974. *Fed. Proc.* 33:1505 (Abstr.)
111. Henry, Y. A., Puget, K., Michelson, A. M. 1974. *Fed. Proc.* 33:1321 (Abstr.)
112. Keele, B. B. Jr., McCord, J. M., Fridovich, I. 1971. *J. Biol. Chem.* 246: 2875–80
113. Finazzi Agro, A., Giovagnoli, C., De-Sole, P., Calabrese, L., Rotilio, G., Mondovi, B. 1972. *FEBS Lett.* 21: 183–85
114. Weser, U., Paschen, W. 1972. *FEBS Lett.* 27:248–50
115. Joester, K. E., Jung, G., Weber, U., Weser, U. 1972. *FEBS Lett.* 25:25–28
116. Zimmermann, R., Flohé, L., Weser, U., Hartmann, H. J. 1973. *FEBS Lett.* 29:117–20
117. Paschen, W., Weser, U. 1973. *Biochim. Biophys. Acta* 327:217–22
118. Weser, U. 1973. *Struct. Bonding* 17: 1–66
119. Kearns, D. R. 1971. *Chem. Rev.* 71: 395–427
120. Foote, C. S. 1975. In *Free Radicals and Biological Systems.* New York: Academic. In preparation.
121. Eigen, M., Hammes, G. G. 1963. *Advan. Enzymol.* 25:1–38
122. Hodgson, E. K., Fridovich, I. 1974. *Biochemistry* 13:3811–15
123. Schapp, A. P., Thayer, A. L., Faler, G. R., Goda, K., Kimura, T. 1974. *J. Am. Chem. Soc.* 96:4025–26
123a.Mayeda, E. A., Bard, A. J. 1974. *J. Am. Chem. Soc.* 96:4023–24
124. Hodgson, E. K., Fridovich, I. 1973. *Photochem. Photobiol.* 18:451–55

124a.Puget, K., Michelson, A. M. 1974. *Biochem. Biophys. Res. Commun.* 58: 830–38
125. Miller, R. W. 1970. *Can. J. Biochem.* 48:935–39
126. Miller, R. W., Rapp, U. 1973. *J. Biol. Chem.* 248:6084–90
127. Chan, P. C., Bielski, B. H. J. 1974. *J. Biol. Chem.* 249:1317–19
128. Land, E. J., Swallow, A. J. 1971. *Arch. Biochem. Biophys.* 145:365–72
129. Williams, D. M., Lee, G. R., Cartwright, G. E. 1974. *J. Clin. Invest.* 53:665–67
130. Petkau, A., Chelak, W. S. 1974. *Fed. Proc.* 33:1505 (Abstr.)
131. Pederson, T. C., Aust, S. D. 1973. *Biochem. Biophys. Res. Commun.* 52: 1071–78
132. Fong, K. L., McKay, P. B., Foyer, J. L., Keele, B. B. Jr., Misra, H. P. 1973. *J. Biol. Chem.* 248:7792–97
133. Fee, J. A., Teitelbaum, D. 1972. *Biochem. Biophys. Res. Commun.* 49:150–57
134. Lavelle, F., Michelson, A. M., Dimitrijevic, L. 1973. *Biochem. Biophys. Res. Commun.* 55:350–57
135. Haber, F., Weiss, J. 1934. *Proc. Roy. Soc. London A* 147:332–51
136. Beauchamp, C., Fridovich, I. 1970. *J. Biol. Chem.* 245:4641–46
137. Goscin, S. A., Fridovich, I. 1972. *Arch. Biochem. Biophys.* 153:778–83
138. Heikkila, R. E., Cohen, G., Manian, A. A. 1975. *Biochem. Pharmacol.* In press
139. Cohen, G., Heikkila, R. 1974. *J. Biol. Chem.* 249:2447–52
140. McCord, J. M. 1974. *Science* 185:529–31

STRUCTURAL ANALYSIS OF   ×878
MACROMOLECULAR ASSEMBLIES BY
IMAGE RECONSTRUCTION FROM
ELECTRON MICROGRAPHS

R. A. Crowther and A. Klug   •

Medical Research Council Laboratory of Molecular Biology, Cambridge CB2 2QH, England

CONTENTS

INTRODUCTION

Electron microscopy and X-ray diffraction have revealed a wealth of information about biological structures. The two techniques are frequently complementary. X-ray diffraction is particularly applicable to specimens that exhibit a high degree of regularity. The intensity distribution of the diffraction pattern gives information about the periodicities present in the specimen, which can frequently be preserved in its native state in an aqueous medium. Furthermore, if the phases of the diffraction pattern can be determined, for example, by using isomorphous heavy atom derivatives as in protein crystallography, a three-dimensional image of the specimen can be computed by Fourier synthesis. This may have atomic resolution if the order in the specimen is sufficiently good. It is the phase determination that is generally limiting in the laborious process of converting the X-ray pattern to an image of the structure.

By contrast the electron microscope provides a direct image of the specimen and

161

can be used on materials unsuitable for X-ray diffraction. The objective lens of the microscope combines the diffracted beams from the specimen while preserving their relative phases. This phase information is therefore recorded in the intensity distribution of the image and may be retrieved from it optically or computationally. It would be lost if only the electron diffraction pattern were recorded. The image given by the microscope, however, suffers from a number of limitations, including distortion of the specimen during preparation and damage during observation. In addition, artificial means of contrast enhancement have to be used, as the majority of atoms in biological specimens have an atomic number too low to give sufficient contrast on their own. Furthermore, the depth of focus of the conventional microscope is several thousand angstroms, so that features at different levels in the specimen are superimposed in the two-dimensional image, which is thus essentially a plane projection of the specimen in the direction of view (1). All these factors limit the structural information that can be obtained by direct inspection of the image. Some of these drawbacks may be overcome by various forms of image processing that exploit the spatial symmetries that are frequently exhibited by the native biological structure or that can be induced by crystallization or by aggregation. More reliable structural information can then be retrieved from the image by techniques analogous to those used in X-ray diffraction analysis.

Broadly, there are two kinds of analysis performed, namely two-dimensional spatial filtering and three-dimensional image reconstruction. The former may be applied to translationally or rotationally symmetric images and provides an average two-dimensional image of the repeating unit of the structure by combining the many images present in the array. This averaging may be realized by direct optical superposition or more reliably by filtering of the optical or computed diffraction patterns. For three-dimensional image reconstruction a series of two-dimensional images of the specimen, viewed from different directions and therefore giving different projections, must be combined to generate a three-dimensional image of the specimen, again showing an average of the repeating unit.

An important feature of these methods is that the initial step consists of a quantitative assessment of the reliability of a particular image based on its degree of symmetry. Images of well-preserved specimens can then be cross-correlated to check their degree of reproducibility. Finally, the best images that exhibit the maximum degree of cross correlation may be combined to produce a best two- or three-dimensional image of the specimen. The unaided eye cannot perform such assessment and combination of different images. For example, stereoscopy cannot be used to disentangle the continuously varying density even in a perfect three-dimensional specimen viewed in transmission. In such a case it can be shown mathematically that a pair of images is not sufficient to allow unambiguous reconstruction, and perceptually we are not accustomed to viewing images of spatially varying translucent objects.

The application of these reconstruction methods has in recent years helped to reveal the molecular architecture of various biological assemblies, such as muscle, viruses, and enzyme complexes. We describe a limited number of examples, which serve to demonstrate the power of the various techniques and the nature of the results they can give. We do not go deeply into the theory of reconstruction nor consider the merits of alternative approaches to the Fourier methods by which most results have so far been obtained. First of all, however, we consider briefly the crucial question of specimen preservation.

Specimen Preservation and Attainable Resolution

In order to make the specimen visible in the conventional microscope, it is necessary to add heavy metal as an electron dense contrasting medium. Bulk specimens such as muscle may be embedded, sectioned, and stained. In such specimens the degree of preservation seldom extends much beyond a resolution of 100 Å, although in one favorable case of insect flight muscle a 48 Å spacing has been recorded (2). Sectioned crystals can give higher resolution. Alternatively the specimen may be frozen, fractured, and possibly etched, the resulting surface topography then being contrasted by heavy metal shadowing. Details in the range 50–100 Å can be visualized, possibly even down to 35 Å, as in polyheads for example (3). Finally, specimens may be deposited from solution and embedded on the grid in a heavy metal salt such as uranyl acetate. Such contrasting is known as negative staining (4), as the presence of biological material is inferred from the absence of stain. The degree of detail revealed is limited by the granularity of the stain and the fidelity with which it follows the surface of the specimen. Although limited to showing the envelope of the structure and accessible internal cavities, this form of contrasting is the most successful so far devised and typically details down to about 20 Å can be observed. As we shall see later, the stain serves not only to contrast the biological material but also to stabilize it, at least partially, during the dehydration and intense irradiation that it suffers in the microscope.

In talking about the resolution present in an image, we actually mean a number of rather different things. First of all, there is the point-to-point resolution, which is set by the electron optical properties and stability of the microscope. For modern instruments correctly operated this should be better than 3 Å and is not limiting in the type of work considered here. Then, there is the degree to which individual macromolecules are preserved during dehydration and irradiation and the fidelity with which their shape is mapped out by the stain. Finally, when one is working with symmetrical structures, one has the problem of distortions in the long range order of the specimen. When one is combining data from the different subunits, it is this variation within individual subunits and the departures from their ideal symmetry-related positions which set a limit to the maximum spatial frequency that can usefully be recovered from the image. There will of course be higher spatial frequencies present, arising from the granularity of the stain and the support film. Thus when one speaks, for example, of an image containing information to 20 Å, one means that spatial frequencies arising from the regular part of the image are detectable above the background noise only for spacings out to 20 Å but not beyond. This spatial frequency cutoff can be converted to the classical Rayleigh criterion for resolving two equal pointlike features by multiplying by 0.61 in two dimensions or 0.72 in three. It is also important to note that the positions or changes in position of well-defined features in the image can be estimated much more accurately than this, possibly to about 0.1 of the spatial frequency cutoff.

It is frequently observed for negatively stained specimens that the regular part of the information extends only to spacings of about the diameter D of the structure units in the array. This is not unexpected, since the stain is only mapping out the envelope of the subunit. If the subunit were spherical, its diffraction pattern (or "Airy sphere") would have its first zero at a radius $1/D$ and would be weak beyond this. Thus, in negatively stained images of structures built from globular subunits, one would not expect to obtain much information about spacings shorter than D. This is nevertheless sufficient to discover the packing geometry of the subunits and to map their gross shape. The limitation on the results arises from the specimen preparation and not from the reconstruction methods themselves.

BACTERIOPHAGE T4

The various parts of bacteriophage T4 display the applicability of a number of image processing techniques. The DNA-containing head has a contractile tail with a complex baseplate structure. The latter serves to attach the phage to the bacterium during infection. Changes in the baseplate trigger contraction of the tail and injection of the phage DNA into the bacterium.

Polyheads: Translational Filtering

The native head structure of T4 is difficult to study directly as it is rather smooth and its features are not strongly contrasted in negative stain. Moreover, the bulk of the tightly folded DNA is superimposed on the protein coat. Ghosts, that is heads from which the DNA has been ejected, are generally poorly preserved and difficult to analyze. Some progress has been made with freeze etching on the closely related T2 phage, but the degree of detail in these images is limiting (3). There are, however, a number of aberrant tubular structures, known as polyheads, which are related to the native head structure and undergo transformations believed to mimic the complicated process of maturation associated with DNA packaging in the native head (5). These structures are easier to analyze than the native head.

Figure 1a shows a negatively stained "coarse" polyhead (6). The tubular structure has flattened so that the image is of two superimposed, approximately planar layers of protein molecules. This superposition plus the granularity of the support film obscures the individual subunits and their arrangement in the structure. Figure 1b shows its optical diffraction pattern. A coherent optical processing system is used (7), in which the micrograph is illuminated by a laser and its Fraunhofer diffraction pattern recorded in the back focal plane of a lens. The spots occur in pairs, symmetrically related about the center of the pattern, and each pair arises by diffraction from a particular strong sinusoidal wave in the image. Pairs of spots near the center of the pattern arise from low spatial frequencies while spots further out come from increasingly higher spatial frequencies. For images with two-dimensional translational periodicity the spots in the diffraction pattern lie on a regular lattice. The case shown here closely approximates this situation, and the spots lie on two regular lattices related by an axial mirror line. The spots from the two layers are spatially separated, so by placing in the diffraction plane of the optical diffractometer an opaque mask with appropriate holes cut in it, the diffracted beams from one of the layers can be selectively recombined by a further lens to produce a filtered image of a single layer (8), as shown in Figure 1c. Besides

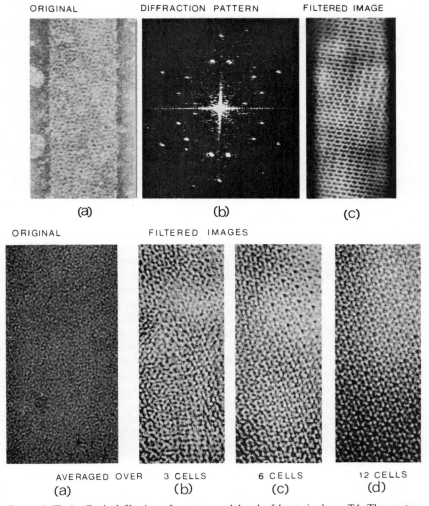

Figure 1 (Top) Optical filtering of a coarse polyhead of bacteriophage T4. The centers of the hexagonal rings of subunits seen in the filtered image are approximately 100 Å apart. *Figure 2 (Bottom)* Computer filtering of a "fine" polyhead of bacteriophage T4, showing different ranges of averaging. The hexagonal rings of subunits seen in (*d*) are approximately 100 Å apart.

removing one of the layers, a large proportion of the aperiodic noise (that is the irregular part of the image, which gives the speckle between the main diffraction spots) has also been removed. In the filtered image the approximately hexagonal arrangement of individual protein subunits can be clearly seen.

It is possible to perform the equivalent operations computationally (9, 10) by digitizing the image and calculating its diffraction pattern or Fourier transform in a computer. This is useful in cases where the diffraction spots are close together or where the signal-to-noise ratio is particularly low. This is the case with so-called fine grained polyheads; an example is shown in Figure 2a. The filtered image is generated by setting the whole transform to zero, except in small regions surrounding the diffraction maxima arising from one of the two layers, and then computing the inverse transform. The range of averaging that takes place in the filtering process is controlled by the size of these "apertures"; the smaller the apertures the greater the range of averaging. If the apertures were reduced to single sample points, a perfectly periodic filtered image would result.

Figure 2b–d (9) shows a series of filtered images of Figure 2a in which the respective range of averaging is 3, 6, and 12 unit cells in the longitudinal direction and about half this laterally. Figure 2b resembles the best that could be achieved readily by optical filtering and is uninterpretable. In Figure 2d, however, a sufficient number of different unit cells have been averaged to enable the fine hexagonal pattern of subunits to be discerned. The patterns are most readily seen at the bottom of the image, and once perceived, can be traced right along this image and also in Figure 2c. It is believed that the protein subunits in the coarse polyhead (Figure 1) have undergone cleavage and rearrangement in forming the fine polyhead of Figure 2. Similar changes occur during the packaging of the DNA inside the native phage head (5), which is thought to have a final structure resembling that of a fine polyhead.

Because this final structure consists of small units rather uniformly distributed, the low spatial frequencies are very weak compared with corresponding terms for the coarse polyhead. The principal contributions therefore come from the region of high spatial frequencies, where the principal noise contributions also occur. Thus, extensive averaging is necessary if the underlying regular pattern is to be seen. Phage ghosts themselves do not provide a sufficiently large and regular specimen for this averaging to be satisfactorily performed.

Baseplates: Rotational Filtering

The baseplate of T4 is a complex structure containing multiple copies of about a dozen different proteins (11, 12) and, as can be seen in Figure 3a, it appears to possess sixfold rotational symmetry. It is not possible to analyze and filter images with rotational symmetry by optical diffraction, as the wanted and unwanted components of the image are not spatially separated in the diffraction pattern. However, it can be done fairly easily by computer (13). The image is again digitized and then numerically decomposed into a series of functions representing harmonics of increasing angular and radial frequency. (These are analogous to the plane sinusoidal waves that give rise to the pairs of spots in the diffraction pattern of a translationally periodic object.) By plotting the strength or power of the angular harmonics as a function of increasing angular frequency, we obtain a rotational power spectrum of the image which is analogous to the distribution of intensities in the optical diffraction pattern of a translationally periodic image and summarizes the results of the analysis in a convenient form.

The power spectrum of the baseplate in Figure 3a is shown in Figure 3c (13). As expected, the harmonics that are multiples of 6 are much stronger than the other harmonics and it is reasonable to say that the image is predominantly six-fold, as is clear from direct inspection in this case. However, we can now recombine just those harmonics whose angular frequencies are multiples of 6 to resynthesize a rotationally filtered image (Figure 3b). The assumption here is that all those harmonics whose angular frequencies are not multiples of 6 arise from distortions, variable staining, and contributions from the support film. The total noise contribution will be reduced by a factor of 6 in this case.

Rotational filtering has a number of advantages over the more straightforward technique of rotational superposition using photographic methods (14). In the latter a compound image is produced by superimposing m copies of the original image, each rotated successively by an angle $2\pi/m$, about the supposed symmetry axis. Features related by m-fold or some multiple of m-fold symmetry are preserved, while features with all other rotational symmetries are suppressed. The disadvantage of this method is that the determination of the symmetry number m and the assessment

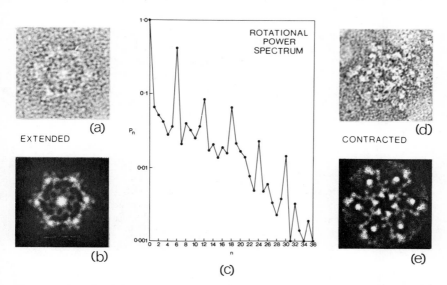

Figure 3 Rotational filtering of baseplates of bacteriophage T4. (*a*) Extended baseplate, approximate diameter 500 Å. (*b*) Sixfold filtered image of (*a*). (*c*) Rotational power spectrum of baseplate shown in (*a*). (*d*) Contracted baseplate, approximate diameter 550 Å. (*e*) Sixfold filtered image of (*d*). Micrographs provided by Dr. J. King.

of the quality of the image are not separated from the production of the final averaged image; only a subjective assessment is possible based on the appearance of the final image. On the other hand, the power spectrum such as that shown in Figure 3c enables us to measure the strength of the sixfold components and to see to what order or resolution they extend. A quantitative comparison of different images can then be made and the best chosen. Similar remarks apply to the comparison of filtering (8) as against linear integration (15) of translationally periodic images.

The filtered image (Figure 3b) shows that the baseplate has a strong hexagonal periphery with spikes at the corners. The central core is surrounded by hexagonal tracery which is in turn connected to the periphery by a series of fine bridges. This complexity of structure is certainly consistent with the fact that it contains the dozen or so components found by genetic and biochemical analysis (11, 12). Preliminary image analysis of a number of structural mutants (J. King and R. A. Crowther, unpublished results) suggests that it may be possible to locate the positions of some of the gene products in the baseplate. It is also possible to get images of the baseplates in their contracted state (Figure 3d and e) from which one may be able to deduce what molecular rearrangement takes place on contraction and how this initiates the contraction of the tail structure, which is discussed in the next section.

Phage Tail: Three-Dimensional Reconstruction

The tail of the T-even phages consists of a hollow core surrounded by a contractile sheath (Figure 4a), which is attached to the baseplate at one end and by a collar structure to the head at the other. The sheath contains a single type of protein subunit arranged in successive annuli which are rotated with respect to one another, so that the subunits also lie along oblique helical lines. On attachment to the host cell the baseplate triggers a contraction of the sheath which rides up the core, causing the core to penetrate the host cell wall and leading to the injection of the phage DNA.

The symmetry of the sheath can be analyzed (16, 17) by using optical diffraction (Figure 4b). The indexing enables the various strong features in the diffraction pattern to be ascribed to the front or back of the helical structure, in much the same way as the contributions from the two planar layers were separated in the case of polyheads. However, filtering of the helical tail structure (Figure 4c) (16) does not lead to a simply interpretable image, as did the polyheads, because there is staining at more than one radius in the structure. The filtered image of a single side of the structure still shows a complicated superposition of different features and it is therefore necessary to undertake a three-dimensional image reconstruction (16).

The optical diffraction pattern records only the intensity of the diffracted beams from the various strong spatial periodicities in the image and not their relative phases, which fix the relative positions of these periodicities in the image. The relative phase information is of course used in optical filtering because the selected diffracted rays themselves are refocused by a lens to form the filtered image. However, it is possible to determine the relative phases by computing the diffraction pattern as a complex Fourier transform. The complex numbers give the amplitudes of the various components and also their phases which are lost in optical recording of the transform.

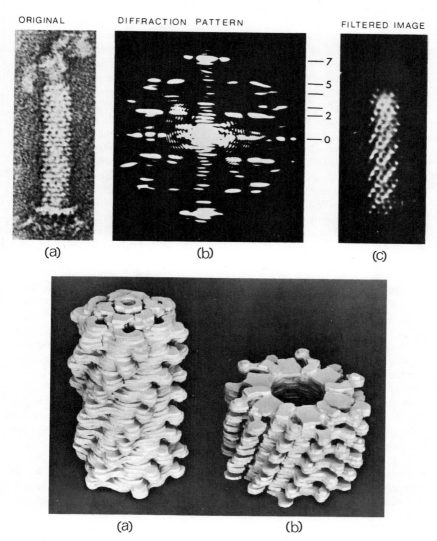

Figure 4 (Top)   Optical filtering of the tail of bacteriophage T4. The axial spacing of the annuli in the tail (a) is approximately 38 Å, corresponding to the layer line marked 7 in (b).
Figure 5 (Bottom)   Models of three-dimensional reconstructions of (a) the extended tail of bacteriophage T4, diameter approximately 240 Å. (b) Polysheath which closely resembles the contracted sheath, diameter approximately 300 Å.

The diffraction pattern from a helical structure consists of a series of horizontal layer lines, each arising from a particular set of families of helical grooves in the structure. Indexing of the pattern consists of determining the number and pitch of the helices in each family. If there is a sufficiently large number of subunits in the helical repeat, there will be only a single helical family contributing to each layer line in the neighborhood of the meridian. This not only simplifies the indexing but also the process of three-dimensional reconstruction. For a helical structure with such high symmetry a single image suffices for the reconstruction. In physical terms, this is because a single image of the specimen presents many different views of the repeating subunit. These views are, moreover, at known relative positions and equally spaced in angle. In the phage tail there are 42 subunits in the axial repeat, since each annulus contains six subunits and the whole structure repeats after seven annuli. There are, however, only 21 distinct views of the subunit, because two views of a subunit from opposite directions give the same projection.

The two-dimensional computed diffraction pattern or Fourier transform of the image is a central section of the three-dimensional transform of the three-dimensional specimen. We may use the known helical symmetry, established by indexing the diffraction pattern, to generate uniquely from this single section the complete three-dimensional transform out to a limit set by the number of distinct views of the subunit. A three-dimensional image of the specimen can then be computed by Fourier synthesis from this filled-in transform. In the computer implementation (18) the filling in of the three-dimensional transform is implicit in the mathematical formulation as a Fourier-Bessel transform, which links the amplitude and phase of each layer line contribution to the strength and position of a particular helical family in the three-dimensional structure. The three-dimensional image is built from a superposition of a number of these helical families. Close to the meridian on each layer line there is in general only a single helical family contributing, and the necessary information about it can be retrieved from a single view of the structure. As we move away from the meridian, other families start to contribute and it is not possible to sort out from a single view the various contributions on each layer line. It is this overlapping that limits the resolution attainable from a single view: the more subunits in the helical repeat, the further out the overlapping starts and the higher the resolution attainable. With more views this limitation can be overcome in a way similar to that described later for spherical viruses.

A reconstruction of the extended phage tail (16; L. A. Amos, unpublished), combining data from several particles to a cutoff of about 30 Å, is shown in Figure 5a. There is a hole of about 15 Å radius along the axis of the particle and then more or less continuous density representing the tail core which extends to about 45 Å radius. Separating this from the main bulk of the sheath is a set of six helical tunnels with bridges between, which link the sheath with the core. At outer radii from about 90–120 Å, the surface is divided up apparently into subunits by two strong families of helical grooves.

The contracted sheath itself is too short to make a good reconstruction. However, an aberrant structure, polysheath, made from sheath subunits assembled in a way believed to resemble closely the normal contracted sheath, has been analyzed

(L. A. Amos, unpublished results) (Figure 5b). A comparison of the two structures taken together with Moody's analysis (19) of the geometrical path of contraction enables the subunit to be tentatively dissected out and its change of configuration to be followed (L. A. Amos, in preparation).

TOBACCO MOSAIC VIRUS: SEPARATE IMAGING OF PROTEIN AND STAIN

The coat protein of tobacco mosaic virus (TMV) can be polymerized in a number of polymorphic forms. One of these, the stacked disc aggregate (20), is a long rod-shaped particle made from discs of protein, successive discs being rotated by 3/10 of $2\pi/17$ radians. The disc consists of two rings, each containing 17 protein molecules, arranged in a polar manner. Three-dimensional image reconstruction shows that the subunits in the two rings have different conformations. The stacked disc rods have been used for testing a new phase contrast imaging technique, using an electrostatic phase plate (21) that enables the contributions of stain and protein to the image to be separated. This has in turn provided important new information on the nature of negative staining and the behavior of the stain when exposed to the beam (22, 23).

The electrostatic phase plate (21) consists of an aperture spanned by a thin, poorly conducting thread, which is placed in the diffraction plane of the microscope. Its effect is analogous to that of an absorbing phase plate in light microscopy. The thread cuts out a large proportion of the unscattered beam, and the charge distribution on it gives an electric field which imparts a more or less uniform phase shift to the various scattered beams. In addition the aperture cuts out most of the electrons scattered by the heavy metal stain. The net effect of all these factors is that the principal contrast in the phase plate image arises from the biological material (24), unlike the conventional bright field image in which the contrast arises predominantly from the stain. Figure 6a and b shows the appearance of the two types of image.

We consider first the phase plate images (22). A number of good stretches of particles, selected by optical diffraction, were further analyzed by computer. After the relative positions and orientations had been determined by searching for the best correlations between the various sets of layer line data, an average transform was computed, in which the consistent parts of the data from five different transforms were averaged. This average transform closely resembled the X-ray diffraction pattern from oriented sols of stacked discs (25), suggesting that the structure imaged by the microscope is very similar to the native structure existing in solution. A cylindrically averaged structure was computed (Figure 6c) from just those parts of the transform that do not involve azimuthal variations, including data out to spacings of about 8.5 Å. There are striking differences between the two layers comprising the disc, the upper running radially while the lower is bent into a zigzag. This gives rise to a local pairing of the layers within a disc at outer radii but between discs at inner radii. There are significant departures from mirror symmetry between the two layers, confirming the polar nature of the disc.

A model of a full three-dimensional reconstruction is shown in Figure 7 (22).

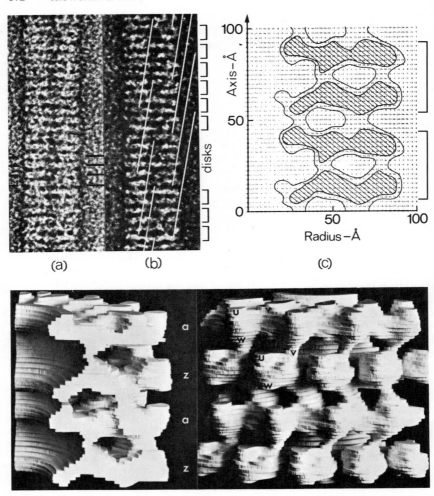

Figure 6 (Top)   Stacked disc rods of tobacco mosaic virus protein, approximate diameter 180 Å. (*a*) Bright field image in which contrast comes mainly from the stain. (*b*) Phase plate image in which contrast comes mainly from the protein. Note that the oblique helical families corresponding to azimuthally varying harmonics are much more visible in (*b*) than (*a*). (*c*) Section through a cylindrically averaged reconstruction from phase plate images, showing differences between the two layers of subunits forming the disc. The axis of the particle is at the left.

Figure 7 (Bottom)   Model of a three-dimensional reconstruction of the stacked disc rod of tobacco mosaic virus protein. The pictures show a cross section and a view of the outside. There are clear differences in conformation between the subunits in the two layers forming a disc (see text).

The shapes of the subunits in the two layers are again rather different, the layer marked *a* running approximately radially while the one marked *z* has a pronounced zigzag. At outer radii the tips of the subunits in the two layers are separated by grooves of different shape; those in the *a* layer are broad and shallow while those in the *z* layer are narrow and angled. By studying various contour maps of the reconstruction sectioned in different ways, it is possible to follow the course of the subunits in the two layers and to make a tentative correlation between features common to the two conformations (22). Three such features, two knobs marked *u* and *v* and a ridge marked *w*, are shown in Figure 7. If these identifications are correct, it appears that the conformation in the *a* layer could be generated by a rotation and slewing of the heads of the subunits in the *z* layer accompanied by a reduction of the axial zigzag. Comparing the average computed transform on which this reconstruction is based with the X-ray diffraction pattern from stacked disc rods in solution, it appears that the conformational differences between the two layers of subunits, although existing in solution, may have been somewhat exaggerated in the negatively stained preparation. It is the conformation of the subunits on the *a* layer that seems most similar to the layer closest to the dyad axis in the structure determined by X-ray diffraction of crystals of discs (26). This in turn is most like the structure of subunits in the virus itself, again as determined by X-ray diffraction (27).

There are two significant discrepancies between the reconstruction from phase plate images just described and the reconstruction made earlier (28) from conventional bright field images which show the stain rather than the protein. There is a much weaker axial modulation at the outer surface of the bright field reconstruction, and the slewing of the outer parts of the subunits in the two different rings is much more similar in the bright field reconstruction. By analyzing the transforms of bright field images of negatively stained specimens exposed to different electron doses, Unwin (23) has shown that, although the bright field image taken with minimal exposure is similar to the phase plate image, increasing irradiation produces consistent and reproducible changes. A difference reconstruction between the images of a weakly and strongly irradiated specimen shows that the changes occur mainly at the inner and outer surfaces of the particle and are accompanied by an increase in the volume unpenetrated by stain. This is explained by a contraction and migration of the stain. On irradiation the uranyl acetate is converted to the considerably more dense UO_2, causing a linear shrinkage of about 15%. Shrinkage alone does not account for all the observed changes, however, and Unwin suggests that there may be surface energy effects which tend to cause the stain to round up. Together these effects can explain the difference between the phase plate reconstruction and the strongly irradiated bright field reconstruction. It is important to note that the contraction and migration of the stain is an orderly affair not accompanied by significant loss of resolution, so that the biological material must be morphologically preserved to act as a template for the movements of stain. What is being seen is probably some highly crosslinked derivative of the original biological material but this must nevertheless bear a close morphological similarity to the native material because of the close resemblance between the transform of the phase plate image and the X-ray diffraction pattern. The stain supports the carbonaceous material while this initial crosslinking to form a more stable radiation product takes place. The latter then serves as a template for the stain movement that occurs on further irradiation. It is clear that caution is necessary in interpreting bright field images of negatively stained specimens and that minimal exposure (29) should be used whenever possible.

HEMOCYANIN

Hemocyanins are copper-containing respiratory proteins which occur dissolved in the hemolymph of many invertebrates. Gastropod hemocyanins have molecular weights of about $7-8 \times 10^6$ and undergo a series of association-dissociation reactions, depending on pH and ionic strength. Electron microscopy shows the particle to be cylindrical with a diameter of about 300 Å and a length of about 360 Å. The first dissociation into half particles occurs in a plane perpendicular to the cylindrical axis.

Rotational filtering of half particles (30) shows that at an inner radius there is a strong fivefold modulation in density, whereas the outer part is dominated by a tenfold modulation (Figure 8). In whole particles the central fivefold modulation, though still present, is considerably reduced in strength compared with the half particles, and the outer part remains tenfold in character (Figure 8). The reduction in the strength of the fivefold components occurs because the two half particles comprising a whole particle are not in rotational register.

The whole particle probably therefore has point group symmetry 52 but this would only allow three-dimensional reconstruction to a cutoff of about 100 Å from a single view. However, taking advantage of the tendency of some hemocyanins to aggregate end-to-end, linear polymers were formed (Figure 9a) (30). The transform of the image showed a set of layer lines characteristic of a helical arrangement with an axial repeat of about 1150 Å, which corresponds to the length of three particles, successive particles being rotated by 120°. There are 15 independent projections of the asymmetric unit within the repeat distance of the polymer, which is therefore suitable for three-dimensional reconstruction to a cutoff of 50 Å. Tilting experiments showed that central rows of particles lying within arrays, such as those shown in Figure 9a, were better preserved than either isolated rows or rows on the edges of arrays. This could be judged both by measuring the apparent widths of the particles in a tilt series and by following the phase variations in the Fourier components on tilting. Four independent reconstructions were made, which were very similar to one another, so an average reconstruction was computed after the relative orientations of the individual polymers had been determined by comparing the relative phases in the transforms. In all reconstructions the halves of the particle lying on either side of the equator appeared to be related by twofold axes normal to the cylindrical axis, as would be expected from the association-dissociation reactions. The presence of twofold axes was therefore tested for in each of the four transforms individually and also in the average. It was found that the averaged data were more nearly twofold related than any of the individual polymers, leaving little doubt that horizontal twofold axes are present to the resolution of the micrograph. This twofold symmetry was therefore imposed and a final map computed having a cutoff of about 30 Å axially and 50 Å radially.

Figure 9b shows a wooden model of a single particle and Figure 9c a section through the cylindrically averaged structure (30). The structure consists of a hollow cylindrical drum closed by fivefold material forming a collar and a central cap. The wall of the drum consists of six layers, each with approximate tenfold rotational symmetry, so that there are 60 morphological units which are of six crystallographically distinct types (two in each layer of the half particle). They are, however, of similar size, shape, and orientation and appear to be quasi-equivalent. The relative rotation between successive layers, whether within or between half particles, is approximately the same, suggesting that there are approximate twofold axes relating the various

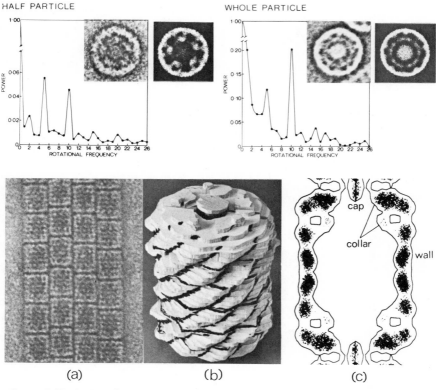

Figure 8 (Top) Rotational power spectra and rotationally filtered images of half and whole particles of gastropod hemocyanin. Particle diameter approximately 300 Å.

Figure 9 (Bottom) Gastropod hemocyanin. (*a*) Negatively stained array of particles. (*b*) Model of a three-dimensional reconstruction of a particle. (*c*) Section through a cylindrically averaged reconstruction on the same scale as (*b*).

layers, in addition to the strict twofolds relating half particles. This would in turn imply twofold axes within each layer, suggesting that the 60 observed morphological units are themselves dimers. This would agree well with chemical evidence that the number of O_2 binding sites per particle is approximately 130. The reconstruction suggests that there are actually 120 functional units arranged in pairs to form 60 dimers, which are assembled as shown in the model. The roles of the extra cap and collar are unknown but they may play a part in the assembly of the structure.

THE ACTIN-TROPOMYOSIN-MYOSIN COMPLEX FROM MUSCLE

The basic structural constituents of vertebrate striated muscle are thick and thin filaments, containing myosin and actin respectively, which are interspersed in a parallel and regular manner. When the muscle contracts, the two sets of filaments slide past one another under the influence of cyclically acting crossbridges (31). The myosin molecule consists of a long tail embedded in the thick filament joined by a flexible hinge to a protruding globular head. When the muscle is relaxed, the heads lie close to the thick filaments and their attachment to the thin filaments is inhibited. Upon electrical stimulation by the motor nerve, calcium ions are released from the sarcoplasmic reticulum and relieve the inhibition of the thin filaments. The myosin heads can then form crossbridges to the actin in the thin filaments. This attachment is followed by a change of the head conformation, which generates tension in the muscle, followed in turn by the release of the crossbridge from the actin. This repeated cycle of events is associated with the splitting of ATP, which provides a source of energy.

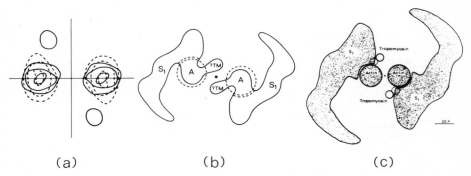

(a)   (b)   (c)

Figure 10   The actin-tropomyosin-myosin interaction. (*a*) Helical projections of reconstructions of actin (dotted contours) and actin plus tropomyosin (solid contours). (*b*) End-on view of a slice of a reconstruction of thin filament decorated with the S_1 subfragment of myosin. (*c*) Diagram showing positions relative to the actin of the tropomyosin in the active (solid circle) and the inhibited (dotted circle) state. The S_1 subfragment is superimposed in profile, showing how its binding may be blocked by the tropomyosin in the inhibited state.

The globular actin monomers are arranged in a staggered double helix in the thin filament. On its own the actin activates the myosin ATPase activity independently of the calcium concentration. Calcium sensitivity is conferred by the presence in the thin filaments of the tropomyosin-troponin complex. The troponin is a globular molecule responsible for binding calcium and is located at approximately 400 Å intervals along the thin filament. The tropomyosin however is a two-chain coiled coil which forms two continuous strands in the grooves of actin double helix and is present in sufficient amounts to make contact with each actin monomer. Since the presence of tropomyosin is necessary for the regulation mechanism to work, it is natural to suppose that the influence of the troponin is transmitted to the actin monomers by the tropomyosin strands.

Evidence for this type of model comes from X-ray diffraction and from three-dimensional image reconstruction from electron micrographs (32). Figure 10*a* shows helical projections of three-dimensional reconstructions of pure actin and of actin plus tropomyosin (33). The two large common features correspond to the double strand of actin while the two smaller features correspond to the tropomyosin lying in the grooves between the actins. Figure 10*b* shows an end-on view of a reconstruction (34) of a thin filament containing tropomyosin-troponin, which has been "decorated" with subfragment 1 (S_1), a proteolytic fragment of myosin containing the globular head but with the long tail removed. The S_1 fragment, besides being tilted and skewed, is attached to the actin monomer in a tangential fashion so that the end of the subfragment extends round into the groove in the actin structure, towards the position of the tropomyosin. This suggests that the inhibitory effect of the tropomyosin-troponin complex may be steric and that in the absence of calcium the troponin holds the tropomyosin in a position where it prevents the binding of the myosin heads to the actin. When the troponin binds calcium it undergoes a structural rearrangement, which allows movement of the tropomyosin and the formation of crossbridges. Further evidence for this comes from studies of the relative positions of the tropomyosin in the active and inhibited complex (T. Wakabayashi, H. E. Huxley, A. Klug, and L. A. Amos, in preparation). The results, summarized in Figure 10*c*, show that in the inhibited complex the tropomyosin is closely bound to the actin in a position that would prevent the myosin head from binding, but that in the active complex the tropomyosin moves away from the actin by about 10 Å to a position that would permit the binding of myosin.

SPHERICAL VIRUSES

The protein coats of all small spherical viruses so far investigated have icosahedral symmetry, point group *532*. They contain multiple copies of one or more proteins. The largest number of identical subunits that can be arranged in identical environments in a spherical shell is 60. However, by relaxing the requirement for strict equivalence of all subunits, it is possible to build larger shells containing 60*T* subunits, in which the subunits are now only quasi-equivalent (35). The triangulation number, *T*, can take only certain integer values, of which the smallest are 1, 3, 4, and 7. The subunits may lie in special positions in the surface lattice, giving rise to dimer, trimer, or hexamer-pentamer clustering. The resulting morphological units give rise to characteristic features in negatively stained preparations, even though the individual subunits may not be distinguishable if the stain does not penetrate between them.

The first successful attempts at disentangling the complicated patterns that arise from superposition of features on opposite sides of the particle were by building wire models and using them to cast shadow graphs (36). However, this gives only a crude simulation of the image and the models are difficult to make and alter. A more flexible and sophisticated technique was therefore developed using a computer with a cathode ray tube display (37). However the process is still inductive, that is, production of a hypothetical model, comparison with observed views of the virus, and modification of model where a discrepancy is found. While this may suffice for simple structures (38), it is likely that for a complex structure no single model can be invented by the simulator to account for all observed views. Even for simple structures, although it may be easy to explain how the gross features arise, no single simulated model may account for the fine details of the images. We therefore need a more powerful and direct method, such as that provided by three-dimensional image reconstruction.

For helical structures, as already described, three-dimensional reconstruction can often be performed in a fairly straightforward way from a single view of the structure. For other types of symmetry, such as icosahedral, the situation is more complicated and more than one view is necessary (39). The two-dimensional Fourier transform of each image represents a central section through the three-dimensional transform of the object. The different views give different sections and so the three-dimensional transform can be filled in plane by plane. When the specimen possesses symmetry, each view gives not only one plane but a whole set of equivalent planes (60 in the case of icosahedral symmetry) generated by the appropriate symmetry operations. When a sufficient number of different views has been included to fill in the three-dimensional transform out to the limiting spatial frequency set by the preservation of the specimen, we perform a three-dimensional Fourier inversion to give a three-dimensional image of the specimen. Since the particles lie in arbitrary orientations, the three-dimensional transform is not filled in uniformly and it is therefore necessary to perform interpolation in the transform prior to inversion. This interpolation involves the solution of sets of linear equations, which will be solvable only if sufficient views have been included. Tests for the solvability of the equations (in the form of eigenvalue spectra) provide a check that for a particular specimen the selected views do uniquely determine the three-dimensional reconstruction to a fineness of detail set by the specimen preservation. Typically for small spherical viruses three or four views are sufficient.

Before including any particular view, its orientation relative to the icosahedral symmetry axes must be found and its preservation assessed. This can be done by searching its two-dimensional transform for the position of the best set of pairs of so-called common lines (40). These are pairs of lines along which the transform should have equal values because of the symmetry of the particle. With ideal data the values along the common lines would agree exactly if the particles were icosahedral, so the symmetry can be tested. Because of distortions of the specimen, nonuniform staining, and contributions from the supporting grid, the values will not agree exactly and the resulting discrepancy may be used as a measure of the quality of the image. It can moreover be computed as a function of increasing spatial frequency, so that it is possible to tell at what scale of detail features in the image cease to be icosahedrally correlated. Typically for small spherical viruses in negatively stained preparations, some degree of icosahedral correlation extends to spatial periodicities of about 25 Å.

The structures of a number of spherical viruses of different T classes and clustering types have been solved (41, 42), demonstrating how the requirement for quasi-equivalence is realized in different ways. Here we consider two more recent and interesting examples.

The first is an insect virus, *Nudaurelia capensis β* virus, which is the first established case of a $T = 4$ surface lattice and also the first to exhibit clear trimer clustering (43). Chemical data show that the capsid consists of about 240 copies of a single protein species. The three-dimensional reconstruction (Figure 11) shows that at outer radii the protein is confined within the triangular faces of the circumscribing icosahedron, leaving clear grooves along the icosahedral edges. Each icosahedral face contains four Y-shaped features, each of which in conjunction with the chemical evidence may be interpreted as a trimer. These trimers follow the local symmetry of the $T = 4$ surface lattice at inner radii. At outer radii, however, the packing of subunits within a face is closer than the packing of subunits between two adjacent faces across an icosahedral edge, the respective center-to-center distances being 35 Å and 50 Å at a radius of 170 Å. These are quite large differences for quasi-equivalent

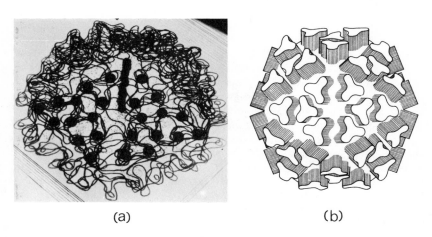

(a)   (b)

Figure 11   Nudaurelia capensis β virus, approximate diameter 400 Å. (*a*) Contour map of a three-dimensional image reconstruction viewed approximately along a twofold axis of symmetry. (*b*) Proposed model for the structure, consisting of trimers of subunits arranged on a $T = 4$ surface lattice, also viewed along a twofold axis.

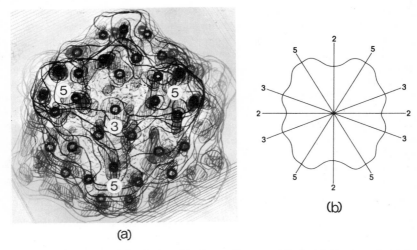

(a)

(b)

Figure 12   Cowpea mosaic virus, approximate diameter 240 Å. (*a*) Contour map of a three-dimensional image reconstruction, viewed approximately along a threefold axis. (*b*) Trace of the equator of the image reconstruction in a plane normal to a twofold axis, showing large bumps at the fivefold positions and smaller ones at the threefold positions.

units and it may be that the capsid of this virus is more like a surface crystal with sharp edges than simple quasi-equivalence theory would allow.

The second example is a plant virus, cowpea mosaic virus. Chemical evidence shows that the capsid of this virus contains two different proteins with molecular weights of about 49,000 and 27,500, each present in about 60 copies per particle. The three-dimensional image reconstruction (Figure 12*a*) (44) suggests how this more complicated capsid is organized. The capsid has a rather smooth but modulated surface, not easily described in terms of an arrangement of discrete morphological units. However, there are strong projecting knobs at the fivefold positions with ridges running from the fivefold positions to additional smaller bumps at the three-fold positions. This modulation is most easily appreciated by looking at the outer contour of the particle around an equator normal to a twofold axis (Figure 12*b*). Just under the surface there are five peaks of density around each fivefold position and also three peaks around each threefold position. A plausible interpretation of these features is that the larger of the two proteins forms 12 pentamers at the five-fold positions while the smaller forms 20 trimers at the fivefold positions. However the asymmetric unit of the reconstructed density may be divided in some more complicated way between the large and small chemical subunits.

CONCLUDING REMARKS

Three-dimensional image reconstruction is in principle applicable to any object, whether it possesses symmetry or not. All the examples discussed above are symmetrical assemblies of molecules. Not only are many such structures biologically important, they also possess enormous advantages for analysis of their images. The symmetrical particle in effect provides many regularly repeated images of the asymmetric unit from which it is built. Thus once the symmetry has been determined, the relative disposition of these various images is known and may be used in two-dimensional filtering or three-dimensional reconstruction. The degree of symmetry exhibited by the image also enables the preservation of the specimen to be judged.

If one wished to study a particle with no symmetry, such as a ribosome, none of these advantages would accrue. Three-dimensional reconstruction would require a complete range of about 25 evenly spaced views to reconstruct to about 20 Å. Because of radiation damage, only a very few useful images can be obtained by tilting a single particle. It is difficult to be sure whether the particle is well preserved even initially, since observations suggest that isolated particles are generally distorted. Data that can be collected by sets of limited tilts then have to be combined, which requires that the relative positions and orientations of the various views be determined—no easy matter when the particle does not contain a natural frame of reference such as is provided by symmetry axes. Symmetrical arrays of ribosomes (45, 46) are likely to provide a more fruitful approach.

The reconstructions presented above demonstrate how structural information can be reliably retrieved by processing electron micrographs and how the results can be interpreted. An important general point emerges. The assessment and selection of images can be quantitatively performed by using their diffraction patterns to see which specimens possess the greatest degree of regularity. Similarly the range of averaging in the filtering process can be precisely controlled: it must be sufficiently great for noise to be reduced but not so great that patches of specimen that differ significantly in preservation or staining become merged. However, the final intepretation of the reconstructed image is much more difficult to quantitate and is more akin to a pattern recognition problem in which prior chemical and biological information must play a part.

NOTE ADDED IN PROOF   A very important advance in specimen preparation has recently been made by Unwin & Henderson (submitted to *J. Mol. Biol.*), who have used glucose to preserve otherwise unstained specimens of purple membrane and catalase crystals, which were then photographed with extremely low electron doses to avoid specimen damage. The resulting micrographs exhibit very low contrast and high electron noise, and computer averaging is crucial for extracting an image of the structure. The small amount of contrast present is produced by underfocusing, which is then compensated for in the computer reconstruction (1). The resolution so far obtained is better than 10 Å and appears to be limited by the performance of the microscope.

Literature Cited

1. Erickson, H. P., Klug, A. 1971. *Phil. Trans. Roy. Soc. London B* 261:105–18
2. Reedy, M. K., Bahr, G. F., Fischman, D. A. 1972. *Cold Spring Harbor Symp.*
   Quant. Biol. 37:397–421
3. Branton, D., Klug, A. 1975. *J. Mol. Biol.* Submitted
4. Brenner, S., Horne, R. W. 1959. *Biochim.*

Biophys. Acta 34 : 103–10
5. Laemmli, U. K. 1970. *Nature* 227 : 680–85
6. DeRosier, D. J., Klug, A. 1972. *J. Mol. Biol.* 65 : 469–88
7. Klug, A., Berger, J. E. 1964. *J. Mol. Biol.* 10 : 565–69
8. Klug, A., DeRosier, D. J. 1966. *Nature* 212 : 29–32
9. Amos, L. A., Klug, A. 1972. *Proc. 5th Eur. Congr. Electron Microscopy*, 580–81
10. Aebi, U., Smith, P. R., Dubochet, J., Henry, C., Kellenberger, E. 1973. *J. Supramolecular Struct.* 1 : 498–522
11. King, J., Mykolajewycz, N. 1973. *J. Mol. Biol.* 75 : 339–58
12. King, J., Laemmli, U. K. 1973. *J. Mol. Biol.* 75 : 315–37
13. Crowther, R. A., Amos, L. A. 1971. *J. Mol. Biol.* 60 : 123–30
14. Markham, R., Frey, S., Hills, G. J. 1963. *Virology* 20 : 88–102
15. Markham, R., Hitchborn, J. H., Hills, G. J., Frey, S. 1964. *Virology* 22 : 342–59
16. DeRosier, D. J., Klug, A. 1968. *Nature* 217 : 130–34
17. Moody, M. F. 1971. *Phil. Trans. Roy. Soc. London B* 261 : 181–95
18. DeRosier, D. J., Moore, P. B. 1970. *J. Mol. Biol.* 52 : 355–69
19. Moody, M. F. 1973. *J. Mol. Biol.* 80 : 613–35
20. Klug, A., Caspar, D. L. D. 1960. *Advan. Virus Res.* 7 : 225–325
21. Unwin, P. N. T. 1972. *Proc. Roy. Soc. London A* 329 : 327–59
22. Unwin, P. N. T., Klug, A. 1974. *J. Mol. Biol.* 87 : 641–56
23. Unwin, P. N. T. 1974. *J. Mol. Biol.* 87 : 657–70
24. Unwin, P. N. T. 1973. *J. Microsc.* 98 : 299–312
25. Finch, J. T., Klug, A. 1974. *J. Mol. Biol.* 87 : 633–40
26. Gilbert, P., Klug, A. 1974. *J. Mol. Biol.* 86 : 193–207
27. Barrett, A. N. et al 1971. *Cold Spring Harbor Symp. Quant. Biol.* 36 : 433–48
28. Finch, J. T., Klug, A. 1971. *Phil. Trans. Roy. Soc. London B* 261 : 211–19
29. Williams, R. C., Fisher, H. W. 1970. *J. Mol. Biol.* 52 : 121–23
30. Mellema, J. E., Klug, A. 1972. *Nature* 239 : 146–50
31. Huxley, H. E. 1969. *Science* 164 : 1356–66
32. Huxley, H. E. 1972. *Cold Spring Harbor Symp. Quant. Biol.* 37 : 361–76
33. Spudich, J. A., Huxley, H. E., Finch, J. T. 1972. *J. Mol. Biol.* 72 : 619–32
34. Moore, P. B., Huxley, H. E., DeRosier, D. J. 1970. *J. Mol. Biol.* 50 : 279–95
35. Caspar, D. L. D., Klug, A. 1962. *Cold Spring Harbor Symp. Quant. Biol.* 37 : 1–24
36. Klug, A., Finch, J. T. 1965. *J. Mol. Biol.* 11 : 403–23
37. Finch, J. T., Klug, A. 1967. *J. Mol. Biol.* 24 : 289–302
38. Josephs, R. 1971. *J. Mol. Biol.* 55 : 147–53
39. Crowther, R. A., DeRosier, D. J., Klug, A. 1970. *Proc. Roy. Soc. London A* 317 : 319–40
40. Crowther, R. A. 1971. *Phil. Trans. Roy. Soc. London B* 261 : 221–30
41. Crowther, R. A., Amos, L. A. 1971. *Cold Spring Harbor Symp. Quant. Biol.* 36 : 489–94
42. Mellema, J. E., Amos, L. A. 1972. *J. Mol. Biol.* 72 : 819–22
43. Finch, J. T., Crowther, R. A., Hendry, D. A., Struthers, J. K. 1974. *J. Gen. Virol.* 24 : 191–200
44. Crowther, R. A., Geelen, J. L. M. C., Mellema, J. E. 1974. *Virology* 57 : 20–27
45. Byers, B. 1967. *J. Mol. Biol.* 26 : 155–67
46. Lake, J. A., Slayter, H. S. 1972. *J. Mol. Biol.* 66 : 271–82

LIPOPROTEINS: ×879
STRUCTURE AND FUNCTION [1,2,3]

Joel D. Morrisett,[4] *Richard L. Jackson*,[4] *and Antonio M. Gotto, Jr.*
Department of Medicine, Baylor College of Medicine and The Methodist Hospital,
Houston, Texas 77025

CONTENTS

[1] Abbreviations: VLDL, very low density lipoprotein; LDL, low density lipoprotein; HDL, high density lipoprotein; VHDL, very high density lipoprotein; apoprotein, a lipid-free protein from a lipoprotein; apoC-I (apoLP-Ser), apoC-II (apoLP-Glu), and apoC-III (apoLP-Ala), three apoproteins from human VLDL with carboxyl-terminal serine, glutamic acid, and alanine, respectively; apoA-I (apoLP-Gln-I) and apoA-II (apoLP-Gln-II), two apoproteins from human HDL, both with carboxyl-terminal glutamine; LP-lipase, lipoprotein lipase; LCAT, lecithin-cholesterol acyltransferase; CO, cholesteryl oleate; PC, phosphatidylcholine; SM, sphingomyelin; SDS, sodium dodecylsulfate; CD, circular dichroism; EPR, electron paramagnetic resonance; NMR, nuclear magnetic resonance; IR, infrared; PAGE. polyacrylamide gel electrophoresis; CNBr, cyanogen bromide; TEMPO-OH, 1-oxyl-2,2,6,6-tetramethylpiperidin-4-ol.

[2] This review was completed August 1, 1974. The authors wish to express appreciation to their colleagues Drs. O. David Taunton, James T. Sparrow, H. Nordean Baker, Henry J. Pownall, and Louis C. Smith for their helpful criticism of this review. The authors are also indebted to Ms. Debbie Mason and Mrs. Jean Atkisson for their assistance in the preparation of the manuscript.

[3] Work from the authors' laboratory described in this review was supported in part by Health, Education, and Welfare Grant No. HL-14194, by NIH Contract No. NIH 71-2156 from The National Heart and Lung Institute Lipid Research Clinic Program, and by a grant from The John A. Hartford Foundation, Inc.

[4] Established Investigators of The American Heart Association.

183

INTRODUCTION

Since the plasma lipoproteins were last reviewed in this series by Scanu & Wisdom (1), knowledge of the structure and function of these macromolecular complexes has greatly increased. The sequences of four plasma lipoprotein-proteins or apo-proteins from man are now known, whereas none were known at the time of the last review. This new structural information has renewed interest in the mechanism of lipid transport by the plasma lipoproteins and has emphasized the protein heterogeneity within individual families of lipoproteins. Thus, the four major classes of plasma lipoproteins, the chylomicrons, VLDL, LDL, and HDL, share several apoprotein components in common. In this review, we have emphasized recent structural, and to a lesser extent functional, studies. Due to limitation of space, we have narrowed this review to human lipoproteins. In addition, we have omitted a discussion of chylomicrons because of the relatively limited knowledge of their apoprotein structure. The reader is referred to other reviews for a discussion of these topics and of lipoprotein metabolism (2–6).

In this review, we have adopted the A, B, C terminology of Alaupovic et al (7). In this system the major protein components of HDL are apoA and those of LDL are apoB. A third group of apoproteins, first detected in VLDL, is designated apoC. ApoC also occurs in HDL, while apoB is found in both VLDL and LDL. Each group of apoproteins is heterogeneous, containing more than one polypeptide. These are described in detail below. A separate question is whether the apoprotein components exist in the lipidated state as separate, discrete entities within the VLDL, LDL, and HDL classes. For example, do apoA and apoC occur in HDL as separate lipoprotein particles rather than as part of a common particle (8)? This question cannot be answered satisfactorily at present.

VERY LOW DENSITY LIPOPROTEINS

General Properties

The VLDL span a wide spectrum of particle sizes (280–750 Å), hydrated densities (0.95–1.006), and flotation rates (20–400) (9). At the lower end of their density range, the VLDL overlap with the chylomicrons. Subfractions of VLDL and chylomicrons have been obtained by density gradient ultracentrifugation (10), chromatography on concanavalin A-Sepharose (11), isoelectric focusing (12), and gel filtration (13–15).

The size of VLDL particles is directly proportional to their triglyceride content and inversely proportional to their phospholipid and protein content. Average values for the composition of VLDL by weight are 55% triglyceride, 20% phospholipid, 15% cholesterol (30% of which is esterified), and 10% protein (9).

Apo VLDL

The apoprotein heterogeneity of VLDL may be shown by affinity chromatography on concanavalin A-Sepharose (16), by ion exchange chromatography (16, 17), by fractionation on hydroxylapatite (8), and by isoelectric focusing (18). After delipidation of VLDL, the protein constituents are soluble in sodium decylsulfate and are partially resolved by chromatography on Sephadex G-100 (17). The protein that elutes at the void volume of the column has been shown by Gotto et al (19) to be identical to the major apoprotein of LDL. There are no detectable differences between this apoprotein and apoLDL as determined by amino acid analysis CD, or immunochemical analysis. The remainder of the VLDL apoproteins are retained by Sephadex and have been referred to collectively as the D-peptides (17), apoC (7), or fraction V (20). These components may be isolated in a homogeneous state by DEAE-cellulose chromatography (16, 17) or isoelectric focusing (18), and include apoC-I (apoLP-Ser), apoC-II (apoLP-Glu), apoC-III (apoLP-Ala), and the "arginine-rich" apoprotein. These components are relatively abundant, but other minor components have also been described and may have important physiologic functions. The ratio of apoB to apoC within VLDL increases as the particle size decreases (15). Shore & Shore (16) have shown that the proportion of the various apoC components in human plasma varies with the individual, his degree of hyperlipidemia, and his age.

Sonication of native VLDL irreversibly destroys the structure of the particle (21), as has also been shown for LDL (21a). Forte et al (22) have reported the isolation of particles (150–1200 Å in diameter) similar in size to VLDL from a sonicated mixture of apoC, PC, and triolein. However, reconstitution of VLDL from its total protein (including apoB) and lipid components has not been described.

ApoC-I

ApoC-I, the first of the apoC components to elute from a DEAE-cellulose column (17), has been reported to activate the enzyme LCAT (23). There is conflicting evidence concerning its activation of LP-lipase from different sources (24–26). ApoC-I is a single polypeptide chain of 57 amino acids and tends to aggregate in the absence of a denaturant (27). Shulman et al (28, 29), initially, and Jackson et al (30), subsequently, have reported the following amino acid sequence

NH_2-Thr-Pro-Asp-Val-Ser-Ser-Ala-Leu-Asp-Lys-Leu-Lys-Glu-Phe-Gly-
        5   10   15

Asn-Thr-Leu-Glu-Asp-Lys-Ala-Arg-Glu-Leu-Ile-Ser-Arg-Ile-Lys-
        20   25   30

Gln-Ser-Glu-Leu-Ser-Ala-Lys-Met-Arg-Glu-Trp-Phe-Ser-Glu-Thr-
        35   40   45

Phe-Gln-Lys-Val-Lys-Glu-Lys-Leu-Lys-Ile-Asp-Ser-COOH
        50   55

This protein was initially called apoLP-Val (17), based on an incorrect identification of the C-terminal amino acid. The molecule contains one residue of methionine and tryptophan, but no cystine, cysteine, histidine, or tyrosine. ApoC-I forms complexes with egg PC which can be isolated by ultracentrifugation (22, 31). The binding of PC results in an increase in α-helix from 56 to 73%, a 6 nm shift in tryptophan fluorescence (31), and a transformation of PC vesicles into rouleaux structures as viewed by electron microscopy (22, 31). Failure by Middelhoff & Brown (32) to observe PC binding by apoC-I may have been due to differences in methodology. The two CNBr fragments of apoC-I differ in their ability to bind phospholipid (31). CNBr I, the N-terminal 38 residue fragment, binds PC to a greater extent than CNBr II, the C-terminal 19 residue fragment, but to a lesser extent than apoC-I (31).

ApoC-I-PC complexes bind the neutral lipids, CO and triolein (22), forming heterogeneous particles. Most of the protein, however, is associated with a relatively lipid-poor complex. The major structures in the d 1.006–1.063 fraction are heterogeneous round particles, while the minor structures are myelin forms (22).

ApoC-II

ApoC-II, the second major protein eluted from DEAE-cellulose chromatography of apoC, has a molecular weight of 12,630 by sedimentation equilibrium and 10,519 from its amino acid composition (27). It is a single polypeptide of approximately 95 residues and is a potent activator of LP-lipase from human and rat postheparin plasma and from cows' milk (24–26, 33). ApoC-II forms a stable complex with PC (22, 32) and also forms a stable surface film with LP-lipase in the absence of PC (26). According to Forte et al (22), apoC-II forms lipid-protein complexes which float predominantly in the d 1.063–1.21 range when sonicated with PC, PC-CO, or PC-triolein.

ApoC-III

ApoC-III is the most abundant of the C-proteins in VLDL and has been reported to inhibit LP-lipase (34). The sequence has been determined (35, 35a) and is shown below

NH$_2$-Ser-Glu-Ala-Glu-Asp-Ala-Ser-Leu-Leu-Ser-Phe-Met-Gln-Gly-Tyr-
　　　　　　　5　　　　　　　　　　　10　　　　　　　　　　　　　15

Met-Lys-His-Ala-Thr-Lys-Thr-Ala-Lys-Asp-Ala-Leu-Ser-Ser-Val-
　　　　　20　　　　　　　　　25　　　　　　　　　　30

Gln-Ser-Gln-Gln-Val-Ala-Ala-Gln-Gln-Arg-Gly-Trp-Val-Thr-Asp-
　　　　35　　　　　　　　　40　　　　　　　　　45

Gly-Phe-Ser-Ser-Leu-Lys-Asp-Tyr-Trp-Ser-Thr-Val-Lys-Asp-Lys-
　　　　50　　　　　　　　　55　　　　　　　　　60

Phe-Ser-Glu-Phe-Trp-Asp-Leu-Asp-Pro-Glu-Val-Arg-Pro-Thr-Ser-
　　　　65　　　　　　　　　70　　　　　　　　｜　75

Ala-Val-Ala-Ala-COOH　　　　　　　　　　　　　　CHO

The molecule contains no cysteine, cystine, or isoleucine. To threonine-74 is attached an oligosaccharide containing one residue each of galactose and galactosamine and differing amounts of sialic acid (27, 35). Three polymorphic forms of apoC-III correspond to a content of 0, 1, or 2 residues of sialic acid (27). The role of the carbohydrate is unknown at the present time.

When apoC-III is sonicated with egg PC, the resulting complex isolated by sucrose density gradient centrifugation has a 46:1 molar ratio of lipid to protein (36). However, when the complex is isolated on a KBr gradient, the ratio is reduced to 30:1, suggesting that salt may dissociate some of the bound PC and that ionic interactions may be involved in the binding. Studies from several laboratories show (22, 32, 36, 37) that apoC-III can bind from 18 to 80 PC molecules and that the binding of lipid is associated with an increase in the α-helix of the protein and a shift of one or more of the three tryptophanyl residues to a more hydrophobic environment.

The effect of apoC-III on the morphology of PC vesicles has been examined by electron microscopy (22, 38). When fractionated bilamellar vesicles of uniform diameter (250 Å) are titrated with increasing amounts of the apoprotein and the negatively stained complexes are examined, the vesicles appear to undergo a series of morphological changes from free vesicles, to stacked discs, to myelin figures (38). The formation of stacked discs or rouleaux as observed by electron microscopy raises the following questions: what is the structure of the apoprotein-phospholipid complex in solution; and do the rouleaux represent artifacts of negative staining and/or drying? To answer these questions, the major complex formed by interaction of apoC-III and PC vesicles was studied by gel filtration, analytical ultracentrifugation, and laser light scattering (39). The complex elutes from a Sepharose-6B column at approximately the same volume as PC vesicles alone, but well ahead of apoC-III. Upon titration with the apoprotein, the observed sedimentation coefficient of the PC vesicles increases from 1.19S to a limiting value of 4.93S, reached when the protein:lipid ratio is 0.23 g/g. This weight ratio corresponds to a molar ratio of about 1:53 or about 46 apoC-III molecules per PC vesicle. As determined by light scattering, the average translational diffusion coefficient for a vesicle-apoprotein complex of this stoichiometry is $D_{20,w} = (2.08 \pm 0.03) \times 10^{-7}$ cm^2 sec^{-1}. These values of S and $D_{20,w}$ exclude the possibility that large multivesicular aggregates such as rouleaux exist in solution. Since the calculated size of the vesicles does not greatly change despite the association with the apoprotein, these hydrodynamic data lead to the conclusion that the protein probably displaces water from the hydration shell around the vesicle (39).

In order to locate the phospholipid binding region(s) of apoC-III, four fragments corresponding to sequence positions 41–79 (I), 48–79 (II), 55–79 (III), and 61–79 (IV) have been synthesized (40). Only the longer fragments I and II bind significant quantities of PC, indicating that residues 48–79 contain the minimal determinants required for binding, but that residues 55–79 do not. These results are in accord with those of Shulman et al (41), who found that when apoC-III combines with PC and the resulting complex is digested with trypsin, cleavage of the N-terminal half proceeds normally while the C-terminal half is protected from cleavage by bound phospholipid.

Most phospholipid binding studies have used egg PC, a heterogeneous mixture of different acyl chains, all of which are in a liquid crystalline state at room temperature. Using dimyristoyl PC, which has a phase transition at 23°C, Pownall et al (42) have shown by CD, fluorescence, and ultracentrifugal methods, that apoC-III interacts with this phospholipid more efficiently in the liquid crystalline than in the gel state. This finding demonstrates the importance of the physical state of the phospholipid in its interaction with protein.

The binding of egg PC and apoC-III has been investigated by quenching the protein fluorescence with iodide and pyridinium ions (43). The steric accessibility of these quenchers to the tryptophanyl residues is reduced by the binding of PC. The results show that the tryptophanyl residues are located in a relatively negatively charged environment and that apoC-III binds PC with an efficiency at least three times greater than apoA-I.

"Arginine-Rich" Apoprotein

Shore & Shore (16) first reported a minor protein component of normal VLDL that is rich in arginine and found preferentially in cholesteryl ester-rich VLDL. High levels of the protein have been found in subjects with type III hyperlipo-

Table 1   Molecular weights of the protein subunits of human plasma LDL as determined by various methods.

Protein	Method	Molecular Weight	Reference
Succinylated apoB	Analytical ultracentrifugation	36,000–38,000	60
ApoB	Gel filtration	80,000	156
LDL	Electron microscopy	27,000	82
ApoB	Analytical ultracentrifugation, pH 11.5	27,500	82
ApoB	Analytical ultracentrifugation, pH 8.6	80,000–100,000	82
Maleylated apoB	Gel filtration	26,000 and 194,000	59
Reduced, carboxymethylated apoB	Gel filtration	275,000	58
	Analytical ultracentrifugation	250,000	58
ApoB	SDS-PAGE	255,000	58
ApoB	SDS-PAGE	230,000	61
ApoB	SDS-PAGE	10,000	56
ApoB	SDS-PAGE	250,000 and 270,000	57
ApoB	SDS-PAGE	9,500 and 13,000	62
LDL	X-ray scattering	8,000	84

proteinemia (44, 45) or with hypothyroidism (45). A similar apoprotein has been found in cholesterol-fed rabbits (46). Shelburne & Quarfordt (47) have reported the preparative isolation of this water-insoluble human protein by gel filtration of apoVLDL on Sepharose-6B in guanidine-HCl or by DEAE-cellulose chromatography in urea. The molecular weight of this protein is about 33,000. The N- and C-terminal residues are lysine and alanine, respectively. Since this apoprotein is preferentially associated with the cholesterol-rich, triglyceride-poor VLDL, it may represent an intermediate in the conversion of VLDL to LDL, possibly related to the "remnant" lipoprotein (48, 49). Further studies of the structure and metabolism of the "arginine-rich" protein may be of great importance in understanding the metabolism of VLDL and its relationship to cholesterol transport.

Studies on Intact VLDL Particles

There is little structural information about intact VLDL due to the instability and variability of the lipid content of these particles. Electron micrographs of negatively stained VLDL show particle diameters of 300 to 900 Å (50). This variation may reflect different stages of degradation by LP-lipase. The particles show no substructure but have electron transparent centers with grayish halos near the periphery. This halo has been interpreted as an outer membrane layer but more likely is due to artifactual flattening of an easily deformed particle. Recent ^{13}C-NMR studies indicate that the spin-lattice relaxation times of fatty acyl chain nuclei in VLDL are slightly longer than the corresponding values for LDL and HDL. These results suggest somewhat greater motion of the lipids in VLDL (50a).

LOW DENSITY LIPOPROTEINS

General Properties

LDL are the principal vehicles for the transport of plasma cholesterol in man. They are usually isolated between d 1.019 to 1.063 and contain about 25% protein and 75% lipid (9). The major lipid constituents by weight are cholesteryl esters (47–55%), phospholipids (28–30%), unesterified cholesterol (10–11%), and triglycerides (8–10%). Although LDL particles are more homogeneous in size than VLDL, the molecular weight may vary by a factor of two (51–53). Since the mass of protein per LDL particle has been found to remain constant at about 510,000 daltons, it is thought that the variation in LDL density and molecular weight is due chiefly to the degree of lipidation of the apoproteins. LDL (d 1.006–1.063) has been ultracentrifugally separated into six subfractions, each containing mainly apoB but also varying quantities of apoA plus apoC (54, 55). The lower the density, the lesser the proportion of apoB present in these particles (54, 55).

ApoLDL

There is little agreement concerning the exact composition and properties of apoLDL (56–63). Molecular weights for apoLDL range from 8,000 to 275,000 (Table 1). Experiments by Smith et al (58) have been interpreted as showing two subunits, each of molecular weight 250,000. This interpretation is based on the assumption that apoLDL is completely dissociated into its subunits by guanidine-HCl. This assumption may not be valid for lipoprotein-proteins, which are notorious for their tendency to aggregate. Subunit heterogeneity has been suggested by immuno-chemical analysis (64), by separation on DEAE-cellulose of several components differing in amino acid composition (65), and by the isolation from Sephadex columns of two fractions of maleylated apoLDL with different amino acid compositions (59). A recent report suggesting that the differences in molecular weight are due to digestion by a protease-like enzyme (57) has been disputed (62, 63). These variations in molecular weights may reflect association-dissociation phenomena or differences in the method of delipidation (62). One of the major hindrances to the characterization of apoLDL is its water insolubility in the lipid free state. Some of the solubilization procedures have used detergents such as sodium deoxycholate (62, 66), Nonidet P40 (66), sodium decylsulfate (19), or cetyltrimethyl ammonium bromide (66); denaturants such as urea (65) or guanidine (58, 59); chemical modification by succinylation (60, 67) or maleylation (59); and high pH solutions (60). The carbohydrate of apoLDL (68) allows selective adsorption of native LDL to a concanavalin A-Sepharose affinity column (11), permitting subsequent fractionation of particles containing only apoB from those containing other apoproteins.

Studies on Intact LDL Particles

The structure of LDL has been studied by several physical techniques including CD (69–73), IR (74), EPR (75, 76), NMR (50a, 77–79), fluorescence (73, 80), laser light scattering (81), electron microscopy (50, 82, 83), and small angle X-ray scattering (84). As determined by CD, the protein in native LDL contains an average of 25% α-helix, 37% β-structure, and 37% disordered structure at 23°C (70). However, the secondary structure is temperature dependent (71) and the CD analysis is further affected by the presence of carotenoids (85). Complete delipidation causes a small but significant decrease in β-structure and an increase in disordered structure (70). The secondary structures of native and delipidated protein are similar as determined by CD. However, apoLDL is more sensitive to environmental perturbation (72) and to chemical modification (70) than native LDL. Dearborn & Wetlaufer (71) found that β-structure increases whereas the α-helix and disordered structures decrease as the lipid content and particle size of LDL increase. Removal of neutral lipids induces a redshift from 330 to 333 nm in the intrinsic (tryptophan) fluorescence maximum of LDL at pH 9.1. In apoLDL, the maximum is at 334 nm. Based on fluorescence experiments, Pollard & Chen (73) have suggested that delipidation exposes the tryptophan residues of the protein to a more polar environment. It should be noted, however, that the fluorescence maximum of apoLDL is still characteristic of an environment significantly more hydrophobic than fully exposed tryptophan, which would fluoresce maximally at longer wave-lengths.

Smith & Green (80) have taken advantage of the fact that the emission band of tryptophan overlaps the absorption band of the conjugated cholesterol analog cholesta-5,7,9(11)-trien-3β-ol and its oleate ester to study the proximity of cholesterol and cholesteryl esters to the apoprotein in LDL. At the level of 2–3 molecules per

LDL particle, these analogs quench tryptophan fluorescence. The efficiency of energy transfer is 0.11 and <0.02 for the sterol and sterol ester, respectively. The investigators conclude from these results that the free sterol is more closely associated with the apoprotein than is the sterol ester. However, this interpretation must be viewed with caution because the observed effect represents the sum of the energy transfer between about 20 donor tryptophan fluorophores and 2 to 3 acceptor analog molecules within a single LDL particle.

The mobility of the lipid constituents of LDL has been studied by NMR spectroscopy. The NMR spectra of HDL and LDL are quite similar and closely resemble those of the lipoprotein-lipids sonically dispersed in water (50a, 77–79). These results indicate that the bulk of the lipids are not tightly bound to the apoprotein but are highly mobile, suggesting a micellar structure for LDL and HDL. Hamilton et al (50a, 78) have recently recorded the natural abundance, proton decoupled ^{13}C-NMR spectra of the human lipoproteins and assigned 22 of the protein and lipid resolved resonances of the LDL spectrum. The assignment of these resonances now makes it possible to monitor structural changes in LDL at a great number and variety of sites within the particle. In addition to affording spectra with highly resolved resonances which are readily assignable, ^{13}C-NMR makes it relatively easy to study the motional properties of specific carbon atoms by spin-lattice relaxation techniques.

The question of substructure in LDL is unsettled. By electron microscopy with negative staining, LDL appears as nearly spherical and highly deformable particles with diameters of 210 to 250 Å (50). LDL from rat, whale, and guinea pig appear to have the same size and shape as the human counterpart (50). Edges, substructures latitudinally circumscribing the particle, and other fine structural details have been described by Forte & Nichols (50). Pollard et al (82) have reported globular subunits of about 40 Å diameter and about 27,000 daltons from analysis of electron micrographs of LDL. The authors concluded that the globular subunits are proteins, because similar structures are present in LDL preparations from which neutral lipids have been removed or which have been treated with phospholipase A and C. Based on these data and ultracentrifugation experiments, these authors proposed an LDL model in which 20 protein subunits are arranged in a dodecahedral pattern with icosahedral symmetry. The surface of the particle is occupied by both protein and phospholipid, while neutral lipids are contained in the hydrophobic core. Further analysis of isodensity maps by Pollard & Devi (83) has been interpreted as supporting this model. However, if the subunit molecular weight of apoLDL is 250,000, this model would be precluded, as there could be only two subunits per particle.

Mateu et al (84) have used small angle X-ray scattering to study the structure of LDL particles. Their results were interpreted as showing spherically symmetrical particles which contain a lipid bilayer and a radius of 65 Å. The outer surface of LDL is described as covered by a loose two-dimensional network of about 60 protein subunits, each with an estimated molecular weight of 8000. The overall symmetry of the particle is described as icosahedral. This symmetry is consistent with the interpretation of Pollard et al (82), although the number of subunits is not.

Phospholipids, cholesterol, cholesteryl esters, and triglycerides are thought to be part of the bilayer. In this model, protein would be distributed on both the outer and inner bilayer surfaces. The authors suggest that the center of the particle is occupied by a protein core. The dimensions of this model do not allow the core to be occupied by neutral lipids. Owing to the limitation of resolution of small angle X-ray scattering, this model should be considered a provisional one (5).

Variant LDL Particles

Lp(a) is a variant of the human LDL. Although both particles have similar lipid compositions, contain the apoB antigen, and have similar electron microscopic appearances, they differ in molecular weight, ultracentrifugal behavior, electrophoretic mobility, and protein-to-lipid ratio (86–88). The protein moiety of Lp(a) consists of about 65% apoB, 15% albumin, and 20% of an apoprotein, apoLp(a), which is found only in Lp(a) (87). In contrast to apoLDL, apoLp(a) has a high content of carbohydrate (0.26 mg/mg of protein). Utermann et al (88) have described the spontaneous or detergent-induced disaggregation of Lp(a) into several components, one of which is a lipid free protein that reacts immunologically with anti-Lp(a). Characterization of the different components of this complex may clarify the physiologic significance of Lp(a). While occurring in the LDL density range, Lp(a) has the same pre-β electrophoretic mobility as VLDL, probably due to its high content of sialic acid (87). Normal subjects may have an increased pre-β-lipoprotein electrophoretic band with normal concentrations of triglycerides and VLDL. This finding, termed the sinking pre-β lipoprotein phenomenon, presumably reflects a relatively high concentration of Lp(a). There is disagreement as to whether Lp(a) is a qualitative or quantitative trait, although evidence favors the latter (89, 90). The regulatory mechanisms affecting the plasma concentration of Lp(a) are not known.

There are several circumstances in which lipoproteins other than LDL float in the density range assigned to this family. An abnormal lipoprotein, LP-X, is isolated in the same density range as LDL in subjects who lack the enzyme LCAT (91, 92) or have obstructive liver disease (93–95). The serum lipid composition is altered such that the HDL class is almost completely absent and a new abnormal family appears. LP-X has the same electrophoretic mobility as LDL, but does not react with antisera to LDL and has a different chemical composition. Its lipid composition is characterized by the presence of cholesterol almost entirely in the unesterified form and by a high phospholipid and low protein content. The protein composition of LP-X is unique, consisting of a mixture of approximately 40% albumin (95) and 60% apoC (apoC-I, -II, -III) (92, 96, 97). LP-X contains an additional constituent referred to variously as the "thin line" protein (97), apoA-III (98), or apolipoprotein D (99). The antigenic sites of albumin, apoC-II, and apoC-III in LP-X are normally masked but are uncovered by partial or total delipidation (100). However, antisera to apoC-I and "thin line" protein react readily with intact LP-X (100). When viewed with the electron microscope, negatively stained LP-X appears as spherical particles of 300–700 Å diameter, which tend to aggregate to form rouleaux or stacks of disclike structures (50, 101). Immunochemical and electron microscopic data suggest that the LP-X particle is structured such that the albumin, apoC-II, and apoC-III constituents are largely buried on the inside, while the phospholipids, apoC-I, and "thin line" protein are on the outer surface. Model systems with the same lipid composition as LP-X have been studied extensively. For example, aqueous lipid mixtures containing 67% PC and 38% cholesterol (molar ratio of 1 : 1) have been shown to form rouleaux, especially in the presence of small amounts of surface active agents such as lysoPC (102). When a mixture of apoA-I, cholesterol, and PC is sonicated, the resulting particles also appear as stacked discs. When these particles are incubated with a plasma $d > 1.21$ fraction containing LCAT, they are transformed into spherical structures resembling HDL in both size and shape (103). A similar morphologic transformation occurs when cholesteryl ester is incorporated into apoHDL-PC mixtures by sonication.

According to Wengeler & Seidel (104), LP-X does not serve as a substrate for LCAT even though it contains cholesterol and PC. This finding likely reflects the absence of an LCAT activator protein such as apoA-I from the LP-X particle. Determination of the structure and physiologic significance of LP-X represents a challenging problem for future study.

HIGH DENSITY LIPOPROTEINS

General Properties

HDL are ultracentrifugally isolated between d 1.063 and 1.21. Within this density range, there are two subclasses, HDL_2 (1.063–1.120) and HDL_3 (1.120–1.210). The HDL class contains approximately equal amounts of lipid and protein by weight (9). The major lipids, PC and cholesteryl ester, represent about 40 and 30%, respectively, of the total lipid mass. The cholesteryl esters are primarily of linoleic, oleic, and palmitic acids. Phosphatidylserine, phosphatidylinositol, SM, and triglycerides occur as minor lipid constituents.

Independently, Shore & Shore (16) and Scanu et al (20) established the protein heterogeneity of human apoHDL by fractionation on DEAE-cellulose and Sephadex G-200 in urea. Despite initial confusion about the C-terminal amino acid, it is now firmly established that apoA consists of two proteins, each with C-terminal glutamine; they are designated apoA-I and -II (105), apoLP-Gln-I and -II (6), or Fractions III and IV (20).

ApoHDL

Earlier work on the reconstitution of HDL from its protein and lipid components has been reviewed by Scanu (2, 3). Water-soluble protein components and aqueous dispersions of lipids have typically been used in the recombination experiments. Complex formation requires sonication of the lipid-protein mixture at temperatures above that of the gel → liquid crystalline transition of the lipid (106).

In a comprehensive study using CD, Lux et al (107) examined the influence of lipid on the conformation of apoHDL. It was estimated that native HDL contains about 70% α-helix, 11% β-structure, and 19% disordered structure. Delipidation decreases α-helix to 52% and increases the disordered structure to 38%. Relipidation of apoHDL with egg PC increases the α-helicity from 52% to 64% and lowers the disordered structure to 28%. Relipidation with both PC and CO further increases the α-helix to about 70%. Chemical analysis of a reconstituted apoHDL:PC:CO complex after ultracentrifugal isolation reveals a weight ratio of 48:38:15 as compared to 48:27:21 in native HDL. ApoHDL also forms a complex with SM, this being somewhat richer in lipid than those containing PC (108). Addition of SM to the PC-apoHDL complex or of PC to the SM-apoHDL complex produces no further change in the CD spectrum.

Microcalorimetry has been used to study the interaction of apoHDL and various phospholipids (109). For lyso-, dicaproyl-, dilauryl-, and dimyristoyl-PC, the interaction is rapid and exothermic, and depends on whether the lipid is in micellar or vesicular form. The lipid-protein ratio of an ultracentrifugally isolated complex was less than the ratio determined by calorimetry. This difference was attributed to the high salt concentration required in the centrifugation experiment (109).

The effect of apoHDL on the motion of the polar head groups of phospholipids in a reconstituted complex has been studied with ^{31}P-NMR (110). The phosphorous resonance linewidth at half-height ($v_{1/2}$) is greater for phospholipid vesicles than for recombined phospholipid-apoprotein complexes. This effect is the reverse of what might have been expected intuitively since it implies that the motion of the phosphorous nuclei in the vesicles is less than that in the complex. Apparently, strong interactions between the polar head groups of adjacent phospholipid molecules are disrupted by apoprotein binding. The authors' suggestion that the effect may be due to preferential interaction by the apoprotein for the vesicles of a specific size seems unlikely, since it has been shown that PC nuclear resonance linewidths cannot be correlated with vesicle size (111, 112). Accurate measurement of spin-lattice relaxation times (T_1) for several different nuclei in the polar head region as well as in the hydrophobic tail region may be quite useful in assessing the relative magnitudes of these interactions.

ApoA-I

ApoA-I, the major protein constituent of HDL, has a single polypeptide chain of 245 amino acids with a calculated molecular weight of 28,331. The sequence of apoA-I has been determined (113–115) and is shown below

Asp-Glu-Pro-Pro-Gln-Ser-Pro-Trp-Asp-Arg-Val-Lys-Asp-Leu-Ala-Thr-Val-Tyr-Val-Asp-
10   20

Val-Leu-Lys-Asp-Ser-Gly-Arg-Asp-Tyr-Val-Ser-Gln-Phe-Gln-Gly-Ser-Ala-Leu-Gly-Lys-
30   40

Gln-Leu-Asn-Leu-Lys-Leu-Leu-Trp-Asp-Asp-Val-Thr-Ser-Thr-Phe-Ser-Lys-Leu-Arg-Gln-
50   60

Glu-Leu-Gly-Pro-Val-Thr-Glu-Glu-Trp-Phe-Asn-Asp-Leu-Gln-Glu-Lys-Leu-Asn-Leu-Glu-
70   80

Lys-Glu-Thr-Gly-Glu-Leu-Arg-Gln-Glu-Met-Ser-Lys-Asp-Leu-Glu-Glu-Val-Lys-Ala-Lys-
90   100

Val-Gln-Pro-Tyr-Leu-Asp-Asp-Phe-Gln-Lys-Lys-Trp-Gln-Glu-Met-Glu-Leu-Tyr-Arg-Gln-
110   120

Lys-Val-Glu-Pro-Leu-Arg-Ala-Glu-Leu-Gln-Glu-Gly-Ala-Arg-Gln-Lys-Leu-His-Glu-Leu-
130   140

Gln-Glu-Lys-Leu-Ser-Pro-Leu-Gly-Glu-Glu-Met-Arg-Asp-Arg-Ala-Arg-Ala-His-Val-Asp-
150   160

Ala-Leu-Arg-Thr-His-Leu-Ala-Pro-Tyr-Ser-Asp-Glu-Leu-Arg-Gln-Arg-Leu-Ala-Ala-Arg-
170   180

Leu-Glu-Ala-Leu-Lys-Glu-Asn-Gly-Ala-Gly-Arg-Leu-Ala-Glu-Tyr-His-Ala-Lys-Ala-Thr-
190   200

Glu-His-Leu-Ser-Thr-Leu-Ser-Glu-Lys-Ala-Lys-Pro-Ala-Leu-Glu-Asp-Leu-Arg-Gln-Gly-
210   220

Leu-Leu-Pro-Val-Leu-Glu-Ser-Phe-Lys-Val-Ser-Phe-Leu-Ser-Ala-Leu-Glu-Glu-Tyr-Thr-
230   240

Lys-Leu-Asn-Thr-Gln
245

It contains three residues of methionine and four of tryptophan, but no cysteine, cystine, or isoleucine. The protein appears to be microchemically or microphysically heterogeneous. Edelstein et al (116) have isolated two polymorphic forms (Fractions III$_a$ and III$_b$) with slight differences in electrophoretic mobility and amino acid composition. Lux & John (117) have observed heterogeneity of the "purified" protein, but its cause is still obscure.

ApoA-I has limited water solubility between pH 4 and 7 with a minimum at about 4.8 (118). Its fluorescent properties indicate that at pH 7.4 the four tryptophan residues are in a very hydrophobic environment (118). In a fluorescence quenching study of apoA-I with iodide and pyridinium ions, it was concluded that the greater proportion of the four tryptophan residues is in a positively charged region of the protein (43). Lowering the pH to 2.3 partially exposes these residues to the polar solution, whereas raising the pH to 12.0 or adding 6 M guanidine exposes the tryptophans completely (118). Molecular weight determination of apoA-I by ultra-centrifugation or by PAGE in SDS gives a value of 26,000–29,000 (116, 119, 120). These measurements are consistent with the molecular weight determined from the amino acid sequence (113). It is important to note that the ultracentrifugal measurements were made at a protein concentration ≤ 1 mg/ml. At higher concentrations significant self association occurs (120), a phenomenon that might well affect the lipid binding capacity of apoA-I.

After isolation in 6 M urea and dialysis, apoA-I contains 55% α-helix, 8% β-structure, and 37% disordered structure as determined by CD (107). A valid concern is the possible effect of the denaturing conditions of isolation on the structure of the protein. It should be noted that the effects of guanidine and urea on the CD spectrum of apoA-I are completely reversed by dialysis (107, 118). When native HDL is dehydrated and rehydrated, apoA-I is dissociated from the remainder of the lipoprotein particle and may be isolated in a lipid free form (121). This material, which has not been exposed to denaturing solvents, exhibits a CD spectrum identical to that of apoA-I isolated by the conventional procedure (107). Relipidation of apoA-I with PC (1 : 1) by sonic irradiation followed by ultracentrifugal isolation at d 1.063–1.210 yields a complex containing 43% protein and 57%

phospholipid (107). This complex contains 69% α-helix, 9% β-structure, and 22% disordered structure. When CO and PC are used for the relipidation, these values are changed to 83%, 2%, and 15%, respectively.

When apoA-I is incubated with small, fractionated bilamellar vesicles (122), complexes are formed having a structure similar (40 Å by 100–200 Å) to those formed by sonic irradiation (107) but containing a smaller proportion of phospholipid. Nichols et al (123, 124) have shown that the stoichiometry of the complex formed from PC vesicles and apoA-I is highly dependent on the initial lipid : protein ratio in the incubation mixture.

Whereas lysoPC is not a major lipid component of HDL, its interaction with HDL apoproteins is of interest because this lipid is formed by the action of LCAT on the HDL particle (91). LysoPC stabilizes the structure of apoA-I to denaturing agents and produces an increase in α-helicity (51 → 67%) of the same magnitude as induced by PC (124). The composition of ultracentrifugally isolated complexes of lysoPC–apoA-I depends on the initial lipid : protein ratio of the mixture. LysoPC may enhance the binding of PC to apoA-I by dissociating the protein or by increasing the fluidity of the PC bilayers. ApoA-I also binds 3–4 molecules of SDS with an association constant of 2×10^4 M^{-1} (120).

Forte et al (50, 125) have studied the morphology of reconstituted complexes of HDL lipids with electron microscopy and negative staining. When the apoprotein is recombined with PC or with a PC-cholesterol mixture, disc-shaped structures arranged as linear arrays are observed. The discs have a periodicity of 50–55 Å and a diameter of 100–200 Å. When cholesteryl ester is included in the lipid mixture, the stacked discs are replaced by spherical particles of 50–100 Å diameter, which resemble native HDL in size and shape. Similar particles can be produced by the action of LCAT on the apoHDL-PC-cholesterol complex (125). This enzyme-mediated transformation demonstrates the importance of cholesteryl ester in influencing the overall structure of the HDL particle.

ApoA-II

The second most abundant apoprotein of HDL is apoA-II. Its amino acid sequence is as follows (126–128)

PCA-Ala-Lys-Glu-Pro-Cys-Val-Glu-Ser-Leu-Val-Ser-Gln-Tyr-Phe-
                5   10   15

Gln-Thr-Val-Thr-Asp-Tyr-Gly-Lys-Asp-Leu-Met-Glu-Lys-Val-Lys-
              20   25   30

Ser-Pro-Glu-Leu-Gln-Ala-Gln-Ala-Lys-Ser-Tyr-Phe-Glu-Lys-Ser-
              35   40   45

Lys-Glu-Gln-Leu-Thr-Pro-Leu-Ile-Lys-Lys-Ala-Gly-Thr-Glu-Leu-
              50   55   60

Val-Asn-Phe-Leu-Ser-Tyr-Phe-Val-Glu-Leu-Gly-Thr-Gln-Pro-Ala-
            65   70   75

Thr-Gln-COOH

ApoA-II contains two identical polypeptide chains of 77 amino acids each. These are linked by a symmetrical disulfide at cystine-6. The molecule contains no histidine, arginine, tryptophan, or carbohydrate.

The sequence of apoA-II from *Maccacus rhesus* monkey is highly homologous to the human protein (129–132). The most significant difference is the substitution of cystine-6 (human) with serine (132), suggesting that the intact disulfide is not required for the lipid binding functions of the protein. Additional human-to-monkey substitutions are Lys-3 → Glu-3, Ser-40 → Ala-40, Ile-53 → Val-53, Gly-71 → Asp-71, and the insertion of arginine between positions 71 and 72 (132). Absence of the disulfide linkage has also been found in the baboon, dog, rabbit, and cow (130). In the chimpanzee, however, apoA-II exists as the dimer (133, 134) as in man. Knowledge of the amino acid sequence from other species may help clarify the physiologic functions of this apoprotein.

Extensive studies of the structural and lipid binding properties of human apoA-II and several of its derivatives and fragments have been carried out. The intact protein contains 35% α-helix, 13% β-structure, and 52% disordered structure as determined by CD (107). Upon sonication with an equal weight of PC and ultra-centrifugal isolation at d 1.063–1.210, a complex is obtained which contains 59% protein and 41% PC and exhibits 50% α-helix, 11% β-structure, and 39% disordered structure. When the apoprotein is relipidated with a 1:1 (by weight) PC-CO mixture, the isolated complex contains 40% protein, 38% PC, and 21% CO. CD analysis indicates 61% α-helix, 6% β-structure, and 33% disordered structure (107). Reduction of the single disulfide of apoA-II to form monomers decreases the α-helix by about 13% (135, 136) but causes little change in the phospholipid binding properties. The conformational changes can be reversed by reoxidation to form the dimer. When a spin label or fluorescent label is attached to the thiol of reduced apoA-II, the motion of the spin label is very rapid ($\tau_c < 10^{-9}$ sec) and the environment of the extrinsic fluorophore is highly polar ($\lambda_{max} = 492$) (136). Each of these derivatives retains the ability to bind phospholipid. Relipidation of these derivatives with PC or PC-CO brings about no detectable change in spin label motion or in polarity of the fluorophore environment (136). These results suggest that the disulfide is not intimately involved in phospholipid binding.

These observations have been extended by examining the phospholipid binding properties of the separated N- and C-terminal CNBr fragments of apoA-II. Cleavage of the single methionine at position 26 with cyanogen bromide yields an N-terminal fragment (CNBr IV)$_2$ containing two chains (residues 1–26) joined at cystine-6, and a C-terminal fragment (CNBr III) spanning residues 27–77 (127, 137, 138). While apoA-II, performic acid oxidized apoA-II, and the C-terminal CNBr fragment (CNBr III) bind PC, the N-terminal fragment, as either the monomer (CNBr IV) or dimer (CNBr IV)$_2$, does not bind (139, 140). CNBr III binds to a lesser extent than does the intact apoprotein. The conclusion from these studies is that the principal phospholipid binding determinants of apoA-II reside in the C-terminal two thirds of the molecule.

ApoA-I and ApoA-II Interaction

An important aspect of lipoprotein structure which has been largely neglected is the interaction between like and unlike protein components and the effects of such interactions on lipid binding. Ritter et al (141) have reported that apoA-I binds less total HDL lipid than does apoA-II. When these apoproteins are mixed, however, apoA-II increases the binding capacity of apoA-I for the lipid mixture. These workers conclude that each of the two major HDL apoproteins has a different affinity for HDL lipids and is dependent on the apoA-I:apoA-II ratio, the lipid:protein ratio, and the conditions of mixing. The overall significance of these experiments might have been extended by studying a single lipid type (e.g. PC or SM) rather than whole HDL lipids.

When present together in equimolar quantities, apoA-I and apoA-II behave as a single species of molecular weight 50,000 as determined by sedimentation equilibrium experiments (120). The calculated molecular weight for a 1:1 molar complex is 45,711 based on the sequence data of the component proteins. The binding of SDS to this complex is significantly less than that calculated on the basis of a simple linear combination of the binding data for the separate proteins. This finding suggests that some of the binding sites available on one or both of the separate apoproteins are blocked by association between apoA-I and apoA-II (120).

Other Apoproteins of HDL

In addition to apoA-I and apoA-II, HDL also contains several minor apoproteins found in VLDL as major components. As a group, these apoproteins are called apoC (7) and are discussed in the section on VLDL. Another protein in HDL has been variously called the "thin line" protein (97), apoD (99), and apoA-III (98). This protein is also found in LDL and VHDL (99). McConathy & Alaupovic (99) have reported that the "thin line" protein can be separated from apoA and apoC by chromatography of intact HDL on hydroxylapatite. These workers designate the protein as part of a fourth lipoprotein family and refer to it as apoLP-D. On the other hand, Kostner (98) has shown by BioGel and DEAE-cellulose chromatography, immunoadsorption studies, and analytical isoelectric focusing that the "thin line" protein is always associated with the apoA group of proteins in intact lipoproteins and has designated it apoA-III. The "thin line" protein is immunologically different from any other apoproteins of apoA, apoB, or apoC and gives no reaction with antibodies to these proteins (98). Its molecular weight by sedimentation equilibrium experiments and amino acid analysis is 19,000 to 20,000 (98). It contains D-glucosamine and all the common amino acids, with the possible exception of half-cystine. Its C-terminal amino acid is serine (98). The function of the "thin line" protein has not been elucidated. Significantly, it is one of the constituents of LP-X (96), an abnormal lipoprotein described in an earlier section.

Studies on Intact HDL Particles

By sedimentation equilibrium, the molecular weights of HDL_2 and HDL_3 are 320,000 and 175,000, respectively (9). Values of 360,000 and 184,000 have been obtained by small angle X-ray scattering studies (142, 143). More than 90% of the lysine ε-amino groups in HDL can be succinylated (144). Succinylated HDL exhibits chemical, ultracentrifugal, and morphological properties similar to native HDL. Whereas these experiments suggest the surface location of the apoproteins, they do not exclude the possibility that succinic anhydride may penetrate the shell-core interface of HDL and react with proteins that might be located at the center of the particle. An appropriate control would involve determining the relative labeling rates of apoprotein in HDL and of a lysine peptide encapsulated in phospholipid vesicles. Based on their electron microscopic data, Forte & Nichols (50) have postulated that HDL contains 4–5 subunits with diameters of 35–50 Å each. Calculated theoretical X-ray scattering curves do not support this subunit model (145). Other workers have suggested that the electron micrographic features interpreted as subunit structures might have resulted from defocus interference effects (146).

Laggner et al (147) have used X-ray techniques to show that both HDL_2 and HDL_3 particles have two regions of significantly different electron density. HDL_2 (HDL_3) consists of a central core with radius 43 Å (37 Å) and electron density 0.312 (0.318) $e\text{Å}^{-3}$, and an outer shell of radius 14 Å (11 Å) and electron density 0.386 (0.386) $e\text{Å}^{-3}$. In a similar study of HDL_2, Shipley et al (148) observed values of 86 Å for the diameter of the central core and 11 Å for the thickness of the outer shell to give a particle diameter of 108 Å. These observed electron densities for the outer shell are significantly less than the value of 0.45 $e\text{Å}^{-3}$ (149) which has been calculated for anhydrous phospholipid polar head groups and protein. This finding together with the fact that the length of the fully extended head group is 11 Å (150) has led to the view that the outer shell region of the particle contains both the polar head groups of phospholipids in micellar arrangement and the protein components. The structure of porcine HDL_3 has been examined by Atkinson et al (149), who report values of 55 Å for the particle radius of gyration, 42 Å for the radius of the electron poor core, and 12 Å for the width of the electron dense outer shell. Based on the low electron density of the central core, the phospholipid acyl chains and the highly hydrophobic cholesteryl esters probably occupy this inner region.

Assmann et al (110) have also assigned a peripheral location for the phospholipid polar head groups in HDL based on the broadening of ^{31}P nuclear resonance linewidth by paramagnetic Eu^{3+} ions. However, the specificity of this effect is subject to question. Indeed, Glonek et al (151) have demonstrated changes in the phosphorus resonance chemical shift and linewidth of HDL phospholipids even with non-paramagnetic ions. Ideally, such an experiment should be performed using a species that is highly polar so it does not penetrate the shell-core interface, and that is uncharged so electrostatic interaction with protein side chains and phospholipid polar head groups is precluded. Several spin labels possess such properties (e.g. TEMPO-OH). Earlier studies using phospholipase A and D have also suggested the surface location of the phospholipid polar head groups of HDL (152, 153). However, caution should be used in structural interpretations of such experiments, unless it is shown that during the course of the enzymic action, the structure of the HDL particle does not change. Considering the demonstrated mobility of phospholipids (154, 155) in membranes and HDL (50a), it is doubtful that HDL could be subjected to such treatment and still retain its native structure.

MODELS OF LIPOPROTEIN STRUCTURE

Models for describing the relative location of the lipid and protein components of the plasma lipoproteins (69, 156–160) have been influenced by existing knowledge about the structure of micelles. In an aqueous environment the more polar components, which include the protein, phospholipids, and cholesterol, would be located at or near the outer surface of the particle, whereas the apolar triglycerides and cholesteryl esters would be in a nonpolar central core. Schneider et al (161) have analyzed this general type of model with a rigorous quantitative treatment, giving special attention to the packing of lipids at curved surfaces. Important parameters for determining the structure of a lipoprotein include the dimensions of the lipids at the shell-water interface, the composition of the interface region, and the size, mass, and lipid composition of the total particle. There are three extreme possibilities for the location of the protein in the outer surface shell: (*a*) interdigitating completely between the surface lipids; (*b*) occupying a monolayer completely covering the outer polar phospholipid, or (*c*) occupying autonomous domains in the matrix of polar lipid. Various combinations of these extreme cases are also possible. An alternative model to the micellar one has been suggested for LDL, in which the protein is at least partially located in the interior of the particle and the phospholipid has bilayer structure (84). This LDL model was derived from an interpretation of low angle X-ray scattering data. None of these models describes on a molecular level how the individual lipid components interact with the apoproteins.

A major lipid-protein interaction in HDL is between the phospholipid and the apoproteins. Several lines of experimental evidence support this view. Delipidation of lipoproteins with organic solvents removes the neutral lipids more easily than the polar lipids. For example, extraction with solvents such as heptane (162) or ether (73) removes most of the neutral lipids but little, if any, of the phospholipids. Further treatment with chloroform-methanol or ether-ethanol gives a more complete delipidation (163), yielding an apoprotein with about 1% residual phospholipid by weight. Sodhi & Gould (164) have shown that HDL lipids recombine with apoHDL to give complexes that contain phospholipids and cholesterol in amounts comparable to those present in native HDL. On the other hand, cholesterol alone combines with apoHDL in much smaller amounts. Generally, triglycerides and cholesteryl esters recombine in the presence, but not in the absence, of phospholipids. Similar results have been observed in reconstitution of HDL using sonication (165). Numerous recent studies have demonstrated the formation of lipid-protein complexes from phospholipids and apoC-I, apoC-II, apoC-III, apoA-I, and apoA-II (22, 31, 32, 36, 38–43, 106–110, 121–125, 139–141, 166). An important consequence of the interaction of these apoproteins with phospholipid is a significant increase in their α-helical structure. The availability of sequence data for four of these apoproteins together with the concepts outlined above has permitted the development of an hypothesis (160) that describes the molecular basis for lipid-apoprotein interaction in plasma lipoproteins.

ApoA-I, apoA-II, apoC-I, and apoC-III each contains four or more pairs of amino acid residues which are in a 1,2 or 1,4 sequence relationship and have oppositely charged side chains (e.g. Asp-9, Lys-10 in apoC-I). The construction of space-filling models of these proteins allows an examination of the steric relationships between these oppositely charged residues and the α-helical regions of the protein. Such an examination (Figure 1) has revealed that in certain helical segments these oppositely charged side chains are oriented in a manner that allows their electrostatic interaction with the complementary polar head groups of zwitterionic phospholipids. This type of ionic interaction could stabilize the hydrophobic bonding between the fatty acyl chains of the phospholipid and the apolar residues of the apoprotein, although this latter interaction is probably of greater magnitude. An important feature of these helical peptide segments is that one of their sides is polar or hydrophilic while the other side is nonpolar or hydrophobic. Thus, they have been called "amphipathic" (two sided) helices (160). Whereas other helical proteins such as myoglobin and hemoglobin (167) contain polar and nonpolar *edges*, the apolipoprotein amphipathic region exhibits a cylindrical surface, a *full half* of which is polar and contains the zwitterionic pairs, while the opposite surface is almost completely hydrophobic. Examination of the polar face reveals a unique distribution of the charged residues. The negatively charged acidic amino acids are clustered in the middle of the surface, whereas the positively charged basic groups are oriented toward the lateral edges. These structural orientations not only allow close steric contact between charged amino acid side chains and oppositely charged groups of the phospholipid, but also permit the segment containing $\sim C_2-C_4$ of the fatty acyl chains to interact with the hydrophobic side of the helix and possibly with the hydrocarbon side chains of lysine and arginine, since the latter are to the sides of the polar surface. Those segments of the acyl chains not associated with the apoprotein ($\sim C_5-C_{18}$) would be available for interaction with cholesterol and other neutral lipids. This amphipathic model is consistent with the experimental data from phospholipid binding studies on apoA-I (108, 166), apoA-II (108, 136, 139, 140), apoC-I (31), and apoC-III (40, 41).

On the other hand, Assmann & Brewer (159) have claimed that ionic binding is not significant in lipoprotein structure and have proposed a model for HDL in which the proteins are distributed as "icebergs in a sea of lipid" and the hydrophobic helical surface is parallel to the fatty acyl chains. This model is made less attractive by the fact that the proteins must extend a significant distance into the hydrophobic domain of the particle. Such a condition would not be thermodynamically favored unless most of the ion pairs and other polar side chains on the polar faces of the amphipathic helices are in close proximity and facing each other, thereby shielding each other from the nearby hydrophobic surfaces; nor is this model consistent with the low angle X-ray scattering measurements of HDL, which locate the protein in an 11 Å outer shell of uniform thickness (142, 143).

In a different model of HDL structure consistent with the peripheral location of

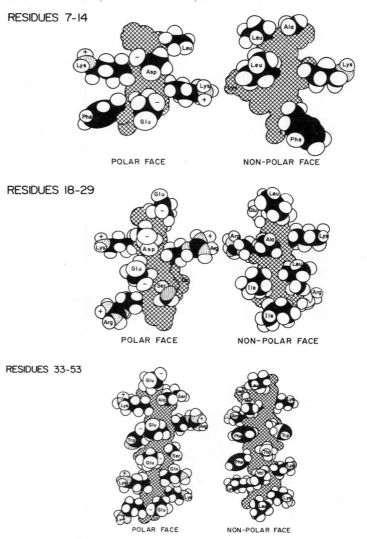

Figure 1 The α-helical surface topography of amphipathic regions of ApoC-I: residues 7 to 14, residues 18 to 29, residues 33 to 53. Each helix was built with Ealing CPK space-filling models. A right handed α-helical backbond with 3.6 residues per turn was constructed. For each amphipathic region amino acid residues were added to the α-carbons in their proper order. Each helix is shown with its axis oriented parallel to the plane of the page and its NH_2-terminal end toward the top of the page. The polar and apolar faces of each helix, rotated around the helix axis by 180° relative to one another, are shown. [From Jackson et al (31) with permission.]

the protein (168), the phospholipids assume a micellar arrangement with the polar head groups oriented outward toward the aqueous bulk phase and the fatty acyl chains directed inward toward the hydrophobic core. In this model, the fatty acyl chains are oriented perpendicular to the long axis of the peptide helix, in contrast to that of Assmann & Brewer (159). Unesterified cholesterol may be either intercalated between the PC acyl chains or located in the hydrophobic core with triglyceride and cholesteryl ester. The apoprotein contains helical regions located between the polar head groups of the phospholipids. It may be calculated that the quantity of protein in HDL can be fully accommodated on the surface of the particle (169). An attractive feature of this model is that it is highly amenable to experimental testing.

SUMMARY

The new information that has evolved in the past five years concerning the chemical and physical properties of the human plasma lipoprotein-proteins is reviewed here. The major families of plasma lipoproteins are heterogeneous with respect to apoprotein composition. One of these families, the HDL, has been reconstituted from its lipid and apoprotein constituents to yield a product similar to the native one. Conditions for recombination of the apoproteins and phospholipids have been determined. Four of the apoproteins have been sequenced and a portion of one synthesized. Phospholipid binding segments have been identified by studies with enzymatic, chemical, and synthetic fragments of the apoproteins. Knowledge of the protein sequences and the phospholipid binding studies have led to an hypothesis explaining lipid binding on the basis of unique spatial orientations identified in three-dimensional models.

Further testing of lipoprotein models is needed. Crystallization of the apoproteins and study of their structure with X-ray crystallography may produce valuable information. The use of isotopic enrichment and Fourier transform may enhance the utility of NMR spectroscopy studies and lead to a better understanding of lipid-protein interaction at the molecular level. Major unsolved problems are the structure of apoLDL, apoC-II, the "arginine-rich" protein, and LP-X.

Literature Cited

1. Scanu, A. M., Wisdom, C. 1972. *Ann. Rev. Biochem.* 41:703–30
2. Scanu, A. M. 1972. *Biochim. Biophys. Acta* 265:471–508
3. Scanu, A. M. 1972. *Ann. NY Acad. Sci.* 195:390–406
4. Hamilton, R. L. 1972. *Advan. Exp. Med. Biol.* 26:7–24
5. Scanu, A. M., Vitello, L., Deganello, S. 1974. *Crit. Rev. Biochem.* 2:175–96
6. Fredrickson, D. S., Lux, S. E., Herbert, P. N. 1972. *Advan. Exp. Med. Biol.* 26:25–56
7. Alaupovic, P., Lee, D. M., McConathy, W. J. 1972. *Biochim. Biophys. Acta* 260:689–707
8. Kostner, G., Alaupovic, P. 1972. *Biochemistry* 11:3419–28
9. Skipski, V. P. 1972. In *Blood Lipids and Lipoproteins: Quantitation, Composition and Metabolism,* ed. G. J. Nelson, 471–583. New York: Wiley-Interscience
10. Lindgren, F. T., Jensen, L. C., Hatch, F. T. Ibid, 181–274
11. McConathy, W. J., Alaupovic, P. 1974. *FEBS Lett.* 41:174–77

12. Pearlstein, E., Aladjem, F. 1972. *Biochemistry* 11:2553–58
13. Quarfordt, S. H., Nathans, A., Dowdee, M., Hilderman, H. L. 1972. *J. Lipid Res.* 13:435–44
14. Sata, T., Havel, R. J., Jones, A. L. 1972. *J. Lipid Res.* 13:757–68
15. Eisenberg, S., Bilheimer, D., Lindgren, F., Levy, R. I. 1972. *Biochim. Biophys. Acta* 260:329–33
16. Shore, V. G., Shore, B. 1973. *Biochemistry* 12:502–7
17. Brown, W. V., Levy, R. I., Fredrickson, D. S. 1969. *J. Biol. Chem.* 244:5687–94
18. Albers, J. J., Scanu, A. M. 1971. *Biochim. Biophys. Acta* 236:29–37
19. Gotto, A. M., Brown, W. V., Levy, R. I., Birnbaumer, M. E., Fredrickson, D. S. 1972. *J. Clin. Invest.* 51:1486–94
20. Scanu, A., Toth, J., Edelstein, C., Koga, S., Stiller, E. 1969. *Biochemistry* 8:3309–16
21. Morrisett, J. D., Gotto, A. M. Unpublished experiments
21a. Sato, J., Shimasaki, H., Hara, I. 1972. *Lipids* 7:404–8
22. Forte, T., Gong, E., Nichols, A. V. 1974. *Biochim. Biophys. Acta* 337:169–83
23. Garner, C. W. Jr., Smith, L. C., Jackson, R. L., Gotto, A. M. Jr. 1972. *Circulation* 45/46 (Suppl. II):958 (Abstr.)
24. Ganesan, D., Bradford, R. H., Alaupovic, P., McConathy, W. J. 1971. *FEBS Lett.* 15:205–8
25. Havel, R. J., Fielding, C. J., Olivercrona, T., Shore, V. G., Fielding, P. E., Egelrud, T. 1973. *Biochemistry* 12:1828–33
26. Miller, A. L., Smith, L. C. 1973. *J. Biol. Chem.* 248:3359–62
27. Brown, W. V., Levy, R. I., Fredrickson, D. S. 1970. *J. Biol. Chem.* 245:6588–94
28. Shulman, R., Herbert, P., Wehrly, K., Chesebro, B., Levy, R. I., Fredrickson, D. S. 1972. *Circulation* 45/46 (Suppl. II):955 (Abstr.)
29. Shulman, R., Herbert, P., Wehrly, K., Chesebro, B., Levy, R. I., Fredrickson, D. S. *J. Biol. Chem.* In press
30. Jackson, R. L., Sparrow, J. T., Baker, H. N., Morrisett, J. D., Taunton, O. D., Gotto, A. M. Jr. 1974. *J. Biol. Chem.* 249:5308–13
31. Jackson, R. L. et al. 1974. *J. Biol. Chem.* 249:5314–20
32. Middelhoff, G., Brown, V. 1973. *Circulation* 7/8 (Suppl. IV):310 (Abstr.)
33. LaRosa, J. C., Levy, R. I., Herbert, P., Lux, S. E., Fredrickson, D. S. 1970. *Biochem. Biophys. Res. Commun.* 41:57–62

34. Brown, W. V., Baginsky, M. L. 1972. *Biochem. Biophys. Res. Commun.* 46:375–81
35. Shulman, R. S., Herbert, P. N., Fredrickson, D. S., Wehrly, K., Brewer, H. B. Jr. 1974. *J. Biol. Chem.* 249:4969–74
35a. Brewer, H. B. Jr., Shulman, R., Herbert, P., Ronan, R., Wehrly, K. 1974. *J. Biol. Chem.* 249:4975–84
36. Morrisett, J. D., David, J. S. K., Pownall, H. J., Gotto, A. M. Jr. 1973. *Biochemistry* 12:1290–99
37. Shulman, R., Herbert, P., Brewer, H. B. 1971. *Fed. Proc.* 30:787 (Abstr.)
38. Hoff, H. F., Morrisett, J. D., Gotto, A. M. Jr. 1973. *Biochim. Biophys. Acta* 296:653–60
39. Morrisett, J. D., Gallagher, J., Aune, K. C., Gotto, A. M. Jr. 1974. *Biochemistry* 13:4765–71
40. Sparrow, J. T., Gotto, A. M. Jr., Morrisett, J. D. 1973. *Proc. Nat. Acad. Sci. USA* 70:2124–28
41. Shulman, R. S., Herbert, P. N., Witters, L. A., Quicker, T., Wehrly, K. A., Fredrickson, D. S., Levy, R. I. 1973. *Fed. Proc.* 32:1858 (Abstr.)
42. Pownall, H. J., Morrisett, J. D., Sparrow, J. T., Gotto, A. M. 1974. *Biochem. Biophys. Res. Commun.* 60:779–86
43. Pownall, H. J., Smith, L. C. 1974. *Biochemistry* 13:2590–93
44. Havel, R. J., Kane, J. P. 1973. *Proc. Nat. Acad. Sci. USA* 70:2015–19
45. Shore, B., Shore, V. 1974. *Biochem. Biophys. Res. Commun.* 58:1–7
46. Shore, V. G., Shore, B., Hart, R. G. 1974. *Biochemistry* 13:1579–85
47. Shelburne, F. A., Quarfordt, S. H. 1974. *J. Biol. Chem.* 249:1428–33
48. Redgrave, T. G. 1970. *J. Clin. Invest.* 49:465–71
49. Hazzard, W. R., Bierman, E. L. 1971. *Clin. Res.* 19:476
50. Forte, T., Nichols, A. V. 1972. *Advan. Lipid Res.* 10:1–41
50a. Hamilton, J. A., Talkowski, C., Childers, R. F., Williams, E., Allerhand, A., Cordes, E. H. 1974. *J. Biol. Chem.* 249:4872–78
51. Fisher, W. R., Hammond, M. G., Warmke, G. L. 1972. *Biochemistry* 11:519–25
52. Fisher, W. R. 1972. *Ann. Clin. Lab. Sci.* 2:198–208
53. Schumaker, V. N. 1973. *Accounts Chem. Res.* 6:398–403
54. Lee, D. M., Alaupovic, P. 1974. *Atherosclerosis* 19:501–20
55. Lee, D. M., Alaupovic, P. 1974. *Biochem. J.* 137:155–67

56. Lipp, K., Wiegandt, H. 1973. *Z. Physiol. Chem.* 354:262–66
57. Krishnaiah, K. V., Wiegandt, H. 1974. *FEBS Lett.* 40:265–68
58. Smith, R., Dawson, J. R., Tanford, C. 1972. *J. Biol. Chem.* 247:3376–81
59. Kane, J. P., Richards, E. G., Havel, R. J. 1970. *Proc. Nat. Acad. Sci. USA* 66:1075–82
60. Scanu, A., Pollard, H., Reader, W. 1968. *J. Lipid Res.* 9:342–48
61. Simons, K., Helenius, A. 1970. *FEBS Lett.* 7:59–63
62. Chen, C-H., Aladjem, F. 1974. *Biochem. Biophys. Res. Commun.* 60:549–54
63. Kane, J. Personal communication
64. Albers, J. J., Chen, C-H., Aladjem, F. 1972. *Biochemistry* 11:57–63
65. Shore, B., Shore, V. 1969. *Biochemistry* 8:4510–16
66. Helenius, A., Simons, K. 1971. *Biochemistry* 10:2542–47
67. Gotto, A. M., Levy, R. I., Fredrickson, D. S. 1968. *Biochem. Biophys. Res. Commun.* 31:151–57
68. Swaminathan, N., Aladjem, F. 1974. *Fed. Proc.* 33:2042 (Abstr.)
69. Gotto, A. M. Jr. 1969. *Proc. Nat. Acad. Sci. USA* 64:1119–27
70. Gotto, A. M. Jr., Levy, R. I., Lux, S. E., Birnbaumer, M. E., Fredrickson, D. S. 1973. *Biochem. J.* 133:369–82
71. Dearborn, D. G., Wetlaufer, D. B. 1969. *Proc. Nat. Acad. Sci. USA* 62:179–85
72. Scanu, A., Pollard, H., Hirz, R., Kothary, K. 1969. *Proc. Nat. Acad. Sci. USA* 62:171–78
73. Pollard, H. B., Chen, R. F. 1973. *J. Supramol. Struct.* 2:177–84
74. Gotto, A. M., Levy, R. I., Fredrickson, D. S. 1968. *Proc. Nat. Acad. Sci. USA* 60:1436–41
75. Keith, A. D., Mehlhorn, R. J., Freeman, N. K., Nichols, A. V. 1973. *Chem. Phys. Lipids* 10:223–36
76. Gotto, A. M., Kon, H., Birnbaumer, M. E. 1970. *Proc. Nat. Acad. Sci. USA* 65:145–51
77. Leslie, R. B., Chapman, D., Scanu, A. M. 1969. *Chem. Phys. Lipids* 3:152–58
78. Hamilton, J. A., Talkowski, C., Williams, E., Avila, E. M., Allerhand, A., Cordes, E. H., Camejo, G. 1973. *Science* 180:193–95
79. Steim, J. M., Edner, O. J., Bargoot, F. G. 1968. *Science* 162:909–11
80. Smith, R. J. M., Green, C. 1974. *Biochem. J.* 137:413–15
81. DeBlois, R. W., Uzgiris, E. E., Devi, S. K., Gotto, A. M. Jr. 1973. *Biochemistry* 12:2645–49
82. Pollard, H., Scanu, A. M., Taylor, E. W. 1969. *Proc. Nat. Acad. Sci. USA* 64:304–10
83. Pollard, H. B., Devi, S. K. 1971. *Biochem. Biophys. Res. Commun.* 44:593–99
84. Mateu, L., Tardieu, A., Luzzati, V., Aggerbeck, L., Scanu, A. M. 1972. *J. Mol. Biol.* 70:105–16
85. Chen, G. C., Kane, J. P. 1974. *Biochemistry* 13:3330–35
86. Simons, K., Ehnholm, C., Renkonen, O., Bloth, B. 1970. *Acta Pathol. Microbiol. Scand.* B78:459–66
87. Ehnholm, C., Garoff, H., Renkonen, O., Simons, K. 1972. *Biochemistry* 11:3229–32
88. Utermann, G., Lipp, K., Wiegandt, H. 1972. *Humangenetik* 14:142–50
89. Albers, J. J., Hazzard, W. R. 1974. *Lipids* 9:15–26
90. Harvie, N. R., Schutz, J. S. 1970. *Proc. Nat. Acad. Sci. USA* 66:99–103
91. Glomset, J. A. See Ref. 9, 745–87
92. Norum, K. R., Glomset, J. A., Gjone, E. 1972. In *The Metabolic Basis of Inherited Disease*, ed. J. B. Stanbury, J. B. Wyngaarden, D. S. Fredrickson, 531–44. New York: McGraw
93. Seidel, D., Agostini, B., Müller, P. 1972. *Biochim. Biophys. Acta* 260:146–52
94. Seidel, D., Alaupovic, P., Furman, R. H. 1969. *J. Clin. Invest.* 48:1211–23
95. Seidel, D., Alaupovic, P., Furman, R. H., McConathy, W. J. 1970. *J. Clin. Invest.* 49:2396–2407
96. Torsvik, H., Berg, K., Magnani, H. N., McConathy, W. J., Alaupovic, P., Gjone, E. 1972. *FEBS Lett.* 24:165–68
97. McConathy, W. J., Alaupovic, P., Curry, M. D., Magnani, H. N., Torsvik, H., Berg, K., Gjone, E. 1973. *Biochim. Biophys. Acta* 326:406–18
98. Kostner, G. M. 1974. *Biochim. Biophys. Acta* 336:383–95
99. McConathy, W. J., Alaupovic, P. 1973. *FEBS Lett.* 37:178–82
100. Magnani, H., Alaupovic, P., Seidel, D. 1973. *Conference on Serum Lipoproteins*, Graz. Oct. 21–23. B16 (Abstr.)
101. Forte, T., Norum, K. R., Glomset, J. A., Nichols, A. V. 1971. *J. Clin. Invest.* 50:1141–48
102. Bangham, A. D., Horne, R. W. 1964. *J. Mol. Biol.* 8:660–68
103. Forte, T. M., Nichols, A. V., Gong, E. L., Lux, S., Levy, R. I. 1971. *Biochim. Biophys. Acta* 248:381–86
104. Wengeler, H., Seidel, D. 1973. *Clin. Chim. Acta* 45:429–32
105. Kostner, G., Alaupovic, P. 1971. *FEBS Lett.* 15:320–24

106. Scanu, A., Cump, E., Toth, J., Koga, S., Stiller, E., Albers, L. 1970. *Biochemistry* 9:1327–35
107. Lux, S. E., Hirz, R., Shrager, R. I., Gotto, A. M. 1972. *J. Biol. Chem.* 247:2598–2606
108. Assmann, G., Brewer, H. B. Jr. 1974. *Proc. Nat. Acad. Sci. USA* 71:989–93
109. Rosseneu, M. Y., Soetewey, F., Blaton, V., Lievens, J., Peeters, H. *Chem. Phys. Lipids.* In press
110. Assmann, G., Sokoloski, E. A., Brewer, H. B. Jr. 1974. *Proc. Nat. Acad. Sci. USA* 71:549–53
111. Horwitz, A. F., Michaelson, D., Klein, M. P. 1973. *Biochim. Biophys. Acta* 298:1–7
112. Seiter, C. H. A., Chan, S. I. 1973. *J. Am. Chem. Soc.* 95:7541–53
113. Baker, H. N., Delahunty, T., Gotto, A. M. Jr., Jackson, R. L. 1974. *Proc. Nat. Acad. Sci. USA* 71:3631–34
114. Baker, H. N., Gotto, A. M., Jackson, R. L. 1975. *J. Biol. Chem.* In press
115. Delahunty, T., Baker, H. N., Gotto, A. M., Jackson, R. L. 1975. *J. Biol. Chem.* In press
116. Edelstein, C., Lim, C. T., Scanu, A. M. 1972. *J. Biol. Chem.* 247:5842–49
117. Lux, S. E., John, K. M. 1972. *Biochim. Biophys. Acta* 278:266–70
118. Gwynne, J., Brewer, H. B. Jr., Edelhoch, H. 1974. *J. Biol. Chem.* 249:2411–16
119. Baker, H. N., Jackson, R. L., Gotto, A. M. Jr. 1973. *Biochemistry* 12:3866–71
120. Reynolds, J. A., Simon, R. H. 1974. *J. Biol. Chem.* 249:3937–40
121. Nichols, A. V., Lux, S., Forte, T., Gong, E., Levy, R. I. 1972. *Biochim. Biophys. Acta* 270:132–48
122. Nichols, A., Forte, T., Gong, E., Blanche, P. 1973. *Conference on Serum Lipoproteins, Graz, Oct. 21–23.* B8 (Abstr.)
123. Nichols, A. V., Forte, T., Gong, E., Blanche, P., Verdery, R. B. 1974. *Scand. J. Clin. Lab. Invest.* 33 (Suppl. 137):147–56
124. Verdery, R. B. III, Nichols, A. V. 1974. *Biochem. Biophys. Res. Commun.* 57:1271–78
125. Forte, T. M., Nichols, A. V., Gong, E. L., Lux, S., Levy, R. I. 1971. *Biochim. Biophys. Acta* 248:381–86
126. Lux, S. E., John, K. M., Brewer, H. B. Jr. 1972. *J. Biol. Chem.* 247:7510–18
127. Brewer, H. B. Jr., Lux, S. E., Ronan, R., John, K. M. 1972. *Proc. Nat. Acad. Sci. USA* 69:1304–8
128. Lux, S. E., John, K. M., Ronan, R.,

Brewer, H. B. 1972. *J. Biol. Chem.* 247:7519–27
129. Scanu, A. M., Edelstein, C., Vitello, L., Jones, R., Wissler, R. 1973. *J. Biol. Chem.* 248:7648–52
130. Edelstein, C., Lim, C. T., Scanu, A. M. 1973. *J. Biol. Chem.* 248:7653–60
131. Edelstein, C., Noyes, C., Scanu, A. M. 1974. *FEBS Lett.* 38:166–70
132. Edelstein, C., Scanu, A. M. Personal communication
133. Scanu, A. M., Edelstein, C., Wolf, R. H. 1974. *Biochim. Biophys. Acta* 351:341–47
134. Blaton, V., Vercaemst, R., Vandecasteele, N., Caster, H., Peeters, H. 1974. *Biochemistry* 13:1127–35
135. Scanu, A. M. 1970. *Biochim. Biophys. Acta* 200:570–72
136. Jackson, R. L., Morrisett, J. D., Pownall, H. J., Gotto, A. M. Jr. 1973. *J. Biol. Chem.* 248:5218–24
137. Jackson, R. L., Gotto, A. M. 1972. *Biochim. Biophys. Acta* 285:36–47
138. Scanu, A. M., Lim, C. T., Edelstein, C. 1972. *J. Biol. Chem.* 247:5850–55
139. Lux, S. E., John, K. M., Fleischer, S., Jackson, R. L., Gotto, A. M. Jr. 1972. *Biochem. Biophys. Res. Commun.* 49:23–29
140. Jackson, R. L., Gotto, A. M. Jr., Lux, S. E., John, K. M., Fleischer, S. 1973. *J. Biol. Chem.* 248:8449–56
141. Ritter, M. C., Kruski, A. W., Scanu, A. M. 1974. *Fed. Proc.* 33:2046 (Abstr.)
142. Müller, K., Laggner, P., Kratky, O., Kostner, G., Holasek, A., Glatter, O. 1974. *FEBS Lett.* 40:213–18
143. Laggner, P., Kratky, O., Kostner, G., Sattler, J. Holasek, A. 1972. *FEBS Lett.* 27:53–57
144. Scanu, A., Reader, W., Edelstein, C. 1968. *Biochim. Biophys. Acta* 160:32–45
145. Laggner, P., Müller, K., Kratky, O., Kostner, G., Holasek, A. 1973. *FEBS Lett.* 33:77–80
146. Haschemeyer, R. H., Fairclough, G. F. 1974. *Fed. Proc.* 33:2041 (Abstr.)
147. Laggner, P., Müller, K., Kratky, O., Kostner, G., Holasek, A. 1973. *Conference on Serum Lipoproteins, Graz, Oct. 21–23.* B7 (Abstr.)
148. Shipley, G. G., Atkinson, D., Scanu, A. M. 1972. *J. Supramol. Struct.* 1:98–104
149. Atkinson, D., Davis, M. A. F., Leslie, R. B. Personal communication
150. Phillips, M. C., Finer, E. G., Hauser, H. 1972. *Biochim. Biophys. Acta* 290:397–402

151. Glonek, T., Henderson, T. O., Kruski, A. W., Scanu, A. M. 1974. *Biochim. Biophys. Acta* 348:155–61
152. Camejo, G. 1969. *Biochim. Biophys. Acta* 175:290–300
153. Ashworth, L. A. E., Green, C. 1963. *Biochem. J.* 89:561–64
154. Kornberg, R. D., McConnell, H. M. 1971. *Biochemistry* 10:1111–20
155. Scandella, C. J., Devaux, P., McConnell, H. M. 1972. *Proc. Nat. Acad. Sci. USA* 69:2056–60
156. Day, C. E., Levy, R. S. 1969. *J. Theor. Biol.* 23:387–99
157. Kamat, V. B., Lawrence, G. A., Barratt, M. D., Darke, A., Leslie, R. B., Shipley, G. G., Stubbs, J. M. 1972. *Chem. Phys. Lipids* 9:1–25
158. Margolis, S., Langdon, R. G. 1966. *J. Biol. Chem.* 241:485–93
159. Assmann, G., Brewer, H. B. Jr. 1974. *Proc. Nat. Acad. Sci. USA* 71:1534–38
160. Segrest, J. P., Jackson, R. L., Morrisett, J. D., Gotto, A. M. 1974. *FEBS Lett.* 38:247–53
161. Schneider, H., Morrod, R. S., Colvin, J. R., Tattrie, N. H. 1973. *Chem. Phys. Lipids* 10:328–53
162. Gustafson, A., Alaupovic, P., Furman, R. H. 1966. *Biochemistry* 5:632–40
163. Scanu, A. M., Edelstein, C. 1971. *Anal. Biochem.* 44:576–88
164. Sodhi, H. S., Gould, R. G. 1967. *J. Biol. Chem.* 242:1205–10
165. Hirz, R., Scanu, A. M. 1970. *Biochim. Biophys. Acta* 207:364–67
166. Jackson, R. L., Baker, H. N., David, J. S. K., Gotto, A. M. 1972. *Biochem. Biophys. Res. Commun.* 49:1444–51
167. Perutz, M. F., Kendrew, J. C., Watson, H. C. 1965. *J. Mol. Biol.* 13:669–78
168. Jackson, R. L., Morrisett, J. D., Gotto, A. M., Segrest, J. P. *Mol. Cell. Biochem.* In press
169. Baker, H. N., Gotto, A. M., Jackson, R. L. Unpublished results

COOPERATIVE INTERACTIONS OF HEMOGLOBIN

×880

Stuart J. Edelstein

Section of Biochemistry, Molecular, and Cell Biology, Cornell University, Ithaca, New York 14850

CONTENTS

INTRODUCTION

Hemoglobin has been studied actively since the 19th century and reviewed on numerous occasions, including an article in this series in 1970 (1). Progress in understanding the molecule continues at a rapid pace and a number of important developments have occurred in recent years which warrant review at this time, including: (*a*) a more detailed understanding of the molecular structure from X-ray crystallographic studies, with direct implications for the mechanism of cooperative binding of oxygen by hemoglobin; (*b*) a recognition of significant differences in the functional properties of the individual α and β chains of hemoglobin; (*c*) the demonstration that isolated $\alpha\beta$ dimers, products of dissociation of the tetrameric $\alpha_2\beta_2$ hemoglobin molecule, are actually devoid of cooperativity and not highly cooperative as believed at the time of the last review; and (*d*) advances in the application of quantitative models, based on new physical-chemical studies, to describe the various features of hemoglobin reactions. The restrictions in length imposed on a review in these volumes prohibit a full exposition of all relevant topics. Therefore, this review focuses on the central and long-standing

209

question that dominates hemoglobin research: What is the structural basis for "heme-heme" interactions, i.e. the cooperative binding of oxygen and other ligands by hemoglobin? Following a discussion of the major conceptual formulations of cooperative interactions, those recent studies that bear on elucidating the mechanism of cooperativity in mammalian hemoglobin will be emphasized.

More complete descriptions of earlier work and background information can be found in existing reviews (1–9), including an excellent historical treatment that has recently appeared (10). An important role has been played by mutants of human hemoglobin in elaborating structure-function relationships, and selected examples are drawn from this source as they relate to individual topics. However, no attempt is made to summarize the vast amount of work that has been done on mutant forms, much of which has been amply reviewed (1, 6, 11, 12). The special problems associated with sickle cell hemoglobin were summarized in June 1974 at the first National Symposium on Sickle Cell disease, and the proceedings of this meeting are to be published.

FORMULATIONS OF THE PROBLEM OF COOPERATIVE LIGAND BINDING

Cooperativity is manifested most directly by the sigmoid curve for the fractional saturation of hemoglobin as a function of the partial pressure of oxygen. In contrast to the sigmoidal behavior, noncooperative binding is reflected by a hyperbolic curve. It is this physiologically important feature, the sigmoid curve, that investigators have for so long attempted to understand; other important properties, such as the Bohr effect and organic phosphate effects, can be viewed as adjustments of the basic phenomenon.

Early Models

Early explanations of cooperative oxygen binding by hemoglobin anticipated the mechanistic alternatives that still confront researchers. The first important mechanism was presented by Hill (13), who postulated that cooperativity arose by a concerted reaction of n binding sites on hemoglobin and could be described by the equation

$$\frac{Y}{1-Y} = Kp^n \qquad\qquad 1.$$

where Y is the fractional saturation, p is the partial pressure of oxygen, and K is an equilibrium constant. Modern data can generally be fit with this equation for values of $Y = 0.1–0.9$ to yield a Hill constant, $n \approx 3$. However, the equation is now only a convenient index of cooperativity; the implications in terms of a mechanism of concerted reaction at n sites cannot be correct, since (a) the number of oxygen binding sites is four (14); (b) the value of n approaches unity as Y approaches 0 or 1 (15); and (c) the kinetics of ligand binding are not consistent with a high order reaction (16).

Another early model of cooperativity was proposed by Douglas, Haldane & Haldane (17) based on a higher degree of aggregation for deoxyhemoglobin than oxyhemoglobin. The finding of Adair (14) of molecular weights ($\sim 65{,}000$) corresponding to four heme-containing units (1 iron/16,000) for both oxy- and deoxyhemoglobin also disqualified this theory. However, the variable aggregation idea has found application for lamprey hemoglobin (18–20) and anticipated the relaxed and constrained states in the theory of Monod et al (21), in which differences in aggregation equilibrium constants are present, although not necessarily revealed in ligand-dependent dissociation, when ligand binding occurs at protein concentrations well above the subunit dissociation constants (see below).

The Adair Scheme

In addition to disqualifying the theories of Hill & Douglas et al, the work of Adair laid the foundation for the modern formulation of the problem of cooperativity by describing a phenomenological equation (22). Since four oxygen molecules are bound per heme, the oxygen binding equilibria can be described by

$$Y = \frac{K_1 p + 2K_1 K_2 p^2 + 3K_1 K_2 K_3 p^3 + 4K_1 K_2 K_3 K_4 p^4}{4(1 + K_1 p + K_1 K_2 p^2 + K_1 K_2 K_3 p^3 + K_1 K_2 K_3 K_4 p^4)} \qquad 2.$$

where the individual equilibrium constants K_i are related to the reaction

$$Hb(O_2)_{i-1} + O_2 \rightleftharpoons Hb(O_2)_i \qquad 3.$$

for $i = 1$–4 and p is the partial pressure of oxygen. This formulation merely describes the system in the broadest terms. The individual binding constants may be determined and efforts along these lines, principally by Roughton and his co-workers (23), revealed that cooperative binding of oxygen is expressed by a value of K_4 some 300 times greater than K_1, when corrected for statistical factors. However, in itself the Adair formation provides no indication of a structural mechanism.

The Sequential Models

The first effort to provide a structural mechanism was a model proposed by Pauling in 1935 (24) which is still relevant with today's knowledge. He argued that the four Adair constants could be expressed in terms of a fundamental oxygen binding equilibrium constant, K', and an interaction constant, α, which describes the stabilization resulting from adjacent oxygen-containing heme units. For a tetrahedral arrangement of hemes, cooperativity arises because the binding of the first oxygen molecule is accompanied by no interheme stabilization, with interactions increasing with each oxygen bound until binding of the fourth oxygen molecule, which is accompanied by three stabilizing interactions. Therefore, the four Adair constants can be expressed as the product of the appropriate statistical factor, the fundamental constant, K', and the interaction constants

$$K_1 = 4K'$$
$$K_2 = \tfrac{3}{2}K'\alpha$$
$$K_3 = \tfrac{2}{3}K'\alpha^2 \qquad 4.$$
$$K_4 = \tfrac{1}{4}K'\alpha^3$$

In this formulation the value of K_4 is greatly enhanced over K' by the added stabilization of heme units. A second related formulation could be derived in which the interactions between oxygen-containing heme units lead to a destabilization. The greatest destabilization occurs with the binding of the first oxygen molecule; the smallest destabilization accompanies the binding of the last oxygen molecule, and cooperativity occurs. In this case, the four Adair constants would be given by the series listed above, each divided by α^3. Thus, mechanisms in which structural factors influence binding energy can be formulated in two alternative ways, which may be called "structural promotion" and "structural constraint." In the case of structural promotion, oxygen binding is facilitated by stronger structural interactions with successive binding steps to give cooperativity. In the case of structural constraint, oxygen is bound while overcoming structural stabilization that diminishes with successive binding steps.

The original formulation of Pauling assumed structural promotion, and this idea was carried through to the modern version of the Pauling model by Koshland, Nemethy & Filmer (25). In contrast, the opposite premise, structural constraint, is an essential feature of the model of Monod et al (21). We now know the structural constraint alternative prevails for hemoglobin. Wyman (2) was the first to conclude that Pauling was incorrect in using structural promotion, on the basis of increased oxygen affinity in the presence of urea, a structure-disrupting agent. More convincing is the finding that α and β chains of hemoglobin have an affinity for oxygen approximately 30 times higher than intact hemoglobin (26). Moreover, the binding constant for chains is in close agreement with the value for the fourth Adair constant. Therefore, the "fundamental" binding properties of hemoglobin sites are exhibited only in the last step. The early steps are reduced in affinity by structural constraint.

Koshland et al (25) extended the Pauling model to other geometric arrangements and other categories of stabilizing interactions. Two distinct conformations, A and B, were assumed with an isomerization equilibrium $K_t = (B)/(A)$. Ligand (S) binding to A induces conformation B, so that the fundamental affinity constant K' of Pauling (24) is given by $K' = K_t K_s$, where $K_s = (BS)/(B)(S)$. The most important addition to the Pauling model was the inclusion of a factor related to stabilizing interactions between mixed pairs. In addition to the relative stabilization of liganded protein over unliganded, K_{BB} (equivalent to the α of Pauling), Koshland et al defined a parameter K_{AB} which gives the relative stabilization arising from type A subunits in contact with type B subunits. This term is very critical to predictions by the model since cooperativity is highly sensitive to its value (27). In the extreme, as K_{AB} goes to zero, the Pauling-Koshland formulation becomes equivalent to the Hill equation. With the terms defined by Koshland et al (25), the Adair constants take the form

$$K_1 = 4(K_s K_t)K_{AB}^3$$
$$K_2 = \tfrac{3}{2}(K_s K_t)K_{BB}K_{AB}$$
$$K_3 = \tfrac{2}{3}(K_s K_t)K_{BB}^2 K_{AB}^{-1}$$
$$K_4 = \tfrac{1}{4}(K_s K_t)K_{BB}^3 K_{AB}^{-3}$$

5.

Since Koshland et al followed the examples of Pauling and continued to define the parameters for hemoglobin in terms of structural promotion, their stabilization constants must be inverted to apply the values to hemoglobin in a physically meaningful way.

The Two-State Model

In 1965 a new formulation of cooperativity was proposed by Monod, Wyman & Changeux (21) which has had a powerful impact on hemoglobin research. The model departed from the earlier sequential ideas and proposed that cooperativity in ligand binding for hemoglobin (or any cooperative proteins) might arise from the existence of just two conformational states. Reasoning from the already identified distinct crystal forms for deoxyhemoglobin and oxyhemoglobin (28), Monod et al noted that the existence of these two conformations was sufficient to generate cooperativity, so long as (a) the two states were freely in equilibrium and differed in affinity for ligand, and (b) the system was governed by structural constraint. Because of the equilibrium between states, structural constraint is required to maintain the protein in the unliganded conformation in the absence of ligand. To reflect this condition Monod et al designated the predominant state for unliganded proteins as T for tense to emphasize structural constraint and the predominant state for liganded proteins as R for relaxed to emphasize the release of constraint during ligand binding (which facilitates binding in the later states of saturation and leads to cooperativity). Since the structural constraint of the T state will be reflected in interactions between subunits, it was designated "quaternary constraint."

The cooperative binding of oxygen by hemoglobin at pH 7.0 in phosphate was described by an intrinsic equilibrium between the T and R states (defined by $L = [T]/[R]$) where $L = 9054$ (21), with the added assumption that the intrinsic affinity of the T state for oxygen is 0.014 times the intrinsic affinity of the R state ($c = K_R/K_T = 0.014$, where K_R and K_T are dissociation constants for the R and T states, respectively). Descriptively, the model indicates that hemoglobin in the absence of ligand is a mixture of molecules in the T and R states, with T predominating about 10,000 to 1. As oxygen is added both the R and T states are populated, but the binding to the R state is more favorable and the $T \rightleftharpoons R$ equilibrium swings toward the R state. The two states are equally populated when the number of ligand molecules bound is equal to $- \log L/\log c$, or about 2 with the parameters of Monod et al (21). When three molecules of ligand are bound, the R state will predominate and the fourth molecule will be bound with an affinity close to K_R. When the population is saturated the R state will predominate, since $L_4 = (T_4)/(R_4)$, which is equivalent to $L_4 = Lc^4 = 4 \times 10^{-4}$, where the subscript indicates moles of ligand bound. As in the case of the sequential models, the two-state model specifies the Adair constants in terms of structural parameters

$$K_1 = 4(1 + Lc)/K_R(1 + L)$$
$$K_2 = \tfrac{3}{2}(1 + Lc^2)/K_R(1 + Lc)$$
$$K_3 = \tfrac{2}{3}(1 + Lc^3)/K_R(1 + Lc^2)$$
$$K_4 = \tfrac{1}{4}(1 + Lc^4)/K_R(1 + Lc^3)$$

6.

and ligand binding can be described by using the Adair equation with the Adair constants defined in this way. However, the two-state formulation leads directly to a simple equation for Y

$$Y = \frac{\alpha(1+\alpha)^3 + Lc(1+c\alpha)^3}{(1+\alpha)^4 + L(1+c\alpha)^4} \qquad 7.$$

where α is the ligand concentration normalized to the dissociation constant of the R state $[\alpha = (X)/K_R]$.

The nomenclature and physical formulation of the two-state model have been widely adopted by hemoglobin researchers in part because of their convenience and in part because the general formulation is consistent with the major observations on hemoglobin, at least to a level of first approximation. These developments arise from certain extensions of the two-state model to the particular properties of hemoglobin. Because of the equilibrium between two states, cooperativity, expressed by the Hill constant n, will be a bell-shaped function of L on a logarithmic scale, as first pointed out by Rubin & Changeux (29). At very low values of L, the R state is present in the absence of ligands, and the $T \to R$ transition, essential for cooperativity, cannot occur. At very high values of L, the T state is so overstabilized that even the addition of ligand does not provide sufficient energy to favor the R state, and again the $T \to R$ transition cannot occur. Thus only at optimal values of L will cooperativity occur, with the maximum cooperativity at $L = c^{-i/2}$ where i is the number of binding sites. The bell curve provides a convenient explanation of the Bohr effect absent in the earlier models. The Bohr effect (alkaline) refers to the increase in affinity with pH (from 7–9) with little or no change in cooperativity. In terms of the two-state model, a decrease in L values in the range corresponding to the top of the bell curve would give just such behavior. However, in order to relate L values and experimental observations, one additional extension was required to fix the value of L from oxygen binding data.

As noted by Edelstein (30), since the isolated chains share many of the physical properties ascribed to the R state (31), the binding of oxygen to hemoglobin can be represented by the reaction

$$T + 4(O_2) \rightleftharpoons R(O_2)_4 \qquad 8.$$

with an apparent overall binding constant that can be expressed as $p_{1/2}$. Similarly the reaction of chains may be represented by

$$R + 4(O_2) \rightleftharpoons R(O_2)_4 \qquad 9.$$

where the apparent binding constant can be expressed by $(p_{1/2})_{\text{chains}}$ and is equivalent to K_R. Solving both equations for L yields

$$L = \alpha_{1/2}^4 \qquad 10.$$

where α is $p_{1/2}/K_R$. From this relationship the L value for any oxygen binding curve can be determined where $1 \ll L \ll c^{-4}$. For pH 7 the value of L is deduced to be at least 3×10^5 (30), considerably higher than the original estimate of Monod et al (21); in addition the correct value of c is probably lower than the original estimate (30). Thus the numbers of molecules in the R and T states become equal only when three molecules of ligand are bound. When various points are located on the bell curve (see Figure 1A), the values of L for pH 7–9 do indeed fall in the region of the top of the bell curve, accounting for the relative invariance of cooperativity in the Bohr effect. Moreover, the behavior of certain affinity mutants can be explained by their positions on the bell curve. The high affinity mutant, hemoglobin Chesapeake (32), and the low affinity mutant, hemoglobin Kansas (33), both of which show low cooperativity, can be explained by their positions on the bell curve. Hemoglobin Chesapeake lies on the left side and is weakly cooperative due to a lack of quaternary constraint. In contrast, hemoglobin Kansas lies on the right side of the bell curve and is only weakly cooperative because it is overconstrained. The decreases in affinity caused by binding of organic phosphates can also be explained by their preferential binding to the T state, which increases L and causes a movement to the right along the bell curve. A further consequence of the two-state formulation is that the value of Y at which the Hill n is a maximum varies between 0.25 and 0.75 in the range of the bell curve (see Figure 1B). Because they lack the "buffering" of cooperativity in the midrange of L values (30) the sequential models (25) are less successful in describing the properties of hemoglobin under a wide range of conditions without arbitrary adjustments of parameters.

While the bell curve at a constant value of c is successful for explaining the general form of hemoglobin behavior under many conditions, a precise descrip-

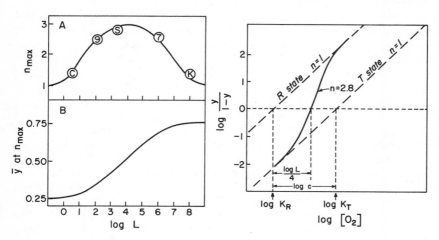

Figure 1 (left)   (*A*) Bell curve of n_{max} vs log L according to (30). (*B*) Variation in Y at n_{max} vs log L according to (167). Symbols refer to values of n_{max} and log L for hemoglobin Chesapeake (C), hemoglobin Kansas (K), normal hemoglobin at pH 7 (7) and at pH 9 (9), and hemoglobin at pH 7 stripped of organic phosphates (S).
Figure 2 (right)   Relationship of the Hill plot to parameters of the two-state model.

tion may require slight changes in c with different conditions. Quaternary constraint leads to stabilization of the T state ($L > 1$) and diminished binding ($c < 1$). The two parameters may be linked in some way so that decreases in L are accompanied by increases in c and vice versa, as some recent data would tend to indicate (34–37). A mechanism for linkage of L and c has been proposed by Szabo & Karplus (38). The distinct effects of L and c can be visualized by considering a Hill plot (Figure 2). The behavior at the extremes relates to the T state at low levels of saturation and the R state at high levels of saturation. If the linear portions at the extremes (with $n = 1$) are extrapolated to a horizontal line at $\log [Y/(1 - Y)] = 0$, the intercepts define $\log K_T$ and $\log K_R$; the distance between the intercepts defines $\log c$ (39). The distance from $\log K_R$ to $\log p_{1/2}$ (observed for the experiment) defines $\frac{1}{4} \log L$. Thus c can be estimated directly along with L if data covering a wide range of Y values are available. In effect the "interaction energy" discussed by Wyman (40) is given by $RT \ln c$. It should be distinguished from the "energy of quaternary constraint" which would be given by $RT \ln L$. Values of L and c can also be determined by kinetic methods (41, 42), and values of L can be determined from subunit dissociation data (43). The relationship of the Hill plot to other models has also been discussed (44).

THE MINIMUM UNIT OF COOPERATIVITY

While the equations cited in the preceding section to describe cooperativity were all based on four oxygen binding sites after the work of Adair, models with fewer sites also give cooperative behavior. Indeed, a dominant idea of the 1960s was the action of heme sites in pairs, such that two-chain dimers were the principal functional units, either as free dimers in solution or as pairs of dimers in intact tetrameric hemoglobin (1, 5, 45, 46). This idea, known as the "dimer hypothesis," arose principally from observations with solutions of salts and other agents that cause dissociation of hemoglobin into subunits. For example, Rossi-Fanelli et al (47) reported dissociation of both oxy- and deoxyhemoglobin largely to dimers in 2 M NaCl on the basis of light scattering and sedimentation data. However, 2 M NaCl had only a slight effect on oxygen binding properties (48). Therefore, the hypothesis was advanced that interactions within dimers are highly cooperative, with a Hill constant near 2 and with only slight cooperativity arising from further interactions between dimers. The fact that the oxygen binding maintained a Hill coefficient near 3, although the hemoglobin was believed to be dissociated into dimeric units containing only two oxygen binding sites, created a complication referred to as the "salt paradox," because the Hill constant can never exceed the number of binding sites (49). The paradoxical aspects of the dimer hypothesis were removed, however, with the observation that dissociation of deoxyhemoglobin is less pronounced in 2 M NaCl than oxyhemoglobin (50, 51), and the dimer hypothesis prevailed.

The first major challenge to the dimer hypothesis arose from kinetic studies. The cooperative binding of ligands to hemoglobin revealed by the sigmoidal oxygen binding curve can be expressed in kinetic terms. For example, the ligand carbon monoxide (CO) binds to deoxyhemoglobin "slowly" ($k \approx 1.5 \times 10^5$ M^{-1} sec^{-1}) in rapid mixing studies; however, if CO-hemoglobin receives a flash of light that photodissociates one CO molecule, it recombines "rapidly" ($k \approx 6 \times 10^6$ M^{-1} sec^{-1}) (4). The flash experiment presumably traps the partially saturated molecule in the R state, whereas the mixing experiment reflects the T state. Cooperativity is also reflected in an acceleration of the apparent rate of CO binding in the mixing experiment (4). With the isolation of individual α and β chains by the p-mercuri-benzoate (PMB) method (52), their high affinity and rapid rate of binding of CO, as well as their spectral properties (26, 53), indicated that the isolated chains are in the R state.

Rapidly reacting material could also be detected in CO-hemoglobin subjected to a large flash that fully dissociated the CO (54). In this case, slowly reacting hemoglobin was the principal product, but a small fraction of rapidly reacting material persisted which was concentration dependent. At concentrations near μM (heme) the rapid component approached 50% (55). Since dimers were believed to be cooperative units (and presumably slowly reacting) and since early studies on hemoglobin with the ultracentrifuge scanner system gave evidence of dissociation of hemoglobin into monomers in very dilute solutions (56), Antonini and co-workers (45, 57, 58) attributed the rapidly reacting material in dilute hemoglobin solutions to free monomers. However, certain difficulties attended this interpretation. Combined flash and flow experiments on hemoglobin and NO-hybrids (59) were incompatible with interactions of subunits within pairs. Also, ultracentrifuge results indicated that dissociation to monomers occurred at concentrations at least an order of magnitude below the concentrations at which rapidly reacting material appeared (27).

The difficulties were resolved in 1969, when Edelstein & Gibson (60) measured the fraction of rapidly reacting material in parallel with dissociation using the ultracentrifuge scanner. The appearance of rapidly reacting hemoglobin was found to correlate closely with dissociation of tetramers into dimers, not formation of monomers. Thus the rapid component was identified as dimeric hemoglobin, presumably $\alpha\beta$ dimers (28), and serious doubt was cast on the dimer hypothesis. In fact, monomers are not produced to any measurable extent under the experimental conditions used (60–63). Identification of dimers as noncooperative raised additional questions about the salt effect. Evidently the extent of dissociation in 2 M NaCl had been overestimated by failure to take into account preferential interactions between the hemoglobin and the water (60, 61, 64). The results of Kellett indicated that after the proper corrections, no dissociation was detectable for deoxyhemoglobin in 2 M NaCl even at concentrations in the μM (heme) range (64). The possibility of dissociation of deoxyhemoglobin at a different plane than for oxyhemoglobin to produce cooperative dimers was briefly considered (62, 65), but trapping of the dimer formed from deoxyhemoglobin by pH jump experiments revealed that it too was rapidly reacting with CO (66). A similar conclusion was reached in studies on the dithionite reaction at low concentrations (67). Hybridization experiments supported the view that oxy- and deoxyhemoglobin dissociate at the same plane (68). Similar experiments have been used to measure the rate constant for dissociation of deoxyhemoglobin tetramers into dimers (69). Finally, it was possible to find conditions where deoxyhemoglobin dissociated into dimers, and noncooperative oxygen binding equilibria were found (70, 71).

The conclusion that tetrameric hemoglobin dissociates to dimers with spectral, kinetic, and equilibrium properties resembling isolated chains or the R state means that as a thermodynamically linked system, hemoglobin and its subunits could be described simply by the scheme

$$T \rightleftharpoons^{L} R \qquad K_{4,2}^{T} \diagdown \quad \Big\Vert K_{4,2}^{R} \qquad 2r \qquad\qquad 11.$$

where $K_{4,2}^{T}$ and $K_{4,2}^{R}$ refer to tetramer-dimer dissociation constants for the T and R states, respectively; L describes the $T \rightleftharpoons R$ equilibrium, $L = (T)/(R)$; and r refers to $\alpha\beta$ dimers in the R state. According to this scheme, quaternary constraint arises entirely from interactions between dimeric units. Since $K_{4,2}^{R}$ can be readily measured and has a value near μM for CO-hemoglobin (60–63, 72, 73), a method of determining $K_{4,2}^{T}$ would permit an independent estimate of L, since

$$L = K_{4,2}^{R}/K_{4,2}^{T} \qquad\qquad 12.$$

A method for determining $K_{4,2}^{T}$ was developed by Thomas & Edelstein (43, 74) based on the variation of the affinity of hemoglobin for CO with hemoglobin concentration. Plots of $\log (CO)_{1/2}$ vs $\log$ (heme) are linear with a slope of 0.25 and yield $K_{4,2}^{T}$ from the intercept. A value of $K_{4,2}^{T} = 3 \times 10^{-12}$ M was obtained for 0.1 M phosphate, pH 7. When combined with values for $K_{4,2}^{R}$, a value of $L = 6.7 \times 10^{5}$ was deduced. Addition of 2 M NaCl produced an increase in $K_{4,2}^{T}$ of about twentyfold, roughly the same increase as is produced in CO-hemoglobin by NaCl (51, 62, 64, 75). Thus the reason 2 M NaCl has little effect on oxygenation, in spite of some increase in the dissociation constant of oxyhemoglobin, is that a similar increase in the dissociation constant of deoxyhemoglobin compensates to leave L largely unaltered. The method of measuring $K_{4,2}^{T}$ has also revealed that the Bohr effect coincides with changes in $K_{4,2}^{T}$ (74). Oxygen affinity increases with pH by destabilizing the T state to decrease L, presumably by breaking salt bridges identified by Perutz (76) and his co-workers (see following section).

Since hemoglobin consists of tetramers in equilibrium with noncooperative dimers (in the R state), the complete description of ligand binding in the formulation of the two-state model takes the form (43)

$$Y = \frac{\alpha(1+\alpha)^{3} + Lc(1+\alpha)^{3} + LK_{4,2}^{T}\alpha(1+\alpha)/2[Hb_{2}]}{(1+\alpha)^{4} + L(1+c\alpha)^{4} + LK_{4,2}^{T}(1+\alpha)^{2}/2[Hb_{2}]} \qquad 13.$$

where $[Hb_{2}]$ refers to the concentration of dimers and can be obtained for any total concentration (heme) from the quadratic equation

$$4[Hb_{2}]^{2}\left\{\frac{(1+\alpha)^{4}}{LK_{4,2}^{T}} + \frac{(1+c\alpha)^{4}}{K_{4,2}^{T}}\right\} + 2[Hb_{2}](1+\alpha)^{2} = (heme) \qquad 14.$$

Evaluation of this equation reveals that for certain systems (high L) cooperativity should increase with dilution of hemoglobin (43), and results on hemoglobin Kansas support this prediction (J. O. Thomas and S. J. Edelstein, unpublished results). This formulation also demonstrates that L can be altered by changes in either $K_{4,2}^R$ or $K_{4,2}^T$. For example, the Bohr effect involves changes in $K_{4,2}^T$, while several other systems involve changes in $K_{4,2}^R$. Cat hemoglobin has a value of L about tenfold higher than most other mammalian hemoglobins due to a tenfold higher value of $K_{4,2}^R$ (37, 77). Similarly, hemoglobin Kansas has an elevated L value due to a high $K_{4,2}^R$, about 10^{-4} M (33). In contrast, hemoglobin Chesapeake has an unusually low L value due to a low $K_{4,2}^R$ (78). These differences in dissociation can also be related to the binding of haptoglobin (79) since haptoglobin reacts only with $\alpha\beta$ dimers (80).

DEDUCTIONS FROM STRUCTURAL STUDIES

The work of Perutz and his associates has now revealed a great deal of information on the precise arrangement of atoms in horse and human deoxyhemoglobin and oxyhemoglobin (actually methemoglobin, since the crystals are oxidized during the measurements) as well as many mutant forms and chemically altered variants (see 76 and 81–83, which review many aspects of this large body of work). Since these papers are synoptic and interpretive, only the main features of the deductions will be reviewed here. However, it should be noted that since the conclusions are based on comparisons of the structures of deoxyhemoglobin and methemoglobin, whose interconversion is only weakly cooperative, some important modifications of the conclusions may be required when the comparison of deoxyhemoglobin to an Fe^{2+} liganded form, such as CO-hemoglobin, is available. The basic difference between the T (deoxy) and R (oxy-met) structures deduced from low resolution data (84) has been supported by data at higher resolution (85, 86). The $T \leftrightarrow R$ transition involves a rotation of $\alpha\beta$ units with respect to one another that maintains the α_1–β_1 interface but alters the α_1–β_2 interface, with the most marked rearrangement involving an increase in distance between the β chains upon deoxygenation (see also 87). It is the separation of β chains that accommodates organic phosphates (88, 89) to depress oxygen affinity by stabilizing the T state (8, 90). Interaction of 2,3-diphosphoglycerate (DPG) with β chains had been proposed from studies on chemically modified hemoglobins (91, 92). Some additional Bohr effect is also caused by DPG binding (93–96).

The conformation of the T state permits formation of several salt bridges (76) with the C-terminal residues, Arg 141α and His 146β, participating. These salt bridges provide the proton binding sites in the T state that produce the Bohr effect. The involvement of the residues has been demonstrated by the measurements of Kilmartin and his co-workers on the oxygen binding properties and Bohr effect of carboxypeptidase-digested hemoglobins (97–99) and carbamylated hemoglobin (100). The later derivatives have also permitted a description of the mechanisms of CO_2 binding (101) that has greatly clarified the physiological processes involved (9). Since the salt bridges in the T state stabilize the protonated forms, transition to the R state breaks the salt bridges, the relevant pK values are lowered, and protons are released. Similarly at high pH the salt bridges are broken by deprotonation even in the T state, accounting for the increased affinity for oxygen with increasing pH (alkaline Bohr effect).

Among other important structural features of hemoglobin is the two-state arrangement of the $\alpha_1-\beta_2$ interface. The contact is "dovetailed" (82) so that a hydrogen bond linking Tyr 42α and Asp 99β in the T state is replaced by a bond between Asp 94α and Asn 102β in the R state. The position of the iron with respect to the heme plane has also come under close scrutiny. Work of Hoard and his associates on model compounds (102, 103) led them to propose that high spin iron, as occurs in deoxyhemoglobin and aquo- and fluoro-methemoglobin, would be too large in radius to be accommodated within the heme plane; low spin iron, as occurs in oxyhemoglobin and cyanide- and azide-methemoglobin would be reduced in radius and could be accommodated within the heme plane. This situation has been partially verified for the iron atoms in aquo-met hemoglobin, which are 0.3 Å from the heme plane (76), and measurements on deoxyhemoglobin indicate a displacement of the iron from the heme plane of 0.75 Å (76).

Apart from the structural data, Perutz (76, 81) has made a number of interpretive proposals on the detailed mechanisms of oxygen binding. While they are of heuristic value, not all are in agreement with experimental evidence. The proposals can be divided into two areas: (a) the sequence of events in the binding of ligands to the α and β chains and their relation to the breaking of the salt bridges and the $T \rightarrow R$ transition, and (b) the role of out-of-the-plane to planar transition of the iron upon oxygenation as a trigger in the $T \rightarrow R$ transition.

Concerning the first area, Perutz (76) argues that the sequential events in oxygenation are connected by a structural interaction involving the four iron atoms and the four penultimate tyrosine residues. The binding of oxygen is postulated to occur first to the hemes of the α chains and coincides with expulsion of the penultimate α chains' tyrosines from pockets between the F and H helices with the attendant breaking of the salt bridges of α chains and release of Bohr protons. According to Perutz, binding occurs first to the hemes of the α chains because access to the hemes of the β chains in deoxyhemoglobin is restricted by the γ methyl group of Val $\beta67$. As each subunit binds ligand, a tertiary conformational change (designated $t \rightarrow r$) is postulated to occur to give an R-like subunit in the T quaternary structure. When several subunits have bound ligand, a transition to the R quaternary structure occurs with a release of organic phosphates and additional Bohr protons. The binding of ligands to the β chain occurs with breakage of the internal salt bridges of the β chain. It now appears highly unlikely that all ligand binding occurs with precisely this sequence of events, although some elements of the proposal may be applicable, as will be apparent from the physical-chemical studies discussed in the remainder of this review. In addition Szabo & Karplus (38) start with the same assumptions as Perutz but arrive at different conclusions concerning the sequence of binding events.

Concerning the role of the iron, Perutz (81) argues that the diminished affinity of the T state for ligands is related to a reciprocal effect between spin state of the iron and interactions with the globin mediated by the imidazole ring of His F8. The basic argument has been related to experiments on methemoglobin (104–107) in which an $R \to T$ transition may occur upon addition of inositol hexaphosphate (IHP), the organic phosphate most effective in stabilizing the T state (8). Spectral changes accompanying addition of IHP to methemoglobin are interpreted as reflecting an increase in the high spin character of the iron that coincides with the $R \to T$ transition. Thus the conformation of the T state is visualized as applying a tension at the heme irons which heightens their high spin character and contributes to the diminished affinity of the T state for ligands. When a ligand is bound, the iron is transformed to a low spin planar form which, Perutz argues, triggers a $t \to r$ conformational change in that subunit. This hypothesis is also supported by the studies of Banerjee and his co-workers on ligand binding to certain valence hybrids (108) and methemoglobin (109); a quantitative analysis of these studies has been presented by Szabo & Karplus (110). However, recent results by Hensley et al (111) are at variance with these conclusions. Many indices of conformation such as tetramer-dimer dissociation and SH reactivity are found to be perturbed by IHP, but in a manner largely independent of spin state. Moreover studies by Edelstein & Gibson (112) on the effect of IHP on the redox reaction of hemoglobin indicate that a major consequence of binding IHP is the enhancement of α–β chain nonequivalence in redox potential. In addition, a quantitative analysis of the effects of the redox reaction which incorporates chain nonequivalence indicates that only a fraction of the molecules of methemoglobin in the presence of IHP are in the T state (112). On the basis of these results, the movement of the iron is not considered to be a major factor in the $T \to R$ transition, so there is at least some degree of uncertainty concerning the validity of the Perutz hypothesis on the role of spin state changes in the iron as a trigger of the conformational transition. Because work on this subject is not yet sufficiently definitive to permit a succinct summary and space limitations preclude a discussion of all the experimental evidence and interpretations, a full exposition of this topic must await a future review. The papers already cited (104–112) discuss the issues in considerable detail and little further clarification can be added at this time.

DETECTION OF THE $T \leftrightarrow R$ TRANSITION

Evidence for a $T \leftrightarrow R$ equilibrium in hemoglobin began accumulating with Gibson's observation (54) of the fast reacting form, referred to as Hb*, in partial flash experiments on carboxyhemoglobin. The rate of CO binding to Hb* is 30–40 times higher than the rate of CO combination with deoxyhemoglobin in rapid mixing experiments, and the spectrum of Hb* is depressed and broadened in the Soret region. Spectral and kinetic properties similar to Hb* were obtained for isolated α and β chains of hemoglobin (26), suggesting that the Hb*-type properties represented an unliganded or deoxy R state. This line of reasoning was supported by studies on carboxypeptidase-treated hemoglobins which also share the spectral and functional properties of the R state in the deoxy form (in the absence of organic phosphates), but which can be switched towards a normal T state deoxy form (typical Soret spectrum, reduced combination velocity with CO, cooperativity in oxygen binding) by addition of IHP (104, 113–117). While a deoxy T state can be produced in these variants in the presence of organic phosphates, the loss of a salt bridge lowers the L value, as seen by higher than normal oxygen affinity. A higher than normal rate of CO combination is also observed and cooperativity is pH dependent, decreasing with increasing pH (96, 98, 113), corresponding to the properties associated with the left side of the bell curve of n vs log L (Figure 1). Broadly similar findings are obtained with mutant forms of hemoglobin such as hemoglobin Hiroshima (118), hemoglobin Bethesda (119), hemoglobin Kempsey (120), and hemoglobin Chesapeake (121).

In the case of hemoglobin Kansas, a low affinity variant, addition of IHP appears to maintain the molecule in the T state, as revealed by NMR studies of Ogawa et al (122). Hopfield et al (123) attempted to explain its ligand binding kinetics on the basis of this observation, although the situation appears to be complicated by specific effects of the mutation on chain heterogeneity reported by Gibson et al (124). In addition, organic phosphates would be expected to perturb the tetramer-dimer equilibrium of hemoglobin according to the hypothesis of Ogawa and Hopfield et al (122, 123), although Gibson et al (124) reported that no such effect was observed. A thirtyfold reduction in $K_{4,2}$ was observed upon addition of IHP ($K_{4,2}$ changes from 4×10^{-4} M to 1.5×10^{-5} M) by Hensley and Edelstein (unpublished results) in sedimentation equilibrium experiments with an online computer system (73, 125), but only in 0.3 M bis-tris buffer, the conditions of the NMR experiments. In buffers used by Gibson et al (124) the effect of IHP is negligible. Thus the various results may simply reflect different conditions and not a fundamental discrepancy.

Work with valence hybrids in which either the α or β chains are prepared in the ferric form (126, 127) has also supported the view of an organic phosphate-dependent T-R equilibrium. Ogawa & Shulman (128) identified NMR peaks characteristic of either oxy- or deoxyhemoglobin that predominated in the valence hybrids in the absence and presence of phosphates, respectively (see also 129–131). Moreover, the addition of organic phosphates produced a slow change from a mixture of slowly and rapidly reacting components in CO combination experiments to slow material only, indicating a $R \rightarrow T$-type transition (132, 133). Similar conclusions have been reached by Lindstrom et al (134) with M-type hemoglobin variants in which one chain is naturally in the ferric form [see review by Ranney et al (135)]. Evidence for a T-R equilibrium permits an explanation of the effect of organic phosphates on valence hybrids without recourse to sequential models as had been proposed (136).

Thus the presence of an $R \leftrightarrow T$ transition has been demonstrated in many hemoglobin variants, and evidence has also been obtained in normal hemoglobin. Cassoly & Gibson (137) have recently demonstrated the production of a form that is fast reacting with CO in experiments on binding of NO and CO mixtures to hemoglobin. Earlier, combined flash photolysis and stopped flow mixing experiments of Gibson & Parkhurst (59) revealed that the fast reacting form produced with a partial flash appears after about three molecules of CO are bound. Development of a kinetic formulation by Hopfield et al (41) of the two-state model indicated that this behavior is in good agreement with the predictions of the model. [Other deductions concerning oxygen binding are complicated by chain heterogeneity (138) discussed below.] In addition, the release of organic phosphate (which binds preferentially to the T state) upon ligand binding occurs with a lag with respect to CO binding, as determined in studies with fluorescent analogs (139, 140). Release is earlier in variants with higher affinity, where lower L and earlier $T \rightarrow R$ transitions are expected (140). The binding of p-mercuribenzoate to the reactive SH group of hemoglobin, β93, also appears to coincide with the $T \rightarrow R$ transition (141). Earlier workers who concluded a coincidence of p-mercuribenzoate binding and ligand binding (142) were misled by the narrow range of conditions examined.

Since salt bridges are present in the T state and absent in the R state, a release of protons associated with the salt bridges (the origin of the alkaline Bohr effect) could be expected to accompany the $T \rightarrow R$ transition. However, pivotal kinetic studies of Antonini et al (143) and Gray (144) indicate a linear release of protons with CO binding, not a lag as would be expected for a parameter that reflects the $T \rightarrow R$ transition. These findings have had a major impact on formulations of cooperativity and are at least partially responsible for Perutz's (76) proposition of a $t \rightarrow r$ transition for each subunit as ligand is bound or a salt bridge is broken. Recently a new light has been cast on this subject by Olson & Gibson (145), who find that proton release does lag behind CO binding when IHP is present.

EVIDENCE FOR NONEQUIVALENCE OF THE α AND β CHAINS

Basic Observations

An important recent development in hemoglobin research is the discovery of significant differences in the ligand binding properties of the α and β chains, principally by Gibson and his co-workers. The first evidence for chain non-equivalence came in studies on methemoglobin (146, 147). For example, the reaction of methemoglobin with azide is about sixfold more rapid with β chains than α chains, with the faster reacting chain identified by spectral studies on isolated chains and intact hemoglobin. Similar results are obtained for the dithionite reaction. Nitrite and thiocyanate also react more rapidly with β chains, although α and β rates are about the same for cyanide and fluoride (146). Chain non-equivalence for imidazole has also been reported (148, 149). The first major α–β difference for deoxyhemoglobin was found for n-butylisocyanide (BIC) by Olson & Gibson (150). This ligand binds to and dissociates from β chains more rapidly than α chains (151, 152). Identification of the chains was made principally by (a) a comparison of properties with isolated chains and p-mercuribenzoate-reacted β chains (151), (b) NMR studies on partially saturated mixtures (153) which indicated preferential disappearance of a resonance identified with β chains in deoxyhemoglobin (154), and (c) binding to valence hybrids (133, 152). While α and β properties are almost equal at pH 9 or at pH 7 in low ionic strength, addition of salt or organic phosphates leads to the preferential binding of BIC to β chains, principally through a reduction in the rate of binding to α chains (155). In the presence of IHP, chain differences are so great that cooperativity is abolished. Excellent agreement with these kinetic studies was obtained by temperature jump relaxation methods (156). The conclusion of preferential binding of BIC to β chains has been challenged by Huestis & Raftery (157) on the basis of NMR measurements with a ^{19}F-trifluoroacetone derivative at the β93 position, but it may be possible to reconcile the disparate observations (see below).

Concerning more traditional ligands, some chain differences in CO binding in the presence of organophosphates were detected by Gray & Gibson (158). Studies on the valence hybrids suggest that in this case α chains represent the rapid component (133). Recently Gibson (138) has demonstrated that oxygen binding kinetics closely parallel those of BIC. For example, in oxygen pulse experiments in which a solution of oxygen is mixed with a solution containing hemoglobin and dithionite, the oxygen binds to hemoglobin but dissociates from partially saturated intermediates due to combination with dithionite. Under these conditions oxygen binds to and dissociates from β chains rapidly. The rate constants for the α chains are so low as to suggest that oxygen binds almost exclusively to β chains in the T state, although the affinity of α chains for oxygen may actually be higher than for β chains due to a compensating dissociation rate (138). This scheme should also bear an important relationship to temperature jump relaxation studies on the kinetics of oxygen binding to hemoglobin (159). Chain differences in oxygen binding affinity were also suggested by Ogata & McConnell (160–163) on the basis of spin-label studies. NMR studies of Lindstrom & Ho (164) indicate a preferential binding of oxygen to α chains. Studies by Henry & Cassoly (165) on NO binding indicate a chain heterogeneity in which NO binds preferentially to α chains (see also 166).

The discovery of conditions under which chain differences are readily apparent has also permitted a more detailed examination of the release of Bohr protons (145). When BIC binding and proton release are compared under conditions where binding to β chains is much more rapid than binding to α chains, about 20% of the proton release is associated with rapid binding to β chains, whereas 80% of the proton release is associated with the slower binding to α chains. This observation appears to be in conflict with the finding of Kilmartin et al (97, 100) that 50% of the Bohr effect is associated with the imidazole groups of His 146β. However, greater association of protons with the α chains is consistent with a larger Bohr effect in $\alpha_2^{II} \beta_2^{III}$ valence hybrids (126), although Brunori et al (127) did not observe this difference. In addition, a larger Bohr effect for M-type variants with a ferrous α chain is observed than for the variants with a ferrous β chain (135). However, since experiments with solutions containing IHP indicate that proton release is coupled to the $T \rightarrow R$ transition (145), the 20% proton release associated with rapid binding to the β chains may simply reflect the extent to which a $T \rightarrow R$ transition occurs with the binding of two molecules of BIC (167). Although the proton release appears to be accommodated by this explanation, there is still some discrepancy concerning release of a fluorescent 2,3-diphosphoglycerate analog, which lags behind ligand binding to a greater extent than proton release. Thus a completely self-consistent explanation of the time course of proton release and organic phosphate release is not yet apparent.

Consequences of Chain Nonequivalence

With the recognition that distinct properties of the α and β chains must be taken into account for a complete description of ligand binding, the binding expression of the two-state model, equation 7, must be expanded to include contributions of each chain. If a, a normalized binding parameter for the R state, is reserved for α chains $[a = (X)/K_R^\alpha]$, a companion term, b, can be defined for the β chains in the R state $[b = (X)/K_R^\beta]$. The properties of the α and β chains in the T state can then be expressed in terms of c_α and c_β, where $c_\alpha = K_R^\alpha/K_T^\alpha$ and $c_\beta = K_R^\beta/K_T^\beta$. With these terms it is possible to derive distinct binding expressions for the α and β chains, Y_α and Y_β, respectively (112, 167)

$$Y_\alpha = \frac{a(1+a)(1+b)^2 + Lc_\alpha a(1+c_\alpha a)(1+c_\beta b)^2}{(1+a)^2(1+b)^2 + L(1+c_\alpha a)^2(1+c_\beta b)^2} \qquad 15.$$

$$Y_\beta = \frac{b(1+b)(1+a)^2 + Lc_\beta b(1+c_\beta b)(1+c_\alpha a)^2}{(1+a)^2(1+b)^2 + L(1+c_\alpha a)^2(1+c_\beta b)^2} \qquad 16.$$

The complete binding properties are then described by

$$Y_{\text{total}} = (Y_\alpha + Y_\beta)/2 \qquad 17.$$

Equation 17 is formally equivalent to an equation derived by Ogata & McConnell (160) and a generating function of Szabo & Karplus (38) but has the advantage that α and β saturation are separated so that each can be evaluated independently, as is required in certain types of analysis (112, 147, 167). Introduction of α–β non-equivalence also alters the interpretation of the asymptotes of the Hill plot (Figure 2) and the equation for L (equation 10).

Since α–β differences in affinity for ligand can occur in either the T or the R state, it is important to determine which state is responsible for any given observation of chain differences. For conditions where a high value of L applies (such as in the presence of organic phosphates for hemoglobin), only chain differences in the T state can give rise to preferential binding to one of the chains in partially saturated solutions (167). This situation arises from the fact that at high L values the T state exists in various degrees of saturation, and chain differences can be revealed. Since the R state only predominates after about three of the four ligand molecules are bound, the possibility of a range of degrees of saturation for the R state in which α–β differences could be revealed does not exist. (At the other extreme, low values of L, only chain differences in the R state can be revealed; depending on the nature of the experiment, kinetic measurements may reveal chain differences in either state.) Thus, the preferential binding of oxygen to α chains in the presence of IHP as determined by NMR measurements (164) must reflect a higher affinity for oxygen of α chains compared to β chains in the T state.

In the case of BIC, although the combination velocities are similar to oxygen, the NMR data indicating preferential binding to β chains (153) must be interpreted as preferential binding to the β chains in the T state. This view is supported by studies of McDonald, Hoffman & Gibson (168) on manganese-iron hybrids in which the manganese chain is effectively locked in the T state (169). With $\alpha^{Fe^{2+}}\beta^{Mn^{2+}}$ only extremely slow binding occurs with BIC. Thus, hemes of the α chains in the T state appear to be relatively inaccessible to BIC. Even the behavior of stripped hemoglobin in BIC binding can be explained without recourse to binding of α chains in the T state. In this case the binding data are highly cooperative and the kinetics relatively homogeneous (155) due to the fact that the transition to the R state occurs at low levels of saturation and the properties of the R state (in which chain differences are relatively minor) dominate the behavior (167). These explanations of chain differences can also be extended to the observations of Huestis & Raftery (157) which were interpreted in terms of preferential binding of BIC to α chains. Examination of the parameters of the two-state model corresponding to the conditions of the measurements (167) indicates that what was interpreted as an indicator of binding to β chains (changes in the NMR signal of ^{19}F-trifluoroacetone attached to the β93 position) is more likely a monitor of the $T \rightarrow R$ transition, i.e. an indicator of the state function, $\bar{R}$, (167) where

$$\bar{R} = \frac{(1+a)^2(1+b)^2}{(1+a)^2(1+b)^2 + L(1+c_\alpha a)^2(1+c_\beta b)^2} \qquad 18.$$

Therefore, the fact that changes in the NMR signal are not linear with saturation is simply an indication that the L value is high and the transition to the R state occurs when about three molecules of ligand are bound.

GENERAL CONCLUSIONS AND CURRENT ISSUES

The availability of structural models for both deoxyhemoglobin and liganded hemoglobin, together with evidence that indicates a two-state model with α–β differences as the basic mechanism for cooperative oxygen binding, have brought hemoglobin research to the point where questions are now phrased in precise physical-chemical terms. One current concern is the extent to which energies reflected in the values of L and c can be identified with particular structural elements in the hemoglobin molecule. Hopfield (170) has suggested three ways in which the affinity of the T state for ligand could be reduced compared to the R state: (*a*) a "direct bond" model in which a chemical interaction of the iron in the T state opposes ligand binding; (*b*) an "indirect bond" model in which energy at a bond at some distance from the iron is set in opposition to ligand binding; and (*c*) a "distributed model" in which many low energy contacts influence binding. Hopfield favors the third alternative and extends it to include a formulation in which interaction energy may vary linearly with the displacement of the iron from the heme. In the mechanism detailed by Perutz (76, 81) involving the breaking of salt bridges accompanying ligand binding, the salt bridges would represent an indirect bond, the second alternative of Hopfield (170). Perutz proposed this mechanism partly to account for the linear release of Bohr protons with ligand binding (143, 144), but the studies already discussed on solutions containing IHP (145) in which a lag is observed cast doubt on the general validity of the coupling of ligand binding and proton release.

Szabo & Karplus (38) have extended the Perutz point of view to a quantitative analysis of ligand binding. The model is based on a set of intrinsic ligand binding constants for α and β chains which are potentiated by coupling to salt bridges to give rise to the low affinity of deoxyhemoglobin. The model assumes a tertiary transition that includes release of Bohr protons with ligand binding and thus may be difficult to reconcile with data indicating a lag in proton release in the presence of IHP (145). These kinetic experiments thus take on major significance. Several other interpretive papers and new formulations of cooperativity have also appeared recently (171–175).

Although the kinetics of CO binding can be accommodated by a kinetic version of the two-state model to a first approximation (41), small deviations in the constants for the first to third binding events have consistently appeared (59, 139, 140) and warrant further investigation. The two-state model therefore provides a more nearly perfect representation for mutant forms with a low L value where high binding rates of the R state apply after the slow CO binding rate for the first site of the T state (118–121). Kinetic formulations of the two-state model for oxygen binding where α–β differences must be included have not yet been reported. In the treatment by Hopfield et al (41), the $T \leftrightarrow R$ transition was assumed to be fast compared to ligand binding. However, studies on variants with a low value indicate that the $T \leftrightarrow R$ transition is relatively slow (117, 133). Since a value of L near unity applies to the $T_3 \leftrightarrow R_3$ transition, a rate-limiting conformational transition might be involved for binding of certain ligands to normal hemoglobin.

The structural reason for the disparate behavior of different ligands remains obscure and deductions from structural models do not appear adequate to explain the phenomena. One intriguing pattern that has emerged is that for the ligands NO, CO, O_2, and BIC, the higher the affinity of the ligand for hemoglobin, the greater the discrimination in favor of α chains over β chains in the binding rates (133, 138, 145, 150, 158, 165).

Ligand Affinity	Relative Binding Rates
(highest to lowest)	(for deoxyhemoglobin)
NO	$\alpha \gg \beta$
CO	$\alpha > \beta$
O_2	$\alpha < \beta$
BIC	$\alpha \ll \beta$

Some understanding of the details of ligand binding may be realized in the mechanistic studies of Austin et al (176). Interest in the mechanism of ligand binding has also been stimulated by recent studies of photosensitivity of CO binding (177–180). The work of Alpert et al (181) with a laser flash system has revealed a very rapid conformational change in the nanosecond range that may require incorporation in any complete kinetic scheme.

Hemoglobin also provides an ideal system for the development and testing of structural principles with broad implications for protein chemistry. A precise description of the energetics of the interfaces should be an ideal testing ground for potential functions related to amino acid side chain interactions. At present, there is not even a satisfactory description of why isolated β chains associate to the tetramer level while isolated α chains remain unassociated. However, Tainsky & Edelstein (182) have provided a tentative explanation for the unusual stability of β_4 tetramers in strong salt solutions (183) based on particular interactions at the subunit interfaces. Perutz (184) has recently offered an attractive explanation for species differences in alkaline denaturation rates based on differences at the inter- faces. Marked differences in the aggregation properties of individual α and β globins exist (185) which have not yet been explained in structural terms. Molecules with heme on only one type of chain, semihemoglobins, provide a rich repertoire of properties, depending on which chain contains the heme and the method of preparation [see review by Cassoly & Banerjee (186)]. The structural principles responsible for these varieties have not yet been revealed. Studies on porphyrin- globin preparations (187–190) and modified porphyrins (6) also provide interesting structural parameters that must be incorporated in any comprehensive description of hemoglobin. The spin-labeled heme of Asakura (191) may be particularly interest- ing in this regard. Recent studies on porphyrins in which the iron is replaced by other metals, including manganese (192–194), cobalt (195–200), and zinc (201), exhibit distinctive features that are likely to contribute important information for a full description of the properties of hemoglobin. Improved physical-chemical approaches also promise to open new avenues of investigation for hemoglobin. Tritium exchange studies by Englander and his colleagues (202, 203) are providing a unique perspective on the conformational dynamics of hemoglobin (see also 204). Improvements are occurring in calorimetric measurements during ligand binding (205, 206). Resonance Raman spectroscopy is emerging as a very powerful method for heme proteins (207–209). Yamamato et al (209) have interpreted their Raman spectra of oxyhemoglobin in terms of iron that is formally in a low spin ferric state. Important variations and new principles may also emerge from studies on non- mammalian hemoglobins. What is certain about the trends in improved methodology is that in spite of the large body of information already accumulated, the ramifications of hemoglobin research are likely to continue to actively engage investigators for some time.

ACKNOWLEDGMENTS

I wish to thank Drs. Q. H. Gibson, J. K. Moffat, C. P. Hensley, Jr., and M. Karplus for their helpful discussions and comments on this review and Ms. Jean Shriro for valuable assistance in preparation of the review. This work was supported by an Alfred P. Sloan Research Fellowship and grants from the National Institute of Health (HL-13591-04) and the National Science Foundation (GB-40529).

Literature Cited

1. Antonini, E., Brunori, M. 1970. *Ann. Rev. Biochem.* 39:977–1042
2. Wyman, J. 1948. *Advan. Protein Chem.* 4:407–531
3. Wyman, J. 1964. *Advan. Protein Chem.* 19:223–86
4. Gibson, Q. H. 1959. *Progr. Biophys. Biophys. Chem.* 9:1–54

5. Rossi-Fanelli, A., Antonini, E., Caputo, A. 1964. *Advan. Protein Chem.* 19 : 73–222
6. Antonini, E., Brunori, M. 1971. *Hemoglobin and Myoglobin in their Reactions with Ligands.* Amsterdam: North-Holland. 436 pp.
7. McConnell, H. M., 1971. *Ann. Rev. Biochem.* 40 : 227–36
8. Benesch, R. E., Benesch, R. 1974. *Advan. Protein Chem.* 28 : 211–37
9. Kilmartin, J. V., Rossi-Bernardi, L. 1973. *Physiol. Rev.* 53 : 836–90
10. Edsall, J. T. 1972. *J. Hist. Biol.* 5 : 205–57
11. Perutz, M. F., Lehmann, H. 1968. *Nature* 219 : 902–9
12. Morimoto, H., Lehmann, H., Perutz, M. F. 1971. *Nature* 232 : 408–13
13. Hill, A. V. 1910. *J. Physiol.* 40 : iv–vii
14. Adair, G. S. 1924. *J. Physiol.* 58 : xxxix–xxxx
15. Roughton, F. J. W. 1965. *J. Gen. Physiol.* 49 : 105–26
16. Roughton, F. J. W. 1934. *Proc. Roy. Soc. B* 115 : 451–503
17. Douglas, C. G., Haldane, J. S., Haldane, J. B. S. 1912. *J. Physiol.* 44 : 275–304
18. Briehl, R. W. 1963. *J. Biol. Chem.* 238 : 2361–66
19. Behlke, J., Scheler, W. 1970. *Eur. J. Biochem.* 15 : 245–49
20. Andersen, M. E. 1971. *J. Biol. Chem.* 246 : 4800–6
21. Monod, J., Wyman, J., Changeux, J.-P. 1965. *J. Mol. Biol.* 12 : 88–118
22. Adair, G. S. 1925. *J. Biol. Chem.* 63 : 529–45
23. Roughton, F. J. W. 1949. *Haemoglobin*, 83. London: Butterworth
24. Pauling, L. 1935. *Proc. Nat. Acad. Sci. USA* 21 : 186–91
25. Koshland, D. E., Nemethy, G., Filmer, D. 1966. *Biochemistry* 5 : 365–85
26. Antonini, E., Bucci, E., Fronticelli, C., Wyman, J., Rossi-Fanelli, A. 1965. *J. Mol. Biol.* 12 : 375–84
27. Edelstein, S. J. 1967. *Functional aspects of the quaternary structure of proteins. Studies on hemoglobin and aldolase.* PhD thesis. Univ. Calif., Berkeley. 218 pp.
28. Perutz, M. F. 1969. *Proc. Roy. Soc. B* 173 : 113–40
29. Rubin, M. M., Changeux, J.-P. 1966. *J. Mol. Biol.* 21 : 265–74
30. Edelstein, S. J. 1971. *Nature* 230 : 224–27
31. Noble, R. W. 1969. *J. Mol. Biol.* 39 : 479–91
32. Nagel, R. L., Gibson, Q. H., Charache, S. 1967. *Biochemistry* 6 : 2395–2402
33. Bonaventura, J., Riggs, A. 1968. *J. Biol. Chem.* 243 : 980–91
34. Bunn, H. F., Guidotti, G. 1972. *J. Biol. Chem.* 247 : 2345–50
35. Tyuma, I., Imai, K., Shimizu, K. 1973. *Biochemistry* 12 : 1491–98
36. Imai, K. 1973. *Biochemistry* 12 : 798–808
37. Hamilton, M. N., Edelstein, S. J. 1974. *J. Biol. Chem.* 249 : 1323–29
38. Szabo, A., Karplus, M. 1972. *J. Mol. Biol.* 72 : 163–97
39. Edelstein, S. J. 1973. *Introductory Biochemistry,* 160. San Francisco: Holden-Day
40. Wyman, J. 1965. *J. Mol. Biol.* 11 : 631–44
41. Hopfield, J. J., Shulman, R. G., Ogawa, S. 1971. *J. Mol. Biol.* 61 : 425–43
42. Shulman, R. G., Ogawa, S., Hopfield, J. J. 1972. *Arch. Biochem. Biophys.* 151 : 68–74
43. Thomas, J. O., Edelstein, S. J. 1972. *J. Biol. Chem.* 247 : 7840–74
44. Saroff, H. A., Minton, A. P. 1974. *Science* 175 : 1253–55
45. Antonini, E. 1967. *Science* 158 : 1417–25
46. Guidotti, G. 1967. *J. Biol. Chem.* 242 : 3704–12
47. Rossi-Fanelli, A., Antonini, E., Caputo, A. 1961. *J. Biol. Chem.* 236 : 391–96
48. Ibid, 397–400
49. Edsall, J. T., Wyman, J. 1958. *Biophysical Chemistry: Thermodynamics, electrostatics, and the biological significance of the properties of matter,* 1 : 591–662. New York: Academic
50. Benesch, R. E., Benesch, R., Williamson, M. E. 1962. *Proc. Nat. Acad. Sci. USA* 48 : 2071–75
51. Guidotti, G. 1967. *J. Biol. Chem.* 242 : 3685–93
52. Bucci, E. et al 1965. *J. Mol. Biol.* 12 : 183–92
53. Antonini, E. et al 1966. *J. Mol. Biol.* 17 : 29–46
54. Gibson, Q. H. 1959. *Biochem. J.* 71 : 293–303
55. Gibson, Q. H., Antonini, E. 1967. *J. Biol. Chem.* 242 : 4678–81
56. Schachman, H. K., Edelstein, S. J. 1966. *Biochemistry* 5 : 2681–2705
57. Antonini, E., Chiancone, E., Brunori, M. 1967. *J. Biol. Chem.* 242 : 4360–66
58. Antonini, E., Brunori, M., Anderson, S. 1968. *J. Biol. Chem.* 234 : 1816–22
59. Gibson, Q. H., Parkhurst, L. J. 1968. *J. Biol. Chem.* 243 : 5521–24
60. Edelstein, S. J., Gibson, Q. H. 1971. *Probes of Structure and Function of Macromolecules and Membranes, Probes of Enzymes and Hemoproteins,* ed. B. Chance, T. Yonetani, A. S. Mildvan, 2 : 417–29. New York: Academic
61. Edelstein, S. J., Gibson, Q. H. 1971.

First Inter-American Symposium on Hemoglobin, ed. R. Nagel, 160–67. Basel: Karger
62. Edelstein, S. J., Rehmar, M. J., Olson, J. S., Gibson, Q. H. 1970. *J. Biol. Chem.* 245:4372–81
63. Kellett, G. L., Schachman, H. K. 1971. *J. Mol. Biol.* 59:387–99
64. Kellett, G. L. 1971. *J. Mol. Biol.* 59:401–24
65. Antonini, E., Brunori, M., Chiancone, E., Wyman, J. See Ref. 60, 431–37
66. Andersen, M. E., Moffat, J. K., Gibson, Q. H. 1971. *J. Biol. Chem.* 246:2796–2807
67. Kellett, G. L., Gutfreund, H. 1970. *Nature* 227:921–26
68. Park, C. M. 1970. *J. Biol. Chem.* 245:5390–94
69. Bunn, H. F., McDonough, M. 1974. *Biochemistry* 13:988–93
70. Kellett, G. L. 1971. *Nature New Biol.* 234:189–91
71. Hewitt, J. A., Kilmartin, J. V., TenEyck, L. F., Perutz, M. F. 1972. *Proc. Nat. Acad. Sci. USA* 69:203–7
72. Chiancone, E., Gilbert, L. M., Gilbert, G. A., Kellett, G. L. 1968. *J. Biol. Chem.* 243:1212–19
73. Crepeau, R. H., Hensley, C. P., Edelstein, S. J. 1974. *Biochemistry* 13:4860–65
74. Thomas, J. O., Edelstein, S. J. 1973. *J. Biol. Chem.* 248:2901–5
75. Kirshner, A. G., Tanford, C. 1964. *Biochemistry* 3:291–96
76. Perutz, M. F. 1970. *Nature* 228:726–39
77. Hamilton, M. N., Edelstein, S. J. 1972. *Science* 178:1104–6
78. Bunn, H. F. 1970. *Nature* 227:839–40
79. Nagel, R. L., Gibson, Q. H. 1972. *Biochem. Biophys. Res. Commun.* 48:959–66
80. Nagel, R. L., Gibson, Q. H. 1971. *J. Biol. Chem.* 246:69–73
81. Perutz, M. F. 1972. *Nature* 237:495–99
82. Perutz, M. F., TenEyck, L. F. 1971. *Cold Spring Harbor Symp. Quant. Biol.* 36:295–310
83. Greer, J. 1971. *Cold Spring Harbor Symp. Quant. Biol.* 36:315–23
84. Muirhead, H., Cox, J. M., Mazzarella, L., Perutz, M. F. 1967. *J. Mol. Biol.* 23:117–56
85. Bolton, W., Perutz, M. F. 1970. *Nature* 228:551–52
86. Muirhead, H., Greer, J. 1970. *Nature* 228:516–19
87. Anderson, L. 1973. *J. Mol. Biol.* 79:495–506
88. Arnone, A. 1972. *Nature* 237:146–49
89. Arnone, A., Perutz, M. F. 1974. *Nature* 249:34–36

90. Benesch, R., Benesch, R. E. 1969. *Nature* 221:618–22
91. Bunn, H. F., Briehl, R. W., Larrabee, P., Hobart, V. 1970. *J. Clin. Invest.* 49:1088–95
92. Benesch, R. E., Benesch, R., Renthal, R., Maeda, N. 1972. *Biochemistry* 11:3576–82
93. Bailey, J. E., Beetlestone, J. G., Irvine, D. H. 1970. *J. Chem. Soc. A* 1970:756–62
94. Riggs, A. 1971. *Proc. Nat. Acad. Sci USA* 68:2062–65
95. Tomita, S., Riggs, A. 1971. *J. Biol. Chem.* 246:547–54
96. Kilmartin, J. V. 1974. *FEBS Lett.* 38:147–48
97. Kilmartin, J. V., Wootton, J. F. 1970. *Nature* 228:766–67
98. Kilmartin, J. V., Hewitt, J. A. 1971. *Cold Spring Harbor Symp. Quant. Biol.* 36:311–14
99. Kilmartin, J. V., Breen, J. J., Roberts, G. C. K., Ho, C. 1973. *Proc. Nat. Acad. Sci. USA* 70:1246–49
100. Kilmartin, J. V., Rossi-Bernardi, L. 1969. *Nature* 222:1243–46
101. Kilmartin, J. V., Fogg, J., Luzzano, M., Rossi-Bernardi, L. 1973. *J. Biol. Chem.* 248:7039–44
102. Hoard, J. L., Hamor, M. J., Hamor, T. A., Caughey, W. S. 1965. *J. Am. Chem. Soc.* 87:2312–19
103. Hoard, J. L. 1971. *Science* 174:1295–1302
104. Perutz, M. F., Ladner, J. F., Simon, S. R., Ho, C. 1974. *Biochemistry* 13:2163–73
105. Perutz, M. F., Fersht, A. R., Simon, S. R., Roberts, G. C. K. 1974. *Biochemistry* 13:2174–86
106. Perutz, M. F. et al 1974. *Biochemistry* 13:2187–2200
107. Kilmartin, J. V. 1973. *Biochem. J.* 133:725–33
108. Banerjee, R., Stetzkowski, F., Henry, Y. 1973. *J. Mol. Biol.* 73:455–67
109. Banerjee, R., Henry, Y., Cassoly, R. 1973. *Eur. J. Biochem.* 32:173–77
110. Szabo, A., Karplus, M. 1974. *Biochemistry.* Submitted
111. Hensley, C. P., Edelstein, S. J., Wharton, D. C., Gibson, Q. H. 1975. *J. Biol. Chem.* 250:952–60
112. Edelstein, S. J., Gibson, Q. H. 1975. *J. Biol. Chem.* 250:961–65
113. Bonaventura, J. et al 1972. *Proc. Nat. Acad. Sci. USA* 69:2174–78
114. Bonaventura, J. et al 1974. *J. Mol. Biol.* 82:499–511
115. Hewitt, J. A., Gibson, Q. H. 1973. *J. Mol. Biol.* 74:489–98
116. Moffat, K., Olson, J. S., Gibson, Q. H.,

Kilmartin, J. V. 1973. *J. Biol. Chem.* 248:6387–93
117. Ogawa, S., Patel, D. J., Simon, S. R. 1974. *Biochemistry* 13:2001–6
118. Olson, J. S., Gibson, Q. H., Nagel, R. L., Hamilton, H. B. 1972. *J. Biol. Chem.* 247:7485–93
119. Olson, J. S., Gibson, Q. H. 1972. *J. Biol. Chem.* 247:1713–26
120. Bunn, H. F., Wohl, R., Bradley, T. B., Cooley, M., Gibson, Q. H. 1974. *J. Biol. Chem.* 249:7402–9
121. Gibson, Q. H., Nagel, R. L. 1974. *J. Biol. Chem.* 249:7255–59
122. Ogawa, S., Mayer, A., Shulman, R. G. 1972. *Biochem. Biophys. Res. Commun.* 49:1485–91
123. Hopfield, J. J., Ogawa, S., Shulman, R. G. 1972. *Biochem. Biophys. Res. Commun.* 49:1480–84
124. Gibson, Q., Riggs, A., Imamura, T. 1973. *J. Biol. Chem.* 248:5976–86
125. Crepeau, R. H., Edelstein, S. J., Rehmar, M. J. 1972. *Anal. Biochem.* 50:213–33
126. Banerjee, R., Cassoly, R. 1969. *J. Mol. Biol.* 42:337–49
127. Brunori, M., Amiconi, G., Antonini, E., Wyman, J. 1970. *J. Mol. Biol.* 49:461–71
128. Ogawa, S., Shulman, R. G. 1971. *Biochem. Biophys. Res. Commun.* 42:9–15
129. Ogawa, S., Shulman, R. G., Fujiwara, M., Yamane, T. 1972. *J. Mol. Biol.* 70:301–13
130. Ogawa, S., Shulman, R. G., Yamane, T. 1972. *J. Mol. Biol.* 70:291–300
131. Ogawa, S., Shulman, R. G. 1972. *J. Mol. Biol.* 70:215–336
132. Cassoly, R., Gibson, Q. H., Ogawa, S., Shulman, R. G. 1971. *Biochem. Biophys. Res. Commun.* 44:1015–21
133. Cassoly, R., Gibson, Q. H. 1972. *J. Biol. Chem.* 247:7332–41
134. Lindstrom, T. R., Ho, C., Pisciotta, A. V. 1972. *Nature New Biol.* 237:263–64
135. Ranney, H. M., Nagel, R. L., Udem, L. See Ref. 61, 143–51
136. Haber, J. E., Koshland, D. E. 1971. *J. Biol. Chem.* 246:7790–93
137. **Cassoly, R., Gibson, Q. H. 1975. *J. Mol. Biol.* 91:301–13**
138. Gibson, Q. H. 1973. *Proc. Nat. Acad. Sci. USA* 70:1–4
139. MacQuarrie, R., Gibson, Q. H. 1971. *J. Biol. Chem.* 246:5832–35
140. MacQuarrie, R., Gibson, Q. H. 1972. *J. Biol. Chem.* 247:5686–94
141. Gibson, Q. H. 1973. *J. Biol. Chem.* 148:1281–84
142. Antonini, E., Brunori, M. 1969. *J. Biol. Chem.* 244:3909–12
143. Antonini, E., Schuster, T. M., Brunori,

M., Wyman, J. 1965. *J. Biol Chem.* 240:PC2262–64
144. Gray, R. D. 1970. *J. Biol. Chem.* 245:2914–21
145. Olson, J. S., Gibson, Q. H. 1973. *J. Biol. Chem.* 248:1623–30
146. Gibson, Q. H., Parkhurst, L. J., Giuseppa, G. 1969. *J. Biol. Chem.* 244:4668–76
147. MacQuarrie, R. A., Gibson, Q. H. 1971. *J. Biol. Chem.* 246:517–22
148. Uchida, H., Heystek, J., Klapper, M. H. 1971. *J. Biol. Chem.* 246:2031–34
149. Klapper, M. H., Uchida, H. 1971. *J. Biol. Chem.* 246:6849–54
150. Olson, J. S., Gibson, Q. H. 1970. *Biochem. Biophys. Res. Commun.* 41:421–26
151. Olson, J. S., Gibson, Q. H. 1971. *J. Biol. Chem.* 246:5241–53
152. Olson, J. S., Gibson, Q. H. 1972. *J. Biol. Chem.* 247:3662–70
153. Lindstrom, T. R., Olson, J. S., Mock, N. H., Gibson, Q. H., Ho, C. 1971. *Biochem. Biophys. Res. Commun.* 45:22–26
154. Davis, D. G. et al 1971. *J. Mol. Biol.* 60:101–11
155. Olson, J. S., Gibson, Q. H. 1973. *J. Biol. Chem.* 248:1616–22
156. Cole, F. X., Gibson, Q. H. 1973. *J. Biol. Chem.* 248:4998–5004
157. Huestis, W. H., Raftery, M. A. 1972. *Biochem. Biophys. Res. Commun.* 48:678–83
158. Gray, R. D., Gibson, Q. H. 1971. *J. Biol. Chem.* 246:7168–74
159. Ilgenfritz, G., Schuster, T. M. 1971. *J. Biol. Chem.* 249:2959–73
160. Ogata, R., McConnell, H. M. 1971. *Cold Spring Harbor Symp. Quant. Biol.* 36:325–35
161. Ogata, R. T., McConnell, H. M. 1972. *Proc. Nat. Acad. Sci. USA* 69:335–39
162. Ogata, R. T., McConnell, H. M., Jones, R. T. 1972. *Biochem. Biophys. Res. Commun.* 47:155–65
163. Ogata, R. T., McConnell, H. M. 1972. *Biochemistry* 11:4792–99
164. Lindstrom, T. R., Ho, C. 1972. *Proc. Nat. Acad. Sci. USA* 69:1707–10
165. Henry, Y., Cassoly, R. 1973. *Biochem. Biophys. Res. Commun.* 51:659–65
166. Henry, Y., Banerjee, R. 1973. *J. Mol. Biol.* 73:469–82
167. Edelstein, S. J. 1974. *Biochemistry* 13:4998–5002
168. McDonald, M. J., Hoffman, B. M., Gibson, Q. H., Bull, C. 1974. *Fed. Proc.* 33:440 (Abstr.)
169. Hoffman, B. M. et al 1974. *Ann. NY Acad. Sci.* In press

170. Hopfield, J. J. 1973. *J. Mol. Biol.* 77: 207–22
171. Deal, W. J. 1973. *Biopolymers* 12: 2057–73
172. Herzfield, J., Stanley, H. E. 1974. *J. Mol Biol.* 82: 231–65
173. Saroff, H. A., Yap, W. T. 1972. *Biopolymers* 11: 957–71
174. Minton, A. P. 1974. *Science* 184: 577–79
175. Minton, A. P., Imai, K. 1974. *Proc. Nat. Acad. Sci. USA* 71: 1418–21
176. Austin, R. H. et al 1973. *Science* 181: 541–43
177. Szabo, A., Karplus, M. 1973. *Proc. Nat. Acad. Sci. USA* 70: 673–74
178. Brunori, M., Bonaventura, J., Bonaventura, C., Antonini, E., Wyman, J. 1972. *Proc. Nat. Acad. Sci. USA* 69: 868–71
179. Brunori, M., Giacometti, G. M., Antonini, E., Wyman, J. 1973. *Proc. Nat. Acad. Sci. USA* 70: 3141–44
180. Phillipson, P. E., Ackerson, B. J., Wyman, J. 1973. *Proc. Nat. Acad. Sci. USA* 70: 1550–53
181. Alpert, B., Banerjee, R., Lindquist, L. 1974. *Proc. Nat. Acad. Sci. USA* 71: 558–62
182. Tainsky, M., Edelstein, S. J. 1973. *J. Mol. Biol.* 75: 735–39
183. Benesch, R. E., Benesch, R., MacDuff, G. 1964. *Biochemistry* 3: 1132–35
184. Perutz, M. F. 1974. *Nature* 247: 341–44
185. Yip, K. Y., Waks, M., Beychok, S. 1972. *J. Biol. Chem.* 247: 7237–44
186. Cassoly, R., Banerjee, R. 1971. *Eur. J. Biochem.* 19: 514–22
187. Noble, R. W., Rossi, G., Berni, R. 1972. *J. Mol. Biol.* 70: 689–96
188. Treffry, A., Ainsworth, S. 1974. *Biochem. J.* 137: 319–29
189. Ainsworth, S., Treffry, A. 1974. *Biochem. J.* 137: 331–37
190. Treffry, A., Ainsworth, S. 1974. *Biochem.*

191. Asakura, T. 1973. *Ann. NY Acad. Sci.* 222: 68–85
192. Yonetani, T. et al 1970. *J. Biol. Chem.* 245: 2998–3003
193. Gibson, Q. H. et al 1974. *Biochem. Biophys. Res. Commun.* 59: 146–51
194. Moffat, K., Loe, R. S., Hoffman, B. M. 1974. *J. Am. Chem. Soc.* 96: 5259–60
195. Hoffman, B. M., Petering, D. H. 1970. *Proc. Nat. Acad. Sci. USA* 67: 637–43
196. Hoffman, B. M., Spilburg, C. A., Petering, D. H. 1971. *Cold Spring Harbor Symp. Quant. Biol.* 36: 343–48
197. Dickinson, L. C., Chien, J. C. W. 1973. *J. Biol. Chem.* 248: 5005–11
198. Yonetani, T., Yamamoto, H., Woodrow, G. V. III. 1974. *J. Biol. Chem.* 249: 682–90
199. Yamamoto, H., Kayne, F. J., Yonetani, T. 1974. *J. Biol. Chem.* 249: 691–98
200. Yonetani, T., Yamamoto, H., Iizuka, T. 1974. *J. Biol. Chem.* 249: 2168–74
201. Leonard, J. J., Yonetani, T., Callis, J. B. 1974. *Biochemistry* 13: 1460–64
202. Englander, S. W., Mauel, C. 1972. *J. Biol. Chem.* 247: 2387–94
203. Englander, S. W., Rolfe, A. 1973. *J. Biol. Chem.* 248: 4852–61
204. Benson, E. S., Rossi-Fanelli, M. R., Giacometti, G. M., Rosenberg, A., Antonini, E. 1973. *Biochemistry* 12: 2699–2706
205. Rudolph, S. A., Gill, S. J. 1974. *Biochemistry* 13: 2451–54
206. Atha, D. H., Ackers, G. K. 1974. *Biochemistry* 13: 2376–82
207. Spiro, T. G., Strekas, T. C. 1974. *J. Am. Chem. Soc.* 96: 338–45
208. Sussner, H., Mayer, A., Brunner, H., Fasold, H. 1974. *Eur. J. Biochem.* 41: 465–69
209. Yamamoto, T., Palmer, G., Gill, D., Salmeen, T. T., Rimai, L. 1973. *J. Biol. Chem.* 248: 5211–13

J. 137: 339–48

BILE ACID METABOLISM ✻881

Henry Danielsson and Jan Sjövall

Department of Chemistry, Karolinska Institutet, Stockholm, Sweden

CONTENTS

INTRODUCTION

Bile acid metabolism has not been reviewed in the *Annual Review of Biochemistry* since 1959. Much of the present knowledge in this area stems from work carried out since that time. A number of reviews on the subject have appeared since 1959 in other publications (1–7). Recently, a two volume treatise on the chemistry, physiology, and metabolism of bile acids, *The Bile Acids* (8), was published, and *Archives of Internal Medicine* devoted its October 1972 issue entirely to a symposium on bile acids. These publications cover the literature up to 1970–1971 and the present review will deal primarily with literature published since that time. The authors' interest as well as the space allotted limit the review to only part of the large volume of literature on bile acids. The review concentrates on the formation and metabolism of bile acids in mammals, and discusses briefly the increasing number of clinical studies of bile acid metabolism.

233

FORMATION OF BILE ACIDS

5β Bile Acids

The principal bile acids formed from cholesterol in mammals are cholic and chenodeoxycholic acids. Several other bile acids, of both the 5β and the 5α series, are also formed and there are considerable variations between species, even to the

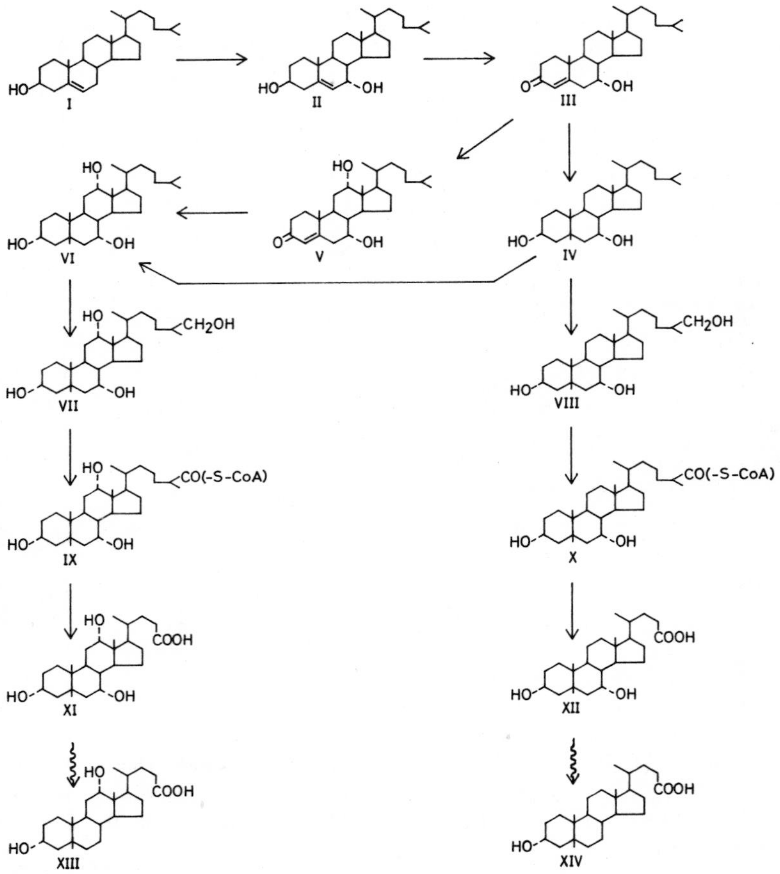

Figure 1 Pathways for formation of bile acids. I, cholesterol; II, 7α-hydroxycholesterol (5-cholestene-3β, 7α-diol); III, 7α-hydroxy-4-cholesten-3-one; IV, 5β-cholestane-3α, 7α-diol; V, 7α, 12α-dihydroxy-4-cholesten-3-one; VI, 5β-cholestane-3α, 7α, 12α-triol; VII, 5β-cholestane-3α, 7α, 12α, 26-tetrol; VIII, 5β-cholestane-3α, 7α, 26-triol; IX, 3α, 7α, 12α-trihydroxy-5β-cholestanoic acid; X, 3α, 7α-dihydroxy-5β-cholestanoic acid; XI, cholic acid; XII, chenodeoxycholic acid; XIII, deoxycholic acid; XIV, lithocholic acid. →, reactions catalyzed by liver enzymes; ↝, reactions catalyzed by microbial enzymes.

extent that some bile acids may be considered species specific, e.g. the muricholic acids in rat and mouse. The major pathways for the biosynthesis of cholic acid and chenodeoxycholic acid in rats were established by the late 1960s (Figure 1). In the biosynthesis of cholic acid, all nuclear changes appear to precede oxidation of the side chain. The initial reaction in the latter oxidation is a 26 hydroxylation, and the substrate for this reaction in cholic acid formation is 5β-cholestane-3α,7α,12α-triol (cf also 9). Introduction of the 7α-hydroxyl group is the first step in the conversion of cholesterol into cholic acid, and the 12α-hydroxyl group appears to be introduced primarily into 7α-hydroxy-4-cholesten-3-one. The further oxidation of 5β-cholestane-3α,7α,12α,26-tetrol leads to formation of 3α,7α,12α-trihydroxy-5β-cholestanoic acid, which in turn is converted into cholic acid (or cholyl coenzyme A) and propionic acid (or propionyl coenzyme A). One pathway for the formation of chenodeoxycholic acid is analogous to that for cholic acid formation. The first step is thus 7α hydroxylation of cholesterol and the major substrate for the 26-hydroxylase is 5β-cholestane-3α,7α-diol. Another pathway with all reactions catalyzed by the mitochondrial fraction fortified with the 100,000 $\times$ g supernatant fraction was described originally by Mitropoulos & Myant in 1967 (10). It involves an initial 26 hydroxylation of cholesterol followed by oxidation of 26-hydroxycholesterol into 3β-hydroxy-5-cholenoic acid. The unsaturated bile acid is converted into lithocholic acid which in turn is 7α-hydroxylated to chenodeoxycholic acid. Further evidence for the intermediary formation of 26-hydroxycholesterol in this pathway was reported recently by the same authors (11). In a series of recent reports, Yamasaki and collaborators (12–15) have provided evidence for an additional pathway for chenodeoxycholic acid formation in the rat involving conversion of 7α-hydroxycholesterol into 3β,7α-dihydroxy-5-cholenoic acid and subsequently 7α-hydroxy-3-keto-4-cholenoic acid. The quantitative importance of these different pathways leading to chenodeoxycholic acid has not been established but some relevant facts may be considered. There is convincing evidence that the major rate-limiting step in the overall conversion of cholesterol into bile acids is the 7α hydroxylation of cholesterol. One might expect the major pathways for biosynthesis of cholic acid and chenodeoxycholic acid to have this regulatory step in common under normal conditions. Hence, a pathway to chenodeoxycholic acid involving an initial 26 hydroxylation may be quantitatively less important under normal conditions. This is supported by the fact that in biliary drainage, which leads to a large increase in the biosynthesis of bile acids, the rate of 26 hydroxylation of cholesterol is unaffected, whereas 7α hydroxylation of cholesterol is increased manyfold (16). Further, if chenodeoxycholic acid were formed to a large extent through an initial 26 hydroxylation, the precursor cholesterol for chenodeoxycholic acid may not be equilibrated with that for cholic acid, because the 26 hydroxylation of cholesterol is a mitochondrial reaction and the 7α hydroxylation of cholesterol is a microsomal reaction. However, several studies have shown that after administration of labeled cholesterol and/or labeled mevalonate, cholic acid and chenodeoxycholic acid attain the same specific radioactivities (e.g. 17). The possibility that cholesterol esters rather than cholesterol might be the immediate precursors of bile acids has been refuted by the results of Ogura et al (17) and Mathé et al (18), who followed specific radioactivities of bile acids, cholesterol, and cholesterol esters after administration of labeled cholesterol or labeled cholesterol esters.

Pathways for biosynthesis of cholic acid and chenodeoxycholic acid in species other than the rat have been studied in only a few instances. Some studies in vivo and in vitro indicate that the pathways in man are similar to or the same as in rat. Thus, the stepwise conversion of cholesterol into 5β-cholestane-$3\alpha,7\alpha,12\alpha$-triol has been shown in liver homogenates (19), and Hanson et al (20) have found that 7α-hydroxy-4-cholesten-3-one is converted in vivo into cholic acid and chenodeoxycholic acid. $3\alpha,7\alpha$-Dihydroxy-5β-cholestanoic acid and $3\alpha,7\alpha,12\alpha$-trihydroxy-5β-cholestanoic acid have been isolated from human bile, and the conversion of cholesterol into these acids as well as the subsequent conversion of these acids into chenodeoxycholic acid and cholic acid, respectively, have been shown (21–24). It is of interest to note that $3\alpha,7\alpha$-dihydroxy-5β-cholestanoic acid was converted to a small extent into cholic acid (24). Javitt and collaborators have found that 26-hydroxycholesterol is converted to an appreciable extent into cholic acid in man (and hamster) in contrast to the rat (25, 26). These studies lead to the conclusion that the same reactions occur in man and rat, but the importance of various pathways may differ quantitatively. Of considerable interest in this connection are the recent reports by Eyssen et al (27) and Setoguchi et al (28). Eyssen et al found that $3\alpha,7\alpha,12\alpha$-trihydroxy-5β-cholestanoic acid was an important or the dominating bile acid in bile of two children with intrahepatic bile duct anomalies. Another possible type of deficiency in side chain oxidation has been detected by Setoguchi et al (28) in patients with the rare disease cerebrotendinous xanthomatosis. Bile from these patients contains large amounts of 5β-cholestane-$3\alpha,7\alpha,12\alpha,25$-tetrol and 5β-cholestane-$3\alpha,7\alpha,12\alpha,24,25$-pentol. These sterols compensate for the decrease in bile acid biosynthesis seen in the patients (29). It is not known whether these patients have a defective oxidation of these sterols into bile acids or a defective 26 hydroxylation. The former case would mean that the mechanisms of side chain oxidation are different from those generally accepted. In the latter case, accumulation of the different side chain hydroxylated sterols would be the result of a side reaction not normally seen when 26 hydroxylation is efficient. Support for this contention comes from the findings that the microsomal fraction of rat liver homogenate catalyzes hydroxylation of 5β-cholestane-$3\alpha,7\alpha,12\alpha$-triol in 23, 24, 25, and 26 positions, and that of these tetrols 5β-cholestane-$3\alpha,7\alpha,12\alpha,26$-tetrol is by far the most efficient precursor of cholic acid (30). On the other hand, preliminary experiments indicate that 5β-cholestane-$3\alpha,7\alpha,12\alpha,25$-tetrol is converted efficiently into cholic acid in man (31).

5α Bile Acids

Bile acids with a 5α configuration (allo bile acids) occur in small amounts in bile from mammals including man. In lower animals, 5α bile acids and/or 5α bile alcohols may be major components in bile. Bile acids and bile alcohols in non-mammalian species will not be discussed in this review; the reader is referred to reviews on this subject by Haslewood (32) and by Hoshita & Kazuno (33) and to papers published by these authors and their collaborators and by Yamasaki and colleagues. In the synthesis of 5α bile acids from cholesterol, the 5α configuration can be introduced at three different stages: conversion of cholesterol into cholestanol; reduction of the Δ^4 double bond of 7α-hydroxy-4-cholesten-3-one or 7α,12α-dihydroxycholest-4-en-3-one; or conversion of a 5β bile acid to the corresponding 5α bile acid. Cholestanol has been shown to be converted into allocholic acid and allochenodeoxycholic acid in the rat (34, 35). The conversion of cholestanol into allocholic acid has also been shown in the gerbil (36). Reactions between cholestanol and the allo bile acids have been studied by Björkhem & Gustafsson (37) and by Elliott, Doisy, and collaborators (38–41). The results so far published indicate considerable similarities between the pathways for biosynthesis of 5α and 5β bile acids. There is, however, a striking difference with respect to 12α hydroxylation (38–41). Whereas in the biosynthesis of cholic acid the presence of a 26-hydroxyl group practically prevents 12α hydroxylation, 5α-cholestane-3β,7α,26-triol is a very efficient precursor of allocholic acid, and allochenodeoxycholic acid is converted into allocholic acid in vivo and in vitro.

The biosynthesis of 5α bile acids through pathways with the intermediary formation of 7α-hydroxycholesterol, 7α-hydroxy-4-cholesten-3-one, and 7α,12α-dihydroxy-4-cholesten-3-one has been studied by Yamasaki and collaborators and by Björkhem & Einarsson. Yamasaki et al (42) found that labeled 7α-hydroxy-cholesterol gave rise to labeled allocholic acid in several species including rat and rabbit. Björkhem & Einarsson (43) showed that microsomal 5α reductase from rat liver catalyzed conversion of 7α-hydroxy-4-cholesten-3-one and 7α,12α-dihydroxy-4-cholesten-3-one into the corresponding 5α steroids. These steroids were precursors of allocholic acid in a rat with biliary fistula. It is likely that the pathways of biosynthesis of 5α bile acids from 7α-hydroxy-4-cholesten-3-one and 7α,12α-dihydroxy-4-cholesten-3-one involve the same intermediates as those from cholestanol.

The formation of 5α bile acids from 5β bile acids has so far been shown only in the case of allodeoxycholic acid (44, 45). The conversion probably occurs by means of the intermediate formation of the 3-keto-5β, the Δ^4-3-keto, and the 3-keto-5α derivatives, and some or all of the reactions are catalyzed by enzymes in intestinal microorganisms. The reactions are reversible and transformation of allocholic, allochenodeoxycholic, allodeoxycholic, and allolithocholic acids into the corresponding 5β bile acids has been shown (46, 47). In most cases the equilibrium appears to favor formation of 5β bile acids.

7α Hydroxylation of Cholesterol

The 7α hydroxylation of cholesterol is the major rate-limiting step in the overall conversion of cholesterol to bile acids and has been extensively studied. The reaction is catalyzed by a microsomal monooxygenase system and there is strong evidence for the participation of cytochrome P-450 and NADPH-cytochrome P-450 reductase (48–50). Recently, the reaction has been shown in a reconstituted system from rat liver microsomes consisting of partially purified cytochrome P-450, NADPH-cytochrome P-450 reductase, and a phospholipid (51). The 7α hydroxylation of cholesterol differs in some respects from most other cytochrome P-450-dependent hydroxylations. The reaction is stimulated severalfold by biliary drainage through a biliary fistula or by administration of the bile acid binding resin cholestyramine (52–55). If it has any effect at all, biliary drainage leads to a decrease in total cytochrome P-450 content, and it has no significant influence on several other microsomal hydroxylations. In rats with adjuvant-induced arthritis, 7α hydroxylation of cholesterol is unaffected, whereas aminopyrine demethylase activity as well as total cytochrome P-450 content are reduced markedly (56). In Sprague-Dawley rats, phenobarbital treatment has no effect on 7α hydroxylation of cholesterol but stimulates most other microsomal hydroxylations (57, 58). However, in Wistar rats 7α hydroxylation is stimulated severalfold by phenobarbital treatment (58). In contrast to several other hydroxylations in the biosynthesis and metabolism of bile acids, the rate of 7α hydroxylation of cholesterol varies markedly with time of day (59–63) and is influenced by dietary bile acids (64, 65). Rat liver microsomes hydroxylate cholesterol essentially only in the 7α position (50, 66), and minor modifications of the cholesterol side chain result in little or no 7α-hydroxylase activity (66–68).

Assay of cholesterol 7α-hydroxylase activity poses several problems. A major one is the presence of large amounts of endogenous cholesterol in microsomes. Most authors have assayed cholesterol 7α-hydroxylase activity by measuring conversion of added labeled cholesterol into 7α-hydroxycholesterol. Mitropoulos & Balasubramaniam (69) have described a method of analyzing for 7α-hydroxycholesterol based on acetylation with tritium labeled acetic anhydride of known specific radioactivity. These authors (69) and Balasubramaniam et al (70) have pointed to the shortcomings of methods involving assay of conversion of labeled cholesterol. A method to assay cholesterol 7α-hydroxylase activity using deuterated carrier and gas chromatography–mass spectrometry has also been recently described (71). Balasubramaniam et al (70) have provided convincing evidence that only part of microsomal cholesterol serves as substrate for cholesterol 7α-hydroxylase and have suggested that the preferred substrate is newly synthesized cholesterol. Their general conclusions have been subsequently confirmed by somewhat differently designed experiments (72).

12α Hydroxylation

The 12α hydroxylation of 7α-hydroxy-4-cholesten-3-one has been shown in reconstituted systems from rat liver microsomes (73). The 12α hydroxylation is stimulated severalfold by starvation (55) and this effect resides in the cytochrome P-450 fraction of the reconstituted system (73). The extent of inhibition of the reaction by carbon monoxide is markedly lower than that of other cytochrome P-450-dependent hydroxylations (73–75), and administration of phenobarbital inhibits 12α hydroxylation (74, 75). In the hyperthyroid rat, the rate of 12α hydroxylation is significantly lower than in the euthyroid rat (76). Administration of bile acids inhibits 12α hydroxylation (64) and the 12α-hydroxylase system has been ascribed a role in the regulation of bile acid formation.

26 Hydroxylation

The 26 hydroxylation is the initial step in the major pathway for oxidation of the side chain. The reaction is catalyzed by the mitochondrial and the microsomal fractions of rat liver homogenate. The microsomal 26-hydroxylase appears to be cytochrome P-450 dependent (30), and the reaction is catalyzed by a reconstituted system from rat liver microsomes (77). Björkhem & Gustafsson (78) have compared the rates of 26 hydroxylation of various C_{27} steroids and found the rate to be considerably faster with 5β-cholestane-$3\alpha,7\alpha$-diol and 5β-cholestane-$3\alpha,7\alpha,12\alpha$-triol than with several other C_{27} steroids that are probable intermediates in the biosynthesis of bile acids. There was no measurable 26-hydroxylase activity toward cholesterol. With 5β-cholestane-$3\alpha,7\alpha,12\alpha$-triol as substrate the microsomal fraction also catalyzes side chain hydroxylations at positions C-23, C-24α, C-24β, and C-25 (30). With 5β-cholestane-$3\alpha,7\alpha$-diol as substrate the only significant hydroxylation in addition to C-26 is in position C-25 (78). The 26 hydroxylation is stimulated in the hyperthyroid rat and inhibited in the hypothyroid as compared with the euthyroid rat (79).

The mitochondrial fraction catalyzes 26 hydroxylation of a number of C_{27} steroids including cholesterol. The mitochondrial 26-hydroxylase system requires NADPH, preferably generated intramitochondrially, and is stimulated markedly by Mg^{2+} (78, 80, 81). The rate of 26 hydroxylation with 5β-cholestane-$3\alpha,7\alpha$-diol, 5β-cholestane-$3\alpha,7\alpha,12\alpha$-triol, or 7α-hydroxy-4-cholesten-3-one as substrate is about twice that with cholesterol (78). The reaction involves incorporation of molecular oxygen and is inhibited by carbon monoxide (78, 80, 81). However, the presence of cytochrome P-450 in liver mitochondria has not yet been described. The 26 hydroxylation of cholesterol is accompanied by a small but significant 25 hydroxylation (81). The 25 hydroxylation also involves incorporation of molecular oxygen (81).

The relative role of mitochondrial and microsomal 26-hydroxylases in bile acid biosynthesis has not been conclusively established. In the rat, the rates of 26 hydroxylation of 5β-cholestane-$3\alpha,7\alpha$-diol and 5β-cholestane-$3\alpha,7\alpha,12\alpha$-triol are similar with mitochondrial and microsomal fraction (78). In man, microsomal 26-hydroxylase activity is, however, very low compared to mitochondrial 26-hydroxylase activity (82). The marked stimulation of this activity by thyroid hormone indicates a role for microsomal 26-hydroxylase in the rat (79). This stimulation coupled with the concomitant inhibition of 12α-hydroxylase activity (76) corresponds well to the reversal of the normal ratio between cholic acid and chenodeoxycholic acid seen in the hyperthyroid rat.

Oxidation of 5β-Cholestane-$3\alpha,7\alpha,12\alpha,26$-tetrol

Convincing evidence has been presented that the oxidation of 5β-cholestane-$3\alpha,7\alpha,12\alpha,26$-tetrol into the corresponding C_{27} acid involves the successive action of liver ethanol and acetaldehyde dehydrogenases (83–87). With horse liver alcohol dehydrogenase both the EE- and the SS-isoenzymes are active (83–85) and the SS-isoenzyme has considerably higher activity than the EE-isoenzyme (85).

METABOLISM OF BILE ACIDS

The bile acids synthesized in the liver are excreted in bile as conjugates with glycine or taurine. Glycine conjugation occurs only in mammals. Some conjugated bile acids may be present to a small extent as sulfate esters. During enterohepatic circulation, the bile acids are subjected to the action of intestinal microorganisms. Major reactions are (a) removal of the 7α-hydroxyl group in cholic acid and chenodeoxycholic acid as well as in the corresponding 5α bile acids and (b) splitting the conjugates and oxidoreductions of the hydroxyl groups. Space does not allow a review of the many reports concerning the microbial metabolism of bile acids, including studies of the metabolic activity of different strains of bacteria. The reader is referred to reviews by Rosenberg (88), Tabaqchali (89), Kellogg (90, 91), and Lewis & Gorbach (92). Most of the bile acids in the intestine are reabsorbed during each enterohepatic cycle. Some less polar bile acids, e.g. lithocholic acid, are bound to intestinal microorganisms and reabsorbed only to a small extent. Upon returning to the liver with the portal blood, the unconjugated bile acids are reconjugated with glycine or taurine, followed in a few instances by sulfation and possibly glucuronidation; ketonic bile acids may be reduced and bile acids may be hydroxylated.

Conjugation with Glycine and Taurine

The metabolism of doubly labeled conjugated bile acids in man has been studied by several groups of workers. Norman (93) reported that the cholyl moiety of glycocholic acid was conserved more than the glycine moiety during the entero-hepatic circulation. Hofmann and collaborators (94, 95) have since studied the metabolism of glycocholic, glycochenodeoxycholic, and glycodeoxycholic acids. The glycine moiety turned over about three times more rapidly than the cholanoyl moiety in all the glycine conjugated bile acids. The labeled glycine was oxidized rapidly and appeared in expired carbon dioxide. A small part was excreted in feces. The rate of taurocholic acid deconjugation is considerably slower than that of glycine conjugated bile acids, and reconjugation occurs preferentially with glycine (96–98).

Esterification with Sulfate and Glucuronic Acid

Palmer (99) first described the occurrence of sulfate esters of bile acids. He found that most of the small amounts of glyco- and taurolithocholic acids in human bile were present as sulfate esters (99, 100). In the rat with a biliary fistula, part of the administered lithocholic acid and almost all the allolithocholic acid are excreted in bile as sulfate ester (101, 102). Subsequently, it has been shown that the small amounts of various bile acids excreted in urine are largely sulfated (103–105). In different hepatobiliary diseases, the excretion of bile acids in urine increases and the bile acids are again largely sulfated. Bile acids excreted in urine may occur to some extent as conjugates with glucuronic acid (106). Sulfation occurs in the liver (101, 107), and possibly also in the kidney.

Reduction of Bile Acids in the Liver

Bile contains little, if any, ketonic bile acids. Any ketonic bile acid returning from the intestine to the liver is apparently reduced. With the exception of ursodeoxycholic acid, the reductions yield primarily the α-hydroxy derivatives. The enzymes catalyzing these reductions have not been studied in any greater detail. The presence of soluble 3α-hydroxysteroid dehydrogenase(s) acting on 5α as well as 5β bile acids is well documented (108–110). The microsomal fraction contains 3α- as well as 3β-hydroxysteroid dehydrogenases (111). Ethanol dehydrogenase catalyzes reduction of 3-keto-5β-cholanoic acid to the corresponding 3β-hydroxy derivative (112–114). No studies of 7- and 12-hydroxysteroid dehydrogenases have been reported. Also in this connection, 3α- and 3β-hydroxy bile acids reaching the liver may be subjected to oxidoreduction before being excreted in bile (45, 102, 115).

Hydroxylation of Bile Acids in the Liver

The 7α hydroxylation of conjugated deoxycholic acid is very efficient in mice and rats, a fact that explains the low concentrations of deoxycholic acid in bile from these species. The reaction is catalyzed by the microsomal fraction and requires NADPH, molecular oxygen, and cytochrome P-450 (116–118). Prager et al (118) have found that the reaction is stimulated by cyclic AMP and saturated fatty acids but not by unsaturated fatty acids. The mechanisms of these stimulatory effects are unknown at present. The reaction has also been shown in a reconstituted system from rat liver microsomes consisting of partially purified cytochrome P-450, NADPH-cytochrome P-450 reductase, and a phospholipid (117, 119). In man, deoxycholic acid is not metabolized appreciably by the liver except for reconjugation (120); however, a very low degree of 7α hydroxylation has since been detected both in vivo (121) and in vitro (122).

The 6β hydroxylation of lithocholic or taurolithocholic acid is very efficient in the mouse and rat. There is strong evidence that the same enzyme system catalyzes 6β hydroxylation of (tauro)lithocholic acid and taurochenodeoxycholic acid (123). The reaction is cytochrome P-450 dependent (124, 125) and has been shown in a reconstituted system from rat liver microsomes (117, 119). In man, administered labeled lithocholic acid is not hydroxylated to any significant extent. With microsomal fraction of human liver homogenate, Trülzsch et al (126) and Czygan et al (127) have reported that taurolithocholic acid is 6α-hydroxylated to yield taurohyodeoxycholic acid in a cytochrome P-450-dependent reaction. In their study of the metabolism of bile acids in homogenates of human liver, Björkhem et al (122) were unable to detect hydroxylation of conjugated or unconjugated lithocholic acid.

The 7α and 6β hydroxylations of bile acids resemble microsomal drug hydroxylations in many respects and thus differ from the 7α hydroxylation of cholesterol, the 12α hydroxylation and the 26 hydroxylation (123).

Metabolism of Bile Acids in Extrahepatic Tissues

Nicholas and collaborators reported in 1969 the presence of lithocholic acid in the brain of guinea pigs with experimental encephalomyelitis (128). The same group has since shown that homogenates of rat brain catalyze oxidoreduction of several 3α-hydroxy bile acids including lithocholic acid as well as 6β hydroxylation of lithocholic acid (129, 130). In vivo, intracerebrally injected lithocholic acid is oxidized to a small extent to the 3-keto acid (131). There is yet no definite information concerning the source of the lithocholic acid detected in brain.

REGULATION OF BILE ACID FORMATION

There is now convincing evidence that the biosynthesis of bile acids is regulated homeostatically by the concentration of bile acids in the enterohepatic circulation. Drainage of bile or administration of bile acid binding resins such as cholestyramine leads to increases in bile acid formation as does decreased absorption, as seen for instance in ileal resections. Feeding of bile acids leads to a decrease in bile acid formation in both normal animals and those subjected to drainage of bile. Generally, changes in the rate of bile acid biosynthesis are parallelled by changes in the rate of cholesterol biosynthesis in the liver. The major rate-limiting step in the biosynthesis of bile acids is the 7α hydroxylation of cholesterol, and in the biosynthesis of cholesterol it is the reduction of hydroxymethylglutaryl coenzyme A (52–54, 65, 132–134). The activities of cholesterol 7α-hydroxylase and hydroxymethylglutaryl coenzyme A (HMG CoA) reductase often change in parallel. The pattern of diurnal variations of these activities in the rat is the same (61, 135, 136). Biliary drainage or feeding of cholestyramine leads to marked increases in both activities (52–55, 135, 138) and feeding of taurocholic acid leads to decreases (64, 65, 137). In addition to those of cholesterol and tomatine feeding, one case where there might be a dissociation between effects on the activities of cholesterol 7α-hydroxylase and HMG CoA reductase has been reported by Shefer et al (65), who found inhibition of HMG CoA reductase and unchanged levels of cholesterol 7α-hydroxylase upon feeding taurochenodeoxycholic acid to rats. In the hamster both activities are inhibited by taurochenodeoxycholic acid (137), whereas other investigations in the rat indicate that cholesterol 7α-hydroxylase activity is inhibited (64, 139).

The mechanisms by which these different factors influence the activities of HMG CoA reductase and cholesterol 7α-hydroxylase and consequently cholesterol and bile acid biosynthesis have not been established. There is also a question of the interrelationship between the two enzyme activities. Is it primarily HGM CoA reductase or cholesterol 7α-hydroxylase activity that is influenced, or are both activities influenced simultaneously? Available information indicates that bile acids do not regulate the two activities directly in an allosteric fashion (54, 140, 141). There are conflicting opinions concerning the site of bile acid action. Weis & Dietschy (142) consider that the bile acids act primarily in the intestine. Bile acids are necessary for absorption of cholesterol, and their level in the intestine will influence the concentration of cholesterol in the lymph, which in turn determines the rate of cholesterol biosynthesis in the liver. On the other hand, Hamprecht et al (143) have found that bile acids act primarily in the liver. The results of Shefer et al (65) also indicate that bile acids influence liver HMG CoA reductase activity directly and not indirectly by influencing cholesterol absorption and enterolymphatic cholesterol flux.

There is little information concerning the relative roles of HMG CoA reductase and cholesterol 7α-hydroxylase in regulating bile acid formation. Early experiments by Myant & Eder (144) showed that in rats subjected to biliary drainage, the increase in cholesterol synthesis measured in vitro preceded the increase in bile acid synthesis measured by excretion of bile acids in bile. The experiments of Weis & Dietschy (142) also indicate a primary effect on HMG CoA reductase. On the other hand, the parallelism in the diurnal variations indicates that HMG CoA reductase and cholesterol 7α-hydroxylase are influenced simultaneously.

The specificity of the effect of bile acids on bile acid formation has been studied to some extent. Taurocholic, taurochenodeoxycholic, and taurodeoxycholic acids, fed at the 1% level in the diet for 3–7 days, were found to markedly inhibit cholesterol 7α-hydroxylase activity in the rat, whereas taurohyodeoxycholic and taurolithocholic acid had no effect (64). Analyses of bile from rats fed various conjugated bile acids show that (a) taurocholic acid inhibits endogenous synthesis of both taurochenodeoxycholic acid and taurocholic acid (145), (b) taurochenodeoxycholic acid inhibits taurocholic acid synthesis (65, 139), and (c) taurodeoxycholic acid inhibits taurochenodeoxycholic acid synthesis (65). Experiments have not been carried out in rats to determine whether taurochenodeoxycholic acid inhibits its own synthesis, or whether taurodeoxycholic acid inhibits synthesis of taurocholic acid. In man, chenodeoxycholic acid inhibits cholic acid synthesis (146) and cholic acid inhibits chenodeoxycholic acid synthesis (147). There is no information on whether cholic acid and chenodeoxycholic acid suppress their own synthesis. With respect to deoxycholic acid there are conflicting reports that might be ascribed to differences in dose administered. Einarsson et al (148) found that deoxycholic acid at 0.5 g/day inhibited cholic acid synthesis but had no consistent effect on chenodeoxycholic acid synthesis. Low-Beer et al (149, 150) found inhibition of chenodeoxycholic acid synthesis and no effect on cholic acid synthesis with 150 mg of deoxycholic acid per day.

Another aspect of the regulation of bile acid formation by bile acids is the relationship between rate of synthesis and concentration of bile acids in the portal blood and/or liver. The studies of Dowling, Small, and collaborators (151–153) on the enterohepatic circulation have shown that in the monkey the response of the liver to bile acids is of an all-or-none type rather than the graded response one might expect. The results of Shefer et al (145) indicate a similar situation in the rat. However, the experimental models used to study regulation of bile acid formation may conceivably not reflect normal conditions in vivo.

The previous discussion has emphasized the relationships between bile acid biosynthesis and cholesterol synthesis in the liver. Chevallier, Lutton, and collaborators (154, 155) have concluded from studies of cholesterol turnover in rats that the intestine may be the major site of cholesterol synthesis and that the rate of cholesterol conversion into bile acids may be adaptively controlled by the dynamic equilibrium of cholesterol. If intestinal synthesis is the major source of cholesterol and consequently of bile acids in the rat, then regulation of bile acid formation by bile acids also might occur in the intestine. There is evidence that, in addition to their role in cholesterol absorption, bile acids influence the rate of cholesterol synthesis in the intestine (156, 157).

EFFECT OF VITAMINS AND HORMONES

Avery & Lupien (158–160) have found that vitamin B-6-deficient rats exhibit an increase in the capacity to conjugate cholic acid with glycine and taurine as measured in vitro. In vivo, there is a proportional increase in glycine conjugated bile acids, as found earlier by Bergeret & Chatagner (161), and an increase in the daily synthesis of cholic acid, which might reflect the increase in cholesterol synthesis seen in vitamin B-6-deficient rats.

In vitamin C-deficient guinea pigs. Ginter and collaborators (162–165) found a significant decrease in the rate of bile acid biosynthesis, which is normalized by administration of vitamin C. The mechanism of action of the vitamin is unclear. In deficient guinea pigs there is a subnormal concentration of cytochrome P-450. Upon administration of vitamin C, cytochrome P-450 concentration and oxidation in vivo of $[26-^{14}C]$ cholesterol increase in parallel (163). Kritchevsky et al (166) found no significant effect on 7α-hydroxylation of cholesterol or oxidation of the side chain of cholesterol by addition of vitamin C to liver homogenate fractions from normal guinea pigs and rats. No experiments with liver homogenates from deficient animals have been reported.

Eskelson et al (167) reported that liver homogenates from vitamin E-deficient rabbits have a decreased capacity to catabolize cholesterol, and Kikuchi et al (168) found that administration of vitamins K-1 and K-2 to rats results in an increased excretion of bile acids in feces. The mechanisms of these effects are unknown.

Recent studies on the influence of hormones on bile acid metabolism have focused on the regulation of the diurnal rhythm of cholesterol 7α-hydroxylase. It has been shown that in the rat, hypophysectomy or adrenalectomy results in abolition of the diurnal rhythm and that the rhythm returns upon treatment of adrenalectomized animals with corticosterone (60, 169). The role of glucocorticoids has been further established by van Cantfort (170).

EFFECT OF DIET

There are a number of recent studies concerning the effect on cholesterol and bile acid metabolism of different diets and drugs, such as anion exchange resins and inhibitors of cholesterol synthesis. Space does not allow a review of the whole subject. The selection has been made to summarize some recent studies on the effect of dietary cholesterol. These studies point to the possibility of species differences in the regulation of bile acid formation. The differing responses toward cholesterol feeding between man and several experimental animals have been confirmed in several detailed studies. Quintão et al (171) have extended the findings of Grundy et al (172) on the effect of a high cholesterol diet on cholesterol balance in man. In contrast to dogs and rats, for instance, man appears unable to increase bile acid production in response to excess cholesterol but can increase excretion of neutral steroids in feces and suppress endogenous cholesterol synthesis. The ability to depress synthesis is rather variable between subjects and ranges from no suppression to almost complete suppression. The rhesus monkey, squirrel monkey, and baboon in particular resemble man in their response to excess cholesterol (173). In long term experiments with dogs on a high cholesterol diet, Pertsemlidis et al (174, 175) found that bile acid formation and excretion increased and cholesterol synthesis decreased to the extent that no marked expansion of tissue cholesterol pools occurred.

The effect of dietary cholesterol on the rate-limiting enzymes of cholesterol and bile acid formation has been the subject of several reports. Boyd et al (53), van Cantfort (176), and Mitropoulos et al (63) found an increase in cholesterol 7α-hydroxylase upon feeding 1–4% cholesterol in the diet to rats, and Mitropoulos et al (63) observed an increase 6 hr after the start of feeding 1% cholesterol in the diet. Shefer et al (65) observed no effect after one week but an increase after two weeks of feeding a diet with 2% cholesterol. The differing results may reflect the problems with assay of cholesterol 7α-hydroxylase activity, and Mitropoulos et al (63) have presented evidence that cholesterol feeding results in an increase in substrate supply rather than in enzyme capacity. Shefer et al (65) followed HMG CoA reductase activity simultaneously and observed the expected decrease upon cholesterol feeding. The opposite response—stimulation of cholesterol synthesis and unchanged bile acid biosynthesis—is observed upon feeding tomatine, a saponin that binds cholesterol (177; G. S. Boyd, personal communication).

CLINICAL ASPECTS OF BILE ACID METABOLISM

Interest in the biosynthesis and metabolism of bile acids in various clinical conditions involving disturbances of lipid metabolism and liver function has increased markedly in the last five years. Areas of particular interest have been gallstone formation, intra- and extrahepatic cholestasis, liver cirrhosis, hyper-lipidemias, and intestinal diseases. These subjects have been treated in excellent reviews in *The Bile Acids* (8) and in the October 1972 issue of *Archives of Internal Medicine,* so only a few problems related to pathways and regulation of bile acid metabolism will be discussed here.

Gallstone Disease

Cholesterol gallstones are formed in bile supersaturated with cholesterol (178). Although the concept of a bile acid-phospholipid-cholesterol micelle in bile (179) has been questioned (180), bile acids play the most important role in keeping cholesterol in solution both directly and by controlling phospholipid secretion into bile (181, 182).

The pools of cholic and chenodeoxycholic acids are smaller in subjects with gallstones than in those without (183–187). Hepatic secretion of bile acids, which depends on synthesis in the liver and reabsorption of the pool during entero-hepatic circulation, is lower in gallstone patients (184, 188), and the half-lives of the primary bile acids are shorter. The rate of bacterial conversion of cholate into deoxycholate is higher (93, 186, 187, 189). The question is whether these changes are the cause or result of the gallstone disease. Impairment of the storage capacity of the gallbladder with increased recycling of bile acids has been suggested as a partial explanation (186), but other factors are probably more important. Thus, American Indian women with a high incidence of gallstones and lithogenic bile (190) have a small bile acid pool with a short half-life, whether gallstones are present or not (191). There is no abnormal loss of bile acids (185, 186, 188) as in patients with ileal dysfunction, who have lithogenic bile (192) and an increased incidence of gallstones (192–194). American Indian women with lithogenic bile may have a defective homeostatic regulation of bile acid synthesis (185, 188), but experimental evidence on this point is lacking.

Pools of bile acids have been reported to increase (189), remain unchanged (195), or decrease (196) upon removing the gallbladder in gallstone patients. Increased fractional turnover of primary bile acids (195, 196) and increased formation of secondary bile acids (189, 195, 197) indicate accelerated recycling of bile acids, which may help to make the bile less saturated with cholesterol. In coeliac disease, which is associated with gallbladder inertia, the opposite is observed, i.e. the pool size and the half-life of cholic acid are markedly increased (198).

When chenodeoxycholic acid is administered to gallstone patients, the total bile acid pool is increased in spite of a marked decrease of the cholic and deoxycholic acid pools (146, 199). Cholesterol solubilization is increased (200). It is not known whether ursodeoxycholic acid, whose concentration in bile increases markedly (201), plays a role in this connection. The decreased pool of cholic acid is explained by inhibition of its synthesis combined with decreased efficiency of intestinal absorption (146). The decreased lithogenicity of bile during chenodeoxycholic acid administration is not due simply to the increased pool size. Administration of cholic acid also leads to an increased bile acid pool, but the cholesterol saturation of bile is not decreased (200).

Several commonly used drugs may be important in gallstone formation. Clofibrate and an estrogen-progestin combination decreased cholic acid synthesis and pool size and increased the lithogenicity of bile (202). The effect of the steroids may be related to both the increased incidence of gallstones in women taking contraceptive steroids and to gallstone formation in pregnancy. In contrast, phenobarbital increases production and pool size of cholic acid in man (203) and decreases cholesterol saturation in bile of monkeys (153).

Liver Cirrhosis and Cholestasis

It is well known that bile acid levels in serum increase in liver disease (204, 205). Patients with cirrhosis have smaller pools of cholic and deoxycholic acids than normal subjects, whereas the chenodeoxycholic acid pool is normal (206, 207). This agrees with the earlier finding that bile of these patients has a high percentage of chenodeoxycholic acid and very little deoxycholic acid. Cholic acid production and fecal excretion of bile acids are reduced (206, 208). Cirrhotic patients have a decreased capacity for bile acid synthesis since cholestyramine increases fecal excretion of bile acids much less than in normal subjects (208). A small fraction of the bile acids is eliminated with urine (208, 209); a major part of such bile acids, particularly the di- and monohydroxycholanoates, is present as sulfate esters (104, 105) and perhaps glucuronides (106).

In patients with extra- or intrahepatic cholestasis, elimination of bile acids with urine may be considerable (104, 105, 210–217). A large number of less common bile acids including 6-hydroxylated ones and epimers of 5α-cholanoates can be found (218). The existence of these compounds, which may be quantitatively important, has been neglected in many studies. In infants with extrahepatic biliary atresia, most of the conjugated metabolites of lithocholic and chenodeoxycholic acids are monosulfates, whereas cholic acid conjugates are sulfated to only a minor extent (214, 219). The pool size and excretion rate of cholic acid are lower in infants with biliary atresia and intrahepatic cholestasis (211, 213) than in normal newborn infants (220). In a child with intrahepatic bile duct hypoplasia, administration of phenobarbital reduced pool sizes of cholic and chenodeoxycholic acids, decreased excretion of bile acids in urine, and increased excretion of bile acids in feces (217).

Subjects with cholestasis excrete sulfated 3β-hydroxy-5-cholenoic acid in urine (215, 221). In cholestatic infants, administered [^{14}C] cholesterol did not appear to be the direct precursor (221). This bile acid is also found in meconium (222) and in trace amounts in bile (223). Other unsaturated and saturated monohydroxycholanoic acids have also been detected (221, 224). The role of these and other hepatotoxic bile acids in cholestatic syndromes is not clear. In rats with cholestasis, produced by bile duct ligation or administration of α-naphthylisothiocyanate, the hepatic concentration of β-muricholic acid increases markedly (225, 226). This is a result of increased 6β hydroxylation of chenodeoxycholic acid (227), which may be a protective mechanism (225). An interesting case where increased production of lithocholic acid could cause cholestatic cirrhosis has been described (228).

Hyperlipoproteinemias

The discovery (229) of a difference in bile acid formation between patients with primary hypercholesterolemia (hyperlipoproteinemia type II) and those with hypercholesterolemia combined with hypertriglyceridemia (hyperlipoproteinemia type IV) has been confirmed and extended. Synthesis of cholic acid is lower than normal in hyperlipoproteinemia type II and much higher in type IV (230–232). Chenodeoxycholic acid turnover tends to be slightly increased in both types of hyperlipoproteinemia (231, 232). These changes are reflected in a lower fecal bile acid excretion in patients with hyperlipoproteinemia type II than in those with type IV (205, 233, 234) and in an increased proportion of deoxycholic acid in feces in the latter group of patients (234). However, the actual amounts of bile acids measured by the isotope dilution technique or by analysis of bile acids in feces are quite different and the old question of the validity of the two methods under different conditions requires further study.

Administration of cholestyramine to patients with hyperlipoproteinemia type II increases bile acid synthesis and excretion (233, 235, 236). The abnormally low ratio between cholic acid production and chenodeoxycholic acid production is normalized by a significant increase in cholic acid pool size and synthesis (237). Cholestyramine does not change cholic acid kinetics in patients with hyperlipoproteinemia type IV who already have a high cholic acid production. Thus, it appears that the capacity of the liver to synthesize bile acids is the same in the two diseases. In patients with hyperlipoproteinemia type II, cholic acid ingestion increases the pools of cholic and deoxycholic acids markedly and depresses chenodeoxycholic acid pool and production (238). In hyperlipoproteinemia type IV, the increase, if any, of the cholic acid pool is small, and chenodeoxycholic acid synthesis is less depressed than normal. This may indicate a defect in cholic acid absorption (238). Chenodeoxycholic acid feeding reduces cholic acid synthesis more in patients with hyperlipoproteinemia type II than in those with type IV, which may indicate a relative insensitivity in the control of cholic acid synthesis in hyperlipoproteinemia type IV (239).

NEW DEVELOPMENTS IN METHODOLOGY

Numerous modifications of conventional analytical methods using thin layer and gas chromatography have been described. Proofs of their specificity are not always satisfactory. Enzymatic determinations with 3-hydroxysteroid dehydrogenase have often been poorly validated for analyses of small amounts of bile acids (cf 240) and comparisons with radioimmunoassay indicate a lack of specificity (241). The combined use of 3α- and 7α-hydroxysteroid dehydrogenases (242, 243) may be useful in routine analyses of bile samples. The development of radioimmunoassay methods for specific types of bile acids is an important advance in clinical bile acid research (241, 244).

Most methods presently used are based on the assumptions that the bile acid composition of the sample is known and that the appropriate bile acid is being determined. These assumptions have often proven to be incorrect. Gas chromatography–mass spectrometry may be used for unbiased bile acid analysis if repetitive scanning and computer evaluation of data are used (245, 246). Fragment ion current chromatograms are then obtained of all m/e values covered by the scan, and computer programs permit location (247) and quantitative determination (246) from peak areas in chromatograms of m/e values specific for individual or groups of bile acid derivatives.

Convenient new methods for extraction and chromatography of bile acids have been described. 2,2-Dimethoxypropane gives a single phase extraction of bile acids from plasma (248). The neutral resin Amberlite XAD-2 is very useful for liquid-solid extraction of bile acids from plasma and urine (211, 249). Neutral (102, 250, 251) and ion exchanging (218) lipophilic-hydrophobic derivatives of Sephadex provide simple means of obtaining straight or reversed phase liquid-gel chromatographic systems for separating bile acids and their conjugates.

Improvements in methods for studying bile acid metabolism include a refined technique for kinetic analysis of data obtained after administration of $[^{14}C]$

cholesterol (252). The progress in gas chromatography–mass spectrometry has revived the use of stable isotopes in both analytical (71) and metabolic applications (220, 253). Ethanol labeled with stable isotopes has been used as precursor in studies of compartmentation and kinetics of bile acid synthesis (254, 255). ^{13}C-NMR permits studies not only of ^{13}C labeling of individual carbon atoms but also of ^{2}H labeling at specific carbon atoms (256) and of intramolecular associations in bile acid solutions (257). Proton NMR has been used to study micellar solutions (258) and also to identify bile acid derivatives (259).

Literature Cited

1. Bergström, S., Danielsson, H., Samuelsson, B. 1960. *Lipide Metabolism,* ed. K. Bloch, 291–336. New York: Wiley
2. Danielsson, H. 1963. *Advan. Lipid Res.* 1:335–85
3. Danielsson, H., Tchen, T. T. 1968. *Metab. Pathways* 2:117–68
4. Danielsson, H., Einarsson, K. 1969. *Biol. Basis Med.* 5:279–315
5. Elliott, W. H., Hyde, P. M. 1971. *Am. J. Med.* 51:568–79
6. Boyd, G. S., Percy-Robb, I. W. 1971. *Am. J. Med.* 51:580–87
7. Tyor, M. P., Garbutt, J. T., Lack, L. 1971. *Am. J. Med.* 51:614–26
8. Nair, P. P., Kritchevsky, D., Eds. 1971/1973. *The Bile Acids,* Vols. 1/2. New York: Plenum
9. Mendelsohn, D., Mendelsohn, L. 1972. *S. Afr. J. Med. Sci.* 37:61–67
10. Mitropoulos, K. A., Myant, N. B. 1967. *Biochem. J.* 103:472–79
11. Mitropoulos, K. A., Avery, M. D., Myant, N. B., Gibbons, G. F. 1972. *Biochem. J.* 130:363–71
12. Ayaki, Y., Yamasaki, K. 1970. *J. Biochem.* 68:341–46
13. Yamasaki, K., Ayaki, Y., Yamasaki, H. 1971. *J. Biochem.* 70:715–18
14. Ayaki, Y., Yamasaki, K. 1972. *J. Biochem.* 71:85–89
15. Ikawa, S., Ayaki, Y., Ogura, M., Yamasaki, K. 1972. *J. Biochem.* 71:579–87
16. Björkhem, I., Gustafsson, J. 1973. *Eur. J. Biochem.* 36:201–12
17. Ogura, M., Shiga, J., Yamasaki, K. 1971. *J. Biochem.* 70:967–72
18. Mathé, D., D'Hollander, F., Chevallier, F. 1972. *Biochimie* 54:1479–81
19. Björkhem, I., Danielsson, H., Einarsson, K., Johansson, G. 1968. *J. Clin. Invest.* 47:1573–82
20. Hanson, R. F., Klein, P. D., Williams, G. C. 1973. *J. Lipid Res.* 14:50–53
21. Staple, E., Rabinowitz, J. L. 1962. *Biochim. Biophys. Acta* 59:735–36
22. Carey, J. B. Jr. 1964. *J. Clin. Invest.* 43:1443–48
23. Hanson, R. F., Williams, G. 1971. *Biochem. J.* 121:863–64
24. Hanson, R. F. 1971. *J. Clin. Invest.* 50:2051–55
25. Anderson, K. E., Kok, E., Javitt, N. B. 1972. *J. Clin. Invest.* 51:112–17
26. Wachtel, N., Emerman, S., Javitt, N. B. 1968. *J. Biol. Chem.* 243:5207–12
27. Eyssen, H., Parmentier, G., Compernolle, F., Boon, J., Eggermont, E. 1972. *Biochim. Biophys. Acta* 273:212–21
28. Setoguchi, T., Salen, G., Tint, G. S., Mosbach, E. H. 1974. *J. Clin. Invest.* 53:1393–1401
29. Salen, G., Grundy, G. S. 1973. *J. Clin. Invest.* 52:2822–35
30. Cronholm, T., Johansson, G. 1970. *Eur. J. Biochem.* 16:373–81
31. Mosbach, E. H., Salen, G. 1975. *Proc. 3rd Bile Acid Meeting in Freiburg, June 13–15, 1974.* Stuttgart: F. K. Schattauer. In press
32. Haslewood, G. A. D. 1967. *Bile Salts.* London: Methuen & Co. 116 pp.
33. Hoshita, T., Kazuno, T. 1968. *Advan. Lipid Res.* 6:207–54
34. Karavolas, H. J., Elliott, W. H., Hsia, S. L., Doisy, E. A., Jr., Matschiner, J. T., Thayer, S. A., Doisy, E. A. 1965. *J. Biol. Chem.* 240:1568–72
35. Ziller, S. A. Jr., Doisy, E. A. Jr., Elliott, W. H. 1968. *J. Biol. Chem.* 243:5280–88
36. Noll, B. W., Walsh, L. B., Doisy, E. A. Jr., Elliott, W. H. 1972. *J. Lipid Res.* 13:71–77
37. Björkhem, I., Gustafsson, J. 1971. *Eur. J. Biochem.* 18:207–13
38. Mui, M. M., Elliott, W. H. 1971. *J. Biol. Chem.* 246:302–4
39. Noll, B. W., Doisy, E. A. Jr., Elliott, W. H. 1973. *J. Lipid Res.* 14:385–90
40. Noll, B. W., Doisy, E. A. Jr., Elliott, W. H. 1973. *J. Lipid Res.* 14:391–99
41. Blaskiewicz, R. J., O'Neil, G. J. Jr., Elliott, W. H. 1974. *Proc. Soc. Exp.*

Biol. Med. 146:92–95
42. Yamasaki, K., Ayaki, Y., Yamasaki, G. 1972. *J. Biochem.* 71:927–30
43. Björkhem, I., Einarsson, K. 1970. *Eur. J. Biochem.* 13:174–79
44. Danielsson, H., Kallner, A., Sjövall, J. 1963. *J. Biol. Chem.* 238:3846–52
45. Kallner, A. 1967. *Acta Chem. Scand.* 21:87–92
46. Kallner, A. 1967. *Acta Chem. Scand.* 21:315–21
47. Kallner, A. 1967. *Ark. Kemi* 26:567–76
48. Wada, F., Hirata, K., Nakano, K., Sakamoto, Y. 1969. *J. Biochem. Tokyo* 66:699–703
49. Boyd, G. S., Grimwade, A. M., Lawson, M. E. 1973. *Eur. J. Biochem.* 37:334–40
50. Johansson, G. 1971. *Eur. J. Biochem.* 21:68–79
51. Björkhem, I., Danielsson, H., Wikvall, K. 1975. *Biochem. Biophys. Res. Commun.* In press
52. Danielsson, H., Einarsson, K., Johansson, G. 1967. *Eur. J. Biochem.* 2:44–49
53. Boyd, G. S., Scholan, N. A., Mitton, J. R. 1969. *Advan. Exp. Med. Biol.* 4:443–56
54. Shefer, S., Hauser, S., Mosbach, E. H. 1968. *J. Lipid Res.* 9:328–33
55. Johansson, G. 1970. *Eur. J. Biochem.* 17:292–95
56. Atkin, S. D., Palmer, E. D., English, P. D., Morgan, B., Cawthorne, M. A., Green, J. 1972. *Biochem. J.* 128:237–42
57. Einarsson, K., Johansson, G. 1968. *Eur. J. Biochem.* 6:293–98
58. Shefer, S., Hauser, S., Mosbach, E. H. 1972. *J. Lipid Res.* 13:69–70
59. Gielen, J., van Cantfort, J., Robaye, B., Renson, J. 1969. *C.R. Acad. Sci. Paris* 269:731–32
60. Mayer, D., Voges, A. 1972. *Z. Physiol. Chem.* 353:1187–88
61. Danielsson, H. 1972. *Steroids* 20:63–72
62. Mitropoulos, K. A., Balasubramaniam, S., Gibbons, G. F., Reeves, B. E. A. 1972. *FEBS Lett.* 27:203–6
63. Mitropoulos, K. A., Balasubramaniam, S., Myant, N. B. 1973. *Biochim. Biophys. Acta* 326:428–38
64. Danielsson, H. 1973. *Steroids* 22:667–76
65. Shefer, S., Hauser, S., Lapar, V., Mosbach, E. H. 1973. *J. Lipid Res.* 14:573–80
66. Brown, M. J. G., Boyd, G. S. 1974. *Eur. J. Biochem.* 44:37–47
67. Aringer, L., Eneroth, P. 1973. *J. Lipid Res.* 14:563–72
68. Boyd, G. S., Brown, M. J. G., Hattersley, N. G., Suckling, K. E. 1974. *Biochim. Biophys. Acta* 337:132–35

69. Mitropoulos, K. A., Balasubramaniam, S. 1972. *Biochem. J.* 128:1–9
70. Balasubramaniam, S., Mitropoulos, K. A., Myant, N. B. 1973. *Eur. J. Biochem.* 34:77–83
71. Björkhem, I., Danielsson, H. 1974. *Anal. Biochem.* 59:508–16
72. Björkhem, I., Danielsson, H. 1975. *Eur. J. Biochem.* In press
73. Bernhardsson, C., Björkhem, I., Danielsson, H., Wikvall, K. 1973. *Biochem. Biophys. Res. Commun.* 54:1030–38
74. Einarsson, K. 1968. *Eur. J. Biochem.* 5:101–8
75. Suzuki, M., Mitropoulos, K. A., Myant, N. B. 1968. *Biochem. Biophys. Res. Commun.* 30:516–21
76. Mitropoulos, K. A., Suzuki, M., Myant, N. B., Danielsson, H. 1968. *FEBS Lett.* 1:13–15
77. Björkhem, I., Danielsson, H., Wikvall, K. *Biochem. Biophys. Res. Commun.* Submitted
78. Björkhem, I., Gustafsson, J. 1973. *Eur. J. Biochem.* 36:201–12
79. Björkhem, I., Danielsson, H., Gustafsson, J. 1973. *FEBS Lett.* 31:20–22
80. Taniguchi, S., Hoshita, N., Okuda, K. 1973. *Eur. J. Biochem.* 40:607–17
81. Björkhem, I., Gustafsson, J. 1974. *J. Biol. Chem.* 249:2528–35
82. Björkhem, I., Gustafsson, J., Johansson, G., Persson, B. 1975. *J. Clin. Invest.* In press
83. Okuda, T., Takigawa, N. 1970. *Biochim. Biophys. Acta* 220:141–48
84. Björkhem, I., Jörnvall, H., Zeppezauer, E. 1973. *Biochem. Biophys. Res. Commun.* 52:413–20
85. Björkhem, I., Jörnvall, H., Åkeson, Å. 1974. *Biochem. Biophys. Res. Commun.* 57:870–75
86. Okuda, K., Higuchi, E., Fukuba, R. 1973. *Biochim. Biophys. Acta* 293:15–25
87. Fukuba, R. 1974. *Biochim. Biophys. Acta* 341:48–55
88. Rosenberg, I. H. 1969. *Am. J. Clin. Nutr.* 22:248–91
89. Tabaqchali, S. 1970. *Scand. J. Gastroenterol.* 5(Suppl. 6):139–63
90. Kellogg, T. F. 1971. *Fed. Proc.* 30:1808–14
91. Kellogg, T. F. 1973. *Bile Acids* 2:283–304
92. Lewis, R., Gorbach, S. 1972. *Arch. Intern. Med.* 130:545–49
93. Norman, A. 1970. *Scand. J. Gastroenterol.* 5:231–36
94. Hepner, G. W., Hofmann, A. F., Thomas, P. J. 1972. *J. Clin. Invest.* 51:1889–97
95. Hepner, G. W., Hofmann, A. F., Thomas, P. J. 1972. *J. Clin. Invest.* 51:1898–1905
96. Garbutt, J. T., Wilkins, R. M., Lack, L.,

Tyor, M. P. 1970. *Gastroenterology* 59: 553–66
97. Ståhl, E., Arnesjö, B. 1972. *Scand. J. Gastroenterol.* 7: 559–66
98. Hepner, G. W., Sturman, J. A., Hofmann, A. F., Thomas, P. J. 1973. *J. Clin. Invest.* 52: 433–40
99. Palmer, R. H. 1967. *Proc. Nat. Acad. Sci. USA* 58: 1047–50
100. Palmer, R. H., Bolt, M. G. 1971. *J. Lipid Res.* 12: 671–79
101. Palmer, R. H. 1971. *J. Lipid Res.* 12: 680–87
102. Cronholm, T., Makino, I., Sjövall, J. 1972. *Eur. J. Biochem.* 26: 251–58
103. Makino, I., Shinozaki, K., Nakagawa, S. 1973. *Lipids* 8: 47–49
104. Stiehl, A. 1974. *Eur. J. Clin. Invest.* 4: 59–63
105. Makino, I., Shinozaki, K., Nakagawa, S., Mashimo, K. 1974. *J. Lipid Res.* 15: 132–38
106. Back, P., Spaczynski, K., Gerok, W. 1974. *Z. Physiol. Chem.* 355: 749–52
107. Liersch, M., Stiehl, A. 1974. *Z. Gastroenterol.* 12: 131–34
108. Kallner, A. 1967. *Ark. Kemi* 26: 553–65
109. Berséus, O. 1967. *Eur. J. Biochem.* 2: 493–502
110. Björkhem, I., Danielsson, H. 1970. *Eur. J. Biochem.* 12: 80–84
111. Björkhem, I., Danielsson, H., Wikvall, K. 1973. *Eur. J. Biochem.* 36: 8–15
112. Waller, G., Theorell, H., Sjövall, J. 1965. *Arch. Biochem. Biophys.* 111: 671–84
113. Reynier, M., Theorell, H., Sjövall, J. 1969. *Acta Chem. Scand.* 23: 1130–36
114. Hoshita, N. 1972. *Hiroshima J. Med. Sci.* 21: 49–58
115. Usui, T., Oshio, R., Kawamoto, M., Yamasaki, K. 1966. *Yonago Acta Med.* 10: 252–59
116. Trülzsch, D. et al 1973. *Biochemistry* 12: 76–79
117. Björkhem, I., Danielsson, H., Wikvall, K. 1973. *Biochem. Biophys. Res. Commun.* 53: 609–16
118. Prager, G. N., Voigt, W., Hsia, S. L. 1973. *J. Biol. Chem.* 248: 8442–48
119. Björkhem, I., Danielsson, H., Wikvall, K. 1974. *J. Biol. Chem.* 249: 6439–45
120. Hanson, R. F., Williams, G. 1971. *J. Lipid Res.* 12: 688–91
121. Einarsson, K., Hellström, K. 1974. *Clin. Sci. Mol. Med.* 46: 183–90
122. Björkhem, I., Einarsson, K., Hellers, G. 1973. *Eur. J. Clin. Invest.* 3: 459–65
123. Björkhem, I., Danielsson, H. 1974. *Mol. Cell. Biochem.* 4: 79–95
124. Voigt, W., Hsia, S. L., Cooper, D. Y., Rosenthal, O. 1968. *FEBS Lett.* 2: 124–26

125. Einarsson, K., Johansson, G. 1968. *FEBS Lett.* 4: 177–80
126. Trülzsch, D. et al 1974. *Biochem. Med.* 9: 158–66
127. Czygan, P. et al 1974. *Biochim. Biophys. Acta* 354: 168–71
128. Naqvi, S. H. M. et al 1969. *J. Lipid Res.* 10: 115–20
129. Martin, C. W., Nicholas, H. J. 1972. *Steroids* 19: 549–65
130. Martin, C. W., Nicholas, H. J. 1973. *Steroids* 21: 633–46
131. Naqvi, S. H. M., Nicholas, H. J. 1973. *Lipids* 8: 651–53
132. Shefer, S., Hauser, S., Bekersky, I., Mosbach, E. H. 1970. *J. Lipid Res.* 11: 404–11
133. Mosbach, E. H., Rothschild, M. A., Bekersky, I., Oratz, M., Mongelli, J. 1971. *J. Clin. Invest.* 50: 1720–30
134. Rodwell, V. W., McNamara, D. J., Shapiro, D. J. 1973. *Advan. Enzymol.* 38: 373–412
135. Back, P., Hamprecht, B., Lynen, F. 1969. *Arch. Biochem. Biophys.* 133: 11–21
136. Hamprecht, B., Nüssler, C., Lynen, F. 1969. *FEBS Lett.* 4: 117–21
137. Schoenfield, L. J., Bonorris, G. G., Ganz, P. 1973. *J. Lab. Clin. Med.* 82: 858–68
138. Shefer, S., Hauser, S., Lapar, V., Mosbach, E. H. 1972. *J. Lipid Res.* 13: 402–12
139. Danielsson, H., Johansson, G. 1974. *Gastroenterology* 67: 126–34
140. Hamprecht, B., Nüssler, C., Waltinger, G., Lynen, F. 1970. *Eur. J. Biochem.* 18: 10–14
141. Liersch, M. E. A., Barth, A., Hackenschmidt, H. J., Ullmann, H. L., Decker, K. F. A. 1973. *Eur. J. Biochem.* 32: 365–71
142. Weis, H. J., Dietschy, J. M. 1969. *J. Clin. Invest.* 48: 2398–2408
143. Hamprecht, B., Roscher, R., Waltinger, G., Nüssler, C. 1971. *Eur. J. Biochem.* 18: 15–19
144. Myant, N. B., Eder, H. A. 1961. *J. Lipid Res.* 2: 363–68
145. Shefer, S., Hauser, S., Bekersky, I., Mosbach, E. H. 1969. *J. Lipid Res.* 10: 646–55
146. Danzinger, R. G., Hofmann, A. F., Thistle, J. L., Schoenfield, L. J. 1973. *J. Clin. Invest.* 52: 2809–21
147. Einarsson, K., Hellström, K., Kallner, M. 1973. *Metabolism* 22: 1477–83
148. Einarsson, K., Hellström, K., Kallner, M. 1974. *Clin. Sci. Mol. Med.* 47: 425–33
149. Low-Beer, T. S., Pomare, E. W., Morris, J. S. 1972. *Nature New Biol.* 238: 215–16

150. Low-Beer, T. S., Pomare, E. W. 1973. *Brit. Med. J.* 338–40
151. Dowling, R. H., Mack, E., Small, D. M. 1970. *J. Clin. Invest.* 49:232–42
152. Dowling, R. H. 1972. *Gastroenterology* 62:122–40
153. Small, D. M., Dowling, R. H., Redinger, R. N. 1972. *Arch. Intern. Med.* 130: 552–73
154. Chevallier, F., Lutton, C. 1973. *Nature New Biol.* 242:61–62
155. Lutton, C., Mathé, D., Chevallier, F. 1973. *Biochim. Biophys. Acta* 306:483–96
156. Dietschy, J. M. 1968. *J. Clin. Invest.* 47:286–300
157. Shefer, S., Hauser, S., Lapar, V., Mosbach, E. H. 1973. *J. Lipid Res.* 14:400–5
158. Avery, M. D., Lupien, P. J. 1971. *Can. J. Biochem.* 49:1026–30
159. Avery, M. D., Lupien, P. J. 1972. *Rev. Can. Biol.* 31:269–75
160. Avery, M. D., Lupien, P. J. 1972. *Rev. Can. Biol.* 31:277–86
161. Bergeret, B., Chatagner, F. 1956. *Biochim. Biophys. Acta* 22:273–77
162. Ginter, E., Nemec, R., Bobek, P. 1972. *Brit. J. Nutr.* 28:205–11
163. Ginter, E., Nemec, R. 1972. *Physiol. Bohemoslov.* 21:539–45
164. Ginter, E. 1973. *Science* 179:702–4
165. Ginter, E., Nemec, R., Cěrveň, J., Mikuš, L. 1973. *Lipids* 8:135–41
166. Kritchevsky, D., Tepper, S. A., Story, J. A. 1973. *Lipids* 482–84
167. Eskelson, C. D., Jacobi, H. P., Fitch, D. M. 1973. *Physiol. Chem. Phys.* 5: 319–29
168. Kikuchi, H., Kuramoto, T., Hoshita, T., Yamamoto, S. 1973. *Life Sci.* 13:933–43
169. Gielen, J., Robaye, B., van Cantfort, J., Renson, J. 1970. *Arch. Int. Pharmacodyn.* 183:403–5
170. van Cantfort, J. 1973. *Biochimie* 55: 1171–73
171. Quintão, E., Grundy, S. M., Ahrens, E. H. Jr. 1971. *J. Lipid Res.* 12:233–47
172. Grundy, S. M., Ahrens, E. H. Jr., Davignon, J. 1969. *J. Lipid Res.* 10: 304–15
173. Eggen, D. A. 1974. *J. Lipid Res.* 15: 139–45
174. Pertsemlidis, D., Kirchman, E. H., Ahrens, E. H. Jr. 1973. *J. Clin. Invest.* 52:2353–67
175. Pertsemlidis, D., Kirchman, E. H., Ahrens, E. H. Jr. 1973. *J. Clin. Invest.* 52:2368–78
176. van Cantfort, J. 1972. *C.R. Acad. Sci. Paris* 275:1015–17
177. Cayen, M. N. 1971. *J. Lipid Res.* 12:

178. Redinger, R. N., Small, D. M. 1972. *Arch. Intern. Med.* 130:618–30
179. Admirand, W. H., Small, D. M. 1968. *J. Clin. Invest.* 47:1043–52
180. Lairon, D., Lafont, H., Hauton, J.-C. 1972. *Biochimie* 54:529–30
181. Swell, L., Bell, C. C. Jr., Entenman, C. 1968. *Biochim. Biophys. Acta* 164:278–84
182. Nilsson, S., Scherstén, T. 1969. *Gastroenterology* 57:525–32
183. Vlahcevic, Z. R., Bell, C. C. Jr., Buhac, I., Farrar, J. T., Swell, L. 1970. *Gastroenterology* 59:165–73
184. Swell, L., Bell, C. C. Jr., Vlahcevic, Z. R. 1971. *Gastroenterology* 61:716–22
185. Vlahcevic, Z. R. et al 1972. *Gastroenterology* 62:73–83
186. Pomare, E. W., Heaton, K. W. 1973. *Gut* 14:885–90
187. Arnesjö, B., Ståhl, E. 1973. *Scand. J. Gastroenterol.* 8:369–75
188. Grundy, S. M., Metzger, A. L., Adler, R. D. 1972. *J. Clin. Invest.* 51:3026–43
189. Hepner, G. W., Hofmann, A. F., Malagelada, J. R., Szczepanik, P. A., Klein, P. D. 1974. *Gastroenterology* 66: 556–64
190. Thistle, J. L., Schoenfield, L. J. 1971. *N. Engl. J. Med.* 284:177–81
191. Bell, C. C. Jr., Vlahcevic, Z. R., Prazich, J., Swell, L. 1973. *Surg. Gynecol. Obstet.* 136:961–65
192. Hofmann, A. F. 1972. *Arch. Intern. Med.* 130:597–605
193. Tyor, M. P. 1973. *Bile Acids* 2:83–101
194. Dowling, R. H., Bell, G. D., White, J. 1972. *Gut* 13:415–20
195. Almond, H. R., Vlahcevic, Z. R., Bell, C. C. Jr., Gregory, D. H., Swell, L. 1973. *N. Engl. J. Med.* 289:1213–16
196. Pomare, E. W., Heaton, K. W. 1973. *Gut* 14:753–62
197. Malagelada, J. R., Go, V. L. W., Summerskill, W. H. J., Gamble, W. S. 1973. *Am. J. Dig. Dis.* 18:455–59
198. Low-Beer, T. S., Heaton, K. W., Pomare, E. W., Read, A. E. 1973. *Gut* 14:204–8
199. Danzinger, R. G., Hofmann, A. F., Schoenfield, L. J., Thistle, J. L. 1972. *N. Engl. J. Med.* 286:1–8
200. Thistle, J. L., Schoenfield, L. J. 1971. *Gastroenterology* 61:488–96
201. Salen, G., Tint, G. S., Eliav, B., Deering, N., Mosbach, E. H. 1974. *J. Clin. Invest.* 53:612–21
202. Pertsemlidis, D., Panveliwalla, D., Ahrens, E. H. Jr. 1974. *Gastroenterology* 66:565–73
203. Miller, N. E., Nestel, P. J. 1973. *Clin. Sci. Mol. Med.* 45:257–62

204. Carey, J. B. Jr. 1973. *Bile Acids* 2:55–82
205. Miettinen, T. A. 1973. *Bile Acids* 2:191–247
206. Vlahcevic, Z. R., Buhac, I., Farrar, J. T., Bell, C. C. Jr., Swell, L. 1971. *Gastroenterology* 60:491–98
207. Vlahcevic, Z. R., Juttijudata, P., Bell, C. C. Jr., Swell, L. 1972. *Gastroenterology* 62:1174–81
208. Miettinen, T. A. 1972. *Gut* 13:682–89
209. Alström, T., Norman, A. 1972. *Acta Med. Scand.* 191:521–28
210. Norman, A., Strandvik, B., Zetterström, R. 1969. *Acta Paediat. Scand.* 58:59–72
211. Norman, A., Strandvik, B. 1971. *J. Lab. Clin. Med.* 78:181–93
212. Norman, A., Strandvik, B. 1973. *Acta Paediat. Scand.* 62:264–68
213. Norman, A., Strandvik, B. 1973. *Acta Paediat. Scand.* 62:253–63
214. Norman, A., Strandvik, B. 1974. *Acta Paediat. Scand.* 63:92–96
215. Back, P. 1973. *Clin. Chim. Acta* 44:199–207
216. Stiehl, A., Thaler, M. M., Admirand, W. H. 1972. *N. Engl. J. Med.* 286:858–61
217. Stiehl, A., Thaler, M. M., Admirand, W. H. 1973. *Pediatrics* 51:992–97
218. Almé, B., Bremmelgaard, A., Sjövall, J., Thomassen, P. *FEBS Lett.* Submitted
219. Norman, A., Strandvik, B., Ojamäe, Ö. 1974. *Acta Paediat. Scand.* 63:97–102
220. Watkins, J. B., Ingall, D., Szczepanik, P., Klein, P. D., Lester, R. 1973. *N. Engl. J. Med.* 288:431–34
221. Makino, I., Sjövall, J., Norman, A., Strandvik, B. 1971. *FEBS Lett.* 15:161–64
222. Back, P., Ross, K. 1973. *Z. Physiol. Chem.* 354:83–89
223. Laatikainen, T., Perheentupa, J., Vihko, R., Makino, I., Sjövall, J. 1972. *J. Steroid Biochem.* 3:715–19
224. Murphy, G. M., Jansen, F. H., Billing, B. H. 1972. *Biochem. J.* 129:491–94
225. Greim, H. et al 1972. *Gastroenterology* 63:837–45
226. Schaffner, F. et al 1973. *Lab. Invest.* 28:321–31
227. Danielsson, H. 1973. *Steroids* 22:567–76
228. Williams, C. N., Kaye, R., Baker, L., Hurwitz, R., Senior, J. R. 1972. *J. Pediat.* 81:493–500
229. Kottke, B. 1969. *Circulation* 40:13–20
230. Einarsson, K., Hellström, K. 1972. *Eur. J. Clin. Invest.* 2:225–30
231. Wollenweber, J., Stiehl, A. 1972. *Klin. Wochenschr.* 50:33–38
232. Einarsson, K., Hellström, K., Kallner, M. 1974. *J. Clin. Invest.* 54:1301–11
233. Miettinen, T. 1970. *Ann. Clin. Res.* 2:300–20
234. Sodhi, H. S., Kudchodkar, B. J. 1973. *Clin. Chim. Acta* 46:161–71
235. Moutafis, C. D., Myant, N. B. 1969. *Clin. Sci.* 37:443–54
236. Grundy, S. M., Ahrens, E. H. Jr., Salen, G. 1971. *J. Lab. Clin. Med.* 78:94–121
237. Einarsson, K., Hellström, K., Kallner, M. 1974. *Eur. J. Clin. Invest.* 4:405–10
238. Einarsson, K., Hellström, K., Kallner, M. 1974. *Metabolism* 23:863–73
239. Kallner, M. *J. Lab. Clin. Med.* Submitted
240. Engert, R., Turner, M. D. 1973. *Anal. Biochem.* 51:399–407
241. Murphy, G. M., Edkins, S. M., Williams, J. W., Catty, D. 1974. *Clin. Chim. Acta* 54:81–89
242. Haslewood, G. A. D., Murphy, G. M., Richardson, J. M. 1973. *Clin. Sci.* 44:95–98
243. MacDonald, I. A., Williams, C. N., Mahony, D. E. 1974. *Anal. Biochem.* 57:127–36
244. Simmonds, W. J., Korman, M. G., Go, V. L. W., Hofmann, A. F. 1973. *Gastroenterology* 65:705–11
245. Reimendal, R., Sjövall, J. 1972. *Anal. Chem.* 44:21–29
246. Axelson, M., Cronholm, T., Curstedt, T., Reimendal, R., Sjövall, J. 1974. *Chromatographia* 7:502–9
247. Back, P., Sjövall, J., Sjövall, K. 1974. *Med. Biol.* 52:31–38
248. Ali, S. S., Javitt, N. B. 1970. *Can. J. Biochem.* 48:1054–57
249. Makino, I., Sjövall, J. 1972. *Anal. Lett.* 5:341–49
250. Sjövall, J., Nyström, E., Haahti, E. 1968. *Advan. Chromatogr.* 6:119–70
251. Ellingboe, J., Nyström, E., Sjövall, J. 1970. *J. Lipid Res.* 11:266–73
252. Quarfordt, S. H., Greenfield, M. F. 1973. *J. Clin. Invest.* 52:1937–45
253. Klein, P. D., Haumann, J. R., Eisler, W. J. 1972. *Anal. Chem.* 44:490–93
254. Cronholm, T., Makino, I., Sjövall, J. 1972. *Eur. J. Biochem.* 24:507–19
255. Cronholm, T., Burlingame, A., Sjövall, J. 1975. *Eur. J. Biochem.* In press
256. Wilson, D. M., Burlingame, A. L., Cronholm, T., Sjövall, J. 1974. *Biochem. Biophys. Res. Commun.* 56:828–35
257. Leibfritz, D., Roberts, J. D. 1973. *J. Am. Chem. Soc.* 95:4996–5003
258. Small, D. M., Penkett, S. A., Chapman, D. 1969. *Biochim. Biophys. Acta* 176:178–89
259. Shalon, Y., Elliott, W. H. 1974. *FEBS Lett.* 41:223–26

BIOLUMINESCENCE: RECENT ADVANCES

×882

Milton J. Cormier, John Lee, and John E. Wampler

Bioluminescence Laboratory, Department of Biochemistry, University of Georgia, Athens, Georgia 30602 and The University of Georgia Marine Institute, Sapelo Island, Georgia 31327

CONTENTS

INTRODUCTION

This review is not intended as a comprehensive one in the field of bioluminescence. A number of topics are not mentioned but most of these have been covered in reviews and popular articles published elsewhere (1–14). The coverage represents relatively recent advances in selected areas which the authors feel are among the more significant ones made during the past five years.

Bioluminescence involves an oxidative event, and the fact that light is a product of the reaction results from the formation of a product molecule in its first electronic singlet excited state. This product molecule may be efficiently fluorescent or, if not, may transfer its excitation energy to an efficiently fluorescent acceptor, thus representing a case of sensitized bioluminescence. The enzymatic mechanisms involved in these oxidations appear to represent variations of the oxygenase and peroxidase ones (2). The terms used in this field may be foreign to the nonspecialist; therefore, the following list of terms and their definitions is provided:

(*a*) Luciferase: a generic term referring to an enzyme that catalyzes the oxidation of a substrate, luciferin, with light emission.

255

(b) Luciferin: a generic term referring to a reduced compound which can be oxidized in an appropriate environment to produce an electronically excited singlet state. This excited singlet state product may be fluorescent and thus may be the emitter. Alternatively, the energy of this excited state product may be transferred to a fluorescent acceptor which functions as the emitter. This would become a case of sensitized bioluminescence as mentioned above.

(c) Photoprotein: a protein-chromophore complex that reacts with certain ions, such as calcium, to produce light. It is believed to be a stabilized oxygenated intermediate of a protein-luciferin chromophore complex (see section on coelenterate bioluminescence).

(d) Bioluminescence or chemiluminescence quantum yield (Q_B or Q_{CL}): the ratio of the total number of photons produced to the total number of molecules of substrate utilized; this relationship is frequently expressed as the number of einsteins per mole. Q_B or Q_{CL} is the product of several terms, thus

$$Q_B \quad \text{or} \quad Q_{CL} = Q_C \times Q_F \times Q_{EX}$$

where Q_C is the chemical yield of product, Q_F is the fluorescence quantum yield of the emitter, and Q_{EX} is the fraction of the product molecules produced in an electronically excited state as opposed to the ground state of the molecule. For some bioluminescent and chemiluminescent reactions, Q_C approaches 1 and this term, therefore, can be ignored in those cases.

For the theoretical organic chemists in this field, one of the central questions relates to how electronic excited states of singlet multiplicity are efficiently generated during bioluminescence. Wherever possible this problem is given attention in this review. Beyond this fundamental question the reader should gain an appreciation of the potential usefulness of these bioluminescent systems in the study of important biological problems, such as the control of nerve-linked calcium transients, mechanisms of flavoprotein catalysis, and mechanisms of the copper-containing peroxidases. In addition, there are numerous analytical applications of these systems reviewed elsewhere (15, 74, 75).

COELENTERATE BIOLUMINESCENCE

Coelenterates are a diverse group of marine animals noted for their beauty and brilliant displays of bioluminescence. The molecular basis for bioluminescence has been investigated in a number of these creatures over the past decade. These data are summarized in this section and emphasis is placed on chemical similarities of the bioluminescent systems among the coelenterates.

The Anthozoans

The bioluminescence of several Anthozoans has been examined at the biochemical level. These include the sea pens (*Stylatula, Renilla,* and *Acanthoptilum*), the sea feather (*Ptilosarcus*), *Parazoanthus,* and *Cavernularia.* Sea pansy (*Renilla*) bioluminescence has been studied in greater detail and consequently discussion is limited to it. However, recent data suggest that the chemical requirements for light emission and its control are similar, if not identical, in all the Anthozoans studied (1, 16–25).

REQUIREMENTS FOR BLUE LIGHT EMISSION   Structural elucidation and chemical synthesis of a partially active analog of *Renilla* luciferin were achieved recently by Hori & Cormier (26). Subsequently, a fully active luciferin (II, Figure 1) was synthesized (27). The imidazolopyrazine ring system which occurs in *Renilla* luciferin also exists as part of the structure of several luciferins of marine origin, which include the crustacean *Cypridina,* a number of fish, numerous coelenterates, and possibly squid (1, 2, 28). The only differences in the structure of *Renilla* and *Cypridina* luciferins are found in the side chains, which are apparently determined by the amino acids utilized in their synthesis. Chemical evidence is available for this since a number of luciferin analogs have been prepared by condensations of any three amino acids selected to yield the desired analog (29).

Synthetic *Renilla* luciferin, as well as native luciferin (I, Figure 1), reacts with luciferase and molecular oxygen to produce blue light ($\lambda_P^1 = 490$ nm). The bioluminescence quantum yield in each case is approximately 5% (27).

The products of bioluminescent oxidation of *Renilla* luciferin are oxyluciferin (IV, Figure 1) and CO_2 (27, 30). Approximately one mole of each is produced per mole of luciferin oxidized. The characteristic visible absorption of luciferin at 435 nm is lost after bioluminescence ceases, and a new absorption is found at 335 nm characteristic of the product oxyluciferin (27).

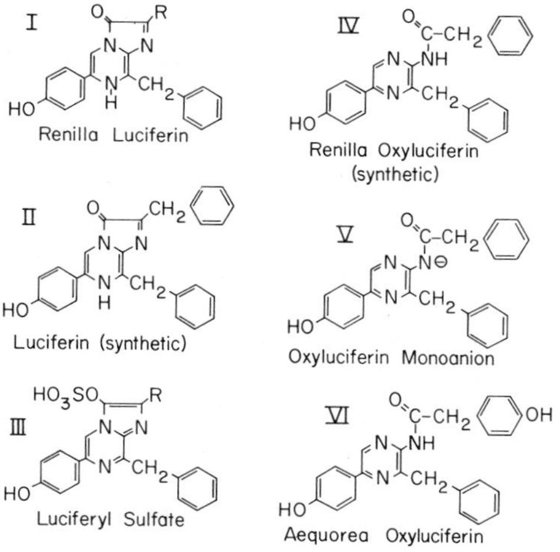

Figure 1   Structures of *Renilla* luciferin, oxyluciferin, and some of their derivatives. Additionally, a comparison of the structure of synthetic *Renilla* oxyluciferin with *Aequorea* oxyluciferin.

[1] λ_B, λ_{CL}, and λ_F refer to the wavelength maxima of bioluminescence, chemiluminescence, and fluorescence emissions, respectively.

Luciferin is oxidized in aprotic solvents, such as dimethylformamide (DMF), via a chemiluminescent pathway (31) to yield oxyluciferin, CO_2, and a bluish luminescence ($\lambda_{CL}^1 = 480$ nm). Oxygen is required for light production and approximately one mole of oxyluciferin is produced per mole of luciferin utilized (27). A study of this chemiluminescent reaction (27, 31) revealed that the blue emission is due to the electronic excited state of the monoanion of oxyluciferin (V, Figure 1) rather than the neutral species (IV, Figure 1) or the dianion, which produces yellow-green light. The evidence suggests that the monoanion is formed directly and that its fluorescence decay rate is faster than the rate of protonation in DMF. Similar conclusions have been reached using model compounds (29, 32, 33).

The electronic excited state of the monoanion of oxyluciferin (V, Figure 1) also appears to be responsible for the bluish emission observed in bioluminescence (27). Of interest here is the observation that neither oxyluciferin nor its monoanion are fluorescent in an aqueous environment (27). In DMF, however, these species are highly fluorescent both for oxyluciferin and several of its analogs (27). For example, oxyluciferin ($\lambda_F^1 = 402$ nm) has a fluorescence quantum yield of 23% in DMF, whereas its monoanion ($\lambda_F = 480$ nm) should have a fluorescence quantum yield of 6%, as deduced from studies on synthetic derivatives (27). Since the bioluminescence quantum yield is 5% and since its λ_B is at 490 nm, it appears that bioluminescence is derived from the electronic excited state of a luciferase-oxyluciferin monoanion complex. Apparently DMF and luciferase provide similar environments for

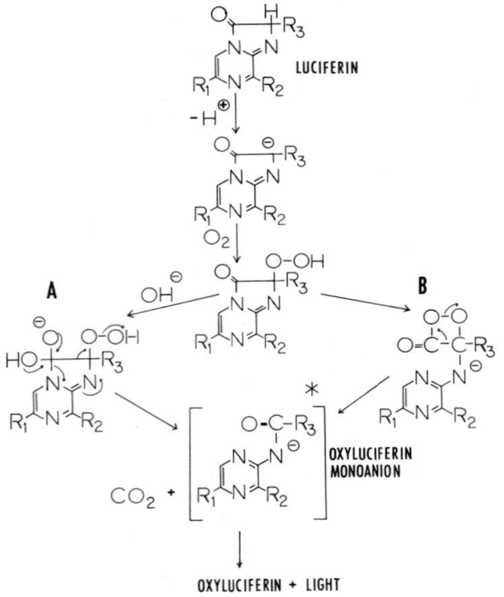

Figure 2 Proposed alternate mechanisms for the bioluminescent oxidation of *Renilla* luciferin.

fluorescence with a quantum yield of 5–6% for the emitting species in each case. An additional reason for suggesting a Q_F of approximately 5% for the luciferase-oxyluciferin complex is due to the observed fivefold increase in Q_B in the presence of the green fluorescent protein (1) whose Q_F is 30% (see next section). The chemiluminescence quantum yield of luciferin in DMF, however, is only 0.1% (27). Thus, during bioluminescence *Renilla* luciferase appears to be generating electronic excited states in yields approaching 100% as opposed to the chemiluminescent path in which the excitation yield is only 2%.

Mechanism studies using ^{18}O labeled water and oxygen suggest the scheme shown in Figure 2 (pathway A) for the bioluminescent oxidation of luciferin (30). The CO_2 labeling patterns were consistent with this scheme since oxygen in the CO_2 produced was derived from H_2O, not O_2. Similar labelings were shown in studies on the bioluminescent oxidation of firefly luciferin (35). Using the same techniques and aprotic solvents, the mechanisms involved in the chemiluminescent oxidations of firefly luciferyl adenylate and *Renilla* luciferin have recently been examined. As in bioluminescence, the results are in accord with pathway A in Figure 2 (unpublished results). Pathway B (Figure 2), which suggests a dioxetanone intermediate, is attractive theoretically (36). Nevertheless, it does not appear to be involved in *Renilla* and firefly luminescence since it is inconsistent with the ^{18}O studies. Furthermore, dioxetanone derivatives have been synthesized but their decompositions yield triplet, not singlet excited states (37–42), and in bioluminescence we are dealing with excited states of singlet multiplicity.

Analogous ^{18}O studies (43) on the mechanism of the bioluminescent oxidation of *Cypridina* luciferin are surprising since the results suggest pathway B (Figure 2) but the structures of *Cypridina* and *Renilla* luciferins are similar (1). However, it is not inconceivable that two different pathways could give rise to an excited state product.

Further work is obviously needed to clearly understand the oxidative mechanisms involved in bioluminescence and chemiluminescence. A mechanism is required that provides a theoretical explanation for the generation of electronic excited states of singlet multiplicity and also takes into account the ^{18}O labeling patterns.

REQUIREMENTS FOR GREEN LIGHT EMISSION   Spectral measurements on the in vitro and in vivo bioluminescence emissions in *Renilla* and a number of coelenterates examined have shown that the in vitro emissions ($\lambda_B = 470$–490 nm) are blue, whereas the in vivo emissions ($\lambda_B = 509$ nm) are green (19–22, 25, 44, 45). Precise spectral comparisons have shown that in *Renilla* and several other coelenterates, the green in vivo emission is due to a highly fluorescent protein which has been isolated from these animals and is referred to as the "green fluorescent protein" (22, 25, 44, 45). In *Renilla* this protein (mol wt $\simeq 40,000$) contains a bound chromophore of unknown structure whose fluorescent quantum yield is 30% (1, 44). The fluorescence emission of this protein matches precisely the green in vivo emission (44).

The possible involvement of energy transfer has been invoked to account for the differences in color between the in vitro and in vivo emissions in *Renilla* (19, 21, 44), and subsequent studies have tended to confirm this suggestion (2, 44, 46). For example, the green fluorescent protein will not catalyze the bioluminescent oxidation

PATHWAY TO BIOLUMINESCENCE IN THE ANTHOZOANS:

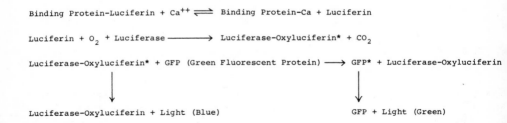

PATHWAY TO BIOLUMINESCENCE IN THE HYDROZOANS:

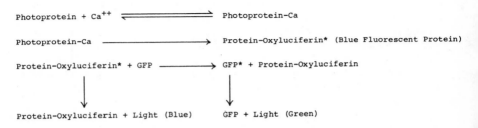

Figure 3   A comparison of the bioluminescent systems in the Anthozoans and the Hydrozoans.

of luciferin in the absence of luciferase, but its addition to an in vitro bioluminescent reaction will change the color of light from blue to the characteristic green in vivo emission (44, 46). Because the concentrations of luciferase and green fluorescent protein were low ($\sim 10^{-6}$ to 10^{-5} M) under conditions in which efficient energy transfer was observed, protein-protein interaction was suggested as a means of achieving a calculated critical transfer distance of 28 Å (44, 46). The phenomenon is highly protein concentration dependent (46), and the quantum yield relative to luciferin increases from 5% for the luciferase-catalyzed reaction to about 25% in the presence of the green fluorescent protein (1). Thus, the production of both blue and green light may be viewed as illustrated in Figure 3. The phenomenon is thought to involve nonradiative energy transfer from the electronic excited state of an oxyluciferin monoanion-luciferase complex to the chromophore on the green fluorescent protein (1). This was the first clear case of a "sensitized bioluminescence" (see Introduction).

CONTROL OF THE BIOLUMINESCENT FLASH   A protein important in the control of bioluminescence in *Renilla* and other Anthozoans has been recently isolated (24). It is a luciferin binding protein (mol wt $\simeq 24,000$), which reversibly discharges its bound luciferin in the presence of calcium ions and exhibits a deep yellow color when highly purified due to its associated luciferin (24). It now appears likely that luciferin sulfokinase, which converts luciferyl sulfate (III, Figure 1) to luciferin in the presence of $3',5'$-diphosphoadenosine (47), may play an important role in controlling the charging of the luciferin binding protein.

The addition of calcium ions to a solution mixture of luciferase, oxygen, and the charged luciferin binding protein results in a bluish luminescence typical of the in vitro reaction described above. This calcium-induced luminescence can be viewed as illustrated in the upper part of Figure 3. The rate of light intensity decay in this reaction is much slower than the rapid flash from the in vivo luminescence (1). Thus, it appears that calcium access to the luciferin binding protein in vivo must be under very fine control. An advance in understanding the mechanisms involved in such control was made with the discovery and isolation of lumisomes from *Renilla* by Anderson & Cormier (23). Lumisomes are membrane-bounded vesicles (average diameter = 0.2 μm) that produce an oxygen-dependent flash of green light ($\lambda_B =$ 509 nm) when exposed to a hypotonic solution of calcium ions (23). These vesicles contain all the proteins necessary for bioluminescence and its control, i.e. luciferase, the luciferin binding protein, and the green fluorescent protein (1, 23). All these proteins appear to be membrane bound (23), providing an excellent environment for the orderly arrangement of proteins necessary for the control and energy transfer processes outlined above.

A recent electron microscopy study has revealed the location and substructural features of *Renilla* photocytes (48). The photocytes are highly specialized cells containing large membrane-bounded vesicles (diameter = 4–6 μm) which have been termed luminelles (48) Luminelles house hundreds of smaller vesicles which are thought to be lumisomes since they are the same size and shape as the lumisomes isolated earlier by Anderson & Cormier (23). In partially purified preparations, luminelles appear as large (4–6 μm) green fluorescent structures which produce flashes of green light when exposed to a hypotonic solution of calcium ions (J. M. Anderson and M. J. Cormier, unpublished). Both the luminelle and lumisome membranes may play important roles in regulating calcium transients to and from the calcium binding sites on the luciferin binding protein. Just how a nerve impulse might trigger the release of calcium in *Renilla* is unknown.

The Hydrozoans and Ctenophores

The bioluminescent systems of the jellyfish, *Aequorea,* and the ctenophores, *Mnemiopsis* and *Beroë*, have been the most thoroughly investigated among these two groups. Protein-chromophore complexes, termed photoproteins, have been isolated and react with calcium ion to produce a bluish luminescence in vitro (49–55). As illustrated in Figure 3, this in vitro luminescence is independent of dissolved oxygen. The photoprotein isolated from *Aequorea* has been termed aequorin while the ones isolated from *Mnemiopsis* and *Beroë* have been termed mnemiopsin and berovin, respectively (49, 53, 55). The protein components of photoproteins apparently are species-specific single polypeptide chains ranging in molecular weights from about 24,000 to 31,000 (54, 56).

Cormier and associates (1, 25–27, 57) have provided evidence that the native chromophore component of photoproteins is similar, if not identical, to *Renilla* luciferin, although Johnson and associates (3, 58) object to this idea. The suggestion that the native chromophores are similar is strengthened by the finding of Shimomura & Johnson (58) that the aequorin bioluminescence reaction forms a product isolated and identified as VI in Figure 1. We have labeled it *Aequorea* oxyluciferin because of its near identity with *Renilla* oxyluciferin (IV, Figure 1). The emitter during aequorin bioluminescence is a protein-oxyluciferin complex (58), again analogous to *Renilla* bioluminescence. Luciferyl sulfates, indistinguishable from *Renilla* luciferyl sulfate (III, Figure 1), have also been isolated from numerous coelenterates including *Aequorea* and *Mnemiopsis* (17, 25). Furthermore, the visible absorption bands of aequorin (1, 59) and mnemiopsin (54) can now be explained on the basis of spectral perturbations of a *Renilla*-like luciferin chromophore (1, 57). Finally, an oxidative path similar to that illustrated in Figure 2 is a reasonable possibility for the formation of *Aequorea* oxyluciferin (VI, Figure 1). Apparently, the oxygen-requiring step has previously been incorporated into these photoproteins, resulting in an oxygenated intermediate that is stable in the absence of calcium. The nature of this intermediate and the lack of CO_2 production during the bioluminescence reaction of aequorin (60) remains to be explained.[2]

Whereas the chemical paths to light emission in *Renilla* and *Aequorea* (and perhaps ctenophores) appear to be similar, there are basic differences reflected at the protein level in control of the luminescence in each case. As shown in Figure 3, two proteins, a luciferin binding protein and luciferase, are involved in calcium-induced bioluminescence in the Anthozoans, whereas only a single protein, a photoprotein, is involved in the Hydrozoans and ctenophores. Photoproteins appear to be playing a dual role; they apparently bind a *Renilla*-like luciferin chromophore in the absence of calcium ions thus acting like a binding protein, while converting this chromophore to oxyluciferin in the presence of calcium ions thus acting like a luciferase.

Like the Anthozoans, the bioluminescent Hydrozoans that have been examined contain a green fluorescent protein (21, 25, 45, 49, 61). The fluorescence emission of the green fluorescent protein isolated from *Aequorea* is identical with the one isolated from *Renilla* (25, 45), although their excitation spectra are quite different (1, 45). This suggests differences in either the proteins, the chromophores, or both. Since the in vivo emission is green ($\lambda_B = 509$ nm) as opposed to the blue in vitro emission of aequorin ($\lambda_B = 469$ nm), a sensitization by the process of energy transfer, such as proposed for *Renilla,* has been suggested as an explanation (2, 21, 45, 62).

Using purified preparations of the *Aequorea* green fluorescent protein, which contained six green fluorescent protein components, a sensitization of the in vitro reaction by energy transfer has been demonstrated (45). However, it is difficult to suggest a mechanism for this sensitization because of the reported quantum yield data. Efficient nonradiative energy transfer should give an increase in quantum yield

---

[2] Since completion of this manuscript, a paper by Shimomura et al demonstrates that CO_2 is produced during the calcium-induced bioluminescence of aequorin (*Biochemistry* 1974. 13:3278). Furthermore, they present data which can be explained by the presence of a *Renilla*-like luciferin chromophore in aequorin.

if the acceptor (green fluorescent protein) has a significantly higher quantum yield than the donor (oxyluciferin-protein complex excited state). Since the bioluminescence quantum yield of aequorin alone is 23% and the fluorescence quantum yield of the product (oxyluciferin-protein complex) is probably not much greater than 23% (45), an increase in quantum yield in the presence of green fluorescent protein ($Q_F = 72\%$) would be expected. However, the bioluminescence quantum yield was only 23% under conditions of efficient energy transfer (45).

The regulation of calcium transients to the photoproteins in vivo probably occurs in a manner analogous to that described above for the Anthozoans, since *Renilla*-like lumisomes have been isolated from several species of bioluminescent Hydrozoans (1, 23). The Hydrozoan lumisomes contain a photoprotein and a green fluorescent protein but no luciferase or luciferin binding protein.

BACTERIAL BIOLUMINESCENCE

The bioluminescence of marine bacteria has been studied mostly on purified extracts and consequently a lot more is known about the in vitro reaction than the in vivo one. The in vitro reaction is a luciferase-catalyzed, molecular oxygen oxidation of reduced flavin mononucleotide ($FMNH_2$) and an aliphatic aldehyde (RCHO) with a carbon chain length longer than heptanal (63–66). It is likely that free $FMNH_2$ is not available within the cell and, in fact, what appear to be flavoproteins that have activity in the light reaction have been isolated from cell extracts (67, 68). There is as yet no strong evidence for RCHO involvement in vivo (69, 70).

It was first thought that reduced nicotine adenine dinucleotide (NADH) was the substrate for the reaction (71), but further purification of the enzyme removed this activity and it was then shown that the true substrate was $FMNH_2$ (63). NADH probably reduces FMN in the crude preparation via an FMN-linked NADH dehydrogenase. Advantage can be taken of this dehydrogenase in applying the bacterial bioluminescence reaction to analytical work, since many enzymes can be readily coupled to NADH (72–75).

The total light produced in the reaction is proportional to the amount of each of the substrates (O_2, $FMNH_2$, and RCHO) when they are present in limiting quantities (63–66). The same can be said of the luciferase since excess $FMNH_2$ is auto-oxidized so rapidly in comparison to the rate of light emission that it is gone when the luciferase finishes its catalytic cycle (76). The luciferase, therefore, acts only once in the in vitro reaction like the other reactants, but it appears capable of turnover on repeated addition of $FMNH_2$ (77, 78). Although the role of RCHO was not understood for a long time (7, 79), it is now established that it is a true substrate and is oxidized to the corresponding carboxylic acid (80–83), as originally suggested by McElroy & Green (84).

Characteristics of the bacterial reaction, such as kinetics, absolute quantum yields, and emission spectra, vary with the type of bacterium from which the luciferase is purified (11, 68, 85, 86). Extensive studies of the luminous bacteria taxonomy are recently available (87–89), and mutants with altered reaction and spectral properties have been demonstrated (90, 142). The fact that this luciferase is of bacterial origin means that large quantities may be produced by fermentation and purified by relatively straightforward methods (76, 91, 92).

The Chemical Reaction

Luciferase is a protein of molecular weight about 80,000 (86) and contains no metals or other cofactors (76, 91, 93). It may be reversibly dissociated into two subunits which differ slightly in molecular weights (86, 93, 94).

The first step in the reaction is a one-to-one association of $FMNH_2$ with the luciferase (E) with an equilibrium constant of 3×10^4 M^{-1} at 23°C (95, 96). As the concentration of E increases above that of $FMNH_2$, the $Q_B(FMNH_2)^3$ also increases and reaches a constant value when $E > 30\,\mu M$ (for $FMNH_2 > 16\,\mu M$) (97). This results from a competition between autooxidation and E for the available $FMNH_2$

$$FMNH_2 + E \rightleftharpoons E - FMNH_2 \xrightarrow{O_2} X$$
$$\downarrow O_2$$
$$FMN + H_2O_2$$

where X represents intermediates in the light path. Stopped-flow observations show no contribution to the rate of FMN appearance from the more rapid auto-oxidation when $E > 30\ \mu M$ (96).

Two approaches to determining the reaction stoichiometry have been made, the first a direct measurement of substrate utilization (97) and the second by steady-state kinetic analysis (98, 99). Using a luciferase concentration of greater than 30 μM to eliminate auto-oxidation, the utilization of each substrate determined on the basis of its quantum yield gives

$$2FMNH_2 + 2O_2 + RCHO \rightarrow 2FMN + H_2O_2 + [H_2O] + RCOOH$$

Both H_2O_2 and RCOOH have been identified as products (80–83, 97) but not H_2O, indicated by the square brackets. A certain ambiguity exists in this approach in that, in common with many hydroxylase enzymes, bound $FMNH_2$ might have a certain probability of reacting in an alternate pathway that does not lead to light production. In the hydroxylases this is evidenced by the variation in utilization ratios when the substrate concentrations are varied (100, 101). With luciferase this does not happen and neither is there any change with variation of pH (6–8.5), temperature (0–35°C), type of bacteria from which the luciferase is isolated, or the number of carbons in the aldehyde from C_7 to C_{14} (96, 97, 102).

An apparently different conclusion results from the kinetic analysis. The initial steady-state rate of light emission is first order in both $FMNH_2$ and luciferase over a broad concentration range (98, 99, 103). The stationary state analysis has a formal relationship to reaction stoichiometry, and with this approach Meighen & Hastings (98) and Watanabe & Nakamura (99) have concluded that the reaction involves one flavin instead of two.

---

[3] $Q_B(FMNH_2)$ refers to the bioluminescence quantum yield relative to $FMNH_2$, while $Q_B(X_2)$ is that relative to the intermediate X_2, etc. See Introduction.

Clearly then, neither a one flavin nor a two flavin simple reaction scheme adequately accounts for the data. Lee & Murphy (96), therefore, propose a more complex scheme which requires two moles of flavin per mole of enzyme but which could still give overall first order kinetics. The oxidation of $FMNH_2$ and the utilization of oxygen via luciferase ($E > 30 \ \mu M$) are both biphasic in rate (96, 104). Although Yoshida et al (105, 106) noted a biphasic oxidation of $FMNH_2$, a significant contribution from auto-oxidation is evident in their data, since insufficient luciferase was present to outcompete this. From studies of the transient kinetics of the reaction at 5°C with luciferase from the bacterium *Beneckea harveyi*, Lee & Murphy (104) proposed the following sequence of events

$$2FMNH_2 + E + O_2 \rightarrow X_1 + FMN + H_2O_2 \qquad \text{fast}$$

$$X_1 + O_2 \rightarrow X_2 \qquad\qquad\qquad t_{1/2} \approx 1.3 \ \text{min}$$

where X_2 and X_1 are viewed as the equivalent of EH_2O_2 FMN and EH_2FMN, respectively.

The first step is complete in 30 sec, at which time all the FMN is completely oxidized. If X_2 is allowed to stand around it eventually decays ($t_{1/2} \approx 7$ min) to FMN and H_2O_2 with the release of native luciferase. If RCHO is added in excess over X_2, even when its concentration is rate limiting, the full quantum yield of X_2 is equal to $2Q_B(FMNH_2)$.[3] X_2 is believed to be the same as the long lived intermediate (II) of Hastings & Gibson (107). Hastings et al (108) attempted to separate this from the reaction mixture by low temperature chromatography, but they allowed only a few seconds for oxygen uptake at 5°C before quenching to low temperature ($-20°C$) in an ethylene glycol-water mixture, and it was not established that the separated intermediate did not continue to react with oxygen before reacting with RCHO. Indeed the absorption spectrum presented for intermediate II appears to have a contribution from $FMNH_2$.

Murphy et al (109) prepared X_2 by converting luciferase quantitatively at 0°C where $t_{1/2} \approx 14$ min. The reaction mixture was then subjected to rapid chromatography on a short G-25 Sephadex column. They found hardly detectable amounts of FMN in the chromatographed preparation even after warming, yet the preparation reacted with RCHO to give the same $Q_B(X_2)$ as for the complete reaction in the presence of FMN. Based on the trace FMN detectable in the chromatographed X_2, the $Q_B(FMN)$ was shown to be greater than unity. Therefore, X_2 that seems to be a flavoprotein can be rapidly dissociated

$$\text{apo-}X_2 + FMN \rightleftarrows X_2$$

and the apo-X_2 is separated by chromatography.

As with most flavoproteins, FMN fluorescence is quenched when bound in holo-X_2. The fluorescence quenching was used to estimate the equilibrium binding constant as $7 \times 10^5 \ M^{-1}$ at 0°C (109).

The light reaction proceeds by

$$X_2 + RCHO \rightarrow FMN + RCOOH + H_2O + E + \text{light}$$

In the complete absence of FMN the light is quenched. Since only a trace of FMN is required for the full quantum yield, the FMN must recycle in what is apparently another example of a sensitized bioluminescence reaction.

The Emitting Chromophore

It is generally believed that the emitting chromophore is some form of the FMN molecule, although the only basis for this is an observed 20 nm blueshift in $\lambda_B{}^1$ when iso-$FMNH_2$ is used to initiate the reaction rather than $FMNH_2$ (110). The radiative transition is almost certainly of singlet character since the presence of molecular oxygen in the reaction would completely quench a triplet, even if the chromophore was buried in the protein (111).

That the emitting chromophore is FMN itself is excluded by the facts that iso-FMN has its $\lambda_F{}^1$ redshifted over FMN and that there is a distinct difference between the λ_F for FMN (535 nm) and the species-dependent bioluminescence emission maxima (λ_B), which covers the range 478–505 nm (11, 85, 110). A perturbation of FMN fluorescence is unlikely since most flavoproteins either quench this fluorescence or, if they are fluorescent, are very slightly blueshifted over free FMN (143). Of course, luciferase could be unique in providing a strongly perturbing environment for FMN, but it is found that even strongly perturbing solvent conditions do not blueshift the fluorescence more than about 25 nm and, more important, such a shift is accompanied by a marked vibronic splitting (102, 112, 113). In contrast, the bioluminescence spectra are all smooth structureless curves (102).

Eley et al (114) suggested that the emitter is a luciferase-bound flavin cation, $FMNH^+$, based on the similarity in fluorescence emission spectra of $FMNH^+$ and iso-$FMNH^+$ in rigid solvents to the bioluminescence spectra produced from $FMNH_2$ and iso-$FMNH_2$, respectively. Although this is still a reasonable suggestion, recent experiments make the requirement that the emitter be some form of FMN less compelling (109). From what we now know, a reaction product of $FMNH_2$ formed in only trace amounts could conceivably sensitize the process; for example, degradation of FMN to a bluer fluorescing product, such as lumichrome, occurs with a yield less than 0.02, based on quantum yield measurements (115). Furthermore, interaction of FMN with luciferase forming a new chromophore, which was thought to have absorption and fluorescence properties of $FMNH^+$ (95, 114), more likely arises from a small amount of light-induced reduction of FMN and subsequent turnover of luciferase to form some X_2, whose typical flavoprotein absorption spectrum would give the same effect (104, 109).

McCapra & Hysert (82) suggested that the emitting chromophore is a complex of aldehyde and flavin with the isoalloxazine ring ruptured at the 2,3 position. Model compounds have fluorescence spectra similar to the bioluminescence. The mechanism proposed for its formation is incorrect, however, since it requires RCHO addition to $FMNH_2$, and the flavin is fully oxidized in X_2 (104).

The identity of the emitting chromophore, therefore, still remains unknown, although $FMNH^+$ remains a good candidate, and a mechanism for its formation by one electron transfer from acyloxy radical provides a sufficiently energetic and physically realistic excitation process (102)

$$FH\cdot + RC\overset{\overset{\displaystyle O}{\|}}{-}O\cdot \rightarrow FH^{+} + RC\overset{\overset{\displaystyle O}{\|}}{-}O^{-} + 70\ kcal$$

FIREFLY BIOLUMINESCENCE

Advances in the understanding of firefly bioluminescence have been reviewed by McElroy & DeLuca (116) and Cormier et al (2). The chemiluminescence reactions of firefly luciferin and its analogs have been reviewed by White et al (117).

The efficient firefly reaction (34) involves a two step conversion of luciferin, VII (Figure 4) to luciferyl adenylate, VIII (118) followed by oxidation to oxyluciferin, IX (119). The conversion to luciferyl adenylate occurs via a reversible reaction (120) as follows

$$E + LH_2 + ATP\text{-}Mg^{2+} \rightleftarrows E(LH_2\text{-}AMP) + P\text{-}P \tag{1.}$$

where E is luciferase, LH_2 is luciferin, P-P is pyrophosphate, and the parentheses indicate an enzyme-bound product. The subsequent reaction of the enzyme-bound luciferyl adenylate involves several postulated steps as indicated by the following reaction scheme

$$E(LH_2\text{-}AMP) \rightleftharpoons E(LH^-\text{-}AMP) + H^+ \tag{2.}$$

$$E(LH^-\text{-}AMP) \rightleftharpoons E'(LH^-\text{-}AMP) \tag{3.}$$

$$E'(LH^-\text{-}AMP) + O_2 \overset{H^+}{\rightarrow} \begin{matrix} E(LHOO^-\text{-}AMP) \\ or \\ E(LHOOH\text{-}AMP) \end{matrix} \tag{4.}$$

$$E(LHOO^-\text{-}AMP)\ or\ E(LHOOH\text{-}AMP) \rightarrow E(oxyluciferin\ dianion\ AMP)^* + CO_2 \tag{5.}$$

$$E(oxyluciferin\ dianion\text{-}AMP)^* \rightarrow E(oxyluciferin,\ AMP) + h\nu \tag{6.}$$

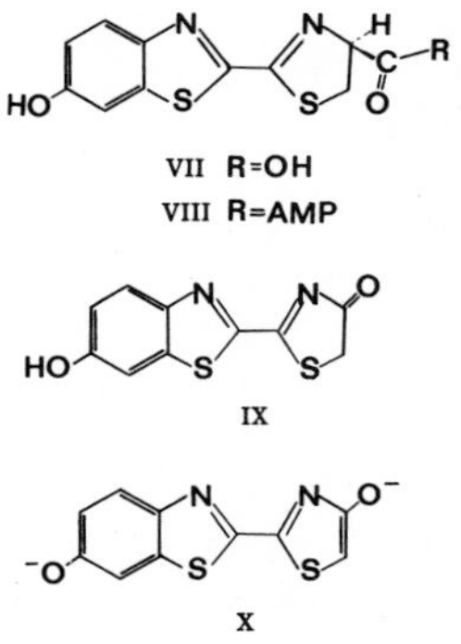

VII R=OH
VIII R=AMP

IX

X

Figure 4 Structures of firefly luciferin (VII), luciferyl adenylate (VIII), oxyluciferin (IX), and the oxyluciferin dianion (X).

The risetime of luminescence in the firefly in vitro reaction is slow (half risetime, 0.2 sec) (121). A recent investigation of the fast kinetics of this step by DeLuca & McElroy (122) reveals at least two slow steps, both occurring after formation of luciferyl adenylate. Both these steps (2 and 3) appear to be intramolecular changes occurring before the addition of oxygen, since preincubation of luciferin, enzyme, ATP, and Mg^{2+} anaerobically gives a rapid flash upon the introduction of oxygen. Initiating the light reaction with preformed LH_2-AMP gives the slow kinetics. Binding of substrate to enzyme does not account for these slow kinetics, since the kinetics are not dependent on enzyme concentration and the spectral changes occurring during binding of a substrate analog dehydroluciferyl adenylate are rapid. Based on this evidence DeLuca & McElroy (122) postulate steps 2 and 3 as changes in the enzyme-luciferyl adenylate complex; first an abstraction of a proton from luciferyl adenylate to form the carbanion for subsequent attack on oxygen and then conformational changes in the enzyme complex.

Details of the mechanism that generates the excited state product (reactions 4 and 5) are not clear. Two mechanisms have been postulated (35, 123) similar to those suggested for *Renilla* (30) and *Cypridina* (43). As mentioned, $^{18}O_2$ labeling studies favor the *Renilla*-like mechanism.

Evidence from many chemiluminescence studies predicts emission (step 6) from the dianion (X) of the oxyluciferin product (117). The structure of oxyluciferin has been confirmed by synthesis (124) and by purification from spent chemiluminescence and bioluminescence reactions (119). The monoanion fluoresces red and accounts for the red emission of the in vitro bioluminescence at acid pHs. In reaction 6, AMP is shown still bound to luciferase. This is strongly suggested by the effect of two ATP analogs, 3-iso-ATP (125) and 1,N^6-etheno-ATP (126), on the reaction. 3-iso-ATP reacts with luciferin to give red light. 1,N^6-etheno-ATP does not react with luciferin, but if the adenylate analog is formed chemically it is a substrate for red bioluminescence. Since the phenol hydroxyl of oxyluciferin appears to be ionized even in the ground state (117, 127), these color changes may be due to competition in the excited state between emission and either enol-keto tautomerization or protonation of the enolate ion.

PHOLAD BIOLUMINESCENCE AND THE PEROXIDE SYSTEMS

The *Pholas dactylus* (clam) bioluminescence system, studied by Michelson and co-workers, involves the oxidation of a protein-bound luciferin (50,000 mol wt) by molecular oxygen (128). This luciferin differs from photoproteins requiring a separate enzyme (310,000 mol wt) for the reaction to occur (129). Recent evidence suggests that *Pholad* luciferase is a copper glycoprotein (130). The in vitro reaction is stimulated by a variety of chemical agents: ferrous ion and $FMNH_2$ (128), pyrogallol, ascorbic acid, dihydroxyfumaric acid, and catechol (129). Each of these effects can be explained by a mechanism involving attack on luciferin by superoxide ions produced enzymically and by contributions to propagation of the light reaction by radical chain reaction mechanisms (129). In fact, an aerobic solution of iron II or

$FMNH_2$ will cause *Pholad* luciferin to chemiluminesce. A detailed investigation of these reactions and other metal-catalyzed oxidations of *Pholad* luciferin and luminol support a mechanism for both chemi- and bioluminescence involving primary attack on luciferin by superoxide ion and a radical chain reaction in the presence of high concentrations of certain reducing agents (131, 132). Additional support for this mechanism is demonstrated by the fact that horseradish peroxidase can substitute for luciferase in producing bioluminescence with molecular oxygen (133). Conversely, *Pholad* luciferase can serve as a peroxidase in the oxidation of several typical peroxidase substrates (133). This is not necessarily unusual since other copper proteins react with peroxide (134, 135).

Some interesting comparisons can be drawn between the *Pholas* system and the few peroxide-requiring bioluminescence systems so far investigated. For instance, the *Chaetopterus variopedatus* photoprotein requires only ferrous ion, oxygen, and a peroxide for light (136), similar to the ferrous ion-catalyzed chemiluminescence of *Pholad* luciferin. The system from the marine worm, *Balanoglossis biminiensis*, cross-reacts with the model bioluminescence system of horseradish peroxidase and luminol (137). And finally, luciferase from the luminescent earthworm, *Diplocardia longa*, may be a copper protein (138, 139). Do these similarities arise from similarities in the mechanism of these reactions? A metal cofactor has not been shown for the *Balanoglossus* system but certainly is indicated for the other systems, and cyanide and azide do inhibit *Balanoglossid* luciferase (140). By analogy to the *Pholas* system, a radical mechanism might then be expected. The *Balanoglossid* reaction is inhibited by cysteine and mercaptoethanol, both of which can act as radical scavengers (137). However, the Diplocardia system does not crossreact with the others and does not oxidize typical peroxidase substrates (141).

Literature Cited

1. Cormier, M. J., Hori, K., Anderson, J. M. 1974. *Rev. Bioenerg.* 346:137–64
2. Cormier, M. J., Wampler, J. E., Hori, K. 1973. *Progr. Chem. Org. Natur. Prod.* 30:1–54
3. Johnson, F. H., Shimomura, O. 1972. *Photophysiology* 7:275–334
4. McElroy, W. D., Seliger, H. 1969. *Photochem. Photobiol.* 10:153–70
5. Hastings, J. W. 1968. *Ann. Rev. Biochem.* 37:597–630
6. Goto, T., Kishi, Y. 1968. *Angew. Chem.* 7:407–14
7. Cormier, M. J., Totter, J. R. 1968. *Photophysiology* 4:315–53
8. Goto, T. 1968. *Pure Appl. Chem.* 17:421–41
9. McCapra, F. 1973. *Endeavour* 32:139–45
10. Adam, W. 1973. *Chem. Unserer Zeit* 6:182–91
11. Seliger, H. H., Morton, R. A. 1968. *Photophysiology* 4:253–314
12. Airth, R., Foerster, G. E., Hinde, R. 1970.

Photobiology of Microorganisms, ed. Per Halldal, 479–94. New York: Wiley-Interscience
13. Cormier, M. J. 1974. *Natur. Hist.* 83:26–34
14. Lee, J. 1974. *Photochem. Photobiol.* 20:535–39
15. Seliger, H. H. 1973. *Chemiluminescence and Bioluminescence,* ed. M. J. Cormier, D. M. Hercules, J. Lee, 461–78. New York: Plenum
16. Cormier, M. J., Eckroade, C. B. 1962. *Biochim. Biophys. Acta* 64:340–44
17. Hori, K., Nakano, Y., Cormier, M. J. 1972. *Biochim. Biophys. Acta* 256:638–44
18. Cormier, M. J., Crane, J. M., Nakano, Y. 1967. *Biochem. Biophys. Res. Commun.* 29:747–53
19. Hastings, J. W., Morin, J. G. 1969. *Biol. Bull.* 137:402
20. Morin, J. G., Hastings, J. W. 1971. *J. Cell. Physiol.* 77:305–12
21. Morin, J. G., Hastings, J. W. 1971. *J. Cell.*

Physiol. 77 : 313–18
22. Wampler, J. E., Karkhanis, Y. D., Morin, J. G., Cormier, M. J. 1973. *Biochim. Biophys. Acta* 314 : 104–9
23. Anderson, J. M., Cormier, M. J. 1973. *J. Biol. Chem.* 248 : 2937–43
24. Anderson, J. M., Charbonneau, H., Cormier, M. J. 1974. *Biochemistry* 13 : 1195–1201
25. Cormier, M. J., Hori, K., Karkhanis, Y. D., Anderson, J. M., Wampler, J. E., Morin, J. G., Hastings, J. W. 1973. *J. Cell. Physiol.* 81 : 291–98
26. Hori, K., Cormier, M. J. 1973. *Proc. Nat. Acad. Sci. USA* 70 : 120–23
27. Hori, K., Wampler, J. E., Matthews, J. C., Cormier, M. J. 1973. *Biochemistry* 12 : 4463–69
28. Goto, T., Iio, H., Inoue, S., Kakoi, H. 1974. *Tetrahedron Lett.* 26 : 2321–24
29. McCapra, F., Manning, M. J. 1973. *Chem. Commun.* 467–68
30. DeLuca, M., Dempsey, M. E., Hori, K., Wampler, J. E., Cormier, M. J. 1971. *Proc. Nat. Acad. Sci. USA* 68 : 1658–60
31. Hori, K., Wampler, J. E., Cormier, M. J. 1973. *Chem. Commun.* 492–94
32. McCapra, F., Chang, Y. C. 1967. *Chem. Commun.* 1011–12
33. Goto, T. 1973. *Tetrahedron Lett.* 29 : 2035–39
34. Seliger, H. H., McElroy, W. D. 1960. *Arch. Biochem. Biophys.* 88 : 136–41
35. DeLuca, M., Dempsey, M. E. 1970. *Biochem. Biophys. Res. Commun.* 40 : 117–22
36. McCapra, F. 1968. *Chem. Commun.* 155–56
37. Kopecky, K. R., Mumford, C. 1969. *Can. J. Chem.* 47 : 709–11
38. White, E. H., Wiecko, J., Wei, C. C. 1970. *J. Am. Chem. Soc.* 92 : 2167–68
39. Lee, D. C-S., Wilson, T. See Ref. 15, 265–83
40. Adam, W., Liu, J. C., Simpson, G., Steinmetzer, H. C. See Ref. 15, 493–94
41. Turro, N. J., Lechtken, P. 1972. *J. Am. Chem. Soc.* 94 : 2886–88
42. Turro, N. J., Lechtken, P. 1973. *J. Am. Chem. Soc.* 95 : 264–66
43. Shimomura, O., Johnson, F. H. 1971. *Biochem. Biophys. Res. Commun.* 44 : 340–46
44. Wampler, J. E., Hori, K., Lee, J. W., Cormier, M. J. 1971. *Biochemistry* 10 : 2903–10
45. Morise, H., Shimomura, O., Johnson, F. H., Winant, J. 1974. *Biochemistry* 13 : 2656–62
46. Wampler, J. E., Karkhanis, Y. D., Hori, K., Cormier, M. J. 1972. *Fed. Proc.* 31 : 419

47. Cormier, M. J., Hori, K., Karkhanis, Y. D. 1970. *Biochemistry* 9 : 1184–90
48. Spurlock, B. O., Cormier, M. J. 1975. *J. Cell Biol.* 64 : 15–28
49. Shimomura, O., Johnson, F. H., Saiga, Y. 1962. *J. Cell. Comp. Physiol.* 59 : 223–40
50. Shimomura, O., Johnson, F. H., Saiga, Y. 1963. *J. Cell. Comp. Physiol.* 62 : 1–8
51. Shimomura, O., Johnson, F. H., Saiga, Y. 1963. *J. Cell. Comp. Physiol.* 62 : 9–16
52. Ward, W. W., Seliger, H. H. 1973. *Fed. Proc.* 32 : 661
53. Girsch, S. J., Hastings, J. W. 1973. *Am. Soc. Photobiol.* 183
54. Ward, W. W., Seliger, H. H. 1974. *Biochemistry* 13 : 1500–10
55. Ward, W. W., Seliger, H. H. 1974. *Biochemistry* 13 : 1491–99
56. Kohama, Y., Shimomura, O., Johnson, F. H. 1971. *Biochemistry* 10 : 4149–52
57. Hori, K., Ward, W. W., Anderson, J. M., Cormier, M. J. 1974. *Am. Soc. Photobiol.* 57
58. Shimomura, O., Johnson, F. H. 1973. *Tetrahedron Lett.* 31 : 2963–66
59. Shimomura, O., Johnson, F. H. 1969. *Biochemistry* 8 : 3991–97
60. Shimomura, O., Johnson, F. H. See Ref. 15, 337–45
61. Morin, J. G., Reynolds, G. T. 1970. *Biol. Bull. Mar. Biol. Lab. Woods Hole* 139 : 430–31
62. Johnson, F. H. 1967. *Comp. Biochem.* 27 : 79–136
63. McElroy, W. D., Hastings, J. W., Coulombre, J., Sonnefeld, V. 1953. *Arch. Biochem. Biophys.* 46 : 399–416
64. Cormier, M. J., Strehler, B. L. 1953. *J. Am. Chem. Soc.* 75 : 4864
65. Strehler, B. L., Cormier, M. J. 1954. *J. Biol. Chem.* 211 : 213–25
66. Hastings, J. W., Spudich, J., Malnic, G. 1963. *J. Biol. Chem.* 238 : 3100–5
67. Mitchell, G. W., Hastings, J. W. 1970. *Biochemistry* 9 : 2699–2708
68. Lee, J., Murphy, C. L., Faini, G., Baucom, T. L. 1974. *Liquid Scintillation Counting: Recent Developments,* ed. P. Stanley, B. Scoggins, 403–21. New York : Academic
69. Eberhard, A., Rouser, G. 1971. *Lipids* 6 : 410–15
70. Ferrell, W. J., Kessler, R. J., Drouillard, M. 1971. *Chem. Phys. Lipids* 6 : 131–34
71. Strehler, B. L. 1953. *Arch. Biochem. Biophys.* 43 : 67–80
72. Stanley, P. E. 1971. *Anal. Biochem.* 39 : 441–53
73. Brolin, S. E., Borglund, E., Tegnei, L., Wettermark, G. 1971. *Anal. Biochem.* 42 : 124–35
74. Schram, E. See Ref. 68, 383–403

75. Stanley, P. 1974. *Liquid Scintillation Counting,* ed. M. A. Crook, P. Johnson. London: Heyden and Son
76. Hastings, J. W., Riley, W. H., Massa, J. 1965. *J. Biol. Chem.* 240:1473–81
77. Erlanger, B. F., Isambert, M. F., Michelson, A. M. 1970. *Biochem. Biophys. Res. Commun.* 40:70–76
78. Lee, J. See Ref. 15, 386
79. Hastings, J. W. 1968. *Ann. Rev. Biochem.* 37:597–631
80. Shimomura, O., Johnson, F. H., Kohama, Y. 1972. *Proc. Nat. Acad. Sci. USA* 69:2086–89
81. Dunn, D. K., Michaliszyn, G. A., Bogacki, I. G., Meighen, E. A. 1973. *Biochemistry* 12:4911–18
82. McCapra, F., Hysert, D. W. 1973. *Biochem. Biophys. Res. Commun.* 52:298–304
83. Vigny, A., Michelson, A. M. 1974. *Biochemie* 56:171–76
84. McElroy, W. D., Green, A. 1955. *Arch. Biochem. Biophys.* 56:240–55
85. Seliger, H. H., McElroy, W. D. 1965. *Light: Physical and Biological Action.* New York: Academic
86. Hastings, J. W., Weber, K., Friedland, J., Eberhard, A., Mitchell, G. W., Gunsalus, A. 1969. *Biochemistry* 8:4681–89
87. Hendrie, M. S., Hodgkiss, W., Shewan, J. M. 1970. *J. Gen. Microbiol.* 64:151–69
88. Chumakova, R. I., Vanyushin, B. F., Kokurina, N. A., Vorob'eva, T. I., Medvedeva, S. E. 1973. *Microbiology* 41:539–46
89. Reicheit, J. L., Baumann, P. 1973. *Arch. Mikrobiol.* 94:283–330
90. Cline, T. W., Hastings, J. W. 1972. *Biochemistry* 11:3359–70
91. Kuwabara, S., Cormier, M. J., Dure, L. S., Kreiss, P., Pfuderer, P. 1965. *Proc. Nat. Acad. Sci. USA* 53:822–28
92. Gunsalus-Miguel, A., Meighen, E. A., Ziegler-Nicoli, M., Nealson, K., Hastings, J. W. 1972. *J. Biol. Chem.* 247:398–404
93. Cormier, M. J., Kuwabara, S. 1965. *Photochem. Photobiol.* 4:1217–25
94. Friedland, J., Hastings, J. W. 1967. *Proc. Nat. Acad. Sci. USA* 58:2336–42
95. Lee, J., Murphy, C. L. 1973. *Biophys. J.* 13:274a
96. Lee, J., Murphy, C. L. 1975. *Biochemistry.* In press
97. Lee, J. 1972. *Biochemistry* 11:3350–60
98. Meighen, E. A., Hastings, J. W. 1971. *J. Biol. Chem.* 246:7666–74
99. Watanabe, T., Nakamura, T. 1972. *J. Biochem.* 72:647–53
100. Walsh, C. T., Schonbrunn, A., Abeles,

R. H. 1971. *J. Biol. Chem.* 246:6855–67
101. White-Stevens, R. H., Kamin, H., Gibson, Q. H. 1972. *J. Biol. Chem.* 247:2371–82
102. Lee, J., Murphy, C. L. See Ref. 15, 381–86
103. Chappelle, E. W., Picciolo, G. L., Altland, R. H. 1967. *Biochem. Med.* 1:252–60
104. Lee, J., Murphy, C. L. 1973. *Biochem. Biophys. Res. Commun.* 53:157–63
105. Yoshida, K., Takahashi, M., Nakamura, T. 1973. *Biochem. Biophys. Res Commun.* 42:1470–74
106. Yoshida, K., Takahashi, M., Nakamura, T. 1974. *J. Biochem.* 75:583–90
107. Hastings, J. W., Gibson, Q. H. 1963. *J. Biol. Chem.* 238:2537–54
108. Hastings, J. W., Balny, C., Le Peuch, C., Douzou, P. 1973. *Proc. Nat. Acad. Sci. USA* 70:3468–72
109. Murphy, C. L., Faini, G., Lee, J. 1974. *Biochem. Biophys. Res. Commun.* 58:119–25
110. Mitchell, G. W., Hastings, J. W. 1969. *J. Biol. Chem.* 244:2572–76
111. Lakowicz, J R., Weber, G. 1973. *Biochemistry* 12:4171–80
112. Palmer, G., Massey, V. 1968. *Biological Oxidations,* ed. T. P. Singer, 263–98. New York: Interscience
113. Sun, M., Moore, T. A., Song, P. S. 1972. *J. Am. Chem. Soc.* 94:1730–40
114. Eley, M., Lee, J., Lhoste, J-M., Lee, C. Y., Cormier, M. J., Hemmerich, P. 1970. *Biochemistry* 9:2902–8
115. Lee, J., Seliger, H. H. 1965. *Photochem. Photobiol.* 4:1015–48
116. McElroy, W. D., DeLuca, M. See Ref. 15, 285–311
117. White, E. H., Rapaport, E., Seliger, H. H., Hopkins, T. A. 1971. *Bioorg. Chem.* 1:92–122
118. Seliger, H. H., McElroy, W. D., White, E. H., Field, G. F. 1961. *Proc. Nat. Acad. Sci. USA* 47:1129–34
119. Suzuki, N., Goto, T. 1972. *Tetrahedron* 28:4075–82
120. Rhodes, W. C., McElroy, W. D. 1958. *J. Biol. Chem.* 233:1528–37
121 McElroy, W. D., Seliger, H. H. 1961. *Light and Life,* ed. W. D. McElroy, B. Glass, 219–57. Baltimore: Johns Hopkins Univ. Press
122. DeLuca, M., McElroy, W. D. 1974. *Biochemistry* 13:921–25
123. Hopkins, T. A., Seliger, H. H., White, E. H., Case, M. W. 1967. *J. Am. Chem. Soc.* 89:7148–50
124. Suzuki, N., Goto, T. 1971. *Tetrahedron Lett.* 22:2021–24

125. McElroy, W. D., Seliger, H. H. 1966. *Molecular Architecture in Cell Physiology,* ed. O. Hayashi, A. Szent-Gyorgi, 63–79. Englewood Cliffs, NJ: Prentice-Hall

126. DeLuca, M., Leonard, N. J., Gates, B. J., McElroy, W. D. 1973. *Proc. Nat. Acad. Sci. USA* 70:1664–66

127. Bowie, L. J., Irwin, R., Loken, M., DeLuca, M., Brand, L. 1973. *Biochemistry* 12:1852–57

128. Henry, J. P., Isambert, M. F., Michelson, A. M. 1970. *Biochim. Biophys. Acta* 205:437–50

129. Michelson, A. M., Isambert, M. F. 1973. *Biochimie* 55:619–34

130. Michelson, A. M. 1974. *Abstr. Am. Soc. Photobiol. Meet. Vancouver,* 62–63

131. Michelson, A. M. 1973. *Biochimie* 55:465–79

132. Michelson, A. M. 1973. *Biochimie* 55:925–42

133. Henry, J. P., Isambert, M. F., Michelson, A. M. 1973. *Biochimie* 55:83–93

134. Jolley, R. L., Evans, L. H., Makino, N., Mason, H. S. 1974. *J. Biol. Chem.* 249:335–45

135. Felsenfeld, G., Printz, M. P. 1959. *J. Am. Chem. Soc.* 81:6259–64

136. Shimomura, O., Johnson, F. H. 1968. *Science* 159:1239–40

137. Dure, L. S., Cormier, M. J. 1964. *J. Biol. Chem.* 239:2351–59

138. Bellisario, R., Cormier, M. J. 1971. *Biochem. Biophys. Res. Commun.* 43:800–5

139. Bellisario, R., Spencer, T. E., Cormier, M. J. 1972. *Biochemistry* 11:2256–66

140. Dure, L. S., Cormier, M. J. 1963. *J. Biol. Chem.* 238:790–93

141. Bellisario, R. 1971. Thesis. Univ. of Georgia, Athens, Ga.

142. Cline, T. W., Hastings, J. W. 1974. *J. Biol. Chem.* 249:4668–69

143. Casola, L., Brumby, P. E., Massey, V. 1966. *J. Biol. Chem.* 241:4977–84

RESTRICTION ENDONUCLEASES IN THE ANALYSIS AND RESTRUCTURING OF DNA MOLECULES   ×883

Daniel Nathans and Hamilton O. Smith
Department of Microbiology, Johns Hopkins University School of Medicine,
Baltimore, Maryland 21205

CONTENTS

INTRODUCTION

Restriction enzymes are endodeoxyribonucleases that recognize specific nucleotide sequences in double stranded DNA and cleave both strands of the duplex. In the cell of origin each restriction enzyme is part of a restriction-modification (R-M) system, consisting of the restriction endonuclease and a matched modification

273

enzyme which recognizes and modifies (generally by methylation) the same nucleotide sequence in DNA recognized by the restriction enzyme. Modification thus protects cellular DNA from restriction; however, foreign (unmodified) DNA is cleaved by the restriction endonuclease and further degraded by other enzymes. Such R-M systems, first detected by phage restriction and modification (1, 2), are widespread in bacteria and are thought to play a role in eliminating foreign DNA that gains entrance to the cell via viruses or as naked DNA. The biochemistry and genetics of R-M systems have recently been reviewed (3).

The usefulness of restriction endonucleases in the analysis and restructuring of DNA, which is the topic of this review, rests on the fact that some of the restriction enzymes cleave DNA at specific nucleotide sequences. These cleavage site-specific endonucleases are thus analogous to specific proteolytic enzymes and are proving as useful in the study of DNA structure and function as trypsin and chymotrypsin have been in protein analysis. After the first characterization of a restriction endonuclease from *Escherichia coli* strain K in 1968 (4), there was a curious lag in the application of restriction enzymes as analytical tools. In a sense this delay was fortunate, since the enzymes isolated initially are in the class now known to be nonspecific in their cleavage sites (5–7). However, after the discovery of cleavage site-specific endonucleases (8, 9), there was immediate application of these enzymes to the analysis of viral genomes (10). In the past three years there has been an almost explosive rate of discovery of new site-specific restriction enzymes and rapid application to physical mapping of chromosomes, nucleotide sequence analysis of DNA, isolation of genes, and restructuring of DNA molecules. Our purpose is to review these recent developments. Other relevant reviews are found in (3, 11–13).

GENERAL PROPERTIES OF RESTRICTION ENDONUCLEASES

Classes of Restriction Enzymes

The usefulness of restriction endonucleases resides in their cleavage specificity. However, as noted, not all restriction enzymes cleave DNA at specific sites, although all appear to recognize specific nucleotide sequences. Those that are nonspecific in their cleavage have been called Class I enzymes (3, 11) and have the following properties: molecular weight of about 300,000 (4, 14), nonidentical subunits (12, 14), and requirement for ATP, Mg^{2+}, and generally S-adenosylmethionine as cofactors (4). Examples are $EcoB^1$ (14) and EcoK (4). EcoPl is similar, being stimulated by S-adenosylmethionine but not requiring it for activity (15, 16). Cleavage site-specific restriction enzymes have been designated Class II enzymes (11); these are

---

[1] The current enzyme nomenclature is detailed in (17). It is based on the use of R-M system names which use a three letter abbreviation for the host organism followed by a strain designation where required (e.g. EcoB for *E. coli* strain B; Hind for *Haemophilus influenzae* strain d). Different R-M systems in a strain are indicated by roman numerals. Restriction enzymes are given the general name endonuclease R or endo R followed by the system name. In this review, for brevity we will generally use only the system name to identify enzymes.

lower in molecular weight than Class I enzymes (about 20,000 to 100,000) (18, 19), require only Mg^{2+} as cofactor (8), and cleave DNA at the enzyme recognition sequence to produce specific fragments (see below). In the best studied case (*Eco*RI), the enzyme contains two identical subunits (20). Only Class II enzymes will be discussed in this review. For a review of Class I enzymes, see (3, 11, 12).

Detection and Assay

The presence of a restriction enzyme in a bacterial strain may be inferred from the demonstration of phage restriction phenomena (1, 2, 4). However, these enzymes can in many cases be detected in crude extracts by virtue of their selective endonuclease activity on foreign as opposed to host DNA (8).

The rationale for many of the assays is simply the discrimination between endo- and exonuclease activity. Some can be used quantitatively and some are qualitative. One useful quantitative assay measures reduction of viscosity of a solution of unmodified DNA (8, 21). This method is relatively insensitive to exonuclease activity and is applicable to crude extracts low in nonspecific endonuclease activity. Activity may also be quantitated from the conversion rate of covalently closed, circular DNA to a denaturable form (6, 22), conversion to a form susceptible to exonuclease (23), or by the inactivation rate of infectious DNA (4, 24, 155). Recently, a solid phase assay has been developed (25) that uses DNA trapped in polyacrylamide beads (26); the rate of release of nonacid soluble fragments from the gel is proportional to enzyme concentration (25).

Qualitative assays are most useful during purification to locate fractions containing activity. A simple clot test (21), based on the observation that restricted DNA forms a fine trichloroacetic acid (TCA) precipitate compared to the fibrous precipitate of high molecular weight substrate, has had limited use because of its lack of specificity. Sucrose gradient assay (4, 27) is most valuable for Class I enzymes but is cumbersome and has little application for Class II enzymes. Assays using gel electrophoresis have proven vastly more useful (18). Restricted DNA yields fragments that form in the gels a spectrum of sharp bands visible in most cases even when the digest is produced with crude extracts containing significant exonuclease activity. Using slab gels with multiple slots (28), numerous assays can be run simultaneously. Furthermore, different restriction specificities in the extract can be identified by the DNA banding pattern as purification proceeds. Because of this, it is often possible to purify more than one enzyme from a given strain (18).

Specific Enzymes and their Cleavage Sites

A relatively large number of site-specific (Class II) restriction endonucleases are now available (Table 1). The cleavage site base sequences have been determined for a number of the enzymes. Usually this has been accomplished by labeling the 5′ termini of the cleavage products with [32P] phosphoryl groups using polynucleotide kinase (9, 29). It is then possible to digest the DNA with a variety of nucleases to yield a collection of labeled oligonucleotides which may be analyzed by chromatographic and electrophoretic procedures. The overlapping terminal oligo-

Table 1 Class II site-specific restriction endonucleases

Strain	Enzyme[a]	Sequence (5'→3')	Number of cleavage sites[b]			Isoschizomeric enzymes[c]
			λ	Ad2	SV40	
Escherichia coli (end I⁻, R⁺, RI)	*Eco*RI (20, 32)	G↓AATTC (39)	5	5	1	
Escherichia coli (end I⁻, R⁺, RII)	*Eco*RII (32)	↓CCTGG (30, 40)	>35	>35	16	
Haemophilus aegyptius (ATCC 11116)	*Hae*III (33)	GG↓CC (38)	>50	>50	17	*Bsu*X5 (38), *Hhg*I (42)
Haemophilus influenzae serotype d	*Hin*dII (8, 21, 34)	GTPy↓PuAC (9)	34	>20	7	*Hinc*II (43)
	*Hin*dIII (21, 35)	A↓AGCTT (38)	6	11	6	*Hinb*III (42), *Hsu*I (42), *Bbr*I (42)
Haemophilus parainfluenzae	*Hpa*I (18, 36, 37)	GTT↓AAC (25, 41)	11	7	4	*Apo*I (44)
	*Hpa*II (18, 36)	C↓CGG (41)	>50	>50	1	*Hap*II (45, 155), *Mno*I (42)
Anabaena variabilis	*Ava*I (38)	CGPu↓PyCG (38)	—	—	—	

a Additional, less well-characterized enzymes and the strains from which they are derived are as follows: *Ava*II, *Anabaena variabilis* (38); *Alu*I, *Arthrobacter luteus*, ATCC 21606 (38, 42); *Hae*II, *Haemophilus aegyptius*, ATCC 11116 (46); *Hga*I, *Haemophilus gallinarium*, ATCC 14385 (47, 155); *Hha*I, *Haemophilus hemolyticus*, ATCC 10014 (38, 42); *Hin*H-1, *Haemophilus influenzae* H-1, (24, 155); *Hph*I, *Haemophilus parahaemolyticus* (48); *Mbo*I, *Mbo*II, *Moraxella bovis*, ATCC 10900 (42); *Sma*I, *Serratia marcescens* (44); *Sac*I, *Sac*II, *Streptomyces achromogenes* ATCC 12767 (42); *Sal*I, *Sal*II, *Streptomyces albus* ATCC 3004 (42); *Xam*I, *Xanthomonas amaranthicola* ATCC 11645 (42), *Hap*I, *Haemophilus aphrophilus*, ATCC 19415 (42).

b See sections on cleavage of DNA and cleavage maps for references.

c The isoschizomeric enzymes are derived from the following strains: *Apo*I, *Arthrobacter polychromogenes*, ATCC 15216; *Bsu*X5, *Bacillus subtilis* X5 (a transformable strain); *Bbr*I, *Bordetella bronchioseptica*, ATCC 19395; *Hap*II, *Haemophilus aphrophilus*, ATCC 19415; *Hhg*I, *Haemophilus hemoglobinophilus*, ATCC 19416; *Hinb*III, *Haemophilus influenzae*, serotype b; *Hinc*II, *Haemophilus influenzae*, serotype c; *Hsu*I, *Haemophilus suis*, ATCC 19417; *Mno*I, *Moraxella nonliquefaciens*, ATCC 19975.

nucleotides uniquely determine the site. In the case of *Eco*RII, the enzyme produced single stranded 5′ termini so that it was possible to fill in the 3′ end with $[\alpha\text{-}^{32}P]$ deoxynucleoside triphosphates using reverse transcriptase (30). By varying the labeled nucleoside triphosphates used and by nearest neighbor analysis, the terminal nucleotide sequence was deduced. With some exceptions (9, 30, 31), the 3′ termini of restriction fragments have not been sequenced.

All known endo R cleavage sites are 4–6 base pairs long and possess twofold rotational symmetry (Table 1). This has led to speculation that restriction enzymes will in general consist of subunits arranged with twofold symmetry (9, 49). As pointed out, the *Eco*RI enzyme is composed of two identical subunits (20). The structures of other Type II enzymes have not yet been reported; however, it is hoped that future work in this area will clarify the nature of the site recognition process. Cleavage positions, marked by arrows in Table 1, are symmetrically placed within the sites and these are either "staggered," as with *Eco*RI cleavage, or "even" as with *Hin*dII. Staggered cleavage results in formation of identical self-complementary cohesive termini (39, 50), whose implications are discussed later. The *Eco*RII site is unique in containing an odd number of base pairs. In this case the dyad axis of symmetry passes through the central A/T base pair. Consequently, two types of 5′ termini are generated so that not all fragment ends are necessarily mutually cohesive (30, 40).

It is striking that among 17 enzymes with known recognition sites listed in Table 1, only eight different sequences are found. Even apparently unrelated organisms share common specificities. For example, restriction endonucleases with GGCC specificity have been isolated from *Bacillus subtilis, Haemcnhilus aegyptius,* and *H. hemoglobinophilus*. R. J. Roberts has coined the term *isoschizomers* to describe such enzymes. Since any enzyme in an isoschizomer group will yield the same cleavage pattern for a given DNA, the most stable and easily purified enzyme can be selected for use. *Hin*dII endonuclease recognizes the degenerate sequence GTPyPuAC and consequently should cleave a set of four possible sites; *Hpa*I cleaves one of these fours sites, GTTAAC, but is not a true isoschizomer of *Hin*dII. So far *Hin*dII has not been subfractionated into component enzymes.

Comparison with Other Endonucleases

Restriction endonucleases are not the only specific endodeoxyribonucleases found among viruses and bacteria. Even so-called general or nonspecific endonucleases usually yield DNA cleavage products terminating with nucleotides that deviate in frequency from the overall base composition (51), but as a rule, the specificity is insufficient to be analytically useful. An exception is phage T4 endonuclease IV which cleaves preferentially between T and C in pyrimidine tracts of single stranded DNA (52) and is valuable in yielding fragments for sequencing (53, 54). At the opposite extreme are several highly specialized site-specific endonucleases. Examples of these are the postulated phage-specific *Ter* enzymes that produce the cohesive ends of the lambdoid phages (55–58) and the gene *A* product of coliphage ϕX174 that breaks a single phosphodiester bond to initiate phage DNA synthesis (59). Single-strand specific nucleases such as S1 from *Aspergillus* (60) and the nuclease from *Neurospora crassa* (61) have been used to cleave supercoiled DNA at specific, presumably A-T rich regions (62) and to cleave single stranded loops (63, 64) or possibly mismatched base pairs (64) in heteroduplex DNA molecules.

CLEAVAGE OF DNA AND SEPARATION OF FRAGMENTS

A first step in the restriction analysis of DNA molecules is to produce a limit digest and fractionate the resulting fragments. The appropriate enzyme and optimal conditions must be determined empirically. Generally, it is possible to use relatively small amounts of enzyme in extended (overnight) incubations because many of the restriction endonucleases are relatively stable and sufficiently free of contaminating nucleases. A limit digest is obtained when the fragment pattern produced by gel electrophoresis is invariant with respect to additional treatment and in the case of homogeneous starting DNA, when the fragments are present in molar amounts (10).

Fractionation of cleavage products is currently achieved almost exclusively by the use of gel electrophoresis. Methodology is simple and equipment is inexpensive (28). By choosing the appropriate gel it is possible to fractionate DNA segments ranging from several nucleotide pairs (65) up to at least 30,000 nucleotide pairs (66). Polyacrylamide gels are suitable for fractionation in the low molecular weight range, whereas the larger pore-sized agarose gels permit separation of large molecules. In general, for a given gel concentration there is a working range over which electrophoretic mobility of the DNA is inversely proportional to log molecular weight; above this range there is little change in mobility with increasing molecular weight (66). However, electrophoretic mobility may also be affected by base composition (67).

To separate fragments of less than 10^6 daltons, 4 to 15% polyacrylamide slab gels of the type first introduced for tRNA fractionation are commonly used (10, 68). Composite polyacrylamide (2.2%) agarose (0.7%) gels or 1–2% agarose gels are useful for separating fragments up to 2–3 × 10^6 mol wt (18). DNA molecules between 3 and 25 × 10^6 daltons may be resolved in 0.3–0.7% agarose gels (66). Slab gels containing a linear gradient of 2.5–7.5% polyacrylamide (69) are also useful for separation of DNA fragments in the molecular weight range of $7 × 10^4$ to $14 × 10^6$. Below $5 × 10^6$, an approximately linear relationship is found between the logarithms of the molecular weights and logarithms of the distances migrated. Advantages of the gradient system are the wide spectrum of sizes separable on a single gel and the extraordinary sharpness of the bands (69, 70).

Both autoradiography and staining have been used to visualize DNA bands in gels. Autoradiography of radioactive DNA in dried gels is feasible for as little as 100 cpm per band (10, 71), and quantitation is possible by densitometer tracing of the film. For more precise determinations, the gels must be sliced and the radio-activity ascertained.

Some commonly used stains are methylene blue (70), toluidine blue (72), and "stains-all" (72). Ethidium bromide is now widely used as a fluorescent stain sensitive to the level of a few nanograms (18, 66). The ethidium may be excited with either short or long wave length ultraviolet light; the latter is preferable for preserving the DNA. Stained gels may be optically scanned for quantitation. DNA fragments are commonly recovered by electrophoresis into dialysis sacs.

MAPPING CHROMOSOMES

Cleavage Maps

One use of restriction endonucleases is in the physical mapping of chromosomes. This has been especially exploited in the case of viral chromosomes consisting of duplex DNA or where a duplex DNA intermediate in replication can be isolated from infected cells. Specific endo R cleavage sites or fragments serve as physical references in the map, hence the terms "cleavage map" or "fragment map." Once constructed, the map can serve as a framework for localizing template functions and genes and for relating nucleotide sequences of fragments to the entire genome. To construct such a map one must isolate individual endo R fragments, determine their sizes, and order the fragments in the molecule. Fragments have been isolated by agarose or acrylamide gel electrophoresis as described above, and their sizes have been determined by electron microscopic measurement of length relative to a reference DNA molecule on the same grid (10, 18, 73, 80b), by determination

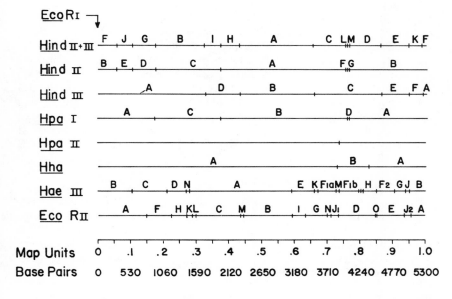

Figure 1 Cleavage map of the Simian Virus 40 genome. The zero point of the map is the *Eco*RI site. For clarity, the circular genome is shown opened at the RI site, and the cleavage sites (and resulting fragments) for each restriction enzyme are indicated on a separate line. Data from refs. 84, 85 (*Eco*RI); 35, 68, 73 (*Hind*); 18, 68, 73 (*Hpa*I and II); 86, (*Hha*); 68, 78, 79 (*Hae*); and 79 (*Eco*RII).

of fragment yield relative to the original DNA (10, 73, 74), or by electrophoretic mobility (see above). The order of fragments in the chromosome has been determined by analysis of partial digest products (75–77), by successive cleavage with multiple restriction endonucleases (68, 75–79), by hybridization of individual fragments (or radioactive transcripts prepared from them with RNA polymerase) to reference fragments previously ordered (78), by pulse-labeling of synthetic DNA (77, 79a), by end-labeling of DNA prior to cleavage (80a), by electron microscopic mapping (10, 18, 81), or by using previously mapped deletion or other mutants (74, 79a, 80). Among the viral genomes dissected in this way are SV40 (10, 18, 68, 73, 78, 79, 84, 85), adenovirus (81a, 82), polyoma (76), phage λ (80, 80b), ϕX174 (77), and filamentous coliphages (79a, 83, 141). A cleavage map of SV40 is shown in Figure 1.

Comparison of Related DNAs: DNA Fingerprints

In analogy with tryptic or chymotryptic maps of proteins, the electrophoretic pattern of endo R fragments of DNA can serve as a fingerprint of the DNA (87). Owing to the sensitivity of electrophoretic separation of fragments in appropriate gels, even relatively minor differences between similar DNAs may be detected, for example, small deletions, additions, rearrangements, or substitutions undetectable by electron microscopic mapping or by tests for nucleotide sequence homology. In this way differences in the DNA from various strains of SV40 (87–90), polyoma (91), human papovaviruses (89, 92), λ and related phages (93), ϕX174 and related phages (94), adenoviruses (95), and herpesviruses (96, 97) have been demonstrated. Also, evolutionary changes in the genomes of SV40 (85, 98–104), polyoma (105), and herpesviruses (97, 106) during serial passage have been followed in this way. Recombination between related viral strains with distinct digest products has also been detected by changes in endo R cleavage patterns (35, 107). Particularly striking applications of this procedure have been the demonstration of maternal inheritance of mitochondrial DNA by the mule and hinny (108) and a detailed comparison of the two ends of adenovirus DNA (109), which is known to have terminal repetitions of nucleotide sequences.

Localization of Chromosomal Functions

Once the relative positions of endo R cleavage sites in a chromosome have been established, these sites can serve as references for localizing or mapping physiologically important chromosomal functions: initiation and termination sites for DNA replication and transcription, templates for specific transcripts or classes of transcripts, binding sites of regulatory proteins, and structural genes. Most progress has been made with small viral chromosomes.

SITES OF INITIATION AND TERMINATION OF DNA REPLICATION   Two general procedures involving restriction endonucleases have been used to map the sites of initiation and termination of viral DNA replication: electron microscopic mapping of cleaved replicating molecules and determination of the temporal sequence of synthesis of different parts of the molecule. The first method is an extension of that used by Schnös & Inman (110) to map replicating linear phage DNAs. Cleavage of circular replicating molecules by a restriction endonuclease, which makes one break, introduces a reference point (the end of the molecule) for determining the position of replication fork(s) relative to the cleavage site (111–113). By repeating the experiment with an enzyme cleaving at a second site, the map position of the origin and termination can be precisely localized (112). Replicating SV40 DNA (111), polyoma DNA (112), and colE1 DNA (113) have been analyzed by this general approach. The second method, analysis of pulse-labeled DNA, is based on Dintzis's procedure for determining the direction of polypeptide elongation (114). Asynchronously replicating molecules that are pulse-labeled (e.g. with [^{3}H] thymidine) generate daughter molecules containing a gradient of label reflecting the temporal order of synthesis of its different parts. By determining the amount of pulse-label present in individual endo R fragments of known map position, one can deduce the sequence of synthesis, including the origin of replication and the termination site. Such studies have been carried out with DNA of SV40 (115, 116), polyoma (112), and ϕX174 replicative form (117).

TRANSCRIPTIONAL MAPPING   RNA transcripts of viral DNA have been mapped by hybridizing RNA to separated strands of individual fragments of DNA, thus localizing in the cleavage map the template from which the transcripts arise. In a widely used procedure, different concentrations of unlabeled RNA from virus-infected cells are incubated with ^{32}P-labeled individual strands of a given fragment under renaturing conditions (80a, 118), and any DNA-RNA hybrid formed is adsorbed to hydroxyapatite under conditions where only duplex structures are retained (119). Hydroxyapatite adsorbable radioactivity is then a measure of sequence homology. To determine the percentage of a given DNA fragment strand with sequence homology to the RNA, the percentage of ^{32}P strand that becomes resistant to the single strand-specific nuclease S1 is measured (120, 121). An alternative procedure for mapping transcripts is the hybridization of labeled RNA to excess unlabeled DNA fragments immobilized on filters (122, 123). These procedures have yielded transcription maps of SV40 (80a, 118, 122, 123), adenovirus (124–126), and polyoma (127) RNA found in productively infected cells, and in the case of SV40 and polyoma also those present in a series of transformed cell lines (127, 128). The most precise mapping of viral mRNA is being done by comparing nucleotide sequences of DNA fragments with nucleotide sequences of viral mRNA isolated from the cytoplasm of infected cells (122). For example, nucleotide sequences at the 3′ end of a class of SV40 mRNA were precisely aligned on the map by establishing their identity to sequences found in specific transcripts of a DNA Hin fragment, thus pinpointing the terminus of this mRNA on the map. Restriction endonucleases have also been used to determine direction of transcription in viral chromosomes by establishing the 5′ → 3′ orientation of each DNA strand relative to the cleavage map (80a, 118).

MAPPING GENES   Restriction endonucleases have been applied to mapping viral genes in several ways: 1. by comparing cleavage maps of certain mutant genomes and wild-type DNA (35, 100, 103, 105); 2. by providing a reference point(s) for electron microscopic mapping of mutants (35, 81, 84, 100, 103, 105, 129); 3. by marker rescue of mutants with endo R fragments (79a, 90, 130–132); and 4. by biological tests of activity of specific fragments (133).

Deletion mutants, insertion mutants, or other types of variant viral genomes may have changes in the mobility of specific endo R fragments compared to wild-type DNA (see above). These alterations can often be localized in the cleavage map (35, 100, 103, 105). Electron microscopic heteroduplex mapping (134) involves visualization and measurement of heteroduplex DNA molecules consisting of one strand of reference DNA (generally wild-type DNA) and one strand of an homologous DNA that has a deletion or substitution to be mapped. The method requires visual reference points in the molecules. Restriction endonucleases simply provide termini at known positions; in the case of a circular DNA molecule, this is an essential prerequisite for the method. This approach has been applied to mapping the SV40 DNA segments present in adeno-SV40 hybrid viruses (81, 84, 129), thus relating the SV40 function expressed by these hybrids to the map positions of their SV40 DNA segments (135, 136). The procedure has also been applied to localizing protein binding sites (137), evolutionary variants of SV40 (100–103, 138) and polyoma (105), and to constructed deletion mutants of SV40 (35, 103; see below).

In the marker rescue procedure (130), denatured individual fragments of wild-type viral DNA are annealed with a single stranded circle of mutant DNA to form a partial heteroduplex molecule consisting of the intact circle and a bound fragment strand held by base pairing. The heteroduplex is then tested for infectivity under conditions where mutant DNA is not infective but wild-type DNA is. If a given fragment corrects (or rescues) the mutant, one can infer that the mutation is present in a part of the genome included in the active fragment. This procedure has been used to map coliphage *amber* mutants and restriction sites (79a, 130–132) (or conversely, to map fragments where the mutational sites had been localized by recombinational analysis), and SV40 *ts* mutants (90, 139).

Another way in which restriction endonucleases have been used to map genes is in the preparation of biologically active fragments of DNA. In molecular cloning experiments (see below) endo R fragments of DNA are linked to bacterial plasmids, and the recombinant molecules are used to infect bacteria in which they replicate and express their genes. Although this procedure has been used primarily to clone and amplify DNA fragments, it also provides information on localization of genes in specific fragments of DNA (140, 181). Another notable example is the localization of genes responsible for transformation of cells by adenovirus or by SV40 (133). In this case, fragments of these viral DNAs were used directly to transform appropriate cells. Activity of DNA was enhanced by adsorption to calcium phosphate during exposure of the cells to DNA. These important experiments have localized the adenovirus gene(s) required for transformation to a small segment near one end of the molecule (133), and in the case of SV40, have so far demonstrated the activity of linear DNA missing a specific segment of the genome (133).

PROTEIN BINDING SITES   Regulatory proteins and polymerases bind to DNA at specific regions that may lie outside structural genes. These regions are often difficult to map genetically. Restriction enzymes offer a useful alternative approach to locating and defining these regions. By cleavage of the genome with a particular enzyme the regions can be isolated (usually attached to adjacent structural gene DNA) on fragments identified by their ability to bind the protein and stick to a membrane filter. Successive cleavage of the fragments with additional enzymes then allows dissection of the binding sites. This approach has been used to isolate and subdivide the λ operators (142, 143), to obtain ϕX174 promoter sites (144), and to isolate *lac* operator (145).

A direct analysis is possible when the protein binding site contains specific restriction endonuclease sites. For example, a series of studies in phage λ shows that the left operator, the leftward promoter, and two leftward promoter mutants intersect a single *Hin*dII site and hence are mutually overlapping (146, 147). Studies of DNA promoter sites further illustrate the possible analyses. The specific binding of *E. coli* RNA polymerase (holoenzyme) to the DNA of SV40, adenovirus 2, and phages λ and T7 blocks cleavage at certain *Hpa* I or *Hin*dII sites (148). Thus, some promoters contain the sequence GTTAAC while some contain members of the degenerate set GTPyPuAC. Other polymerase binding sites exist that do not involve a cleavage site. Additional promoter sequence differences are demonstrated by experiments that show the successive protection of binding sites with increasing levels of polymerase (148). The SV40 *E. coli* RNA polymerase binding site contains an *Hpa* I site, which has been located, by direct sequencing, 30 base pairs in front of the transcription start site (149), suggesting that polymerase starts adjacent to its binding site and need not drift along the DNA prior to initiation (148).

NUCLEOTIDE SEQUENCE ANALYSIS OF DNA

The availability of specific DNA fragments has greatly simplified nucleotide sequence analysis of DNA molecules. By sequential cleavage with a series of enzymes, starting with those that cleave at few sites, even a rather large molecule can be reduced to small, often overlapping fragments amenable to sequencing procedures. This general strategy is exemplified by recent analyses of SV40 DNA (see Figure 1) in which this molecule has been systematically dissected into a series of fragments varying from <20 to ~750 nucleotide pairs; by cleavage of the operator-promoter region of λ DNA beginning with a fragment of ~3500 nucleotide pairs into a series of fragments ranging from 150 to ~1125 nucleotide pairs (142), and by enzymatic dissection of the *lac* promoter-operator region (145). Although it is not our purpose to review DNA sequencing methodology or results (reviewed recently in 150), we describe two general approaches that make use of restriction endonucleases.

One approach is the use of DNA fragments as templates for the synthesis of [^{32}P] RNA by *E. coli* RNA polymerase (for review, see 151); the electrophoretically purified transcripts can then be sequenced by standard methods (152, 153). Transcription of fragments varies in efficiency and in the relative amount of transcript from each strand; in some cases there is predominantly asymmetric transcription (68, 154, 156). However, as a rule all or nearly all sequences in each strand are represented in the product (68, 151, 156), thus providing RNA segments with complementary sequences, often useful for sequence verification and determination of overlaps. Further limitation of the transcript has been achieved by using small ribo-oligonucleotide primers together with low nucleotide concentrations (65, 68, 157). In the case of SV40 DNA fragments, unit length RNA as well as smaller RNA transcripts are produced (68, 149, 154, 156, 158), indicating variable start and/or end points for transcription. A variation of this approach is the transcription of intact viral DNA or DNA segment and subsequent hybridization of ^{32}P transcript to a single fragment immobilized on a filter, followed by RNase removal of nonhybridized RNA and recovery of the bound RNA for sequence analysis (151, 159). These procedures have been used quite effectively to sequence portions of SV40 DNA (68, 149, 154, 155, 158, 159) and *lac* operon DNA (145), starting with fragments up to $\sim$350 nucleotide pairs long.

Another general approach is the use of a single strand of endo R fragment as a primer for [^{32}P] DNA synthesis by *E. coli* DNA polymerase I (150, 160–163). In the presence of Mn^{2+} and a single ribonucleoside triphosphate, ribo substitution occurs (164), giving DNA with one type of ribonucleotide. Subsequent cleavage with restriction enzyme followed by denaturation yields single stranded radioactive products of variable length, whose 5′ termini originate at the enzyme cleavage site. Products are separated into size classes by electrophoresis and homochromatography (153), split at the 3′-ribonucleoside phosphodiester bond by an appropriate RNase, and then sequenced by Sanger's methods (162). These procedures have been used to sequence operator-promoter regions of coliphage λ (163).

ISOLATION OF GENES

General procedures for gene isolation have been recently reviewed (165). Restriction enzymes provide a powerful new approach to this problem. In theory a complex DNA genome can be cleaved into a number of fragments from which a specific gene-bearing fragment may be isolated. If the fragment contains a protein binding site, it can be retrieved as a protein-DNA complex on a membrane filter as seen in the previous section for operator and promoter elements. A more general approach involves fractionation of DNA cleavage products by gel electrophoresis. Specific DNA bands are then assayed for the gene in question by hybridization to radioactive transcripts of the gene purified from cells, binding to specific proteins, or other suitable means. The complete gene or parts of it may be obtained on even smaller fragments by successive cleavage with additional enzymes. An example of this approach is the isolation of the *lac* regulatory region (166). Starting with *lac* transducing phage, a *Hin*d digest yields 40–50 fragments of which only one, separable on gel electrophoresis, binds to *lac* repressor. This fragment is 660 bases long and contains the entire regulatory region between the end of the *i* gene and the beginning of the z gene and extends at least 210 bases into the z gene. This fragment is active in a coupled transcription-translation system, and responds at least partially to control by isopropylthiogalactoside and cyclic AMP. The 660 nucleotide Hind fragment can be further cleaved with Hae enzyme to a piece 174 base pairs long which still contains the entire regulatory region. The isolation of tRNA$_1^{\text{ry}}$ genes from $\phi80$ transducing phages has also been reported (167). In this case the genes were assayed by hybridization to tRNA$_1^{\text{ry}}$.

Retrieval of specific genes from eukaryotic chromosomes presents new challenges. A typical gene represents as little as $10^{-5}\%$ of the total DNA in human cells (165). A restriction enzyme digest may contain over a million fragments that, except for the repetitive segments (see below), produce a continuous banding profile on gel electrophoresis. However, repeated fractionation and enrichment are possible (165); with a suitable assay (e.g. hybridization to a specific RNA or its transcript) considerable purification appears feasible. As a special example of this approach, SV40 sequences possibly linked to host sequences have been detected in transformed mouse cells by digesting the cell DNA with EcoRI enzyme, fractionating the fragments by gel electrophoresis, and then locating the SV40 sequences within the gel by hybridization with radioactive SV40 complementary RNA (168). Most promising are molecular cloning techniques that result in the propagation of DNA segments in bacteria and hence amplification of specific genes (see below).

ORGANIZATION OF EUKARYOTIC DNA

Eukaryotic genomes carry both unique and repeated sequences, as shown by reassociation kinetics of the sheared DNA (169). Restriction enzymes are proving particularly valuable for analyzing the repetitious DNA complement which contains the satellite DNA as well as families of repeated sequences of greater complexity. Restriction sites appear to occur almost randomly in the unique DNA, giving rise on gel electrophoresis to an unresolved continuum of perhaps a million or more bands for the average mammalian genome (170). However, repetitious DNA containing regularly spaced restriction sites gives rise to discrete bands superimposed on the unique DNA profile (171, 173). Among several eukaryotic species analyzed with restriction enzymes, calf thymus DNA has been most carefully examined (171, 174). Using Hind enzyme, 12 distinct bands (representing about 5% of the total DNA) can be seen in the gel profile. The repeat frequency of the DNA in these bands, determined from their amount and molecular weight, varies between 2,000 and 140,000. Cleavage with Hae enzyme uncovers additional orders of repeat within the 12 Hind fragments. Sequential digestion of individual fragments with the two enzymes has permitted tentative ordering of a few fragments (174). Extension of this type of analysis should facilitate determination of the arrangement of repeated sequences in the genome.

The highly repetitious satellite DNAs are particularly amenable to restriction analysis. Purified mouse satellite DNA is cleaved by EcoRII enzyme into a remarkable series of fragments which on gel electrophoresis yield 20 or more regularly spaced bands representing integral multiples of a 240 base pair repeat unit (175). Cleavage of satellite DNA from guinea pig (172) with *Hin*d enzyme or from mouse with *Hin*d or *Hae* enzyme (172, 175) yields regular series of minor bands superimposed on the gel profile of largely undigested DNA. The series of bands are postulated to reflect evolutionary loss or addition of restriction sites in the DNA repeat units (175). Similar analysis of amplified ribosomal DNA of *Xenopus laevis* using *Eco*RI enzyme has revealed unsuspected heterogeneity in the lengths of spacer DNA (176, 179).

Detailed analysis of the unique fraction of eukaryotic DNA will require amplification of these sequences artificially, e.g. by molecular cloning (see below). Such DNA segments might then be amenable to the same type of analyses applied to viral chromosomes and satellite DNA.

RESTRUCTURING AND CLONING OF DNA MOLECULES

One of the most far-reaching applications of restriction endonucleases is the in vitro construction of DNA molecules with novel biological activities. With appropriate enzymes segments of DNA can be excised, added, or rearranged in a given genome (35, 103), or DNA from diverse sources can be joined to yield artificial recombinant molecules (177, 181, 184). Such restructured molecules can then be cloned in suitable cells (178, 180, 181, 183) and subsequently propagated to yield large quantities of the new DNA and, in some cases, specific gene products (180, 181).

Construction and Cloning of Mutant Genomes

Deletion mutants of SV40 and phage λ have been generated by excision of DNA segments from the viral genome with multicut restriction endonucleases (35, 103, 180) followed by cloning the resulting short linear molecules in the presence of helper virus in the case of SV40 (35, 100, 103). As expected, the limits of many of the deletions corresponded to enzyme cleavage sites when enzymes producing cohesive termini were used (35, 180). Prior ligase closure of DNA circles was not required. In addition to these excisional deletions, a second class of deletions was generated by cell-mediated intramolecular recombination near the termini of linear molecules (35, 180). Thus, cohesive termini are not required for cyclization within cells, and a variety of specific and nonspecific endonucleases can be used to generate deletion mutants (35, 103, 139).

Insertion mutants of SV40 have also been constructed (103). In this case, poly (dA·dT) was inserted (see below) at the *Hap*II cleavage site and the resulting infectious mutant DNA cloned by direct plaquing. By using appropriate helper virus it should be possible to clone defective mutants with insertions at different positions in the genome. Indeed, a variety of constructed genomes could be cloned in this way as long as the initiation site for DNA replication is present (100, 102).

Construction and Cloning of Artificial Recombinants

There are basically three steps in this procedure: 1. cleavage of DNA and isolation of products with the desired genes or regulatory signals; 2. end-to-end joining of these molecules; and 3. cloning in a suitable cell with or without prior purification by physical-chemical methods. Cloning requires inclusion in the restructured molecule of genetic information for *cis* functions needed for replication of the recombinant DNA.

PREPARATION OF FRAGMENTS TO BE JOINED   Isolation of endo R fragments of DNA with specific genes or regulatory elements depends on prior mapping of the genome from which the fragments are derived. Appropriate fragments can then be separated from other digest products by gel electrophoresis or by sedimentation procedures. Alternatively, digests of complex DNA can be fractionated and fractions enriched for specific genes as described earlier and detailed in (165). Final purification can be effected by cloning (see below). This procedure has been used successfully for cloning and amplifying rRNA genes of *Xenopus* (179). As will be described, a suitably sensitive selective cloning procedure may obviate the need for preliminary isolation of a specific gene segment (180–182).

END-TO-END JOINING OF DNA SEGMENTS   Procedures for joining DNA molecules are of two types. In the first approach, complementary single stranded termini are added enzymatically to the 3' ends of the linear molecules to be joined (177, 184). After preliminary removal of a small number of nucleotides from the 5' end of each strand with λ exonuclease, the overhanging 3' ends are extended by adding dAMP residues to one molecular species and dTMP residues to the other by means of the terminal deoxynucleotidyl transferase reaction (185). Molecules with oligo (dA) termini are then joined to molecules with oligo (dT) termini by base pairing, and the joined molecules are covalently linked by action of exonuclease III, DNA polymerase I, and DNA ligase. The products include circular DNA duplexes containing both parental molecules (177, 184). However, other more complex products are also produced (184). A more direct and facile means of joining DNA fragments is via cohesive termini produced by restriction endonucleases which make staggered breaks in DNA (see Table 1). Any pair of fragments resulting from cleavage by a given enzyme of this type will have complementary single stranded ends and under appropriate conditions can join end-to-end and/or cyclize via these cohesive termini (50). Adjacent 5' phosphoryl-3' hydroxyl ends can then be linked covalently by the DNA ligase reaction (186). This procedure has been used to form recombinants between DNA from bacterial plasmids and DNA from diverse sources (178, 179, 181, 182), and between λ DNA and bacterial DNA containing the tryptophan operon of *E. coli* (180).

In the procedure utilizing complementary oligonucleotide extensions, parental molecules can join head-to-head, tail-to-tail, or head-to-tail as well as form multimeric circular structures, but a given parental molecule cannot cyclize or polymerize with other like molecules (177, 184). In contrast, joining parental fragments by cohesive termini produced by restriction endonucleases permits coupling in all orientations owing to the twofold symmetry of the nucleotide sequence at the cleavage site and also cyclization or polymerization of each parental type. [In the case of *Eco*RII ends, however, coupling may be restricted because of the asymmetrical central A/T base pair (Table 1).]

CLONING RECOMBINANT MOLECULES   Although it has been possible to partially purify desired recombinant molecules by physical-chemical means (177, 184), the recombinant products may be exceedingly heterogeneous. In these instances purification has been effected by molecular cloning in a suitable biological system; at the same time the recombinant molecule is greatly amplified in amount. A widely applicable procedure entails the use of *E. coli* plasmids as vehicles for cloning DNA fragments from diverse sources (178, 179). The plasmid contributes *cis* functions needed for autonomous replication and also genetic markers useful in selecting bacterial colonies that contain recombinant molecules. In some instances the parental molecule coupled with the vehicular DNA may also contain genes (e.g. for biosynthetic enzymes or antibiotic resistance) which can be selected for (178, 179, 182). Once the recombinant has been cloned and purified DNA isolated, the original parental fragments can be generated by suitable endo R cleavage (178, 179). An example of the cloning procedure is outlined in Figure 2. In this instance, *Eco*RI-cleaved *Xenopus* DNA enriched for rRNA genes was joined to *Eco*RI-cleaved *E. coli* plasmid (pSC101) DNA containing tetracycline resistance genes (179). Unfractionated recombinant molecules were then used to transform sensitive *E. coli* to tetracycline resistance. Many of the resulting tet^r colonies contained plasmids that were recombinants between pSC101 DNA and *Xenopus* DNA. In cells containing recombinant plasmids *Xenopus* rDNA sequences were transcribed (179), although the precise structure of the transcript is not known.

Other vehicles for cloning DNA segments in *E. coli* have also been used. RI-cleaved *col*E1 DNA, an *E. coli* colicinogenic factor (187, 188), when coupled to other RI-cleaved DNA, retains its ability to replicate in the presence of chloramphenicol and eventually comprises over 40% of the total cellular DNA (181).

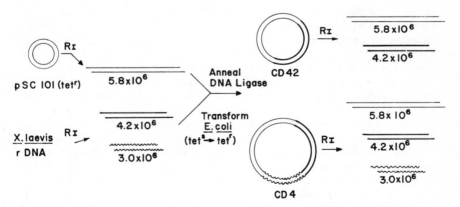

Figure 2   Scheme for construction and cloning of recombinants between *E. coli* plasmid DNA (pSC 101) and ribosomal DNA (rDNA) from *X. laevis,* adapted from data in (179). For simplicity, minor components of rDNA resulting from cleavage by *Eco*RI have been omitted. Abbreviations used: tet^s, sensitive to tetracycline; tet^r, resistant to tetracycline or containing genes conferring resistance.

This plasmid therefore provides a high yield of cloned DNA. It has been used successfully to clone and amplify the tryptophan operon of *E. coli* (181). Another versatile and high yield vehicle is a mutant of coliphage λ, missing a nonessential segment of DNA between *EcoRI* sites (180). Inserted DNA carrying tryptophan genes has been cloned and greatly amplified in the λ vehicle (180). In these instances of tryptophan gene amplification, the enzyme products of these genes comprised some 25 to 50% of the soluble protein of cells containing the plasmid or phage DNA (180, 181). No doubt other DNA viral genomes containing extensive deletions could likewise serve as effective vehicles for cloning segments of DNA.

The methodology recently developed for artificially recombining DNA fragments and subsequently cloning recombinant molecules provides a powerful tool for studying the structure and function of DNA. One should be able to clone any DNA segment of appropriate size and end sequence and prepare it in large quantities. Moreover, by proper choice of the vehicle it should be possible to construct and clone specific transducing viruses for animal cells (100, 102). These approaches will be of particular value in analyzing the organization and expression of eukaryotic chromosomes. The DNA amplification procedures may also prove valuable in preparing large amounts of gene products. By taking advantage of the detailed knowledge of the genetics and regulation of *E. coli* and its phages, rather precise molecules could be designed for specific purposes. An important unanswered question is whether eukaryotic DNA present in bacteria will be correctly transcribed and the mRNA correctly translated.

Possible Biohazards

In regard to the construction and cloning of artificial recombinant DNA molecules, we call the reader's attention to possible biohazards in the creation of new replicating DNA molecules, some of which might be infectious and pathogenic. The basis of these concerns has been outlined in recently published letters (189, 190), in which postponement of certain types of DNA cloning experiments has been suggested until biohazards can be better evaluated (190).

ACKNOWLEDGMENTS
We are grateful to many colleagues for supplying reprints and preprints of manuscripts in press, and particularly to Richard J. Roberts and Kenneth Murray for providing the data on isoschizomers and additional enzymes shown in Table 1. The authors' research included in this review was supported by grants from the National Cancer Institute (CA 11895), the National Science Foundation (GB 35434), the American Cancer Society, and the Whitehall Foundation.

Literature Cited

1. Luria, S. E., Human, S. L. 1952. *J. Bacteriol.* 64:557
2. Bertani, G., Weigle, J. J. 1953. *J. Bacteriol.* 65:113
3. Arber, W. 1974. *Progr. Nucl. Acid Res. Mol. Biol.* 14:1
4. Meselson, M., Yuan, R. 1968. *Nature* 217:1110
5. Horiuchi, K., Zinder, N. D. 1972. *Proc. Nat. Acad. Sci. USA* 69:3220
6. Adler, S. P., Nathans, D. 1973. *Biochim. Biophys. Acta* 299:177

7. Murray, N. E., Batten, P. L., Murray, K. 1973. *J. Mol. Biol.* 81:395
8. Smith, H. O., Wilcox, K. W. 1970. *J. Mol. Biol.* 51:379
9. Kelly, T. J. Jr., Smith, H. O. 1970. *J. Mol. Biol.* 51:393
10. Danna, K. J., Nathans, D. 1971. *Proc. Nat. Acad. Sci. USA* 68:2913
11. Boyer, H. W. 1971. *Ann. Rev. Microbiol.* 25:153
12. Meselson, M., Yuan, R., Heywood, J. 1972. *Ann. Rev. Biochem.* 41:447
13. Murray, K., Old, R. W. 1974. *Progr. Nucl. Acid Res. Mol. Biol.* 14:117
14. Linn, S., Lautenberger, J. A., Eskin, B., Lackey, D. 1974. *Fed. Proc.* 33:1128
15. Haberman, A., Heywood, J., Meselson, M. 1972. *Proc. Nat. Acad. Sci. USA* 69:3138
16. Haberman, A. 1974. *J. Mol. Biol.* 89:545
17. Smith, H. O., Nathans, D. 1973. *J. Mol. Biol.* 81:419
18. Sharp, P. A., Sugden, B., Sambrook, J. 1973. *Biochemistry* 12:3055
19. Boyer, H. W. 1974. *Fed. Proc.* 33:1125
20. Greene, P. J., Betlach, M. C., Boyer, H. W., Goodman, H. M. 1974. *Methods Mol. Biol.* 7:87
21. Smith, H. O. 1974. *Methods Mol. Biol.* 7:71
22. DeFilippes, F. M. 1973. *Anal. Biochem.* 52:637
23. Eskin, B., Linn, S. 1972. *J. Biol. Chem.* 247:6183
24. Takanami, M., Kojo, H. 1973. *FEBS Lett.* 29:267–70
25. Sack, G. H. Jr. 1974. PhD thesis. Johns Hopkins Univ., Baltimore
26. Melgar, E., Goldthwaite, D. A. 1968. *J. Biol. Chem.* 243:4401
27. Meselson, M., Yuan, R. 1971. *Proc. Nucl. Acid. Res.* 2:889
28. Studier, F. W. 1973. *J. Mol. Biol.* 79:237
29. Weiss, B., Richardson, C. C. 1967. *J. Mol. Biol.* 23:405
30. Boyer, H. W., Chow, L. T., Dugaiczyk, A., Hedgpeth, J., Goodman, H. M. *Nature New Biol.* 244:40
31. Englund, P. T. 1971. *J. Biol. Chem.* 246:3269
32. Yoshimori, R. 1971. PhD thesis. Univ. of California, San Francisco
33. Middleton, J. H., Edgell, M. H., Hutchison, C. A. III. 1972. *J. Virol.* 10:42
34. Humphries, P., Gordon, R. L., McConnell, D. J., Connolly, P. 1974. *Virology* 58:25
35. Lai, C-J., Nathans, D. 1974. *J. Mol. Biol.* 89:179
36. Gromkova, R., Goodgal, S. H. 1972. *J. Bacteriol.* 109:987
37. DeFilippes, F. M. 1974. *Biochem. Biophys. Res. Commun.* 58:586
38. Murray, K. Personal communication
39. Hedgpeth, J., Goodman, H. M., Boyer, H. W. 1972. *Proc. Nat. Acad. Sci. USA* 69:3448
40. Bigger, C. H., Murray, K., Murray, N. E. 1973. *Nature New Biol.* 244:7
41. Garfin, D. E., Goodman, H. M. 1974. *Biochem. Biophys. Res. Commun.* 59:108
42. Roberts, R. J. Personal communication
43. Landy, A., Ruedisueli, E., Robinson, L., Foeller, C., Ross, W. 1974. *Biochemistry* 13:2134
44. Green, R., Mulder, C. Personal communication
45. Sugisaki, H., Takanami, M. 1973. *Nature New Biol.* 246:138
46. Roberts, R. J., Breitmeyer, J. B., Tabachnik, N. F., Myers, P. A. 1975. *J. Mol. Biol.* In press
47. Takanami, M. 1973. *FEBS Lett.* 34:318
48. Middleton, J. H., Stankus, P. V., Edgell, M. H., Hutchison, C. A. III. Personal communication
49. Sobell, H. M. 1973. *Advan. Genet.* 17:411
50. Mertz, J. E., Davis, R. W. 1972. *Proc. Nat. Acad. Sci. USA* 69:3370
51. Barnardi, G., Ehrlich, S. D., Thiery, J-P. 1973. *Nature New Biol.* 246:36
52. Sadowski, P. D., Hurwitz, J. 1969. *J. Biol. Chem.* 244:6192
53. Ziff, E. B., Sedat, J. W., Galibert, F. 1973. *Nature New Biol.* 241:34
54. Galibert, F., Sedat, J., Ziff, E. 1974. *J. Mol. Biol.* 87:377
55. Wu, R., Taylor, E. 1971. *J. Mol. Biol.* 57:491
56. Weigel, P. H., Englund, P. T., Murray, K., Old, R. W. 1973. *Proc. Nat. Acad. Sci. USA* 70:1151
57. Murray, K., Murray, N. E. 1973. *Nature New Biol.* 243:134
58. Wang, J. C., Brezinski, D. P. 1973. *Proc. Nat. Acad. Sci. USA* 70:2667
59. Henry, T. J., Knippers, R. 1974. *Proc. Nat. Acad. Sci. USA* 71:1549
60. Ando, T. 1966. *Biochim. Biophys. Acta* 114:158
61. Rabin, E. Z., Preiss, B., Fraser, M. J. 1971. *Prep. Biochem.* 1:283
62. Beard, P., Morrow, J. F., Berg, P. 1973. *J. Virol.* 12:1303
63. Bartok, K., Garon, C. F., Berry, K. W., Fraser, M. J., Rose, J. A. 1974. *J. Mol. Biol.* 87:437
64. Shenk, T. E., Rhodes, C., Rigby, P. W. J., Berg, P. 1974. *Cold Spring Harbor*

Symp. Quant. Biol. 39: In press
65. Gilbert, W., Maxam, A. 1973. *Proc. Nat. Acad. Sci. USA* 70: 3581
66. Helling, R. B., Goodman, H. M., Boyer, H. W. 1974. *J. Virol.* 14: 1235
67. Zeiger, R. S., Salomon, R., Dinginan, C. W., Peacock, A. C. 1972. *Nature New Biol.* 238: 65
68. Fiers, W., Danna, K., Roglers, R., Van de Voorde, A., Van Herreweghe, J., Van Heuverswyn, H., Volckaert, G., Yang, R. 1974. *Cold Spring Harbor Symp. Quant. Biol.* 39: In press
69. Jeppesen, P. G. N. 1974. *Anal. Biochem.* 58: 195
70. Allet, B. 1973. *Biochemistry* 12: 3972
71. Bonner, W. M., Laskey, R. A. 1974. *Eur. J. Biochem.* 46: 83
72. Streeck, R. E., Philippsen, P., Zachau, H. G. 1974. *Eur. J. Biochem.* 45: 489
73. Sack, G. H. Jr., Nathans, D. 1973. *Virology* 51: 517
74. Hutchison, C. A. III, Edgell, M. H. 1972. *J. Virol.* 9: 574
75. Danna, K. J., Sack, G. H. Jr., Nathans, D. 1973. *J. Mol. Biol.* 78: 363
76. Griffin, B. E., Fried, M., Cowie, A. 1974. *Proc. Nat. Acad. Sci. USA* 71: 2077
77. Lee, A., Sinsheimer, R. L. 1974. *Proc. Nat. Acad. Sci. USA* 71: 2882
78. Lebowitz, P., Siegel, W., Sklar, J. 1974. *J. Mol. Biol.* 88: 105
79. Subramanian, K. N., Pan, J., Zain, B. S., Weissman, S. M. 1974. *Nucl. Acid Res.* 1: 727
79a. Seeburg, P. H., Schaller, H. 1975. *J. Mol. Biol.* In press
80. Allet, B., Jeppesen, P. G. N., Katagiri, K. J., Delius, H. 1973. *Nature* 241: 120
80a. Sambrook, J., Sugden, B., Keller, W., Sharp, P. A. 1973. *Proc. Nat. Acad. Sci. USA* 70: 3711
80b. Thomas, M., Davis, R. W. 1975. *J. Mol. Biol.* In press
81. Lebowitz, P., Kelly, T. J. Jr., Nathans, D., Lee, T. N. H., Lewis, A. M. Jr. 1974. *Proc. Nat. Acad. Sci. USA* 71: 441
81a. Pettersson, U., Mulder, C., Delius, H., Sharp, P. A. 1973. *Proc. Nat. Acad. Sci. USA* 70: 200
82. Sharp, P. A., Pettersson, U., Sambrook, J. 1974. *J. Mol. Biol.* 86: 709
83. Takanami, M., Okamoto, T., Sugimoto, K., Sugisaki, H. 1975. *J. Mol. Biol.* In press
84. Morrow, J. F., Berg, P. 1972. *Proc. Nat. Acad. Sci. USA* 69: 3365
85. Mulder, C., Delius, H. 1972. *Proc. Nat. Acad. Sci. USA* 69: 3215
86. Weissman, S. M. Personal communication
87. Nathans, D., Danna, K. J. 1972. *J. Mol. Biol.* 64: 515
88. Huang, E-S., Newbold, J. E., Pagano, J. S. 1973. *J. Virol.* 11: 508
89. Sack, G. H. Jr., Narayan, O., Danna, K. J., Weiner, L. P., Nathans, D. 1973. *Virology* 51: 345
90. Lai, C-J., Nathans, D. 1974. *Virology* 60: 466
91. Fried, M., Griffin, B. E., Land, E., Robberson, D. L. 1974. *Cold Spring Harbor Symp. Quant. Biol.* 39: In press
92. Osborn, J. E., Robertson, S. M., Padgett, B. L., Zurhein, G., Walker, D. L., Weisblum, B. 1974. *J. Virol.* 13: 614
93. Landy, A., Ruedisueli, E., Robinson, L., Foeller, C., Ross, W. 1974. *Biochemistry* 13: 2134
94. Godson, G. N., Boyer, H. 1974. *Virology.* 62: 270
95. Mulder, C., Sharp, P. A., Delius, H., Pettersson, U. 1974. *J. Virol.* 14: 68
96. Skare, J., Summers, W. P., Summers, W. C. 1975. *J. Virol.* In press
97. Hayward, G., Roizman, B. Personal communication
98. Brockman, W. W., Lee, T. N. H., Nathans, D. 1973. *Virology* 54: 384
99. Rozenblatt, S., Lavi, S., Singer, M. F., Winocour, E. 1973. *J. Virol.* 12: 501
100. Brockman, W. W., Nathans, D. 1974. *Proc. Nat. Acad. Sci. USA* 71: 942
101. Khoury, G., Fareed, G. C., Berry, K., Martin, M. A., Lee, T. N. H., Nathans, D. 1974. *J. Mol. Biol.* 87: 289
102. Brockman, W. W., Lee, T. N. H., Nathans, D. 1974. *Cold Spring Harbor Symp. Quant. Biol.* 39: In press
103. Mertz, J. E., Carbon, J., Herzberg, M., Davis, R. W., Berg, P. 1974. *Cold Spring Harbor Symp. Quant. Biol.* 39: In press
104. Fareed, G. C., Byrne, J. C., Martin, M. A. 1974. *J. Mol. Biol.* 87: 275
105. Robberson, D. L., Fried, M. 1974. *Proc. Nat. Acad. Sci. USA* 71: 3497
106. Wagner, M., Skare, J., Summers, W. C. 1974. *Cold Spring Harbor Symp. Quant. Biol.* 39: In press
107. Grodzicker, T., Williams, J. F., Sharp, P. A., Sambrook, J. 1974. *Cold Spring Harbor Symp. Quant. Biol.* 39: In press
108. Hutchison, C. A. III, Newbold, J. E., Potter, S. S., Edgell, M. H. 1974. *Nature* 251: 536
109. Roberts, R. J., Arrand, J. F., Keller, W. 1974. *Proc. Nat. Acad. Sci. USA* 71: 3829
110. Schnös, M., Inman, R. B. 1970. *J. Mol. Biol.* 51: 61
111. Fareed, G. C., Garon, C. F., Salzman, N. P. 1972. *J. Virol.* 10: 484

112. Crawford, L. V., Robbins, A. K., Nicklin, P. M., Osborne, K. 1974. *Cold Spring Harbor Symp. Quant. Biol.* 39: In press
113. Inselberg, J. 1974. *Proc. Nat. Acad. Sci. USA* 71: 2256
114. Dintzis, H. M. 1961. *Proc. Nat. Acad. Sci. USA* 47: 247
115. Nathans, D., Danna, K. J. 1972. *Nature New Biol.* 236: 200
116. Danna, K. J., Nathans, D. 1972. *Proc. Nat. Acad. Sci. USA* 69: 3097
117. Godson, G. N. 1974. *J. Mol. Biol.* 90: 127
118. Khoury, G., Martin, M. A., Lee, T. N. H., Danna, K. J., Nathans, D. 1973. *J. Mol. Biol.* 78: 377
119. Britten, R. J., Kohne, D. E. 1968. *Science* 161: 529
120. Sutton, W. D. 1971. *Biochim. Biophys. Acta* 240: 522
121. Leong, J., Garapin, A., Jackson, N., Fanshier, L., Levinson, W., Bishop, J. M. 1972. *J. Virol.* 9: 891
122. Dhar, R., Zain, B. S., Weissman, S. M., Pan, J., Subramanian, K. 1974. *Proc. Nat. Acad. Sci. USA* 71: 371
123. Weinberg, R. A., Ben-Ishai, Z., Newbold, J. E. 1974. *J. Virol.* 13: 1263
124. Tibbetts, C., Pettersson, U., Philipson, L. 1974. *J. Mol. Biol.* 88: 767
125. Sharp, P. A., Gallimore, P. H., Flint, S. J. 1974. *Cold Spring Harbor Symp. Quant. Biol.* 39: In press
126. Tal, J., Craig, E. A., Zimmer, S., Raskas, H. J. 1974. *Proc. Nat. Acad. Sci. USA* 71: 4057
127. Kamen, R., Lindstrom, D. M., Shure, H. 1974. *Cold Spring Harbor Symp. Quant. Biol.* 39: In press
128. Khoury, G., Martin, M., Lee, T. N. H., Nathans, D. 1975. *Virology.* In press
129. Morrow, J. F., Berg, P., Kelly, T. J. Jr., Lewis, A. M. Jr. 1973. *J. Virol.* 12: 653
130. Hutchison, C. A. III, Edgell, M. H. 1971. *J. Virol.* 8: 181
131. Sclair, M., Edgell, M. H., Hutchison, C. A. III. 1973. *J. Virol.* 11: 378
132. van den Hondel, C. A., Weijers, A., Konings, R. N. H., Schoenmakers, J. G. G. 1975. *Eur. J. Biochem.* In press
133. Graham, F. L., Abrahams, P. J., Mulder, C., Heijneker, H. L., Warnaar, S. O., de Vries, F. A. J., Fiers, W., van der Eb, A. J. 1974. *Cold Spring Harbor Symp. Quant. Biol.* 39: In press
134. Davis, R. W., Simon, J., Davidson, N. 1971. *Methods Enzymol.* 21: 413
135. Lewis, A. M. Jr., Levine, A. S., Crumpacker, C. S., Levin, M. J., Samaha,

136. R. J., Henry, P. H. 1973. *J. Virol.* 11: 655
136. Kelly, T. J. Jr., Lewis, A. M. Jr. 1973. *J. Virol.* 12: 643
137. Morrow, J. F., Berg, P. 1973. *J. Virol.* 12: 1631
138. Risser, R., Mulder, C. 1974. *Virology* 58: 424
139. Lai, C-J., Nathans, D. 1974. *Cold Spring Harbor Symp. Quant. Biol.* 39: In press
140. Cohen, S. N., Chang, A. C. Y., Hsu, C. L. 1972. *Proc. Nat. Acad. Sci. USA* 69: 2110
141. van den Hondel, C. A., Schoenmakers, J. G. G. 1975. *Eur. J. Biochem.* In press
142. Maniatis, T., Ptashne, M. 1973. *Nature* 246: 133
143. Maniatis, T., Ptashne, M., Maurer, R. 1973. *Cold Spring Harbor Symp. Quant. Biol.* 38: 857
144. Chen, C-Y., Hutchison, C. A. III, Edgell, M. H. 1973. *Nature New Biol.* 243: 233
145. Gilbert, W., Gralla, J., Majors, J., Maxam, A. 1975. In *Symp. Protein-Ligand Interactions.* Berlin: Walter de Gruyter
146. Allet, B., Solem, R. 1974. *J. Mol. Biol.* 85: 475
147. Maurer, R., Maniatis, T., Ptashne, M. 1974. *Nature* 249: 221
148. Allet, B., Roberts, R. J., Gesteland, R. F., Solem, R. 1974. *Nature* 249: 217
149. Dhar, R., Weissman, S. M., Zain, B. S., Pan, J., Lewis, A. M. Jr. 1974. *Nucl. Acid Res.* 1: 595
150. Salser, W. A. 1974. *Ann. Rev. Biochem.* 43: 923
151. Marotta, C. A., Lebowitz, P., Dhar, R., Zain, B. S., Weissman, S. M. 1974. *Methods Enzymol.* 29: 254
152. Sanger, F., Brownlee, G. G. 1967. *Methods Enzymol.* 12A: 361
153. Brownlee, G. G., Sanger, F. 1969. *Eur. J. Biochem.* 11: 395
154. Zain, B. S., Weissman, S. M., Dhar, R., Pan, J. 1974. *Nucl. Acid Res.* 1: 577
155. Takanami, M. 1974. *Methods Mol. Biol.* 7: 113
156. Lebowitz, P., Bloodgood, R. 1975. *J. Mol. Biol.* In press
157. Doroney, K. M., So, A. G. 1970. *Biochemistry* 9: 2520
158. Dhar, R., Subramanian, K., Zain, B. S., Pan, J., Weissman, S. M. 1974. *Cold Spring Harbor Symp. Quant. Biol.* 39: In press
159. Zain, B. S., Dhar, R., Weissman, S. M., Lebowitz, P., Lewis, A. M. Jr. 1973. *J. Virol.* 11: 683
160. Salser, W. 1972. *Proc. Nat. Acad. Sci.*

USA 69:238

161. van de Sande, J. H., Loewen, P. C., Khorana, H. G. 1972. *J. Biol. Chem.* 247:6140

162. Sanger, F., Donelson, J. E., Coulson, A. R., Kössel, H., Fischer, D. 1973. *Proc. Nat. Acad. Sci. USA* 70:1209

163. Maniatis, T., Ptashne, M., Barrell, B. G., Donelson, J. 1974. *Nature* 250:394

164. Berg, P., Fancher, H., Chamberlin, M. 1963. In *Symposium on Informational Macromolecules,* ed. H. Vogel, V. Bryson, J. O. Lampen, 467. New York/London: Academic

165. Brown, D. D., Stern, R. 1974. *Ann. Rev. Biochem.* 43:667

166. Landy, A., Olchowski, E., Ross, W., Reiness, G. 1974. *Mol. Gen. Genet.* In press

167. Landy, A., Foeller, C., Ross, W. 1974. *Nature* 249:738

168. Botchan, M., McKenna, G., Sharp, P. A. 1974. *Cold Spring Harbor Symp. Quant. Biol.* 38:383

169. Britten, R. J., Davidson, E. H. 1971. *Quant. Rev. Biol.* 46:111

170. Southern, E. M., Roizes, G. 1974. *Cold Spring Harbor Symp. Quant. Biol.* 38:429

171. Mowbray, S. L., Landy, A. 1974. *Proc. Nat. Acad. Sci. USA* 71:1920

172. Hörz, W., Hess, I., Zachau, H. G. 1974. *Eur. J. Biochem.* 45:501

173. Philippsen, P., Streeck, R. E., Zachau, H. G. 1974. *Eur. J. Biochem.* 45:479

174. Mowbray, S. L., Gerbi, S. A., Landy, A. *Nature.* In press

175. Southern, E. M. *J. Mol. Biol.* In press

176. Wellauer, P. K., Reeder, R. H., Carroll, D., Brown, D. D., Deutch, A., Higashinakagawa, T. 1974. *Proc. Nat. Acad. Sci. USA* 71:2823

177. Jackson, D. A., Symons, R. H., Berg, P. 1972. *Proc. Nat. Acad. Sci. USA* 69:2904

178. Cohen, S. N., Chang, A. C. Y., Boyer, H. W., Helling, R. B. 1973. *Proc. Nat. Acad. Sci. USA* 70:3240

179. Morrow, J. F., Cohen, S. N., Chang, A. C. Y., Boyer, H. W., Goodman, H. M., Helling, R. B. 1974. *Proc. Nat. Acad. Sci. USA* 71:1743

180. Murray, N. E., Murray, K. 1974. *Nature* 251:476

181. Hershfield, V., Boyer, H. W., Yanofsy, C., Lovett, M. A., Helinski, D. R. 1974. *Proc. Nat. Acad. Sci. USA* 71:3455

182. Chang, A. C. Y., Cohen, S. N. 1974. *Proc. Nat. Acad. Sci. USA* 71:1030

183. Cohen, S. N., Chang, A. C. Y. 1973. *Proc. Nat. Acad. Sci. USA* 70:1293

184. Lobban, P. E., Kaiser, A. D. 1973. *J. Mol. Biol.* 78:453

185. Chang, L. M. S., Bollum, F. J. 1971. *Biochemistry* 10:536

186. Richardson, C. C. 1969. *Ann. Rev. Biochem.* 38:795

187. Helinski, D. R., Clewell, D. B. 1971. *Ann. Rev. Biochem.* 40:899

188. Clewell, D. B., Helinski, D. R. 1969. *Proc. Nat. Acad. Sci. USA* 62:1159

189. Singer, M., Soll, D. 1973. *Science* 181:1114

190. Berg, P., Baltimore, D., Boyer, H. W., Cohen, S. N., Davis, R. W., Hogness, D. S., Nathans, D., Roblin, R., Watson, J. D., Weissman, S., Zinder, N. D. 1974. *Science* 185:303

PROTEIN COMPLEMENTATION ×884

Irving Zabin[1] *and Merna R. Villarejo*

Department of Biological Chemistry, School of Medicine and Molecular
Biology Institute, University of California, Los Angeles 90024

CONTENTS

INTRODUCTION

Protein complementation has not been treated previously as a separate topic in the *Annual Review of Biochemistry*. This review defines and illustrates complementation using selected enzyme systems, with emphasis on work of the last five or six years. In addition to Fincham's monograph, *Genetic Complementation* (1), several reviews have appeared (2, 3). Other authors have discussed possible mechanisms (4–7).

For this review, complementation is defined as the restoration of a biological activity by noncovalent interaction of different proteins or polypeptides. Complementation experiments were first performed in vivo. Genes specifying a particular enzyme and carrying different mutations were introduced into heterocaryons (cells with nuclei from different parents) of *Neurospora* (8, 9). Enzyme activity was restored under conditions in which recombination of DNA did not occur. Comple-

[1] The preparation of this review and the work described in the authors' laboratory were supported by a grant (AI 04181) from the National Institute of Allergy and Infectious Diseases.

295

mentation experiments are also performed in vitro. The earliest observation was that of Woodward (10), who obtained adenylosuccinase activity by mixing cell free extracts of two different mutant strains of *Neurospora*. In many systems, complementation is performed by simply mixing the protein components. In others, the mixture of proteins is subjected to denaturing and renaturing conditions. The conditions necessary to achieve complementation no doubt can provide insight into the mechanism of interaction.

It is convenient to divide complementation into two types: intercistronic and intracistronic. Intercistronic complementation involves two different polypeptide chains, each specified by a different cistron. A simple example is the tryptophan synthetase of *Escherichia coli*, $\alpha_2\beta_2$, which catalyzes the formation of tryptophan from indoleglycerol phosphate and L-serine. Mutant strains that produce normal α- but defective β-subunits or normal β- but defective α-subunits cannot catalyze this reaction. When partial diploids are formed of the two mutant strains, each of which contributes one wild-type subunit, activity is restored. Activity is also restored when extracts of the two strains are mixed. In intercistronic complementation, any strain with a mutation in one cistron is expected to complement any other with a mutation in the second cistron. The active oligomer formed is the wild-type enzyme. The test for intercistronic complementation [the *cis-trans* test in genetic terminology (11)] may not always give a positive result, however. In many bacterial operons, a mutation in one cistron may exert a polar effect on the expression of an operator-distal cistron, so that the polypeptide chain corresponding to the latter is not produced (12). An additional complication is that complementing pairs that map in two groups are not always intercistronic but may be intracistronic.

Intracistronic complementation involves proteins specified by a single cistron and may be considerably more complex. It may involve the interaction of two partial chains, each derived from a different mutant gene, as in β-galactosidase. The classical chemical analogy is that of ribonuclease treated with subtilisin to produce the S peptide, residues 1–20, and S protein, residues 21–124. Neither has activity by itself, but full activity is restored on mixing the two fractions (13). A number of elegant experiments of a similar nature have been carried out on *Staphylococcal nuclease* by Anfinsen and his colleagues (14).

Intracistronic complementation can also occur with complete polypeptides specified by the same cistron. This is known as interallelic complementation; a polypeptide chain with an amino acid substitution resulting from a mutation in one allele may complement a polypeptide with an amino acid substitution elsewhere in the sequence. The mechanism of such complementation was discussed by Crick & Orgel (5), who suggested that restoration of function might be due to the correction of misfolding of one monomer by some unaltered part of the other.

It might be expected that a subunit from a wild-type organism would correct the misfolding of a protein from a mutant. In such a case, the biological activity of the enzyme complex would be greater than that due to the native subunit alone; this is positive complementation. The reverse case, where there is a decrease in activity when a mutant subunit is combined with wild-type, is known as negative complementation.

Complementation has been used to determine the nature of a gene product, i.e. the number and identity of subunits in an oligomeric enzyme; to determine the function of different regions of a polypeptide chain; and to investigate tertiary and quaternary structure. Most experiments on complementation have been carried out using organisms of the same species. However, interspecies and intergeneric combinations have also been tested in both intracistronic and intercistronic systems in order to decipher evolutionary relationships. Examples of the use of complementation in each of these areas will be presented in the systems discussed below.

ALKALINE PHOSPHATASE

Intracistronic Complementation

E. coli alkaline phosphatase is a dimeric protein with a molecular weight of 86,000 (15, 16). Early experiments on the molecular basis of intracistronic complementation were carried out by Garen & Garen, who observed that certain pairs of inactive structural mutants of alkaline phosphatase would complement in vivo to produce enzyme activity (17). Schlesinger & Levinthal purified inactive proteins from alkaline phosphatase negative mutants of E. coli and dissociated the inactive dimeric proteins to monomers using mild acid treatment or reduction with thioglycollate in urea. When appropriate pairs of inactive monomers were allowed to reassociate, a partially active, heat labile enzyme was formed. These results suggested that the active enzyme was a hybrid molecule composed of a monomer from each of the mutant proteins used in the reaction (18, 19).

In order to prove that this was the case, complemented enzymes were isolated and characterized. Hybrid dimers were formed in vitro with one subunit from each of various mutants. One subunit was labeled for density (D, ^{15}N) so that the complemented enzymes could be isolated by equilibrium density gradient centrifugation. It was found that hybrids formed from different mutant subunits had different specific enzymic activities and that affinities between subunits were also a function of the particular structural alteration involved. There was no obvious correlation between complementation activity and position of mutations on the genetic map. This is not surprising, since it is likely that complementation is dependent not only on the position of the mutation, but also on the particular amino acid substitution and the way it affects the entire tertiary structure (20).

In more recent work, density labeled dimers have been used to study the symmetry of subunit association in the wild-type enzyme (21). A hybrid enzyme consisting of a mutant phosphatase subunit and a wild-type subunit was used to examine the role of functional subunit interactions in the catalytic mechanism. In this case, the defective subunit displayed neither positive nor negative complementation; that is, it had no effect on the catalytic properties of the wild-type subunit (22).

Intergeneric Complementation

The alkaline phosphatases from E. coli and Serratia marcescens have similar catalytic properties although they are quite different in amino acid composition and immunological properties. Nonetheless, active hybrid enzyme containing one subunit of each kind can be formed both in vivo and in vitro, indicating that the structure involved in subunit interaction has been conserved in the evolution of these species (23).

β-GALACTOSIDASE

The structure and properties of β-galactosidase have been reviewed (24, 25). It is a tetramer of molecular weight 540,000, and each of four identical chains contains approximately 1170 amino acids. Despite the extraordinary length of the polypeptide chain, the enzyme can be renatured easily from a denatured state. Active enzyme was obtained after treatment with 8 M urea (26), and about 30% of the original activity was recovered after 10 min of boiling (27). Under certain conditions quantitative recovery of enzyme activity was obtained after denaturation with urea or guanidine (28). The possibilities of error in refolding might be expected to increase with size of the protein; clearly the forces determining the native conformation of β-galactosidase are powerful and specific. Perhaps, then, it is not surprising that certain fragments of β-galactosidase may interact with each other in a specific manner to produce complemented enzyme.

β-Galactosidase is specified by the first structural gene, the Z gene, of the lac operon in E. coli and has been at the center of a large number of studies concerned with gene expression and its control (29). Thousands of bacterial strains carrying mutations in the Z gene have been isolated; hence, many kinds of enzymically inactive β-galactosidase can be obtained. By the preparation of partial diploid cells, Perrin (30, 31) demonstrated that the combination of many pairs of point mutants would result in the in vivo formation of active enzyme. In vitro experiments, carried out by mixing certain pairs of extracts or partially purified proteins, also resulted in the restoration of enzyme activity. The complemented enzymes appeared to be similar to wild-type in immunological response and in sedimentation behavior, but differed in thermal stability. Further, the pairs of complementing mutants did not map in two genetically distinct groups, as would be expected for intercistronic complementation, but instead showed a complicated pattern. The results gave no reason to believe that the Z gene is anything but a single cistron producing a single polypeptide chain, albeit an unusually long one.

Additional complementation studies with genetically well-characterized mutant strains raised the possibility that β-galactosidase might contain more than one kind

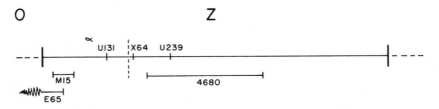

Figure 1   Mutant strains of E. coli illustrating α-complementation of β-galactosidase.

of subunit. These experiments utilized deletion mutants which produced one portion of the polypeptide chain in combination with other mutants supplying the remainder of the sequence (32, 33).

However, thorough investigation of the properties of complemented enzymes showed that they were different from wild-type β-galactosidase (34, 35). Detailed examination of these complementing systems has revealed some interesting aspects of protein-protein interaction.

α-Complementation

The mutant strain M15 (Figure 1) contains an operator-proximal Z deletion of a length that was roughly approximated by genetic recombination data (36). Since the operator end of the Z gene corresponds to the amino terminus of the polypeptide chain (37), the β-galactosidase protein produced by M15 differs from the wild-type protein by the absence of a chain segment near the amino terminus. The M15 protein has no enzyme activity. Peptide fragments from mutant strains that contained an intact operator-proximal (α) segment of the gene were found by Ullmann et al (33) to restore β-galactosidase enzyme activity to extracts of M15. The peptide fragments were called α-donors; the M15 was the α-acceptor. Certain other mutants with deletions extending from another operon into the α-segment, such as E65, could, like M15, serve as α-acceptors. Extracts of termination mutant strains, such as X64 and U239 but not U131, served as α-donors. The fact that these mutant strains mapped in two distinct groups suggested intercistronic complementation. One group consisted of those deletion mutants that map to the left of the "α-barrier" (dotted line, Figure 1); these are α-acceptors. The second group included termination and deletion mutants that map past the α-barrier; these are α-donors. Further, strain 4680 produced an α-donor peptide whose size on Sephadex columns and in the centrifuge, perhaps 30,000 daltons, corresponded to the size expected for the α-region (33).

But was the complemented enzyme the same as wild-type? Further experiments showed it to be more heat labile and it could not be renatured following treatment with 8 M urea (34). Although the α-donor peptide could be dissociated from complemented enzyme, only traces could be obtained from native β-galactosidase. This clearly demonstrated that complemented enzyme was different in structure from wild-type enzyme.

Later work showed that the α-barrier is not an essential feature of α-complementation. Morrison & Zipser (38) discovered that when extracts of strains containing β-galactosidase or certain fragments of β-galactosidase were autoclaved, a soluble peptide fraction of molecular weight 7400 was produced which was an effective α-donor to extracts of M15. Lin et al (39) found that treatment of pure β-galactosidase with cyanogen bromide yielded a peptide of about 80 residues with considerable α-donor activity. The peptide was isolated, and its amino acid sequence was shown to correspond to residues 3–92 of the complete protein (40). These experiments demonstrated that considerably less than the NH_2-terminal one fourth or one fifth of the protein could serve as a complementing partner for M15.

If only the first 92 residues of the β-galactosidase sequence are necessary for α-complementing activity, why did short chain termination mutants, such as U131, which mapped within the α-region of the gene, produce no α-donor activity? The explanation is that such fragments were degraded rapidly by proteolysis during growth, and extracts prepared for the test no longer contained the gene product. Goldschmidt (41) characterized fragments produced by Z gene termination mutants by autoradiography after separation by sodium dodecylsulfate (SDS) polyacrylamide gel electrophoresis and demonstrated by pulse-chase experiments that they were highly unstable. Lin & Zabin (42) used the α-complementation reaction following cyanogen bromide treatment of extracts to measure the half-life of fragments. Incomplete chains produced by termination mutants were found to disappear rapidly; those that were very short, i.e. from mutants mapping within the α-region, had particularly short half-lives. Interestingly, no direct correlation between length of fragment and half-life was found, but in general, larger chains had longer half-lives. Hence, extracts containing the longer chains would complement.

α-Complementation was also studied in vivo by preparation of partial diploids. Polypeptide chains that were degraded rapidly in haploid cells were also degraded rapidly in the diploid cells, but were protected to some extent in the latter (43, 44). Protein degradation has been reviewed recently (45–47).

These experiments have concerned the subunit structure of β-galactosidase. The question of how tertiary structure is modified by the complementation reaction is of considerable interest. The α-donor activity of the short chain survives autoclave treatment and even boiling in guanidine for 3 hr (35). On the other hand, the M15 protein loses α-acceptor activity readily. So far, attempts to renature the protein from urea or guanidine have not succeeded. It is reactive to anti-β-galactosidase serum (48) and contains a binding site for β-galactosides (49). Possibly, the α-donor acts by modifying the conformation of the larger fragment. However, a contribution of the smaller chain directly to the active site cannot be ruled out.

α-Complementation with pure components results in a complemented enzyme which has at least 50% of the specific activity of native β-galactosidase (40). The interaction is slow; it requires 3 hr at 28°C. Once formed, the complemented enzyme is not readily dissociated and it is stable in growing cells (43).

ω-Complementation

ω-Complementation involves mutants at the operator-distal end, comprising approximately the last third or ω portion of the Z gene (Figure 2). ω-Acceptor proteins are those produced by strains with termination or missense mutations, such as S908, 366, or X90. Deletion mutants that mapped within Z (4680) or extended from another operon to various distances within Z (O5, B9) produced ω-donors. In the ω-complemented enzyme the complete sequence of β-galactosidase is present. However this is not sufficient, because if the deletion extended into the ω-segment, as in D34, complementation with an ω-acceptor such as X90 would not occur. These results suggested, in analogy to reasons discussed above in α-complementation, that an "ω-barrier" exists and that the wild-type protein might contain different subunits (32).

The prediction that ω-donors supplied an "ω peptide" appeared to be true, because extracts of a number of different deletion mutants and point mutants were found to contain a peptide fraction active as ω-donor with a sedimentation velocity of 3.15. This is compatible with a molecular weight of about 35,000–40,000, about one third to one fourth the minimum molecular weight of β-galactosidase. Filtration of extracts through Sephadex columns also indicated that the active fraction had the predicted size, although in some cases activity was also associated with larger protein fractions (32).

However, ω-complemented and native enzyme were not identical. Complemented enzyme released the ω peptide quantitatively by treatment with urea or guanidine but only traces of the peptide were released from native enzyme under the same conditions. ω-Complemented enzyme is more heat labile than native, and renaturation experiments have yielded only a few percent of the initial activity (34).

The structural nature of ω-complemented β-galactosidase was elucidated by isolation of pure material from several partial diploid strains (50–52). The catalytic properties and the specific activity of complemented enzyme from the diploid strain 366/B9 (Figure 2) were found to be essentially the same as those of the native protein. But the molecular weight was 595,000 as compared to 540,000 for the wild-type enzyme. Sedimentation equilibrium of the wild-type enzyme in 6 M guanidine showed that it contained a single chain of molecular weight 135,000. Two chains were derived from the complemented enzyme, one of molecular weight 110,000, which must be derived from the ω-acceptor 366, and a second of molecular weight 40,000 from the ω-donor B9. These results suggested that an overlapping portion of the sequence was present on each protomer, or four in each tetramer, which did not interfere with proper association within the molecule. Since it was likely that such extra segments would be located on the enzyme surface, treatment with papain was carried out under conditions that do not result in loss of native or complemented enzyme activity. The complemented enzyme was reduced in molecular weight from 595,000 to 540,000. Although it was not directly demonstrated, presumably the extra segment was removed from the carboxyl terminus of the longer chain by the mild proteolysis. Complemented enzyme was also isolated from the diploid strain X90/B9 and was found to have a molecular weight of 670,000.

The ω-peptide is inactivated by urea and guanidine but it can be renatured, in

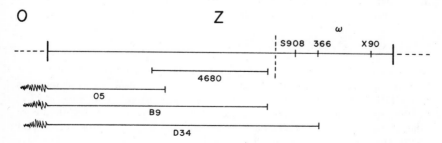

Figure 2   Mutant strains of *E. coli* illustrating ω-complementation of β-galactosidase.

contrast to ω-acceptors. It is not a random coil; its frictional coefficient is compatible with a globular structure. It is highly heat labile. Goldberg has suggested that ω-globules are present in both ω-complemented enzyme and native enzyme (52). In the latter, the terminal third of the polypeptide chain, folded into its normal three-dimensional structure, is joined to the remainder of the molecule through a continuous polypeptide chain. In complemented enzyme, on the other hand, the same association of globular structures must occur, but the joining part of the polypeptide chain is not continuous. If this model is correct, it offers a reasonable explanation for the apparent requirement for an ω-complementing peptide of about 40,000 daltons. Such a peptide may be the minimum size required to form the ω-globule, and without the complete globule, association of the two components could not occur. It should be emphasized that although this is an attractive hypothesis, there is no direct evidence to support it. The negative result obtained with strain D34 might be due to rapid proteolysis of the gene product. Also, it has been claimed that a peptide fraction of 18,000 mol wt present in extracts of a β-galactosidase mutant strain had ω-donor activity. However, the peptide was characterized only by gel filtration (53).

Complementation between ω-donor and ω-acceptor is complete in 2 hr at 28°C. The kinetics of interaction are complex (54).

Complementation of Wild-Type and Mutant Polypeptides

Positive complementation has been reported with a $lacZ^-/lacZ^+$ heterogenote of *E. coli* which contained 60% more enzyme activity than could be accounted for on the basis of its wild-type β-galactosidase subunits alone (55). The Z^- mutant chosen was particularly favorable for this experiment. It had been shown earlier by Rotman & Celada (56) that although the protein itself had quite low activity, it was activated 550-fold by anti-β-galactosidase antibody. However, positive complementation was not demonstrated by in vitro experiments using similar antibody activatable mutants (57). Hybrid molecules were prepared by renaturation from urea of mixtures of native and mutant protein. With the aid of an assay sensitive enough to measure the activity of an individual molecule, it was shown that each wild-type subunit was independently active and the mutant subunit inactive (58). These contrasting results suggest that assembly in vivo might be a critical factor in the partial restoration of activity. There is no evidence, however, to show that complemented enzyme produced in vivo is different from that made in vitro.

GLUTAMIC DEHYDROGENASE

Neurospora crassa glutamic dehydrogenase (GDH), the gene product of the *am* locus, is an NADP-dependent allosteric enzyme with a central role in metabolism (59). GDH is composed of six identical subunits, with a total molecular weight of 300,000. At neutral pH, the enzyme is activated by glutamate, α-ketoglutarate, or other dicarboxylic acids. At alkaline pH, the activation is more rapid and can occur in the absence of substrate or effectors (60). Fincham and his colleagues have identified and characterized many mutant forms of GDH (1), several of which are ineffective in vivo because they do not readily convert to the active form (61). The mutant enzymes can, however, be activated in vitro by higher concentrations of substrates and by more alkaline pH than that required to activate wild-type enzyme (62). Mutants of this class, which include am^2, am^3, and am^{19}, apparently produce a GDH in which the inactive form of the enzyme is favored. Another mutant protein (am^1) appears to have an amino acid substitution in the catalytic site and has no enzymatic activity (63). Complementation between mutant proteins of altered conformation, such as am^2, am^3, or am^{19}, with am^1 produces enzymatically activated hybrid multimers (64). Hybridization has been demonstrated in vivo in hetero-caryons, and in vitro by freezing and thawing in the presence of NaCl.

The activation of am^{19} protein is accompanied by an alteration in net surface charge as shown by a change from an abnormal to normal electrophoretic mobility (65). Complementation of an excess of am^1 with am^{19} results in an activated hybrid enzyme with normal electrophoretic properties. If the proportions of subunits are varied so that am^{19} is in excess, the hybrid enzyme is electro-phoretically abnormal and is only gradually activated (66). Recently, the mutation in am^{19} was shown to be the substitution of a methionine for lysine (67, 68). This might account for the increase in net negative surface charge at pH 8.5 observed in the inactive form of the mutant enzyme.

Hybridization between mutant forms of GDH and wild-type subunits has also been used to study conformational interactions. The combination of mutant subunits from am^2, am^3, or am^{19} with wild-type subunits results in a destabilization of wild-type structure and a net loss of activity. This is an example of negative complementation (69). Mutant am^1 protein is neutral when complemented with wild-type, neither interfering with nor enhancing wild-type activity. This situation is quite similar to that observed in hybrids of wild-type and mutant forms of alkaline phosphatase (22).

Mutant am^{14} seems to be an extreme case of a protein with a defective conformation. This strain produces no active GDH and a very low level of immunological crossreactivity (70). Nevertheless, it can complement with wild-type and several other am mutants in vivo. The complementation between wild-type and am^{14} has an extremely detrimental effect on both the stability and activity of the wild-type subunits (71).

The GDH formed in vivo from am^{14} and am^3 was partially purified. This complemented enzyme was also exceptionally thermolabile and largely in the in-active conformation (72). These data suggest that the am^{14} protein is in an unstable, rather disordered conformation, unable to aggregate by itself to form a GDH-like protein, and subject, therefore, to in vivo proteolytic degradation. This phenomenon has been observed with mutant forms of β-galactosidase, which also can be protected from degradation by in vivo complementation (43). The fact that am^{14} is able to complement and thus form active enzyme with a large number of other mutants is somewhat surprising. The mechanism of complementation with am^{14} is likely to be very different from the conformational correction provided by interaction with am^1. In cases where am^{14} is hybridized with am^2, am^3, or am^{19}, it would be interesting to determine which catalytic site is being activated or whether a new site is formed from portions of each mutant chain.

TRYPTOPHAN SYNTHETASE

Intracistronic Complementation

Intracistronic complementation of the α chain of E. coli tryptophan synthetase (TS) has been reviewed by Yanofsky & Crawford (73). Active dimers have been reconstituted by denaturation-renaturation of mutant monomers containing different amino acid substitutions (74). In general, pairs of α chains that form functional heterologous dimers have amino acid substitutions at opposite ends of the molecule. One such pair has been shown to refold to form a single, fully functional active site.

A cyanogen bromide fragment containing residues 2–84 of the E. coli α chain can complement a mutant with an amino acid substitution at residue 49. The homologous fragment derived from the α chain of Salmonella typhimurium is also effective, indicating that these regions of the enzymes from E. coli and S. typhimurium are functionally equivalent despite some amino acid differences (75, 76).

Fragments of the α chain produced by termination mutations in the structural gene should also complement if they overlap the altered region. This has been used as a specific assay for the presence of α chain fragments; the level of functional enzyme formed is only slightly affected by nonspecific proteins (75). However, two ochre mutants which were expected to produce complementing fragments did not have any detectable activity. This may have been due to in vivo proteolysis as observed with β-galactosidase fragments (41, 42).

Recently, intracistronic complementation has been observed between mutationally altered β_2 subunits of E. coli TS (77). Modest increases in activity were observed when suitable pairs of mutant proteins interacted to form hybrids.

Evolutionary Relationships

The extent of the interaction between tryptophan synthetase α- and β-subunits from different organisms has been used to measure structural homology at the subunit binding site. Relationships among the tryptophan synthetases of enteric bacteria were examined using interspecies intercistronic complementation. The subunit affinities of α and β from E. coli, Shigella dysenteriae, S. typhimurium, Aerobacter aerogenes, and S. marcescens were remarkably similar (78–80). However, immunological comparison of the α-subunits of these organisms showed considerable antigenic diversity among them (80).

The pattern of relationships within the Bacillus genus was more complex. β-Subunits from Bacillus subtilis formed active enzyme with α-subunits from Bacillus pumilis but virtually no activity was obtained when Bacillus alvei was the α-subunit donor (81).

Nonetheless, the β-subunit of B. subtilis could be activated by association with the α-subunits of E. coli and Pseudomonas putida to a level of activity 30% as great as that obtained with the α-subunit of *B. subtilis* itself. The activity specific to the α-subunit, however, was not stimulated by the interaction (82, 83).

Functional binding between α- and β-subunits of *E. coli* and *P. putida* has not been detected despite repeated attempts to do so (82, 84, 85). *E. coli* and *S. typhimurium* α-subunits have been shown to form active enzyme with β_2-subunits from two algae, *Anabaena variabilis* and *Chlorella ellipsoidea* (86).

The high degree of interspecies and intergeneric homology of the α-β binding site detected by complementation can be compared with the extent of evolutionary divergence indicated by protein primary sequence. The complete amino acid sequences of the α-subunits of *E. coli* (87), *S. typhimurium* (88), and *A. aerogenes* (89) have been determined. The sequence of the amino terminal region of the α chains from *S. dysenteriae* (76), *S. marcescens* (90), *P. putida* (91), and *B. subtilis* (92) has been reported. The degree of similarity of the amino-terminal regions to that of *E. coli* is in the following order: *S. dysenteriae* > *A. aerogenes* > *S. typhimurium* > *S. marcescens* > *P. putida* > *B. subtilis* (90, 93). Yet, despite considerable differences in amino-terminal sequence between α-subunits of *E. coli* and *P. putida*, each α-subunit is equally effective in activating β_2-subunits of *B. subtilis* (see above). These results and the contrast between the immunological diversity vs complementation similarity of the α chains from the *Enterobacteria* suggest that the intersubunit binding site has been conserved relative to the rest of the molecule during evolution.

ANTHRANILATE SYNTHETASE

The first step in tryptophan biosynthesis is catalyzed by anthranilate synthetase (AS), which has been reviewed in an excellent paper by Zalkin (94). Intercistronic complementation has been used to assign function to a particular polypeptide chain or to a portion of a polypeptide in a multifunctional aggregate. Evolutionary relationships have been examined by interspecies and intergeneric complementation.

Anthranilate synthetase catalyzes the conversion of chorismic acid to anthranilic acid, using glutamine as amido donor (reaction 1). In several enteric bacteria, AS

$$\text{L-glutamine} + \text{chorismate} \underset{}{\overset{Mg^{2+}}{\rightleftharpoons}} \text{anthranilate} + \text{pyruvate} + \text{L-glutamate} \qquad 1.$$

activity is associated with the second enzyme of tryptophan biosynthesis, anthranilate-5-phosphoribosylpyrophosphate phosphoribosyl transferase (PR transferase). The enzyme complexes from *E. coli* and *S. typhimurium* were characterized as oligomers containing two types of nonidentical subunits of approximately equal size (95, 96). The normally tightly associated subunits of 60,000 mol wt were obtained from mutants in which one component was defective or absent. Component I (CoI) alone could catalyze the NH_3-dependent formation of anthranilate from chorismate (reaction 2) and contained a binding site for the feedback inhibitor,

$$\text{ammonia} + \text{chorismate} \underset{}{\overset{Mg^{2+}}{\rightleftharpoons}} \text{anthranilate} + \text{pyruvate} + H_2O \qquad 2.$$

tryptophan (97). Component II (CoII) could independently catalyze the PR transferase reaction and had a glutamine binding site. When combined with CoI, the CoII polypeptide contributed the glutamine amido-transferase function of the complex and restored reaction 1 activity. Thus CoII has two distinct activities (98, 99).

Function of Different Regions of CoII Polypeptide

Evidence from complementation has helped to localize these two activities to nonoverlapping regions of the CoII polypeptide chain. Termination and deletion mutant proteins containing only the amino-terminal third of the CoII polypeptide from *E. coli* are still able to associate with and activate CoI, restoring glutamine amido-transferase function. PR transferase activity is not present in these fragments (100, 101). Conversely, mutant polypeptides containing only the carboxyl-terminal two thirds of the CoII sequence will catalyze the PR transferase reaction (102).

Fragments of CoII, produced by termination mutant strains of *S. typhimurium*, which possess the NH_2-terminal 40% of the sequence, could be detected in cell extracts by their ability to complement purified CoI in the glutamine-dependent AS reaction. These fragments are degraded in vivo. The end product is a polypeptide of 24,000 mol wt which possesses full activity in complementing CoI, but has no PR transferase activity (103).

Limited proteolysis also indicated that the two activities are on different portions of the polypeptide chain. Trypsin treatment of intact anthranilate synthetase-PR transferase aggregate reduced the molecular weight of the CoII polypeptide from 62,000 to 15,000–19,000. This fragment, presumably the amino-terminal region of CoII, retained a functional glutamine binding site and could complement CoI to form glutamine-dependent AS. The AS activity equaled that of the native enzyme (104).

Evidence from complementation and proteolysis experiments on CoII of *E. coli* and *S. typhimurium* indicates that the 60,000 mol wt polypeptide has two distinct regions of activity. The NH_2-terminal third, which contains a glutamine amido-transferase site, complexes with and activates CoI of anthranilate synthetase. The remainder of the polypeptide forms the PR transferase. This is analogous to DNA polymerase I (105–107) and the *TrpC* gene product (108) which also have two independent regions of activity. It has been suggested that the bifunctional CoII polypeptide could have arisen through the fusion of contiguous genes coding for distinct proteins (102, 103, 109, 110). Mutations leading to such gene fusions have been obtained in the *his* operon of *S. typhimurium* (111) and the *lac* operon of *E. coli* (112).

AS isolated from *S. marcescens* is not associated with PR transferase activity as it is in the related enteric bacteria *E. coli*, *S. typhimurium*, and *A. aerogenes* (113). Instead, the enzyme contains subunits of dissimilar molecular weights 60,000 and 21,000, with the smaller CoII contributing only a glutamine binding site (110). The enzyme from *S. marcescens* is in fact remarkably similar to trypsin-treated AS from *S. typhimurium* (104).

AS from nonenteric bacteria is structurally and functionally analogous to the enzyme from *S. marcescens*. *P. putida* (114), *B. subtilis* (115), *Acinetobacter calcoaceticus* (116), *Chromobacterium violaceum* (117), and *Clostridium butyricum*

(118) produce AS enzymes which consist of a dissociable complex of two different subunits of unequal size. The larger subunit (CoI) can be isolated by dissociation of the complex. It can also be obtained from mutant strains lacking the second component (CoII). CoI alone is active in the synthesis of anthranilate using ammonia as amide donor (reaction 2). In vitro reassociation of the two subunits results in regeneration of glutamine-dependent anthranilate synthesis (reaction 1) (116).

In *Neurospora*, AS activity is associated with a more complex multifunctional aggregate of enzymes of the tryptophan biosynthetic pathway. Complementation with extracts of mutant strains has been used to assign enzymatic functions to the various polypeptides in this system (119).

Evolutionary Relationships

Interspecies and intergeneric complementation has provided insight into evolutionary relationships. The data obtained by such complementation experiments are summarized in Table 1. Hybrid molecules containing CoI from *E. coli* with CoII from *E. coli*, *S. typhimurium*, and *A. aerogenes* have been constructed (113, 120). The complemented enzymes were identical with respect to glutamine-dependent anthranilate synthetase activity and sensitivity to tryptophan inhibition, suggesting that the sites for interaction are the same in these closely related genera. Comparable compatibility was observed in complementation of the tryptophan synthetase of these organisms (see above).

Although *Serratia* is the most divergent genus within the *Enterobacteria*, an active hybrid enzyme has been produced from the *S. marcescens* CoII subunit and *E. coli* CoI. Subunit binding in this complemented enzyme is, however, reversible and glutamine dependent in contrast to the irreversible association found in the homologous enzymes from both species (121). Thus, intergeneric complementation reinforces the idea that the glutamine subunit from *S. marcescens* is functionally and structurally similar to the amino-terminal region of the *E. coli* CoII protein. The amino acid sequences of the amino-terminal portions of CoI and CoII from *E. coli*, *S. typhimurium*, and *S. marcescens* reveal the basis of this relationship. The sequences

Table 1   Relative activity of AS complemented enzymes with subunits derived from different organisms

Component I	Component II				
	P. putida	*P. aerugenosa*	*B. subtilis*	*S. marcescens*	*S. typhimurium*
P. putida	100	100	—	—	0
P. aerugenosa	100	100	22	—	0
B. subtilis	—	93	100	—	—
A. calcoaceticus	50	—	5	0	—
E. coli	—	—	—	100	100

are all homologous, but *E. coli* and *S. typhimurium* are more closely related to each other than to *S. marcescens* (122).

Interspecies complementation among the *Bacilli* indicated that the dissociable AS subunits from *B. subtilis*, *B. licheniformis*, *B. pumilis*, *B. mascerans*, and *B. coagulans* are readily interchangeable. Hybrid complexes formed between subunits of these species were as catalytically active as the homologous complexes. Hybrid complexes utilizing CoI from *B. alvei* and CoII from the other species were considerably less active (123). The uniqueness of *B. alvei* enzymes has also been observed in complementation experiments of tryptophan synthetase (see above).

Anthranilate synthetases from *B. subtilis* and *A. calcoaceticus* are functionally closely related. In both cases, the glutamine-amido-transferase subunit has been shown to be part of two enzymes, AS and *p*-aminobenzoate synthetase, of the folate biosynthetic pathway (115, 116, 124). However, preliminary attempts to form hybrid enzymes between *A. calcoaceticus* CoI and *B. subtilis* CoII yielded only a low level of enzyme activity (116).

The subunits of *B. subtilis* and *A. calcoaceticus* will hybridize successfully, though not perfectly, with those of *Pseudomonads*. Complementation between *P. putida* and *A. calcoaceticus* resulted in hybrid enzyme which had approximately half the specific activity of native *A. calcoaceticus* enzyme (116). Dissociated subunits from *B. subtilis* and *P. aerugenosa* have been recombined. The activities of the hybrid complexes are shown in Table 1 (125).

Interspecies complementation among *Pseudomonads* indicates that two groups can be distinguished by the activity of the enzyme hybrids, the *putida-aeruginosa* group and the *acidovorans-testosteroni* group. Within each group the hybrid enzymes are equivalent in activity to the native enzyme; between groups the complemented enzymes are lower in activity (126).

Attempts to form hybrid anthranilate synthetase by combining subunits from an enteric bacterium with those from a nonenteric bacterium have been unsuccessful. No complementation was observed between *A. calcoaceticus* CoI and *S. marcescens* CoII (116) or between *P. aeruginosa* and *S. typhimurium* (114). This is reminiscent of the failure to form active tryptophan synthetase from subunits of *P. putida* and *E. coli* (see above).

OTHER COMPLEMENTING SYSTEMS

Intracistronic complementation has been used to study several other enzyme systems. Hybrid forms of isopropylmalate synthetase from heterocaryons of *Neurospora* were similar in size and shape to the wild-type enzyme, but different from each other and from the normal enzyme in sensitivity to thermal inactivation, catalysis, and allosteric interactions (127). Complemented forms of threonine dehydratase consisting of pairs of defective polypeptides resulting from missense mutations were isolated from diploid cells of *Saccharomyces cerevisiae*. The complemented enzymes were resistant to feedback inhibition by isoleucine in contrast to wild-type (128). Mutants of the *GuaB* locus in *S. typhimurium* produce defective IMP dehydrogenase. Restoration of enzyme activity was achieved in vitro by denaturation and renaturation of complementary mutant polypeptides. The hybrid enzymes had the same binding constants for substrates as the wild-type, but the maximum velocity was considerably lower (129). Complementation has been detected in a DNA restriction enzyme system, and is being used to assign functions to the three component polypeptides (130).

Complementation has been observed in proteins that are not enzymes. Detection of complementation between mutants in the *i* gene of *E. coli* suggested that the *lac* repressor was an oligomeric protein (131). The *i*[s] mutation results in a *lac* repressor protein with reduced affinity for inducer. When this defective protein combines in vivo with subunits of the wild-type repressor, it exhibits negative complementation, so that the cells are phenotypically Lac⁻. This effect of the *i*[s] mutation can be partially overcome by combining it with a large excess of functional repressor protein such as *i*[q] (132, 133).

Negative complementation can result in severe disability for the cell. A mutation in the structural gene for flagellin in *Salmonella* results in nonmotile cells with "straight" flagella. Hybrid flagellar filaments have been assembled in vitro using varying proportions of the straight mutant and normal flagellin subunits. The hybrids are intermediate in their physical wave properties between the normal and straight (134). If both types of subunits are present in vivo, a homogeneous copolymer is formed which is only slightly helical. However, flagellar function has been lost through negative complementation and the cells are nonmotile (135).

CONCLUDING REMARKS

Complementation was originally a genetic tool for studying the nature of the gene and its product. The classical studies of Benzer (136), which led to the definition of a cistron, were carried out by intercistronic complementation with mutants of bacteriophage T4 mapping in two distinct groups. Intracistronic complementation, which may also occur between two groups of mutants, is more complex. Complementation is of considerable interest to the biochemist, because such studies serve as useful models of noncovalent, protein-protein interacting systems.

Each of the oligomeric enzymes discussed above contributed uniquely to the development of information in this area. The formation of active alkaline phosphatase from isolated inactive monomers was the first direct demonstration that differently altered polypeptide chains can regain enzyme activity. Determination of amino acid sequence (137) and three-dimensional structure (138) of this enzyme is now in progress. When completed, it should be possible to correlate complementation behavior with specific alterations in conformation.

Complementation with β-galactosidase fragments has some similarities to the association between ribonuclease S peptide and S protein. In α-complementation a cyanogen bromide peptide no more than one thirteenth of the total chain (and perhaps considerably less) restores enzyme activity to a large fragment lacking a segment of the chain near the amino terminus. Sequence studies in progress (40, 139–141) have determined the exact position in the polypeptide of the complementing peptide. It should be rewarding to explore the nature of the interaction between the small and the large polypeptide fragments. In ω-complementation a large segment comprising the carboxyl-terminal third of the polypeptide chain may add to the first two thirds to restore enzyme activity. This has suggested the presence in the native molecule of at least two folded areas or globules, each containing independent centers of nucleation.

A complication in complementing studies using crude systems is that there may be considerable proteolytic destruction of proteins from mutant strains. Therefore, if complementation is the only test used for the presence of a gene product, a negative result can be misleading. This has become clearer recently from studies with β-galactosidase but is of general applicability. This also means that comparison of complemented enzyme formed in vivo to enzyme made in vitro cannot be valid unless both enzymes are shown to have the same primary structure (142).

Studies of complementation in the allosteric enzyme, glutamic dehydrogenase of N. crassa, focused attention on conformational interactions in multimeric enzymes. The evidence suggests a conformational correction of mutant proteins with altered tertiary structure when hybridized with an inactive mutant protein. The latter is inactive apparently because of an amino acid substitution in the active site. Therefore amino acid sequence analysis of this protein should help to delineate the active site by determining the position of the substitution. Clearly, determining the position of an amino acid substitution in any enzyme in relation to its effect on conformation or activity is a special kind of site-specific protein modification which can be very informative.

For the determination of evolutionary relationships, proteins from different species have been compared by investigating primary structure and by immunological relatedness. Intercistronic complementation is an additional important measure, since it tests for the conservation of intersubunit binding sites dependent on tertiary structure. Such studies with tryptophan synthetase and anthranilate synthetase have contributed a new look at an old problem of classification of enteric and nonenteric bacteria.

For complex enzymes with several activities or binding sites, complementation has been useful in assigning functions to various portions of polypeptide chains. The CoII polypeptide of anthranilate synthetase has a glutamine binding site in its amino-terminal third, while the carboxyl-terminal two thirds of this chain has PR transferase activity.

It can be expected that complementation studies will be applied in still other areas, for example, to test models of subunit interaction (143). As in experiments already performed with β-galactosidase (40), the judicious use of complementing fragments can aid amino acid sequence determination. No doubt some surprises are also in store, such as the discovery that a pair of β-galactosidase complementing fragments were converted in vivo to a single polypeptide chain by a splicing mechanism whose nature is unknown at present (144).

ACKNOWLEDGMENTS

We are indebted to Dr. A. V. Fowler for thoughtful and stimulating comments.

Literature Cited

1. Fincham, J. R. S. 1966. *Genetic Complementation.* New York : Benjamin. 143 pp.
2. Schlesinger, M. J., Levinthal, C. 1965. *Ann. Rev. Microbiol.* 19 : 267–84
3. Schlesinger, M. J. 1970. *Enzymes* 1: 241–66
4. Kapular, A. M., Bernstein, H. 1963. *J. Mol. Biol.* 6 : 443–51
5. Crick, F. H., Orgel, L. E. 1964. *J. Mol. Biol.* 8 : 161–65
6. Gillie, O. J. 1966. *Genet. Res.* 8 : 9–31
7. McGavin, S. 1968. *J. Mol. Biol.* 37 : 239–42
8. Woodward, D. O., Partridge, C. W. H., Giles, N. H. 1958. *Proc. Nat. Acad. Sci. USA* 44 : 1237–44
9. Fincham, J. R. S. 1959. *J. Gen. Microbiol.* 21 : 600–11
10. Woodward, D. O. 1959. *Proc. Nat. Acad. Sci. USA* 45 : 846–50
11. Jacob, F., Wollman, E. L. 1961. *Sexuality and the Genetics of Bacteria,* 249–50. New York : Academic. 374 pp.
12. Zipser, D. 1969. *Nature* 221 : 21–25
13. Richards, F. M., Vithayathil, P. J. 1959. *J. Biol. Chem.* 234 : 1459–65
14. Anfinsen, C. B. 1973. *Science* 181 : 223–30
15. Rothman, F., Byrne, R. 1963. *J. Mol. Biol.* 6 : 330–40
16. Reid, T. W., Wilson, I. B. 1971. *Enzymes* 4 : 373–416
17. Garen, A., Garen, S. 1963. *J. Mol. Biol.* 7 : 13–22
18. Schlesinger, M. J., Levinthal, C. 1963. *J. Mol. Biol.* 7 : 1–12
19. Schlesinger, M. J., Torriani, A., Levinthal, C. 1963. *Cold Spring Harbor Symp. Quant. Biol.* 28 : 539–42
20. Fan, D. P., Schlesinger, M. J., Torriani, A., Barrett, K. J., Levinthal, C. 1966. *J. Mol. Biol.* 15 : 32–48
21. Halford, S. E., Schlesinger, M. J. 1973. *J. Mol. Biol.* 81 : 261–66
22. Bloch, W., Schlesinger, M. J. 1974. *J. Biol. Chem.* 249 : 1760–68
23. Levinthal, C., Signer, E., Fetherolf, K. 1962. *Proc. Nat. Acad. Sci. USA* 48 : 1230–37
24. Zabin, I., Fowler, A. V. 1970. *The Lactose Operon,* ed. J. R. Beckwith, D. Zipser, 27–48. Cold Spring Harbor, NY : Cold Spring Harbor Lab. 437 pp.
25. Wallenfels, K., Weil, R. 1972. *Enzymes* 7 : 617–63
26. Zipser, D. 1963. *J. Mol. Biol.* 7 : 113–21
27. Perrin, D., Monod, J. 1963. *Biochem. Biophys. Res. Commun.* 12 : 425–28
28. Ullmann, A., Monod, J. 1969. *Biochem.*

Biophys. Res. Commun. 35 : 35–42
29. Beckwith, J. R., Zipser, D., Eds. 1970. *The Lactose Operon.* Cold Spring Harbor, NY : Cold Spring Harbor Lab. 437 pp.
30. Perrin, D. 1963. *Ann. NY Acad. Sci.* 103 : 1058–66
31. Perrin, D. 1963. *Cold Spring Harbor Symp. Quant. Biol.* 28 : 529–32
32. Ullmann, A., Perrin, D., Jacob, F., Monod, J. 1965. *J. Mol. Biol.* 12 : 918–23
33. Ullmann, A., Jacob, F., Monod, J. 1967. *J. Mol. Biol.* 24 : 339–43
34. Ullmann, A., Jacob, F., Monod, J. 1968. *J. Mol. Biol.* 32 : 1–13
35. Ullmann, A., Perrin, D. See Ref. 29, 143–72
36. Beckwith, J. R. 1964. *J. Mol. Biol.* 8 : 427–30
37. Fowler, A. V., Zabin, I. 1966. *Science* 154 : 1027–29
38. Morrison, S. L., Zipser, D. 1970. *J. Mol. Biol.* 50 : 359–71
39. Lin, S., Villarejo, M., Zabin, I. 1970. *Biochem. Biophys. Res. Commun.* 40 : 249–54
40. Langley, K. E., Fowler, A. V., Zabin, I. 1975. *J. Biol. Chem.* In press
41. Goldschmidt, R. 1970. *Nature* 228 : 1151–54
42. Lin, S., Zabin, I. 1972. *J. Biol. Chem.* 247 : 2205–11
43. Villarejo, M., Zamenhof, P. J., Zabin, I. 1972. *J. Biol. Chem.* 247 : 2212–16
44. Zamenhof, P. J., Villarejo, M. 1972. *J. Bacteriol.* 110 : 171–78
45. Pine, M. J. 1972. *Ann. Rev. Microbiol.* 26 : 103–26
46. Goldberg, A. L., Howell, E. M., Li, J. B., Martel, S. B., Prouty, W. F. 1974. *Fed. Proc.* 33 : 1112–20
47. Goldberg, A. L., Dice, J. F. 1974. *Ann. Rev. Biochem.* 43 : 835–69
48. Fowler, A. V., Zabin, I. 1968. *J. Mol. Biol.* 33 : 35–47
49. Villarejo, M. R., Zabin, I. 1973. *Nature New Biol.* 242 : 50–52
50. Goldberg, M. E., Edelstein, S. J. 1969. *J. Mol. Biol.* 46 : 431–40
51. Goldberg, M. E. 1969. *J. Mol. Biol.* 46 : 441–46
52. Goldberg, M. E. See Ref. 29, 27–48
53. Marinkovic, D. V., Tang, J. 1971. *Biochem. Biophys. Res. Commun.* 45 : 1288–93
54. Ullmann, A., Monod, J. See Ref. 29, 265–72
55. Hall, B. G. 1973. *J. Bacteriol.* 114 : 448–50

56. Rotman, M. B., Celada, F. 1968. *Proc. Nat. Acad. Sci. USA* 60:660–66
57. Melchers, F., Messer, W. 1971. *J. Mol. Biol.* 61:401–7
58. Melchers, F., Messer, W. 1973. *Eur. J. Biochem.* 34:228–31
59. Smith, E. L., Austen, B. M., Blumenthal, K. M., Nyc, J. F. 1975. *Enzymes* 11: In press
60. West, D. J., Tuveson, R. W., Barratt, R. W., Fincham, J. R. S. 1967. *J. Biol. Chem.* 242:2134–38
61. Fincham, J. R. S., Coddington, A. 1963. *Cold Spring Harbor Symp. Quant. Biol.* 28:517–28
62. Fincham, J. R. S. 1962. *J. Mol. Biol.* 4:257–74
63. Fincham, J. R. S., Stadler, D. R. 1965. *Genet. Res.* 6:121–29
64. Coddington, A., Fincham, J. R. S. 1965. *J. Mol. Biol.* 12:152–61
65. Sundaram, T. K., Fincham, J. R. S. 1964. *J. Mol. Biol.* 10:423–37
66. Coddington, A., Fincham, J. R. S., Sundaram, T. K. 1966. *J. Mol. Biol.* 17:503–12
67. Wootton, J. C., Chambers, G. K., Taylor, J. G., Fincham, J. R. S. 1973. *Nature New Biol.* 241:42–43
68. Wootton, J. C. 1973. *Biochem. Soc. Trans.* 1:1250–52
69. Fincham, J. R. S. 1967. *Mol. Genet., Wiss. Konf. Ges. Deut. Naturforsch. Aerzte,* 4th:1–5
70. Roberts, D. B., Pateman, J. A. 1964. *J. Gen. Microbiol.* 34:295–305
71. Sundaram, T. K., Fincham, J. R. S. 1967. *J. Mol. Biol.* 29:433–39
72. Sundaram, T. K., Fincham, J. R. S. 1968. *J. Bacteriol.* 95:787–92
73. Yanofsky, C., Crawford, I. P. 1972. *Enzymes* 7:1–31
74. Jackson, D. A., Yanofsky, C. 1969. *J. Biol. Chem.* 244:4526–38
75. Jackson, D. A., Yanofsky, C. 1969. *J. Biol. Chem.* 244:4539–46
76. Li, S. L., Yanofsky, C. 1972. *J. Biol. Chem.* 247:1031–37
77. Kida, S., Crawford, I. P. 1974. *J. Bacteriol.* 118:551–59
78. Balbinder, E. 1964. *Biochem. Biophys. Res. Commun.* 17:770–74
79. Creighton, T. E., Helsinki, D. R., Somerville, R. L., Yanofsky, C. 1966. *J. Bacteriol.* 91:1819–26
80. Murphy, T. M., Mills, S. E. 1969. *J. Bacteriol.* 97:1310–20
81. Hoch, S. O., Crawford, I. P. 1973. *J. Bacteriol.* 116:685–93
82. Hoch, S. O. 1973. *J. Biol. Chem.* 248:2992–98
83. Hoch, S. O. 1973. *J. Biol. Chem.* 248:2999–3003
84. Maurer, R., Crawford, I. P. 1971. *Arch. Biochem. Biophys.* 144:193–203
85. Enatsu, T., Crawford, I. P. 1971. *J. Bacteriol.* 108:431–38
86. Sakaguchi, K. 1970. *Biochim. Biophys. Acta* 220:580–93
87. Guest, J. R., Drapeau, G. R., Carlton, B. C., Yanofsky, C. 1967. *J. Biol. Chem.* 242:5442–46
88. Li, S. L., Yanofsky, C. 1973. *J. Biol. Chem.* 248:1830–36
89. Li, S. L., Yanofsky, C. 1973. *J. Biol. Chem.* 248:1837–43
90. Li, S. L., Drapeau, G. R., Yanofsky, C. 1973. *J. Bacteriol.* 113:1507–8
91. Crawford, I. P., Yanofsky, C. 1971. *J. Bacteriol.* 108:248–53
92. Li, S. L., Hoch, S. O. 1974. *J. Bacteriol.* 118:187–91
93. Hoch, S. O., Li, S. L. 1974. *J. Bacteriol.* 118:187–91
94. Zalkin, H. 1973. *Advan. Enzymol.* 38:1–39
95. Ito, J., Yanofsky, C. 1966. *J. Biol. Chem.* 241:4112–14
96. Henderson, E. J., Nagano, H., Zalkin, H., Hwang, L. H. 1970. *J. Biol. Chem.* 245:1416–23
97. Ito, J., Cox, E. C., Yanofsky, C. 1969. *J. Bacteriol.* 97:725–33
98. Bauerle, R. H., Margolin, P. 1966. *Cold Spring Harbor Symp. Quant. Biol.* 31:203–14
99. Nagano, H., Zalkin, H. 1970. *J. Biol. Chem.* 245:3097–3103
100. Ito, J., Yanofsky, C. 1969. *J. Bacteriol.* 97:734–42
101. Yanofsky, C., Horn, V., Bonner, M., Stasiowski, S. 1971. *Genetics* 69:409–33
102. Jackson, E. N., Yanofsky, C. 1974. *J. Bacteriol.* 117:502–8
103. Grieshaber, M., Bauerle, R. 1972. *Nature New Biol.* 236:232–35
104. Hwang, L. H., Zalkin, H. 1971. *J. Biol. Chem.* 246:2338–45
105. Setlow, P., Brutlag, D., Kornberg, A. 1972. *J. Biol. Chem.* 247:224–31
106. Setlow, P., Kornberg, A. 1972. *J. Biol. Chem.* 247:232–40
107. Jacobsen, H., Klenow, H., Overgaard-Hansen, K. 1974. *Eur. J. Biochem.* 45:623–27
108. Creighton, T. E. 1970. *Biochem. J.* 120:699–707
109. Li, S. L., Hanlon, J., Yanofsky, C. 1974. *Nature* 248:48–50
110. Zalkin, H., Hwang, L. H. 1971. *J. Biol. Chem.* 246:6899–6907

111. Yourno, J., Kohno, T., Roth, J. R. 1970. *Nature* 228:820–24
112. Müller-Hill, B., Kania, J. 1974. *Nature* 249:561–62
113. Egan, A. F., Gibson, F. 1972. *Biochem. J.* 130:847–59
114. Queener, S. W., Queener, S. F., Meeks, J. R., Gunsalus, I. C. 1973. *J. Biol. Chem.* 248:151–61
115. Kane, J. F., Holmes, W. M., Smiley, K. L. Jr., Jensen, R. A. 1973. *J. Bacteriol.* 113:224–32
116. Sawula, R. V., Crawford, I. P. 1973. *J. Biol. Chem.* 248:3573–81
117. Wegman, J., Crawford, I. P. 1968. *J. Bacteriol.* 95:2325–35
118. Baskerville, E., Twarog, R. 1974. *J. Bacteriol.* 117:1184–94
119. Arroyo-Begovich, A., DeMoss, J. A. 1973. *J. Biol. Chem.* 248:1262–67
120. Ito, J. 1969. *Nature* 223:57–59
121. Robb, F., Belser, W. L. 1972. *Biochim. Biophys. Acta* 285:243–52
122. Li, S. L., Hanlon, J., Yanofsky, C. 1974. *Biochemistry* 13:1736–44
123. Patel, N., Holmes, W. M., Kane, J. F. 1974. *J. Bacteriol.* 119:220–27
124. Kane, J. F., Holmes, W. M., Jensen, R. A. 1972. *J. Biol. Chem.* 247:1587–96
125. Patel, N., Holmes, W. M., Kane, J. F. 1973. *J. Bacteriol.* 114:600–2
126. Queener, S. F., Gunsalus, I. C. 1970. *Proc. Nat. Acad. Sci. USA* 67:1225–32
127. Webster, R. E., Nelson, C. A., Gross, S. R. 1965. *Biochemistry* 4:2319–27
128. Zimmermann, F. K., Gundelach, E. 1969. *Mol. Gen. Genet.* 103:348–62
129. Schafer, M. P., Hannon, W. H., Levin, A. P. 1974. *J. Bacteriol.* 117:1270–79
130. Hadi, S. M., Yuan, R. 1974. *J. Biol. Chem.* 249:4580–86
131. Bourgeois, S., Cohn, M., Orgel, L. E. 1965. *J. Mol. Biol.* 14:300–2
132. Müller-Hill, B., Crapo, L., Gilbert, W. 1968. *Proc. Nat. Acad. Sci. USA* 59:1259–64
133. Davies, J., Jacob, F. 1968. *J. Mol. Biol.* 36:413–17
134. Asakura, S., Iino, T. 1972. *J. Mol. Biol.* 64:251–68
135. Silverman, M., Simon, M. 1974. *J. Bacteriol.* 118:750–52
136. Benzer, S. 1957. *The Chemical Basis of Heredity*, ed. W. D. McElroy, B. Glass, 70–93. Baltimore: Johns Hopkins Univ. Press. 848 pp.
137. Kelley, P. M., Neumann, P. A., Shriefer, K., Cancedda, F., Schlesinger, M. J., Bradshaw, R. A. 1973. *Biochemistry* 12:3499–3503
138. Knox, J. R., Wyckoff, H. W. 1973. *J. Mol. Biol.* 74:533–45
139. Fowler, A. V., Zabin, I. 1970. *J. Biol. Chem.* 245:5032–41
140. Fowler, A. V. 1972. *J. Biol. Chem.* 247:5425–31
141. Zabin, I., Fowler, A. V. 1972. *J. Biol. Chem.* 247:5432–35
142. Suyama, Y., Bonner, D. M. 1964. *Biochim. Biophys. Acta* 81:565–75
143. Cornish-Bowden, A. J., Koshland, D. E. Jr. 1971. *J. Biol. Chem.* 246:3092–3102
144. Apte, B. N., Zipser, D. 1973. *Proc. Nat. Acad. Sci. USA* 70:2969–73

GENETIC MODIFICATION OF ×885
MEMBRANE LIPID

David F. Silbert[1]

Department of Biological Chemistry, Washington University,
St. Louis, Missouri 63110

CONTENTS

INTRODUCTION

At the present time, the basis for the heterogeneity and biological specificity of membrane lipids is poorly understood. As one general approach to this problem, attempts have been made to simplify, influence, or control the lipid composition of the membranes of cultured cells by genetic and environmental means. These efforts have led to the identification of requirements for specific lipids in biological functions. An important and concurrent aspect of this research has been the development of physical methods for studying the properties of lipids in membranes. Using appropriate biological systems, it should now be possible to characterize specific lipid requirements for structure and function in terms of the physical behavior of the component lipids.

This chapter is concerned with recent work on modification of membrane lipids and associated physical, biochemical, and physiological studies. Most of the current literature in this area deals with one of four systems: bacteria (primarily *Escherichia*

[1] The preparation of this review and research in the author's laboratory were supported by grants from the U.S. Public Health Service GM-16292 and 1-K4-GM 70,654.

315

coli), yeast (*Saccharomyces cerevisiae*), *Mycoplasma*, and cultured mammalian cells. The reader's attention is drawn to several recent reviews covering bacterial phospholipid metabolism (1), fatty acid biosynthesis (2), bacterial membranes (3), bacterial cell surfaces (4), genetic modification of membrane lipid (5), dynamics of lipids in natural and model membranes (6), and lipid mutants and physical properties of membrane lipids in yeast and *Neurospora* (7). The present review will focus on publications in the two year period ending in mid-1974.

ISOLATION OF MUTANTS IN LIPID BIOSYNTHESIS

In view of the considerable body of information available on their genetics and biochemistry, it is not surprising that the two organisms most extensively utilized for isolating mutants in membrane lipid biosynthesis are *E. coli* and *S. cerevisiae*, prototypes for prokaryotic and eukaryotic cells, respectively. In *E. coli*, over 90% of the lipids are phospholipids and these are located exclusively in the cytoplasmic and outer membranes of the cell envelope. In contrast, yeasts have a more differentiated system of membranes and the composition of their surface membranes, for example, includes phospholipids, triglycerides, and sterols (8). In discussing the isolation of mutants, the biochemical and genetic characterizations of these strains will be presented in detail as a foundation for an intelligent application of mutants to the study of the synthesis and properties of membrane lipid.

Escherichia coli Lipid Mutants

The pathway of fatty acid biosynthesis in *E. coli* (9–11) involves the following enzymes and partial reactions

1. acetyl CoA carboxylase:
   acetyl-S-CoA + HCO_3^- + ATP $\rightleftharpoons$ malonyl-S-CoA + ADP + P_i

2. acetyl CoA-ACP transacylase:
   acetyl-S-CoA + HS-ACP $\rightleftharpoons$ acetyl-S-ACP + CoA-SH

3. malonyl CoA-ACP transacylase:
   malonyl-S-CoA + HS-ACP $\rightleftharpoons$ malonyl-S-ACP + CoA-SH

4. β-ketoacyl synthetase:
   acetyl-S-ACP + malonyl-S-ACP $\rightleftharpoons$ acetoacetyl-S-ACP + CO_2 + HS-ACP

5. β-ketoacyl reductase:
   acetoacetyl-S-ACP + NADPH + H^+ $\rightleftharpoons$ D(−)-β-hydroxybutyryl-S-ACP + $NADP^+$

6. β-hydroxyacyl dehydrase:
   D(−)-β-hydroxybutyryl-S-ACP $\rightleftharpoons$ crotonyl-S-ACP + H_2O

7. enoyl reductase:
   crotonyl-S-ACP + NADPH [or NADH (11)] + H^+ $\rightleftharpoons$ butyryl-S-ACP + $NADP^+$ [or NAD (11)]

Reactions 4–7 are reutilized in a cyclic fashion with each successive addition of two carbon units derived from malonyl ACP. An important feature of fatty acid synthesis in *E. coli,* distinguishing it from this process in higher organisms, is the mechanism of unsaturated fatty acid synthesis. As depicted below, *cis* monoenoic acids are formed in the pathway by the action of a β,γ dehydrase competing with the α,β dehydrase (reaction 6) for the β-hydroxydecanoyl ACP intermediate (12) followed elongation of the *cis*-Δ^3-decenoyl ACP in the manner shown for acetyl ACP (reactions 4–7)

acetyl-S-ACP → β-hydroxydecanoyl-S-ACP → saturated fatty acids

↓

cis β,γ-decenoyl-S-ACP → *cis* unsaturated fatty acids

The fatty acids synthesized by *E. coli* and incorporated into the lipid-containing macromolecules of the cell include: 14:0, 16:0, *cis*-Δ^9-16:1, and *cis*-Δ^{11}-18:1 in phospholipid and lipoprotein (13); 12:0, 14:0, trace 16:0, and β-hydroxy 14:0 in lipid A of lipopolysaccharide. Investigations concerned with the response of *E. coli* to exogenous fatty acid supplementation (14–17), changes in growth temperature (18), glycerol deprivation (19) and mutations in glycerol 3-phosphate acyltransferase activity (20), and amino acid starvation (21) indicate that mechanisms exist for regulating the type and amount of fatty acids synthesized. In general, these are not yet well understood.

There are now a number of different mutants in fatty acid biosynthesis. Table 1 lists those that have been most adequately studied: *fabA, fabB, fabD,* and *cvc*. *FabA* and *fabB* are unsaturated fatty acid auxotrophs, and extracts from these cells cannot form unsaturated fatty acid products. *FabA* strains are defective in the β,γ dehydrase, the enzyme that introduces the *cis* ethylenic bond at the C_{10} level during chain elongation. This activity is low in auxotrophs (22) and the enzyme is thermolabile in *fabA* temperature-sensitive mutants (23). Hence, it is the structural gene for this enzyme. Furthermore, when merodiploid strains are constructed for the region of the chromosome carrying the *fabA* gene, the specific activity of the β,γ dehydrase is increased twofold, indicating that synthesis of this enzyme is not under transcriptional or translational control (24). *FabB* mutants are defective in an activity of β-ketoacyl ACP synthetase necessary for unsaturated fatty acid synthesis (25). Recent studies reveal that there are two β-ketoacyl ACP synthetases in *E. coli* differing in molecular weight, pH optimum, and heat stability (G. D'Agnolo, I. S. Rosenfeld, and P. R. Vagelos, submitted for publication). Synthetase I is absent in *fabB* mutants, and unsaturated fatty acid synthesis is restored by adding synthetase I from wild-type cells to mutant extracts. Although synthetase II has little if any correcting activity, it is catalytically active like synthetase I in condensation reactions involving fatty acyl ACPs that are intermediates in the synthesis of unsaturated as well as saturated fatty acids. Synthetases I and II are further distinguished by the fact that synthetase I has a higher K_m and a lower V_{max} than synthetase II with palmitoleyl ACP. Gelmann & Cronan (26) have reported another mutant (*cvc*) in unsaturated fatty acid biosynthesis that results in vitro in a deficiency in the conversion of *cis*-Δ^9-16:1 to *cis*-Δ^{11}-18:1. *Cvc* strains are not unsaturated fatty acid auxotrophs and have normal total amounts of unsaturated fatty acids. They have increased *cis*-Δ^9-16:1 and very little *cis*-Δ^{11}-18:1 in their phospholipid. Furthermore, in contrast to *cvc*$^+$ strains, the mutants do not increase their content of

Table 1 Lipid biosynthetic mutants of *Escherichia coli*

Mutant	Phenotype	Biochemical Defect	Map Position	References
Fatty acid biosynthesis				
fabA	Unsaturated fatty acid auxotroph	β-Hydroxydecanoyl thioester dehydrase	22	22–24, 30
fabB	Unsaturated fatty acid auxotroph	β-Ketoacyl ACP synthetase I	44	25, 31, 32, see text
cvc	Reduced cis-Δ^{11}-18:1 synthesis	β-Ketoacyl ACP synthetase II	?	26, see text
fabD	Temperature-sensitive total fatty acid synthesis	Malonyl CoA-ACP transacylase	24	28, 29, see text
Phospholipid biosynthesis				
gpsA	L-Glycerol-3-phosphate auxotroph	L-Glycerol-3-phosphate dehydrogenase	71	39–41
plsB	L-Glycerol-3-phosphate auxotroph	L-Glycerol-3-phosphate acyltransferase (altered K_m for L-glycerol-3-phosphate)	69	38, 40, 42, 52
plsA	Temperature-sensitive phospholipid synthesis	L-Glycerol-3-phosphate acyltransferase	13	43–45, 56, see text
pls	Temperature-sensitive phosphatidylethanolamine synthesis	Phosphatidylserine synthetase	?	54, 55

cis-Δ^{11}-18:1 in the phospholipid in response to growth at 15° as compared to 37°C. Preliminary experiments suggest that a reduction in β-ketoacyl ACP synthetase II may account for the cvc phenotype (P. R. Vagelos, personal communication). The cvc mutation, however, has no known physiological effect and was discovered accidentally as a silent mutation in a $fabA$ strain derived by nitrosoguanidine mutagenesis (26).

The counterpart for cvc is a mutation (vtr) described by Broekman (27) which also has no known physiological effect. Vtr strains have increased cis-Δ^{11}-18:1 and reduced cis-Δ^{9}-16:1 in their phospholipid, and growth at 42 vs 30°C does not result in a reduction of cis-Δ^{11}-18:1, as seen in vtr^{+} strains. No in vitro fatty acid synthesis studies with this type of strain have been reported. A possible enzymatic basis for this mutation is discussed below.

Temperature-sensitive mutants affecting total fatty acid synthesis have been isolated using a [^{3}H] acetate radiation suicide procedure which killed cells capable of fatty acid synthesis at 40°C (28). The first of these mutants to be well characterized biochemically are temperature-sensitive $fabD$ strains which possess a thermo-labile malonyl CoA-ACP transacylase (29). Saturated and unsaturated fatty acid synthesis is temperature-sensitive both in vivo and in vitro, and heat-inactivated extracts are restored to normal activity by malonyl transacylase purified from wild-type cells. In vivo, a decrease in malonyl transacylase causes a preferential reduction in the long chain fatty acyl groups (16:0 and 18:1) of the phospholipid. At temperatures that result in a reduction in fatty acid synthesis just sufficient to arrest growth, the addition of either long chain saturated or long chain unsaturated fatty acids to the culture medium restores normal fatty acid composition and normal growth to these mutants. On the other hand, at restrictive temperatures that arrest fatty acid synthesis, single or multiple fatty acid supplementation of these and other temperature-sensitive general fab mutants (28) does not sustain growth. Hence, essential products of fatty acid synthesis must be lacking in the supplement or are not effectively utilized by the cell from the exogenous source. As suggested, the required components may be the acyl groups of lipid A (28).

The location of lipid biosynthetic mutants on the genetic map of E. coli is shown in Figure 1. Although most of the unsaturated fatty acid auxotrophs belong to $fabA$ and $fabB$ classes that map at minutes 22 and 44, respectively (30–32), recently an unsaturated fatty acid-requiring strain has been reported to cotransduce with $nalA$ but not with $pyrD$ like $fabA$ or with $aroC$ like $fabB$ (33). Hence, it represents a third class called $fabC$. Three factor recombination studies based on mating experiments, together with the cotransduction data, favor locating $fabC$ at minute 41 between $nalA$ and his (33) but do not strongly rule out the order $fabC$–$nalA$–his. This mutant has not been characterized by any in vitro studies.

Preliminary mapping, based on the proximity of the affected genes to the origin of specific Hfr strains (and in some cases more definitive mapping), shows that the temperature-sensitive fab mutants affecting total fatty acid synthesis and the un-saturated fatty acid auxotrophs are widely distributed on the chromosome. Although many of the temperature-sensitive fab mutants included on the map (Figure 1) have not been fully characterized (hence designated by numbers in parentheses),

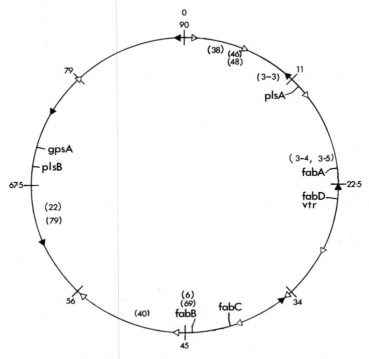

Figure 1   Location of lipid biosynthetic mutants on the genetic map of *E. coli*.

all show selective reduction in fatty acid synthesis in vivo and fatty acid growth requirements distinct from those of unsaturated fatty acid auxotrophs (5). For several of these mutants, in vitro studies have revealed reduced fatty acid synthetase or acetyl CoA carboxylase activity (D. F. Silbert, unpublished observations).

Two exceptions to the conclusion that the *fab* genes are generally not clustered on the chromosome may be noted. *FabD* and *vtr* give similar cotransduction frequencies with *pyrC* and *purB* (K. Semple and D. F. Silbert, in preparation; 27) and thus must be close together at minute 24. This finding, together with what is known about the nature of these two defects (see above), suggest that *vtr* may be a second type of mutation located in the *fabD* gene and compatible with normal physiology. For example, it is conceivable that other structural alterations in malonyl transacylase might lead to higher steady-state levels of malonyl ACP (and less free ACP) and, as a consequence, enhance rather than reduce the extent of chain elongation. The second instance of apparent clustering of different types of *fab* mutations occurs in the *fabB* region. Mutants that cotransduce with *aroC* are not all unsaturated fatty acid auxotrophs. For example, *fab6* requires both saturated and unsaturated fatty acids, and *fab69* is a leaky mutant that grows normally with either long chain saturated or unsaturated fatty acid. By P_1 transduction these two mutants map close to but clearly to the *purF* side of two different *fabB* mutants (5; D. F. Silbert, S. Ross, and K. Semple, in preparation). These observations do not distinguish between two different types of mutations in the same or closely linked genes. If in vivo and in vitro studies reveal no complementation between *fabB* and *fab6*, the former interpretation would be more likely and should encourage comparative studies on the β-ketoacyl synthetases from these two types of mutants as an extension of the biochemical characterization of the *fabB* defect (25; see above).

The pathway of phospholipid biosynthesis in *E. coli* (34) is depicted in Figure 2. The principal glycerophosphatides are formed from a common liponucleotide intermediate, CDP (and dCDP) diglyceride (35). Although phosphatidylethanolamine arises from phosphatidylserine and cardiolipin derives from phosphatidylglycerol (36), there is no known mechanism for modification of the membrane phospholipids through interconversion of the glycerophosphatides rather than through de novo

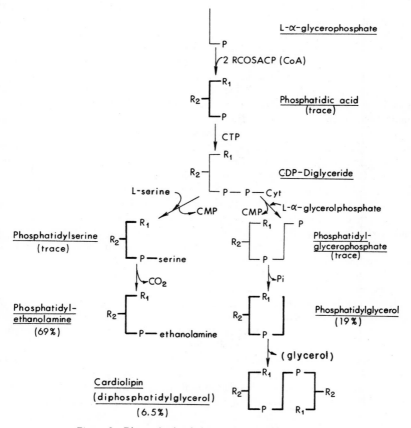

Figure 2   Biosynthesis of glycerophosphatides in *E. coli.*

synthesis. The isolation of mutants affecting lipid biosynthesis at this level has two principal objectives with respect to the modification of membrane lipid. On one hand, cellular mechanisms for coordinating macromolecular synthesis and assembly can be examined by controlling the overall production of phospholipid. On the other hand, mutations that selectively reduce the formation of one type of glycerophosphatide make it possible to investigate the requirement for different polar head groups in the membrane lipid. Hence, mutations in the common portion of the pathway leading to CDP (or dCDP) diglyceride, including those affecting the availability of fatty acids or glycerol 3-phosphate, would have particular application to the former problem. Strains with mutations beyond the liponucleotide level would be most suitable for the second type of study, although a reduced level of intermediates in the common pathway or increased activity of enzymes in the anionic glycerophosphatide branch (37) causes a preferential reduction in the zwitterionic glycerophosphatides (21, 35, 38). Glycerol 3-phosphate auxotrophs have been isolated and belong to two groups. GpsA strains are defective in the biosynthetic L-glycerol 3-phosphate dehydrogenase (39–41). This activity is absent in the auxotrophs and thermolabile in a temperature-sensitive auxotroph (41). Hence, it is the structural gene for this enzyme. PlsB strains are characterized by a K_m defect in the first enzyme in the phospholipid biosynthetic pathway (38, 40, 42). The apparent K_m for glycerol 3-phosphate of the glycerol 3-phosphate acyltransferase from these strains is severalfold higher than the K_m of the enzyme from the normal cell. By means of ^{3}H-glycerol 3-phosphate radiation suicide, which kills cells capable of phospholipid synthesis at 40°C (hence enriching for temperature-sensitive phospholipid mutants), a second class of mutations that affect the acyltransferase activity but map in another locus called plsA has been isolated (43, 44). In the plsA strains, the activity catalyzing transfer of a fatty acyl group from a saturated or an unsaturated fatty acyl thioester to glycerol 3-phosphate is thermolabile in vitro and lost when cells are placed at the restrictive temperature prior to conducting the in vitro measurements. The acylation of 1- or 2-acylglycerol 3-phosphates (lysophosphatidic acids), however, is not affected (45). Although it is likely that plsA and plsB are structural genes for two different subunits of glycerol 3-phosphate acyltransferase, this question cannot be resolved until the purified enzyme has been obtained and studied. All the enzymes of the phospholipid pathway, except phosphatidylserine synthetase (46), are localized in the cytoplasmic membrane (47–49), and methods for purifying them are only now being developed (50, 51). It is possible that the properties of the membrane-bound glycerol 3-phosphate acyltransferase in the plsA or plsB mutants are due to secondary effects of an alteration in another membrane component which is closely associated with the acyltransferase (52). In this connection, it has been found recently that plsA strains also possess an adenylate kinase that is thermolabile in vivo and in vitro (M. Glaser, W. Nulty, and P. R. Vagelos, submitted for publication). Revertants that are no longer temperature sensitive for growth arise from plsA strains with a frequency consistent with a single mutation. Several of these have been examined and have normal acyltransferase and adenylate kinase activity. Although the molecular basis for these changes has yet to be established, apparently a single mutation has led to thermolability of two activities, one (acyltransferase) that is tightly and the other (adenylate kinase) loosely associated with the membrane.

The radiation suicide experiments using glycerol 3-phosphate of high specific activity might be expected to give rise primarily to defects early in the phospho-lipid biosynthetic pathway, although mutations beyond the liponucleotide inter-mediate would reduce in part the incorporation of damaging radiation and theoretically might also be recovered. To date, however, other mutations isolated by this method have not been adequately characterized. As a more direct approach to finding mutants specifically defective in the synthesis of phosphatidylethanolamine, a procedure was devised to selectively incorporate serine of very high specific radio-activity into phosphatidylethanolamine (53, 54). It was then possible to select for temperature sensitive mutants that survived the exposure to the radioisotope because they were defective in phosphatidylethanolamine synthesis. Phosphatidyl-serine synthetase has been identified as the site of the defect in one of the strains isolated by this approach (55).

With respect to mutations affecting phospholipid biosynthesis, the genetic locations of $gpsA$, $plsB$, and $plsA$ have been placed at minutes 71.5, 69, and 12, respec-tively (41, 52, 56; see Figure 1). One of the mutants that is defective in phospha-tidylethanolamine synthesis also maps at minute 12 (53), suggesting a possible cluster of genes in this region associated with membrane lipid biosynthesis. However, other temperature-sensitive mutants selected by the glycerol 3-phosphate radiation suicide procedure and tentatively identified as phospholipid mutants have been mapped in a number of regions on the chromosome (44). Considerably more biochemical infor-mation is needed before the genetic grouping of these mutants can be interpreted meaningfully.

Regulation of membrane lipid content may result from the combined activity of biosynthetic and degradative activities. The enzymes catalyzing β oxidation have been demonstrated in E. coli (57). Mutants in many steps in the pathway have been isolated and designated fad for fatty acid degradation (58–60). Fad mutations do not appear to affect the fatty acid composition of E. coli growing without fatty acid supplement. However, they do prevent partial degradation of exogenously supplied fatty acids prior to incorporation into the phospholipid. For this reason, fab fad double mutants are more useful than fab mutants for nutritional modifica-tion of the membrane lipid (61).

A number of enzymes capable of participating in the turnover of phospholipids have been identified in E. coli. Recent studies by Albright et al (62) have led to a clearer picture of the possible sequences of reactions involved. Furthermore, localiza-tion of the enzymatic activities with respect to the outer wall, inner membrane, and cytoplasmic fraction suggests that the initial steps in catabolism occur in the outer wall through the action of a phospholipase A_1 and/or phospholipase A_2. Although this enzyme may also have lysophospholipase activity, there is another lysophospho-lipase in the inner membrane and a glycerophosphorylethanolamine phosphodi-esterase in the cytoplasm. Since phospholipid synthesis occurs on the cytoplasmic membrane (47, 48) and phospholipids are translocated to the outer envelope (63), a compartmentalization of synthetic and degradative activities would serve to remove degradative intermediates such as lysophospholipids from the inner membrane where they could disrupt structure and vital functions. The finding of another phospholipase A in the supernatant (62, 64), specific for phosphatidylglycerol but of unknown positional specificity, indicates that this scheme may not fully account for the metabolism of phosphatidylglycerol.

Ohki et al (65) have isolated a phospholipase A mutant in *E. coli* by means of a procedure that reveals colonies unable to undergo autolysis and crossfeed an unsaturated fatty acid auxotroph. This mutant has lost detergent-resistant phospholipase A_1, phospholipase A_2, and lysophospholipase activities, and the affected gene has been designated *pldA* for phospholipid degradation (66). The loss in activity is not due to an inhibitor (65). It seems likely that this mutant is defective in a single enzyme, the phospholipase found in the outer envelope (47, 48, 62), and that this is the phospholipase extensively purified by Scandella & Kornberg (67). However, further biochemical work is necessary to establish this point, and the availability of a *pldA* mutant with a temperature-sensitive phospholipase A would help. A second type of phospholipase A mutant has been found (68) which is defective in the detergent-sensitive, phosphatidylglycerol-specific, cytoplasmic phospholipase A (62, 64). It has been suggested that phospholipase A is activated by genetic or environmental factors affecting the integrity of the cell (69). However, no known physiological abnormalities have been detected thus far in either type of phospholipase A mutant (68).

In addition to glycerophosphatides, the cell envelope of gram-negative enteric bacteria contains another type of complex lipid, known as lipid A, which accounts for approximately 20% of the fatty acyl groups of the membrane. This molecule is a disaccharide of glucosamine in which the two amino groups are in amide linkage with β-hydroxymyristate (a fatty acyl residue unique to lipid A), three of the six available sugar hydroxyls and one of the two β-hydroxy acid hydroxyls are esterified with fatty acids (β-hydroxymyristate, laurate, myristate, and palmitate), two of the six sugar hydroxyls are esterified by phosphate residues that are thought to crosslink lipid A molecules, and the last of the available sugar hydroxyls forms a ketosidic linkage with 2-keto-3-deoxyoctonate (KDO) which is the innermost residue of a complex polysaccharide chain anchored on the lipid A molecule (63). Of the sugars extending outward from the lipid A moiety, the three residues of KDO and the two residues of L-glycero-D-mannoheptose are characteristic of the lipopolysaccharide of the gram-negative enteric organisms. Recently, mutants defective in the synthesis of the innermost portion of the polysaccharide have been shown to have very profound effects on the synthesis and/or assembly of membrane components. Conditional mutants have been obtained in *Salmonella typhimurium* which are defective in the synthesis of lipid A as a consequence of a primary defect in the synthesis of KDO (63). One of these mutants is a D-arabinose 5-phosphate auxotroph in which KDO-8-phosphate synthetase has a K_m for D-arabinose 5-phosphate that is manyfold higher than that of the enzyme from the parental strain (63, 70). In the absence of D-arabinose 5-phosphate, lipopolysaccharide synthesis ceases in the mutant, while phospholipid synthesis continues at a reduced rate for about one generation before stopping along with all macromolecular synthesis and growth (63, 71). Under these conditions, a lipid A precursor accumulates which is a glucosamine disaccharide lacking ester-linked palmitic, myristic, and lauric acid residues and one of the two amide-linked β-hydroxymyristic acid groups as well as the ketosidic-linked KDO (63, 72). Thus, it appears that during lipid A biosynthesis the incorporation of KDO in general precedes the acylation of the disaccharide with fatty acid residues. That synthesis of the inner portion of lipopolysaccharide is important to the assembly of the outer membrane is further supported by studies on the lipid and protein composition of this membrane in heptose-deficient mutants of *E. coli* (73) and *S. typhimurium* (74). In these mutants, chemical analyses and gel electrophoresis show a marked reduction in total protein and in most major polypeptides normally present in the outer membrane. In one of these investigations more lipid A appeared in the cytoplasmic than in the outer membrane, a distribution opposite to that found in normal cells (73).

Before concluding this section, brief mention should be made of other approaches that can turn up lipid mutants but are designed more broadly to find strains with defective membranes. These efforts focus on the selection of mutants with altered permeability or increased susceptibility to lysis in response to environmental stress, such as low or high growth temperature (75), pH (76), or ionic strength (77). A moderate fraction of temperature-sensitive lysis mutants was found to be defective in unsaturated fatty acid synthesis or to have altered patterns of phospholipid synthesis (75). Isolation of the former type of defect was anticipated because unsaturated fatty acid auxotrophs were known to lyse when grown without supplement (78). Another study has produced a mutant with an active membrane phospholipase that renders protoplasts of this strain osmotically very fragile (77). Although these examples represent novel approaches to the isolation of membrane mutants, identification of primary defects in general will require devising additional methods for subgrouping the mutants with respect to the broad biochemical nature of the lesion. As for any type of mutant, full characterization at the biochemical and genetic level is an important prerequisite to adequate evaluation of physiological properties.

Saccharomyces cerevisiae Lipid Mutants

Fatty acid synthesis in yeast is catalyzed by a very stable multienzyme complex with a molecular weight of 2.3×10^6 (79). This property of the fatty acid synthetase is common to all eukaryotic systems that have been studied but contrasts to the unassociated or potentially loosely associated nature of the enzymes catalyzing these reactions in bacteria and plants (2). As discussed below, the genetic information coding for the yeast fatty acid synthetase, unlike that for *E. coli* system (see above), is highly clustered. It may be anticipated that the close physical association of the components of the yeast fatty acid synthetase and the clustering of the genetic information for this pathway will give rise to particular problems in the characterization of mutants. For example, mutations in one component can lead to the loss of activity of other components through a failure to synthesize and/or assemble them into a complex or through conformational effects exerted by the mutation on the whole complex. Alternately, a modification in the structure of one component may be expressed only during synthesis of that component but not after incorporation into the complex (80).

Although the pathway of fatty acid synthesis in yeast is very similar to that in *E. coli,* there are a few notable differences. First, the products of fatty acid synthesis are released from the complex as CoA thioesters by the action of fatty acyltransferase (81). Second, yeast derives unsaturated fatty acids from the long chain saturated fatty acyl CoA products of the synthetase by an aerobic desaturase which acts on methylene groups in the middle of the hydrocarbon chain (82). Although yeast prototrophic for fatty acid synthesis does not avidly assimilate exogenous fatty acids (83), the discovery of mutants in fatty acid synthesis showed that this organism can utilize the supplement extensively for complex lipid synthesis and, furthermore, can elongate fatty acyl chains of 12 carbons or more (83–85). The principal fatty acids found in *S. cerevisiae* are 16:0, 8:0, *cis*-Δ^9-16:1, and *cis*-Δ^9-18:1. These are found largely as ester groups in triglycerides and the diverse glycerophosphatides of yeast (86).

Fatty acid (*ole* and *fas*) and sterol (*erg*) mutants have been isolated in yeast, and Table 2 lists those types most adequately studied. Unsaturated (*ole*) and saturated (*fas*) fatty acid auxotrophs were obtained by screening for organisms requiring the respective fatty acid supplement. *Ole1* mutants (87) are defective in the Δ^9-desaturase (88). The nature of the defect is inferred from in vivo studies revealing impaired incorporation of acetate into unsaturated but not saturated fatty acids. I am unaware of any work in vitro concerning the biochemical nature of this defect, which could be in one of several structural components of the enzyme (82) or in a regulatory element required for this activity. Since the syntheses of *cis*-Δ^9-16:1 and *cis*-Δ^9-18:1 are affected in the *ole1* mutants, the same enzyme seems to be involved in both desaturations (88). In some revertant strains, the two unsaturated fatty acids are synthesized in different proportions than that found in the parental strain, suggesting that the revertant enzyme may have an altered substrate specificity. Other mutations leading to unsaturated fatty acid auxotrophy have been described but are associated with additional defects such as respiratory deficiency and requirements for sterols or other undefined nutrients (88). In these mutants desaturation as well as other processes are affected through a block in the synthesis of a common type of component, for example, the heme group of various cytochromes (89).

The *fas* mutants have been assigned by in vivo complementation to a number of groups (90, 91). Measurements of the individual enzymes of fatty acid synthesis reveal that mutants in one group had normal fatty acid synthetase activity as well as fully active component enzymes, mutants in several groups (e.g. II, V, VI, VII, VIII) were lacking activity in one component and sometimes had reduced activity in other enzymatic steps (80, 90, 92, 93), and mutants in other groups (III, IV, IX) were missing both the activities for all the component reactions and an isolatable fatty acid synthetase complex (93). The complementation data together with the enzymatic measurements suggested six distinct *fas* cistrons, five of which could be associated with a specific component activity: dehydratase, β-ketoacyl reductase, enoyl reductase, and β-ketoacyl synthetase (two cistrons). Based on evidence of

Table 2 Lipid biosynthetic mutants in *Saccharomyces cerevisiae*

Mutant	Phenotype	Biochemical Defect	References
Fatty acid biosynthesis			
ole1	Unsaturated fatty acid auxotroph	Fatty acyl CoA Δ^9-desaturase	87, 88
fas1			
Complementation Group			
II	Saturated fatty acid auxotroph	Enoyl reductase	90, 91, 93, 96, 97
V	Saturated fatty acid auxotroph	Dehydratase	90, 91, 93, 96, 97
fas2			
Complementation Group			
VI	Saturated fatty acid auxotroph	β-Ketoacyl synthetase	80, 90–93, 96, 97
VII	Saturated fatty acid auxotroph	Pantetheine-deficient fatty acid synthetase	90, 91, 93, 94, 96, 97
VIII	Saturated fatty acid auxotroph	β-Ketoacyl reductase	90, 91, 93, 96, 97
fas1 ole1	Saturated and unsaturated fatty acid auxotroph	*fas1* uncharacterized *ole1* see above	98
Sterol biosynthesis			
erg1	Sterol auxotroph	Conversion of squalene to lanosterol	100

linkage obtained from induced mitotic recombination and tetrad analysis, Schweizer and co-workers (90, 93) and Henry & Fogel (91) have proposed that these cistrons are collected into three distinct regions on the yeast genome containing one, two, and three cistrons, respectively (see Figure 3). The two clusters have been called *fas1* and *fas2,* and the third region is no longer thought to be involved in fatty acid biosynthesis (94, 95). *Fas1* has been localized to fragment V of yeast chromosome XI (96, 97). Very recently, mutants in one of the cistrons of *fas2* affecting β-ketoacyl synthetase were found to produce a fatty acid synthetase complex of normal molecular weight but lacking bound pantetheine, which is essential for the β-ketoacyl synthetase reaction (94). It has not been determined whether this deficiency results from a lesion in the apoprotein portion of the complex or in the enzyme that transfers the prosthetic group to the complex. In this study attempts were made to characterize the subunit structure of the yeast fatty acid synthetase complex. Under conditions retarding proteolysis, two high molecular weight (185,000 and 180,000) components were observed on sodium dodecylsulfate polyacrylamide gel electrophoresis, and only one of these contained pantothenic acid. Should these components prove to be fully dissociated subunits and, furthermore, to be two different polypeptide chains rather than two collections of different polypeptide chains that are unresolved by the electrophoresis, then these authors suggest that the acyl carrier protein of yeast is part of a larger multifunctional polypeptide and that the two *fas* regions are not gene clusters but rather encode for two large polyfunctional subunits which are each present several times in the fatty acid synthetase complex.

Since *fas* mutants are still capable of synthesizing unsaturated fatty acids, strains combining *fas* and *ole1* mutations would be most desirable for studies on genetic modification of membrane lipid. This double mutant has been constructed (98) as shown in Table 2. As noted earlier, *fas* mutants can elongate fatty acid of 12 carbons or longer. Although this elongation system has not been characterized (83, 85), it is presumably distinct from the fatty acid synthetase complex. No mutants defective in this capacity have been reported. The combination of *fas, ole1,* and an elongation defect should render *S. cerevisiae* amenable to precise control of its fatty acid composition.

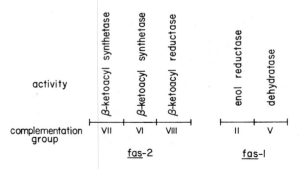

Figure 3   Gene-enzyme relationships for fatty acid synthesis in *S. cerevisiae.*

The second category of membrane lipid mutants in yeast involved sterol bio-synthesis. In plasma membranes of *S. cerevisiae* sterols account for about 6% of the dry weight (8, 86). A number of sterol auxotrophs have been isolated (89) which have additional nutritional requirements (e.g. unsaturated fatty acids) and are respiratory deficient. The primary defect is in porphyrin synthesis (89, 99) leading to the loss of heme-containing enzymes required for various processes. In the case of sterol synthesis, a block occurs in the conversion of lanosterol to ergosterol (99). A very recent report describes the isolation of a mutant (*erg1*) in *S. cerevisiae* which requires only sterol for growth (100) and incorporates acetate into squalene but forms virtually no lanosterol or ergosterol. Genetic characterization suggests that there may be more than one mutation in this strain. Nevertheless, *erg1* should be very useful for biochemical and physical studies on the properties of membrane lipids under conditions regulating the type and availability of sterols. Furthermore, a *fas ole1 erg1* triple mutant can now be constructed and would provide a system for very extensive control of the composition of the membrane lipid and for the study of hydrocarbon chain-sterol interactions.

MODIFICATION OF MEMBRANE LIPID

When synthesis of lipid is affected by mutation or the action of a chemical inhibitor (e.g. 101–103), two principal means for modifying membrane lipid structure become available. If the component normally obtained through synthesis can be derived from an exogenous source, manipulation of the membrane composition can be achieved by controlling the type and relative availability of replacements. Alternatively, a condition of starvation results when the membrane component is not or cannot be provided as a supplement. This section discusses recent studies on qualitative and quantitative alterations in membrane phospholipid and sterol composition and on attendant changes in the physical and biochemical properties of the membrane. Modifications in membrane lipid may be grouped with respect to those affecting 1. the hydrocarbon chains, 2. the head groups of the phospholipids, or 3. the type of sterol.

Acyl Group

REPLACEMENTS AND ATTEMPTS TO QUANTITATE REQUIREMENTS   Although unique acyl groups may be required for certain biochemical processes, early work with unsaturated fatty acid auxotrophs revealed that the requirement for *cis* unsaturated fatty acids in *E. coli* could be satisfied by a wide variety of fatty acid analogs of the monoenoic acids synthesized by the cell (5). Recently, certain branched chain fatty acids have been added to this list (104), demonstrating that a special electronic configuration in the acyl chain is not critical to the role of unsaturated fatty acids in membranes. Although the initial studies with unsaturated fatty acid auxotrophs of *S. cerevisiae* suggested much more stringent structural requirements (105), recent findings indicate that an extensive range of modifications in unsaturated fatty acyl group structure is also possible in this organism (7, 106, 107). Included in the group of suitable analogs are 9, 10, or 12 hydroxy fatty acids and polyunsaturated fatty acids containing as many as five and six *cis* double bonds (106–108) ranging from the Δ^4 to the Δ^{19} position relative to the carboxyl terminus. Such striking modifications in the acyl group structure of the phospholipids may only be possible in membranes containing sterols where the localized disruptive effect of hydroxyl groups or the ethylenic bonds on the packing of acyl chains in the bilayer may not be as great as in membranes not containing sterols (109).

Various estimates indicate that the minimum amount of unsaturated fatty acids required for normal growth of unsaturated fatty acid auxotrophs of *E. coli* at 37°C is about 20% of the total fatty acid content (23, 78, 104, 110). This value is greatly influenced by temperature (104) and very likely also by the nature and arrangement of all the acyl groups present in the phospholipid (111). In an analogous mutant of *S. cerevisiae,* the minimum amount of monoenoic fatty acids required for normal mitochondrial function at 28°C is about 80% of the total fatty acid content (106). Depletion of unsaturated fatty acids leads progressively to less efficient oxidative phosphorylation. When the level is only 20% of the mitochondrial fatty acid composition, the ability of yeast to grow on nonfermentable substrates is lost and the mitochondria are found to have increased passive permeability to protons, defective potassium transport, and completely uncoupled oxidative phosphorylation (112, 113). If no further unsaturated fatty acid depletion occurs, full function is regained on restoration of unsaturated fatty acid composition to normal.

The requirement for saturated fatty acid in the phospholipids of *E. coli* has been examined by growing strains defective in total fatty acid synthesis on various supplements at 37°C. Of the *n*-saturated fatty acids, only 16:0 and 17:0 are fully effective (111). Also, anteisoheptadecanoic (14-me-16:0) and *trans*-hexadecenoic (*trans*-Δ^9-16:1) acids are suitable replacements (28, 104). When these strains are grown on *cis*-octadecenoic acid (*cis*-Δ^{11}-18:1), the saturated fatty acid content can be progressively decreased. When it is reduced below approximately 15% of the total, a minimum level approached by wild-type strains grown under the same conditions, passive permeability to small molecules increases abruptly (111). However, the deficiency in saturated rather than unsaturated fatty acids (78) does not lead to cell lysis. Saturated fatty acid mutants of *S. cerevisiae* grow at 30°C on synthetic media with *n*-saturated fatty acids 14:0, 15:0, or 16:0 but not with 12:0, 17:0, 18:0, or *trans* unsaturated fatty acids (84). *Cis*-Δ^9-18:1 was observed to support growth for a few generations, but the fatty acid composition and the physiological properties of the cell were not evaluated at the time growth ceased (84). Differences in the fatty acid requirements of saturated fatty acid mutants of *E. coli* vs *S. cerevisiae* may be related to differences in the growth temperatures (37° vs 30°C) and in the metabolic capabilities of the two organisms (e.g. *S. cerevisiae* but not *E. coli* can elongate saturated fatty acid supplements and/or convert them into *cis* unsaturated fatty acids).

As a consequence of the success in modifying the membrane lipid of microorganisms, techniques for manipulating the fatty acid composition in the phospholipids of animal cells cultured in a defined medium have been developed recently (114–116). A large variety of saturated and unsaturated fatty acids can be utilized for phospholipid synthesis when provided in the culture medium as sorbitan (Tween) esters (114, 115). If fatty acid synthesis is partially inhibited by the addition of a biotin antagonist to the medium, the extent of incorporation of the fatty acid supplement is enhanced (114). Additional complications with different fatty acid supplements include elongation, desaturation, and/or β oxidation of the acyl group prior to incorporation into the phospholipids (115). The most marked changes with respect to the content of saturated fatty acid and unsaturated sites can be achieved with 19:0 (115) and *cis* polyunsaturated fatty acids (115, 116), respectively. The latter type of supplement can be effectively provided as the fatty acid complexed to bovine serum albumin (116), and under these circumstances, there is extensive incorporation of the polyunsaturated fatty acid into the phospholipids.

FATTY ACID STRUCTURE AND PHYSICAL PROPERTIES OF MEMBRANE LIPID   Although unique acyl groups may be required for certain biochemical processes in the membrane, the bulk of the evidence indicates that the physical attributes of the hydrocarbon residues in the phospholipids are the important determinants. With respect to these properties, the major thrust has been to determine the precise structure of the lipids in membranes under physiological conditions. Recently, Oldfield & Chapman (6) have reviewed the application of a number of physical techniques to the study of lipids in biological membranes. Investigations have focused on a number of structural features. For example, paramagnetic resonance spectroscopic studies, using spin labels placed at varying distances from the carboxyl terminus of a fatty acyl group, indicate that the region of the lipid bilayer close to the head group is less mobile than that in the interior where the methyl terminal portion of the hydrocarbon chain lies (117). It has been suggested that the position of the ethylenic bond in membranes of organisms utilizing monounsaturated fatty acids determines the depth from the membrane surface at which the packing of acyl chains becomes less ordered (109). In addition to this difference in mobility along the fatty acyl chains, there may exist *within the plane* of the membrane distinct physical phases in equilibrium with one another, creating regions that are relatively solid (gel-like) or relatively fluid (liquid-crystalline) (6). The temperature range over which this equilibrium exists is markedly influenced by the configuration and chain length of the fatty acyl residues of the phospholipid molecules. Finally, to mention another structural feature of the membrane, the thickness of the bilayer at any given point may be affected by the phase state of the lipid and the chain length of the acyl groups in that region (118), and this consideration may receive renewed attention in the future.

By differential thermal analysis, McElhaney (119) has reinvestigated the relationship between fatty acid composition, the temperature range of the phase separation, and the temperature range over which *Acholeplasma laidlawii* B can grow. In this system it is possible to vary widely the temperature over which the phase separation occurs by manipulating the fatty acid composition of the cells. McElhaney found that the range between minimum and maximum growth temperatures may lie completely above or may overlap that associated with the phase transition. In the latter instance, normal growth can occur with as much as half of the lipid molecules in the gel-like state, and growth stops altogether only when less than 10% of the lipid remains in a fluid state. When the cells are enriched with unsaturated fatty acid and growth temperatures lie above that of the phase separation, the maximum growth temperatures are slightly reduced below those observed under other conditions of supplementation. This observation suggests that the liquid-crystalline phase associated with different lipid compositions may not be identical and that a state that is too fluid might exist.

In wild-type or mutant *E. coli* strains, the fatty acid composition can be varied by supplementation from 20 to 60% saturated fatty acids, at the least, without noticeably affecting growth at 37°C (17, 111). These changes in composition resemble those found when *E. coli* is grown at increasing temperatures from 10 to 43°C (18). Using the information obtained from paramagnetic resonance spectroscopic measurements, Sinensky computed the viscosity of lipid regions in membranes and of isolated lipids from *E. coli* grown at different temperatures (120). When the temperature of measurement corresponded to the temperature of growth, the viscosity remained constant over the range of 15 to 43°C, suggesting that the membrane lipids of *E. coli* are modified to maintain a constant viscosity regardless of growth conditions. Furthermore, the membrane lipids are in the liquid-crystalline state at the growth temperature and about 15° above the temperature at which phase separation would begin (120). Studies on the relation of fatty acid composition to growth using fatty acid supplementation at 37°C (17, 111) or growth at different temperatures (121), however, indicate that "homeoviscous" adaptation (120) is not essential. Hence, *E. coli,* like *A. laidlawii,* may be able to function normally with its lipids in a mixed liquid-crystalline and gel-like condition.

LIPID STRUCTURE, TOPOGRAPHY OF MEMBRANE PROTEINS, AND ACTIVITY OF SPECIFIC FUNCTIONS   Several recent studies have examined the effect of phase separation of membrane lipid promoted by lowering temperature on the topography of membrane proteins as revealed by freeze-fracture electron microscopy (122–125). Findings in the different investigations are not in complete agreement as to whether the intramembranous protein particles aggregate at a temperature above, coincident with, or below the onset of lateral phase separation. These differences arise at least in part from the difficulty connected with determining precisely the temperature of onset of the phase change. Membranes prepared from an unsaturated fatty acid auxotroph of *E. coli* grown on *trans*-Δ^9-18:1 have a relatively simple lipid composition, and under these conditions, the whole phase separation process can be followed by electron paramagnetic resonance measurements (125, 126). The conclusions from this study are that the intramembranous particles are randomly distributed, beginning to aggregate, or extensively aggregated at temperatures above, coincident with, and below the onset of phase separation, respectively.

The recent literature contains numerous reports concerned with the effect of fatty acid composition of membrane lipid on specific biochemical functions. Due to the space limitation, only a few of these studies can be considered here. The initial observation (127) that the induction of sugar transport requires the simultaneous synthesis of lipids containing unsaturated fatty acids has been reinvestigated by several laboratories (128–130). It is generally agreed that the original claim of a specific lipid requirement is not warranted. In the most recent study, normal induction of transport was observed even after the synthesis of phospholipids containing unsaturated fatty acids had fallen to 10% of the normal rates (130). Furthermore, when loss of transport did occur, it was accompanied also by loss of other lactose operon functions.

The activity of a number of membrane-associated processes, such as transport, phospholipid and lipopolysaccharide synthesis, and electron transfer (see references in 5), are affected by the fatty acid composition of the membrane. Furthermore, Arrhenius plots of some of these functions show sharp discontinuities occurring at temperatures characteristic of the fatty acid composition. Recent work has documented that these discontinuities arise as a result of lateral phase separations of the membrane phospholipids (126, 131–137). Under certain conditions, two breaks in the Arrhenius plot are observed and have been correlated with the onset and completion of the lipid phase change (133–136). In most instances, only single discontinuities have been detected, and although these are associated with changes in the physical state of the lipid, they have not generally been correlated precisely with a given part of the phase separation (e.g. 137). Recent observations on membrane bound ATPase of *A. laidlawii* are an exception (138). In this example, the single discontinuity in the Arrhenius plot occurred at a temperature corresponding to almost complete conversion of the lipid to the gel state. Although the studies described in this section indicate that membrane functions can be affected by the phase behavior of membrane lipid, there is limited evidence (119) to conclude that changes in fatty acid composition or temperatures occurring under normal physiological conditions promote phase separations. Furthermore, other explanations must be found to explain differences in biochemical activities associated with membrane lipid structure that are observed at temperatures above the liquid-gel phase separation (5, 139).

Polar Head Group

In comparison to the extensive studies on modification of the fatty acyl groups, very little progress has been made in altering the polar head groups of the phospholipids. As noted in the section on isolation of mutants, conditional lethal mutants of *E. coli* which do not synthesize phosphatidylethanolamine have been found (53–55). Although no detailed studies on changes in composition and function under the nonpermissive conditions have been made, the authors demonstrate that growth continues for one to two generations in the virtual absence of phosphatidylethanolamine synthesis but that viable cell number begins to fall long before growth stops (54). Attempts to block glycerophosphatide synthesis by the action of inhibitors have met with some success (140) but this approach has not been fully developed. Several years ago *Neurospora crassa* strains that were defective in lecithin synthesis were isolated and shown to require monomethylethanolamine, dimethylethanolamine, or choline supplementation (141, 142). One strain had subnormal phosphatidylethanolamine N-methyltransferase activity and a second was deficient in the activities required to convert phosphatidylmonomethylethanolamine to lecithin (143). The growth response to supplements and the compositional changes observed in these strains indicated that phosphatidyldimethylethanolamine and possibly phosphatidylmonomethylethanolamine could substitute for lecithin (144). Recently, it has been shown that cultured animal cells that require choline for growth can utilize one of several choline analogs. The substitution results in a marked reduction in lecithin synthesis and a considerable modification in the composition of the polar head groups of the phospholipids (116). As in the case of *Neurospora,* dimethylethanolamine is most comparable to choline as a growth factor and, in the cells grown with this supplement, lecithin is replaced extensively by phosphatidyldimethylethanolamine. Some interesting physical and biochemical studies should now be possible with membranes containing phospholipids modified with respect to the polar head groups.

Sterol Component

The effect of sterols in biological membranes has been investigated extensively by Van Deenen and co-workers using *A. laidlawii* B. Although the organism does not require sterols, there is good reason to believe that these studies are pertinent to understanding the requirement for sterols in other systems. A wide variety of sterols can be incorporated into the membrane of *A. laidlawii,* and in most studies, the sterol taken up accounts for 10% of the total membrane lipid (145, 146). Sterol incorporation is not influenced by fatty acid composition (146) and, conversely, has little or no effect on the fatty acid composition of the phospholipids (145–147). Furthermore, monolayer and calorimetric measurements of phospholipid-cholesterol mixtures suggest that the nature of the head groups of the lipid molecules does not control the lipid-sterol interaction (148). Although many sterols can be incorporated into the membrane, only those containing a 3-β-hydroxyl group and a relatively planar nucleus influence membrane function (146). For example, cholesterol causes a decrease in the permeability of *A. laidlawii* (as well as model membranes) toward nonelectrolytes (146, 147). Analogs of cholesterol, such as ergosterol, which have an additional alkyl group in the side chain or additional ethylenic bonds in the nucleus and side chain, are also active. Differential scanning calorimetry demonstrates that cholesterol reduces the energy content associated with the thermotropic phase separation of the phospholipid and extends the transition to lower temperatures (an effect at the higher end of the transition is not certain) (138, 145, 148). These physical changes are not observed with 3-α-hydroxysterols (145). Arrhenius plots of ATPase activity in membranes from cells grown without sterol show that a discontinuity occurs at a temperature corresponding to the lower end of the lateral phase separation. Cholesterol, but not epicholesterol (3-α-hydroxy isomer of cholesterol), lowers the temperature of this break by about 6°C, which is consistent with the change in the physical properties of the membrane lipid (138). A similar effect on the Arrhenius plot of ATPase from yeast mitochondria has also been observed following an increase in mitochondrial ergosterol levels (149).

A strain of *Mycoplasma* that requires a sterol for growth has been adapted to grow in low cholesterol concentrations (150). By adjusting the concentration of the cholesterol supplement, these investigators could vary the cholesterol content in the membrane between a normal level and one tenth the normal content. In most respects, their observations (150, 151) on physical and functional properties of membranes containing normal vs low sterol contents resemble and confirm those described for *A. laidlawii,* an organism that does not require cholesterol. The difference between the two systems is the discovery that in the *Mycoplasma* strain but not in *A. laidlawii* the saturated fatty acid content of the polar lipids increased in response to a diminished source of cholesterol, suggesting a possible adaptive mechanism to maintain certain physical properties of the membrane (150). The newly isolated sterol-requiring mutant in *S. cerevisiae* (100) described earlier should afford another system to examine the role of sterols in biological membranes.

INTERRELATIONSHIP OF LIPID AND OTHER MACROMOLECULAR SYNTHESES

To conclude this review, we consider some observations pertinent to the problem of mutual regulation of macromolecular synthesis. A number of recent studies have established that lipid synthesis in *E. coli* is coupled to protein synthesis by the stringent control mechanism (21, 152–157). Furthermore, there is excellent correlation between the accumulation of intracellular guanosine tetraphosphate (ppGpp), a mediator in this control process, and the reduction in phospholipid synthesis (156). Several specific enzymatic steps in lipid biosynthesis have been shown to be inhibited 50 to 60% more or less specifically by ppGpp at concentrations comparable to those in stringent (*rel*+) strains under conditions of amino acid starvation: the carboxyltransferase component of acetyl CoA carboxylase (21), glycerol-3-P-acyltransferase (156), and phosphatidylglycerolphosphate synthetase (156). There is evidence that the control mechanism in vivo acts at the level of fatty acid as well as phospholipid synthesis. In the first instance, fatty acid synthesis decreases about 60% in *rel*+ but not in *rel*− bacteria during amino acid starvation (21). Secondly, when *rel*+ but not *rel*− strains are made dependent on exogenous fatty acids for phospholipid synthesis because of a *fab* mutation or the action of an inhibitor of fatty acid biosynthesis, phospholipid synthesis is still under stringent control (157).

The availability of lipid biosynthetic mutants has provided a means to examine the effect of limited phospholipid synthesis on other macromolecular processes in *E. coli.* Results from several studies are somewhat in conflict, particularly with respect to how tightly protein and nucleic acid synthesis are coupled to lipid synthesis. Using a temperature-sensitive glycerol 3-phosphate acyltransferase mutant (*plsA*), Glaser et al (20) showed that growth and all macromolecular synthesis ceased abruptly on shifting the cells to the restrictive temperature. All other studies utilized glycerol 3-phosphate auxotrophs, either of the *gpsA* (39, 40) or *plsB* (38, 40) types. When deprived of glycerol 3-phosphate, growth of these strains continued for at least one generation even though net phospholipid synthesis decreased to a very low rate in at least one study (40). The short term persistence of protein and nucleic acid synthesis is best documented in the paper by Bell (40) in which the K_m of the acyltransferase for glycerol 3-phosphate in the *plsB* strain was 10- to 15-fold above that of a normal enzyme. Recent observations by Pizer et al (38) and by M. Glaser, W. Nulty, and P. R. Vagelos (submitted for publication) reveal that intracellular concentrations of ATP decrease very abruptly and drastically in temperature-sensitive *plsA* strains when they are placed at the restrictive temperature. Since the adenylate kinase as well as the acyltransferase is thermolabile in these strains as noted in an earlier section, the tight coupling of macromolecular syntheses in *plsA* mutants can be explained by the simultaneous thermolability of phospholipid and ATP synthesis. A less dramatic reduction in ATP levels has also been observed in a *plsB* strain when glycerol 3-phosphate deprivation caused a 50% reduction in phospholipid synthesis (38). Although the change in ATP levels following glycerol 3-phosphate removal needs to be documented better kinetically in *plsB* strains, the present findings suggest a possible mechanism for coupling protein and nucleic acid synthesis to phospholipid synthesis. Perhaps the physiologically significant conditions for this type of coupling occur during cell elongation associated with the cell cycle or with the availability of a more favorable carbon source (158, 159). During this period, the surface-to-volume ratio and the envelope content relative to total dry weight decrease. Since phospholipid, unlike protein and nucleic acid, is located exclusively in the cell envelope of *E. coli*, these changes in cell shape must reflect an adjustment in the relative content of lipid and other macromolecules. It will be of interest to see if the observations of Huzyk & Clark (160) that nucleoside and deoxynucleoside triphosphate pools fluctuate during the cell cycle of *E. coli* can be tied to a control mechanism coordinating the synthesis of macromolecules in the cell envelope and cytoplasm.

Literature Cited

1. Cronan, J. E. Jr., Vagelos, P. R. 1972. *Biochim. Biophys. Acta* 265:25–60
2. Volpe, J. J., Vagelos, P. R. 1973. *Ann. Rev. Biochem.* 42:21–60
3. Machtiger, N. A., Fox, C. F. 1973. *Ann. Rev. Biochem.* 42:575–600
4. Glaser, L. 1973. *Ann. Rev. Biochem.* 42:91–112
5. Silbert, D. F., Cronan, J. E. Jr., Beacham, I. R., Harder, M. E. 1974. *Fed. Proc.* 33:1725–32
6. Oldfield, E., Chapman, D. 1972. *FEBS Lett.* 23:285–97
7. Keith, A. D., Wisnieski, B. J., Henry, S., Williams, J. C. 1973. *Lipids and Biomembranes of Eukaryotic Microorganisms,* ed. J. A. Erwin, 259–321. New York: Academic
8. Longley, R. P., Rose, A. H., Knights, B. A. 1968. *Biochem. J.* 108:401–12
9. Bloch, K. et al 1961. *Fed. Proc.* 20:921–27
10. Prescott, D. J., Vagelos, P. R. 1972. *Advan. Enzymol.* 36:269–309
11. Weeks, G., Wakil, S. J. 1968. *J. Biol. Chem.* 243:1180–89
12. Helmkamp, G. M., Bloch, K. 1969. *J. Biol. Chem.* 244:6014–22
13. Hantke, K., Braun, V. 1973. *Eur. J. Biochem.* 34:284–96
14. Sinensky, M. 1971. *J. Bacteriol.* 106:449–55
15. Esfahani, M., Ioneda, T., Wakil, S. J. 1971. *J. Biol. Chem.* 246:50–56
16. Silbert, D. F., Cohen, M., Harder, M. E. 1972. *J. Biol. Chem.* 247:1699–1707
17. Silbert, D. F., Ulbright, T. M., Honegger, J. L. 1973. *Biochemistry* 12:164–71
18. Marr, A. G., Ingraham, J. L. 1967. *J. Bacteriol.* 84:1260–67
19. Mindich, L. 1972. *J. Bacteriol.* 110:96–102
20. Glaser, M., Bayer, W. H., Bell, R. M., Vagelos, P. R. 1973. *Proc. Nat. Acad. Sci. USA* 70:385–89
21. Polakis, S. E., Guchhait, R. B., Lane, M. D. 1973. *J. Biol. Chem.* 248:7957–66
22. Silbert, D. F., Vagelos, P. R. 1967. *Proc. Nat. Acad. Sci. USA* 58:1579–86
23. Cronan, J. E. Jr., Gelmann, E. P. 1973. *J. Biol. Chem.* 248:1188–95
24. Cronan, J. E. Jr. 1974. *Proc. Nat. Acad. Sci. USA* 71:3758–62
25. Rosenfeld, I. S., D'Agnolo, G., Vagelos, P. R. 1973. *J. Biol. Chem.* 248:2452–60
26. Gelmann, E. P., Cronan, J. E. Jr. 1972.

J. Bacteriol. 112:381–87
27. Broekman, J. H. F. F. 1973. *Mutants of Escherichia coli K-12 impaired in the biosynthesis of the unsaturated fatty acids.* PhD thesis. Univ. of Utrecht, Utrecht. 84 pp.
28. Harder, M. E., Beacham, I. R., Cronan, J. E. Jr., Beacham, K., Honegger, J. L., Silbert, D. F. 1972. *Proc. Nat. Acad. Sci. USA* 69:3105–9
29. Harder, M. E., Ladenson, R. C., Schimmel, S. D., Silbert, D. F. 1974. *J. Biol. Chem.* 249:7468–75
30. Cronan, J. E. Jr., Silbert, D. F., Wulff, D. L. 1972. *J. Bacteriol* 112:206–11
31. Schairer, H. U., Overath, P. 1969. *J. Mol. Biol.* 44:209–14
32. Epstein, W., Fox, C. F. 1970. *J. Bacteriol.* 103:274–75
33. Broekman, J. H. F. F., Hoekstra, W. P. M. 1973. *Mol. Gen. Genet.* 124:65–67
34. Chang, Y. Y., Kennedy, E. P. 1967. *J. Biol. Chem.* 242:516–19
35. Raetz, C. R. H., Kennedy, E. P. 1973. *J. Biol. Chem.* 248:1098–1105
36. Hirschberg, C. B., Kennedy, E. P. 1972. *Proc. Nat. Acad. Sci. USA* 69:648–51
37. Bell, R. M., Mavis, R. D., Vagelos, P. R. 1972. *Biochim. Biophys. Acta* 270:504–12
38. Pizer, L. I., Merlie, J. P., Ponce deLeon, M. 1974. *J. Biol. Chem.* 249:3212–24
39. Hsu, C. C., Fox, C. F. 1970. *J. Bacteriol.* 103:410–16
40. Bell, R. M. 1974. *J. Bacteriol.* 117:1065–76
41. Cronan, J. E. Jr., Bell, R. M. 1974. *J. Bacteriol.* 118:598–620
42. Kito, M., Lubin, M., Pizer, L. I. 1969. *Biochem. Biophys. Res. Commun.* 34:454–58
43. Cronan, J. E. Jr., Ray, T. K., Vagelos, P. R. 1970. *Proc. Nat. Acad. Sci. USA* 65:737–44
44. Godson, G. N. 1973. *J. Bacteriol.* 113:813–24
45. Ray, T. K., Cronan, J. E. Jr., Mavis, R. D., Vagelos, P. R. 1970. *J. Biol. Chem.* 245:6442–48
46. Raetz, C. R. H., Kennedy, E. P. 1973. *J. Biol. Chem.* 247:2008–14
47. Bell, R. M., Mavis, R. D., Osborn, M. J., Vagelos, P. R. 1971. *Biochim. Biophys. Acta* 249:628–35
48. White, D. A., Albright, F. R., Lennarz, W. J., Schnaitman, C. A. 1971. *Biochim. Biophys. Acta* 249:636–42
49. Machtiger, N. A., Fox, C. F. 1973. *J. Supramol. Struct.* 1:545–64
50. Raetz, C. R. H., Hirschberg, C. B., Dowhan, W., Wickner, W. T., Kennedy, E. P. 1972. *J. Biol. Chem.* 247:2245–47
51. Dowhan, W., Wickner, W. T., Kennedy,

E. P. 1974. *J. Biol. Chem.* 249:3079–84
52. Cronan, J. E. Jr., Bell, R. M. 1974. *J. Bacteriol.* 120:227–33
53. Cronan, J. E. Jr. 1972. *Nature New Biol.* 240:21–22
54. Ohta, A., Okonogi, K., Shibuya, I., Maruo, B. 1974. *J. Gen. Appl. Microbiol.* 20:21–32
55. Ohta, A., Shibuya, I., Maruo, B., Ishinaga, M., Kito, M. 1974. *Biochim. Biophys. Acta* 348:449–54
56. Cronan, J. E. Jr., Godson, G. N. 1972. *Mol. Gen. Genet.* 116:199–210
57. Overath, P., Raufuss, E. 1967. *Biochem. Biophys. Res. Commun.* 29:28–33
58. Overath, P., Pauli, G., Schairer, H. U. 1969. *Eur. J. Biochem.* 7:559–74
59. Weeks, G., Shapiro, M., Burns, R. O., Wakil, S. J. 1969. *J. Bacteriol.* 97:827–36
60. Klein, K., Steinberg, R., Fiethen, B., Overath, P. 1971. *Eur. J. Biochem.* 19:442–56
61. Overath, P., Schairer, H. U., Stoffel, W. 1970. *Proc. Nat. Acad. Sci. USA* 67:606–12
62. Albright, F. R., White, D. A., Lennarz, W. J. 1973. *J. Biol. Chem.* 248:3968–77
63. Osborn, M. J., Rick, P. D., Lehmann, V., Rupprecht, E., Singh, M. 1974. *Ann. NY Acad. Sci.* 235:52–65
64. Doi, O., Ohki, M., Nojima, S. 1972. *Biochim. Biophys. Acta* 260:244–58
65. Ohki, M., Doi, O., Nojima, S. 1972. *J. Bacteriol.* 110:864–69
66. Doi, O., Nojima, S. 1973. *J. Biochem. Tokyo* 74:667–74
67. Scandella, C. J., Kornberg, A. 1971. *Biochemistry* 10:4447–56
68. Nojima, S., Doi, O., Okamoto, N., Abe, M. 1972. *Membrane Research,* ed. C. F. Fox, 135–44. New York: Academic
69. Audet, A., Nantel, G., Proulx, P. 1974. *Biochim. Biophys. Acta* 348:334–43
70. Rick, P. D., Osborn, M. J. 1972. *Proc. Nat. Acad. Sci. USA* 69:3756–60
71. Rick, P. D., Osborn, M. J. 1973. *Fed. Proc.* 32(3):1457 (Abstr.)
72. Rick, P. D., Osborn, M. J. 1974. *Fed. Proc.* 33(5):282 (Abstr.)
73. Koplow, J., Goldfine, H. 1974. *J. Bacteriol.* 117:527–50
74. Ames, G. F., Spudich, E. N., Nikaido, H. 1974. *J. Bacteriol.* 117:406–16
75. Steenbakkers, J. F., Broekman, J., Kerkenaar, A., deHaan, P. G. 1973. *J. Bacteriol.* 116:535–40
76. Kent, C., Lennarz, W. J. 1972. *Biochim. Biophys. Acta* 288:225–30
77. Kent, C., Krag, S. S., Lennarz, W. J. 1973. *J. Bacteriol.* 113:874–83
78. Henning, U., Dennert, G., Rehn, K.,

338   SILBERT

Deppe, G. 1969. *J. Bacteriol.* 98:784–96
79. Oesterhelt, D., Bauer, H., Lynen, F. 1969. *Proc. Nat. Acad. Sci. USA* 63:1377–82
80. Meyer, K. H., Schweizer, E. 1972. *Biochem. Biophys. Res. Commun.* 46:1674–80
81. Schweizer, E., Lerch, I., Kroeplin-Rueff, L., Lynen, F. 1970. *Eur. J. Biochem.* 15:472–82
82. Bloomfield, D. K., Bloch, K. 1960. *J. Biol. Chem.* 235:337–45
83. Orme, T. W., McIntyre, J., Lynen, F., Kuhn, L., Schweizer, E. 1972. *Eur. J. Biochem.* 24:407–15
84. Henry, S. A., Keith, A. D. 1971. *Chem. Phys. Lipids* 7:245–65
85. Southwell-Keely, P. T., Lynen, F. 1974. *Biochim. Biophys. Acta* 337:22–28
86. Hunter, K., Rose, A. H. 1971. *The Yeasts*, ed. A. H. Rose, J. S. Harrison, 2:211–70. New York: Academic
87. Resnick, M. A., Mortimer, R. K. 1966. *J. Bacteriol.* 92:597–600
88. Keith, A. D., Resnick, M. A., Haley, A. B. 1969. *J. Bacteriol.* 98:415–20
89. Bard, M., Woods, R. A., Haslam, J. M. 1974. *Biochem. Biophys. Res. Commun.* 56:324–30
90. Schweizer, E., Kuhn, L., Castorph, H. 1971. *Z. Physiol. Chem.* 352:377–84
91. Henry, S. A., Fogel, S. 1971. *Mol. Gen. Genet.* 113:1–19
92. Schweizer, E., Bolling, H. 1970. *Proc. Nat. Acad. Sci. USA* 67:660–66
93. Kuhn, L., Castorph, H., Schweizer, E. 1972. *Eur. J. Biochem.* 24:492–97
94. Schweizer, E., Kniep, B., Castorph, H., Holzner, U. 1973. *Eur. J. Biochem.* 39:353–62
95. Meyer, K. H., Schweizer, E. 1974. *J. Bacteriol.* 117:345–50
96. Burkl, G., Castorph, H., Schweizer, E. 1972. *Mol. Gen. Genet.* 119:315–22
97. Culbertson, M. R., Henry, S. A. 1973. *Genetics* 75:441–58
98. Henry, S. A. 1973. *J. Bacteriol.* 116:1293–1313
99. Gollub, E. G., Trocha, P., Liu, P. K., Sprinson, D. B. 1973. *Biochem. Biophys. Res. Commun.* 56:471–77
100. Karst, F., Lacroute, F. 1974. *Biochem. Biophys. Res. Commun.* 59:370–76
101. Nomura, S., Horiuchi, T., Omura, S., Hata, T. 1972. *J. Biochem. Tokyo* 71:783–96
102. Vance, D., Goldberg, I., Mitsuhashi, O., Bloch, K., Omura, S., Nomura, S. 1972. *Biochem. Biophys. Res. Commun.* 48:649–56
103. D'Agnolo, G., Rosenfeld, I. S., Awaya, J., Omura, S., Vagelos, P. R. 1973. *Biochem. Biophys. Acta* 326:155–66

104. Silbert, D. F., Ladenson, R. C., Honegger, J. L. 1973. *Biochim. Biophys. Acta* 311:349–61
105. Wisnieski, B., Keith, A. D., Resnick, M. R. 1970. *J. Bacteriol.* 101:160–65
106. Proudlock, J. W., Haslam, J. M., Linnane, A. W. 1971. *J. Bioenerg.* 2:327–49
107. Williams, M. A., Taylor, D. W., Tinoco, J., Ojakian, M. A., Keith, A. D. 1973. *Biochem. Biophys. Res. Commun.* 54:1560–66
108. Barber, E. D., Lands, W. E. M. 1973. *J. Bacteriol.* 115:543–51
109. Eletr, S., Keith, A. D. 1972. *Proc. Nat. Acad. Sci. USA* 69:1353–57
110. Silbert, D. F. 1970. *Biochemistry* 9:3631–40
111. Davis, M., Silbert, D. F. 1974. *Biochim. Biophys. Acta* 373:224–41
112. Haslam, J. M., Proudlock, J. W., Linnane, A. W. 1971. *J. Bioenerg.* 2:351–70
113. Haslam, J. M., Spothill, T. W., Linnane, A. W. 1973. *Biochem. J.* 134:949–57
114. Wisnieski, B. J., Williams, R. E., Fox, C. F. 1973. *Proc. Nat. Acad. Sci. USA* 70:3669–73
115. Williams, R. E., Wisnieski, B. J., Rittenhouse, H. G., Fox, C. F. 1974. *Biochemistry* 13:1969–77
116. Glaser, M., Ferguson, K. A., Bayer, W. H. 1974. *Fed. Proc.* 33(5):421 (Abstr.)
117. Hubbell, W. L., McConnell, H. M. 1971. *J. Am. Chem. Soc.* 93:314–26
118. Engelman, D. 1971. *J. Mol. Biol.* 58:153–65
119. McElhaney, R. N. 1974. *J. Mol. Biol.* 84:145–58
120. Sinensky, M. 1974. *Proc. Nat. Acad. Sci. USA* 71:522–25
121. Shaw, M. K., Ingraham, J. L. 1965. *J. Bacteriol.* 90:141–46
122. Verkleij, A. J., Ververgaert, P. H. J., Van Deenen, L. L. M., Elbers, P. F. 1972. *Biochim. Biophys. Acta* 288:326–32
123. Speth, V., Wunderlich, F. 1973. *Biochim. Biophys. Acta* 291:621–28
124. James, R., Branton, D. 1973. *Biochim. Biophys. Acta* 323:378–90
125. Kleemann, W., McConnell, H. M. 1974. *Biochim. Biophys. Acta* 345:220–30
126. Linden, C. D., Keith, A. D., Fox, C. F. 1973. *J. Supramol. Struct.* 1:523–34
127. Fox, C. F. 1969. *Proc. Nat. Acad. Sci. USA* 63:850–55
128. Overath, P., Hill, F. F., Lamnek-Hirsch, I. 1971. *Nature New Biol.* 234:264–67
129. Robbins, A. R., Rotman, B. 1972. *Proc.*

Nat. Acad. Sci. USA 69:2124–29
130. Nunn, W. D., Cronan, J. E. Jr. 1974. *J. Biol. Chem.* 249:724–31
131. Overath, P., Trauble, H. 1973. *Biochemistry* 12:2625–34
132. Trauble, H., Overath, P. 1973. *Biochim. Biophys. Acta* 307:491–512
133. Linden, C. D., Wright, K. L., McConnell, H. M., Fox, C. F. 1973. *Proc. Nat. Acad. Sci. USA* 70:2271–75
134. Shimshick, E. J., McConnell, H. M. 1973. *Biochemistry* 12:2351–60
135. Shimshick, E. J., McConnell, H. M. 1973. *Biochem. Biophys. Res. Commun.* 53:446–51
136. Shimshick, E. J., Kleemann, W., Hubbell, W. L., McConnell, H. M. 1973. *J. Supramol. Struct.* 1:285–94
137. Eletr, S., Williams, M. A., Watkins, T., Keith, A. D. 1974. *Biochim. Biophys. Acta* 339:190–201
138. DeKruyff, B., Van Dijck, P. W. M., Goldbach, R. W., Demel, R. A., Van Deenen, L. L. M. 1973. *Biochim. Biophys. Acta* 330:269–82
139. Beacham, I. R., Silbert, D. F. 1973. *J. Biol. Chem.* 248:5310–18
140. Shopsis, C. S., Engel, R., Tropp, B. E. 1974. *J. Biol. Chem.* 249:2473–77
141. Hall, M. O., Nyc, J. F. 1959. *J. Am. Chem. Soc.* 81:2275
142. Hall, M. O., Nyc, J. F. 1961. *J. Lipid Res.* 2:321–27
143. Scarborough, G. A., Nyc, J. F. 1967. *J. Biol. Chem.* 242:238–42
144. Crocken, B. J., Nyc, J. F. 1964. *J. Biol. Chem.* 239:1727–30
145. DeKruyff, B., Demel, R. A., Van Deenen, L. L. M. 1972. *Biochim. Biophys. Acta* 255:331–47

146. DeKruyff, B., DeGreef, W. J., Van Eyk, R. V. W., Demel, R. A., Van Deenen, L. L. M. 1973. *Biochim. Biophys. Acta* 298:479–99
147. McElhaney, R. N., DeGier, J., Van der Neut-Kok, E. C. M. 1973. *Biochim. Biophys. Acta* 298:500–12
148. DeKruyff, B., Demel, R. A., Slotboom, A. J., Van Deenen, L. L. M., Rosenthal, A. F. 1973. *Biochim. Biophys. Acta* 307:1–19
149. Cobon, G. S., Haslam, J. M. 1973. *Biochem. Biophys. Res. Commun.* 52:320–26
150. Rottem, S., Yashouv, J., Ne'eman, Z., Razin, S. 1973. *Biochim. Biophys. Acta* 323:495–508
151. Rottem, S., Cirillo, V. P., DeKruyff, B., Shinitzky, M., Razin, S. 1973. *Biochim. Biophys. Acta* 323:509–19
152. Sokawa, Y., Nakao, E., Kaziro, Y. 1968. *Biochem. Biophys. Res. Commun.* 38:108–12
153. Sokawa, Y., Nakao, E., Kaziro, Y. 1970. *Biochim. Biophys. Acta* 199:256–64
154. Golden, N. G., Powell, G. L. 1972. *J. Biol. Chem.* 247:6651–58
155. Pizer, L. I., Merlie, J. P. 1973. *J. Bacteriol.* 114:980–87
156. Merlie, J. P., Pizer, L. I. 1973. *J. Bacteriol.* 116:355–66
157. Nunn, W. D., Cronan, J. E. Jr. 1974. *J. Biol. Chem.* 249:3994–96
158. Schaechter, M. 1968. *Biochemistry of Bacterial Cell Growth,* ed. J. Mandelstam, K., McQuillen, 144–46. New York: Wiley
159. Ballesta, J. P. G., Schaechter, M. 1972. *J. Bacteriol.* 110:452–53
160. Huzyk, L., Clark, D. J. 1971. *J. Bacteriol.* 108:74–81

CHEMOTAXIS IN BACTERIA ×886

Julius Adler

Departments of Biochemistry and Genetics, University of Wisconsin,
Madison, Wisconsin 53706

CONTENTS

"I am not entirely happy about my diet of flies and bugs, but it's the way I'm made.
A spider has to pick up a living somehow or other, and I happen to be a trapper.
I just naturally build a web and trap flies and other insects. My mother was a trapper
before me. Her mother was a trapper before her. All our family have been trappers.
Way back for thousands and thousands of years we spiders have been laying for flies
and bugs."

"It's a miserable inheritance," said Wilbur, gloomily. He was sad because his new
friend was so bloodthirsty.

"Yes, it is," agreed Charlotte. "But I can't help it. I don't know how the first spider in
the early days of the world happened to think up this fancy idea of spinning a web,
but she did, and it was clever of her, too. And since then, all of us spiders have had
to work the same trick. It's not a bad pitch, on the whole."

E. B. White, *Charlotte's Web*

Bacterial chemotaxis, the movement toward or away from chemicals, was
discovered nearly a century ago by Engelmann (1) and Pfeffer (2, 3). The subject was
actively studied for about fifty years, but then there were very few reports until quite

341

recently. For reviews of the literature up to about 1960, see Berg (4), Weibull (5), and Ziegler (6). This review will restrict itself to the recent work on chemotaxis in *Escherichia coli* and *Salmonella typhimurium*. Some of this work is also covered in Berg's review (4), and a review by Parkinson (7) should be consulted for a more complete treatment of the genetic aspects.

OVERVIEW

Motile bacteria are attracted by certain chemicals and repelled by others; this is positive and negative chemotaxis. Chemotaxis can be dissected by means of the following questions:
1. How do individual bacteria move in a gradient of attractant or repellent?
2. How do bacteria detect the chemicals?
3. How is the sensory information communicated to the flagella?
4. How do bacterial flagella produce motion?
5. How do flagella respond to the sensory information in order to bring about the appropriate change in direction?
6. In the case of multiple or conflicting sensory data, how is the information integrated?

DEMONSTRATION AND MEASUREMENT OF CHEMOTAXIS IN BACTERIA

Work before 1965, although valuable (4–6), was carried out in complex media and was largely of a subjective nature. It was therefore necessary to develop conditions for obtaining motility and chemotaxis in defined media (8–11) and to find objective, quantitative methods for demonstrating chemotaxis.

(*a*) Plate method: For positive chemotaxis, a petri dish containing metabolizable attractant, salts needed for growth, and soft agar (a low enough concentration so that the bacteria can swim) is inoculated in the center with the bacteria. As the bacteria grow, they consume the local supply of attractant, thus creating a gradient, which they follow to form a ring of bacteria surrounding the inoculum (8). For negative chemotaxis, a plug of hard agar containing repellent is planted in a petri dish containing soft agar and bacteria concentrated enough to be visibly turbid; the bacteria soon vacate the area around the plug (12). By searching in the area of the plate traversed by wild-type bacteria, one can isolate mutants in positive or negative chemotaxis (for example, 10, 12–15).

(*b*) Capillary method: In the 1880s Pfeffer observed bacterial chemotaxis by inserting a capillary containing a solution of test chemical into a bacterial suspension and then looking microscopically for accumulation of bacteria at the mouth of and inside the capillary (positive chemotaxis) or movement of bacteria away from the capillary (negative chemotaxis) (2, 3). For positive chemotaxis this procedure has been converted into an objective, quantitative assay by measuring the number of bacteria accumulating inside a capillary containing attractant solution (10, 11). For negative chemotaxis, repellent in the capillary decreases the number of cells that will enter (12). Alternatively, repellent is placed with the bacteria but not in the capillary; the number of bacteria fleeing into the capillary for refuge is then measured (12). Unlike in the plate method, where bacteria make the gradient of attractant by metabolizing the chemical, here the experimenter provides the gradient; hence nonmetabolizable chemicals can be studied.

(c) Defined gradients: Quantitative analysis of bacterial migration has been achieved by making defined gradients of attractant (16) or repellent (17), and then determining the distribution of bacteria in the gradient by measuring scattering of a laser beam by the bacteria. The method allows the experimenter to vary the shape of the gradient.

(d) A change in the bacterium's tumbling frequency in response to a chemical gradient, described next, is also to be regarded as a demonstration and a measurement of chemotaxis.

THE MOVEMENT OF INDIVIDUAL BACTERIA IN A CHEMICAL GRADIENT

The motion of bacteria can of course be observed microscopically by eye, recorded by microcinematography, or followed as tracks that form on photographic film after time exposure (18, 19). Owing to the very rapid movement of bacteria, however, significant progress was not made until the invention of an automatic tracking microscope, which allowed objective, quantitative, and much faster observations (20). A slower, manual tracking microscope has also been used (21). A combination of these methods has led to the following conclusions.

In the absence of a stimulus (i.e. no attractant or repellent present, or else a constant, uniform concentration—no gradient) a bacterium such as E. coli or S. typhimurium swims in a smooth, straight line for a number of seconds—a "run," then it thrashes around for a fraction of a second—a "tumble" (or abruptly changes its direction—a "twiddle"); and then it again swims in a straight line, but in a new, randomly chosen direction (22). (A tumble is probably a series of very brief runs and twiddles.)

Compared to this unstimulated state, cells tumble less frequently (i.e. they swim in longer runs) when they encounter increasing concentrations of attractant (22, 23) and they tumble more frequently when the concentration decreases (23). For repellents, the opposite is true: bacteria encountering an increasing concentration tumble more often, while a decreasing concentration suppresses tumbling (17). (See Figure 1.) [Much smaller concentration changes are needed to bring about suppression of tumbling than stimulation of tumbling (22, 24).]

All this applies not only to *spatial* gradients (for example, a higher concentration of chemical to the right than to the left) but also to *temporal* gradients (a higher concentration of chemical now than earlier). The important discovery that bacteria can be stimulated by temporal gradients of chemicals was made by mixing bacteria quickly with increasing or decreasing concentrations of attractant (23) or repellent (17) and then immediately observing the alteration of tumbling frequency. After a short while (depending on the extent and the direction of the concentration change), the tumbling frequency returns to the unstimulated state (17, 23). A different way to provide temporal gradients is to destroy or synthesize an attractant enzymatically; as the concentration of attractant changes, the tumbling frequency is measured (25). [For a history of the use of temporal stimulation in the study of bacterial behavior, see the introduction to (25).] The fact that bacteria can "remember" that there is a different concentration now than before has led to the proposal that bacteria have a kind of "memory" (23, 24).

The possibility that a bacterium in a spatial gradient compares the concentration at each end of its cell has not been ruled out, but it is not necessary to invoke it now, and in addition, the concentration difference at the two ends would be too small to be effective for instantaneous comparison (23, 24).

These crucial studies (17, 22, 23, 25) point to the regulation of tumbling frequency as a central feature of chemotaxis. The results are summarized in Figure 1.

By varying the tumbling frequency in this manner, the bacteria migrate in a "biased random walk" (24) toward attractants and away from repellents: motion in a favorable direction is prolonged, and motion in an unfavorable direction is terminated.

[Bacteria that have one or more flagella located at the pole ("polar flagellation," as in *Spirillum* or *Pseudomonas*) back up instead of tumbling (26, 27). Even bacteria that have flagella distributed all over ("peritrichous flagellation," as in *E. coli* or *S. typhimurium*) will go back and forth instead of tumbling if the medium is sufficiently viscous (28).]

THE DETECTION OF CHEMICALS BY BACTERIA: CHEMOSENSORS

What is Detected?

Until 1969 it was not known if bacteria detected the attractants themselves or instead measured some product of metabolism of the attractants, for example ATP. The latter idea was eliminated and the former established by the following results (10). (*a*) Some extensively metabolized chemicals are not attractants. This includes chemicals that are the first products in the metabolism of chemicals that do attract. (*b*) Some essentially nonmetabolizable chemicals attract bacteria: nonmetabolizable analogs of metabolizable attractants attract bacteria, and mutants blocked in the metabolism of an attractant are still attracted to it. (*c*) Chemicals attract bacteria even in the presence of a metabolizable, nonattracting chemical. (*d*) Attractants that are closely related in structure compete with each other but not

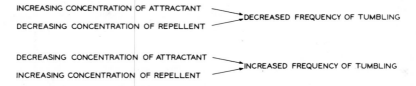

Figure 1   Effect of change of chemical concentration on tumbling frequency.

with structurally unrelated attractants. (*e*) Mutants lacking the detection mechanism but normal in metabolism can be isolated.(*f*) Transport of a chemical into the cells is neither sufficient nor necessary for it to attract.

Thus, bacteria can sense attractants per se: these cells are equipped with sensory devices, "chemoreceptors," that measure changes in concentration of certain chemicals and report the changes to the flagella (10). It is a characteristic feature of this and many other sensory functions that when the stimulus intensity changes, there is a response for a brief period only, i.e. the response is *transient* (17, 23). In contrast, all other responses of bacteria to changes in concentration of a chemical *persist* as long as the new concentration is maintained. For example, when the concentration of lactose is increased (over a certain range), there is a persisting increase in the rate of lactose transport, the rate of lactose metabolism, or the rate of β-galactosidase synthesis. To emphasize this unique feature, the sensory devices for chemotaxis will now be called "chemosensors."

"Chemoreceptor" will now be used for that part of the chemosensor that "receives" the chemicals—a component that recognizes or binds the chemicals detected. A chemosensor must in addition have a component—the "signaller"—that signals to the flagella the change in the fraction of chemoreceptor occupied by the chemical. Further, a chemosensor may contain transport components for the sensed chemical, needed either directly or to place the chemoreceptor in a proper conformation. Bacteria have "receptors" for other chemicals—phage receptors, bacteriocin receptors, enzymes, repressors, etc, but these do not serve a *sensory* function in the manner just defined.

Since metabolism of the attractants is not involved in sensing them (10), the mechanism of positive chemotaxis does not rely upon the attractant's value to the cell. Similarly, negative chemotaxis is not mediated by the harmful effects of a repellent (12): (*a*) repellents are detected at concentrations too low to be harmful; (*b*) not all harmful chemicals are repellents; and (*c*) not all repellents are harmful. Nevertheless, the survival value of chemotaxis must lie in bringing the bacteria into a nutritious environment (the attractants might signal the presence of other undetected nutrients) and away from a noxious one.

The chemosensors serve to alert the bacterium to changes in its environment.

The Number of Different Chemosensors

For both positive and negative chemotaxis the following criteria have been used to divide the chemicals into chemosensor classes. (*a*) For a number of chemosensors, mutants lacking the corresponding taxis, "specifically nonchemotactic mutants," have been isolated (10, 12–15, 29–31). (*b*) Competition experiments: chemical A, present in high enough concentration to saturate its chemoreceptor, will completely block the response to B if the two are detected by the same chemoreceptor but not if they are detected by different chemoreceptors (10, 12, 17, 30–33). (*c*) Many of the chemosensors are inducible, each being separately induced by a chemical it can detect (10, 15).

Table 1 lists chemosensors identified so far for positive chemotaxis in *E. coli,* and Table 2 for negative chemotaxis in *E. coli.* Altogether evidence exists for about

Table 1   Partial list of chemosensors for positive chemotaxis in *Escherichia coli*

Attractant		Threshold Molarity[a]
	N-acetyl-glucosamine sensor	
N-Acetyl-D-glucosamine		1×10^{-5}
	Fructose sensor	
D-Fructose		1×10^{-5}
	Galactose sensor	
D-Galactose		1×10^{-6}
D-Glucose		1×10^{-6}
D-Fucose		2×10^{-5}
	Glucose sensor	
D-Glucose		3×10^{-6}
	Mannose sensor	
D-Glucose		3×10^{-6}
D-Mannose		3×10^{-6}
	Maltose sensor	
Maltose		3×10^{-6}
	Mannitol sensor	
D-Mannitol		7×10^{-6}
	Ribose sensor	
D-Ribose		7×10^{-6}
	Sorbitol sensor	
D-Sorbitol		1×10^{-5}
	Trehalose sensor	
Trehalose		6×10^{-6}
	Aspartate sensor	
L-Aspartate		6×10^{-8}
L-Glutamate		5×10^{-6}
	Serine sensor	
L-Serine		3×10^{-7}
L-Cysteine		4×10^{-6}
L-Alanine		7×10^{-5}
Glycine		3×10^{-5}

Data from (15, 29, 33); see more complete listing of specificities there. O_2 is also attractive to *E. coli* (8, 34), as are certain inorganic ions (unpublished data).

[a] The threshold values are lower in mutants unable to take up or metabolize a chemical. For example, the threshold for D-galactose is 100 times lower in a mutant unable to take up and metabolize this sugar (15).

20 different chemosensors in *E. coli*, but the evidence for each of them is not equally strong. Oxygen taxis (8, 34) has not yet been studied from the point of view of a chemosensor. *S. typhimurium,* insofar as its repertoire has been investigated, shows some of the same responses as *E. coli* (16, 17, 23, 30, 35).

Nature of the Chemosensors

Protein components of some of the chemosensors have been identified by a combination of biochemical and genetic techniques. Each chemosensor, it is believed,

Table 2   Partial list of chemosensors for negative chemotaxis in *Escherichia coli* [a]

Repellent	Threshold Molarity
Fatty acid sensor	
Acetate (C2)	3×10^{-4}
Propionate (C3)	2×10^{-4}
n-Butyrate, isobutyrate (C4)	1×10^{-4}
n-Valerate, isovalerate (C5)	1×10^{-4}
n-Caproate (C6)	1×10^{-4}
n-Heptanoate (C7)	6×10^{-3}
n-Caprylate (C8)	3×10^{-2}
Alcohol sensor	
Methanol (C1)	1×10^{-1}
Ethanol (C2)	1×10^{-3}
n-Propanol (C3)	4×10^{-3}
iso-Propanol (C3)	6×10^{-4}
iso-Butanol (C4)	1×10^{-3}
iso-Amylalcohol (C5)	7×10^{-3}
Hydrophobic amino acid sensor	
L-Leucine	1×1^{-4}
L-Isoleucine	1.5×10^{-4}
L-Valine	2.5×10^{-4}
L-Tryptophan	1×10^{-3}
L-Phenylalanine	3×10^{-3}
L-Glutamine	3×10^{-3}
L-Histidine	5×10^{-3}
Indole sensor	
Indole	1×10^{-6}
Skatole	1×10^{-6}
Aromatic sensors	
Benzoate	1×10^{-4}
Salicylate	1×10^{-4}
H^+ *sensor*	
Low pH	pH 6.5
OH^- *sensor*	
High pH	pH 7.5
Sulfide sensor	
Na_2S	3×10^{-3}
2-Propanethiol	3×10^{-3}
Metallic cation sensor	
$CoSO_4$	2×10^{-4}
$NiSO_4$	2×10^{-5}

[a] Data from (12). See more complete listing of specificities there.

has a protein that recognizes the chemicals detected by that chemosensor—the "chemoreceptor" (or "receptor") or "recognition component" or "binding protein." Wherever this protein has been identified, it has also been shown to function in a transport system for which the attractants of the chemosensor class are substrates. Yet both the transport and chemotaxis systems have other, independent components, and transport is not required for chemotaxis. These relationships are diagrammed in Figure 2.

Transport and chemotaxis are thus very closely related; but not all substances that are transported, or for which there are binding proteins, are attractants or repellents (12, 15).

The first binding protein shown to be required (14, 36) for chemoreception was the galactose binding protein (37). This protein is known to function in the β-methylgalactoside transport system (36, 38, 39), one of several by which D-galactose enters the $E.$ $coli$ cell (40). This is one of the proteins released from the cell envelope of bacteria—presumably from the periplasmic space, the region between the cytoplasmic membrane and the cell wall—by an osmotic shock procedure (41).

The evidence that the galactose binding protein serves as the recognition component for the galactose sensor is the following: (a) Mutants (Type 1 in Figure 2) lacking binding protein activity also lack the corresponding taxis (14), and they are defective in the corresponding transport (38). Following reversion of a point mutation in the structural gene for the binding protein (39), there is recovery of the chemotactic response (M. Goy, unpublished) and of the ability to bind and transport galactose (39). (b) For a series of analogs, the ability of the analog to inhibit taxis towards galactose is directly correlated to its strength of binding to the protein (14). (c) The threshold concentration and the saturating concentration for chemotaxis toward galactose and its analogs are consistent with the values expected from the dissociation constant of the binding protein (42). (d) Osmotically shocked bacteria exhibit a greatly reduced taxis towards galactose, while taxis towards some other attractants is little affected (14). (e) Galactose taxis could be restored by mixing the shocked bacteria with concentrated binding protein (14), but this phenomenon requires further investigation and confirmation.

Binding activities for maltose and ribose were revealed by a survey for binding proteins released by osmotic shock, which might function for chemosensors other

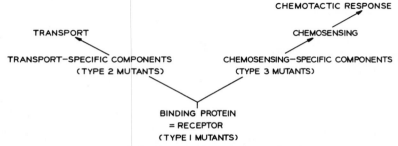

Figure 2   Relation between chemosensing and transport.

than galactose (14). The binding protein for maltose has now been purified (43), and mutants lacking it fail to carry out maltose taxis (31), as well as being defective in the transport of maltose (31, 43). That for ribose has been purified from *S. typhimurium* and serves as ribose chemoreceptor by criteria *a* to *c* above (30, 35).

Mutants of Type 2 (Figure 2) are defective in transport but not necessarily in chemotaxis, even though the two share a common binding protein. Thus, certain components of the transport system, and the process of transport itself, are not required for chemotaxis (at least for certain chemosensors). This has been studied extensively in the case of galactose, where transport is clearly not required (10, 14, 44). Two genes of Type 2 were found for the β-methylgalactoside transport system for galactose (44). Some of the mutations in these genes abolished transport without affecting chemotaxis; other mutations in these genes affected chemotaxis as well (44). Such chemotactic defects may reflect interactions, direct or indirect, that these components normally have with the chemosensing machinery, or some kind of unusual interaction of the mutated component with the binding protein that would hinder its normal function in chemosensing. Two genes whose products are involved in the transport system for maltose (45) can be mutated without affecting taxis toward that sugar (15, 31). Type 2 gene products are most likely located in the cytoplasmic membrane, since they function in transport.

Mutants of Type 3 are defective in chemosensing but not in transport. Presumably they have defects in gene products—"signallers" (44)—which signal information from the binding protein to the rest of the chemotaxis machinery without having a role in the transport mechanism. Such mutants, defective only for galactose taxis or jointly for galactose and ribose taxis, are known (44). The chemistry and location of Type 3 gene products are as yet undescribed.

A mutant in the binding protein gene (by the criterion of complementation) is known that binds and transports galactose normally but fails to carry out galactose taxis, presumably because this binding protein is altered at a site for interaction with the Type 3 gene product (44). Conversely, some mutations mapping in the gene for the galactose (44) or maltose (31) binding proteins affect transport but not binding or chemotaxis. The binding protein thus appears to have three sites—one for binding the ligand, one for interacting with the next transport components, and one for interacting with the next chemotaxis components.

Whereas the binding proteins mentioned above can be removed from the cell envelope by osmotic shock, other binding proteins exist that are tightly bound to the cytoplasmic membrane. Examples of such are the enzymes II of the phosphotransferase system, a phosphoenolpyruvate-dependent mechanism for the transport of certain sugars (46, 47). A number of sugar sensors utilize enzymes II as recognition components: for example, the glucose and mannose sensors are serviced by glucose enzyme II and mannose enzyme II, respectively (29). In these cases, enzyme I and HPr (a phosphate-carrier protein) of the phosphotransferase system (46) are also required for optimum chemotaxis (29). This could mean that phosphorylation and transport of the sugars are required for chemotaxis in these cases; that enzyme I and HPr must be present for interaction of enzyme II with subsequent chemosensing components; or, as seems most likely, that the enzyme II binds sugars more effectively after it has been phosphorylated by phosphoenolpyruvate under the influence of enzyme I and HPr. The products of the phosphotransferase system, the phosphorylated sugars, are not attractants, even when they can be transported by a hexose phosphate transport system (15, 29). This rules out the idea that the phosphotransferase system is required to transport and phosphorylate the sugars so that they will be available to an *internal* chemoreceptor, and indicates instead that interaction of the sugar specifically with the phosphotransferase system somehow leads to chemotaxis (29). Certainly it is not the metabolism of the phosphorylated sugars that brings about chemotaxis: several cases of non-metabolizable phosphorylated sugars are known, yet the corresponding free sugars are attractants (29).

Bacteria detect changes over time in the concentration of attractant or repellent (17, 23, 25), and experiments with whole cells indicate that it is the time rate of change of the binding protein fraction occupied by ligand that the chemotactic machinery appears to detect (25, 42). How this is achieved remains unknown. A conformational change occurs when ligand (galactose) interacts with its purified binding protein (48, 49), and possibly this change is sensed by the next component in the system, but nothing is known about this linkage.

COMMUNICATION OF SENSORY INFORMATION FROM CHEMOSENSORS TO THE FLAGELLA

Somehow the chemosensors must signal to the flagella that a change in chemical concentration has been encountered. The nature of this system of transmitting information to the flagella is entirely unknown, but several mechanisms have been suggested (10, 23, 50).

(*a*) The membrane potential alters, either increasing or decreasing for attractants, with the opposite effect for repellents. The change propagates along the cell membrane to the base of the flagellum. The cause of the change in membrane potential is a change in the rate of influx or efflux of some ion(s) when the concentration of attractant or repellent is changed.

(*b*) The level of a low molecular weight transmitter changes, increasing or decreasing with attractant or repellent. The transmitter diffuses to the base of the flagella. Calculations (4) indicate that diffusion of a substance of low molecular weight is much too slow to account for the practically synchronous reversal of flagella at the two ends of *Spirillum volutans,* which occurs in response to chemotactic stimuli (26, 27). Thus for this organism, at least, a change in membrane potential appears to be the more likely of the two mechanisms.

Although the binding protein of chemosensors is probably distributed all around the cell because it is shared with transport, possibly only those protein molecules at the base of the flagellum serve for chemoreception. In that case, communication between the chemosensors and flagella could be less elaborate, taking place by means of direct protein-protein interaction.

Several tools are available for exploring the transmission system. One is the study of mutants that may be defective in this system; these are the "generally nonchemotactic mutants," strains unable (fully or partly) to respond to any attractant or repellent (10, 12, 51). Some of these mutants swim smoothly, never tumbling (51–53), while others—"tumbling mutants"—tumble most of the time (28, 53–55). Genetic studies (56–58) have revealed that the generally nonchemotactic mutants map in four genes (53, 58). One of these gene products must be located in the flagellum, presumably at the base, since some mutations lead to motile, nonchemotactic cells while other mutations in the same gene lead to absence of flagella (59). The location of the other three gene products is unknown. The function of the four gene products is also unknown, but it has been suggested (7, 53) that they play a role in the generation and control of tumbling at the level of a "twiddle generator" (22).

A second tool comes from the discovery that methionine is required for chemotaxis (60), perhaps at the level of the transmission system. Without methionine, chemotactically wild-type bacteria do not carry out chemotaxis (11, 55, 60, 61) or tumble (55, 60, 62). This is not the case for tumbling mutants (55), unless they are first "aged" in the absence of methionine (62), presumably to remove a store of methionine or a product formed from it. There is evidence that methionine functions via S-adenosylmethionine (55, 61–64), but the mechanism of action of methionine in chemotaxis remains to be discovered.

THE FUNCTIONING OF FLAGELLA TO PRODUCE BACTERIAL MOTION

For reviews of bacterial flagella and how they function, see (5, 65–70).

For many years it was considered that bacterial flagella work either by means of a wave that propagates down the flagellum, as is known to be the case for eucaryotic flagella, or by rotating as rigid or semirigid helices [for a review of the history, see (71)]. Recently it was argued from existing evidence that the latter view is correct (71), and this was firmly established by the following experiment (72). E. coli cells with only one flagellum (obtained by growth on D-glucose, a catabolite repressor of flagella synthesis) (9) were tethered to a glass slide by means of antibody to the filaments. (The antibody of course reacts with the filament and just happens to stick to glass.) Now that the filament is no longer free to rotate, the cell instead rotates, usually counterclockwise, sometimes clockwise (72). By using such tethered cells, the dynamics of the flagellar motor were then characterized (73).

Energy for this rotation comes from the intermediate of oxidative phosphorylation (the proton gradient in the Mitchell hypothesis), not from ATP directly (64, 74), unlike the case of eucaryotic flagella or muscle; this is true for both counterclockwise and clockwise rotation (64).

In S. typhimurium, light having the action spectrum of flavins brings about tumbling, and this might in some way be caused by interruption of the energy flow from electron transport (75).

It is now possible to isolate "intact" flagella from bacteria, i.e. flagella with the basal structure still attached (Figure 3) (76–78). There is the helical filament, a hook, and a rod. In the case of E. coli four rings are mounted on the rod (76), whereas flagella from gram-positive bacteria have only the two inner rings (76, 78). For

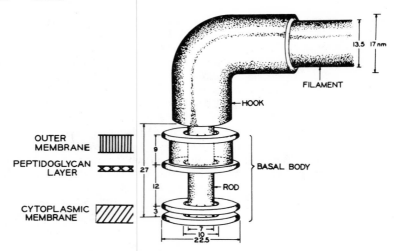

Figure 3   Model of the flagellar base of *E. coli* (76, 77). Dimensions are in nanometers.

E. coli it has been established that the outer ring is attached to the outer membrane and the inner ring to the cytoplasmic membrane (Figure 3) (77). The basal body thus (*a*) anchors the flagellum into the cell envelope; (*b*) provides contact with the cytoplasmic membrane, the place where the energy originates; and (*c*) very likely constitutes the motor (or a part of it) that drives the rotation.

The genetics of synthesis of bacterial flagella is being vigorously pursued in *E. coli* and *S. typhimurium* (28, 79, 80). It is consistent with such a complex structure that at least 20 genes are required for the assembly and function of an *E. coli* flagellum (79) and many of these are homologous to those described in *Salmonella* (28, 80).

THE RESPONSE OF FLAGELLA TO SENSORY INFORMATION

Addition of attractants to *E. coli* cells, tethered to glass by means of antibody to flagella, causes counterclockwise rotation to the cells as viewed from above (52). (Were the flagellum free to rotate, this would correspond to clockwise rotation of the flagellum and swimming toward the observer, as viewed from above. But since a convention of physics demands that the direction of rotation be defined as the object is viewed moving away from the observer, the defined direction of the flagellar rotation is counterclockwise.) On the other hand, addition of repellents causes clockwise rotation of the cells (52). These responses last for a short time, depending on the strength of the stimulus; then the rotation returns to the unstimulated state, mostly counterclockwise (52).

Mutants of *E. coli* that swim smoothly and never tumble always rotate counterclockwise, while mutants that almost always tumble rotate mostly clockwise (52).

From these results and from the prior knowledge that increase of attractant concentration causes smooth swimming (i.e. suppressed tumbling) (22, 23, 25) while addition of repellents causes tumbling (17), it was concluded that smooth swimming results from counterclockwise rotation of flagella and tumbling from clockwise rotation (52).

When there are several flagella originating from various places around the cell, as in *E. coli* or *S. typhimurium,* the flagella function together as a bundle propelling the bacterium from behind (75, 81, 82). Apparently the bundle of flagella survives counterclockwise rotation of the individual flagella to bring about smooth swimming (no tumbling), but comes apart as a result of clockwise rotation of individual flagella to produce tumbling. That tumbling occurs concomitantly with the flagellar bundle flying apart has actually been observed by use of such high intensity light that individual flagella could be seen (75).

Presumably less than a second of clockwise rotation can bring about a tumble, and the long periods of clockwise rotation reported (52) result from the use of unnaturally large repellent stimuli. (The corresponding statement can be made for the large attractant stimuli used.) Some kind of a recovery process is required for return to the unstimulated tumbling frequency. The mechanism of recovery is as yet unknown, but it appears that methionine is somehow involved (55, 62).

The information developed so far is summarized in Figure 4.

The reversal frequency of the flagellum of a *Pseudomonad* is also altered by gradients of attractant or repellent, and reversal of flagellar rotation can explain the backing up of polarly flagellated bacteria (27).

INTEGRATION OF MULTIPLE SENSORY DATA BY BACTERIA

Bacteria are capable of integrating multiple sensory inputs, apparently by algebraically adding the stimuli (17). For example, the response to a decrease in repellent concentration could be overcome by superimposing a decrease in concentration of attractant (17). Whether bacteria will "decide" on attraction or repulsion in a "conflict" situation (a capillary containing both attractant and repellent) depends on the relative effective concentration of the two chemicals, i.e. how far each is present above its threshold concentration (3, 83). The mechanism for summing the opposing signals is unknown.

ROLE OF THE CYTOPLASMIC MEMBRANE

There is increasing evidence that the cytoplasmic membrane plays a crucial role in chemotaxis. (*a*) Some of the binding proteins that serve in chemoreception—the enzymes II of the phosphotransferase system (29)—are firmly bound to the

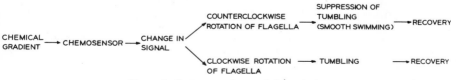

Figure 4   Summary scheme of chemotaxis.

cytoplasmic membrane (46, 47). Binding proteins that can be released by osmotic shock are located in the periplasmic space (41), perhaps loosely in contact with the cytoplasmic membrane. (b) The base of the flagellum has a ring embedded in the cytoplasmic membrane (77). (c) The energy source for motility comes from oxidative phosphorylation (64, 74), a process that along with electron transport is membrane-associated (84). (d) Chemotaxis, but not motility, is unusually highly dependent on temperature, which suggested a requirement for fluidity in the membrane lipids (11). This requirement for a fluid membrane was actually established by measuring the temperature dependence of chemotaxis in an unsaturated fatty acid auxotroph that had various fatty acids incorporated (85). (e) A number of reagents (for example, ether or chloroform) that affect membrane properties inhibit chemotaxis at concentrations that do not inhibit motility (26, 86–88).

This involvement of the cytoplasmic membrane in chemotaxis, especially the location there of the chemoreceptors and flagella, makes the membrane potential hypothesis for transmission of information from chemoreceptors to flagella plausible, but of course by no means proves it.

UNANSWERED QUESTIONS

While the broad outlines of bacterial chemotaxis have perhaps been sketched, the biochemical mechanisms involved remain to be elucidated: How do chemosensors work? By what means do they communicate with the flagella? What is the mechanism that drives the motor for rotating the flagella? What is the mechanism of the gear that shifts the direction of flagellar rotation? How does the cell recover from the stimulus? How are multiple sensory data processed? What are the functions of the cytoplasmic membrane in chemotaxis?

RELATION OF BACTERIAL CHEMOTAXIS TO BEHAVIORAL BIOLOGY AND NEUROBIOLOGY

The inheritance of behavior (see opening quotation) and its underlying biochemical mechanisms are nowhere more amenable to genetic and biochemical investigation than in the bacteria. From the earliest studies of bacterial behavior (2, 3, 89–91) to the present (8, 10, 24, 42, 50, 92, 93) people have hoped that this relatively simple system could tell us something about the mechanisms of behavior of animals and man. Certainly, striking similarities exist between sensory reception in bacteria and in higher organisms (16, 24, 42, 92, 93).

Already in 1889 Alfred Binet wrote in *The Psychic Life of Micro-organisms* (89)

If the existence of psychological phenomena in lower organisms is denied, it will be necessary to assume that these phenomena can be superadded in the course of evolution, in proportion as an organism grows more perfect and complex. Nothing could be more inconsistent with the teachings of general physiology, which shows us that all vital phenomena are previously present in non-differentiated cells.

Literature Cited

1. Engelmann, T. W. 1881. *Pflüger's Arch. Gesamte Physiol. Menschen Tiere* 25: 285–92
2. Pfeffer, W. 1884. *Untersuch. Bot. Inst. Tübingen* 1: 363–482
3. Pfeffer, W. 1888. *Untersuch. Bot. Inst. Tübingen* 2: 582–661
4. Berg, H. C. 1975. *Ann. Rev. Biophys. Bioeng.* 4: 119–36
5. Weibull, C. 1960. *Bacteria* 1: 153–205
6. Ziegler, H. 1962. *Encyclopedia of Plant Physiology*, ed. W. Ruhland, 17-II: 484–532. Berlin: Springer. (In German)
7. Parkinson, J. S. 1975. *Cell.* In press
8. Adler, J. 1966. *Science* 153: 708–16
9. Adler, J., Templeton, B. 1967. *J. Gen. Microbiol.* 46: 175–84
10. Adler, J. 1969. *Science* 166: 1588–97
11. Adler, J. 1973. *J. Gen. Microbiol.* 74: 77–91
12. Tso, W.-W., Adler, J. 1974. *J. Bacteriol.* 118: 560–76
13. Hazelbauer, G. L., Mesibov, R. E., Adler, J. 1969. *Proc. Nat. Acad. Sci. USA* 64: 1300–7
14. Hazelbauer, G. L., Adler, J. 1971. *Nature New Biol.* 230(12): 101–4
15. Adler, J., Hazelbauer, G. L., Dahl, M. M. 1973. *J. Bacteriol.* 115: 824–47
16. Dahlquist, F. W., Lovely, P., Koshland, D. E. Jr. 1972. *Nature New Biol.* 236: 120–23
17. Tsang, N., Macnab, R., Koshland, D. E. Jr. 1973. *Science* 181: 60–63
18. Dryl, S. 1958. *Bull. Acad. Pol. Sci.* 6: 429–32
19. Vaituzis, Z., Doetsch, R. N. 1969. *Appl. Microbiol.* 17: 584–88
20. Berg, H. C. 1971. *Rev. Sci. Instrum.* 42: 868–71
21. Lovely, P., Macnab, R., Dahlquist, F. W., Koshland, D. E. Jr. 1974. *Rev. Sci. Instrum.* 45: 683–86
22. Berg, H. C., Brown, D. A. 1972. *Nature* 239: 500–4
23. Macnab, R. M., Koshland, D. E. Jr. 1972. *Proc. Nat. Acad. Sci. USA* 69: 2509–12
24. Koshland, D. E. Jr. 1974. *FEBS Lett.* 40 (Suppl.): S3–S9
25. Brown, D. A., Berg, H. C. 1974. *Proc. Nat. Acad. Sci. USA* 71: 1388–92
26. Caraway, B. H., Krieg, N. R. 1972. *Can. J. Microbiol.* 18: 1749–59
27. Taylor, B. L., Koshland, D. E. Jr. 1974. *J. Bacteriol.* 119: 640–42
28. Vary, P. S., Stocker, B. A. D. 1973. *Genetics* 73: 229–45
29. Adler, J., Epstein, W. 1974. *Proc. Nat. Acad. Sci. USA.* 71: 2895–99
30. Aksamit, R. R., Koshland, D. E. Jr. 1974. *Biochemistry* 13: 4473–78
31. Hazelbauer, G. L. 1975. *J. Bacteriol.* In press
32. Rothert, W. 1901. *Flora* 88: 371–421
33. Mesibov, R., Adler, J. 1972. *J. Bacteriol.* 112: 315–26
34. Baracchini, O., Sherris, J. C. 1959. *J. Pathol. Bacteriol.* 77: 565–74
35. Aksamit, R., Koshland, D. E. Jr. 1972. *Biochem. Biophys. Res. Commun.* 48: 1348–53
36. Kalckar, H. M. 1971. *Science* 174: 557–65
37. Anraku, Y. 1968. *J. Biol. Chem.* 243: 3116–22
38. Boos, W. 1969. *Eur. J. Biochem.* 10: 66–73
39. Boos, W. 1972. *J. Biol. Chem.* 247: 5414–24
40. Rotman, B., Ganesan, A. K., Guzman, R. 1968. *J. Mol. Biol.* 36: 247–60
41. Heppel, L. A. 1967. *Science* 156: 1451–55
42. Mesibov, R., Ordal, G. W., Adler, J. 1973. *J. Gen. Physiol.* 62: 203–23
43. Kellerman, O., Szmelman, S. 1974. *Eur. J. Biochem.* 47: 139–49
44. Ordal, G. W., Adler, J. 1974. *J. Bacteriol.* 117: 517–26
45. Hofnung, M., Hatfield, D., Schwartz, M. 1974. *J. Bacteriol.* 117: 40–47
46. Roseman, S. 1972. *Metab. Pathways* 6: 41–89
47. Kundig, W., Roseman, S. 1971. *J. Biol. Chem.* 246: 1407–18
48. Boos, W., Gordon, A. S., Hall, R. E., Price, H. D. 1972. *J. Biol. Chem.* 247: 917–24
49. Rotman, B., Ellis, J. H. Jr. 1972. *J. Bacteriol.* 111: 791–96
50. Doetsch, R. N. 1972. *J. Theor. Biol.* 35: 55–66
51. Armstrong, J. B., Adler, J., Dahl, M. M. 1967. *J. Bacteriol.* 93: 390–98
52. Larsen, S. H., Reader, R. W., Kort, E. N., Tso, W.-W., Adler, J. 1974. *Nature* 249: 74–77
53. Parkinson, J. S. 1975. *Nature* 252: 317–19
54. Armstrong, J. B. 1968. Chemotaxis in *Escherichia coli*. PhD thesis, Univ. of Wisconsin, Madison, 43–45; 85–86
55. Aswad, D., Koshland, D. E. Jr. 1974. *J. Bacteriol.* 118: 640–45
56. Armstrong, J. B., Adler, J. 1969. *Genetics* 61: 61–66
57. Armstrong, J. B., Adler, J. 1969. *J. Bacteriol.* 97: 156–61

58. Parkinson, J. S. 1975. *J. Bacteriol.* In press
59. Silverman, M., Simon, M. 1973. *J. Bacteriol.* 116:114–22
60. Adler, J., Dahl, M. M. 1967. *J. Gen. Microbiol.* 46:161–73
61. Armstrong, J. B. 1972. *Can. J. Microbiol.* 18:591–96
62. Springer, M. S., Kort, E. N., Larsen, S. H., Ordal, G. W., Reader, R. W., Tso, W.-W., Adler, J. 1975. *J. Bacteriol.* In press
63. Armstrong, J. B. 1972. *Can. J. Microbiol.* 18:1695–1701
64. Larsen, S. H., Adler, J., Gargus, J. J., Hogg, R. W. 1974. *Proc. Nat. Acad. Sci. USA* 71:1239–43
65. Newton, B. A., Kerridge, D. 1965. *Symp. Soc. Gen. Microbiol.* 15:220–49
66. Doetsch, R. N., Hageage, G. J. 1968. *Biol. Rev.* 43:317–62
67. Iino, T. 1969. *Bacteriol. Rev.* 33:454–75
68. Asakura, S. 1970. *Advan. Biophys.* 1:99–155
69. Smith, R. W., Koffler, H. 1971. *Advan. Microb. Physiol.* 6:219–339
70. Doetsch, R. N. 1971. *Crit. Rev. Microbiol.* 1:73–103
71. Berg, H. C., Anderson, R. A. 1973. *Nature* 245:380–82
72. Silverman, M., Simon, M. 1974. *Nature* 249:73–74
73. Berg, H. C. 1974. *Nature* 249:77–79
74. Thipayathasana, P., Valentine, R. C. 1974. *Biochim. Biophys. Acta* 347:464–68
75. Macnab, R., Koshland, D. E. Jr. 1974. *J. Mol. Biol.* 84:399–406

76. DePamphilis, M. L., Adler, J. 1971. *J. Bacteriol.* 105:384–95
77. DePamphilis, M. L., Adler, J. 1971. *J. Bacteriol.* 105:396–407
78. Dimmitt, K., Simon, M. 1971. *J. Bacteriol.* 105:369–75
79. Hilemn, M., Silverman, M., Simon, M. 1975. *J. Supramol. Struct.* 2: In press
80. Yamaguchi, S., Iino, T., Horiguchi, T., Ohta, K. 1972. *J. Gen. Microbiol.* 70:59–75
81. Pijper, A. 1957. *Ergeb. Mikrobiol. Immunitaetsforsch. Exp. Ther.* 30:37–91
82. Pijper, A.; Nunn, A. J. 1949. *J. Roy. Microsc. Soc.* 69:138–42
83. Adler, J., Tso, W.-W. 1974. *Science* 184:1292–94
84. Harold, F. M. 1972. *Bacteriol. Rev.* 36:172–230
85. Lofgren, K. W., Fox, C. F. 1974. *J. Bacteriol.* 118:1181–82
86. Rothert, W. 1904. *Jahrb. Wiss. Bot.* 39:1–70
87. Chet, I., Fogel, S., Mitchell, R. 1971. *J. Bacteriol.* 106:863–67
88. Faust, M. A., Doetsch, R. N. 1971. *Can. J. Microbiol.* 17:191–96
89. Binet, A. 1889. *The Psychic Life of Micro-organisms,* iv–v. Chicago: Open Court.
90. Verworn, M. 1889. *Psycho-Physiologische Protisten-Studien.* Jena: Fischer
91. Jennings, H. S. 1906. *Behavior of the Lower Organisms.* Republished by Indiana Univ. Press, Bloomington, 1962
92. Clayton, R. K. 1953. *Arch. Mikrobiol.* 19:141–65
93. Clayton, R. K. 1959. See Ref. 6, 17-I:371–87

INHERITED DISORDERS OF LYSOSOMAL METABOLISM

Elizabeth F. Neufeld, Timple W. Lim, and Larry J. Shapiro

National Institute of Arthritis, Metabolism, and Digestive Diseases,
National Institutes of Health, Bethesda, Maryland 20014

CONTENTS

INTRODUCTION

Inherited lysosomal diseases were discussed by Raivio & Seegmiller in this series in 1972 (1) as part of the broader subject of genetic diseases of metabolism. In the intervening years, the lysosomal subgroup has been the focus of so much intensive and fruitful research that, regretfully, we have had to limit the discussion to a few selected topics within the scope of this already restricted subject matter.

What are heritable lysosomal disorders? There is general agreement that the designation is appropriate when there is a documented deficiency of a lysosomal enzyme or when such deficiency can be reasonably suspected because of lysosomal storage of some undegraded metabolite (Table 1); eventually the group will probably include defects of lysosomal membrane function. This definition of lysosomal disorders does not invalidate the criteria established by Hers (2) at a time when only one lysosomal enzyme deficiency was established.

Table 1 Summary of lysosomal disorders

Disorder	Enzyme Deficiency	Metabolite Primarily Affected
Mucopolysaccharidoses		
Hurler and Scheie syndromes	α-L-Iduronidase	Dermatan sulfate, heparan sulfate
Hunter syndrome	Iduronate sulfatase	Dermatan sulfate, heparan sulfate
Sanfilippo syndrome		
A subtype	Heparan N-sulfatase	Heparan sulfate
B subtype	N-Acetyl-α-glucosaminidase	Heparan sulfate
Maroteaux-Lamy syndrome	N-Acetylgalactosamine sulfatase (arylsulfatase B)	Dermatan sulfate
β-Glucuronidase deficiency	β-glucuronidase	Dermatan sulfate, heparan sulfate
Morquio syndrome	Uncertain	Keratan sulfate
Sphingolipidoses		
GM$_1$ gangliosidosis	β-Galactosidase	GM$_1$ ganglioside, fragments from glycoproteins
Krabbe's disease	β-Galactosidase	Galactosylceramide
Lactosylceramidosis	β-Galactosidase	Lactosylceramide
Tay-Sachs disease	Hexosaminidase A	GM$_2$ ganglioside
Sandhoff's disease	Hexosaminidases A and B	GM$_2$ ganglioside, globoside
Gaucher's disease	β-Glucosidase	Glucosylceramide
Fabry's disease	α-Galactosidase	Trihexosylceramide
Metachromatic leukodystrophy	Arylsulfatase A	Sulfatide
Niemann-Pick disease	Sphingomyelinase	Sphingomyelin
Farber's disease	Ceramidase	Ceramide

Disorders of Glycoprotein Metabolism

Disease	Enzyme Defect	Stored Material
Fucosidosis	α-L-Fucosidase	Fragments from glycoproteins, glycolipids
Mannosidosis	α-Mannosidase	Fragments from glycoproteins
Aspartylglycosaminuria	Amidase	Aspartyl-2-deoxy-2-acetamido glucosylamine

Other Disorders with Single Enzyme Defect

Disease	Enzyme Defect	Stored Material
Pompe's disease	α-Glucosidase	Glycogen
Wolman's disease	Acid lipase	Cholesterol esters, triglyceride
Acid phosphatase deficiency	Acid phosphatase	Phosphate esters

Multiple Enzyme Deficiencies

Disease	Enzyme Defect	Stored Material
Multiple sulfatase deficiency	Sulfatases (arylsulfatase A, B, C: steroid sulfatases; iduronate sulfatase; heparan N-sulfatase)	Sulfatide, steroid sulfate, mucopolysaccharide
I cell disease and pseudo-Hurler polydystrophy	Almost all lysosomal enzymes deficient in cultured fibroblasts; present extracellularly	Mucopolysaccharide and glycolipids

Disorders of Unknown Etiology

Disease	Enzyme Defect	Stored Material
Cystinosis	Accumulation of cystine in lysosomes	Cystine
Mucolipidoses I, IV	Ultrastructural evidence of lysosomal storage	Unknown

Any classification involves some arbitrary choices. For example, in Table 1 we have avoided the category of "mucolipidosis" (3) because it combines deficiencies of single enzymes with broad substrate specificity with the fundamentally different multiple enzyme deficiencies. Disorders of the phagocytic or bactericidal system of leucocytes and of ceroid lipofuscin metabolism have been considered outside the scope of the review, even though malfunction of lysosomes may be directly or indirectly involved.

The survey of the literature ended in July 1974. Extensive clinical and biochemical background is provided in recent books (4–6).

MUCOPOLYSACCHARIDOSES

Enzyme Deficiencies

The enzymatic defects of most mucopolysaccharide storage disorders have been elucidated within the past two years. With the exception of the Morquio syndrome, which is a disturbance of keratan sulfate metabolism, the classified mucopolysaccharidoses involve the accumulation and excretion of dermatan sulfate and/or heparan sulfate (7). As shown schematically in Figure 1, these polymers are composed of repeating units of uronic acid and sulfated hexosamine joined by a variety of linkages, each requiring a specialized lysosomal exoenzyme for cleavage.

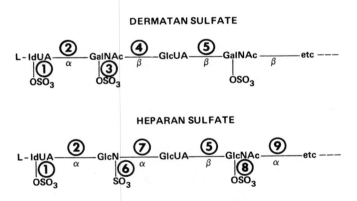

Figure 1 Schematic representation of the catabolism of dermatan sulfate and of heparan sulfate. Sugars are of the D configuration unless designated L. Numbers represent hydrolytic enzymes and associated deficiency diseases discussed in the text. All the linkages known to be present in the two polymers are illustrated within the tetrasaccharides, but this should not be construed to mean that the tetrasaccharides are regular repeating units. Dermatan sulfate is composed of alternating uronic acid and sulfated β linked N-acetylgalactosamine residues; heparan sulfate, of alternating uronic acid and either N-sulfated or N-acetylated α linked glucosamine residues (8). However, there may be considerable variation in the ratio of iduronic to glucuronic acid, of N-sulfated to N-acetylated glucosamine, and of sulfated to unsulfated iduronic acid residues, as well as in their position along the chain. For a discussion of microheterogeneity within mucopolysaccharides, see (9).

When the stepwise degradation is interrupted by the malfunction or deficiency of an enzyme, the undigested chain accumulates intralysosomally.

In some tissues, endoglycosidases (hyaluronidase for dermatan sulfate and either hyaluronidase or some unknown enzyme for heparan sulfate) are presumably able to bypass the block and generate fragments, some of which may be small enough for release from the tissue and for urinary excretion. It is not known whether this auxiliary pathway operates in normal tissues or only comes into play when there is abnormal accumulation of mucopolysaccharide. In any case, no hyaluronidase pathway has been demonstrated in cultured fibroblasts (8).

The enzyme reactions and associated disorders, listed by the number assigned in Figure 1, are as follows: 1. iduronate sulfatase, deficient in the Hunter syndrome (10–12); 2. α-L-iduronidase, deficient in the Hurler and Scheie syndromes (13, 14); 3. N-acetylgalactosamine sulfatase, an enzyme activity which is deficient in the Maroteaux-Lamy syndrome (14a); 4. N-acetyl-β-galactosaminidase, an activity that may be related to the β-hexosaminidases hydrolyzing sphingolipids (15, 16); 5. β-glucuronidase, the absence of which results in a recently described mucopolysaccharidosis (17); 6. heparan N-sulfatase, deficient in the A subtype of the Sanfilippo syndrome (18–20); 7. α-glucosaminidase, a presumptive activity; 8. N-acetylglucosamine sulfatase, likewise a presumptive activity; and 9. N-acetyl-α-glucosaminidase, the enzyme deficient in the B subtype of the Sanfilippo syndrome (21, 22). Reactions 7 and 8 are not associated with any known enzyme or disease, but are deduced from the chemistry of the polymer. Since both involve sulfated glucosamine, a structure peculiar to heparan sulfate, further subtypes of heparan sulfate storage disease (Sanfilippo syndrome) might be expected.

Patients with the Maroteaux-Lamy syndrome have been shown, on the one hand, to have a generalized deficiency of arylsulfatase B (23–25), and on the other hand, to have a deficiency of N-acetylgalactosamine sulfatase in their fibroblasts (reaction 3, 14a). Possibly these sulfatase activities are but two aspects of the same enzyme, observed with the use of an artificial and a natural substrate, respectively.

Although disorders of chondroitin sulfate metabolism have been reported (26), they seem much rarer than disorders of dermatan sulfate and heparan sulfate metabolism. Neither the normal pathway of chondroitin sulfate degradation nor the relevant enzyme deficiencies are known. Some defect of chondroitin sulfate metabolism occurs in β-glucuronidase deficiency (17) and in Sandhoff's disease (16).

Although a defect in keratan sulfate degradation may be the basis of the Morquio syndrome, the only deficiency reported thus far is, surprisingly, that of a sulfatase for N-acetyl β-galactosamine 6-sulfate in oligosaccharides derived from chondroitin sulfate (27).

In the conceptual framework of degradation by exoenzymes, a block in the degradative pathway should lead to the accumulation of macromolecules with a nonreducing terminal residue that would have been the substrate for the missing enzyme. This prediction has been experimentally verified for the mucopolysaccharides that accumulate in fibroblasts derived from patients with β-glucuronidase deficiency (28), the Hunter syndrome (11), or the Maroteaux-Lamy syndrome (14a).

These stored polymers have terminal β-linked glucuronic acid, sulfated iduronic acid, and sulfated N-acetylgalactosamine residues, respectively.

Corrective Factors

Identification of the enzyme deficiencies in the mucopolysaccharidoses proceeded by an unorthodox route, necessitated by ignorance of the fine structure and enzymology of dermatan sulfate and heparan sulfate at the time the studies were undertaken. By using $^{35}SO_4^{2-}$ as a convenient precursor (29), it was shown that radioactive mucopolysaccharide accumulated excessively in fibroblasts from patients with any mucopolysaccharidosis other than the Morquio syndrome. The abnormal accumulation could be reduced, or "corrected," by adding to the medium secretions of cells or concentrates of urine, provided the donor of cells or urine was not himself affected by the genetic condition to be corrected (30). The active substances effecting normalization of mucopolysaccharide catabolism, designated "corrective factors," were extensively purified from normal urine (18, 22, 31, 32) and eventually shown to be the missing enzyme in each disorder. Thus, the corrective factor for Hurler and Scheie cells is the enzyme α-L-iduronidase (13); the Hunter corrective factor, iduronate sulfatase (10); the Sanfilippo A and B corrective factors, heparan N-sulfatase (18) and N-acetyl-α-glucosaminidase (22, 33), respectively. β-Glucuronidase of human or bovine origin serves as a corrective factor for cells deficient in that enzyme (34–36). Identity of the Maroteaux-Lamy factor (37) with arylsulfatase B has been tentatively suggested (25) but must await the use of more purified preparations for definitive proof.

Correction is accompanied by uptake of enzyme from the medium into fibroblasts, presumably into lysosomes, so that the catabolic pathway is reconstituted (13, 33–36, 38). Only a small fraction of the normal complement of enzyme needs to be incorporated in order to give essentially normal mucopolysaccharide catabolism (13, 33).

The uptake of lysosomal enzymes, originally thought to take place by pinocytic imbibition of medium with all its macromolecular contents (9), turned out to be a highly selective process requiring a recognition marker on the protein (39) and, by implication, a receptor on the fibroblast surface. Given a constant catalytic activity, ability of an enzyme to correct is correlated with effectiveness of uptake (35, 36, 40). Those enzymes that were originally purified as corrective factors were later found to be taken up efficiently and selectively (13, 38); that is not surprising, since the corrective factor assay would have selected for a combination of high uptake as well as high catalytic activity.

Correlation with Clinical Phenotype

Classification of mucopolysaccharidoses by factor, and subsequently by enzyme deficiency, has revealed a complicated relationship between biochemical defect and clinical manifestations.

There is a wide clinical spectrum within the framework of one enzyme deficiency. Although the greatest contrast occurs between the Hurler (very severe) and Scheie (mild) syndromes, both due to deficiency of α-L-iduronidase, there also exist mild and severe forms of the Hunter and Maroteaux-Lamy syndromes (7) as well as of β-glucuronidase deficiency (41). By analogy with the well-studied variants of glucose 6-phosphate dehydrogenase (42), residual activity of the affected enzyme under physiological conditions is usually invoked as an explanation for the mild forms of mucopolysaccharidoses, but has not yet been demonstrated in vitro. The likelihood that some mutational alterations of one enzyme would be allelic implies the existence of "genetic compounds"—individuals with a different mutation of the same gene on the maternal and paternal chromosome, and a resulting intermediate phenotype (43).

Since dermatan sulfate and heparan sulfate have a different tissue distribution, a different sort of clinical variability must arise from the existence of some enzymatic deficiencies that are specific to one or the other polymer, and of other deficiencies common to both (see Figure 1). Microheterogeneity of dermatan sulfate and heparan sulfate may account for yet a third kind of clinical variability; for instance, the consistent differences between α-L-iduronidase and iduronate sulfatase deficiency diseases suggest that in some tissues, L-iduronic acid residues close to the nonreducing end may not be sulfated.

SPHINGOLIPIDOSES

With the sole exception of ceramidase deficiency (44), the enzyme defects of the sphingolipidoses have been known for some years and have been extensively reviewed (1, 4–6, 45, 46). Although the discovery of each enzyme defect listed in Table 1 followed its own course, the usual sequence was a logical progression of 1. chemical identification of the stored substance; 2. discovery of an enzyme that would hydrolyze this substance in the normal situation; and 3. finding of the enzymatic deficiency in patients and of an intermediate level in heterozygotes. Recent studies have been aimed at understanding the mutational event that produced the deficiency and have of necessity required the isolation and intensive study of the normal enzymes as a first step.

β-Galactosidase Deficiencies

Three distinct inherited disorders are presently recognized as caused by the deficiency of some particular β-galactosidase activity.

GM_1 gangliosidosis results from the deficiency of a β-galactosidase, which is the most readily detected of the three, as it is the only one that reacts with synthetic substrates such as p-nitrophenyl or methylumbelliferyl β-galactoside (47, 48). Its major natural substrate is GM_1[1] ganglioside, but it can also cleave the β-galactosyl residues of some glycoproteins (49). As a result of this relative lack of

[1] Abbreviations: GM_1, Gal-($\beta1 \rightarrow 3$)-GalNAc-($\beta1 \rightarrow 4$)-[NANA-($\alpha2 \rightarrow 3$)]-Gal-($\beta1 \rightarrow 4$)-Glc-($\beta1$–1')-ceramide; GM_2, GalNAc-($\beta1 \rightarrow 4$)-[NANA-($\alpha2 \rightarrow 3$)]-Gal-($\beta1 \rightarrow 4$)-Glc-($\beta1$–1')-ceramide; globoside, GalNAc-($\beta1 \rightarrow 3$)-Gal-($\alpha1 \rightarrow 4$)-Gal-($\beta1 \rightarrow 4$)-Glc-($\beta1$–1')-ceramide; ceramide trihexoside, Gal-($\alpha1 \rightarrow 4$)-Gal-($\beta1 \rightarrow 4$)-Glc-($\beta1$–1')-ceramide; lactosylceramide, Gal-($\beta1 \rightarrow 4$)-Glc-($\beta1$–1')-ceramide; digalactosylceramide, Gal-($\alpha1 \rightarrow 4$)-Gal-($\beta1$–1')-ceramide; sulfatide, Gal-3-sulfate ($\beta1$–1')-ceramide; ceramide, N-acylsphingosine.

specificity, patients may show not only massive accumulation of GM_1 in the brain, but also storage of material resembling partially degraded glycoprotein (50) or desulfated keratan sulfate (51) in visceral organs. There is clinical heterogeneity, in that at least two types of GM_1 gangliosidosis can be distinguished by age of onset, clinical course, and extent of visceral involvement (46)—though without clear correlation with the degree of residual β-galactosidase activity (52–54).

The major form of GM_1 β-galactosidase (A form) has been purified to apparent homogeneity from normal human liver (49). The B form of the enzyme may be structurally related (a multimer?) to the A form, because it is precipitable by anti-A antibody, and because both forms are concurrently absent from tissues of GM_1 gangliosidosis patients.

Krabbe's disease is characterized by the absence of galactosylceramide β-galactosidase. It is unique among lysosomal disorders in that there is no massive accumulation of the affected metabolite, galactosylceramide, probably because of the nearly total disappearance of oligodendroglial (myelin-forming) cells which normally synthesize this glycolipid (46). What storage there is occurs in globoid cells; these are presumably macrophages that invade the white matter in response to the extensive demyelination.

The loss of galactosylceramide β-galactosidase activity in Krabbe's disease is accompanied by loss of activity towards monogalactosyldiglyceride (55) and β-galactosylsphingosine (56), suggesting that the three lipids are degraded by the same enzyme. It may be galactosylsphingosine, a cytotoxic compound, that is responsible for the loss of oligodendroglial cells.

Although activity towards GM_1 ganglioside is unaffected in Krabbe's disease, activity towards synthetic substrates is quantitatively unchanged but qualitatively altered, as demonstrated by isoelectric focusing (57). This might be an indication of some previously unsuspected relationship between β-galactosidase activities for galactosylceramide and for methylumbelliferyl galactoside, although indirect effects cannot be ruled out without purification of the enzyme.

Lactosylceramidosis is considered a separate disease entity, characterized by accumulation of lactosylceramide and deficiency of lactosylceramide β-galactosidase in cells and tissues of the one known patient (58). Presence of that activity in tissues of GM_1 gangliosidosis (54) and of most Krabbe's patients (57) has been considered strong evidence for the existence of a distinct β-galactosidase for lactosylceramide. The recent report that this activity is missing in tissues and cells of two Krabbe's patients (59) may shed some doubt on the earlier view. This is probably another question that will be resolved only by study of purified enzymes.

As a reminder of the possibility of secondary enzyme deficiencies in lysosomal disorders, it should be noted that until a few years ago the mucopolysaccharidoses were considered by some investigators to be β-galactosidase deficiency diseases because of a striking deficit of β-galactosidase activity in patients' tissues (60). The apparent loss of β-galactosidase in the mucopolysaccharidoses has turned out, in fact, to be an inhibition or a change in electrophoretic pattern due to binding of stored mucopolysaccharide to the enzyme (61).

There are well-characterized animal models for β-galactosidase deficiency diseases: canine Krabbe's disease (62) and feline (63, 64) as well as bovine (65) GM_1 gangliosidoses, probably similar to the juvenile form of the human disease.

β-Hexosaminidase Deficiencies

There are two major hereditary disorders to be considered in this group. In Sandhoff-Jatzkewitz disease (also called Sandhoff's disease or Tay-Sachs O variant) total inability to degrade lipid hexosaminides leads to storage of GM_2 ganglioside, asialo GM_2 ganglioside, and globoside. Both hexosaminidase A and hexosaminidase B are absent from all tissues (the two forms of the enzyme are defined in terms of hydrolysis of synthetic substrates, such as p-nitrophenyl or methylumbelliferyl N-acetyl β-glucosaminide and β-galactosaminide). In classical Tay-Sachs disease (Tay-Sachs B variant) the enzyme deficiency is not as complete; inability to degrade GM_2 ganglioside and absence of hexosaminidase A occur with retention of globoside hydrolysis, and a normal or even elevated level of hexosaminidase B. We shall examine the hypotheses that have been proposed to explain these biochemical alterations, keeping in mind that each disease must be explained as the effect of one mutant gene and the resultant malfunction of the polypeptide encoded in that gene.

Studies of the enzyme derived from normal tissues reveal that the A and B forms are closely related. They have the same molecular weight (66–68) and number of subunits (67, 68), similar though not identical amino acid compositions (67), and the same kinetic constants for the hydrolysis of synthetic substrates (66–70). In addition to antigenic determinants common to both forms, hexosaminidase A has determinants not shared by B (71–73). A small crossreacting protein, perhaps a subunit, occurs in liver (74).

The most obvious differences between the two hexosaminidases are in charge and heat stability, and these are used in most procedures designed to differentiate between them (70). Hexosaminidase A can be converted to a form similar to hexosaminidase B by a variety of procedures, such as storage (66), controlled heating (68), or treatment with a mercurial (75). The finding (70, 76) of a conversion of hexosaminidase A into hexosaminidase B by incubation with neuraminidase should be interpreted in light of recent reports that it is not the removal of sialic acid that effects the conversion but the merthiolate used as preservative in certain commercial lots of neuraminidase (75, 77). The asialo derivative of hexosaminidase A is not hexosaminidase B (68).

Somatic cell genetics has supplied interesting, if not always clear-cut information. Heterokaryons resulting from the fusion of Sandhoff and Tay-Sachs fibroblasts produce hexosaminidase A (78–80). Study of human-hamster hybrids suggests that hexosaminidase A and hexosaminidase B are synthetized independently of each other (81). On the other hand, study of human-mouse hybrids, in which the human enzymes were identified more rigorously by their antigenicity as well as by their electrophoretic positions, showed that whereas hexosaminidase B was frequently found in the absence of hexosaminidase A, the reverse combination was not found once in 60 clones; this result implies that the biosynthesis of hexosaminidase A is contingent on that of hexosaminidase B (82). Though apparently inconsistent with each other, the hybridization studies indicate that at least two genes are required for the production of hexosaminidases A and B; they probably invalidate one-gene hypotheses such as the recent proposal that the two forms are conformers of each other (68).

A two-gene theory (which was in fact the first suggested model) proposed that hexosaminidase B, active towards globoside and synthetic substrates but not towards GM_2 ganglioside, would be converted to hexosaminidase A, which has a broader substrate specificity (70, 83). The enzyme catalyzing the conversion would be absent in Tay-Sachs disease, whereas the very formation of hexosaminidase B would be defective in Sandhoff's disease. Another two-gene hypothesis would have hexosaminidases A and B composed of different subunits, $(\alpha\beta)n$ and $(\beta\beta)n$, respectively (84); a three-gene model would assign a unique subunit to each hexosaminidase, as well as one subunit common to both forms, $(\alpha\beta)n$ and $(\beta\gamma)n$ (73). In the subunit models, the Sandhoff mutation would affect the common subunit, and the Tay-Sachs mutation, the subunit specific to hexosaminidase A. All three models are consistent with immunological data and with analyses that show the two hexosaminidases to be similar proteins; the two-gene or three-gene model is favored by the simplest interpretation of the mouse-human or the hamster-human hybridization studies, respectively.

However, the above models are incompatible with the surprising recent finding that hexosaminidase B from normal tissue is as active as hexosaminidase A in catalyzing the hydrolysis of GM_2 ganglioside (68). The mutation in Tay-Sachs disease is therefore pleiotropic, resulting in the absence of hexosaminidase A and the concomitant formation of a defective hexosaminidase B. To accommodate this new evidence, it becomes necessary to postulate that the Tay-Sachs gene affects hexosaminidase B or a subunit thereof in such a way as to alter catalytic activity as well as to prevent the eventual formation of hexosaminidase A. The Sandhoff mutation would also affect hexosaminidase B or a subunit thereof, but in a different way: all catalytic activity would be lost, but formation of antigenically reactive proteins, A and/or B, could occur (73). In this view, the Tay-Sachs and Sandhoff mutations might be allelic (i.e. affect the same subunit) and the formation of functional hexosaminidase A observed in Tay-Sachs: Sandhoff heterokaryons would occur by assembly of nonidentically defective subunits.

There is little to be gained by further elaboration of this hypothesis, which clearly calls for experimental evidence such as peptide mapping of normal and mutant hexosaminidase subunits. It is sufficiently flexible to easily accommodate three other mutant conditions: juvenile GM_2 gangliosidosis, a variant of Tay-Sachs disease with later onset and with partial hexosaminidase A deficiency (85); the AB variant of Tay-Sachs disease, in which both hexosaminidase A and B are active towards synthetic substrates presumably inactive towards GM_2 ganglioside (83); and a benign deficiency of hexosaminidase A (86). The hypothesis does not take into account the recently discovered glycoprotein "activator" of hexosaminidase, which is discussed below.

In addition to the A and B forms, there exist several minor forms of β-hexosaminidase: hexosaminidase C, which has a higher pH optimum (87);

hexosaminidases I_1, I_2, and P (the last two perhaps identical) which are found in serum, particularly of pregnant women (88, 89); hexosaminidase M, found in male urine (90); and yet other forms seen in fibroblasts from patients with I cell disease (91). These minor forms could represent intermediates in the synthesis or degradation of hexosaminidases A and B, or be unrelated. Although their biological function is unknown, the minor forms should not be ignored, if only because they may interfere with analytical determinations of hexosaminidases A and B and with interpretation of the results.

Other Glycosphingolipidoses

The relationship of the various forms of α-galactosidase to each other and to the deficiency state seems at least as complicated as that just discussed for β-hexosaminidases, and a coherent model has yet to emerge. The accumulation of lipids with a terminal α-galactosyl residue in Fabry's disease [i.e. ceramide trihexoside, digalactosylceramide, and a complex lipid with blood group B specificity (92)] is associated with loss of activity towards the di- and trihexosylceramides, but with residual activity towards methylumbelliferyl and p-nitrophenyl α-galactosides. Of the major electrophoretic forms of α-galactosidase, A and B (determined by hydrolysis of synthetic substrates), only the A form is deficient in Fabry's disease (93, 94). A third form may, in fact, be an N-acetyl α-galactosaminidase (95). Superficially, the A and B α-galactosidases appear related to Fabry's disease in the same way as hexosaminidase A and B are to Tay-Sachs disease. However, the analogy may be spurious, because in contrast to the β-hexosaminidases, the α-galactosidases are not antigenically related (96). A family of α-galactosidases that catalyzes only the hydrolysis of natural substrates has been isolated from plasma (97–99); on the other hand, highly purified placental trihexosylceramide α-galactosidase retains activity towards synthetic substrates (J. F. Tallman and R. O. Brady, personal communication). In the two Fabry cases examined so far, there was no protein crossreactive with α-galactosidase A (100, 100a).

The interrelationship of enzyme activities is clearer in the case of sphingolipid sulfatases. Cerebroside sulfatase and arylsulfatase A activities, measured by hydrolysis of cerebroside sulfate and of nitrocatechol sulfate, respectively, are two properties of the same enzyme protein (101, 102). Accumulation of sulfatides in metachromatic leukodystrophy is associated with deficiency of that enzyme, though with the presence, in two cases examined, of crossreactive protein (103). For technical convenience, it is the arylsulfatase A activity that is generally measured for diagnostic purposes. Arylsulfatase B probably functions in mucopolysaccharide catabolism (see above); although they share the ability to hydrolyze nitrocatechol sulfate, the two arylsulfatases are neither genetically nor antigenically related.

As most lysosomal disorders, metachromatic leukodystrophy occurs in clinical forms that differ in time of onset and rate of progression. Although there is no clear relationship between clinical phenotype and enzyme deficit (104), biochemical distinction has been made with intact fibroblasts in culture (105). These cells ingest ^{35}S-sulfatide from the medium; the rate of degration thereof is reduced in proportion to the severity of the disease.

The β-glucosidase relevant to Gaucher's disease is most active towards the natural substrate, β-glucosylceramide, but also catalyzes the hydrolysis of synthetic (p-nitrophenyl and methylumbelliferyl) β-glucosides and of β-glucosylsphingosine (106). In addition, the enzyme shows transglucosylating activity (107). Gaucher's disease is the only one of the conditions discussed in this chapter in which clinical severity has been successfully correlated with the degree of enzyme deficiency (108).

Activators of Glycosphingolipid Hydrolases

The enzymatic hydrolysis of sphingolipids must generally be performed in the presence of a detergent, usually a bile salt such as taurocholate. Since there is no such requirement for the hydrolysis of synthetic substrates, the detergent is thought to act, at least in part, by forming mixed micelles with the lipid substrate (e.g. 102, 109). Another type of activator has been described for certain hydrolases. It is in each case a naturally occurring, heat-stable material that may or may not eliminate the requirement for a detergent.

The first activator to be reported is a substance that enhances the activity of cerebroside sulfatase (101). Originally, it was considered to be "complementary" to the heat-labile protein, the two combining to form an active enzyme. It is now interpreted to be an activator and has been found in the liver of metachromatic leukodystrophy patients as well as of normal individuals (110). Its chemical nature remains to be elucidated.

The next report concerned the marked stimulation of acid β-glucosidase (measured with synthetic substrate) by a substance from human spleen (111), particularly that of Gaucher's patients. This same activator, a glycoprotein of molecular weight about 20,000, also stimulates the hydrolysis of glucosylcerebroside (112), but this may be primarily due to a broadening of the pH curve, so that activation is only observed below pH 6 (106). Activation depends on the presence of phospholipids (111a).

Most recently, a glycoprotein of molecular weight 25,000 has been purified from human liver and found to stimulate the hydrolysis of GM_2 ganglioside by hexosaminidase A, of ceramide trihexoside by α-galactosidases A and B, and of GM_1 ganglioside by β-galactosidase (113, 114).

The relationship of these various activators to each other, their physiological role, and the possibility of activator deficiency diseases constitute most interesting problems. As a first step in resolving them, it would be useful to determine whether the activators are in fact located within lysosomes, otherwise an activating function in vivo would seem dubious.

MULTIPLE ENZYME DEFICIENCY DISEASES

Included in the category of multiple enzyme deficiencies are disorders that are inherited as single gene defects, yet manifest themselves by the simultaneous deficiency of several enzymes. They are particularly instructive since they point to a common feature of these enzymes, which might otherwise not have been suspected.

Multiple Sulfatase Deficiency

Accumulation of sulfatide, mucopolysaccharide, and sulfated steroids is accompanied by a deficit of arylsulfatases A, B, and C, steroid sulfatases (115), as well as heparan N-sulfatase and iduronate sulfatase (116). The disease, inherited in autosomal recessive fashion, is clinically similar to metachromatic leuko-dystrophy, probably because the neurological consequences of arylsulfatase A (cerebroside sulfatase) deficiency are the earliest and most prominent clinical problems. There is little experimental data to explain the simultaneous loss of all sulfatase activities. It is unlikely that all these enzymes are coded by neighboring genes which might be regulated or deleted as a unit, since one of the enzymes, iduronate sulfatase, is subject to X-linked inheritance (10). A defect specific to an interaction of sulfatase with lysosomes is also ruled out, since some of the deficient sulfatases (arylsulfatase C and steroid sulfatases) are microsomal. A protein crossreacting with arylsulfatase A has been found in liver of a multiple sulfatase deficient patient (117). It may be postulated that the arylsulfatase A requires some component (subunit or modifier) for catalytic activity and, by extension, that the other sulfatases require the same component.

Disorders of Lysosomal Enzyme Localization: Mucolipidoses II and III

Although these conditions, inherited in an autosomal recessive manner, are clinically different [I cell disease or mucolipidosis II is a very severe disorder reminiscent of the Hurler syndrome, whereas pseudo-Hurler polydystrophy or mucolipidosis III is a relatively mild condition (7)], they are similar enough biochemically to be considered together.

Several apparently unrelated lysosomal enzymes are markedly deficient in cultured fibroblasts: α-L-fucosidase, β-galactosidase, arylsulfatase A, β-glucuronidase, α-galactosidase and α-mannosidase (118–122), α-L-iduronidase (13), and iduronate sulfatase (10). The degree of deficiency depends on the enzyme, the patient, and the culture conditions; it may be as complete as for the one-enzyme deficiency diseases or leave up to half the normal activity. Only two enzymes are found at normal levels: acid phosphatase and β-glucosidase (119, 121).

The intracellular deficiencies are not due to lack of synthesis, since the enzymes can be found in the culture medium (121–123). A very marked elevation of several of these enzymes is observed in body fluids (120, 122, 124). Thus, in contrast to the diseases discussed previously, I cell disease and pseudo-Hurler polydystrophy are not due to enzyme deficiency but rather to inappropriate enzyme localization: extracellular rather than intralysosomal.

The intracellular enzyme deficiencies are also unsystemic; for example, enzyme activities in liver and leucocytes are normal (122, 125). Electron microscopy shows lysosomal storage primarily in cells of connective tissue and in a very few other types of cells (125, 126). In spite of the nearly complete deficit of arylsulfatase A and β-galactosidase in cultured fibroblasts, I cell disease patients do not show the characteristic neurological signs of metachromatic leukodystrophy or of GM_1

gangliosidosis, presumably because the relevant cells in the central nervous system are adequately supplied with these enzymes.

The disorders must involve interaction of hydrolytic enzymes with lysosomes of the affected cells, and the question as to which is defective has been resolved by crossfeeding experiments (127). Since fibroblasts from I cell disease or pseudo-Hurler polydystrophy patients take in normal enzyme (presumably by adsorptive pinocytosis) and retain it in normal fashion, whereas enzymes elaborated by these mutant cells are not taken in by other fibroblasts, it must be that the enzymes are defective. Though possessing normal catalytic activity, the enzymes produced by these patients' fibroblasts are believed to lack a recognition marker for adsorptive pinocytosis and for entry into lysosomes. The marker would be common to all the enzymes involved.

What is the nature of the recognition marker? The suggestion that it might be a carbohydrate chain or residue on the enzymes is backed by the following arguments: 1. lysosomal enzymes are glycoproteins (48, 67, 76, 128–134); 2. carbohydrate residues are known to function as recognition markers to bind circulating glycoproteins to hepatocyte membranes and thereby ensure their entry into hepatocyte lysosomes (135, 136); 3. a high uptake form of β-hexosaminidase can be converted to a low uptake form by periodate oxidation under conditions suitable for destruction of carbohydrate residues (39); and 4. high and low uptake forms of α-L-iduronidase show a difference in lectin binding (40). However, definitive identification of the recognition marker is not yet available.

A modification required to ensure the localization of an enzyme in a specific organelle has been previously proposed with respect to β-glucuronidase (137). In mouse liver, part of the enzyme activity is found in microsomes. Although one structural gene controls the synthesis of both lysosomal and microsomal forms of the enzyme (138), a second gene is required for embedding the enzyme into microsomes. In this situation, the second gene is thought to control the addition of an auxiliary polypeptide, rather than of a carbohydrate, to anchor the β-glucuronidase to the microsomal membrane.

PRACTICAL APPLICATIONS

Prenatal Diagnosis

Since so many of the lysosomal disorders are devastating and untreatable, it is fortunate that they are generally amenable to prenatal diagnosis. Cells of fetal origin, obtained from amniotic fluid and established in culture, have a content of lysosomal enzymes that is surprisingly similar to that of fibroblasts grown from the skin of children and adults, and can therefore be used to detect lysosomal enzyme deficiencies. In some cases, determination of the stored product may be easier or more reliable. The following disorders listed in Table 1 have been detected prenatally (139, 140, 140a): Hurler, Hunter, Sanfilippo A, GM$_1$ gangliosidosis, Krabbe's, Tay-Sachs, Sandhoff's, Gaucher's, Fabry's, metachromatic leukodystrophy, Niemann-Pick, Pompe's, acid phosphatase deficiency, I cell disease, and cystinosis. The remaining disorders listed in the table should in theory be diagnosable prenatally by biochemical methods, except for mucolipidoses I (3) and IV (141), for which neither enzyme deficiency nor storage material is known. However, until an instance of enzyme deficiency in fetal cells has actually been documented, the possibility that it might be masked by fetal isoenzymes should not be overlooked.

Heterozygote Detection

The desirability of detecting families at risk before the birth of an affected child has spurred research for biochemical identification of the heterozygous state. The gene dosage effect predicts that heterozygotes for recessive disorders should have a level of enzyme intermediate between that of the affected and normal individual. This has been verified for a number of lysosomal disorders, but usually on a statistical basis, because of the overlap in enzyme level between normal and heterozygous groups (142). To minimize that overlap, it is customary to express the level of the enzyme of interest in terms of some marker enzyme that is likely to show concurrent variation (e.g. 143, 144). The choice of marker enzyme is empirical since there seems to be no obligatory coupling in the levels of different lysosomal enzymes.

The β-hexosaminidases are unique in that the synthesis of hexosaminidase A is related to that of B (see above), and the ratio of the two hexosaminidases is therefore a reliable discriminant for Tay-Sachs heterozygotes (145). Because the Tay-Sachs gene is also unusual for its concentration in a small ethnic group, there have been developed in many large cities of the U. S. heterozygote screening programs which fulfill the criteria for diagnostic reliability, social usefulness and acceptability, and economic feasibility (146, 147).

An interesting and extensive veterinary screening program is carried out in New Zealand for heterozygotes of mannosidosis, which may number up to 25% of some herds of Angus cattle (148). Automated measurements of plasma α-mannosidase, followed by withdrawal of affected bulls from the breeding population, should effectively and inexpensively eliminate the disease, though not the gene. Biochemical variability causes few problems here, since doubtful cases can be resolved by breeding and examining the progeny.

Enzyme Replacement

The subject of enzyme replacement in vitro has already been discussed in detail in connection with corrective factors (p. 362) and with disorders of enzyme localization (p. 370). In addition to shedding light on the biochemical basis of some lysosomal disorders and providing a model for possible in vivo therapy, the uptake of lysosomal enzymes by deficient fibroblasts is useful for determining the fate of normal enzymes within lysosomes. For example, the intracellular stability of a number of lysosomal enzymes has been determined in that manner, and the half-lives were found impressively long, ranging from a few days to several weeks (33, 36, 38, 39, 127, 149, 149a).

The original discovery of lysosomal deficiency disease was accompanied by the hopeful expectation that such disorders would be amenable to replacement therapy

(150). Attempts of the past ten years [see (151) for experience up to 1972] have clarified the many problems that must be overcome and have suggested some general approaches.

One possibility would be to supply the patient with cells or tissues that would serve as a continuous source of enzyme. Infusion of leucocytes into patients with mucopolysaccharidoses (152, 153) and Fabry's disease (154) has been tried with variable short term effects. In an interesting situation involving an animal model, a chimeric male calf (with circulating lymphocytes presumably acquired as a fetal transplant from his female co-twin) had clinically expressed mannosidosis. The discouraging conclusion is that the lymphocytes, normal with respect to the mannosidosis genotype, had apparently little protective value against the disease (R. D. Jolly, personal communication).

In Fabry's disease, the uremia, which is a consequence of renal storage of ceramide trihexoside, has been treated by renal transplantation (155). Whether the grafted kidney, in addition to its primary function, can also alter the course of the disease by supplying other organs with ceramide trihexosidase is a subject that has been debated without clear resolution (156).

The present emphasis is on testing highly purified enzymes. Encouraging aspects of single injection trials (expected to yield biochemical information rather than clinical benefit) are the short term decrease of ceramide trihexoside in the plasma of a patient with Fabry's disease following an injection of ceramide trihexosidase (154), and a relatively long lasting reduction of β-glucosylceramide in erythrocytes of a patient with the adult form of Gaucher's disease following an injection of glucosylceramide β-glucosidase (157). In anticipation that such an approach will eventually prove beneficial, at least in selected cases, procedures are being developed for the stabilization of lysosomal enzymes by chemical alteration (158); protection against immunogenicity by encapsulation in non-antigenic material such as the recipient's own erythrocyte ghosts (159, 160); and targeting to affected cells by attachment of recognition markers (161) or by injecting high uptake forms previously selected with the aid of cultured cells (162).

Literature Cited

1. Raivio, K. O., Seegmiller, J. E. 1972. *Ann. Rev. Biochem.* 41:543–76
2. Hers, H. G. 1965. *Gastroenterology* 48:625–33
3. Spranger, J. W., Wiedemann, H. R. 1970. *Humangenetik* 9:113–39
4. Stanbury, J. B., Wyngaarden, J. B., Fredrickson, D. S., Eds. 1972. *Biochemical Basis of Inherited Disease.* New York: McGraw-Hill. 3rd ed. 1778 pp.
5. Volk, B. W., Aronson, S. M., Eds. 1972. *Sphingolipids, Sphingolipidoses and Allied Disorders.* New York: Plenum. 691 pp.
6. Hers, H. G., van Hoof, F., Eds. 1972. *Lysosomes and Storage Disease.* New York: Academic. 666 pp.
7. McKusick, V. A. 1972. *Heritable Disorders of Connective Tissue,* 521–686. St. Louis: Mosby. 4th ed. 878 pp.
8. Dorfman, A., Matalon, R., Cifonelli, J. A., Thompson, J., Dawson, G. See Ref. 6, 195–210
9. Neufeld, E. F., Barton, R. W. 1973. *Biology of Brain Dysfunction,* ed. G. E. Gaull, I:1–30. New York: Plenum. 403 pp.
10. Bach, G., Eisenberg, F. Jr., Cantz, M., Neufeld, E. F. 1972. *Proc. Nat. Acad. Sci. USA* 70:2134–38
11. Sjöberg, I., Fransson, L. A., Matalon, R., Dorfman, A. 1973. *Biochem. Biophys. Res. Commun.* 54:1125–32

12. Coppa, G. V., Singh, J., Nichols, B. L., DiFerrante, N. 1973. *Anal. Lett.* 6:225–33
13. Bach, G., Friedman, R., Weissmann, B., Neufeld, E. F. 1972. *Proc. Nat. Acad. Sci. USA* 69:2048–51
14. Matalon, R., Dorfman, A. 1972. *Biochem. Biophys. Res. Commun.* 47:959–64
14a. O'Brien, J. F., Cantz, M., Spranger, J. 1974. *Biochem. Biophys. Res. Commun.* 60:1170–77
15. Thompson, J. N., Stoolmiller, A. C., Matalon, R., Dorfman, A. 1973. *Science* 181:866–67
16. Cantz, M., Kresse, H. 1974. *Eur. J. Biochem.* 47:581–90
17. Sly, W. S., Quinton, B. A., McAlister, W. H., Rimoin, D. L. 1973. *J. Pediat.* 82:249–57
18. Kresse, H., Neufeld, E. F. 1972. *J. Biol. Chem.* 247:2164–70
19. Matalon, R., Dorfman, A. 1974. *J. Clin. Invest.* 54:907–12
20. Kresse, H. 1973. *Biochem. Biophys. Res. Commun.* 54:1111–18
21. O'Brien, J. S. 1972. *Proc. Nat. Acad. Sci. USA* 69:1720–22
22. von Figura, K., Kresse, H. 1972. *Biochem. Biophys. Res. Commun.* 48:262–69
23. Stumpf, D. A., Austin, J. H. 1972. *Trans. Am. Neurol. Assoc.* 97:29–32
24. Stumpf, D. A., Austin, J. H., Crocker, A. C., LaFrance, M. 1973. *Am. J. Dis. Child.* 126:747–55
25. Fluharty, A. L., Stevens, R. L., Sanders, D. L., Kihara, H. 1974. *Biochem. Biophys. Res. Commun.* 59:455–61
26. Freitag, F., Küchemann, K., Schuster, W., Spranger, J. 1971. *Virchows Arch. B* 8:1–18
27. Matalon, R., Arbogast, B., Dorfman, A. 1974. *Pediat. Res.* 8:436/162 (Abstr.)
28. Eisenberg, F. Jr. 1974. *Anal. Biochem.* 60:181–87
29. Fratantoni, J. C., Hall, C. W., Neufeld, E. F. 1968. *Proc. Nat. Acad. Sci. USA* 60:699–706
30. Neufeld, E. F., Cantz, M. J. 1971. *Ann. NY Acad. Sci.* 179:580–87
31. Barton, R. W., Neufeld, E. F. 1971. *J. Biol. Chem.* 246:7773–79
32. Cantz, M., Chrambach, A., Bach, G., Neufeld, E. F. 1972. *J. Biol. Chem.* 247:5456–62
33. O'Brien, J. S., Miller, A. L., Loverde, A. W., Veath, M. L. 1973. *Science* 181:753–55
34. Hall, C. W., Cantz, M., Neufeld, E. F. 1973. *Arch. Biochem. Biophys.* 155:32–38
35. Brot, F. E., Glaser, J. H., Roozen, K. J., Sly, W. S. 1974. *Biochem. Biophys. Res. Commun.* 57:1–8
36. Lagunoff, D., Nicol, D. M., Pritzl, P. 1973. *Lab. Invest.* 29:449–53
37. Barton, R. W., Neufeld, E. J. 1972. *J. Pediat.* 80:114–16
38. von Figura, K., Kresse, H. 1974. *J. Clin. Invest.* 53:85–90
39. Hickman, S., Shapiro, L. J., Neufeld, E. F. 1974. *Biochem. Biophys. Res. Commun.* 57:55–61
40. Shapiro, L. J., Hickman, S., Hall, C. W., Neufeld, E. F. 1974. *Am. J. Hum. Genet.* 26:79a
41. Beaudet, A. L., DiFerrante, N. M., Ferry, G. D., Nichols, B. L., Mullins, C. E. 1974. *J. Pediat.* In press
42. Yoshida, A. 1973. *Science* 179:532–37
43. McKusick, V. A., Hussels, I. E., Howell, R. R., Neufeld, E. F., Stevenson, R. E. 1972. *Lancet* 1:993–96
44. Sugita, M., Dulaney, J. T., Moser, H. W. 1972. *Science* 178:1100–2
45. Brady, R. O. 1973. *Advan. Enzymol.* 38:293–315
46. Suzuki, K., Suzuki, K. 1973. *Biology of Brain Dysfunction,* ed. G. E. Gaull, 2:1–73. New York: Plenum. 422 pp.
47. Ho, M. W., Cheetham, P., Robinson, D. 1973. *Biochem. J.* 136:351–59
48. Norden, A. G. W., O'Brien, J. S. 1973. *Arch. Biochem. Biophys.* 159:383–92
49. Norden, A. G. W., Tennant, L. L., O'Brien, J. S. 1974. *J. Biol. Chem.* 249:7969–76
50. Tsay, G. C., Dawson, G. 1973. *Biochem. Biophys. Res. Commun.* 52:759–66
51. Wolfe, L. S., Senior, R. G., Ng Ying Kin, N. M. K. 1974. *J. Biol. Chem.* 249:1828–38
52. Singer, H. S., Schafer, I. A. 1972. *Am. J. Hum. Genet.* 24:454–63
53. Chou, L., Kaye, C. I., Nadler, H. L. 1974. *Pediat. Res.* 8:120–25
54. Suzuki, Y., Suzuki, K. 1974. *J. Biol. Chem.* 249:2113–17
55. Wenger, D. A., Sattler, M., Markey, S. P. 1973. *Biochem. Biophys. Res. Commun.* 53:680–85
56. Miyatake, T., Suzuki, K. 1972. *Biochem. Biophys. Res. Commun.* 48:538–43
57. Suzuki, Y., Suzuki, K. 1974. *J. Biol. Chem.* 249:2105–8
58. Dawson, G., Matalon, R., Stein, A. O. 1971. *J. Pediat.* 79:423–29
59. Wenger, D. A., Sattler, M., Hiatt, W. 1974. *Proc. Nat. Acad. Sci. USA* 71:854–57
60. van Hoof, F., Hers, H. G. 1968. *Eur. J. Biochem.* 7:34–44
61. Kint, J. A., Dacremont, G., Carton, D., Orye, E., Hooft, C. 1973. *Science* 181:

352–54
62. Suzuki, Y., Miyatake, T., Fletcher, T. F., Suzuki, K. 1974. *J. Biol. Chem.* 249: 2109–12
63. Handa, S., Yamakawa, T. 1971. *J. Neurochem.* 18: 1275–80
64. Farrell, D. F., Baker, H. J., Herndon, R. M., Lindsey, J. R., McKhann, G. M. 1973. *J. Neuropathol. Exp. Neurol.* 32: 1–18
65. Donnelly, W. J. C., Sheahan, B. J., Kelly, M. 1973. *Res. Vet. Sci.* 15: 139–41
66. Wassle, W., Sandhoff, K. 1971. *Z. Physiol. Chem.* 352: 1119–33
67. Srivastava, S. K., Yoshida, A., Awasthi, Y. C., Beutler, E. 1974. *J. Biol. Chem.* 249: 2049–53
68. Tallman, J. F., Brady, R. O., Quirk, M. V., Villalba, M., Gal, A. E. 1974. *J. Biol. Chem.* 249: 3489–99
69. Wenger, D. A., Okada, S., O'Brien, J. S. 1972. *Arch. Biochem. Biophys.* 153: 116–29
70. Robinson, D., Stirling, J. L. 1968. *Biochem. J.* 107: 321–27
71. Carroll, M., Robinson, D. 1973. *Biochem. J.* 131: 91–96
72. Bartholemew, W. R., Rattazzi, M. C. 1974. *Int. Arch. Allergy Appl. Immunol.* 46: 512–24
73. Srivastava, S. K., Beutler, E. 1974. *J. Biol. Chem.* 249: 2054–57
74. Carroll, M., Robinson, D. 1974. *Biochem. J.* 137: 218–21
75. Carmody, P. J., Rattazzi, M. C. 1974. *Biochim. Biophys. Acta* 371: 117–20
76. Goldstone, A., Konecny, P., Koenig, H. 1971. *FEBS Lett.* 13: 68–72
77. Beutler, E., Villacorte, D., Srivastava, S. K. 1974. *Int. Res. Commun. Syst.* 2: 1090
78. Rattazzi, M. C., Brown, J. A., Davidson, R. G., Shows, T. B. 1974. *Am. J. Hum. Genet.* 26: 71a
79. Galjaard, H. et al 1974. *Exp. Cell. Res.* 187: 444–48
80. Thomas, G. H., Taylor, H. A., Miller, C. S., Axelman, J., Migeon, B. R. 1974. *Nature* 250: 580–82
81. van Someren, H., van Henegouwen, H. B. 1973. *Humangenetik* 18: 171–74
82. Lalley, P. A., Rattazzi, M. C., Shows, T. B. 1974. *Proc. Nat. Acad. Sci. USA* 71: 1569–73
83. Sandhoff, K., Harzer, K., Wassle, W., Jatzkewitz, H. 1971. *J. Neurochem.* 18: 2469–89
84. Srivastava, S. K., Beutler, E. 1973. *Nature New Biol.* 241: 463–64
85. O'Brien, J. S. 1973. See Ref. 6, 323–44
86. Navon, R., Padek, B., Adam, A. 1973.

Am. J. Hum. Genet. 25: 287–93
87. Braidman, I. et al 1974. *FEBS Lett.* 41: 181–84
88. Price, R. G., Dance, N. 1972. *Biochim. Biophys. Acta* 271: 145–53
89. Stirling, J. L. 1972. *Biochim. Biophys. Acta* 271: 154–62
90. Grebner, E. E., Tucker, J. 1973. *Biochim. Biophys. Acta* 321: 228–33
91. Lie, K. K., Thomas, G. H., Taylor, H. A., Sensenbrenner, J. A. 1973. *Clin. Chim. Acta* 45: 243–48
92. Wherrett, J. R., Hakomori, S. I. 1973. *J. Biol. Chem.* 248: 3046–51
93. Beutler, E., Kuhl, W. 1972. *Am. J. Hum. Genet.* 24: 237–49
94. Wood, S., Nadler, H. L. 1972. *Am. J. Hum. Genet.* 24: 250–55
95. Tallman, J. F., Pentchev, P. G., Brady, R. O. 1974. *Enzymes* 18: 136–49
96. Beutler, E., Kuhl, W. 1972. *J. Biol. Chem.* 247: 7195–7200
97. Mapes, C. A., Sweeley, C. C. 1972. *FEBS Lett.* 25: 279–81
98. Mapes, C. A., Suelter, C. H., Sweeley, C. C. 1973. *J. Biol. Chem.* 248: 2471–79
99. Mapes, C. A., Sweeley, C. C. 1973. *J. Biol. Chem.* 248: 2461–70
100. Rietra, J. G. M., Molenaar, J. L., Hamers, M. N., Tager, J. M., Borst, P. 1974. *Eur. J. Biochem.* 46: 89–98
100a. Beutler, E., Kuhl, W. 1973. *N. Engl. J. Med.* 289: 694–95
101. Mehl, E., Jatzkewitz, H. 1968. *Biochim. Biophys. Acta* 151: 619–27
102. Jerfy, A., Roy, A. B. 1973. *Biochim. Biophys. Acta* 293: 178–90
103. Neuwelt, E., Stumpf, D., Austin, J., Kohler, P. 1971. *Biochim. Biophys. Acta* 236: 333–46
104. Stumpf, D., Austin, J. 1971. *Arch. Neurol.* 24: 117–24
105. Porter, M. T., Fluharty, A. L., Trammell, J., Kihara, H. 1971. *Biochem. Biophys. Res. Commun.* 44: 660–66
106. Pentchev, P. G., Brady, R. O., Hibbert, S. R., Gal, A. E., Shapiro, D. 1973. *J. Biol. Chem.* 248: 5256–61
107. Raghavan, S. S., Mumford, R. H., Kanfer, J. N. 1974. *Biochem. Biophys. Res. Commun.* 58: 99–106
108. Brady, R. O., Kanfer, J. N., Bradley, R. M., Shapiro, D. 1966. *J. Clin. Invest.* 45: 1112–15
109. Ho, M. W. 1973. *Biochem. J.* 133: 1–10
110. Jatzkewitz, H., Stinshoff, K. 1973. *FEBS Lett.* 32: 129–31
111. Ho, M. W., O'Brien, J. S. 1971. *Proc. Nat. Acad. Sci. USA* 68: 2810–13
111a. Ho, M. W., Light, N. D. 1973. *Biochem. J.* 136: 821–23

112. Ho, M. W., O'Brien, J. S., Radin, N. S., Erickson, J. S. 1973. *Biochem. J.* 131: 173–76
113. Li, Y. T., Mazzotta, M. Y., Wan, C. C., Orth, R., Li, S. C. 1973. *J. Biol. Chem.* 248: 7512–15
114. Li, S. C., Wan, C. C., Mazzotta, M. Y., Li, Y. T. 1974. *Carbohyd. Res.* 34: 180–93
115. Murphy, J. V., Wolfe, H. J., Balasz, E. A., Moser, H. W. 1971. *Lipid Storage Diseases,* ed. J. Bernsohn, H. Grossman, 67–110. New York: Academic. 316 pp.
116. Eto, Y., Wiesmann, U. N., Carson, J. H., Herschkowitz, N. N. 1974. *Arch. Neurol.* 30: 153–56
117. Stumpf, D., Neuwelt, E., Austin, J., Kohler, P. 1971. *Arch. Neurol.* 25: 427–31
118. Lightbody, J., Wiesmann, U., Hadorn, B., Herschkowitz, N. 1971. *Lancet* 1: 451
119. Leroy, J. G., Ho, M. W., MacBrinn, M. C., Zielke, K., Jacob, J., O'Brien, J. S. 1972. *Pediat. Res.* 6: 752–57
120. Thomas, G. H., Taylor, H. A., Reynolds, L. W., Miller, C. S. 1973. *Pediat. Res.* 7: 751–56
121. Berman, E. R., Kohn, G., Yatziv, S., Stein, H. 1974. *Clin. Chim. Acta* 52: 115–24
122. Glaser, J. H., McAlister, W. H., Sly, W. S. 1974. *J. Pediat.* 85: 192–98
123. Wiesmann, U. N., Lightbody, J., Vassella, F., Herschkowitz, N. N. 1971. *N. Engl. J. Med.* 284: 109–10
124. Wiesmann, U., Vassella, F., Herschkowitz, N. 1971. *N. Engl. J. Med.* 285: 1090–91
125. Tondeur, M. et al 1971. *J. Pediat.* 79: 366–78
126. Tondeur, M., Neufeld, E. F. 1975. *Molecular Pathology,* ed. S. Day, R. A. Good. New York: Thomas. In press
127. Hickman, S., Neufeld, E. F. 1972. *Biochem. Biophys. Res. Commun.* 49: 992–99
128. Verpoorte, J. A. 1972. *J. Biol. Chem.* 247: 4787–93
129. Stahl, P. D., Touster, O. 1971. *J. Biol. Chem.* 246: 5398–5406
130. Graham, E. R. B., Roy, A. B. 1973. *Biochim. Biophys. Acta* 329: 88–92
131. Ahmad, A., Bishayee, S., Bacchawat, B. K. 1973. *Biochem. Biophys. Res. Commun.* 53: 730–35
132. Keller, K., Touster, O. 1973. *Fed. Proc.* 32: 638 (Abstr.)
133. Mapes, C. A., Sweeley, C. C. 1973. *Arch. Biochem. Biophys.* 158: 297–304
134. Norden, A. G. W., O'Brien, J. S. 1974. *Biochem. Biophys. Res. Commun.* 56:

135. Gregoriadis, G., Morell, A. G., Sternlieb, I., Scheinberg, I. H. 1970. *J. Biol. Chem.* 245: 5833–37
136. Pricer, W. E. Jr., Ashwell, G. 1971. *J. Biol. Chem.* 246: 4825–33
137. Swank, R. T., Paigen, K. 1973. *J. Mol. Biol.* 77: 371–89
138. Paigen, K. 1971. *Exp. Cell. Res.* 25: 286–301
139. Milunsky, A. et al 1970. *N. Engl. J. Med.* 283: 1370–81; 1441–47; 1498–1504
140. Burton, B. K., Gerbie, A. B., Nadler, H. L. 1974. *Am. J. Obstet. Gynecol.* 118: 718–46
140a. Harper, P. S. et al 1974. *J. Med. Genet.* 11: 123–32
141. Berman, E. R., Livni, N., Shapira, E., Merin, S., Levij, I. S. 1974. *J. Pediat.* 84: 519–26
142. Kihara, H. et al 1973. *Am. J. Ment. Defic.* 77: 389–94
143. Zielke, K., Veath, M. L., O'Brien, J. S. 1972. *J. Exp. Med.* 136: 197–99
144. Hall, C. W., Neufeld, E. F. 1973. *Arch. Biochem. Biophys.* 158: 817–21
145. Okada, S., Veath, M. L., Leroy, J. G., O'Brien, J. S. 1971. *Am. J. Hum. Genet.* 23: 55–61
146. O'Brien, J. S. 1973. *Fed. Proc.* 32: 191–99
147. Kaback, M. M., O'Brien, J. S. 1973. *Hosp. Pract.* 8: 107–16
148. Jolly, R. D. 1975. *Recent Advances in Veterinary Science and Comparative Medicine,* ed. C. A. Brandly, C. R. Cornelius, Vol. 19. New York: Academic. In press
149. Porter, M. T., Fluharty, A. L., Kihara, H. 1971. *Science* 172: 1253–65
149a. von Figura, K., Kresse, H. 1974. *J. Clin. Invest.* 53: 85–90
150. Baudhuin, P., Hers, H. G., Loeb, H. 1964. *Lab. Invest.* 13: 1139–52
151. Desnick, R. J., Bernlohr, R. W., Krivit, W., Eds. 1973. *Enzyme Therapy in Genetic Diseases,* Birth Defects Original Articles Series, National Foundation—March of Dimes, Vol. 9. Baltimore: Williams & Wilkins. 236 pp.
152. Knudson, A. G., DiFerrante, N., Curtis, J. E. 1971. *Proc. Nat. Acad. Sci. USA* 68: 1738–41
153. Moser, H. W. et al 1974. *Arch. Neurol.* 31: 329–37
154. Brady, R. O. et al 1973. *N. Engl. J. Med.* 289: 9–14
155. Desnick, R. J. et al 1972. *Surgery* 72: 203–10
156. Clarke, J. T. R., Guttmann, R. D.,

Wolfe, L. S., Beaudoin, J. G., Morehouse, D. D. 1972. *N. Engl. J. Med.* 287:1215–18

157. Brady, R. O., Pentchev, P. G., Gal, A. E., Hibbert, S. R., Dekaban, A. S. 1974. *N. Engl. J. Med.* 291:989–93

158. Snyder, P. D. Jr. et al 1974. *Biochim. Biophys. Acta* 350:432–36

159. Ihler, G. M., Glew, R. H., Schnure, F. W. 1973. *Proc. Nat. Acad. Sci. USA* 70:2663–66

160. Fiddler, M. B., Thorpe, S. R., Desnick, R. J. 1974. *Am. J. Hum. Genet.* 26:30a

161. Rogers, J. C., Kornfeld, S. 1971. *Biochem. Biophys. Res. Commun.* 45:622–29

162. Sly, W. S., Glaser, J. H., Roozen, K., Brot, F., Stahl, P. 1974. *Enzyme Therapy in Lysosomal Storage Diseases,* ed. J. M. Tager, G. J. M. Hooghwinkel, W. T. Daems, 288–89. Amsterdam: North-Holland. 308 pp.

OXYGENASE-CATALYZED BIOLOGICAL HYDROXYLATIONS[1]

I. C. Gunsalus, T. C. Pederson, and S. G. Sligar

Department of Biochemistry, University of Illinois, Urbana, Illinois 61801

CONTENTS

INTRODUCTION

The year 1974 marks the 200th anniversary of the discovery of oxygen by Priestley (1), 100 years since Pflüger (2) reported O_2 use by all animal cells, and a mere 20 years since Mason (3) and Hayaishi (4) demonstrated independently the direct substrate incorporation of molecular oxygen bypassing an equilibration with the oxygen of water. This review is devoted to a subsection of the *oxygenase* area defined by the latter discovery as a molecular problem in biocatalysis.

Dioxygen reductive cleavage proceeds in several ways in biological systems. *Dioxygenase* is the term applied to systems that incorporate both atoms from a

[1] This work was supported in part by grants from the National Science Foundation GB 41629X and the United States Public Health Service AM00562.

single O_2 molecule into one substrate; *mixed function oxidase* or *monooxygenase* denotes the two electron reductive O_2 cleavage to yield one H_2O with the second atom incorporated into an organic substrate. The two electron reduction of O_2 to H_2O_2 and the four electron reduction to 2 H_2O are not considered among the oxygenases; the latter is quantitatively preponderant and accompanies most energy coupling for cellular work. The two electron reductant for hydroxylases is generally an external donor, frequently a reduced pyridine nucleotide molecule. A variation in reductant includes an abundant and regeneratable essential metabolite, e.g. an α-keto acid, alcohol, or another functional group on the same substrate. In the latter, an O_2 atom may be incorporated also into the reductant, which could have mechanistic implications. Examples of each appear among the hydroxylases described in this review. For a general description of the known systems and a simple pattern of grouping them, the reader is referred to the recent review by Hayaishi (5) or to an earlier one by Mason (6).

Over the past 20 years, several prosthetic groups have been identified in oxygenase proteins and shown to participate in the binding of O_2 and organic substrate and in the catalytic steps of oxygenation. Today, the two largest identified classes are metalloproteins, containing iron or copper, and a group of flavoproteins. Among the iron-containing hydroxylases are heme in *b*-type cytochromes of the P-450 class and one or more dissociable forms of iron with as yet unidentified ligands. The iron-linked monoxygenases commonly appear in multienzyme systems where they are coupled to short electron transfer chains. A flavoprotein usually initiates electron transport by mobilizing the two reducing equivalents required for eventual reduction of O_2. The substrate and oxygen reactive cytochromes may be coupled directly to the flavoproteins, e.g. TPNH or DPNH, dehydrogenases, or may be linked through an iron-sulfur protein of the $FeCys_4$ or $Fe_2S_2^*Cys_4$ type. The P-450 cytochromes of steroid 11β-hydroxylase in the adrenal cortex and the camphor 5-*exo*-hydroxylase in *Pseudomonas putida* are coupled via $Fe_2S_2^*Cys_4$ redoxins, whereas the "iron-containing" ω-hydroxylase for normal C_6-C_{14} hydrocarbons found in *Pseudomonas oleovorans* is coupled via a $FeCys_4$ rubredoxin.

The flavoprotein hydroxylases are generally single component systems with both hydroxylation and reductase sites on a single polypeptide chain, usually bearing a molecule of FAD, or occasionally FMN, as the reactive center. Oxygenated intermediates have been detected in both the cytochrome and flavin-mediated hydroxylases by spectroscopic, subzero temperature, and kinetic techniques. The ferrous P-450 cytochromes oxygenate reversibly to intermediates resembling the oxy forms of myoglobin and hemoglobin; they also autoxidize more rapidly to the ferric *met*-like forms with release of superoxide anion. The reduced forms of the flavin-coupled monoxygenases have recently also been shown to oxygenate to intermediates stable only at subzero temperatures (7). Evidence has recently been obtained by Kaufman and co-workers (8) for an intermediate in the iron-linked pteridine (tetrahydrobiopterin) aromatic hydroxylase (phenylalanine → tyrosine), which would appear to follow O_2 cleavage.

Oxidative energy coupling, for example by oxidative phosphorylation by the monoxygenases or monoxygenase systems, has not been observed and is believed, without formal proof, not to occur. Thus the concept is generally held that the regulation of monoxygenase activity by energy charge would likely prevent essential biosynthesis on the entry of "inert" molecules into energy and carbon, yielding pathways with consequences deleterious to the organism. Energy supply for cellular work would thus appear to derive from subsequent dehydrogenases in a reaction pathway rather than from the initial oxygenase steps.

Enzyme induction and electron flow among the monoxygenases, including hydroxylase systems, appear to be modulated by potential substrates or substrate analogs. In the cytochromes of the P-450 class, the modulation of electron flow by substrates appears to occur by a modification of the redox potential of the ferric-ferrous couple on binding the carbon substrate (9). An alternate mode of regulation, observed in the hepatic microsomal P-450 and bacterial ω-hydroxylase systems, depends on a phospholipid component (10, 11). Among the flavoprotein oxygenases, reduction of the prosthetic group ($FAD \rightarrow FADH_2$) has been shown in many cases to depend upon prior substrate binding (12).

Any sequential discussion of the various hydroxylase systems necessitates resorting to artificial separations. This review has been guided by the original suggestions of Mason (6), modified in the light of accumulating information, and discussed most recently by Hayaishi (5) in the first chapter of his second ten-year summary of oxygenases under the title *Molecular Mechanisms of Oxygen Activation*. A comprehensive statement of knowledge concerning most of the hydroxylases discussed in this manuscript through mid-1973 appeared in the latest Hayaishi volume. Reviews concerning enzymatic hydroxylation last appeared in the *Annual Review of Biochemistry* in 1969 by Hayaishi (13) and in earlier volumes by Mason (14) and Massart & Vercanteren (15).

HEME-COUPLED MONOXYGENASES: CYTOCHROME P-450

Proteins with reduced CO adduct showing pronounced absorbances in the 450 nm region, called P-450 cytochromes (16, 17), were recognized over a decade ago as a special class of heme proteins (18) and were implicated in microsomal hydroxylations with monoxygenase stoichiometry (19). Their study over the past 20 years has opened a new area of oxidative metabolism and led to the first clear understanding in molecular and physical terms of enzymatic dioxygen reduction. Such proteins, now recognized as widely distributed in animals and plants, including single pro- and eucaryotic cells, were found to play significant roles in the biogenesis of essential metabolites, particularly lipophilic and regulatory steroid hormones, in the oxygenation and excretion of foreign or synthetic molecules in the mammalian liver or kidney, and in the recycling processes in the biosphere carried out by many microorganisms (20).

The P-450 systems best understood at present are those freed from cellular organelles with recovery of separate activity components in homogeneous form. Although these have furnished insight at the molecular level, much remains to be completed in defining essential components, enzyme properties, reactions, and mechanisms. Encouraging progress has occurred with the selective conversion of cholesterol to steroid hormones in the adrenal cortex, particularly in the study of 11β hydroxylation and the side chain cleavage system. The early evidence of NADPH-dependent metabolism was supplemented by evidence of CO sensitivity and an action spectrum for its photoreversibility corresponding to the cytochrome P-450 absorption (21). More recently, enzyme isolations have revealed for the 11β hydroxylation a three component system composed of a flavoprotein TPNH dehydrogenase, an $Fe_2S_2^*Cys_4$ iron sulfide redoxin, and a P-450 cytochrome (22).

Perhaps the P-450 system obtained in the purest form and the best characterized in molecular terms is the 5-exo-hydroxylase of the bicyclic monoterpene, camphor, which has been isolated from $P. putida$ (23). Much conclusive data on the catalytically active sites and chemical intermediates of P-450 cytochrome hydroxylases in general have been provided by a detailed study of this microbial system.

Hepatic microsomes have now yielded several individual cytochromes of the P-450 class. The phenobarbital-induced system has been isolated in essentially homogeneous form, free of other oxidative activities (24, 25). These hepatic systems have resisted solubilization until recently and have often yielded a variety of enzymatically inactive forms of the cytochrome, termed P-420. In a definition of the components and activities of the most purified preparations, TPNH-coupled flavoprotein reductases furnish the electrons, apparently directly to the cytochrome, with the electron flow dependent on a phospholipid replaceable by synthetic phosphatidylcholine.

Camphor Methylene Hydroxylases

Although the steroids are terpenoids indispensable to the structure and regulation of animal cells, biosynthesis of the overwhelming preponderance of terpenoids occurs in plants, and the role of recycling their carbon residues falls to the microbes. Through microorganisms whose growth is supported by these bicyclic largely hydrocarbon skeletons, we are afforded a vehicle for the elucidation of the enzymatic mechanisms converting these branched chain compounds to essential metabolites. The oxygenated bicyclic monoterpene, D or L camphor, is metabolized by organisms that accumulate the 5-exo or 6-$endo$ hydroxy camphor together with a variety of more substituted derivatives. These pathways have been termed, respectively, the 2,5 and 2,6 hydroxylating systems (26). In $Mycobacterium$ $rhodochrous$, the 2,6 pathway has been explored but the enzymology of hydroxylation has not been studied in detail (27). The oxidation of camphor, however, by the 2,5 pathway via isobutyrate in the soil bacterium $P. putida$ has been investigated in great detail and has yielded much valuable information toward understanding energy transfer and segregation during mixed function oxidation.

The methylene 5-exo-hydroxylase consists of three protein components, a DPNH specific FAD flavoprotein dehydrogenase, an $Fe_2S_2^*Cys_4$ iron sulfide redoxin (putidaredoxin, abbreviated Pd), and a b-type cytochrome, termed P-450$_{cam}$ or cytochrome m (28). This system, perhaps one of the most intensively studied hydroxylases from genetic, chemical, and physical approaches, has contributed significantly to progress in elucidating the general molecular mechanism of mixed function oxygenative hydroxylation. Of prime importance in this undertaking has been the homogeneity of the enzymes obtained by purification procedures using gentle sizing and ion exchange chromatography (29). The reductase and redoxin have been prepared devoid of contaminating proteins, and the cytochrome has been crystallized (29, 30) with recrystallization as a final step in the bulk purification procedure. The purity of the isolated components has allowed a study of several questions of prime biochemical importance. Two relate to the molecular mechanism of redox transfer and the details of intermediates involved in O_2 reduction and hydroxylation. A general account can be obtained from several recent reviews (9, 23, 31, 32) and the details in the original papers cited here and in the reviews.

The reductase, molecular weight 45,000 (33), bears a single FAD prosthetic group that undergoes a two electron reduction by DPNH at a potential of -285 mV. On chemical reduction, the protein binds DPN^+ and exhibits a charge-transfer band in the 650 nm region; evidence for a stable semiquinone has not been observed (34).

Putidaredoxin (Pd), an iron-sulfide protein of molecular weight $\sim 12,000$ (33), contains 106 amino acid residues of known primary sequence (35) and two atoms each of acid-labile iron and sulfide attached to the protein through the thiol groups of four of the six cysteine residues in the primary structure. Pd accepts a single electron at a potential of -240 mV (36) with conversion of one of the iron atoms to the ferrous state. The physical properties have been studied by optical and fluorescence (23, 45) spectroscopy and the state of the iron by electron paramagnetic resonance (EPR) (38) and Mössbauer (39, 42) spectroscopy. Both reduction and oxidation of the flavoprotein, Pd, and P-450$_{cam}$ pose interesting and important questions when one considers the coupling of a two electron reductant to a one electron acceptor and the crucial reactions in each of two single electron reductions of the cytochrome. The reconstituted system and the redox reactions with product formation will be described.

Cytochrome P-450$_{cam}$ (cyt m) has received by far the most intensive study, with investigation centered on the stable states and on the reactions of the oxygen and carbon substrates. The overall scheme to emerge from research over the past five years is

The Soret maxima of the stable intermediate states of the cytochrome are indicated. Usually isolated in the more stable ferric substrate form, the cytochrome can be completely freed of the substrate which will rebind rapidly in a second order reaction with the blueshift in the strong Soret maximum. The first reduction of this intermediate occurs at a potential of -170 mV with a ferric-ferrous transition of the heme iron. This first electron, normally supplied in the native system by reduced Pd, can be supplied to the pure cytochrome in free or substrate form by chemical reduction. In the camphor-free P-450, the iron redox potential is -340 mV, well below that of the substrate form, -170 mV (36). Thus, substrate binding regulates the flow of energy to the cytochrome in the sequence cyt $m^o \rightarrow m^{os} \rightarrow m^{rs}$. Free energy changes on substrate binding and reduction are constant and offer a comprehensive view of the regulatory role of camphor from fundamental thermodynamic principles. Again, the presence of iron in the active center has allowed detailed studies using EPR (40), Mössbauer (41, 42), fluorescence (43), and optical (23) spectroscopy.

Chemical modifications of the P-450$_{cam}$ were undertaken in an effort to delineate specific reaction roles of certain amino acids. In a detailed study of the six cysteinyl residues using N-ethylmaleimide (NEM), Lipscomb et al (44; unpublished) identified one sulfhydryl on the exterior and three in the substrate binding region of the cytochrome. The first redox reaction and substrate-modulated spectral shift have been suppressed in the NEM-treated cytochrome. The redox potential of the modified protein is -266 mV, in contrast to the -170 mV substrate-bound native form. However, the chemically reduced modified cytochrome still binds oxygen and forms product in the presence of putidaredoxin.

Multienzyme complexes of the cytochrome and redoxin play a crucial role in camphor hydroxylation. A cyt m-Pd 1:1 stoichiometric complex with dissociation constant 3 μM was directly demonstrated by the quenching of the fluorescence associated with a fluorescein moiety covalently linked to the cytochrome (43, 45). The carboxyl terminus of the redoxin has been identified as a critical locus for this inter-protein binding (45, 46). The terminal tryptophan, shown by fluorescence measurements to be in a highly hydrophilic environment, was removed together with the penultimate glutamine to form a new protein with an arginine carboxyl terminus. This protein, termed des-Trp-Gln-Pd, retains the optical and EPR spectra of the native redoxin, indicative of no crucial modification of the $Fe_2S_2^*Cys_4$ active center. However, des-Trp-Gln-Pd exhibits a drastically altered binding constant for the cytochrome, $K_D = 150$ μM, a fiftyfold loss in specific activity in the effector role requiring the second electron, and in the complete hydroxylase system resulting from the three components a fiftyfold decreased catalytic activity. Thus, the formation and breakdown of the multienzyme complex are extremely critical to the hydroxylation reaction. The thermodynamic parameters of the interactions relating to the hydroxylation mechanism appear well established by recent data. Detailed analysis of the time course for both the decay of $m^{rs}_{O_2}$ and the generation of hydroxylated substrate show saturation kinetics indicative of a Pd-$m^{rs}_{O_2}$ multienzyme complex as the active state prior to the oxygenase reaction. Individual rate constants for the breakdown of the Pd-$m^{rs}_{O_2}$ complex by autoxidation or product formation (7 and 18 sec^{-1}, respectively) are consistent with the turnover number in the complete reconstituted hydroxylase (17 sec^{-1}) and the total normalized yield of product at saturating levels of effector (0.72). In addition, the regenerated oxidized cytochrome (m^{os}) competes with $m_{\beta_2}^{rs}$, for the effector molecule, creating a product inhibition ($K_I = 3\ \mu M$).

The effector role has provided a second focal point of recent work in the product-forming reaction from Pd-$m_{\beta_2}^{rs}$. In details of cytochrome turnover, Pd is by far the best effector; other proteins carrying effector activity include mammalian cytochrome b_5 and several bacterial rubredoxins (23, 32). Small molecules have recently been demonstrated to replace the native effector, including organic thiols, notably lipoic acid (37, 44). The molecular mechanism following the formation of the Pd-$m_{\beta_2}^{rs}$ complex and the role of the effector in product release clearly require further elucidation.

Adrenal Cortex Hydroxylases

Triterpenoids, including cholesterol, are precursors of an array of mammalian regulatory steroid hormones formed in the adrenal cortex, corpus luteum, ovary, placenta, and testis (47). Specific hydroxylase systems involving a P-450 cytochrome have been implicated in many of the accompanying conversions of hydrocarbon groups to oxygenated and aromatic residues. The adrenal cortex is the richest source of cytochrome P-450-catalyzed steroid hydroxylases, which are reviewed in terms of their specific hydroxylation activities in this section.

20α, 22R HYDROXYLATION AND LYASE   Removal of the cholesterol side chain carbons 22 to 27 leading to pregnenolone occurs in the adrenal mitochondria via hydroxylation to 20α, 22R-dihydroxycholesterol. In addition to isolation from bovine adrenal glands

CHOLESTEROL   20 α-,22R-HYDROXYLASE   PREGNENOLONE

(48), this intermediate is formed from radioactively labeled cholesterol on incubation with an acetone powder of adrenal mitochondria (49, 50). Subsequent action of the C20-22 lyase removes isocaproic aldehyde to yield pregnenolone (50, 51).

Some debate persists on the nature and order of the individual steps leading to 20α,22R-dihydroxycholesterol, arising from the disagreement of several investigators as to the isolation of proposed intermediates. N. Orme-Johnson (personal communication) has recently indicated that the first step consists of the 22R hydroxylation, followed by the 20α, whereas others have held the converse view (52–54). The accumulation of 20α-hydroxycholesterol, a presumed intermediate in pregnenolone formation, is indicated in (55), whereas others have failed to observe this compound (56–59). Several suggestions have been made that the 20α hydroxylation is the rate-limiting step and the site of ACTH action in the overall conversion (60, 61). A concerted attack on both carbons by dioxygen without a monohydroxylated intermediate has also been suggested (46, 62, 63). However, the random incorporation of $^{15}O_2$ and $^{16}O_2$ into 20,22-dihydroxycholesterol does not support such a mechanism (64). Determination of the reaction sequence and enzyme specificities is hampered by the presence of multiple activities in currently available preparations. Bryson & Sweat (65) demonstrated that side chain cleavage is dependent on TPNH, a flavoprotein dehydrogenase, an iron-sulfur protein, and a P-450 cytochrome. The flavoprotein-iron-sulfur protein electron transport chain appears to be the same as for the 11β-hydroxylase, whereas the P-450s are different (66–69); 11β-hydroxylase system reconstituted by a number of investigators from "purified" subunits is found to retain marked side chain cleavage activity (70, 71). Steroid hydroxylations by intact adrenal mitochondria have received the attention of numerous investigators by multiple techniques, including incubation with proposed pathway intermediates, observation of spectral and spin state changes, etc (72–74). Attempts to clarify enzyme and electron transport selectivity of cholesterol side chain hydroxylation and removal have used a variety of procedures for solubilization and separation, although these studies have not reached a common basis of agreement at this time.

11β HYDROXYLATION   The steroid hydroxylase subjected to the most successful fractionation and examination of enzyme components is the system responsible for the hydroxylation of 11-deoxycorticosterone and 11-deoxycortisol. Earlier studies on this hydroxylation system were reviewed by Schleyer et al (75). The system is

located in the inner mitochondrial membrane of adrenal cortex cells (76), contains three enzyme components, and uses NADPH as a two electron reductant. The system and reaction steps, partially reconstructed by analogy of their remarkable similarity to the purified microbial system for the 5-exo-methylene hydroxylation of the bicyclic monoterpene camphor (23), is composed of FAD flavoprotein reductase (22), an iron-sulfide protein, adrenodoxin (77, 78), and a selective cytochrome P-450 (79). Although Cooper et al (80) have found that isolated components promote a maximum rate of hydroxylation with the ratio of reductase : adrenodoxin : cytochrome = 1 : 50 : 1, debate continues on the ratio of the proteins in the intact organ.

The flavoprotein, NADPH-adrenodoxin reductase, isolated from bovine adrenal mitochondria (22) was reported to be immunologically distinct from the NADPH P-450 reductase from liver microsomes isolated by Masters et al (81).

The iron-sulfur protein, adrenodoxin, isolated in homogeneous crystalline form from both bovine and porcine adrenal cortex (82) bears a remarkable homology in sequence to the microbial putidaredoxin (33, 35, 83). Comprehensive reviews of the physical and chemical properties of the protein, including the redox potential, have appeared recently (84, 85). Although there are many physical and chemical similarities between adrenodoxin and putidaredoxin, they fail to replace one another in the heterologous systems (9), which may be related to the formation of a redoxin-cytochrome complex and the consequent regulatory effects (45).

The role of cytochrome P-450 in 11β hydroxylation by bovine adrenal cortex mitochondria was demonstrated in 1967 by Cooper and co-workers (79), who subsequently described the purification of the cytochrome (86). Details of solubilization appear to be critical, and the best preparation of the cytochrome, isolated in the ferric form, bears EPR, optical, and resonance spectra indicative of a low spin ferric iron (87). It also appears that enzymatic reduction of the cytochrome requires the binding of substrate, as demonstrated with the microbial P-450$_{cam}$-putidaredoxin couple (23).

17α-HYDROXYLASE   Steroids hydroxylated at the apex of the D-ring, for example the conversion of pregnenolone to testosterone, are precursors of androgens (88). The 17α-hydroxylase activity has been reported in the microsomal fraction of the

adrenal glands and testis. So far, efforts to purify this activity have been partially successful only with adrenal preparations, and it appears that the presence of the 11β hydroxylating system inhibits 17α-hydroxylase activity (89).

19-HYDROXYLASE   Steroids hydroxylated in the 19-methyl position are obligatory intermediates to the biogenesis of estrogens (90, 91). An NADPH-dependent system with 19-methyl hydroxylation activity has been identified in adrenal mitochondria and in the microsomal fraction from human placenta. The system that converts 4-androstene-3,17-dione to the 19-hydroxy compound has been reported to lack carbon monoxide sensitivity, thus casting doubt on the role of a cytochrome P-450 active center (92, 93).

21-HYDROXYLASE   In general, hydroxylation of the 21-methyl carbon adjacent to a C20 ketone leads to the androgen generation. The nature of the enzyme system(s) and the oxygenase prosthetic group remains clouded, due to the variety of

cytochromes and hydroxylase activities found in the adrenal cortex and the lack of success in freeing active 21-hydroxylase preparations from other enzymatic activities. The adrenal cortex microsomal fraction is known to catalyze the 21 hydroxylation. Ryan & Engel (94) reported carbon monoxide inhibition reversible by light, and thus suggested a P-450 cytochrome oxygen binding pigment, but Matthijssen et al (95, 96), with enzymatically active preparations, were unable to obtain spectral evidence for a P-450 cytochrome. Mackler et al (97) purified a P-450 system from adrenal microsomes that lost activity towards progesterone and pregnenolone as purification proceeded. The formation from a P-450 substrate complex, based on a Type I (blue) spectral shift, has been both reported (98, 99) and contested (100), the latter with the added suggestion that the spectral changes occur only in proportion to loss of enzymatic activity.

Hepatic Microsomal Cytochrome P-450 Monoxygenase

Orrenius & Ernster (101) have reviewed the state of progress with mammalian microsomal cytochrome P-450 hydroxylases obtained from the endoplasmic reticulum of several organs. The most extensively characterized system, found in hepatic microsomes, catalyzes oxygenation of both endogenous substrates, such as steroids and fatty acids, and a large variety of substances foreign to the natural environment of the animal body (102–107). The monoxygenase stoichiometry is preserved with O_2 incorporation (108, 109) at the expense of one NADPH per O_2 and substrate converted to product (110, 111). Among the reactive groups, both aryl and alkyl compounds are hydroxylated and substituted N-, O-, and S-residues are dealkylated or undergo deamination, N-oxidation, desulfurization, and S oxidation. Aromatic hydroxylation is characterized by ortho migration and retention of substituents at the position of hydroxylation (NIH shift) (112). In some aromatic hydroxylations, an epoxide has been identified as the actual enzymatic product that undergoes nonenzymatic rearrangement to a phenolic compound simultaneous with the migration reaction (113).

A breakthrough in the difficulties encountered in solubilizing and identifying catalytically active microsomal hydroxylase components occurred in 1968 with the advances of Lu & Coon (114), who, by using detergents in the presence of glycerol and sulfhydryl reagents, were able to separate three soluble fractions from liver microsomes, which on recombination possessed monoxygenase activity toward fatty acids. The recent review by Lu & Levin (115) summarizes the subsequent use of these methods. Such studies have demonstrated a direct reduction of liver microsomal cytochrome P-450 by a NADPH-specific flavoprotein without the intervention of an $Fe_2S_2^*Cys_4$ iron-sulfur protein, as observed in the adrenal mitochondrial and microbial cytochrome P-450 hydroxylases. Coon and his co-workers (24) and Imai & Sato (25) have made a number of advances, culminating in the purification of cytochrome P-450 preparations from phenobarbital-induced rabbit liver microsomes which migrate as a single band in sodium dodecylsulfate (SDS)-polyacrylamide electrophoresis with apparent minimum molecular weights of 45,000 to 50,000. The purified hemoproteins retain their catalytic activity. The reductant is a microsomal flavoprotein, NADPH-cytochrome P-450 reductase, which has also been purified to homogeneity by both Sato and co-workers (116) and Vermilion & Coon (117). For activity, the cytochrome preparation of Coon et al requires a synthetic phospholipid, phosphatidylcholine, whereas that of Sato et al does not. This dissimilarity in phospholipid requirement may result from an activating effect of the detergent used in isolation of the components (118) and from differences in the maximal turnover number of the purified proteins. This dissimilarity is certain to be resolved shortly.

Although the highly purified preparation of cytochrome P-450 appears to be homogeneous, it is quite evident that intact microsomes contain multiple species, perhaps less than ten but certainly more than one, of the cytochrome. The differential effect in the induction of cytochrome P-450 synthesis and hydroxylase activity following pretreatment of animals with barbiturates vs polycyclic aromatic hydrocarbons has been the most significant form of evidence for multiple cytochromes (101). Barbiturate induction increases activity toward a large variety of substrates, but activity induced by the aromatic hydrocarbons, such as 3-methylcholanthrene, is highly selective for hydroxylation of aromatic substrates and is associated with the synthesis of a new spectral form of the cytochrome, called cytochrome P-448, characterized by an absorption maximum in the CO difference spectrum at 448 (or 446) nm. Lu et al (119) demonstrated that partially purified preparations of cytochromes P-450 and P-448 obtained from phenobarbital- or 3-methyl-cholanthrene-treated animals retained the substrate specificity observed with intact microsomes. Nebert and co-workers (120) have similarly found that genetic differences in the hydroxylase activity of inbred strains of mice are associated with the cytochrome fraction. Additional evidence for multiple cytochromes has been obtained with SDS-polyacrylamide electrophoresis of microsomal membrane proteins (121, 122). Welton & Aust (123) utilized the peroxidase activity retained by SDS-solubilized hemoproteins to show that rat liver microsomes have three separable hemoproteins in the 50,000 mol wt range. Pretreating the animals with pheno-barbital induced the hemoprotein with molecular weight 44,000, while 3-methylcholanthrene induced a hemoprotein of molecular weight 53,000. Coon and co-workers (124) similarly found that rabbit liver microsomes contain four separable hemoproteins. Phenobarbital pretreatment induces the hemoprotein of molecular weight 50,000, which they have purified to homogeneity (25).

The NADPH-cytochrome P-450 reductase, commonly called NADPH-cytochrome c reductase because of its capacity to reduce exogenous cytochrome c, can be purified to homogeneity following proteolytic digestion (101). Iyangi & Mason (125), in detailed studies of the pig and rabbit liver flavoprotein purified following proteolytic digestion, reported the presence of one molecule each of FAD and FMN. Similar flavin content has also been determined by investigating preparations from other animal species as well as other organs (126–128). The reductases purified following proteolytic digestion promote little or no hydroxyla-tion activity when substituted for the detergent-solubilized reductase (129, 130). SDS-polyacrylamide electrophoresis indicated that the detergent-solubilized preparation from rat liver microsomes is larger by about 8,000 daltons than the 71,000 mol wt protein isolated after proteolytic release (131). The reductase appears similar to cytochrome b_5 and the NADH-cytochrome b_5 reductase, which are termed "amphipathic" because they are attached to the microsomal membrane by terminal sequence of predominantly hydrophobic amino acid residues and can be readily liberated from this hydrophobic tail by proteolytic cleavage (132). Other studies also indicate that the hydroxylase activity promoted by the detergent-solubilized preparations may involve hydrophobic interactions, as indicated by the formation of dissociable complexes with the detergent-solubilized cytochrome under reaction conditions promoting hydroxylation (133).

Microsomal drug hydroxylation is dependent on NADPH. NADH is one or two orders of magnitude less active when added alone, but when added to saturating levels of NADPH, the rate of hydroxylation is in many instances stimulated appreciably (101). Changes in the redox state of microsomal cytochrome b_5 accompanying hydroxylation led Hildebrandt & Estabrook (134) to propose that cytochrome b_5 is involved in the transfer of the second reducing equivalent to cytochrome P-450. The reduction of cytochrome b_5 by NADH-cytochrome b_5 reductase would account for the NADH enhancement of cytochrome P-450-catalyzed activity. Subsequent studies suggest that cytochrome b_5 is unlikely to be an obligatory component but does promote the hydroxylation activity in the intact microsome (135, 136). Both the increase in activity observed with NADH plus NADPH and the slow hydroxylation activity observed with NADH alone apparently depend on electron transfer by cytochrome b_5 (137–139). West et al (140) have recently shown that hydroxylation can be promoted by NADH with detergent-solubilized NADH-cytochrome b_5 reductase, cytochrome b_5, cytochrome P-448, and lipid.

The uncoupling of oxygen reduction from hydroxylation activity by microsomal cytochrome P-450 has led to an alternative explanation of the role of cytochrome b_5. Narasimhula (141) observed that adrostenedione binds to cytochrome P-450 in adrenal cortex microsomes and promotes oxygen uptake and disappearance of NADPH without producing hydroxylated products. Ullrich and co-workers (142) found a similar reaction in liver microsomes promoted by perfluorinated hydrocarbons lacking hydroxylatable C–H bonds. In the absence of substrate, liver microsomes catalyze a slow oxidation of NADPH with formation of H_2O_2, but the oxidase activity promoted by these uncouplers results in the reduction of oxygen to water. Liver microsomes will hydroxylate both n-hexane and cyclohexane; the latter hydroxylation is tightly coupled stoichiometrically in NADPH/product, but hydroxylation of n-hexane is accompanied by oxidation of additional pyridine nucleotide (111). The rate of n-hexane but not cyclohexane hydroxylation is enhanced by NADH through the reduction of cytochrome b_5. Thus the suggestion was made that cytochrome b_5 reduction by NADH frees the NADPH oxidase activity, thereby enhancing the hydroxylation rate.

There are several unresolved problems of P-450 cytochrome action both in intact microsomes and with the purified preparations. For example, in intact systems, substrate interactions are often predicted on the characteristic change in Soret absorbance maxima, either to the blue, Type I, or to the red, Type II (101). In some reports, the addition of substrate enhances the rate of P-450 reduction by NADPH (111, 143, 144). The data are unclear on the transfer route of the second electron for hydroxylation by microsomal cytochrome P-450. Estabrook and co-workers (145, 146) observed a new spectral species which appears similar to the well-characterized oxygenated intermediate of purified cytochrome P-450 from *P. putida* (23). However, with the purified microsomal P-450 preparations, the spectrum of an oxygenated intermediate has not been observed. The reduction of purified microsomal cytochrome P-450 was found by Coon and co-workers (147) to have a reduction equivalence of two, using either dithionite or NADPH, whereas reduction of pure cytochrome P-450$_{cam}$ gave an equivalence of one. However, the redox center reduced by the additional reducing equivalent and its role in the hydroxylation activity remain unknown.

OTHER IRON-LINKED HYDROXYLASES

The Pteridine-Dependent Hydroxylases

More than ten years ago Kaufman (148) reported the pteridine-coupled hydroxylation of phenylalanine to tyrosine by a two component system recovered from mammalian liver and kidney. The excellent series of precisely documented advances included identification of the liver enzyme as a hydroxylase with the new prosthetic group, biopterin (149), that incorporates an atom from O_2 into the aromatic 4-hydroxyl group (150) when coupled to the FAD-containing NADPH reductase isolated from kidney. Using stoichiometric concentrations of tetra-hydropteridine, hydroxylation proceeds as indicated to form the quinoid form of dihydropteridine. With catalytic concentrations of biopterin or other pteridines, the flavin-coupled NADPH dehydrogenase serves to keep the cofactor in the reduced state. Similar roles of reduced pteridines have been demonstrated for the phenylalanine hydroxylase of a *Pseudomonas* species (151), mammalian tyrosine hydroxylase (152), and trytophan hydroxylase (153). For detail, the reader is referred to the review by Kaufman & Fisher (154).

The role of iron in pterin-dependent hydroxylation was first suggested by the inhibitory effect of ferrous ion chelators (152, 155–159). The addition of ferrous ion also stimulates activity, but possibly by accelerating the breakdown of hydrogen peroxide, since catalase will usually replace the iron requirement (157). The pure phenylalanine hydroxylase contains nearly two atoms of iron (158). Their removal is accompanied by loss of activity and a restoration on the readdition of ferrous iron. The native holoenzyme has a minor absorption band in the visible region near 410 nm and electron spin resonance signal at $g = 4.28$, indicative of high spin ferric iron, which disappears upon substrate addition. The role of iron in other pteridine-dependent oxygenase has not been demonstrated as thoroughly, although Petrack et al (159) reported that bovine adrenal medulla tyrosine hydroxylase is activated by preincubation with ferrous ion. Tyrosine hydroxylase from the adrenal medulla is primarily a particulate enzyme and purification involves an initial proteolytic digestion to solubilize the enzyme. In brain, tyrosine hydroxylase is found both in a soluble form and attached to synaptic membranes, with the membrane-bound form having a greater affinity for tetrahydropterin (160). Treatment of the soluble form of the enzyme with either heparin or trypsin "activates" the enzyme by increasing the affinity for the substrates (161).

With the natural substrate and cofactor, NADPH oxidation via phenylalanine hydroxylase is tightly coupled to hydroxylation, but in the presence of 4-fluorophenylalanine, oxidation of excess NADPH was found to occur (162). The use of 7-methyl-tetrahydropteridine in place of tetrahydrobiopterin also promoted oxidation of excess NADPH with the rest of the reducing equivalents being recovered as hydrogen peroxide (163). Tyrosine hydroxylase from adrenal medulla is also partially uncoupled in the presence of some substitute cofactors (153). The uncoupled oxidase activities observed using other tetrahydropteridines also depend on the presence of substrate, indicating the probable relationship of oxidase activity to the oxygen activation process. In the presence of tyrosine, phenylalanine hydroxylase is partially uncoupled by the addition of lysolecithin or α-chymotrypsin, which stimulate activity with the natural substrate (164). A protein isolated from liver, called phenylalanine hydroxylase-stimulating protein, also increases the activity of this enzyme but apparently by a different mechanism, because the stimulation by lysolecithin and α-chymotrypsin was synergistic with the stimulation by this protein (165).

Initial kinetic studies with phenylalanine hydroxylase suggested that catalysis involved a cyclic process commonly referred to as the ping-pong mechanism, but later work failed to confirm these findings and indicated that the reaction involved the formation of a quarternary complex without evidence of ordered binding (166). In addition, hydroxylation was accompanied by migration of the substituent at position 4 to position 3 on the ring (NIH shift) (167). Tyrosine hydroxylation by the tyrosine hydroxylase does not involve any shift of ring substituents, but the hydroxylation of phenylalanine by this enzyme does (168). Additional insight into the mechanism of phenylalanine hydroxylase has come from an examination by Kaufman and co-workers (165) of the increase in activity observed in the presence of the stimulating protein. Kinetic analysis is consistent with a mechanism of product precursor release that either breaks down to form product or rebinds to the enzyme with the stimulating protein effecting the product-forming reaction (8). Since the kinetics indicate that binding of this species competes with binding of tetra-hydropteridine, the intermediate probably retains the pterin moiety. S. Kaufman (personal communication) has recently found that the sum of intermediate plus dihydrobiopterin that accumulates is equal to the tyrosine formed at any given interval, strongly implying that the intermediate is a precursor to the dihydropteridine, possibly a hydrated form. Further characterization of this intermediate may be anticipated.

Decarboxylation-Coupled Hydroxylases

A specific requirement for α-ketoglutarate in the hydroxylation of collagen proline residues was discovered in 1966 by Hutton et al (169). Subsequent studies have revealed a similar cosubstrate requirement for a number of hydroxylases, described in detail in recent reviews (170, 171). These include the prolyl and lysyl hydroxylases, which hydroxylate the indicated residues in protocollagen and related peptides; plant prolyl hydroxylase, which acts upon proteins of the cell wall; γ-butyrobetaine hydroxylase from both mammalian and bacterial sources, which converts γ-butyrobetaine to ornithine; and several enzymes from *Neurospora crassa,* which catalyze the multistep oxidation of thymine to uracil-5-carboxylic acid and the 2′ hydroxylation of the deoxyribonucleosides of thymidine and deoxyuridine. Ferrous ion chelators inhibit the activity of these enzymes in cell extracts, and the purified hydroxylases commonly require addition of ferrous ion. Maintenance or restoration of enzymatic activity also requires a reducing agent, commonly ascorbic acid.

Rhoads & Udenfriend (172) were able to purify prolyl hydroxylase from chick embryos and show that during hydroxylation, α-ketoglutarate is oxidized stoichio-metrically to succinate plus CO_2. The incorporation of dioxygen into hydroxy

$$R-H + \begin{array}{c} COOH \\ | \\ C=O \\ | \\ CH_2 \\ | \\ CH_2 \\ | \\ COOH \end{array} + O_2 \xrightarrow[\text{reducing agent}]{Fe^{2+}} R-OH + \begin{array}{c} COOH \\ | \\ CH_2 \\ | \\ CH_2 \\ | \\ COOH \end{array} + CO_2$$

proline had also been demonstrated (173, 174). However, Lindblad et al (175) discovered that during the hydroxylation of γ-butyrobetaine, one atom of dioxygen is incorporated into ornithine and the other into the carboxyl group of succinic acid. Similar oxygen incorporations were shown for other α-ketoglutarate-coupled systems (176, 177). Thus, Hayaishi (5) has used the more selective term "intermolecular dioxygenase" for this subclass.

Ferrous iron must be added to the majority of purified ketoglutarate-dependent hydroxylases because endogenous iron is lost during purification. In the cases of prolyl hydroxylase, some preparations reportedly retain tightly bound iron (170), but Pänkäläinen & Kivirikko (178) found that purified preparations contained much less than 1 g iron per mole of enzyme. The replacement of ascorbate by other reducing agents suggests that the ascorbate is not directly involved in catalysis of hydroxylation (169, 179). Ascorbate is also required for normal collagen synthesis in vivo, but other observations, such as the failure of protocollagen to accumulate in tissues of scorbutic animals, leave the role of ascorbate in dispute (180). It has been suggested that the reducing agents are required to maintain the ferrous form of the iron atom or perhaps sulfhydryl groups in the reduced state. Sulfhydryl reagents, including mercurials, have been shown to inhibit these hydroxylases, and α-ketoglutarate exhibits a substrate-type protective effect (179, 181). The generation of hydrogen peroxide will also inactivate these enzymes and the addition of catalase is commonly required (182). The highly purified preparation of prolyl hydroxylase also requires the addition of other proteins, such as serum albumin, for optimal activity (171).

Hydroxylation coupled to oxidative decarboxylation is also catalyzed by p-hydroxyphenylpyruvate hydroxylase from liver or kidney, but this reaction is promoted by decarboxylation of the pyruvate side chain accompanied by migration of the carboxymethyl group to the adjacent carbon to form homogentisic acid (183,

184). As with α-ketoglutarate-dependent hydroxylases, ferrous iron and a reducing agent are required by this enzyme (185). Initial studies indicated that a net incorporation of one atom from dioxygen occurred in this reaction (186), but more recent studies by Lindblad et al (187), who considered the extent of oxygen exchange with water, demonstrate that dioxygen is incorporated into both the hydroxyl and carboxyl groups of homogentisate, suggesting that this enzyme should be classified as a dioxygenase.

To account for the γ-butyrobetaine hydroxylase-catalyzed incorporation of dioxygen into both succinate and the hydroxylated product, Lindstedt & Lindstedt (179) proposed that an enzyme-bound ferrous ion-oxygen complex attacks the substrate to form a peroxide which then oxidizes α-ketoglutarate, liberating CO_2, succinate, and the hydroxylated product. A similar mechanism was proposed by Cardinale et al (176) for prolyl hydroxylase. However, Hamilton (188) postulated that an oxyferrous complex attacks the α position on α-ketoglutarate to form persuccinic acid which then hydroxylates the substrate. He noted that experiments of Rhoads & Udenfriend (172) showed that prolyl hydroxylase catalyzes a small release of carbon dioxide from α-ketoglutarate in the absence of the prolyl substrate. Abbott & Udenfriend (170) have recently reported that experiments with substrate levels of prolyl hydroxylase confirm the previous finding. An initial attack by oxygen on α-ketoglutarate followed by an electrophilic hydroxylation reaction has also been suggested by Hobza et al (189) on the basis of molecular orbital calculations. Additional insight into the hydroxylation mechanism comes from the study of p-hydroxyphenylpyruvate hydroxylase. Goodwin & Witkop (190) first proposed a reaction mechanism involving a cyclic peroxide, which was initiated by deprotonization of the aromatic hydroxyl group followed by addition of oxygen, but the enzyme will also catalyze hydroxylation of phenylpyruvate to form o-hydroxyphenylacetate. The hydroxylation of p-fluorophenylpyruvate to form 5-fluoro-2-hydroxyphenylacetate also demonstrated that migration of the ring substituent still occurred (191). Hamilton (188) therefore suggested the involvement of a perbenzylic acid intermediate analogous to his

earlier interpretation of the mechanism of α-ketoglutarate-dependent hydroxylase and the nonenzymatic hydroxylation by peracids of aromatic compounds, which involves formation of an epoxide followed by rearrangement to the phenolic compound described by Daly et al (113).

The ω-Hydroxylase and 4-Methoxybenzoate Monoxygenase Systems

A nonheme iron prosthetic group is involved in the hydroxylations catalyzed by both α-ketoglutarate- and pteridine-dependent hydroxylases, and mechanisms in which oxygenation of the cofactor precedes the hydroxylation of the substrate have been postulated for both types of oxygenases, leaving the role of iron in the actual hydroxylating reaction in doubt. However, two hydroxylase systems have been characterized in which nonheme irons are apparently the only prosthetic groups directly involved in the hydroxylation reactions. These are the ω-hydroxylation system isolated by Coon and co-workers (192) from *P. oleovorans*, which hydroxylates the terminal carbon of long chain fatty acids and hydrocarbons, and the 4-methoxybenzoate monooxygenase isolated by Bernhardt et al (193) from *P. putida*, which catalyzes both oxidative *o*-demethylation and hydroxylation of the aromatic ring. Stoichiometric studies showed that the ω-hydroxylation system consumes one mole of oxygen and oxidizes one mole of NADH per mole of substrate hydroxylated (194), and 4-methoxybenzoate monooxygenase, in a reaction requiring NADH, consumes a mole of oxygen per mole of product formed (195). However, direct evidence for incorporation of molecular oxygen has not been established for either enzyme system.

The ω-hydroxylation system was found to be separable into three components: a nonheme iron protein similar to the rubredoxins, a flavoprotein, and a final component required for hydroxylase activity called the ω-hydroxylase (196). The third component has recently been characterized as an iron-containing protein of molecular weight 42,000 with one atom of iron and no labile sulfide (11). Its visible absorption spectrum is characterized by only a single weak absorbance band at about 415 nm, similar to that of liver phenylalanine hydroxylase (158). When the iron was removed by chelating agents, the apo-ω-hydroxylase was catalytically inactive, but incubation with ferrous ion restored activity. The purified hydroxylase also contains a large amount of bound phospholipid; hydroxylation activity is lost in parallel with the removal of phospholipid, but the activity is recovered when phospholipids are added back. The rubredoxin from this system is a nonheme iron protein with iron bound to four of the protein's ten cysteine residues, and it contains no labile sulfide (197). The oxidized protein has an EPR spectra with a complex resonance at $g = 4.3$, characteristic of the rubredoxins. The reduction of this rubredoxin is catalyzed by the NADH-specific flavoprotein containing one mole of FAD per mole of enzyme (198). It will also readily reduce the rubredoxins isolated from other bacteria, but these rubredoxins will not promote any hydroxylation activity when substituted for the rubredoxin from *P. oleovorans* (199). The isolated rubredoxin, a single polypeptide with a molecular weight of 19,000, contains one atom of iron, but the protein will also bind a second atom of iron at a similar binding site, doubling the intensities of both the visible absorption bands and the EPR signal (197). Both irons can be enzymatically reduced, but the addition of a second iron atom does not increase the catalytic effectiveness. It was also found that cyanogen bromide cleaves the rubredoxin into two peptides with each peptide containing one of the iron binding sites. The peptide from the carboxyl terminal, which contains the iron present in the one iron protein, will still function in the hydroxylation system, but the iron-containing peptide from the NH_2 terminal promoted negligible hydroxylation.

The 4-methoxybenzoate monoxygenase was first characterized as a three component system containing a flavoprotein, an iron-sulfur protein, and a third unidentified component (193). The flavoprotein catalyzed the NADH-dependent reduction of the iron-sulfur protein, but substrate hydroxylation required the addition of the third component. It was later reported that the system could be fractionated into two components, the iron-sulfur protein and an iron-containing flavoprotein (195). The iron-containing flavoprotein is a homogeneous protein as determined by polyacrylamide disc electrophoresis and sedimentation studies, but SDS dissociates the protein into several peptides separable by polyacrylamide electrophoresis. The visible absorption spectrum of the iron-containing flavoprotein suggests that an iron-sulfur chromophor is present (200). The second component in this system was found to contain acid-labile sulfide and an electron spin resonance exhibiting an anisotropic signal with g values at 2.01, 1.91, and 1.78 when reduced. The reduced protein also exhibits absorbance bands at 416 and 516 nm. The iron-sulfur center is reduced by the iron-containing flavoprotein and both components are required for hydroxylation. Sensitivity to oxidative loss of activity necessitated the use of anaerobic conditions during purification. Addition of NADH stabilized the iron-containing flavoprotein, whereas the iron-sulfur protein was stabilized by 4-methoxybenzoate, indicating that the iron-sulfur protein contains the catalytic site.

The catalytic roles of the prosthetic groups in either the ω-hydroxylation system or the 4-methoxybenzoate monoxygenase are not well defined, but other characteristics indicate that the mechanism of oxygen activation and substrate hydroxylation is similar to that of other monoxygenases. Oxygen consumption by the 4-methoxybenzoate monoxygenase system depends on the addition of substrate which is hydroxylated stoichiometrically with the consumption of oxygen. However, when a substrate analog lacking a hydroxylatable C–H bond is added, NADH-dependent reduction of oxygen to H_2O_2 occurs (195). Other substitute substrates such as 4-methylbenzoate or 3-methoxybenzoate are hydroxylated but account for only a fraction of the oxygen consumed, and the apparent Michaelis constant for oxygen is increased. The ω-hydroxylase system will also catalyze the epoxidation of olefins such as 1,7-octadiene in a reaction that still requires all three protein components plus NADH and oxygen (201). This activity competes with the ω hydroxylation of octane, suggesting that epoxidation is promoted by the same activated complex that promotes ω hydroxylation. It has also been shown that the ω-hydroxylation system will catalyze the reduction of octyl-hydroperoxide to octanol (202), suggesting that hydroxylation may occur via formation of a peroxide, but there has been no other evidence for the formation of hydroperoxides during catalysis of hydroxylation.

COPPER-CONTAINING HYDROXYLASES

Dopamine-β-Hydroxylase and the Phenolases

The hydroxylation of dopamine to norepinephrine catalyzed by the enzyme dopamine-β-hydroxylase, which is associated with the chromaffin granule in the adrenal medulla (203), is the most thoroughly characterized enzymatic hydroxylation involving a copper prosthetic group. Kaufman and co-workers (204) isolated the enzyme and found that hydroxylation was coupled to the oxidation of ascorbic acid with the overall stoichiometry indicated. The subsequently demonstrated incorporation of molecular oxygen into product confirmed the classification of this enzyme as a monoxygenase (150). There are also a number of copper-containing oxygenases found in vertebrates, arthropods, and diverse forms of plant life, which are known collectively as phenolases or tyrosinases (205, 206). These proteins do not have any strong absorbance bands in the visible region and are classified as nonblue copper enzymes (207). Phenolases catalyze the oxygen-dependent conversions, of either phenols or catechols to *o*-quinones, referred to as the monophenolase and diphenolase activities. The monophenolase activity is a complex reaction as evidenced by an induction or lag period which can be eliminated by adding an *o*-dihydroxyphenol. *o*-Dihydroxyphenols can also be isolated from reaction mixtures during the oxidation of monophenols, which suggests that the monophenolase activity is actually a hydroxylation of a monophenol to an *o*-diphenol coupled to the oxidation of a diphenol to an *o*-quinone (205, 206). Although this hydroxylation of

monophenols has been regarded by some investigators as nonenzymatic (206), the incorporation of an atom of oxygen from molecular oxygen into one of the hydroxyl groups of the *o*-diphenol (3) and other evidence have led most investigators to conclude that the reaction is enzyme mediated and should be classified as a monoxygenase. A more detailed account of work on the properties of phenolases and dopamine-*β*-hydroxylase can be found in the recent review of the copper-containing oxygenases by Vanneste & Zuberbuhler (208).

The involvement of copper in phenolases was implicated by the historic experiments of Kubowitz (209), demonstrating that activity was lost upon removal of copper and could be restored by adding back cupric ions. After dopamine-*β*-hydroxylase was first purified to homogeneity and characterized as a nonblue copper-containing enzyme, the involvement of the copper in catalysis was indicated by the inhibitory effect of copper chelating agents (210). The inhibition of both dopamine-*β*-hydroxylase and phenolases by carbon monoxide was indicative of cuprous copper involvement in catalysis. With substrate quantities of dopamine-*β*-hydroxylase, the enzyme was shown to oxidize two equivalents of ascorbate, and the reduced form of the enzyme could be combined with substrate and oxygen to form an equivalent amount of product (210). The oxidation of ascorbate was shown to reduce approximately two equivalents of Cu^{2+} per mole of enzyme, and EPR studies of the reaction showed that the Cu^+ was reoxidized during the subsequent hydroxylation reaction (211, 212). On the basis of these results, a mechanism involving cyclic reduction by ascorbate and oxidation by substrate and oxygen at a catalytic center containing two copper ions was proposed (212)

$$E(Cu^{2+})_2 + ascorbate \rightarrow E(Cu^+)_2 + dehydroascorbate$$

$$E(Cu^+)_2 + O_2 + dopamine \rightarrow E(Cu^{2+})_2 + norepinephrine + H_2O$$

This mechanism was supported by the studies of Goldstein et al (213), who showed that the kinetics of the steady-state reaction were consistent with a ping-pong mechanism in which the enzyme first interacts with ascorbate followed by a reaction with O_2 and substrate to form a ternary complex before products are released. Much of the difficulty in characterizing the active site of dopamine-β-hydroxylase stems from variations in the measure of total copper content. The original purified preparations were reported to contain from three to seven equivalents of copper per mole of enzyme whose molecular weight is approximately 290,000 (210). Recent preparations of the enzyme, characterized as a glycoprotein composed of four 75,000 mol wt peptides, have been found to contain 4 to 5 equivalents of copper per mole (214). The actual number of copper ions associated with an active center is not known with certainty. A mechanism involving the ascorbate radical at an active center containing a single cuprous ion has been proposed (211). It has recently been suggested that a histidine-mediated proton transfer is involved in catalysis of hydroxylation, as evidenced by the pH dependence of the enzymatic reaction, inactivation of the enzyme by the histidine reagent, diethylpyrocarbonate, and a pH-dependent deuterium isotope effect (215). It was also shown that the transferred proton apparently originates from ascorbate rather than water. Other aspects of this hydroxylation reaction, such as the stimulation by carboxylic compounds (204, 213) and the postulated involvement of an oxygen adduct of the reduced enzyme, remain to be resolved (216).

The reaction mechanism of the phenolase reactions is much less clearly defined. In contrast to dopamine-β-hydroxylase, the copper in phenolases shows no EPR signal, suggesting a cuprous state. Chemical determinations by some investigators have indicated that essentially all the copper is in the form of Cu^+ (206). However, other evidence, such as the reaction of phenolase with HOOH to form a product that reversibly binds oxygen and has an absorbance spectrum like oxyhemocyanin (217), indicates that the resting enzyme contains an active center with two Cu^{2+} ions. The existence of Cu^{2+} in spin-coupled pairs could also explain the lack of electron paramagnetic resonance. The spectrum of the oxygenated enzyme disappeared rapidly in the presence of phenolic substrates, suggesting that it may be an intermediate in the enzymatic reaction. A reaction mechanism originally proposed by Mason (205)

$$E(Cu^{2+})_2 + o\text{-diphenol} \rightarrow E(Cu^+)_2 + o\text{-quinone} + 2H^+$$

$$E(Cu^+)_2 + O_2 \rightleftarrows E(Cu^+)_2 O_2$$

$$E(Cu^+)_2 O_2 + monophenol \rightarrow E(Cu^{2+})_2 + o\text{-diphenol} + 2H_2O$$

$$E(Cu^+)_2 O_2 + o\text{-diphenol} \rightarrow E(Cu^{2+})_2 + o\text{-quinone} + 2OH^-$$

is accepted as the most probable reaction mechanism, although a reaction mechanism involving a single copper ion per catalytic site has also been proposed for these enzymes (208).

FLAVOPROTEIN HYDROXYLASES

The flavoprotein hydroxylases reviewed in this section are monoxygenases incorporating one atom of dioxygen into substrate and the other into water, which requires two reducing equivalents obtained from reduced pyridine nucleotides. The enzymes discussed are primarily of bacterial origin and many have been highly purified, allowing elegant studies on the reaction mechanisms involved. These enzymes are usually a single polypeptide containing one molecule of FAD which participates in both the oxidation of reduced pyridine nucleotide and the oxygenation reactions leading to product formation.

Similarity in catalytic properties has provided a basic mechanistic sequence applicable to the flavoprotein hydroxylases. In almost all cases the binding of substrate markedly affects the rate of coupled flavin reduction and pyridine nucleotide oxidation. In many cases distinct intermediates have been suggested in the $NAD(P)H/FAD \rightarrow NAD(P)^+/FADH_2$ reaction. In addition reduction of oxygen is uncoupled from the oxygenase activity by substrate analogs known as effectors, which stimulate the rate of pyridine nucleotide oxidation but are not hydroxylated.

Perhaps the most interesting aspect of these flavoprotein enzymes is the elucidation of the molecular mechanism of oxygen activation and attack on the carbon chain of the substrate. During the oxygenation reaction, intermediates including flavin oxygen adducts have been indicated and, in some cases, characterized. Further study of these important reactions will undoubtedly contribute much to an overall understanding of hydroxylation mehanisms. Several reviews of the flavoprotein oxygenases have recently appeared (12, 217). The excellent review by Flashner & Massey (12) stresses the elucidation of the molecular mechanisms through detailed kinetic studies.

p-Hydroxybenzoate Hydroxylase

The most carefully studied FAD hydroxylases and perhaps the best characterized in terms of molecular mechanism are the bacterial enzymes that catalyze the hydroxylation of p-hydroxybenzoates. These enzymes, which are all very similar in terms of molecular weight ($\simeq 65,000$), prosthetic group (FAD), and catalytic properties, have been obtained in crystalline form from four species of pseudomonads (218). The hydroxylation activity of all these enzymes is coupled to the oxidation of NADPH.

Howell, Spector & Massey have isolated and crystallized a hydroxybenzoate hydroxylase from *Pseudomonas fluorescens* (218). The rate of reduction of the oxidized enzyme by NADPH is markedly affected by the presence or absence of substrate, which is bound with 1:1 stoichiometry by the oxidized enzyme. The substrate-bound species is reduced with a pseudo first order rate constant of 253 sec^{-1}, whereas in the absence of substrate the rate is reduced to 0.0068 sec^{-1}. The reduction of the enzyme proceeds through an NADPH-enzyme complex and in the absence of *p*-hydroxybenzoate, the dissociation constant for NADPH is increased 13-fold. Rapid reaction studies by Massey and co-workers (218–220) allowed the identification of several intermediates during anaerobic reduction of the enzyme by NADPH in the presence of *p*-hydroxybenzoate or 2,4-dihydroxybenzoate. These intermediates are characterized by long wavelength charge-transfer bands.

The stable precursor to the hydroxylation reaction appears to be a reduced enzyme-substrate complex. Upon admission of oxygen, stoichiometric formation of hydroxylated substrate is observed. Further examination of this reaction with rapid reaction techniques by Massey and co-workers (219, 221) indicated the involvement of three intermediates

$$
\begin{array}{c}
O_2 \\
\underset{\substack{| \\ S}}{E}\!-\!FADH_2 + O_2 \xrightarrow{k_1} \underset{\substack{| \\ S}}{E}\ FADH_2 \xrightarrow{k_2} II \xrightarrow{k_3} \underset{\substack{| \\ S-OH}}{E}\!-\!FAD_2(H_2) \xrightarrow{k_4} E\!-\!FAD + product
\end{array}
$$

The reaction of O_2 with the reduced enzyme-substrate complex proceeds very fast, forming an oxygenated flavin adduct. The molecular structure of this oxygenated flavin intermediate was suggested by Spector & Massey (219) to have an –OOH attached to the 10-N position of the isoalloxazine ring. Müller, Hartmann & Hemmerich (222) believe the –OOH group is attached at the 1a position, whereas Hamilton (188) proposes a detailed oxenoid mechanism involving –OOH substitution at the 4a position. However, Orf & Dolphin (223) have proposed, on the basis of molecular orbital calculations, that the reduced flavin and oxygen form a 1a-4a-dioxetane, which rearranges to give an oxaziridine, and that the oxaziridine serves as the hydroxylating agent. Rapid acid quenching experiments indicate that an oxygen atom is transferred to a substrate-forming intermediate (II), perhaps a hydroxydihydroflavin product precursor complex, with a rate constant $k_2 = 5300$ min^{-1}. This product precursor might be a quinoid-type intermediate shown below. Deprotonation then yields a compound (III) with $k_3 = 21$ min^{-1}, perhaps a trihydroxybenzoate-hydroxydihydroflavin complex. Elimination of H_2O and release of product would then regenerate oxidized enzyme in a limiting rate of

$k_4 = 8.3$ min^{-1}. Similar intermediates may be involved in the reaction with p-hydroxybenzoate as substrate and indeed may be a general mechanism of flavoprotein hydroxylations. The overall rates for the individual steps are a function of the substrate used.

The reduction of oxygen is uncoupled from substrate hydroxylation by substrate analogs called effectors (224). 3,4-Dihydroxybenzoate and 6-hydroxynicotinate stimulate the NADPH oxidase activity, but are not themselves metabolized. Benzoate is an effector but is metabolized to m-hydroxybenzoate with less than 5% efficiency. 2,4-Dihydroxybenzoate is also both effector and substrate, being rapidly converted to 2,3,4-trihydroxybenzoate. The substrate, p-hydroxybenzoate, is an inhibitor of the enzyme at high concentrations (224–226), which may indicate the presence of two substrate binding sites. However, 2,4-dihydroxybenzoate, although a substrate, is not inhibitory at high concentrations (224). The substrate-free enzyme or complexes of reduced enzyme with the nonsubstrate effectors show no evidence for an intermediate in the reoxidation process, even though the rate of reoxidation is faster than that of the unliganded enzyme.

Salicylate Hydroxylase

The first flavoprotein hydroxylase shown to be coupled to an external source of reducing equivalents was salicylate hydroxylase from P. putida (227). A detailed description of the studies with this enzyme can be found in several recent reviews (217, 228). The hydroxylation is accompanied by a decarboxylation, resulting in the conversion of salicylate to catechol. The purified NADH-specific flavoprotein is a monomer of 57,000 mol wt which binds one molecule of FAD (229, 230).

The binding of salicylate causes a small shift in the absorption bands of the enzyme and concomitantly quenches the fluorescence of the prosthetic group. Both techniques yield a $1:1$ binding stoichiometry with a dissociation constant of about 3.5 μM (230). The apoprotein can also bind salicylate with a dissociation constant measured by fluorescence techniques to be 1.8 μM. 3-Methylsalicylate has been shown to be a substrate for the enzyme, being hydroxylated to 3-methylcatechol (231). Other substrate analogs, such as 2,3-, 2,4-, and 2,5-dihydroxybenzoate, p-amino-salicylate, and 1-hydroxy-2-naphthoate, stimulate NADH oxidation, but as no product analyses were made it is impossible to classify them as either substrate or simply effectors for NADH oxidase activity (229). The compounds m- and p-hydroxybenzoate and 3,4-dihydroxybenzoate are not substrates for the enzyme and do not cause any observable shift in the absorption spectra of the enzyme (231).

Anaerobic reductive titration of the salicylate-bound enzyme with NADH shows two electron stoichiometry with no evidence for a semiquinone intermediate. The reduced E-FADH$_2$ substrate complex rapidly undergoes stoichiometric conversion of salicylate to catechol upon admission of oxygen. Katagiri and co-workers proposed the following model of the salicylate hydroxylase reaction (228)

$$E\text{—}FAD + S \rightleftarrows E\text{—}\overset{\overset{\displaystyle S}{\diagup}}{FAD} \xrightarrow{+NADH} X_1 \rightarrow E\text{—}\overset{\overset{\displaystyle S}{\diagup}}{FADH_2} + O_2 \rightarrow E\text{—}FAD + catechol + CO_2 + H_2O$$

Although these authors indicate the presence of intermediates in both reduction and reoxidation steps, no such intermediates have been found. In addition, Flashner & Massey (12) have criticized the interpretation of the kinetic data indicating X_1 and X_2. Clearly, more careful work is needed to unambiguously identify these intermediates which are, in all probability, general features of the flavoprotein hydroxylation mechanism.

Kamin and co-workers (232–234) have isolated a different salicylate hydroxylase from a soil microorganism. This enzyme has a total molecular weight of 91,000 with two subunits, each containing one FAD moiety as prosthetic group. This enzyme shows the substrate-dependent kinetics for the reduction of the enzyme-bound flavin by pyridine nucleotide, which appears to be a common feature of flavoprotein hydroxylases. There are also a group of effector molecules that stimulate pyridine nucleotide oxidation but are not themselves metabolized (233). Among these effectors are benzoate, o-nitrobenzoate, p- and m-hydroxybenzoate, and salicylamide. The benzoate-stimulated NADH oxidase activity was shown to stoichiometrically produce H_2O_2 (234). p-Aminosalicylate, 3-methylsalicylate, 2,3-, 2,4-, and 2,6-dihydroxybenzoates are both substrates and effectors.

Melilotate Hydroxylase

Melilotate, β-(2-hydroxyphenyl)propionate, is converted to β-(2,3-dihydroxyphenyl)propionate by a flavoprotein hydroxylase isolated from an *Arthrobacter* species (235). The purified enzyme was found to have FAD as a prosthetic group and a molecular weight of 65,000 (236). A similar melilotate hydroxylase has been recently isolated from *Pseudomonas* and kinetically characterized. In a systematic study, Strickland & Massey (237) measured the steady-state kinetics of the catalytic reaction and proposed the following reaction scheme

$$E\text{—}FAD + S \underset{k_2}{\overset{k_1}{\rightleftharpoons}} E\text{—}\underset{\underset{\displaystyle NADH}{|}}{\overset{\overset{\displaystyle S}{|}}{FAD}} \xrightarrow{k_3} E\text{—}\underset{\underset{\displaystyle S}{|}}{\overset{\overset{\displaystyle NAD^+}{|}}{FADH_2}} \underset{k_6}{\overset{k_5}{\rightleftharpoons}} E\text{—}\underset{\underset{\displaystyle S}{|}}{FADH_2} + NAD^+$$

$$E\text{—}\underset{\underset{\displaystyle S}{|}}{FADH_2} + O_2 \overset{k_7}{\rightarrow} E\text{—}\underset{\underset{\displaystyle S}{|}}{\overset{\overset{\displaystyle FADH_2}{|}}{O_2}} \xrightarrow{k_9} E\text{—}\underset{\underset{\displaystyle H_2O \quad P}{|}}{FAD} \underset{k_{12}}{\overset{k_{11}}{\rightleftharpoons}} E\text{—}FAD + P$$

Here S refers to melilotate and P to the hydroxylated product. Rapid reaction spectrophotometry was used to determine the individual rate constants in the reaction. When the corresponding steady-state parameters were calculated using the individual rate constants, excellent agreement was noted, strongly supporting the proposed reaction scheme. In the absence of melilotate, the enzyme-bound flavin is reduced through the formation of an E-FAD-NADH complex, whereas in the presence of substrate the reduction reaction proceeds at a much greater rate without any detectable Michaelis intermediate (238). However, the reduced species produced appears to retain NAD^+ in a charge-transfer interaction. This NAD^+ moiety must dissociate prior to oxygen addition (237). The rate of oxygen addition is increased 16-fold by the presence of melilotate. It appears that the same type of intermediates observed in the reoxidation pathway with p-hydroxybenzoate hydroxylase also occurs in the case of melilotate hydroxylase.

Other Flavoprotein Hydroxylases

Although not studied in as great detail, several other flavoprotein hydroxylases have been isolated. In 1968 Ribbons & Chapman (239) isolated a strain of P. fluorescens which could use orcinol as a sole carbon source. Subsequently an orcinol hydroxylase from P. putida was crystallized (240) and shown to be an FAD flavoprotein of 60,000 to 70,000 mol wt that catalyzed the conversion of orcinol to trihydroxytoluene.

Pseudomonas aeruginosa contains a m-hydroxybenzoate-6-hydroxylase that catalyzes the conversion of m-hydroxybenzoate to gentisic acid (241). The enzyme is

a flavoprotein using either NADH or NADPH as electron donor. An FAD-containing hydroxybenzoate 4-hydroxylase has been isolated from Aspergillus niger (242) and from Pseudomonas testosteroni (12).

Several hydroxylases reported to be flavoproteins have been isolated from liver microsomal and mitochondrial preparations. Ziegler et al (243) report a 500,000 mol wt enzyme with seven molecules of FAD bound which catalyzes an NADPH and O_2-dependent N-oxidation of secondary and tertiary amines. The outer membrane of rat liver mitochondria contains a kynurenine-3-hydroxylase activity (244–246) coupled to oxidation of either NADH or NADPH (247).

Although almost all flavin hydroxylases requiring an external source of reducing equivalents have aromatic substrates, the breakdown of histidine by pseudomonads proceeds through a hydroxylation reaction (248, 249). The enzyme has been crystallized (248, 250) and found to contain one FAD prosthetic group per 87,000 to 90,000 mol wt. Okamoto et al (251) found, with either p-chloromercuribenzoate or silver nitrate as reagent, that two sulfhydryl groups are titratable in the holoenzyme, but that in the presence of substrate only one mercaptide group is available. In addition, the substrate protected the enzyme from inactivation by the mercurial. In general, it was found that the sulfhydryl reagents caused a stimulation in the NADH oxidase activity of the enzyme, with the concomitant production of hydrogen peroxide. One unusual aspect of this enzyme, as opposed to the other flavoprotein hydroxylases, is that substrate binding causes no observable shift in either the absorption or circular dichroism/optical rotary dispersion (CD/ORD) spectra.

Literature Cited

1. Priestley, J. 1775. Republished 1961. In *Alembic Club Reprints,* No. 7. Edinburgh: Livingstone; see also Leicester, H. M., Klickstein, H. S. 1952. *A Source Book in Chemistry,* 101. New York: McGraw-Hill
2. Pflüger, E. F. W. 1877. *Pfluegers Arch.* 14:1
3. Mason, H. S., Fowlks, W. B., Peterson, E. W. 1955. *J. Am. Chem. Soc.* 77:2914
4. Hayaishi, O., Katagiri, M., Rothberg, S. 1955. *J. Am. Chem. Soc.* 77:5450
5. Hayaishi, O. 1974. *Molecular Mechanisms of Oxygen Activation,* ed. O. Hayaishi, 1. New York: Academic
6. Mason, H. S. 1957. *Science* 125:1185
7. Hastings, J. W., Balny, C., LePeuch, C., Douzou, P. 1974. *Proc. Nat. Acad. Sci. USA* 71:4389
8. Huang, C. Y., Kaufman, S. 1973. *J. Biol. Chem.* 248:4242
9. Gunsalus, I. C., Tyson, C. A., Lipscomb, J. D. 1973. *Oxidases Relat. Redox Syst.* 2:583
10. Strobel, H. W., Lu, A. Y. H., Heidema, J., Coon, M. J. 1970. *J. Biol. Chem.* 245:4851
11. Ruettinger, R. T., Olson, S. T., Boyer, R. F., Coon, M. J. 1974. *Biochem. Biophys. Res. Commun.* 57:1011
12. Flashner, M., Massey, V. See Ref. 5, 245
13. Hayaishi, O. 1969. *Ann. Rev. Biochem.* 38:21
14. Mason, H. S. 1965. *Ann. Rev. Biochem.* 34:595
15. Massart, L., Vercanteren, R. 1959. *Ann. Rev. Biochem.* 28:527
16. Kingenberg, M. 1958. *Arch. Biochem. Biophys.* 75:376
17. Garfinkel, D. 1958. *Arch. Biochem. Biophys.* 27:493
18. Omura, T., Sato, R. 1963. *Biochim. Biophys. Acta* 71:224
19. Ernster, L., Orrenius, S. 1965. *Fed. Proc.* 24:1190
20. Dagley, S. 1972. *Degradation of Synthetic Organic Molecules in the Biosphere,* 1. Washington, DC: Nat. Acad. Sci.
21. Cooper, D. Y., Levin, S., Narasimhulu, S., Rosenthal, O., Estabrook, R. W. 1965. *Science* 147:400
22. Omura, T., Sanders, E., Estabrook, R. W., Cooper, D. Y., Rosenthal, O. 1966. *Arch. Biochem. Biophys.* 117:660
23. Gunsalus, I. C., Meeks, J. R., Lipscomb, J. D., Debrunner, P., Munck, E. See Ref. 5, 559
24. Van der Hoeven, T. A., Haugen, D. A., Coon, M. J. 1974. *Biochem. Biophys. Res. Commun.* 60:569
25. Imai, Y., Sato, R. 1974. *Biochem. Biophys. Res. Commun.* 60:8
26. Gunsalus, I. C., Chapman, P. J., Kuo, J. F. 1965. *Biochem. Biophys. Res. Commun.* 18:924
27. Chapman, P. J., Meerman, G., Gunsalus, I. C., Srinivasan, R., Rinehardt, K. L. 1966. *J. Am. Chem. Soc.* 88:618
28. Katagiri, M., Ganguli, B. N., Gunsalus, I. C. 1968. *J. Biol. Chem.* 243:3543
29. Yu, C. A., Katagiri, M., Suhara, K., Takemori, S., Gunsalus, I. C. 1974. *J. Biol. Chem.* 249:94
30. Yu, C. A., Gunsalus, I. C. 1970. *Biochem. Biophys. Res. Commun.* 40:1431
31. Gunsalus, I. C., Lipscomb, J. D. 1973. *Iron-Sulfur Proteins,* ed. W. Lovenberg, 1:151. New York: Academic
32. Gunsalus, I. C., Lipscomb, J. D. 1972. *The Molecular Basis of Electron Transport,* ed. J. Schultz, B. F. Cameron, 179. New York: Academic
33. Tsai, R. L., Gunsalus, I. C., Dus, K. 1971. *Biochem. Biophys. Res. Commun.* 45:1300
34. Marbach, W. 1973. *Putidaredoxin reductase: purification and characterization.* Senior thesis. Univ. of Illinois, Urbana
35. Tanaka, M., Haniu, M., Yasunobu, K., Dus, K., Gunsalus, I. C. 1974. *J. Biol. Chem.* 249:3689
36. Wilson, D. S., Tsibris, J. C. M., Gunsalus, I. C. 1973. *J. Biol. Chem.* 248:6059
37. Gunsalus, I. C., Marshall, V. P., Meeks, J. R., Lipscomb, J. D. 1973. *Proc. Int. Congr. Biochem. 9th* (Abstr.)
38. Der Vartanian, D. V. et al 1967. *Biochem. Biophys. Res. Commun.* 26:569
39. Münck, E., Debrunner, P., Tsibris, J.,

Gunsalus, I. C. 1972. *Biochemistry* 11:855

40. Tsai, R. L. et al 1970. *Proc. Nat. Acad. Sci. USA* 66:1157

41. Sharrock, M. 1973. *Mössbauer studies of cytochrome P450 from Pseudomonas putida.* PhD thesis. Univ. of Illinois, Urbana

42. Sharrock, M. et al 1973. *Biochemistry* 12:258

43. Sligar, S. G., Debrunner, P. G., Lipscomb, J. D., Gunsalus, I. C. 1973. *Int. Congr. Biochem. 9th,* 339 (Abstr.)

44. Lipscomb, J. D., Dus, K., Gunsalus, I. C. 1974. *Fed. Proc.* 33:1291 (Abstr.)

45. Sligar, S., Debrunner, P. G., Lipscomb, J. D., Namtvedt, M. J., Gunsalus, I. C. 1974. *Proc. Nat. Acad. Sci. USA* 71:3906

46. Sligar, S., Debrunner, P., Lipscomb, J., Gunsalus, I. C. 1973. *Int. U. Biochem. Symp. 61*

47. Hamberg, M., Samuelsson, B., Björkhem, I., Danielsson, H. See Ref. 5, 29

48. Dixon, R., Furutachi, T., Lieberman, S. 1970. *Biochem. Biophys. Res. Commun.* 40:161

49. Burstein, S., Kimball, H. L., Gut, M. 1970. *Steroids* 15:809

50. Constantopoulos, G., Satoh, P. S., Tchen, T. T. 1962. *Biochem. Biophys. Res. Commun.* 8:50

51. Staple, E., Lynn, W. S. Jr., Gurin, S. 1956. *J. Biol. Chem.* 219:845

52. Shimizu, K., Hayano, M., Gut, M., Dorfman, R. I. 1961. *J. Biol. Chem.* 236:695

53. Constantopoulos, G., Tchen, T. T. 1961. *J. Biol. Chem.* 236:65

54. Shimizu, K., Gut, M., Dorfman, R. I. 1962. *J. Biol. Chem.* 237:699

55. Ichii, S., Omata, S., Kobayashi, S. 1967. *Biochim. Biophys. Acta* 139:308

56. Sulimovici, S., Boyd, G. S. 1968. *Eur. J. Biochem.* 3:332

57. Koritz, S. B., Hall, P. F. 1964. *Biochemistry* 3:1298

58. Hall, P. F., Koritz, S. B. 1964. *Biochim. Biophys. Acta* 93:441

59. Simpson, E. R., Boyd, G. S. 1967. *Eur. J. Biochem.* 2:275

60. Koritz, S. B. 1963. *Biochim. Biophys. Acta* 56:63

61. Hall, P. F., Young, D. G. 1968. *Endocrinology* 82:559

62. Van Lier, J. E., Smith, L. L. 1970. *Biochem. Biophys. Res. Commun.* 40:510

63. Van Lier, J. E., Smith, L. L. 1970. *Biochim. Biophys. Acta* 210:153

64. Burstein, S., Middleditch, B. S., Gut, M. 1974. *Biochem. Biophys. Res. Commun.* 61:692

65. Bryson, M. J., Sweat, M. L. 1968. *J. Biol. Chem.* 243:2799

66. Wilson, L. D., Harding, B. W. 1970. *Biochemistry* 9:1621

67. Jefcoate, C. R., Hume, R., Boyd, G. S. 1970. *FEBS Lett.* 9:41

68. Young, D. G., Holroyd, J. D., Hall, P. F. 1970. *Biochem. Biophys. Res. Commun.* 40:184

69. Isaka, S., Hall, P. F. 1971. *Biochem. Biophys. Res. Commun.* 43:747

70. Huang, J. J., Kimura, T. 1971. *Biochem. Biophys. Res. Commun.* 43:737

71. Mitani, F., Horie, S. 1970. *J. Biochem. Tokyo* 68:529

72. Whyner, J. A., Harding, B. W. 1968. *Biochem. Biophys. Res. Commun.* 32:921

73. Van Lier, J. E., Smith, L. L. 1970. *Biochim. Biophys. Acta* 218:320

74. Burstein, S. et al 1972. *Biochemistry* 11:4

75. Schleyer, H., Cooper, D. Y., Levin, S. S., Rosenthal, O. 1972. *Biological Hydroxylation Mechanisms,* ed. G. S. Boyd, R. M. S. Smellie, 187. New York: Academic

76. Satre, M., Vignais, P. V., Idelman, S. 1969. *FEBS Lett.* 5:135

77. Suzuki, K., Kimura, T. 1965. *Biochem. Biophys. Res. Commun.* 19:340

78. Omura, T., Sato, R., Cooper, D. Y., Rosenthal, O., Estabrook, R. W. 1965. *Fed. Proc.* 24:1181

79. Cooper, D. Y. et al 1967. *Fed. Proc.* 26:431

80. Cooper, D. Y., Narasimhulu, S., Rosenthal, O. 1968. *Advan. Chem. Ser.* 77:220

81. Masters, B. S. S., Baron, J., Taylor, W. E., Isaacson, E. L., LoSpalluto, J. 1971. *J. Biol. Chem.* 246:4143

82. Katagiri, M., Takemori, S. 1973. *Proc. Int. Congr. Biochem. 9th,* 327 (Abstr.)

83. Tanaka, M., Haniu, M., Yasunobu, K. T. 1973. *J. Biol. Chem.* 248:1141

84. Huang, J. J., Kimura, T. 1973. *Biochemistry* 12:406

85. Estabrook, R. W. et al. See Ref. 31, 193

86. Cooper, D. Y., Schleyer, H., Rosenthal, O. 1968. *Z. Physiol. Chem.* 349:1592

87. Schleyer, H., Cooper, D. Y., Levin, S. S., Rosenthal, O. 1971. *Biochem. J.* 125:10

88. Eik-Nes, K. B. 1970. *The Androgens of the Testis,* ed. K. B. Eik-Nes, 1. New York: Dekker

89. Young, R. B., Sweat, M. L. 1967. *Arch. Biochem. Biophys.* 121:576

90. Longchamp, J. E., Gual, C., Ehrenstein, M., Dorfman, R. I. 1960. *Endocrinology* 66:416

91. Wilcox, B. R., Engel, L. L. 1964. *Steroids*

I : 49 (Suppl.)
92. Meigs, R. A., Ryan, K. J. 1968. *Biochim. Biophys. Acta* 165:476
93. Meigs, R. A., Ryan, K. J. 1971. *J. Biol. Chem.* 246:83
94. Ryan, K., Engel, L. L. 1957. *J. Biol. Chem.* 225:103
95. Matthijssen, C., Mandel, J. E. 1967. *Biochim. Biophys. Acta* 146:613
96. Matthijssen, C., Mandel, J. E. 1970. *Steroids* 15:541
97. Mackler, B., Haynes, B., Tattoni, D. S., Tippit, D. F., Kelly, V.-C. 1971. *Arch. Biochem. Biophys.* 145:194
98. Lewis, A. M., Bryan, G. T. 1971. *Life Sci.* 10:901
99. Cooper, D. Y., Narasimhulu, S., Rosenthal, O., Estabrook, R. W. 1968. *Functions of the Adrenal Cortex*, ed. K. McKerns, 2:897. New York: Appleton
100. Narasimhulu, S. 1971. *Arch. Biochem. Biophys.* 147:391
101. Orrenius, S., Ernster, L. See Ref. 5, 215
102. Brodie, B. B., Gillette, J. R., LaDu, B. N. 1958. *Ann. Rev. Biochem.* 27:427
103. Shuster, L. 1964. *Ann. Rev. Biochem.* 33:571
104. Remmer, H. A. 1965. *Ann. Rev. Pharmacol.* 5:405
105. Gillette, J. R. 1966. *Advan. Pharmacol.* 5:219
106. Conney, A. H. 1967. *Pharmacol. Rev.* 19:317
107. Gillette, J. R., Davis, D. C., Sasame, H. A. 1972. *Ann. Rev. Pharmacol.* 12:57
108. Posher, H. S., Mitoma, C., Rothberd, S., Udenfriend, S. 1961. *Arch. Biochem. Biophys.* 94:280
109. Baker, J. R., Chaykin, S. 1962. *J. Biol. Chem.* 237:1309
110. Orrenius, S. 1965. *J. Cell Biol.* 26:712
111. Staudt, H., Lichtenberger, F., Ullrich, V. 1974. *Eur. J. Biochem.* 46:99
112. Guroff, G. et al 1967. *Science* 157:1524
113. Daly, J. W., Jerina, D. M., Witkop, B. 1972. *Experientia* 28:1129
114. Lu, A. Y. H., Coon, M. J. 1968. *J. Biol. Chem.* 243:1331
115. Lu, A. Y. H., Levin, W. 1974. *Biochim. Biophys. Acta* 344:205
116. Satake, H., Imai, Y., Sato, R. 1972. *Seikagaku* 44:765
117. Vermilion, J. L., Coon, M. J. 1974. *Biochem. Biophys. Res. Commun.* 60:1315
118. Vore, M., Hamilton, J. G., Lu, A. Y. H. 1974. *Biochem. Biophys. Res. Commun.* 56:1038
119. Lu, A. Y. et al 1973. *J. Biol. Chem.* 248:456
120. Nebert, D. W., Heidema, J. K., Strobel, H. W., Coon, M. J. 1973. *J. Biol. Chem.* 248:7631
121. Welton, A. F., Aust, S. D. 1973. *Fed. Proc.* 32:665 (Abstr.)
122. Alvares, A. P., Siekevitz, P. 1973. *Biochem. Biophys. Res. Commun.* 54:923
123. Welton, A. F., Aust, S. D. 1974. *Biochem. Biophys. Res. Commun.* 56:898
124. van der Hoeven, T. A., Haugen, D. A., Coon, M. J. 1974. *Pharmacologist* 16:321 (Abstr.)
125. Iyangi, T., Mason, H. S. 1973. *Biochemistry* 12:2297
126. Prough, R. A., Masters, B. S. S., Kamin, H. 1974. *Fed. Proc.* 33:1397 (Abstr.)
127. Fan, L. L., Masters, B. S. S. 1974. *Fed. Proc.* 33:1397 (Abstr.)
128. van der Hoeven, T. A., Coon, M. J. 1974. *J. Biol. Chem.* 249:6302
129. Coon, M. J., Strobel, H. W., Boyer, R. F. 1973. *Drug Metab. Disposition* 1:92
130. Satake, J., Imai, Y., Sato, R. 1972. *Abstr. Ann. Meet. Jap. Biochem. Soc., Nov. 23–26, 1972*
131. Welton, A. F., Pederson, T. C., Buege, J. A., Aust, S. D. 1973. *Biochem. Biophys. Res. Commun.* 54:161
132. Spatz, L., Strittmatter, P. 1973. *J. Biol. Chem.* 248:793
133. Autor, A. P., Kaschnitz, R. M., Heidema, J. K., Coon, M. J. 1973. *Mol. Pharmacol.* 9:93
134. Hildebrandt, A. G., Estabrook, R. W. 1971. *Arch. Biochem. Biophys.* 143:66
135. Correia, M. A., Mannering, G. J. 1972. *Drug Metab. Disposition* 1:139
136. Sasame, H. A., Mitchell, J. R., Thorgeirsson, S., Gillette, J. R. 1972. *Drug Metab. Disposition* 1:150
137. Mannering, G. J., Kuwahara, S., Omura, T. 1974. *Biochem. Biophys. Res. Commun.* 57:476
138. Sasame, H. A., Thorgeirsson, S. S., Mitchell, J. R., Gillette, J. R. 1974. *Life Sci.* 14:35
139. Hrycay, E., Estabrook, R. W. 1974. *Biochem. Biophys. Res. Commun.* 60:771
140. West, S. B., Levin, W., Ryan, D., Vore, M., Lu, A. Y. H. 1974. *Biochem. Biophys. Res. Commun.* 58:516
141. Narasimhulu, S. 1971. *Arch. Biochem. Biophys.* 147:384
142. Ullrich, V., Diehl, H. 1971. *Eur. J. Biochem.* 20:509
143. Schenkman, J. B. 1968. *Z. Physiol. Chem.* 349:1624
144. Gigon, P. L., Gram, T. E., Gillette, J. R. 1969. *Mol. Pharmacol.* 5:109
145. Estabrook, R. W., Hildebrandt, A. G., Baron, J., Netter, K. J., Leibman, K.

1971. *Biochem. Biophys. Res. Commun.* 42:132
146. Baron, J., Hildebrandt, A. G., Peterson, J. A., Estabrook, R. W. 1973. *Drug Metab. Disposition* 1:129
147. Coon, M. J. et al 1975. *Proc. 2nd Philadelphia Conf. Heme Protein P450.* In press
148. Kaufman, S. 1959. *J. Biol. Chem.* 234:2677
149. Kaufman, S. 1963. *Proc. Nat. Acad. Sci. USA* 50:1085
150. Kaufman, S., Bridgers, W. F., Eisenberg, F., Friedman, S. 1962. *Biochem. Biophys. Res. Commun.* 9:497
151. Guroff, G., Rhoads, C. A. 1967. *J. Biol. Chem.* 242:3641
152. Nagatsu, T., Levitt, M., Udenfriend, S. 1964. *J. Biol. Chem.* 239:2910
153. Shiman, R., Akino, M., Kaufman, S. 1971. *J. Biol. Chem.* 246:1330
154. Kaufman, S., Fisher, D. B. See Ref. 5, 285
155. Kaufman, S. 1970. *Methods Enzymol.* 17A:603
156. Ichiyama, A., Nakamura, S., Nishizuka, Y., Hayaishi, O. 1970. *J. Biol. Chem.* 247:4165
157. Friedman, R. A., Kappelman, A. H., Kaufman, S. 1972. *J. Biol. Chem.* 247:4165
158. Fisher, D. B., Kirkwood, R., Kaufman, S. 1972. *J. Biol. Chem.* 247:5161
159. Petrack, B., Sheppy, F., Fetzer, F., Manning, T., Chertock, H. 1972. *J. Biol. Chem.* 247:4872
160. Kuczenski, R. T., Mandell, A. J. 1972. *J. Biol. Chem.* 247:3114
161. Kuczenski, R. T. 1973. *J. Biol. Chem.* 248:2261
162. Kaufman, S. 1961. *Biochim. Biophys. Acta* 51:619
163. Storm, C. B., Kaufman, S. 1968. *Biochem. Biophys. Res. Commun.* 32:788
164. Fisher, D. B., Kaufman, S. 1973. *J. Biol. Chem.* 248:4300
165. Huang, C. Y., Max, E. E., Kaufman, S. 1973. *J. Biol. Chem.* 248:4235
166. Kaufman, S. 1971. *Advan. Enzymol.* 35:245
167. Guroff, G., Reifsnyder, C. A., Daly, J. W. 1966. *Biochem. Biophys. Res. Commun.* 24:720
168. Daly, J. W., Levitt, M., Guroff, G., Udenfriend, S. 1968. *Arch. Biochem. Biophys.* 126:593
169. Hutton, J. J., Tappel, A. L., Udenfriend, S. 1966. *Biochem. Biophys. Res. Commun.* 24:179
170. Abbott, M. T., Udenfriend, S. See Ref. 5, 168
171. Cardinale, G. J., Udenfriend, S. 1973.

Oxidases Relat. Redox Syst. 1:195
172. Rhoads, R. E., Udenfriend, S. 1968. *Proc. Nat. Acad. Sci. USA* 60:1473
173. Fujimoto, D., Tamiya, N. 1962. *Biochem. J.* 84:333
174. Prockop, D., Kaplan, A., Udenfriend, S. 1962. *Biochem. Biophys. Res. Commun.* 9:162
175. Lindblad, B., Lindstedt, G., Tofft, M., Lindstedt, S. 1969. *J. Am. Chem. Soc.* 91:4604
176. Cardinale, G. J., Rhoads, R. E., Udenfriend, S. 1971. *Biochem. Biophys. Res. Commun.* 43:537
177. Holme, E., Lindstedt, G., Lindstedt, S., Tofft, M. 1971. *J. Biol. Chem.* 246:3314
178. Pänkäläinen, M., Kivirikko, K. I. 1971. *Biochim. Biophys. Acta* 229:504
179. Lindstedt, G., Lindstedt, S. 1970. *J. Biol. Chem.* 245:4178
180. Barnes, M. J., Kodicek, E. 1972. *Vitam. Horm.* 30:1
181. Popenoe, E. A., Aronson, R. B., VanSlyke, D. D. 1969. *Arch. Biochem. Biophys.* 133:286
182. Rhoads, R. E., Udenfriend, S. 1970. *Arch. Biochem. Biophys.* 139:329
183. LaDu, B. N., Zannoni, V. G. 1955. *J. Biol. Chem.* 217:777
184. Hager, S. E., Gregerman, R. I., Knox, W. E. 1957. *J. Biol. Chem.* 225:935
185. Goswami, M. N. D. 1964. *Biochim. Biophys. Acta* 85:390
186. Yasunobu, K., Tanaka, T., Knox, W. E., Mason, H. S. 1958. *Fed. Proc.* 17:340
187. Lindblad, B., Lindstedt, G., Lindstedt, S. 1970. *J. Am. Chem. Soc.* 92:7446
188. Hamilton, G. A. 1971. *Progr. Bioorg. Chem.* 1:83
189. Hobza, P., Hurych, J., Zahradnik, R. 1973. *Biochim. Biophys. Acta* 304:466
190. Goodwin, S., Witkop, B. 1956. *J. Am. Chem. Soc.* 79:179
191. Taniguchi, K., Kappe, T., Armstrong, M. D. 1964. *J. Biol. Chem.* 239:3389
192. Kusunose, M., Kusunose, E., Coon, M. J. 1964. *J. Biol. Chem.* 239:2135
193. Bernhardt, F. H., Ruf, H. H., Staudinger, H., Ullrich, V. 1971. *Z. Physiol. Chem.* 352:109k
194. McKenna, E. J., Coon, M. J. 1970. *J. Biol. Chem.* 245:3882
195. Bernhardt, F. H., Erdin, N., Staudinger, H., Ullrich, V. 1973. *Eur. J. Biochem.* 35:126
196. Peterson, J. A., Basu, D., Coon, M. J. 1966. *J. Biol. Chem.* 241:5162
197. Lode, E. T., Coon, M. J. 1971. *J. Biol. Chem.* 246:791
198. Ueda, T., Lode, E. T., Coon, M. J.

1972. *J. Biol. Chem.* 247:2109
199. Ueda, T., Coon, M. J. 1972. *J. Biol. Chem.* 247:5010
200. Bernhardt, F. H., Staudinger, H. 1973. *Z. Physiol. Chem.* 354:217
201. May, S. W., Abbott, B. J. 1973. *J. Biol. Chem.* 248:1725
202. Boyer, R. F., Lode, E. T., Coon, M. J. 1971. *Biochem. Biophys. Res. Commun.* 44:925
203. Molinoth, P. B., Axelrod, J. 1971. *Ann. Rev. Biochem.* 40:465
204. Levin, E. Y., Levenberg, B., Kaufman, S. 1960. *J. Biol. Chem.* 235:2080
205. Mason, H. S. 1957. *Advan. Enzymol.* 19:79
206. Kertesz, D., Zito, R. 1962. *Oxygenases,* ed. O. Hayaishi, 307. New York: Academic
207. Malkin, R., Malmstrom, B. G. 1970. *Advan. Enzymol.* 32:177
208. Vanneste, W. H., Zuberbuhler, A. See Ref. 5, 371
209. Kubowitz, F. 1938. *Biochem. Z.* 299:32
210. Friedman, S., Kaufman, S. 1965. *J. Biol. Chem.* 240:4763
211. Blumberg, W. E., Goldstein, M., Lauber, E., Peisach, J. 1965. *Biochim. Biophys. Acta* 99:187
212. Friedman, S., Kaufman, S. 1966. *J. Biol. Chem.* 241:2256
213. Goldstein, M., Joh, T. H., Garvey, T. Q. 1968. *Biochemistry* 7:2724
214. Wallace, E. F., Krant, M. J., Lovenberg, W. 1973. *Proc. Nat. Acad. Sci. USA* 70:2253
215. Aunis, D., Miras-Portugal, M. T., Mandel, P. 1974. *Biochem. Biophys. Res. Commun.* 57:1192
216. Jolley, R. L. Jr., Evans, L. H., Mason, H. S. 1972. *Biochem. Biophys. Res. Commun.* 46:878
217. Katagiri, M., Takemori, S. 1971. *Flavins Flavoproteins* 3:447
218. Howell, L. G., Spector, T., Massey, V. 1972. *J. Biol. Chem.* 247:4340
219. Spector, T., Massey, V. 1972. *J. Biol. Chem.* 247:5632
220. Howell, L. G., Massey, V. 1971. *Flavins Flavoproteins* 3:499
221. Entsch, B., Massey, V., Ballou, D. 1974. *Biochem. Biophys. Res. Commun.* 57:1018
222. Müller, M., Hartmann, U., Hemmerich, P. 1973. *Z. Physiol. Chem.* 354:215
223. Orf, H. W., Dolphin, D. 1974. *Proc. Nat. Acad. Sci. USA* 71:2646
224. Spector, T., Massey, V. 1972. *J. Biol. Chem.* 247:4679
225. Nakamura, S., Ogura, Y., Yano, K., Higashi, N., Arima, K. 1970. *Biochemistry* 9:3235

226. Hosokawa, K., Stanier, R. Y. 1966. *J. Biol. Chem.* 244:5644
227. Katagiri, M., Yamamoto, S., Hayaishi, O. 1962. *J. Biol. Chem.* 237:2413
228. Takemori, S., Nakamura, M., Katagiri, M. Nakamura, T. 1971. *Flavins Flavoproteins* 463
229. Yamamoto, S., Katagiri, M., Maeno, H., Hayaishi, O. 1965. *J. Biol. Chem.* 240:3408
230. Suzuki, K., Takemori, S., Katagiri, M. 1969. *Biochim. Biophys. Acta* 191:77
231. Takemori, S., Yasuda, H., Mihara, K., Suzuki, K., Katagiri, M. 1969. *Biochim. Biophys. Acta* 191:58
232. White-Stevens, R. H., Kamin, H. 1970. *Biochem. Biophys. Res. Commun.* 38:882
233. White-Stevens, R. H., Kamin, H. 1972. *J. Biol. Chem.* 247:2358
234. White-Stevens, R. H., Kamin, H. Gibson, Q. H. 1972. *J. Biol. Chem.* 247:2371
235. Levy, C. C., Frost, P. 1966. *J. Biol. Chem.* 241:997
236. Levy, C. C. 1967. *J. Biol. Chem.* 242:747
237. Strickland, S., Massey, V. 1973. *J. Biol. Chem.* 248:2953
238. Strickland, S., Massey, V. 1973. *J. Biol. Chem.* 248:2944
239. Ribbons, D. W., Chapman, P. J. 1968. *Biochem. J.* 106:44P
240. Ohta, Y., Ribbons, D. W. 1970. *FEBS Lett.* 11:189
241. Groseclose, E. E., Ribbons, D. W. 1972. *Bacteriol. Proc.* 273
242. Premkumar, R., Rao, P. V. Subba, Sreeleela, N. S., Vaidyanathan, C. S. 1969. *Can. J. Biochem.* 47:825
243. Ziegler, D. M., Mitchell, C. H. 1972. *Arch. Biochem. Biophys.* 150:116
244. Okamoto, H. 1970. *Methods Enzymol.* 17:460
245. Okamoto, H., Yamamoto, S., Nozaki, M., Hayaishi, O. 1967. *Biochem. Biophys. Res. Commun.* 26:309
246. Saito, Y., Hayaishi, O., Rothberg, S. 1957. *J. Biol. Chem.* 229:921
247. Hayaishi, O., Okamoto, H. 1971. *Am. J. Clin. Nutr.* 24:805
248. Maki, Y., Yamamoto, S., Nozaki, M., Hayaishi, O. 1969. *J. Biol. Chem.* 244:2942
249. Rothberg, S., Hayaishi, O. 1957. *J. Biol. Chem.* 229:897
250. Maki, Y., Yamamoto, S., Nozaki, M., Hayaishi, O. 1966. *Biochem. Biophys. Res. Commun.* 25:609
251. Okamoto, H., Nozaki, M., Hayaishi, O. 1968. *Biochem. Biophys. Res. Commun.* 32:30

ENERGY CAPTURE IN   ×889
PHOTOSYNTHESIS: PHOTOSYSTEM II[1]

Richard Radmer and Bessel Kok

Martin Marietta Corporation, Martin Marietta Laboratories,
Baltimore, Maryland 21227

CONTENTS

The photochemical processes of green plant photosynthesis produce molecular O_2, reduced NADP, and ATP. According to current thinking electrons are promoted from the level of H_2O to NADPH by two photoacts connected in series. Photosystem II forms a weak reductant Q^- and a strong oxidant $Z^+(P_{690}{}^+)$ which indirectly oxidizes H_2O. Photosystem I forms a weak oxidant $P_{700}{}^+$ and a strong reductant X^- which reduces NADP. P_{700} and Q are regenerated (from $P_{700}{}^+$ and Q^-) by dark reactions connecting the two photoacts.

We will limit this review to some of the more recent information pertaining to System II and will only tangentially mention artificial donor and acceptor reactions and the effects of inhibitors. In the first section, we will describe the results of kinetic experiments related to the O_2 evolution system. We will then consider results related to the identity and function of the components involved in the overall process.

---

[1] Abbreviations used: A pools, A_1 and A_2, secondary acceptor pools of System II; CCCP, carbonylcyanide-*m*-chlorophenylhydrazone; DBMIB, dibromothymoquinone; DCMU, dichlorophenyldimethylurea; DCPIP, dichlorophenolindophenol; EPR, electron paramagnetic resonance; FCCP, carbonylcyanide-*p*-trifluoromethoxyphenylhydrazone; PMS, phenazine methosulfate; and ϕ, quantum yield.

Our aim is not to be encyclopedic, but rather to describe and evaluate some of the data and hypotheses presented since the last comprehensive review of the topic by Cheniae (1). Other information germane to this subject can be found in reviews by Trebst (2), Bishop (3), Witt (4), and Kok & Cheniae (5).

SOME ASPECTS OF O_2 EVOLUTION KINETICS

Kinetic observations and proposed model   The formation of a molecule of O_2 from water requires the removal of four electrons; the average redox potential of these four equivalents is about $+0.81$ V. However, the four individual reaction steps in the chemical oxidation of H_2O vary from -0.45 to $+2.33$ V (6). Since the energy of a red photon is about 1.8 eV, some of these oxidation steps might be energetically unfavorable without a "moderating" mechanism.

The mechanism by which the O_2 evolving system effects the collaboration of four oxidizing equivalents produced by four different photoacts remains obscure. The development of sensitive polarographic techniques by Joliot (7, 8) allowed the precise measurement of small quantities of O_2 evolved by weak illumination or by single short (1–10 μsec) flashes. Studies using this technique have resulted in a clearer understanding of the O_2 evolving system.

Figure 1 shows the oscillating behavior of the O_2 flash yields after long dark in spinach chloroplasts, reported by Forbush et al (9). Flash yields one and two (Y_1 and Y_2) are very low and the O_2 yields show a damped oscillation with a period of four. The oscillations shown here are more extreme than those reported earlier (cf 1, 10, 11), presumably due to improvements in the experimental technique and, in contrast to previous reports, Y_3 is greater than $2Y_{ss}$.[2] This observation is not consistent with the binary (two-step) mechanisms that have been proposed (11, 13) and instead suggests the operation of a four step mechanism.

A similar oscillation pattern has been reported for the release of protons by a series of flashes (14). Although these data are not as convincing as the comparable O_2 data, they point to a concerted reaction in which protons are released in synchrony with oxygen.

In continuous light the same basic pattern of O_2 evolution is observed. In this case, however, the low Y_1 and Y_2 are reflected by an S-shaped time course, the "activation" first described by Joliot (15). Preflashes before the continuous illumination alter the subsequent time course of O_2 evolution: after one flash the sigmoidal character of the rate at the onset of continuous illumination is no longer present; after two flashes the initial rate is twice that of the steady state (11).

According to the generally accepted model for this system proposed by Kok et al (11), each O_2 evolving center undergoes the following reactions

$$S_0 \xrightarrow{hv} S_0^* \xrightarrow{k_0} S_1 \xrightarrow{hv} S_1^* \xrightarrow{k_1} S_2 \xrightarrow{hv} S_2^* \xrightarrow{k_2} S_3 \xrightarrow{hv} S_3^* \xrightarrow{k_3} S_4 \xrightarrow{k_4} S_0 \qquad 1.$$
$$\underbrace{}_{2H_2O} \qquad \underbrace{}_{O_2+4H^+}$$

[2] In the presence of low concentrations of NH_2OH the maximum O_2 yield is obtained on the fifth or sixth flash, apparently due to the reduction of a component near the O_2 evolving system (12).

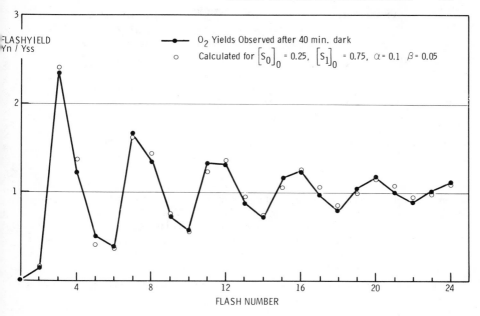

Figure 1   Observed and calculated O_2 flash yields after long dark (9). $Y_n = O_2$ yield of nth flash, Y_{ss} = steady-state flash yield. See text for other details.

In this model, the S_n states differ by (at least) the number of oxidizing equivalents in the system (indicated by n); the dark reactions $S_n^* \rightarrow S_{n+1}$ include the reoxidation of the photochemical electron acceptor ($Q^- + A_2 \rightarrow Q + A_2^-$). To be consistent with the experimental observations the following assumptions are required:

1. Each O_2 evolving system operates independently, and no exchange of oxidizing equivalents occurs between centers.

2. Each photostep involves a one quantum process and is sensitized by the same pigment system.

3. The S_1 and S_0 states are both stable in the dark. The lifetimes of the S_2 and S_3 states are long compared to the dark transitions ($S_n^* \rightarrow S_{n+1}$) but less than about 15 min; they decay to S_1. Consequently, after about 15 min dark, the distribution of the S states is $S_0 : S_1 : S_2 : S_3 = 0.25 : 0.75 : 0 : 0$.

4. During each flash a fraction (α) of the O_2 centers does not undergo a transition ("misses") and another fraction (β) undergoes a double transition ($S_n \rightarrow S_{n+2}$ "double hits"). The aberrant transitions occur randomly in the O_2 centers; it is this randomization that accounts for the damping in the oscillation of the flash yield.

In the following sections we will consider some aspects of O_2 evolution within the framework of this model.

Independence of O_2 evolving centers   If O_2 evolution involves the cooperation of neighboring O_2 centers, one might expect that the kinetic pattern would be altered by partial inhibition (assuming that the inhibition is a random process). Kok et al (11) reported that the inhibition of a fraction of the O_2 centers by DCMU, UV radiation, or Mn^{2+} deficiency resulted in a proportionate loss in the amplitude of the O_2 flash yields, but did not alter the flash yield pattern. From these results the authors concluded that the O_2 evolving centers operate independently.

A similar result under (possibly) more physiological conditions was obtained using mutants with a low concentration of O_2 centers (P. Bennoun, personal communication). Again, however, it was not possible to prove the absence of "islands" of active centers which could cooperate.

The significance of "misses (α)" and "double hits (β)"   The origin of misses is probably biological rather than technical, since it was observed that the fractional miss factor (α) remained constant even when the flash intensity was increased far above the saturation level (11). Misses presumably reflect the fact that in any given flash a percentage of the traps randomly fails to undergo a photoact because either 1. a fraction of Q is in the reduced state before the flash or 2. a fraction of the traps undergoes a backreaction. The ramifications of the first mechanism are obvious, provided there is interaction between Qs in the system (16). Joliot et al (17) invoked the second mechanism and interpreted the decay of the fluorescence yield in the range of a few tenths of a second as being due to backreactions. From the

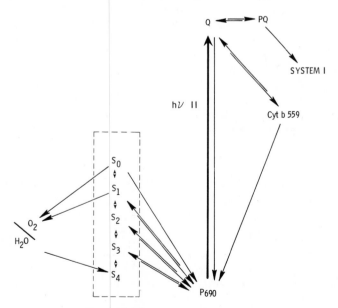

Figure 2   A proposed electron flow scheme for Photosystem II reactions (33). In this scheme the deactivator R is cytochrome b_{559}; see text for details.

fluorescence decay in this time range, they computed α values for the four states varying between 0.08 and 0.28. We prefer to restrict the term "misses" to the events more closely associated with the photoacts (~ 1 msec) and apply the term "deactivation" for the slower losses of oxidizing equivalents from the S states. In the scheme of Figure 2, however, these distinctions are rather academic.

For convenience in the interpretation of O_2 flash yield sequences, it is often assumed that misses are distributed uniformly among the four transitions. However, most workers agree that this distribution is probably unequal; an extreme assumption (18) that misses occur exclusively in the transition $S_2 \rightarrow S_3$ appears unlikely.

Unlike misses, the occurrence of double hits (β) is probably due to technical limitations in the flash measurements: double hits will occur when the duration of the flash is not negligible compared to the dark relaxation time of the traps (k_0 to k_3 in equation 1), so that a fraction of the centers can undergo two transitions during a single flash (e.g. $S_1 \rightarrow S_2 \rightarrow S_3$) (5, 17). Double hits do not occur when very short xenon flashes (17) or laser flashes (19) are used.

Relaxation times of the steps $S_n^* \rightarrow S_{n+1}$ After a trapping center is excited, both the electron donor and electron acceptor must recover before the next excitation can be effective; i.e. the two reactions $S_n P_{690}{}^+ \rightarrow S_{n+1} P_{690}$ and $Q^- \rightarrow Q$ must both take place. The common notation $S_n^* \rightarrow S_{n+1}$ includes both processes, and without additional information either process could be rate limiting.

Joliot (7) measured the phase delay of O_2 evolved in weak modulated light as a function of modulation frequency, and computed that the time lapse between the photochemical act and the appearance of O_2 from S_3 states was 1.2 msec. An "overall rate constant" of System II turnover determined by O_2 evolution in long sequences of paired flashes (20) yielded a time constant of 0.6 msec [this value might be a composite of two half-times of 0.2 and 2 msec, respectively (21)].

The reaction rates of the specific transitions $S_1^* \rightarrow S_2$ and $S_2^* \rightarrow S_3$ were first measured by Kok et al (11), who found relaxation times of 200 and 400 μsec, respectively, at room temperature.[3] Bouges-Bocquet (22) performed similar experiments and reported that 1. the first two transitions were nonexponential with half-times of 400 μsec, 2. the transition $S_2^* \rightarrow S_3$ showed a lag (suggesting the participation of an intermediate step), and 3. the transition $S_3^* \rightarrow S_4 \rightarrow S_0 + O_2$ was significantly slower than the others and appeared to be first order, exhibiting a half-time very similar (~ 1 msec) to that obtained by phase shift measurement (see above). The reason for the discrepancy between (11) and (22) is not apparent.

Deactivation   Forbush et al (9) reported that deactivation proceeds primarily by one step processes; i.e. $S_3 \rightarrow S_2 \rightarrow S_1$. One feature of this scheme is that it predicts that S_2 should transiently increase in the dark; this is usually observed.

---

[3] These measurements were made using variably spaced sequences of saturating flashes. For example, to determine the rate of the $S_1^* \rightarrow S_2$ transition, the spacing between the first two flashes was varied and the yield of the third flash (given a long constant interval later) was measured.

Joliot and co-workers (17) explored the possible occurrence of two step deactivation; i.e. $S_3 \rightarrow S_1$ and $S_2 \rightarrow S_0$. Although a strict two step mechanism is probably not consistent with the observed phenomena, it may contribute as a minor process to some of the observed second order deviations from the one step model.

The rate of deactivation varies widely, depending on the experimental material and its pretreatment, and is much slower in isolated chloroplasts than in whole algae. For example, the S_3 state is reported to have a first half-life of about 60 sec in chloroplasts (9, 17) and about 5 sec in *Chlorella* (17), a difference that can probably be ascribed to the more reduced milieu in the whole cell. Chloroplasts preilluminated in the absence of an electron acceptor (to reduce Q and the redox pools between the photosystems) deactivate even more rapidly; the reported first half-time of S_3 decay is about 0.6 sec (16), presumably reflecting a backflow of electrons from Q^- to S_3. Similar but less dramatic changes in the lifetime of S_3 have also been reported when the redox state of the A pools was changed by the color of the preillumination (23).

The deactivation rate can also be accelerated by chemicals such as CCCP, anilinothiophenes (24), and indophenols (25). The mode of action of these agents is not clear; they may operate in a cycle between the oxidizing side of System II and a chloroplast reductant other than Q^- (26). In the hypothesis shown in Figure 2, one could assume that the above agents convert cytochrome b_{559} to its low potential form, which could then rapidly donate electrons to the S states. The reported accelerating effect of DCMU (which presumably blocks the transfer of electrons from Q to the A pools) on deactivation in chloroplasts (24, 27) is probably related to the fast deactivation described above, reflecting an enhanced backreaction from Q to the higher S states.[4]

In the analysis of deactivation under normal conditions (in which the system is largely oxidized), Forbush et al (9) assumed the existence of a deactivator (R) of limited concentration. Since this deactivator reacts with the S_2 and S_3 states in a second order reaction, one must conclude that R is diffusible or that Rs of different chains are interconnected. Ultimately, the reducing equivalents for deactivation are probably derived from undefined constituents of the chloroplasts or their suspension medium. Even after far red light, which oxidizes all intermediates of the electron chain, P_{700} still becomes slowly reduced in the dark (31). Deactivation under oxidized conditions could be similar; it might proceed via a deactivator, but the ultimate source of electrons would be unrelated to light-generated reducing power.

The S_1 state   The nature and apparent stability of the S_1 state have received a good deal of scrutiny. There are reports that the first flash after a long dark period might move two equivalents (32), which may imply that the O_2 centers could

---

[4] Duysens (28) reported that *Chlorella* still evolved O_2 when DCMU was added to cells in the S_3 state (i.e. two preflashes), whereas Rosenberg et al (29) and Etienne (30) reported the absence of O_2 evolution under similar conditions in chloroplasts. These results were ascribed to the different effect of DCMU on deactivation in whole cells vs chloroplasts (27); however, on close examination the argument is not very convincing.

undergo two transitions in the first flash. However, such a transition could not be sequential (i.e. $S_0 \rightarrow S_1 \rightarrow S_2$), since O_2 is produced on the third flash even with short (20–40 nsec) laser flashes (19), while the relaxation time of the two transitions is >0.1 msec. The strongest argument for the relative stability of the S_1 state (over a period of a few hours) is that the ratio S_1/S_0 can be varied by appropriate flash preilluminations; after one preflash the ratio Y_3/Y_4 is found to be higher than after dark, and after three preflashes the ratio is lower (9).

There is evidence that the S_1 state can convert to the S_0 state (and vice versa) in the dark [17, see also Figure 7 in (9)]; during prolonged dark periods the S_1/S_0 ratio tends to approach a rather constant value (about 3) independent of the preillumination, as though S_1 and S_0 can interconvert. This would imply a low equilibrium constant for the reaction $S_0 \rightarrow S_1$, suggesting that this transition conserves virtually none of the energy of the photon that sensitized it, and that energetically O_2 evolution is a three-quantum-four electron process.

The distribution of states can also be altered in the dark by chemical means: in the presence of ferricyanide the ratio S_1/S_0 becomes very high, whereas in the presence of ascorbate-DCPIP this ratio approaches unity (22). This might indicate that the midpoint potential of the system is correspondingly low (22). However, we prefer to ascribe these observations to an indirect effect of the added redox reagents: if the S_0 state (and possibly the S_1 state) is slowly oxidized by molecular O_2 and reduced by R, the ratio S_0/S_1 will reflect the balance between the two reactions $(R \rightarrow S_1/S_0 \rightarrow O_2)$. The effect of mild oxidants and reductants would then be due to changes in the redox state of R (see Figure 2).[5]

IDENTITY AND FUNCTION OF SYSTEM II COMPONENTS

Primary Donor

One of the fundamental tenets in photosynthesis is that of a reaction center. This reaction center is usually envisioned as a special chlorophyll complex associated with a primary electron donor (D) and acceptor (A). Upon excitation the following reaction sequence presumably occurs

$$D \; Chl \; A \xrightarrow{h\nu} D \; Chl^* \; A \rightarrow D \; Chl^+ \; A^- \rightarrow D^+ \; Chl \; A^- \qquad 2.$$

In System I of green plants the reaction center P_{700} has an absorption maximum at about 700 nm which is lost upon photo-oxidation. Plastocyanin and ferredoxin presumably function as the donor and acceptor, respectively. Although there has been some disagreement (35–37), it is now generally accepted that the System II reaction center acts analogously. It had long been suspected that the System II

---

[5] Bennoun (34) reported that NH_2OH inhibited the backreaction between P_{690}^+ and Q^- in the presence of DCMU (determined from the return of the fluorescence yield in darkness; see below). This suggests that the reduction of P_{690}^+ by NH_2OH is irreversible, in contrast to the reduction of P_{690}^+ by the O_2 system. Other compounds, such as CCCP, have a similar effect. However, a detailed discussion of these studies is beyond the scope of this review.

reaction center would have a bleachable absorption band 10 to 20 nm to the blue side of P_{700}. However, measurements in this spectral region are difficult because of high chlorophyll absorption and possible interferences by changes in the chlorophyll fluorescence yield (see e.g. 38). In addition the System II trap seems to turn over much faster than P_{700} under most conditions, requiring wider bandwidth detection equipment and thus resulting in "noisier" measurements.

Spectral observations   Döring et al (39), using repetitive flashes, reported the observation of a flash-induced absorption change at about 690 nm, which recovered with a half-time of 200 μsec. They attributed this absorption change to the bleaching of the System II reaction center chlorophyll molecule "Chl a_{II}." [Other laboratories label this pigment according to its absorption maximum; the most recent data (40) would then suggest the nomenclature P_{690}, which we will use in the following discussions.] Later experiments using System II "particles" showed a maximum absorption change at about 682 nm (41) and spectral changes in the Soret band (435 nm) with the same kinetics as those at 682 nm and a similar response to heat treatment. This suggested that the changes attributed to Chl a_{II} are indeed due to a special chlorophyll molecule.

Recently, Glaser et al (40), again using repetitive flashes, reported that the relaxation of the absorption change was biphasic, having a 35 and a 200 μsec component, and that the maximal change at 690 nm was about equal to that of P_{700}. Additional arguments in favor of P_{690} being the reaction center of System II were its 1. insensitivity to far red light, 2. insensitivity to DBMIB (which inhibits oxidation of the System II acceptor by System I), and 3. sensitivity to DCMU.

Floyd et al (42) have reported experiments at $-196°C$ in which they monitored the redox state of "P_{680}" and cytochrome b_{559} after a laser flash. They correlated the oxidation of cytochrome b_{559} with a slow (5 msec) kinetic component ascribed to P_{690} reduction. Spectral changes attributed to P_{690} were also reported by Ke et al (43) at $-196°C$ in Triton-treated Photosystem II-enriched particles. However, Lozier & Butler (44) reported that they were unable to detect $P_{690}{}^{+}$ spectroscopically at this temperature.

Electron paramagnetic resonance signal   Oxidized P_{700}, like other chlorophyll radicals, has a characteristic EPR signal. Recently Malkin & Bearden (45) observed at $-196°C$ that after oxidizing P_{700} with ferricyanide and thus maximizing its EPR signal, a further increase could be observed in 645 nm light. In the presence of ferricyanide, this additional signal could also be observed in System II particles but not in System I particles. The spin concentration of this signal ascribed to "$P_{690}{}^{+}$" [not to be confused with "Signal II" of Commoner et al (46)] is about one per 300 chlorophylls. [About ten years ago, Beinert & Kok (47) reported that the ratio of EPR spins per P_{700} was about two, but gave no cogent explanation for this finding.] However, Lozier & Butler (44), who could not spectroscopically detect $P_{690}{}^{+}$, ascribed the EPR signal (which they *did* detect) to a secondary donor oxidized by P_{690}^+ when the normal donor at $-196°C$ (cytochrome b_{559}, see below) is oxidized by pretreatment with ferricyanide.

In Photosystem II particles isolated by Triton treatment, Ke et al (43) found a good correlation between the EPR signal and the optical determinations of P_{690}^+ at $-196°C$. In deoxycholate-treated particles, Van Gorkom et al (48) reported that a light-induced bleaching as well as an EPR signal which they ascribed to P_{690} were readily observable at room temperature (in the presence of ferricyanide to oxidize P_{700}) and had a lifetime of several milliseconds. (These particles seem to uniquely reveal several "isolated" partial reactions; however, the physiological significance is not always evident.)

The Primary Acceptor of System II

FLUORESCENCE   The primary electron acceptor of Photosystem II has generally been equated with Q, the notation used by Duysens & Sweers (49) for the quencher of the variable fluorescence (F_v) emanating from System II. (No comparable variable fluorescence component has been observed that correlates with the redox state of System I.) Even with Q oxidized and P_{690} reduced, a residual fluorescence yield ("dead fluorescence" F_0) is observed, indicating that not all excitations are trapped. One might expect that in general the variable fluorescence is indicative of the state of the traps; i.e. the fluorescence from the light harvesting chlorophyll increases when the excitations do not find an open trapping center (Q reduced, P_{690} oxidized, or both). However, more recent studies indicate that this picture is oversimplified, and that phenomena unrelated to the "openness" of the traps can also affect the variable fluorescence yield. [For a discussion of other aspects of fluorescence in photosynthetic systems, see (50).]

Quenching by carotenoids   The behavior of the fluorescence yield during and immediately after strong microsecond flashes is complex (51); during the flash the yield first rises and then declines (52, 53). The rise is ascribed to a closing of the traps and the decline to the photo formation of new quenchers formed with high efficiency after the traps are closed (52, 53). The computed lifetimes of these quenchers (a few microseconds) and their response to oxygen are similar to those of carotenoid radicals or triplets complexed with chlorophyll (54–56). The 3 μsec rise of the fluorescence yield observed by Mauzerall (57) after a laser flash probably reflects the decay of this quencher. It is now generally agreed that this process protects the photosynthetic apparatus by draining excess energy from the system.

Quenching by oxidized quinones   It has been recognized for some time (58, 59) that quinones in their oxidized (but not reduced) forms quench the fluorescence of chlorophyll in solution as well as F_0 and F_v in chloroplasts (without affecting electron transport). This effect may be significant in experiments involving a change in the redox state of added quinones; a recent example has been reported using DBMIB (60). Plastoquinone, the secondary electron acceptor of System II, can presumably exhibit a similar "nonphotochemical" quenching effect (61, 62). In normal chloroplasts, its contribution [defined "Q_R" by Joliot & Joliot (63)] is relatively small. In deoxycholate-treated chloroplasts, however, most of the variable fluorescence yield correlates with the redox state of the plastoquinone pool (48) (rather than with "C550" which the authors assumed to reflect the primary electron acceptor).

Quenching related to the System II donor   Several experiments have been reported that suggest the fluorescence yield is a function not only of acceptor Q but also of the primary donor of System II. Delosme (64, 65) and Joliot et al (17, 66) observed oscillations in the fluorescence yield in sequences of light flashes which correlated with the changes in the distribution of S states. Zankel (53) reported that the extent of the 35 μsec rise in the fluorescence yield after a flash varied with the flash number (high after 2 and 3), and apparently reflected a conversion at the donor side of System II (see also 64). Okayama & Butler (68) observed that the maximum variable fluorescence yield at $-196°C$ was fourfold lower in the presence of ferricyanide, which presumably oxidized cytochrome b_{559}, the secondary System II donor at these temperatures. They suggested that oxidized P_{690} (or another compound oxidized by $P_{690}{}^+$) was a quencher of fluorescence (however, see below and 69, 70).

A similar type of fluorescence quenching has been reported at $-50°C$ when the traps are in the S_2-S_3 state (two flashes given before cooling). The removal of this quenching by light is inefficient (a saturating flash has little effect) and does not occur in the presence of ferricyanide (cytochrome b_{559} oxidized?). Joliot & Joliot (63) denoted this quenching "Q_S." As we discuss later, the nature of this quenching is not understood.

Quenching by Q   Although the above enumeration seems rather gloomy, the variable fluorescence yield (F_v) still can serve as a useful indication of Q under most conditions; i.e. in active chloroplasts at room temperature where P_{690} is mostly reduced and Q_S and Q_R are relatively small. Adhering to Joliot's nomenclature (if only because it seems operationally convenient) the criterion for quenching due to Q("Q_F") is that it can be removed by a single microsecond flash. Again there are complications:

1. The complement of $F_v(F_{max}-F_v)$ is not quite proportional to the fraction of Q in the oxidized state, due to the fact that excitations can migrate to more than one trap. For a given redox state of Q, this transfer between units tends to increase the rate of photochemistry and decrease the fluorescence (71).

2. The yield of this fluorescence ($\emptyset F_v$) is not invariant; for instance, the maximum fluorescence change observed at $-50°C$ is much smaller when traps are in the S_2 and S_3 states (although the same number of quanta are required to attain this maximum as in the S_0-S_1 case).

3. Homann (72) and Murata (73) have shown that the ionic environment of chloroplasts (e.g. the concentration of Mg^{2+}) can influence the variable component of the fluorescence yield. Two effects may be involved: 1. a change of the fluorescence yield ($\emptyset F_v$) due to changes in the membrane structure (74–76) and

2. a true effect on the redox state of Q which would imply a redistribution of excitation energy between the photosystems (73). A related phenomenon is the reported quenching of fluorescence by the light-generated "high energy state" (77–79) (e.g. the proton gradient formed in the light by System I in the presence of DCMU and PMS) which may be related to the associated efflux of cations (80). [The reverse effect, an increase of the yield upon addition of ATP to a system in which the ATPase is activated, is presumably due to a true reduction of Q (81).]

Fluorescence yield changes in whole cells reflect an even more complex set of phenomena, which can include effects due to carbon metabolism and changes in the state of the pigment systems, such as the state 1–state 2 phenomenon described by Bonaventura & Myers (82) and Murata et al (83). (For a review of this topic, see 84. For a more general review of in vivo fluorescence changes, see 85.) Again, our intention is not to discourage the reader; used with caution, fluorescence measurements provide a convenient method for monitoring the redox state of Q.

"C550"   Knaff & Arnon (86) discovered a light-induced absorbance change near 550 nm which they attributed to an electron transport component "C550." This change is sensitized by System II light and occurs even at liquid nitrogen temperatures with the concomitant oxidation of cytochrome b_{559} (87) (see below), suggesting that these components are located close to the primary photoact. The low temperature observations were corroborated and extended by Bendall & Sofrova (88) and by Butler and co-workers (89–93). The 550 nm absorption change could also be induced by dithionite, suggesting that the bleaching is indicative of a photoreduction (89). The extent of the C550 change and fluorescence yield change observed at liquid nitrogen temperature followed the same redox titration curve (preset at room temperature) (90). These changes also behaved in parallel in mutants (91), when samples were heated (90), irradiated with UV light (92), or extracted with lipase (93). Butler and co-workers concluded from these correlations that C550 is identical to Q, the primary electron acceptor of System II.

However, the correspondence between C550 and Q was found to break down at higher temperature. Ben-Hayyim & Malkin (94) reported that C550 and Q (i.e. fluorescence yield) responded similarly to short wavelength light but not to long wave light. The response had two phases, only one of which was purely photochemical. Butler (95) subsequently reported a contribution to the 550 nm change at room temperature which he attributed to light-induced changes in membrane potential, at least partially due to System I. [This interpretation has been disputed, however (96).]

Noncorrelative behavior between C550 and other Photosystem II activities was also observed in other studies. Katoh & Kimimura (96) found that O_2 evolution and fluorescence yield had the same sensitivity to UV radiation, whereas C550 was much more resistant. Malkin & Knaff (97) reported that all C550 could be destroyed by chemical oxidation without affecting the rate of electron transport or the EPR signal which (presumably) reflects P_{690}^+.

Studies of C550 in various detergent-treated preparations in which System II donor reactions (but not O_2 evolution) are retained or enriched led to conflicting conclusions: Boardman (98), Van Gorkom et al (48), and Ke et al (43) found C550 in the reduced state in the dark while the fluorescence yield was low; under similar conditions, Kitajama & Butler (99) and Wessels et al (100) found that C550 was largely oxidized. Van Gorkom et al (48) found in their preparation that the oxidation of C550 by low (but not high) concentrations of ferricyanide was inhibited by DCMU, suggesting identity with Q. On the other hand, its state of reduction had only a small effect upon the fluorescence yield, which was primarily determined by the redox state of the plastoquinone.

Evidently the significance of C550 is still not clear. Experiments in which plastoquinone and β-carotene were extracted with organic solvents suggested that β-carotene was required to observe the C550 change (101, 102). The change is not due to a cytochrome, since there is no related absorbance in the Soret region (86, 88). There is good evidence that C550 is actually due to a bandshift from 547 to 543 nm at liquid nitrogen temperature (89, 103). Unlike other changes attributed to the primary acceptor of System II (i.e. fluorescence and "X-320"), there is no report in the literature of a C550 change induced by a single brief flash at room temperature, which would imply that it is not a primary event. Thus, although C550 seems to be a good indicator of the state of Q under some circumstances (e.g. at $-196°C$), it is rather unlikely that it is an actual participant in the redox reaction.

ROLE OF PLASTOQUINONE   The chemical identity of the primary acceptor is not known. One prime candidate is plastoquinone, a chloroplast constituent that has been shown to undergo light-induced changes consistent with the position of Q in the electron transport chain (59, 104–110). However, although its function as a secondary acceptor seems reasonably well established (the "A pool," see 59), its identity as the primary acceptor has not been demonstrated (see e.g. 111).

Stiehl & Witt (112, 113) reported the occurrence of a light-induced absorption change in the UV (called "X-320" or "X-335"), which they attributed to the reduction of the primary System II acceptor. On the basis of the observations that 1. the X-320 reaction was not sensitized by System I light, and 2. the lifetime of X-320 corresponded to that of Q [as determined by fluorescence measurements (114)], they proposed that X-320 was the System II primary acceptor and that it was the semiquinone of plastoquinone. A similar suggestion was later made on the basis of light-induced difference spectra in subchloroplast fragments (115). Bensasson & Land (116) reported that the semiquinone of plastoquinone has an absorption maximum at 320 nm and a spectrum that is generally equivalent to that reported for X-320.[6] Additional support for the identity of X-320 and Q was obtained when it was shown that X-320 is formed at 140°K, and is therefore presumably a participant in a primary photoreaction (117). Like that of Q, the recovery of X-320 at room temperature is inhibited by DCMU (117).

---

[6] Diner & Mauzerall (21) speculated that the primary acceptor is a plastochromonol, a compound that Kohl and associates proposed as the source of "EPR Signal II" (118, 119). Although the significance of "Signal II" is unresolved (120–124), its very sluggish decay kinetics makes a close association with Q doubtful.

Kinetic studies of fluorescence (114, 125) and O_2 evolution (16, 126) indicated that Q interacts with the A_2 pool as if the equilibrium constant for the reactions is near unity. This suggests (16) that Q is a "special" equivalent of the A_2 pool (which is probably plastoquinone). In any event, if Q is a plastoquinone molecule it must have properties much different from those of the A pools in order to be compatible with its unique characteristics (e.g. the effect of DCMU).

Cytochrome b_{559}

Cytochrome b_{559} was first described by Lundegardh (127, 128) in 1962. Despite nearly 100 publications devoted to the subject, its function in photosynthesis remains to be clarified. [For a description of some of the techniques used in these studies, see (129).]

Redox potential   The first problem encountered in the study of cytochrome b_{559} was its apparently variable redox potential. Bendall (130) reported the redox potential to be $+370$ mV, but hinted that there might also be cytochromes with similar spectral characteristics at a lower potential. Subsequent determinations fell into two groups centering around $+350$ mV (131–136) and $+70$ mV (137–139). Wada & Arnon (140) suggested that cytochrome b_{559} could exist in three forms (of differing redox potentials) and that the high potential form could be converted to the lower potential forms by aging and other disruptive treatments. In disrupted chloroplasts, Erixon et al (135) observed broad titration curves and ascribed this to a continuum of lower redox potentials. Extraction of chloroplasts with organic solvents (which removes most of the plastoquinone and β-carotene) has been found to change the high potential form (cytochrome b_{559} HP) to its low potential form, a change that could be partially reversed by the addition of either the extract or plastoquinone (101, 102, 141). Many other treatments have been found that convert the high potential to the low potential form. These include rather specific reagents, such as FCCP and antimycin A (142), and treatments that tend to destroy O_2 evolution, such as hydroxylamine (143), tris-washing, heat, and detergent (140).

Cytochrome b_{559} associated with System II   Studies to date indicate that there is at least one, and possibly two, molecules of cytochrome b_{559} per 400 chlorophylls (a photosynthetic unit) (137, 138, 144–146). [For a compilation of some of these studies, see (139).] Systematic disruption of chloroplasts to separate the two photosystems has established that there is at least one cytochrome b_{559} molecule closely associated with each System II unit (43, 100, 137, 139, 144, 145, 147, 148).

The physiological significance (if any) of the low potential form of the cytochrome is unclear. Cramer et al (142) have ascribed a regulatory function to a switch from the high potential to the low potential form. Erixon et al (135) suggest, however, that the low potential cytochrome b_{559} observed in fresh chloroplasts is a different cytochrome and not "modified cytochrome b_{559}." This view has received support from reports that cytochrome b_{559} is present in its low potential form as part of a System I complex with cytochromes b_6 and f (149) and in System I fragments (150). In the following discussions we will assume that the low potential form, if present in vivo, is associated with System I, and that only the high potential form is associated with Photosystem II.

Requirement of cytochrome b_{559} HP for O_2 evolution?   Studies with mutant algae (151) showed that strains lacking 50% of their cytochrome b_{559} HP were incapable of evolving O_2, suggesting that cytochrome b_{559} HP existed in two distinct pools, one of which is required for O_2 evolution. A similar hypothesis was proposed (152) and later withdrawn (153) by Bendall and co-workers, who found that treatments that converted cytochrome b_{559} HP to its low potential form did not have a parallel effect on O_2 evolution. Cramer & Böhme (143), on the basis of studies using FCCP (see above), also concluded that cytochrome b_{559} HP was not on the pathway of H_2O oxidation.

Studies using greening etiolated tissue also gave mixed results: some workers reported that the development of cytochrome b_{559} HP correlated with that of O_2 evolution (154), whereas others (155, 156) observed that the O_2 evolving system was operative in the absence of cytochrome b_{559} HP, which did not appear until much later in the greening process. Although the overall picture is still ambiguous, the bulk of the evidence seems to indicate that cytochrome b_{559} HP does not directly participate in the O_2 evolution process.

Photo-oxidation by System II   Cytochrome b_{559} is photo-oxidized by System II at liquid nitrogen temperatures (42, 87, 89, 90, 133, 135, 152). This strongly suggests a close association with P_{690}, the primary oxidant. At room temperature, photo-oxidation of cytochrome b_{559} by System II is also observed in Tris-extracted chloroplasts (87, 132) or at high pH (134), where O_2 evolution is inhibited.[7]

These observations have suggested to some that cytochrome b_{559} functions in a cycle or side path around System II, donating electrons to the primary System II oxidant in parallel with the "real" donor of the O_2 system (38, 89, 133, 139, 143, 157). The significance of this process under physiological conditions is hard to assess; it might be an alternate pathway when the normal electron donor to P_{690} is unavailable. Other aspects of this pathway will be discussed below.

Turnover mediated by Photosystems I and II   There is little agreement among the numerous reports in the literature concerning the redox changes of cytochrome b_{559} mediated by the two photosystems in intact systems at room temperature (which presumably reflect bonafide physiological processes). The photo-oxidation of cytochrome b_{559} by System I and its reduction by System II (136, 158–163) have been reported by several laboratories, whereas other groups have reported the absence of this push-and-pull process in untreated tissue (i.e. no CCCP, etc, see below) (87, 110, 143, 164–167). [There is one report of System I photoreduction (168).] Although this work has resulted in proposals that cytochrome b_{559} is located in the electron

---

[7] There is one report (157) of cytochrome b_{559} oxidation by System II at room temperature in a chloroplast preparation which had significant O_2 evolution activity (about one half that of intact chloroplasts). In this preparation it may be that the observed cytochrome oxidation occurs in the inactive fraction.

transport chain connecting the two photosystems (136, 158–161, 169, 170), its exceedingly sluggish turnover under physiological conditions (e.g. see data in 157, 158, 170, 171) and the nonstoichiometric behavior of the assumed reaction partners [e.g. see (158)] make such a scheme doubtful.

The addition of agents such as CCCP and FCCP (133, 142, 160, 166, 167, 169, 172, 173) results in a much more pronounced light-induced turnover of the cytochrome, probably due to a change of the high potential to the low potential form. Although most workers observed reduction by System II and oxidation by System I (i.e. push-and-pull), reports to the contrary have appeared (165, 171).

The scheme presented in Figure 2 attempts to account for some of the conflicting data. According to this hypothesis cytochrome b_{559} has a low leak rate to System I so that it can be either oxidized or reduced by long wave light depending on the ratio of the rate of incidence of Photosystem II quanta to the leak rate. Some of the reported diversity observed in supposedly identical experiments could then be ascribed to variations in the intensity and wavelength composition of the System I light between different experiments.

RELATED TOPICS

Role of Manganese-Photoactivation

To date, little is known about the enzymatic system that mediates the flow of electrons from water to P_{690}. Undoubtedly this is partly due to the fragility of the system and its apparent inaccessibility—most substances that can replace water as an electron donor tend to destroy the O_2 system.

The suggestion that manganese is an active component of the O_2 evolving system has received a good deal of support, particularly from the work of Cheniae et al. [For a detailed review of earlier work related to the function of Mn in photosynthesis, see (1).] The active Mn is apparently associated with a protein; Cheniae & Martin (174) reported that the Mn-containing entity was water soluble and could be isolated and partially purified. However, the O_2 evolving system could not be reconstituted by the readdition of this protein, and thus its function in System II is still open to question.

The Mn associated with System II seems to be located in two discrete pools of about two and four Mn per O_2 center, respectively. Experiments in which the Mn content of the cells or chloroplasts was altered during growth or by extraction suggested a stoichiometric relationship between the larger, more readily removable pool of bound manganese and O_2 evolution activity (175–179). The removal of this pool, however, does not affect the photo-oxidation of artificial System II donors (180). [There are several reports (181–183) suggesting that the Mn pool associated with O_2 evolution is a substantially smaller fraction of the total chloroplast Mn pool ($\leq 30\%$). The reason for the discrepancy is unclear; the outer chloroplast membrane can be quite impermeable to Mn^{2+} (179), and it may be that some extraction procedures cause the release of Mn into the intrachloroplast space rather than the surrounding medium, so that it is difficult to distinguish free vs bound Mn^{2+} (179).]

At present there is no information concerning the redox state of Mn in situ or direct evidence for its participation in the process of O_2 evolution. To date there has been no report of an EPR signal attributable to bound Mn. Lozier et al (184) found that a signal due to soluble Mn^{2+} is observed when bound Mn is released by Tris extraction; this represents about 60% of the total bound Mn pool (179). The presence of bound Mn is also related to EPR signal II; Chen & Wang (177) observed that the magnitude of this signal correlates with the loss of bound Mn, suggesting that the signal is generated between the Mn complex and P_{690}.

An intriguing aspect of the O_2 evolving system is the process of photoactivation. Cells grown in dark cannot immediately evolve O_2 upon illumination; the observed lag apparently reflects the activation of O_2 centers and involves the light-driven transformation of inactive, soluble Mn^{2+} to a bound, active form (174, 185, 186). In this process Mn is incorporated directly into the O_2 evolving system without the synthesis of protein or chlorophyll (174). Apparently a photo-oxidation of Mn^{2+} is involved, since compounds known to reduce higher oxidation states of manganese (or the light-induced System II oxidant) also inhibit photoactivation (187).

Photoactivation is strictly a System II reaction and is specific for Mn^{2+} salts (188). It appears to be a multiquantum process of low overall efficiency which is rate saturated in very weak light (189). Analysis of data obtained in flashing light (190) yielded as a minimal hypothesis a mechanism containing two light-driven steps as shown in equation 3 (191)

$$Mn + E \rightleftarrows MnE \xrightarrow{h\nu} MnE' \xrightarrow{h\nu} MnE_{act} \qquad \qquad 3.$$

$$\underset{k_{-1}}{\overset{\big\uparrow \underline{\hspace{4cm}} \big\downarrow}{}}$$

According to this hypothesis at any time only a small fraction of the enzyme E is associated with Mn (MnE) so that overall quantum yield is low. The backreaction (k_{-1}) might be related to deactivation (see above).

To date, attempts to induce photoactivation in isolated chloroplasts have not been successful. In vivo the process has been demonstrated in algae (blue-green as well as green) that were 1. grown in darkness (192) or in Mn-free medium (189), 2. extracted with NH_2OH (187), and 3. depleted by aging (193) or prolonged anaerobiosis (188). Photoactivation thus seems to be a universal process in the development of the O_2 evolving apparatus.[8]

Interaction of Q with Plastoquinone

The oxidation of Q by the electron transport chain to System I presumably involves the (direct?) interaction of a one electron donor Q (195) with a two-electron acceptor plastoquinone (PQ). Stiehl & Witt (113) studied the kinetics of the photoreduction of plastoquinone by System II using a repetitive flash technique. From the second order dependence of the reduction on the redox state of PQ, they concluded that two System II units act in tandem to furnish the two electrons required to reduce PQ. More recently, Bouges-Bocquet (196) and Velthuys &

[8] The one apparent exception reported in the literature (194) has been shown to be in error (G. M. Cheniae, personal communication).

Amesz (197) reported the observation of a binary oscillation of the redox state of P_{700} (196) and the magnitude of the dithionite-induced fluorescence, as a function of flash number (197). Both reports suggested that an equivalent is stored by an intermediate acceptor so that each System II unit independently reduces a PQ molecule. The reactions that cause these oscillations are probably independent of the S states because they occur in the absence of O_2 evolution if appropriate electron donors are present (197). The nature of this intermediate "charge accumulator" is unknown.

Luminescence, Fluorescence, and Correlation of Rate Constants

Luminescence   All light emitted from chlorophyll later than 10^{-9} sec after excitation reflects energy stored in forms other than the singlet state. Since the spectrum of delayed light resembles that of prompt fluorescence, these emissions are probably due to a backreaction of the primary photoact (198). However, the largest potential span in System II $(Q-P_{690})$ is only about one volt, and a simple backreaction would not release enough energy to emit a red photon (~ 1.8 eV). The source of the additional energy is unknown; it may be derived in part from the high energy state of the thylakoid (199).

Measurement of luminescence is simpler than that of fluorescence or spectral changes, and thus it provides a unique method for monitoring rapid ($< \sim 1$ msec) processes close to the photoact of System II. The ensuing discussion will be restricted to the fast components of luminescence decay observed after short single flashes. For a more detailed account, the reader is referred to the review by Lavorel (200).

Zankel (199) reported the observation of three kinetic components of chloroplast luminescence after short (microsecond) saturating flashes; one was very rapid (< 10 μsec) and the other two were slower. One of the slower components had a first order half-life of 35 μsec at 25°C (60 μsec at 0°C) and was not sensitive to DCMU. The total emission of this component after a saturating flash was about 10^{-4} photons per System II trap. This emission could be induced by a single flash after 10 min in the dark. After subsequent flashes, it oscillated like the O_2 yield; i.e. it was highest when traps in the S_3 state were excited.

The other slow emission component had a half-life of 200 μsec, was annihilated by DCMU, absent after the first flash after long dark, and also oscillated in phase with the O_2 yield [Barbieri et al (201) found that emissions occurring at longer times (> 10 msec) after the flashes oscillated out of phase, with a maximum after Y_2. This change of phase occurs at about 5 msec, the faster components being "in phase" with O_2 evolution, and the slower ones "a step ahead." A similar phase delay was also observed in the oscillations of so-called triggered luminescence (202).]

Using higher time resolution measurements, Lavorel (203) generally confirmed these observations with whole cells. He reported that the fast (5–10 μsec) component was a substantial portion (70–80%) of the total signal. NH_2OH eliminated this fast emission, while enhancing the medium component (50–70 μsec, probably corresponding to the 35 μsec component of Zankel). His data suggested that the lifetimes of these components were variable and depended on pretreatment.

Fluorescence   Similar time constants have been observed in the fluorescence yield after a brief flash. Zankel (53) ascribed the rise of the yield with a half-time of 35 μsec to reactions on the donor side (53); it was more pronounced after the second and third flashes than after the first and fourth. This kinetic component was also observed by Joliot (67) but is not apparent in the data of Mauzerall (57). Following the rise, the fluorescence decays with 200 μsec and slower (millisecond) components (53, 57).

Rate constants   The values derived for the rate constants in these and other kinetic studies of (presumably) the same system tend to fall into a few well-defined ranges (see Table 1), and it is tempting—albeit a bit dangerous—to ascribe these similar kinetics components to the same reactions. With this caveat in mind, we will consider the rate constants shown in Table 1 as follows:

1. Since luminescence is probably a result of the backreaction of the primary photoproducts P_{690}^+ and Q^- (see above), its kinetic behavior should reflect their disappearance. The recovery of P_{690} after a flash, determined by spectral measurements, is reported to have two kinetic phases, a large 35 μsec component and a smaller ($\sim 1/3$) 200 μsec component (40); both may be reflected in the luminescence emission.[9] The 200 μsec luminescence component (reported to be sensitive to DCMU) probably also reflects the recovery of Q.

2. If we accept the interpretation that P_{690}^+ is a quencher (see above), the 35 μsec rise in fluorescence reported by Zankel (but not by Mauzerall) may reflect the recovery of P_{690}^+. In this case, however, the 200 μsec components of fluorescence decay and P_{690} recovery must be unrelated, since the predicted effects are antiparallel. The slower fluorescence decay components (i.e. 200 μsec and ~ 2 msec) probably reflect exclusively the recovery of Q.

3. If X-320 is the primary acceptor, and thus identical to Q, the reported monophasic 600 μsec X-320 decay curve (113) does not agree with the observed two component (0.2 and 2 msec) fluorescence decay. This disparity may or may not be significant.

4. Measurements of the O_2 system relaxation rate in paired flashes yielded the same values (and disparity between laboratories) as the rate of recovery of the primary acceptor. This would imply that these measurements also reflect the recovery of Q. (The final O_2 evolution dark reaction, which has a half-time of about 1 msec, may not show up in paired flash measurements.)

Photoreactions at Low Temperatures

When dark adapted chloroplasts are illuminated at $-196°C$, the yield of the System II fluorescence emission bands ("F_{685} and F_{696}") rises from a low initial value (F_0) to a 2–5 times higher final value F_{max} (204). A number of absorption changes

---

[9] The 35 μsec P_{690} component was reported to be abolished by DCMU, whereas the corresponding luminescence component was insensitive to this treatment. However, the half-time for recovery of Q in the presence of DCMU is about 1 sec, and the faster flashing rate used in the P_{690} experiments (4 flashes/sec vs 1 flash/sec) probably did not allow a significant recovery.

Table 1 Half-times of various kinetic components at room temperature

		References
1. Luminescence decay (single flashes)	a. <10 μsec, 35 μsec, and 200 μsec	199
	b. 5–10 μsec, 50–70 μsec, and 200 μsec	203
2. Fluorescence (single flashes)	a. rise, 35 μsec; decay, 200 μsec and ~ 2 msec	53
	b. rise, 3 μsec; decay, 200 μsec and 2 msec	57
	c. ~ 35 μsec rise; decay data not presented	67
3. P_{690} recovery (repetitive flashes)	35 μsec and 200 μsec	40
4. X-320 recovery (repetitive flashes)	600 μsec	113
5. O_2 evolution relaxation (repetitive paired flashes)	a. 600 μsec	20
	b. 200 μsec and 2 msec	21

also occur at $-196°$C, some of which correlate with System II. In addition to the C550 and cytochrome b_{559} changes discussed earlier, a relatively large (and broad) change at 518 nm induced by System II (called P518) has been described by Amesz et al (205) and correlated with the similar absorption changes at room temperature. Although several studies of these various parameters have been reported, no consistent picture has emerged (70, 205–215).

The rise of fluorescence at $-196°$C is slower than the analogous rise observed at $-50°$C or at room temperature in the presence of DCMU (under all these conditions the oxidation of Q^- by the A pools is prevented) (204, 213, 214). At $-196°$C Murata et al (215) observed a half risetime of 35 msec in strong saturating light, which implies that at this temperature dark reactions are involved in the rise. In addition, the photoreaction must be inefficient (or involve a backreaction) because a single flash does not effect more than about 20% of F_{max} and a comparably small fraction of cytochrome b_{559}, C550, and P518 (106, 205). The apparent efficiency of the photoact (determined as the fluorescence risetime) undergoes a rather sudden transition at about $-100°$C (70, 205, 214) in dark adapted samples. Peculiarly, chloroplasts given two preflashes before cooling (so that the traps are predominantly in the S_2 and S_3 states) show a slow fluorescence rise at $-50°$C, which is comparable to that observed in dark adapted samples at $-196°$C (205). Two questions arise: 1. what causes the slow rise at $-196°$C, and 2. what underlies the similar slow rise at $-50°$C when the S_2 and S_3 states are predominant?

The first problem was addressed by Butler et al (70) and Murata et al (215), who proposed a scheme in which a relatively fast backreaction competed with the final transformation as shown in equation 4

$$\text{DPA} \underset{k_{-1}}{\overset{h\nu}{\rightleftharpoons}} \text{DP}^+\text{A}^- \overset{k_2}{\rightarrow} \text{D}^+\text{PA}^- \qquad \qquad 4.$$

According to Butler et al (70), $D \equiv$ cytochrome b_{559}, $P \equiv P_{690}$, and $A \equiv C550$. Mathis & Vermeglio (207) observed a ~ 5 msec backreaction of the (complete) flash-induced C550 change concurrent with a partial oxidation of cytochrome b_{559}, and concluded that the half-times of k_{-1} and k_2 were about 5 and 20 msec, respectively, at $-170°C$. These reaction times agree with the data of Floyd et al (42).

Although this scheme has appeared in several reports, it seems to be inconsistent with properties ascribed to its components. For example 1. if, as described earlier, P_{690}^+ is reduced by cytochrome b_{559} at low temperatures, equation 4 predicts that there should be no fluorescence rise in the presence of ferricyanide; 2. two reports on the intensity dependence of the rates of $\Delta 550$ and $\Delta 559$ were at variance, but neither was compatible with equation 4;[10] and 3. Den Haan et al (69) could not detect a change of the fluorescence yield in the millisecond range following a flash, which is not consistent with the assumptions that the P_{690}^+ is a quencher and is reduced in 4 msec by cytochrome b_{559}.

A more consistent picture might be obtained if this scheme is amended to include two secondary System II donors, one of which is cytochrome b_{559} [see (209)]. Such a scheme might be compatible with the reported redox behavior of the electron spin resonance (ESR) signal attributed to P_{690}^+ (216).

Experiments in which the behavior of System II components was studied at temperatures between -100 and $-50°C$ have yielded a rather incoherent picture. Butler et al (70) and Amesz et al (205) observed no photo-oxidation of cytochrome b_{559} at these temperatures, which might imply that now P_{690}^+ is rapidly reduced by its normal electron donor. [Vermeglio & Mathis (210) reported the conversion of a small amount at $-50°C$.] At $-40°C$, Amesz et al (205), using dark adapted chloroplasts, observed efficient (one flash) conversion of C550, P518, and fluorescence (which was unaffected by ferricyanide). However, after two preflashes before cooling (traps in the S_2 and S_3 states) a single flash elicited no C550 change, a smaller P518 change (70%), 30% of the fluorescence rise (which took many flashes for completion), and a lower F_{max} which was further ($\sim 20\%$) depressed by ferricyanide. These fluorescence data confirmed those of Joliot & Joliot discussed above (63). By way of contrast, Vermeglio & Mathis (210) reported that the kinetics of the transformations was unaffected by two preflashes, and that compared to dark adapted samples, preilluminated ($S_2 S_3$) samples showed the same C550 change, half the P518 change, and twice the amount of cytochrome b_{559} oxidation.

Although it is evident that System II behaves differently when the traps have accumulated positive charges, this behavior is not well described. Attempts to rationalize these observations show recurrent referral to a proposal of Joliot & Joliot (66) that each System II center contains two acceptor-donor couples, one operative in states S_0 and S_1 and the other in states S_2 and S_3.

The significance of observations made at low temperatures to physiological

---

[10] In one study (70) both processes saturated in very weak light (half-times of 0.3 and 1 min, respectively) which is not consistent with the role of C550 as a primary acceptor. In the other study (208) the conversion half-time of C550 (but not of cytochrome b_{559}) was linear with intensity over a 100-fold range.

processes remains to be determined. At room temperature, the fluorescence oscillation is relatively small and can be destroyed by 100 mM methylamine without affecting the O_2 oscillation (53). This might imply that the low temperature pathways, although instructive, play only a secondary role under normal conditions.

ACKNOWLEDGMENTS

The authors would like to thank Drs. G. Cheniae, C. F. Fowler, and K. L. Zankel for many helpful discussions.

This work was supported in part by grants from the Atomic Energy Commission, Contract AT(11-1)-3326, and the National Science Foundation, Contract NSF-C705.

Literature Cited

1. Cheniae, G. M. 1970. *Ann. Rev. Plant Physiol.* 21:467–98
2. Trebst, A. 1974. *Ann. Rev. Plant Physiol.* 25:423–58
3. Bishop, N. I. 1971. *Ann. Rev. Biochem.* 40:197–226
4. Witt, H. T. 1971. *Quart. Rev. Biophys.* 4:365–477
5. Kok, B., Cheniae, G. 1966. *Curr. Top. Bioenerg.* 1:1–47
6. George, P. 1964. *Oxidases Related Redox Syst. Proc. Symp.* 3–36
7. Joliot, P. 1967. *Brookhaven Symp. Biol.* 19:418–33
8. Joliot, P., Joliot, A. 1967. *Biochim. Biophys. Acta* 153:625–34
9. Forbush, B., Kok, B., McGloin, M. 1971. *Photochem. Photobiol.* 14:307–21
10. Joliot, P., Barbieri, G., Chabaud, R. 1969. *Photochem. Photobiol.* 10:309–29
11. Kok, B., Forbush, B., McGloin, M. 1970. *Photochem. Photobiol.* 11:457–75
12. Bouges, B. 1971. *Biochim. Biophys. Acta* 234:103–12
13. Joliot, P. 1968. *Photochem. Photobiol.* 8:451–63
14. Fowler, C., Kok, B. 1974. *Biochim. Biophys. Acta* 357:299–307
15. Joliot, P. 1961. *J. Chim. Phys. Physico-chim. Biol.* 58:584
16. Radmer, R., Kok, B. 1973. *Biochim. Biophys. Acta* 324:28–41
17. Joliot, P., Joliot, A., Bouges, B., Barbieri, G. 1971. *Photochem. Photobiol.* 14:287–305
18. Delrieu, M. 1973. *C. R. Acad. Sci. Paris* 277:2809–12
19. Weiss, C. Jr., Sauer, K. 1970. *Photochem. Photobiol.* 11:495–501
20. Vater, J., Renger, G., Stiehl, H., Witt, H. 1968. *Naturwissenschaften* 55:220–21
21. Diner, B., Mauzerall, D. 1973. *Biochim. Biophys. Acta* 305:353–63
22. Bouges-Bocquet, B. 1973. *Biochim. Biophys. Acta* 292:772–85
23. Lemasson, C., Barbieri, G. 1971. *Biochim. Biophys. Acta* 245:386–97
24. Renger, G. 1973. *Biochim. Biophys. Acta* 314:113–16
25. Vater, J. 1973. *Biochim. Biophys. Acta* 292:786–95
26. Renger, G., Bouges-Bocquet, B., Delosme, R. 1973. *Biochim. Biophys. Acta* 292:796–807
27. Bouges-Bocquet, B., Bennoun, P., Taboury, J. 1973. *Biochim. Biophys. Acta* 325:247–54
28. Duysens, L. 1972. *Biophys. J.* 12:858–63
29. Rosenberg, J. L., Sahu, S., Bigat, T. K. 1972. *Biophys. J.* 12:839–49
30. Etienne, A. L. 1974. *Biochim. Biophys. Acta* 333:320–30
31. Marsho, T., Kok, B. 1970. *Biochim. Biophys. Acta* 223:240–50
32. Doschek, W., Kok, B. 1972. *Biophys. J.* 12:832–38
33. Kok, B., Radmer, R., Fowler, C. 1974. Presented at 3rd Int. Congr. Photosynthesis, Rehovot, Israel
34. Bennoun, P. 1970. *Biochim. Biophys. Acta* 216:357–63
35. Döring, G., Renger, G., Vater, J., Witt, H. 1969. *Z. Naturforsch. B* 24:1139–43
36. Döring, G., Witt, H. 1971. Presented at 2nd Int. Congr. Photosynthesis, Stresa, Italy
37. Govindjee, Döring, G., Govindjee, R. 1970. *Biochim. Biophys. Acta* 205:303–6
38. Butler, W. L. 1972. *Biophys. J.* 12:851–57
39. Döring, G., Stiehl, H., Witt, H. 1967. *Z. Naturforsch. B* 22:639–44

40. Glaser, M., Wolff, C., Buchwald, H., Witt, H. 1974. *FEBS Lett.* 42:81–85
41. Döring, G., Bailey, J. Kreutz, W., Witt, H. 1968. *Naturwissenschaften* 55:220
42. Floyd, R., Chance, B., Devault, D. 1971. *Biochim. Biophys. Acta* 226:103–12
43. Ke, B., Sahu, S., Shaw, E., Beinert, H. 1974. *Biochim. Biophys. Acta* 347:36–48
44. Lozier, R., Butler, W. 1974. *Biochim. Biophys. Acta* 333:465–80
45. Malkin, R., Bearden, A. 1973. *Proc. Nat. Acad. Sci. USA* 70:294–97
46. Commoner, B. et al 1957. *Science* 126: 57–63
47. Beinert, H., Kok, B. 1964. *Biochim. Biophys. Acta* 88:278–88
48. Van Gorkom, H., Tamminga, J., Haveman, J. 1974. *Biochim. Biophys. Acta* 347:417–38
49. Duysens, L., Sweers, H. 1963. In *Studies on Microalgae and Photosynthetic Bacteria,* ed. Japanese Soc. Plant Physiologists, 353–72. Tokyo: Univ. Tokyo Press
50. Goedheer, J. 1972. *Ann. Rev. Plant Physiol.* 23:87–112
51. Duysens, L. M. N. 1963. In *Photosynthetic Mechanisms of Green Plants,* 1–17. Washington, DC: Nat. Acad. Sci. Nat. Res. Council
52. Duysens, L. N. M., Van der Schatte, T., Den Haan, G. 1972. Presented at 6th Int. Congr. Photobiol., Bochum, Germany, 1972. Abstr. 277
53. Zankel, K. 1973. *Biochim. Biophys. Acta* 325:138–48
54. Chessin, M., Livingston, R., Truscott, T. G. 1966. *Trans. Faraday Soc.* 62: 1519–24
55. Mathis, P., Galmiche, J. 1967. *C. R. Acad. Sci. Paris* 264:1903–6
56. Wolff, C., Witt, H. 1969. *Z. Naturforsch. B* 24:1031–37
57. Mauzerall, D. 1972. *Proc. Nat. Acad. Sci. USA* 69:1358–62
58. Amesz, J., Fork, D. 1967. *Biochim. Biophys. Acta* 143:97–107
59. Amesz, J. 1973. *Biochim. Biophys. Acta* 301:35–52
60. Felker, P., Izawa, S., Good, N., Haug, A. 1973. *Biochim. Biophys. Acta* 325: 193–96
61. Morin, P. 1964. *J. Chim. Phys. Physico-chim. Biol.* 61:674–80
62. Delosme, R. 1967. *Biochim. Biophys. Acta* 143:108–28
63. Joliot, P., Joliot, A. 1973. *Biochim. Biophys. Acta* 305:302–16
64. Delosme, R. 1971. *C. R. Acad. Sci. Paris* 272:2828–31
65. Delosme, R. 1972. In *Proc. 2nd Int.*

Congr. Photosynthesis Res., Stresa, Italy, 1971, ed. G. Forti, M. Avron, A. Melandri, 187–96. The Hague: W. Junk
66. Joliot, P., Joliot, A. See Ref. 65, 26–48
67. Joliot, A. 1974. Presented at 3rd Int. Congr. Photosynthesis, Rehovot, Israel
68. Okayama, S., Butler, W. 1972. *Biochim. Biophys. Acta* 267:523–29
69. Den Haan, G., Warden, J., Duysens, L. 1973. *Biochim. Biophys. Acta* 325:120–25
70. Butler, W., Visser, J., Simons, H. 1972. *Biochim. Biophys. Acta* 292:140–51
71. Joliot, A., Joliot, P. 1964. *C. R. Acad. Sci. Paris* 258:4622–25
72. Homann, P. 1969. *Plant Physiol.* 44: 932–36
73. Murata, N. 1969. *Biochim. Biophys. Acta* 189:171–81
74. Mohanty, P., Braun, B., Govindjee. 1973. *Biochim. Biophys. Acta* 292:459–76
75. Clemet-Metral, J., Lefort-Tran, M. 1974. *Biochim. Biophys. Acta* 333:560–69
76. Jennings, R., Forti, G. 1974. *Biochim. Biophys. Acta* 347:299–310
77. Murata, N., Sugahara, K. 1969. *Biochim. Biophys. Acta* 189:182–92
78. Wraight, C., Crofts, A. 1970. *Eur. J. Biochem.* 17:319–27
79. Cohen, W., Sherman, L. 1971. *FEBS Lett.* 16:319–23
80. Krause, G. 1974. *Biochim. Biophys. Acta* 333:301–13
81. Rienits, K., Hardt, H., Avron, M. 1973. *FEBS Lett.* 33:28–32
82. Bonaventura, C., Myers, J. 1969. *Biochim. Biophys. Acta* 189:366–83
83. Murata, N., Tashiro, H., Takamiya, A. 1970. *Biochim. Biophys. Acta* 197:250–56
84. Myers, J. 1971. *Ann. Rev. Plant Physiol.* 22:289–312
85. Govindjee, Papageorgiou, G. 1971. *Photophysiology* 6:1–46
86. Knaff, D., Arnon, D. 1969. *Proc. Nat. Acad. Sci. USA* 63:963–69
87. Knaff, D., Arnon, D. 1969. *Proc. Nat. Acad. Sci. USA* 63:956–62
88. Bendall, D., Sofrova, D. 1971. *Biochim. Biophys. Acta* 234:371–80
89. Erixon, K., Butler, W. 1971. *Photochem. Photobiol.* 14:427–33
90. Erixon, K., Butler, W. 1971. *Biochim. Biophys. Acta* 234:381–89
91. Epel, B., Butler, W. 1972. *Biophys. J.* 12: 922–29
92. Erixon, K., Butler, W. 1971. *Biochim. Biophys. Acta* 253:483–86
93. Okayama, S. et al 1971. *Biochim. Biophys. Acta* 253:476–82

94. Ben-Hayyim, G., Malkin, S. See Ref. 65, 61–72
95. Butler, W. 1972. FEBS Lett. 20:333–37
96. Katoh, S., Kimimura, M. 1974. Biochim. Biophys. Acta 333:71–84
97. Malkin, R., Knaff, D. 1973. Biochim. Biophys. Acta 325:336–40
98. Boardman, N. 1972. Biochim. Biophys. Acta 283:469–82
99. Kitajama, M., Butler, W. 1973. Biochim. Biophys. Acta 325:558–64
100. Wessels, J., Von Alphen-Van Waveren, O., Voorn, G. 1973. Biochim. Biophys. Acta 292:741–52
101. Okayama, S., Butler, W. 1972. Plant Physiol. 49:769–44
102. Cox, R., Bendall, D. 1974. Biochim. Biophys. Acta 347:49–59
103. Butler, W., Okayama, S. 1971. Biochim. Biophys. Acta 245:237–39
104. Klingenberg, M., Muller, A., Schmidt-Mende, P., Witt, H. 1962. Nature 194:379–80
105. Rumberg, B., Schmidt-Mende, P., Weikard, J., Witt, H. See Ref. 51, 18–34
106. Amesz, J. 1964. Biochim. Biophys. Acta 79:257–65
107. Rumberg, B., Schmidt-Mende, P., Witt, H. 1964. Nature 201:466–68
108. Schmidt-Mende, P., Rumberg, B. 1968. Z. Naturforsch. B 23:225–28
109. Schmidt-Mende, P., Witt, H. 1968. Z. Naturforsch. B 23:228–35
110. Amesz, J., Visser, J., Van den Engh, G. J., Dirks, M. 1972. Biochim. Biophys. Acta 256:370–80
111. Trebst, A. See Ref. 65, 399–418
112. Stiehl, H., Witt, H. 1968. Z. Naturforsch. B 23:220–24
113. Stiehl, H., Witt, H. 1969. Z. Naturforsch. B 24:1588–98
114. Forbush, B., Kok, B. 1968. Biochim. Biophys. Acta 162:243–53
115. Van Gorkom, H. 1974. Biochim. Biophys. Acta 347:439–42
116. Bensasson, R., Land, E. 1973. Biochim. Biophys. Acta 325:175–81
117. Witt, K. 1973. FEBS Lett. 38:116–18
118. Kohl, D., Wood, P. 1969. Plant Physiol. 44:1439–45
119. Kohl, D., Wright, J., Weissman, M. 1969. Biochim. Biophys. Acta 180:536–44
120. Warden, J., Bolton, J. 1974. Photochem. Photobiol. 20:245–50
121. Esser, A. 1974. Photochem. Photobiol. 20:167–72
122. Esser, A. 1974. Photochem. Photobiol. 20:173–81
123. Babcock, G., Sauer, K. 1973. Biochim.
Biophys. Acta 325:483–503
124. Babcock, G., Sauer, K. 1973. Biochim. Biophys. Acta 325:504–19
125. Joliot, P. 1965. Biochim. Biophys. Acta 102:135–48
126. Joliot, P. 1965. Biochim. Biophys. Acta 102:116–34
127. Lundegardh, H. 1965. Proc. Nat. Acad. Sci. USA 53:703–10
128. Lundegardh, H. 1962. Physiol. Plant. 15:390–98
129. Bendall, D., Davenport, H., Hill, R. 1971. Methods Enzymol. 23A:327–44
130. Bendall, D. 1968. Biochem. J. 109:46–47P
131. Ikegami, I., Katoh, S., Takamiya, A. 1968. Biochim. Biophys. Acta 162:604–6
132. Knaff, D., Arnon, D. 1969. Proc. Nat. Acad. Sci. USA 64:715–22
133. Boardman, N., Anderson, J., Hiller, R. 1971. Biochim. Biophys. Acta 234:126–36
134. Knaff, D., Arnon, D. 1971. Biochim. Biophys. Acta 226:400–8
135. Erixon, K., Lozier, R., Butler, W. 1972. Biochim. Biophys. Acta 267:375–82
136. Knaff, D. 1973. Biochim. Biophys. Acta 325:284–96
137. Hind, G., Nakatani, H. 1970. Biochim. Biophys. Acta 216:223–25
138. Fan, H., Cramer, W. 1970. Biochim. Biophys. Acta 216:200–7
139. Ke, B., Vernon, L., Chaney, T. 1972. Biochim. Biophys. Acta 256:345–57
140. Wada, K., Arnon, D. 1971. Proc. Nat. Acad. Sci. USA 68:3064–68
141. Sofrova, D., Bendall, D. See Ref. 65, 561–68
142. Cramer, W., Fan, H., Böhme, H. 1971. Bioenergetics 2:289–303
143. Cramer, W., Böhme, H. 1972. Biochim. Biophys. Acta 256:358–69
144. Boardman, N., Anderson, J. 1967. Biochim. Biophys. Acta 143:187–203
145. Arnon, D., Tsujimoto, H., McSwain, B., Chain, R. 1968. In Comparative Biochemistry and Biophysics of Photosynthesis, ed. K. Shibata, A. Takamiya, A. Jagendorf, R. Fuller, 113–32. Tokyo: Univ. of Tokyo Press
146. McEvoy, F., Lynn, W. 1972. Arch. Biochem. Biophys. 150:624–31
147. Vernon, L., Ke, B., Mollenhauer, H., Shaw, E. 1969. Progr. Photosyn. Res. 1:137–48
148. Vernon, L., Shaw, E., Ogawa, T., Raveed, D. 1971. Photochem. Photobiol. 14:343–57
149. Anderson, J., Boardman, N. 1973. FEBS Lett. 32:157–60

150. Knaff, D., Malkin, R. 1973. *Arch. Biochem. Biophys.* 159:555–62
151. Epel, B., Butler, W., Levine, R. 1972. *Biochim. Biophys. Acta* 275:395–400
152. Bendall, D., Sofrova, D. 1971. *Biochim. Biophys. Acta* 234:371–80
153. Cox, R., Bendall, D. 1972. *Biochim. Biophys. Acta* 283:124–35
154. Plesnicar, M., Bendall, D. See Ref. 65, 2367–74
155. Whatley, F., Gregory, P., Haslett, B., Bradbeer, J. See Ref. 65, 2375–82
156. Henningsen, K., Boardman, N. 1973. *Plant Physiol.* 51:1117–26
157. Fork, D. 1972. *Biophys. J.* 12:909–21
158. Levine, R., Gorman, D., Avron, M., Butler, W. 1967. *Brookhaven Symp. Biol.* 19:143–48
159. Levine, R., Gorman, D. 1966. *Plant Physiol.* 41:1293–1300
160. Cramer, W., Butler, W. 1967. *Biochim. Biophys. Acta* 143:332–39
161. Ben-Hayyim, G., Avron, M. 1970. *Eur. J. Biochem.* 14:205–13
162. Malkin, S. 1969. *Progr. Photosyn. Res.* 2:845–56
163. Satoh, K., Katoh, S. 1972. *Plant Cell Physiol.* 13:807–20
164. Hind, G. 1968. *Biochim. Biophys. Acta* 153:235–40
165. Amesz, J., Pulles, M., Visser, J., Sibbing, F. 1972. *Biochim. Biophys. Acta* 275:442–52
166. Anderson, J., Than-Nyunt, Boardman, N. 1973. *Arch. Biochem. Biophys.* 155:436–44
167. Hiller, R., Anderson, J., Boardman, N. 1971. *Biochim. Biophys. Acta* 245:439–52
168. Satoh, K., Yakushiji, E., Katoh, S. 1973. *Plant Physiol.* 14:763–67
169. Hind, G. 1968. *Photochem. Photobiol.* 7:369–75
170. Ikegami, I., Katoh, S., Takamiya, A. 1970. *Plant Cell Physiol.* 11:777–91
171. Ben-Hayyim, G. 1972. In *6th Int. Congr. Photobiol., Bochum, 1972*, ed. G. O. Schenck, 255 (Abstr.)
172. Hiller, R., Anderson, J., Boardman, N. See Ref. 65, 547–60
173. Böhme, H., Cramer, W. 1971. *FEBS Lett.* 15:349–51
174. Cheniae, G., Martin, I. See Ref. 158, 406–17
175. Cheniae, G., Martin, I. 1970. *Biochim. Biophys. Acta* 197:219–39
176. Cheniae, G., Martin, I. 1971. *Plant Physiol.* 47:568–75
177. Chen, K.-Y., Wang, J. 1974. *Bioinorg. Chem.* 3:367–80
178. Homann, P. 1968. *Biochem. Biophys. Res. Commun.* 33:229–34
179. Blankenship, R., Sauer, K. 1974. *Biochim. Biophys. Acta* 357:252–66
180. Cheniae, G., Martin, I. 1968. *Biochim. Biophys. Acta* 153:819–37
181. Itoh, M. et al 1969. *Biochim. Biophys. Acta* 180:509–19
182. Selman, B., Bannister, T., Dilley, R. 1973. *Biochim. Biophys. Acta* 292:566–81
183. Yamashita, T., Tsuji-Kaneko, J., Yamada, Y., Tomita, G. 1972. *Plant Cell Physiol.* 13:353–64
184. Lozier, R., Baginsky, M., Butler, W. 1971. *Photochem. Photobiol.* 14:323–28
185. Anderson, J., Pyliotis, N. 1969. *Biochim. Biophys. Acta* 189:280–93
186. Homann, P. 1967. *Plant Physiol.* 42:997–1007
187. Cheniae, G., Martin, I. 1972. *Plant Physiol.* 50:87–94
188. Cheniae, G., Martin, I. 1967. *Biochem. Biophys. Res. Commun.* 28:89–95
189. Cheniae, G., Martin, I. 1969. *Plant Physiol.* 44:351–60
190. Cheniae, G., Martin, I. 1971. *Biochim. Biophys. Acta* 253:167–81
191. Radmer, R., Cheniae, G. 1971. *Biochim. Biophys. Acta* 253:182–86
192. Cheniae, G., Martin, I. 1973. *Photochem. Photobiol.* 17:441–59
193. Margulies, M. 1972. *Biochim. Biophys. Acta* 267:96–103
194. Teichler-Zallen, D. 1969. *Plant Physiol.* 44:701–10
195. Cramer, W., Butler, W. 1969. *Biochim. Biophys. Acta* 172:503–10
196. Bouges-Bocquet, B. 1973. *Biochim. Biophys. Acta* 324:250–56
197. Velthuys, B., Amesz, J. 1974. *Biochim. Biophys. Acta* 333:85–94
198. Arnold, W., Davidson, J. 1954. *J. Gen. Physiol.* 37:677–84
199. Zankel, K. 1971. *Biochim. Biophys. Acta* 245:373–85
200. Lavorel, J. In *Bioenergetics of Photosynthesis*, ed. Govindjee. New York: Academic. In press
201. Barbieri, G., Delosme, R., Joliot, P. 1970. *Photochem. Photobiol.* 12:197–206
202. Hardt, H., Malkin, S. 1973. *Photochem. Photobiol.* 17:433–40
203. Lavorel, J. 1973. *Biochim. Biophys. Acta* 325:213–29
204. Kok, B. See Ref. 51, 45–55
205. Amesz, J., Pulles, M. P., Velthuys, B. 1973. *Biochim. Biophys. Acta* 325:472–82
206. Butler, W., Visser, J., Simons, H. 1973. *Biochim. Biophys. Acta* 325:539–45

207. Mathis, P., Vermeglio, A. *Biochim. Biophys. Acta.* In press
208. Mathis, P., Villaz, M., Vermeglio, A. 1974. *Biochem. Biophys. Res. Commun.* 56:682–88
209. Visser, J., Butler, W. See Ref. 171, 237 (Abstr.)
210. Vermeglio, A., Mathis, P. 1974. Presented at 3rd Int. Congr. Photosynthesis, Rehovot, Israel, 1–12
211. Vermeglio, A., Mathis, P. 1973. *Biochim. Biophys. Acta* 292:763–71
212. Vermeglio, A., Mathis, P. 1973. *Biochim. Biophys. Acta* 314:57–65
213. Thorne, S., Boardman, N. 1971. *Biochim. Biophys. Acta* 234:113–25
214. Malkin, S., Michaeli, G. See Ref. 65, 149–67
215. Murata, N., Itoh, S., Okada, M. 1973. *Biochim. Biophys. Acta* 325:463–71
216. Bearden, A., Malkin, R. 1973. *Biochim. Biophys. Acta* 325:266–74

BIOLOGICAL METHYLATION: SELECTED ASPECTS[1]

×890

Giulio L. Cantoni

The Laboratory of General and Comparative Biochemistry,
The National Institute of Mental Health, Bethesda, Maryland 20014

CONTENTS

Biological methylation is a broad subject that spans two principal areas of metabolism. First is a series of reactions leading to the biosynthesis of methionine (methylation of homocysteine) and then the multitude of reactions that utilize the methyl group of methionine after its activation to S-adenosylmethionine (A-Met). These, in fact, are not completely separate areas, and a link between them is provided by the highly specific requirement for A-Met in the terminal reaction of the methionine biosynthetic pathway (1–4). Moreover, A-Met plays a role as an inhibitor or a repressor of several enzymes that catalyze earlier steps in the methionine biosynthetic pathway (5–13). It would be impossible to cover this large and varied section of biochemistry in any depth in the space allotted, so I have chosen to limit very severely and quite arbitrarily the topics covered in this review. In this choice, I have been guided principally by my own biased curiosity, which has led me to examine in some detail areas relatively less familiar, not fully understood, or under rapid development. I will cover the methylation of amino acids and proteins, the methylation of carbohydrates and polysaccharides, and present views on the role of A-Met as the sole biological methyl donor compound to substrates other than homocysteine. Finally, I will list a number of methods that may be useful in the study of biological methylation. I am fully aware that this partial coverage will result in the complete omission of many significant areas of biological methylation. But, fortunately, some of these areas have been covered expertly by recent reviews:

[1] The abbreviations used are: A-Met, S-adenosylmethionine; ASR, S-adenosylhomocysteine; NG, NG dimethylarginine, $(CH_3)_2\text{-N-C}(=NH)\text{-NH-}(CH_2)_3\text{-CH}(NH_2)\text{-COOH}$; NG, N'G dimethylarginine, $(CH_3)\text{-NH-C}(=NCH_3)\text{-NH-}(CH_2\text{-})_3\text{-}(CHNH_2)\text{-COOH}$.

435

DNA methylation and the modification and restriction phenomena (14–16), tRNA methylation (17–20), methionine biosynthesis (3, 4, 21, 22), adenosyltransferases (23–25), the formation of adenosylethionine by adenosyltransferase and its biochemical consequences (26), biological mechanisms in methyl group transfer (27, 28), and polyamine metabolism (29–31). The proceedings of a recent symposium on the biochemistry of A-Met contain many excellent reviews and reports on a number of new developments (32).

IS S-ADENOSYLMETHIONINE THE ONLY BIOLOGICAL METHYL DONOR?

Over twenty years ago, I proposed as a working hypothesis (33, 34) the concept that A-Met is the sole methyl donor in all methyl transfer reactions, except those resulting in the biosynthesis of methionine. No serious experimental exception was presented to challenge this concept, and we were ready to change the "working hypothesis" into an "established general principle"; recently, however, a series of interesting reports suggested that N^5-methyltetrahydrofolate (CH_3-THF) may be the donor of a methyl group for the biosynthesis of epinine and other methylated catechol or methylated indoleamines (35–38).

This possibility is especially interesting from two points of view: From a biochemical standpoint, whereas the role of CH_3-THF in the biosynthesis of methionine is well established in bacteria, plants, and vertebrates, including man (3, 4, 39), the possible role of CH_3-THF in the methylation of biogenic amines would be novel. The energetics of these methyl transfer reactions is clearly different from those proceeding from an onium pole. Coward (in 32; personal communication) estimates that methyl ammonium compounds such as 5-CH_3-THF are $\sim 10^3$ times less reactive toward "soft" polarizable nucleophiles than the corresponding sulfonium compounds. Even protonated or partially oxidized (quarternized) 5-CH_3-THF would be considerably less reactive than A-Met with a soft nucleophile such as homocysteine and presumably oxyanions and amines. Therefore in order to catalyze methyl transfer reactions, 5-CH_3-THF enzymes must utilize other mechanisms, such as reaction of the apoenzymes with vitamin B-12, taking advantage of the fact that B-12 is one of the most powerful nucleophiles known (3, 40). Alternative mechanisms would require the protonation or quarternization of the nitrogen bearing the methyl group in the nonpolar environment on the enzyme surface.

Laduron's and Snyder's observations are exciting for a second reason, namely, the possibility that a decrease in the availability of a methylated amine caused by deficiency of CH_3-THF or of specific methylating enzymes might affect behavior and mentation. A working hypothesis along these lines has been advanced by my colleague, Harvey Mudd, in collaboration with J. M. Freeman (41). Based on the observation that when the supply of CH_3-THF is low because of deficiency in 5–10 methyltetrahydrofolate reductase, normal methylation of homocysteine is impaired (42), Mudd & Freeman proposed that a lowered level of CH_3-THF could also interfere with the methylation of biogenic amines. The K_m values of these two enzymes for CH_3-THF are of similar magnitude: the K_m for homocysteine methyltransferase is 6×10^{-5} M and that for dopamine is 6×10^{-5} M (36) or 1×10^{-5} M (38). As Mudd & Freeman (41) have pointed out

> Several lines of evidence have recently converged to provide tentative support for the hypothesis that one or more abnormalities of dopamine metabolism, turnover, or function may contribute significantly to the etiology of schizophrenia (43, 44). Among the specific models proposed are those involving (a) the overactivity of dopaminergic transmission (44), or (b) a genetically determined error of metabolism which might lead to increased formation of a dopamine metabolite. Failure of dopamine removal due to impaired flow through the step catalyzed by N^5-methyltetrahydrofolate-dopamine methyltransferase might be envisioned as a significant underlying factor in such models.

Furthermore, it is now apparent that one or more CH_3-THF-dependent methyltransferases may be present in a number of tissues besides brain (38), and that not only dopamine but also a number of other biogenic amines might be active as methyl acceptors. Substrates of interest include tryptamine, norepinephrine, and serotonin, the latter being predominately O-methylated (36, 38, 45, 46). S. H. Snyder (personal communication) identified two distinct enzymes capable of using CH_3-THF as a methyl donor. One N-methylates a variety of amines. The other, which is quantitatively much more impressive, methylates primarily serotonin. The product of this reaction, 5-methoxytryptamine, occurs in significant concentrations in the brain. Moreover, it behaves neurophysiologically very much like a psychedelic drug. It could be a normally occurring psychedelic agent which may function to facilitate perceptual integration. Thus, it is difficult at present to specify exactly the physiological role(s) of CH_3-THF methyl transfer reactions. For example, interference with serotonin metabolism might play a role in the precipitation of schizophrenic manifestations (47). A deficiency rather than an overproduction of 5-methoxytryptamine may be associated with emotional disturbances. The intent of Mudd & Freeman's hypothesis is not to maintain that deficient activity of $N^{5,10}$-methylenetetrahydrofolate reductase per se is the cause of a significant portion of those cases now diagnosed as schizophrenia (48). Rather, it is to point out the possibility that CH_3-THF-dependent transmethylations may be a crucial area where pathophysiological sequences converge, starting from one or another primary lesion, and ultimately lead to a common chemical aberration. A deficiency of $N^{5,10}$-methylenetetrahydrofolate reductase may be one example of a primary lesion that leads to a decrease in CH_3-THF-dependent methylations, and hence to schizophrenia. The psychotogenic effect of methionine (49) may be another example. This effect, often interpreted in terms of increased methylation leading to schizophrenia, may just as well be related to the demonstrated capacity (6, 12) of methionine to prevent the accumulation of CH_3-THF. To evaluate this working hypothesis, it becomes of obvious interest to further define the distribution, molecular basis, and physiological role of CH_3-THF-dependent methyl transfer reactions. A deficiency of an enzyme catalyzing one such methyl transfer reaction might be the primary cause of some cases of schizophrenia.

The working hypothesis set forth by Mudd & Freeman differs fundamentally from that recently suggested by Laduron (36). In order to reconcile his discovery of

CH_3-THF-dependent methylation of biogenic amines (35) with earlier theories postulating overmethylation in schizophrenia, Laduron postulated that an increase in CH_3-THF-dependent methylation might be responsible for schizophrenic disorders (35). With respect to CH_3-THF-dependent methylation, this observation appears to favor *over*methylation rather than *under*methylation as a factor in psychotic symptoms. However, it is just as possible that the deficiency of a methylated compound, caused by either lack of CH_3-THF or deficiency of a CH_3-THF-methyltransferase, is the predominant factor in the etiology of schizophrenia.

AMINO ACID AND PROTEIN METHYLATION

The existence of methylated amino acids has been known for almost fifty years, ever since 1-methylhistidine was recognized as a product of anserine hydrolysis (50). Later a number of methylated amino acids, such as 1-methylhistidine (51); 3-methylhistidine (52); and three different methylated arginine derivatives (53–54), CH_3-NH-C($=$NH)-NH-$(CH_2)_3$-CH(NH_2)-COOH, CH_3-NH-C($=$NCH_3$)-NH-$(CH_2)_3$-CH($NH_2$)-COOH, and $(CH_3)_2$-N-C($=$NH)-NH-$(CH_2)_3$-CH(NH_2)-COOH, have been isolated from human urine.

It has become clear only in recent years that these amino acids are the product of the catabolism of proteins (55) of which the different methylated amino acids are components, and that the methylation step occurred at the polypeptide level. This was first demonstrated by Comb et al (56), who found that methyllysine, a methylated amino acid previously identified as a product of the hydrolysis of the flagellar protein of *Salmonella typhimurium* (55), is formed in vitro upon incubation of washed ribosomes from *Blastocladiella emersonii* with A-Met (56). In vitro methylation of histones with A-Met and a partially purified extract of calf thymus nuclei led Paik & Kim (57) and Kaye & Sheratzky (58) to identify methylarginine as the methylation product.

The presence of methylarginine in proteins was confirmed by Eylar (59), Brostoff & Eylar (60), and Carnegie (61, 62). These investigators established that methylarginine occurs specifically at residue 107 in a basic protein, called A1 protein, which is a component of myelin from the brain of many vertebrates, including man. As noted above, there are three different N-methylarginines, and the identification of the molecular species found in the A1 protein has been rather tortuous. Baldwin & Carnegie (63) at first reported that NG, N′G dimethylarginine was the methylated component in myelin. Later Brostoff & Eylar reported that both monomethylarginine and NG, NG dimethylarginine (also called unsymmetrical dimethylarginine) are present in the A1 protein, and finally Deibler & Martenson (64), using a highly sensitive column chromatographic method, found that myelin from different species varies in methylarginine content: most species contain monomethylarginine and NG, N′G dimethylarginine in widely different molar ratios, different total amounts, and no NG, NG dimethylarginine. However, the preparation from turtle brain contains (64) both symmetrical and unsymmetrical dimethylarginine. The heterogeneity of the basic encephalitogenic brain proteins noted by Martenson et al (65–67) might therefore be ascribed to the presence of both the mono and dimethyl derivatives of arginine at position 107. It is not yet entirely clear whether the discrepancies in the values for the various methylated arginines in the A1 protein from different species are due to variation in the degree and type of methylation or to differences in the methods of analysis. The microheterogeneity of myelin basic protein is in all probability a real and not a trivial phenomenon and might well be an indication that methylation of arginine 107 is a dynamic process, possibly related to the structural role and metabolism of this unusual protein.

In vitro methylation of arginine 107 in myelin basic protein from beef brain has been reported (68). The products of methylation have been identified as N-monomethylarginines and NG, N'G dimethylarginine. The enzyme has a higher specific activity toward myelin basic proteins than toward histones.

The A1 protein (mol wt 18,400) is a major component of myelin, representing up to 30% of its total protein (69), and it has been suggested that it participates in the interaction of myelin with lipid components in the membrane. The A1 basic protein of myelin is of great biological interest because of its ability to induce allergic encephalomyelitis when injected into animals (70, 71). However, methylation of arginine 107 is not directly related to encephalitogenic activity of the A1 proteins, since this function is determined by the presence or absence of tryptophan 116 (72). This has been demonstrated by the determination of the encephalitogenic activity of various tryptic peptides derived from A1 proteins. It has been found that peptide Phe-Ser-Try-Gly-Ala-Gly-Arg (or Lys), corresponding to residues 114–121 in the A1 protein from beef in human brain, is the encephalitogenic determinant, and this has been confirmed by synthesis of the encephalitogenic peptide (72, 73). The methylated arginine in A1 protein is at position 107; Arg 121 terminates the encephalitogenic sequence and is not methylated. It can be argued that the presence of methylated residue at position 107 changes the protein conformation to expose tryptophan 117, but until further work on the native conformation of the protein becomes available, the relationship of methylation to the conformation of the encephalitogenic determinant remains unclear. (See also 72, 74.)

Reporter & Corbin (75) identified dimethylarginine in the hydrolysate of myosin derived from embryonic muscle or cultured myoblast, an important finding that will be discussed below.

The enzymology of protein methylation has advanced considerably in the last few years, thanks primarily to the efforts of Kim and Paik. They have characterized three different protein methylases: protein methylase I (A-Met protein arginine methyltransferase) (76), protein methylase II (A-Met protein carboxyl methyltransferase) (77), and protein methylase III (A-Met protein lysine methyltransferase) (78). The first two enzymes are soluble and found in the cytosol, whereas lysine methylase is nuclear bound (56, 79).

Protein arginine methyltransferase has so far been only partially purified (76) and little information is available concerning its properties.

Protein methylase II has been extensively purified (2500-fold) (77) from calf thymus and from human erythrocytes (80). The protein from erythrocytes has a molecular weight of 25,000 and exhibits three different peaks by electrofocusing.

Some evidence suggests that it may be composed of subunits with a molecular weight of 8000. Protein methylase II has also been purified from bovine pituitary and its substrate specificity investigated: it can carboxymethylate all anterior pituitary hormones, whereas posterior pituitary hormones are essentially without effect as substrates (81). The high activity of the enzyme in the pituitary and its activity on anterior pituitary protein hormones suggest that the enzyme might function in the storage and/or transport of these hormones. A-Met-protein carboxyl methylpherase had been originally discovered by Liss and colleagues (82, 83) and, as is now realized, is identical with the so-called methanol-forming enzyme found in posterior pituitary extracts by Axelrod & Daly (84), because the methyl ester formed by this reaction hydrolyzes at an alkine pH to yield methanol (85). While the unstable nature of the protein methyl ester provides a stumbling block in the identification of the natural substrate for this enzyme, the nature of the reaction and the change in net protein charge that it brings about suggest that this methylation reaction could be a significant mechanism for varying the conformation of certain proteins, thereby controlling their biological activity.

The existence of more than one arginine protein methylase can be surmised from the work of Gallwitz (86) and of Sundarraj & Pfeiffer (68). In fact, the presence of a variety of methylated proteins and the specificity of the methylation pattern suggest greater complexity in the specificity and distribution of the protein methylases than has been discovered so far.

Protein methylases use A-Met as the sole methyl donor, and the K_m for A-Met is of the order of 1×10^{-6} M, more than one order of magnitude lower than the intracellular concentration of A-Met. It therefore would seem a priori that the activity of these methylases is not controlled by the methyl donor concentration, although the distribution of A-Met intracellularly is not really known (see also 87). The protein methylase in human erythrocytes, however, is probably not saturated with respect to A-Met which occurs in serum at $\sim 1 \times 10^{-6}$ M. Less is known about protein methylase III (79), but its distribution provides a problem: as noted above the enzyme is chromatin bound; while this is consistent with the presence of methyllysine in histones, it is difficult to reconcile with the presence of methylated lysine residues in non-nuclear proteins.

In spite of the notable advances in this area of polymer methylation, a number of critical questions are still unanswered or only partially resolved by experimentation. These are the identification and characterization of the various proteins that function as substrates for the different methylases, the timing of methylation in relation to the synthesis of the polypeptide chain (88), and the biological significance of the reactions. Most of the proteins where methylated amino acids have been detected are structural proteins or carriers such as histones, flagellar proteins, encephalitogenic brain protein, cytochrome c, or proteins endowed with complex biochemical or biophysical properties such as myosin, actin, or rhodopsin (89, 90). Methylation of what we might call a "simple" enzyme is not yet known, with one possible exception discussed below. Consequently, it has not been possible so far to identify with certainty the biological significance of protein methylation or to assign a specific difference in the functions of methylated and unmethylated proteins.

The possible exception referred to earlier is methylation of the protein component of *Escherichia coli* ribosomes. Methylation of proteins of *E. coli* ribosomes has been known for some time (56). More recently, proteins containing ε-N-methyllysine have been characterized. Terhorst et al (91, 92) have shown that about half the molecules of L7 and L12, two closely related 50S ribosomal proteins of *E. coli* MRE600, contain ε-N-methyllysine in specific positions in the peptide chain. Moreover, Beaud & Hayes (93, 94) have shown that if *met⁻* strains of *E. coli* are grown in the presence of ethionine, their ribosomes are functionally altered, and the 23 and 16S RNAs as well as the proteins of the 50S ribosomal proteins are submethylated and can act in vitro as methyl acceptors (94). A very recent study (95, 96) extends these findings and focuses attention on L11 protein as one of the submethylated ribosomal proteins. Methylation of the methyl-deficient ribosomes in vivo (achieved by incubation of ethionine-treated *E. coli* in a methionine-containing medium) leads to rapid recovery of normal physiological properties. Similar results are seen upon incubation of crude extract of submethylated *E. coli* with A-Met or methionine. Upon methylation in vitro with A-Met as the methyl donor, a single ε-N-trimethyllysine is found in L11, a protein identified by Nierhaus & Montejo (97) as a polypeptide transferase enzyme. Since L11 protein is also responsible for the ribosomal binding of chloroamphemical, the suggestion of Alix & Hayes (95) of a correlation between methylation and ribosomal function becomes particularly intriguing and might provide the first demonstration that protein methylation, like protein phosphorylation, is used as a biological control mechanism.

Cytochrome *c* methylation might provide another interesting area for further studies on the biological significance of methylation. Cytochrome *c* is one of the proteins most extensively investigated: the amino acid sequences of more than 30 different cytochrome *c*s from many different species have been determined. Cytochrome *c* from several ascomycetes (*Neurospora crassa, Saccharomyces cerevisiae, Candida krusei*) (98, 99) contains a single residue of ε-N-trimethyllysine located at residue 78. Cytochrome *c* from wheat germ and other higher plants (100) contains two residues of ε-N-trimethyllysine at positions 78 and 86, respectively. Methylated amino acids, on the other hand, are absent from animal cytochromes (98). From our knowledge about the evolutionary relationships of cytochrome *c* from fungi, plants, and vertebrates, it can be surmised that the methylated cytochromes branched from the phylogenetic tree some 1 to 2 billion years ago (101). Over this very long period of evolution, the amino acid sequence comprising residues 76 to 87

Asn-Pro-Lys-Lys-Tyr-Ile-Pro-Gly-Thr-Lys-Met

which contains the two methylatable lysines (78 and 86), has remained entirely free of mutations in all of the more than 30 cytochrome *c*s sequenced so far from different species, whereas the rest of the polypeptide chain on both sides of this sequence is variable and the mutation frequency increases with the distance from this key sequence. The three-dimensional structure of equine cytochrome *c* (98) shows that there is "a hole or channel" leading from the back of the molecule to the heme, and that the side chains of lysine residues 78 and 86 are on the exterior of the molecule and located near the hole or channel, thus providing relatively easy exposure to the methylase.

Insofar as the relationship between methylation and biological function is concerned, there are unfortunately no data that allow comparison of the physicochemical, biochemical, and functional characteristics of the three types of cytochrome c, namely, animal cytochromes that are not methylated, those from fungi that contain one trimethyllysine residue, and those from higher plants that contain two trimethyllysine residues. Such studies might be very revealing and help assign a physiological function to the methylation of cytochrome c.

As noted, protein-lysine methylase activity in mammalian tissues is chromatin bound and therefore might not be able to act on cytochrome c. It would be interesting to explore the subcellular distribution of lysine protein methylase in plants and see if plant extracts could methylate mammalian cytochrome c. This approach might begin to throw light on the nature of the recognition site or sites for protein methylation. In other amino acid side chain modifications, for example in the hydroxylation of proline and lysine in collagen or in the biosynthesis of certain glycoproteins, the recognition site is a short specific segment around the amino acid to be modified. It is not clear, however, whether the amino acids that form the recognition site must be adjacent to the amino acid residue that becomes modified or, in our specific case, methylated (histidine in myosin, lysine in cytochrome c), or whether they are near it in the tertiary structure but, in fact, far apart in the unfolded peptide chain.

Another biological system that might afford interesting opportunities to correlate protein methylation with function is the developing muscle cell. As noted above, Reporter & Corbin (75) and Reporter (88) found unsymmetrical dimethylarginine in myosin from developing leg muscle (four residues of dimethylarginine/5×10^5 g protein). This amino acid is absent in the adult chicken or rat myosins, as well as in cardiac myosin and actin. Thus, in agreement with earlier data of Kuehl & Adelstein (102) and of Hardy et al (103), it was shown that myosin from differentiating striated muscle is different from adult myosin. Myosin methylation is influenced by hormonal action (104), which supports the idea that the modification of protein via methylation is somehow related to regulation of the specific function catalyzed by the methylated protein.

Further studies on the differences in the methylation patterns in fetal and adult muscle have been carried out by Huszar (105) using rabbit muscle. Myosin from white skeletal muscle of adult rabbits contains one residue of methylhistidine and two residues of ε-trimethyllysine. All these residues are in subfragment 1, the globular head of myosin that carries the active sites for ATP hydrolysis and actin combination (105–111). In myosin from newborn rabbits 3-methylhistidine is absent (102, 105, 112). Myosin methylation can be demonstrated in cell-free systems and A-Met is the methyl donor (103, 113); the process is clearly highly specific because only one histidine out of the 35 found in the heavy chain of myosin is methylated. The amino acid sequence of the peptide from adult rabbit myosin that contains a methylatable histidine residue is

Leu-Leu-*Gly*-Ser-Ile-Asp-*Val*-His-Gln-Thr-Tyr-Lys

In fetal muscle this sequence is changed in two residues, with alanine replacing glycine and isoleucine replacing valine; this proves that two different genes direct the synthesis of myosin in rabbits. There is a correlation between the 3-methyl-histidine content of myosin and muscle contraction speed. The full implications of this correlation are, however, not yet clear.

As to timing, protein methylation occurs after synthesis of cytochrome c (114). Reporter (88) also studied this question and proposed that selective methylation of nascent myosin can begin at the polyribosomal level and be completed in the cytosol. Methylation of the nascent ribosomal bound protein chains would be directed toward the more hydrophobic region of the molecule, whereas the peripheral residue on the outside of the protein could interact more freely with methylases in the cytosol. As noted above, however, our present knowledge of the subcellular distribution of protein methylases is inadequate.

The discovery of a protein demethylase has been reported by Paik & Kim (115). This finding suggests at first glance that protein methylation, like protein phosphorylation, might emerge as a mechanism for the regulation and control of enzyme activity. Unfortunately, the enzyme discovered by Paik & Kim is an oxidative lysine demethylase, and it appears very unlikely on structural grounds that this system can be involved in regulation; rather it is more likely to be a catabolic system, although firm evidence is not available on this question. This, then, is an area that needs urgent clarification and that can be framed by a question: Do methyl groups in proteins turn over repeatedly, independent of the amino acid residue to which they are attached? Experimental approaches to this question can be formulated but they are frought with difficulties. [See Paik & Kim (79) for full discussion.]

As a result of recent advances, carnitine biosynthesis can now be considered a special subchapter of lysine-protein methylation. It has been known for more than a decade that the methyl groups of carnitine are derived from methionine, but the nature of the methyl acceptor remained elusive until recently. The key observation that unraveled the biosynthetic pathway was the finding of Horne & Broquist (116) that the biosynthesis proceeds through successive methylation of either free or, more likely, protein-bound lysine to give ε-N-trimethyllysine which would then be metabolized by a series of reactions and cleaved with the loss of carbons 1 and 2 to yield γ-butyrobetaine. The latter compound is then hydroxylated to form carnitine by an enzyme, γ-butyrobetaine hydroxylase, found in liver cytosol (117). In agreement with this formulation, it has been established conclusively by Broquist and his colleagues (116, 118) and by Cox & Hoppel (119, 120) that ε-N-trimethyllysine fed to *Neurospora* or rats is efficiently converted to carnitine. The evidence on the origin of trimethyllysine is not yet conclusive. As discussed above, Paik and his colleagues (89) have identified an enzyme fraction, protein methylase III, that will generate peptide-bound ε-trimethyllysine, but the subcellular tissue distribution of this enzyme is probably inadequate to account for carnitine biosynthesis. An enzyme capable of methylating free lysine has not been described. Moreover, free lysine as such is unlikely to be a direct precursor of carnitine, since carnitine synthesis in liver was observed after trimethyllysine administration, but was not seen when lysine was the precursor. These data confirm the pathway from trimethyllysine but do not establish the metabolic state of lysine when methylation takes place (121). The lysine residue destined to become carnitine differs from other protein lysine residues in that it appears to be in equilibrium with an environment such that the ε-nitrogen atom can exchange with nitrogen sources in the medium. This conclusion is based on an experiment in which the N^{15} contents of carnitine and protein-bound lysine were compared after administration of ε-N^{15}-lysine to a lysine-dependent *Neurospora* mutant (*St*15069) grown on a carnitine-free diet. The ε-N^{15} atom excess of protein-bound lysine was identical with that of the lysine in the growth medium, whereas in carnitine the ε-N^{15} was diluted by about 50%. It is important to point out, however, that this comparison was made between N^{15} atom excess in carnitine and in protein lysine and not between that in carnitine and in ε-N-trimethyllysine from proteins (possibly there is none in *Neurospora*). This observation must be correlated mechanistically with the fact that carbon atom 6 of lysine, when labeled with C^{14}, is incorporated into carnitine without dilution (118). The mechanism underlying the equilibration of the ε-nitrogen atom from protein-bound lysine with a nitrogen source in the medium might be analogous to that involved in the synthesis of desmosine (122), where protein-bound lysine is oxidatively deaminated, yielding protein-bound aminoadipic semialdehyde. This mechanism, it should be emphasized, is offered only speculatively. As to the loss of carbons 1 and 2 of N^{ε}-trimethyllysine, a good analogy is provided by the catabolism of lysine in *Clostridium sticklandii,* where Morley & Stadtman (123) have shown that the C^1 and C^2 of lysine yield acetate and C^3-C^4C^5-C^6 of lysine yields butyrate.

CARBOHYDRATE AND POLYSACCHARIDE METHYLATION[2]

It has been found in the last few years that A-Met is the methyl donor to complex branched chain deoxy sugars, such as vinelose (124) and micarose (125). In both cases, sugars bound to pyrimidine nucleotides are the substrate for the methylation reaction.

In confirmation of the earlier structural data of Okuda et al (126), the biosynthetic pathway for vinelose synthesis in *Azotobacter vinelandii* was found to consist of the reduction of carbon 6 of CDP-D-glucose to form CDP 4-keto-6-deoxy-D-glucose, an obligatory intermediate, which then tautomerizes to the Δ-3,4-enediol derivative; next, carbon atom 3 is methylated by A-Met to give CDP-3-C-methyl-4-keto-6-deoxyhexose. The subsequent steps in the synthesis of CDP-vinelose are less well understood: CDP-vinelose contains a second O-methyl group at carbon atom 2, but so far an A-Met requirement for this methylation has not been demonstrated. The reaction has, however, been studied with only crude extracts, and further work in progress will hopefully elucidate the biosynthetic reactions involved.

An analogous reaction underlies the formation of thymidine diphosphate micarose (2-6-di-deoxy-3-C-methyl-L-ribohexose) (125), a component of the macrolide antibiotic tylosin produced by *Streptomyces rimosus* and of other macrolide antibiotics such as erythromycin. In tylosin, micarose is the terminal sugar of a

---

[2] The methylation of ribose in tRNA by A-Met is not reviewed here.

disaccharide whereas in erythromycin it is linked directly to the 12-membered lactone ring (127). Pape & Brillinger (125) have investigated the reaction in some detail and found that it is specific with regard to thymidine diphosphate-D-glucose as the methyl acceptor. However the requirement for NADP$^+$ in the overall system suggested a 6-deoxy compound as a probable obligatory intermediate, and in agreement with this expectation it was found that thymidine diphosphate-4-keto-6-deoxy-D-glucose (a byproduct in the biosynthesis of thymidine diphosphate-L-rhamnose) (128) could substitute for thymidine diphosphate-glucose as the methyl acceptor. It is presumed that the reaction proceeds from this intermediate through the Δ-3,4-enediol derivative as in the CDP vinelose synthesis.

The addition of the methyl group of A-Met across the 3–4 double bond of Δ-3,4-enediol derivatives is new but formally identical with other double bonded carbon methylations recently reviewed by Mudd (27) and Lederer (129).

Carbohydrate methylation at the macromolecular level was discovered by Hassid and his collaborators. Kauss & Hassid (130) found that the biosynthesis of the 4-O-methyl-D-glucuronic acid of hemicellulose B (an ill-defined group of heteropoly-saccharides of plant cell walls) does not proceed from a nucleotide-bound 4-O-methyl-D-glucuronic acid. They showed, moreover, that particulate preparations of corncob are able to transfer the methyl group of A-Met into hemicellulose B. It was assumed that the macromolecular acceptor was a glucuronoxylan rich in D-glucuronic acid and present in the particulate enzyme preparation. Likewise, the O-methyl ester groups of pectin are introduced at the macromolecular level (131). The enzyme responsible for the transfer of methyl groups into pectins has different properties from the enzyme that introduces the 4-O-methylether group into glucuronoxylans. The latter has a pH optimum of 8 and requires a divalent cation such as Co^{2+} or Mn^{2+}; the former is not stimulated by cations and has a pH optimum of 6.7.

Methylation of the carboxyl groups of pectins plays an important role in their biological function because the ion exchange and water binding capacity of pectins and their possible involvement in crosslinkages through cations, esters, or hydrogen bonds depend to a large extent on the degree of methylation (132, 133). Regardless of whether or not the carboxyl groups of pectins play a direct role in plant growth, their partial or full methylation appears to be part of the complex process of cell wall formation. Growing tissues always contain a pectin methyl esterase (132), a hydrolytic enzyme that reverses the action of the enzyme that transfers the methyl of A-Met into pectins.

Ballou and his collaborators (134–141) have discovered a new type of poly-saccharide methylation and remarkably advanced our understanding of its biological features. A polysaccharide methyltransferase that catalyzes the transfer of the methyl group from A-Met to position 6 of α(1 → 4)-D-glucooligosaccharides has been demonstrated in extracts of *Mycobacterium phlei* (135, 136). A similar enzyme is presumably responsible for the transfer of the methyl group of A-Met to the methyl hexoses found in *M. tuberculosis* St. 37 Ra (139).

The discovery of polysaccharide methylation stemmed from Lee & Ballou's earlier identification of 6-O-methyl-D-glucose as a hydrolysis product of a *M. phlei* lipid extract (137). O-Methylglucose, as was shown later, originates from a unique liposaccharide with 18 hexose units, 10 of which are 6-O-methylglucose units, while others are acylated with D-glyceric acid and with acetyl, propionyl, isobutyryl, octanoyl, and succinyl groups (138). The 6-O-methylglucose units are in a row or adjacent to each other. In order to study the properties of the polysaccharide methylase, Ballou and his collaborators examined a number of plausible exogenous carbohydrates as potential substrates. The recognition that detailed analysis in vitro of polysaccharide methylation would require use of an exogenous substrate was based on Ballou's acute perception that the formation of the 6-O-methylglucose units would involve a reaction at the polymeric level analogous to that seen in the field of protein or nucleic acid biochemistry. As is well known, nucleic acid methylation in vitro can be studied using as substrates either undermethylated homologous nucleic acids or heterologous nucleic acids having a different pattern of methylation. Thus, *E. coli* tRNA methylases can be studied by use of (*a*) methyl-poor tRNA derived from methionine-starved *E. coli rc* mutants, or (*b*) tRNA from different bacterial or eukaryotic cells. In their search for suitable substrates, Ballou and colleagues found that amylo-oligosaccharides from various sources containing seven to ten glucose units in α-(1 → 4) linkage acted as methyl acceptors, and on hydrolysis after in vitro methylation, yielded 6-O-methylglucose as the only methylated compound. Corresponding β-(1 → 4)-linked oligosaccharides were inactive and, in fact, acted as inhibitors.

Several additional features make in vitro methylation of exogenous amylo-oligosaccharides a particularly interesting system: in the first place methylation occurred only with an acceptor that had been partially acylated, reflecting a requirement for nonpolar sites for the binding of the methyltransferase (136). As noted above, the natural lipopolysaccharide from *M. phlei* contains several acyl hexoses in addition to methylated hexoses, and investigation of the enzymatic acyl transfer from acyl-CoA to the natural lipopolysaccharide, as well as to artificial acceptors, revealed that the requirement for partial acylation of the polysaccharide unit is common to methyl transfer and acyl transfer systems (140). This observation may be explained by assuming that during its biosynthesis the lipopolysaccharide is attached to a lipid carrier where presumably methylation and acylation take place. The incorporation of acyl, propionyl, or isobutyryl groups is probably catalyzed by a single enzyme, but succinyl groups are transferred by a different enzyme (140). This concept is supported by in vivo studies of Narumi et al (142) that show competition between acetate, propionate, and isobutyrate and the non-competitiveness of succinate.

In addition to the 10 6-O-methylglucose residues, the natural lipopolysaccharide from *M. phlei* contains one 3-O-methylglucose residue that occupies the terminal position at the nonreducing end of the polysaccharide. The methylation reaction that generates this 3-O-methylglucose unit is also dependent on the degree of acetylation of the substrate.

Finally and most interesting, the 10 O-methylglucose groups occupy a specific location in the liposaccharide unit starting precisely at the fifth glucose residue (from the nonreducing end). The specificity of the 6-O-methylglucose forming enzyme is such that if an octasaccharide is used as substrate, methylation starts at position 5 from the nonreducing end and *only* the four remaining hexose groups are methylated. This specificity is consistent with the structure of the lipopolysaccharide and suggests a mechanism by which the sequence of the methylated section of the polysaccharide could be generated, assuming that the polysaccharide is made first by the stepwise polymerization of glucose units. Once the glucose chain is formed, Ballou postulates that the methyltransferase could recognize, as a start or stop signal for methylation, the site five units in from the nonreducing end of the polysaccharide. The corresponding signal at the other end of the polymer could be the side chain attached to the fourth glucose unit from the glyceric acid end (134). One cannot say whether the methylation occurring within these two limits is directional or random; this would be interesting to study. A precedent for directional methylation at the polymer level is provided by the beautiful studies of Drahovsky & Morris on methylation of the calf thymus DNA (143, 144). In determining these signals the structure of the four hexose units near the nonreducing end of the polysaccharide must be of particular importance: and it is noteworthy that three of the four hexose units bear an acyl group (141).

The requirement for substrate acylation that is common to the methyl transfer and acyl transfer reactions is interesting and unusual. It suggests that the biosynthesis of the fully modified mycobacterial polysaccharide should follow an autocatalytic course with the first acyl transfer favoring subsequent acylation and/or methyl transfer reactions.

As a result of all these modifications, it has been proposed (141) that the methyl-glucose-containing lipopolyssacharide of M. *phlei* can assume a helical conformation and can have a hydrophilic side with a heavy concentration of hydroxyl groups and a lipophilic side where the methyl and acyl groups would be concentrated. This molecular arrangement would permit the methylated lipopolysaccharide to function in the cell "at the interface between membranes and cytosols or between hydrophobic and hydrophilic surfaces of proteins." Here then is a beautiful and most satisfying interpretation of an elegant and complex chemical structure in functional biological terms, surely an ultimate aim of biochemists.

METHODS

A variety of methods relevant to biological methylation studies have been published in the last few years. A new assay for adenosylmethionine transferase is based on determining the amount of A-Met formed. The latter is measured by means of a double isotope derivative technique by coupling adenosylmethionine transferase to serotonin methylpherase (145). The main advantage of this method is that it can estimate the loss of A-Met during the assay, which might be of value in certain cases (146).

Chou & Lombardini (147) proposed that cellulose phosphate ion exchange discs might usefully replace ion exchange techniques used earlier (148, 149) for the assay of adenosylmethionine transferase. Whether this technique represents an advance or merely offers a choice between equally valid methods is not clear.

The determination of a number of compounds after enzymatic methylation by specific enzymes can be used (*a*) to determine the tissue concentration of the methyl acceptor compound (in the presence of excess A-Met and methyltransferase) or (*b*) to assay the methyltransferase activity. Many such methods have been reported; most depend on the separation of the labeled or doubly labeled methylated products using ion exchange chromatography (150, for S-methylmethionine) or extraction into organic solvents (151, for metanephrine; 152, for amphetamine; 153, for puromycin; 154–156, for methylhistamine; 157, for phospholipids; 158, for cyclopropane fatty acids; 159, for methylated purines; 160, for methyltryptamine; and 161, 162, for dopamine). Spectrophotometric measurements permit a rapid assay of catechol O-methylpherase (163). All these methods are discontinuous and, depending on the purpose of the assay (determination of enzyme activity or the amount of substrate), are performed as a function of time and enzyme concentration or at equilibrium. These methods and their refinements provide the combined advantage of specificity and high sensitivity (picamole range).

A continuous method capable in principle of monitoring *any* methyl transfer reaction has been proposed. This method (164–166) is based on the fact that enzymatic deamination of ASR, a product of methyl transfer from A-Met, results in a spectrophotometric change at 265 nm; it has the advantage of overcoming errors due to accumulation of ASR, an inhibitor of most methyl transfer reactions. The method is limited by both a relatively low sensitivity (10–20 nmol range) and the requirement that the substrate or the enzymes involved have low absorbance at 265 nm.

ACKNOWLEDGMENTS

In spite of my diligent efforts, I am sure I have failed to quote a number of valuable contributions even in the restricted areas covered in this review. I apologize sincerely to these authors and will be grateful if they will drop me a note for my personal benefit and so that in the future, if the opportunity should arise, I may not fail to mention them.

I am grateful to my colleague, S. H. Mudd, not only for many interesting discussions and for a valuable exchange of ideas, but also for allowing me to quote from his paper with J. M. Freeman, currently in press; my thanks are also due to him and P. Chiang for reviewing this manuscript.

Literature Cited

1. Rudiger, H., Jaenicke, L. 1969. *Eur. J. Biochem.* 10:557–60
2. Taylor, R. T., Hanna, M. L. 1970. *Arch. Biochem. Biophys. Acta* 137:453–59
3. Taylor, R. T., Weissbach, H. 1973. *Enzymes,* 9:121–65
4. Blakley, R. L. 1969. *The Biochemistry of Folic Acid and Related Pteridines.* Amsterdam: North-Holland
5. Wijesundera, S., Woods, D. D. 1960. *J. Gen. Microbiol.* 22:229–41
6. Buehring, K. U., Batra, K. K., Stokstad, E. L. R. 1972. *Biochim. Biophys. Acta* 279:498–512
7. Kerr, D. S., Flavin, M. 1970. *J. Biol. Chem.* 245:1842–55
8. Cherest, H., Surdin-Kerjan, Y., Antoniewski, J., De Robichon-Szulmajster, H. 1973. *J. Bacteriol.* 114:928–33
9. Masselot, M., De Robichon-Szulmajster, H. 1973. *Mol. Gen. Genet.* 129:349–62

10. Ahmed, A. 1973. *Mol. Genet.* 123:299–324
11. Hobson, A. C., Smith, D. A. 1973. *Mol. Gen. Genet.* 126:7–18
12. Kutzbach, C., Stokstad, E. L. R. 1967. *Biochim. Biophys. Acta* 139:217–20
13. Greene, R. C., Su, C.-H., Holloway, C. T. 1970. *Biochem. Biophys. Res. Commun.* 38:1120–26
14. Boyer, H. W. 1971. *Ann. Rev. Microbiol.* 25:153–76
15. Marx, J. L. 1973. *Science* 180:482–85
16. Meselson, M., Yuan, R., Heywood, J. 1972. *Ann. Rev. Biochem.* 41:447–66
17. Craddock, V. M., 1970. *Nature* 228:1264–68
18. Borek, E., Srinivasan, P. R. 1966. *Ann. Rev. Biochem.* 35:275–98
19. Kerr, S. J., Borek, E., 1973. *Enzymes* 9:167–95
20. Kuchino, Y. 1972. *Protein Nucl. Acid Enzymes* 17:252–64
21. Weissbach, H., Taylor, R. T. 1970. *Vitam. Horm.* 28:415–40
22. Flavin, M. 1975. *Metabolism of Sulfur Compounds.* New York: Academic. In press
23. Savitskii, I. V., Zelinskii, V. G. 1972. *Ukr. Biokhim. Zh.* 44:662–76
24. Lombardini, J. B., Talalay, P. 1971. *Advan. Enzyme Regul.* 9:349–84
25. Mudd, S. H. 1973. *Enzymes* 8:121–54
26. Farber, E. 1971. *Ann. Rev. Pharmacol.* 11:71–96
27. Mudd, S. H. 1973. *Metabolic Conjugation and Metabolic Hydrolysis,* ed. W. Fishman, 3:297–350
28. Jencks, W. P. 1969. *Catalysis in Chemistry and Enzymology.* New York: McGraw-Hill
29. Tabor, H., Tabor, C. W. 1972. *Advan. Enzymol.* 36:203–68
30. Williams-Ashman, H. G., Janne, J., Coppoc, G. L., Geroch, M. E., Schenone, A. 1972. *Advan. Enzyme Regul.* 10:225–45
31. Cohen, S. 1971. *Introduction to the Polyamines.* Englewood Cliffs, NJ: Prentice-Hall. 179 pp.
32. Salvatore, F. et al 1975. *The Biochemistry of Adenosylmethionine,* New York: Columbia Univ. Press. In press
33. Cantoni, G. L. 1952. *Phosphorus Metab.* 2:129–50
34. Cantoni, G. L. 1960. *Handbook of Comparative Biochemistry,* ed. M. Florkin, H. Mason, 1:181–241. New York: Academic.
35. Laduron, P. 1972. *Nature New Biol.* 238:212–13
36. Laduron, P. 1973. *Advan. Neuropsychopharmacol.* 3:233–38
37. Leysen, J., Laduron, P. 1973. *Arch. Int. Physiol. Biochim.* 81:978
38. Banerjee, S. P., Snyder, S. H. 1973. *Science* 182:74–75
39. Mudd, S. H. 1974. *Heritable Disorders of Amino Acid Metabolism,* ed. W. L. Nyhan, 429–51. New York: Wiley
40. Schrauzer, G. N., Deutsch, E. 1969. *J. Am. Chem. Soc.* 91:3341–50
41. Mudd, S. H., Freeman, J. M. 1975. *J. Psychiat. Res.* In press
42. Mudd, S. H., Levy, H. L., Morrow, G. III. 1970. *Biochem. Med.* 4:193–214
43. Kety, S. S., 1972. *Semin. Psychiat.* 4:233–38
44. Matthysse, S. 1973. *Fed. Proc.* 32:200–5
45. Korevaar, W. C., Geyer, M. A., Knapp, S. S., Hsu, L. L., Mandell, A. J. 1973. *Nature New Biol.* 245:244–45
46. Hsu, L. L., Mandell, A. J. 1973. *Life Sci.* 13:847–58
47. Weil-Malherbe, H., Szara, S. I. 1971. *The Biochemistry of Functional and Experimental Psychoses,* Springfield, Ill.: Thomas. 406 pp.
48. Gershon, E. S., Shader, R. I. 1969. *Arch. Gen. Psychiat.* 21:82–88
49. Pollin, W., Cardon, P. V. Jr., Kety, S. S. 1961. *Science* 133:104–5
50. Ackermann, D., Timpe, G., Poller, K. 1929. *Z. Physiol. Chem.* 183:1–10
51. Westall, R. G. 1952. *Biochem. J.* 52:638–42
52. Tallan, H. H., Stein, W. H., Moore, S. 1954. *J. Biol. Chem.* 206:825–34
53. Kakimoto, Y., Akazawa, S. 1970. *J. Biol. Chem.* 245:5751
54. Nakajima, T., Matsuoka, Y., Kakimoto, Y. 1971. *Biochim. Biophys. Acta* 230:212–22
55. Ambler, R. P., Rees, M. W. 1959. *Nature* 184:56–57
56. Comb, D. G., Sarkar, N., Pinzino, C. J. 1966. *J. Biol. Chem.* 241:1857–62
57. Paik, W. K., Kim, S. 1970. *J. Biol. Chem.* 245:88–92
58. Kaye, A. M., Sheratzky, D. 1969. *Biochim. Biophys. Acta* 190:527–38
59. Eylar, E. H. 1970. *Proc. Nat. Acad. Sci. USA* 67:1425–31
60. Brostoff, S., Eylar, E. H. 1971. *Proc. Nat. Acad. Sci. USA* 68:765–69
61. Carnegie, P. R. 1971. *Biochem. J.* 123:57–65
62. Carnegie, P. R. 1971. *Nature* 229:25–28
63. Baldwin, G. S., Carnegie, P. R. 1971. *Science* 171:579–81
64. Deibler, G. E., Martenson, R. E. 1973. *J. Biol. Chem.* 248:2392–96
65. Martenson, R. E., Deibler, G. E., Kies,

M. W. 1969. *J. Biol. Chem.* 244:4268–72
66. Martenson, R. E., Deibler, G. E., Kies, M. W. 1971. *J. Neurochem.* 18:2427–33
67. Martenson, R. E., Deibler, G. E., Kies, M. W. 1971. *Immunological Disorder of the Nervous System,* ed. L. P. Rowland, 49:76–93. Baltimore: Williams & Wilkins
68. Sundarraj, N., Pfeiffer, S. E. 1973. *Biochem. Biophys. Res. Commun.* 52: 1039–45
69. Eylar, E. H., Salk, J., Beveridge, G., Brown, L. 1969. *Arch. Biochem. Biophys.* 132:34–48
70. Kies, M. W. 1965. *Ann. NY Acad. Sci.* 122:161
71. Paterson, P. 1966. *Advan. Immunol.* 5: 131
72. Eylar, E. H., Brostoff, S., Jackson, J., Carter, H. 1972. *Proc. Nat. Acad. Sci. USA* 69:617–19
73. Eylar, E. H. See Ref. 67, 50–75
74. Carnegie, P. R., Caspary, E. A., Smythies, J. R., Field, E. J. 1972. *Nature* 240:561–63
75. Reporter, M., Corbin, J. L. 1971. *Biochem. Biophys. Res. Commun.* 43: 644–50
76. Paik, W. K., Kim, S. 1968. *J. Biol. Chem.* 243:2108–14
77. Kim, S. 1973. *Arch. Biochem. Biophys.* 157:476–84
78. Paik, W. K., Kim, S. 1970. *J. Biol. Chem.* 245:6010–15
79. Paik, W. K., Kim, S. 1975. *Advan. Enzymol.* In press
80. Kim, S. 1973. *Arch. Biochem. Biophys.* 161:652–57
81. Diliberto, E. J., Axelrod, J. 1974. *Proc. Nat. Acad. Sci USA* 71:1701–4
82. Liss, M., Edelstein, L. M. 1967. *Biochem. Biophys. Res. Commun.* 26:497–504
83. Liss, M., Maxam, A. M. 1967. *Biochim. Biophys. Acta* 140:555–57
84. Axelrod, J., Daly, J. 1965. *Science* 150: 892–93
85. Morin, A. M., Liss, M. 1973. *Biochem. Biophys. Res. Commun.* 52:373–78
86. Gallwitz, D. 1971. *Arch. Biochem. Biophys.* 145:650–57
87. Judes, C., Jacob, M. 1972. *FEBS Lett.* 27:289–92
88. Reporter, M. 1973. *Arch. Biochem. Biophys.* 158:577–85
89. Paik, W. K., Kim, S. 1971. *Science* 174:114–19
90. Reporter, M., Reed, D. W. 1972. *Nature New Biol.* 239:201–3
91. Terhorst, C., Wittmann-Liebold, B., Moeller, W. 1972. *Eur. J. Biochem.* 25: 13–19

92. Terhorst, C., Moeller, W., Laursen, R., Wittmann-Liebold, B. 1972. *FEBS Lett.* 28:325–28
93. Beaud, G., Hayes, D. H. 1971. *Eur. J. Biochem.* 19:323–39
94. Beaud, G., Hayes, D. H. 1971. *Eur. J. Biochem.* 20:525–34
95. Alix, J.-H., Hayes, D. 1974. *J. Mol. Biol.* 86:139–59
96. Alix, J.-H., Hayes, D. H. 1974. *C. R. Acad. Sci. Paris* 278:951–53
97. Nierhaus, K. H., Montejo, V. 1973. *Proc. Nat. Acad. Sci. USA* 70:1931–35
98. DeLange, R. J., Glazer, A. N., Smith, E. L. 1970. *J. Biol. Chem.* 245:3325–27
99. DeLange, R. J., Glazer, A. N., Smith, E. L. 1969. *J. Biol. Chem.* 244:1385–88
100. Boulter, D., Laycock, M. V., Ranshaw, J. A. M., Thompson, E. W. 1970. *Phytochemical Phylogeny,* ed. J. B. Harborne, 179–86. New York: Academic. 335 pp.
101. Dayhoff, M. O., Park, C. M., McLaughlin, P. J. 1972. *Atlas Protein Sequence Struct.* 5:7–16
102. Kuehl, W. M., Adelstein, R. S. 1970. *Biochem. Biophys. Res. Commun.* 39: 956–64
103. Hardy, M. F., Harris, C. I., Perry, S. V., Stone, D. 1970. *Biochem. J.* 120:653–60
104. Reporter, M. 1971. *J. Gen. Physiol.* 57:244
105. Huszar, G. 1972. *Nature New Biol.* 240:260–64
106. Trayer, I. P., Harris, C. I., Perry, S. V. 1968. *Nature* 217:452–53
107. Hardy, M. F., Perry, S. V. 1969. *Nature* 223:300–2
108. Huszar, G., Elzinga, M. 1969. *Nature* 223:834–35
109. Kuehl, W. M., Adelstein, R. S., 1969. *Biochem. Biophys. Res. Commun.* 37: 59–65
110. Huszar, G., Elzinga, M. 1971. *Biochemistry* 10:229–36
111. Huszar, G. 1972. *J. Biol. Chem.* 247: 4057–62
112. Huszar, G., Elzinga, M. 1972. *J. Biol. Chem.* 247:745–53
113. Krzysik, B., Vergnes, J. P., McManus, I. R., 1971. *Arch. Biochem. Biophys.* 146: 34–55
114. Scott, W. A., Mitchell, H. K. 1969. *Biochemistry* 8:4282–89
115. Paik, W. K., Kim, S. 1973. *Biochem. Biophys. Res. Commun.* 51:781–88
116. Horne, D. W., Broquist, H. P. 1973. *J. Biol. Chem.* 248:2170–75
117. Lindstedt, G., Lindstedt, S. 1965. *J. Biol. Chem.* 240:316–21
118. Tanphaichitr, V., Broquist, H. P. 1973.

J. Biol. Chem. 248:2176–81

119. Cox, R. A., Hoppel, C. L. 1973. *Biochem. J.* 136:1075–82
120. Cox, R. A., Hoppel, C. L. 1973. *Biochem. J.* 136:1083–90
121. Haigler, H. T., Broquist, H. P. 1974. *Biochem. Biophys. Res. Commun.* 56: 676–81
122. Anwar, R. A., Oda, G. 1967. *Biochim. Biophys. Acta* 133:151–56
123. Morley, C. G. D., Stadtman, T. 1970. *Biochemistry* 9:4890–4900
124. Eguchi, Y. et al 1973. *J. Biol. Chem.* 248:3341–52
125. Pape, H., Brillinger, G. U. 1973. *Arch. Mikrobiol.* 88:25–35
126. Okuda, S., Suzuki, N., Suzuki, S. 1967. *J. Biol. Chem.* 242:958–66
127. Martin, S. R. 1970. *Progr. Antimicrob. Anticancer Chemother. Proc. Int. Congr. Chemother. 1969.* 2:1112–16
128. Okazaki, R., Okazaki, T., Strominger, J. L., Michelson, A. M. 1962. *J. Biol. Chem.* 237:3014–26
129. Lederer, E. 1969. *Quant. Rev. London* 23:453
130. Kauss, H., Hassid, W. Z. 1967. *J. Biol. Chem.* 242:1680–84
131. Kauss, H., Hassid, W. Z. 1967. *J. Biol. Chem.* 242:3449–53
132. Deul, H., Stutz, E. 1958. *Advan. Enzymol.* 20:341–82
133. Henglein, F. A. 1958. *Encyclopedia of Plant Physiology,* ed. W. Ruhland, 6: 405–65. New York: Springer. 1444 pp.
134. Saier, M. H. Jr., Ballou, C. E. 1968. *J. Biol. Chem.* 243:4332–41
135. Ferguson, J. A., Ballou, C. E. 1970. *J. Biol. Chem.* 245:4213–23
136. Grellert, E., Ballou, C. E. 1972. *J. Biol. Chem.* 247:3236–41
137. Lee, Y. C., Ballou, C. E. 1964. *J. Biol. Chem.* 239:PC3602–3
138. Ballou, C. E. 1968. *Accounts Chem. Res.* 1:366–73
139. Keller, J. M., Ballou, C. E. 1968. *J. Biol. Chem.* 243:2905–10
140. Tung, K.-K., Ballou, C. E. 1973. *J. Biol. Chem.* 248:7126–33
141. Smith, W. L., Ballou, C. E. 1973. *J. Biol. Chem.* 248:7118–25
142. Narumi, K., Keller, J. M., Ballou, C. E.

1973. *Biochem. J.* 132:329–40
143. Drahovsky, D., Morris, N. R. 1971. *J. Mol. Biol.* 57:475–89
144. Drahovsky, D., Morris, N. R. 1971. *J. Mol. Biol.* 61:343–56
145. Matthyse, S., Baldessarini, R. J., Vogt, M. 1972. *Anal. Biochem.* 48:410–21
146. Hobson, A. C. 1974. *Mol. Gen. Genet.* 13:263–73
147. Chou, R. C., Lombardini, J. B. 1972. *Biochim. Biophys. Acta* 276:399–406
148. Cantoni, G. L., Durell, J. 1957. *J. Biol. Chem.* 225:1033–48
149. Mudd, S. H., Finkelstein, J. D., Irreverre, F., Laster, L. 1965. *J. Biol. Chem.* 240:4382
150. Allamong, B. D., Abrahamson, L. 1973. *Anal. Biochem.* 53:343–349
151. Passon, P. F., Peuler, J. D. 1973. *Anal. Biochem.* 51:618–31
152. Kreuz, D. S., Axelrod, J. 1974. *Science* 183:420–21
153. Sankaran, L., Pogell, B. M. 1973. *Anal. Biochem.* 54:146–52
154. Axelrod, J. 1971. *Methods Enzymol.* 17: 761
155. Taylor, K. M., Snyder, S. H. 1972. *J. Neurochem.* 19:1343–58
156. Kobayashi, Y., Maudsley, D. V. 1972. *Anal. Biochem.* 46:85–90
157. Akamatsu, Y., Law, J. H. 1970. *J. Biol. Chem.* 245:701–8
158. Akamatsu, Y., Law, J. H. 1970. *J. Biol. Chem.* 245:707–13
159. Gold, M., Hurwitz, J. 1968. *Methods Enzymol.* 12:491
160. Saavedra, J. M., Axelrod, J. 1972. *J. Pharmacol. Exp. Ther.* 182:363–69
161. Cuello, A. C., Hiley, R., Iversen, L. L. 1973. *J. Neurochem.* 21:1337–40
162. Christensen, N. J. 1973. *Scand. J. Clin. Lab. Invest.* 31:343–46
163. Herblin, W. F. 1973. *Anal. Biochem.* 51:19–22
164. Coward, J. K., Wu, F. Y. H. 1973. *Anal. Biochem.* 55:406–10
165. Schlenk, F., Zydek-Cwick, C. R., Hutson, N. K. 1971. *Arch. Biochem. Biophys.* 142:144–49
166. Zappia, V., Galletti, P., Carteni-Farina, M., Servillo, L. 1974. *Anal. Biochem.* 58:130–38

INTERMEDIATES IN PROTEIN FOLDING REACTIONS AND THE MECHANISM OF PROTEIN FOLDING[1]

×891

Robert L. Baldwin

Department of Biochemistry, Stanford University School of Medicine,
Stanford, California 94305

CONTENTS

[1] Abbreviations used for proteins are: nuclease, staphylococcal nuclease; metMb, sperm whale metmyoglobin; RNase A, bovine pancreatic ribonuclease A; lysozyme, hen egg white lysozyme; cyt c, horse heart ferricytochrome c; CGN-A, bovine pancreatic chymotrypsinogen A; BPTI, bovine pancreatic trypsin inhibitor; and PIR: des 121–124 RNase A. Physicochemical abbreviations are: k, rate constant; τ, time constant for one phase of a reaction (or relaxation time in studies of relaxation kinetics, with very small perturbations); ΔH, standard enthalpy change of a reaction; ΔC_p, change in heat capacity at constant pressure; T_m, temperature midpoint of an unfolding transition; T jump, temperature jump; and NMR, nuclear magnetic resonance.

453

INTRODUCTION

Scope of this Review

The direct study of the mechanism of protein folding by the classical procedure of detecting and characterizing intermediates has been a goal of protein chemists for two decades. Until recently, most analyses of reversible unfolding transitions have reached the negative conclusion that intermediates are present at concentrations too low to detect. In the last three years, fast reaction techniques have demonstrated that intermediates are present at least transiently. The nature of these intermediates and the information they provide about the mechanism of protein folding are the chief subjects of this review. However, work on their characterization is only beginning. A different kind of intermediate can also be studied now: covalent intermediates in disulfide bond formation can be trapped and the locations of the first −SS− bonds determined. The relationship between these noncovalent folding intermediates and covalent disulfide bond intermediates is beginning to be evaluated.

Limitations of space prevent the review of other approaches, which are listed below to supply perspective. The long range goal of all approaches is to find the stereochemical determinants of folding; the ultimate goal is to predict the tertiary structure from the amino acid sequence. Rapid progress is being made in correlating sequence with local secondary structure: α-helices, pleated sheets, and hairpin turns (1). Sophisticated statistical techniques are being used to evaluate the reliability of correlations between sequence and secondary structure derived from X-ray structures of proteins (2, 3). Also the correlations are being studied experimentally with amino acid copolymers (4, 5), and the results are used to test semi-empirical calculations of helix stability based on potential functions (6). A "blind test" of several predictive schemes has been made recently (1): predictions of the secondary structure of adenyl kinase were based on its sequence and later compared with the known structure when the X-ray analysis was released. A different approach to the search for stereochemical determinants uses chemically modified or mutant proteins: amino acid substitutions and deletions, as well as complementation between fragments. Much of this work has been done with staphylococcal nuclease and has been briefly reviewed recently (7–10). A third approach is to study the dynamic accessibility, or fluctuations in folding, of native proteins by hydrogen isotope exchange (11) or by susceptibility to proteolytic enzymes (12, 13). The relation between dynamic accessibility and the pathway of unfolding is a challenging current problem.

Other Reviews

The analysis of protein unfolding transitions was reviewed definitively by Tanford in 1968 and 1970 (14, 15); discussion of the material in his review will not be repeated here. Proposals about the nature of unfolding transitions are given in recent chapters by Brandts (16), by Hermans et al (17), and by Lumry & Biltonen (18). Kauzmann's 1954 review (19) remains informative and stimulating. A critical review of all approaches to the study of protein folding has been provided recently by Wetlaufer & Ristow (20). Ptitsyn et al have discussed probable mechanisms involved in protein folding (21). New reviews are forthcoming from Anfinsen & Scheraga (22) and Pain (23).

Background

The modern study of reversible unfolding transitions began in the 1950s (24–27) as protein chemists sought methods of detecting the α-helix (28) in globular proteins and realized the possibility of using optical rotation to measure changes in secondary structure (29, 30). Earlier studies of unfolding transitions were aimed principally at understanding heat denaturation, which today is equated to irreversible unfolding (usually precipitation of the heat-unfolded protein). Denaturation was both a mystery and an extreme annoyance, especially to enzyme chemists. Several workers suggested in the 1930s that denaturation is the unfolding of an ordered three-dimensional structure but this view was not accepted generally until the persuasive kinetic analyses of Kauzmann and co-workers (19) in the mid 1950s. Even at that time many believed that denaturation was always irreversible, and protein folding was thought to be a process under cellular control. Although reversible denaturation of the proteolytic enzymes and proenzymes had been studied as early as 1930 by Northrop, Kunitz, Anson, and colleagues (31–33), these were not accepted as actual denaturation reactions by many [see the 1954 discussion by Eyring and Neurath at the end of Kauzmann's review (19)] probably because these proteins contain –SS– bonds that were left intact during unfolding. Probability calculations showed that if the –SS– bonds form by random pairing, then the yield of native protein with correct –SS– bonds will be small even for a protein with only three or four –SS– bonds (34, 35). Formation of the correct –SS– bonds was believed to be under cellular control until Anfinsen and co-workers showed that the gain in thermodynamic stability upon folding provides a driving force for correct in vitro folding and –SS– bond formation (36).

Protein Folding in vitro and in vivo

Although one aim of studying protein folding in vitro is to understand the in vivo pathway, it should be recognized that the in vitro and in vivo pathways may not be identical. In vivo a protein may begin to fold as it is being synthesized on a ribosome because in vitro folding reactions are fast for some small proteins without –SS– bonds [about 1 sec for nuclease (37) and about 10 sec for metmyoglobin (metMb) (38)] and folding may proceed more rapidly than biosynthesis. Many experiments suggest that protein folding is under thermodynamic, not kinetic, control in the sense that the same product can be formed by different pathways. For example, reduced, unfolded ribonuclease A (RNase A) forms a scrambled set of –SS– bonds upon re-oxidation in 8 M urea where native RNase A is not stable, but removal of the urea and addition of an agent to reshuffle the –SS– bonds cause the gradual formation of a folded protein with the correct –SS– bonds and the native structure of RNase A (36). It seems unlikely that the scrambled –SS– bonds formed in 8 M urea are also formed in one stage of the normal re-oxidation reaction. If the same product can be formed by different pathways, then refolding in vitro to give the native structure is not evidence that the in vitro and in vivo pathways are identical. Presumably certain basic properties of the mechanism dominate the folding process both in vitro and in vivo and a major goal of in vitro studies is to discover these properties. The question of kinetic control vs thermodynamic control of folding is considered briefly in a later section of this review.

THE TWO STATE APPROXIMATION FOR EQUILIBRIUM STUDIES OF UNFOLDING

Earlier Work

Equilibrium studies reviewed previously (14–16) have established that the thermal and pH-induced unfolding transitions of many small proteins are highly cooperative. The failure to detect intermediates in these studies has suggested the use of the two state approximation, $N \rightleftharpoons U$, to represent unfolding (31, 39, 40).[2] Two types of experiments have been used. (a) Thermodynamic properties of the transition (ΔH, ΔS, ΔC_p, dependence of equilibrium constants on guanidine concentration,[3] etc) have been estimated from equilibrium constants based on the two state transition. In very careful studies (14, 15, 41, 42), these results have been found to be self-consistent and in reasonable agreement with data for model compounds. (b) Transition curves measured by different properties (intrinsic viscosity, optical rotation, absorbance or fluorescence of buried chromophores) have been found to be superimposable (43, 44). This behavior is characteristic of a two state reaction because the same fraction of native protein, f_N, must be given by any property X

$$f_N = \frac{X - X_U}{X_N - X_U} \qquad \qquad 1.$$

There are two problems in using equation 1. First, the value of X for the native and unfolded forms, X_N and X_U, at a particular temperature or pH inside the transition zone must be obtained by extrapolation. Second, the measured property X is assumed to be a weighted average of X_N and X_U. If intermediates

[2] N = native, U = unfolded, I = intermediate in folding, as defined by the conditions in which these species are stable (i.e. populated at equilibrium). N (or N_1, N_2, . . .) is used for species stable below the transition zone, U (or U_1, U_2 . . .) is used for species stable above this zone, and I (or I_1, I_2 . . .) is used for species that are stable either only inside the transition zone or else not at all (i.e. I then refers to a transient kinetic intermediate).

[3] The term "guanidine" refers here to a strongly denaturing salt such as guanidinium chloride; anions such as sulfate largely compensate for the denaturing properties of the guanidinium cation (155).

are present they will affect the value of X but the comparison of transition curves is not a definitive test for intermediates unless it gives a positive result (noncoincident transition curves). Not all small proteins show cooperative unfolding by this test: well-resolved stages of unfolding are observed in the guanidine-induced unfolding of bovine carbonic anhydrase B (45, 46) and of staphylococcal penicillinase (47), as well as in the urea-induced unfolding of a large globular protein, rabbit muscle phosphorylase b (48), whose pyridoxal phosphate has been bound covalently by reduction.

Calorimetric Test

Direct measurements of $\Delta H (\Delta H_{cal})$ provide a quantitative test of the two state approximation, because (14, 49)

$$\Delta H_{cal} \geqq \Delta H_{vH} \qquad\qquad 2.$$

Here ΔH_{vH} is the value obtained from the two state approximation by using the van't Hoff equation for the temperature dependence of the equilibrium constant. The equality holds for a two state reaction; if any intermediates differing in enthalpy from N and U are populated the inequality holds. For example, the α-helix yields very different values for ΔH_{cal} and ΔH_{vH}, according to the relation $\Delta H_{cal}/\Delta H_{vH} = 1/\sigma^{1/2}$, where σ is the Zimm-Bragg (50) nucleation parameter; typically $\sigma \approx 10^{-4}$ (51, 52). Very sensitive and precise calorimetric measurements of thermal transition curves are needed to test equation 2. The first reports of successful measurements in Sturtevant's laboratory (49, 53) were followed quickly by reports from other laboratories (54, 55). Calorimetric transition curves for protein unfolding are complex: substantial heat evolution occurs both below and above, as well as inside, the transition zone measured by optical properties. Above the transition zone, the measurements confirm Brandt & Hunt's estimates of ΔC_p, based on two state equilibrium constants (42). Interpretation of the heat evolved at low temperatures is controversial (49, 54) and strongly affects the evaluation of equation 2. Privalov and co-workers have proposed the following interpretation (55–57). First, the transition zone for cooperative unfolding is defined by optical measurements of the transition at a wavelength where X_N in equation 1 is independent of temperature. Then the calorimetric curve is divided into a "pretransition" zone and a cooperative unfolding zone that matches the one defined optically. The value of ΔH_{cal} is taken from the area under the peak of heat evolved vs temperature (plotted as a differential curve), whereas ΔH_{vH} is taken from the temperature dependence of the equilibrium constant, based on values of f_N (equation 1). Values of ΔH_{cal} and ΔH_{vH} obtained in this way for the cooperative unfolding transition agree closely [e.g. $\Delta H_{cal}/\Delta H_{vH} = 1.05 \pm 0.03$ for RNase A (58)]. Whether or not events in the pretransition zone are connected to the mechanism of unfolding remains to be decided. Volume changes on unfolding are reported to be complex (59) and an abrupt pretransition zone can be discerned, which however is not matched by any abrupt change in latent heat, according to the calorimetric curves (57).

NMR Measurements of Transition Curves for Individual Protons

If the equilibrium unfolding curves for individual protons could be resolved from each other, they could be used to establish the pathway of unfolding when the proton resonances are assigned and the X-ray structure is known. Study of these transition curves also provides a sensitive test of whether intermediates are populated at equilibrium.

A thorough, early NMR study of the thermal unfolding of lysozyme, using continuous wave spectroscopy, concluded that equilibrium measurements of the transition are completely described by the two state approximation (60). More recent studies of other proteins confirm that the two state approximation is a useful first approximation but offer promise of detecting some intermediates at equilibrium: this is true for the acid transition of nuclease (61), the thermal unfolding of RNase A at low pH (62), and the guanidine-induced unfolding of RNase A (63, 64). Three basic features of these NMR studies are as follows:

(a) *Slow exchange between native and unfolded protein*—a given proton typically shows separate resonances in the native and unfolded forms of a protein. The peaks are often well resolved from each other, and no intermediate peaks are observed. Interconversion of the native and unfolded forms is slow on the NMR time scale: $\tau > 20$ msec is a typical result at 100 MHz. No exception to the rule of slow exchange in a cooperative unfolding reaction has been observed. However, evidence of fast exchange between different native conformations, in addition to slow exchange between native and unfolded, has been observed in the thermal unfolding of RNase A (62): the chemical shifts of the four His protons studied become strongly temperature dependent inside the transition zone. This suggests that intermediates in unfolding, whose properties are close to those of native RNase A, are present in temperature-dependent equilibria and that they re-equilibrate rapidly on the NMR time scale ($\tau \ll 20$ msec).

(b) *Closely superimposable transition curves for different protons, with some exceptions*—with Fourier transform NMR, the area under the peak for a well-resolved proton can be determined with sufficient accuracy ($\pm 10\%$, in favorable cases) to measure the transition curve from native to unfolded for that proton. This is a more direct method of determining a transition curve than the use of equation 1 to estimate the curve from the change in some average property X. In the native conformation the resonances of a given type (e.g. the C2 protons of the His residues) are often well resolved from each other; the problem is to separate them from the thousand-odd other proton resonances and to assign the resonances to known protons. Selective deuteration can decrease greatly the complexity of the spectrum (65). Individual transition curves from native to unfolded have been measured for the C2 protons of four His residues in both the acid transition of nuclease (61) and the thermal unfolding of RNase A (62). In each case, three out of four His transition curves are superimposable and the fourth is resolved. The nonconforming His residue in nuclease clearly participates in a subtle conformational change caused by titration of neighboring groups in the native enzyme. Thus the same question arises as in interpreting the calorimetric results: are subtle changes in the predenaturation zone linked directly to the cooperative unfolding reaction? The nonconforming His residue of RNase A probably reveals the existence of one or more true intermediates in unfolding: this residue displays an apparently normal transition curve whose T_m is displaced by about 1° from the other three (62). There is a change in mechanism of unfolding for RNase A between pH 1.3 and 4.0, where all four His transition curves are superimposable (66). NMR evidence of multistate behavior has been reported also for the guanidine-induced unfolding of RNase A (63, 64).

(c) *Unfolded proteins give relatively simple NMR spectra*—compared to spectra for native proteins, those of the unfolded forms are relatively simple, and the proton resonances of a given type are closely superimposed. This has been interpreted to mean that all parts of a thermally unfolded protein are exposed to solvent and have similar magnetic environments (60). Fourier transform NMR spectra of RNase A unfolded by heat (67) and by urea or guanidine (64) show that differences between these unfolded forms can be detected by NMR. Moreover, two different pK values are observed for the four His residues of thermally unfolded RNase A: two residues show a pK of 5.75 and two show a pK of 5.96 at 69°C (67).

Local unfolding also can be studied by NMR. A subtle conformational change at low pH, involving one His residue of nuclease, has been mentioned above. At alkaline pH, complex changes in chemical shift with pH are observed for the four His residues, the seven Tyr residues, and the single Trp residue of nuclease (68). The unfolding is not very cooperative: the pH midpoints vary from 8.8 for one His residue to 10.8 for the single Trp residue. Only fast exchange is observed, in contrast to the cooperative unfolding of nuclease at acid pH which shows slow exchange (61, 69). The acid transition is 90% complete within 0.5 pH units and three out of four His residues studied show a pH midpoint of 3.90 ± 0.05.

Pressure-Induced Unfolding

Equilibrium studies of pressure-induced unfolding are usually represented by the two state approximation, but little effort has been made to use these studies to test it. Instead, interest has centered on the pressure-temperature relation governing the temperature of maximum stability of a protein (70, 71) and on comparison of ΔV_{vH}, the volume change for unfolding obtained by the van't Hoff equation, with the volume changes accompanying hydrophobic interactions in model compounds. The two do not agree, which may indicate that present estimates of the role of hydrophobic interactions in determining protein stability are not in a satisfactory state (71, 72). Regarding the two state approximation, ΔV_{vH} is typically much smaller than the volume change measured directly by dilatometry, but the discrepancy has been attributed (71, 72) to problems in the direct measurement of volume changes rather than to failure of the two state approximation. Preliminary kinetic studies of the pressure-induced unfolding of metMb (71) indicate that the two state approximation is satisfactory within the transition zone but fails outside this zone. The kinetics of protein unfolding become very slow at high pressures, and the native form has been separated from the unfolded form of chymotrypsinogen A (CGN-A) in an equilibrium mixture of the two at low pH by high pressure gel electrophoresis at 3250 kg/cm^2 (73).

NATURE OF THE INTERMEDIATES OBSERVED BY KINETICS

Multistate Kinetic Behavior

The kinetic criterion for a two state reaction (N ⇌ U) is that its time course should be a single exponential whose time constant $\tau(\tau = 1/k)$ is the same no matter what property is studied. It is a standard result of kinetics that if the time course of a reaction is a sum of two exponentials, then at least three species must participate in the reaction; conversely three reacting species also must produce biphasic kinetics except in rare cases where the amplitude of one kinetic phase is zero (74). In principle, the number of species reacting is one greater than the number of kinetic phases (75, 76), but this method of counting intermediates is impractical for complex reactions: typically some kinetic phases have negligible amplitudes and others are poorly resolved (76).

The first kinetic studies of reversible unfolding were reported to show two state behavior, both for thermal or pH-induced unfolding (25, 77–80) [with one report to the contrary (81)] and for guanidine-induced unfolding (82). The methods used to initiate the reaction in these early studies often had dead times of 0.1 or even 5 sec. Changes in the absorbance, or sometimes in the fluorescence, of buried tyrosine or tryptophan groups have been used commonly to follow unfolding. These early results typically suggested that only a single unfolding or refolding reaction could be observed, since its amplitude accounted, approximately, for the entire difference between the equilibrium values in the initial and final states. Any disagreement was attributed to uncertainty in the extrapolated values of the molar absorbancy of the native or of the unfolded species.

However, when fast reaction techniques (stopped-flow and T-jump methods) were used, additional faster reactions were found, both in thermal (83, 84) and pH-induced transitions (37, 38) and sometimes also for guanidine-induced unfolding (15, 85, 86). Certain patterns of behavior are observed, but a general understanding of this behavior has not yet been achieved. Inside or below the transition zone the kinetics of thermal unfolding or refolding are typically biphasic with a minor fast phase in milliseconds and a major slow phase in seconds. In the case of cyt c, changes in the heme spectrum also demonstrate the presence of a third, much faster phase in microseconds for thermal unfolding (87). In two respects the biphasic kinetics often approach two state behavior within the transition zone, for thermal (88), pH-induced (38, 89), guanidine-induced (85, 86), and pressure-induced (71) unfolding. Near the midpoint of the transition, the amplitude of the fast phase drops to a small value and the τs of the two kinetic phases approach each other. In the pH-induced unfolding of metMb, the fast and slow phases actually merge at pH_m (38). For the guanidine-induced unfolding of lysozyme, strict two state behavior is reported inside the transition zone, whereas the kinetics are biphasic both below and above the transition zone (86). Another common observation in pH-induced (89) and thermal unfolding (88) is that in unfolding the fast phase becomes the major phase outside the transition zone.

Finally, it appears to be a rule that intermediates that are observed readily in thermal or pH-induced unfolding are seen only with difficulty in denaturant-induced unfolding. For example the thermal unfolding or refolding of RNase A shows biphasic kinetics at all pHs, whereas the guanidine-induced transition shows two state kinetics (90). The urea transition of RNase A is reported to show complex slow kinetics (91), but the fast phase characteristic of pH-induced or thermal unfolding is absent. It is probable that denaturants such as urea and guanidine destabilize partly folded intermediates. The kinetics suggest this, and equilibrium measurements show that the addition of guanidine to thermally unfolded proteins reduces or eliminates secondary structure that remains after thermal unfolding (92, 93). One aim in studying guanidine-induced unfolding is to measure refolding after all nucleation centers for refolding have been eliminated (14, 15). The physical properties of guanidine-unfolded proteins have been studied carefully by Tanford and co-workers (94), and the results are consistent with a random coil conformation constrained by any –SS– bridges present. However, it is possible that some elements of secondary structure remain: the fluorescence of the single Trp residue of cyt c is reported to monitor some residual secondary structure present after unfolding in 4 M guanidine (95).

Relation to Physiological Conditions

Although extremes of temperature, pH, or denaturant concentration are used to cause unfolding, refolding can be studied in closely physiological conditions (pH 7, 37°C, no denaturant present). Thus the intermediates observed in refolding are of particular interest. When unfolding is reversible and measurements of unfolding and refolding can be made in the same final conditions, inside the transition zone, one can relate the intermediates observed in unfolding to those observed in refolding and characterize the pathway of folding from both ends. This is one reason for studying reversible unfolding transitions. Another reason is that kinetic mechanisms can be tested by comparing measurements of unfolding and refolding in the same final conditions (96–98). If the aim is to study refolding in physiological conditions, then it is awkward to cause unfolding by a denaturant such as guanidine because 6 M guanidine often is needed to unfold, and in refolding the guanidine concentration cannot be reduced to a negligible level by dilution in a stopped-flow apparatus.

Possible Artifacts

Particularly because of the equilibrium evidence supporting the two state approxima-tion, it is important to check for artifacts in the kinetics. Instrumental artifacts can be checked by comparing different methods: both stopped-flow and T-jump measurements of unfolding show biphasic kinetics for RNase A (83, 88) and also for CGN-A (84). There is, however, a serious potential artifact in stopped-flow studies at temperatures above room temperature. Unless care is taken to ensure that the temperatures of the drive syringes and observation chamber are identical, a temperature gradient will cause a time-dependent loss of light that can appear as a spurious reaction, especially in the time range of 1 sec (99). This effect can be eliminated by controlling independently the thermostating of the drive syringes and observation chamber (100). An artifact more difficult to evaluate is the possible confusion of subtle changes in native conformation with reactions that are part of the cooperative unfolding process. This problem arises in all methods of studying unfolding. It has been studied carefully for the thermal unfolding of RNase A. One approach is to use a series of T jumps of fixed size to study thermal unfolding, in order to find out where the fast and slow reactions first appear and later disappear. For RNase A (83), both reactions first appear at the same temperature when the final temperature of the T jump enters the transition zone, and both reactions disappear together when the initial temperature of the T jump goes above the transition zone. Thus both fast and slow reactions appear to be part of the unfolding transition. Stopped-flow studies of the pH-induced refolding of RNase A confirm that the fast reaction is part of the refolding process, and not an isomerization of N or U (101). It cannot be an isomerization of N as in

$$\overset{\text{slow}}{}\quad\overset{\text{fast}}{}$$
$$U \rightleftharpoons N_1 \rightleftharpoons N_2 \tag{3.}$$

because it occurs when no N is present initially, and it cannot be an isomerization of U as in

$$\overset{\text{fast}}{}\quad\overset{\text{slow}}{}$$
$$U_1 \rightleftharpoons U_2 \rightleftharpoons N \tag{4.}$$

because the product of the fast refolding reaction is N (see below).

Two Procedures for Characterizing Kinetic Intermediates

One can try to study the kinetic properties of intermediates by solving the kinetic mechanism, or one can use fast kinetics as a separation method analogous to chromatography and characterize the intermediates by fast response spectral methods. There are three steps in the latter procedure: (a) find conditions in which intermediates are present, (b) measure their concentrations on a weight or molar basis, and (c) find adequate spectral methods of characterizing them in times as short as a few milliseconds. An ideal kinetic experiment would be stopped-flow NMR (102), and it should eventually become feasible even though present NMR experiments on proteins require signal averaging over times that are long compared to milliseconds. Using the second approach, one might hope to rely only on spectral characterization of intermediates and to dodge the problem of determining the kinetic mechanism. However, it may be necessary to understand the kinetic mechanism even to know when an intermediate is being observed. It had been supposed at first that the fast refolding reaction of RNase A must yield an intermediate; in fact, it yields the native enzyme (101).

Different Approaches to Analysis of the Kinetics

The reader interested in this problem should read the discussions by Tanford (15, 97, 98). Basically four approaches have been tried; each has some merit, and all have the defect of being valid only in special circumstances.

(a) *Location of steps in folding by known chromophores*—the aim is to identify steps involving folding of a given segment by the changes in a chromophore located in that segment. For illustration, the unfolding of RNase A shows both fast and slow reactions when followed by the exposure to solvent of three buried Tyr groups. A possible explanation is that a segment with one Tyr group unfolds in a fast reaction, and the slow unfolding of a larger segment exposes the other two. This suggestion was tested (103) by measuring the unfolding kinetics of a chemically modified RNase A that has a single dinitrophenyl chromophore, 41-DNP RNase A (104). The results show that the fast and slow unfolding kinetics are the same when measured by the single DNP group as by the three buried Tyr groups. Whatever the nature of the fast and slow unfolding reactions, both are monitored efficiently by a single chromophore. Analysis of simple kinetic models indicates that this approach gives useful results only when a given step in folding goes to completion in a single kinetic phase. Consider the case in which two segments, a and b, fold separately. Each has a known chromophore: A changes only when a folds to a', B changes only when b folds to b'. Let the two steps be well resolved in time.

$$\overset{\text{fast}}{ab} \overset{\text{slow}}{\rightleftharpoons a'b} \rightleftharpoons a'b' \qquad\qquad 5.$$
$$(1) \quad (2) \quad (3)$$

The exact kinetic analysis of such reactions is well known (75). Unless the fast reaction goes to completion, chromophore A will change in both the fast and slow kinetic phases. At the end of the fast phase, the concentration ratio C_2/C_1 is close to its final value but the conversion of species 2 to 3 pulls the reaction from 1 to 2 in the slow phase. Observing a change in chromophore A in both kinetic phases does not mean that there are two steps in the folding of a to a'.

(b) *Direct analysis*—simple kinetic mechanisms can be tested by solving for the kinetic parameters and applying consistency checks. As is usual in kinetics, this procedure serves to eliminate mechanisms, not to prove them. If only biphasic kinetics are observed, only three species mechanisms can be tested. A rigorous framework for direct analysis has been worked out by Ikai & Tanford (96–98); diagnostic tests for different mechanisms are illustrated and consistency checks are given in analytical form. Application to the guanidine-induced unfolding of cyt c and lysozyme (85, 86, 105) indicates that at least four species are present in both cases. Since only two kinetic phases are observed, the consistency tests cannot be carried through to a quantitative test of these mechanisms.

(c) *Computer analysis of models*—protein folding must proceed via a complex series of intermediates. Even if the intermediates are only transient and poorly populated, they can have striking effects on the observed kinetics. The simplest way of studying these effects is often by analyzing the predicted properties of models. For example, the rate of formation of a polynucleotide double helix from two complementary strands shows a striking dependence on temperature, decreasing towards zero roughly in proportion to $(T_m - T)$ (106). In the absence of complications caused by forming three stranded helices, this is a second order reaction without observable intermediates (107). The temperature dependence of the rate can be understood by analyzing a sequential model of nucleation followed by "zipping-up" of the helix (108, 109). The activation enthalpies computed by the two state approximation for the rates of folding and unfolding of small proteins (77–80) resemble those of double helix formation by oligonucleotides (110). An attempt was made to apply a similar, nucleation-dependent, sequential model to the kinetics of protein folding (111, 112). The model did successfully predict the biphasic kinetics of unfolding observed soon afterwards in the thermal unfolding of RNase A (83) and CGN-A (84). However, this interpretation of the biphasic kinetics found in the RNase A thermal transition (88) has been disproven by a later study of RNase A refolding (101), since the fast refolding reaction does not produce one or more intermediates, as required by this model, but rather yields native enzyme.

(d) *Ask specific questions about the mechanism*—this approach works only if positive answers can be found to specific questions. Typically negative answers give little information. For example, if the question is whether active enzyme is formed in the fast phase of refolding, a positive answer provides a lead to the next set of experiments but a negative answer gives no clue. Some results obtained by this approach for RNase A are discussed in a following section.

Possibility of Abortive Intermediates

Ikai & Tanford (96–98) have shown that there is a qualitative difference between the kinetic behavior shown by each of these three species mechanisms: the first has an intermediate I on the direct pathway of folding

$$U \leftrightharpoons I \leftrightharpoons N \qquad\qquad\qquad 6.$$

and in the second I is off this pathway

$$I \leftrightharpoons U \leftrightharpoons N \qquad\qquad\qquad 7.$$

When the kinetics of unfolding and refolding are compared in the same final conditions, mechanism 6 requires a lag during the fast phase of either unfolding or refolding, whereas mechanism 7 does not. A lag is another way of saying that the amplitudes of the fast and slow phases have opposite signs. Since no instance of a lag in a reversible unfolding reaction in either unfolding or refolding has yet been reported, this observation raises the possibility that the intermediates detected by fast kinetics are only abortive intermediates. In the derivation by Ikai & Tanford (96–98), it was assumed that the intermediate I is never populated at equilibrium outside the transition zone. In the case of the thermal unfolding of RNase A, a different three species mechanism

$$U_1 \overset{\text{slow}}{\rightleftharpoons} U_2 \overset{\text{fast}}{\rightleftharpoons} N \qquad\qquad 8.$$

describes most of the kinetics, both qualitatively and quantitatively (100, 101, 113), and it does not predict a lag in either unfolding or refolding although the intermediate U_2 is on the direct pathway of folding. It differs from mechanisms 6 and 7 in that both U_1 and U_2 are populated at equilibrium at temperatures above the thermal transition zone. Application of the equations of Ikai & Tanford to the guanidine transitions of cyt c and lysozyme suggests that abortive intermediates are prominent in the former (85) but not in the latter (86) case. In the case of cyt c, stable intermediate(s) have been detected inside the transition zone (85), i.e. intermediate(s) are present at equilibrium.

Hijazi & Laidler point out (114) that a different mechanism, not considered by Ikai & Tanford, shows a kinetic behavior similar to mechanism 7

$$U \rightleftharpoons N \rightleftharpoons I \qquad\qquad 7a.$$

Here I is an abortive intermediate in unfolding.

Nature of the Fast and Slow Refolding Reactions of RNase A

By asking specific questions, some basic properties of the RNase A mechanism have been learned before attempting to solve the mechanism. The basic approach is to use fast kinetics as a separation method. The refolding of RNase A is a favorable case for doing this because the fast and slow refolding reactions are well resolved in time (roughly 100-fold at temperatures sufficiently below the transition zone). (a) Is the product of the fast refolding reaction sufficiently folded to bind a specific ligand, 2'CMP? Binding of 2'CMP is known to be fast ($\tau \approx 1$ msec) compared to the fast refolding reaction ($\tau \approx 0.1$ sec). The answer is yes; in fact, the species formed in the fast reaction shows the same binding constant for 2'CMP as native RNase A, within experimental error. Moreover, this species hydrolyzes the substrate CpA (101). Thus, because RNase A is a small protein, in which the substrate binding and catalytic sites account for a large fraction of the whole enzyme, the product of the fast refolding reaction must be native RNase A or some isomeric form. (b) Are the fast and slow refolding reactions produced by different starting materials? If different fast and slow refolding species exist, and if the equilibrium between them is slow compared to the stopped-flow mixing time, their presence can be demonstrated by varying the initial conditions to change this equilibrium. The ratio of the two species then would be measured by the amounts of RNase formed in the fast and slow refolding reactions, in fixed final conditions. Refolding experiments as a function of the initial pH show that the fraction of fast refolding material varies from 20% at pH 2 to 0% at pH 5, and follows a pH titration curve with a midpoint of pH 4.0 at 50°C (113). We conclude that fast and slow refolding reactions arise from different species present in the initial conditions; these species are not chemically different because they are interconvertible by changing pH. (c) Is the fast refolding species an intermediate present at equilibrium only inside the transition zone? Refolding experiments as a function of temperature show that it is present also above the transition zone; the apparent equilibrium between fast and slow refolding species is not measurably temperature dependent (113). Therefore heat-unfolded RNase A contains an equilibrium mixture of fast and slow refolding species.

These observations are all consistent with a simple three-species mechanism

$$\overset{\text{slow}}{U_1} \overset{\text{fast}}{\rightleftharpoons} U_2 \rightleftharpoons N \qquad\qquad 9.$$

or:

$$\overset{\text{slow}}{U_1} \overset{\text{fast}}{\rightleftharpoons} N \rightleftharpoons U_2 \qquad\qquad 9a.$$

A physical difference between U_1 and U_2 has been found in studying a derivative of RNase A in which the three freely ionizing Tyr groups are converted to mononitrotyrosyl groups: the pKs of these groups are different in U_1 and U_2 (115). Mechanisms 9 and 9a predict a fast reaction in unfolding, as observed. Also they are consistent with the two state description of equilibrium studies of the thermal transition, since the ratio $(U_2)/(U_1)$ is not temperature dependent. Because 9 and 9a are three-species mechanisms, they can be solved and tested quantitatively. In fact, mechanism 9 does describe the kinetics fairly quantitatively (100), but the temperature dependence of τ_2 (for the fast phase) indicates that a series of intermediates must be involved in the actual unfolding transition between U_2 and N. Also a third, faster kinetic phase is observed in and above the upper half of the transition zone, indicating that an additional intermediate is present at these temperatures. The structures of U_1 and U_2 remain to be characterized. At present one might assign the 100-fold difference in their rates of refolding either to "correct" structure in U_2 or "incorrect" structure in U_1. The negligible difference in enthalpy between U_1 and U_2 suggests that a topologically difficult change in conformation connects them.

DISULFIDE INTERMEDIATES

Mechanism of Disulfide Bond Formation in Relation to the Mechanism of Folding

Early work had shown that the –SS– bonds of RNase A stabilize the native enzyme: with –SS– bonds intact, unfolded RNase A refolds rapidly, but after breaking the –SS– bonds with performic acid, the oxidized RNase A remains unfolded (26). After the discovery that reduction and re-oxidation give native enzyme (116), one could hope that trapping and characterizing intermediates with 1, 2, or 3 disulfide bonds formed would give direct information about intermediates in folding, since the –SS– bonds evidently confer stability on intermediates in folding. However, further work showed that scrambled –SS– bonds are formed rapidly upon re-oxidation (117) and that native RNase A is recovered only slowly via the disulfide interchange reaction. The discovery of an enzyme catalyzing these disulfide interchange reactions (36) suggests that this mechanism may be prevalent in vivo. In a system where the wrong disulfide bonds are formed initially, characterization of intermediates with 1, 2, or 3 –SS– bonds can give only indirect clues about intermediates in folding. A recent, very thorough characterization of –SS– bond formation by re-oxidation of reduced RNase A has confirmed this picture (118): diagonal mapping of the peptides at early times shows only a mixture of small amounts of native RNase A with large amounts of highly scrambled RNase A. Analogous studies with lysozyme show that certain –SS– bonds are formed preferentially at early stages of re-oxidation (119). When the number of scrambled species is large, diagonal mapping of the peptides shows them only as a blur in the background, and the yield of scrambled species, when there are many such species, has to be computed from the percentage yield of peptides observed on the chromatogram.

The small protein BPTI (bovine pancreatic trypsin inhibitor), which has 3 disulfide bridges and 58 amino acid residues, offers a more favorable system. Creighton's recent work with this protein (120–122) promises to yield the detailed mechanism relating folding to disulfide bond formation in BPTI. When a disulfide such as hydroxyethyl disulfide or oxidized dithiothreitol (DTT) is used to re-oxidize reduced BPTI, formation of the –SS– bonds follows a direct pathway by the criterion that no external thiol need be added to break incorrect –SS– bonds. The reaction proceeds by formation of a mixed disulfide between oxidized DTT and reduced BPTI, followed by a disulfide interchange reaction within the protein to form an internal –SS– bridge and to eliminate oxidized DTT. The reaction follows second order kinetics in oxidized DTT, so that formation of the mixed disulfide is rate limiting. At 1×10^{-2} M oxidized DTT, the half-time of complete re-oxidation is 5 min. Scrambled BPTI accumulates when the reaction is carried out at higher concentrations of oxidized DTT. Significant concentrations of partially re-oxidized intermediates are present before the reaction goes to completion. The single disulfide intermediates have been trapped and characterized. Four out of fifteen possible species containing single –SS– bonds are detected. The most prominent species (Cys 30-Cys 51) accounts for half of the total and contains an –SS– bond present in native BPTI. The other three species contain bonds not found in native BPTI. The ratio of the concentrations of these species is independent of reaction time, trapping agent, and oxidizing agent. This finding suggests (121) that the single disulfide intermediates are in mobile equilibrium via internal disulfide interchange reactions.

These results suggest a possible generalization for understanding the relation between –SS– bond formation and the mechanism of folding: the concentrations of the possible one disulfide intermediates are governed by the ratio of their thermodynamic stabilities, similarly for the two disulfide intermediates, etc. If this is correct, the single disulfide intermediates with "wrong" –SS– bonds disappear when the second disulfide bridge is formed because they are unstable. This mechanism requires that free –SH groups be available in the protein to break wrong –SS– bonds, a function otherwise provided (in vitro) by an added thiol.

It would not be surprising if the postulated mobile equilibrium becomes sluggish when only two free –SH groups remain, and rather stable abortive intermediates with two disulfide bonds have been observed (121) in the re-oxidation of BPTI. In other words, the generalization that now appears possible from Creighton's work may be an oversimplification, but nevertheless it may prove to be the only useful generalization relating the mechanism of –SS– bond formation to the mechanism of folding. If it is correct, then intermediates in –SS– bond formation give no information about kinetic intermediates in folding, because the concentrations of different one disulfide intermediates are determined by their thermodynamic stabilities. There is other evidence that the kinetics of folding are not likely to be rate limiting in –SS– bond formation. Small proteins without –SS– bonds such as nuclease (37) and metMb (38) can refold in times of 1 to 10 sec.

What structural features account for the formation of these particular one disulfide species? The "correct" bond connects Cys 30, located inside an antiparallel pleated sheet (residues 16–36), to Cys 51, in the midst of the only sizable α-helix (residues 47–56). The incorrect bonds also have either Cys 30 or Cys 51 as one member, and the other two participating residues Cys 5 and Cys 55 are in α-helices, Cys 5 being in a very short helix. No other obvious correlation between structure and formation of these particular –SS– bonds has been found. In native BPTI, the polypeptide backbone has to be threaded through a loop formed by the Cys 30–Cys 51 bond, and so it is interesting that this is the first correct –SS– bond to form.

Formation of Disulfide Bonds as a Test for Critical Interactions in Folding

It is possible to delete or modify certain residues and to ask if the modified protein is still capable of forming the correct –SS– bonds on refolding. Early studies of this kind by Anfinsen and co-workers (123) showed that the proteolytic enzyme α-chymotrypsin, which is activated by peptide bond cleavage, forms scrambled –SS– bonds after reduction and re-oxidation, although the proenzyme chymotrypsinogen is able to form the correct –SS– bonds on re-oxidation. The experiment may be carried out by adding an agent to promote disulfide interchange and asking if the native protein is converted to a scrambled form. Experiments of this type with insulin (123) led to the discovery of proinsulin (124). This approach has been used to ask whether the N-terminal and C-terminal ends of the polypeptide are needed for the correct re-oxidation of RNase A. A 20 peptide fragment can be removed from the N-terminal end without preventing correct re-oxidation in moderate yield (125, 126), whereas deletion of a C-terminal tetrapeptide by peptic cleavage blocks correct re-oxidation (127). This type of experiment shows the influence of the entire polypeptide chain in determining correct folding and pairing of Cys residues to form –SS– bonds (7, 8, 127). The case of the pepsin-inactivated ribonuclease (PIR) is particularly intriguing because the protein unfolds and refolds reversibly as long as its –SS– bonds are left intact. The T_m for thermal unfolding of PIR is only 16° below that of native RNase A (128, 129). Is there a kinetic block in the correct re-oxidation of PIR, or is native PIR thermodynamically unstable relative to scrambled PIR? The question has been answered by using a thiol to promote disulfide interchange among the –SS– bonds of native PIR: at 37°C, where RNase A is stable, the native PIR disappears and is replaced by a precipitate containing, presumably, scrambled PIR (127). Thus native PIR is thermodynamically unstable relative to scrambled PIR but stable relative to unfolded PIR. There probably is a significant increase in entropy when unfolded PIR is converted to scrambled PIR because the number of possible conformations is greatly increased. This type of experiment apparently depends on, and gives information about, the relative thermodynamic stabilities of the different disulfide intermediates much as in the case of BPTI.

Other Observations

The porphyrin ring of cyt c is needed for correct folding, although the heme iron is not needed (130). When the re-oxidation of –SS– bonds in lysozyme is allowed to proceed at high temperatures, the recovery of enzymatic activity is reported to pass through a maximum and then to decline with time (131). This has been interpreted as showing the importance of kinetic factors in determining the products of –SS– bond formation.

RELATED TOPICS

Limitations of space prevent any review of these topics, but we wish to call attention to their close relation to the problem of intermediates in protein folding reactions.

Dynamic Accessibility

Native proteins exhibit dynamic accessibility: parts of the protein that are not accessible to solvent, as shown by the X-ray structure, nevertheless contact the solvent for some fraction of the time. This behavior is most clearly demonstrated by the exchange of tritium or deuterium between water and internal protons or the protein (11). It may also be studied by chemical reactivity of buried groups or by proteolytic enzymes that attack preferentially the unfolded form of a protein. In the latter two cases especially, an uncertainty principle operates: the reagent used to study accessibility can change the equilibria and kinetics of the reactions that provide accessibility. Regarding unfolding, the basic question is: what is the relation between dynamic accessibility and cooperative unfolding? It is commonly accepted, but not proven, that local unfolding reactions provide accessibility. However, there is no general agreement as to whether a network of local unfolding reactions constitutes part of the cooperative unfolding process (15, 132). That this is a serious possibility was pointed out by Hermans and co-workers (133) with computer calculations for a three-dimensional Ising model. There is general agreement, on the basis of their pH dependence, that the rates of local unfolding reactions are fast compared to the rate of exchange itself (for peptide H atoms) and that therefore H exchange measurements give apparent equilibrium constants for local unfolding reactions (11). We would like to note here three points about dynamic accessibility and unfolding. 1. Measurements of dynamic accessibility often give direct structural information. The site at which a proteolytic enzyme attacks can be determined by peptide mapping. Such measurements have been used to propose a pathway of unfolding for RNase A (13). Sites of H exchange also are beginning to be determined. The dynamic accessibility of five of the six Trp residues of lysozyme has been determined by NMR (134). 2. The standard reason given in the past for supposing that dynamic accessibility and local unfolding are not connected to the cooperative unfolding process has been that cooperative unfolding is a two state process, but it is clear now that the kinetics of thermal unfolding do not show two state behavior. Even the three state approximation used to describe the thermal unfolding of RNase A is only a first approximation (100). 3. The relation between local unfolding and cooperative unfolding will probably be worked out by exchange measurements on individual protons. Thus it is significant that the first exchange measurements on single protons have recently been made by NMR spectroscopy of $H \leftrightarrow D$ exchange for the Trp residues of lysozyme (134).

Antigenic Detection of Stages in Folding

Complementation between overlapping polypeptide fragments has been shown to be a powerful technique for evaluating critical interactions in folding. The technique is best suited to small proteins without –SS– bonds, such as nuclease. Recent experiments have shown that conformational equilibria in the separate, unfolded fragments can be probed by sensitive antigen-antibody techniques (8, 135, 136). The approach is to immunize a goat with intact nuclease, then to fractionate the resultant antibodies on an immunoabsorbent column containing the covalently bound fragment. In this way antibodies directed against nuclease but capable of reacting with fragment are obtained. The antibodies react with the native rather than the unfolded form of nuclease, because ligands that stabilize the native conformation (Ca^{2+} and pdTp) do not inhibit nuclease precipitation by these antibodies. These ligands do inhibit precipitation of nuclease by other antibodies directed against the unfolded fragment (135). To evaluate the affinity of fragment for antibodies directed against nuclease, the ratio of their equilibrium constants for combination with antibody is measured by determining the enzymatic activity of nuclease remaining at equilibrium after adding varying amounts of neutralizing antibody and competing fragment. This ratio of equilibrium constants is of the order of 10^{-4} (135). It has been interpreted in terms of a conformational equilibrium $U \leftrightarrow N$ in the fragment. By hypothesis, the affinity of those fragment molecules combining with antibody (the N subfraction) is said to be the same as that of native nuclease. Then one calculates that 10^{-4} of the fragment molecules are in the N subfraction, or in "the native format." Alternatively, the affinity of fragment for antibody might be interpreted in terms of a $U \leftrightarrow N^*$ conformational equilibrium, where the affinity of N^* for antibody is less than that of nuclease because N^* has an incomplete antigenic determinant. If the affinity for antibody of N^* is 100-fold less than that of nuclease, then one calculates from the ratio of equilibrium constants that 10^{-2} of the fragment molecules are in the N^* subfraction.

Further analysis depends in part on obtaining antibodies with sharply defined specificities for the N and U conformations of nuclease. Recent progress in this work has been reported (137, 138). The Scatchard plots for combination of antibody with fragment are strongly curved. This is caused by the presence of two antibody populations with 100-fold different affinities for nuclease (137).

Kinetics of Formation of the α-Helix and the β-Pleated Sheet

A theoretical framework for interpreting the kinetics of α-helix formation has been developed by Schwarz and co-workers (139, 140), who use a formalism consistent with the Zimm-Bragg model for describing the equilibrium between helix and random coil. The theory predicts that an average relaxation time τ^* increases inside the transition zone and passes through a maximum at the midpoint of the transition. This prediction has been verified experimentally for different α-helix-forming polypeptides. The maximum τ^* for poly L-ornithine in an aqueous solvent (0.2 M NaCl: methanol, 85:15, v/v) is 2×10^{-8} sec (141). The formation of α-helices in proteins should occur some 10^8 times faster than the observed rates of protein folding. Thus α-helices may serve as nuclei in protein folding, but the nucleation of α-helical segments is too fast to be a rate-limiting process in protein folding. NMR studies of the α-helix ⇌ random coil transition frequently show separate resonances for helix and coil as if there were slow exchange between them on the NMR time scale (142). It is difficult to reconcile this view with the kinetic results. An alternative interpretation of the NMR results has been suggested: the two resonances are the result of polydispersity, the coil resonance being given by short polypeptides and the helix resonance by long ones (143, 144).

Formation of the β-structure of poly L-lysine follows nucleation-dependent kinetics. Although the rate is independent of concentration, indicating that nucleation is intramolecular, large aggregates are formed—about 100 molecules per aggregate (145). Rate enhancement by seeding has been observed. The half-times are in the range of minutes, suggesting that nucleation of β-structures could be slow enough to be a rate-limiting process in protein folding (145, 146).

Refolding of Oligomeric Proteins

When an oligomeric enzyme is dissociated into protomers, they usually are found to be inactive. The "quaternary constraint" imposed by association controls the conformation and function of the assembled protomers (147). When the refolding kinetics of oligomeric enzymes are studied (148–151), physical measurements of refolding often show rapid refolding (e.g. in seconds), while the recovery of enzymatic activity is a slow, first order process (in minutes or even hours). Addition of a specific ligand may catalyze the recovery of enzymatic activity, as in the addition of NAD to yeast glyceraldehyde 3-phosphate dehydrogenase (152). In some cases "maturation" by specific ligands may be an important biological control mechanism: the allosteric effectors L-isoleucine and L-valine, as well as the substrate L-threonine, catalyze the recovery of the active tetramer of threonine deaminase from *Salmonella typhimurium* (153). This subject, which appears to be more closely related to conformation and control of function in allosteric proteins than to the control of folding in small, monomeric proteins, has been reviewed by Paulus & Alpers (154).

FOLDING INTERMEDIATES AND MODELS FOR PROTEIN FOLDING

The significant conclusion at this time is that folding intermediates exist and can be characterized, and future results will give valuable information about the mechanism of folding. This comes from four independent lines of work: (*a*) the discovery of unfolding transitions that are not highly cooperative (45, 47), in which intermediates can be characterized by equilibrium methods; (*b*) the finding that NMR can sometimes detect folding intermediates at equilibrium even in highly cooperative transitions (61–64); (*c*) the finding that intermediates can be observed by fast kinetics in cooperative unfolding transitions for thermal or pH-induced unfolding (38, 83, 84, 87) and sometimes for refolding (37, 38, 88); and (*d*) the successful characterization of the single disulfide intermediates in the re-oxidation of BPTI (122).

A second conclusion is that stable intermediates exist at very early stages in folding or, to put it another way, that polypeptides that appear to be unfolded actually contain functional elements of structure. Earlier work (93) had shown the presence of some structure in heat-unfolded proteins, without indicating whether it is nonspecific or functional structure. Evidence for specific structure is provided by the discovery of conformational equilibria in unfolded fragments of nuclease that produce native antigenic determinants a small fraction of the time (135, 136), and by finding a conformational equilibrium between fast and slow refolding species in heat-unfolded RNase A (113). The nature of these primal structures will have to be established by future work; the current guess is that α-helices and β-pleated sheets play a prominent role. After completion of the first stage in its refolding from guanidine, the optical rotatory properties of penicillinase indicate that α-helical structure is present (47). The early formation of α-helices and β-pleated sheets is suggested for BPTI by finding that the four actual, out of the fifteen possible, single disulfide intermediates involve residues located in α-helices and in a β-pleated sheet (122).

ACKNOWLEDGMENTS

The people who have worked with me on the study of kinetic intermediates in protein folding have contributed much to the discussion presented here. I would like to mention in particular correspondence with Elliot Elson, Peter McPhie, and Tian Tsong, and discussions with Jean-Renaud Garel, Paul Hagerman, Robert Matthews, and Barry Nall. The writing of this review has been supported by research grants from the National Science Foundation (Grant GB 354, 32X) and from the National Institutes of Health (Grant GM AM 19, 983–13).

Literature Cited

1. Schulz, G. E., Barry, C. D., Friedman, J., Chou, P. Y., Fasman, G. D., Finkelstein, A. V., Lim, V. I., Ptitsyn, O. B., Kabat, E. A., Wu, T. T., Levitt, M., Robson, B., Nagano, K. 1974. *Nature* 250:140–42
2. Finkelstein, A. V., Ptitsyn, O. B. 1971. *J. Mol. Biol.* 62:613–24
3. Robson, B., Pain, R. H. 1971. *J. Mol. Biol.* 58:237–59
4. Lewis, P. N., Go, N., Go, M., Kotelchuck, D., Scheraga, H. A. 1970. *Proc. Nat. Acad. Sci. USA* 65:810–15
5. Chou, P. Y., Fasman, G. D. 1974. *Biochemistry* 13:211–21
6. Kotelchuck, D., Scheraga, H. A. 1968. *Proc. Nat. Acad. Sci. USA* 61:1163–70
7. Anfinsen, C. B. 1972. *Biochem. J.* 128:737–49
8. Anfinsen, C. B. 1973. *Science* 181:223–30
9. Taniuchi, H. 1972. *PAABS Symp.* 1:419–23
10. Schechter, A. N. 1972. *PAABS Symp.* 1:461–64
11. Hvidt, A., Nielsen, S. O. 1966. *Advan. Protein Chem.* 21:287–386
12. Klee, W. A. 1967. *Biochemistry* 12:3736–42
13. Burgess, A. W., Weinstein, L. I., Gabel, D., Scheraga, H. A. 1975. *Biochemistry* 14:197–200
13a. Burgess, A. W., Scheraga, H. A. 1975. *J. Theor. Biol.* In press
14. Tanford, C. 1968. *Advan. Protein Chem.* 23:121–282
15. Tanford, C. 1970. *Advan. Protein Chem.* 24:1–95
16. Brandts, J. F. 1969. *Structure and Stability of Biological Macromolecules,* ed. S. N. Timasheff, G. E. Fasman, 213–90. New York: Dekker, 694 pp.
17. Hermans, J. Jr., Lohr, D., Ferro, D. 1972. *Advan. Polym. Sci.* 9:230–83
18. Lumry, R., Biltonen, R. See Ref. 16, 65–212
19. Kauzmann, W. 1954. *The Mechanism of Enzyme Action,* ed. W. D. McElroy, B. Glass, 70–120. Baltimore: The Johns Hopkins Press. 819 pp.
20. Wetlaufer, D. B., Ristow, S. 1973. *Ann. Rev. Biochem.* 42:135–58
21. Ptitsyn, O. B., Lim, V. I., Finkelstein, A. V. 1972. *Fed. Eur. Biochem. Soc.* 25:421–29
22. Anfinsen, C. B., Scheraga, H. A. 1975. *Advan. Protein Chem.* In preparation
23. Pain, R. H. 1975. *Progr. Biophys.* In preparation

24. Kunitz, M. 1948. *J. Gen. Physiol.* 32:241–63
25. Eisenberg, M. A., Schwert, G. W. 1951. *J. Gen. Physiol.* 34:583–606
26. Harrington, W. F., Schellman, J. A. 1956. *C. R. Trav. Lab. Carlsberg* 30:21–43
27. Foss, J. G., Schellman, J. A. 1959. *J. Phys. Chem.* 63:2007–12
28. Pauling, L., Corey, R. B., Branson, H. R. 1951. *Proc. Nat. Acad. Sci. USA* 37:205–11
29. Yang, J. T., Doty, P. 1957. *J. Am. Chem. Soc.* 79:761–75
30. Schellman, J. A. 1958. *C. R. Trav. Lab. Carlsberg* 30:363–461
31. Northrop, J. H. 1930. *J. Gen. Physiol.* 13:739–66
32. Anson, M. L., Mirsky, A. E. 1934. *J. Gen. Physiol.* 17:393–98
33. Kunitz, M., Northrop, J. H. 1934. *J. Gen. Physiol.* 17:591–615
34. Kauzmann, W. 1959. *Symposium on Sulfur in Proteins,* ed. R. Benesch, 93–108. New York: Academic. 469 pp.
35. Sela, M., Lifson, S. 1959. *Biochim. Biophys. Acta* 36:471–78
36. Epstein, C. J., Goldberger, R. F., Anfinsen, C. B. 1963. *Cold Spring Harbor Symp. Quant. Biol.* 27:439–49
37. Epstein, H. F., Schechter, A. N., Chen, R. F., Anfinsen, C. B. 1971. *J. Mol. Biol.* 60:499–508
38. Shen, L. L., Hermans, J. Jr. 1972. *Biochemistry* 11:1836–41
39. Brandts, J. F. 1964. *J. Am. Chem. Soc.* 86:4291–4301
40. Lumry, R., Biltonen, R., Brandts, J. F. 1966. *Biopolymers* 4:917–44
41. Brandts, J. F. 1964. *J. Am. Chem. Soc.* 86:4302–14
42. Brandts, J. F., Hunt, L. 1967. *J. Am. Chem. Soc.* 89:4826–38
43. Ginsburg, A., Carroll, W. R. 1965. *Biochemistry* 4:2159–74
44. Anfinsen, C. B., Schechter, A. N., Taniuchi, H. 1972. *Cold Spring Harbor Symp. Quant. Biol.* 36:249–55
45. Wong, K.-P., Tanford, C. 1973. *J. Biol. Chem.* 248:8518–23
46. Yazgan, A., Henkens, R. W. 1972. *Biochemistry* 11:1314–18
47. Robson, B., Pain, R. H. 1973. *Conformation of Biological Molecules and Polymers,* ed. E. D. Bergmann, B. Pullman, 161–72. Jerusalem: Israel Acad. Sci. Hum.
48. Chignell, D. A., Azhir, A., Gratzer, W. B. 1972. *Eur. J. Biochem.* 26:37–42
49. Tsong, T. Y., Hearn, R. F., Wrathall, D. P., Sturtevant, J. M. 1970. *Biochemistry*

9:2666–77
50. Zimm, B. H., Bragg, J. K. 1958. *J. Chem. Phys.* 28:1246–47
51. Karasz, F. E., O'Reilly, J. M. 1966. *Biopolymers* 4:1015–23
52. Hayashi, Y., Teramoto, A., Kawahara, K., Fujita, H. 1969. *Biopolymers* 8:403–20
53. Danforth, R., Krakauer, H., Sturtevant, J. M. 1967. *Rev. Sci. Instrum.* 38:484–87
54. Jackson, W. M., Brandts, J. F. 1970. *Biochemistry* 9:2294–2301
55. Privalov, P. L., Khechinashvili, N. N., Atanasov, B. P. 1971. *Biopolymers* 10:1865–90
56. Khechinashvili, N. N., Privalov, P. L., Tiktopulo, E. I. 1973. *FEBS Lett.* 30:57–60
57. Privalov, P. L. 1974. *FEBS Lett.* 40:S140–53
58. Tiktopulo, E. I., Privalov, P. L. 1974. *Biophys. Chem.* 1:349–57
59. Bull, H. B., Breese, K. 1973. *Biopolymers* 12:2351–58
60. McDonald, C. C., Phillips, W. D., Glickson, J. D. 1969. *J. Am. Chem. Soc.* 93:235–46
61. Epstein, H. F., Schechter, A. N., Cohen, J. S. 1971. *Proc. Nat. Acad. Sci. USA* 68:2042–46
62. Westmoreland, D. G., Matthews, C. R. 1973. *Proc. Nat. Acad. Sci. USA* 70:914–18
63. Benz, F. W., Roberts, G. C. K. 1973. *FEBS Lett.* 29:263–66
64. Roberts, G. C. K., Benz, F. W. 1973. *Ann. NY Acad. Sci.* 222:130–48
65. Markley, J. L., Putter, I., Jardetzky, O. 1968. *Science* 161:1249–51
66. Matthews, C. R., Westmoreland, D. G. 1973. *Ann. NY Acad. Sci.* 222:240–54
67. Matthews, C. R., Westmoreland, D. G. 1975. *Biochemistry.* To be submitted
68. Jardetzky, O., Thielmann, H., Arata, Y., Markley, J. L., Williams, M. N. 1972. *Cold Spring Harbor Symp. Quant. Biol.* 36:257–61
69. Arata, Y., Khalifah, R., Jardetzky, O. 1973. *Ann. NY Acad. Sci.* 222:230–39
70. Hawley, S. A. 1971. *Biochemistry* 10:2436–42
71. Zipp, A., Kauzmann, W. 1973. *Biochemistry* 12:4217–28
72. Brandts, J. F., Oliveira, R. J., Westort, C. 1970. *Biochemistry* 9:1038–47
73. Hawley, S. A. 1973. *Biochim. Biophys. Acta* 317:236–39
74. Cerf, R. 1973. *Dynamic Aspects of Conformations, Changes in Biological Macromolecules,* ed. C. Sadron, 247–69. Dordrecht and Boston: D. Reidel. 519 pp.
75. Eigen, M., DeMaeyer, L. 1963. *Tech.*

Org. Chem. 8(2):895–1054
76. Elson, E. L. 1972. *Biopolymers* 11:1499–1520
77. Pohl, F. M. 1968. *Eur. J. Biochem.* 4:373–77
78. Pohl, F. M. 1968. *Eur. J. Biochem.* 7:146–52
79. Pohl, F. M. 1969. *FEBS Lett.* 3:60–64
80. Pohl, F. M. 1972. *Angew. Chem. Int. Ed.* 11:894–906
81. Scott, R. A., Scheraga, H. A. 1963. *J. Am. Chem. Soc.* 85:3866–73
82. Tanford, C., Pain, R. H., Otchin, N. S. 1966. *J. Mol. Biol.* 15:489–504
83. Tsong, T. Y., Baldwin, R. L., Elson, E. L. 1971. *Proc. Nat. Acad. Sci. USA* 68:2712–15
84. Tsong, T. Y., Baldwin, R. L. 1972. *J. Mol. Biol.* 69:145–48
85. Ikai, A., Fish, W. F., Tanford, C. 1973. *J. Mol. Biol.* 73:167–84
86. Tanford, C., Aune, K. C., Ikai, A. 1973. *J. Mol. Biol.* 73:185–97
87. Tsong, T. Y. 1973. *Biochemistry* 12:2209–14
88. Tsong, T. Y., Baldwin, R. L., Elson, E. L. 1972. *Proc. Nat. Acad. Sci. USA* 69:1809–12
89. Summers, M. R., McPhie, P. 1972. *Biochem. Biophys. Res. Commun.* 47:831–37
90. Salahuddin, A., Tanford, C. 1970. *Biochemistry* 9:1342–47
91. Barnard, E. A. 1964. *J. Mol. Biol.* 10:235–62
92. Bigelow, C. G. 1964. *J. Mol. Biol.* 8:696–701
93. Aune, K. C., Salahuddin, A., Zarlengo, M. H., Tanford, C. 1967. *J. Biol. Chem.* 242:4486–89
94. Tanford, C., Kawahara, K., Lapanje, S. 1967. *J. Am. Chem. Soc.* 89:729–49
95. Tsong, T. Y. 1974. *J. Biol. Chem.* 249:1988–90
96. Ikai, A., Tanford, C. 1971. *Nature* 230:100–2
97. Tanford, C. 1972. *Ciba Found. Symp.* 7:125–46
98. Ikai, A., Tanford, C. 1973. *J. Mol. Biol.* 73:145–64
99. Gibson, Q. H. 1964. *Rapid Mixing and Sampling Techniques in Biochemistry,* ed. B. Chance, Q. H. Gibson, R. H. Eisenhardt, K. K. Lonberg-Holm, 115–17. New York: Academic. 400 pp.
100. Hagerman, P. J., Baldwin, R. L. 1975. *Biochemistry.* To be submitted
101. Garel, J.-R., Baldwin, R. L. 1973. *Proc. Nat. Acad. Sci. USA* 70:3347–51
102. Grimaldi, J., Baldo, J., McMurray, C., Sykes, B. D. 1972. *J. Am. Chem. Soc.* 94:7641–45

103. Tsong, T. Y., Baldwin, R. L. 1972. *J. Mol. Biol.* 69:149–53
104. Ettinger, M. J., Hirs, C. H. W. 1968. *Biochemistry* 7:3374–80
105. Henkens, R. W., Turner, S. R. 1973. *Biochemistry* 12:1618–21
106. Ross, P. D., Sturtevant, J. M. 1960. *Proc. Nat. Acad. Sci. USA* 46:1360–65
107. Blake, R. D., Klotz, L. C., Fresco, J. R. 1968. *J. Am. Chem. Soc.* 90:3556–62
108. Saunders, M., Ross, P. D. 1960. *Biochem. Biophys. Res. Commun.* 3:314–18
109. Wetmur, J. G., Davidson, N. 1968. *J. Mol. Biol.* 31:349–70
110. Pörschke, D., Eigen, M. 1971. *J. Mol. Biol.* 62:361–81
111. Tsong, T. Y., Baldwin, R. L., McPhie, P. 1972. *J. Mol. Biol.* 63:453–69
112. Elson, E. L. 1972. *J. Mol. Biol.* 63:469–75
113. Garel, J.-R., Baldwin, R. L. 1975. *J. Mol. Biol.* In press
114. Hijazi, N. H., Laidler, K. J. 1972. *J. Chem. Soc.* 68:1235–42
115. Garel, J.-R., Baldwin, R. L. 1975. *J. Mol. Biol.* In press
116. Sela, M., White, F. H., Anfinsen, C. B. 1957. *Science* 125:691–92
117. Anfinsen, C. B., Haber, E., Sela, M., White, F. H. Jr. 1961. *Proc. Nat. Acad. Sci. USA* 47:1309–14
118. Hantgan, R. R., Hammes, G. G., Scheraga, H. A. 1974. *Biochemistry* 13:3421–31
119. Ristow, S., Wetlaufer, D. B. 1973. *Biochem. Biophys. Res. Commun.* 50:544–50
120. Creighton, T. E. 1974. *J. Mol. Biol.* 87:563–77
121. Creighton, T. E. 1974. *J. Mol. Biol.* 87:579–602
122. Creighton, T. E. 1974. *J. Mol. Biol.* 87:603–24
123. Givol, D., DeLorenzo, F., Goldberger, R. F., Anfinsen, C. B. 1965. *Proc. Nat. Acad. Sci. USA* 53:676–84
124. Steiner, D. F., Oyer, P. 1967. *Proc. Nat. Acad. Sci. USA* 57:473–80
125. Haber, E., Anfinsen, C. B. 1961. *J. Biol. Chem.* 236:422–24
126. Kato, I., Anfinsen, C. B. 1969. *J. Biol. Chem.* 244:1004–7
127. Taniuchi, H. 1970. *J. Biol. Chem.* 245:5459–68
128. Lin, M. C. 1970. *J. Biol. Chem.* 245:6726–31
129. Puett, D. 1972. *Biochemistry* 11:1980–90
130. Fisher, W. R., Taniuchi, H., Anfinsen, C. B. 1973. *J. Biol. Chem.* 248:3188–95
131. Wetlaufer, D., Kwok, D., Anderson, W. L., Johnson, E. R. 1974. *Biochem. Biophys. Res. Commun.* 56:380–85
132. Woodward, C. K., Rosenberg, A. 1971. *J. Biol. Chem.* 246:4105–13
133. Hermans, J. Jr., Lohr, D., Ferro, D. 1969. *Nature* 224:175–77
134. Glickson, J. D., Phillips, W. D., Rupley, J. A. 1971. *J. Am. Chem. Soc.* 93:4031–38
135. Sachs, D. H., Schechter, A. N., Eastlake, A., Anfinsen, C. B. 1972. *Proc. Nat. Acad. Sci. USA* 69:3790–94
136. Sachs, D. H., Schechter, A. N., Eastlake, A., Anfinsen, C. B. 1972. *Biochemistry* 11:4268–73
137. Eastlake, A., Sachs, D. H., Schechter, A. N., Anfinsen, C. B. 1974. *Biochemistry* 13:1567–71
138. Furie, B., Schechter, A. N., Sachs, D. H., Anfinsen, C. B. 1974. *Biochemistry* 13:1561–66
139. Schwarz, G. 1965. *J. Mol. Biol.* 11:64–77
140. Schwarz, G. 1968. *Biopolymers* 6:873–97
141. Hammes, G. G., Roberts, P. B. 1969. *J. Am. Chem. Soc.* 91:1812–16
142. Ferretti, J. A., Ninham, B. W., Parsegian, V. A. 1973. *Macromolecules* 3:34–42
143. Ullman, R. 1973. *Biopolymers* 9:471–87
144. Nagayama, K., Wada, A. 1973. *Biopolymers* 12:2443–58
145. Hartman, R., Schwaner, R. C., Hermans, J. Jr. *J. Mol. Biol.* In press
146. Snell, C. R., Fasman, G. D. 1973. *Biochemistry* 12:1017–25
147. Monod, J., Wyman, J., Changeux, J.-P. 1965. *J. Mol. Biol.* 12:88–118
148. Teipel, J. W., Koshland, D. E. Jr. 1971. *Biochemistry* 10:792–98
149. Ibid, 798–805
150. Teipel, J. W. 1972. *Biochemistry* 11:4100–7
151. Waley, S. G. 1973. *Biochem. J.* 135:165–72
152. Deal, W. C. 1969. *Biochemistry* 8:2795–2805
153. Hatfield, G. W., Burns, R. O. 1970. *Proc. Nat. Acad. Sci. USA* 66:75–76
154. Paulus, H., Alpers, J. B. 1971. *Enzymes* 12:385–401
155. VonHippel, P. H., Wong, K.-Y. 1964. *Science* 145:577–80

NOTE ADDED IN PROOF   Two recent papers on the kinetics of unfolding of human serum albumin were unfortunately overlooked when this review was written: Steinhardt, J., Stocker, N. 1973. *Biochemistry* 12:1789–97; 2798–2802.

METHODS FOR THE STUDY OF   ✳892
THE CONFORMATION OF SMALL
PEPTIDE HORMONES AND ANTIBIOTICS
IN SOLUTION

Lyman C. Craig,[1] *David Cowburn, and Hermann Bleich*[2]
The Rockefeller University, New York, New York 10021

CONTENTS

This review considers techniques for determining the conformation of peptides in solution. It concentrates on the uses and limitations of such techniques. Several recent reports and reviews have appeared in which the determined structures are extensively discussed (1, 2).

A peptide molecule in solution may have a relatively rigid structure in which only one or a very narrow spectrum of conformations is allowed or, if the constraints are weaker, a broad distribution averaged in time tending towards the theoretical "random coil."

The parameters that determine these constraints are considered to be the partial double bond character of the peptide bond, steric hindrance restricting allowed

---

[1] Dr. Craig died while this manuscript was in preparation.

[2] Present address: Department of Biochemistry, University of Connecticut Health Center, Farmington, Connecticut 06032.

values of backbone and side chain dihedral angles added to the normal torsional restrictions on such angles, coulombic interactions, hydrogen bonding, and hydrophobic and van der Waals interactions. Because some of these determining factors are of relatively low energy compared to Boltzmann fluctuation in normal solvents at room temperature, it is obvious that unless there are particularly strong reinforcing interactions there will always be a spectrum of time-dependent conformations, rather than a single one as would be determined by X-ray diffraction from a crystal.

Some of the methods used for determining conformation of peptides are: 1. hydrodynamic methods for the determination of Stokes radius, including free diffusion, gel filtration, thin film dialysis, and uses of the analytical ultracentrifuge (these methods rest on a relatively firm thermodynamic basis, and although a determination of Stokes radius alone provides no detail of conformation, it does considerably restrict the number of possible conformations and aids the interpretation of more detailed information available from other methods); 2. spectroscopic methods: circular dichroism, high resolution nuclear magnetic resonance (NMR), fluorescence and phosphorescence, and ultraviolet and infrared absorption; 3. hydrogen exchange rates determined by exchange with deuterium or tritium; 4. simple model building by hand or computer assisted, a method not to be ignored for its ability to exclude possible conformations; and 5. calculation of energies for a conformation, and energy minimization techniques. At this point it seems unlikely that any single approach can adequately define conformational states; a combination is most fruitful. Information from the above techniques may be increased by studies of conformation as a function of solvent, pH, and temperature, and by studies with carefully chosen synthetic analogs, which may allow variation of configuration in a single residue, a specific incorporation of isotope (for specific assignment of overlapping NMR signals), or as a "conformational" perturbation, either the addition of residue allowing increasing flexibility, e.g. glycine or D- or L-alanine, or restriction of the backbone conformation by substitution of proline or ·bulkily substituted residues like nor-leucine.

This review concentrates on those techniques that have been used by the authors and seem to them most useful. In addition to specialist reviews of the individual techniques, general reviews have appeared (3–6).

THIN FILM DIALYSIS

The details and theory of this technique have been previously summarized (7–10).

If certain precautions and care are exercised, the thin film dialysis method can reliably determine relative diffusional size when suitable models of known dimensions are available for comparison. In addition, it provides an excellent approach to the study of molecular interactions, self association, and binding (11, 12).

In order to correlate thin film data unambiguously with the overall shape of a solute, it must be demonstrated that the rate of dialysis reflects primarily diffusional activity. It has been proposed that the controlling factor in dialysis rate is the probability of a solute molecule entering a membrane pore (10). The solute must behave ideally, i.e. 1. it must not interact with the membrane by adsorption or electrostatic interaction, and 2. its chemical potential must be linearly related to analytical concentration, and there may be no diminution of chemical potential by self association and no change in chemical potential by other factors, such as gross flow under pressure or changes in electrical potential across the membrane associated with the movement of counterions (an equivalent of the Donnan membrane equilibrium).

Problems connected with membrane interaction have been largely eliminated.

Membrane Preparation

Peptides that are apparently "random coil-like" with high charge densities, such as ACTH, have stimulated research aimed at eliminating all residual fixed charge from "Visking" membranes. Treatment with glycine amide and water-soluble coupling agents has proven to be completely satisfactory in the preparation of inert membranes (13). The abolition of the effects of sodium chloride on diffusion of charged solutes through treated membranes has been considered to reflect the elimination of charged groups on the membrane. Derivatization similarly increased the inertness of Sephadex, and the inertness of these derivatized materials may recommend their use in certain preparative procedures.

Interactions of solutes with membranes can usually be detected by recovery (8). Control of pore size by stretching, zinc chloride treatment, and acetylation has produced membranes that have useful pore sizes for studying the rates of dialysis for molecules in the molecular weight range 100–135,000 (7, 14).

Potential Artifacts and Interpretation of Results

The effects of interconverting dialyzable monomers, concentration-dependent aggregation, and problems of technique connected with sampling frequency have been calculated by numerical simulation (15). The effect of salt on the dialysis rate of small synthetic peptides (16) and angiotensinamide II (17, 18; L. C. Craig and D. Cowburn, unpublished) shows that strongly charged basic solutes generally dialyze faster in low salt concentrations at neutral pH.

When measured in acetylated membranes, such changes in rate imply either that low salt concentration favors a more compact form of these peptides, or that there is some change in chemical potential difference across the membrane unrelated to conformation. The effect of changes in ionic strength on rates of free diffusion of charged species has been discussed previously (19). Essentially, the more freely dialyzing counterions of the charged peptide introduce a change in the electrochemical potential of the peptide, and hence a change in apparent dialysis rate. That such effects are significant at low ionic strength has been confirmed (L. C. Craig and D. Cowburn, unpublished). It is important to establish the strengths of ionic intramolecular interactions in model peptides (16) and to examine the possibility that biological potentiation of angiotensin and vasopressin (18, 20) by sodium chloride is conformationally mediated. For thin film dialysis studies, it is necessary to measure quantitatively the relative contributions of changes in conformation and changes in electrochemical potential from coupled ionic flow to the perturbation of dialysis rates.

The study by thin film dialysis of a number of analogs of oxytocin and vasopressin (9, 21) demonstrated a clear variation in molecular size, nearly all oxytocins studied being significantly smaller than vasopressins or vasotocin. Because it was considered that only small changes in size would result from conformational differences in the ring, it was originally suggested that the apparent size differences resulted from the folding of the tripeptide tail moiety into a more compact fashion in oxytocin. This suggestion has been confirmed by NMR studies, although the details of such folding are controversial (9, 22–24).

The use of dialysis for the measurement of binding is illustrated in the study of the interaction of a fluorescent probe with several peptides (12). Measurement of the escape rate provides a convenient and rapid method of estimating an apparent association constant, and it does not require that there be any modulation of fluorescence on binding as does conventional fluorimetric titration.

The possibility of intermolecular association of peptides must be considered when assigning spectroscopic effects, particularly at the high concentrations used in magnetic resonance spectroscopy. Linearity of escape rate in thin film dialysis provides a clear and simple demonstration of lack of association.

Noncovalent association of the tyrocidines, previously investigated by dialysis (25), has also been studied by ultracentrifugation (26, 27). A scheme involving two sizes of aggregate with variations in relative proportion and micelle number with concentration and temperature fits the data, but perhaps not exclusively.

FLUORESCENCE

The use of fluorescence probes was recently reviewed in this series (28). Application of 2-p-toluidinylnaphthalene-6-sulfonate to investigate binding properties to small antibiotics (12) demonstrated a relatively complex picture in a small peptide system. Whereas binding requires a positively charged molecule, fluorescent enhancement requires that there be incorporation of the probe into aggregates and presumably transfer into a less polar environment. Intrinsic phosphorescence of tryptophan in tyrocidines was also found to reflect aggregation state (29), with clearly differential components of both emission spectra and phosphorescence decay being associated with exposed residues in the monomer and with internal residues in the aggregate.

The utility of energy transfer measurements in small peptides is limited by the relative infrequency of naturally suitable peptides and the probability that synthetic substitution of relatively bulky aromatic groups represents a major perturbation. The occurrence of tyrosine and tryptophan in the same molecule and in relatively close sequential proximity in tyrocidines B and C allowed estimates for inter-residue distances (30). Absence of triplet transfer between tyrosinate and tryptophan in tyrocidine B and extensive triplet transfer in tyrocidine C indicate that D-tryptophan[7] and tyrosine[10] lie about 8 Å apart, consistent with the conformation derived from NMR studies (31).

The synthesis of a biologically active analog of ACTH, the N^ε-dansyllysine[21] 1–24 peptide is described (32) and the distance between dansyllysine[21] and tryptophan[9] estimated at about 23 Å. Such a distance seems compatible with previously reported intramolecular distances between aromatics (33–35) in this part of the molecule. Thus energy transfer measurements would seem a feasible conformational tool in the favorable cases of tyrosine and tryptophan, both being present and close within a molecule, or where substitution of bulky groups in a synthesis does not interfere with biological activity.

HYDROGEN EXCHANGE

Variation of the exchange rate of hydrogen from peptides to solvent contains many factors, only some of which are conformationally related. The simple observation of classes of fast and slow exchanging protons may perhaps, under suitable circumstances, be interpreted by equating slow exchange with steric restriction or presence of a hydrogen bond. The interpretation of intermediate exchange rates requires extensive experimental work. The chemistry of the exchange process has been reviewed in detail in this series (36); that review summarized experimental technique, and NMR methods are reviewed here later. For small peptides, backbone amide protons are the only class containing potential conformational information. The key to applying this method to the study of small peptides resides in the use of data for model compounds. The controlling factors for exchange rate are temperature, primary structure and/or substituent effects (37, 38), displacement of pH of observation from pH of minimum exchange, and conformational restriction. Choice of suitable conditions of temperature and pH is critical. For example, for $\partial \log k / \partial T \simeq 0.05$ and $\partial \log k / \partial \text{ pH} \simeq 1$ (where k is the apparent first order rate constant for exchange and pH is well away from the pH of minimum rate) there is a predicted and observed difference of about 10^5 in the rate of exchange of hydrogen bonded protons in gramicidin SA, between 20°C, pH 7.0 and 0°C, pH 3.0.

The primary structure substituent problem has been carefully approached by Molday et al (37, 38).

The proposed additivity rule for a first order neighbor effect is shown to effectively explain considerable variations in exchange of peptide amides normally regarded as "intermediate." The use of their data for analysis of peptides exchange requires that the experiment covers the range of the pH of minimum exchange rate. Then it is possible to distinguish between, on one hand, the separate modification of the acid- and base-catalyzed components of the rate (39) and, on the other, a conformational factor, which should slow the rate of exchange throughout the pH range, assuming a pH-independent conformation, which is clearly not always the case (40). For example, the pH profile for HD exchange of the C-terminal hexapeptide of angiotensin II shows two clearly anomalous exchanging protons assigned to His^6 (numbering as in angiotensin) and Val^5 (41). The first shows an anomaly of solely the base-catalyzed rate, reflecting either on the unsuitability of the model data to calculate rate profiles for this residue or on specific restrictions of imidazole in this analog of angiotensin II as suggested (42). On the other hand, the clear displacement of rate through the pH range covering the pH minimum of Val^5 argues for steric hindrance of exchange independent of exchange mechanism, and thus related to conformation. Note that this rate of exchange is still 10 times faster than those observed for protons in hydrogen bonds in gramicidin SA (43, 44). It is obviously misleading to refer to this peptide proton as potentially taking part in a hydrogen bond, the data merely implying the unavailability of the proton to normal solvent access.

Although very considerable precision may be obtained in determining exchange rate by isotope methods (36, 45, 46), the NMR methods of measuring exchange clearly contain greater detail when individual protons can be examined, even though absolute rates may be less precise. Greater chemical shift separation of amides from aromatic protons and from each other at frequencies higher than 220 MHz probably will allow greater use of this technique in larger peptides (47), and larger model compounds will permit more subtle discrimination of primary structure effects from conformational restriction of exchange. The identification of specific amide protons with restricted exchange properties will greatly aid the deduction of time-averaged conformational sets and permit a more discriminating analysis of the role of hydrogen bonding. A major difficulty may be encountered in the case of pH-dependent conformations, however.

OPTICAL ACTIVITY

Measurements of circular dichroic dispersion (CD) and, prior to current instrument development, optical rotatory dispersion have constituted a major area of research in synthetic polypeptides and in studies of globular and fibrous proteins in solution. This was largely due to the apparently successful correlation of spectral parameters with secondary structure, as has been extensively reviewed elsewhere (48–51). The use of such spectral analyses has many theoretical and practical pitfalls in large peptides and proteins. It is difficult to find correct models of a particular conformation (e.g. the α helix or the so-called random coil). The effect of short sections of a conformation having different mean residue optical activities from the same conformation in a model homopolypeptide is poorly understood. Large optical activities intrinsic to a side chain or generated from interactions of aromatic (or disulfide) chromophores with each other and with the peptide chromophore are observed and cannot easily be removed from the analysis. Some of these problems have been mitigated by the use of reference spectra deduced from the measured CD and the known mole fractions of α, β, and irregular regions in proteins whose structure has been deduced by X-ray crystallography (52, 53).

It has been demonstrated many times that the objections to spectral analyses of amide optical activity for small peptides are very great, particularly in the case of cyclic molecules (25, 54, 55). Therefore, approaches to the conformation of small peptides by optical activity measurements are more effectively made by either careful use of the method phenomenologically, assessing spectral variation as a function of perturbations such as temperature, pH, solvent, and sequence, or application of a more thorough theory of optical activity by, for example, comparison of calculated and observed spectra for a proposed model.

The effect of temperature on CD of angiotensin II in aqueous solution has been interpreted as a stabilization of secondary structure with increasing temperature

(56, 57). Changes in CD on transfer to fluorinated ketones interpreted as the introduction of new intramolecular interaction(s) possibly involving a cross β structure were also described. A similar stabilization of structure with increasing temperature has been proposed in bradykinin and an elegant method for the interpretation of solvent perturbation suggested (58, 59). Based on the study of model compounds, particularly of proline (60, 61), the difference CD on transfer from water to 90% dioxan may be interpreted as indicating the number of new intra-molecular hydrogen bonds. Thus the data on a number of analogs of bradykinin might indicate one *trans*-proline bond in the whole molecule in aqueous solution.

Total calculation of optical activity for a conformation has benefited from advances in formulation (62, 63). Using simple extensions of methods for dipeptides, Bayley (64) calculated complete peptide CD for gramicidin SA. In particular, it may be noted that the summation of nearest neighbor interactions is inadequate, and that the optical activities of prolyl residues are particularly sensitive to conformation [in accordance with (60, 61)]. The extension of such calculations to slightly larger peptides and to include other chromophores than the amide would be an interesting development.

Analysis of the CD of aromatic and disulfide chromophores in small peptides has become more feasible. Model studies of phenylalanyl, tyrosinyl, and trypto-phanyl derivative (65–68) and theoretical and model studies of cystine derivatives (69, 70) extending previous work (71–73) should allow more detailed spectral interpretation of the effects of perturbants. The optical activities or oxytocin analogs, of which the disulfide is partially or totally substituted by methylene groups (74), appear to resolve some details of CD band assignment in this system.

HIGH RESOLUTION NUCLEAR MAGNETIC RESONANCE

High resolution nuclear magnetic resonance (NMR) has the advantage over other spectroscopic techniques in that information concerning individual nuclei can be derived. A variety of information resides in the fine structure of the resonance lines, and several methods have been developed that provide some insight into the dynamic behavior of peptides in solution. The relatively low sensitivity of NMR limits most investigations to solutions that are above physiological concentrations. For this reason the higher intrinsic sensitivity of proton NMR has made it the most commonly used NMR technique, and will thus be emphasized.

The Assignment of Resonances

The assignment of resonances makes extensive use of the spectra that have been recorded and analyzed for amino acids and peptides (75, 76). All proton resonances are generally located 1 to 10 ppm downfield from the accepted standard tetra-methylsilane and fall into well separated regions: peptide amide resonances from 6.5 to 8.5 ppm, α-proton resonances from 4 to 5 ppm, etc, and the resonance patterns are to a useful extent characteristic for each amino acid. Whenever the resonances of adjacent protons have insufficiently differing chemical shifts, the fingerprints of resonances due to each residue have to be ascertained by spin decoupling. This task is made difficult by the lack of coupling between aromatic ring protons and nearby β protons, the overlap of resonances, and the reoccurrence of identical residues. In favorable cases the overlap of resonances can be overcome by changes in temperature and pH and by solvent perturbation (77). The increasing availability of Fourier transform and time-shared decoupling techniques will in the future also help in such cases, since overlapping resonances can sometimes be resolved through differences in spin-lattice relaxation rates.

Small changes in overlapping resonances that occur during spin decoupling have been elegantly amplified by the internuclear double resonance (INDOR) method (78). Other methods, which are particularly helpful in the case of identical residues, are the study of chemical analogs (41) and the incorporation of isotopes during synthesis (79–81). The use of difference spectroscopy has been suggested as a means to further magnify slight differences in NMR spectra that might be obtained by isotopic labeling or by chemical analogs (81, 82). The results of energy minimization calculations have been applied to the assignment of resonances (83), but this method might be criticized for its lack of experimental foundation.

The Assignment of Amide Resonances

Even with some prior knowledge of the conformation of peptides, prediction of the chemical shifts of amide resonances has not been possible. The study of closely related model peptides provides information of only a general nature, which has to be substantiated by independent methods (41, 84). The extension of spin decoupling to amide resonances has proven to be difficult due to their absence in most deuterated solvents and to the close proximity of the α-proton resonances to the overwhelming solvent peak of many nondeuterated solvents. Dry deuterated dimethylsulfoxide has been a preferred solvent for proton NMR studies because amides exchange only slowly in this solvent (43). By following the chemical shifts of amide resonances in gradual mixtures of nondeuterated solvents, the assignments determined by spin decoupling in deuterated dimethylsulfoxide can provide a basis for amide resonance assignments in other solvents (85, 86). This method, however, has to be limited to cases where no crossover or merger of amide resonances occurs during this solvent transition.

Isotopic labeling of the amide nitrogen or replacement of the α protons with deuterium is an unambiguous method for identifying amide resonances because characteristic spectral changes occur due to heteronuclear spin couplings. For assignment of amide resonances in aqueous solution, spin decoupling during hydrogen-deuterium exchange in deuterium oxide has been shown to be possible if care is taken to perform this experiment near the pH minimum for exchange catalysis and at low temperature (41). The technical difficulties of performing spin decoupling in nondeuterated solvents have recently been analyzed (87), and the routine application of spin decoupling techniques in such cases has been shown to be possible (86, 88).

Conformation and Coupling Constants

A detailed description of the solution conformation of a peptide should include a complete listing of all its dihedral angles (89). The coupling constants between adjacent protons determine the fine structure of the high resolution NMR spectrum, and a relationship between dihedral angle and coupling constants should therefore provide some measure of conformational information. However, in practice only the coupling constants between amide protons and α protons, J_{NH-CH}, and between α protons and β protons, $J_{\alpha\beta}$, are deduced from the proton spectrum. Furthermore, the semi-empirical relationships used in their interpretation are not unique functions of the dihedral angle, and rapidly interconverting rotamers (i.e. at a rate faster than the observed linewidths) would provide only coupling constants consistent with a rotationally averaged conformation. The analysis of the fine structure of proton NMR spectra has hence tended to be performed with specific conformational models in mind, whether based on energy minimization calculations or on independent experiments.

The measurement of coupling constants is straightforward in the case of well-resolved amide doublets. The analysis of resolved β-proton resonances is fairly simple when only three spins forming an ABX system have to be considered (90). For more complex spin systems and in the case of overlapping resonances, the INDOR method might be used (91), which provides not only the magnitude but also the sign of the coupling constants. The two constants for coupling between one α proton and two vicinal β protons, J_{AX} and J_{BX}, do not in general provide unique dihedral angles, as determined from the Karplus relationship of Abraham & McLauchlan (92). A set of rotationally averaged rotamers should bring the Karplus relationship and the measured coupling constants into agreement. Two approaches are possible: if various models based on energy minimization calculations are available, averages of pairs or triplets of conformational models might be considered; otherwise the preferred *gauche* and *trans* configurations are assumed (93) and averaged populations of rotamers are used to satisfy the observed coupling constants (94).

The Bystrow-Karplus relationship (95) between backbone dihedral angle and coupling of vicinal amide and α protons is acknowledged to be of limited accuracy (96). However, the fact that this relationship is a multivalued function of the dihedral angle poses the major problem for the interpretation of measured coupling constants. The observed correlation between the results of energy minimization calculations and the conformation of proteins as deduced by X-ray diffraction has therefore been coupled to the Bystrow-Karplus relationship (97) to provide some guidance when a choice among possible dihedral angles has to be made. Here again proposed conformations are generally the basis for spectral interpretation (98, 99).

Hydrogen Bonding

Several NMR methods are available for studying the accessibility of amide protons to the solvent. While they are unable to differentiate between amide protons that are either hydrogen bonded or simply unaccessible to the solvent, the information they provide is valuable because it restricts the choice among possible models and is often indicative of conformational stability. Four methods available to measure the exchange rate of amide protons with the solvent differ in the time scale on which this exchange occurs. For exchange times longer than 10 min the disappearance of amide resonances can be followed as a function of time after dissolving the peptide in a deuterated solvent (41, 43). On the time scale of 10 msec to 1 sec the exchange in protonated solvents affects both the lineshape and the spin-lattice relaxation time T_1 of the amide resonance, and transfer of saturation (88), lineshape analysis (77, 100), and T_1 measurements (101, 102) can be used. Any of these three methods alone provides only an indication of relative changes in exchange rates with changes in solvent composition or temperature. However, transfer of saturation coupled to lineshape analysis or to T_1 measurements allows an absolute determination of the exchange rate (102).

Since the above methods do not cover the whole time scale over which an amide can become accessible to the solvent, several methods depending on solvent perturbation have been proposed. The information they provide has to be used with caution since a perturbation of the solvent might also cause a perturbation of the conformation of the peptide. A correlation between slow exchange and a reduced temperature dependence of the chemical shift of amide resonances was an early observation (43) which has been applied successfully to the more rigid cyclic peptides (96, 103). In more flexible peptides (104) and in some linear peptides in aqueous solution (41), such a correlation has been found less reliable. Perturbation of dimethylsulfoxide solutions by trifluoroethanol (104–106) has the same limitations as perturbation by changes in temperature. The use of the line broadening produced by paramagnetic probes (107) might be a more rigorous approach to solvent perturbation because binding of the probe to the peptide can be assessed by titration of the probe, and the conformational perturbation assessed.

Our knowledge of the interaction of peptides with their solvent environment is still limited, and the interpretation of data resulting from the use of the above methods is often ambiguous. Catalysis of the exchange by the solvent has been studied in aqueous solutions (108) for model peptides which were assumed to be in a random chain conformation. This study and the investigation of the inductive effects of neighboring side chains (38) have provided the basis of interpretation for some exchange studies in aqueous solution by nuclear magnetic resonance (40, 41, 109).

Technical Developments

The emphasis on proton NMR has been due not only to its intrinsic higher sensitivity but also to the wealth of information residing in spin-spin couplings. Carbon-carbon coupling constants have recently been reported for carbon-13 enriched amino acids (110), and high resolution NMR of peptides has also been performed with the isotopes of deuterium (111), fluorine-19 (112), and nitrogen-15 (113, 114). The use of carbon-13 NMR in the study of peptides has become routinely possible by the application of the pulsed Fourier transform method (115–117). The implications of larger chemical shifts of carbon-13 resonances have been recognized (118) and some model studies of amino acids and peptides (119–121) have laid the foundations for later applications. Among these, the studies of gramicidin SA (81, 118), angiotensin II (122), and oxytocin (123–125) are noteworthy.

Another application of pulsed Fourier transform NMR has been the measurement of spin-lattice relaxation times both of protons (126, 127) and of carbon-13 nuclei (23, 128–130). These data promise to furnish information on segmental motion of peptides in solution, but their interpretation is presently ambiguous because the detailed features of the spin-lattice relaxation process are still not well understood. The cases of a rigid ellipsoid undergoing Brownian motion (131) and of methyl protons attached to such a body (132) have been treated theoretically and used in experimental interpretation. The conformational stability of gramicidin S has made it the subject of the best model study to date (129). Effects of the strength of the magnetic field (126) and of paramagnetic impurities (127) have been investigated experimentally.

In pulsed Fourier transform all resonances are recorded simultaneously and an overwhelming solvent peak often drowns all features of interest, due to the limited dynamic range of most NMR spectrometers. The application of long, weak radio-frequency pulses has made possible the use of pulsed Fourier transform proton NMR in nondeuterated aqueous solution (133). Another approach to this problem has been a method that is called either rapid scan Fourier transform or correlation spectroscopy (134–136). This method makes use of computer programs similar to those used in Fourier transform NMR but does not require the high radiofrequency power normally used in pulsed NMR.

CONCLUSION

The substantial description of the conformational properties of a small peptide requires a multifaceted approach, continuously bearing in mind the possibility of existence of conformational averaging. Uses of nuclear resonance have greatly extended the detail that may be ascertained, but other techniques provide very necessary complementary information. It may be assumed that extension of detailed studies up to the level of insulin is feasible.

As more details of conformation emerge, studies of structure/activity relationships will move from the primary structural level; sequence substitutions will be directed at alterations of either in-solution or on-receptor conformation. It is possible to look forward to a most interesting testing of hypotheses of hormone action (137, 138).

ACKNOWLEDGMENTS

Partial support from NIH Grants AM-02493, RR-00639, and the University of Connecticut Research Foundation is gratefully acknowledged. We are indebted to R. Galardy for discussion on hydrogen exchange and to Ms. Goggin for secretarial assistance.

Literature Cited

1. Hanson, H., Jakubke, H.-D. 1973. *Peptides 1972. Proc. 12th Eur. Peptide Symp., Reinjardbrunn Castle, GDR, 1972.* Amsterdam: North-Holland

2. Law, H. D. 1972. See Ref. 3, 384–447
3. Sheppard, R. C., Ed. 1972. *Amino-acids, Peptides, and Proteins, 5. A Specialist Periodical Report.* London: Chem. Soc.

4. Meinhofer, J., Ed. 1972. *Chemistry and Biology of Peptides.* Ann Arbor, Mich.: Ann Arbor Sci.
5. Chignell, C. F., Ed. 1972. *Methods Pharmacol.* Vol. 2
6. Ovchinnikov, Yu. A. 1972. See Ref. 1, 3–37
7. Craig, L. C. 1967. *Methods in Enzymol.* 11:870–905
8. Craig, L. C. 1970. *Membrane Science and Technology,* ed. J. E. Flinn, 1–15. New York: Plenum
9. Craig, L. C., Chen, H. C., Gibbons, W. A. 1973. *Advan. Chem.* 125:286–97
10. Craig, L. C., Chen, H. C. 1972. *Proc. Nat. Acad. Sci. USA* 69:702–5
11. Chen, H. C., Craig, L. C. 1971. *Bioorg. Chem.* 1:51–65
12. Beyer, C. F., Craig, L. C., Gibbons, W. A. 1972. *Biochemistry* 11:4920–26
13. Chen, H. C., Craig, L. C., Stoner, E. 1972. *Biochemistry* 11:3559–64
14. Stewart, K. K., Craig, L. C. 1970. *Anal. Chem.* 42:1257–60
15. Stewart, K. K., Craig, L. C., Williams, R. C. 1970. *Anal. Chem.* 42:1252–56
16. Harris, M. J., Craig, L. C. 1974. *Biochemistry* 13:1510–15
17. Ferreira, A. T., Hampe, O. G., Paiva, A. C. M. 1969. *Biochemistry* 8:3483–87
18. Walter, R., Schaechtelin, G., Craig, L. C. 1974. *Experimentia* 30:306–7
19. Gosting, L. J. 1956. *Advan. Protein Chem.* 11:429–554
20. Schaechtelin, G., Walter, R., Salomon, H., Jelínek, J., Karen, P., Cort, J. H. 1974. *Mol. Pharmacol.* 10:57–67
21. Craig, L. C., Harfenist, E. J., Paladini, A. C. 1964. *Biochemistry* 3:764–69
22. Walter, R., Ballardin, A., Schwartz, I. L., Gibbons, W. A., Wyssbrod, H. R. 1975. *Proc. Nat. Acad. Sci. USA* 71:4528–33
23. Deslauriers, R., Smith, I. C. P., Walter, R. 1974. *J. Am. Chem. Soc.* 96:2289–91
24. Honig, B., Kabat, E. A., Katz, L., Levinthal, C., Wu, T. T. 1973. *J. Mol. Biol.* 80:277–95
25. Ruttenberg, M. A., King, T. P., Craig, L. C. 1966. *Biochemistry* 5:2857–64
26. Laiken, S. L., Printz, M. P., Craig, L. C. 1971. *Biochem. Biophys. Res. Commun.* 43:595–600
27. Williams, R. C., Yphantis, D. A., Craig, L. C. 1972. *Biochemistry* 11:70–77
28. Brand, L., Gohlke, J. R. 1972. *Ann. Rev. Biochem.* 41:843–68
29. Beyer, C. F., Craig, L. C., Gibbons, W. A., Longworth, J. W. 1975. *Proc. Symp. Excited States of Biological Molecules (Lisbon).* New York: Wiley
30. Beyer, C. F., Gibbons, W. A., Craig, L. C.,

Longworth, J. W. 1974. *J. Biol. Chem.* 249:3204–11
31. Wyssbrod, H., Fein, M., Balaram, P., Bothner-By, A. A., Sogn, J. A., Ziegler, P., Gibbons, W. A. 1975. *J. Biol. Chem.* In press
32. Schiller, P. W. 1972. *Proc. Nat. Acad. Sci. USA* 69:975–79
33. Eisinger, J. 1969. *Biochemistry* 8:3902–8
34. Eisinger, J., Fener, B., Lamola, A. A. 1969. *Biochemistry* 8:3908–15
35. Edelhoch, H., Lippoldt, R. E. 1969. *J. Biol. Chem.* 244:3876–3900
36. Englander, S. W., Downer, N. W., Teitelbaum, H. 1972. *Ann. Rev. Biochem.* 41:903–24
37. Molday, R. S., Kallen, R. G. 1972. *J. Am. Chem. Soc.* 94:6738–45
38. Molday, R. S., Englander, S. W., Kallen, R. G. 1972. *Biochemistry* 11:150–58
39. Galardy, R. E., Craig, L. C., Printz, M. P. 1974. *Biochemistry* 13:1674–77
40. Llinas, M., Klein, M. P., Neilands, J. B. 1973. *J. Biol. Chem.* 248:915–23
41. Bleich, H. E., Galardy, R. E., Printz, M. P., Craig, L. C. 1973. *Biochemistry* 12:4950–57
42. Weinkam, R. J., Jorgensen, E. C. 1971. *J. Am. Chem. Soc.* 93:7038–44
43. Stern, A., Gibbons, W. A., Craig, L. C. 1968. *Proc. Nat. Acad. Sci. USA* 61:734–41
44. Laiken, S. L., Printz, M. P., Craig, L. C. 1969. *Biochemistry* 8:519–26
45. Printz, M. P., Williams, H. P., Craig, L. C. 1972. *Proc. Nat. Acad. Sci. USA* 69:378–82
46. Laiken, S. L., Printz, M. P. 1970. *Biochemistry* 9:1547–53
47. Masson, A., Wüthrich, K. 1973. *FEBS Lett.* 31:114–18
48. Gratzer, W. B., Cowburn, D. A. 1969. *Nature* 222:426–31
49. Lotan, N., Berger, A., Katchalski, E. 1972. *Ann. Rev. Biochem.* 41:869–902
50. Bayley, P. M. 1972. See Ref. 3, 237–57
51. Madison, V., Schellman, J. 1972. *Biopolymers* 11:1041–76
52. Chen, Y. H., Yang, J. T., Martinez, H. M. 1972. *Biochemistry* 11:4120–4200
53. Saxena, V. P., Wetlaufer, D. B. 1971. *Proc. Nat. Acad. Sci. USA* 68:969–72
54. Urry, D. W., Ruiter, A. 1970. *Biochem. Biophys. Res. Commun.* 38:800–6
55. Laiken, S. L., Printz, M. P., Craig, L. C. 1969. *J. Biol. Chem.* 244:4454–56
56. Fermandjian, S., Morgat, J.-L., Fromageot, P. 1971. *Eur. J. Biochem.* 24:252–58
57. Fermandjian, S., Fromageot, P. 1973. See Ref. 1, 303–10

58. Cann, J. R., Stewart, J. M., Matsueda, G. R. 1973. *Biochemistry* 12:3780–88
59. Cann, J. R. 1972. *Biochemistry* 11:2654–59
60. Madison, V. S. 1969. *The optical activity of diamides, peptides & proteins.* PhD thesis. Univ. of Oregon. 70-2523, University Microfilms Inc., Ann Arbor, Mich.
61. Madison, V., Schellman, J. 1970. *Biopolymers* 9:511–67
62. Bayley, P. M., Nielsen, E. B., Schellman, J. A. 1969. *J. Phys. Chem.* 73:228–43
63. Johnson, W. C. Jr., Tinoco, I. Jr. 1969. *Biopolymers* 7:727–49
64. Bayley, P. M. 1971. *Biochem. J.* 125:90P–91P
65. Andrews, L. J., Forster, L. S. 1972. *Biochemistry* 11:1875–79
66. Horwitz, J., Strickland, E. H., Billups, C. 1970. *J. Am. Chem. Soc.* 92:2119–29
67. Horwitz, J., Strickland, E. H. 1971. *J. Biol. Chem.* 246:3749–52
68. Strickland, E. H., Wilchek, M., Horwitz, J., Billups, C. 1972. *J. Biol. Chem.* 247:572–80
69. Casey, J. P., Martin, R. B. 1972. *J. Am. Chem. Soc.* 94:6141–51
70. Strickland, R. W., Webb, J., Richardson, F. S. 1974. *Biopolymers* 13:1269–90
71. Beychok, S., Breslow, E. 1968. *J. Biol. Chem.* 243:151–56
72. Kahn, P., Beychok, S. 1968. *J. Am. Chem. Soc.* 90:4168–70
73. Imanishi, A., Isemura, T. 1969. *J. Biochem.* 65:309–12
74. Fric, I., Kodíček, M., Jost, V., Bláha, K., 1972. See Ref. 1, 318–25
75. McDonald, C. C., Phillips, W. D. 1969. *J. Am. Chem. Soc.* 91:1513–21
76. Roberts, G. C. K., Jardetzky, O. 1970. *Advan. Protein Chem.* 25:447–45
77. Glickson, J. D., Cunningham, W. D., Marshall, G. R. 1972. See Ref. 4, 563–69
78. Gibbons, W. A., Alms, H., Bockman, R. S., Wyssbrod, H. R. 1972. *Biochemistry* 11:1721–25
79. Brewster, A. I. R., Hruby, V. J. 1973. *Proc. Nat. Acad. Sci. USA* 70:3806–9
80. Bradbury, A. F., Burgen, A. S. V., Feeney, J., Roberts, G. C. K., Smyth, D. G. 1974. *FEBS Lett.* 42:179–82
81. Sogn, J. A., Craig, L. C., Gibbons, W. A. 1974. *J. Am. Chem. Soc.* 96:3306–9
82. Patel, D. J. 1971. *Macromolecules* 4:251–54
83. Patel, D. J., Tonelli, A. E., Pfaender, P., Faulstich, H., Wieland, Th. 1973. *J. Mol. Biol.* 79:185–96
84. Feeney, J., Roberts, G. C. K., Rockey, J. H., Burgen, A. S. V. 1971. *Nature New Biol.* 232:108–10
85. Glickson, J. D., Urry, D. W., Walter, R. 1972. *Proc. Nat. Acad. Sci. USA* 69:2566–69
86. Von Dreele, P. H., Brewster, A. I., Dadok, J., Scheraga, H. A., Bovey, F. A., Ferger, M. F., du Vigneaud, V. 1972. *Proc. Nat. Acad. Sci. USA* 69:2169–73
87. Dadok, J., Von Dreele, P. H., Scheraga, H. A. 1972. *Chem. Commun.* 1055–56
88. Glickson, J. D., Dadok, J., Marshall, G. R. 1974. *Biochemistry* 13:11–14
89. IUPAC-IUB Commission on Biochemical Nomenclature. 1970. *J. Mol. Biol.* 52:1–17
90. Pople, J. A., Schneider, W. G., Bernstein, J. 1959. *High Resolution Nuclear Magnetic Resonance,* 132–38. New York: McGraw-Hill
91. Gibbons, W. A., Alms, H., Sogn, J., Wyssbrod, H. R. 1972. *Proc. Nat. Acad. Sci. USA* 69:1261–65
92. Abraham, R. J., McLauchlan, K. A. 1962. *Mol. Phys.* 5:513–23
93. Pachler, K. G. R. 1964. *Spectrochim. Acta* 20:581–87
94. Feeney, J., Roberts, G. C. K., Brown, J. P., Burgen, A. S. V., Gregory, H. 1972. *J. Chem. Soc. Perkin Trans.* 2 601–4
95. Bystrow, V. F., Portnova, S. L., Tsetlin, V. I., Ivanov, V. T., Ovchinnikov, Yu. A. 1969. *Tetrahedron* 25:493–515
96. Urry, D. W., Ohnishi, M. 1970. In *Spectroscopic Approaches to Biomolecular Conformation,* ed. D. W. Urry, 263–300. Chicago: Am. Med. Assoc.
97. Gibbons, W. A., Nemethy, G., Stern, A., Craig, L. C. 1970. *Proc. Nat. Acad. Sci. USA* 67:239–46
98. Patel, D. J. 1973. *Biochemistry* 12:677–88
99. Patel, D. J., Tonelli, A. E. 1974. *Biochemistry* 13:788–92
100. Torchia, D. A., Wong, S. C. K., Deber, C. M., Blout, E. R. 1972. *J. Am. Chem. Soc.* 94:616–20
101. Redfield, A. G., Gupta, R. K. 1971. *Advan. Magn. Resonance* 5:81–116
102. Bockman, R. S. 1971. *The conformation of bacitracin,* PhD thesis. The Rockefeller Univ., New York
103. Bovey, F. A., Brewster, A. I., Patel, D. J., Tonelli, A. E., Torchia, D. A. 1972. *Accounts Chem. Res.* 5:193–200
104. Pitner, T. P., Urry, D. W. 1972. *Biochemistry* 11:4132–37
105. Pitner, T. P., Urry, D. W. 1972. *J. Am. Chem. Soc.* 94:1399–1400
106. Walter, R., Glickson, J. D. 1973. *Proc. Nat. Acad. Sci. USA* 70:1199–

1203

107. Kopple, K. D., Schamper, T. J. 1972. *J. Am. Chem. Soc.* 94:3644–46

108. Englander, S. W., Poulsen, A. 1969. *Biopolymers* 7:379–93

109. Llinas, M., Klein, M. P., Neilands, J. B. 1973. *J. Biol. Chem.* 248:924–31

110. Sogn, J. A., Craig, L. C., Gibbons, W. A. 1974. *J. Am. Chem. Soc.* 96:4694–96

111. Glasel, J. A., Hruby, V. J., McKelvy, J. F., Spatola, A. F. 1973. *J. Mol. Biol.* 79:555–75

112. Vine, W. H., Brueckner, D. A., Needleman, P., Marshall, G. R. 1973. *Biochemistry* 12:1630–37

113. Sogn, J. A., Gibbons, W. A., Randall, E. W. 1973. *Biochemistry* 12:2100–5

114. Mehlis, B., Muller, H. G., Hintsche, R., Niedrich, H. 1972. See Ref. 1, 328–32

115. Farrar, T. C., Becker, E. D. 1971. *Pulse and Fourier Transform NMR.* New York: Academic

116. Becker, E. D., Farrar, T. C. 1972. *Science* 178:361–68

117. Anet, F. A. L., Levy, G. C. 1973. *Science* 180:141–48

118. Gibbons, W. A., Sogn, J. A., Stern, A., Craig, L. C., Johnson, L. F. 1970. *Nature* 227:840–42

119. Horsley, W. J., Sternlicht, H. 1968. *J. Am. Chem. Soc.* 90:3738–48

120. Horsley, W., Sternlicht, H., Cohen, J. S. 1970. *J. Am. Chem. Soc.* 92:680–86

121. Christl, M., Roberts, J. D. 1972. *J. Am. Chem. Soc.* 94:4565–73

122. Zimmer, S., Haar, W., Maurer, W., Rueterjans, H., Fermandjian, S.,

123. Fromageot, P. 1972. *Eur. J. Biochem.* 29:80–87

123. Deslauriers, R., Walter, R., Smith, I. C. P. 1972. *Biochem. Biophys. Res. Commun.* 48:854–59

124. Brewster, A. I. R., Hruby, V. J., Spatola, A. F., Bovey, F. A. 1973. *Biochemistry* 12:1643–49

125. Deslauriers, R., Walter, R., Smith, I. C. P. 1974. *Proc. Nat. Acad. Sci. USA* 71:265–68

126. Coates, H. B., McLauchlan, K. A., Campbell, I. D., McColl, C. E. 1973. *Biochim. Biophys. Acta* 310:1–10

127. Wasylishen, R. E., Cohen, J. S. 1974. *Nature* 249:847–50

128. Allerhand, A., Oldfield, E. 1973. *Biochemistry* 12:3428–33

129. Allerhand, A., Komoroski, R. A. 1973. *J. Am. Chem. Soc.* 95:8228–31

130. Deslauriers, R., Walter, R., Smith, I. C. P. 1973. *FEBS Lett.* 37:27–32

131. Woessner, D. E. 1962. *J. Chem. Phys.* 37:647–54

132. Hubbard, P. S. 1970. *J. Chem. Phys.* 52:563–68

133. Redfield, A. G., Gupta, R. K. 1971. *J. Chem. Phys.* 54:1418–19

134. Petersson, G. A. 1970. *Filtering NMR spectra,* PhD thesis. Calif. Inst. Tech., Pasadena

135. Dadok, J., Sprecher, R. F. 1974. *J. Magn. Resonance* 13:243–48

136. Gupta, R. K., Ferretti, J. A., Becker, E. D. 1974. *J. Magn. Resonance* 13:275–90

137. Printz, M. P., Némethy, G., Bleich, H. 1972. *Nature New Biol.* 237:135–40

138. Schwyzer, R. 1972. See Ref. 1, 424–36

ROLE OF CYCLIC NUCLEOTIDES IN GROWTH CONTROL[1]

Ira H. Pastan, George S. Johnson, and Wayne B. Anderson
Laboratory of Molecular Biology, National Cancer Institute,
National Institutes of Health, Bethesda, Maryland 20014

CONTENTS

[1] Abbreviations used: cyclic AMP, adenosine 3′, 5′-monophosphate; Bt$_2$cAMP, N^6-2′-O-dibutyryl adenosine 3′, 5′-monophosphate; PGE$_1$, prostaglandin E$_1$; cyclic GMP, guanosine 3′, 5′-monophosphate; cyclic CMP, cytidine 3′, 5′-monophosphate; PHA, phytohemagglutinin; Con A, concanavalin A; and FGF, fibroblast growth factor.

491

INTRODUCTION

How is cellular growth regulated? The answer to this question has been at the center of biological investigation since its inception. The development of animal cell culture systems has enabled investigators to study this question under conditions where many of the variables that can not be controlled in intact animals do not operate. Two types of cells have been widely used for such studies. One type is the cultured fibroblastic cell. Normal and transformed fibroblastic cells have been prepared from a variety of animals. The second cell type falls in the general category of lymphocyte. Again studies have been done on lymphocytes from different animals and tissue sources. This review focuses on the role of cyclic nucleotides in controlling the growth of these two cell types, but information on other types of cells is also included. The role of cyclic AMP[1] in controlling growth as well as other aspects of the behavior of cultured cells has been reviewed elsewhere (1). This review will concentrate on information not covered previously.

FIBROBLASTIC CELLS

Effect of Cyclic AMP Treatment

Fibroblastic cells prepared from the embryos of mice, rats, chickens, and man will grow readily in culture. Normal untransformed cells will grow only when attached to a supporting glass or plastic surface and require a variety of undefined growth factors contained in serum. Generally such cells grow logarithmically until they crowd together; then growth markedly slows or stops. This phenomenon has a variety of names, each implying a different mechanism of growth control. For purposes of this review the term "density-dependent inhibition of growth" (2) will be used. A permanent cell line which shows particularly striking growth control has been isolated by Todaro et al (3) from mouse embryo cells. This cell line, 3T3, has been widely used in studies on growth control.

Transformed cells and cancer cells maintained in culture differ from normal cells in their growth requirements and behavior. Generally transformed and cancer cells require less serum for growth, grow rapidly, and have lost density-dependent inhibition of growth.

The first clue that cyclic AMP might have a role in controlling growth of cultured cells came from two studies. Burk observed that two drugs, caffeine and theophylline, slowed the growth of normal and transformed baby hamster kidney (BHK) cells (4). Caffeine and theophylline were used because they inhibit the activity of cyclic AMP phosphodiesterases and raise cyclic AMP levels in cells. About the same time, Ryan & Heidrick reported that cyclic AMP itself inhibited the growth of HeLa cells (5). Because caffeine and theophylline have other effects on cells besides raising cyclic AMP levels, and because cyclic AMP is rapidly converted to 5' AMP, a compound also known to slow cellular growth, it was necessary to perform a variety of other experiments to establish the physiological role of cyclic AMP as an inhibitor of growth. The following observations support such a role.

Cyclic AMP analogs known to mimic cyclic AMP's action on well-defined systems were found to also slow cellular growth. The most prominent of these is Bt_2cAMP (reviewed in 1). Prostaglandin E_1, an agent that activates adenylate cyclase and raises cyclic AMP levels in fibroblastic cells, also slowed cellular growth (6). The identification of a cell line in which PGE_1 failed to activate adenylate cyclase and raise cyclic AMP levels permitted an important control experiment to be performed. In this cell line PGE_1 did not inhibit growth (7). In addition, in PGE_1 responsive cells, prostaglandin B_1, a structurally similar compound which cannot elevate cyclic AMP levels, had no effect on the rate of cell growth (6).

Analogs of cyclic AMP are effective at much lower concentrations than cyclic AMP itself. The reasons for the relative effectiveness of cyclic AMP are not clear, but may in part be due to more rapid degradation of cyclic AMP in the growth medium, poorer penetration of cyclic AMP into the cell, and rapid degradation of cyclic AMP inside the cell. It also is not clear whether the inhibition of growth by added cyclic AMP represents a specific growth regulating function. Cyclic AMP, 5'AMP, and adenosine inhibit growth of cells apparently by a depletion of pyrimidine precursors for DNA and RNA synthesis (8, 9). This depletion is not observed in growth inhibition by Bt_2cAMP (9).

By now the addition of analogs of cyclic AMP, cyclic AMP itself, or agents that raise the intracellular levels of cyclic AMP has been observed to decrease the growth rate of numerous cell lines of fibroblastic origin. Included in this list are: mouse L cells (5, 6, 10–16), transformed mouse 3T3 cells (11, 17–21), BHK and transformed BHK cells (4, 11, 22, 23), Nil cells (24), embryonic rat cells (25, 26), Rous sarcoma virus-transformed mouse, rat, and hamster cells (12, 13, 27a), normal rat kidney (NRK) and transformed NRK cells (27), human diploid fibroblasts (28), human rhabdomyosarcoma (29), and Chinese hamster ovary cells (30, 31).

Two bacterial agents that inhibit DNA synthesis or growth appear to do so by an effect on cyclic AMP mechanism. One of these is cholera toxin, which activates adenylate cyclase (32). The other class consists of glycolipids from *Salmonella minnesota R* mutants. These glycolipids bind to the plasma membranes of normal and transformed rat embryo cells and inhibit their growth. Accompanying this inhibition of growth is an elevation of cyclic AMP (26).

Cyclic AMP Levels in Logarithmic Growth

All in all, these various experiments strongly suggest that cyclic AMP has a physiological role in regulating the rate of cell division. This suggestion has been verified by direct measurement of the cyclic AMP levels in different cell lines dividing at different rates under constant conditions (10, 33, 34), in chick embryo cells dividing at different rates in varying concentrations of serum (35), and in BHK cells stimulated to grow more rapidly by the addition of insulin to their growth medium (22).

Cyclic AMP Levels in Relation to Cessation of Growth in Confluent Cells

In our own laboratory, we have observed that cyclic AMP levels rise as normal cells cease growth at confluency. In these studies we have used 3T3 cells (33, 36), human diploid fibroblasts, MA308 (36) and WI38 (37), and NRK cells (27). Transformed cells that do not cease growth do not elevate their cyclic AMP levels (33). NRK cells transformed with a temperature-sensitive Kirsten sarcoma virus show density-dependent inhibition of growth and also increase their cyclic AMP levels at 39°C, the nonpermissive temperature, but not at 32°C, where they behave like transformed cells in both properties (27).

Sheppard at first did not find a rise in cyclic AMP levels at confluency in 3T3 cells (34), but in a later study he observed a rise in cyclic AMP levels (38). On the other hand, both Oey et al (39) and Burstein et al (40) have been unable to show a rise in cyclic AMP levels in confluent 3T3 cells. Oey et al (39) did observe a rise in cyclic AMP with serum starvation and suggested that the reason others observed a rise in cyclic AMP was due to inadvertent serum starvation. While it has been reported that serum starvation will raise cyclic AMP levels in cells (41, 42), serum deprivation does not explain the difference between the results of Oey et al (39) and those in our own laboratory. We believe there are unsolved technical problems in measuring cyclic AMP levels in cultured cells. These problems have been discussed elsewhere (1).

Cyclic Nucleotide Levels and Initiation of DNA Synthesis

One of the early events following addition of agents that stimulate quiescent cells to make DNA and divide is a rapid fall in cyclic AMP levels. This is observed following serum or insulin addition to quiescent mouse 3T3 cells (34, 36, 39, 42–44) or human diploid fibroblasts (28). Also treatment with some proteases rapidly lowers cyclic AMP levels and induces cell division (45). Abortive transformation by certain transformation viruses also induces DNA synthesis; this induction is preceded by a fall in cyclic AMP (43, 46). On the other hand, using chick embryo cells, Humphreys (35) and Hovi et al (47) did not observe a decrease in cyclic AMP levels with serum stimulation.

A fall in cyclic AMP is apparently important for the initiation of growth because growth stimulation is prevented if cyclic AMP is maintained at high levels following addition of growth-promoting agents. Bt$_2$cAMP will prevent this serum-induced DNA synthesis in quiescent 3T3 cells (42, 48), rat embryo fibroblasts (25), chick embryo fibroblasts (35), human diploid fibroblasts (28, 49), and BHK cells (50). Bt$_2$cAMP will prevent DNA synthesis following protease stimulation (45, 51, 52), and the DNA synthesis in BHK cells produced by adenovirus is also prevented by Bt$_2$cAMP (53).

Cyclic GMP

Cyclic AMP is not the only cyclic nucleotide found in cells. Cyclic GMP (reviewed in 54) and more recently cyclic CMP (55) have also been identified. Since the time of its discovery the role of cyclic GMP has been obscure.

Goldberg and his collaborators have suggested that cyclic AMP and cyclic GMP function together in a yin-yang fashion to control growth and other cellular functions, and they have emphasized that it is the ratio of these two nucleotides that is important (56). This suggestion is based in part on the fact that fibroblasts in which DNA synthesis has been initiated have high levels of cyclic GMP and low levels of cyclic AMP, whereas the opposite is true in resting cells (56).

If cyclic GMP has an important role in promoting growth, one might expect that the addition of cyclic GMP or an analog of cyclic GMP would promote cell growth. Carchman et al (27) found no effect of cyclic GMP analogs on the growth of NRK cells. Seifert & Rudland (44) did observe a small stimulatory effect of cyclic GMP on thymidine incorporation into DNA in serum-starved 3T3 cells; however, this effect was not specific for cyclic GMP. Clearly in the next few years we should see a lot of work in this area.

Restoration of Normal Growth Properties of Fibroblastic Cells

A variety of experiments have been done to determine if elevation of cyclic AMP levels would restore proper growth control or "contact inhibition of growth" to transformed cells. Initially it was claimed that polyoma transformed and spontaneously transformed 3T3 cells treated with Bt_2cAMP plus theophylline ceased growth at the same density as normal untreated 3T3 cells (19). From these studies it was concluded that contact inhibition of growth was restored (19). Grimes & Schroeder (18), however, found that with these same cells, the final saturation density of the treated cells is dependent upon the initial plating density and that the earlier observations (19) may have resulted from a fortuitous plating density.

We have tested the effect of Bt_2cAMP and theophylline on the growth of numerous transformed cell lines (11, 27). In all cell lines but one, the treated cells continue to grow to high densities, at which point they begin to come off the substratum. The one exception, KNRK C132 (Kirsten sarcoma virus-transformed NRK cells), did not increase in cell density beyond 1×10^5 cells/cm^2 when grown in the presence of Bt_2cAMP (1, 27). However, many holes were observed in the monolayer and dying cells were apparent. Thus the final cell density apparently represented an equilibrium between dying and dividing cells and not restoration of normal growth control. Blat et al (23) also reported that Bt_2cAMP plus theophylline did not prevent polyoma transformed BHK cells from growing to very high densities.

Where in the Cell Cycle Does Cyclic AMP Act?

The point in the cell cycle at which cell growth is slowed or arrested varies, depending on both the cell type used and the type of experiment performed. Three experimental designs have been used in such studies. In one the cells have been treated with a cyclic AMP analog and the point of growth arrest assessed by measuring the DNA content of each cell. In another the cyclic AMP analog has been removed and the subsequent sequence in the increase in DNA synthesis or in cell number used to determine where in the cycle the cells were arrested. In the third type of experiment, cells have been synchronized by a

Table 1 Site of growth inhibition by cyclic AMP

Cell Type	Authors	Block in[a] G1	Block in[a] G2
3T3	Willingham et al (48)	+	+
3T3	Kram, Mamont & Tomkins (42)	+	n
SV 3T3	Smets (20)	o	+
Human skin fibroblast	Froehlich & Rachmeler (49)	+	n
Rat fibroblast	Frank (25)	+	n
HeLa	Zeilig (57)	o	+
CHO	Remington & Klevecz (58)	(+)	+
Lymphoma S49	Bourne et al (59)	+	o
Lymphocytic leukemia	Smets (59a)	+	o
Lymphocytic leukemia	Millis et al (60)	n	+

[a] +, inhibition found; o, no inhibition; n, not examined.

variety of techniques, the synchronizing agent removed, cyclic AMP added, and its effect on blocking process through the cell cycle observed. Table 1 summarizes the data from only a few of many such experiments. It is clear that cyclic AMP almost always arrests cell growth in G_1 or G_2. In a simple study with Reuber hepatoma cells, growth inhibition in S phase was observed (61). Other data suggest that cyclic AMP may actually have a stimulatory role if added to

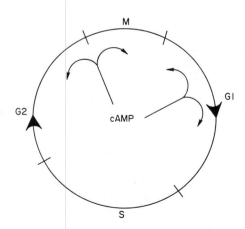

Figure 1 Diagrammatic representation of possible points of cyclic AMP action in the cell cycle. Large arrows on circle indicate the progression of cells through the cell cycle. Small arrows in clockwise direction indicate stimulation by cyclic AMP; small arrows in counterclockwise direction indicate inhibition by cyclic AMP.

cells late in G_1 or in mitosis (48, 57). Figure 1 illustrates these points of growth control.

Pardee (62) has observed that if the growth of BHK cells is arrested by Bt_2cAMP treatment or serum, isoleucine, or glutamine starvation, the time required for the cells to begin DNA synthesis after the arresting treatment is removed is about 8 hr. On the basis of these experiments, he has suggested that there is a common restriction point in the G_1 phase of the cell cycle on which all these treatments operate. In this regard it is interesting that serum starvation has been reported to raise cyclic AMP levels (41, 42). The effect of isoleucine or glutamine starvation on cyclic nucleotide levels in BHK cells needs to be investigated. In related experiments, Seifert & Rudland (63) have found that Balb 3T3 cells become arrested in G_1 following glutamine and histidine starvation, and that upon readdition of the amino acids cyclic AMP levels fall.

It is frequently observed that cells showing normal growth control are arrested in the G_1 (G_0) phase of the cell cycle. As noted above, cyclic AMP levels are elevated in G_1 arrested cells in agreement with the finding that Bt_2cAMP treatment arrests cells in G_1.

Stimulation of the Growth of Fibroblasts by Cyclic Nucleotides

In lymphoblasts (64) cyclic AMP at low concentrations has a growth-promoting effect, whereas it is inhibitory at higher concentrations. Also in fibroblasts, cyclic nucleotides at about 10^{-5} M have been reported to stimulate growth. Cyclic AMP, cyclic GMP, and their dibutyryl derivatives, cyclic IMP, adenosine, 5'IMP, and ATP, stimulate the growth rate of chick embryo fibroblasts (65) but do not reinitiate cell growth in quiescent cultures. Schor & Rozengurt (66) found that several nucleotides at 10^{-5} M, including cyclic AMP, cyclic GMP, cyclic IMP, adenosine, AMP, ADP, and ATP, enhanced the induction of DNA synthesis by serum in 3T3 cells. Furmanski & Lubin (50) reported that low concentrations of Bt_2cAMP increased serum-induced DNA synthesis in BHK cells, whereas higher concentrations were inhibitory. We have made similar observations in Balb 3T3 cells with Bt_2cAMP, PGE_1, and 1-methyl-3-isobutyl xanthine (67). The mechanism of this growth-promoting effect is unknown but, due to its lack of specificity, it probably does not represent a specific action of cyclic nucleotides.

LYMPHOCYTES

The lymphocyte is a cell that normally exists in a resting (nongrowing) state. However, if lymphocytes are exposed to antigen or to certain other mitogenic substances, such as the plant lectins phytohemagglutinin (PHA) or concanavalin A (Con A), they are changed from quiescent into proliferative, blastlike cells (68), and 50–70 hr later a peak of DNA synthesis is observed (69, 70). Investigators have taken advantage of this mitogen activation property of lymphocytes to study the biological changes that occur when a resting cell is converted to a cell type capable of DNA synthesis and mitotic division, and to study the possible involvement of the cyclic nucleotides in regulating these proliferative events.

Cyclic AMP and Lymphoblast Proliferation

MacManus and Whitfield and their colleagues have completed numerous studies on the effects of cyclic AMP on the proliferation of thymic cells, which are a mixture of lymphocytes and lymphoblasts. Initial results from their laboratory indicated that physiological concentrations (10^{-8}–10^{-6} M) of cyclic AMP or Bt_2cAMP were able to stimulate the proliferation of thymic lymphocytes in culture (71). In addition, agents that produce an elevation in endogenous cyclic AMP levels, such as epinephrine or PGE_1, also were shown to induce thymic lymphocytes to initiate DNA synthesis and cell division (72–74). Yet, concentrations of cyclic AMP above 10^{-6} M were found to inhibit cell proliferation (71). These results were initially interpreted to indicate that cyclic AMP was acting as a mitogenic signal to convert the resting lymphocyte to a proliferative lymphoblast. However, since DNA synthesis ($[^3H]$-thymidine incorporation) occurs within 1 hr after cyclic AMP treatment and cell division begins within 3 to 4 hr, it seems more likely that cyclic AMP rather promotes DNA synthesis and mitosis of a subpopulation of already activated lymphoblasts (64).

Studies with synchronized HeLa cells (75), Chinese hamster ovary cells (76), 3T3 cells (45), and human lymphoid cells (60, 77) show that the intracellular concentration of cyclic AMP varies in different stages of the cell cycle, and these changes in cyclic AMP levels are thought to regulate the passage of cells through the cell cycle. These cell cycle studies also suggest that cyclic AMP could have either a stimulatory or an inhibitory effect on cellular proliferation, depending upon the period in the cell cycle when the cyclic nucleotide was added. Thus, an increase in endogenous cyclic AMP levels in late G_1 of the cell cycle might be required to initiate DNA synthesis (S phase); this would explain the stimulatory effects of cyclic AMP on lymphoblast proliferation observed by MacManus & Whitfield (71). Willingham et al (48) observed a similar situation with fibroblasts in culture, where the addition of Bt_2cAMP in G_1 caused DNA synthesis to begin at an earlier time.

From other evidence, which has been summarized by Whitfield et al (64), it appears that a complicated interrelationship may exist between calcium and cyclic AMP to regulate lymphoblast proliferation. They believe that calcium ion is the principal regulator of cell proliferation both in vivo and in vitro and in some cells calcium may act via cyclic AMP. Their studies indicate that calcium promotes the initiation of DNA synthesis by increasing endogenous cyclic AMP levels (78). The calcium concentration in the growth medium also can change cyclic AMP from a stimulator to an inhibitor of thymic lymphoblast proliferation (64, 79). Low concentrations (10^{-7}–10^{-5} M) of exogenous cyclic AMP increase lymphoblast proliferation in low calcium medium. Conversely, the growth of lymphoblasts maintained in a high calcium (1.5 mM) medium is inhibited at all exogenous cyclic AMP concentrations between 10^{-7} and 10^{-3} M.

The mechanism by which cyclic AMP at low concentrations (10^{-7}–10^{-6} M) promotes cell proliferation is somewhat obscure because the cyclic nucleotide does not enter the cell at this extracellular concentration or cause an elevation in intracellular cyclic AMP levels (64, 80). MacManus & Whitfield envision a binding site for cyclic AMP on the cell surface which is an "activation site" for initiating DNA synthesis. If calcium does play a regulatory role in the lymphoblast proliferative process, it is equally likely that low levels of cyclic AMP are able to either mobilize intracellular calcium or enhance calcium uptake from the low calcium medium to elevate calcium ion sufficiently to induce a proliferative response.

These studies seem to indicate that cyclic AMP is not acting as a mitogen to activate lymphocytes. Rather, depending upon the calcium concentration, cyclic AMP appears to serve as either a positive or negative regulator of the initiation of DNA synthesis in cycling lymphoblasts. For a more detailed discussion of this area the reader is referred to a review on the calcium–cyclic nucleotide inter-relationship in regulating cell proliferation (64).

Cyclic AMP and Lymphocyte Proliferation

Cultures of peripheral blood lymphocytes are predominantly in a resting state, with low [^{3}H]-thymidine incorporation into DNA. When incubated with a mitogen such as PHA, however, the majority of these cells undergo transformation to a proliferatively active blast cell. Cyclic AMP has been implicated as a possible regulator of this transformation process.

Several reports have noted small increases in [^{3}H]-thymidine incorporation when low concentrations of cyclic AMP are added to unstimulated peripheral lymphocytes (81–86). Cyclic AMP does not exhibit nearly the stimulatory properties of other mitogens such as the plant lectins; for example, stimulation by the cyclic nucleotide is only about 2% that observed with PHA. Krishnaraj & Talwar (86) found that whereas cyclic AMP could cause a slight enhancement in [^{3}H]-thymidine incorporation, Bt$_2$cAMP was ineffective in promoting DNA synthesis. Others have also observed that the addition of cyclic AMP or Bt$_2$cAMP to unstimulated lymphocytes does not initiate DNA synthesis at all (87–90).

In the absence of mitogens, the addition of cyclic AMP to peripheral lympho-cytes in culture can have a biphasic effect; in some instances where low doses of cyclic AMP are reported to stimulate DNA synthesis, higher concentrations (10^{-4}–10^{-2} M) inhibit [^{3}H]-thymidine incorporation (81, 84, 85, 87). Mendelsohn (91) observed that PGE$_1$, Bt$_2$cAMP, or cortisol treatment of unstimulated human peripheral blood lymphocytes causes inhibition of DNA synthesis.

The interpretation of [^{3}H]-thymidine incorporation studies with unstimulated lymphocytes in culture is complicated by the fact that only a small percentage of the cell population is actively cycling. Thus, it has been suggested that the effects noted with cyclic AMP could be influencing only a small subpopulation of lymphocytes; or, perhaps the effects are on a cell type in the culture other than lymphocytes (92). All in all these studies do indicate that cyclic AMP is not an effective mitogen in the lymphocyte system; the slight enhancement of DNA synthesis reported in some cases is possibly due to stimulation of previously activated cells as discussed earlier.

Cyclic AMP and Mitogen-Activated Lymphocytes

A clearer indication of a possible regulatory role for the cyclic nucleotides on lymphocyte proliferation has been obtained from studies with cell populations stimulated with PHA or Con A. When peripheral lymphocytes are exposed to high concentrations of PHA, a slight rise in intracellular cyclic AMP levels has been observed within a few minutes of mitogen treatment (86, 93, 94). However, cyclic AMP levels in PHA-activated lymphocytes fall within 6 hr to below the levels found in control cultures (93). Novogrodsky & Katchalski (87) found that PHA did not elevate cyclic AMP levels in rat lymph node lymphocytes. Although Hadden et al (95) did note a slight increase in intracellular cyclic AMP levels in lymphocytes stimulated with an impure PHA preparation, no change in cyclic AMP levels was observed after stimulation with purified PHA. The activation of lymphocytes with another potent mitogen, Con A, does not increase intracellular cyclic AMP levels even though DNA synthesis and cell division are induced (86, 94, 95). The fact that cyclic AMP, Bt$_2$cAMP, and agents that elevate endogenous cyclic AMP levels are not effective in initiating DNA synthesis, and that the mitogens PHA and Con A do not provoke a marked elevation in cyclic AMP levels seems to indicate that cyclic AMP does not mediate the effect of mitogens in the transformation of lymphocytes.

What effect then do elevated concentrations of cyclic AMP have on the PHA-induced activation of lymphocytes? Johnson & Abell (88) have demonstrated that isoproterenol, cyclic AMP, and Bt$_2$cAMP inhibit DNA synthesis in PHA-stimulated lymphocytes from patients with chronic lymphocytic leukemia. Studies by Smith et al (96) have shown that Bt$_2$cAMP produced maximal inhibition of PHA-induced DNA synthesis if present during the first hour after initial exposure to PHA. Several agents that elevate intracellular cyclic AMP levels (aminophylline, isoproterenol, and prostaglandins) also inhibited [^{3}H]-thymidine incorporation (96). Other studies have shown that Bt$_2$cAMP inhibits the PHA-induced transformation of human peripheral lymphocytes (81, 83). Gallo & Whang-Peng (89) found that low concentrations of Bt$_2$cAMP (10^{-7}–10^{-6} M) enhanced PHA transformation and DNA synthesis, while concentrations above 10^{-6} M were inhibitory. Cholera toxin can inhibit lymphocyte stimulation by the mitogenic plant lectins PHA and Con A (97, 98). The inhibitory action of cholera toxin was due to activation of adenylate cyclase and elevation of cyclic AMP levels. Hirschhorn (69) has reported that high doses of cyclic AMP will inhibit PHA, pokeweed mitogen, and Con A stimulation of lymphocytes to the same extent. Mendelsohn et al (91) have shown that incubation of human peripheral lymphocytes with PGE$_1$, Bt$_2$cAMP, or cortisol results in a concentration-dependent inhibition of PHA-stimulated DNA synthesis. These investigators found that although cortisol alone does not significantly alter cyclic AMP concentrations, it does potentiate the PGE$_1$ response to produce greater enhancement of intracellular cyclic AMP levels. They suggest that cortisol may act synergistically with PGE$_1$ to elevate lymphocyte cyclic AMP levels and regulate [^{3}H]-thymidine incorporation and transformation.

When S49 lymphosarcoma cells are exposed to Bt$_2$cAMP or to stimulators of adenylate cyclase, such as cholera toxin, the logarithmically growing cells complete one cycle of replication and then are arrested in the G_1 phase of the cell cycle (59). Millis et al (60, 77) found that in thymidine-synchronized human lymphoid cells the onset of mitosis is prevented by increasing or maintaining high levels of cyclic AMP during the G_2 period of the cell cycle. These results, as well as those showing the effect of cyclic AMP on mitogen-activated lymphocytes, clearly indicate that elevated levels of cyclic AMP inhibit lymphocyte DNA synthesis and prevent mitogen-induced proliferation.

Monahan et al (99) have measured intracellular cyclic AMP levels in L5178Y lymphoblasts, a malignant lymphoblastic cell line. Cyclic AMP levels did not vary significantly throughout the growth period and did not increase when these cells became stationary. Further, altered serum concentrations in the growth medium did not change either the cyclic AMP levels or the growth rate of these neoplastic cells. The authors suggest that the unrestricted growth of these lymphoblasts may be due to an aberration in the cyclic AMP system.

Cyclic GMP and Lymphocyte Proliferation

There are indications that cyclic GMP also is a regulatory agent in lymphocyte proliferation, although confusion exists as to its biological effect on this process. Whitfield et al (100) initially reported that exposure of previously programmed thymic lymphoblasts to low $(10^{-11}–10^{-10}$ M) or high $(10^{-6}–5 \times 10^{-6}$ M) concentrations of cyclic GMP raised the intracellular cyclic AMP content and thereby stimulated proliferation. Intermediate cyclic GMP concentrations $(10^{-9}–10^{-7}$ M) had no effect on this system. Further study, however, revealed that these intermediate concentrations of cyclic GMP can either promote or inhibit the progression of DNA synthesizing cells into mitosis, depending on the intracellular cyclic AMP content and extracellular calcium concentration (101). If calcium is not present, cyclic GMP does not arrest the cells in G_2 of the cell cycle (80). These investigators have also reported cyclic GMP $(10^{-7}$ M) to prevent completely calcium from initiating DNA synthesis and cell division (64). Other work from their laboratory has shown that a low concentration of cyclic GMP $(5 \times 10^{-11}$ M) initiates DNA synthesis in lymphoblasts by a calcium-independent process (102).

Hirschhorn (69) found cyclic GMP to be more effective than cyclic AMP in inhibiting the increase in DNA synthesis induced by PHA in lymphocytes in culture. However, this inhibition was not specific for the cyclic nucleotides because 5'-AMP, ATP, and ADP also inhibited the PHA-induced DNA synthesis. Parker (103), however, found that the addition of cyclic GMP produced little stimulation of PHA-induced cells, and no change was noted in the cyclic GMP levels in PHA-stimulated cells.

Hadden et al (95) have shown that PHA and Con A treatment of lymphocytes produced at least a tenfold increase in intracellular cyclic GMP within the first 20 min of exposure to the mitogens. The presence of acetylcholine, which elevates cyclic GMP levels in peripheral blood lymphocytes, has been reported to enhance PHA-induced [^{3}H]-thymidine incorporation (56, 104). These findings tend to support the proposal of Goldberg and his colleagues that at the time of initiation of proliferation (early G_1), a temporal increase in cellular cyclic GMP concentration serves as the initial mitogenic signal to induce the proliferative process (56, 95). They further suggest that cyclic GMP may act by increasing the translocation or transport of calcium into the cell (54).

Cyclic CMP

Recently a novel cyclic nucleotide, cyclic CMP, has been detected in leukemia L1210 cells by Bloch (55). Although the precise concentration of this nucleotide in cells is not apparent, it is present in substantial quantities. When leukemia L1210 cells are grown in vitro to a stationary phase and cooled to 4°C for 1 hr, a lag period of 2 hr is observed before growth ensues. The addition of cyclic CMP, but not cyclic GMP or cyclic AMP, abolishes this lag (105).

Interpretations of Cyclic Nucleotide Effects on Lymphocytes

A number of problems are encountered when studying the proliferative process of lymphocytes. In studies with already activated lymphoblasts, only a small population of cells is activated, and it is not known if these cells are similar to those initiated by PHA or Con A. Further, the initial mitogenic event has been missed in these cells. In studies with peripheral lymphocytes in culture, one encounters a heterogeneous cell population because most cultures contain variable numbers of erythrocytes, monocytes, platelets, and polymorphonuclear leukocytes (103). The small amount of [^{3}H]-thymidine incorporation noted in unstimulated peripheral lymphocyte preparations could be due to DNA synthesis in one of these contaminating cell types or it conceivably could represent repair synthesis.

Problems also are encountered in trying to interpret the relevance of results obtained from studies on plant lectin-induced proliferation. DNA synthesis does not begin until at least 24 hr after PHA exposure, and the period of peak DNA synthesis is 50 to 70 hr after PHA treatment (69, 70). Thus, in studies with PHA induction it is conceivable that the mitogenic signal is not initially expressed but rather follows a series of other steps that occur over the first 24 hr period. Weber et al (106) recently studied the kinetics of lymphocyte stimulation in vitro by mitogens. They found the activation of lymphocytes in vitro to be a complex process which they divided into four distinct steps, three of which are mitogen dependent and the fourth mitogen independent. It is evident from their studies that a short period of contact between the lymphocytes and the mitogen is not enough to push the cells into the DNA synthetic phase. In fact, it is possible that the plant lectin initially induces an immune response in these cells and that the mitogenic event required to convert resting cells to proceed into the G_1 phase of the cell cycle may follow some hours later. Teh & Paetkau (107) have reported that cyclic AMP plays a positive role in the first phase of an immune response, whereas it is inhibitory in the proliferative phase which begins 24 to 36 hr after mitogen activation. With these studies in mind one might ask if the initial changes noted in PHA-stimulated lymphocytes, such as the rapid elevation in cyclic GMP levels found by Hadden et al (95), have any relationship to the actual initiation of proliferative events which follow several hours later.

In spite of these difficulties it seems likely that the cyclic nucleotides are involved in the regulation of lymphocyte proliferation. The calcium ion also has been implicated as a potential regulator, although the complex interrelationship that probably exists between calcium and the cyclic nucleotides remains to be determined. The predominant effect of sustained elevation of lymphocyte cyclic AMP levels is an inhibition of PHA-stimulated DNA synthesis and cell division. This suggests that a fall in cyclic AMP levels, perhaps associated with an increase in cellular cyclic GMP, may be required as an activation signal to initiate the proliferative process in lymphocytes.

CYCLIC NUCLEOTIDE EFFECT ON GROWTH OF OTHER CELL TYPES

Skin

It has long been known that epinephrine inhibits the growth of epidermal skin cells (108) and that proliferative skin diseases are accompanied by accumulation of glycogen (109). With the discovery of the role of cyclic AMP in the actions of epinephrine in the liver, it seemed likely that these two seemingly unrelated facts could be tied together by cyclic AMP.

Bronstad et al (110) and Powell et al (111) reported that epinephrine raised cyclic AMP levels in skin. Later it was shown that Bt_2cAMP (112) and cyclic AMP plus theophylline (113) inhibit skin growth in vitro. Thus high levels of cyclic AMP are associated with inhibition of growth; the blockage appears to be in G_2.

The observation that in mouse epidermis isoproterenol is less effective in elevating cyclic AMP levels following injection of the carcinogen, 3, 4-benzpyrene, has led the authors to suggest that the growth stimulatory effects of 3, 4-benzpyrene may result from an inability of adenylate cyclase to respond to hormonal stimulation (114). On the other hand, Curtis et al (115) observed that injection of cyclic AMP increased the incidence of skin papillomas and squamous cell carcinomas in 7, 12-dimethylbenzanthracene-treated mice.

In the rapidly proliferating psoriatic skin, cyclic AMP levels are lower than in uninvolved areas (116, 117). Rapidly proliferating areas also have higher cyclic GMP levels (118).

Melanoma Cells

Cyclic AMP, Bt_2cAMP, theophylline, caffeine, and melanocyte stimulating hormone (MSH) inhibit the growth of cultured melanoma cells (119–121). Since MSH can only elevate cyclic AMP levels in G_2 (122), it is likely that the cells are inhibited by cyclic AMP in the G_2 phase of the cell cycle. Cholera toxin, which is known to raise cyclic AMP levels, also has been found to inhibit the growth of melanoma cells (123).

Hepatoma and Liver Cells

STUDIES IN VITRO Bt_2cAMP (0.5 mM) increases the doubling time of Reuber H35 hepatoma cells from 40 to 56 hr (124). Contrary to the G_1 and G_2 blocks found in other cell types, this effect was completely due to an extension of the S phase (61). The inhibition is overcome by the addition of deoxyribopyrimidine nucleosides, suggesting that Bt$_2$cAMP interferes with the synthesis of these DNA precursors (125). The effects of cyclic AMP analogs and theophylline in other hepatoma and liver cells have been tested, but the site of inhibition in the cell cycle has not been determined. Bt$_2$cAMP inhibits growth of MHL (hepatoma) and two Morris hepatoma lines, but not normal rat liver cells (125, 126). 8-SH cyclic AMP and 8-SCH cyclic AMP are inhibitory, but 5'AMP or cyclic AMP itself is ineffective. Theophylline (127) but not Bt$_2$cAMP (124) inhibits the growth of HTC (hepatoma) cells. Butyrate, a frequent contaminant in Bt$_2$cAMP solutions, inhibits the growth of some (126) but not all (125) of these cell lines.

STUDIES IN VIVO   Butcher et al (128) measured the cyclic AMP levels in normal livers and in hepatomas of 14 hepatoma-bearing rats. They found no differences in the content of cyclic AMP in the hepatomas compared to normal livers. Thomas et al (129) compared the cyclic AMP and cyclic GMP levels in eight Morris hepatoma lines of varying doubling times. They could detect no correlation between the growth rate and either the cyclic AMP or cyclic GMP levels. Furthermore, they found that all eight livers had higher cyclic AMP and cyclic GMP levels than normal livers in tumor-bearing rats. On the other hand, Hickie et al (130) recently reported that one derivative of a Morris hepatoma line, 5123T.C. (h), has significantly lower levels of cyclic AMP than either its parent or normal liver. Chayoth et al (131) measured cyclic AMP levels in livers of ethionine-treated rats. Ethionine is a carcinogen that induces hepatoma formation. They found that uninvolved liver and benign hyperplastic nodules have similar cyclic AMP levels and malignant hepatomas have much higher levels. All three have significantly higher values than normal liver. Thus in vivo assays of cyclic AMP in hepatoma cells have failed to show any consistent role of cyclic nucleotides in cell growth.

Miscellaneous Animal Cells

A variety of brain and nervous system cells are also under cyclic AMP control. Bt$_2$cAMP has been reported to inhibit the growth of mouse neuroblastoma cells (132–135), astrocytoma cells (134), glial cells (135), human tumor astrocyte and neuroblast cell lines (136), and dispersed cell cultures of fetal rat brain (137). However, Richelson (138) observed only a very small inhibition of growth with a mouse neuroblastoma line.

There is evidence that cyclic AMP inhibits the growth of many other cell lines. These include KB cells (139), HeLa cells (9, 57), normal chick retinal pigmented epithelium (140), adrenal tumor cells (141), rat anterior pituitary cells (line GH$_1$) (142), an established myogenic cell line L6 (143), murine mastocytoma cells (144, 145), and cultured bone marrow cells (146).

Plants

Early studies suggested that cyclic AMP may stimulate the growth of plants, although the evidence supporting this role for cyclic AMP was indirect. Cytokinesin I, a potent growth-promoting agent isolated from crown gall tumor tissue of Vinca Rosea L or from normal plant tissues induced to proliferate by the addition of cytokinins, is an inhibitor of cyclic nucleotide phosphodiesterase (147). Also, theophylline, a common inhibitor of phosphodiesterase activity in cell homogenates, stimulates plant growth (147). It is difficult to conclude from these studies alone that cyclic AMP promotes growth of plants. It is well known that phosphodiesterase inhibitors have additional actions besides inhibiting phosphodiesterase, and many substances inhibit phosphodiesterase activity in broken cell homogenates which do not elevate cyclic AMP in the intact cell (148).

More recent studies cast some doubt on the hypothesis that cyclic AMP stimulates plant growth. The addition of cyclic AMP and its analogs has failed to show any specific growth-promoting activity. Cyclic AMP itself has no effect on growth, whereas its dibutyryl and 8-Br derivatives are active (149). But these experiments must be interpreted with caution, since 8-Br-5'AMP and N^6-butyryl adenine are equally effective (149, 150). Furthermore, Drlica et al (151) found no detectable differences in cyclic AMP levels between several normal and transformed plant cells (crown gall) growing in culture, even if the cells were stimulated to divide by the addition of hormone-supplemented medium.

Recent studies with the Jerusalem artichoke tubors suggest that cyclic AMP may have an inhibitory function in plant cell growth. In a series of experiments which, at least superficially, resemble experiments on the stimulation of growth in quiescent fibroblasts described earlier in this review, Giannattasio et al (152) observed that cyclic AMP levels fall rapidly when dormant tubors are activated by incubation in aerated water; the fall precedes an induction of RNA and protein synthesis.

One problem in studying plant growth regulation is the lack of stable cell culture systems where growth of cloned cell types can be studied under carefully controlled conditions. The indirect studies described above have failed to show a specific action for cyclic AMP, and the direct measurements of cyclic AMP have been done with mixed cell populations where it is difficult to show that changes in cyclic AMP are directly related to changes in growth.

ENZYMES OF CYCLIC AMP METABOLISM

The fact that cyclic AMP appears to be an important mediator of cellular proliferation dictates that the intracellular concentration of this cyclic nucleotide be well controlled. The level of cyclic AMP in the cell is determined in large measure by the activity of the enzyme adenylate cyclase, which catalyzes the formation of cyclic AMP from ATP, and cyclic AMP phosphodiesterases, which degrade cyclic AMP into 5'-AMP and PP_i. To understand the role of cyclic AMP in mediating cellular proliferation, it is essential that we also understand how the activities of these enzymes are regulated and influenced by both intra- and extracellular agents and events.

Adenylate Cyclase Activity and Cell Proliferation

In studies with several fibroblast lines which exhibit density-dependent inhibition of growth it has been demonstrated that cyclic AMP levels rise as the cells approach confluency and growth is arrested (27, 33, 36, 153). The elevation of cyclic AMP levels in normal cells at confluency appears to be due to an increase in adenylate cyclase activity. Makman (154) observed that adenylate cyclase activity is increased as cell density is increased in 3T3 and Chang liver cells. In a short communication, Zacchello et al (155) showed that cyclase activity increased at the time of confluency in human diploid fibroblasts. We have measured adenylate cyclase and cyclic AMP phosphodiesterase activites as a function of increasing cell population density in contact-inhibited NRK cells (153, 156). In lightly seeded growing cells, cyclic AMP levels were low; both enzyme activities increased with increasing cell population. As NRK cells reached confluency and growth slowed, adenylate cyclase activity continued to rise, but the cyclic AMP phosphodiesterase activity decreased somewhat and then remained constant. This divergence in enzymatic activities coincided with an elevation in cyclic AMP levels, a declining growth rate, and the eventual onset of the stationary growth phase. It is felt that the elevated cyclase activity reflects an increased population of cells arrested in the G_1 (G_0) phase of the cell cycle, apparently caused by cell-to-cell contact.

Results from a number of studies have clearly shown that the intracellular concentration of cyclic AMP varies during the cell cycle (45, 60, 75–77). Thus, it is important to determine the enzymatic basis for the alterations in the concentration of this cyclic nucleotide as the cell progresses from the resting stage (G_1 or G_0) to the DNA synthetic phase (S) and eventual mitosis.

Makman & Klein (157) observed that after release of Chang's liver cells from a thymidine blockade, adenylate cyclase activity dropped in the S phase of the cell cycle. As the cells progressed through the cycle, enzyme activity increased and reached maximal levels at the peak of mitosis.

Infection of BHK21 cells, arrested in the G_1 phase of the cell cycle, with adenovirus type 12 results in an abortive infection, but does lead to a release from contact inhibition and to a progression through the mitotic cycle. Raska (46) found that cyclase activity is high during the G_1 period of adenovirus-infected BHK cells, and then drops dramatically during S phase.

Millis et al (60) measured cyclase activity in human lymphoid cells synchronized by either excess thymidine or colcemid treatment. In thymidine-synchronized lymphoid cells, the levels of adenylate cyclase activity are high in G_2 as compared with S. Cells synchronized with colcemid were found to have high cyclase activity in early G_1.

Regulation of Adenylate Cyclase Activity

The activity of adenylate cyclase does vary with the rate or phase of cell growth, although the mechanism by which the activity of this enzyme is controlled has not been elucidated. Certain polypeptide hormones and catecholamines increase the activity of this enzyme in appropriate target cells (158). In vitro studies of the adenylate cyclase systems from the thyroid (159), liver (160), platelets (161), lymphoblasts (162), and macrophages (163) show a stimulatory response with prostaglandins. Prostaglandins of the E type (mainly PGE_1 and PGE_2) also stimulate cyclase activity from a number of fibroblast cell lines in tissue culture (154, 156, 164, 165). Nucleotides, particularly GTP, potentiate the activation of adenylate cyclase by prostaglandins (156, 159–161), thyroid-stimulating hormone (159), glucagon (166, 167), isoproterenol (168), oxytocin (169), and epinephrine (170).

The ability of certain hormones to stimulate adenylate cyclase is influenced by the conditions of incubation and cell growth. Makman & Klein (157) found the adenylate cyclase in Chang's liver cells to be more sensitive to hormone stimulation during mitosis and decreased during the S phase of the cell cycle. Synchronized mouse melanoma cells respond to melanocyte-stimulating hormone only in the G_2 phase of the cell cycle (171).

Clark & Perkins (172) observed that the response of cultured astrocytoma cells to norepinephrine differed in various stages of the growth cycle; a good response was found in low density, logarithmically growing cells. However, hormonal stimulation of cyclic AMP levels in Chang's liver cells, WI38 fibroblasts, HeLa cells, and lens epithelial cells increases with an increase in cell density (173). In studies with glial tumor cells, Schwartz et al (174) similarly found that norepinephrine was able to elevate cyclic AMP levels only after the cells reached confluency. Kelly & Butcher (165) also observed an increase in PGE_1 response with increasing cell population density with WI38 human fibroblasts. In contrast, WI38 fibroblasts grown to high population density respond poorly or not at all to epinephrine, whereas cells at low density exhibit a good response with this hormone (165).

Serum is required for the growth of cells in culture (3); insulin can replace some of the factors in serum required for growth of fibroblasts (175). Such treatments result in rapidly decreased levels of cyclic AMP and increased growth (22, 34, 36). The insulin effect is apparently due to inhibition of adenylate cyclase. Physiological concentrations of insulin inhibit cyclase activity in liver and fat cell crude membrane preparations (176–178) as well as the adenylate cyclase activity of fibroblasts (22, 179).

Modulation of Hormonal Response—Inhibition

Another property that may contribute to the regulation of adenylate cyclase is hormone-induced desensitization. The hormone-induced increase in cyclic AMP levels in a number of tissues is transitory; the levels of the cyclic nucleotide reach a peak and then decline even in the continued presence of hormone (180–182). The decrease in cyclic AMP concentration appears to result from a loss of the hormonal response.

A time- and temperature-dependent inactivation of the catecholamine and prostaglandin response has been demonstrated in lymphoid, HeLa, lens epithelial, and Chang's liver cells after exposure to the hormone (173, 183). Manganiello & Vaughan (184) observed an apparent loss in the response of L cells in culture to PGE_1. In studies with human diploid fibroblasts in culture, Franklin & Foster (185) also have shown a desensitization of the ability of isoproterenol and PGE_1 to elevate cyclic AMP levels following a preincubation of the cells with the appropriate hormone. Exposure of intact macrophages to prostaglandin E_2 for 2 hr markedly reduced the ability of this hormone to stimulate adenylate cyclase; similar desensitization was noted with epinephrine (163).

Ho & Sutherland (186) observed a refractory hormonal response in fat cells and found a hormone antagonist which was formed in response to elevated cyclic AMP levels and then prevented further stimulation of the enzyme by epinephrine, ACTH, or glucagon. The identification of this antagonist and an understanding of its synthesis and mode of action could be of considerable importance in elucidating how adenylate cyclase activity is altered in various stages of growth and the cell cycle.

Modulation of Hormonal Response—Activation

As opposed to the prevention of hormonal responses, we have recently obtained evidence for a macromolecular factor that potentiates the hormonal responsiveness of fibroblast adenylate cyclase. When washed NRK membrane preparations were used as the source of enzyme, very little stimulation of adenylate cyclase was noted with PGE_1 unless GTP was also present (156). An activator of the cyclase system is present in the 100,000 x g supernatant from fibroblasts, which increases basal activity as well as supports a prostaglandin response (187). This modulatory activity has not been fully characterized so the possibility remains that some, or all, of the stimulatory activity noted may be due to the presence of guanyl nucleotides and the generation of GTP.

The changes noted in hormonal responsiveness may result from altered membrane structure induced during the division cycle or as a result of contact between cells. It is not clear if this leads to modification of hormone receptor sites or if the activity of hormone antagonists or modulators may be affected. In any event, it is conceivable that the potentiation or prevention of hormonal regulation of adenylate cyclase could be responsible for the altered cyclic AMP levels noted in specific periods of the proliferative process.

Agents That Alter Adenylate Cyclase Activity and Cellular Proliferation

Usually, factors that decrease cyclic AMP levels stimulate the proliferation of cells growing in culture, while factors that increase cyclic AMP levels cause a decrease in cellular growth rate. As described in the section on lymphocytes, certain plant lectins act as mitogens to induce proliferation of this cell type, possibly by altering the intracellular concentration of cyclic AMP.

Smith et al (93) found that PHA has a stimulatory effect on adenylate cyclase in broken cell preparations. However, Novogrodsky & Katchalski (87) reported that lymphocyte adenylate cyclase was not altered by PHA at 50 μg per ml, although higher concentrations of the agglutinin (125–500 μg per ml) reduced enzymatic activity. Low concentrations of Con A inhibit fat cell adenylate cyclase, whereas higher concentrations of the lectin increased enzyme activity (188). Certain phytohemagglutinins also have been shown to inhibit platelet cyclase activity (189). Makman (183), however, found that PHA added directly to the assay had little effect on the activity of the enzyme in homogenates. Rather, if intact thymocytes were preincubated with PHA or Con A, a definite decrease in cyclase activity was observed in lysates of these cells.

From these studies it has been suggested that perhaps the plant lectins interact with cell surface constituents to either directly or indirectly alter the adenylate cyclase system. Con A and wheat germ agglutinin can prevent the binding of insulin to the insulin receptor of liver and fat cell membranes (188). These investigators have suggested that lymphocytes may have "incomplete" insulin receptors that are nevertheless capable of being activated by the plant lectins. If so, these mitogens could initiate insulin-like effects in cells lacking receptors for the hormone, i.e. altering cyclic AMP levels and initiating cell replication.

Cholera toxin is another agent that can be used to alter adenylate cyclase activity and increase cyclic AMP levels. This protein stimulates cyclase activity of plasma membranes from a variety of tissues (190–192). Apparently, cholera toxin interacts with a membrane ganglioside, GM_1, to promote its effects (193–195). Hollenberg & Cuatrecasas (32) found that cholera toxin would suppress the stimulation of DNA synthesis provoked by epidermal growth factor and serum when added to human fibroblasts, presumably by altering the levels of intracellular cyclic AMP through the activation of adenylate cyclase.

Influence of Surface Membrane Changes on Adenylate Cyclase

The enzyme adenylate cyclase is localized within the plasma membrane of the cell where it is susceptible to the influence of both intra- and extracellular agents. A common feature of the possible growth regulating factors discussed in this section is that they all appear to alter adenylate cyclase by some interaction with the membrane. Thus, it seems likely that alterations in membrane structure or composition which might be induced at different stages of the cell cycle or with cell-to-cell contact could dramatically change the catalytic and regulatory properties of this enzyme.

Cell cycle-dependent changes in the surface membrane have been detected by variations in [^{3}H] Con A binding in different periods of the cycle (196, 197), as well as by increased synthesis of membrane components in specific cycle phases (198–200). When surface membranes were labeled by a procedure utilizing galactose oxidase and [^{3}H] borohydride, labeling occurred maximally during the G_1 phase and minimally during S phase (201). Also, various cell surface antigens may be maximally expressed during mitosis or early G_1 phase and in some instances greatly decreased during the S phase (202, 203).

Several studies have shown that the concentration of certain glycolipids increases on cell-to-cell contact of normal cells (204–206). A specific high molecular weight glycoprotein has been detected in confluent cells; this substance is present in low amounts in growing cell populations (207, 208). However, it has not been established that these membrane changes directly influence the rate of cellular proliferation; they could be a result of the cell's proliferative state.

Adenylate Cyclase and Malignant Transformation

A characteristic property of neoplastic cells is that they appear to have lost their normal growth control mechanisms; they often grow faster than normal cells and do not exhibit contact inhibition of growth when grown in cell culture. Studies discussed elsewhere in this review have demonstrated that altered cyclic AMP metabolism may, in part, be responsible for some of the abnormal properties of transformed cells, including uncontrolled growth. Adenylate cyclase activity has been measured in tumors and transformed cell lines to determine if an alteration in the activity of this enzyme might be particularly responsible for the aberrant behavior of these cells.

Granner et al (209) initially reported that the adenylate cyclase activity of cultured rat hepatoma cells was low and unresponsive to glucagon. Emmelot & Bos (210) also found that the cyclase activity of plasma membranes isolated from a rat hepatoma cell line was less than half that of normal liver membranes and exhibited little stimulation by glucagon or epinephrine. However, membranes isolated from mouse liver and mouse hepatoma cells showed no significant differences in basal cyclase activity or hormonal stimulation.

Higher adenylate cyclase activity has been reported by Brown et al (211) in studies with several rat hepatoma cell lines, and Pennington et al (212) found an increased epinephrine sensitivity in hepatoma cells. These studies were carried out in the presence of NaF and high activities were reported. Thus, some question arises as to the interpretation of these experiments.

Yet, Makman (154, 213) also found the activity of this enzyme to be elevated in Chang's liver cells, a cultured liver tumor cell line, although the cyclase from these tumor cells was not stimulated by glucagon. However, in agreement with Granner et al (209), hepatoma (HTC) cells grown in suspension culture showed no detectable adenylate cyclase activity, and low activity when grown in stationary culture (213). The hormonal response of adenylate cyclase has been determined in several other hepatoma cells; the enzyme from some of these malignant lines is unresponsive to glucagon, whereas in others the cyclase system appears to behave in a normal manner (128, 214).

In fully developed hepatocarcinomas induced by the chemical carcinogen 2-acetylaminofluorene, basal adenylate cyclase activity is reported to be increased but shows only slight responses to hormones (215). In a related study, Boyd et al (216) have determined cyclase activity in rat liver tumors induced by the carcinogen 3'-methyl-4-dimethylaminoazobenzene. They found that both basal activity and isoproterenol responsiveness of adenylate cyclase are markedly increased in premalignant liver. However, it was observed that basal activity was low and hormonal stimulation variable in established tumors.

Another means of altering adenylate cyclase activity has been suggested from subcellular distribution studies of adenylate cyclase in Yoshida hepatoma cells (217) and SV40-transformed hamster fibroblasts and Rauscher leukemic spleen cells of mice (218). It is reported that adenylate cyclase is found in the cytosol, as well as in the plasma membrane, in these transformed cells. They suggest that shifting adenylate cyclase to the cell interior would reduce or eliminate regulation by hormones, which could contribute to tumor development. These results should be viewed with caution, however, for adenylate cyclase is a difficult enzyme to solubilize and maintain in an active state, let alone be able to retain hormonal responsiveness. It seems possible that in such experiments the authors were working with very small membrane fragments resulting from their disruption procedure. Since the membranes from malignant cells clearly differ from normal membranes, varied degrees of fragmentation could result when disrupting each type of membrane.

The ability of ACTH and epinephrine to increase intracellular cyclic AMP levels and stimulate adenylate cyclase has been determined in a number of studies with adrenocortical carcinoma tissue and cells in culture. While hormonal stimulation does vary in the different cell lines, a number of the cultured tumor cell lines have lost their responsiveness to glucagon (219–224). Schorr et al (220) measured adenylate cyclase derived from rat adrenal tumor and found that multiple hormone receptors had appeared so that ACTH, TSH, LH, and FSH stimulated the enzyme's activity. Abnormal hormonal responsiveness has also been observed in several human endocrine tumors (225) and in an islet tumor (226).

Changes in the adenylate cyclase activity of thymus cells during the development of leukemia in ARK mice have also been determined (227). Investigators found that thymus cyclase was elevated in both preleukemic and leukemic ARK mice when compared to nonleukemic mice. However, there was a decrease in the enzyme's response to epinephrine in preleukemic and leukemic thymus preparations. In lymphocytes of patients with chronic lymphatic leukemia, however, both basal adenylate cyclase activity and the enzyme's response to prostaglandins are decreased when compared with the cyclase from normal human lymphocytes (228).

Brown et al (229) have reported that a chemically induced mammary carcinoma in rats has both increased cyclase activity and increased sensitivity to epinephrine. Yet, mouse epidermis treated with the chemical carcinogen 3,4-benzpyrene shows a decreased response to isoproterenol (114).

Psoriasis is a common skin disorder characterized by accelerated proliferation of skin epithelium. Although it is not a malignant disease, it does present a good model for determining if altered adenylate cyclase activity might account for the decreased cyclic AMP levels and uncontrolled growth found in involved tissue. Hsia et al (230) and Wright et al (231) have presented evidence suggesting that adenylate cyclase activity is low in psoriasis tissue and does not respond well to epinephrine. Voorhees and colleagues (232, 233), however, are not convinced that adenylate cyclase activity is decreased; they cite a number of experimental problems encountered in working with this tissue which make it difficult to evaluate results.

Altered Adenylate Cyclase Activity with in vitro Virus Transformation

Cells growing in culture provide a good system for studying the effect of viral transformation on adenylate cyclase activity. Numerous reports have appeared which indicate that cells transformed by viruses have low intracellular concentrations of cyclic AMP. Paradoxically, there are reports that BHK cells transformed by the Bryan strain of Rous sarcoma virus (4), spontaneously transformed 3T6 mouse embryo fibroblasts (213, 234), and polyoma virus (234), MSV/MuLV (164), and SVT2 (164) transformed 3T3 mouse embryo cells have elevated basal adenylate cyclase levels. SV40-transformed 3T3 cells have fluoride-stimulated adenylate cyclase levels much higher than observed with the enzyme from 3T3 cells (164).

The predominant finding, however, is that adenylate cyclase activity is low, or has lost hormonal responsiveness, in transformed cells. Burk (4) noted decreased cyclase activity in polyoma virus-transformed BHK cells, while Peery et al (164) found 3T3 cells transformed by polyoma virus to have slightly decreased levels of this enzyme. Further, transformation of hamster astrocytes (235) or 3T3 cells (236) by SV40 results in a substantial decrease in the specific activity of adenylate cyclase. No attempt was made, however, to characterize these changes in cyclase activity with respect to possible alterations in the kinetic parameters of the enzyme.

Studies from our laboratory have shown that chick embryo fibroblasts (CEF) transformed by either the Bryan high titer (RSV-BH) (237) or Schmidt-Ruppin (RSV-SR) (238) strains of Rous sarcoma virus have decreased adenylate cyclase activity. The mechanism for lowering cyclase activity differs with these two related viruses. With chick cells transformed by RSV-BH, the enzyme's apparent K_m (ATP) and response to increasing Mg^{2+} are altered (237). The enzyme from CEF transformed by RSV-SR shows little change in either its affinity for ATP or in its response to increasing Mg^{2+} concentration, although its V_{max} is decreased (238). As described earlier, we have evidence for an activator of fibroblastic adenylate cyclase present in the $100,000 \times g$ supernatant (187). It appears that RSV-SR transformed CEF have decreased amounts of the cyclase activator or modulator.

Anderson et al (156) also have shown that adenylate cyclase activity is decreased in NRK cells transformed by either the Kirsten sarcoma virus (Ki-SV) or the Moloney sarcoma virus (Mo-SV). The enzyme from Ki-SV transformed (KNRK) and Mo-SV transformed (MNRK) rat kidney fibroblasts exhibits no change in its affinity for ATP, although differences are apparent in its responsiveness to increasing Mg^{2+}

Of particular interest is the observation that the enzyme from KNRK and MNRK cells is unresponsive to PGE_1 in either the presence or absence of GTP (156). This is similar to results with 3T3-SV40 (Cl-x) cells, which have an adenylate cyclase that is unresponsive to prostaglandins (164). Other studies in our laboratory indicate that KNRK and MNRK cells do contain supernatant modulators, but that the enzymes from these transformed cells do not respond to activator (239). Burk (4) has described a similar lack of response to a supernatant activator with the enzyme from polyoma-transformed BHK cells. With regard to the control of cellular proliferation, it is of interest that NRK cells transformed by Ki-SV have lost their ability to elevate cyclase activity with increasing cell density (156); these transformed cells exhibit rapid uncontrolled growth.

Viral mutants that are temperature sensitive in their transformation function have been used to establish that altered cyclase activity is due to the transformation event. It was found that in CEF infected with the Rous sarcoma virus temperature-sensitive mutants, RSV-BH-Ta and RSV-SR-T5, altered adenylate cyclase activity was among the earliest events to be discerned following the shift from non-permissive to permissive temperature (237, 238). The rapid decrease in adenylate cyclase activity which follows the temperature shift of RSV-BH-Ta infected CEF from the nonpermissive to the permissive temperature is not blocked by the inhibition of protein synthesis (cycloheximide) or RNA synthesis (actinomycin D) (240). Thus, the rapid change in adenylate cyclase activity seems directly attributable to the temperature-sensitive viral gene product (transformation factor).

Adenylate cyclase appears to be an integral component of the plasma membrane in all animal cells. Thus, modification of its membranous environment could alter the properties and activity of this enzyme. Numerous studies have shown that transformation produces a number of chemical changes in the plasma membrane. These modifications in membrane structure probably account for the altered properties of adenylate cyclase.

Apparently, modification of the membrane can alter the properties of the catalytic unit of the enzyme, or it can alter some of the regulatory properties of the cyclase system (240). In any event the decrease in enzyme activity leads to lower cyclic AMP levels, which in turn seem to cause many of the abnormal properties of transformed cells, including an altered rate of proliferation.

Interpretations of Adenylate Cyclase Studies

Any conclusions drawn from these studies must be tempered by taking into account the inherent difficulties encountered when trying to determine and compare adenylate cyclase activities in different stages of growth or in different cell types. The activity and regulatory properties of adenylate cyclase are dependent on membrane structure, so it is conceivable that the activity measured in broken cell preparations does not necessarily bear any relationship to the level of cyclic AMP formation in the intact cell. Adenylate cyclase may exist in an activated or inhibited state within the cell, but escapes this regulated state with homogenization. Enzyme levels determined in homogenates or membrane preparations may only reflect an elevated or decreased activity, and may not be a measure of the full catalytic potential available to the cell at that time. The enzyme's response to hormones or other modulators also might be altered in broken cell preparations. For example, it has been difficult to show an insulin effect on adenylate cyclase activity in cell-free extracts. This could negate, or perhaps introduce, regulatory parameters that may or may not be functionally important in the intact cell.

Caution is particularly advisable in studies with the enzyme from malignant systems. Determining cyclase activity in tumor preparations presents a problem of obtaining good normal control tissue for comparative purposes. Also, a tumor can be severely contaminated with other cell types, which may contribute significantly to the measured cyclase activity and make interpretation difficult. Artifacts, i.e. altered hormonal response, also may be introduced in going from tissue preparations to in vitro cell culture systems.

In a number of studies care has not been taken to thoroughly investigate all the parameters of the cyclase system that might be altered as a result of transformation or the proliferative state of the cell. From our studies it is evident that cyclase activities can be meaningfully compared only after careful kinetic analysis.

Nevertheless, taken together these studies indicate that altered adenylate cyclase activity may be responsible for some of the growth characteristics and transformed properties of the cell. Cyclase activity does vary at different periods in the cell cycle and with increasing cell population density. Alterations also are apparent in the enzyme's properties and activity in solid tumors and virus-transformed cells.

A common feature of altered cyclase activity involves a change in its response to certain hormones, particularly glucagon and prostaglandins. Fluctuations in hormonal response may arise from alterations in membrane composition which occur during the cell cycle, during cell-cell contact, or are induced by malignant transformation. It will be of interest to determine if the changes in enzyme activity result from some modification of hormonal receptor sites, or if they are due to the presence or absence (activation or inactivation) of modulators that may be required to elicit a hormonal response.

CYCLIC AMP PHOSPHODIESTERASES

Although easier to measure than adenylate cyclase, much less work has been done on cyclic AMP phosphodiesterases than on adenylate cyclase. Extracts of cultured cells contain enzymatic activities that degrade cyclic AMP. In extracts of chick embryo cells, phosphodiesterase activity is found in both the soluble and membranous fractions. The soluble activity has been subjected to DEAE-cellulose column chromatography and resolved into two major components. One of these has a high K_m for cyclic AMP and also hydrolyzes cyclic GMP. Indeed, cyclic GMP is the preferred substrate (241, 242). The other activity appears identical to the membrane-bound enzyme. This enzyme is associated with the plasma membrane, and apparently during homogenization variable amounts are solubilized. The enzyme can be almost completely solubilized by sonication. After solubilization and purification, it shows negatively cooperative kinetics of hydrolysis of cyclic AMP. Studies on changes in phosphodiesterase activity must be interpreted in relation to these observations.

One situation in which total phosphodiesterase activity rises is after cells are exposed to agents that raise their cyclic AMP levels (184, 243). This change is not due to enzyme activation but represents new enzyme synthesis caused by an action of cyclic AMP at the transcriptional level (242). Presumably such a mechanism protects the cell from long periods of high cyclic AMP levels. Another situation in which phosphodiesterase changes is after trypsin treatment. Trypsin, as described previously, affects cyclic AMP levels and cell growth. One action of trypsin on intact cells is to alter the kinetic properties of the membrane phosphodiesterase (241). This appears to be a case in which an action of trypsin on the cell exterior affects a catalytic activity on the inner surface of the membrane.

Rapidly growing chick cells at low cell densities have low levels of phosphodiesterase. As the cells crowd together, the amount of enzyme increases (153). In chick cells the rise occurs in concert with a rise in adenylate cyclase, and cyclic AMP levels do not rise. In contrast, the phosphodiesterase of NRK cells stops rising as the cells crowd together and then cyclic AMP levels do rise (153). Why NRK cells fail to increase their phosphodiesterase levels in the face of rising cyclic AMP levels needs to be clarified.

As mentioned above, transformed fibroblastic cells fail to raise their cyclic AMP levels at confluency. To date this failure has been chiefly associated with changes in adenylate cyclase (see above) and not phosphodiesterase. Sheppard has reported a small increase in phosphodiesterase activity in a line of polyoma-transformed 3T3 cells (234). However, D'Armiento et al (243) found low phospho-diesterase activity in SV40-transformed 3T3 cells. In studies on chick cells transformed by RSV-BH and RSV-SR, Russell & Pastan have not observed marked changes in phosphodiesterase activity (244). In NRK cells transformed by Kirsten or Moloney sarcoma viruses, the levels of cyclic AMP phosphodiesterases tend to be diminished (244), and it is mainly the membrane-bound enzyme that is affected.

Rhoads et al (245) have measured the activity of phosphodiesterase in the soluble portion of liver tissue comparing hepatoma with normal liver. They found that the tumors had a generally lower activity than normal tissue. Unfortunately in most of their studies activity was measured at a relatively high concentration of cyclic AMP and the activity of the membranous fraction was ignored. Clark et al (246) have performed similar studies on soluble phosphodiesterase activity and found that the activity measured at low substrate tended to be increased in more rapidly growing cells and the activity at high substrate increased. The possibility that some of these changes were due to selective solubilization of the membrane-bound enzyme in tumor cells needs to be evaluated.

GROWTH FACTORS

Although it was emphasized at the beginning of this review that cultured cells are usually propagated in medium containing serum (normally 10%), some factors have been identified that will partially replace factors in serum. The most interesting of these has been called fibroblast growth factor (FGF) by Gospodarowicz and co-workers (247). This polypeptide, isolated from brain, will support the growth of many cell types when supplemented with a small amount of serum (0.6–0.8%). In 3T3 cells but not other cell types, FGF also requires the presence of hydrocortisone to stimulate growth (247). Another factor that will promote growth of resting fibroblasts is insulin. In BHK cells maintained in low serum the growth-promoting effect of insulin is particularly striking, but rather large amounts (8 μg/ml) are required (22). In contrast, FGF is effective in 3T3 cells at around 1 ng/ml. A third factor found to promote DNA synthesis in cultured cells is epidermal growth factor (32). However, this material has not been shown to promote cell division.

A principal action of FGF is to raise cyclic GMP levels in resting cells (179). FGF produces only a small fall in cyclic AMP levels. The rise in cyclic GMP levels promoted by FGF appears to be due to increased synthesis, since FGF activates guanylate cyclase in plasma membranes prepared from broken cell extracts (248). FGF also has a small inhibitory effect on adenylate cyclase but at much higher concentrations than those used to activate guanylate cyclase. Insulin, on the other hand, is an effective inhibitor of adenylate cyclase, and only at very high concentrations does it activate guanylate cyclase and then just slightly (179).

Because insulin is a poor growth factor for 3T3 cells, it has been argued that inhibition of adenylate cyclase and a fall in cyclic AMP levels are not sufficient to promote growth (179). Since insulin is a good growth factor for BHK cells (22) this argument is not entirely convincing. Certainly the action of FGF and insulin on cyclic GMP metabolism needs to be investigated. Further, it is probable that not all the growth-promoting effects of both FGF and insulin are mediated by cyclic nucleotides.

It is essential to bear in mind the state of the cell when evaluating the effects of growth factors. If nongrowing cells are studied at a time when cyclic AMP levels are high, then it would seem probable that a fall in cyclic AMP levels would be needed to initiate cell growth. It seems reasonably well established that treating cells with cyclic AMP analogs inhibits growth. On the other hand, if the growth of cells is arrested by a mechanism that does not raise cyclic AMP levels, then a fall in cyclic AMP would not be expected.

EFFECT OF CYCLIC AMP ON GROWTH OF TUMOR CELLS IN VIVO

The observation that cyclic AMP inhibits the growth of many types of cells in vitro has obvious implications for the chemotherapy of cancer. To date, there have been few publications on the effects of cyclic AMP in vivo. However, the results of the few experiments that have been published are quite interesting.

Prior treatment of CELO virus-transformed hamster cells (249), mouse neuroblastoma cells (250), and human KB tumor cells (251) with Bt_2cAMP and theophylline in culture reduces the ability of these cells to produce tumors when injected into animals. Injection of theophylline into animals infected with untreated transformed hamster cells also significantly inhibits the growth of some tumors (249).

Administration of cyclic AMP together with aminophylline to mice curbs the growth of Ehrlich tumor cells in both the solid and ascites form (252). Cyclic AMP, and also cyclic UMP and cyclic IMP to lesser degrees, slow the growth of a transplanted lymphosarcoma in mice (253).

Bt_2cAMP and theophylline as well as epinephrine and isoproterenol inhibit the growth of Walter 256 carcinoma cells and also block the enhanced tumor growth elicited by the addition of *Nippostrongylus brasiliensis* antiserum (254). (Antiserum to this nematode contains a blocking factor which inhibits some macrophage functions, thus potentiating tumor growth.)

Cho-Chung & Gullino have observed that Bt_2cAMP inhibits the growth of two estrogen-dependent mammary tumors in rats (255) and also a hormone-independent rat mammary carcinoma (256). The latter tumor was found to be composed of two cell types, Bt_2cAMP sensitive and insensitive populations (256).

It seems likely that a number of studies have been performed in which Bt_2cAMP has been ineffective and that these have not been reported. For example, in collaboration with others in the National Cancer Institute we have investigated the effect of Bt_2cAMP on the growth of a number of tumors in mice. We have seen no response with a polyoma-induced sarcoma (Py89), L1210 leukemia, Lewis lung carcinoma, or B1 melanoma cells (257).

The mechanism of inhibition of tumor growth in vivo by cyclic AMP analogs is not clear. Is it a direct or indirect effect on the tumor cells? In some cases an interaction of cyclic AMP with the immune system may be involved. For example, Rigby (258) reported that cyclic AMP inhibits tumor growth in mice, but only if mice are immunized with X-irradiated tumor cells prior to challenge with live tumor cells. Poly(AU), poly(IC), and theophylline retard the intradermal growth of Rauscher leukemia virus-induced tumors (259). Since these agents enhance the immune response, part of their actions may be due to an augmentation of the immune response. But part of their action apparently is not mediated by the immune system because significant inhibition of growth is also observed in irradiated mice and after treatment of cells with these agents prior to implantation.

MUTANTS IN CYCLIC AMP METABOLISM

Very little is known about how cyclic nucleotides act to inhibit growth, although some data are available on the stage of the cell cycle that is affected. In bacteria the isolation of mutants defective in cyclic AMP synthesis or in their response to cyclic AMP has been instrumental in elucidating how cyclic AMP controls gene expression (260, 261). A similar approach has been initiated with cultured cells. Willingham et al (262) have isolated a mutant of 3T3 cells in which the intracellular concentration of cyclic AMP changes after a change in temperature. Although these cells show rapid phenotype changes characteristic of cells with low cyclic AMP levels (decreased adhesiveness, retraction of cell processes and rounding, increased agglutinability), the growth of these cells is not affected. This is due to the fact that cyclic AMP levels are depressed transiently, for about 1 hr, and then overshoot to very high levels for 1 to 2 hr before returning to normal.

Tompkins and co-workers have taken advantage of the fact that cyclic AMP inhibits the growth and subsequently kills lymphoma S49 cells to isolate some very interesting mutant cell lines (59). Some of these have the ability to grow in the presence of Bt_2cAMP and theophylline. In one type of mutant the failure of cyclic AMP to inhibit growth is associated with a marked diminution in the amount of cyclic AMP-dependent protein kinase in extracts of these cells (263). Both the regulatory and catalytic subunit are diminished. This finding suggests that cyclic AMP acts to control growth by increasing the phosphorylation of one or more regulatory proteins.

A second type of mutant has been identified in which the amount of cyclic AMP-dependent protein kinase is normal (59). Apparently in these cells the mutation effects some step beyond the initial phosphorylation step.

The isolation of other mutants both in lymphoma cells and in fibroblastic cells and the characterization of the biochemical defects possessed by these cells will be very helpful in elucidating how cyclic AMP acts. Mutants will also be useful in determining a role for cyclic GMP. To date no mutants in cyclic GMP metabolism have been identified.

Literature Cited

1. Pastan, I., Johnson, G. S. 1974. *Advances in Cancer Research,* ed. G. Klein, S. Weinhouse, A. Haddow, 303–29. New York: Academic

2. Stoker, M. G. P., Rubin, H. 1967. *Nature* 215:171–72
3. Todaro, G. J., Lazar, G. K., Green, H. 1965. *J. Cell. Comp. Physiol.* 66:325–33

4. Burk, R. R. 1968. *Nature* 219:1272–75
5. Ryan, W. L., Heidrick, M. L. 1968. *Science* 162:1484–85
6. Johnson, G. S., Pastan, I. 1971. *J. Nat. Cancer Inst.* 47:1357–64
7. Johnson, G. S., Pastan, I., Peery, C. V., Otten, J., Willingham, M. C. 1972. *Prostaglandins in Cellular Biology and the Inflammatory Process,* ed. P. W. Ramwell, B. B. Pharriss, 195–219. New York: Plenum
8. Ishii, K., Green, H. 1973. *J. Cell Sci.* 13:429–39
9. Hilz, H., Kaukel, E. 1973. *Mol. Cell. Biochem.* 1:1–11
10. Heidrick, M. L., Ryan, W. L. 1971. *Cancer Res.* 31:1313–15
11. Johnson, G. S., Pastan, I. 1972. *J. Nat. Cancer Inst.* 48:1377–87
12. Johnson,G.S.,Friedman,R.M.,Pastan,I. 1971. *Proc. Nat. Acad. Sci. USA* 68:425–29
13. Johnson,G.S.,Friedman,R.M.,Pastan,I. 1971. *Ann. NY Acad. Sci.* 185:413–16
14. Taylor-Papadimitriou, J. 1974. *Int. J. Cancer* 13:404–11
15. Thomas, D. R., Philpott, G. W., Jaffe, B. M. 1974. *Exp. Cell Res.* 84:40–46
16. Oler, A., Iannaccone, P. M., Gordon, G. B. 1974. *In Vitro* 9:35–38
17. Gazdar, A., Hatanaka, M., Herberman, R., Russell, E., Ikawa, Y. 1972. *Proc. Soc. Exp. Biol. Med.* 139:1044–50
18. Grimes, W. J., Schroeder, J. L. 1973. *J. Cell Biol.* 56:487–91
19. Sheppard, J. R. 1971. *Proc. Nat. Acad. Sci. USA* 68:1316–20
20. Smets, L. A. 1972. *Nature New Biol.* 239:123–24
21. Paul, D. 1972. *Nature New Biol.* 240:179–81
22. Jimenez de Asua, L., Surian, E. S., Flawia, M. M., Torres, H. N. 1973. *Proc. Nat. Acad. Sci. USA* 70:1388–92
23. Blat, C., Boix, N., Harel, L. 1973. *Cancer Res.* 33:2104–8
24. Sakiyama, H., Robbins, P. W. 1973. *Arch. Biochem. Biophys.* 154:407–14
25. Frank, W. 1972. *Exp. Cell Res.* 71:238–41
26. Brailovsky, C., Trudel, M., Lallier, R., Nigam, V. N. 1973. *J. Cell Biol.* 57:124–32
27. Carchman, R. A., Johnson, G. S., Pastan, I., Scolnick, E. M. 1974. *Cell* 1:59–64
27a. Kurth, R., Bauer, H. 1973. *Nature New Biol.* 243:243–45
28. Froehlich, J. E., Rachmeler, M. 1972. *J. Cell Biol.* 55:19–31
29. Sandor, R. 1973. *J. Nat. Cancer Inst.* 51:257–59
30. Rozengurt, E., Pardee, A. B. 1972. *J. Cell. Physiol.* 80:273–79
31. Nagyvary, J., Gohill, R. N., Kirchner, C. R., Stevens, J. D. 1973. *Biochem. Biophys. Res. Commun.* 55:1072–77
32. Hollenberg, M. D., Cuatrecasas, P. 1973. *Proc. Nat. Acad. Sci. USA* 70:2964–68
33. Otten, J., Johnson, G., Pastan, I. 1971. *Biochem. Biophys. Res. Commun.* 44:1192–98
34. Sheppard, J. R. 1972. *Nature New Biol.* 236:14–16
35. Humphreys, T. 1972. *Cell Interaction,* ed. L. G. Silvestri, 264–76. Amsterdam: North-Holland
36. Otten, J., Johnson, G., Pastan, I. 1972. *J. Biol. Chem.* 247:7082–87
37. D'Armiento,M.,Johnson,G.S.,Pastan,I. 1973. *Nature New Biol.* 242:78–80
38. Bannai, S., Sheppard, J. R. 1974. *Nature* 250:62–64
39. Oey, J., Vogel, A., Pollack, R. 1974. *Proc. Nat. Acad. Sci. USA* 71:694–98
40. Burstein, S. J., Renger, H. C., Basilico, C. 1974. *J. Cell. Physiol.* 84:69–74
41. Seifert, W., Paul, D. 1972. *Nature New Biol.* 240:281–83
42. Kram, R., Mamont, P., Tomkins, G. M. 1973. *Proc. Nat. Acad. Sci. USA* 70:1432–36
43. Rein,A.,Carchman,R.A.,Johnson,G.S., Pastan, I. 1973. *Biochem. Biophys. Res. Commun.* 52:899–904
44. Seifert, W. E., Rudland, P. S. 1974. *Nature* 248:138–40
45. Burger, M. M., Bombik, B. M., Breckenridge, B. M., Sheppard, J. R. 1972. *Nature New Biol.* 239:161–63
46. Raska, K. 1973. *Biochem. Biophys. Res. Commun.* 50:35–41
47. Hovi, T., Keski-Oja, J., Vaheri, A. 1974. *Cell* 2:235–40
48. Willingham, M. C., Johnson, G. S., Pastan, I. 1972. *Biochem. Biophys. Res. Commun.* 48:743–48
49. Froehlich, J. E., Rachmeler, M. 1974. *J. Cell Biol.* 60:249–57
50. Furmanski, P., Lubin, M. 1973. *The Role of Cyclic Nucleotides in Carcinogenesis,* ed. J. Schultz, H. G. Gratzer, 239–54. New York: Academic
51. Bombik, B. M., Burger, M. M. 1973. *Exp. Cell Res.* 80:88–94
52. Noonan, K. D., Burger, M. M. 1973. *Exp. Cell Res.* 80:405–15
53. Zimmerman, J. E., Raska, K. 1972. *Nature New Biol.* 239:145–47
54. Goldberg, N. D., O'Dea, R. F., Haddox, M. K. 1973. *Advan. Cyclic Nucleotide Res.* 3:155–223
55. Bloch, A. 1974. *Biochem. Biophys. Res. Commun.* 58:652–59

56. Goldberg, N. D., Haddox, M. K., Dunham, E., Lopez, C., Hadden, J. W. 1974. *Control of Proliferation in Animal Cells,* ed. B. Clarkson, R. Baserga, 609–25. New York: Cold Spring Harbor Lab.
57. Zeilig, C. 1974. PhD thesis. Vanderbilt Univ., Nashville, Tenn.
58. Remington, J. A., Klevecz, R. R. 1973. *Biochem. Biophys. Res. Commun.* 50: 140–46
59. Bourne, H. R., Coffino, R., Melmon, K. L., Tomkins, G. M., Weinstein, Y. *Advan. Cyclic Nucleotide Res.* In press
59a. Smets, L. A. 1973. *Nature New Biol.* 243: 113–15
60. Millis, A. J. T., Forrest, G. A., Pious, D. A. 1974. *Exp. Cell Res.* 83: 335–43
61. Van Wijk, R., Wicks, W. D., Bevers, M. M., Van Rijn, J. 1973. *Cancer Res.* 33: 1331–38
62. Pardee, A. B. 1974. *Proc. Nat. Acad. Sci. USA* 71: 1286–90
63. Seifert, W., Rudland, P. S. 1975. *Proc. Nat. Acad. Sci. USA.* In press
64. Whitfield, J. F., Rixon, R. H., MacManus, J. P., Balk, S. D. 1973. *In Vitro* 8: 257–78
65. Hovi, T., Vaheri, A. 1973. *Nature New Biol.* 245: 175–77
66. Schor, S., Rozengurt, E. 1974. *J. Cell. Physiol.* 81: 339–46
67. Johnson, G. S. Unpublished data
68. Nowell, P. C. 1960. *Cancer Res.* 20: 462–66
69. Hirschhorn, R. 1974. *Cyclic AMP, Cell Growth, and the Immune Response,* ed. W. Braun, L. M. Lichtenstein, C. W. Parker, 45–54. New York: Springer
70. Hadden, J. W., Hadden, E. M., Good, R. A. 1971. *Biochim. Biophys. Acta* 237: 339–47
71. MacManus, J. P., Whitfield, J. F. 1969. *Exp. Cell Res.* 58: 188–91
72. Franks, D. J., MacManus, J. P., Whitfield, J. F. 1971. *Biochem. Biophys. Res. Commun.* 44: 1177–83
73. MacManus, J. P., Whitfield, J. F., Youdale, T. 1971. *J. Cell. Physiol.* 77: 103–16
74. Whitfield, J. F., MacManus, J. P., Braceland, B. M., Gillan, D. J. 1972. *J. Cell Physiol.* 79: 353–62
75. Zeilig, C. E., Johnson, R. A., Friedman, D. L., Sutherland, E. W. 1972. *J. Cell Biol.* 55(2, Pt. 2): 296a (Abstr.)
76. Sheppard, J. R., Prescott, D. M. 1972. *Exp. Cell Res.* 75: 293–96
77. Millis, A. J. T., Forrest, G., Pious, D. A. 1972. *Biochem. Biophys. Res. Commun.* 49: 1645–49
78. MacManus, J. P., Whitfield, J. F. 1971. *Exp. Cell Res.* 69: 281–88

79. Whitfield, J. F., MacManus, J. P., Gillan, D. J. 1973. *J. Cell. Physiol.* 81: 241–50
80. MacManus, J. P., Whitfield, J. F., Rixon, R. H. See Ref. 69, 302–16
81. Hirschhorn, R., Grossman, J., Weissmann, G. 1970. *Proc. Soc. Exp. Biol. Med.* 133: 1361–65
82. Cross, M. E., Ord, M. G. 1971. *Biochem. J.* 124: 241–48
83. Rigby, P. G., Ryan, W. L. 1970. *Eur. J. Clin. Biol. Res.* 15: 774–77
84. McCrery, J. E., Rigby, P. G. 1972. *Proc. Soc. Exp. Biol. Med.* 140: 1456–59
85. Whitney, R. B., Sutherland, R. M. 1972. *J. Immunol.* 108: 1179–83
86. Krishnaraj, R., Talwar, G. P. 1973. *J. Immunol.* 111: 1010–17
87. Novogrodsky, A., Katchalski, E. 1970. *Biochim. Biophys. Acta* 215: 291–96
88. Johnson, L. D., Abell, C. W. 1970. *Cancer Res.* 30: 2718–23
89. Gallo, R. C., Whang-Peng, J. 1971. *J. Nat. Cancer Inst.* 47: 91–94
90. Averner, M. J., Brock, M. L., Jost, J. P. 1972. *J. Biol. Chem.* 247: 413–17
91. Mendelsohn, J., Multer, M. M., Bome, R. F. 1973. *J. Clin. Invest.* 5: 2129–37
92. Abell, C. W., Monahan, T. M. 1973. *J. Cell Biol.* 59: 549–58
93. Smith, J. W., Steiner, A. L., Newberry, W. M. Jr., Parker, C. W. 1971. *J. Clin. Invest.* 50: 432–41
94. Webb, D. R., Stites, D. P., Perlman, J., Austin, K. E., Fudenberg, H. H. See Ref. 69, 55–76
95. Hadden, J. W., Hadden, E. M., Haddox, M. K., Goldberg, N. D. 1972. *Proc. Nat. Acad. Sci. USA* 69: 3024–27
96. Smith, J. W., Steiner, A. L., Parker, C. W. 1971. *J. Clin. Invest.* 50: 442–48
97. Sulzer, B. M., Craig, J. P. 1973. *Nature New Biol.* 244: 178–80
98. Holmgren, J., Lindholm, L., Lonnroth, I. 1974. *J. Exp. Med.* 139: 801–19
99. Monahan, T. M., Fritz, R. R., Abell, C. W. 1973. *Biochem. Biophys. Res. Commun.* 55: 642–46
100. Whitfield, J. F., MacManus, J. P., Franks, D. J., Gillan, D. J., Youdale, T. 1971. *Proc. Soc. Exp. Biol. Med.* 137: 453–57
101. Whitfield, J. F., MacManus, J. P. 1972. *Proc. Soc. Exp. Biol. Med.* 139: 818–24
102. Whitfield, J. F., MacManus, J. P., Rixon, R. H., Gillan, D. J. 1973. *Proc. Soc. Exp. Biol. Med.* 144: 808–12
103. Parker, C. W. See Ref. 69, 35–44
104. Hadden, J. W., Hadden, E. M., Meetz, C., Good, R. A., Haddox, M. K., Goldberg, N. D. 1973. *Fed. Proc.* 32: 1022
105. Bloch, A., Dutschman, G., Maue, R.

1974. *Biochem. Biophys. Res. Commun.* 59:955–59

106. Weber, T. H., Skoog, V. T., Mattsson, A., Lindahl-Kiessling, K. 1974. *Exp. Cell Res.* 85:351–61
107. Teh, H., Paetkau, V. 1974. *Nature* 250:505–7
108. Bullough, W. S., Laurence, E. B. 1961. *Proc. Roy. Soc. London* 154:540–56
109. Goltz, R. W., Fusaro, R. M., Jarvis, J. 1958. *J. Invest. Dermatol.* 31:331–41
110. Bronstad, G. O., Elgjo, K., Oye, I. 1971. *Nature New Biol.* 233:78–79
111. Powell, J. A., Duell, E. A., Voorhees, J. J. 1971. *Arch. Dermatol.* 104:354–65
112. Voorhees, J. J., Duell, E. A., Kelsey, W. H. 1972. *Arch. Dermatol.* 105:384–86
113. Marks, F., Rebien, W. 1972. *Naturwissenschaften* 59:41–42
114. Murray, A. W., Verma, A. K. 1973. *Biochem. Biophys. Res. Commun.* 54:69–74
115. Curtis, G. L., Stenbeck, F., Ryan, W. L. 1974. *Cancer Res.* 34:2192–95
116. Voorhees, J. J., Duell, E. A., Bass, L. J., Powell, J. A., Harrell, E. R. 1972. *Arch. Dermatol.* 105:695–701
117. Hsia, S. L., Wright, R. K., Halprin, K. M. See Ref. 50, 303–23
118. Voorhees, J. J., Starviski, M., Duell, E. A., Haddox, M. K., Goldberg, N. D. 1973. *Life Sci.* 13:639–53
119. Johnson, G. S., Pastan, I. 1972. *Nature New Biol.* 237:267–68
120. Wong, G., Pawelek, J. 1973. *Nature New Biol.* 241:213–15
121. Kreider, J. W., Rosenthal, M., Lengle, N. 1973. *J. Nat. Cancer Inst.* 50:555–58
122. Wong, G., Pawelek, J., Sansone, M., Morowitz, J. 1974. *Nature* 248:351–54
123. O'Keefe, E., Cuatrecasas. P. 1974. *Proc. Nat. Acad. Sci. USA* 71:2500–4
124. Van Wijk, R., Wicks, W. D., Clay, K. 1972. *Cancer Res.* 32:1905–11
125. Wicks, W. D. et al. See Ref. 50, 103–24
126. Weber, G. See Ref. 50, 57–102
127. Stellwagen, R. H. 1974. *Biochim. Biophys. Acta* 338:428–39
128. Butcher, F. R., Scott, D. F., Potter, V. R., Morris, H. P. 1972. *Cancer Res.* 32:2135–40
129. Thomas, E. W., Murad, F., Looney, W. B., Morris, H. P. 1973. *Biochim. Biophys. Acta* 297:564–67
130. Hickie, R. A., Walker, C. M., Croll, G. A. 1974. *Biochem. Biophys. Res. Commun.* 59:167–73
131. Chayoth, R., Epstein, S. M., Field, J. B. 1973. *Cancer Res.* 33:1970–74
132. Prasad, K. N., Kumar, S. See Ref. 56, 581–94

133. Furmanski, P., Silverman, D. J., Lubin, M. 1971. *Nature* 233:413–15
134. Lim, R., Mitsunobu, K. 1972. *Life Sci.* 11:1063–70
135. Jaffe, B. M., Philpott, G. W., Hamprecht, B., Parker, C. W. 1972. *Advan. Biosci.* 9:179–82
136. MacIntyre, E. H., Wintersgill, C. J., Perkins, J. P., Vatter, A. E. 1972. *J. Cell Sci.* 11:639–67
137. Shapiro, D. L. 1973. *Nature* 241:203–4
138. Richelson, E. 1973. *Nature New Biol.* 242:175–77
139. Teel, R. W., Hall, R. G. 1973. *Exp. Cell Res.* 76:390–94
140. Newsome, D. A., Fletcher, R. T., Robison, W. G., Kenyon, K. R., Chader, G. J. 1974. *J. Cell Biol.* 61:369–82
141. Masui, H., Garren, L. D. 1971. *Proc. Nat. Acad. Sci. USA* 68:3206–10
142. Hertelendy, F., Keay, L. 1974. *Prostaglandins* 6:217–25
143. Wahrmann, J. P., Winand, R., Luzzati, D. 1973. *Nature New Biol.* 245:112–13
144. Keller, R., Keist, R. 1973. *Life Sci.* 12:97–105
145. Thomas, D. B., Midley, G., Lingwood, C. A. 1973. *J. Cell Biol.* 57:397–405
146. Tisman, G., Herbert, V. 1973. *In Vitro* 9:86–91
147. Wood, H. N., Lin, M. C., Braun, A. C. 1972. *Proc. Nat. Acad. Sci. USA* 69:403–6
148. Johnson, G. S., D'Armiento, M., Carchman, R. A. 1974. *Exp. Cell Res.* 85:47–56
149. Wood, H. N., Braun, A. C. 1973. *Proc. Nat. Acad. Sci. USA* 70:447–50
150. Dekhuijzen, H. M., Overeem, J. C. 1972. *Phytochemistry* 11:1669–72
151. Drlica, K. A., Gardner, J. M., Kado, C. I., Vijay, I. K., Troy, F. A. 1974. *Biochem. Biophys. Res. Commun.* 56:753–59
152. Giannattasio, M., Mandato, E., Macchia, V. 1974. *Biochem. Biophys. Res. Commun.* 57:365–71
153. Anderson, W. B., Russell, T. R., Carchman, R. A., Pastan, I. 1973. *Proc. Nat. Acad. Sci. USA* 70 (Part II):3802–5
154. Makman, M. H. 1971. *Proc. Nat. Acad. Sci. USA* 68:2127–30
155. Zacchello, F., Bensen, P. F., Giannelli, F., McGuire, M. 1972. *Biochem. J.* 126:27p
156. Anderson, W. B., Gallo, M., Pastan, I. 1974. *J. Biol. Chem.* 249:7041–48
157. Makman, M. H., Klein, M. I. 1972. *Proc. Nat. Acad. Sci. USA* 69:456–58
158. Robison, G. A., Butcher, R. W., Sutherland, E. W. 1968. *Ann. Rev. Biochem.* 37:149–74

159. Wolff, J., Cook, G. H. 1973. *J. Biol. Chem.* 248:350–55
160. Sweat, F. W., Wincek, T. J. 1973. *Biochem. Biophys. Res. Commun.* 55:522–29
161. Krishna, G., Harwood, J. P., Barber, A. J., Jamieson, G. A. 1972. *J. Biol. Chem.* 247:2253–54
162. MacManus, J. P., Whitfield, J. F. 1974. *Prostaglandins* 6:475–87
163. Remold-O'Donnell, E. 1974. *J. Biol. Chem.* 249:3615–21
164. Peery, C. V., Johnson, G. S., Pastan, I. 1971. *J. Biol. Chem.* 246:5785–90
165. Kelly, L. A., Butcher, R. W. 1974. *J. Biol. Chem.* 249:3098–3102
166. Rodbell, M., Birnbaumer, L., Pohl, S. L., Krans, H. M. J. 1971. *J. Biol. Chem.* 246:1877–82
167. Rodbell, M., Lin, M. C., Salomon, Y. 1974. *J. Biol. Chem.* 249:59–65
168. Bilezikian, J. P., Aurbach, G. D. 1974. *J. Biol. Chem.* 249:157–61
169. Bockaert, J., Roy, C., Jard, S. 1972. *J. Biol. Chem.* 247:7073–81
170. Leray, F., Chambaut, A. M., Hanoune, J. 1972. *Biochem. Biophys. Res. Commun.* 48:1385–91
171. Varga, J. M., DiPasquale, A., Pawelek, J., McGuire, J. S., Lerner, A. B. 1974. *Proc. Nat. Acad. Sci. USA* 71:1590–93
172. Clark, R. B., Perkins, J. P. 1971. *Proc. Nat. Acad. Sci. USA* 68:2757–60
173. Makman, M. H., Dvorkin, B., Keehn, E. See Ref. 56, 649–63
174. Schwartz, J. P., Morris, N. R., Breckenridge, B. M. 1973. *J. Biol. Chem.* 248:2699–2704
175. Temin, H. 1967. *J. Cell. Physiol.* 69:377–84
176. Hepp, K. D., Renner, R. 1972. *FEBS Lett.* 20:191–94
177. Hepp, K. D. 1972. *Eur. J. Biochem.* 51:266–76
178. Illiano, G., Cuatrecasas, P. 1972. *Science* 175:906–8
179. Rudland, P. S., Gospodarowicz, D., Seifert, W. 1974. *Nature* 250:741–42, 773–74
180. Robison, G. A., Butcher, R. W., Oye, I., Morgan, H. W., Sutherland, E. W. 1965. *Mol. Pharmacol.* 1:168–77
181. Butcher, R. W., Ho, R. J., Meng, H. C., Sutherland, E. W. 1965. *J. Biol. Chem.* 240:4515–23
182. Exton, J. H., Robison, G. A., Sutherland, E. W., Park, C. R. 1971. *J. Biol. Chem.* 246:6166–77
183. Makman, M. H. 1971. *Proc. Nat. Acad. Sci. USA* 68:885–89
184. Manganiello, V., Vaughan, M. 1972. *Proc. Nat. Acad. Sci. USA* 69:269–73
185. Franklin, T. J., Foster, S. J. 1973. *Nature New Biol.* 246:146–48
186. Ho, R. J., Sutherland, E. R. 1971. *J. Biol. Chem.* 246:6822–27
187. Anderson, W. B., Gallo, M., Pastan, I. Unpublished data
188. Cuatrecasas, P., Tell, G. P. E. 1973. *Proc. Nat. Acad. Sci. USA* 70:495–98
189. Majerus, P. W., Brodie, G. N. 1972. *J. Biol. Chem.* 247:4253–57
190. Field, M. 1971. *N. Engl. J. Med.* 284:1137–44
191. Pierce, N. F., Greenough, W. B., Carpenter, C. C. J. 1971. *Bacteriol. Rev.* 35:1–13
192. Sharp, G. W. G. 1973. *Ann. Rev. Med.* 24:19–28
193. Holmgren, J., Lonnroth, I., Svennerholm, L. 1973. *Scand. J. Infect. Dis.* 5:77–78
194. Cuatrecasas, P. 1973. *Biochemistry* 12:3558–66
195. van Heyningen, S. 1974. *Science* 183:656–57
196. Noonan, K. D., Burger, M. M. 1973. *J. Biol. Chem.* 248:4286–92
197. Noonan, K. D., Levine, A. J., Burger, M. M. 1973. *J. Cell Biol.* 58:491–97
198. Bosmann, H. B., Winston, R. A. 1970. *J. Cell Biol.* 45:23–33
199. Glick, M. C., Buck, C. A. 1973. *Biochemistry* 12:85–89
200. Chatterjee, S., Sweeley, C. C., Velicer, L. F. 1973. *Biochem. Biophys. Res. Commun.* 54:585–92
201. Gahmberg, C. G., Hakomori, S. 1974. *Biochem. Biophys. Res. Commun.* 59:283–91
202. Cikes, M., Friberg, S. 1971. *Proc. Nat. Acad. Sci. USA* 68:566–69
203. Thomas, D. B. 1971. *Nature* 233:317–21
204. Hakomori, S. 1970. *Proc. Nat. Acad. Sci. USA* 67:1741–47
205. Sakiyama, H., Gross, S. K., Robbins, P. W. 1972. *Proc. Nat. Acad. Sci. USA* 69:872–76
206. Critchley, D., MacPherson, I. 1973. *Biochim. Biophys. Acta* 246:145–59
207. Gahmberg, C. G., Hakomori, S. 1973. *Proc. Nat. Acad. Sci. USA* 70:3329–33
208. Gahmberg, C. G., Kiehn, D., Hakomori, S. 1974. *Nature* 248:413–15
209. Granner, D., Chase, L. R., Aurbach, G. D., Tomkins, G. M. 1968. *Science* 162:1018–20
210. Emmelot, P., Bos, C. J. 1971. *Biochim. Biophys. Acta* 249:285–92
211. Brown, H. D., Chattopadhyay, S. K., Morris, H. P., Pennington, S. N. 1970. *Cancer Res.* 30:123–26
212. Pennington, S. N., Brown, H. D.,

Chattopadhyay, S. K., Conaway, C., Morris, H. P. 1970. *Experimentia* 26: 139

213. Makman, M. H. 1970. *Science* 170: 1421–23
214. Allen, D. O., Munshower, J., Morris, H. P. Weber, G. 1971. *Cancer Res.* 31:557–60
215. Christoffersen, T., Moerland, J., Osnes, J. B., Elgjo, K. 1972. *Biochim. Biophys. Acta* 279:363–66
216. Boyd, H., Louis, C. J., Martin, T. J. 1974. *Cancer Res.* 34:1720–25
217. Tomasi, V., Rethy, A., Trevisani, A. See Ref. 50, 127–52
218. Rethy, A., Varzi, L., Toth, F. D., Boldogh, I. See Ref. 50, 153–79
219. Ney, R. L., Hochella, N. J., Grahame-Smith, J. G., Dexter, R. N., Butcher, R. W. 1969. *J. Clin. Invest.* 48:1733–39
220. Schorr, I., Rathnam, P., Saxena, B. B., Ney, R. L. 1971. *J. Biol. Chem.* 246: 5806–11
221. Schimmer, B. P. 1972. *J. Biol. Chem.* 247:3134–38
222. Sharma, R. K., Hashimoto, K. E. 1973. *Cancer Res.* 32:666–74
223. Sharma, R. K. 1973. *Eur. J. Biochem.* 32:506–12
224. Brush, J. S., Sutliff, L. S., Sharma, R. K. 1974. *Cancer Res.* 34:1495–1502
225. Schorr, I., Hinshaw, H. T., Cooper, M. A., Mahaffee, D., Ney, R. L. 1972. *J. Clin. Endocrinol.* 34:447–51
226. Goldfine, I. D., Roth, J., Birnbaumer, L. 1972. *J. Biol. Chem.* 247:1211–18
227. Kemp, R. G., Duquesnoy, R. J. 1973. *Science* 183:218–19
228. Polgar, P., Vera, J. C., Kelley, P. R., Rutenburg, A. M. 1973. *Biochim. Biophys. Acta* 297:378–83
229. Brown, H. D., Chattopadhyay, S. K., Spjut, H. J., Spratt, J. S. Jr., Pennington, S. N. 1969. *Biochim. Biophys. Acta* 192:372–75
230. Hsia, S. L., Wright, R., Mandy, S. H., Halprin, K. M. 1972. *J. Invest. Dermatol.* 59:109–13
231. Wright, R. K., Mandy, S. H., Halprin, K. M., Hsia, S. L. 1973. *Arch. Dermatol.* 107:47–53
232. Voorhees, J. et al. See Ref. 50, 325–73
233. Voorhees, J. J., Duell, E. A., Stawiski, M., Harrell, E. R. 1974. *Advan. Cyclic Nucleotide Res.* 4:117–62
234. Sheppard, J. R., Bannai, S. See Ref. 56, 571–79
235. Weiss, B., Shein, H. M., Snyder, R. 1971. *Life Sci.* 10 (Part 1): 1253–60
236. Yoshikawa-Fukada, M., Nojima, T.

1972. *J. Cell. Physiol.* 80:421–30
237. Anderson, W. B., Johnson, G. S., Pastan, I. 1973. *Proc. Nat. Acad. Sci. USA* 70:1055–59
238. Anderson, W. B., Lovelace, E., Pastan, I. 1973. *Biochem. Biophys. Res. Commun.* 52:1293–99
239. Anderson, W. B., Gallo, M., Pastan, I. Unpublished data
240. Anderson, W. B., Pastan, I. 1975. *Advan. Cyclic Nucleotide Res.* 5: In press
241. Russell, T., Pastan, I. 1973. *J. Biol. Chem.* 248:5835–40
242. Russell, T. R., Pastan, I. 1975. *J. Biol. Chem.* In press
243. D'Armiento, M., Johnson, G. S., Pastan, I. 1972. *Proc. Nat. Acad. Sci. USA* 69:459–62
244. Russell, T. R., Pastan, I. Unpublished results
245. Rhoads, A. R., Morris, H. P., West, W. L. 1973. *Cancer Res.* 32:2651–55
246. Clark, J. F., Morris, H. P., Weber, G. 1973. *Cancer Res.* 33:356–61
247. Gospodarowicz, D. 1974. *Nature* 249: 123–27
248. Rudland, P. S., Hamilton, M., Hamilton, R., Gospodarowicz, D., Seifert, W. 1975. *J. Biol. Chem.* Submitted
249. Reddi, P. K., Constantinides, S. M. 1972. *Nature* 238:286–87
250. Prasad, K. N. 1972. *Cytobios* 6:163
251. Smith, E. E., Handler, A. H. 1973. *Res. Commun. Chem. Pathol. Pharmacol.* 5:863–66
252. Seller, M. J., Benson, P. F. 1973. *Eur. J. Cancer* 9:525–26
253. Gericke, D., Chandra, P. 1971. *Z. Physiol. Chem.* 350:1469–71
254. Keller, R. 1972. *Life Sci.* 11:485–91
255. Cho-Chung, Y. S., Gullino, P. M. 1974. *Science* 183:87–88
256. Cho-Chung, Y. S., Gullino, P. M. 1974. *J. Nat. Cancer Inst.* 52:995–96
257. Pastan, I., Johnson, G. S., Oyer, D. Unpublished data
258. Rigby, P. 1972. *Cancer Res.* 32:455–57
259. Webb, D., Braun, W., Plescia, O. J. 1972. *Cancer Res.* 32:1814–19
260. Pastan, I., Perlman, R. 1970. *Science* 169:339–44
261. Perlman, R. L., Pastan, I. 1971. *Curr. Top. Cell. Regul.* 3:117–34
262. Willingham, M., Carchman, R., Pastan, I. 1973. *Proc. Nat. Acad. Sci. USA* 70: 2906–10
263. Daniel, V., Litwack, G., Tompkins, G. M. 1973. *Proc. Nat. Acad. Sci. USA* 70:76–79

THE ENERGETICS OF BACTERIAL ACTIVE TRANSPORT ✖894

Robert D. Simoni

Department of Biological Sciences, Stanford University,
Stanford, California 94305

Pieter W. Postma

Laboratory of Biochemistry, BCP Jansen Institute,
University of Amsterdam, Amsterdam, The Netherlands

CONTENTS

INTRODUCTION TO TRANSPORT MECHANISM[1]

The general topic of solute transport is enormously broad. This review will concentrate on a few selected aspects concerning the mechanism of coupling metabolic energy to active transport in bacterial cells. Such studies are particularly relevant to both transport phenomena and energy transduction in biological systems. The scope will be restricted to bacterial systems even though most of the ideas discussed in this field originate from studies on mitochondria and chloroplasts.

In the simplest sense, one can consider solute translocation systems as composed of two distinct elements; a solute-specific membrane carrier and a system for energy coupling. The solute-specific membrane carrier provides solute recognition and translocation. When considered alone, such a component or components act as a facilitated diffusion system. The characteristics are a saturable, solute-specific carrier which moves solute down a concentration gradient at rates far more rapid than would be predicted for diffusion of a hydrophilic molecule, such as a sugar, amino acid, or ion, across a hydrophobic barrier. This mechanism involves a symmetrical carrier system that will transport solute across the membrane in both directions with equivalent kinetic properties. Obviously, an imposed solute gradient provides the only energy necessary, and no metabolic coupling is required. There are many examples of facilitated diffusion; glucose is transported into most animal cells (the red blood cell is the classical example) by this mechanism.

It is conceptually convenient, although somewhat inaccurate, to consider that systems with a preferential direction of solute movement consist of a facilitated diffusion carrier to which an energetic component has been added. The introduction of energy results in an asymmetry such that solute gradients are established.

Unfortunately, little information is available on the translocation mechanisms at the molecular level. In contrast, a variety of mechanisms have been described to explain solute accumulation in many systems. These are reviewed below.

Solute Modification (Group Translocation)

One possibility for assuring unidirectional solute flow is to modify the solute as it is transported, thus preventing exit via the same solute-specific carrier. Such a mechanism does not fit the strict definition of active transport, which requires the solute to be accumulated unaltered; however, the consequences are the same. This type of mechanism is best exemplified by the phosphoenolpyruvate (PEP)-sugar phosphotransferase system, originally described by Kundig et al (1) for *Escherichia coli*. The system is widespread in the bacterial world, having been demonstrated in anaerobes and facultative organisms. Space prohibits a description of this system; however, the reader is directed to other reviews (2–5).

---

[1] Abbreviations used in this review are: DCCD, N,N'-dicyclohexylcarbodiimide; DDA$^+$, dibenzyldimethyl ammonium; CCCP, carbonylcyanide m-chlorophenylhydrazone; FCCP, carbonylcyanide p-trifluoromethoxyphenylhydrazone; OSCP, oligomycin sensitivity-conferring protein; PCB$^-$, phenyldicarbaundecaborane; PMS, phenazine methosulfate; TMG, thiomethyl galactoside; and TPB$^-$, tetraphenylboron.

Carrier Modification

Several systems have been described in which the energy component is directed toward modification of the solute carrier itself. The best studied examples are the ion-specific ATPases, particularly the Na^+-K^+-dependent ATPase described for a variety of animal cell systems, and the Ca^{2+}-ATPase from sarcoplasmic reticulum. Specific ion movements in these systems occur only as the result of the reversible phosphorylation of the carrier component by ATP. Thus the carrier exists in either the phosphorylated or dephosphorylated state, each of which has differential ion binding properties. The reader is directed to other reviews (6, 7, 72) for additional information on this area of active transport.

Indirect Coupling (Cotransport)

Solute modification or carrier modification mechanisms affect the transport system directly in that energy input such as the high energy phosphate bond is directly applied to that system. There is, however, considerable evidence to suggest that energy and transport can be coupled via indirect mechanisms in both animal and bacterial systems, and these involve neither solute nor carrier modification. Certainly one of the most familiar systems is Na^+ cotransport of amino acids and sugars in animal cells. The active transport of glucose in intestinal epithelium can be represented by a model originally proposed by Crane (8). The brush border side of the intestinal epithelial cell contains a facilitated diffusion carrier for glucose which requires concomitant binding of Na^+ in order to translocate efficiently. Therefore glucose and Na^+ are transported simultaneously by the same carrier. The accumulation of glucose is dependent on a component that pumps Na^+ out of the cell, thus keeping the level of intracellular Na^+ well below that of the outside. This results in the accumulation of intracellular glucose. While ATP is required for the Na^+ extrusion, the only link to the glucose carrier is the Na^+ concentration gradient itself. Since concomitant Na^+ and glucose binding to the carrier are required for maximal transport and the Na^+ (out)/Na^+(in) is high, glucose entry will be greater than glucose exit, resulting in $glucose_{in}/glucose_{out} > 1$.

Thus glucose accumulation is dependent on the generation of an electrochemical Na^+ gradient in the opposite direction, and in this case neither the carrier nor the solute need be directly involved in the energy-coupling process.

Although Na^+-melibiose cotransport system has been proposed by Stock & Roseman in Salmonella typhimurium (9), most ion-solute symporters in bacteria seem to be dependent on a proton gradient. The remainder of this review will be concerned primarily with these transport systems.

Because, as we hope to show, there exists a close connection between the energetics of active transport and the conservation of energy during the process of oxidative phosphorylation, we will first present the biochemical and physiological data that bear on the question of energy coupling and then the evidence obtained from the analysis of various mutants that have been altered in components involved in energy transduction.

ENERGY COUPLING TO TRANSPORT IN CELLS
AND MEMBRANE VESICLES

Most transport studies have been performed with intact cells that have the advantage of physiological relevance, but the disadvantage that results is sometimes difficult to interpret. One step toward overcoming the latter has been the development and use of isolated membrane vesicles, as originated by Kaback and his collaborators (for review, see 10). Some controversy has developed in attempts to explain the conflicting results obtained in these systems, and the discussion in this section is intended to help resolve these problems. For this reason, it is convenient to place the discussion in a conceptual framework. It now becomes apparent that energy-coupling mechanisms can best be interpreted by the chemiosmotic theory of transport and metabolism first formulated by Mitchell, which is briefly outlined below.

Chemiosmosis

The basic postulate in Mitchell's concept of energy transduction, whether it be for oxidative phosphorylation or solute transport, is that the hydrogen and electron carriers of the respiratory chain are arranged in loops across the cytoplasmic membrane in such a way that electron flow via the chain results in translocation of protons from one side of the membrane to the other. Since the membrane is virtually impermeable to protons, the result is generation of a proton motive force (Δp), which is composed of a membrane potential ($\Delta\Psi$) and a pH gradient (ΔpH) such that $\Delta p = \Delta\Psi - Z\Delta pH$, where $Z = 2.3\ RT/F$ and under usual conditions has a value of about 60 mV. The relevance of this proton motive force to transport processes becomes clear from Mitchell's proposal that solute movement is coupled to the movement of protons. For those interested in mitochondria and chemiosmosis, the reader is referred to a number of excellent reviews (11–15). Here only a few observations with bacterial systems will be discussed.

Much of the evidence to support chemiosmosis depends on the ability to measure a membrane potential. Measurement of the sign and magnitude of the membrane potential has been attempted in a number of bacterial systems by measuring the distribution of lipid-soluble anions and cations across the membrane, a technique introduced by Skulachev, Liberman, and their co-workers (16–19). Using a lipid-soluble cation, DDA^+, Harold & Papineau (20, 21) determined a membrane potential of 150 to 200 mV, interior negative, for cells of *Streptococcus faecalis*. Using a fluorescent probe, 1,1'-dihexyl-2,2'-oxycarbocyanine (22), Laris & Perhadsingh (23) calculated a membrane potential of 140 mV in the same bacteria. This probe has also been used by Kashket & Wilson (24) to estimate the potential in *Streptococcus lactis* to be about 40 mV in the presence of glucose. Hirata et al (25) found a value of 100 mV, interior negative, in membrane vesicles of *E. coli*. Using DDA^+ uptake, Griniuviene et al (26) calculated a value of 140 mV, interior negative, for *E. coli* cells. Grinius et al (27) showed that sonicated vesicles of *Micrococcus lysodeikticus* take up PCB^- or TPB^-, indicating an interior positive potential and presumably inside-out vesicles. Scholes & Mitchell (28) have calculated from proton conductance in *Micrococcus denitrificans* a membrane potential of about 250 mV, interior negative. Jeacocke et al (29) have demonstrated a similar potential in *Staphylococcus aureus* based on the distribution of potassium, a method introduced for mitochondria by Mitchell & Moyle (30). This evidence can leave little doubt that bacteria generate a membrane potential, interior negative, of the order of 50 to 250 mV.

In related studies, a more direct measurement of a membrane potential was reported recently by Skulachev and co-workers (31), who found that illumination of bacteriorhodopsin, incorporated in a planar phospholipid bilayer, resulted in a potential of about 50 mV. It was dependent on light and sensitive to uncouplers. Kayushin & Skulachev (32), using bacteriorhodopsin in phospholipid vesicles, have shown that illumination results in uptake of PCB⁻ and the quenching of atebrine fluorescence. The authors suggest that both a membrane potential and a pH gradient are generated. In a recent study, Racker & Stoeckenius (33) demonstrated that a system composed of lipid vesicles, bacteriorhodopsin, ATPase (F_1), oligomycin sensitivity-conferring protein (OSCP), and mitochondrial hydrophobic proteins can carry out light-dependent ATP synthesis. Thus it appears that the proton motive force is capable of driving ATP synthesis.

Since the demonstration of respiration-driven electrogenic proton extrusion in *M. denitrificans* by Scholes & Mitchell (34), this phenomenon has been established in a number of bacteria (26, 29, 35–37) and in *E. coli* membrane vesicles (38, 39). Respiration-driven uptake of protons is observed in membrane vesicles in which the orientation of the membrane has been inverted (40, 41). An alternative to oxidation-dependent proton extrusion exists in the proton-translocating ATPase (11, 42). In

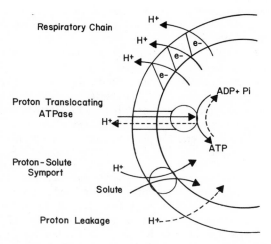

Figure 1   Chemiosmotic coupling. Proton motive force $\Delta p = \Delta \Psi - Z\Delta pH$. From Mitchell (179).

mitochondria, this system is usually considered to facilitate proton influx with concomitant ATP synthesis. Since the proton-translocating ATPase is reversible, ATP hydrolysis should catalyze proton extrusion (or in inverted vesicles proton influx). This property has also been demonstrated by Hertzberg & Hinkle (41) and by West & Mitchell (43) in E. coli. These results are similar to earlier findings with mitochondria and submitochondrial particles.

Critical to our discussion of active transport is the electrogenic proton imbalance which can be generated either by flow of reducing equivalents through the respiratory chain or by ATP hydrolysis by the Mg^{2+}-Ca^{2+}-ATPase complex. Mitchell (44) proposes that a number of solutes are carried across the membrane by so-called proton-solute symporters which are driven by the proton motive foce. This basic scheme is illustrated in Figure 1. One of the implications of such a proposal is that artificially generated membrane potentials should be able to drive a variety of energy-dependent processes. It has been demonstrated that ion movements down an electrochemical gradient can lead to ATP synthesis in mitochondria (45), red cells (46), and sarcoplasmic reticulum (47). Experiments that bear on the interaction between solute transport and the proton motive force are discussed below.

Energy Coupling in Intact Cells

By definition, energy is required for the accumulation of a solute against its electrochemical potential in a process called active transport. A major goal during the last few years has been to describe the mechanism coupling metabolism and transport (for reviews, see 10, 12, 48, 49). Most models proposed for this coupling have implicated the process of oxidative phosphorylation to explain this aspect of the transport phenomenon. [Curiously, Boyer and co-workers (50) have recently proposed that ATP is synthesized on the mitochondrial ATPase molecule and subsequently released by energization from electron flow much in the way Winkler & Wilson (51) suggested that energization caused solute to be released at the internal face of the membrane.]

Both ATP and a high energy intermediate of oxidative phosphorylation have been implicated by a number of workers (52–55). Boyer & Klein (49) have proposed a specific conformational change model. In contrast, Mitchell (44) has argued convincingly for chemiosmosis. The important consideration is that all these models propose that in principle both oxidation and ATP hydrolysis can energize transport. This is in contrast to the oxidation-reduction model proposed by Kaback (10), in which only electron flow through the electron transport chain can energize solute accumulation. This model will be discussed subsequently.

ENERGY FROM ELECTRON TRANSPORT AND ATP   There is now a great deal of evidence to support the notion that energy can be coupled to transport from either electron flow or ATP hydrolysis, both of which lead to the same "high energy state" (56–61). It has been shown by Harold and co-workers (56, 57) that transport of some amino acids and cations (i.e. K^+) in S. faecalis (an anaerobe that lacks a functional respiratory chain) is dependent on ATP hydrolysis via the membrane-bound Mg^{2+}-Ca^{2+}-ATPase. Dicyclohexylcarbodiimide (DCCD) and Dio-9, inhibitors of the membrane-bound ATPase, inhibit active transport. In a mutant that is DCCD resistant and possesses a DCCD-resistant ATPase, transport of K^+ and isoleucine is insensitive to this inhibitor (62). Asghar et al (63) extended these results to show that both DCCD and uncouplers inhibit net uptake of a number of amino acids. They were also able to drive transport with an artificially generated potential, which will be discussed below.

In *E. coli*, Pavlasova & Harold (64) showed that β-galactosides are accumulated under anaerobic conditions. This process was sensitive to uncouplers, although the ATP levels were not affected. It was suggested that a proton gradient generated by the hydrolysis of glycolytic ATP was involved in the accumulation of sugars. Klein & Boyer (58) showed that transport of a number of solutes in *E. coli* can be energized both by electron flow and by ATP hydrolysis via the ATPase. The same conclusion has been reached using mutants in oxidative phosphorylation and will be discussed below.

ENERGY FROM A MEMBRANE POTENTIAL   In Mitchell's concept, accumulation of solutes is brought about by coupling solute movement to proton movement down the electrochemical gradient. The electrochemical gradient or proton motive force can be generated by either oxidation or ATP hydrolysis. A third way to create a membrane potential consists of making the membrane selectively permeable for a cation or anion or using an ion that is readily permeant through a biological membrane. In the first case, potassium plus valinomycin can be used, whereas in the second case chaotropic anions such as SCN^- of NO_3^- serve this purpose. Consequently, generation of a membrane potential using these artificial systems should lead to solute accumulations.

Kashket & Wilson (65), using *Streptococcus lactis* 7962 [a strain of *S. lactis* that takes up β-galactosides by active transport rather than by group translocation (66)], showed transient accumulation of β-galactosides upon addition of valinomycin to cells in a potassium-free medium. Since *S. lactis* cells have a high intracellular potassium concentration (~ 400 mM), valinomycin induces a potassium efflux, generating a membrane potential interior negative. The accumulation was sensitive to uncouplers but not to inhibitors of the Mg^{2+}-Ca^{2+}-ATPase, like DCCD. In an extension of this study (67) the authors measured the ΔpH and the internal and external potassium concentrations and found that the calculated proton motive force agreed reasonably well with the solute gradient. Although they did not measure the actual proton movement, it can be calculated that the stoichiometry of H^+ movement with respect to sugar movement is much greater than 1. It is not clear why net thiomethylgalactoside (TMG) movement stops almost immediately after the addition of valinomycin, while the potassium gradient is still about 1000. Kashket & Wilson (67) also showed that acidification of the medium leads to transient TMG accumulation. Surprisingly, no permeant anion such as SCN^- was required to compensate for the proton influx. Such a requirement is shown in *E. coli* for lactose transport by West & Mitchell (35). Since the experiment was conducted at low external galactoside levels, the relative small flux (about 0.5 nmol TMG/mg dry weight) may be the explanation.

Similar experiments were conducted by Asghar et al (63) using cells of S. faecalis. The cells are somewhat simpler than E. coli in that, being anaerobes, they have no functional electron transport chain. Starved cells were treated with valinomycin to induce electrogenic efflux of K^+, and concomitant uptake of glycine and threonine occurred. The amounts of amino acid taken up were quite low, but inhibitor and ionophore studies supported the interpretation of potential-dependent uptake. An imposed pH gradient could also stimulate amino acid uptake to a low extent, but a ΔpH in the presence of valinomycin was much more effective.

Hamilton and co-workers (68, 69), using S. aureus, have shown that lysine distributes itself according to the membrane potential. Comparison of the lysine gradient and the membrane potential calculated from the distribution of potassium shows that both are in equilibrium. In an extension of these studies (69), the authors showed that cationic, neutral, and anionic amino acids are accumulated in response to the membrane potential ($\Delta\Psi$), the proton motive force (Δp), and the pH gradient (ΔpH), respectively.

Although not strictly active transport, West and Mitchell's experiments should be mentioned here. West (70) showed that lactose movement in E. coli is accompanied by proton movement. In a later paper, West & Mitchell (71) measured the stoichiometry of proton and lactose movement and concluded that they move in a one-to-one ratio. It was essential to include a permeant anion, SCN^-, to compensate for the proton movement in these experiments. West & Mitchell (35) also showed that addition of TMG to anaerobic suspensions of E. coli (in the presence of iodoacetate to inhibit glycolysis) elicits a small proton influx which is increased severalfold by the addition of SCN^-. A similar stimulation can be obtained upon addition of TMG or lactose by making the cells permeable to potassium by the addition of valinomycin. However, the rate of H^+ influx under both conditions, SCN^- and valinomycin, is very different, being approximately 22 and 3 nmol H^+/min/mg dry weight, respectively. It would be interesting to know whether the sugar movement follows the same pattern.

Energy Coupling in Membrane Vesicles

As has been discussed above, transport studies in whole cells have provided useful information on the coupling between energy and transport. However, interpretation of the data is not completely unambiguous because of lack of control over endogenous metabolism and pool sizes. In order to circumvent these difficulties, Kaback and his collaborators have exploited the use of isolated membrane vesicles much in the way red blood cell ghosts were used previously (73). The preparation of membrane vesicles by lysis of spheroplasts that have been prepared by treatment with lysozyme and EDTA is described elsewhere (10, 73). The main feature of these vesicles is that they accumulate a variety of solutes only when supplied with an exogenous electron donor such as D-lactate. In sharp contrast to whole cells, these vesicles could not use exogenous ATP to drive transport, nor could they synthesize ATP from ADP and phosphate in response to an electron donor (58). Based mainly on these results, a model was proposed by Kaback in which the various solute carriers were located in the respiratory chain between the primary dehydrogenases and cytochrome *b*. Energy coupling was viewed as alternate oxidation reduction of sulfhydryl groups in the carriers due to electron flow. Later results with electron transport uncoupled mutants (*etc* mutants) led to a modification of the original model, the solute carriers now being on a shunt pathway (74).

Space prohibits extensive criticism of this model, and arguments for and against it can be found in a number of reviews (2, 5, 10, 12, 75–78). The proponents seem to now consider chemiosmosis or some aspect of it to be a more likely mechanism (79).

As discussed at the outset, there are a number of discrepancies between results obtained with cells and vesicles; the following discussion is designed to clarify these differences.

ORIENTATION OF MEMBRANE VESICLES   Great emphasis has been placed by Kaback and his collaborators on the finding that some electron donors are much more efficient in supporting transport than others. D-Lactate, for example, is supposedly the best energy source in *E. coli* vesicles, whereas NADH and ATP are unable to support transport.

As has been pointed out by several authors (5, 12, 54, 61, 75), the efficiency of coupling substrate oxidation to transport is dependent on the accessibility of the various compounds to the proper site of metabolism. At the risk of belaboring the obvious, it must be clear that the problem of accessibility is unique to vesicles, because cells generate the necessary energy from compounds that are present internally. Moreover, it is equally clear that membrane enzymes are distributed asymmetrically across the membrane in much the same way as components of the electron transport chain, and that the preparation of vesicles may alter the gross sidedness, i.e. invert some membrane vesicles, and/or reorganize components so that the original distribution is altered. Such complications, unless thoroughly understood, make interpretation of energy-dependent uptake virtually impossible.

Kaback and his collaborators have claimed that vesicles prepared from *E. coli* ML308-225 by the procedure described in (73) were strictly right-side-out, i.e. with the same orientation as the cell from which they were derived. They cited various lines of evidence for this orientation, including freeze-fracture electron microscopy. In a recent study, Altendorf & Staehelin (80) have confirmed this observation with fresh vesicles prepared from this strain. When vesicles were frozen for storage, as described by Kaback, about 25% of the vesicles became inverted. The inverted vesicles were apparently very small and in terms of transport represent an insignificant internal volume. They can, however, contribute appreciably to the enzymatic activities that are normally latent. Similar results have been obtained by Konings et al (81) with vesicles from *Bacillus subtilus*. Although there still remains some uncertainty, it appears that the original contention of Kaback is correct, at least as far as gross orientation of fresh vesicles is concerned.

The question remains, however, whether components within the membrane can be reorganized. Oppenheim & Salton (82) studied vesicles from *M. lysodeikticus* which were prepared by osmotic lysis of spheroplasts. Ferritin-labeled antibody was prepared against the Mg^{2+}-Ca^{2+}-ATPase and was shown to be bound to the outside of some vesicles, whereas no binding could be demonstrated in intact spheroplasts. Thus it appears that in this case, a portion of what is normally an internal enzyme has managed to get to the outside surface. Gorneva & Ryabova (83) claim that vesicle preparations of the same organism, obtained by osmotic lysis, contain both inside-in and inside-out vesicles, as judged by ion uptake (see below) and electron microscopy using the ATPase as a marker. Some vesicles contained no ATPase (inside-in), whereas others had the characteristic knobs (inside-out).

Extensions of similar observations to E. coli membrane vesicles have led to the conclusion that orientation of some marker enzymes is not the same as in cells. Studies with the Mg^{2+}-Ca^{2+}-ATPase (61, 84), glycerol phosphate dehydrogenase (85), NADH dehydrogenase (84), and succinic dehydrogenase (85) as marker enzymes have supported this view.

It has been reported by van Thienen & Postma (61) that low concentrations of the detergent Triton X-100 stimulate the Mg^{2+}-Ca^{2+}-ATPase activity of lysozyme-EDTA vesicles about twofold. The detergent supposedly disrupts the permeability barrier for ATP. The nonstimulated value is the same as reported by Prezioso et al (86) and by Futai for such vesicles. Membranes prepared by sonication, however, do not show any detergent stimulation (or inhibition, for that matter). Furthermore, the ATPase activity detectable in the absence of detergent can be removed from lysozyme-EDTA vesicles by a low ionic strength wash as in sonicated particles, but the detergent-stimulated activity remains. This suggests that the ATPase activity measured in the absence of detergent is at the outside of the vesicle where the enzyme is accessible to the substrate. ATP cannot permeate the membrane and reach the enzyme present at the inner surface unless the permeability barrier is disrupted with detergent. Sonicated membranes, on the other hand, have most or all of the membrane ATPase on the outside, suggesting that these vesicles are inverted. Similar results have been obtained by Weiner (85) and Futai (84) in more extensive studies. Weiner (85) found that a portion of the glycerol phosphate dehydrogenase and succinate dehydrogenase in E. coli vesicles is accessible to ferricyanide as an electron acceptor. Spheroplasts show no glycerol phosphate or succinate ferricyanide reductase activity, because the enzymes are internal and ferricyanide is impermeant. Futai (84) studied the ATPase in spheroplasts and vesicles in the presence or absence of toluene, Triton X-100, and deoxycholate. While the activity of spheroplasts is stimulated around 15-fold, the ATPase activity of vesicles is increased about 25%. The inhibition of ATPase activity by antibody was also studied in vesicles and spheroplasts and gave results consistent with those obtained from studies with detergent, although the quantitation leaves something to be desired. For instance, after the toluene treatment, antibody against the ATPase inhibited the ATPase activity in spheroplasts 95%, but inhibited the ATPase in vesicles only 60% in the presence or absence of toluene. Addition of Triton X-100 or deoxycholate gave somewhat higher inhibition.

Futai (84) observed that osmotic vesicles as well as sonicated particles made from the unc^-ATPase$^-$ mutant DL54 (59) can bind purified ATPase (F_1) in a DCCD-sensitive way, whereas spheroplasts were unable to bind F_1. It can be concluded that components other than some "marker enzymes" are moving during lysis. This finding, together with the observation that ATP can energize membranes, as measured by 9-amino-6-chloro-2-methoxyacridine (ACMA) fluorescence quenching (see later) (61) via this ATPase, suggests that the complete ATPase complex has changed its orientation in some vesicles. Asano et al (87) have also observed that *Mycobacterium phlei* ghosts prepared by osmotic lysis are unable to carry out oxidative phosphorylation unless purified ATPase is added.

Hampton & Freese (88) have concluded that 20% of the vesicles prepared from *B. subtilis* are inside-out. Matin & Konings (89) have measured both the rate of uptake and oxidation of various electron donors such as D-lactate, L-lactate, and succinate in *E. coli* vesicles. From their data it can be calculated that oxidation is 10- to 30-fold faster than uptake of the electron donor, indicating that only a small portion of the total oxygen consumption is by inside-in vesicles, since transport of the electron donor is rate limiting in such vesicles.

Gorneva & Ryabova (83) have used the uptake of the lipid-soluble anion PCB$^-$ and potassium plus valinomycin to determine the orientation of osmotic vesicles and sonicated particles of *M. lysodeikticus*. Whereas PCB$^-$ is taken up by osmotic vesicles and sonicated particles, potassium in the presence of valinomycin is taken up mainly by osmotic vesicles. This suggests again that osmotic vesicles contain a population with the orientation of sonicated particles, inside-out.

Measurements, such as the direction of proton movement (38) or anilinonaphthalene sulfonate (ANS) fluorescence (90) upon energization of the membrane, cannot be used to determine the orientation of membrane vesicles, since the extent to which inside-in and inside-out membranes contribute to these processes is not known.

Several possibilities exist to explain the external localization of enzymes that are normally on the internal surface. 1. Vesicles are a mixed population of inside-out and outside-out vesicles (61, 83, 85, 88). This seems unlikely if one accepts the freeze-fracture evidence. 2. Each vesicle is a patchwork with portions oriented in different directions. Such vesicles could arise from fusion or partial invagination during preparation. 3. Portions of the vesicles are simply leaky, and they contribute to the enzymatic activity measured but do not contribute to transport measurements. This seems unlikely, since data on the energy-dependent quenching of 9-amino-6-chloro-2-methoxyacridine (ACMA) fluorescence (61) suggest that at least some of the lysozyme-EDTA vesicles are inverted and closed, similar to sonicated vesicles. 4. The gross orientation of the membrane is correct, but some of the enzymes have now become oriented on the wrong side. Although this seems to be the most probable explanation, it is not without objections. Clearly, experiments must be designed to differentiate between these possibilities if one hopes to interpret transport data in any meaningful way. One way to separate inside-in and inside-out bacterial vesicles, if distinct populations exist, is indicated by the recent reports using free flow electrophoresis (91) and concanavalin A binding (92) to separate such populations of inside-in and inside-out vesicles from erythrocyte ghosts and plasma membranes.

ENERGY FROM ELECTRON TRANSPORT AND ATP   The main feature of the oxidation-reduction model is the proposal that electron transport through the respiratory chain is sufficient and necessary to energize solute accumulation (10).

It has been reported repeatedly by Kaback and co-workers (10) that ATP is not able to energize transport in vesicles. This, coupled to the observation that lysozyme-EDTA vesicles are incapable of ATP synthesis, led to the conclusion that oxidative phosphorylation is not involved in energizing transport. In order to avoid a semantic problem, oxidative phosphorylation refers to the synthesis of ATP coupled to the oxidation of an electron donor, and in this regard vesicle transport is certainly not dependent on oxidative phosphorylation. However, it is equally clear from studies with mutants defective in oxidative phosphorylation that the coupling of solute transport to electron transport is dependent upon the generation of the high energy intermediate of oxidative phosphorylation. Konings & Kaback (93) have shown that E. coli, grown under special anaerobic conditions using nitrate or formate as electron acceptor, is able to couple anaerobic electron flow to transport. This work, however, simply demonstrates an alternative to normally grown cells, which undoubtedly use glycolytic ATP under anaerobic conditions. Furthermore, van Thienen & Postma (61) have recently shown that ATP can drive active transport of serine in vesicles if ATP is shocked into the vesicles at high concentration. The objection by Kaback (79) that these results represent incorporation of serine into phospholipids ignores the facts that uptake is sensitive to uncouplers and DCCD, that it was not demonstrable in mutants lacking the Mg^{2+}-Ca^{2+}-ATPase, and that the accumulated solute leaked from the vesicles when the ATP was depleted. In addition, H. Hirata (personal communication) has recently demonstrated similar results for uptake of alanine in E. coli vesicles. It is therefore clear that ATP can serve as an energy source for transport in vesicles if the experiments are done under the correct conditions. Previous failures must be attributed both to impermeability of ATP (94) and to the low concentrations added to the vesicles. Although it has been claimed by Weisbach et al (95) that ATP can enter vesicles, the rate is less than 0.1 nmol/min/mg. This rate is insufficient to keep up with the rate of hydrolysis of ATP by the ATPase. Similar studies bearing on this problem have been conducted by Asano et al (96), who have shown that ghosts of M. phlei do not perform oxidative phosphorylation. This is similar to the work of Klein & Boyer with E. coli vesicles (58). However, Asano et al (96) showed that oxidative phosphorylation occurred normally if ADP was shocked into the ghosts. Sonicated ghosts were shown to be competent for oxidative phosphorylation, so that the problem with vesicles was simply sidedness. This is also the case with E. coli vesicles (97).

Similar considerations can be raised to explain the apparent inefficiency of NADH in energizing transport. Futai (98) showed that when spheroplasts were lysed in the presence of NAD^+ and alcohol dehydrogenase, transport in the resulting vesicles upon oxidation of internal NADH was perfectly normal. NADH was almost as effective as D-lactate.

An interesting series of experiments has been conducted by Kaback and his collaborators (99, 100) in the reconstitution of energy coupling. They can restore D-lactate-dependent transport to vesicles prepared from a mutant strain lacking D-lactate dehydrogenase by adding back purified enzyme at the outside. Futai (101) has confirmed this result and extended it to show similar results with α-glycerol-phosphate dehydrogenase. Furthermore, Futai demonstrated that ferricyanide inhibits transport completely in vesicles reconstituted in this way, whereas sphero-plasts are not affected. This indicates that the enzyme is indeed located at the outside. Short et al (100) have also verified the external location of the enzyme. In "normal" vesicles only 50% of the dehydrogenase is accessible to ferricyanide. Thus some interesting questions are posed. Is the enzyme facing the outside in normal vesicles coupled to the electron transport chain in the same way as the reconstituted enzymes? Does it contribute to a normal membrane potential by circumventing one loop of the electron transport chain, or does it generate a potential opposite to that generated by the correctly oriented enzyme? This is also a critical question with regard to the proton-translocating ATPase, which also appears to exist on both sides of the vesicle surface, because ATP hydrolysis at the outside can possibly lead to a potential interior positive [as suggested by ACMA fluorescence quenching (61)] and promote solute efflux. This simply emphasizes the need to clarify the problem of vesicle orientation.

ENERGY FROM A MEMBRANE POTENTIAL   As with intact cells, it can be shown also in membrane vesicles that a membrane potential, created by the efflux of potassium in the presence of valinomycin, is sufficient to energize solute accumulation. Harold and co-workers (25, 102) were able to demonstrate that oxidation of D-lactate by E. coli membrane vesicles results in the formation of a membrane potential as measured by the uptake of the lipid-soluble cation DDA^+. This observation is an extension of an earlier observation by Reeves that protons are extruded by the vesicles under these conditions (38). The potential generated was about 100 mV, interior negative, and the potential could be dissipated by the proton conductor CCCP or by valinomycin-induced K^+ influx. Ionophores that facilitated electro-neutral exchanges, such as nigericin, did not dissipate the gradient. Therefore it seems likely that D-lactate-driven respiration creates a membrane potential that is due to an electrogenic extrusion of protons.

The key experiment, however, is the generation of a membrane potential independent of the metabolic machinery of the cell. This was achieved by pre-loading E. coli vesicles with high levels of K^+ and then suspending these cells in media low in K^+ (25, 102). When valinomycin is added to such vesicles, a rapid efflux of K^+ occurs, and the concomitant uptake of DDA^+ verifies the membrane potential. Significantly, proline uptake is markedly increased as K^+ efflux occurs. This potential-induced proline uptake appears to be free of the metabolic machinery of the vesicle, since neither DCCD (an inhibitor of the ATPase) nor 2-n-heptyl-4-hydroxyquinoline-N-oxide (HOQNO) and CN^- (respiration inhibi-tors) have any effect. Proton conductors, however, abolish the potential and proline uptake as predicted. Surprisingly, nigericin inhibits. Although nigericin does not change the membrane potential, catalyzing an electroneutral H^+-K^+ exchange, it may create a ΔpH in the wrong direction. This is supported by the observation (102) that transport is low in vesicles in which the internal pH is lower than the external pH. Hirata et al (102) also found, as reported earlier by Kashket & Wilson (67), that potassium added outside at low concentrations inhibits the transient solute accumulation, although an appreciable K^+ gradient still exists. A similar observation has been made during the cation-induced ATP synthesis in mitochondria (103).

Role of Energy in Facilitated Diffusion

The discussion has so far centered on active transport and how energy is coupled to the accumulation of solutes. The same carrier can equilibrate internal and external solutes, a process called facilitated diffusion. The equilibration proceeds supposedly in the absence of energy.

In the past, the transport of solutes in bacterial cells has been studied with emphasis on the kinetics of the carrier process itself and the effect of energy coupling on those kinetics. The purpose was to describe how a facilitated diffusion system changed to an active transport system.

Since most studies have been conducted with the β-galactoside transport system of E. coli, we will concentrate on this transport system. Koch (53) and Winkler & Wilson (51) suggested that the addition of energy to this system changes the affinity of the carrier for its solute at the inner face of the membrane, resulting in solute release. Thus no energy is required for entry. A note of caution is required here, since such a statement implies that the change in the solute-carrier affinity is a consequence of a change in the carrier. In terms of chemiosmosis, the change is considered to be a lowered internal proton concentration which results in lowered β-galactoside affinity, since solute binding is dependent on a concomitant proton binding. The above results stem largely from the observation that the influx kinetics do not change appreciably when the cells are treated with energy poisons. Manno & Schachter (104) and Scarborough et al (55), on the other hand, claim that energization changes the K_m for influx. The various possibilities and results have been discussed by Wong et al (105), who conclude that the most likely site for energy coupling is at the inner surface of the membrane. These and other results have been interpreted as showing that in the absence of energy the carrier can shuttle across the membrane, equilibrating the inside and outside solute concentration.

Koch (106) has shown, however, that E. coli cells completely devoid of endogenous energy sources cannot carry out facilitated diffusion as measured by o-nitrophenylgalactoside (ONPG) hydrolysis. Addition of an exogenous energy source such as glucose or succinate restores transport. Cecchini & Koch (107) report that an uncoupler like CCCP can restore facilitated diffusion in these energy-depleted cells. Although Koch originally gave a rather complex explanation of these results, they can more readily be explained in chemiosmotic terms. In the normal energized state, solute and protons enter the cell via the symporter in response to the proton motive force, and the protons are extruded in the energy-dependent step, thus preventing solute exit via the carrier and restoring electroneutrality. However, in the energy-depleted state there is no mechanism for proton extrusion, so that as symport begins there is the establishment of a membrane potential interior positive due to the H^+ accumulation. This accumulation prevents further symport unless a substitute for metabolic proton extrusion is provided, such as the movement of a permeant anion like SCN^- to compensate for charge. Alternatively, the membranes can be made leaky for protons by the addition of an uncoupler like CCCP, which then prevents the buildup of a membrane potential and allows solute to equilibrate.

Although discussed in an earlier section, the experiments of West & Mitchell (35, 71) should be mentioned here because they are complementary to those of Koch. West & Mitchell (35) showed that addition of β-galactosides to metabolically inactive *E. coli* cells induced an influx of protons when either a permeant anion such as SCN^- is present or the cells are made permeable to K^+ with valinomycin. Although they measured the actual uptake of lactose in the presence of SCN^- (71), they did not measure whether facilitated diffusion was arrested in the absence of SCN^-.

In recent studies, Kaback and collaborators (79, 108) added the interesting observation that the fluorescence of dansylgalactoside is changed upon energization of *E. coli* membrane vesicles. From both fluorescence enhancement and polarization studies, they concluded that solute is bound to the carrier protein (M protein) only after energization. They have suggested that energization either increases the affinity for solute at the outside or promotes the movement of carriers from the inside to the outside. In contrast to these results, Kennedy et al (109) observe that direct binding of β-galactosides can be measured without any energization. These conflicting results could be due to the use of different preparations. Kaback and co-workers used lysozyme-EDTA vesicles which are supposedly right-side-out, whereas Kennedy and co-workers used sonicated membranes which have been shown to be inside-out. It would be most interesting to compare the direct binding assay with the fluorescence method with both types of vesicles. It should also be pointed out that the amount of binding measured with the two techniques differs greatly. Using the same strain, *E. coli* ML308-225, Kennedy et al (109) reported 0.11 nmol of sugar bound/mg of membrane protein, similar to the data obtained with binding of labeled N-ethylmaleimide to the M protein, whereas Reeves et al (108) obtained 1.14 nmol of sugar bound/mg of protein. Clearly, several basic problems, such a membrane sidedness and the quantitative interpretation of fluorescence enhancement data, must be resolved before any meaningful interpretation can be attempted.

Additional observations on the requirement of "energy" for facilitated diffusion have been reported. Abrams (110) observed that protoplasts of *S. faecalis* do not swell in a sucrose solution unless supplied with an oxidizable substrate such as glucose. Asghar et al (63) reported that the same protoplasts do not swell in 0.5 M threonine unless glucose is added. Lysis in the presence of glucose is prevented by uncouplers like CCCP or inhibitors like DCCD, both of which prevent the generation of the high energy state formed as a result of the hydrolysis of glycolytic ATP.

Höfer (111) found that the yeast *Rodotorula gracilis* does not equilibrate monosaccharides unless energy is present, and suggested a model in which the empty carrier cannot move across the membrane unless it is energized (112). A similar model has been proposed by Tanner and co-workers for sugar transport in *Chlorella* (113, 114). They observed that uncouplers do not induce sugar efflux in the steady state but inhibit steady-state influx. In *Azotobacter vinelandii* it has been found that uncouplers and inhibitors of oxidative phosphorylation prevent the movement of anions such as succinate, citrate, and malate down their gradients (115, 116). Finally, although not an active transport mechanism, mutants missing enzyme I of the PEP-sugar phosphotransferase system in *S. aureus* are virtually unable to equilibrate sugars (117).

It should be apparent from this discussion that the seemingly simple question of an energy requirement for facilitated diffusion is not yet resolved. It is tempting to suggest that in the β-galactoside system of *E. coli*, energy is required for solute movement even down a gradient. In fact, what one means by energy in this case is somewhat confusing, since facilitated diffusion can apparently be restored in energy-depleted cells by the addition of an uncoupler. In Mitchell's scheme for solute transport the "energy requirement" for facilitated diffusion is simply the need of the cell to dissipate the proton gradient.

Transport Systems That Can Use Only ATP

It should not be concluded from the previous sections that transport of all solutes can be energized by either electron transport or ATP hydrolysis. Berger & Heppel (118, 119) have recently described a number of transport systems that seem to derive their energy directly and solely from ATP. These authors have compared the transport of amino acids via "shockable" and "nonshockable" systems. For the former class, a periplasmic binding protein released upon osmotic shock of intact cells has been demonstrated (for reviews, see 120 and 121). Berger & Heppel (119) and Berger (118) have systematically compared the energetics of transport of the shockable to the nonshockable systems with the use of unc^-ATPase$^-$ mutants (see next section) and metabolic inhibitors. The transport assays involved the use of cells that had been depleted of endogenous energy supply by incubation in the presence of dinitrophenol, followed by washing to remove the inhibitor. Cells treated in this way require an added energy source for transport. Such cells may offer an alternative to normal cells and membrane vesicles.

E. coli strain DL54, an unc^-ATPase$^-$ mutant, shows about a 50% decrease in proline transport with glucose as an energy source. When D-lactate was used, however, normal transport of proline was observed. In contrast, whereas glutamine uptake in this strain in the presence of glucose is normal, in the presence of lactate it is less than 5% of the parental level. CN^- inhibited proline uptake completely, yet it only partly inhibited glutamine uptake with glucose as an energy source. The authors (118, 119) concluded that the shockable transport systems are energized directly by ATP or by some high energy derivative of ATP and do not require the high energy state. This is supported by additional studies showing that the uncoupler FCCP will completely inhibit proline uptake while only inhibiting glutamine uptake about 30%. They attribute this decrease in glutamine uptake to a 50% decrease in the ATP pool. One of the clearest differences between the two types of energy coupling is the effect of arsenate. Proline transport is not inhibited by arsenate as had been observed previously (58), whereas glutamine transport is abolished. Again, the direct role of ATP or some high energy phosphate intermediate is implicated. There are, however, some uncertainties, such as the inhibition of the shockable systems by uncouplers, that must be clarified in a more quantitative way.

Recent work has extended these observations, namely, energization of transport by ATP alone, to ribose (122) and glycylglycine (123). Galactose transport is also considered to be energized by ATP directly (124). This is in contrast to the conclusion of Parnes & Boos (125). However, in the experiments reported by these authors, an inhibition by arsenate was also observed.

Conclusions: Sources of Energy for Active Transport

Models for active transport in bacteria should apply equally well to both intact cells and membrane vesicles derived from them. From the previous section it is clear that results obtained with both systems are consistent. Energization of the transport of a number of solutes can be achieved by oxidation via the respiratory chain, by ATP hydrolysis via the membrane-bound Mg^{2+}-Ca^{2+}-ATPase, or by ion gradients. Each of these pathways leads to the same "high energy state" of the membrane, which can be utilized for energy-requiring processes such as ATP synthesis, energy-linked transhydrogenase, reversed electron transport, and solute transport.

Mitchell's chemiosmotic theory ties all these processes together, identifying the high energy state of the membrane as a proton motive force generated by the electrogenic transport of protons across the membrane. Solute carriers are thought to be proton-solute symporters which respond to the proton motive force. This proposal can also explain why facilitated diffusion is not operating in the "absence of energy": obligate coupling between H^+ and solute movement creates an electrochemical potential which has to be dissipated.

It is equally clear that Kaback's oxidation-reduction model cannot be correct. Transport can proceed in the absence of electron transport.

We have discussed extensively only one class of transport systems, the proton-solute symporters. Other systems derive their energy from other sources such as PEP, ATP, Na^+ gradients, or possibly other energy sources.

GENETIC ANALYSIS OF ENERGY COUPLING

While previous sections have examined the mechanism of coupling between transport and energy, this section will provide information on how the macromolecular components involved in oxidative phosphorylation actually affect active transport.

As in the initial stages of describing bacterial transport systems, the use of genetics is proving to be a most valuable approach. It has many obvious advantages, one of the most important for this type of study being that it reduces the reliance upon metabolic inhibitors, the specificities of which are always questionable. While genetic alterations overcome these objections, they have their own set of difficulties that must be eliminated before any given phenotype can be assigned to the alteration of a single polypeptide. This is particularly true when considering any multicomponent system where a mutation may affect not only a single peptide but also the interaction of an entire complex. Needless to say, the problems of double mutations, polarity, deletions, and regulatory mutations necessitate cautious interpretation of phenotypes and comparison of different mutants.

As discussed above, the most simple interpretation of available data is that energy coupling involves generation of a proton motive force. It can be generated by respiratory activity or by ATP hydrolysis via the membrane-bound ATPase complex. In facultative aerobic organisms such as *E. coli* both mechanisms appear to be operative. In anaerobic organisms such as the *Streptococci* only the latter is present. Analysis of mutants impaired in the energy-transducing apparatus owes much to studies with mitochondria and chloroplasts. It is through the use of genetics, however, that microbial systems (and eukaryotic systems such as yeast) offer a real opportunity to extend our knowledge beyond that obtained with the more complex systems. Although an extensive genetic analysis of energy transduction is only beginning, the potential is obvious. The reader is directed to a recent review of Cox & Gibson (126) for a discussion of the initial stages of this investigation.

The Mg^{2+}-Ca^{2+}-ATPase Complex

A major area of recent research has been the genetic and biochemical analysis of the ATPase complex of *E. coli*. To avoid confusion it is first necessary to define the complex. We will refer to the entire functional unit as the ATPase complex. This complex is most familiar as the knoblike structures seen on the outer surface of the membrane of submitochondrial particles when viewed in the electron microscope after negative staining. What is visualized in this way is only a portion of the complex commonly referred to as the headpiece or F1. The F1 is the most easily removed portion and contains probably five distinct polypeptides. The second portion is the stalk or the peptide or peptides that join the F1 to the membrane. In mitochondria the so-called oligomycin sensitivity-conferring protein (OSCP) is presumably in the stalk. The protein nectin isolated from *S. faecalis* by Baron & Abrams (127) also performs this function. The remaining portion of the complex, the membrane sector, is the least well understood and may contain four or five distinct polypeptides. One of the interesting features of this ATPase complex is the amazing phylogenetic conservation of structure. The structure is basically the same in mitochondria, chloroplasts, yeast, and bacteria. The review by Senior (128) is an excellent discussion of the structural features of this complex, while Abrams & Smith (129) have compared the bacterial ATPases. In functional terms, the complex was initially described as an ATPase activity and considered to operate in the final step in oxidative phosphorylation, namely the transphorylation reaction. The complex has, however, a bidirectional function. It can use the high energy state generated by respiration for the synthesis of ATP. It can also generate that same high energy state by the reverse reaction, hydrolysis of ATP. In Mitchell's view, this is accomplished by a complex that serves as a reversible proton translocator, while the high energy state is a proton gradient. Although there is considerable evidence for this general function, the precise mechanism is still a matter of conjecture (for a recent proposal by Mitchell, see 130).

The following general features of the *E. coli* ATPase complex are important for our discussion:

1. The *E. coli* complex has an F1 component quite similar to that of mitochondria and chloroplasts. It is easily removed from the membrane by washing with buffer of low ionic strength in the presence of EDTA (stripping procedure). It probably has five distinct polypeptides, as reported by Bragg & Hou (131), and the molecular weights of the peptides appear to be approximately α, 56,000; β, 52,000; γ, 30,000; δ, 20,000; and ε, 10,000. Although it has been concluded from gel staining intensity (132) that the subunits in the mitochondrial ATPase may be present in the ratio of $3:3:1:1:1$, no such data have been reported yet for the bacterial ATPase. There is some disagreement on the number of different peptides, however, since Hanson & Kennedy (133) and Nelson et al (134) have isolated a soluble ATPase containing only four types of peptides. The δ (20,000) peptide is not present. Bragg et al (135) have reported, however, that when the molecule containing five peptides was subjected to gel electrophoresis and reisolated, the δ peptide was missing. Further, the molecule, lacking the δ peptide, is unable to reconstitute the energy-linked transhydrogenase activity in stripped membranes. This observation suggests that the 20,000 peptide may be analogous to the OSCP or nectin or is needed to bind the F1 to the stalk protein. Furthermore, it suggests that this component may be lost during certain purification procedures. Additional reconstitution experiments are required to resolve this point, which is obviously crucial to our understanding of the F1 in *E. coli*. Recently Futai et al (136) have verified that the procedure of Nelson et al (134) yields a complex containing four peptides, and that a modification [which in essence is simply the Bragg & Hou procedure (131)] yields a complex consisting of five peptides. Although Futai et al suggest that the product is dependent upon the procedural modifications for *E. coli* K12, *E. coli* ML308-225 yields an F1 containing five peptides regardless of the procedure. In addition, examination of another *E. coli* K12 strain, 1100, yields a four peptide complex regardless of the procedure (137). It would thus appear that the nature of the product depends upon both the procedure and the strain.

2. The ATPase complex is activated by Mg^{2+}, which also plays a role in attaching the F1 to the membrane portion of the complex.

3. The ATPase activity of the complex is inhibited by DCCD and Dio-9, but the ATPase activity of the soluble F1 is not sensitive (138). Therefore a DCCD sensitivity-conferring protein (DSCP) has been implicated as a component of the membrane portion of the complex. In contrast, azide inhibits both the membrane-bound and soluble ATPase (138).

4. Removal (stripping) of the F1 from the membrane results in energetic uncoupling that appears to be due to a proton leak through the membrane portion of the complex, the proton channel (42, 139). This uncoupled state can be recoupled by treatment with DCCD or by reconstitution with F1 (39, 61, 140–142).

Uncoupled (unc) Mutants

A host of mutants have now been isolated, particularly in *E. coli*, that have alterations in the ATPase complex. The first such mutants were reported by Butlin et al (143); these have been designated *unc* for uncoupled. Such mutants were incapable of coupling electron transport to ATP synthesis (oxidative phosphoryl-ation); the original mutant, AN120, was also lacking ATPase activity and was designated *uncA*. This same group subsequently isolated a strain that was *unc⁻* but had ATPase activity, which they designated *uncB* (144). There has been some tendency to designate all subsequent mutants of the *unc⁻* ATPase⁻ class as *uncA* and all of the *unc⁻* ATPase⁺ class as *uncB* mutants. This designation is somewhat misleading. Although the ATPase activity resides in the F1, it does not follow that all mutants lacking ATPase activity have mutations in the F1 part of the complex. Conversely, not all mutations in the F1 part of the ATPase complex lead to an enzymatically inactive molecule. Consequently, neither of these designations is generally applicable, and the designation *uncA* or *uncB* should not be used until polypeptide assignments can be made. For this reason, we shall refer to *unc* mutants as either ATPase⁺ or ATPase⁻ (presence or absence of enzymatic activity) with no assumption as to the nature of the mutation unless additional information is available.

As mentioned, the first such mutants of *E. coli* were isolated by Gibson and his collaborators. Subsequently, a number of ATPase⁻ and ATPase⁺ mutants have been isolated by other workers (59, 60, 141, 145–148). The general phenotype of all *unc* mutants reported up to now can be described as follows: 1. *unc* mutants are able to utilize primarily fermentable carbon sources such as glucose or glycerol for growth but are unable to use carbon sources that yield energy primarily as a result of ATP synthesis via oxidative phosphorylation such as succinate, malate, lactate, etc; 2. the mutants give low aerobic growth yields when grown on limiting amounts of glucose, or more simply the amount of cell mass produced per amount of glucose utilized is low; 3. *unc* mutants have normal respiratory activity; 4. the mutants have little or no detectable oxidative phosphorylation; 5. ATPase activity can be present or missing. [A word of caution is needed here because Gunther & Maris (149) have recently shown that AN120, an *unc⁻* ATPase⁻ mutants isolated first by Gibson and co-workers (143), can exhibit as much as 50% of the parental ATPase activity under certain conditions, such as high salt concentrations.] 6. All *unc* mutations so far described in *E. coli* are located at about 73.5 min on the chromosome and are 20 to 50% cotransducible with the *ilvC* locus and about 50% cotransducible with the *asn* locus (147).

With this basic phenotype one can begin to compare the various properties of the mutants that have been studied. Work on these mutants has been expanded in three general directions: 1. correlation of *unc* mutations with alterations in other energy-dependent functions such as active transport, the energy-linked transhydro-genase, motility, etc; 2. use of *unc* mutants to examine the general mechanism of energy transduction, for example, to see if mutants lacking the F1 component are really leaky to protons as demonstrated in mitochondria (150); and 3. correlation of a specific mutation with alteration in one peptide and a description of the role of that peptide in the overall function of the complex. This may be complicated by the possibility of phenotypic changes in *unc* mutants. Gunther & Maris, for instance (149), claim that in strain AN120, an *unc⁻* ATPase⁻ point mutant, the lipid composition is changed. Phosphatidylglycerol is increased while cardiolipin is decreased. Curiously enough, Santiago et al (151) have found that degradation of cardiolipin in rat liver mitochondria leads to a decrease in ATPase activity. Racker (152) has shown that in reconstituted cytochrome oxidase vesicles the $^{32}P_i$-ATP exchange and P/O ratio are increased by cardiolipin. As will become apparent, a great deal of additional work will be required before any of these goals are satisfied.

ASSIGNMENT OF MUTATIONS TO SPECIFIC POLYPEPTIDES   Comparison of mutants that have changes in a complex having perhaps as many as ten polypeptides is extremely difficult and will be largely dependent on the assignment of mutational events to specific polypeptides. Unfortunately, little such information is available as yet. This problem is being approached in three ways: 1. in vitro complementation of various mutant extracts in order to localize the defect in either the F1 fraction or the membrane fraction, 2. isolation of the mutant proteins and analysis of the peptides, and 3. genetic complementation analysis. In vitro complementation tests have shown that the unc^- ATPase$^-$ strain DL54 has an altered F1 component by virtue of the fact that both transhydrogenase activity (142) and active transport (153) can be restored by reconstitution with parental F1. Cox et al (154) have been able to reconstitute oxidative phosphorylation by mixing membrane fragments from strain AN249, unc^- ATPase$^-$, with the low ionic strength wash of an unc^- ATPase$^+$ mutant. Washing was required presumably to remove the residual defective F1. This had previously been noted by Bragg & Hou (142). Thus the assignment of ATPase negativity to a mutation in the F1 seems appropriate in these cases. It has also been demonstrated by Nelson et al (134) that the α and β subunits of F1 are apparently responsible for ATPase activity. Treatment of F1 with trypsin destroyed the two smaller subunits, but ATPase activity was relatively unaffected by this digestion. These results emphasize the danger of assuming that unc^- ATPase$^+$ mutants have an intact F1.

A more definitive approach to mutant assignment will come from analysis of the mutant proteins. This is a complex problem when dealing with the F1 component, since it is difficult to demonstrate a single amino acid substitution in a protein that has a molecular weight of nearly 4×10^5. In addition, since the most convenient method of peptide analysis is sodium dodecylsulfate (SDS)-gel electrophoresis, which discriminates solely on the basis of size, it is impossible to detect missense mutations. Therefore, one must concentrate on nonsense mutations or deletions for this sort of analysis. It is possible, however, that a major alteration in one of the peptides of F1, for example, will prevent the assembly of the complex or promote its degradation and prohibit analysis. Moreover, assignment of genetic loci to the membrane components is greatly hindered by the fact that they have been neither isolated nor identified. In spite of these problems one mutant has been analyzed in this manner. Bragg et al (135) have demonstrated that the etc -15, unc^- ATPase$^+$, strain has a mutation in the γ subunit of F1. This sort of observation requires caution, since on the basis of normal ATPase activity one would have assumed that this mutation affected a peptide in the membrane portion of the complex. This is supported by the finding of Futai et al (136) that part of the F1 molecule, the δ subunit, is involved in binding.

Comparison with mitochondrial, chloroplast, and other bacterial ATPase can be useful. Nelson et al (155) have shown that in chloroplast F1 (CF1) the α and γ subunits are probably involved in the enzymatic function of the coupling factor, whereas the ε subunit is a regulatory subunit [possibly the inhibitor (156)]. The three largest subunits, α, β, and γ, together still give coupling activity. Kozlov & Mikelsaar (157) have shown that the mitochondrial ATPase missing the γ and δ subunits has the same K_m and V_{max} with ATP as substrate as the native F1, and they suggest that γ and δ are involved in binding. Salton & Schor (158, 159) found that *M. lysodeikticus* ATPase isolated by a shock wash contained two major and one to three minor subunits. The coupling factor was able to bind to membranes. However, a fraction extracted with n-butanol contained the two major subunits and was unable to bind. Both fractions had about the same specific activity. Unquestionably many more analyses must be conducted before a clear picture of the function of each peptide in the complex will emerge. Genetic complementation tests, which will help define the number of components in the complex, should be equally informative. These approaches give the bacterial systems added attractiveness and will permit extension of our understanding of this complex beyond that possible with systems not amenable to genetic manipulation.

ENERGY-LINKED TRANSHYDROGENASE   The *unc* mutants have been examined for their ability to carry out another energy-dependent reaction in addition to oxidative phosphorylation: the reduction of $NADP^+$ by NADH. This reaction can be driven either by respiration or by ATP hydrolysis via the ATPase complex, as in mitochondria. There is agreement that the ATP-driven reaction is missing in all *unc* mutants. In the unc^- $ATPase^-$ strains this is, of course, due to inability to hydrolyze ATP, and in the unc^- $ATPase^+$ strains it is due to the inability to couple hydrolysis to generation of the high energy state. In any case, examination of this process gives further insight into the effects of such mutations. Bragg & Hou (142) examined strains DL54 and NI44, both unc^- $ATPase^-$. Isolated membranes [in this procedure membranes are prepared by sonication, presumably inverted (84), and are not to be confused with those prepared by lysozyme-EDTA] of strain NI44 exhibited normal levels of respiration-driven transhydrogenase while totally lacking ATP-driven activity as reported earlier by Kanner & Gutnick (160). DL54 also had no ATP-driven activity, but respiration-driven activity was reduced by about 50%. More interestingly, these workers were able to restore the respiration-driven transhydrogenase activity to the parental level by treating membranes of DL54 with DCCD or by incubation with F1 isolated from the parental strain. In this latter case, reconstitution of ATP-driven transhydrogenase was also achieved. It was concluded on the basis of analogous work with mitochondria (150, 161, 162) that DCCD could "repair" membranes from which the F1 had been removed. Furthermore, it has been suggested that the F1 serves a dual role, functioning as both a catalytic and a structural complex. The action of DCCD appears to be fairly specific for a polypeptide in the membrane portion of the ATPase complex. Abrams and his collaborators (62) have isolated a DCCD-resistant mutant of *S. faecalis* and showed that the mutation had affected the membrane-bound part of the ATPase complex. The DCCD binding protein has been isolated from mitochondria by Cattell et al (163) and Stekhoven et al (164). For our purposes, DCCD has proven a useful diagnostic reagent for different *unc* defects. Bragg & Hou (142) suggested that the defect in the ATPase complex of DL54 resulted in reduced affinity of the F1 for the membrane and increased dissociation of the defective complex from the membrane during isolation. Although this interpretation is reasonable, no difference could be detected between DL54 and NI44 in the distribution of residual ATPase activity in the membrane and soluble fraction of extracts.

The general picture that emerges from a variety of strains is the following: whereas all *unc* mutants are unable to couple ATP hydrolysis to the energy-linked transhydrogenase, some strains have normal and some have defective respiration-driven transhydrogenase activity. The differences in these two classes probably reflect the leakiness of their membranes to protons.

ENERGY-DEPENDENT QUENCHING OF 9-AMINO-6-CHLORO-2-METHOXYACRIDINE (ACMA) FLUORESCENCE   Another parameter that has been used to examine the energetics of *unc* mutants is the energy-dependent quenching of the fluorescence of the acridine dye ACMA. When this dye is added to a suspension of membranes, the observed fluorescence can be quenched by the addition of ATP or an oxidizable substrate, such as succinate. This quenching is taken as a measure of the energized state of the membrane. While this technique has the advantage of convenience and un-questionably relates to the generation of a high energy state, it has the disadvantage that it is not known exactly what is being measured. It does provide an additional parameter to compare to transhydrogenase and transport activities (to be discussed later).

Nieuwenhuis et al (140) have reported extensive studies on four strains, NI44, *unc*⁻ ATPase⁻, and BV4, KI1, and AI44, all *unc*⁻ ATPase⁺. NI44 shows some succinate-dependent ACMA fluorescence quenching, but both the rate and extent are considerably less than in the parental strain. This is curious in that the oxidation-driven quenching of ACMA fluorescence is defective, yet the respiration-driven transhydrogenase is normal. The difference found with these assays may be due to different affinities of each process for the high energy state. That is, transhydrogenase requires less energy for maximal coupling than does fluorescence quenching. It was also demonstrated that DCCD restored the capacity of NI44 membranes to quench ACMA fluorescence. Several observations on the three *unc*⁻ ATPase⁺ strains are also pertinent and shed some light on the nature of the mutations. 1. The ATPase activity in these strains is resistant to inhibition by DCCD. The simplest interpretation of this observation is that the mutation has altered the peptide in the membrane portion of the complex that binds DCCD. This would be analogous to the DCCD-resistant mutants described earlier by Abrams and his collaborators in *S. faecalis* (62). 2. The three *unc*⁻ ATPase⁺ strains can, however, be subdivided into two classes: (*a*) BV4 has normal respiration-driven transhydrogenase, but poor ACMA fluorescence quenching, and it is not repaired by treatment with DCCD. This phenotype, then, is still consistent with a defective DSCP; and (*b*) KI1 and AI44, however, have poor respiration-driven transhydrogenase and poor ACMA fluorescence quenching but can be "repaired" by treatment with DCCD. Thus in these strains DCCD-resistant ATPase activity is apparently not due to a defective DSCP, but to some other component of the complex. These authors suggest a stalk protein defect, but it seems equally likely that a peptide of F1 has been altered. This question could be answered by in vitro complementation experiments but, for some unknown reason, this set of strains does not reconstitute.

Role of the ATPase Complex in Active Transport

Schairer & Haddock (60) reported that cells of an unc^- ATPase$^-$ strain, unc A103c, were able to transport β-galactosides normally under aerobic conditions. However, if the cells were treated with KCN to inhibit respiration, transport in the mutant was abolished, whereas the parent was only slightly affected. The interpretation seems fairly straightforward. The parental strain can couple energy to transport from either respiration or ATP hydrolysis via the ATPase. However, the mutant has lost the latter alternative, and when treated with KCN it has lost both pathways. Furthermore, this evidence suggests that the ATPase complex is not required for respiratory coupling. Schairer & Gruber (147) have subsequently isolated an unc^- ATPase$^+$ strain, unc 253, which has essentially the same transport phenotype. Or et al (165) and van Thienen & Postma (61) have examined a similar set of strains, NI44, unc^- ATPase$^-$, and BV4, unc^- ATPase$^+$, for proline and serine transport and obtained normal transport under aerobic conditions and low transport under anaerobic conditions. The interpretation of these data is the same as that discussed above. It would appear from these mutants that unc mutations affect ATP-driven but not respiration-driven transport. Prezioso et al (86) have examined strain AN120, unc^- ATPase$^-$, isolated by Butlin et al (143) and found also normal aerobic transport.

In contrast to the above, other unc^- ATPase$^-$ mutants behave somewhat differently. Simoni & Shallenberger (59) have reported an unc^- ATPase$^-$ strain, DL54, which has about a 50% reduction in transport of proline and alanine by intact cells with glucose as an energy source. Furthermore, when transport was tested in vesicles with D-lactate as the energy source, the respiratory defect was more pronounced. It should be added that van Thienen & Postma (61) examined transport in vesicles of strains NI44, unc^- ATPase$^-$, and KI1, unc^- ATPase$^+$, which exhibit normal aerobic transport in cells, and they demonstrated greatly reduced respiration-driven uptake of serine. The increased defect seen in vesicle preparations is possibly due to increased proton leakiness in vesicles resulting from the preparation procedure. Defects are thus magnified in vesicles when compared with cells. Rosen (141) has reported that an unc^- ATPase$^-$ strain, NR70, has a severe defect in aerobic transport of several solutes in the presence of glucose as an energy source. Yamamoto et al (148) have reported similar results with a strain that they isolated. These results indicate that some unc^- ATPase$^-$ strains have defective respiratory coupling.

Results with mutants of the unc^- ATPase$^+$ class are also dependent on the particular mutant strain studied. As discussed above, several mutants of this class show a defect in only the ATP-coupled reaction. Van Thienen & Postma (61), however, demonstrated a defect in respiration-driven transport in vesicles of KI1 that had normal transport in cells. Hong & Kaback (74) reported that an unc^-

ATPase$^+$ strain of *S. typhimurium* had a marked defect for aerobic transport of a variety of solutes in cells. (They have called this strain *etc* for electron transport coupling, and suggested that the defect was in a component involved in a shunt off the respiratory chain that was unique in coupling of oxidation to transport. This was the first evidence by this group that contradicted their original proposal that the solute carriers were intermediates of the respiratory chain. In light of subsequent observations it is likely that the *etc* mutation is in the ATPase complex.) Green & Simoni (166) have examined transport in vesicles of an *unc*$^-$ ATPase$^+$ strain, BG31, and found that respiration-driven proline transport was defective.

Van Thienen & Postma (61) and Rosen (167) have reported that transport in cells and vesicles of ATPase mutants, when defective, could be restored by DCCD. This is similar to restoration of the energy-linked transhydrogenase (142) and quenching of ACMA fluorescence (140), discussed in an earlier section. Van Thienen & Postma (61) showed that vesicles of *unc*$^-$ ATPase$^-$ and *unc*$^-$ ATPase$^+$ mutants were unable to transport serine in the presence of D-lactate or phenazine methosulfate (PMS)-ascorbate unless DCCD was added. DCCD had no effect on wild-type transport.

Rosen reported the same results in intact cells of NR70, *unc*$^-$ ATPase, and extended these results to the proton permeability of wild type and mutant (141, 167). As discussed above, strain NR70, *unc*$^-$ ATPase$^-$, has a marked defect in respiration-driven transport in cells. The nature of the mutation is not clear; this strain possesses no crossreacting material toward antibody prepared against parental F1 and does not revert. This suggests that it may possess a deletion of a portion of the ATPase complex, which may explain the severity of the respiratory defect. Rosen demonstrated that this strain had an increased rate of passive proton movement and that this rate was reduced by treatment with DCCD. Active transport could also be restored by this treatment. It would thus appear that the defect in the ATPase complex results in a membrane that is leaky to protons, as suggested above, and consequently is unable to maintain the high energy state generated by respiration. DCCD repairs that leakiness and restores respiration-driven functions. Altendorf et al (39) have made basically the same observation with strain DL54, *unc*$^-$ ATPase$^-$, which appears to be a single point mutation. This work demonstrated that vesicles of this strain are leaky for protons. In fact, they are as proton leaky as the parental membranes treated with the uncoupler CCCP. The following pertinent observations were made with this strain: 1. respiration-driven proline transport is defective; 2. upon addition of valinomycin, mutant vesicles showed a rapid uptake of protons, in contrast to wild-type vesicles; 3. vesicles were unable to generate a proton gradient or membrane potential in response to D-lactate; 4. vesicles were unable to generate a membrane potential in response to valinomycin which induces potassium efflux; and, most importantly, 5. all the above defects can be repaired by treatment with DCCD.

This information explains the respiratory defect in this strain, as well as others of the same phenotype but, more importantly, provides evidence that the proton motive force is inextricably linked to active transport.

It has recently been possible to restore transport in this strain by reconstitution with parental F1 (153).

Conclusions on unc Mutants

As will be clear from the results discussed earlier, no simple explanation is available for apparent differences in various *unc* mutants. As a case in point, the results with NI44, *unc⁻* ATPase⁻, provide a good demonstration of the difficulty in equating various parameters from mutants derived in one laboratory. NI44 shows no defect in aerobically driven active transport in whole cells, but has a severe defect in transport in vesicles (61). Membranes prepared by sonication show no defect in aerobically driven transhydrogenase (160), but are defective in respiration-dependent ACMA quenching (140). It should be clear that comparison of different energy-dependent processes in a single strain can give different views of the defect.

These results, however, are not really conflicting, but rather extend the role of the ATPase complex. Thus we can view the complex as having a dual role: coupling ATP hydrolysis to the generation of the high energy state and, in addition, functioning to stabilize that same high energy state when it is generated by respiration.

Since the ATPase complex is so complicated, mutations can occur in many polypeptides, each leading to the same general phenotype. More detailed studies reveal many different specific defects. It is clear that loss of the catalytic activity of the F1 results in all cases in loss of ATP-driven reactions. This need not be true for the oxidation-driven reaction.

Much more work needs to be done before direct correlation can be made between changes in specific polypeptides and alteration in energy-linked functions.

Mutations Affecting the Electron Transport Chain

Although the electron transport chain of *E. coli* is not as well defined as that of mitochondria, the genetic approach should provide further insight into this problem as well.

QUINONE-DEFICIENT MUTANTS   Gibson and his collaborators have described mutants of *E. coli* that are defective in ubiquinone biosynthesis (168, 169). Two classes have been described, *ubiB* and *ubiD*, which accumulate the ubiquinone precursors 2-octaprenylphenol and 3-octaprenyl-4-hydroxybenzoate, respectively. The *ubiB* class is of particular interest because a transport defect has been demonstrated. Membrane preparations are defective in the ability to oxidize NADH or D-lactate. Transport of phosphate and of serine in cells of a *ubiB* strain under aerobic conditions with glucose as an energy source was reduced to 10 and 5% of the parental levels, respectively (126). Anaerobic transport in cells grown anaerobically was normal. The severity of the aerobic defect is somewhat surprising. Most *E. coli* strains in which respiration is inhibited due to anaerobiosis or the presence of CN⁻ show only small reduction in transport capacity because there is an alternative pathway to energize the membrane via ATP hydrolysis. In addition, an electron transport mutant isolated by Simoni & Shallenberger (59) shows no apparent defect in proline or alanine uptake. One possibility is that the quinone deficiency results in proton-leaky membranes that make it difficult to generate a high energy state from ATP hydrolysis. Parnes & Boos (125) examined galactose uptake in the *ubiB* strain and found it to be defective.

D-LACTATE DEHYDROGENASE MUTANTS   The work of Kaback and his collaborators has clearly demonstrated that amino acid and sugar uptake in vesicles is preferentially driven by D-lactate as energy source (10). This work prompted an examination of D-lactate dehydrogenase-deficient mutants in order to evaluate the suggested uniqueness of this energy source in cells. Simoni & Shallenberger (59) demonstrated that such a mutant has completely normal aerobic uptake of proline and alanine. This observation is hardly surprising in light of our understanding of energy-coupling mechanisms. As predicted, transport of these amino acids in vesicles showed no D-lactate stimulation. Hong & Kaback (74) have subsequently reported essentially the same result. Although these mutants have eliminated any unique role for D-lactate oxidation in supplying energy for active transport, as originally suggested, they have provided a system for studying reconstitution of the solubilized D-lactate dehydrogenase with membranes lacking this enzyme and its role in energizing transport, which was discussed in an earlier section.

MUTATIONS AFFECTING HEME BIOSYNTHESIS   Strains of *E. coli* have been isolated that have specific defects in the biosynthesis of heme (170, 171). One class of *heme A* mutants contains no detectable cytochromes unless the cells are grown in the presence of 5-aminolevulinic acid. In an interesting set of experiments, Haddock & Schairer (172) were able to reconstitute the cytochromes by incubation of membranes of this strain prepared from cells grown in the absence of 5-amino-levulinic acid with hematin, plus ATP. This indicates that the cytochrome apoproteins are present in the membrane. Such strains, grown in the absence of 5-aminolevulinic acid, have virtually no electron transport capacity. Devor et al (173) have shown that cultures supplemented with 5-aminolevulinic acid transport β-galactosides as the parental strain, i.e. transport is inhibited 50% by either KCN or DCCD. In contrast, however, the unsupplemented cells exhibited transport that was insensitive to KCN but 80% inhibited by DCCD. Singh & Bragg (174) have studied the transport of phenylalanine in *heme A* mutants grown in the presence or absence of 5-aminolevulinic acid. The deficient culture is virtually unable to transport phenylalanine with D-lactate as an energy source, since this compound yields energy chiefly through respiration. However, when glucose was used as energy source, transport was equivalent to the supplemented culture as expected because energy could now be obtained through hydrolysis of glycolytically generated ATP via the ATPase complex.

Energy-Coupling Mutations in Specific Transport Systems

We have thus far discussed mutations in the general components of the energy transduction apparatus of the cell. There are, however, mutations that apparently affect the energetics of only a single solute transport system. Again, mutations in the β-galactoside system of *E. coli* have been most extensively studied by Wilson and his collaborators (175, 176). One such mutant, X7154, is defective in the accumulation of β-galactosides, although the rate of entry appears to be normal.

The defect seems to result in an increased exit rate due to an increased affinity of the carrier for the solute at the inner surface of the membrane. An apparently similar mutation has been described by Hechtman & Scriver (177) in *Pseudomonas fluorescens*. The mutant is unable to accumulate alanine and proline, which share the same transport system. Transport of other amino acids is normal. It may be recalled that the energy requirement for facilitated diffusion is open to question. The isolation of such mutants lends weight to the suggestion of Winkler & Wilson (51) that energy is not required for entry but rather affects exit. Genetic analysis of the mutation in *E. coli* indicates that the defect occurs in or near the gene coding for the β-galactoside carrier, the M protein.

West & Wilson (178) have shown that addition of TMG to anaerobic mutant cells in the presence of a permeant anion, SCN^-, does not lead to H^+ uptake as observed with the parental strain. Separate experiments showed that the mutant is not leaky to protons. The authors conclude that the mutant is defective in the coupling of H^+ influx to β-galactoside influx. The results can be equally well explained, however, if the M protein binds H^+ so strongly that protons are not released at the inside, resulting in a cycling of H^+ in the presence of sugar. Whatever the correct explanation, in both cases the transport system of the mutant is different from the wild type in that either a galactoside-M protein complex moves inward or a H^+-M protein complex moves outward. In both cases, however, SCN^- should not be required for TMG movement, an experiment that is regrettably not reported by the authors.

GENERAL CONCLUSIONS

In summary, it appears that available evidence strongly implicates the proton motive force or some element of it in the indirect coupling of metabolic energy to active transport of a number of solutes. It is equally clear that this force can be generated by respiration or hydrolysis of ATP. The contribution of either or both depends on the organism under study. In facultative bacteria such as *E. coli* both are operative. While this general conclusion seems justified, many questions remain. The quantitative aspects of proton-solute symport remain poorly defined. Elucidation of the mechanism of proton translocation and general function and properties of the Mg^{2+}-Ca^{2+}-ATPase complex will require a great deal of additional information. In this regard the use of genetics will continue to be a fruitful approach.

ACKNOWLEDGMENTS

The authors wish to thank those who sent reprints and preprints. Their cooperation made this review possible.

We also thank J. U. Umbreit, R. Humbert, and Pamela Talalay for helpful criticism during the preparation of this manuscript.

P. Postma was a recipient of a ZWO stipend from the Netherlands Organization for the Advancement of Pure Research (ZWO).

Work conducted in the laboratory of R. Simoni was supported by National Institutes of Health Grant GM 18539.

Literature survey as of September 1, 1974.

Literature Cited

1. Kundig, W., Ghosh, S., Roseman, S., 1964. *Proc. Nat. Acad. Sci. USA* 52: 1067–74
2. Roseman, S. 1972. *Metab. Pathways* 6: 41–89
3. Simoni, R. D. 1972. *Membrane Molecular Biology,* ed. C. F. Fox, A. D. Keith, 284–322. Stamford, Conn.: Sinauer
4. Lin, E. C. C. 1971. *Structure and Function of Biological Membranes,* ed. L. I. Rothfield, 285–341. New York: Academic
5. Hamilton, W. A. 1975. *Advan. Microb. Physiol.* 12: In press
6. Baker, P. F. 1972. *Metab. Pathways* 6: 243–68
7. Martonosi, A. 1972. *Metab. Pathways* 6: 317–49
8. Crane, R. K. 1965. *Fed. Proc.* 24: 1000–6
9. Stock, J., Roseman, S. 1971. *Biochem. Biophys. Res. Commun.* 44: 132–38
10. Kaback, H. R. 1972. *Biochim. Biophys. Acta* 265: 367–417
11. Harold, F. M. 1972. *Bacteriol. Rev.* 36: 172–230
12. Mitchell, P. 1966. *Biol. Rev.* 41: 445–502
13. Greville, G. D. 1969. *Curr. Top. Bioenerg.* 3: 1–78
14. Skulachev, V. P. 1971. *Curr. Top. Bioenerg.* 4: 127–90
15. Slater, E. C. 1971. *Quart. Rev. Biophys.* 4: 35–71
16. Grinius, L. L., Jasaitis, A., Kadziauskas, P., Liberman, E. A., Skulachev, V. P., Topaly, V. P., Tsofina, L. M., Vladimirova, M. A. 1970. *Biochim. Biophys. Acta.* 216: 1–12
17. Bakeeva, L. E., Grinius, L. L., Jasaitis, A., Kuliene, V., Levitskii, D. O., Liberman, E. A., Severina, I. I., Skulachev, V. P. 1970. *Biochim. Biophys. Acta* 216: 13–21
18. Isaev, P. I., Liberman, E. A., Samuilov, V. D., Skulachev, V. P., Tsofina, L. M. 1970. *Biochim. Biophys. Acta* 216: 22–29
19. Liberman, E. A., Skulachev, V. P. 1970. *Biochim. Biophys. Acta* 216: 30–42
20. Harold, F. M., Papineau, D. 1972. *J. Membrane Biol.* 8: 27–44
21. Harold, F. M., Papineau, D. 1972. *J. Membrane Biol.* 8: 45–62
22. Sims, P. J., Waggoner, A. S., Wong, C. H., Hoffman, J. F. 1974. *Biochemistry* 13: 3315–30
23. Laris, P. C., Perhadsingh, W. A. 1972. *Biochem. Biophys. Res. Commun.* 57: 620–26
24. Kashket, E. R., Wilson, T. H. 1974. *Biochem. Biophys. Res. Commun.* 59: 879–86
25. Hirata, H., Altendorf, K., Harold, F. M. 1973. *Proc. Nat. Acad. Sci. USA* 70: 1804–8
26. Griniuviene, B., Chmieliauskaite, V., Grinius, L. 1974. *Biochem. Biophys. Res. Commun.* 56: 206–13
27. Grinius, L. L., Il'ina, M. D., Mileikovskaya, E. I., Skulachev, V. P., Tikhonova, G. V. 1972. *Biochim. Biophys. Acta* 283: 442–55
28. Scholes, P., Mitchell, P. 1970. *Bioenergetics* 1: 61–72
29. Jeacocke, R. E., Niven, D. F., Hamilton, W. A. 1972. *Biochem. J.* 127: 57P
30. Mitchell, P., Moyle, J. 1969. *Eur. J. Biochem.* 7: 471–84
31. Drachev, L. A., Kaulen, A. D., Ostroumov, S. A., Skulachev, V. P. 1974. *FEBS Lett.* 39: 43–45
32. Kayushin, L. P., Skulachev, V. P. 1974. *FEBS Lett.* 39: 39–42
33. Racker, E., Stoeckenius, W. 1974. *J. Biol. Chem.* 249: 662–65
34. Scholes, P., Mitchell, P. 1970. *Bioenergetics* 1: 309–23
35. West, I., Mitchell, P. 1972. *Bioenergetics* 3: 445–62
36. Lawford, H. G., Haddock, B. A. 1974. *Biochem. J.* 136: 217–20
37. van Dam, K., Postma, P. W. 1974. *Dynamics of Energy-Transducing Membranes,* ed. L. Ernster, R. W. Estabrook, E. C. Slater, 433–45. Amsterdam: Elsevier
38. Reeves, J. P. 1971. *Biochem. Biophys. Res. Commun.* 45: 931–36
39. Altendorf, K. H., Harold, F. M., Simoni, R. D. 1974. *J. Biol. Chem.* 249: 4587–93
40. Bening, G. J., Eilermann, L. J. M. 1973. *Biochim. Biophys. Acta* 292: 402–12
41. Hertzberg, E. L., Hinkle, P. C. 1974. *Biochem. Biophys. Res. Commun.* 58: 178–84
42. Mitchell, P. 1973. *FEBS Lett.* 33: 267–74
43. West, I. C., Mitchell, P. 1974. *FEBS Lett.* 40: 1–4
44. Mitchell, P. 1963. *Biochem. Soc. Symp.* 22: 142–68
45. Cockrell, R. S., Harris, E. J., Pressman, B. C. 1967. *Nature* 215: 1487–88
46. Lew, V. L., Glynn, I. M., Ellory, J. C. 1970. *Nature* 225: 865–66
47. Makinose, M., Hasselbach, W. 1971. *FEBS Lett.* 12: 271–72
48. Mitchell, P. 1970. *Membranes and Ion Transport,* ed. E. E. Bittar, 1: 192–256. New York: Wiley
49. Boyer, P. D., Klein, W. L. See Ref. 3, 323–44
50. Boyer, P. D., Cross, R. L., Momsen, W. 1973. *Proc. Nat. Acad. Sci. USA* 70:

2837–39

51. Winkler, H. H., Wilson, T. H. 1966. *J. Biol. Chem.* 241:2200–11
52. Kepes, A. 1960. *Biochim. Biophys. Acta* 40:70–84
53. Koch, A. L. 1964. *Biochim. Biophys. Acta* 79:177–200
54. Kepes, A. 1971. *J. Membrane Biol.* 4:87–112
55. Scarborough, G. A., Rumley, M. K., Kennedy, E. P. 1968. *Proc. Nat. Acad. Sci. USA* 60:951–58
56. Harold, F. M., Baarda, J. R., Baron, C., Abrams, A. 1969. *J. Biol. Chem.* 244:2261–68
57. Harold, F. M., Baarda, J. R., Baron, C., Abrams, A. 1969. *Biochim. Biophys. Acta* 183:129–42
58. Klein, W. L., Boyer, P. D. 1972. *J. Biol. Chem.* 247:7257–65
59. Simoni, R. D., Shallenberger, M. K. 1972. *Proc. Nat. Acad. Sci. USA* 69:2663–67
60. Schairer, H. U., Haddock, B. A. 1972. *Biochem. Biophys. Res. Commun.* 48:544–51
61. van Thienen, G., Postma, P. W. 1973. *Biochim. Biophys. Acta* 323:429–40
62. Abrams, A., Smith, J. B., Baron, C. 1972. *J. Biol. Chem.* 247:1484–88
63. Asghar, S. S., Levin, E., Harold, F. M. 1973. *J. Biol. Chem.* 248:5225–33
64. Pavlasova, E., Harold, F. M. 1969. *J. Bacteriol.* 98:198–204
65. Kashket, E. R., Wilson, T. H. 1972. *Biochem. Biophys. Res. Commun.* 49:615–20
66. Kashket, E. R., Wilson, T. H. 1972. *J. Bacteriol.* 109:784–89
67. Kashket, E. R., Wilson, T. H. 1973. *Proc. Nat. Acad. Sci. USA* 70:2866–69
68. Niven, D. F., Jeacocke, R. E., Hamilton, W. A. 1973. *FEBS Lett.* 29:248–52
69. Niven, D. F., Hamilton, W. A. 1974. *Eur. J. Biochem.* 44:517–22
70. West, I. C. 1970. *Biochem. Biophys. Res. Commun.* 41:655–61
71. West, I. C., Mitchell, P. 1973. *Biochem. J.* 132:587–92
72. Whittam, R. 1967. *The Molecular Mechanism of Active Transport,* ed. G. C. Quarton, T. Melnechuk, F. U. Schmitt, 313–25. New York: Rockefeller Inst.
73. Kaback, H. R. 1971. *Methods Enzymol.* 22:99–120
74. Hong, J. S., Kaback, H. R. 1972. *Proc. Nat. Acad. Sci. USA* 69:3336–40
75. Mitchell, P. 1973. *Bioenergetics* 4:63–91
76. Boos, W. 1974. *Ann. Rev. Biochem.* 43:123–46
77. Harold, F. M. 1974. *Ann. NY Acad. Sci.* 227:297–311
78. Lombardi, F. J., Reeves, J. P., Short, S. A., Kaback, H. R. 1974. *Ann. NY Acad. Sci.* 227:312–27
79. Kaback, H. R. 1974. *Science* 186:882–92
80. Altendorf, K. H., Staehelin, L. A. 1974. *J. Bacteriol.* 117:888–99
81. Konings, W. N., Bisschop, A., Veenhuis, M., Vermeulen, C. A. 1973. *J. Bacteriol.* 116:1456–65
82. Oppenheim, J. D., Salton, M. R. J. 1973. *Biochim. Biophys. Acta* 298:297–322
83. Gorneva, G. A., Ryabova, I. D. 1974. *FEBS Lett.* 42:273–74
84. Futai, M. 1974. *J. Membrane Biol.* 15:15–28
85. Weiner, J. H. 1974. *J. Membrane Biol.* 15:1–14
86. Prezioso, G., Hong, J. S., Kerwar, G., Kaback, H. R. 1973. *Arch. Biochem. Biophys.* 154:575–82
87. Asano, A., Hirata, H., Brodie, A. F. 1972. *Biochem. Biophys. Res. Commun.* 46:1340–46
88. Hampton, M. L., Freese, E. 1974. *J. Bacteriol.* 118:495–504
89. Matin, A., Konings, W. N. 1973. *Eur. J. Biochem.* 34:58–67
90. Reeves, J. P., Lombardi, F. J., Kaback, H. R. 1972. *J. Biol. Chem.* 247:6204–11
91. Heidrich, H. G., Leitner, G. 1974. *Eur. J. Biochem.* 41:37–43
92. Zachowski, A., Paraf, A. 1974. *Biochem. Biophys. Res. Commun.* 57:787–92
93. Konings, W. N., Kaback, H. R. 1973. *Proc. Nat. Acad. Sci. USA* 70:3376–81
94. Brockman, R. W., Heppel, L. A. 1968. *Biochemistry* 7:2554–62
95. Weissbach, H., Thomas, E., Kaback, H. R. 1971. *Arch. Biochem. Biophys.* 147:249–54
96. Asano, A., Cohen, N. S., Baker, R. F., Brodie, A. F. 1973. *J. Biol. Chem.* 248:3386–97
97. Mével-Ninio, M., Yamamoto, T. 1974. *Biochim. Biophys. Acta* 357:63–66
98. Futai, M. 1974. *J. Bacteriol.* 120:861–65
99. Reeves, J. P., Hong, J.-S., Kaback, H. R. 1973. *Proc. Nat. Acad. Sci. USA* 70:1917–21
100. Short, S., Kaback, H. R., Kohn, L. D. 1974. *Proc. Nat. Acad. Sci. USA* 71:1461–63
101. Futai, M. 1974. *Biochemistry* 13:2327–33
102. Hirata, H., Altendorf, K. H., Harold, F. M. 1975. *J. Biol. Chem.* 249:2939–45
103. Haaker, H., Berden, J. A., Kraayenhof, R., Katan, M., van Dam, K. 1972. *Biochemistry and Biophysics of Mitochondrial Membranes,* ed. G. F. Azzone, E. Carafoli, A. L. Lehninger, E. Quag-

liariello, N. Siliprandi, 329–40. New York: Academic
104. Manno, J. A., Schachter, D. 1970. *J. Biol. Chem.* 245:1217–23
105. Wong, J. T. F., Pincock, A., Bronskill, P. M. 1971. *Biochim. Biophys. Acta* 233:176
106. Koch, A. L. 1971. *J. Mol. Biol.* 59: 447–59
107. Cecchini, G., Koch, A. L. 1974. *Abstr. Ann. Meet. Am. Soc. Microbiol.,* 193
108. Reeves, J. P., Schechter, E., Weil, R., Kaback, H. R. 1973. *Proc. Nat. Acad. Sci. USA* 70:2722–26
109. Kennedy, E. P., Rumley, M. K., Armstrong, A. 1974. *J. Biol. Chem.* 249: 33–37
110. Abrams, A. 1959. *J. Biol. Chem.* 234: 383–88
111. Höfer, M. 1971. *Arch. Mikrobiol.* 80: 50–61
112. Höfer, M. 1971. *J. Theor. Biol.* 33: 599–603
113. Komor, E., Haas, D., Tanner, W. 1972. *Biochim. Biophys. Acta* 266:649–60
114. Komor, E., Loos, E., Tanner, W. 1973. *J. Membrane Biol.* 12:89–99
115. Visser, A. S., Postma, P. W. 1973. *Biochim. Biophys. Acta* 298:333–40
116. Postma, P. W., Cools, A., van Dam, K. 1973. *Biochim. Biophys. Acta* 318: 91–104
117. Simoni, R. D., Roseman, S. 1973. *J. Biol. Chem.* 248:966–76
118. Berger, E. A. 1973. *Proc. Nat. Acad. Sci. USA* 70:1514–18
119. Berger, E. A., Heppel, L. A. 1975. *J. Biol. Chem.* In press
120. Oxender, D. L. 1972. *Metab. Pathways* 6:133–85
121. Oxender, D. L. 1972. *Ann. Rev. Biochem.* 41:777–809
122. Curtis, S. J. 1974. *J. Bacteriol.* 120: 295–303
123. Cowell, J. L. 1974. *J. Bacteriol.* 120: 139–46
124. Wilson, D. B. 1974. *J. Bacteriol.* 120: 866–71
125. Parnes, J. R., Boos, W. 1973. *J. Biol. Chem.* 248:4429–35
126. Cox, G. B., Gibson, F. 1974. *Biochim. Biophys. Acta* 346:1–26
127. Baron, C., Abrams, A. 1971. *J. Biol. Chem.* 246:1542–44
128. Senior, A. E. 1973. *Biochim. Biophys. Acta* 301:249–77
129. Abrams, A., Smith, J. B. 1974. *Enzymes* 10:395–429
130. Mitchell, P. 1974. *FEBS Lett.* 43:189–94
131. Bragg, P. D., Hou, C. 1972. *FEBS Lett.* 28:309–12
132. Catterall, W. A., Coty, W. A., Pedersen, P. L. 1973. *J. Biol. Chem.* 248:7427–31
133. Hanson, R. L., Kennedy, E. P. 1973. *J. Bacteriol.* 114:772–81
134. Nelson, N., Kanner, B. I., Gutnick, D. L. 1974. *Proc. Nat. Acad. Sci. USA* 71: 2720–24
135. Bragg, P. D., Davies, P. L., Hou, C. 1973. *Arch. Biochem. Biophys.* 159: 664–70
136. Futai, M., Sternweis, P. C., Heppel, L. A. 1974. *Proc. Nat. Acad. Sci. USA* 71:2725–29
137. Simoni, R. D. In preparation
138. Roisin, M. P., Kepes, A. 1973. *Biochim. Biophys. Acta* 305:249–59
139. Mitchell, P. 1967. *Fed. Proc.* 26:1370–79
140. Nieuwenhuis, F. J. R. M., Kanner, B. I., Gutnick, D. L., Postma, P. W., van Dam, K. 1973. *Biochim. Biophys. Acta* 325:62–71
141. Rosen, B. P. 1973. *J. Bacteriol.* 116: 1124–29
142. Bragg, P. D., Hou, C. 1973. *Biochem. Biophys. Res. Commun.* 50:729–36
143. Butlin, J. D., Cox, G. B., Gibson, F. 1971. *Biochem. J.* 124:75
144. Butlin, J. D., Cox, G. B., Gibson, F. 1973. *Biochim. Biophys. Acta* 292:366–75
145. Kanner, B. I., Gutnick, D. L. 1972. *J. Bacteriol.* 111:287–89
146. Gutnick, D. L., Kanner, B. I., Postma, P. W. 1972. *Biochim. Biophys. Acta* 283:217–22
147. Schairer, H. U., Gruber, D. 1973. *Eur. J. Biochem.* 37:282–86
148. Yamamoto, T. H., Mével-Ninio, M., Valentine, R. C. 1973. *Biochim. Biophys. Acta* 314:267–75
149. Gunther, T., Maris, G. 1974. *Z. Naturforsch. C* 29:60–62
150. Hinkle, P. C., Horstman, L. L. 1971. *J. Biol. Chem.* 246:6024–28
151. Santiago, E., López-Moratalla, N., Segovia, J. L. 1973. *Biochem. Biophys. Res. Commun.* 53:439–45
152. Racker, E. 1972. *J. Membrane Biol.* 10:221–35
153. Humbert, R., Simoni, R. D. In preparation
154. Cox, G. B., Gibson, F., McCann, L. 1973. *Biochem. J.* 134:1015–21
155. Nelson, N., Deters, D., Nelson, H., Racker, E. 1973. *J. Biol. Chem.* 248: 2049–55
156. Nelson, N., Nelson, H., Racker, E. 1972. *J. Biol. Chem.* 247:7657–68
157. Kozlov, I. A., Mikelsaar, H. N. 1974. *FEBS Lett.* 43:212–14
158. Salton, M. R. J., Schor, M. T. 1972.

Biochem. Biophys. Res. Commun. 49: 350–57

159. Salton, M. R. J., Schor, M. T. 1974. *Biochim. Biophys. Acta* 345:74–82

160. Kanner, B. I., Gutnick, D. L. 1972. *FEBS Lett.* 22:197–99

161. Mitchell, P., Moyle, J. 1965. *Nature* 208:1205–6

162. Papa, S., Guerrieri, F., Rossi-Bernardi, L., Tager, J. M. 1970. *Biochim. Biophys. Acta* 197:100–3

163. Cattell, K. J., Lindop, C. R., Knight, I. G., Beechey, R. B. 1971. *Biochem. J.* 125:169–77

164. Stekhoven, F. S., Waitkus, R. F., van Moerkerk, H. Th. B. 1972. *Biochemistry* 11:1144–50

165. Or, A., Kanner, B., Gutnick, D. L. 1973. *FEBS Lett.* 35:217–19

166. Green, B., Simoni, R. D. In preparation

167. Rosen, B. P. 1973. *Biochem. Biophys. Res. Commun.* 53:1289–96

168. Cox, G. B., Gibson, F., Pitlard, J. 1968. *J. Bacteriol.* 95:1591–98

169. Cox, G. B., Young, I. G., McCann, L., Gibson, F. 1969. *J. Bacteriol.* 99: 450–58

170. Wulff, D. L. 1967. *J. Bacteriol.* 93: 1473–74

171. Sasarman, A., Surdeanu, M., Saegli, G., Horodniceanu, T., Greceanu, V., Dumi, A. 1968. *J. Bacteriol.* 96:570–72

172. Haddock, B. A., Schairer, H. U. 1973. *Eur. J. Biochem.* 35:34–45

173. Devor, K. A., Schairer, H. U., Ross, D., Overath, P. 1974. *Eur. J. Biochem.* 45: 451–56

174. Singh, A. P., Bragg, P. D. 1974. *Biochem. Biophys. Res. Commun.* 57:1200–6

175. Wilson, T. H., Kusch, M., Kashket, E. R. 1970. *Biochem. Biophys. Res. Commun.* 40:1409–14

176. Wilson, T. H., Kusch, M. 1972. *Biochim. Biophys. Acta* 255:786–97

177. Hechtman, P., Scriver, C. R. 1970. *Biochim. Biophys. Acta* 219:428–36

178. West, I. C., Wilson, T. H. 1973. *Biochem. Biophys. Res. Commun.* 50:551–58

179. Mitchell, P. 1970. *20th Symp. Soc. Gen. Microbiol.* 121–66

VIRUS ASSEMBLY

Sherwood Casjens[1] and Jonathan King

Department of Biology, Massachusetts Institute of Technology,
Cambridge, Massachusetts 02139

CONTENTS

INTRODUCTION

Despite the importance of virus studies in the development of molecular biology, the study of particle assembly has blossomed only recently. The original reconstitution of tobacco mosaic virus (TMV) by Fraenkel-Conrat & Williams (9), though the root of the self-assembly concept, suggested that assembly was the straightforward binding of macromolecules to each other. In addition, Caspar & Klug's satisfying and simple theory of icosahedral virus structure (47) indicated that virus assembly held no further biological secrets. Subsequently, the discovery by Epstein et al (107) that a very large number of phage T4 genes were involved in assembly and the elucidation

[1] Present address: Dept. of Microbiology, University of Utah Medical School, Salt Lake City, Utah.

of the complicated pathways of T4 morphogenesis (109, 201, 207, 224) indicated that things were not so simple. The discovery of proteolytic processing in poliovirus and T4 (64, 65, 162–165), the analysis of the actual pathway of TMV reconstitution (13, 15), and the numerous studies of the formation of enveloped viruses, to name only a few areas, have made it clear that a great deal is yet to be learned from and about the formation of viruses.

In the last few years we have witnessed an explosion of information about viral assembly. This has depended in large part on technical advances in the dissociation and separation of strongly bonded proteins and on intensive study of cells infected with viral mutants blocked in assembly. The introduction by Laver and by Maizel of the detergent dodecylsulfate as a protein denaturing agent during polyacrylamide gel electrophoresis has made possible the separation and identification of viral proteins according to the particularly meaningful criterion of molecular weight (39). Laemmli's combination of the sodium dodecylsulfate (SDS) method with discontinuously buffered gel systems (162) enables the resolution of even minor proteins in such complex mixtures as cell lysates.

The development by Epstein, Edgar, and Campbell (107, 108) of conditional lethal mutations as a method of identifying essential genes greatly increased our understanding of the genetic control of cellular processes. The *amber* mutants of bacteriophage permit the identification of the protein products of viral genes. To the extent that the defective phenotypes of the mutants are known, the correlation of protein with gene indicates the function of the protein in the assembly process. Analysis of the incomplete structures accumulating in mutant-infected cells often reveals steps in the in vivo pathway of particle formation. Mutant-infected cells also serve as an invaluable source of precursors for in vitro assembly studies, as originally demonstrated by Edgar & Wood (201). These studies, using precursors formed in vivo, should not be confused with reconstitution experiments utilizing degradation products.

Previous general reviews of virus assembly are Levine (1) and Eiserling & Dickson (2) in this series, and Kellenberger (7), and Kushner's (3) review of general assembly processes. A less detailed but very readable general review has recently been published by Russell (4). Immunological studies of viral components have been reviewed by Neurath & Rubin (457). Reviews of individual groups of viruses will be referred to in the relevant sections of this review.

Rather than concentrating on a few well-understood varieties, we have chosen to cover the assembly of most types of viruses, since this may be the last time it is feasible to review the assembly of animal, plant, and bacterial viruses in a single article. Thus, we have tried to gather and summarize material on each group of viruses to make it available to those who work on different viruses. We hope this will stimulate communication among those studying different groups of viruses, and provide a reference source and a guide for assimilating proliferating new developments. Since it is unclear to us which features of assembly processes are general and which particular, we have avoided attempts at grand synthesis.

To insure effective access to the literature we have discussed only a few studies that were not in press at the time we finished the literature survey, August 1974. For the same reason we have sometimes referenced the most recent publication on a subject, rather than the original. Due to space limitations we concentrate explicitly on mechanisms of assembly, at the expense of discussing virus structure per se. In another review in this volume, Crowther & Klug (4a) describe recent advances in electron microscopic techniques for analyzing virus structure. The proteolytic reactions in virus assembly are considered in greater detail in Hershko & Fry's review in this volume (4b).

One must keep in mind the function of viral capsids in discussing their formation. Clearly they have to protect the nucleic acid from degradation. They also may function in adsorption to the host cell, penetration of the cell envelope, and perhaps transport of the nucleic acid to a proper site. Capsids of many viral groups must be designed so that they can be uncoated by the host cell, whereas others, such as the cores of reoviruses, represent enzymatic apparatus that must function in the host cell. In many cases, such as in the large phage, encapsidation of nucleic acid is intimately associated with the maturation of the nucleic acid, for example formation of the sticky ends of phage λ DNA. Thus formation of the capsid represents the last step in the formation of an infectious nucleic acid molecule.

One of the major problems in keeping up with virus studies is that a very large number of different viruses are under active investigation. We have grouped the viruses that we will discuss in Figure 1 according to criteria that are useful for discussing their assembly.[2] Thus, since ssRNA is not rigid and often contains considerable spontaneous secondary structure, nucleic acid condensation is not a problem, and we have grouped these viruses together. The overall grouping has the disadvantage of separating the enveloped viruses, even though they may share common features in the formation of their envelopes.

Our general framework in gathering data pivots around a few questions. How do protein subunits come together to form closed shells? How are nucleic acid molecules packaged into these shells? What other steps are required for the formation of an infectious virus particle? How are assembly processes regulated so that correct structures are formed? How is the information for accurate assembly genetically specified?[3]

The motivation for this review derives not only from the desire to understand the mechanisms and genetic control of morphogenesis, but also from the awareness that no effective therapy exists for most diseases of viral orgin, and that virion assembly may be a target for the development of such therapies.

HELICAL VIRUSES

Rod-shaped viruses with helical symmetry have been isolated from many plants and bacteria (5, 6). The plant viruses, all containing ssRNA genomes, fall into two

[2] ss and ds will be used to abbreviate single stranded and double stranded, respectively.

[3] In the text we use "gp" to mean gene product in describing the protein product of a gene. Thus gp7 refers to the polypeptide chain of gene 7. We recommend this as a general nomenclature which would distinguish genetically identified proteins from genetically unidentified ones called simply "pA," etc for proteins A, etc.

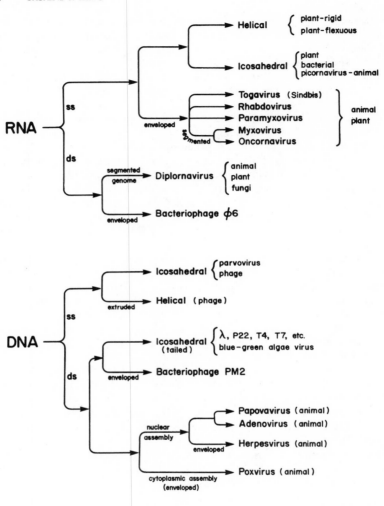

Figure 1 Major groups of viruses according to genome structure, symmetry, site of assembly, and whether or not enveloped. (The groupings would be quite different if the envelope criterion was primary rather than secondary.)

morphological classes: (*a*) rigid rods, including tobacco mosaic virus, barley stripe mosaic virus, and tobacco rattle virus, and (*b*) longer flexuous rods, including potato X virus, potato Y virus, and clover yellow mosaic virus. The assembly of the latter class has not been studied. The helical filamentous bacterial viruses contain circular ssDNA genomes. Their morphogenesis differs substantially from the plant viruses in being intimately associated with cell membranes, and we discuss it in a separate section. Since plant cells have cell walls, the penetration by plant viruses may be quite different from animal viruses, and there is considerable evidence that mechanical or insect mediated wounding of the host tissue is required, perhaps to break the wall.

Assembly of the Rigid Helical Viruses

Tobacco mosaic virus was the first sub-bacterial particle identified as the etiological agent of a plant disease, and its biology has been studied in great detail (8, 458). Its reconstitution in vitro by Fraenkel-Conrat & Williams (9) was a dramatic event in molecular biology, and since that time its structure and reconstitution have been studied more extensively than any other virus. The TMV virion contains one ssRNA molecule 6300 nucleotides long and 2130 helically arranged identical protein subunits of molecular weight 17,500. The RNA molecule is also helically arranged, being sandwiched between the helical layers of protein subunits. Early studies of the reconstitution showed that the RNA molecule normally determines the length of the nucleoprotein helix. However, under certain conditions, the protein alone can polymerize into long virus-like helical structures or long stacked disc structures. Furthermore, two anomalously titrating carboxyl groups are thought to control the type of polymerization the coat protein undergoes (helical or stacked disc). These studies have been extensively reviewed (8, 10–15), so we present only a brief discussion with emphasis on recent results to focus the distinction between the initiation of particles and their elongation.

TMV COAT PROTEIN   All the information necessary for the assembly of the TMV virion resides in the coat protein and RNA molecules. In order to understand how this information is used, Klug, Butler & Durham (13, 16) have extensively studied the oligomeric states which TMV coat protein (derived from disrupted virons) assumes in solution. Three easily distinguishable equilibrium forms have been defined by this work: (a) A protein: a set of small oligomers (up to about pentamers) which sediment at about 4S; (b) 8S protein: a less well-characterized higher oligomer(s) thought to contain about seven monomer subunits; and (c) 20S discs: made up of 34 coat protein subunits arranged in two 17 subunit rings, with the rings being bound head-to-tail [i.e. the polarity of the two rings is the same (17)].

A protein is the major species present at pHs above 7 and ionic strengths below 0.4, and the disc predominates between pH 6 and 7 at similar ionic strengths. Below pH 6, the protein polymerizes into long virus-like helices (17). The interconversion between these species has been studied, and it has generally been concluded that there exists an equilibrium between the various protein oligomers, although there has been some disagreement on the rates of the interconversions (18–21). Coat protein to be used in reconstitution experiments is generally stored at acid pH as protein helices; A protein is prepared by dialysis to pH 7.0 in ionic strength 0.1 solution at 4°C, and discs are prepared in the same manner except at 20°C. A protein prepared in this way exhibits a single 4S sedimentation boundary, and the disc preparations are 60–80% 20S material (discs) and 20–40% 4S material (22, 23).

It is possible that the state of the protein depends on its preparative history, and so different preparative techniques could result in some of the differences of opinion mentioned below.

INITIATION OF TMV ASSEMBLY   There is general agreement among workers in the field that the initial interaction of the RNA molecule with coat protein requires the presence of discs (19, 22, 24–26), as first shown by Butler & Klug (38). The 5' sequence of the TMV RNA molecule is also specifically required for efficient assembly in vitro (27). These results have led to the proposal of a model in which initiation of assembly occurs through the formation of an initiation complex. This consists of one (or two) 20S discs bound to the 5' end of the RNA molecule which are dislocated into a lock-washer-type structure (see 28). Details of the initiation process are not known. Butler (28) has suggested from kinetic data that two discs are required for a successful initiation event, and he has proposed a specific model for initiation which awaits confirmation from experiments of more direct methodology.

Tobacco leaf (host) RNA as well as several synthetic polynucleotides are not encapsidated efficiently by TMV coat protein in vitro, implying that the nucleotide sequence of the RNA is important in successful encapsidation (29, 30). A number of more recent studies have shown that assembly is inhibited if terminal 5' nucleotides are removed from the TMV RNA molecule, whereas if 3' nucleotides are removed, no such effect is observed (reviewed by Atabekov 27). These results led to the belief that the sequence which the coat protein recognizes most strongly resides in the 5' terminal end of the RNA molecule. It is not known whether this recognition is strictly sequence specific or if it is partially dependent on secondary structure. Curiously, Richards et al (31) found that the TMV RNA fragment (from partial T_1 ribonuclease digests) most efficiently protected by coat protein from further nuclease degradation under assembly conditions was not the 5' terminal sequence, but the part of the genome that codes for amino acids 95–130 of the coat protein. This implies that either the sequence at the 5' end is not the most favorable sequence for the coat protein interaction, or that the 5' terminal sequence was so badly damaged by the original T_1 ribonuclease digestion that it could not be used for assembly. Richards et al (31) suggest that this may not represent an interaction directly related to assembly, but could be a translational repression similar to that seen with the RNA bacteriophage coat proteins (see below).

ELONGATION STAGE OF TMV ASSEMBLY   Considerable disagreement exists among workers in the area about which coat protein oligomers are the active species in the elongation of the TMV rod—those steps that occur after the formation of the initiation complex (15). Several authors have argued that the disc structures are needed directly (without breakdown) for the elongation of the TMV nucleoprotein rod (13, 21, 23, 32), whereas other workers suggest that 4S A protein and discs are equally competent for elongation (19, 22) or that only 4S material is competent (25, 26). Since discs and smaller species are in equilibrium and there is disagreement on the rates of interconversion, unambiguous interpretation of elongation from disc preparations is difficult. Our own judgement is that the data support the idea that elongation proceeds by addition of A protein to the growing rod.

Recent studies using tobacco rattle virus (TRV) (33) and cucumber green mottle mosaic virus watermelon strain (CGMMV-W) (34, 35) strongly support the notion that disc-type structures are also necessary for the formation of the initiation complex of these viruses. The disc structure made by the CGMMV-W coat protein is, however, probably a single layer ring of subunits. Conditions are easily found for both of these viruses where initiation occurs but elongation does not; these generally coincide with those conditions where only discs are present. Ohno et al (34) showed that if a CGMMV-W coat protein and RNA are incubated under conditions where only rings are present, no formation of complete virions occurs. However, if low molecular weight (2.5S) coat protein is then added, or if conditions are shifted such that the rings break down into smaller units, infectious virus particles are then formed. The simplest interpretation of these results is that rings are necessary for initiation, but some oligomer smaller than rings is required for the elongation process.

Thus far, all information on the assembly of rigid helical viruses has been obtained from in vitro assembly experiments using protein and RNA from disrupted virions. The recent successful productive infection of tobacco mesophyll protoplasts by TMV should be of great help in defining the intracellular development of these viruses (36, 37).

ICOSAHEDRAL SINGLE STRANDED RNA VIRUSES

Three general classes of icosahedral ssRNA viruses have been identified: 1. the spherical ssRNA plant viruses, 2. the ssRNA phages, and 3. the animal picornaviruses.

Icosahedral ssRNA Plant Viruses

The plant virus, tomato bushy stunt virus, was the first virus for which 5:3:2 symmetry was firmly established (40), and since that time the structure of this virus class has been extensively studied by X-ray diffraction (12, 41, 42) and image analysis of electron micrographs (43, 44). The icosahedral ssRNA plant viruses can be divided into two general structural classes: 1. those with 180 identical subunits in a $T = 3$ surface lattice, represented by cowpea chlorotic mottle virus (CCMV), cucumber mosaic virus (CMV), turnip yellow mosaic virus (TYMV), and many others, some of which may contain one molecule of a minor protein per virion (48); and 2. those with 60 subunits composed of two structural proteins arranged in interpenetrating $T = 1$ lattices[4] [represented by cowpea mosaic virus (CPMV)] (44). Many plant viruses are multicomponent viruses, in that they consist of two or more

[4] Crick & Watson (45, 46) and Caspar & Klug (47) pointed out that there are a limited number of ways in which identical subunits with specific bonding properties can be arranged on a symmetrical closed surface. Caspar & Klug (47) argued that all spherical viruses have 5:3:2 symmetry, and they presented a general model of the possible arrangements of protein subunits in the shells of icosahedral viruses. According to their notation the triangulation number (T) defines the number and arrangement (i.e. the surface lattice) of the structural units; T is the number of small unit triangles into which each triangular face of the icosahedron has been subdivided. $60T$ is the number of protein subunits used to build the structure.

RNA molecules which are apparently packaged separately, and simultaneous infection by all components is required for successful infection (5). Very little is known about the in vivo assembly of plant viruses, but virus particles of the first type (180 identical subunits) have been successfully reconstituted in vitro, and the conditions for reconstitution and products of the reconstitution have been studied in detail. This work is presented in Bancroft's extensive review (50) and will not be discussed in depth here. Despite the apparent simplicity of the experimental system, extensive data on the ionic and pH conditions required for reconstitution have not yielded the mechanism by which coat protein subunits assemble into a shell. This is partly because efforts have not yet been made to identify intermediates in the assembly process. In general, there does not seem to be a specific sequence of nucleotides required for assembly as with TMV (50); the coat protein will polymerize around any RNA-like polyamion. However, Jonard et al (51) have reported that under their conditions TYMV protein shows a preference for TYMV RNA.

A number of attempts have been made to determine whether protein-protein or protein-nucleic acid interactions dominate the stabilization forces within the virion (49). It would be interesting to know whether these bonding properties of the virus proteins reflect the mode of assembly or the mode of uncoating upon infection of a host cell.

ssRNA Bacteriophages

The ssRNA bacteriophages (R17, fr, f2, MS2, and Qβ) resemble the Class 1 (see above) spherical plant viruses in that they contain a linear RNA (messenger strand) molecule about 3000 bases long, 180 capsid subunits ($T = 3$ surface lattice), and one molecule of a minor protein, called A protein. In this case the minor protein is required for infectivity (52) and is transported into the host cell along with the RNA (53). Hohn & Hohn (52) have reviewed in detail the in vivo assembly and reconstitution of these phages, and have reviewed evidence for a model of assembly in which a few (6–8) molecules of coat protein initially complex with the viral RNA strand (called complex I) and then additional molecules of coat protein are added to form the protein shell around the RNA. This would presumably involve interaction of RNA and already assembled coat protein with each newly added subunit (i.e. co-condensation of nucleic acid and protein). Physical measurements and base sequence analysis have shown that the RNA molecule has a very high degree of secondary structure and exists in solution in a very compact form (52). Thus the RNA may actually not be greatly condensed during encapsidation. Kaerner (54) has shown that A protein must act early in the assembly process, probably before encapsidation of the RNA molecule is completed. The relationship, if any, between the A protein and complex I is unknown. The presence of RNA stimulates the in vitro polymerization of coat protein into shell structures (52, 55), and under certain conditions MS2 and Qβ coat proteins specifically encapsidate only homologous RNA (52, 56), however the level of the determination of this specificity (A protein, complex I, etc.) was not determined. The oligomeric state of the polymerizing unit of coat protein is unknown, although Zelanzo & Haschemeyer (57) have suggested that an 11S aggregate may be able to assemble into protein shells directly.

Complex I has also been implicated in the control of the translation of the viral mRNA molecule (reviewed by Kozack & Nathans 58), in that the 6–8 bound coat protein molecules specifically block translation of the replicase gene. In fact, a segment of RNA protected by those few coat protein molecules has been isolated, sequenced, shown to be the initiation region of the replicase gene, and shown to bind coat protein quite specifically (59). This coupling between replication (or more specifically replicase synthesis) and the presence of coat protein may insure the synthesis of viral RNA strands rather than more replication complexes when there is sufficient coat protein present to begin encapsidation. The role of complex I in assembly is not yet clear. Chroboczek et al (60) found that formaldehyde-treated RNA cannot form complex I but can form virus-like structures with coat protein in vitro, suggesting that complex I may not be an obligate intermediate in in vitro assembly. However, infectivity could not be measured and thus the integrity of their virus-like structures is unknown.

Picornavirus Morphogenesis

Picornaviruses [from "pico" (small) plus "RNA virus"] comprise the group of small (260–300 nm) icosahedral animal viruses containing ssRNA genomes. Four sub-groups are distinguished: 1. the human enteroviruses, including polio-, coxsackie, and ECHO viruses, which were isolated from the alimentary tract in the search for poliovirus; 2. the cardioviruses of rodents, including encephalomyocarditis, mouse Elberfeld, and Mengo viruses; 3. the rhinoviruses, associated with human respiratory infections; and 4. foot-and-mouth disease virus, a highly contagious virus of cloven hoofed animals. Despite differences in host range and serology, these viruses are structurally very closely related with essentially identical protein compositions. We will first discuss work on the analysis of precursors from infected cells and the assembly of picornaviruses, and then cover work on the breakdown products of the mature virions and the light it sheds on the assembly process. Further details are given in the valuable reviews of Rueckert (61), Phillips (62), and Baltimore (63). The proteolytic processing reactions are described in detail by Hershko & Fry (4b) in this volume.

Formation of viral proteins by the cleavage of larger precursor chains was discovered in the poliovirus system by Jacobson & Baltimore (64) and Summers & Maizel (65). The entire genome is translated as a single giant polypeptide which is cleaved, prior to the termination of its synthesis, into three smaller proteins. The only other virus group in which the entire genome is known to be translated in one piece is the togavirus, described below. One of the initial cleavage products is the precursor for all four virion proteins (65–67, 74), which are present in the virion in equal numbers (68). This virion precursor, the 125,000 mol wt NCVP1 (noncapsid viral protein), is derived from the 5' end of the genome (69, 70) and is completed prior to being further cleaved (66, 71), suggesting that it may fold before it is further cleaved. The secondary cleavage products of NCVP1, VP0 (40,000 mol wt), VP1 (35,000 mol wt), and VP3 (23,000 mol wt) (71, 74), are found associated with each other within infected cells as 5S and 14S aggregates and empty capsids (72). The 5S complex is a precursor to the 14S structure (73), which in turn is a precursor to virions and empty capsids (72, 75). VP0 is not found in the mature virion, but is the precursor polypeptide to the virion proteins VP2 and VP4 (74). Essentially identical results have been obtained with rhino- and cardioviruses (68, 76–78). Jacobson & Baltimore (74) found that 73S empty capsids containing VP0, VP1, and VP3 accumulated in cells incubated in the presence of guanidine, and that upon removal of the guanidine, radioactive label in empty capsids decreased and appeared in mature virions. They suggested that 73S empty capsids were precursors to virions, and that cleavage of VP0 to form VP2 and VP4 might be coupled with the encapsidation of the RNA by the empty capsid.

The assembly of partially purified 14S precursors into empty capsids proceeds in vitro in extracts of infected cells (72, 75). In the absence of added extract there is some formation of empty capsids, but the reaction is highly concentration dependent (75). Incubation with a rough membrane fraction of infected cells abolishes the concentration dependence of the in vitro reaction (79). Perlin & Phillips (80) have recently shown that the assembly activity of the membranes can be dissected into two components: one is labile and appears to convert inactive 14S complexes into active ones, and the other component is not used up in the reaction. They suggested that this represents a poliovirus specified or induced membrane protein which organizes or concentrates the 14S particles for their assembly into capsids.

Although empty capsids are generally associated with poliovirus infection, they are not found in cells infected with Mengo or encephalomyocarditis virus infection (82, 83). Furthermore, Ghendon et al (73) have found that poliovirus-infected MiO cells (from rhesus monkey tonsils) do not accumulate empty capsids in the presence of guanidine, instead they accumulate 14S particles. Upon removal of the guanidine, the accumulated 14S material chases into virions without the appearance of 73S empty capsids. Interestingly, if the 14S particles are incubated in vitro in the Phillips system, empty capsids do form. These experiments suggest that in these cells empty capsids may not be intermediates in virus assembly. The reconstitution of infectious poliovirus in low yield by dilution of urea denatured virions through phosphate buffer has been reported (81).

Fernandez-Tomas & Baltimore (84) have described a new particle, the 125S provirion, which appears to be the immediate precursor of mature virus. This structure contains RNA in an RNase-resistant form, but in contrast to virions, is sensitive to EDTA and to low concentrations of SDS. These provirions contain uncleaved VP0, making it unlikely that association of the capsid protein with RNA depends on cleavage, as has previously been suggested. The cleavage of VP0 to VP2 .and VP4 probably stabilizes the particle and generates the active sites for cell attachment and infection. Neutralizing antibody against poliovirus is directed against VP4 (85), and particles that have transiently interacted with host cells and desorbed, lose their VP4 and are not infectious (86, 87). Empty capsids containing the precursor VP0 do not adsorb to cells and do not have the VP4 antigen (86, 87). Breindl & Koch (88) have shown that incubation of cells with VP4 (but not with empty capsids) sensitizes them to infection by naked poliovirus RNA. They suggest that VP4 might bring the RNA to its transcription or replication site in the cell.

The relationship of the various protein components to the final structure of picornaviruses has been greatly clarified by the work of Rueckert and colleagues. In a careful study, Dunker & Rueckert (89) showed that mouse Elberfeld virus dissociated into 14S units in the presence of 0.1 M NaCl. Analysis by sedimentation equilibrium showed that the structures had a molecular weight of about 475,000. The 14S structures had equimolar amounts of VP1, VP2, and VP3 but lacked VP4, which was presumably lost during the decomposition of the virus. (We use the poliovirus nomenclature for ease of comparison.) Treatment of the 14S material with urea led to the formation of 5S subunits, although these adsorbed to glass and plastic and were easily lost. These structures were apparently homogeneous and had a molecular weight of 86,000. They contained the same proteins as the 14S complex, also in equimolar amounts. Dunker & Rueckert (89) proposed that the 5S structure was a protomer containing one VP1, VP2, and VP3 chain, and that five of these aggregated to form a 14S pentamer; each 14S pentamer would then represent a vertex of the mature virion. In their model the VP4 chains hold the pentamers together. The virus also contains an average of two uncleaved copies of VP0 (68), suggesting that the pentamers containing the uncleaved VP0 may play some special role in the particle. The protein composition of the Mengo virion is very similar to that of mouse Elberfeld virus, but in addition to 1–2 copies of uncleaved VP0 it also contains 1–2 copies of the uncleaved precursor to VP1 and VP3 (77). The decomposition of Mengo virus proceeds just like mouse Elberfeld virus (82), and electron microscopy of the 14S structure reveals an ellipsoidal disc, 17 × 14 nm and about 7 nm thick. Twelve of these would just cover the viral surface.

Although their protein compositions are very similar to the other picornaviruses, the decomposition of foot-and-mouth disease virus (90) and coxsackie virus (91) requires stronger disruption procedures and breakdown into structures somewhat different from those viruses mentioned above. In the case of coxsackie virus, the bonds broken in the disassembly reaction are probably different from those formed during the assembly of the protein shell.

A number of temperature-sensitive (*ts*) mutants of poliovirus have been isolated and their defective phenotypes characterized. Three classes can be distinguished: 1. RNA and antigens synthesized, 2. antigens made but no RNA, and 3. neither RNA nor antigens made (92, 93). More detailed analysis revealed that a Class 1 mutant was blocked in the conversion of a 10S (14S?) complex to virions, while Classes 2 and 3 accumulated 5S material, which presumably represents the VP0, VP1, VP3 complex. It is not clear how defects in capsid assembly are coupled with defects in RNA synthesis, but it is not surprising since it is the messenger RNA for the capsid proteins that is being encapsidated. Cooper et al (94) have proposed a model in which an assembly intermediate regulates the translation or replication of the genome.

Abortive high temperature infections with two common strains of poliovirus also appear to be due to blocks in the early stages of the capsid protein assembly. That is, these strains appear to carry temperature-sensitive mutations which affect capsid assembly. Cells infected with the Sabin type 1 poliovirus strain fail to form virions, 73S, or 14S particles, even though protein synthesis is normal, and VP0, VP1, and VP3 are formed. Upon shift to permissive temperature the accumulated protein matures into virions, even in the presence of cyclohexamide (95), suggesting that the

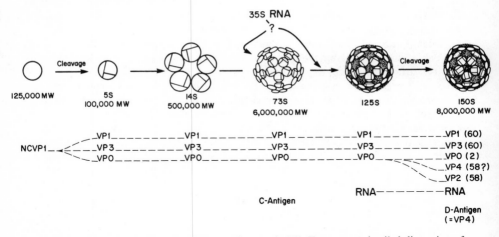

Figure 2   Pathway of picornavirus morphogenesis (62). For a more detailed discussion of how the subunits may be arranged, see Rueckert's review (61).

formation of 14S material from 5S protomers is blocked. Cells infected with the attenuated Lsc1 strain accumulate uncleaved precursors at high temperature (96). These precursors, but not host proteins, are degraded at high temperature, suggesting an altered conformation of the substrate or an altered protease if the protease is virus specified (96).

The pathway of picornavirus morphogenesis is summarized in Figure 2. NCVP1 folds and is cleaved by a yet unidentified activity into three polypeptide chains that remain tightly associated as a 5S structure. Five of these combine into a soluble 14S pentamer whose assembly into capsids is activated and catalyzed by an infected cell membrane fraction. This involves the same protein structure sequentially forming two different sets of bonds. The 14S aggregate either polymerizes together with the RNA or forms an empty shell which then interacts to form a provirion. Only after this interaction is VP0 cleaved to VP2 and VP4, generating the mature infectious virus. The nature and site of the encapsidation are obscure, although Caliguiri & Compans (97) suggest that encapsidation occurs in association with the replicating RNA complex, indicating that newly synthesized strands are encapsidated.

COMPLEX DOUBLE STRANDED DNA BACTERIOPHAGES

The assembly of the complex dsDNA bacteriophages has been studied more intensively than for any other group of viruses. Shell formation, DNA encapsidation, particle completion, and the regulation of assembly are particularly clearly delineated because of mutant availability. Recent reviews have discussed the structure and assembly of the *Escherichia coli* phages T4 (2, 2a, 98), λ (99), P2 and P4 (100, 101), and the *Salmonella typhimurium* phage P22 (102). Other phages whose assembly is under study are the *E. coli* phages T7 (103), T3 (104), and T5 (105), and the *Bacillus subtilus* phage ϕ29 (106). The rapid progress in this area of macromolecular assembly research began with the isolation and characterization of conditional lethal mutants of phages T4 and λ by Epstein et al (107) and Campbell (108) and the characterization of structures that accumulate in mutant-infected cells. The *amber* mutant phage has been particularly valuable, in that the protein product of the mutant gene has a smaller than wild-type molecular weight, and thus can be identified by this criterion in dodecylsulfate electrophoresis gels. Of particular importance is that genetic analysis often permits the identification of the functional role of even very minor proteins in the assembly of the virus. It is perhaps unfortunate that proteins, which are so critical in even small numbers, are called "minor."

Bacteriophage heads and tails generally assemble independently and then join to form complete virions. Head and tail assemblies are often very different, so we discuss them separately below.

Bacteriophage Head Assembly

A large number of gene products (10–20 in the phages studied) are required to assemble a phage head, and functions can be ascribed to many of them. The requirement for many proteins reflects a number of factors: 1. the formation of a large protein shell of accurate dimensions seems to require the action of gene products other than only the coat protein, 2. noncapsid proteins may be required to condense the DNA into a very compact structure within the phage head, 3. in general dsDNA phage chromosomes are cut from longer than phage length replicating DNA by capsid-associated proteins during encapsidation, and 4. these phages have an active mechanism for injecting their DNA into the host cell, which must be built into the virion at the time of assembly.

A central question raised by workers in the field has been the mechanism of DNA encapsidation. Three general classes of packaging models have been proposed: 1. the DNA condenses first and the protein capsid forms around it (110), 2. the capsids form first and the DNA is subsequently brought inside the head [the "headful" encapsidation model (111, 112)], and 3. the DNA and protein shell condense simultaneously. The recent identification of empty capsid precursors to phage has brought strong support to model 2. Such empty capsid precursors have been found during infection by T4 (112–114), λ (115, 116), T7 (117, 118), T3 (104), and P22 (102, 119), and their existence has been inferred for phages P2 and P4 (120). Recent experiments have shown the successful packaging of λ (115, 116, 124), T7 (117), and P4 (120) DNA in vitro. This allowed these workers to show directly that complete exogenously added DNA molecules can be packaged by preformed empty protein capsids. Laemmli et al (125) have shown that partially filled heads can encapsidate at least 50% of the phage chromosome in vivo, thus bringing strong support to the second model for in vivo T4 assembly.

A second central question is the actual mechanism of shell formation. How do capsid subunits come together to form accurately dimensioned closed shells? The experiments of Kellenberger and Laemmli have shown how complex this problem is for phage T4. Although we have only recently discovered how complex this process is with phage P22, we will discuss it first, since its assembly is more easily traced than T4 or λ. The assembly of all three phages depends on proteins which are absent from the complete particle. With T4, the proteins are removed by proteolysis; this is reviewed in the article by Hershko & Fry (4b) in this volume.

P22 HEAD ASSEMBLY   P22 is a temperate phage of *S. typhimurium* which is capable of generalized transduction. Its head is isometric, about 600 nm in diameter, and contains a chromosome of 27×10^6 mol wt.

The products of ten phage genes are required for the formation of infectious particles; six of these are found in the mature particle and three are not (119, 122). Figure 3 shows the overall pathway of P22 morphogenesis; as is also true of λ and T4, a prohead empty of DNA is assembled which then encapsidates a headful of DNA and is converted to an infectious phage (119). About 250 molecules of the 42,000 mol wt gene 8 scaffolding protein catalyze the assembly of about 420 55,000 mol wt gene 5 coat protein molecules into a double shell prohead temporarily containing both proteins (126). Upon encapsidation of the DNA, all the scaffolding molecules depart from the prohead without proteolytic cleavage; these then take part in further rounds of prohead assembly (119, 126). In cells

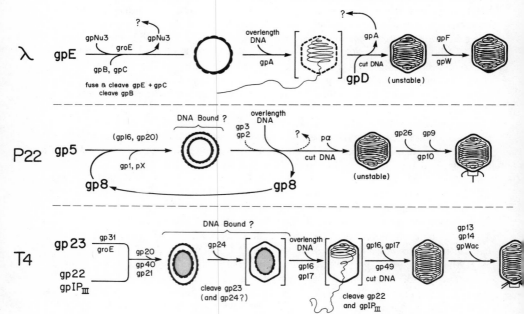

Figure 3   Schematic assembly pathways of λ, P22, and T4 heads as they are currently understood. The pertinent references are given in the text. A "gp" before a gene name (number or letter) refers to the protein product of that gene. Brackets around a structure designate that its existence is inferred, but the structure has not yet been isolated and fully characterized.

infected with *amber* mutants of the scaffolding protein gene, the coat protein accumulates unassembled and as aberrant aggregates (119, 123). In cells infected with *amber* mutants of the coat protein gene, the scaffolding protein accumulates as free subunits (102). Thus, we envision the coat protein subunits forming a complex with scaffolding subunits, and this complex then accurately polymerizing into a prohead shell with the coat protein on the outside and the scaffolding protein on the inside (102, 127).

The encapsidation of DNA by the prohead and the exit of the scaffolding protein require the products of three phage genes (122). One of these, gp1, is incorporated into the prohead prior to packaging, whereas the other two, gp2 and gp3, are not found in association with any particle. We do not know whether all the scaffolding protein exits from the prohead prior to DNA entry or during the course of the process. Either way the process represents a rather extraordinary exchange reaction between a very large DNA molecule and 250 protein molecules. We favor a model in which the coat protein bonding alters, allowing the scaffolding protein molecules to escape between them. As noted above, after exit from the prohead, the scaffolding proteins recycle; they polymerize together with newly synthesized coat protein to form new proheads made from new coat protein and recycled scaffolding protein (126). In a sense, the prohead is a giant enzyme-substrate complex, whose dissociation requires DNA and three other proteins. Since each molecule takes part in multiple rounds of assembly, the scaffolding protein, though a major component of the prohead, is present in only small amounts in the total culture. Furthermore, protein analysis from incorporation of labeled precursors can be very deceiving, since the recycling of gp8 dilutes newly synthesized gp8. Thus either staining or continuous labeling is required to accurately estimate the amount of such a protein in the precursor shell. No protein cleavages have been found to occur during P22 morphogenesis (102, 122).

There are two proteins (gp16 and gp20) incorporated into the prohead during assembly, which are not required for morphogenesis (122). Morphologically normal particles can be assembled in their absence, but these particles are defective in DNA injection. Presumably these proteins have to be placed in the particle at an early stage in order to be in the proper location for their later participation in the penetration of DNA.

After encapsulation the newly filled capsid is unstable and requires the addition of two minor proteins (gp10 and gp26) for conversion to a stable head. In the absence of these gene products the heads lose their DNA both in vivo and in vitro, and convert to empty capsids which can be easily confused with precursor capsids (122, 123). These proteins probably form the head-tail junction. The last step in morphogenesis is the addition of the tailplate proteins to form infectious phage. The head-tail joining reaction was one of the first demonstrations of in vitro phage assembly (128).

P22 replicating DNA does not have the structure of mature phage DNA but instead is found as long molecules thought to be repeated genomes (129, 130). The fast sedimenting DNA molecules accumulate in cells infected with mutants unable to assemble head structures, demonstrating the coupling between head assembly and cutting of mature phage lengths of DNA from the long replicating molecules. This is easily understood in terms of the headful encapsidation model of Streisinger et al (111). Encapsidation has been shown to begin at a specific point on this long DNA and to proceed unidirectionally with sequential encapsidation events, incorporating 2% more than a complete genome at each cutting (131). If all intracellular long replicating DNA molecules have identical ends, then the ends are a likely candidate for the original point at which encapsidation begins. Since proheads and uncut DNA accumulate in the absence of the products of genes 1, 2, and 3, these proteins are likely to be involved in the DNA cutting reaction.

The genetics and protein composition of phage T3 (104) and T7 (103, 118) have also been analyzed in detail. The morphology and protein compositions of T3 and T7 are similar to P22, and the assembly pathways also proceed via prohead shells which encapsulate a chromosome from an overlength DNA molecule (118, 132). Both proheads contain protein species not found in mature phage and not cleaved during assembly, which may be scaffolding proteins for these phages. Since T7 chromosomes have unique terminal sequences (133), the encapsidation process must differ somewhat from P22 and T4 in having sequence specificity, as in phage λ. Kerr & Sadowski (117) have demonstrated the packaging of T7 DNA into proheads in vitro. Serwer (132) has isolated nucleoprotein complexes which appear to represent capsids bound to mature T7 DNA molecules. When isolated from infected cells the capsids are always found bound to one end of the molecule, and may represent the specific binding of prohead and DNA necessary for chromosome encapsidation and cutting.

LAMBDA HEAD ASSEMBLY   The products of ten genes are required to produce complete phage heads, two of which are major coat proteins (gpE and gpD). These are both present on the surface of the head in equimolar amounts, arranged as two superimposed $T = 7$ lattices with one protein (probably gpD) hexamer-pentamer clustered and the other trimer clustered (134, 135). In addition there are several minor proteins (gpF, gpB, modified gpB, and modified gpC) in the phage head, (136), at least one of which (gpF) is located at the head-tail junction (137). The pathway of phage λ head assembly is shown in Figure 3.

Phage λ proheads (called petit λ in the literature) were for many years thought to be an aberrant assembly of the coat protein primarily because it is smaller (about $\frac{3}{4}$ diameter) than the mature phage head. These particles contain only one major protein (gpE) and several less abundant proteins (136, 138) that are required for the proper assembly of the phage head (139). These minor proteins (gpB, gpC, and gpNu3) somehow guide the capsid protein (gpE) in the proper assembly of the capsid protein shell. In their absence the major coat protein polymerizes into a long tubular polyhead, or spiral or other head-related structures. After assembly of a protein shell built of gpE and the minor proteins, additional assembly steps must take place in order for the prohead to become competent to encapsidate a DNA molecule: 1. gpB is cleaved and at least the larger fragment (56,000 mol wt) remains bound to the particle (136, 140); 2. gpNu3 is lost from the particle; and 3. in a complex, not yet fully understood reaction, all the molecules of the gene C protein are covalently bonded to an equal number of the major capsid subunits (gpE), and both the pC and pE portions are cleaved (141). The remaining major capsid subunits are not cleaved. Such a reaction is difficult to recognize by standard dodecylsulfate electrophoretic analysis, because the product of the reaction has been both covalently fused and cleaved; thus the product of the reaction may have either higher or lower molecular weight than the precursors. Whether such protein-joining reactions occur more generally during virus assembly remains to be seen. The functions of these covalent modifications are at present perplexing. It has been suggested that they serve to make assembly irreversible, stabilize structures, insure obligate ordering of certain assembly steps, or change the properties of the head in some manner not possible without covalent changes.

The study of replicating λ DNA has been particularly fruitful because of the existence of a biological infectivity assay for λ DNA which is dependent on the presence of mature molecular ends (142). Mature λ DNA has specific ends with protruding 5' single strands. These two single stranded regions (called cohesive ends) are complementary in sequence. In recent years it has become clear that proper head assembly is necessary for the production of progeny DNA with such cohesive ends (143). More recent studies have shown that λ replicating DNA, like that of P22 and T4, is present as long molecules made up of tandem repeats of the mature DNA sequence, and that these replicating structures accumulate during infection by mutants unable to assemble heads (144, 145). λ Proheads encapsidate and hydrolyze DNA sequentially along these long molecules (146). Both in vivo and in vitro experiments suggest that λ packaging proceeds from left to right along the vegetative genetic map (146–148). In this context it is interesting that the right end of the DNA molecule is specifically susceptible to nuclease in free phage heads but not in complete phage with tails attached (149), suggesting that the right end of the DNA molecule, the last end to be encapsidated, is located close to the head-tail joining site.

There may be two types of hydrolysis events during λ growth. If the end of a long replicating molecule is not the same as a mature DNA molecular end, then the first encapsidation event on each long DNA molecule must create two new ends while subsequent events need create only one. There is clear evidence that the λ packaging machinery is capable of creating two new ends in such a molecule (150). Wang & Brezinski (151) have suggested a model for the recognition of the two potential molecular ends in such a packaging event, which takes into consideration the known base sequence of the DNA sites to be hydrolyzed.

The DNA encapsidation and cutting by a prohead requires functioning gene A and D proteins. Gene A product is thought to be the nuclease that hydrolyzes the long DNA (152), and gene D protein is a major external capsid component. Kaiser and co-workers (116) have presented evidence for a two step model of gpA function in which gpA first aids in the entry of the DNA into the prohead. Then, perhaps with the completion of DNA condensation, the gpD molecules bind to the outside of the head in a manner allowing the gpA to hydrolyze the long DNA molecule, thereby generating the cohesive ends. The protein that hydrolyzes the DNA must have unique properties which allow it to recognize and hydrolyze the DNA only when the DNA is completely encapsidated (i.e. it must recognize a headful of DNA). This "enzyme" [it is not yet known whether this protein(s) acts catalytically] also has sequence specificity since the ends are always exactly the same. Completed phage do not contain gene A protein (153).

Subsequent assembly steps stabilize and prepare the head for tail joining (see below).

T4 HEAD ASSEMBLY   Phage T4, structurally the most elegantly endowed of the phages under study, has an elongated head with a collar and extending "whiskers" attached to one apical vertex (to which the tail attaches). Electron microscopic examination of freeze-etched phage T2 (a very closely related phage) has shown it to have a basic hexamer-pentamer clustered $T = 13$ surface lattice with an extra near-equatorial band of hexamers in the elongated direction (154). This makes assembly more complex since two dimensions of the head must be specified by the proteins involved, rather than one dimension as with the isometric head phage. At least 18 gene products are required for T4 head assembly (156, 201). Laemmli (162) detected 11 protein species in purified heads, with the most abundant being the gene 23 protein. At least four of these are cleavage products of precursor proteins which are generated during assembly. The linear DNA molecule is terminally redundant and circularly permuted (155).

The pathway of T4 head assembly as it is currently understood is shown in Figure 3. One of the earliest steps in the process is the apparent "solubilization" of the major capsid protein by the product of gene 31 (156, 157). Next, the protein shell is assembled from uncleaved major capsid protein (gp23) with the aid of a number of minor components. Showe & Black (161) found that besides the coat protein, the two most abundant prohead structural proteins—gp22 and internal protein III (gpIPIII) —are in the interior of the prohead. Their absence results in the formation of poly-heads of variable diameters, suggesting that they may form an assembly core which aids in the selection of the short diameter of phage heads. On the other hand, gp20 and gp40 may play a role in the selection of the long diameter of the head, because 20^- and 40^- infected cells accumulate polyheads with a properly specified short diameter which contain cores made up of gp22 and gpIPIII (98, 158, 161). Presumably these minor proteins are required to control the proper assembly of the major coat protein; however missense mutations in the major coat protein gene can also cause misshapen phage heads to be formed (159). Thus, the exact shape determination depends on both the major coat protein and the minor components; Kellenberger (160) has reviewed the ability of the coat protein to assemble into structures of different shapes and the control of its polymerization into these structures.

After assembly of the prohead but before DNA encapsidation, the major capsid protein, gp23, is cleaved (162–165). About 10,000 daltons are removed from the N-terminal end of the protein (166). The cleavage does not occur unless gp23 is organized into a proper prohead structure (162). T4 proheads also appear to encapsidate DNA from a membrane-bound "replicating structure" containing over-length DNA (167, 168). Laemmli & co-workers (98, 113) have found that cleavage of the internal components gp22 and gpIPIII parallels DNA encapsidation. The gene 22 protein is cut into small peptides, and 2500 daltons are removed from gpIPIII.

Several small peptides (whose formation is dependent on normal assembly) are present in purified phage heads (169). These are most likely the cleavage products of gp22 (161, 170). The product of gene 24 is also cleaved during assembly, but the assembly stage at which it is cleaved is not yet clear. Cleavage of the T4 proteins has been accomplished in vitro and the phage gene 21 protein is necessary for cleavage to occur (171–174). This will no doubt allow a more detailed characterization of the biochemistry of the cleavages.

The products of genes 49, 16, and 17 are not required for prohead assembly but are required for DNA encapsulation and cutting (112). Frankel et al (175) have shown the gene 49 protein to be a nuclease which hydrolyzes overlength T4 replicating DNA into shorter, but still high molecular weight fragments without the presence of capsids. The known complexity of replicating T4 DNA (176) suggests that gp49 may recognize and hydrolyze at structures other than headfuls, such as branch points, which might impede the encapsidation machinery. Therefore it may not be the normal end generating enzyme. The roles of gp16 and gp17 are not known, but gp17 is likely not in phage heads (177); either of these proteins could be the headful cutting nuclease.

Several other gene products (gp2, gp50, gp64, gp65) are required for the formation of phage heads, but the assembly stage at which they act is not yet known. The products of genes 13 and 14 finish the head after the DNA is packaged and make it competent to join tails (102).

Phage P2 and P4 heads may be assembled in a manner similar to the T4 head since its major protein is also cleaved during assembly. However in this case the sequence of assembly events is not yet clearly defined (121, 178).

CURRENT DNA PACKAGING MODELS   The actual condensing of DNA within the phage head is a complex process requiring the action of several phage-coded proteins as well as the long replicating DNA molecule and the prohead. In each case studied, major changes in the protein components of the capsids accompany DNA encapsidation: 1. Alteration in the major protein composition of the particle: P22, T7, and T3 lose a major internal component; T4 internal proteins are cleaved; and the λ particle adds a major external protein during encapsidation. 2. In each case the prohead expands and becomes more icosahedral at the time of DNA encapsulation [although Dawson et al (179) have recently suggested that the λ head expands at a stage after DNA encapsidation]. Aebi et al (180) have argued from indirect evidence that changes in the conformation of the coat protein accompany gp23 cleavage and/ or DNA encapsidation, while X-ray scattering experiments with phage P22 and P22 proheads have indicated that their surface lattices are very similar even though the shell has expanded (127).

The structure of the extremely compact DNA molecule within phage heads is not clearly understood. Dorman & Maestre (181) have shown that intraphage DNA probably has a structure close to that of C-form DNA. These authors point out that this DNA form has the lowest specific volume of the DNA forms analyzed thus far. Recently, Richards et al (182) have presented electron micrographs of partially disrupted phage heads which suggest that DNA is packed within phage heads in a structure analogous to a ball of yarn. This is consistent with X-ray scattering (183, 184) and birefringence (185, 186) studies of oriented T4 pellets, which indicated a small preferential alignment of DNA with the head-tail axis (which would be true of an elongated yarn ball).

Lerman (187) has shown that various anionic polymers cause DNA to collapse into very compact structures in solution, and Laemmli et al (98) have found that the small negatively charged gene 22 protein fragments purified from phage heads can cause a similar collapse of DNA in solution if they are present in sufficiently high concentration. From this work, Laemmli has proposed a model for DNA condensation within phage heads in which the small peptides created by the proteolytic cleavage of the gene 22 protein cause the DNA to collapse into its compact form within the head. Alternative mechanisms have not yet been ruled out, and in fact may be required to explain condensation of DNA within virions that do not exhibit such extensive proteolytic cleavages during assembly. Serwer (118) has suggested that expansion of the prohead could, if the capsid were impermeable, lower the pressure within the head and thereby literally suck the DNA inside. Equally likely at this point, since an internal protein is removed from T7 and P22 at the time of DNA encapsidation, is a model in which the DNA undergoes a displacement reaction with the internal proteins. Not included in any of these models are the related observations that the T4 internal protein, cleaved gpIPIII, binds tightly to DNA (188, 189) as does the major capsid protein, gpE, of λ (190), and that packaging of T4 DNA appears to require continuous DNA synthesis (191). Thus, details of the encapsidation process remain ill understood, and a general underlying principle has not yet emerged.

HOST INTERACTIONS   Recently a number of workers have isolated bacterial mutants in which T4 and/or λ will not multiply. One group of such host mutants was found in which all known phage functions except head assembly take place normally (summarized in 192 and 193). Given such a restrictive host strain, it is possible to isolate mutants of the phage that overcome the block and thus identify inter-acting viral and host gene products. Because these phage mutations map in genes 31 and 23 of T4 and genes B and E of λ, these proteins probably interact with the host in some undefined manner during morphogenesis. The nature of this host factor and its role in phage assembly are as yet unknown; however, because of its relation-ship with the T4 gene 31 protein, which seems to keep gp23 from accumulating on the inner cell membrane, it has been suggested that the host function may be membrane related (193). Initial experiments measuring electron spin resonance of spin-label probes have shown differences between the inner membrane of wild-type bacteria and at least one such mutant (194). On the other hand, the assembly block by the mutant could be at the level of protein cleavage in λ (193) and T5 (105). Thus, it is unclear whether the primary defect is membrane associated or protease associated.

Chao et al (195) have found that T4, upon infecting a host lacking endonuclease I and polymerase I, produces a large number of short headed particles. Since the bulk of T4 DNA can be packaged into preformed heads (125), it is surprising that a change in DNA metabolism could affect the size of the capsid unless the DNA were also involved in the formation of proheads. Furthermore, small amounts of DNA have been found attached to certain defective structures that have not encapsulated DNA (196). These observations have led Simon (196) to propose a model of T4 prohead assembly in which a small amount of DNA may initially be associated with proheads during assembly, with the bulk of the DNA being pulled into the head at a later stage.

HEAD COMPLETION AND HEAD-TAIL JOINING   After DNA encapsidation, several assembly steps are required that stabilize the heads and make them competent to join with tails. These head completion steps can generally be carried out in vitro (109, 122, 197, 198), but their biochemical characterization has been reported only in studies of phage λ. In this case two proteins, gene W and F_{II} products, prepare the head for tail joining (197). In vitro phenotypic mixing experiments showed that the final protein to act on λ heads, gpF_{II}, most likely forms the site on one vertex of the head with which tails combine (137). The analogous T4 proteins are also found at the head-tail joint (199). This view is supported by the observation that in heads lacking gpF_{II}, the right end of the DNA is probably more nuclease sensitive than DNA within complete heads (200).

How do tails attach to one and only one corner of a structure thought to have $5:3:2$ symmetry? Several hypotheses can explain the observation: 1. the ends of the DNA molecule, which must necessarily disrupt the $5:3:2$ symmetry, select a specific corner for tail union, 2. proheads already have one corner that is unique, perhaps a prohead assembly initiation site, or 3. binding of a head completion protein to any corner causes a conformational change in the capsid which renders the other corners unavailable. Head-tail union has been studied with phages T4 (201) and λ (202). Heads and tails appear to join spontaneously without the aid of other factors. Since the right end of the λ chromosome is thought to be extended into the tail tube in the complete phage (203), head-tail joining must be more complex than a simple binding of tails to heads; the act of tail addition must in some way release the DNA end from its site in the free head. The problem of joining a tail that has sixfold symmetry, such as the T4 tail, with a head corner that has fivefold symmetry has been discussed by Moody (204).

Tail and Tail Fiber Assembly

Although the tail and tail fibers of the large bacterial viruses have no analogs in other viruses, they have proven valuable experimental systems for studying the regulation of protein assembly and the formation of morphologically and functionally complex organelles. Characterization of the structures accumulating in cells infected with *amber* and *ts* mutants blocked in assembly has shown clearly that tail and fiber assemblies are sequential processes; the proteins interact with each other *only* in a particular order. If a protein is missing due to mutation, the precursor accumulates and the subsequent proteins remain functional and unaggregated. Addition of the missing gene product to an extract of mutant-infected cells often results in formation of viable phage. Such an in vitro complementation assay,

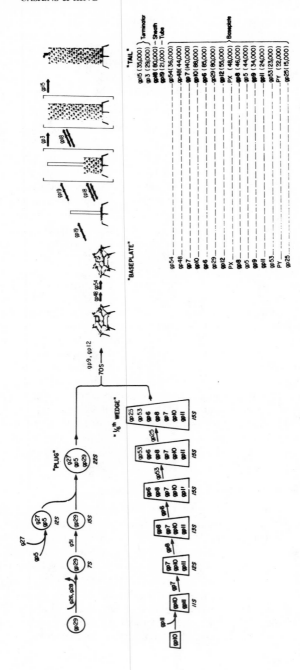

Figure 4 Pathway of T4 tail morphogenesis, from Kikuchi & King (212) and King & Mykolajewycz (210). The gene products interact only in the order shown. When a step is blocked by mutation, the subsequent proteins remain soluble and unassembled. Baseplate assembly represents the aggregation of six 15S complexes around the central plug.

developed by Edgar & Wood (201) and Edgar & Lielausis (109), makes it possible to assay for the assembly activity of the gene products involved in morphogenesis and to determine the sequence of their interaction.

T4 TAIL ASSEMBLY   The T4 tail is a metastable organelle; during infection the hexagonal baseplate rearranges into a hexagram, releasing the tip of the tail tube and triggering the contraction of the sheath which forces the tail tube through the cell envelope where the chromosome is released into the host cell (205, 206).

Assembly of this syringe-like organelle requires the product of at least 21 phage genes (107, 207). The protein products of 17 of these genes have been identified in SDS gels and are found in the complete tail (209, 210). No proteolytic cleavage of any of these tail proteins has been detected. Most of the proteins are needed for baseplate assembly; the central core and surrounding sheath are each formed from a single species of subunit, and the terminal connector is formed of two protein species. The baseplate also contains the enzyme dihydrofolate reductase and a folic acid conjugate (211), which may function in triggering injection.

Tail formation (Figure 4) first involves assembly of the hexagonal baseplate, then polymerization of tube and sheath subunits on the baseplate, followed by termination of the completed tube and sheath (207, 208). Kikuchi & King (212) have shown that the baseplate forms from the interactions of the products of two subassembly pathways. In one subassembly pathway the five major baseplate structural proteins assemble sequentially into a 15S complex that represents morphologically a $\frac{1}{6}$th wedge of the baseplate. This complex is soluble, as are all its precursors. Addition of another protein, the gene 53 product, converts the 15S complex to a form that polymerizes into a 70S hexagon.

The second baseplate subassembly pathway involves a set of minor and catalytic proteins which interact to form a 22S complex, representing the central plug of the baseplate. Six 15S wedges assemble around this 22S plug to form a functional 70S baseplate. In the absence of the plug, the 15S complexes assemble with low efficiency into aberrant hexagonal structures without a central plug. The proteins of the plug are probably involved in triggering baseplate expansion and DNA injection (206, 213, 214) and may include the dihydrofolate reductase (211). A number of gene products appear to act catalytically in plug assembly, and Kozloff & Lute (215) have implicated one of them, the gene 28 product, in the synthesis of the folate.

After the assembly of 70S structures, four more proteins interact with the baseplate: gp9 adds to form the site for tail fiber attachment, and gp12 forms the short fibers that interact with the bacteria (216). These are present in only 70S hexagonal structures. Two other proteins, gp48 and gp54, add sequentially to the baseplate, converting it to a substrate for tail tube polymerization (208, 212).

Polymerization of the tail tube subunits on the baseplate proceeds essentially irreversibly until 24 annuli of six subunits each (217) have formed and then stops (208). The gene 3 product adds to the tip of the tube. Polymerization of the sheath subunits begins only after tube polymerization has started, and then proceeds until the sheath subunits have polymerized to the end of the tube. Sheath subunit polymerization is reversible; stabilization of the sheath requires the action of gp15, which forms a stable bond between the end of the tube and the end of the sheath, at the same time creating the site for head attachment (207, 208, 210).

In the absence of the complete baseplate the tube subunits remain soluble and have no tendency to aggregate. The assembly of the sheath subunits is somewhat less tightly regulated; if the normal substrate is missing, they will eventually polymerize intercellularly into aberrant tubular structures, polysheaths.

In summary the T4 tail proteins appear to be synthesized in a state in which they do not spontaneously assemble; rather they are activated during the assembly process itself by incorporation into a substrate complex. Thus, the regulation of the entire pathway takes place at the level of the assembly process by limiting reactive sites to growing structures.

A distinct mystery in tail assembly is the mechanism that determines tail length so precisely. This cannot be a vernier mechanism between tube and sheath subunits, since in the absence of the sheath protein the tubes are still the precise length (208). No mutants are known that produce either shorter or longer tails. Two models have been proposed to account for the length determination: 1. the length determiner model, which proposes that an extended chain or chains determine the length of tube polymerization (208), and 2. the induced strain model of Kellenberger, which proposes that small perturbations are induced in gp19 during polymerization, eventually leading to the accumulation of sufficient distortion to terminate polymerization (218). Tail tube subunits purified from degraded phage assemble in vitro into tubes in the absence of baseplates, but these are of varying lengths (219). These subunits may differ in conformation and properties from precursor subunits.

LAMBDA TAIL ASSEMBLY   The λ tail is morphologically simpler than the T4 tail, with a single fiber—the gene J protein—at the end, a small basal tip, and then a 180 nm stacked disc tube made of the gene V product (134, 140). The morphological simplicity is probably only superficial, since the products of 11 genes are required for its formation. Katsura & Kühl (220, 221) have studied the structures accumulating in λ mutant-infected cells by methods similar to those used with T4, and they derived a pathway for most of the gene products during assembly. Initiation of the tail begins with the 15S fiber, the product of gene J. Three gene products act on the 15S fiber to convert it to a more advanced 15S complex, which in turn is converted by the action of two other gene products into a 25S complex. This 25S complex probably represents morphologically the fiber and basal part. The V gene product, the major tail protein, then polymerizes on this substrate structure to form a proper length tail. In the absence of a complete 25S structure, gpV does not polymerize into tail. One of the λ tail proteins, gpH, which is probably a component of the 25S structure, is proteolytically cleaved during assembly (153).

The normal polymerization of gpV depends not only on the 25S basal tip, but also on the presence of gpU, a major protein of the infected cell but only a very minor protein of the tail. In the absence of gpU, the tail proteins polymerize into very long polytails. Addition of gpU to polytails results in their attachment to heads to form particles. Thus U product is necessary for the termination of the tail. However U product must also regulate the polymerization process, since in the presence of gpU but the absence of the basal part, tail protein does not form extralong tubes, while it does in the absence of both. Tail subunits purified from degraded phage also polymerize into long tubes (222). Thus, gpU acts both to regulate assembly and to determine tail length. We suggest that these observations can be explained by a length determiner model, in which polymerization acts via interaction of a U-V complex with the unidentified length determiner, and gpU is displaced by this interaction, except at the very end where gpU remains bound, preventing further gpV polymerization and forming the site for head attachment.

OTHER PHAGE   The tail of T5 is morphologically similar to that of λ and a high molecular weight minor protein is cleaved during its formation. As noted in the section on host interactions, this cleavage is blocked in certain bacterial hosts that do not support phage growth (105).

Assembly of the contractile tail of P2 probably proceeds similarly to that of T4, in that a basal structure must be formed prior to the polymerization of the tube and sheath proteins (223).

T4 TAIL FIBERS   The long slender tail fibers of T4 continue to reveal surprising new aspects of molecular assembly processes. The products of six phage genes are required for fiber assembly; five of these (genes 34–38) map contiguously on the T4 chromosome and the other (gene 57) is separated. In addition the products of two other phage genes (63 and wac-whisker antigen control) are required for attachment of the completed fiber to the particle. Fiber assembly has been recently

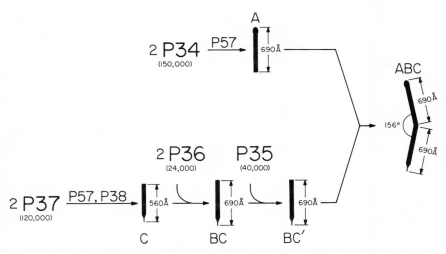

Figure 5   Pathway of T4 tail fiber assembly, from Bishop, Conley & Wood (225), Ward & Dickson (226), and King & Laemmli (227).

reviewed (192, 225) so we briefly summarize only the general features. The complete fiber is composed of two different half fibers, named for their antigenic properties, A and BC (224). The A half fiber (which will be joined to the particle) is a dimer of the 150,000 mol wt gene 34 product (226, 228, 229). The BC half fiber, which has the site for attachment to the host receptor, is more complex (224, 230). Its formation involves the polymerization of two copies of the 120,000 mol wt gene 37 product into a half fiber, the C antigen. Addition of two copies of the gene 36 product to the elbow end lengthens the fiber and results in the appearance of antigen B. Finally, the gene 35 protein also adds at the elbow end, activating this BC′ half fiber for attachment to the A fiber. The two halves then spontaneously assemble into the characteristically kinked 1400 Å complete fiber. All the intermediate structures have been purified and further characterized by Wood and co-workers (192, 226, 228) and Tsugida and co-workers (229, 230).

Fiber assembly requires the protein products of three genes, 38, 57, and 63, which are not incorporated into the completed structure. The gene 38 and 57 proteins are required at the earliest stages of fiber assembly (226, 227). In the absence of gp57, both large fiber proteins are synthesized, but they do not display the antigens found in wild-type fibers, nor are they morphologically recognizable as fibers. Rather, the polypeptides are found with the pellet of cell debris, perhaps precipitated or bound to cell membrane. The protein forming the short fibers of the baseplate (216), gp12, also requires the function of gene 57 product for its assembly into a functional precursor (227). Thus this protein is required for the proper assembly of three fibrous structural proteins of the phage. Assembly of gp37 into a half fiber requires gp38 in addition to gp57. In the absence of gp38, the gp37 polypeptide also behaves as described above. It is not clear whether these nonstructural proteins act to solubilize the fiber proteins prior to their assembly, to catalyze their efficient dimerization, or perhaps to perform some subtle chemical modification.

The tail fiber, once formed, does not spontaneously join to the baseplate. The attachment process requires the product of gene 63, which catalyzes the attachment of the completed fiber to the particle (192). The mechanism of catalysis is unknown; Wood (192) has suggested that the enzyme may help provide a hydrophobic environment for the formation of the flexible but stable (though not covalent) fiber-particle bond. Complete fibers do not join efficiently to free baseplates or tails even in the presence of gp63. This has now been shown to be due to the participation of the whisker proteins in fiber attachment, 400 Å fibers extending from the neck of the phage (165, 461). These whisker proteins may serve to first adsorb the fibers to the particle, or they may serve as an assembly jig to align the fibers and bring the tips of the A fibers into proximity with the apices of the baseplate (462).

The differences between the host range specificity of T2 and T4 reside in the gene 37 and 38 proteins. In a series of ingenious experiments, Beckendorf, Kim & Lielausis (459, 460) have taken advantage of this observation to determine the orientation of the 37 protein in the fiber. Their experiments showed that the 37 protein is linearly arranged in the half fiber, with the N terminus near the elbow and the C terminus at the tip that interacts with the cell. The gene 38 product interacts with the C terminus of gp37. From the dimensions of the half fiber and the molecular weight of the protein it was possible to determine the rise per residue, 0.5 Å, which does not correspond with any known fibrous protein packing. Thus, it is quite possible that the packing of the fiber proteins represents a new fibrous protein structure.

ADENOVIRUSES

Adenoviruses are large (80 nm) dsDNA containing icosahedral viruses, which cause respiratory disease in their mammalian and avian hosts. Adenovirions display particularly distinctive morphology; the morphological units of the icosahedron faces (hexons) are readily visible in negatively stained preparations, and the pentons, morphological units at the vertices, have 10–20 nm fibers extending from them. The 252 total morphological subunits are arranged on a $T = 25$ lattice (47). The virus particles (type 2 has been most extensively studied) contain at least 10–12 protein species (231, 232). Hexon and penton units, which are antigenically distinct, have been purified and their proteins characterized in considerable detail (reviewed by Philipson & Petterson 233).

The chromosome, 12–14% of the particle by weight, is about 35,000 base pairs long and has terminally redundant sequences in which the two redundant sequences have opposite polarity (234, 235). The chromosome can, under certain conditions, be released from the virion as a circular structure with the ends apparently held together by protein (236). Virions also contain enough spermine and spermidine to neutralize 10% of the negative charge of the DNA (237). The infection process is initiated when the virion penetrates the plasma membrane and enters the cytoplasm where the capsid shell is removed, leaving viral core-type structures. These cores then move to the nucleus, where they are further uncoated and DNA replication and virus assembly take place (234).

The capsid proteins and perhaps several minor nonstructural proteins are synthesized in infected cells at late times after infection (> 10 hr postinfection) (238, 239). As with the large phages, replication of the viral DNA is required for synthesis of these late proteins (234). Proteins made in the cytoplasm subsequently move rapidly to the nucleus where virions appear (246). The polypeptide subunits of the hexon and the subunits of the penton appear to be assembled into complete hexons and pentons by the time they are inside the nucleus (240). The assembly of these units also occurs in vitro with proteins synthesized in vitro (241). It is not clear why these do not assemble further before they reach the nucleus.

The more recent work of Sundquist et al (242) and Ishibashi & Maizel (231) has shown that the capsid proteins of type 2 virus assemble into a DNA-free empty shell before the DNA is encapsulated, reminiscent of the dsDNA phages. (Although these particles are isolated as empty structures, it is possible that they contain some DNA in the cell.) The form of the progeny DNA, which is to be encapsidated, is unknown; replicating adenovirus DNA has large single stranded regions and contains no strands longer than viral length (reviewed by Tooze 234). Ishibashi & Maizel (231) also presented evidence for two classes of DNA-containing virions (called young and aged) and showed that young virions are precursors to aged

582   CASJENS & KING

virions. The polypeptide compositions of these particles have been characterized and are summarized in Figure 6. Several proteins of empty capsids and young virions appear to be cleaved in the conversion of young to aged virions. The cleavage that forms protein VII of the virion core has been shown directly by typtic fingerprinting of the precursor and product molecules (243). Similar cleavages may occur with adenovirus types 3, 5, 12, and 16 (242, 244–246). In conjunction with DNA entry and protein cleavage, several protein components are lost, and proteins IVa and V are added during the conversion from empty capsids to aged virions. The fiber polypeptide (IV) of type 2 virions is a glycoprotein containing glucosamine (247), and in some strains the fiber and core polypeptides are phosphorylated (although no quantitation of the extent of phosphorylation was reported) (244, 248).

Arginine deprivation inhibits the production of virus but does not inhibit as greatly the synthesis of the capsid proteins. Recently Rouse & Schlesinger (249) have found that capsid proteins made during arginine deprivation are used only inefficiently in the production of virions upon restoration of arginine. Winters & Russell (250), however, found that in vitro some proteins made under conditions of arginine starvation could be converted to DNase-resistant particles with the density

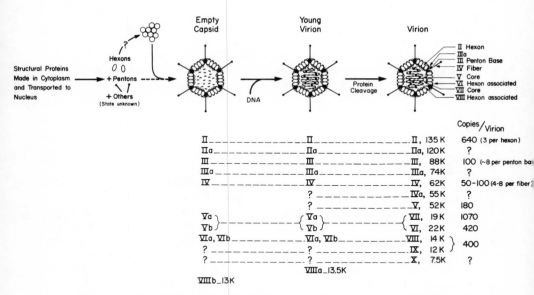

Figure 6 Adenovirus assembly. This diagram summarizes the pathway of adenovirus assembly as delineated by Sundquist et al (242) and Ishibashi & Maizel (231). Beneath each structure are listed the polypeptides present in that structure, and after the polypeptides of the virus are given the molecular weight of each protein and the number of molecules of that protein in one virion [empty capsids may have several additional minor protein species not listed (231)].

of virions when incubated with an extract of productively infected cells. In addition, Ishibashi & Maizel (231) have reported that the protein cleavages associated with the young → aged virion conversion also occur in crude extracts.

Insight into the pathway of shell formation has come from some fascinating experiments by Pereira & Wrigley (251) on the reconstitution of adenovirus. Incubation with trypsin, or sodium deoxycholate and heat treatment of virions, results in the appearance of groups of nine hexons, arranged as three hexon trimers (see Figure 6). Electron microscopic analysis of these structures shows that they have threefold symmetry (252). Pereira & Wrigley (251) have shown that such groups of nine hexons can reassemble in vitro into shell structures in which each of the 20 sides of the icosahedron is apparently made of one group of nine hexons. The structures thus built contain 180 hexons; these lack the 12 pentons and 60 peripentonal hexons (those hexons adjacent to the pentons). They have also proposed an intriguing model for the detailed structure of the hexon nonomer which explains the bonding properties between hexons within and between groups of nine. Although it is not known whether adenovirus shells are assembled via the hexon-nonomer intermediates in vivo, Sundquist et al (242) have seen shell structures lacking pentons (and perhaps the peripentonal hexons) in lysates of types 2 and 3 infected cells.

More detailed knowledge of the assembly pathway of adenovirus may be obtained from analysis of the *ts* mutants of type 5 virus isolated by Williams et al (253) and Ensinger & Ginsberg (254). Preliminary experiments by Russell and co-workers (255, 256) have shown that in a number of these mutants, various structural antigens are not produced normally at nonpermissive temperatures although DNA is replicated.

PAPOVAVIRUS ASSEMBLY

The papovavirus group, which includes polyoma, SV40, and many wart viruses, is tumorigenic and produces latent and chronic infections in mammalian hosts. The virions contain a double stranded, covalently closed, circular genome 6,000 to 10,000 base pairs long. They apparently penetrate the cells in pinocytotic vesicles, from which they make their way to the nucleus where they are uncoated (234). Both replication of the DNA and morphogenesis of the virions take place in the cell nucleus. Although the viral structural proteins are presumably made in the cytoplasm, where the viral mRNA is found bound to polysomes, attempts to demonstrate their presence in the cytoplasm have not been successful; this implies that the virion proteins must be very rapidly transported to the nucleus after their synthesis.

The surface topology of all papovaviruses examined has been found to be a $T = 7$ icosahedral surface lattice—there are 420 molecules of the major coat protein per virion (257). These virions are about 12% DNA, suggesting that the DNA may be considerably less compact than in phage heads. The papovaviruses studied all have one major coat protein of molecular weight 43,000–48,000 and also contain 4–6 minor proteins of lower molecular weight (257, 258). Several of the smallest of these minor proteins appear to be host histones packaged in the virion (259,

260, 261). Tan & Sokol (262) showed that during growth, radioactive phosphorus is incorporated into all SV40 capsid proteins, but no quantitation of the phosphorylation was reported.

Several of the capsid proteins and minor nonstructural proteins of SV40 and polyoma viruses have been detected in infected cells by gel electrophoretic analysis (234, 263). Synthesis of capsid proteins (but not the histones), as in the larger phages, is dependent on viral DNA replication (234, 263). Ozer (264, 265) has found that the major capsid protein of SV40 is present in infected cells as virions, empty capsids, and a small amount of 8S structures. These naturally occurring empty capsids were shown to be precursors to infectious virus particles. This is similar to the phage and adenovirus systems discussed above, although without in vitro assembly studies it is very difficult to prove that such structures are not actually derived from full but unstable capsids within the cell. The encapsidation of DNA into a preformed shell may be analogous to phage encapsulation, but is probably simpler in that encapsidation does not require the cutting of DNA to mature length. The empty capsids do not contain normal amounts of the cellular histones present in virions (264, 266), and under certain conditions disrupted virions release their DNA in association with the histones (267, 268). Newly replicated DNA can also be recognized in extracts of infected cells, and the existing evidence supports a mode of replication in which individual circles are duplicated without the appearance of any structures with strands longer than viral length (234). The newly replicated DNA is found in association with the same histones present in the virion (266).

These results suggest a model for the assembly of papovaviruses in which an empty shell encapsidates a DNA molecule with the virion histones already bound to it.

Several papovaviruses also produce aberrant elongated empty capsids. Kiselev & Klug (269) have analyzed the surface topology of these particles. The existence of such particles may imply that, as with the phage discussed above, the minor virion proteins may be involved in aiding the correct assembly of the coat protein.

Attempts at reconstituting polyoma from particles disrupted into 7–10S material have led to the formation of DNA-free protein shells which contain all the low molecular weight nonhistone virion proteins in addition to the coat protein (267, 270). The particles formed in vitro also contain one of the virion histones, suggesting that it may be bound to the particle by protein-protein interactions as well as bound to the DNA.

As with other animal viruses that replicate in the nucleus, arginine blocks virus assembly. Tan & Sokol (271) have found that during arginine deprivation all the viral proteins are made and transported to the nucleus normally, but are not assembled. Restoration of arginine led to the assembly of the previously made proteins into infectious virions and empty capsids. A number of mutants, including conditional lethal mutants, of SV40 and polyoma have been isolated that seem to affect late stages of virus development, and they should be useful in further analysis of virus morphogenesis (272–274).

HERPESVIRUS

The herpesviruses are large (140–170 nm) enveloped viruses containing dsDNA. They are responsible for a number of persistent infections in humans, including cold sores, shingles, and infectious mononucleosis, and they have been implicated as the etiologic agents of cervical carcinoma and Burkitt's lymphoma. They are widely distributed among vertebrates (Marek's disease of fowl and equine abortion virus). The virus particle is composed of an outer envelope containing a lipid membrane, within which is a spherical nucleocapsid about 100 nm in diameter with a central dense core of DNA. The structure of the virus has been reviewed by Roizman & Spear (275) and the general biology of the virus has recently been reviewed (234). The morphology of the capsid is interpreted as representing 162 hollow capsomers organized into a shell with 5:3:2 symmetry (276). Surrounding the nucleocapsid is a glycoprotein-lipid envelope, whose thickness (about 30 nm) indicates either the presence of matrix material between nucleocapsid and envelope, or an unusual structure of the membrane envelope. The envelope is derived from the nuclear membrane.

The most recently determined molecular weight of the herpesvirus genome is $82 \pm 5 \times 10^6$, 75% the size of the T4 genome (277). The chromosome has been reported to have terminal redundancy (278). The virus contains sufficient polyamines to neutralize 50% of the DNA phosphates, with spermine localized in the nucleocapsid and spermidine in the outer envelope (279). The outer envelope also contains protein kinase (280).

After uptake of herpesvirus the outer envelope is removed in the cytoplasm and the nucleocapsid migrates to the nucleus (281). Messenger RNA synthesis and transport to the cytoplasm may be quite similar to host cell synthesis, involving processing of large transcripts. The viral proteins are synthesized in the cytoplasm and then migrate to the nucleus (282, 283), where the virus DNA is replicated and the viral proteins assemble. Arginine deprivation blocks virus production, though not DNA synthesis, by preventing the migration of the capsid proteins from cytoplasm to nucleus (284, 285).

A number of capsid-related structures can be seen in the nuclei of herpesvirus-infected cells, including 1. empty capsids early in viral development (286), 2. capsids with a lightly staining core, and 3. capsids with a densely staining core which are the mature nuclear particles (286). In a frog herpes-type virus, capsids can be seen with an inner shell (287) rather than an inner lightly staining solid center. Structures resembling free inner shells or cores also occur free in the nucleoplasm and in paracrystalline aggregates. The particles with densely staining cores, and only these particles (288, 289), bud through the nuclear envelope into the perinuclear cisternae (289, 290). This process is associated with thickening of the nuclear membrane (286, 291). The enveloped viruses probably proceed to the cell exterior via cytoplasmic membrane channels formed after infection (289, 292). This may prevent virions from being uncoated by the cytoplasm prior to their exit. At late times after infection, nuclei may lyse, releasing unenveloped particles into the cytoplasm. These particles can apparently be enveloped by budding through cytoplasmic membranes (289, 291).

In the presence of DNA synthesis inhibitors, viral proteins are synthesized and capsids with light cores form, but the dense core particles do not form. The light core particles formed do not pass through the nuclear membrane (288). These experiments suggested that capsids lacking DNA but containing some inner material were precursors to mature nucleocapsids. In a careful study of a frog herpesvirus in which virus growth at low temperatures was extended over a long period, Stackpole (287) showed that double shelled capsids appeared in the nucleus prior to mature nucleocapsids and did not accumulate, suggesting that they were precursors to nucleocapsids. Furthermore, autoradiography experiments showed that the double shelled particles were the first class to be labeled by tritiated thymidine, and subsequently the dense core nucleocapsids were heavily labeled. Early in infection both 100 nm empty capsids and 55 nm core particles appeared, which accumulated in the cells, suggesting that these were aberrant particles formed prior to the full establishment of late protein synthesis. Sydiskis (293), in a study of pseudorabies virus in baby hamster kidney (BHK) cells, isolated five classes of particles in sucrose gradients by fixing in formaldehyde prior to separation. Two classes of particles with variable diameters (45–80 nm) had densities of 1.4 and 1.5 gm/cc, compatible with what one might expect for a nucleic acid-protein complex. In pulse-chase experiments using ^{3}H-thymidine as label, these complexes behave as precursors to mature virions. The label in them was completely released by DNase treatment, in contrast to nucleocapsids which are resistant to DNase.

Herpes infection shuts off host protein synthesis, and a large number of viral proteins are synthesized. The viral proteins are synthesized sequentially in three groups: the first group consists of nonstructural proteins; the second group includes some of the minor structural proteins, perhaps the enzymatic components of the capsid; and the third group contains the major structural proteins (294). Synthesis of the first group is required for synthesis of the second group, and synthesis of the second group for the third group. No proteolytic cleavages of these proteins have been detected, even with protease inhibitors (295). Twenty-four proteins have been detected in mature virus. At least 12 of these are glycoproteins (296) and can be found in membranes purified from infected cells (297). The viral proteins are not glycosylated on polysomes but in the membranes (298). Although viral proteins and host proteins are found together in membrane fragments and host proteins are not displaced from the membrane (299), no host proteins are found in the mature virus (296). Roizman & Heine (300) have written a short review of the modification of the host membranes by herpesvirus.

Purified naturally occurring empty capsids contain four proteins, one of which, p5, is the major capsid protein. Nucleocapsids isolated from disrupted nuclei, and which presumably have not yet budded through the nuclear membrane, contain two additional proteins, p21 and p22a, and nucleocapsids isolated by detergent treatment of mature enveloped nucleocapsid contain three more proteins, although all the glycoproteins are removed (301). Protein 22a, which is found only in association with nucleocapsids that have not been enveloped, has unusual staining properties and also incorporates radioactive phosphate (302). The protein is missing from nucleocapsids derived from particles that have undergone envelopment, but a somewhat smaller protein, p22, appears, which shares the unusual properties of p22a. Gibson & Roizman (302) therefore suggest that p22 may be derived from p22a by cleavage. Alternatively the labeling kinetics of p22a are consistent with it being a recycling protein similar to the scaffolding protein of bacteriophage P22 (102). A schematic depiction of herpesvirus assembly is shown in Figure 7.

Twenty-eight proteins have been described for closely related equine herpesvirus; four are glycoproteins, four lipoproteins, and nine glycolipoproteins, although incorporation of labeled precursors has been the sole criteria for identifying modified proteins (303). Perdue et al (303) reported five major structural proteins, comprising 96% of viral protein in the nucleocapsid, none of which was glyco- or lipoprotein.

A number of *ts* mutants of herpesvirus have been isolated and to date 15 complementation groups have been defined (304, 305). Most groups are defined by a

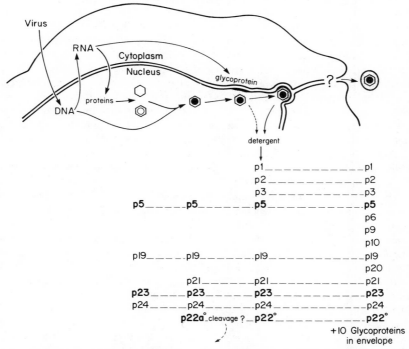

Figure 7   Pathway of herpesvirus morphogenesis. The diagram shows the intracellular states of particles that have been isolated from herpesvirus infected cells, with the polypeptide composition of the particle shown below it. Data are from mainly Roizman and co-workers (296, 301, 302). The figure is modified from Tooze (234). A superscript o represents a phosphoprotein.

single mutant, suggesting that it will be possible to identify a large number of genes by the isolation of more mutants. Some mutants fail to synthesize DNA, but all mutant-infected cells display virion antigens and all accumulate some form of particles. The preliminary data suggested that mutants unable to synthesize viral DNA were also blocked in the envelopment of the nucleocapsid, consistent with the pathway described above (306).

NONBUDDING LIPID-CONTAINING VIRUSES

Four biologically unrelated virus groups have been studied which contain lipid (presumably in bilayer form) but do not appear to bud from cellular membranes. These viruses, which seem to assemble membranes de novo, are PM2 and $\phi6$ (dsDNA and RNA bacteriophages, respectively), the "Iridescent viruses" (icosahedral dsDNA insect viruses), and the poxvirus group of dsDNA animal viruses. The latter two assemble in the cytoplasm of infected cells (307, 308).

The assembly of $\phi6$ (309) has not been studied, and although an Iridescent virus was one of the first viruses shown to be icosahedral (310), it has not yet been studied biochemically beyond the demonstration of lipid in the virions (311, 312). Likewise the assembly of PM2 is not understood. However, its chemical and physical structure has been studied extensively (reviewed by Franklin 313), and low angle X-ray scattering (314) and electron spin resonance studies (315) have proven directly that the lipid in PM2 virions is arranged in a bilayer surrounding the nucleic acid-containing portion of the particle. PM2 is the only lipid containing virus known not to contain any glycoproteins (313). Electron microscopic analysis of infected cells suggests that virions are formed along the inside of the inner membrane (but not by a budding process) and that empty virions may be precursors to complete virions (316). Recent experiments have shown that the virus does alter the phospholipid metabolism of the host upon infection and that two thirds of the viral lipids are made after infection (317).

The poxviruses are the largest and most complex of the viruses. There are a large number of poxviruses that infect mammals, insects, and birds. The smallpox (variola) agents have been of major importance to man. Fowl pox and vaccinia, the only members of the group to be biochemically characterized, contain about 30 different species of polypeptides (318, 319). The dsDNA molecule of fowl poxvirus is about 400,000 base pairs long, the largest of any virus (320). The virus particles are brick shaped with a dumbbell-like DNA containing internal structure or core (321). Electron microscopic studies suggest that a trilaminar membrane (most likely a lipid bilayer) is present in the layer surrounding the core (308). These workers also found that most of the phosphatidylcholine in virus particles is synthesized during infection rather than incorporated from pre-existing cellular phospholipids. The external layer of the virion also contains one or more glycoproteins (322–324).

The morphogenesis of poxviruses has been studied extensively by electron microscopy of infected cells (308, 325, 326). DeHarven & Yohn (326) have defined five types of immature virions in infected cells which are compatible with a morphogenetic sequence as follows: Particles initially appear in thin sections as crescent-shaped objects within specific areas of cytoplasm called "factories" or "viroplasmic matrix"—even at this stage they appear to contain the trilaminar unit membrane. The crescents are then completed into spherical structures, dense material (presumably DNA) appears within the particle and assumes a dumbbell shape, and then the external surface is thickened (especially in the area between the two ends of the dumbbell) to form completed virions. Recently, Sarov & Joklik (327) have begun biochemical characterization of incomplete particles (presumptive assembly intermediates) found in vaccinia virus-infected cells. They have isolated three forms of such incomplete virions, and their polypeptide content was determined. Joklik & Becker (328) found that the replicating DNA sediments more rapidly than mature viral DNA. This replicating DNA has two proteins (presumably viral coded) bound to it (327, 329). The earliest intermediate (IPA) isolated by Sarov & Joklik (327) contains no DNA as isolated and has only two major proteins, one of which, a phosphoprotein, is almost totally lost during the completion of the virion (it is similar in this way to phages P22 and T7 and herpesvirus). Several minor proteins are also present, including the two proteins found in association with the replicating DNA. The significance of these proteins being found in both structures is unknown; there may have been DNA in IPAs in the cell or, since the proteins are known to bind specifically to DNA (329), they could be part of the apparatus that draws the DNA into the particle. The other two incomplete particles (IPB and IPC) isolated by Sarov & Joklik (327) appear to be nearly complete, DNA-containing particles missing only exterior components. They presented no evidence regarding the phospholipid content of any of the incomplete particles.

No conditional lethal mutants have been described for any nonbudding enveloped viruses, but various inhibitors have been used to dissect vaccinia virus development. Of particular interest is the effect of rifampicin, which blocks viral development after the synthesis of all the viral proteins and DNA, but before the assembly of any structures visible in cells by electron microscopy (reviewed by Follet & Pennington 330). Analysis of the proteins made in infected cells in the presence of rifampicin has shown that two virion proteins, 4a and 4b, are missing, suggesting that they are not primary translation products. Pulse-chase labeling experiments have subsequently shown that proteins 4a, 4b, and 10 (a or b) appear during the chase, while two higher molecular weight proteins (105,000 and 74,000) disappear (330). Moss & Rosenblum's (331) tryptic fingerprint analysis establishes that the virion proteins 4a and 4b are derived by proteolytic cleavage from the 105,000 and 74,000 molecular weight proteins, respectively. It is not known whether this cleavage takes place in association with the assembly process, or whether the cleavage is part of some function of these proteins unrelated to assembly. Further study of these viruses should increase our understanding of the mechanism of de novo membrane biosynthesis.

DIPLORNAVIRUS

Diplornaviruses constitute those viruses with segmented dsRNA genomes. Although the best studied members of the group, reovirus and bluetongue virus of sheep, infect vertebrates, they are most frequently associated with insects and insect-borne plant infections. Other well-characterized members of this class are wound tumor virus of plants, maize rough dwarf virus (MRDV), rice dwarf virus, and cytoplasmic poly-hedrosis virus (CPV) of silkworms. The biology of diplornaviruses has recently been thoroughly reviewed by Wood (332), as has the expression of the genome (333). The mature virion has a diameter of about 75 nm and particle weight of about 70×10^6 daltons, of which 15% is RNA.

The reovirion is composed of seven species of proteins and 10 distinct dsRNA segments (334, 335) and appears as a double shelled structure in the electron micro-scope. Virions also contain a large amount of oligo adenylate (about 20% of the total virion nucleotides). Chymotrypsin removes the oligonucleotides and three of the virion proteins, including the most abundant protein, $\sigma 3$, and leaves a stable ribo-nucleoprotein core, about 52 nm in diameter, with a number (probably 12) of projections containing all 10 RNA segments (336, 337). These projections are particularly distinct in cores from MRDV and CPV and appear as hollow cylinders (338, 339). These cores have RNA polymerase activity which complete virions lack, and incubation with sufficient substrate results in the continued initiation and completion of messenger RNA strands from all 10 segments (340). Removal of $\sigma 3$ only is sufficient to unmask the polymerase activity. Bluetongue virus has similar protein composition arrangement, although the outer protein layer shows less distinct surface morphology (341). Recent electron micrographs of reovirus cores in the process of synthesizing RNA show the RNA being extruded from many places on the cores, perhaps coming out through the projections (342). These cores are essentially giant enzyme-template complexes (since the RNA segments and polymerase must be moving cyclically past each other) that are protected by a protein shell. Structural integrity of the core is required for RNA polymerase activity.

Upon infection, reovirus is uncoated in association with lysosomes (343), yielding a subviral particle ($\rho = 1.40$ gm/cc, 57 nm) that is very similar to cores prepared from virions. These parental cores remain intact in the cytoplasm and synthesize messenger RNA. The synthesis of nine proteins has been detected in reovirus-infected cells: the seven virion proteins and two nonstructural proteins (344). Eight of these are primary gene products, but one protein $\mu 2$ is derived from a precursor $\mu 1$, probably by proteolytic cleavage (344). All the capsid proteins except $\mu 2$ have blocked N termini (345).

A number of observations suggest that the proteins can associate into a double shelled capsid independently of the RNA: 1. small amounts of empty capsids are found in reovirus preparations, which look like double shells and have the same protein composition as virions, 2. lysine starvation results in the accumulation of double shelled particles at the density of protein (346), and 3. cells infected with *ts* mutants unable to synthesis dsRNA at high temperature still form double capsids (347).

The association of the virion proteins with the genome is of particular interest, since each of the 10 segments has to be gathered into a single particle. The genetic reassortment studies of Fields & Joklik (348) show that the segments can reassort and behave as if they are segregating independently. Synthesis of dsRNA requires new protein synthesis and is thus separable from single strand synthesizing activity.

The dsRNA appears to be made by synthesizing the complement of the newly made messenger RNA (349) and is never found free in solution. The dsRNA synthesizing activity sediments in a broad peak between 300 and 500S, whereas the product of the reaction sediments in a sharp peak at about 550S. Treatment of the product with chymotrypsin converts it to a denser particle with properties very similar to cores prepared from virions. Apparently this activity represents the completion of complementary strands that were initiated in vivo, and upon completion the dsRNA remains associated with the particle, which is likely to be a precursor to virions (350). Morgan & Zweerink (351) have shown that the proteins of 400S corelike particles found in infected cells are labeled more rapidly than virions, suggesting that they are precursors to virions. The implication is that it is single stranded instead of double stranded molecules that are gathered into virions.

Temperature-sensitive mutants of reovirus have been isolated and so far define seven complementation groups, A–G. At high temperature all those mutants analyzed synthesize ssRNA in the same ratios as wild-type infected cells. All protein species were also synthesized, although a mutant in gene D displayed abnormal cleavage of u1 to u2, suggesting an alteration in the specificity of the cleavage reaction (352). Chymotrypsin cores prepared from mutant particles synthesize ssRNA in vitro, although the tsG cores gave a low yield (353). Mutants in three genes, C, D, and E, fail to synthesize double stranded progeny viral RNA (353, 354).

These results are consistent with a model in which dsRNA synthesis depends on the incorporation of single strands into a core precursor particle. The cores that accumulate in B$^-$ and G$^-$ infected cells presumably contain dsRNA and may represent intermediates in morphogenesis. The empty capsids that accumulate in C$^-$ and D$^-$ infected cells may be either aberrant byproducts of the assembly process or precursor capsids that have not yet encapsidated a chromosome.

All the above findings can be incorporated into the following overall scheme. The parental particle is uncoated and activated to synthesize messenger RNA. These are translated into proteins which sequester 10 single (messenger) strands into a new core. This complex, in association with other proteins, synthesizes the complementary strands and then the outer proteins are added to yield infectious virus. The adenylic acid oligonucleotides found in virions are synthesized in the maturing core, after the synthesis of the complementary strand and probably by the same activity that forms the dsRNA (355).

ICOSAHEDRAL ssDNA VIRUSES

Two classes of icosahedral ssDNA viruses are known; the bacteriophages represented by ϕX174 and S13 and the insect and animal parvoviruses. These are very small viruses (25–30 nm) with genomes 5,000–6,000 bases long.

ϕX174, the best characterized of this group, contains a circular DNA molecule in a capsid containing three major equimolar proteins (60 copies each) and two minor protein species (356–360).

Conditional lethal mutations of ϕX174 and S13 define 9–10 complementation groups (360). Ten or eleven viral-specified proteins have been identified in phage-

Table 1 Phenotypes of φX174 mutants

Gene	D	E	F	G	H	A	B	C	(I)?	(J)?
Product in capsid	no	no	yes	yes	yes	no	no	no	?	yes
Molecular weight	14.5k	17.5k	50k	20k	37k	13k or 67k	25k	?	?	9k
RF synthesis	+	+	+	+	+	−	+	?	?	+
Viral strand accumulation	−	+	−	−	+	−	−	−	?	+
Structures accumulated	12sF+G	Virion	6sG	9sF	Defective virions	12sF+G	9F+6sG	12sF+G	12sF+G	Defective virions
Gene product function	ss synthesis	Lysis	Major capsid	Major capsid spike	Major capsid spike	RF synthesis	Assembly factor	?	?	Capsid

infected cells, and many of these matched with their corresponding genes (although polarity effects confuse some of the interpretations). The products of genes F, G, H, J, and one genetically unidentified protein are found in the capsid. The F gene protein molecules apparently make up the protein shell, whereas the others are in the spikes present at each vertex (356). ϕX174 assembly has not been reviewed previously, so we have tried to synthesize the data available in Table 1. Readers wishing to consult this literature are warned that different research groups have used conflicting genetic nomenclature. Hayashi and co-workers (357, 361) have analyzed the protein structures that accumulate in mutant infected cells. F protein is found as a 9S aggregate, probably a pentamer, and G protein is found as a 6S aggregate, also a pentamer. In the presence of gene B function the two structures aggregate into a 12S complex, presumably a corner of the icosahedral virion. The gene B protein is not found in the mature phage and may function catalytically. Twelve of the 12S corners then assemble into the protein shell of the virion, analogous to the assembly of picornaviruses from 14S precursors. It is not clear at which stage the DNA and the remaining one major and two minor proteins are added. Reconstitution of ϕX174 has been attempted and noninfectious protein shells can be reassembled in vitro, but the reaction has not been further characterized (362).

A particularly interesting feature of ϕX174 morphogenesis is that production of progeny single strands depends upon capsid formation; cells infected with mutants defective in genes F, G, B, C, and D fail to accumulate viral single strands, suggesting that virion assembly or encapsidation is intimately coupled with viral strand synthesis (366).

The parvoviruses contain a linear ssDNA molecule and three structural proteins (reviewed by Tinsley & Longworth 363). The three proteins may have considerable sequence overlap (364). Empty capsids containing all three proteins are found in infected cell extracts (365), but their role in assembly is unknown.

FILAMENTOUS ssDNA BACTERIOPHAGES

Filamentous bacteriophages contain a circular ssDNA genome about 6000 bases in length packaged into a long slender particle 700–900 nm long and 6 nm wide. These phages attach to the tips of bacterial sex pili, and thus are limited to male strains of bacteria harboring a sex factor. Their biology was carefully reviewed by Marvin & Hohn in 1969 (367). Although their genome is similar to the spherical ssDNA phages, they are distinguished by their unusual mode of cell entry and by their extraordinary morphogenesis, which takes place by extrusion of the genome through the cytoplasmic membrane containing precursor capsid subunits (368).

Mature virions are composed of about 2000 molecules of an unusual 5000 mol wt coat protein; the banana-shaped protein has 20 hydrophobic residues in its center and hydrophilic residues at each end and is predominantly alpha helical (369). In addition to the major coat protein, the virions contain 1–4 copies of a 70,000 mol wt protein (370) located at one end of the virion (371, 372). This minor protein is required for attachment of the virus to the sex pilus (371). The genetics of fd has been well characterized by Pratt and co-workers (375) with eight complementation groups defined by *amber* mutations. The major coat protein is the product of gene 8, and the minor particle protein is the product of gene 3 (370). The products of genes 2 and 5 are needed for DNA replication (see below) while the four other gene products, though not incorporated into virions, are required for morphogenesis.

X-ray diffraction studies of these phages by Marvin and co-workers have shown that the coat protein of one set of filamentous phages is organized into a simple helix of pitch 1.5 nm with 4.4 units per turn. The protein subunits are elongated axially and slope radially around the particle, overlapping each other like the scales of a fish (373). The better studied filamentous phage, infecting *E. coli* (fd, f1, M13), represents a slight variant of this basic design, in which the helix is slightly twisted around its axis. The pitch is still 1.5 nm but there are 4.5 subunits every turn (374). Both classes of particles have a 2 nm inner diameter and 6 nm outer diameter, and the two strands of the circular DNA are contained in the central hole. The relationship between the closely opposed but noncomplementary DNA strands is unclear.

Upon attachment of the virion to the sex pilus, the pilus with attached phage is retracted to bring the phage to the cell surface (376), although this is controversial (377). The gene 3 protein enters the cell and is found associated with double stranded replicative form DNA (378). Infection with particles containing a thermosensitive gene 3 protein does not result in the synthesis of replicative form or the uncoating of the parental DNA (378), indicating that gene 3 protein is necessary for entry of the DNA and its replicative function. The coat protein from the infecting phage is found associated with the inner membrane (see below) (379). The virus can be iodinated with retention of infectivity, and Smilowitz has used such labeled particles to show that protein subunits derived from the coat of the parental phage are found dispersed among the progeny particles (379).

The product of gene 2 is required for synthesis of replicative form and also of virus strand (380). Continued synthesis of virus strand also requires the product of gene 5. The gene 5 protein is a nonstructural protein present in roughly 100,000 copies per infected cell (381). Within infected cells, gp5 is associated with single stranded progeny DNA (382). This complex, which does not appear tightly bound to any cellular constituent, appears in the electron microscope as a rod 1100 nm long and about 16 nm wide (383). The complex contains about 1300 molecules of gp5, and almost all the cellular ssDNA is in this coated form. In vitro studies with purified gene 5 product show clearly that gp5 binds specifically to ssDNA and not dsDNA, reminiscent of the gene 32 product of T4, although it does not stimulate in vitro DNA synthesis (384). It also differs from gene 32 product in that the structure formed in vitro is not an extended circle but a rod. The protein appears to bind to the single strands and coalesce them together, although it is not clear whether this is through protein-protein interactions or by each active subunit binding both strands. Careful binding studies show that one molecule of protein binds four nucleotides (387). Gene 5 protein-ssDNA complexes accumulate in cells infected with mutants blocked in morphogenesis (except those in gene 2, which do not make ssDNA). However, in wild-type infected cells the DNA leaves the cells and appears as virions, and the gp5, presumably displaced by the coat protein, is reused on a new viral strand (382). For a long time workers in the area were puzzled by the inability to detect intracellular phage (367). Subsequently, it was shown that the newly synthesized coat protein was associated with the E. coli inner membrane and could not be detected in the cytoplasm or in association with the outer membrane (368). Newly synthesized coat protein appears immediately in the inner membrane and later in the medium as completed phage (368). Coat protein also accumulates in the inner membrane of cells infected with mutants blocked in assembly (382). These results are consistent with a model of virion assembly in which the ssDNA molecule, coated with gene 5 protein, associates with the inner membrane, whereas the gene 5 protein is replaced by the gene 8 coat protein, perhaps simultaneously with extrusion through the membrane into the medium. The bases in the gp5-DNA complex are unstacked, whereas those in the particle are stacked. Day (387) has proposed that the transition occurs in the exchange reaction between the α helical coat protein and the non-α helical DNA binding protein. The gene 3 protein may terminate the particle, since 3$^-$ infected cells extrude extra long phage containing multiple genomes (385). The remaining genes, whose functions have not been defined, may serve in the insertion of the DNA into the membrane or during its extrusion from the cell.

Studies of phospholipid synthesis in phage-infected cells suggest a linkage with incorporation of the coat protein subunits into the membrane (382, 386).

ENVELOPED ssRNA VIRUSES

At least seven groups of animal ssRNA viruses are known that replicate in the cytoplasm and are released from infected cells by budding through the plasma membrane (or cytoplasmic vacuole membrane), thereby becoming surrounded by a portion of that membrane (388). These are listed, along with some of their important characteristics, in Table 2. These viruses generally cause acute infections in their host; for instance in man, a togavirus causes yellow fever, a rhabdovirus causes rabies, a myxovirus causes influenza, a paramyxovirus causes measles, coronaviruses cause acute upper respiratory tract infections, and an arenavirus causes lymphocytic choriomeningitis. The oncornaviruses include the leukemia, sarcoma, and mammary tumor viruses of mice, cats, and subhuman primates. Plant viruses structurally similar to rhabdo- and myxoviruses are known, but their development has not been studied (389, 390). Recently, extremely rapid progress has been made in defining the structural components of these viruses. These findings have been reviewed in a number of excellent articles (391–399) and will not be discussed in detail here. We concentrate on the assembly of these viruses; however, some discussion of their structures is necessary, and the reader is referred to these reviews for further specific references.

General Structure of the Enveloped ssRNA Viruses

Structural models for the five best studied groups of this type of virus are shown in Figure 8. All those studied have a membranous lipid-containing envelope surrounding the particle. The lipid in the envelope has been directly shown to be present in the form of a bilayer by low angle X-ray scattering (400) and by electron spin resonance studies (401, 402) for various toga-, myxo-, paramyxo-, and oncornaviruses.

Table 2 Properties of the enveloped RNA viruses[a]

| Virus group | Nucleic Acid | | | | Virion shape | Nucleocapsid shape | Approx. number of protein components |
	Type	Length in bases	State	Viral strand			
Togavirus	ssRNA	6,000–10,000	single molecule	mRNA	spherical	icosahedral	3–4
Rhabdovirus	ssRNA	12,000–20,000	single molecule	anti-mRNA	bullet	helical	5
Myxovirus	ssRNA	10,000–13,000	segmented	anti-mRNA	spherical or filamentous	helical	5–7
Paramyxovirus	ssRNA	20,000–25,000	single molecule	anti-mRNA	spherical or filamentous	helical	5–9
Oncornavirus	ssRNA	30,000–40,000	segmented(?)	?	spherical	?	~14–26
Coronavirus	?	?	?	?	spherical	?	6
Arenavirus	?	?	segmented(?)	?	spherical	?	4

[a] Adapted from Lenard & Compans (391).

GLYCOPROTEINS OF THE VIRION   There is extensive evidence that these viruses contain one or more glycoproteins, and that these glycoproteins represent the spikes seen on the exterior of the virions in negative stained preparations. The spikes often have hemagglutinating and/or neuraminidase activities, and if they are removed by treatment with proteolytic enzymes or reacted with antiglycoprotein serum, the virus particles are rendered noninfectious. It seems likely that the glycoproteins, at least in some cases, do not control the shape of the virus, since a number of viruses can form hybrid envelopes with each other (403, 404). McSharry et al (404) were able to show that in cells coinfected with bullet-shaped rhabdovirus (vesicular stomatitus virus) and spherical paramyxovirus (SV5), bullet-shaped particles were produced which contained some paramyxovirus glycoprotein, suggesting a lack of shape specificity for these proteins. No other paramyxovirus proteins were found in the bullet-shaped particles.

The chemical structure of the carbohydrate portions of these proteins is not known, although initial progress in this direction is being made (405–407). The geometrical arrangement of the glycoproteins on the surface of the virions is also in general unknown; however, it has been suggested that those of myxo- and togaviruses have icosahedral symmetry (42, 408). The carbohydrate moiety may play a role in the infection process; recent reports show that modification of the carbohydrate results in loss of hemagglutinability and infectivity (409, 410).

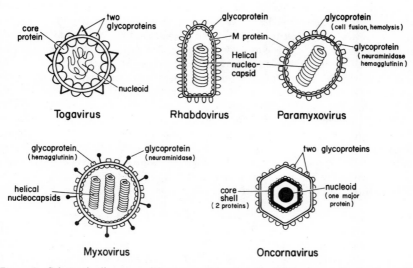

Figure 8   Schematic diagrams of five types of enveloped ssRNA viruses. In the figure M protein stands for matrix protein (see text). The structural models of the viruses are adapted from the following sources: togavirus (42), rhabdovirus (399, 455), oncornavirus (393, 456), myxovirus (391, 394), and paramyxovirus (437). The prototype viruses used in these studies were sindbis virus, vesicular stomatitis virus, murine leukemia virus, influenza A virus, and sendai virus, respectively.

M PROTEIN   The rhabdo-, myxo-, and paramyxoviruses contain a major protein resistant to chemical modification from without (the glycoproteins are not), suggesting that it is internal to the lipid bilayer. It has not been found associated with the nucleocapsid, appears membrane bound in infected cells (see below), and is surmised to be associated with the internal surface of the lipid bilayer in the virion. This protein has been called the M protein for "matrix" or "membrane" protein. Its structural role in the virus particle is unclear, but it may stabilize the lipid bilayer, perhaps by giving the particle a more rigid framework.

NUCLEOCAPSID PROTEIN   If the envelope of these viruses is removed by treatment with various weak or non-ionic detergents, it is usually possible to isolate an apparently undisrupted internal nucleocapsid structure (or structures in the case of viruses with segmented genomes). By electron microscopic examination, these appear to have helical symmetry. The togaviruses, however, are thought to have icosahedral nucleocapsids (with a $T = 4$ surface topology) (42). The togaviruses do not have an M protein and the nucleocapsid protein may perform the functional and structural roles of both nucleocapsid and M proteins. In all cases only one major polypeptide component has been found in the isolated nucleocapsid structure.

Assembly of the Enveloped ssRNA Viruses

CURRENT MODELS   A large number of studies utilizing electron microscopy of ultrathin sections of infected cells have accumulated evidence for a general model for the assembly of the enveloped RNA viruses [see Lenard & Compans (391) for a more detailed review]. Four events leading to the formation of mature virions have been defined by this work: 1. nucleocapsids form in the cytoplasm, 2. patches of cellular plasma membrane accumulate viral glycoproteins, 3. the nucleocapsids become aligned along the inner surface of the modified membrane patches, and 4. the nucleocapsid buds through the modified membrane, being enveloped by the membrane in the process. In some cases steps 1 and 3 seem to occur simultaneously. During the budding process no host membrane proteins are incorporated into the viral particle, although evidence is mounting that the bulk of the viral lipids are derived from the host's normal complement of lipids. Klenk (392) has reviewed these data and concluded that, because these compositions are very similar while the same virus grown in different hosts can have different phospholipid compositions, the host phospholipids are likely to be used by the budding virus.

ENVELOPE FORMATION   Biochemical studies have shown that many proteins made by a number of enveloped RNA viruses can be detected in infected cells, and recent experiments using cell fractionation techniques have begun to localize the viral proteins within the cell. Much of this work has been done with myxoviruses (411, 412) and rhabdoviruses (413–416). The exact site of synthesis is not known for any viral protein. It is possible, for instance, that viral membrane proteins are made on membrane-bound ribosomes; however, it is clear that myxo- (influenza A and fowl plague viruses) and rhabdovirus (vesicular stomatitis virus) envelope proteins are found in membrane fractions soon after their synthesis. The envelope proteins (glycoproteins and M protein) of these viruses accumulate in cytoplasmic membrane fractions (411, 414, 416). However, Cohen et al (414) found that under certain conditions the vesicular stomatitus virus M protein can bind to membranes in vitro, and they cautioned against the assumption that these proteins are always membrane bound in vivo. Pulse-chase experiments have shown that the myxovirus (influenza A) hemagglutinin glycoprotein first appears in the rough membrane fraction and then migrates into the smooth membrane fraction (411, 412, 417). Furthermore, the hemagglutinin in the rough membrane has an incomplete carbohydrate moiety, in that its fucose content is low compared to that found in the smooth membrane fraction (418). The mechanism of migration from the rough to the smooth membranes is not known and could occur by lateral diffusion along cytoplasmic membrane channels, free diffusion through the cytoplasm, or by transport via cytoplasmic membrane vesicles.

Recent in vitro reconstitution experiments of Hosaka & Shimizu (419) have shown that spikes removed from paramyxoviruses (Newcastle disease and Sendai viruses) by solubilization in the mild detergent NP40 do not have hemolytic activity (as do virions), but dialysis with viral, host, or synthetic phospholipids restores the activity. This reconstituted activity is associated with membrane vesicles with visible internal and external spikes. This implies that the spike glycoproteins can associate with membranes in a partially biologically active form after they are completely glycosylated. It also points out the unsolved problem of how the glycoproteins are inserted in the membrane in vivo in such a manner that they are all external in the mature virion. Studies using inhibitors of glycosylation indicate that completely unglycosylated proteins can be inserted into membranes in vivo (412).

Several lines of evidence suggest that the carbohydrates of these glycoproteins are added by host enzymes: 1. the smaller viruses do not have sufficient genetic information to code for such enzymes; 2. Grimes & Burge (420) found that the specific activity of several glycosyltransferases measured did not change after infection; 3. the carbohydrate moiety of the glycoproteins varies with the cell type infected (421, 422); and 4. a carbohydrate host antigen is covalently linked to the hemagglutinin and neuraminidase molecules of myxoviruses (423). It is not certain, however, that parts of the carbohydrate are not viral specified in some cases. The myxo- and paramyxovirus envelopes do not contain neuraminic acid and it has been suggested that this is due to the viral neuraminidase (392), although this has not been shown directly.

NUCLEOCAPSID ASSEMBLY   The nucleocapsid protein of these viruses appears first in the cytoplasm. In the cases of the rhabdovirus and vesicular stomatitis virus, it takes much longer for it to associate with the membrane fraction than it takes the envelope proteins (413, 414, 416). It is bound to viral RNA strands in the cytoplasm, presumably in the form of nucleocapsids. Kiley & Wagner (424) found that there were three different sizes of these intracellular nucleocapsids in cells infected with vesicular stomatitus virus under conditions where there were three sizes of RNA molecules and virions produced. This implies that it is the RNA molecules that determine the size of the nucleocapsid and not the nucleocapsid protein because each size of nucleocapsid contains the same protein. This in turn implies that the size of the nucleocapsid determines the length of the rhabdovirus particle.

Myxovirus nucleocapsid protein was found to migrate from cytoplasmic fractions to the nucleus and to the plasma membrane fraction (425). The significance of its presence in the nucleus of infected cells is unknown because the viral RNA is thought to replicate in the cytoplasm. Pons (426) has shown that viral RNA isolated from infected cells is associated with the nucleocapsid protein that is also found in polysome-containing fractions.

The nucleocapsid protein of a myxovirus (influenza A) binds both viral and antiviral strands of RNA in vitro (427). The protein also binds the anionic polymer polyvinyl sulfate (even tighter than RNA) in a nucleocapsid-like structure (428). It seems likely, therefore, that the nucleocapsid protein subunit carries the information needed to assemble a helical structure (capsid), but alone it may not be able to determine the strand specificity of packaging.

The togavirus (sindbis virus) has an icosahedral nucleocapsid or core which appears to assemble in a slightly different manner. The core protein is found associated with the membrane fraction in infected cells (429), and electron microscopy of infected cells has shown that cores are formed only adjacent to modified membrane patches (430). These observations agree with the hypothesis that the togavirus core protein may function as a sort of combination M protein and nucleocapsid protein.

PROTEIN CLEAVAGES   Cleavage of structural proteins has been reported during the development of many enveloped RNA viruses. The togaviruses, which contain an mRNA genome like the nonenveloped picornaviruses, also appear to translate their entire complement of protein as one giant molecule which is then proteolytically cleaved into the various proteins necessary for successful infection (431–433). There is no direct evidence that any of the cleavages that lead to the formation of the virion proteins are coupled to assembly, although the glycosylation of at least one of the two glycoproteins does begin before the final cleavage to form the viral proteins (431–434).

The hemagglutinin glycoprotein of the myxoviruses is cleaved into two fragments, both of which are found in the virion, after its migration (discussed above) from the rough to smooth membranes (417, 435, 436). This cleavage, which is not required for hemagglutinin activity (417), occurs in the presence of inhibitors of glycosylation (412), occurs only very slowly in some host cell lines, and may occur at different sites along polypeptide chains in different host cells (417). Thus, although the biological function of the cleavage (if any) remains mysterious, the available evidence suggests that host enzymes are responsible for the cleavage.

The paramyxovirus (Sendai virus) envelope contains a number of biological activities; hemagglutinin and neuraminidase are associated with one glycoprotein, and hemolytic and cell fusion activities are associated with a second glycoprotein (437). The recent experiments of Homma & Ohuchi (438) and Scheid & Choppin (437) have shown that the Sendai virus glycoprotein with cell fusion and hemolytic activity is derived from a glycosylated precursor protein without these activities. Furthermore, virus prepared from some cell lines is deficient in these functions, but can be activated in vitro by treatment with trypsin (438) with concomitant cleavage to form the normal viral glycoprotein (437). It is not known whether the small cleavage fragment(s) remains in the virus. A similar cleavage occurs in vivo with another paramyxovirus, Newcastle disease virus (439, 440).

At least two of the virion proteins of the oncornavirus, avian myeloblastosis virus, are derived by cleavage from a larger precursor protein (441), and no cleavages of proteins are known to accompany the development of rhabdoviruses (442).

Another type of post-translational modification, phosphorylation of structural proteins, accompanies the development of a number of the enveloped RNA viruses (415, 443, 454), but the rôle of these phosphorylations in viral assembly and/or function is obscure. One difficulty in giving significance to these reports is the general lack of data concerning the quantitation of the phosphorylation, particularly because the one report containing these data (443) indicates considerably less than one phosphorus atom per polypeptide.

THE BUDDING PROCESS   The biochemical mechanisms by which the nucleocapsids bud through the modified host membrane remain mysterious. A number of electron microscopic studies have suggested that viral envelope antigens (glycoproteins) are present in the plasma membrane in discrete patches (391). The mode of assembly of these patches is unknown, and a number of pathways for arriving at such a point have been discussed by Compans & Caliguiri (444). Finding the M protein in association with cellular membrane fractions rather than with the nucleocapsid implies that these patches of viral-specific membrane may also have M protein bound on the inside of the membrane. Presumably, when a patch is assembled, a nucleocapsid can interact with it to cause the budding process. The actual budding process has not yet been amenable to experimentation; however, it has been shown that bivalent anti-neuraminidase antibodies in the medium block the budding of myxoviruses, whereas univalent antibodies do not (445). Hopefully, the use of conditional lethal mutants will allow further dissection of this process. Temperature-sensitive mutants affecting virus assembly and budding have been isolated for a number of these viruses [toga- (446), rhabdo- (447–449), myxo- (450), and oncorna- (451–453)].

CONCLUSIONS

We summarize by setting out some pertinent questions and answering them where possible. How are shells of spherical viruses formed—often with the help of proteins not found in the mature virion (ϕX174, T4, P22, λ, adenovirus). These proteins, which aid the accurate polymerization of coat proteins, may be incorporated into a precursor capsid and subsequently removed by proteolysis (the assembly core of T4), or they may be removed intact and then recycle, catalyzing assembly (the scaffolding protein of P22). In other cases, they catalyze assembly without forming a long lived complex, such as the B protein of ϕX174 or the membrane assembly factor of poliovirus-infected cells. Formation of the capsids of the small ($T = 1$) viruses, ϕX174 and polio, probably occurs by the aggregation of 12 pentamer corners into an icosahedral shell. Little is known about the course of initiation of this series of reactions. Though only clearly established with TMV-like viruses, the requirement for a special aggregate of the coat protein for initiation may be a general feature of capsid formation, even with spherical viruses.

How is DNA encapsidated into the spherical viruses—generally by a preformed shell, in concert with other proteins, which may be part of either the precursor or the mature virion. In the large phages the condensation of the DNA is associated with the removal of the assembly core proteins. It is unclear whether or not there is enzymatic machinery to drive the DNA into the head independently of its collapse within. Laemmli has in fact shown that T4 DNA can be collapsed into a tight ball in vitro by incubation with the cleavage products of the assembly core proteins. Thus, the capsid provides a mechanism for generating a high concentration of acidic peptides in a localized environment. Lambda DNA encapsidation is associated with the addition of a second major capsid protein, and phage P22 is encapsidated in concert with the exit of a major capsid protein.

What is the relationship between genome structure and encapsidation? The spherical ϕX174 phages and filamentous f1-like phages contain circular ssDNA. Yet the ϕX genome is encapsidated into a small icosahedral shell in the cytoplasm, whereas the f1 chromosome, coated with a DNA binding protein, extrudes through the inner cell membrane out into the medium via replacement of all the binding protein by membrane-associated coat protein. Poliovirus, TMV, and enveloped RNA viruses have ssRNA genomes but unrelated assembly processes, although the assembly of the nucleocapsids of the enveloped RNA viruses might be analogous to TMV polymerization.

In the case of dsDNA viruses, condensation is probably a much more difficult problem, requiring the formation of a complex capsid precursor structure which may undergo substantial reorganization during nucleic acid packaging. In the large phages one can identify gene products involved in cutting the chromosome from the overlength replicating DNA. In phage λ this protein is clearly sequence specific, generating the complementary single stranded ends of the mature chromosome, but with phages T4 and P22 these proteins must not be sequence specific, since the chromosome ends are not unique. The special ends of the adenovirus chromosome may also be generated in coupling with encapsidation.

What is the relationship between the mode of virus penetration and the mode of assembly? A very important function of viral capsids, in addition to the protection of the nucleic acid, is delivery of the nucleic acid into the next host cell. We assume that particle structure has been selected in part for the performance of this function, whether in cell attachment, penetration, uncoating within the cell, or perhaps enzymatic function within the cell, such as reovirus cores or RNA tumor viruses. The membrane of enveloped viruses probably represents a penetration organelle, and the mechanism of the virus coating may not be unrelated to the mechanism of cell penetration. However, for herpesvirus, which is enveloped by the nuclear membrane but uncoated in the cytoplasm, the release of virus from the cell probably must bypass the cytoplasm to prevent premature uncoating. For the filamentous phage, the dissolution of the coat and penetration of the cell via pili could be in part the reverse of its assembly via extrusion through the inner membrane. For the large phage, the

DNA release mechanism has to be built into the particle during morphogenesis, unlike viruses that are uncoated by the host cell.

How is assembly regulated—in general by the reactions of the proteins with each other and not by a temporal sequence in their synthesis. As proteins organize into large aggregates, they generate sites and reactivities which program further steps in assembly. Thus, the 5S protomers of poliovirus first assemble into pentamers and only then into shells, even though the same proteins are involved in making both sets of bonds. This kind of regulation at the level of assembly is particularly clear in T4 tail assembly. Some 20 or more proteins are synthesized simultaneously but only interact with each other in a particular order. That is, the reactive sites for further assembly steps get generated during the previous assembly reactions per se. This is not to say that control of the numbers of protein molecules synthesized is not important, but that the accuracy and efficiency of the assembly process are primarily controlled by protein-protein interactions during assembly, and not by the time of protein synthesis, or by concentration.

What new ideas emerge in terms of general biological interest? (a) Structure: 1. nucleic acid—the detailed study of the mechanisms of DNA condensation and of the existence of large ordered states of condensed DNA opens up a new approach for understanding chromosome structure and the state of inactive DNA; 2. proteins—the tail fibers of T4, for example, represent a fibrous protein organization different from any known ones; the emergence of new kinds of protein structures will certainly follow from further analysis of viral organelles; and 3. membranes—the insertion of viral membrane proteins into preformed envelopes and the formation of de novo viral membranes are probably the areas of most rapid progress for understanding membrane assembly.

(b) New reactions: the dependence of a number of enzymatic reactions on the state of viral structure organization is liable to be a general property of enzymes in eukaryotes. For example, proteins that cut the DNA of P22, T4, and λ do it only in concert with a capsid, and in fact a filled capsid. Somehow these proteins monitor the state of the whole particle. The mechanism is completely mysterious. Similarly the proteases in the T4 head act only in the organized structure. It is not clear whether this represents activation of the substrate protein during their organization or of the protease activity. The exit of a very large number of proteins in an exchange reaction with DNA, as in phage P22, represents a surprising kind of reorganization, as does the exchange reaction between DNA binding protein and membrane-associated coat protein during extrusion of the f1 chromosome through the membrane into the medium.

As shown clearly with phage tail proteins, structural proteins must often be synthesized in a form that does not spontaneously assemble, probably to prevent the formation of either incorrect structures or a structure at the wrong site. Rather, the proteins bind only to sites created on the growing structure by the previous assembly steps.

(c) Genetics: the development of techniques to use temperature- and cold-sensitive mutants in more powerful ways (see, for example, 463) will certainly have general significance.

ACKNOWLEDGMENTS

We thank Roni McCall and Wanda Fischer for their unflagging efforts in the preparation of the manuscript, and Ruth Griffin Shea and Samuel Kayman for proofreading and reference checking. Costs for the preparation of the paper were borne by NIH grant 17,980 to Jonathan King from the Institute of General Medical Sciences and by a grant from the MIT Health Sciences Fund. Sherwood Casjens was supported by a Helen Hay Whitney Fellowship.

Literature Cited

1. Levine, M. 1969. *Ann. Rev. Genet.* 3:323
2. Eiserling, F., Dickson, R. 1972. *Ann. Rev. Biochem.* 41:467
2a. Poglazov, B. F. 1973. *Monogr. Dev. Biol.* Vol. 7
3. Kushner, D. 1969. *Bacteriol. Rev.* 33:302
4. Russell, W. C. 1975. *Progr. Med. Virol.* 16: In press
4a. Crowther, R. A., Klug, A. 1975. *Ann. Rev. Biochem.* 44:161–82
4b. Hershko, A., Fry, M. 1975. *Ann. Rev. Biochem.* 44:775–97
5. Harrison, B. et al, 1971. *Virology* 45:356
6. Marvin, D., Hohn, B. 1969. *Bacteriol. Rev.* 33:172
7. Kellenberger, E. 1972. *Ciba Found. Symp.* 7, 189 pp.
8. Caspar, D. 1963. *Advan. Protein Chem.* 18:37
9. Fraenkel-Conrat, H., Williams, R. 1955. *Proc. Nat. Acad. Sci. USA* 41:690
10. Lauffer, M., Stevens, C. 1968. *Advan. Virus Res.* 13:1
11. Fraenkel-Conrat, H. 1970. *Ann. Rev. Microbiol.* 24:463
12. Finch, J. 1972. *Contemp. Phys.* 13:1
13. Klug, A. 1972. *Fed. Proc.* 31:30
14. Klug, A. 1972. *Ciba Found. Symp.* 7:205
15. Okada, Y. 1975. *Advan. Biophys.* In press
16. Durham, A. 1972. *J. Mol. Biol.* 67:289
17. Finch, J., Klug, A. 1971. *Phil. Trans. Roy. Soc. London B* 261:211
18. Durham, A., Klug, A. 1972. *J. Mol. Biol.* 67:315
19. Richards, K., Williams, R. 1972. *Proc. Nat. Acad. Sci. USA* 69:1121
20. Lebeurier, G., Lonchampt, M., Hirth, L. 1973. *FEBS Lett.* 35:54
21. Rodionova, N., Vesenina, N., Atabekova, T., Yavakhya, V., Atabekov, I. 1973. *Virology* 51:24
22. Richards, K., Williams, R. 1973. *Biochemistry* 12:4574
23. Butler, P. 1974. *J. Mol. Biol.* 82:333
24. Butler, P. 1974. *J. Mol. Biol.* 82:343
25. Okada, Y., Ohno, T. 1972. *Mol. Gen. Genet.* 114:205
26. Ohno, T., Yamaura, R., Kuriyama, K., Inoue, H., Okada, Y. 1972. *Virology* 50:76
27. Atabekov, I. 1973. In *Generation of Subcellular Structures,* ed. R. Markham, R. Home, 77–100
28. Butler, P. 1971. *Cold Spring Harbor Symp. Quant. Biol.* 36:461
29. Fraenkel-Conrat, H., Singer, B. 1964. *Virology* 23:354
30. Fritsch, C., Stussi, C., Witz, J., Hirth, L. 1973. *Virology* 56:33
31. Richards, K., Guilley, H., Jonard, G., Hirth, L. 1975. *FEBS Lett.* In press
32. Lebeurier, G., Hirth, L. 1973. *FEBS Lett.* 34:19
33. Abou, Haidar, M., Pfeiffer, P., Fritsch, C., Hirth, L., 1973. *J. Gen. Virol.* 21:83
34. Ohno, T., Inoue, H., Okada, Y. 1972. *Proc. Nat. Acad. Sci. USA* 69:3680
35. Ohno, T., Okada, Y., Nonomura, Y., Inoue, H. 1975. *J. Biochem. Tokyo* In press
36. Hitchborn, J., Hills, G. 1965. *Virology* 27:528
37. Coutts, R., Cocking, E., Kassanis, B. 1972. *J. Gen. Virol.* 17:289
38. Butler, P., Klug, A. 1971. *Nature* 229:47
39. Shapiro, A., Vinuela, E., Maizel, J. Jr. 1967. *Biochem. Biophys. Res. Commun.* 28:815
40. Caspar, D. 1956. *Nature* 177:477
41. Harrison, S. 1971. *Cold Spring Harbor Symp. Quant. Biol.* 36:495
42. Harrison, S., Jack, A., Goodenough, D., Sefton, B. 1974. *J. Supramol. Struct.* 2:486
43. Crowther, R., Amos, L. 1971. *Cold Spring Harbor Symp. Quant. Biol.* 36:489
44. Crowther, R., Geelen, J., Mellema, J. 1974. *Virology* 57:20
45. Crick, F., Watson, J. 1956. *Nature* 177:473
46. Crick, F., Watson, J. 1957. *Nature of Viruses Ciba Found. Symp.,* 5
47. Caspar, D., Klug, A. 1962. *Cold Spring Harbor Symp. Quant. Biol.* 27:1
48. Zeigler, A., Harrison, S., Leberman, R. 1974. *Virology* 59:509

49. Kaper, J. 1973. *Virology* 55:299
50. Bancroft, J. B. 1970. *Advan. Virus Res.* 16:99
51. Jonard, G., Witz, J., Hirth, L. 1972. *J. Mol. Biol.* 66:165
52. Hohn, T., Hohn, B. 1970. *Advan. Virus Res.* 16:43
53. Krahn, P., O'Callaghan, R., Paranchych, W. 1972. *Virology* 47:628
54. Kaerner, H. 1970. *J. Mol. Biol.* 53:515
55. Matthews, K., Cole, R. 1972. *J. Mol. Biol.* 65:1
56. Ling, C., Hung, P., Overby, L. 1970. *Virology* 40:920
57. Zelanzo, P., Haschemeyer, R. 1970. *Science* 168:1461
58. Kozak, M., Nathans, D. 1972. *Bacteriol. Rev.* 36:109
59. Steitz, J. 1974. *Nature* 248:223
60. Chroboczek, J., Pietrzak, M., Zagorski, W. 1973. *J. Virol.* 12:230
61. Rueckert, R. 1971. In *Comparative Virology,* ed. K. Maramorosch, E. Kustrak, 225–306. New York: Academic
62. Phillips, B. 1972. *Curr. Top. Microbiol.* 58:156
63. Baltimore, D. 1971. *Perspect. Virol.* 7:1
64. Jacobson, M. F., Baltimore, D. 1968. *Proc. Nat. Acad. Sci. USA* 61:77
65. Summers, D., Maizel, J. 1968. *Proc. Nat. Acad. Sci. USA* 59:966
66. Butterworth, B., Rueckert, R. 1972. *Virology* 50:535
67. Oberg, B., Shatkin, A. 1972. *Proc. Nat. Acad. Sci. USA* 69:3589
68. Stoltzfus, C., Rueckert, R. 1972. *J. Virol.* 10:347
69. Butterworth, B., Hall, L., Stoltzfus, C., Rueckert, R. 1971. *Proc. Nat. Acad. Sci. USA* 68:30
70. Rekosh, D. 1972. *J. Virol.* 9:479
71. Jacobson, M., Asso, J., Baltimore, D. 1970. *J. Mol. Biol.* 49:657
72. Phillips, B., Summers, D., Maizel, J. Jr. 1968. *Virology* 35:216
73. Ghendon, Y., Yakobson, E., Mikhejeva, A. 1972. *J. Virol.* 10:261
74. Jacobson, M. F., Baltimore, D. 1968. *J. Mol. Biol.* 33:369
75. Phillips, B. 1971. *Virology* 44:307
76. Butterworth, B. 1973. *Virology* 56:439
77. Ziola, B., Scraba, D. 1974. *Virology* 57:531
78. Korant, B., Lonberg-Holm, K., Noble, J., Stasny, J. 1972. *Virology* 48:71
79. Perlin, M., Phillips, B. 1973. *Virology* 53:107
80. Perlin, M., Phillips, B. 1975. *Virology.* In press
81. Drzeniek, R., Bilello, P. 1972. *Nature New Biol.* 240:118
82. Mak, T., Colter, J., Scraba, D. 1974. *Virology* 57:543
83. Medappa, K., McLean, C., Rueckert, R. 1971. *Virology* 44:259
84. Fernandez-Tomas, C., Baltimore, D. 1973. *J. Virol.* 12:1122
85. Breindl, M. 1971. *Virology* 46:962
86. Breindl, M. 1971. *J. Gen. Virol.* 11:147
87. Crowell, R., Philipson, L. 1971. *J. Virol.* 8:509
88. Breindl, M., Koch, G. 1972. *Virology* 48:136
89. Dunker, A., Rueckert, R. 1971. *J. Mol. Biol.* 58:217
90. Talbot, P., Brown, F. 1972. *J. Gen. Virol.* 15:163
91. Philipson, L., Beatrice, S., Crowell, R. 1973. *Virology* 54:69
92. Mikhejeva, A., Yakobson, E., Soloviev, G. 1970. *J. Virol.* 6:188
93. Cooper, P., Stancek, K., Summers, D. 1970. *Virology* 40:971
94. Cooper, P., Steiner-Pryor, A., Wright, P. 1974. *Intervirology* 1:1
95. Fiszman, M., Reynier, M., Bucchini, D., Girard, M. 1972. *J. Virol.* 10:1143
96. Garfinkle, D., Tershak, D. 1972. *Nature New Biol.* 238:206
97. Caliguiri, L., Compans, R. 1973. *J. Gen. Virol.* 21:99
98. Laemmli, U., Paulson, J., Hitchins, V. 1974. *J. Supramol. Struct.* 2:276
99. Kellenberger, E., Edgar, R. 1971. In *The Bacteriophage Lambda,* 271. Cold Spring Harbor, NY: Cold Spring Harbor Press
100. Barrett, K., Calendar, R., Gibbs, W., Goldstein, R., Lindquist, B., Six, E., 1973. *Progr. Med. Virol.* 15:309
101. Goldstein, R., Lengyel, J., Pruss, G., Barrett, K., Calendar, R., Six, E. 1975. *Curr. Top. Microbiol. Immunol.* In press
102. Casjens, S., King, J. 1974. *J. Supramol. Struct.* 2:202
103. Studier, F. W. 1972. *Science* 176:367
104. Matsuo-Kato, H., Fugisawa, H. 1975. *Virology* 63:105
105. Zweig, M., Cummings, D. 1973. *J. Mol. Biol.* 80:505
106. Carrascosa, J., Camacho, A., Vinuela, E., Salas, M. 1974. *FEBS Lett.* In press
107. Epstein, R., Bolle, A., Steinberg, C., Kellenberger, E., Boy de la Tour, E., Chevalley, R., Edgar, R., Susman, M., Denhardt, G., Lielausis, A. 1963. *Cold Spring Harbor Symp. Quant. Biol.* 28:375
108. Campbell, A. 1961. *Virology* 14:22
109. Edgar, R., Lielausis, I. 1968. *J. Mol. Biol.* 32:263
110. Kellenberger, E., Ryter, A., Sechaud, J.

1958. *J. Biophys. Biochem. Cytol.* 4:671

111. Streisinger, G., Emrich, J., Stahl, M. 1967. *Proc. Nat. Acad. Sci. USA* 57:292

112. Luftig, R., Wood, W., Okinaka, R. 1971. *J. Mol. Biol.* 57:555

113. Laemmli, U., Favre, M. 1973. *J. Mol. Biol.* 80:575

114. Bijlenga, R., van der Broek, R., Kellenberger, E. 1974. *Nature* 249:825

115. Hohn, B., Hohn, T. 1974. *Proc. Nat. Acad. Sci. USA* 71:2372

116. Kaiser, A. D., Syvanen, M., Masuda, T. 1975. *J. Mol. Biol.* 91:175

117. Kerr, C., Sadowski, P. 1974. *Proc. Nat. Acad. Sci. USA* 71:3545

118. Serwer, P. 1974. Personal communication

119. King, J., Lenk, E., Botstein, D. 1973. *J. Mol. Biol.* 80:697

120. Pruss, G., Goldstein, R., Calendar, R. 1974. *Proc. Nat. Acad. Sci. USA* 71:2367

121. Pruss, G., Barrett, K., Lengyel, J., Goldstein, R., Calendar, R. 1974. *J. Supramol. Struct.* 2:337

122. Botstein, D., Waddell, C., King, J. 1973. *J. Mol. Biol.* 80:669

123. Lenk, E., Casjens, S., Weeks, J., King, J. 1975. *Virology.* In press

124. Kaiser, A. D., Masuda, T. 1973. *Proc. Nat. Acad. Sci. USA* 70:260

125. Laemmli, U., Teuff, N., D'Ambrosia, J. 1974. *J. Mol. Biol.* 88:749

126. King, J., Casjens, S. 1974. *Nature* 251:112

127. Earnshaw, W., Casjens, S., Harrison, S. 1975. In preparation

128. Israel, J., Anderson, T., Levine, M. 1967. *Proc. Nat. Acad. Sci. USA* 57:284

129. Botstein, D., Levine, M. 1968. *Cold Spring Harbor Symp. Quant. Biol.* 33:659

130. Tye, B., Chan, R., Botstein, D. 1974. *J. Mol. Biol.* 85:485

131. Tye, B., Huberman, J., Botstein, D. 1974. *J. Mol. Biol.* 85:501

132. Serwer, P. 1974. *Virology* 59:70

133. Ritchie, D., Thomas, C. Jr., MacHattie, L. 1967. *J. Mol. Biol.* 23:365

134. Casjens, S., Hendrix, R. 1974. *J. Mol. Biol.* 88:535

135. Williams, R., Richards, K. 1974. *J. Mol. Biol.* 88:547

136. Hendrix, R., Casjens, S. 1975. *J. Mol. Biol.* 91:187

137. Casjens, S. 1975. *J. Mol. Biol.* 90:1

138. Hohn, B., Wurtz, M., Klein, B., Lustig, A., Hohn, T. 1974. *J. Supramol. Struct.* 2:302

139. Kemp, C., Howatson, A., Siminovitch, L. 1968. *Virology* 36:490

140. Murialdo, H., Siminovitch, L. 1972. *Virology* 48:785

141. Hendrix, R., Casjens, S. 1974. *Proc. Nat. Acad. Sci. USA* 71:1451

142. Kaiser, A., Hogness, D. 1960. *J. Mol. Biol.* 2:392

143. MacKinlay, A., Kaiser, A. D. 1969. *J. Mol. Biol.* 39:679

144. Wake, R., Kaiser, A. D., Inman, R. 1972. *J. Mol. Biol.* 64:519

145. Skalka, A., Poonian, M., Bartl, P. 1972. *J. Mol. Biol.* 64:541

146. Emmons, S. 1974. *J. Mol. Biol.* 83:511

147. Sternberg, N. 1974. Personal communication

148. Syvanen, M. 1975. *J. Mol. Biol.* 91:165

149. Padmanabhan, R., Wu, R., Bode, V. 1972. *J. Mol. Biol.* 69:201

150. Freifelder, D., Chud, L., Levine, E. 1974. *J. Mol. Biol.* 83:503

151. Wang, J., Brezinski, D. 1973. *Proc. Nat. Acad. Sci. USA* 70:2667

152. Wang, J., Kaiser, A. D. 1973. *Nature New Biol.* 241:16

153. Hendrix, R., Casjens, S. 1974. *Virology* 61:156

154. Branton, D., Klug, A. 1975. In preparation

155. MacHattie, L., Ritchie, D., Thomas, C., Richardson, C. 1967. *J. Mol. Biol.* 23:355

156. Laemmli, U., Beguin, F., Gujer-Kellenberger, G. 1970. *J. Mol. Biol.* 47:69

157. Simon, L. 1972. *Proc. Nat. Acad. Sci. USA* 69:907

158. Laemmli, U., Molbert, E., Showe, M., Kellenberger, E. 1970. *J. Mol. Biol.* 49:99

159. Doermann, A., Eiserling, F., Boehner, L. 1973. *Virus Research,* ed. C. F. Fox, W. Robinson. New York: Academic. 243 pp.

160. Kellenberger, E. 1973. In *Generation of Subcellular Structures,* ed. R. Markham, R. Horne, 59–75. New York: American Elsevier

161. Showe, M., Black, L. 1973. *Nature New Biol.* 242:70

162. Laemmli, U. 1970. *Nature* 227:680

163. Hosoda, J., Cone, R. 1970. *Proc. Nat. Acad. Sci. USA* 66:1275

164. Kellenberger, E., Kellenberger-Van der Kamp, C. 1970. *FEBS Lett.* 8:140

165. Dickson, R., Barnes, S., Eiserling, F. 1970. *J. Mol. Biol.* 53:461

166. Celis, J., Smith, J., Brenner, S. 1973. *Nature New Biol.* 241:130

167. Fugisawa, H., Minagawa, T. 1971. *Virology* 45:289

168. Siegel, P., Schaechter, M. 1973. *J. Virol.* 11:359

169. Howard, G., Wolin, M., Champe, S.

1972. *Ann. NY Acad. Sci.* 34:36
170. Sternberg, N., Champe, S. 1969. *J. Mol. Biol.* 46:377
171. Laemmli, U., Quittner, S. 1974. *Virology* 62:483
172. Bachrach, U., Benchetrit, L. 1974. *Virology* 59:51
173. Goldstein, J., Champe, S. 1974. *J. Virol.* 13:419
174. Showe, M., Onotaro, L. 1972. *Ann. Meet. Am. Soc. Microbiol.* 206 (Abstr.)
175. Frankel, F., Batcheler, M., Clark, C. 1971. *J. Mol. Biol.* 62:439
176. Huberman, J. 1968. *Cold Spring Harbor Symp. Quant. Biol.* 33:509
177. Vanderslice, R., Yegian, C. 1974. *Virology* 60:265
178. Lengyel, J., Goldstein, R., Marsh, M., Sunshine, M., Calendar, R. 1973. *Virology* 53:1
179. Dawson, P., Skalka, A., Simon, L. 1975. *J. Mol. Biol.* In press
180. Aebi, U., Bijlenga, R., van der Broek, J., van der Broek, R., Eiserling, F., Kellenberger, C., Kellenberger, E., Mesyanzhinov, V., Muller, L., Showe, M., Smith, R., Steven, A. 1974. *J. Supramol. Struct.* 2:253
181. Dorman, B., Maestre, M. 1973. *Proc. Nat. Acad. Sci. USA* 70:255
182. Richards, K., Williams, R., Calendar, R. 1973. *J. Mol. Biol.* 78:255
183. Maestre, M., Kilkson, R. 1962. *Nature* 193:366
184. North, A., Rich, A. 1961. *Nature* 191:1242
185. Bendet, I., Goldstein, D., Lauffer, M., 1960. *Nature* 187:781
186. Gellert, M., Davies, D. 1964. *J. Mol. Biol.* 8:341
187. Lerman, L. 1973. *Cold Spring Harbor Symp. Quant. Biol.* 38:59
188. Black, L., Ahmad-Zadeh, C. 1971. *J. Mol. Biol.* 57:71
189. Bachrach, U., Benchetrit, L. 1974. *Virology* 59:443
190. Brody, T. 1973. *Virology* 54:441
191. Luftig, R., Lundh, N. 1973. *Virology* 51:432
192. Wood, W., Dickson, R., Bishop, R., Revel, H. 1973. *Proc. 1st John Innes Symp.*, ed. R. Markham, 25–58 Amsterdam: Elsevier
193. Georgopoulos, C., Eisen, H. 1974. *J. Supramol. Struct.* 2:349
194. Simon, L., Snover, D., McLaughlin, T., Grisham, C. 1974. Personal communication
195. Chao, J., Chao, L., Speyer, J. 1974. *J. Mol. Biol.* 85:41
196. Simon, L. 1974. Personal communication

197. Casjens, S., Hohn, T., Kaiser, A. D. 1972. *J. Mol. Biol.* 64:551
198. Studier, F. W. 1969. *Virology* 39:562
199. Coombs, D. 1974. PhD thesis. University of California, Los Angeles
200. Boklage, C., Wong, E., Bode, V. 1974. *Virology* 61:22
201. Edgar, R., Wood, W. 1966. *Proc. Nat. Acad. Sci. USA* 55:498
202. Weigle, J. 1968. *J. Mol. Biol.* 33:483
203. Saigo, K., Uchida, H. 1974. *Virology* 61:524
204. Moody, M. 1965. *Virology* 26:567
205. Simon, L., Anderson, T. 1967. *Virology* 32:298
206. Benz, W. C., Goldberg, E. B. 1973. *Virology* 53:225
207. King, J. 1968. *J. Mol. Biol.* 32:231
208. King, J. 1971. *J. Mol. Biol.* 58:693
209. King, J., Laemmli, U. 1973. *J. Mol. Biol.* 75:315
210. King, J., Mykolajewycz, N. 1973. *J. Mol. Biol.* 75:339
211. Kozloff, L., Verses, C., Lute, M., Crosby, L. 1970. *J. Virol.* 5:740
212. Kikuchi, Y., King, J. 1975. *J. Mol. Biol.* In press
213. Dawes, J., Goldberg, E. B. 1973. *Virology* 55:380
214. Yamamoto, M., Uchida, H. 1975. *J. Mol. Biol.* In press
215. Kozloff, L., Lute, M. 1973. *J. Virol.* 11:630
216. Kells, S., Haselkorn, R. 1974. *J. Mol. Biol.* 83:473
217. Moody, M. 1971. *Phil. Trans. Roy. Soc. London B.* 261:181
218. Kellenberger, E. 1969. *Nobel Symp., 11th,* 349 pp.
219. Poglazov, B., Nikolskaya, T. 1969. *J. Mol. Biol.* 43:231
220. Katsura, I., Kühl, P. W. 1974. *J. Supramol. Struct.* 2:239
221. Katsura, I., Kühl, P. W. 1975. *Virology.* 63:221
222. Bleviss, M., Easterbrook, K. 1971. *Can. J. Microbiol.* 17:947
223. Lengyel, J. A., Goldstein, R. N., Marsh, M., Calendar, R. 1975. *Virology.* In press
224. King, J., Wood, W. B. 1969. *J. Mol. Biol.* 39:583
225. Bishop, R., Conley, M., Wood, W. 1974. *J. Supramol. Struct.* 2:196
226. Ward, S., Dickson, R. 1971. *J. Mol. Biol.* 62:479
227. King, J., Laemmli, U. 1971. *J. Mol. Biol.* 62:465
228. Dickson, R. 1973. *J. Mol. Biol.* 79:633
229. Takata, R., Tsugita, A. 1970. *J. Mol. Biol.* 54:45
230. Imada, S., Tsugita, A. 1972. *Mol. Gen.*

Genet. 119 : 185
231. Ishibashi, M., Maizel, J. V. Jr. 1974. *Virology* 57 : 409
232. Everitt, E., Sundquist, B., Petterson, U., Philipson, L. 1973. *Virology* 52 : 130
233. Philipson, L., Petterson, U. 1973. *Advan. Exp. Tumor Virus Res.* 18 : 1
234. Tooze, J. 1973. *The Molecular Biology of Tumor Viruses.* Cold Spring Harbor, NY : Cold Spring Harbor Lab.
235. Koczot, F., Carter, B., Garon, C., Rose, J. 1973. *Proc. Nat. Acad. Sci. USA* 70 : 215
236. Robinson, A., Younghusband, H., Bellet, A. 1973. *Virology* 56 : 54
237. Shortridge, K., Stevens, L. 1973. *Microbiosis* 7 : 61
238. Russell, W., Skehel, J. 1971. *J. Gen. Virol.* 15 : 45
239. Walter, G., Maizel, J. V. Jr. 1974. *Virology* 57 : 402
240. Velicer, L., Ginsberg, H. 1970. *J. Virol.* 5 : 338
241. Ginsberg, H. In Eiserling, F., Dickson, R. 1972. *Ann. Rev. Biochem.* 41 : 467
242. Sundquist, B., Everitt, E., Philipson, L., Hoglund, S. 1973. *J. Virol.* 11 : 449
243. Anderson, C., Baum, P., Gestland, R. 1973. *J. Virol.* 12 : 241
244. Russell, W., Skehel, J. 1973. *J. Gen. Virol.* 20 : 195
245. Prage, L., Hoglund, S., Philipson, L. 1972. *Virology* 49 : 745
246. Wadell, G., Hammarskjold, M., Varsangi, T. 1973. *J. Gen. Virol.* 20 : 287
247. Ishibashi, M., Maizel, J. V. Jr. 1974. *Virology* 58 : 345
248. Russell, W., Skehel, J., Machado, R., Pereira, H. 1972. *Virology* 50 : 931
249. Rouse, H., Schlesinger, R. 1972. *Virology* 48 : 463
250. Winters, W., Russell, W. 1971. *J. Gen. Virol.* 10 : 181
251. Pereira, H., Wrigley, N. 1974. *J. Mol. Biol.* 85 : 617
252. Crowther, R., Franklin, R. 1972. *J. Mol. Biol.* 68 : 181
253. Williams, J., Gharpure, M., Ustacelebi, S., Macdonald, S. 1971. *J. Gen. Virol.* 11 : 95
254. Ensinger, M., Ginsberg, H. 1972. *J. Virol.* 10 : 328
255. Russell, W., Newman, C., Williams, J. 1972. *J. Gen. Virol.* 17 : 265
256. Wills, E., Russell, W., Williams, J. 1973. *J. Gen. Virol.* 20 : 407
257. Finch, J., Crawford, L. 1975. *Compr. Virol.* In press
258. Mullarkey, M., Hruska, J., Takemoto, K. 1974. *J. Virol.* 13 : 1014
259. Frearson, P., Crawford, L. 1972. *J. Gen. Virol.* 14 : 141
260. Lake, R., Barban, S., Salzman, N. 1973. *Biochem. Biophys. Res. Commun.* 54 : 640
261. Friedmann, T. 1974. *Proc. Nat. Acad. Sci. USA* 71 : 257
262. Tan, K., Sokol, F. 1972. *J. Virol.* 10 : 985
263. Seehafer, J., Weil, R. 1974. *Virology* 58 : 75
264. Ozer, H. 1972. *J. Virol.* 9 : 41
265. Ozer, H., Tegtmeyer, P. 1972. *J. Virol.* 9 : 52
266. Seebeck, T., Weil, R. 1974. *J. Virol.* 13 : 567
267. Friedmann, T., David, D. 1972. *J. Virol.* 10 : 776
268. Estes, M., Huang, E., Pagano, J. 1971. *J. Virol.* 7 : 635
269. Kiselev, N., Klug, A. 1969. *J. Mol. Biol.* 40 : 155
270. Friedmann, T. 1971. *Proc. Nat. Acad. Sci. USA* 68 : 2574
271. Tan, K., Sokol, F. 1974. *J. Gen. Virol.* 25 : 37
272. Kit, S., Tokuno, S., Nakijima, K., Trkula, D., Dubbs, D. 1970. *J. Virol.* 6 : 286
273. Tegtmeyer, P., Ozer, H. 1971. *J. Virol.* 8 : 516
274. Barban, S. 1973. *J. Virol.* 11 : 971
275. Roizman, B., Spear, P. See Ref. 61, 135
276. Wildy, P., Russell, W., Horne, R. 1960. *Virology* 12 : 204
277. Wagner, E., Tewari, K., Kolodner, R., Warner, R. 1974. *Virology* 57 : 436
278. Grafstrom, R., Alwine, J., Steinhart, W., Hill, C. 1974. *Fed. Proc.* 33 : 1543 (Abstr.)
279. Gibson, W., Roizman, B. 1971. *Proc. Nat. Acad. Sci. USA* 68 : 2818
280. Rubenstein, A. S., Gravell, M., Darlington, R. 1972. *Virol.* 50 : 287
281. Morgan, C., Rose, H. M., Mednis, B. 1968. *J. Virol.* 2 : 507
282. Ben-Porat, T., Shimono, H., Kaplan, A. S. 1969. *Virol.* 37 : 56
283. McCombs, R. M. 1974. *Virol.* 57 : 448
284. Courtney, R. J., McCombs, R. M., Benyesh-Melnick, M. 1971. *Virol.* 43 : 350
285. Mark, G. F., Kaplan, A. S. 1971. *Virology* 45 : 53
286. Nii, S., Morgan, C., Rose, H. 1968. *J. Virol.* 2 : 517
287. Stackpole, C. 1969. *J. Virol.* 4 : 75
288. Nii, S., Rosenkranz, H. S., Morgan, C., Rose, H. M. 1968. *J. Virol.* 2 : 1163
289. Fong, C., Tenser, R., Hsiung, G., Gross, P. 1973. *Virology* 52 : 468
290. Darlington, R. W., Moss, L. H. 1959.

Progr. Med. Virol. 11:16
291. Wolf, K., Darlington, R. 1971. *J. Virol.* 8:525
292. Schwartz, J., Roizman, B. 1969. *Virology* 38:42
293. Sydiskis, R. J. 1969. *J. Virol.* 4:283
294. Honess, R., Roizman, B. 1974. *J. Virol.* 14:8
295. Honess, R., Roizman, B. 1973. *J. Virol.* 12:1347
296. Spear, P., Roizman, B. 1972. *J. Virol.* 9:143
297. Heine, J., Spear, P., Roizman, B. 1972. *J. Virol.* 9:431
298. Spear, P., Roizman, B. 1970. *Proc. Nat. Acad. Sci. USA* 66:730
299. Spear, P., Keller, J., Roizman, B. 1970. *J. Virol.* 5:123
300. Roizman, B., Heine, J. 1972. In *Membrane Research,* ed. C. Fred Fox, 202–37. New York: Academic
301. Gibson, W., Roizman, B. 1972. *J. Virol.* 10:1044
302. Gibson, W., Roizman, B. 1974. *J. Virol.* 13:155
303. Perdue, M., Kemp, M., Randall, C., O'Callaghan, D. 1974. *Virology* 59:201
304. Schaffer, P., Aron, G., Biswal, L., Benyesh-Melnick, M. 1973. *Virology* 52:57
305. Halliburton, I., Timbury, M. 1973. *Virology* 54:60
306. Esparza, J., Purifoy, D., Schaffer, P., Benyesh-Melnick, M. 1974. *Virology* 57:554
307. Kelly, D., Tinsley, T. 1974. *Microbios* 9:75
308. Dales, S., Mosbach, E. 1968. *Virology* 35:564
309. VanEtten, J., Vidaver, A., Koski, R., Burnett, J. 1974. *J. Virol.* 13:1254
310. Williams, R., Smith, K. 1958. *Biochim. Biophys. Acta* 28:464
311. Kelly, D., Vance, D. 1973. *J. Gen. Virol.* 21:417
312. Kelly, D., Robertson, J. 1973. *J. Gen. Virol.* 20:17 (Suppl.)
313. Franklin, R. 1975. *Curr. Top. Microbiol. Immunol.* In press
314. Harrison, S., Caspar, D., Camerini-Otero, R., Franklin, R. 1971. *Nature* 229:197
315. Scandella, C., Schindler, H., Franklin, R., Seelig, J. 1975. *Eur. J. Biochem.* Submitted
316. Dahlberg, L., Franklin, R. 1971. *Virology* 42:1073
317. Tsukagoshi, N., Franklin, R. 1974. *Virology* 59:408
318. Sarov, I., Joklik, W. 1972. *Virology* 50:579

319. Obijeski, J., Palmer, E., Gafford, L., Randall, C. 1973. *Virology* 51:512
320. Gafford, L., Randall, C. 1967. *J. Mol. Biol.* 26:303
321. Medzon, E., Bauer, H. 1970. *Virology* 40:860
322. Moss, B., Rosenblum, E., Garon, C. 1973. *Virology* 55:143
323. Sarov, I., Joklik, W. 1972. *Virology* 50:593
324. Katz, E., Margalith, E. 1973. *J. Gen. Virol.* 18:381
325. Morgan, C., Ellison, S., Rose, H., Moore, D. 1964. *J. Exp. Med.* 100:301
326. DeHarven, E., Yohn, D. 1966. *Cancer Res.* 26:995
327. Sarov, I., Joklik, W. 1973. *Virology* 52:223
328. Joklik, W., Becker, Y. 1964. *J. Mol. Biol.* 10:452
329. Polisky, B., Kates, J. 1972. *Virology* 49:168
330. Follett, E., Pennington, T. 1973. *Advan. Virus Res.* 18:105
331. Moss, B., Rosenblum, E. 1973. *J. Mol. Biol.* 81:267
332. Wood, H. 1973. *J. Gen. Virol.* 20:61 (Suppl.)
333. Shatkin, A. 1974. *Ann. Rev. Biochem.* 43:643
334. Shatkin, A., Sipe, J., Loh, P. 1968. *J. Virol.* 2:986
335. Loh, P., Shatkin, A. 1968. *J. Virol.* 2:1353
336. Joklik, W. 1972. *Virology* 49:700
337. Luftig, R., Kilham, S., Hay, A., Zweerink, H., Joklik, W. 1972. *Virology* 48:170
338. Milne, R., Conti, M., Lisa, V. 1973. *Virology* 53:130
339. Lewandowski, L., Traynor, B. 1972. *J. Virol.* 10:1053
340. Skehel, J., Joklik, W. 1969. *Virology* 39:822
341. Martin, S., Pett, D., Zweerink, H. 1973. *J. Virol.* 12:194
342. Bartlett, N. M. Gillies, S. C., Bullivant, S., Bellamy, A. R. 1974. *J. Virol.* 14:315
343. Silverstein, S., Astell, C., Levin, D., Schonberg, M., Acs, G. 1972. *Virology* 47:797
344. Zweerink, H., McDowell, M., Joklik, W. 1971. *Virology* 45:716
345. Pett, D., Vanaman, T., Joklik, W. 1973. *Virology* 52:174
346. Loh, P., Oie, H. 1969. *J. Virol.* 4:890
347. Fields, B., Raine, C., Baum, S. 1971. *Virology* 43:569
348. Fields, B., Joklik, W. 1969. *Virology* 37:335
349. Acs, G., Klett, H., Schonberg, M.,

Christman, J., Levin, D., Silverstein, S. 1971. *J. Virol.* 8:684

350. Sakuma, S., Watanabe, Y. 1972. *J. Virol.* 10:943

351. Morgan, E., Zweerink, H. 1974. *Virology* 59:556

352. Ito, Y., Joklik, W. 1972. *Virology* 50: 282

353. Cross, R., Fields, B. 1972. *Virology* 50:799

354. Ito, Y., Joklik, W. 1972. *Virology* 50:189

355. Silverstein, S. C., Astell, C., Christman, J., Klett, H., Acs, G. 1974. *J. Virol.* 13:740

356. Burgess, A. 1969. *Proc. Nat. Acad. Sci. USA* 64:613

357. Tonegawa, S., Hayashi, M. 1970. *J. Mol. Biol.* 48:219

358. Scruda, A., Poljak, R. 1971. *Virology* 46:164

359. Godson, G. 1971. *J. Mol. Biol.* 57:541

360. Benbow, B., Mayol, R., Picchi, J., Sinsheimer, R. 1972. *J. Virol.* 10:99

361. Siden, E., Hayashi, M. 1974. *J. Mol. Biol.* 89:1

362. Takai, M. 1966. *Biochim. Biophys. Acta* 119:20

363. Tinsley, T., Longworth, J. 1973. *J. Gen. Virol.* 20:7 (Suppl.)

364. Crawford, L., Shatkin, A., Tattersall, P. 1974. Personal communication

365. Johnson, F., Hoggan, M. 1973. *Virology* 51:129

366. Iwaya, M., Denhardt, D. 1971. *J. Mol. Biol.* 57:159

367. Marvin, D., Hohn, B. 1969. *Bacteriol. Rev.* 33:172

368. Smilowitz, H., Carson, J., Robbins, P. 1972. *J. Supramol. Struct.* 1:8

369. Asbeck, F., Beyreuther, K., Koehler, H., vonWettstein, G., Braunitzer, G. 1969. *Z. Physiol. Chem.* 350:1047

370. Henry, T., Pratt, D. 1969. *Proc. Nat. Acad. Sci. USA* 62:800

371. Rossomando, E., Zinder, N. 1968. *J. Mol. Biol.* 36:387

372. Pratt, D., Tzagoloff, H., Beaudoin, J. 1969. *Virology* 30:397

373. Marvin, D., Wiseman, R., Wachtel, E. 1974. *J. Mol. Biol.* 82:121

374. Marvin, D., Pigram, W., Wiseman, R., Wachtel, E., Marvin, F. 1974. *J. Mol. Biol.* 88:581

375. Pratt, D. 1969. *Ann. Rev. Genet.* 3:343

376. Jacobson, A. 1972. *J. Virol.* 10:835

377. Henry, T. J., Brinton, C. C. 1971. *Virology* 46:754

378. Jazwinski, S., Marco, R., Kornberg, A. 1973. *Proc. Nat. Acad. Sci. USA* 70:205

379. Smilowitz, H. 1974. *J. Virol.* 13:94

380. Tseng, B., Marvin, D. 1972. *J. Virol.* 10:384

381. Alberts, B., Frey, L., Delius, H. 1972. *J. Mol. Biol.* 68:139

382. Webster, R., Cashman, J. 1973. *Virology* 55:20

383. Pratt, D., Laws, P., Griffith, J. 1974. *J. Mol. Biol.* 82:425

384. Oey, J., Knippers, R. 1972. *J. Mol. Biol.* 68:125

385. Beaudoin, J., Henry, T., Pratt, D. 1974. *J. Virol.* 13:470

386. Woolford, J., Cashman, J., Webster, R. 1974. *Virology* 58:544

387. Day, L. A. 1973. *Biochemistry* 12:5329

388. Melnick, J. 1973. In *Ultrastructure of Animal and Bacterial Viruses,* ed. A. Dalton, F. Haguenan, 1–20. New York: Academic

389. Francki, R. 1973. *Advan. Virus Res.* 18:257

390. Mohamed, N., Randles, J., Francki, R. 1973. *Virology* 56:12

391. Lenard, J., Compans, R. 1974. *Biochim. Biophys. Acta* 344:51

392. Klenk, H. 1975. *Curr. Top. Microbiol. Immunol.* In press

393. Bolognesi, D. 1975. *Advan. Virus Res.* In press

394. Schulze, I. 1973. *Advan. Virus Res.* 18:1

395. Laver, W. 1973. *Advan. Virus Res.* 18:57

396. Bukrinskaya, A. 1973. *Advan. Virus Res.* 18:195

397. Simons, K., Garoff, H., Helenius, A., Kaarianen, L., Renkonen, O. 1975. In *Perspectives in Membrane Biology,* ed. G. Gitler. In press

398. Blough, H., Tiffany, J. 1973. *Advan. Lipid Res.* 11:267

399. Wagner, R., Prevec, L., Brown, F., Summers, D., Sokol, F., MacLeod, R. 1972. *J. Virol.* 10:1228

400. Harrison, S., David, A., Jumblatt, J., Darnell, J. 1971. *J. Mol. Biol.* 60:523

401. Landsberger, F., Compans, R., Paxton, J., Lenard, J. 1972. *J. Supramol. Struct.* 1:50

402. Landsberger, F., Compans, R., Choppin, P., Lenard, J. 1973. *Biochemistry* 12:4498

403. Zavada, J. 1972. *Nature New Biol.* 240:122

404. McSharry, J., Compans, R., Choppin, P. 1971. *J. Virol.* 8:722

405. Sefton, B., Keegstra, K. 1974. *J. Virol.* 14:522

406. Moyer, S., Summers, D. 1974. *Cell* 2:71

407. Krantz, M., Lee, Y., Hung, P. 1974. *Nature* 248:684

408. Nermut, M., Frank, H. 1971. *J. Gen. Virol.* 10:37

409. Schulze, I. 1975. In *Negative Strand Viruses*, ed. R. Barry, B. Mahy. New York: Academic
410. Schloemer, R., Wagner, R. 1974. *J. Virol.* 14:270
411. Compans, R. 1973. *Virology* 51:56
412. Klenk, H., Woellert, W., Rott, R., Scholtissek, C. 1974. *Virology* 57:28
413. David, A. 1973. *J. Mol. Biol.* 76:135
414. Cohen, G., Atkinson, P., Summers, D. 1971. *Nature New Biol.* 231:121
415. Moyer, S., Summers, D. 1974. *J. Virol.* 13:455
416. Wagner, R., Kiley, M., Snyder, R., Schnaitman, C. 1972. *J. Virol.* 9:672
417. Stanley, P., Gandhi, S., White, D. 1973. *Virology* 53:92
418. Compans, R. 1973. *Virology* 55:541
419. Hosaka, Y., Shimizu, Y. 1972. *Virology* 49:640
420. Grimes, W., Burge, B. 1971. *J. Virol.* 7:309
421. Burge, B., Huang, A. 1970. *J. Virol.* 6:176
422. Lai, M., Duesberg, P. 1972. *Virology* 50:359
423. Laver, W., Webster, R. 1966. *Virology* 30:104
424. Kiley, M., Wagner, R. 1972. *J. Virol.* 10:244
425. Taylor, J., Hampson, A., Layton, J., White, D. 1970. *Virology* 42:744
426. Pons, M. 1972. *Virology* 47:823
427. Scholtissek, C., Becht, H. 1971. *J. Gen. Virol.* 10:11
428. Goldstein, E., Pons, M. 1970. *Virology* 41:382
429. Bose, H., Brundige, M. 1972. *J. Virol.* 9:785
430. Pedersen, C., Sagik, B. 1973. *J. Gen. Virol.* 18:375
431. Schlesinger, M., Schlesinger, S. 1973. *J. Virol.* 11:1013
432. Sefton, B., Burge, B. 1973. *J. Virol.* 12:1366
433. Morser, M., Burke, D. 1974. *J. Gen. Virol.* 22:395
434. Shapiro, D., Kos, K., Russell, P. 1973. *Virology* 56:88
435. Klenk, H., Rott, R. 1973. *J. Virol.* 11:823
436. Lazarowitz, S., Compans, R., Choppin, P. 1973. *Virology* 52:199
437. Scheid, A., Choppin, P. 1974. *Virology* 57:475
438. Homma, M., Ohuchi, M. 1973. *J. Virol.* 12:1457
439. Kaplan, J., Bratt, M. 1973. *Ann. Meet. Am. Soc. Microbiol.* 243 (Abstr.)
440. Samson, A., Fox, C. 1974. *J. Virol.* 13:775
441. Vogt, V., Eisenman, R. 1973. *Proc. Nat. Acad. Sci. USA* 70:1734
442. Stampher, M., Baltimore, D. 1973. *J. Virol.* 11:520
443. Waite, M., Lubin, M., Jones, K., Bose, H. 1974. *J. Virol.* 13:244
444. Compans, R., Caliguiri, L. See Ref. 409
445. Becht, H., Hammerling, U., Rott, R. 1971. *Virology* 46:337
446. Jones, K., Waite, M., Bose, H. 1974. *J. Virol.* 13:809
447. Reichmann, M., Pringle, C., Follett, E. 1971. *J. Virol.* 8:154
448. Printz, P., Wagner, R. 1971. *J. Virol.* 7:651
449. Wunner, W., Pringle, C. 1974. *J. Gen. Virol.* 23:97
450. Ghendon, Y., Markushin, S., Marchenko, A., Sitnikov, B., Ginzburg, V. 1973. *Virology* 55:305
451. Kawai, S., Hanafusa, H. 1973. *Proc. Nat. Acad. Sci. USA* 70:3493
452. Stephenson, J., Aaronson, S. 1973. *Virology* 54:53
453. Wong, P., McCarter, J. 1974. *Virology* 58:396
454. Sokol, F., Clark, H. 1973. *Virology* 52:246
455. Cartwright, B., Smale, C., Brown, F., Hull, R. 1972. *J. Virol.* 10:256
456. Nermut, M., Hermann, F., Schäfer, W. 1972. *Virology* 49:345
457. Neurath, R. A., Rubin, B. A. 1971. *Monogr. Virol.* 4:1
458. Corbett, M., Sisler, H., Eds. 1964. *Plant Virology.* Gainesville, Univ. of Florida Press
459. Beckendorf, S., Kim, J., Lielausis, I. 1973. *J. Mol. Biol.* 73:17
460. Beckendorf, S. 1973. *J. Mol. Biol.* 73:37
461. Terzaghi, E. 1971. *J. Mol. Biol.* 59:319
462. Dewey, M., Wiberg, J., Frankel, F. 1974. *J. Mol. Biol.* 84:625
463. Jarvik, J., Botstein, D. 1973. *Proc. Nat. Acad. Sci. USA* 70:2046

EUKARYOTIC NUCLEAR RNA POLYMERASES

✷896

Pierre Chambon

Institut de Chimie Biologique, Faculté de Médecine,
67085 Strasbourg, France

CONTENTS

INTRODUCTION

The control of transcription in nuclei of eukaryotic cells is undoubtedly a major mechanism of cellular regulation (for review, see 1). Although it was conceivable that, as in prokaryotes (see 2), a single enzyme could be responsible for the synthesis of all types of cellular RNA (3), a major advance has been the discovery of multiple forms of nuclear DNA-dependent RNA polymerase which differ in their structure, localization, and function. While our present picture of the control of transcription in eukaryotic cells is largely drawn from analyses of the RNA synthesized in vivo, our understanding of the regulation of transcription at the molecular level requires studies with cell-free systems. Since space is limited, this review will concentrate primarily on some recent advances in our knowledge of the multiple animal nuclear RNA polymerases. However, some relevant properties of RNA polymerases isolated from lower eukaryotes will also be mentioned. Reviews on mammalian (4), animal

613

(5), and eukaryotic (6) RNA polymerases should be consulted for historical background and for discussions of some important properties of the enzymes that will not be discussed here.

MULTIPLICITY AND NOMENCLATURE

The existence of multiple RNA polymerases is supported by several lines of evidence obtained independently in several laboratories. First, multiple peaks of RNA polymerase activity were eluted by chromatography of solubilized enzyme preparations on DEAE-Sephadex (7, 8) or DEAE-cellulose (9) columns (for other references, see 4–6, 10). Second, several classes of enzymes were distinguished according to the inhibitory effect of amanitin (9, 11–13). Third, structural analysis and immunochemical properties of the purified enzymes firmly established the multiplicity of RNA polymerases (see below).

The nomenclature of the multiple RNA polymerases is complex and confusing since different criteria have been used to classify the enzymes. One terminology (enzymes I, II, and III) was based mainly on the order of elution from DEAE-Sephadex of the three enzymic activities found in a variety of cells (7, 8, 14). Two difficulties are encountered with this nomenclature. First, additional enzyme activities were found that elute at lower salt concentrations than enzyme II. Second, the position of the peak corresponding to a given enzymic activity could change from one type of cell to another (15, 16). Furthermore, the elution of a specific enzyme may differ on DEAE-cellulose and on DEAE-Sephadex. The other terminology of RNA polymerases is based on an unequivocal and easily determined criterion: sensitivity to amanitin (5, 6, 9, 17, 18). Enzymes of class A are not inhibited by amanitin at concentrations up to 10^{-3} M (1 mg/ml), whereas enzymes of class B are inhibited by very low concentrations of amanitin (10^{-9} to 10^{-8} M) and class C enzymes are inhibited only at much higher amanitin concentrations (10^{-5} to 10^{-4} M). Table 1 lists the various animal RNA polymerase activities that have been isolated as distinct peaks after ion-exchange chromatography. The major change from our previous terminology (5, 6, 17) is related to the recent finding that several enzyme activities, including enzyme III, belong to class C.

Enzyme AI (or I) is defined as the class A-type enzyme activity that was highly purified from calf thymus (19, 20), mouse myeloma (21), and rat liver (22, 23). Chromatographic heterogeneity in class A RNA polymerase has been reported in rat liver nuclei (24), ascites tumor (25), *Xenopus laevis* (26), *Drosophila* (27), and HeLa cells (28, 45) after either phosphocellulose, CM-Sephadex, or DEAE-Sephadex column chromatography. Whether these additional peaks correspond to additional forms of class A enzyme can be questioned, since none have been purified and their resistance to high concentrations of amanitin was not tested except in the case of HeLa cells (28, 45). In rat liver, both class A enzyme activities appear to be of nucleolar origin (24).

The mixture of class B enzymes was extensively purified from calf thymus (17, 29–31), rat liver (30, 32, 33), KB cells (34), mouse myeloma (35), chick liver (36), and oviduct (36, 37), and subsequently resolved by repeated DEAE-cellulose

Table 1 Nomenclature and localization of animal DNA-dependent RNA polymerases[a]

Class of Enzyme	Author's Terminology		Other Terminology		Principal Localization
	Enzymes		Enzymes	Class of Enzyme	
A (9, 18) insensitive to amanitin	AI(a+b)	(19, 20, 24, 38)	I (21)	I (7, 8)	nucleolar
	AII	(24, 28, 45)	I_B		nucleolar
B (9, 18) sensitive to low concentrations of amanitin, 10^{-9}–10^{-8} M	B0	(17, 32, 38)	II_0 (35)	II (7, 8)	nucleoplasmic
	BI	(17, 29, 38)	II_A (35)		
	BII(a+b)	(17, 29, 38)	II_B (35)		
C (40, 42, 44, 46) sensitive to high concentrations of amanitin, 10^{-5}–10^{-4} M	CI	(28, 45)			?
	CII	(28, 45)			?
	CIIIa	(28, 45)	III_A (48)	III (7, 8)	nucleoplasmic
	CIIIb	(28, 45)	III_B (48)		cytoplasmic

[a] Numbers in parentheses refer to references. Enzymes that are underlined correspond to those for which the subunit pattern is known.

chromatography (29) or polyacrylamide gel electrophoresis (32, 35) into three forms: B0 (or II_0), BI (or II_A), and BII (or II_B) (38). The BII form was further resolved by polyacrylamide gel electrophoresis into two isoenzymes, BIIa and BIIb (38). An additional enzyme peak belonging to class B (or II) has recently been isolated from *X. laevis* by chromatography on DEAE-Sephadex (26, 39). This enzyme peak, which is eluted before the main peak of activity, is detected in whole cell extracts but not in purified nuclei, and thus may be of cytoplasmic origin, although nuclear leakage has not been ruled out. A similar enzyme peak belonging to class B was also isolated from whole HeLa cell extracts (28, 45).

The first class C enzymes to be characterized were cytoplasmic RNA polymerase activities in rat liver (40) and calf thymus (41). Although the rat liver class C enzyme activity behaves as a single component when subjected to DEAE-cellulose chromatography and elutes at the same place as class A activity, two enzyme peaks were obtained by chromatography on DEAE-Sephadex (42). The minor peak was eluted just after the peak of class A activity whereas the major peak was eluted at the same place as the peak of class B activity. Furthermore, the enzyme peak eluted at the higher ionic strength was also isolated from rat liver nuclei by chromatography on DEAE-Sephadex (42). A class C-type polymerase activity with chromatographic and amanitin sensitivity similar to those of rat liver class C enzyme was also found recently in the cytoplasm of Chinese hamster kidney (CHK) cells (43, 44). In addition, two peaks of class C-type enzyme activity (CI and CII), eluted between class A and B activity, were obtained by chromatography of whole HeLa cell extracts on DEAE-Sephadex (28, 45). In *X. laevis* oocytes (46) a major fraction of the nuclear RNA polymerase activity appears to belong to class C. The enzyme peak was eluted as a single peak from DEAE-cellulose column at the same place as class A activity. When chromatographed on DEAE-Sephadex two broad peaks of activity were obtained (46), one eluted at very low ionic strength and the other at approximately the place of enzyme CIII (see below).

RNA polymerase III was defined as the nuclear enzyme activity eluted after class B (or II) activity from DEAE-Sephadex columns (7, 8). Enzyme III has been found, usually as a minor component, in sea urchin (7, 8), rat liver (7, 14), amphibian oocytes and embryos (26, 39), human KB (47), and HeLa cells (28, 45). Mouse myeloma cells have recently been found to contain high levels of RNA polymerase III (48). Two chromatographic forms (III_A and III_B) have been purified (48, 49). Enzyme III_A appears to be nuclear in origin, whereas enzyme III_B is found mainly in the cytoplasm. Enzyme III was initially characterized as an amanitin-resistant enzyme (13, 26, 39, 47), since it was not inhibited by amanitin concentrations which completely block class B enzyme activity. However, recent studies have shown that mouse myeloma (48) and HeLa cell (28, 45) enzymes III are completely inhibited by high concentrations of the toxin. Furthermore, the inhibition curves of enzymes III are identical to those of the class C enzymes isolated from rat liver (40, 42), *X. laevis* (46), HeLa cells (28, 45), and CHK cells (43, 44) (50% inhibition at approximately 30 μg/ml). Therefore, enzyme(s) III belong to RNA polymerase class C as previously defined (5, 6, 40) and may tentatively be termed enzyme CIII as opposed to the two peaks of class C activity eluted before class B activity from a

DEAE-Sephadex column (28, 45). Although there is no doubt that, in the case of HeLa cells, enzyme CIII is distinct from the two class C enzyme peaks eluted from DEAE-Sephadex column between classes A and B, additional studies are required to assess whether the major peak of class C activity, isolated on DEAE-Sephadex column from rat liver (40, 42) and CHK cell (43, 44) cytoplasms, is in fact identical to enzymes CIII (48).

As first demonstrated by Sergeant & Krsmanovic (47), enzyme CIII is not detected after chromatography on DEAE-cellulose. This absence was explained when it was found that class C enzymes were eluted from DEAE-cellulose at a much lower ionic strength at the same place as enzyme AI (28, 45, 48). This coincidental elution of enzymes CIII and AI on DEAE-cellulose and the much higher activity of enzyme AI certainly explain why enzyme CIII was not previously detected in many tissues. The reason for this unique difference in behavior on DEAE-cellulose and DEAE-Sephadex is unknown. However, the discovery that, at least in some cells, enzymes of class C represent a significant fraction of the RNA polymerase activity and chromatography together with class A enzymes on DEAE-cellulose points to the necessity of further characterizing all enzyme peaks previously identified as class A-type enzymes. More specifically, the sensitivity of class A enzymes to high concentrations of amanitin and their behavior on DEAE-cellulose and DEAE-Sephadex columns should be systematically investigated. Such studies will certainly explain some of the discrepancies between the results published by different groups, for instance those concerning the characterization of the various polymerase activities isolated from X. laevis oocytes (26, 39, 43, 46, 50, 178).

In any case, the appearance of an additional peak of activity after ion-exchange chromatography may not represent a new enzyme. The new peak could correspond to a previously known enzyme, from which a subunit or a nonspecifically bound component has been dissociated. Furthermore, two peaks could also be obtained from the same enzyme if one of the subunits is present in the cell in limiting amounts. The present nomenclatures should therefore be considered merely as working tools, and it will not be possible to propose a fully satisfactory and systematic terminology before the molecular structure and the function of all the enzyme activities corresponding to the various chromatography peaks have been elucidated.

The terminology currently used for the multiple RNA polymerase activities isolated from other eukaryotes was originally based on the same criteria as those used for animal enzymes. Two types of enzymes, A (or I) and B (or II), were distinguished according to their sensitivity to amanitin or their elution pattern during ion-exchange chromatography. However, the concentration of amanitin required for inhibition of class B enzymes is usually higher by one or two orders of magnitude than that required to inhibit the animal class B enzymes (see 6). In some instances a third enzyme activity (class C or III) was isolated from yeast and lentil roots (see 6). It is unknown whether this enzyme activity actually belongs to an enzyme class similar to class C (as defined for animal RNA polymerases), since the possible inhibitory effect of very high concentrations of amanitin (up to 10^{-3} M) was not tested.

PURIFICATION AND PURITY

Detailed study of the various RNA polymerases requires significant quantities of highly purified enzymes. Although on a weight basis the amount of RNA polymerase activity in eukaryotic cells is much lower than in prokaryotes, and despite many difficulties related to solubilization and instability problems (for discussion, see 4–6), RNA polymerases belonging to classes A (or I) and B (or II) have been purified to a considerable extent from a variety of eukaryotic cells. However, even from calf thymus, which is a tissue rich in RNA polymerase (51), only a few milligrams of pure class A or B enzymes were obtained per kilogram of tissue (19, 20, 29, 31). This yield, which is one to two orders of magnitude lower than that of the *Escherichia coli* RNA polymerase, illustrates one of the difficulties encountered in studying the regulation of transcription at the molecular level in eukaryotes.

The purity of the most purified RNA polymerases from animal cells (19–21, 29, 30, 33–36) or yeast (52–55) is above 90%, as judged by electrophoretic analysis on polyacrylamide gels under nondenaturing conditions, and their specific activity is in the order of that of the purified *E. coli* RNA polymerase. Like many other nucleotidyltransferases, the animal RNA polymerases are metalloproteins requiring a tightly bound metal ion, possibly zinc (56). Both pure animal (19, 29, 34) and yeast (53, 55) enzymes lack measurable DNase, RNase, and RNase H activities. In experiments aimed at studying specific transcription in vitro the methods used for measuring the possible DNase contamination should be very sensitive, since initiation of RNA synthesis could readily occur nonspecifically at single strand breaks (57) (see below).

MOLECULAR WEIGHT

Purified AI RNA polymerase [calf thymus (18, 38), mouse myeloma (21)], rat liver AI and AII RNA polymerases (40, 58), purified class B RNA polymerases [calf thymus (18, 30, 38), rat liver (40, 58), and KB cells (34)], and class C RNA polymerase (40) sediment through glycerol or sucrose gradients at about 14–15S, faster than the *E. coli* core enzyme [mol wt 380,000–400,000 (59)]. Class B enzymes sediment slightly faster than A enzymes (38, 40), whereas class B and C enzymes are not resolved by centrifugation through sucrose gradients (40). These observations suggest a molecular weight of about 500,000. No drastic variation of the sedimentation rate was observed at different ionic strengths (34, 38).

Molecular weights of $550,000 \pm 10\%$, $600,000 \pm 10\%$, and $570,000 \pm 10\%$ were found for calf thymus AI, BI, and BIIa or BIIb, respectively, by electrophoresis in polyacrylamide gels of increasing porosity (38).

At high ionic strength, yeast RNA polymerases of classes A and B sediment faster than *E. coli* holoenzyme (52, 60, 61), suggesting molecular weights in the order of 500,000. However, a lower molecular weight (440,000) was found for yeast RNA polymerase B using polyacrylamide gels of graded porosity (60). At low salt concentrations, the sedimentation constant of yeast enzymes A and B increased to 24 and 21S, respectively (52, 60), suggesting that at low ionic strength these enzymes may form dimers like the bacterial enzyme but unlike the animal enzymes.

Sedimentation studies in sucrose or glycerol gradients also indicate that the molecular weights of RNA polymerase B (or II) of *Dictyostelium discoideum* (62) and RNA polymerase A (or I) of *Physarum polycephalum* (16) are in the order of 420,000–500,000. Maize RNA polymerase IIa (amanitin sensitive) (63) and soybean RNA polymerases I (amanitin resistant) and II (amanitin sensitive) (64) also sediment at about 16S, suggesting molecular weights around 500,000.

MOLECULAR STRUCTURE

The molecular structure of purified eukaryotic nuclear RNA polymerase has been investigated by polyacrylamide gel electrophoresis in the presence of sodium dodecyl-sulfate (SDS). In some instances (38), since possible charge differences cannot be detected by SDS polyacrylamide gel electrophoresis, the molecular structure of the various enzymes was further investigated by polyacrylamide gel electrophoresis in the presence of urea.

The molecular weight of the constitutive polypeptide chains was estimated from SDS polyacrylamide gel electrophoresis by comparison with the migration of markers of known molecular weights (65, 66). Although this method is accurate, its accuracy depends greatly on both the number of markers used for the calibration curve and the accuracy of the methods used for determining their molecular weights. It follows in particular that the values obtained for the large polypeptide chains are only indicative, since there are very few polypeptide chains with a molecular weight in the range of 120,000–200,000 which can be used for constructing a calibration curve. This certainly explains most of the discrepancies observed when comparing the molecular weights obtained for the various enzyme components in different laboratories. Such discrepancies, which are only related to differences in the slopes of the calibration curves, result in a systematic increase or decrease of the molecular weight estimates for all the components and should be distinguished from true discrepancies, which are indicated by nonsystematic variations, when comparing the molecular weights of the various polypeptide chains.

For convenience, the polypeptide chains isolated on polyacrylamide gels in the presence of SDS were called subunits of the various enzymes. However, some of them (particularly some of the polypeptide chains of low molecular weight) could correspond to polypeptide chains tightly bound to the enzyme rather than to true subunits. Dissociation-reassociation experiments will be necessary to assess the effective role of these subunits in RNA synthesis.

The molar ratios of the polypeptide chains in the various enzymes were estimated from the amount of dye adsorbed on each of them. For reasons discussed elsewhere (38) this method is not accurate, which certainly accounts for the differences observed when the molar ratios of the various components of the same enzyme were determined in different laboratories.

Subunit Structure of Animal RNA Polymerases

The molecular structure of calf thymus RNA polymerases AI, BI, BIIa, and BIIb has been studied in great detail (19, 29, 38, 67, 68). The overall pattern of the eukaryotic enzymes resembles that of prokaryotic enzymes (59), since each enzyme consists of two subunits of high molecular weight and several molecular subunits of lower molecular weights (Table 2). The molar ratio of the large subunits is one, whereas it is higher than unity for some of the smaller subunits. The data (38) are consistent with subunit models of $(SA1)_1 (SA2)_2 (SA3)_1 (SA4)_1 (SA5)_2 (SA6)_2, (SB1)_1 (SB3)_1 (SB4)_{1-2} (SB5)_2 (SB6)_{3-4}$, and $(SB2)_1 (SB3)_1 (SB4)_{1-2} (SB5)_2 (SB6)_{3-4}$ for calf thymus AI, BI, and BII enzymes, respectively. The molecular weights of enzymes calculated from their subunit composition are $501{,}000 \pm 10\%$, $538{,}000 \pm 10\%$, and $504{,}000 \pm 10\%$ for AI, BI, and BII enzymes, respectively. These values are in the range of the values expected from the sedimentation coefficients of the enzymes and, within the error limits, in rather good agreement with the molecular weight numbers obtained by electrophoresis in acrylamide gel of graded porosity (38).

Gissinger & Chambon have recently shown (20) that it is possible to split by CM-Sephadex chromatography the purified calf thymus enzyme AI into two fractions: a minor one, AIa, and a major one, AIb (in their order of elution). Fraction AIa, which lacks subunit SA3, is still enzymatically active on commercial calf thymus DNA, indicating that the SA3 component is not mandatory for RNA synthesis on this template. Since AIa is only a minor fraction of enzyme AI, it is presently unknown .whether a small fraction of SA3 is actually removed from the full AI enzyme during CM-Sephadex chromatography, resulting in the appearance of AIa, or whether SA3 is in fact present in a limiting amount in vivo, resulting in two populations of AI molecules which would then be separated by CM-Sephadex chromatography.

The calf thymus BII enzyme exists in two forms, BIIa and BIIb, which differ only in the charge of their largest subunits, SB2a and SB2b, respectively (38). The SB6 component of the class B enzymes consists in fact of two charge isomers, SB6a and SB6b (38). An additional component called SB5' was always associated with enzymes BI and BII, and polypeptide chains moving faster than subunits SA6 and SB6 were also seen on SDS gels (38). As pointed out above, reconstruction of the enzymes from their isolated polypeptides is required to ascertain whether these additional polypeptide chains are actually part of the enzyme structure. The structure of purified calf thymus class B RNA polymerases was also analyzed to some extent by Weaver et al (30) and by Ingles (31). Although the latter author found a subunit composition suggesting the presence of the two forms, BI and BII, the former group found only one predominant form similar to enzyme BII (for discussion, see 5 and 38). It is interesting that AI and B enzymes contain two pairs of small subunits of identical molecular weight and charges, SA5 and SB5, SA6 and SB6a, since it raises the possibility that there could be a common pool of low molecular weight subunits for AI and B enzymes. Fingerprint studies are required to ascertain that these subunits are, in fact, identical.

The subunit pattern of the purified nucleolar mouse myeloma (21, 49) and rat

Table 2 Subunits of calf thymus RNA polymerases AI and B (19, 20, 29, 38, 67)

Form AIb		Form AIa		Form BI		Form BIIa or BIIb	
Subunit	Molecular weight[a]	Subunit	Molecular weight[a]	Subunit	Molecular weight[a]	Subunit	Molecular weight[a]
SA1	197 (1)[b]	SA1	197 (1)	SB1	214 (1)	—	—
SA2	126 (1)	SA2	126 (1)	—	—	SB2a ⎫ or SB2b ⎭	180 (1)
SA3	51 (1)	—	—	SB3	140 (1)	SB3	140 (2)
SA4	44 (1)	SA4	44 (1)	SB4	34[b] (1–2)	SB4	34 (1–2)
SA5	25 (2)	SA5	25 (2)	SB5	25 (2)	SB5	25 (2)
				SB5'	20 (1)	SB5'	20 (1)
SA6	16.5 (2)	SA6	16.5 (2)	SB6 (a + b)	16.5 (3–4)	SB6 (a + b)	16.5 (3–4)

[a] Molecular weight: daltons $\times 10^{-3}$

[b] Numbers in parentheses correspond to molar ratios.

liver (22, 23) RNA polymerase AI (or I) is very similar to that of the corresponding calf thymus enzyme. Two large subunits (corresponding to the calf thymus SA1 and SA2 subunits) and several smaller components (analogous to the calf thymus SA3, SA4, SA5, and SA6 subunits) were found. The rat liver enzyme AI (22) was resolved by phosphocellulose chromatography into two enzyme forms which are both enzymatically active and differ only by the absence of subunit SA3 in the enzyme that is eluted first from phosphocellulose. The mouse myeloma enzyme AI (21, 49) was also resolved into two components by polyacrylamide gel electrophoresis under nondenaturating conditions. The slower, enzymatically active band contains the six subunits, whereas the faster, enzymatically inactive band is lacking the subunit corresponding to the calf thymus SA3 subunit and is therefore analogous to the calf thymus AIa enzyme. However, since the yield of enzymic activity eluted from the gel was poor, no definitive conclusion can be drawn about the possible involvement of this subunit in the transcription process.

The rat liver class B enzymes consist of at least three enzyme forms, B0, BI, and BIIb (32, 38, 67). BI and BIIb are identical in every respect with the corresponding calf thymus enzymes, while the only difference between B0 and the other B enzymes lies in the molecular weight of its largest subunit, SB0 (220,000 daltons). The subunit analysis of the rat liver class B enzyme (or II) purified by Weaver et al (30) suggests the presence of two forms of class B activity possibly similar to enzymes BI and BII (see 5, 6, and 38). The mouse myeloma enzymes B were recently purified (35). Their subunit analysis reveals the presence of enzyme forms similar to the calf thymus enzymes BI and BII and rat liver B0 enzyme. It is unknown whether the myeloma enzyme BII can be split into two components corresponding to the BIIa and BIIb forms of calf thymus class B enzymes. Within the experimental error of determining molecular weight, all the components found in calf thymus or rat liver class B enzymes have their counterparts in mouse myeloma class B enzymes. Although the mouse myeloma class B enzymes appear to contain some additional components of low molecular weight, it is interesting that, as in the case of the calf thymus enzymes, two subunits of low molecular weight appear to be common (at least with respect to molecular weight) to mouse myeloma enzymes AI (or I) and B (or II). The analysis of the class B enzymes purified from chick oviduct (36, 37) and liver (36) indicates very little change of these enzymes during evolution. Chick oviduct class B enzymes consist of three enzyme forms, BI, BIIa, and BIIb, whose subunit patterns are identical to the corresponding calf thymus enzymes (36). Unlike rat liver, the purified chick liver enzyme B contains only one enzyme B form corresponding to calf thymus, rat liver, and chick oviduct enzyme BIIb (36). Purified polymerase B from KB cells (34) and human placenta (69) consists also of two major components with molecular weights of about 200,000 and 140,000, a minor component of molecular weight 170,000–180,000, and additional components of lower molecular weight.

Enzyme CIII (or III) was recently purified from *X. laevis* oocytes (49) and mouse myeloma (48, 49). The molecular structures of enzymes CIII from both sources are identical. They contain two large subunits of molecular weight 155,000 and 138,000 (49) and several components of lower molecular weight. Two of these components have molecular weights identical with the two pairs of polypeptide chains (SA5 and SB5, SA6 and SB6a) which are already common to AI and B enzymes, while two others have molecular weights identical with the SB4 component of B enzymes and the SA4 component of AI enzyme, respectively. Although the results concerning the low molecular weight component have to be confirmed (49), they suggest that some subunits of low molecular weight could be common to enzymes of all classes. On the other hand, the 138,000 dalton subunit of enzyme CIII appears to be clearly different from the 140,000 dalton subunit of enzymes B (or II), as shown by mixing the two enzymes prior to electrophoresis in the presence of SDS (49). Although the mouse myeloma CIII (or III) enzyme activity can be resolved by DEAE-Sephadex chromatography into two components, CIIIa (or III_A) and CIIIb (or III_B), their subunit pattern was undistinguishable after gel electrophoresis in the presence of SDS (49), suggesting that the difference between the two CIII forms could lie only in charge differences as previously noted for the calf thymus BIIa and BIIb enzymes (see above). In any case, these results indicate that the molecular structures of the CIII enzymes are very similar irrespective of their intracellular origin, since enzyme CIIIa was found predominantly in the nucleus of myeloma cells, whereas enzyme CIIIb was found mainly in the cytoplasm of the same cells (48). None of the other class C enzyme activities found in the nucleus or in the cytoplasm (28, 40–47) have been sufficiently purified at the present time to permit meaningful structural analysis. Since the class C enzyme activities found in the cytoplasm of various cells present a chromatography behavior very similar to that of the nuclear CIII enzyme (26, 39, 42–48), it would be of particular interest to investigate whether their molecular structures are similar with those of the myeloma CIII enzymes.

Immunological Properties of Animal RNA Polymerases

An antiserum to class B enzyme of calf thymus was obtained in hens (31), although all attempts to obtain such an antiserum in rabbits repeatedly failed (31, 38). This antiserum precipitates and inhibits to the same extent both the calf thymus and the rat liver class B RNA polymerases, in keeping with their striking structural analogies. Partially purified RNA polymerases form B from X. laevis and Tetrahymena pyriformis were also inhibited but to a lower extent, indicating a somewhat reduced crossreactivity (31). In these experiments, calf thymus RNA polymerase AI was also inhibited to some extent (31), whereas in other experiments (C. J. Ingles, personal communication) no inhibition occurred.

Antibodies to calf thymus RNA polymerase AI specifically precipitated calf thymus enzyme AI and failed to give precipitation lines with either E. coli RNA polymerases or partially purified or pure calf thymus class B enzymes (5, 38). In addition these antibodies completely inhibited partially purified or pure calf thymus AI preparations, while the pure calf thymus class B enzymes and E. coli RNA polymerase were not inhibited (38). Since structural studies suggest that there could be a common pool of low molecular weight subunits (see above), one has to consider that these low molecular weight subunits are in fact different, that the anti-AI antibody preparation does not contain antibodies to SA5 and SA6 subunits, or that these low molecular weight polypeptide chains are not required for the transcription of a "common" preparation of calf thymus DNA containing nicks and single stranded regions. Antibodies directed against the various subunits are required to distinguish between these various possibilities. Crude preparations of AI activity from rat liver and chick oviduct are also inhibited by antibodies to calf thymus enzyme AI although to a lower extent than calf thymus RNA polymerase AI, suggesting that RNA polymerases AI from widely different animals are structurally related. This result is not unexpected, since RNA polymerase AI is involved in the synthesis of ribosomal RNA (see below). What is in fact surprising is that there is less similarity between the AI enzymes of two different species than between their class B enzymes, which are supposed to synthesize widely different messenger RNAs. This observation and the striking structural similarities of the class B enzymes argue against the view that the transcription of some genes specifically expressed in some tissues could be regulated by different forms of class B RNA polymerases, which would be tissue specific.

Although pure class B enzymes from calf thymus, rat liver, or chick oviduct are not inhibited by antiserum to calf thymus enzyme AI, crude class B enzyme preparations from the same tissues are inhibited, but to a lower extent than the corresponding AI enzymes. This inhibition, which gradually disappears as the purification progresses, suggests that the class B enzymes could contain a loosely bound component, lost during their purification and structurally related to one of the polypeptide chains present in pure calf thymus AI enzyme. In this respect it is interesting that the three class C enzyme peaks isolated from HeLa cells (28, 45) are also inhibited, and all to the same extent, by the antiserum to calf thymus RNA polymerase AI, suggesting not only that they are structurally related but also that there are some structural similarities between AI and class C enzymes. Studies with antibodies of the various subunits are required to elucidate the basis of these crossreactivities and to assess whether they reflect the effective presence of common antigenic determinants in the various RNA polymerase classes.

Possible Relationship Among Various Animal RNA Polymerases

The studies mentioned above firmly establish on structural ground the multiplicity of nuclear RNA polymerases, which was initially suggested by their catalytic and chromatographic properties (7, 8) and by their different sensitivity to amanitin (9). The striking structural resemblance among class B RNA polymerases purified from a variety of sources and their closely related immunological properties suggest that very little change has occurred in the structure of these enzymes during evolution. In this respect it is striking that all of the class B animal RNA polymerases so far studied have the same affinity for the amatoxins (70, 71).

Although the structural and immunological studies suggest that some subunits could be common to the class A, B, and C RNA polymerases, it is likely that most of the subunits of these three classes of enzymes are coded by distinct genes. An interconversion among the various classes (72) is therefore very unlikely. However, interconvertibility is a real possibility for a given class of enzymes. For instance, the structural difference among the class B enzymes appears to lie in only one subunit, the heavier of the two large subunits. It has been suggested that enzymes BII could be artifacts derived in vitro from enzyme BI by proteolysis of subunit SB1 during purification (30, 68). This hypothesis is unlikely (29, 31, 34, 36, 38). Whether the various class B enzymes are derived in vivo from a common precursor (B0?) by specific proteolysis of the highest molecular weight subunit or whether they have their largest subunit coded by different genes is still to be established. Although BI and BII enzymes are inactivated differently by increasing temperature or urea concentration (73), no distinct functional properties have yet been found for these two enzyme forms (74).

Subunit Structure of RNA Polymerases from Lower Eukaryotes

Yeast RNA polymerases of classes A and B have been purified to apparent homogeneity and their subunit pattern has been analyzed (52, 53, 55, 75). RNA polymerase of class A contains two large subunits (190,000 and 135,000 daltons) in equimolar amount and several smaller subunits (48,000, 44,000, 37,000, 29,000, 24,000, 20,000, 16,000, and 14,000 daltons), while RNA polymerase of class B also contains two large subunits in equimolar amount (180,000 and 150,000 daltons) and a number of components of lower molecular weight (46,000, 41,000, 34,000, 29,000, 24,000, 18,000, 16,000, and 14,000 daltons). Immunological studies (75) have shown that the largest subunits of class A and B enzymes are unrelated. On the other hand, four small polypeptide chains appear to have the same molecular weight in the two enzymes. Fingerprint analysis indicates that the subunits of 29,000, 24,000, and 16,000 daltons are identical, respectively, in the two enzymes, whereas the subunits of 14,000 daltons are clearly different (75). The similarity with the animal enzymes is striking and suggests that in yeast there is a common pool of some of the small molecular weight subunits. The identity of these small subunits in yeast RNA polymerase of classes A and B is in keeping with immunological studies (76) indicating that class A (or I) and class B (or II) of yeast RNA polymerase share some common antigenic determinants. Purification of yeast class A enzyme on a column of Sepharose-bound antibodies shows that the polypeptide chains retained on the column correspond to the subunit pattern of pure class A enzyme with two exceptions corresponding to the polypeptides of 48,000 and 37,000 daltons, which were not retained on the column. Polyacrylamide gel electrophoresis of pure class A enzyme and phosphocellulose chromatography indicate that these two components are in fact reversibly associated to the enzyme (75). They are not required for the transcription of poly(dA-T), but are in some way involved in the transcription of calf thymus DNA. Although these observations are reminiscent of the behavior of the bacterial σ factor, there are no indications at present concerning the function of these proteins. In addition, yeast class A enzyme deprived of these two components is inhibited by a high concentration of amanitin (75). This surprising finding raises the question of a possible relationship between yeast class A and C RNA polymerases.

The molecular structures of purified slime mold class A (or I) (16, 77, 78) and B (or II) (62, 77) enzymes consist also of two subunits of high molecular weight in a 1:1 ratio and several components of lower molecular weight. The large subunits of class A enzyme appear to be different from those of class B enzyme (77). The subunit patterns of the nuclear amanitin-sensitive and -resistant enzymes from higher plants are also similar to those of other eukaryotic nuclear RNA polymerases (79–82).

INTRANUCLEAR LOCALIZATION AND FUNCTION OF THE MULTIPLE POLYMERASES

In animal cells RNA polymerases AI and B0, BI and BII have been detected primarily in nucleolar and nucleoplasmic fractions, respectively (Table 1; for other references, see 4–6), whereas RNA polymerases of the C class have been found in both the nucleoplasmic and cytoplasmic fractions. However, in this latter case, the possibility of nuclear leakage was not ruled out (40, 42–44, 48). Yeast RNA polymerase of class A appears to be located in the yeast nuclear regions (dark crescents) which would correspond to nucleoli (83). *Physarum polycephalum* (77, 84), *Dictyostelium discoideum* (85), and *Acetabularia* (86) enzymes A are also preferentially associated with nucleoli.

The preferential nucleolar localization of the amanitin-resistant RNA polymerase of class A immediately suggested that it could be responsible for the synthesis of rRNAs, while the nucleoplasmic localization of the other enzymes suggested that they might be involved in the synthesis of the other RNA species. This hypothesis has been tested by studying either the effect of administration of amanitin on the synthesis of the various RNA classes in vivo or the effect of amanitin on RNA synthesis catalyzed in vitro by isolated nuclei. This latter approach was very successful and led to the demonstration that each of the enzyme classes has a specific role in the transcription of the genetic information.

The involvement of class B RNA polymerases in the synthesis of heterogeneous nuclear RNA, a part of which is presumably the precursor of messenger RNA, was demonstrated by incubating isolated nuclei in the presence of a concentration of amanitin which completely inhibits the purified class B enzymes (14, 87–89). At present nothing is known about possible distinct functions of the various class B enzymes. Although there has been no demonstration of the direct involvement of a class B RNA polymerase in the synthesis of a specific cellular messenger, several reports indicate that class B RNA polymerases are involved in the synthesis of viral mRNA both early and late during infection with animal DNA viruses (90–95). Furthermore, class B RNA polymerase appears to be responsible for the transcription of the viral DNA in isolated nuclei prepared from cells infected with a RNA tumor virus (96, 97). Direct evidence that a host cellular class B enzyme is involved in the transcription of polyoma virus DNA was obtained by infecting amanitin-resistant mutant cells altered in class B RNA polymerase (98). In addition, using hybrid cells, it was shown that the class B polymerase of a nonpermissive cell line is capable of transcribing the viral genome, suggesting that the limitation of viral growth in unsusceptible cells is not at the polymerase level (98).

The synthesis of rRNAs was not affected by incubating nuclei in the presence of low concentrations of amanitin (87, 89). This observation suggested that the nucleolar class A RNA polymerase was in fact responsible for the synthesis of the rRNAs, but it did not eliminate a possible involvement of the class C (or III) enzyme in rRNA synthesis. The more recent discovery of the inhibition of the class C enzymes by high concentrations of amanitin under conditions where the A class is completely resistant led to the unequivocal demonstration that a nucleolar class A enzyme is responsible for the synthesis of rRNAs (99).

Circumstantial evidence suggested that in isolated nuclei a nucleoplasmic RNA polymerase activity exhibiting some properties of class C (or III) enzyme might catalyze the synthesis of 5S and pre-4S RNAs (the precursor of cytoplasmic tRNA) (87, 88), but only the discovery of the sensitivity of class C enzymes to high amanitin concentrations permitted the conclusive demonstration of the role of class C RNA polymerase in the synthesis of 5S and pre-4S RNA species in isolated nuclei (99). Since the cell appears to contain several class C enzymes, all of which have the same pattern of inhibition by amanitin, it is not yet possible to assign a specific role to the various class C enzymes in the transcription of pre-4S and 5S genes. A class C RNA polymerase is also involved in the transcription of a virus-associated 5.5S RNA late in the productive infection of KB cells with adenovirus 2 (88, 95). There is, in addition, some circumstantial evidence that an enzyme activity which might belong to class C is involved in the synthesis of some heterogeneous nuclear RNA (87, 111, 124).

Several lines of evidence obtained after in vivo administration of amanitin also support the idea that different genes or groups of genes are transcribed by different enzymes. First, biochemical, autoradiographic, and electron microscopic studies (100–114; see 5) indicate that the synthesis of giant heterogeneous nuclear RNA is inhibited when amanitin is administered in vivo or when polytene chromosomes are incubated in its presence. Second, the possible involvement of class B (or II) RNA polymerases in the synthesis of giant heterogeneous nuclear RNA is upheld by the isolation of cell mutants that contain altered class B RNA polymerases resistant to amanitin and growing normally in the presence of amanitin (115, 116). In addition, these studies indicate that the regulation of class B RNA polymerase activity is independent of the regulation of class A (or I) activity and suggest that the level of class B RNA polymerase could be autogenously regulated, as seen for holoenzyme in bacteria (116).

Contrary to expectations, the synthesis of ribosomal RNA precursors and of rRNA is rapidly inhibited in rat and mouse liver after in vivo administration of amanitin (100–102, 107, 117, 118). This inhibition is accompanied by fragmentation of liver nucleoli (107, 113, 114, 119–122). But, at the same time, the in vivo treatment with amanitin does not result in any significant inhibition of the class A activity in isolated nuclei (11, 102, 107, 118), although there is a decrease in the amount of class A activity associated with the nucleoli (100, 123). Several explanations were proposed to account for the inhibition of rat and mouse ribosomal RNA precursor synthesis after administration of amanitin in vivo, despite the lack of inhibition of both the purified AI enzyme and the amanitin-resistant activity measured in isolated nuclei of amanitin-treated rats (4, 5, 107, 123). It appears, however, that in this respect liver could be a special case, possibly related to the very high concentrations of amanitin achieved in this organ after in vivo administration of the poison. Ribosomal RNA synthesis is indeed not inhibited by amanitin in nucleoli of *Triturus* oocytes (112) or in nucleoli of polytene chromosomes of salivary glands (109–111, 120). Furthermore, amanitin does not affect RNA synthesis in growing oocytes of *X. laevis* at stage four, when more than 90% of the synthesized RNA is ribosomal (125, 126). Moreover, when amanitin is added to cells in culture, ribosomal RNA synthesis is much less affected than heterogeneous nuclear RNA synthesis (104, 105, 108). Kedinger & Simard (108) have conclusively shown, using both biochemical and autoradiographic analysis, that at early times after the addition of amanitin to Chinese hamster ovary (CHO) cells, the synthesis of rRNAs and their precursors is not affected, but the synthesis of giant heterogeneous nuclear RNA is completely depressed. In addition, these authors observed that the integrity of the nucleoli is not required for rRNA synthesis, since this synthesis was still very active when the nucleoli were highly fragmented (108). Although these studies were most likely carried out at intracellular concentrations of amanitin too low for inhibiting the class C polymerases, their results support the hypothesis that an RNA polymerase, insensitive to amanitin and most likely belonging to class A, is actually responsible for the in vivo synthesis of the ribosomal RNAs.

Several reports have indicated that the synthesis of the 4 and 5S RNAs is not inhibited when dipteran salivary glands are incubated in the presence of low concentrations of amanitin (109, 111, 120). Similarly, synthesis of the 5S RNA and pre-4S RNAs was only slightly, if at all, affected after in vivo administration of amanitin to mice or rats (107, 127) or after addition of amanitin to cells in culture (108). Although these studies were carried out at concentrations of amanitin too low to differentiate between the A and C class enzymes, they demonstrate that pre-4S and 5S RNAs are not synthesized by class B enzymes. Taken together with the results obtained in isolated nuclei (99), they strongly suggest that enzymes belonging to class C are in fact responsible in vivo for the synthesis of pre-4S and 5S RNAs.

GENERAL PROPERTIES OF EUKARYOTIC RNA POLYMERASES

The general properties of eukaryotic RNA polymerases will not be discussed here, since they have recently been extensively reviewed (5, 6). One important feature of the eukaryotic enzymes should be stressed, however. It is possible, using an excess of commercial calf thymus DNA, to characterize the three classes of eukaryotic RNA polymerases according to some of their catalytic properties (preferential activation by Mn^{2+} or Mg^{2+}, ionic strength optimum, preferential transcription of native or denatured DNA) (7, 8, 48; for other references, see 4–6). However, these catalytic properties should not be considered as invariant characteristics of the three enzyme classes. In particular, the optimum ionic strength and the relative stimulating effects of Mn^{2+} and Mg^{2+} are highly dependent on the nature of the DNA (natural or synthetic), the state of the DNA (native or denatured), and its concentration in the reaction (45, 48, 73, 128–130).

IN VITRO SELECTIVE TRANSCRIPTION BY EUKARYOTIC RNA POLYMERASES

The existence of multiple RNA polymerase forms with distinct subcellular localizations and specific functions suggests that gene expression in eukaryotic cells might be regulated at least in part via distinct enzymes which would specifically transcribe different genes or classes of genes. A possible mechanism to accomplish this consists of the specific recognition by the different polymerases of sequences which would be characteristic of a class of genes, although there is at present no evidence that the eukaryotic genomes contain regulatory sites similar to the prokaryotic promoter and termination regions (2, 131). Nevertheless, the similarity between the overall molecular structure of the prokaryotic and eukaryotic RNA polymerases suggests that some of the regulatory mechanisms operating in the prokaryotic cells might have their counterpart in eukaryotic cells and that the different eukaryotic enzymes could therefore play a positive role in regulating the expression of the genetic information.

In order to investigate in vitro whether the specificity of transcription is governed by mainly DNA-RNA polymerase interactions, one needs, in addition to the purified enzymes, an appropriate DNA template and a method to accurately measure the possible selective transcription. There are two difficulties in performing such experiments. First, eukaryotic genomes are very complex (1). Only a small fraction of the DNA is likely to code for messenger RNA and, for higher eukaryotes, the DNA sequences corresponding to a given "unique" gene represent less than 1×10^{-6} of the total DNA sequences. Since breaking the DNA during its isolation is unavoidable and since the enzymes can bind to and initiate RNA synthesis at these breaks (see below), the problem of the enzyme locating a given gene is enormous. In principle, this situation could be improved by using chromatin instead of DNA as template, since the number of sites for RNA synthesis appears to be restricted in chromatin as compared to deproteinized DNA (132). Thus, chromatin might contain components helping the enzyme to locate a promoter site. The situation is more favorable with reiterated genes (e.g. ribosomal, 5S, and histone genes), particularly when it is possible to obtain a DNA fraction greatly enriched in these reiterated genes (133–136). An alternative approach to the problem of studying the selectivity of in vitro transcription consists of using appropriate DNAs from animal viruses as templates. Transcription studies of bacteriophage DNAs have been extremely useful for elucidating some of the mechanisms involved in the regulation of gene expression in prokaryotes (2, 131, 142), and animal viruses might offer a model system in which the regulation of the expression of a few genes could be accomplished by mechanisms that resemble those of the host cell (129, 137, 138). The second problem lies in the difficulty of assessing the selectivity of transcription (for further discussion of this problem, see 131). One first wants to know whether a given RNA is present in the collection of synthesized RNAs. This can be achieved by qualitative and quantitative DNA-RNA hybridization [with rDNA or 5S DNA when one is exploring the selective transcription of these genes, or with the DNA complementary (cDNA) to a given messenger RNA when one is interested in the transcription of a protein-coding gene]. However, in order to assess whether the transcription is actually selective, one has to investigate whether the synthesized RNA molecules are initiated and terminated at the proper sites (which can be achieved by analyzing the RNA sequence at the 5′ and 3′ ends) and also whether only RNA species found in vivo are transcribed in vitro (which can be explored to some extent by analyzing whether the sense strand of the DNA only is transcribed). Needless to say, in the case of in vitro study of transcription of animal messenger RNAs such a characterization of selective transcription is at present beyond our capability for many reasons, not the least being that in no case is the structure of the primary transcript corresponding to a given messenger known [this is unfortunately true even for viral messenger RNAs (for discussion, see 137)]. Again the situation is much more favorable for the ribosomal and 5S genes, since it is possible in this case to explore the strand selectivity of the in vitro transcription. However, as we will see, and despite a considerable amount of work, there is at present no convincing evidence that the selective transcription operating in vivo is governed by specific RNA polymerases-DNA or -chromatin interactions that can be reproduced in vitro.

The Problem of Initiation of RNA Synthesis on an Intact Double Stranded DNA

Commercial DNA preparations of cellular DNA are often used as templates to study whether the purified eukaryotic RNA polymerases can in fact initiate RNA synthesis on a double stranded DNA or whether additional initiation factor(s) similar to the E. coli σ initiation factor is required. These native DNA preparations usually correspond to DNA molecules of $5–20 \times 10^6$ daltons that contain single strand breaks (nicks) as well as denatured regions. When these DNAs, for instance calf thymus DNA, were used as templates for RNA polymerases isolated from animal cells (8, 10, 24, 34, 57, 73, 139), yeast (130; for other references, see 6), slime molds (16, 62, 78), and higher plants (for references, see 6), it was generally found that enzymes AI (or I) prefer native DNA, whereas class B enzymes (or II) prefer denatured DNA and class C enzymes (or III) transcribe equally efficiently native and denatured DNAs. However, assays in which only the amount of synthesized RNA was measured did not give any information concerning the nature of the DNA regions where RNA chains are initiated. In fact, it was shown for both animal (37, 73) and yeast (140) class A and B RNA polymerases that prior treatment of the DNA with an endonuclease specific for single stranded DNA decreases its template efficiency, suggesting that initiation of RNA chains was occurring mainly at gaps or nicks. Removing the unpaired 3′-hydroxyl ends of DNA molecules also decreases their template efficiency for class A and B polymerases of rat liver (141). Moreover, the lower template efficiency of denatured calf thymus DNA for enzyme AI is related to an inefficient elongation step and not to an inability to initiate on this template, since chain initiation occurs much more readily on denatured than on native calf thymus DNA (73). It is therefore clear that the question of whether the purified eukaryotic RNA polymerases can initiate RNA synthesis on a double stranded DNA cannot be answered with such ill-defined preparations of DNA. Phage DNAs are obvious templates to use in answering this question, since they are relatively easy to prepare in a native and intact state and because their in vivo and in vitro products of transcription by the *E. coli* enzyme are well characterized (2, 131, 142). None of the purified eukaryotic RNA polymerases are able to efficiently transcribe phage DNAs (14, 18, 54, 57, 62, 69, 140, 143). The basis for the restriction of RNA synthesis clearly lies at the initiation level, since the introduction of nicks or gaps converts phage DNAs to efficient templates for calf thymus (18, 57) and yeast (140) class A and B RNA polymerases. The reason the yeast enzymes require gaps while the mammalian enzymes can initiate RNA chains at nicks is unknown. These observations are very important because they illustrate the absolute necessity to use a DNA template as intact as possible and a polymerase preparation completely free of DNase in any study investigating a possible selective in vitro transcription (144).

Since phage DNAs are obviously nonphysiological templates for eukaryotic RNA polymerases, there are two possibilities to account for the above results: either the purified enzymes are lacking a general initiation factor or they are unable to recognize the initiation sites in the DNA of these too remote organisms. In an attempt to distinguish between these two possibilities, an intact linear double stranded animal viral DNA (adenovirus 2 DNA) and a high molecular weight DNA preparation from African monkey cells (CV1) devoid of single stranded breaks (145) were used as templates for purified calf thymus RNA polymerases AI and B (146, 147). Adenovirus 2 DNA was essentially not transcribed by either enzyme and the high molecular weight CV1 DNA was a very poor template. Adenovirus 2 and 5 DNAs were also inefficient templates for purified KB cell class A and B RNA polymerases (34, 128). The inability of the purified calf thymus RNA polymerase AI and B to transcribe CV1 DNA efficiently is not related to a species-specific selectivity, since calf thymus high molecular weight unnicked DNA preparations are also poorly transcribed by the purified calf thymus enzymes (73; M. Gross-Bellard, unpublished observation). In all cases RNA synthesis is blocked at the initiation step (6, 146, 147). Several hypotheses could account for these results: 1. Initiation proteins, similar to the *E. coli* σ protein, could be lost during the purification of the enzymes. In this respect, it is possible that class C enzymes might be different from class A and B enzymes, since some recent results suggest that animal class C RNA polymerases can in fact initiate RNA synthesis on intact linear double stranded DNAs. Further work is required to assess this important observation since it was made with enzymes only partially purified (28, 45, 46); 2. The very limited RNA synthesis on cellular DNA could be physiologically relevant. This synthesis could correspond to initiations at a small number of initiation sites laboriously located by the enzymes; and 3. The eukaryotic RNA polymerases could be intrinsically incapable of initiating RNA synthesis on a naked intact double stranded DNA. In this case initiation would occur only when the secondary structure of DNA is altered. Since the natural template of the enzymes is never a naked DNA but rather a complex deoxyribonucleoprotein structure (chromatin), this alteration could be achieved by the introduction of specific single stranded breaks (148, 149), by the binding of some chromatin regulatory factors [protein or RNA (1, 150–152)], or by the supercoiling of the DNA within the chromatin, which would result in loops of single stranded DNA (153). In this respect, it is worth recalling that the purified animal enzymes can readily initiate RNA synthesis on the superhelical but not on the open circular or linear DNA of Simian viruses 40 (129, 137, 138). I will briefly discuss these three possibilities.

Stimulatory Proteins (Factors)

A number of protein fractions, often called factors, have been found which stimulate the transcription of native DNA by class B (or II) animal enzymes (34, 154–165), class A (or I) animal enzymes (23, 25, 166, 167; A. G. Lezius, personal communication), yeast class A, B, or C (or III) enzymes (168, 169), and plant RNA polymerases (170). Some of these stimulatory proteins were extensively purified (169, 171, 172), free of nucleases, and appear to consist of a family of basic proteins. There was no definite indication that any of these proteins act at the initiation step of transcription. In fact, when the mechanism of action of these proteins was more extensively investigated, it was found that they act primarily by stimulating chain elongation (6, 158, 171, 172). It seems clear that these proteins act rather nonspecifically and that none of them plays a specific role in initiation as does the bacterial σ protein. In this respect, it should be kept in mind that many basic proteins including histones stimulate the bacterial (173) and yeast (169) RNA polymerases when added at appropriate concentrations.

Analysis of Selectivity of Transcription on Deproteinized DNA

The different animal (14, 48, 69, 158) and yeast (52, 54, 82, 140, 174) RNA polymerases transcribe various synthetic templates with widely different efficiencies. For instance, poly(dA-T) is a much more effective template for enzymes CIIIa and CIIIb than for class A and B enzymes (48). It is unknown whether these template preferences correspond to a specific recognition of nucleotide sequences, which would be present at regulatory sites in the natural DNAs, or to much less specific mechanisms related to the particular secondary structure of the various synthetic templates. Asymmetric transcription was observed with some very simple synthetic templates, indicating that mechanisms other than specific recognition of a promoter nucleotide sequence could be responsible for strand selection during in vitro transcription. It is therefore clear that asymmetric transcription per se could suggest, but not prove, that selective transcription had occurred in vitro.

Although there is some indication that under certain conditions the calf thymus AI and B and E. coli holoenzyme RNA polymerase initiate at different sites on calf thymus DNA (74), there is at present no firm evidence that any of the eukaryotic RNA polymerases exhibit template specificity and accurately select the sequences to transcribe on a naked DNA. Polymerase I (or AI), polymerase II (or B) (175), and polymerase III (or CIII) (39) did not show any striking preference for X. laevis rDNA vs bulk DNA, and neither enzyme specifically transcribed the regions of the rDNA that are transcribed in vivo. It is possible, however, that in these experiments initiation at single stranded breaks or at ends of the molecules obscured the pattern of transcription (176). When a high molecular weight *X. laevis* DNA preparation enriched in ribosomal DNA was used as template for *X. laevis* ovarian RNA polymerase A and B, the class A polymerase transcribed rRNA coding sequences, whereas the form B did not (43, 50, 177). However, these experiments cannot be taken as evidence that the solubilized class A enzyme transcribed selectively the ribosomal genes in vitro, since 1. the strand selectivity was not investigated, and 2. the class A enzyme preparation was only partially purified [containing for instance a large amount of RNA polymerase C (178)]; it is therefore impossible to know whether the preferential transcription of the rDNA is related to its specific recognition by class A enzyme or to the presence of contaminating proteins interacting with the DNA. On the other hand, partially purified class C enzyme from *X. laevis* ovaries (46) transcribed only the sense strand (H strand) of purified rDNA of *X. laevis* (179). The physiological relevance of this result is at present unknown, since at least in somatic cells, class C enzymes are not involved in the synthesis of rRNA (see above). There is also some suggestion that yeast class A (or I) enzyme, but not class B (or II) enzyme, could display some template selectivity when transcribing rDNA in vitro, since hybridization experiments suggest that it transcribes with a twofold greater efficiency the L strand, the strand transcribed in vivo, than the H strand of γ DNA (yeast DNA enriched in rDNA) or total nuclear DNA (180). However, since the strand selectivity was not very striking and the actual efficiency of hybridization was not measured, these experiments cannot be taken as evidence for a selective in vitro transcription of the ribosomal genes by yeast class A RNA polymerase. Comments similar to those already made for the *X. laevis* and yeast systems apply to the apparent in vitro selective transcription by coconut RNA polymerases (82, 181).

For reasons already discussed, it was not yet possible, using naked DNA, to study an eventual selective transcription of genes other than the ribosomal genes. However, both mouse and monkey class A and B RNA polymerases actively transcribed mouse and monkey satellite DNAs, which are not transcribed in vivo (182). Moreover, mouse polymerase A transcribed both strands of mouse satellite DNA with equal efficiency, whereas class B polymerase transcribed five times more RNA from the T-rich H strand than from the L strand. Provided RNA synthesis was not initiated at nicks, these observations suggest that promoter site recognition by specific RNA polymerases is not the only mechanism dictating whether a given part of the genome should be transcribed or not in vivo. The asymmetry of the transcription of mouse satellite DNA with class B RNA polymerase is interesting in the context of the comments above concerning asymmetric transcription of synthetic templates.

Analysis of Selectivity of Transcription on Chromatin

Although RNA synthesis appears to be generally restricted on chromatin (when compared to the template efficiency of nicked deproteinized DNA), several observations have indicated that chromatin preparations isolated from various sources are as effectively transcribed by their homologous RNA polymerases as by *E. coli* RNA polymerase (22, 128, 183–192). However, Butterworth et al (183) found that rat liver enzyme AI is unable to transcribe rat liver chromatin, whereas in other systems both class A and B polymerases appear to have roughly the same activity on chromatin (128, 191). The reason for this discrepancy is unknown, but could possibly be related to the number of single strand breaks in the DNA of various chromatin preparations (188). It should indeed be stressed that there is no indication in any study of chromatin transcription whether initiation of RNA synthesis occurred at single stranded breaks or at double stranded regions of the DNA. The question as to whether the animal RNA polymerases can initiate RNA synthesis on intact double stranded regions of the chromatin DNA is therefore still unanswered.

Several investigators have reported that the RNAs synthesized by eukaryotic and bacterial polymerases on chromatin are different (183, 185–187, 189, 190), but their studies did not indicate whether the eukaryotic polymerases were, in fact, more accurate in selecting the sequences that are actually transcribed in vivo. For reasons already discussed, it is impossible at the present time to fully explore a possible selective in vitro transcription of genes coding for messenger RNAs. It was recently reported that a hemoglobin gene was more efficiently transcribed from a chromatin template by a eukaryotic RNA polymerase than by a prokaryotic enzyme (193). However, further studies are required to confirm these experiments, which are inconclusive because of methodological reasons beyond the scope of this review.

When a rat liver nucleolar chromatin preparation was used as a template for *E. coli* holoenzyme and for rat liver RNA polymerases AI and B, hybridization-competition experiments have indicated that the nucleolar RNA polymerase AI might transcribe more efficiently the ribosomal genes than either the *E. coli* holoenzyme or the nucleoplasmic RNA polymerase B (22). No such preference was observed with naked nucleolar DNA (22), suggesting that some chromatin-bound component could be important in helping the polymerase to select the DNA to transcribe. However, the accuracy of the transcription was not tested in these experiments. The exhaustive studies of Honjo & Reeder (191) have conclusively shown that both *X. laevis* RNA polymerases AI (or I) and B (or II) aberrantly transcribe the genes for both ribosomal and 5S RNAs when using homologous chromatin as template. More specifically, both strands of ribosomal and 5S DNAs are transcribed in vitro by AI (or I) and B (or II) polymerases, although the in vivo selective transcription of these genes is asymmetric and carried out by enzyme AI and C (or III) for the ribosomal and 5S genes, respectively. It is clear that in spite of the presence of chromatin proteins and the restricted RNA synthesis on chromatin, the eukaryotic enzymes, which are unable to initiate and terminate at the proper sites, are not more specific than *E. coli* RNA polymerase on the same template (192). It is therefore very likely that components other than chromatin proteins and purified RNA polymerases are involved in selective transcription and that these components must be identified before accurate transcription can be achieved in vitro with eukaryotic systems.

ACKNOWLEDGMENTS

I thank Dr. J. Wilhelm for his critical reading of the manuscript. Work in my laboratory has been supported by grants from the CNRS, the Délégation à la

Recherche Scientifique et Technique, the Commissariat à l'Energie Atomique, the Fondation pour la Recherche Médicale Française, and the INSERM. I am grateful to all those who have provided unpublished information.

Literature Cited

1. Davidson, E. H., Britten, R. J. 1973. *Quart. Rev. Biol.* 48:565–613
2. Losick, R. 1972. *Ann. Rev. Biochem.* 41:409–46
3. Bonner, J., Dahmus, M. E., Fambrough, D., Huang, R. C., Marushige, K., Tuan, D. Y. H. 1968. *Science* 159:47–56
4. Jacob, S. T. 1973. *Progr. Nucl. Acid Res. Mol. Biol.* 13:93–136
5. Chambon, P., Gissinger, F., Kedinger, C., Mandel, J. L., Meilhac, M. 1974. *The Cell Nucleus,* ed. H. Busch, 270–307. New York: Academic
6. Chambon, P. 1974. *The Enzymes X,* ed. P. D. Boyer, 261–331. New York: Academic
7. Roeder, R. G. 1969. *Multiple RNA polymerases and RNA synthesis in eukaryotic systems.* PhD thesis. Univ. of Washington, Seattle
8. Roeder, R. G., Rutter, W. J. 1969. *Nature* 224:234–37
9. Kedinger, C., Gniazdowski, M., Mandel, J. L., Gissinger, F., Chambon, P. 1970. *Biochem. Biophys. Res. Commun.* 38:165–70
10. *Cold Spring Harbor Symp. Quant. Biol.* 1970. 35:641–737
11. Stirpe, F., Fiume, L. 1967. *Biochem. J.* 105:779–82
12. Jacob, S. T., Sajdel, E. M., Munro, H. N. 1970. *Biochem. Biophys. Res. Commun.* 38:765–70
13. Lindell, T. J., Weinberg, F., Morris, P. W., Roeder, R. G., Rutter, W. J. 1970. *Science* 170:447–49
14. Blatti, S. P. et al 1970. *Cold Spring Harbor Symp. Quant. Biol.* 35:649–58
15. Hildebrandt, A., Sauer, H. W. 1973. *FEBS Lett.* 35:41–44
16. Gornicki, S. Z., Vuturo, S. B., West, T. V., Weaver, R. F. 1974. *J. Biol. Chem.* 249:1792–98
17. Kedinger, C., Nuret, P., Chambon, P. 1971. *FEBS Lett.* 15:169–74
18. Chambon, P. et al 1970. *Cold Spring Harbor Symp. Quant. Biol.* 35:693–708
19. Gissinger, F., Chambon, P. 1972. *Eur. J. Biochem.* 28:277–82
20. Gissinger, F., Chambon, P. 1975. *Eur. J. Biochem.* In press
21. Schwartz, L. B., Roeder, R. G. 1974. *J. Biol. Chem.* 249:5898–5906
22. Muramatsu, M., Onishi, T., Matsui, T.,

Kawabata, C., Tokugawa, S. 1975. *Proc. 9th FEBS Meet., Budapest.* In press
23. Goldberg, M., Perriard, J. C., Hager, G., Hallick, R. B., Rutter, W. J. 1974. *Basic Life Sci.* 3:241–56
24. Chesterton, C. J., Butterworth, P. H. W. 1971. *Eur. J. Biochem.* 19:232–41
25. Froehner, S. C., Bonner, J. 1973. *Biochemistry* 12:3064–71
26. Roeder, R. G. 1974. *J. Biol. Chem.* 235:241–48
27. Philips, J. P., Forest, H. S. 1973. *J. Biol. Chem.* 248:265–69
28. Wells, D., Hossenlopp, P., Chambon, P. 1975. See Ref. 22
29. Kedinger, C., Chambon, P. 1972. *Eur. J. Biochem.* 28:283–90
30. Weaver, R. F., Blatti, S. P., Rutter, W. J. 1971. *Proc. Nat. Acad. Sci. USA* 68:2994–99
31. Ingles, C. J. 1973. *Biochem. Biophys. Res. Commun.* 55:364–71
32. Mandel, J. L., Chambon, P. 1971. *FEBS Lett.* 15:175–80
33. Chesterton, C. J., Butterworth, P. H. W. 1971. *FEBS Lett.* 15:181–85
34. Sudgen, B., Keller, W. 1973. *J. Biol. Chem.* 248:3777–88
35. Schwartz, L. B., Roeder, R. G. 1975. *J. Biol. Chem.* In press
36. Krebs, G., Chambon, P. 1975. *Eur. J. Biochem.* In press
37. Houghton, M., Cox, R. F. 1974. *Nucl. Acid. Res.* 1:299–308
38. Kedinger, C., Gissinger, F., Chambon, P. 1974. *Eur. J. Biochem.* 44:421–36
39. Roeder, R. G. 1974. *J. Biol. Chem.* 249:249–56
40. Seifart, K. H., Benecke, B. J., Juhasz, P. P. 1972. *Arch. Biochem. Biophys.* 151:519–32
41. Amalric, F., Nicoloso, M., Zalta, J. P. 1972. *FEBS Lett.* 22:67–72
42. Seifart, K. H., Benecke, B. J. 1975. See Ref. 22
43. Butterworth, P. H. W., Austoker, J. L., Chesterton, C. J., Beebee, T. J. C. 1975. See Ref. 22
44. Austoker, J. L., Beebee, T. J. C., Chesterton, C. J., Butterworth, P. H. W. 1974. *Cell* 3:227–34
45. Hossenlopp, P., Wells, D., Chambon, P. 1975. *Eur. J. Biochem.* In press
46. Wilhelm, J., Dina, D., Crippa, M. 1974.

Biochemistry 13 : 1200–8
47. Sergeant, A., Krsmanovic, V. 1973. *FEBS Lett.* 35 : 331–35
48. Schwartz, L. B., Sklar, V. E. F., Jaehning, J. A., Weinmann, R., Roeder, R. G. 1974. *J. Biol. Chem.* 249 : 5889–97
49. Sklar, V. E. F., Schwartz, L. B., Roeder, R. B. 1975. *Proc. Nat. Acad. Sci. USA* 72 : 348–52
50. Beebee, T. J. C., Butterworth, P. H. W. 1974. *Eur. J. Biochem.* 44 : 115–22
51. Cochet-Meilhac, M., Nuret, P., Courvalin, J. C., Chambon, P. 1974. *Biochim. Biophys. Acta* 353 : 185–92
52. Ponta, H., Ponta, U., Wintersberger, E. 1972. *Eur. J. Biochem.* 29 : 110–18
53. Dezélée, S., Sentenac, A. 1973. *Eur. J. Biochem.* 34 : 41–52
54. Frederick, E. W., Maitra, U., Hurwitz, J. 1969. *J. Biol. Chem.* 244 : 413–24
55. Buhler, J. M., Sentenac, A., Fromageot, P. 1974. *J. Biol. Chem.* 249 : 5963–70
56. Valenzuela, P., Morris, R. W., Faras, A., Levinson, W., Rutter, W. J. 1973. *Biochem. Biophys. Res. Commun.* 53 : 1036–41
57. Gniazdowski, M., Mandel, J. L., Gissinger, F., Kedinger, C., Chambon, P. 1970. *Biochem. Biophys. Res. Commun.* 38 : 1033–40
58. Chesterton, C. J., Humphrey, S. M., Butterworth, P. H. W. 1972. *Biochem. J.* 126 : 675–81
59. Burgess, R. R. 1971. *Ann. Rev. Biochem.* 40 : 711–40
60. Dezélée, S., Sentenac, A., Fromageot, P. 1972. *FEBS Lett.* 21 : 1–6
61. Brogt, T. M., Planta, R. J. 1972. *FEBS Lett.* 20 : 47–52
62. Pony, S. S., Loomis, W. F. 1973. *J. Biol. Chem.* 248 : 3933–39
63. Strain, G. C., Mullinix, K. P., Bogorad, L. 1971. *Proc. Nat. Acad. Sci. USA* 68 : 2647–51
64. Horgen, P. A., Key, J. L. 1973. *Biochim. Biophys. Acta* 294 : 227–35
65. Shapiro, A. L., Vinuela, E., Maizel, J. V. 1967. *Biochem. Biophys. Res. Commun.* 28 : 815–20
66. Weber, K., Osborn, M. 1969. *J. Biol. Chem.* 244 : 4406–12
67. Chambon, P. et al 1972. *Acta Endocrinol.* 168 (Suppl.) : 222–46
68. Chambon, P. et al 1973. *Basic Life Sci.* 1 : 75–90
69. Kaufman, R., Voigt, H. P. 1973. *Z. Physiol. Chem.* 354 : 1432–38
70. Cochet-Meilhac, M., Chambon, P. 1974. *Biochim. Biophys. Acta* 353 : 160–84
71. Cochet-Meilhac, M., Nuret, P., Courvalin, J. C., Chambon, P. 1974. *Biochim. Biophys. Acta* 353 : 185–92
72. Chesterton, C. J., Butterworth, P. H. W. 1971. *FEBS Lett.* 12 : 301–9
73. Gissinger, F., Kedinger, C., Chambon, P. 1974. *Biochimie* 56 : 319–33
74. Cochet-Meilhac, M., Chambon, P. 1973. *Eur. J. Biochem.* 35 : 454–63
75. Buhler, J. M., Dezélée, S., Huet, J., Ibarra, F., Sentenac, A. 1975. See Ref. 22
76. Hildebrandt, A., Sebastian, J., Halvorson, H. O. 1973. *Nature New Biol.* 246 : 73–74
77. Hildebrandt, A., Sauer, H. W. 1973. *FEBS Lett.* 35 : 41–44
78. Burgess, A. B., Burgess, R. R., 1974. *Proc. Nat. Acad. Sci. USA* 71 : 1174–77
79. Mullinix, K. P., Strain, G. C., Bogorad, L. 1973. *Proc. Nat. Acad. Sci. USA* 70 : 2386–90
80. Jendrisak, J. J., Becker, W. M. 1974. *Biochem. J.* 139 : 771–77
81. Ganguly, A., Das, A., Mondal, H., Mandal, R. K., Biswas, B. B. 1972. *Eur. J. Biochem.* 28 : 143–50
82. Biswas, B. B., Mondal, H., Ganguly, A., Das, A., Mandal, R. K. 1974. *Basic Life Sci.* 3 : 279–94
83. Sebastian, J., Bhargava, M. M., Halvorson, H. O. 1973. *J. Bacteriol.* 114 : 1–6
84. Grant, W. D. 1972. *Eur. J. Biochem.* 29 : 94–98
85. Pong, S. S., Loomis, W. F. 1973. *Molecular Techniques and Approaches in Developmental Biology,* ed. M. Chrispeels, 93–116. Palo Alto, Cal : Wiley
86. Brandle, E., Zetsche, K. 1973. *Planta* 111 : 209–17
87. Zylber, E. A., Penman, S. 1971. *Proc. Nat. Acad. Sci. USA* 68 : 2861–65
88. Price, R., Penman, S. 1972. *J. Mol. Biol.* 70 : 435–50
89. Reeder, R. H., Roeder, R. G. 1972. *J. Mol. Biol.* 70 : 433–41
90. Price, R., Penman, S. 1972. *J. Virol.* 9 : 621–26
91. Wallace, R. D., Kates, J. 1972. *J. Virol.* 9 : 627–35
92. Chardonnet, Y., Gazzolo, L., Pogo, B. G. T. 1972. *Virology* 48 : 305–7
93. Jackson, A., Sugden, B. 1972. *J. Virol.* 10 : 1086–89
94. Alwine, J. C., Steinhart, W. L., Hill, C. W. 1974. *Virology* 60 : 302–7
95. Weinmann, R., Raskas, H. J., Roeder, R. G. 1974. *Proc. Nat. Acad. Sci. USA* 71 : 3426–30
96. Rymo, L., Parsons, J. T., Coffin, J. M., Weissmann, C. 1974. *Proc. Nat. Acad. Sci. USA* 71 : 2782–86
97. Jacquet, M., Groner, Y., Monroy, G., Hurwitz, J. 1974. *Proc. Nat. Acad. Sci. USA* 71 : 3045–49
98. Amati, P., Blasi, F., di Porzio, U.,

Riccio, A., Trabeni, C. 1975. *Proc. Nat. Acad. Sci. USA*. Submitted
99. Weinmann, R., Roeder, R. G. 1974. *Proc. Nat. Acad. Sci. USA* 71:1790–94
100. Jacob, S. T., Muecke, W., Sajdel, E. M., Munro, H. N. 1970. *Biochem. Biophys. Res. Commun.* 40:334–42
101. Niessing, J., Schnieders, B., Kunz, W., Seifart, K. H., Sekeris, C. E. 1970. *Z. Naturforsch. B* 25:1119–25
102. Tata, J. R., Hamilton, M. J., Shilds, D. 1972. *Nature* 238:161–64
103. Shraya, E., Clever, V. 1972. *Biochim. Biophys. Acta* 272:373–81
104. Hastie, N. D., Mahy, B. W. J. 1973. *FEBS Lett.* 32:95–99
105. Kuwano, M., Ikehara, T. 1973. *Exp. Cell Res.* 82:454–57
106. Montanaro, L., Novello, F., Stirpe, F. 1973. *Biochim. Biophys. Acta* 319:188–98
107. Hadjiolov, A. A., Dabeva, M. D., Mackedonski, M. D. 1974. *Biochem. J.* 138:321–34
108. Kedinger, C., Simard, R. 1974. *J. Cell Biol.* 63:831–42
109. Beerman, W. 1971. *Chromosoma* 34:152–60
110. Wobus, U., Panitz, R., Serfling, E. 1971. *Experientia* 27:1202–3
111. Egyhazi, E., D'Monte, B., Edström, J. E. 1972. *J. Cell Biol.* 53:523–31
112. Bucci, S., Nordi, I., Mancino, G., Fiume, L. 1971. *Exp. Cell Res.* 69:462–65
113. Marinozzi, V., Fiume, L. 1971. *Exp. Cell Res.* 67:311–22
114. Petrov, P., Sekeris, C. E. 1971. *Exp. Cell Res.* 69:393–401
115. Chan, V. L., Whitmore, G. F., Siminovitch, L. 1972. *Proc. Nat. Acad. Sci. USA* 69:3119–23
116. Somers, D. G., Pearson, M. L., Ingles, C. J. 1975. *Nature* 253:371–74
117. Jacob, S. T., Sajdel, E. M. Muecke, W., Munro, H. N. 1970. *Cold Spring Harbor Symp. Quant. Biol.* 35:681–91
118. Sekeris, C. E., Schmid, W. C. 1972. *FEBS Lett.* 27:41–45
119. Fiume, L., Laschi, L. 1965. *Sperimentale* 115:228–35
120. Fiume, L., Marinozzi, V., Nordi, I. 1969. *Brit. J. Exp. Pathol.* 50:270–76
121. Meyer-Schultz, F., Porte, A. 1971. *Cytobiologie* 3:387–400
122. Emannilov, I., Nicolova, R. C., Dabeva, M. D., Hadjiolov, A. A. 1974. *Exp. Cell Res.* 86:401–3
123. Schmid, W., Sekeris, C. E. 1973. *Biochim. Biophys. Acta* 312:549–54
124. Serfling, E., Wobus, U., Panitz, R. 1972. *FEBS Lett.* 20:148–52

125. Tocchini-Valentini, G. P., Crippa, M. 1970. *Cold Spring Harbor Symp. Quant. Biol.* 35:737–42
126. Tocchini-Valentini, G. P., Crippa, M. 1970. *Nature* 228:993–95
127. Montecuccoli, G., Novello, F., Stirpe, F. 1972. *FEBS Lett.* 25:305–8
128. Austin, G. E., Bello, L. J., Furth, J. J. 1973. *Biochim. Biophys. Acta* 324:488–500
129. Mandel, J. L., Chambon, P. 1974. *Eur. J. Biochem.* 41:367–78
130. Ponta, H., Ponta, U., Kraft, V., Wintersberger, E. 1974. *Eur. J. Biochem.* 46:473–79
131. Chamberlin, M. J. 1974. *Ann. Rev. Biochem.* 43:721–75
132. Cedar, H., Felsenfeld, G. 1973. *J. Mol. Biol.* 77:237–54
133. Birnstiel, M., Speirs, J., Purdom, I., Jones, K., Loening, U. 1968. *Nature* 219:454–63
134. Dawid, I. B., Brown, D. D., Reeder, R. H. 1970. *J. Mol. Biol.* 51:341–60
135. Brown, D. D., Sugimoto, K. 1973. *J. Mol. Biol.* 78:397–415
136. Birnstiel, M., Telford, J., Weinberg, E., Stafford, D. 1974. *Proc. Nat. Acad. Sci. USA* 71:2900–4
137. Mandel, J. L., Chambon, P. 1974. *Eur. J. Biochem.* 41:379–95
138. Hossenlopp, P., Oudet, P., Chambon, P. 1974. *Eur. J. Biochem.* 41:397–411
139. Doenecke, D., Pfeiffer, C., Sekeris, C. E. 1972. *FEBS Lett.* 21:237–43
140. Dezélée, S., Sentenac, A., Fromageot, P. 1974. *J. Biol. Chem.* 249:5971–77
141. Nohara, H., Mizuno, T., Iizuka, K. 1973. *Biochim. Biophys. Acta* 319:55–61
142. Bautz, E. K. F. 1972. *Progr. Nucl. Acid. Res. Mol. Biol.* 12:129–60
143. Furth, J. J., Pizer, L. I., Austin, G. E., Fujii, K. 1972. *Life Sci.* 11:1001–10
144. Flint, S. J., de Pomerai, D. I., Chesterton, C. J., Butterworth, P. H. W. 1974. *Eur. J. Biochem.* 42:567–79
145. Gross-Bellard, M., Oudet, P., Chambon, P. 1973. *Eur. J. Biochem.* 36:32–38
146. Chambon, P. et al 1973. *Regulation of Transcription and Translation in Eukaryotes,* ed. E. K. F. Bautz, 125–38. Berlin: Springer
147. Chambon, P. et al 1974. *Basic Life Sci.* 3:257–68
148. Cascino, A., Riva, S., Geiduschek, E. P. 1970. *Cold Spring Harbor Symp. Quant. Biol.* 35:213–20
149. Chinnadurai, G., McCorquodale, D. J. 1973. *Proc. Nat. Acad. Sci. USA* 70:3502–5
150. Britten, R. J., Davidson, E. H. 1969.

Science 165:349–57

151. Bram, S. 1972. *Biochimie* 54:1005–11
152. Paul, J. 1972. *Nature New Biol.* 238: 444–46
153. Crick, F. 1971. *Nature* 234:25–27
154. Seifart, K. H., Sekeris, C. E. 1969. *Eur. J Biochem.* 7:408–12
155. Seifart, K. H. 1970. *Cold Spring Harbor Symp. Quant. Biol.* 35:719–25
156. Stein, H., Hausen, P. 1970. *Cold Spring Harbor Symp. Quant. Biol.* 35:709–18
157. Stein, H., Hausen, P. 1970. *Eur. J. Biochem.* 14:270–77
158. Lentfer, D., Lezius, A. G. 1972. *Eur. J. Biochem.* 30:278–84
159. Natori, S. 1972. *J. Biochem.* 72:1291–94
160. Natori, S., Takeuchi, K., Mizuno, D. 1973. *J. Biochem.* 73:345–51
161. Natori, S., Takeuchi, K., Takahashi, K., Mizuno, D. 1973. *J. Biochem.* 73: 879–88
162. Natori, S., Takeuchi, K., Mizuno, D. 1974. *J. Biochem.* 74:1177–82
163. Chuang, R., Chuang, L., Laszlo, J. 1974. *Biochem. Biophys. Res. Commun.* 57: 1231–39
164. Shea, M., Kleinsmith, L. J. 1973. *Biochem. Biophys. Res. Commun.* 50: 473–77
165. Lee, S. C., Dahmus, M. E. 1973. *Proc. Nat. Acad. Sci. USA* 70:1383–87
166. Higashinakagawa, T., Muramatsu, M. 1972. *Biochem. Biophys. Res. Commun.* 47:1–6
167. Higashinakagawa, T., Onishi, T., Muramatsu, M. 1972. *Biochem. Biophys. Res. Commun.* 48:937–44
168. DiMauro, E., Hollenberg, C. P., Hall, B. 1972. *Proc. Nat. Acad. Sci. USA* 69:2818–22
169. Huet, J., Sentenac, A., Fromageot, P. 1974. *Eur. J. Biochem.* In press
170. Mondal, H., Mandal, R. K., Biswas, B. B. 1972. *Eur. J. Biochem.* 25:463–70
171. Seifart, K. H., Juhasz, P. P., Benecke, B. J. 1973. *Eur. J. Biochem.* 33:181–91
172. Hameister, H., Stein, H., Kedinger, C. See Ref. 146, 125–38
173. Konishi, G., Koide, S. S. 1971. *Experientia* 27:262–64

174. Adman, R., Schultz, L. D., Hall, B. 1972. *Proc. Nat. Acad. Sci. USA* 69: 1702–6
175. Roeder, R. G., Reeder, R. H., Brown, D. D. 1970. *Cold Spring Harbor Symp. Quant. Biol.* 35:727–35
176. Hecht, R. M., Birnstiel, M. L. 1972. *Eur. J. Biochem.* 29:489–99
177. Beebee, T. J. C., Butterworth, P. H. W. 1974. *Eur. J. Biochem.* 45:395–406
178. Beebee, T. J. C., Butterworth, P. H. W. 1974. *FEBS Lett.* 47:304–6
179. Crippa, M., Dina, D., Long, E., Wilhelm, J. See Ref. 22
180. Cramor, J. H., Sebastian, J., Rownd, R. H., Halvorson, H. O. 1974. *Proc. Nat. Acad. Sci. USA* 71:2188–92
181. Mondal, H., Ganguly, A., Das, A., Mandal, R. K., Biswas, B. B. 1972. *Eur. J. Biochem.* 28:143–50
182. Maio, J. J., Kurnitt, D. M. 1974. *Biochim. Biophys. Acta* 349:305–19
183. Butterworth, P. H. W., Cox, R. F., Chesterton, C. J. 1971. *Eur. J. Biochem.* 23:229–41
184. Farber, J. L., Rovera, G., Baserga, R. 1972. *Biochem. Biophys. Res. Commun.* 49:558–62
185. Keshgegian, A. A., Furth, J. J. 1972. *Biochem. Biophys. Res. Commun.* 48: 757–63
186. Keshgegian, A. A., Garibian, G. S., Furth, J. J. 1973. *Biochemistry* 12: 4337–42
187. Maryanka, D., Gould, H. 1973. *Proc. Nat. Acad. Sci. USA* 70:1161–65
188. Pays, E., Ronsse, A. 1973. *Eur. J. Biochem.* 40:119–31
189. Dupras, M., Bonner, J. 1974. *Mol. Cell. Biochem.* 3:27–33
190. Tsai, M. J., Saunders, G. F. 1973. *Biochem. Biophys. Res. Commun.* 51: 756–65
191. Honjo, T., Reeder, R. H. 1974. *Biochemistry* 13:1896–99
192. Reeder, R. H. 1973. *J. Mol. Biol.* 80: 229–41
193. Steggles, A. W. et al 1974. *Proc. Nat. Acad. Sci. USA* 71:1219–23

THREE-DIMENSIONAL STRUCTURE OF IMMUNOGLOBULINS

×897

David R. Davies and Eduardo A. Padlan

Laboratory of Molecular Biology, National Institute of Arthritis,
Metabolism, and Digestive Disea es, National Institutes of Health,
Bethesda, Maryland 20014

David M. Segal

Immunology Branch, National Cancer Institute,
Bethesda, Maryland 20014

CONTENTS

INTRODUCTION

The immune system of vertebrate species is characterized by the capacity to synthesize humoral antibodies in response to challenge with antigen. These antibody molecules will bind to the antigen and with multivalent antigens will cause precipitation by formation of a crosslinked lattice. In vivo, antibodies can also play a crucial role in a variety of cellular and humoral responses.

Except under very special conditions, the antibodies produced generally form a heterogeneous spectrum of molecules, both in terms of amino acid composition and sequence and with regard to the strength of binding to antigen. Because of this heterogeneity, most structural studies have been carried out on homogeneous myeloma proteins from tumor cells. There is now a considerable body of evidence to show that some of these proteins are immunologically, chemically, and physically indistinguishable from natural antibodies (1–5).

Since the last reviews on this subject (2, 6) there has been tremendous progress in our understanding of the three-dimensional structure of antibodies and the nature of the interaction between antibody and antigen. High resolution X-ray crystallography on a number of immunoglobulin fragments has shed a great deal of light on antibody structure. In this review we shall to a large extent be concerned with these studies and with other investigations that relate either directly or indirectly to these structure determinations. We make no claim to have been comprehensive in these other areas and rather have been quite selective. A number of recent reviews have been written on physical-chemical studies (7–9) and on the amino acid sequences (10–12) of immunoglobulins.

CHEMICAL STRUCTURE

Structural immunology is unusual in that long before the recent crystallographic contributions, many ideas on structure and function had been proposed based on a wealth of chemical data. These include the concept of the basic immunoglobulin structure along with the location of the antigen binding site and the concept of domain structure.

The molecular organization of immunoglobulins has been discussed in many previous reviews (2, 6, 8, 9, 13, 14). The basic structure consists of two identical light (L) chains of molecular weight 22,500 and two heavy (H) chains of molecular weight 50,000–75,000 linked by noncovalent interactions and disulfide bridges to form a structure with twofold symmetry (Figure 1). This basic immunoglobulin structure can join with other like structures through disulfide bridges between the heavy α and μ chains to form IgA molecules (chiefly dimers) and IgM molecules (chiefly pentamers), respectively. Earlier reviews summarize the various classes of immunoglobulins (2, 6, 8).

Porter (15) showed that limited proteolysis results in three fragments of roughly equal weight, two Fabs and one Fc (Figure 1). The antigen binding site was localized on the Fab part of the molecule, whereas the Fc portion plays an important role in complement fixation and other biological functions. Between each Fab and the Fc lies a "hinge" region, some 25 residues long, containing the interchain disulfides and noteworthy for often having an unusually high percentage of proline (16).

Light Chains

The light chains are on average 217 amino acids long (16). They may be divided into two groups, κ and λ, clearly distinguishable by their amino acid sequence.

The relative proportions of κ and λ vary considerably with species, being 65 to 35% in humans and 97 to 3% in mice. Within each group there are several subgroups that may be identified by the degree of sequence homology within the subgroup (16).

Each light chain is divided into two approximately equal parts, the amino-terminal V half which is quite variable between different immunoglobulins, and the carboxyl-terminal C half which is constant or common in different light chains of the same group (17). Light chains, either whole or fragmented, have been observed for over 100 years (18) in the urine of patients with multiple myeloma (Bence-Jones proteins).

Heavy Chains

Depending on the immunoglobulin class, heavy chains contain about 450 or 576 amino acids. Appreciable variation arises from insertions and deletions in the variable region as well as in the region near the hinge residues. The original sequence determination of a human $\gamma 1$ molecule revealed that the γ heavy chain has three C_H domains with homologous sequence in addition to the amino-terminal V_H domain (19, 20). The γ chain C domains are also homologous with the light chain C domain.

No homology exists between the variable and constant domains. Nevertheless, they are of similar length and both possess an internal disulfide loop of about 60 (C) or 67 (V) amino acid residues, starting at about position 23. It has been

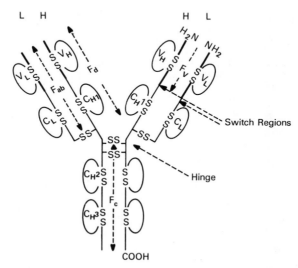

Figure 1   Schematic representation of polypeptide chains in the basic immunoglobulin structure.

proposed that all domains form similar discrete globular units, each specialized for carrying out a particular function (19, 20). From 12 to 18 residues beyond the first cysteine of the internal disulfide is an "invariant" tryptophan. The V_H region sequences, like those of V_L, permit separation into subgroups, distinguishable by very high homology within a subgroup and by having less than 50% homology between members of different subgroups.

The complete amino acid sequence of the 576 residues of the μ chain of a human IgM immunoglobulin (Ou) has recently been reported (21). This chain has five domains of sequence homology, two (V_H and $C_\mu 1$) are associated with light chain, and two ($C_\mu 3$ and $C_\mu 4$) are released in the Fc fragment after limited tryptic digestion. An intervening domain ($C_\mu 2$) and a segment that probably corresponds to the hinge region of γ chains are degraded during this digestion. The intersubunit cysteine used to form the pentameric IgM is located in $C_\mu 3$ at residue 414. Preliminary results indicate that IgE, another class of immunoglobulin, also has five heavy chain domains (22).

J Chain

A third type of polypeptide, known as J chain, has been found in polymeric IgA and IgM (23–28) from a variety of species (29–31). Because J is always found with polymeric immunoglobulin and never with the monomer, it is generally considered to be involved with joining [hence the term "J" (23)] immunoglobulin monomers.

J chain has been isolated by a variety of techniques (24–28, 32, 33). It has a molecular weight of approximately 15,000 (24, 25, 34) and an axial ratio of 17.9 (32).

J chain is bound covalently to polymeric immunoglobulin via disulfide bridges (23). Assuming one mole of J binds to one mole of IgM pentamer (24, 25), it appears that the J chain does not contain sufficient SH groups to bind to every μ chain (32). Alpha chains bind J at the penultimate cysteine (35). In one study, J chain was removed from IgM pentamer without causing the polymer to convert to monomer (36). It thus appears J does not link all monomers together directly (32, 36).

Hypervariability

Comparison of V_L sequences revealed that some regions, notably around residues 24–34, 89–96 (13, 37, 38), and 50–55 (39), showed unusually high variability. Wu & Kabat (40) carried out a statistical analysis of the data that firmly established these three regions as hypervariable. Similar regions (31–35, 50–65, 95–102) were noted in the heavy chains by Kabat & Wu (41) and Capra & Kehoe (12, 42), who also observed hypervariability at positions 81–85. In subsequent discussions the first, second, and third hypervariable regions of the light chain will be designated L1, L2, and L3, respectively; the homologous hypervariable regions in the heavy chain will be designated H1, H2, and H3, and the additional hypervariable region (residues 81–85) will be called He. It was proposed (38, 40) that these hypervariable regions would be found together at the antigen binding site.

The amino acid residues involved in binding were independently determined by the method of affinity labeling (43). In this technique a reagent, structurally similar to the hapten ligand, is covalently bound to the immunoglobulin and the position of the amino acid side chains with which it reacts is determined. In a recent review Givol (44) has listed the affinity-labeled residues and it is apparent that they are all located in the hypervariable regions of both L and H chains. Moreover, all the hypervariable regions, with the exception of that at position 81–85 of the H chain, were labeled in one protein or another.

OVERALL THREE-DIMENSIONAL STRUCTURE

Physical Chemistry and Electron Microscopy

The model of three relatively rigid fragments linked by a flexible hinge region was proposed originally by Noelken et al (45) on the basis of a physical chemical analysis of the fragments and the intact IgG molecule. These authors also concluded that there was relatively little α helix in the structure. This overall model was confirmed by the electron micrographs (46) of rabbit anti-DNP (2, 4-dinitrophenyl) antibodies linked by a bivalent hapten.

The aggregation of the basic immunoglobulin structure to form IgA (dimers) and IgM (mostly pentamers) has been confirmed by many elegant micrographs (47–58). In IgA the molecules join through the Fc region to form end-to-end dimers. In IgM they aggregate again through the Fc to form a pentameric structure with a central disc. The studies of Feinstein and colleagues (55–58) emphasized the flexibility of attachment of the individual units to the central disc.

Preliminary X-Ray Studies on Fc Fragments

Since the Fc portion is identical in a wide variety of otherwise heterogeneous antibodies, large amounts of pure fragment can be obtained by digestion of normal IgG. In fact, Fc from rabbit IgG crystallizes spontaneously from the digestion mixture (15). Preliminary X-ray data have been presented for human and rabbit Fc, but early work (59, 60) was hindered by twinning of the crystals.

Crystals more suitable for high resolution X-ray studies were obtained by Humphrey (61) from papain digests of a human IgG myeloma protein. The symmetry of these crystals changed when they were soaked in dilute solutions of $KAuCl_3$ or o-chloromercuriphenol. The new crystal form produced by this soaking contained only half an Fc fragment per asymmetric unit, indicating that the fragment must have a twofold symmetry axis (62, 63).

Low Resolution X-Ray Analysis of the Intact Molecule

Crystallization of the human γG1 cryoglobulin Dob (64) provided some information about the intact molecule. In particular, it emphasized the dyad axis of the molecule with the Fc part located on a twofold symmetry axis. Similar observations of twofold symmetry have been reported for other intact human IgG molecules (65, 66).

The work with Dob was extended (67–70) to provide a low resolution (6 Å) electron density map of the protein. This exhibited three globular regions of density corresponding to the Fc and two Fabs. Because of the crystal packing it was difficult at this resolution to distinguish between various ways of linking the Fc to the two Fabs, but a T-shaped model was preferred. The dimensions for this model placed the two combining sites 140 Å apart, at the ends of the arms of the T. An accompanying electron microscope investigation of these crystals yielded essentially the same result (69–72). Difficulties with crystal quality have hitherto limited the X-ray analysis to low resolution. It is interesting to note that in Dob and in one other crystalline intact IgG (Mcg), there is a deletion in the hinge region of the molecule (73, 74).

HIGH RESOLUTION CRYSTAL STRUCTURES

All the high resolution results hitherto achieved have dealt with fragments of the intact molecule. This has been due primarily to the difficulty of getting acceptable crystals of the intact molecule, together with the obvious attractions of having to collect and analyze fewer data for the fragments. To date, there are reports of four structures from four different laboratories (Table 1). Two of these, the human light chain dimer and the human Fab fragment, appeared approximately simultaneously and have been closely followed by the structures of the human Bence-Jones V_L dimer and the mouse myeloma Fab with phosphorylcholine binding specificity. Already, then, it is possible to make some comparisons of the structures, although because most of the published results are preliminary and, with one exception (75), do not include atomic coordinates, such comparisons will only be possible at a qualitative level. Where specific amino acid residues are referred to, the sequence numbering is taken from the original publications.

Crystallographic Studies on the Mcg Bence-Jones Protein

Structural studies on the Mcg Bence-Jones protein are particularly interesting since crystals of the whole IgG1 (65, 76) myeloma protein from the same patient have been obtained. This whole molecule presumably contains the same light chains as found in the Bence-Jones dimer, and its crystal structure is being determined.

Table 1   High resolution immunoglobulin structures

Protein	Source	Molecular composition	Class or type	Resolution	Bound ligands	Reference
Mcg	Human	L-chain dimer	λ	2.3 Å	Dnp compounds and many other small molecules	65, 76–82, 88
New	Human	Fab(pepsin)	$\lambda, \gamma 1$	2.8 Å	Vitamin K-1 OH and several others	83, 94–103
REI	Human	V_L dimer	$\kappa 1$	2.8 Å		75, 89, 92, 93
McPC 603	Mouse	Fab(pepsin)	κ, α	3.1 Å	Phosphorylcholine	105–110

The Mcg Bence-Jones protein dimer is made up of two lambda-type light chains joined together by a disulfide bond between the penultimate cysteine residues.

Structure of the Mcg Light Chain Dimer

The crystal structure of the Mcg dimer (77) has been analyzed successively at 6 (78), 3.5 (79), and 2.3 (80) Å resolutions. Even at 6 Å resolution, it was clear that the dimer was represented by two globular modules differing in size and structure. The electron density map at 3.5 Å resolution (78) was sufficiently clear to trace the course of the polypeptide chains. The large and small modules previously observed at low resolution were identified as the V and C regions, respectively, on the basis of the known sequence of the C-terminal half and on the location of the mercury atom inserted into the interchain disulfide bridge (81). The domains were clearly delineated and the two segments of electron density bridging the two regions were identified as the switch regions between the V and C parts of the amino acid sequence.

The analysis at 2.3 Å resolution (80), aided by knowledge of the complete amino acid sequence (82), has led to a complete molecular model. The improved resolution more fully revealed the details regarding the structure and interactions of the various domains.

The relative disposition of the two monomers and the four domains in the Mcg dimer is shown in Figure 2. Each domain is approximately 40 Å long and the dimer is about 77 Å long. The two chemically identical monomers are not equivalent in their three-dimensional structures although approximate twofold axes relate the paired domains in both V and C regions. These axes are not collinear, rather they intersect at an angle of 120°. The spatial relationships between the V and C domains of the same chain are very different in the two monomers. In monomer 1, the distance (center to center) between the intradomain disulfide bonds is 25 Å and the angle between the long axes of the V and C domains is about 70°. In monomer 2, the measured values are 43 Å and about 110°, respectively. These angles are similar to those measured in the human Fab, New (83), with the domains of monomer 1 disposed as in the heavy chain and those of monomer 2 as in the light chain of the Fab. This, and the observation that the additional disulfide bridge found in rabbit light chains (84) could be formed

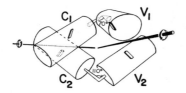

Figure 2   Schematic representation illustrating steric relationship of the four domains in Mcg light chain dimer (79). The appropriate twofold axes relating $V_1:V_2$ and $C_1:C_2$ are indicated. They intersect at an angle of about 120°. Reprinted with permission from Schiffer et al 1973. *Biochemistry* 12: 4620. Copyright by the American Chemical Society.

between the corresponding residues (82, 174 in Mcg) in monomer 2 but not in monomer 1, led the authors to suggest that monomer 2 is a closer homolog to light chains and that monomer 1 fills the structural role of the heavy chain in Fab (79, 80).

Differences in the spatial relationships lead to differences in the longitudinal interactions between the V and C domains of the two Mcg monomers. Thus, in monomer 1, the ring of Pro 8 and the main chain between Ala 147 and Val 148 are very close, as are Ser 9 and Ser 11 and the side chain of Glu 202 (80). In monomer 2, the corresponding residues are widely separated.

The V and C domains were found to have very similar tertiary structures. The basic domain structure consists of two layers of antiparallel segments (Figure 3). There are four segments in one layer and three in the other. The intradomain disulfide bond is located in the center of each domain, connecting the middle segment of the three chain layer with a parallel strand in the other layer. The invariant tryptophan [residues 37 in V and 152 in C (82)] lies close to the disulfide and the interior of the domain is filled with hydrophobic side groups. In both layers, the formation of several hydrogen bonds is possible between adjacent segments. The commonly observed "twisted sheet" of β structure is seen in the Mcg dimer, particularly in the C domains.

The arrangement of the segments in the C domain is more regular, with the antiparallel chains forming two distinct layers. In the V domain, the N-terminal segment lies in the four chain layer but is also close and parallel to the C-terminal segment which is in the three chain layer. In addition to the more irregular arrangement of segments and layers in the V than in the C domain (Figure 3), an

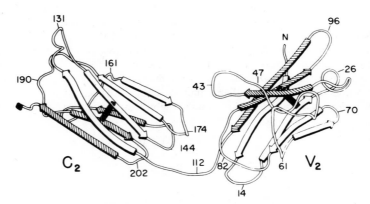

Figure 3   Schematic representation of monomer 2 of the Mcg dimer (79). Arrows indicate direction of polypeptide chain. In both domains, white arrows represent stretches of extended chain lying in one plane, hatched arrows represent stretches lying in the other. Disulfide bonds are indicated as black bars connecting the two planes in each domain. Reprinted with permission from Schiffer et al 1973. *Biochemistry* 12: 4620. Copyright by the American Chemical Society.

irregular loop is found in the former which has no equivalent in the latter. This additional loop contains the second hypervariable region. Two short right-handed helices were found in the C domains involving residues 125–132 and 185–192.

Although the two V domains are related by an approximate twofold axis, distinct differences between them were observed. The first hypervariable loops, for example, were found to be significantly different in the two domains. In addition, the structure of the V domain of monomer 1 appears to be generally less ordered with apparently less hydrogen bonding between segments. The C domains are more nearly equivalent.

The domains in the Mcg λ chain are clearly structurally homologous, yet their association in pairs was found to be very different in the V and C modules. Thus the difference in size of the two regions is due to the greater separation of the V domains, reflected in the distance between the intradomain disulfide bonds, which is 24 Å in the V module but only 18 Å in the C.

The lateral interactions between the paired domains appear to be stronger in the C region. The four chain layers of the C domains form the interface in this half of the dimer. These layers form concave and highly complementary surfaces with hydrophobic residues interdigitating across the boundary between domains.

In marked contrast, it is found that in the V region, the four chain layers are on the outside. The three chain layers face each other with relatively few interactions between domains (Table 2). Among the few points of contact between the V domains are those between the Tyr 51 and Asp 97 side groups, and between the Gln 40 side chains from the two monomers. The side group of Tyr 89 is situated close to Pro 46 and to the main chain between Lys 44 and Ala 45 of the opposite monomer.

The wider separation of the V domains in the Mcg dimer results in the presence of a "solvent channel" between the three chain layers (80). As revealed in the 2.3 Å study, the solvent channel begins at the tip of the molecule as a large conical cavity approximately 15 Å across at the entrance and 17 Å deep. At the base of the cavity is a 5–6 Å opening leading to an ellipsoidal pocket with dimensions of $8 \times 8 \times 10$ Å (80). The main cavity is lined by 21 side chains, most of which emanate from the three hypervariable regions (residues 23–36, 50–56, and 91–100) of each monomer. Tyr 34, 51, and 93 and Glu 52 from both monomers as well as Asp 97 from monomer 2 line the rim of the main cavity. The walls are formed by Ser 36 and Ser 91, Val 48 and Phe 99 from both monomers. Tyr 38 and Phe 101 form the floor of the cavity and the roof of the inner pocket. The walls of this pocket consist of Tyr 89, Pro 46, and Gln 40 from both monomers, which are not hypervariable.

The positions of various substitutions and antigenic markers were located in the Mcg dimer. Position 147, at which Val was substituted for Ala in protein Mz (85), was found in the close contact between V and C domains of monomer 1 and on the surface in monomer 2. The second Mz substitution, Lys for Asn at position 174, was located on an exposed loop in both monomers. The Kern [position 156 (86)] and Oz [position 193 (87)] markers were located on the surface on adjacent loops.

Table 2   Residues in contact across the V region interface in the REI and Mcg Bence-Jones dimers[a]

REI (75)

Tyr 36(—OH) . . . . Gln 89$^+$ (O=C NH$_2$)
Tyr 36 . . . . Phe 98
Pro 44 . . . . Tyr 87
Pro 44 . . . . Phe 98
Ala 43 . . . . Tyr 87
Lys 42(C=O) . . . . Tyr 87(—OH)
Gln 38(O=C—NH$_2$) . . . . Gln 38(O=C—HN$_2$)
Tyr 49(—OH) . . . . Tyr 96$^+$(—OH)
Tyr 49 . . . . Leu 94$^+$
Gln 55$^+$ . . . . Leu 94$^+$
Gln 55$^+$ . . . . Pro 95$^+$

Mcg (79)

Tyr 51 . . . . Asp 97$^+$
Gln 40 . . . . Gln 40
Pro 46 . . . . Tyr 89
Tyr 89 . . . . backbone between Lys 44
                                                                                    and Ala 45

[a] Each interaction occurs twice because of the twofold axis between domains; exact in REI, approximate in Mcg. The exception is the contact between Tyr 51 (monomer 1 only) and Asp 97 (monomer 2 only) in Mcg. The hypervariable residues are indicated by +.

Binding Studies on Mcg Crystals

Binding studies in the crystal (88) revealed that dinitrophenyl (DNP) amino acids, purine and pyrimidine derivatives such as caffeine and 5-acetyluracil, aromatic compounds like menadione (vitamin K-3) and colchicine, and lipid components like triacetin bound in the solvent channel of the Mcg dimer. There were two distinct binding sites, A and B, in the main cavity. Site A, which is more spacious, is situated near the entrance but closer to monomer 2 than monomer 1. The site related to A by the approximate twofold axis is blocked by another molecule in the crystal. Site B is located near the approximate twofold axis farther down the cavity. Sites A and B act as discrete sites in the binding of small molecules. Large molecules like colchicine, on the other hand, bind in sites A, B, and other parts of the cavity. The most probable contact residues for site A are Tyr 34, Tyr 93, and Glu 52 from monomer 2, with Tyr 51 in close proximity. The potential contributors of site B are Phe 99 and 101, Ser 36 and 91, and Tyr 38 from both monomers. An additional binding site was found in the inner pocket. The center of this third site is between the two Pro 46 residues, with Tyr 38 and 89 and Phe 101 from both chains being accessible for binding.

The binding of some of these compounds produces changes in the protein structure around the binding site. 5-Iodo-DNP, which was covalently bound to Tyr 34 of monomer 2, produced a displacement of the phenolic ring of about 3 Å. In addition, the polypeptide backbone from Tyr 34 to Ser 36 was shifted approximately 2.5 Å. Large compounds like colchicine were more disruptive. Menadione (vitamin K-3) binds in the inner pocket quite avidly, cracks the crystals within 2 hr, and causes a displacement of the polypeptide chain between Lys 44 and Ala 45 in monomer 2. The ease with which compounds like menadione could disrupt the crystal structure was attributed by the authors to the flexibility in the protein "which permits an oversized entering group with the proper affinity to induce a fit" in the binding channel (88).

Crystal Structure of Bence-Jones Protein REI

Crystals suitable for X-ray analysis have been obtained from dimers of the variable halves of the κ-type Bence-Jones proteins REI (89), LEN (90), and Au (91). The structure of the REI dimer has been determined to 2.8 Å resolution (75). The REI dimer crystallizes with the two V_L fragments related by a local (non-crystallographic) twofold axis (75, 92, 93). A model of one domain was constructed from an electron density map that had been averaged by rotation about this axis. At 2.8 Å resolution, the polypeptide chain is found to fold into two sheets covering the hydrophobic interior of the domain as found in the Mcg dimer. One sheet is described as being formed by five strands and the other by three strands, with both sheets showing the commonly observed left-handed twist (Figure 9). The hydrogen bonding between strands is found to be characteristic of antiparallel pleated sheets with about 50% of the residues contributing to the β-sheet structure. As in Mcg, the N-terminal strand adds to the three strand layer in an antiparallel fashion and to the other in a parallel fashion. One turn of a distorted α-II helix formed by residues Gln 79 to Ile 83 was the only helical segment found in the structure. Three hairpin turns involving conserved glycine residues (positions 16, 41, 57) were observed.

Although the REI V_L domain was observed to have a close structural resemblance to other light chain V domains, differences existed in the folding of the first hypervariable region. This is an extended chain in V_{REI} and is folded into a helical structure in Fab New and in the Mcg light chain, presumably to accommodate the three extra residues inserted in this region in the latter proteins.

The REI dimer contact is found to include hydrophobic as well as hydrogen-bonding interactions across the $V_L : V_L$ interface (Table 2). Some contact residues are hypervariable (Figure 12). Finding that the $V_L : V_L$ contact in REI is virtually identical with the $V_L : V_L$ contact of the Mcg dimer and apparently similar to the corresponding contact in Fab New, the authors conclude that the mode of association of the V domains does not depend on the presence of the C domains.

The main chain atoms of the hypervariable regions were observed to form a cavity around the diad axis. Side chains, predominantly from the three hypervariable regions of both chains, protrude into the cavity. Tyr 49, 91, and 96 form a "slit pocket" around the diad, while Tyr 36 and Gln 89 form the bottom of the pocket. The walls of the pocket include the polar residues Gln 89 and Asn 34. The potential binding surface involves side chains and/or main chains of the hypervariable segments Tyr 49 to Asn 53, Lys 31 to Leu 33, and in particular Gln 90 to Thr 97. It was predicted that the arrangement of the residues in the cavity can nicely accommodate a hapten consisting of an aromatic ring with polar substituents, with the hapten being oriented in the pocket by the six tyrosine side chains, but no binding studies have yet been reported.

Crystallographic Studies on Human Fab Fragments

The first Fab crystals were obtained from human myeloma proteins by Rossi & Nisonoff (94) using papain digestion and later, pepsin (95). Both digests gave isomorphous crystals, but the pepsin crystals were larger (95–97) and were used in subsequent investigations. Crystals were obtained from three (out of six) myeloma proteins, two of which, New and Hil, were suitable for high resolution studies (96, 97). Since New crystals have one molecule in the asymmetric unit, whereas Hil has two, Fab' New has been investigated first.

Heavy atom derivatives have been obtained and an early 6 Å electron density map (98–100) was sufficiently detailed at this resolution to permit a clean delineation of the molecular boundaries. The Fab' structure consists of two weakly connected, globular structural subunits, each approximately $30 \times 40 \times 50$ Å in size, whereas the entire Fab' is approximately $40 \times 50 \times 80$ Å.

The authors suggest that each globular portion of the Fab' molecule represents a region consisting of two domains, one from the L chain and one from the H. Thus one subunit would be the V region ($V_L : V_H$) and the other the C region ($C_L : C_H 1$). This 6 Å map was consistent with the domain model for immunoglobulin structure originally suggested by Edelman and co-workers (6, 19, 20). The overall Fab structure could be described as a distorted tetrahedron, with the domains of homology forming the vertices. An interesting feature of the Fab' New 6 Å electron density map was the existence of a cavity-like space at one end of the molecule. This cavity was compatible with the dimensional requirements of the antigen binding site, and on this basis the authors suggested that the subunit containing this feature was the V region.

This work has been extended to yield a 2.8 Å electron density map of Fab' New (83). The V_L amino acid sequence has also been determined (101). The C_L and $C_H 1$ sequences were known, but V_H was surmised by homology with other human V_H regions and from the electron density map. Using the known part of the sequence and the electron density map, a model of the Fab New structure was constructed (83). At this resolution it became clear that the subunit previously identified as the V region was in fact the C region.

All four domains are similar in their tertiary structures to the domain structures described above. Fab' New is atypical in that V_L contains a seven residue deletion in the second hypervariable loop. Thus the V_L domain in New is more similar in structure to the constant domains.

In spite of the fact that the Fab contains two chains of differing amino acid sequence, within each region, V or C, the paired domains are related by an approximate local twofold rotational axis. The major axes of domains within each chain are also not collinear, the angle between the axes being 100–110° in the

L chain and 80–85° in the Fd. The C_L and C_H1 domains interact over a wider area than the V_L and V_H domains.

As expected, the serologic λ chain markers, Kern (86) and Oz (87), lie on the outside of the molecule, each accessible to appropriate typing antisera. Disulfide bridges present in other immunoglobulins, notably in human $\gamma2$, $\gamma3$, $\gamma4$, and μ chains, rabbit γ, mouse $\gamma2a$ and $\gamma2b$, guinea pig $\gamma2a$, another human $\gamma1$, and rabbit κ chains, could be formed in the New structure without otherwise distorting the molecule. This is taken as evidence that all of these other classes of immunoglobulin are similar to New in conformation.

In both L and H chains the hypervariable sequences (40) occur at one end of the molecule and are fully exposed to the solvent. The hypervariable regions occur at "adjacent bends of tightly packed, linear polypeptide chains." They form a shallow groove 15×6 Å in area and 6 Å in depth. Residues 55–65 and 30–33 from the heavy chain form the lower area of the site, while 27a–30 and residues close to 50 of the L chain form the upper boundary. The sides are comprised of residues 90–95 (L chain) on the left and 102–107 (H chain) on the right. The heavy chain contributes more residues to this region than the light.

Recently, it has been observed that Fab′ New binds several apolar small molecules in solution and in the crystal (102, 103). Bound in solution with relatively low affinity ($\sim 10^3$ l/mol) were orceine, dichlorophenolindophenol, coenzyme Q50, uridine, folic acid, the N-terminal hexapeptide of MSH-1 and menadione. In contrast, one molecule of IgG New (whole protein) binds two molecules of vitamin K-1 OH in solution with an affinity of 1.7×10^5 l/mol. Both the phytyl tail and the 2-methylnaphthoquinone moieties of the vitamin K-1 OH contribute to the total binding energy.

Difference Fourier maps were computed at 6 Å resolution on Fab′ New crystals treated with dichlorophenolindophenol, orceine, and menadione, and at 3.5 Å for the vitamin K-1 OH complex. All compounds showed one major peak at the same position and several minor peaks. In the vitamin K-1 OH map at 3.5 Å resolution the minor peaks were barely above background, and the major peak could be fitted with the menadione ring and the phytyl tail of the vitamin K-1 OH. The location of this major binding site is on the relatively flat end of the molecule adjacent to the hypervariable regions of both chains (see Figure 13). The menadione moiety lies in the shallow groove between heavy and light chains described previously.

The methylnaphthoquinone ring of vitamin K-1 OH makes close contact with the ring of Tyr 90 of the L chain at the bottom of the crevice, with the backbone and side chain of residue 104 (H chain), and with the backbone of L chain residues 29 and 30. Starting from the 2-methyl-1,4 naphthoquinone moiety, the phytyl tail makes contact with Gly 29 and Asn 30 of the L chain. It then approaches the backbone of L chain residues 93, 94, and residue 104 of the H chain, and terminates at the side chains of residues 54, 57, and 63 of the heavy chain. At least 10–12 residues from both V_L and V_H make contact with vitamin K-1 OH. No conformational change in the protein structure was observed subsequent to vitamin K-1 OH binding.

Crystallographic Studies on Mouse Myeloma Proteins with Antigen Binding Properties

The availability of large amounts of homogeneous proteins from transplantable tumors in the mouse (14), many with known antigen binding properties, has provided a rich source of material for crystallographic investigation. Stimulated by the observation (104) that small crystals could be obtained from MOPC 315 Fab (pepsin) fragments, a systematic attempt was made to obtain crystals of Fab (pepsin) fragments of mouse myeloma proteins with known binding properties (105). Five proteins with phosphorylcholine binding activity, TEPC 15, HOPC 8, MOPC 167, McPC 603, and MOPC 511 (1, 14), were investigated. These proteins precipitate with phosphorylcholine-containing natural antigens from a variety of organisms including *Pneumococcus, Lactobacillus, Trichoderma, Aspergillus, Proteus morganii,* and *Ascaris* (1, 14). Crystals were obtained from MOPC 167 and McPC 603. Of these, one crystal form from McPC 603 was suitable for X-ray diffraction analysis. It was demonstrated by chemical methods that these crystals bound 1 mol phosphorylcholine per mole Fab, with a binding constant of 4×10^3 l/mol. This binding constant is about 45-fold lower than that in free solution, a result assumed to be due to the ammonium sulfate of crystallization.

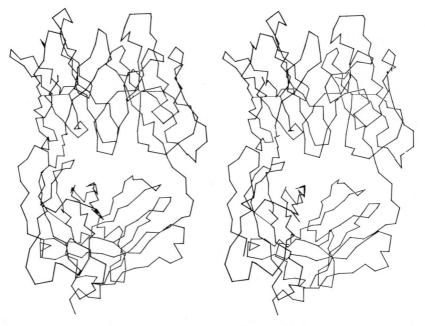

Figure 4 Stereo drawing of α-carbon skeleton of McPC 603 Fab. The V region is at the top, and the light chain on the right.

A subsequent study of an electron density map at 4.5 Å resolution was reported together with the location of the binding site for phosphorylcholine (106). At this resolution the molecule, $40 \times 50 \times 80$ Å in overall dimensions, clearly consisted of four distinct globular regions, which were interpreted to be the four domains. The domains cluster in pairs to form two regions separated by a hole and linked by two thin ribbons of density. At this resolution it was not possible to distinguish H from L, but it was clear that the structure closely resembled that found at 6 Å resolution for Fab New (99).

In order to locate the phosphorylcholine binding site (106), a difference Fourier synthesis was calculated using data from crystals soaked in solutions containing 2-(5′-acetoxymercury-2′-thienyl)-ethylphosphorylcholine (AMTEPC). The difference map showed two sites of incorporation, one on the surface of the molecule in what was called the C region, and the other at the tip of the molecule in the crevice between the two V region domains. When crystals that had been soaked originally in AMTEPC were subsequently soaked in solutions containing AMTEPC plus an excess of p-nitrophenylphosphorylcholine, the peak in the V region disappeared whereas the C region peak persisted. Since the studies in solution and in the crystal showed specific binding of phosphorylcholine to only one site per Fab (105), the displaced peak was taken to represent the hapten binding site. This conclusion was confirmed by the observation of a peak in this position in difference maps between native crystals and those soaked in a saturating amount of phosphorylcholine. The binding studies with AMTEPC illustrate one of the problems of working with crystals, namely that the AMTEPC difference map at the antigen binding site showed two peaks separated by a region of low density where there was a peak in the native map. This was interpreted as being due to the binding in the crystal of a sulfate ion from the ammonium sulfate of crystallization.

The observation of no significant peaks other than those representing hapten binding was interpreted to mean that no significant conformational changes had occurred on ligand binding. However, the possibility could not be ruled out that the sulfate binding might have caused the Fab fragment to adopt a liganded conformation before crystallization.

The subsequent extension of this work to 3.1 Å resolution (107–110) together with complete sequence data for V_H and partially for V_L (110) led to an interpretation of the above results in molecular terms. The overall appearance of the molecule (Figure 4)[1] was very similar to that previously reported for the human IgG (λ) Fab New (83) and the Mcg Bence-Jones dimer (79). The $V_H : V_L$ and $C_H 1 : C_L$ domains are related in pairs to one another by approximate twofold axes of symmetry. These two axes are not collinear, but make an angle of approximately 135° with each other.

The two domains of each chain are oriented approximately at right angles to one another. The long axes of the L domains make an angle of approximately

---

[1] The authors suggest that stereo viewers be used with figures 4–6, 8, 9, and 12. Stereo viewers may be obtained from Hubbard Scientific Company, P.O. Box 105, Northbrook, Illinois 60062, Abrams Instrument Corporation, 606 East Shiwassee Street, Lansing, Michigan 48901, or other sources.

100° with each other, whereas the corresponding angle for the H chain is approximately 80°. This results in a center-to-center distance of about 40 Å between the V_L and C_L intradomain disulfides vs a distance of about 30 Å between V_H and C_H1. A possible explanation of this distortion of symmetry might come from the more extensive interaction between the C and V domains of the H chain vs the L chain. In the $V_H:C_H1$ contact region residues 8–10, Gly-Gly-Gly, and residues 111–113, Gly-Thr-Thr of V_H are close to the C_H1 domain. It appears that these small side chains facilitate the close approach of the two heavy domains.

Within each domain the same sandwich-like structure is observed, illustrated in Figure 5. Adjacent segments within each layer are antiparallel to one another and frequently assume a β-pleated sheet configuration. The interior of the sandwich contains principally hydrophobic residues. The extended segments are linked by bends with varying degrees of sharpness, and the electron density in many of the tight bends can be fitted with β-bend configurations (111).

The quaternary structure of McPC 603 Fab (Figure 4) is similar to those in Fab New (83) and the Mcg dimer (79). The interaction between the C domains occurs principally between the four chain layers. The interactions between the V domains involve the three chain layers as well as a major portion of the additional loops found in the V domains. The C region is more compact, the interacting layers being about 10 Å apart. The distance between the intradomain disulfide bonds in the C region is 18 Å. In the V region, the interacting segments are 12–15 Å apart and the distance between the disulfide bonds is about 25 Å. The residues involved in the V interface span the complete range of variability. Some are quite conserved while others are hypervariable. For example, the large loop containing the first light chain hypervariable region is in intimate contact with most of the residues in the third heavy chain hypervariable region.

The tip of the Fab molecule contains a large wedge-shaped cavity, approximately 12 Å deep, 15 Å wide at the mouth, and 20 Å long, whose walls are lined exclusively with hypervariable residues (109, 110). Hapten binds in this cavity where it is located asymmetrically, being closer to the H chains than to the L chain (Figure 6). The choline end is bound in the interior of the cavity with the

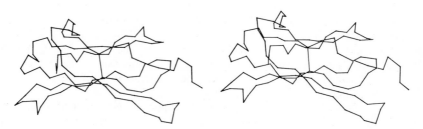

Figure 5   Stereo drawing of α-carbon skeleton of McPC 603 C_L domain. This view depicts the bilayer nature of the domain. The N terminus is on the right, with the three chain layer on top and the four chain layer on the bottom. A disulfide bond in the center of the diagram connects the two layers. Note the concavity of the four chain layer.

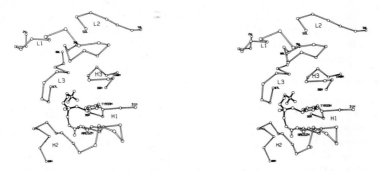

Figure 6   The hapten binding site of McPC 603. Phosphorylcholine is shown in black, liganded to the protein. The complete hypervariable surface is shown. L1, L2, and L3 refer to light chain hypervariable regions, whereas H1, H2, and H3 refer to those of the heavy chain. Numbers of the first and last residues are indicated for each hypervariable region.

phosphate group more towards the exterior. The phosphate appears to be exclusively bound by the H chain and forms specific interactions with Tyr 33 (H) and Arg 52 (H). These side chains are located so that they can form hydrogen bonds with the phosphate, and the positively charged guanidinium group presumably also acts to neutralize partly the negative charge on the phosphate. In the case of a mono-esterified phosphate, as in phosphorylcholine, the phosphate is doubly charged, but in the "true" antigenic determinant which presumably contains phosphorylcholine in a doubly esterified form (112), the phosphate will have only a single charge. The choline group appears to interact with both the L and H chains, and the acidic side chain of Glu 35 (H) is only about 5 Å away from the positively charged nitrogen of the choline. The choline also forms van der Waals interactions with the main chain atoms of residues 102–103 of the H chain and 91–94 of the L chain. The entire hapten is in close contact with the ring atoms of Tyr 33 (H).

The magnitude of the hypervariable cavity in McPC 603 is to a large extent a reflection of the insertions that occur in regions L1, H2, and H3, causing these hypervariable loops to extend farther out, thus increasing the cavity depth. The region L2 does not form a part of the cavity, being screened from it by the large loop containing L1.

COMPARISON OF STRUCTURES

The crystallographic studies discussed above confirm the general concept of antibody structure developed through chemical studies. The regions of sequence homology fold into compact domains which act as building blocks in the formation of the whole immunoglobulin molecule (19, 20), and all 11 chemically distinct domains described above have the same basic tertiary structure. Yet, whereas the domains bear a strong familial resemblance, it is clear that structural changes have occurred during the course of evolution resulting in specialized functions for the various domains of the antibody molecule. In this section we make a comparison of these domain tertiary structures and the manner in which they aggregate into functional units.

Domain Tertiary Structure

Each domain consists of segments of extended polypeptide chains connected by bends of varying lengths and shapes (75, 79, 83, 110). The basic domain structure is shown schematically in Figure 7 for the V (right) and C (left) domains. The extended segments are designated S1, S2 ... S9, and the bends connecting the segments as B12, B23 ... B89. In Figure 7 the segments lying within one layer are designated by heavy lines, and those within the other layer by thin lines. The numbering of the stretches and bends in Figure 7 has been chosen to preserve homology between C and V domains. The intradomain disulfide bond joins S2 and S8 in all the domains (109, 110).

In Figure 8 a comparison of the tertiary structures of C_L and V_H of McPC 603 (109, 110) is made, and in Figure 9 the tertiary structures of V_L REI, V_L and V_H of McPC 603 (75, 109, 110) are compared. In general, whereas the domains have similar tertiary structures, larger differences exist between the V and C domains than between the six V domains or between the five C domains thus far reported. The C and V domains bear essentially no homology in amino acid sequence (16) and it is therefore not surprising that the greatest structural differences exist between these domains.

A major difference between C and V domains lies in the existence of an additional loop in the V which has no equivalent in the C domains (79, 83, 110). This loop can be seen in Figure 8 and is designated as S4, B45, S5 in Figure 7. This additional loop does not lie in either layer and thus the sandwich nature of the V domains is less apparent.

The V and C domains (Figure 8) differ also in the length, shape, and chemical

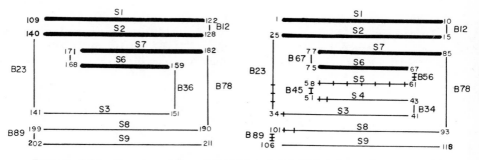

Figure 7   Schematic representation of V (right) and C (left) domain structures. Extended segments are labeled S1, S2, ... S9, and the bends connecting them are B12, B23, ... B89. Dark lines indicate stretches of one layer, while stretches indicated by thin lines lie in the other. Hypervariable regions are indicated by short perpendicular lines in the V domain. Numbering refers to McPC 603 V_H (right) and C_L (left).

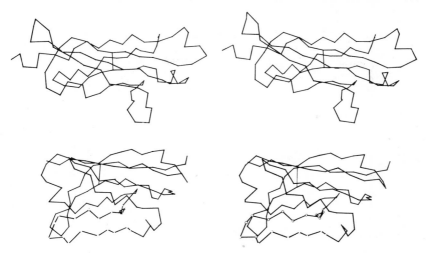

Figure 8 Stereo drawing of α-carbon backbone for McPC 603 C_L (top) and V_H (bottom) domains. Both domains are in approximately the same orientation. The loop that is present in the V domains, but missing from C, is indicated by unjoined bonds and lies on the bottom left part of the domain.

nature of the common stretches and bends. The straight segments in the C domains are apparently longer. Many of the V domains contain large, convoluted bends at the N-terminal end of the domain not found in the constant region. In the C domains, the four chain layer forms a concave surface (Figure 5); in the V domains this layer is curved in the opposite direction. Residues on the external surface of the four chain layers are mainly hydrophobic in the C domains and are involved in the strong interaction between the homologous domains across the C interface. The homologous residues in the V domains are exposed to solvent and are in general hydrophilic.

All the V domains are quite similar to each other in their three-dimensional structure (Figure 9). The most remarkable similarity is between the two V_L domains of McPC 603 and REI. The difference between the two is principally in the greater extension of the loop B23 in McPC 603 V_L (in the first hypervariable region) where McPC 603 has a six residue insertion compared to REI. The V_L REI domain has been superimposed by least square methods on V_L McPC 603. For 94 residues the root-mean-square distance between homologous α-carbon atoms is 1.4 Å (109, 110).

It is apparent from Figure 9 that V_L and V_H are also quite similar in conformation, but not to the same extent as the two V_L domains. The major differences between McPC 603 V_L and V_H lie in the loop B23, where V_L is larger, and in B45, where V_H is more extensive. Comparative amino acid sequence analyses (16) indicate that the V_H domain around B45 usually contains about five more residues than its

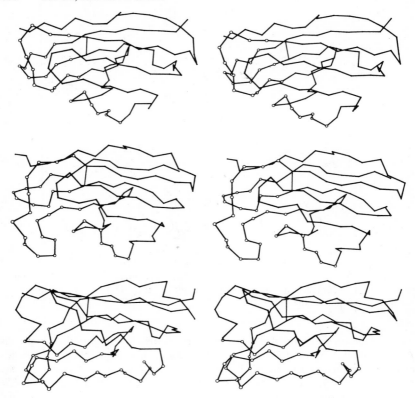

Figure 9 Stereo drawing of α-carbon backbone of three variable region domains, all in similar orientations. (*Top*) REI light chain domain. (*Middle*) McPC 603 light chain domain. (*Bottom*) McPC 603 heavy chain domain. Hypervariable residues are indicated by large circles.

light chain counterpart. When V_H and V_L from McPC 603 are superimposed, the root-mean-square displacement for 74 α-carbons is 1.9 Å. This has been represented in Figure 10 as a plot of displacement vs position in the two chains (109). The displacements observed in this plot can be compared with the variability plot of Kabat & Wu (41) for the amino acid sequence (Figure 11). It can be seen that by and large the sites of greatest displacement correspond to the hypervariable regions. The principal exception occurs in the extra hypervariable region (residues 82–89) of the H chain (42). This analysis lends strong support to the concept that immunoglobulins can incorporate large amounts of variation, i.e. substitutions, insertions, and deletions, within a localized area of the V domain without significantly altering the structure of the nonhypervariable "framework" residues.

Quaternary Structure

Changes in the chemical nature of surface side chains have resulted in a completely different quaternary structure in V regions compared to C (79, 83, 110). In the C region, the four chain layers face each other and form a large surface of interaction between domains. In striking contrast, the four chain layers are on the outside in the V region and the interface involves the three chain layers (Figures 3 and 4). This mode of association of the V domains brings the three hypervariable regions together in space to form a continuous hypervariable surface. In addition, the V region interaction causes hypervariable residues to come into contact with each other across the interface (Figure 12). This could be an important factor in determining the size as well as the shape of the binding cavity.

Since a detailed picture of the Fc has not been obtained, we can only speculate

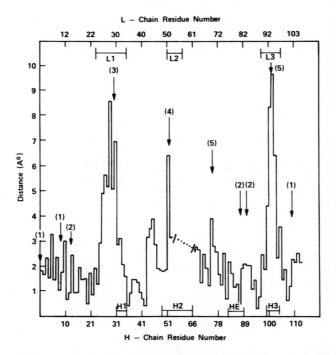

Figure 10 Differences in α-carbon positions (Å) after McPC 603 V_H is superimposed on McPC 603 V_L. Regions of sequence hypervariability are shown at top and bottom; L1, L2, and L3 for the L chain and H1, H2, HE, and H3 for the heavy. HE is the extra hypervariable region residues 81–85 (42). Arrows with numbers refer to the following: (1) One residue in L, not in H; (2) One residue in H, not in L; (3) Four residues in L, not in H; (4) Six residues in H, not in L; and (5) Two residues in H, not in L. Dotted portion represents region of uncertain structure in V_L.

on the tertiary structures of the Fc domains as well as on the nature of the interactions in this region of the molecule. The homology of the sequences of the Fc domains to those of C_L and C_H1 implies that the Fc domains will resemble the C domains more closely than the V domains of the Fab. A closer examination of the sequences, especially around the intradomain disulfide, suggests that the quaternary structure of the C_H2, and probably also of the C_H3 region, will be similar to that of the C region in the Fab.

The high resolution structure analyses of the two Bence-Jones dimers (75, 79) and of the human (83) and murine (110) Fabs have provided some interesting details regarding the domain interactions in the C and V regions of these fragments. It was found, for example (Figure 5), that the four chain layers in the C_L and C_H1 domains in McPC 603 Fab form slightly concave surfaces, cross each other at

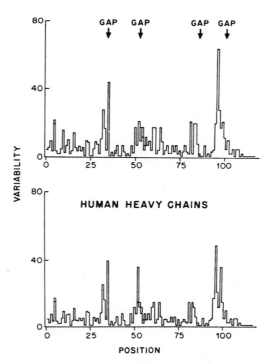

Figure 11 Amino acid variability among heavy chains as a function of position within the sequence (41). Variability is defined as the number of different residues divided by the frequency of the most common residue at a position. Reprinted with permission from Kabat, E. A., Wu, T. T. 1971. *Ann. NY Acad. Sci.* 190: 386.

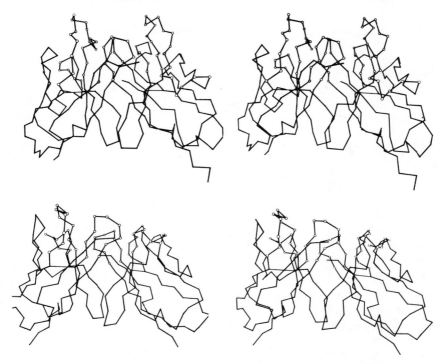

Figure 12 Skeletal representation of the McPC 603 V region (top) and REI V_L dimer (bottom). Both structures are in approximately the same orientation. The hypervariable surfaces are at the top, and the C termini at the bottom. Hypervariable residues are indicated with large circles.

approximately right angles (Figure 4), and together form a rather extensive interface (109, 110). The segments across the C region interface are almost uniformly 10 Å apart, and the C region, in comparison with the V, is more compact. Very similar results were obtained in the Mcg Bence-Jones dimer (79). In McPC 603 Fab, there are fewer aromatic side chains in the C region compared to the V region interface. In the latter, the interacting segments are more widely separated, being 12–15 Å apart. Consequently, the distance between the intradomain disulfides is about 25 Å (center to center) in the V region and 18 Å in the C. Approximately the same distances were observed in the Mcg dimer. Moreover, the distance calculated for the REI protein, where the two V_L domains are related by a twofold axis, is ∼ 24 Å.

Some of the interactions observed in the McPC 603 V region (109, 110) were also found in the REI and Mcg dimers (Table 2), especially those near the carboxyl end of the V domains. In McPC 603 Fab, interactions between many hypervariable residues have been observed. For example, almost the entire third heavy chain hypervariable segment and most of the residues in the first light chain hypervariable region are in close contact. Contacts between hypervariable residues were also found in the REI dimer (Figure 12), but only one such interaction was observed in the Mcg dimer. In this respect, the REI V_K dimer more closely resembles the McPC 603 V region. It is interesting that the two V domains have few points of contact in the Mcg dimer. Indeed, many of the residues that are in contact in both REI and McPC 603 are found to line the walls of the pocket and the main cavity in Mcg. It seems that in Mcg, part of the interface in the V region had opened up and become accessible for binding by aromatic molecules like DNP and menadione (vitamin K-3). Comparing the known λ and κ light chain sequences (16), it is not obvious why the V region interactions in the λ-type Mcg should be different from those in the κ-type REI.

The longitudinal interactions between the V_L and C_L domains and those between the V_H and C_H1 domains in the human and in the murine Fab are similar to those occurring in the two monomers of the Mcg light chain dimer. In the Fabs, the domains of the heavy chain approach each other more closely at the switch region. In McPC 603, this close approach is apparently allowed by the existence of small residues at the interface.

The Antigen Binding Site

Since the binding studies with both phosphorylcholine and vitamin K-1 OH have been performed with myeloma proteins, it is reasonable to question their physiological significance. Although direct evidence is not available, there is considerable indirect evidence to suggest that what is being observed is indeed a satisfactory model for antigen-antibody interaction.

McPC 603 is one of a number of independently induced mouse plasma cell tumor proteins that precipitate with several natural antigens, including *Pneumococcus* C polysaccharide (1, 113). Choline is a common constituent of all of these antigens, and Leon & Young (114, 115) observed that phosphorylcholine would inhibit the precipitation of *Pneumococcus* C polysaccharide by McPC 603. Antibodies to phosphorylcholine-containing antigens have been prepared from Balb/C mice that have been shown to be idiotypically identical with one phosphorylcholine binding myeloma, TEPC 15 (3). Although under these conditions, no idiotypic identity with McPC 603 has been observed, there is no good reason to doubt that phosphorylcholine is the major antigenic determinant in the interaction between *Pneumococcus* C polysaccharide and McPC 603 protein (1). In the case of Fab' New binding to vitamin K-1 OH, evidence has been cited (102, 103) that anti-idiotypic antibodies against IgG New will combine with some of the immunoglobulin species from the serum of a rabbit immunized with a vitamin K-1 bovine gamma-globulin complex.

Secondly, the affinity constants observed in McPC 603 for phosphorylcholine and in New for vitamin K-1 OH are in the right range $[5 \times 10^4 – 1 \times 10^9 \text{ l/mol} (116)]$ to correspond to antigen-antibody complexes, although both are at the lower end of the range.

Thirdly, the location of the binding site on the tip of the Fab in the region between the light and heavy chains is reassuringly reasonable when viewed in the light of existing chemical knowledge. In both, the hapten is to a large extent in contact with hypervariable residues. For McPC 603 the results of affinity labeling are consistent with the interaction being predominantly with the heavy chain (117).

Although the hypervariable end of the molecule in the case of Fab New is described as being rather flat with a narrow crevice between the light and heavy domains, whereas for McPC 603 it is more in the nature of a large cavity, both haptens bind to their respective proteins in approximately the same place (Figure 13). In both cases there are interactions with hypervariable regions of light and heavy chains. In McPC 603 the interactions come principally from the H chain, while in Fab New the extended phytyl chain of vitamin K-1 OH ensures hapten interaction with residues from each of the five hypervariable regions. Despite the disparity in size between vitamin K-1 OH and phosphorylcholine, they both bind to their respective proteins with roughly equal affinity (102, 105). In both proteins the light chain second hypervariable region plays no role in binding—in Fab New because it has been deleted and in McPC 603 because this loop is excluded from the binding site by the extensive first L chain hypervariable region. In neither protein is there any large conformational change on binding of hapten, in agreement with a low angle X-ray diffraction study on Fab/antigen interactions (118). In a recent review, Metzger (7) has dealt comprehensively with the data regarding

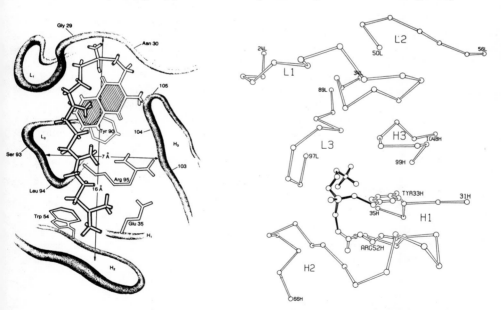

Figure 13   Binding sites of Fab' New (left) and McPC 603 (right) in similar orientations. Fab' New is shown with vitamin K-1 OH (102, reproduced with authors' permission); McPC 603 is shown binding phosphorylcholine.

possible conformational changes in whole molecules as a result of antigen or hapten binding.

The interesting binding studies with the Mcg L chain dimer (88) are more difficult to assess. Three different binding sites are observed, A and B being in the hypervariable cavity in the vicinity of the binding sites observed with Fab New and McPC 603. In contrast to these proteins where the L2 region is either absent (New) or not involved in the binding site (McPC 603), site A of Mcg involves residues 51, 52 from the second hypervariable region. The third site lies within the inner pocket in a region that involves nonhypervariable residues and defines the $V_L : V_H$ interface in the other proteins. These binding studies were performed in the crystal, and it would be interesting to know the binding constants involved.

CONCLUSION

The elucidation of the complete primary structure of an intact human immuno-globulin revealed the existence of repeating regions of sequence homology (19). This observation led to the hypothesis that these regions would fold into compact domains with similar tertiary structures (18–20). The crystallographic studies described above confirm the domain hypothesis conclusively.

While the domains act as the building blocks of immunoglobulin molecules, it is nevertheless clear from the X-ray studies, even at low resolution, that the basic structural units consist of pairs of strongly interacting, homologous domains. The immunoglobulin molecule can be viewed as a series of loosely connected structural units; two in each Fab and two (α, γ) or three (μ, ε) in the Fc.

In the V region, the basic structural unit consisting of the two variable domains is the functional entity responsible for antigenic recognition. The distinct mode of association of the V domains produces a continuous hypervariable surface whose topology can be altered by amino acid substitutions, insertions, and deletions in the hypervariable regions of both the light and heavy chains (109, 110). The hypervariable surfaces present in the total immunoglobulin pool probably generate the entire immunogenic potential of the organism.

The correlation of structure with function in the V region is made possible by the existence of small molecules that bind specifically to the hypervariable surface and demonstrate the antibody-hapten interaction. Similar correlations involving C region structural units require the development of suitable markers to serve as probes for functional specialization. These probes, with the elucidation of the detailed structures of Fc and whole immunoglobulin molecules, should in the future lead to an understanding of the structural basis of effector mechanisms in the immune response.

ACKNOWLEDGMENTS
We wish to acknowledge Drs. E. Kabat, H. Metzger, M. Navia, M. Potter, and S. Rudikoff for their advice and criticisms of the manuscript. We are grateful to Dr. Enid Silverton for permitting us to use her data on comparisons in advance of publication.

Literature Cited

1. Potter, M. 1971. *Ann. NY Acad. Sci.* 190:306–21
2. Metzger, H. 1970. *Ann. Rev. Biochem.* 39:889–928
3. Cosenza, H., Koehler, H. 1972. *Science* 176:1027–29
4. Blomberg, B., Geckler, W. R., Weigert, M. 1972. *Science* 177:178–80
5. Lieberman, R., Potter, M., Mushinski, E. B., Humphrey, W. Jr., Rudikoff, S. 1974. *J. Exp. Med.* 139:983–1001
6. Edelman, G. M., Gall, W. E. 1969. *Ann. Rev. Biochem.* 38:415–66
7. Metzger, H. 1974. *Advan. Immunol.* 18:169–207
8. Cathou, R. E., Dorrington, K. J. 1974. *Biological Macromolecules "Subunits in Biological Systems,"* ed. S. N. Timasheff, G. D. Fasman, 7. New York: Dekker, In press
9. Gally, J. A. 1973. *The Antigens,* ed. M. Sela, 161–298. New York: Academic
10. Hood, L., Prahl, J. 1971. *Advan. Immunol.* 14:291–351
11. Gally, J. A., Edelman, G. M. 1972. *Ann. Rev. Genet.* 6:1–46
12. Capra, J. D., Kehoe, J. M. 1975. *Advan. Immunol.* 19:1
13. Milstein, C., Pink, J. R. L. 1970. *Progr. Biophys. Mol. Biol.* 21:209–63
14. Potter, M. 1972. *Physiol. Rev.* 52:631–719
15. Porter, R. R. 1959. *Biochem. J.* 73:119–26
16. Dayhoff, M. O., Ed. 1972. *Atlas of Protein Sequence and Structure,* Washington DC: Nat. Biomed. Res. Found.
17. Hilschmann, N., Craig, L. C. 1965. *Proc. Nat. Acad. Sci. USA* 53:1403–9
18. Edelman, G. M., Gally, J. A. 1962. *J. Exp. Med.* 116:207–27
19. Edelman, G. M., Cunningham, B. A., Gall, W. E., Gottlieb, P. D., Rutishauser, U., Waxdal, M. J. 1969. *Proc. Nat. Acad. Sci. USA* 63:78–85
20. Edelman, G. M. 1970. *Biochemistry* 9:3197–3205
21. Putnam, F. W., Florent, G., Paul, C., Shinoda, T., Shimizu, A. 1973. *Science* 182:287–91
22. Bennich, H., von Bahr-Lindstroem, H. 1973. *9th Int. Congr. Biochem.* p. 299. (Abstr.)
23. Halpern, M. S., Koshland, M. E. 1970. *Nature* 228:1276–78
24. Mestecky, J., Zikan, J., Butler, W. T. 1971. *Science* 171:1163–65
25. O'Daly, J. A., Cebra, J. J. 1971. *Bio-chemistry* 10:3843–50
26. O'Daly, J. A., Cebra, J. J. 1971. *J. Immunol.* 107:436–48
27. Morrison, S. L., Koshland, M. E. 1972. *Proc. Nat. Acad. Sci. USA* 69:124–28
28. Kownatzki, E. 1971. *Eur. J. Immunol.* 1:486–91
29. Weinheimer, P. F., Mestecky, J., Acton, R. T. 1971. *J. Immunol.* 107:1211–12
30. Kehoe, J. M., Tomasi, T. B., Ellouz, F., Capra, J. D. 1972. *J. Immunol.* 109:59–64
31. Zikan, J. 1973. *Immunochemistry* 10:351–54
32. Wilde, C. E. III, Koshland, M. E. 1973. *Biochemistry* 12:3218–24
33. Kobayashi, K., Vaerman, J. P., Heremans, J. F. 1973. *Biochim. Biophys. Acta* 303:105–17
34. Schrohenloher, R. E., Mestecky, J., Stanton, T. H. 1973. *Biochim. Biophys. Acta* 295:576–81
35. Mestecky, J., Schrohenloher, R. E., Kulhavy, R., Wright, G. P., Tomana, M. 1974. *Proc. Nat. Acad. Sci. USA* 71:544–48
36. Tomasi, T. B. 1973. *Proc. Nat. Acad. Sci. USA* 70:3410–14
37. Milstein, C. 1967. *Nature* 216:330–32
38. Kabat, E. A. 1970. *Ann. NY Acad. Sci.* 169:43–54
39. Franěk, F. 1969. *Developmental Aspects of Antibody Formation and Structure,* ed. J. Sterzl, I. Riha. Prague: Czechoslovak Acad. Sci.
40. Wu, T. T., Kabat, E. A. 1970. *J. Exp. Med.* 132:211–50
41. Kabat, E. A., Wu, T. T. 1971. *Ann. NY Acad. Sci.* 190:382–92
42. Capra, J. D., Kehoe, J. M. 1974. *Proc. Nat. Acad. Sci. USA* 71:845–48
43. Wofsy, L., Metzger, H., Singer, S. J. 1962. *Biochemistry* 1:1031–38
44. Givol, D. 1974. *Essays Biochem.* 10:73–103
45. Noelken, M. E., Nelson, C. A., Buckley, E. C. III, Tanford, C. 1965. *J. Biol. Chem.* 240:218–24
46. Valentine, R. C., Green, N. M. 1967. *J. Mol. Biol.* 27:615–17
47. Svehag, S. E., Chesebro, B., Bloth, B. 1967. *Science* 158:933–36
48. Chesebro, B., Bloth, B., Svehag, S. E. 1968. *J. Exp. Med.* 127:399–410
49. Svehag, S. E., Bloth, B. 1970. *Science* 168:847–49
50. Svehag, S. E. 1972. *3rd Int. Convoc. Immunol. Buffalo, 1973.* 8–91. Basel: Karger

51. Shelton, E., McIntyre, K. R. 1970. *J. Mol. Biol.* 47:595–97
52. Shelton, E., Smith, M. 1970. *J. Mol. Biol.* 54:615–17
53. Dourmashkin, R. R., Virella, G., Parkhouse, R. M. E. 1971. *J. Mol. Biol.* 56:207–8
54. Green, N. M., Dourmashkin, R. R., Parkhouse, R. M. E. 1971. *J. Mol. Biol.* 56:203–6
55. Feinstein, A., Rowe, A. J. 1965. *Nature* 205:147–49
56. Feinstein, A., Munn, E. A. 1966. *J. Physiol. London* 186:64–66P
57. Feinstein, A., Munn, E. A., Munro, A. J. 1971. *Nature* 231:527–29
58. Feinstein, A., Munn, E. A., Richardson, N. E. 1971. *Ann. NY Acad. Sci.* 190:104–21
59. Poljak, R. J., Dintzis, H. M. 1966. *J. Mol. Biol.* 17:546–47
60. Poljak, R. J., Dintzis, H. M., Goldstein, D. J. 1967. *J. Mol. Biol.* 24:351–52
61. Humphrey, R. L. 1967. *J. Mol. Biol.* 29:525–26
62. Poljak, R. J., Goldstein, D. J., Humphrey, R. L., Dintzis, H. M. 1967. *Cold Spring Harbor Symp. Quant. Biol.* 32:95–98
63. Goldstein, D. J., Humphrey, R. L., Poljak, R. J. 1968. *J. Mol. Biol.* 35:247–49
64. Terry, W. D., Matthews, B. W., Davies, D. R. 1968. *Nature* 220:239–41
65. Edmundson, A. B., Wood, M. K., Schiffer, M., Hardman, K. D., Ainsworth, C. F., Ely, K. R., Deutsch, H. F. 1970. *J. Biol. Chem.* 245:2763–64
66. Palm, W., Colman, P. M. 1974. *J. Mol. Biol.* 82:587–88
67. Sarma, V. R., Silverton, E. W., Davies, D. R., Terry, W. D. 1971. *J. Biol. Chem.* 246:3753–59
68. Sarma, V. R., Davies, D. R., Labaw, L. W., Silverton, E. W., Terry, W. D. 1971. *Cold Spring Harbor Symp. Quant. Biol.* 36:413–19
69. Davies, D. R., Sarma, V. R., Labaw, L. W., Silverton, E. W., Segal, D. M., Terry, W. D. 1971. *Ann. NY Acad. Sci.* 190:122–29
70. Davies, D. R., Sarma, V. R., Labaw, L. W., Silverton, E. W., Terry, W. D. 1971. *Proc. First Int. Congr. Immunol.,* 25–32. New York: Academic
71. Labaw, L. W., Davies, D. R. 1971. *J. Biol. Chem.* 247:3760–62
72. Labaw, L. W., Davies, D. R. 1972. *J. Ultrastruct. Res.* 40:349–65
73. Deutsch, H. F., Suzuki, T. 1971. *Ann. NY Acad. Sci.* 190:472–85
74. Lopes, A. D., Steiner, L. A. 1973. *Fed. Proc.* 32:1003 (Abstr.)
75. Epp, O., Colman, P., Fehlhammer, H., Bode, W., Schiffer, M., Huber, R., Palm, W. 1974. *Eur. J. Biochem.* 45:513–24
76. Edmundson, A. B., Schiffer, M., Wood, M. K., Hardman, K. D., Ely, K. R., Ainsworth, C. F. 1971. *Cold Spring Harbor Symp. Quant. Biol.* 36:427–32
77. Schiffer, M., Hardman, K. D., Wood, M. K., Edmundson, A. B., Hook, M. E., Ely, K. R., Deutsch, H. F. 1970. *J. Biol. Chem.* 245:728–30
78. Edmundson, A. B., Schiffer, M., Ely, K. R., Wood, M. K. 1972. *Biochemistry* 11:1822–27
79. Schiffer, M., Girling, R. L., Ely, K. R., Edmundson, A. B. 1973. *Biochemistry* 12:4620–31
80. Edmundson, A. B., Ely, K. R., Girling, R. L., Abola, E. E., Schiffer, M., Westholm, F. A. 1974. *Progress in Immunology II,* ed. L. Brent, J. Holborow, 1:103–13. Amsterdam: North-Holland
81. Ely, K. R., Girling, R. L., Schiffer, M., Cunningham, D. E., Edmundson, A. B. 1973. *Biochemistry* 12:4233–37
82. Fett, J. W., Deutsch, H. F. 1974. *Biochemistry* 13:4102–14
83. Poljak, R. J., Amzel, L. M., Avey, H. P., Chen, B. L., Phizackerly, R. P., Saul, F. 1973. *Proc. Nat. Acad. Sci. USA* 70:3305–10
84. Strosberg, A. D., Fraser, K. J., Margolies, M. N., Haber, E. 1972. *Biochemistry* 11:4978–85
85. Milstein, C., Frangione, B., Pink, J. R. L. 1967. *Cold Spring Harbor Symp. Quant. Biol.* 32:31–36
86. Hess, M., Hilschmann, N., Rivat, L., Rivat, C., Ropartz, C. 1971. *Nature New Biol.* 234:58–60
87. Appella, E., Ein, D. 1967. *Proc. Nat. Acad. Sci. USA* 57:1449–54
88. Edmundson, A. B., Ely, K. R., Girling, R. L., Abola, E. E., Schiffer, M., Westholm, F. A., Fausch, M. D., Deutsch, H. F. 1974. *Biochemistry* 13:3816–27
89. Palm, W. H. 1970. *FEBS Lett.* 10:46–48
90. Solomon, A., McLaughlin, C. L., Wei, C. H., Einstein, J. R. 1970. *J. Biol. Chem.* 245:5289–91
91. Schramm, H. J. 1971. *Z. Physiol. Chem.* 352:1134–38
92. Epp, O., Palm, W., Fehlhammer, H., Ruehlmann, A., Steigemann, W., Schwager, P., Huber, R. 1972. *J. Mol. Biol.* 69:315–18
93. Colman, P. M., Epp, O., Fehlhammer, H., Bode, W., Schiffer, M., Lattman, E. E., Jones, T. A., Palm, W. 1974. *FEBS Lett.* 44:194–99
94. Rossi, G., Nisonoff, A. 1968. *Biochem.*

Biophys. Res. Commun. 31:914–18

95. Rossi, G., Choi, T. K., Nisonoff, A. 1969. *Nature* 223:837–38
96. Avey, H. P., Poljak, R. J., Rossi, G., Nisonoff, A. 1968. *Nature* 220:1248–49
97. Humphrey, R. L., Avey, H. P., Becka, L. N., Poljak, R. J., Rossi, G., Choi, T. K., Nisonoff, A. 1969. *J. Mol. Biol.* 43:223–26
98. Poljak, R. J., Amzel, L. M., Avey, H. P., Becka, L. N., Goldstein, D. J., Humphrey, R. L. 1971. *Cold Spring Harbor Symp. Quant. Biol.* 36:421–25
99. Poljak, R. J., Amzel, L. M., Avey, H. P., Becka, L. N., Nisonoff, A. 1972. *Nature New Biol.* 235:137–40
100. Poljak, R. J. 1973. *Contemp. Top. Mol. Immunol.* 2:1–26
101. Chen, B. L., Poljak, R. J. 1974. *Biochemistry* 13:1295–1302
102. Amzel, L. M., Poljak, R. J., Saul, F., Varga, J. M., Richards, F. F. 1974. *Proc. Nat. Acad. Sci. USA* 71:1427–30
103. Richards, F. F., Amzel, L. M., Konigsberg, W. H., Manjula, B. N., Poljak, R. J., Rosenstein, R. W., Saul, F., Varga, J. M. 1974. *The Immune System,* ed. E. E. Sercarz, A. R. Williamson, C. F. Cox, 53–67. New York: Academic
104. Inbar, D., Rotman, M., Givol, D. 1971. *J. Biol. Chem.* 246:6272–75
105. Rudikoff, S., Potter, M., Segal, D. M., Padlan, E. A., Davies, D. R. 1972. *Proc. Nat. Acad. Sci. USA* 69:3689–92
106. Padlan, E. A., Segal, D. M., Rudikoff, S., Potter, M., Spande, T., Davies, D. R. 1973. *Nature New Biol.* 245:165–67
107. Padlan, E. A., Segal, D. M., Cohen, G. H., Davies, D. R., Rudikoff, S., Potter, M. See Ref. 103, 7–14
108. Padlan, E. A., Segal, D. M., Cohen, G. H., Davies, D. R., Rudikoff, S., Potter, M. 1974. *Transplant. Proc.* In press
109. Segal, D. M., Padlan, E. A., Cohen, G. H., Silverton, E. W., Davies, D. R., Rudikoff, S., Potter, M. 1974. See Ref. 80, 93–102
110. Segal, D. M., Padlan, E. A., Cohen, G. H., Rudikoff, S., Potter, M., Davies, D. R. 1974. *Proc. Nat. Acad. Sci. USA* 71:4298–4302
111. Venkatachalam, C. M. 1968. *Biopolymers* 6:1425–36
112. Brundish, D. E., Baddiley, J. 1968. *Biochem. J.* 110:573–82
113. Cohn, M. 1967. *Cold Spring Harbor Symp. Quant. Biol.* 32:211–21
114. Leon, M. A., Young, N. M. 1970. *Fed. Proc.* 29:437 (Abstr.)
115. Leon, M. A., Young, N. M. 1971. *Biochemistry* 10:1424–29
116. Karush, F. 1962. *Advan. Immunol.* 2:1–40
117. Chesebro, B., Hadler, N., Metzger, H. 1972. *3rd Int. Convoc. Immunol. Buffalo,* 1973. 205–17. Basel: Karger
118. Pilz, I., Kratky, O., Licht, A., Sela, M. 1973. *Biochemistry* 12:4998–5005

PROSTAGLANDINS[1]

×898

B. Samuelsson, E. Granström, K. Green, M. Hamberg, and S. Hammarström
Department of Chemistry, Karolinska Institutet, 104 01 Stockholm, Sweden

CONTENTS

The literature on prostaglandins[2] is expanding at a rapidly increasing rate. This topic was covered three years ago in the *Annual Review of Biochemistry* (1) and in many reviews elsewhere. Therefore and because of the space limitation, we have decided to deal with only a limited number of problems.

[1] Abbreviations: PG: prostaglandin, PGA_1: 15-hydroxy-9-ketoprosta-10,13(*trans*)-dienoic acid, PGA_2: 15-hydroxy-9-ketoprosta-5(*cis*),10,13(*trans*)-trienoic acid, PGB_1: 15-hydroxy-9-ketoprosta-8(12),13(*trans*)-dienoic acid, PGC_1: 15-hydroxy-9-ketoprosta-11,13(*trans*)-dienoic acid, PGD_1: 9α,15-dihydroxy-11-ketoprost-13(*trans*)-enoic acid, PGE_1: 11α,15-dihydroxy-9-ketoprost-13(*trans*)-enoic acid, PGE_2: 11α,15-dihydroxy-9-ketoprosta-5(*cis*),13(*trans*)-dienoic acid, $PGF_{1\alpha}$: 9α,11α,15-trihydroxyprost-13(*trans*)-enoic acid, $PGF_{2\alpha}$: 9α,11α,15-trihydroxyprosta-5(*cis*),13(*trans*)-dienoic acid, PGG_2: 15-hydroperoxy-9α,11α-peroxidoprosta-5(*cis*),13(*trans*)-dienoic acid, and PGH_2: 15-hydroxy-9α,11α-peroxido-prosta-5(*cis*),13(*trans*)-dienoic acid.

[2] A prostaglandin bibliography, prepared by J. E. Pike and J. R. Weeks, is distributed by The Upjohn Company.

BIOSYNTHESIS

Assay Methods

Most work on biosynthesis of prostaglandins in various tissues has been performed with labeled precursor acids and chromatographic identification of the labeled product. During recent years prostaglandin production by tissue homogenates has also been monitored by gas-liquid chromatography with electron capture detection (2) and by radioimmunoassay and multiple ion analysis (see section on analytical methods). In addition the Zimmerman reaction (3), spectrophotometric determination of adrenochrome (4), and the thiobarbituric acid reaction (5) have been used.

With these methods a vast number of tissues are found to produce prostaglandins. Recent examples are prostaglandin synthesis in skin (2, 6–10), platelets (11–14), mouse fibrosarcoma cells (15), virus-transformed baby hamster kidney fibroblasts (16), and a number of tissues of lower animals (17). The gorgonian, *Plexaura homomalla,* was a rich source of 15-epi-PGA_2 and its methyl ester acetate derivative (18) and also contained smaller amounts of 15-epi-PGE_2, PGE_2, PGA_2, and the methyl esters of 15-epi-PGE_2 and 15-epi-PGA_2 (19). Homogenates of *P. homomalla* catalyzed the conversion of labeled arachidonic acid into PGA_2 and other products (20).

The endogenous synthesis of prostaglandins in intact animals has been determined by monitoring specific urinary metabolites (21–26). The following conditions and agents were accompanied by an altered rate of synthesis of prostaglandins: cold stress in the rat (increased formation of $PGE_1 + PGE_2$) (22), scalding injury in the guinea pig (increased formation of $PGE_1 + PGE_2$) (27), anaphylaxis in the guinea pig (increased formation of $PGF_{1\alpha} + PGF_{2\alpha}$) (26), and pregnancy in the human (increased formation of $PGF_{1\alpha} + PGF_{2\alpha}$) (28); indomethacin and other nonsteroidal anti-inflammatory drugs were shown to inhibit synthesis of $PGE_1 + PGE_2$ in the guinea pig (23) and man (24) and of $PGF_{1\alpha} + PGF_{2\alpha}$ in the guinea pig (26). Basal levels in man of the major plasma metabolites of PGE_2 and $PGF_{2\alpha}$ (11α-hydroxy-9,15-diketoprost-5-enoic acid and 9α,11α-dihydroxy-15-ketoprost-5-enoic acid, respectively) have also been determined (29). By measuring the major plasma metabolite, an elevated synthesis rate of $PGF_{2\alpha}$ during parturition in the human was demonstrated (30).

The half-life of the major plasma metabolites of PGE_2 and $PGF_{2\alpha}$ in the circulation is very short (21, 31). Thus, quantitative determination of these metabolites makes it possible to follow acute changes in prostaglandin synthesis. However, if frequent sampling is not made, the short half-life may be a drawback in certain situations, since synthesis and release of prostaglandins have occurred intermittently in a number of studies (cf 32). In such cases determination of urinary metabolites may be preferable.

Enzymes and Cofactors

The prostaglandin synthetase complex is membrane bound and there is one report on the solubilization of the enzyme system (33). Preparation and enzymatic properties of an acetone pentane powder of the sheep vesicular gland enzyme have been described (34).

Bovine seminal vesicle microsomes were reported to possess a prostaglandin synthetase, whose activity was apparent when a heat labile, nondialyzable inhibitor present in the supernatant fraction was removed (3). The conversion of 8,11,14-eicosatrienoic acid into prostaglandins by the bovine synthetase was stimulated by catecholamines and serotonin. L-Epinephrine enhanced formation of $PGF_{1\alpha}$, whereas serotonin favored formation of PGD_1 (35, 36). Catecholamines also stimulated prostaglandin biosynthesis in the vesicular gland of sheep (37) and in homogenates of rat stomach fundus (38).

In the presence of Cu^{2+}, prostaglandin production by sheep vesicular gland tissue shifts from PGE to PGF_α compounds (39, 40). Formation of PGF_α compounds could be further enhanced by addition of dithiols, such as dithiothreitol and dihydrolipoamide (39). The effect of copper ion was suggested to result from pref-

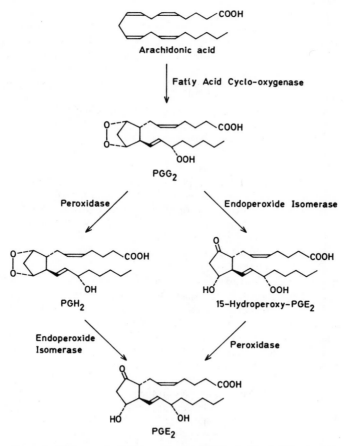

Figure 1   Pathways in the biosynthesis of PGE_2 from arachidonic acid.

erential inactivation of a PGE synthetase isoenzyme (40). However, other experiments have shown that PGE and PGF_α compounds originate in a common endoperoxide intermediate (37, 41). Furthermore, recent work with prostaglandin endoperoxides has shown that the microsomal sheep vesicular gland synthetase consists of fatty acid cyclo-oxygenase(s) and an endoperoxide isomerase (42) (Figure 1). The latter enzyme was readily inactivated by p-mercuribenzoate and other sulfhydryl blockers (42), and it thus seems probable that the observed effect of Cu^{2+} can be ascribed at least partly to inhibition of the endoperoxide isomerase.

Two types of fatty acid dioxygenase activities were present in acetone powder preparations of the sheep vesicular gland (43). One activity (E_a) was stimulated by phenol and suppressed by glutathione peroxidase. The latter fact was taken as evidence for the presence of a separate hydroperoxide (product) binding site as proposed for soybean lipoxygenase (44). The other dioxygenase activity (E_b) was not affected by phenol or glutathione peroxidase. E_a and E_b were both inactivated in the presence of the fatty acid substrate and oxygen. The kinetic formulation of E_a resembled that of soybean lipoxygenase in being product-activated and self-destructive. Further studies (45, 46) showed that diethyldithiocarbamate and other agents capable of complexing Cu^{2+} could reversibly inhibit oxygenation of unsaturated fatty acids by acetone powders of the sheep vesicular gland. This suggested that protein-bound copper ion could play a role in the interaction of the oxygenase with oxygen and the fatty acid substrate. o-Phenanthroline and other aromatic hydrocarbons also inhibited the oxygenation, which was suggested to be due to binding to a hydrophobic site on the oxygenase (46).

Inhibitors

A significant contribution since the previous review (1) was the discovery that indomethacin and aspirin inhibit prostaglandin biosynthesis (47–49). In a homogenate of guinea pig lung, the ID_{50}s for indomethacin and aspirin were 0.75 μM and 35 μM, respectively. Naproxen, another nonsteroidal anti-inflammatory drug, was somewhat less potent than indomethacin (50). Interestingly, a number of psychotropic drugs, such as phenelzine and tranylcypromine, are also potent inhibitors of prostaglandin biosynthesis (51).

Limited information is available on the mode of action of nonsteroidal anti-inflammatory drugs on the prostaglandin synthetase complex. Aspirin and indomethacin did not instantly inhibit the dioxygenase but acted in a time- and concentration-dependent manner; o-phenanthroline protected the enzyme preparation from the inhibitory action of the two drugs (52). Different drugs did not always inhibit formation of PGE, PGF_α, and PGD compounds to the same extent, which suggests that different anti-inflammatory drugs may have different target enzymes in the synthetase complex (5). Furthermore, a given drug did not always inhibit prostaglandin synthesis in different tissues to the same extent (53).

Comprehensive reviews on nonsteroidal anti-inflammatory drugs as inhibitors of prostaglandin biosynthesis are available (54, 55).

Other inhibitors of prostaglandin biosynthesis include 5,8,11,14-eicosatetraynoic acid (56), 8,12,14-eicosatrienoic acid, and 5,8,12,14-eicosatetraenoic acid (57), decanoic acid (34), a bicyclo [2.2.1] heptene derivative (58), and a number of 5-oxa- and 7-oxa-prostanoate derivatives (59). The first mentioned acid was also a potent inhibitor of soybean lipoxygenase (60) and of a novel lipoxygenase in human platelets (61).

Prostaglandin Endoperoxides

The previously postulated endoperoxide intermediate in prostaglandin biosynthesis (for review, see 62) was recently detected (63, 64) and isolated for the first time (42) (PGH_2, Figure 1). In subsequent work an additional endoperoxide carrying a hydroperoxy group at C-15 was isolated (13, 65) (PGG_2, Figure 1). The two endoperoxides were stable for months in dry acetone at $-20°C$ but decomposed rapidly upon addition of hydroxylic solvents. The half-life in aqueous medium at $37°$ was about 5 min (13). The products formed in aqueous medium were predominantly PGE compounds (13, 65), whereas treatment of the endoperoxides with mild chemical reducing agents afforded $PGF_{2\alpha}$ (13, 42, 65). The isolation of PGG_2 for the first time proved that the hydroxyl group at C-15 of the prostaglandins is introduced by a dioxygenase reaction. The recent observation that cis-14,15-epoxy-8,11-eicosadienoic acid did not serve as a precursor of PGE_1 is in agreement with this result (66).

Two pathways in the formation of PGE_2 from PGG_2 seemed possible (65, 67) (Figure 1). The facts that 15-hydroperoxy-PGE_2 had been detected in incubation mixtures of arachidonic acid with preparations of sheep vesicular gland (67) and that the fatty acid cyclo-oxygenase(s) and endoperoxide isomerase are both membrane bound whereas the peroxidase is soluble (65, 67) suggested that the preferred pathway would be $PGG_2 \rightarrow$ 15-hydroperoxy-$PGE_2 \rightarrow PGE_2$ (67). Reduced glutathione was found to stimulate both the endoperoxide isomerase and the peroxidase (65, 67), and the former enzyme was readily inhibited by p-mercuribenzoate and other SH group blocking agents (42).

The presence in a number of tissues of a soluble endoperoxide isomerase that catalyzed the conversion of PGH compounds into PGD compounds has been reported (65). It was also claimed that PGD_1 was the major prostaglandin formed from 8,11,14-eicosatrienoic acid in lung, stomach, intestine, coagulating gland, and skin from rats and that PGD compounds are devoid of biological activity (65). However, PGD_2 was found to be considerably more potent than $PGF_{2\alpha}$ on the isolated guinea pig trachea and in increasing the airway resistance in the anesthetized guinea pig (67, 68). PGD_2 has also been shown to be a very potent inhibitor of platelet aggregation (69).

The pronounced activity of the endoperoxides on a number of smooth muscle organs (65, 67, 68), the effect of PGH_2 on adenyl cyclase in adipocyte ghosts (70), and the fact that PGG_2 and PGH_2 cause aggregation of human blood platelets (13) indicate that the prostaglandin endoperoxides may be biologically important not only as precursors of the classical prostaglandins but also through their own effects.

METABOLISM

Determination of the structures of prostaglandin metabolites, studies of their metabolic pathways, and the development of analytical methods are prerequisites for quantitative studies on the endogenous formation of prostaglandins under physiological and pathological conditions. Metabolic studies have recently been carried out in the human and other species (21, 23, 31, 71–82).

Tritium-labeled PGE_2 or $PGF_{2\alpha}$ administered intravenously to human subjects is rapidly converted into the corresponding 15-keto-13,14-dihydro metabolites by the enzymes prostaglandin dehydrogenase and Δ^{13} reductase (21, 31). Thus, $1\frac{1}{2}$ min after the injection of PGE_2, only about 3% was present as PGE_2 in blood, whereas more than 40% was recovered as the 15-keto-13,14-dihydro metabolite (21). Similar data were obtained for $PGF_{2\alpha}$ (31). In the latter case, 13,14-dihydro-$PGF_{2\alpha}$ was also identified (31, 83). The half-life of the primary prostaglandins is obviously very short, less than 1 min, whereas that of the major metabolite was found to be about 8 min (21, 31). The sequence of these early metabolic steps has been studied. It was found that 13,14-dihydro-PGE_2 was not formed by direct reduction of the Δ^{13} double bond of PGE_2 but via the 15-keto compound (84, 85) and that the reduction of the keto group at C-15 was catalyzed by an enzyme different from prostanoate dehydrogenase (85).

Identification of some of these $PGF_{2\alpha}$ metabolites in the human after administration of PGE_2 is noteworthy (81). The in vivo conversion of PGE_2 into a PGF derivative was first demonstrated in the guinea pig, where the major urinary PGE_2 metabolite was identified as a derivative of $PGF_{2\beta}$ (23). In an in vitro study using the soluble fraction of homogenates of guinea pig liver it was demonstrated, however, that PGE_2, 15-keto-13,14-dihydro-PGE_2, and 13,14-dihydro-PGE_2 were reduced into the corresponding F_α derivatives (86). Later, the enzymatic conversion of PGE_2 into $PGF_{2\alpha}$ in several rat tissues was observed (87). The presence of a 9-keto reductase was demonstrated in the cellular fraction of sheep blood (88). A similar enzyme has also been found in certain microorganisms (89).

The further metabolism of prostaglandins has been extensively studied in several species: man (21, 71–74, 81), rat (75, 76, 80, 82), guinea pig (23, 77, 78), and rabbit (79). Prior to excretion into urine the compounds are degraded by several mechanisms: one or two steps of β oxidation from the carboxyl end to yield di-nor or tetra-nor compounds, ω oxidation to afford $\omega 1$ and $\omega 2$ hydroxy compounds and eventually dicarboxylic acids, and in some species β oxidation also from the ω end of the dioic acids with the formation of C_{14} metabolites. In the case of E prostaglandins, dehydration to B derivatives may also occur. A deoxy-prostaglandin was also identified as a metabolite of $PGF_{2\alpha}$ in man (73). A large number of urinary metabolites, formed by combinations of these reactions, have been identified. Thus, the main metabolites of PGE_2 and $PGF_{2\alpha}$ are 7α-hydroxy-5,11-diketotetranorprostane-1, 16-dioic acid and 5α,7α-dihydroxy-11-ketotetranorprostane-1,16-dioic acid, respectively. Figure 2 shows the urinary metabolites of $PGF_{2\alpha}$ so far identified in the human.

C_{20} ACIDS	C_{18} ACIDS	C_{16} ACIDS	C_{14} ACIDS

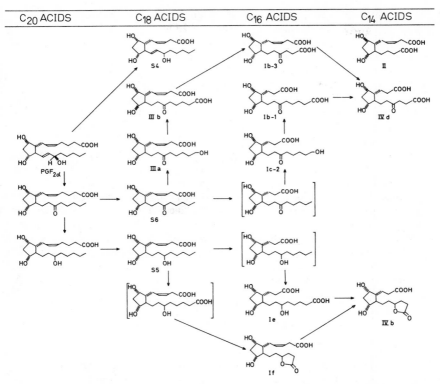

Figure 2   Tentative pathways in the metablism of $PGF_{2\alpha}$ in the human female (71–74). Compounds within brackets have not been identified. The structures of metabolites S4, S5, and S6 have recently been determined (E. Granström, unpublished results).

Methods for quantitative determination of the major urinary metabolites of $PGE_1 + PGE_2$ and $PGF_{1\alpha} + PGF_{2\alpha}$ have been developed (24, 25). Data on excretion of these metabolites have been used together with information on recovery of metabolites from their precursors to calculate the synthesis of prostaglandins of the PGE and PGF groups. For human males, a daily PGE production of 46–333 μg was found; for females 18–38 μg. The PGF production was 42–120 μg and 36–61 μg, respectively (24, 25).

Recently, considerable interest has been taken in metabolism of prostaglandins of the A type. It was earlier demonstrated that the biological activity of PGA survived passage through the lungs in perfusion experiments (90, 91). This finding has led many investigators to assume that prostaglandins of the A type are not metabolized by the lungs but are circulating hormones. However, the identity of the biologically active compound was never established, and the possibility that the product is an active metabolite of PGA must be considered. In the earlier studies of the properties of 15-prostanoate dehydrogenase from swine lung, prostaglandins of the A type were found to be good substrates for the enzyme, particularly PGA_1 (92).

Recently the presence of an isomerase, catalyzing the conversion of PGA to PGC, was demonstrated in the plasma of several species (93, 94). The nature of the isomerase(s) has been studied by two groups (95, 96). Several discrepancies in their results have been reported, concerning both its occurrence in some species and some of its properties. The presence of PGA isomerase has not been demonstrated in any tissues so far (95), nor has it been found in the human. An enzyme that catalyzes the conversion of PGA but not E or F into polar, biologically inactive metabolites has recently been found in human red blood cells (97).

Studies on the metabolism of prostaglandins in the human placenta showed that this tissue was a rich source of a 15-hydroxyprostaglandin dehydrogenase, and a partial purification of the enzyme and some properties were reported (98). Further studies on this enzyme, including the effects of progesterone and estrogen have recently been published (99). The presence of a placental dehydrase, catalyzing the conversion of PGE_1 to PGA_1, has been reported (100). However, the thin layer system used did not distinguish between PGA_1 and 15-keto-13,14-dihydro-PGE_1. The latter product is likely to be formed to a great extent, and additional identification therefore seems justified. The authors compared placentas from normal and eclamptic subjects and found a depressed metabolism in placentas from the latter group.

QUANTITATIVE ANALYSIS

General Considerations

During the past few years highly sensitive techniques based on gas chromatography with electron capture detectors, gas chromatography-mass spectrometry, and radio-immunological methods have been developed for quantitative analysis of prostaglandins.

The prostaglandins are rapidly synthesized from the precursor acids by the prostaglandin synthetase complex, and prostaglandins can also be formed by autoxidative cyclization (101). The rapid biosynthesis of prostaglandins expected after removal of a tissue sample and especially after homogenization makes it doubtful whether it is possible to determine the in situ levels of these compounds, even if there are no objections to the method of analysis. Thus, no PGE_2 could be detected (less than 10 ng/g) in rat tissues immediately homogenized in ethanol, whereas homogenization in saline of 0°C for 5 min gave 200–300 ng PGE_2 per gram tissue (cf 2, 102). It should be stressed that even gentle handling or slight mechanical trauma of a tissue can initiate biosynthesis of prostaglandins.

A few aspects have to be emphasized in connection with analyses of prostaglandins in plasma. Intravenous administration of tracer amounts of tritiated PGE_2 or $PGF_{2\alpha}$ to humans (simulating endogenous release from a tissue) has demonstrated a very short half-life of primary prostaglandins and a significantly longer half-life of the corresponding 15-keto-13,14-dihydro prostaglandins (cf section on metabolism).

During infusion of $PGF_{2\alpha}$ at a rate of 75 $\mu g/min$ the serum levels of 15-keto-13,14-dihydro-$PGF_{2\alpha}$ were 10–70 times higher (55–143 ng/ml) than the levels of $PGF_{2\alpha}$ (1.7–10.4 ng/ml) (103). The amount of $PGF_{2\alpha}$ reaching the blood in this experiment was about 2000 times higher than that calculated to be released under basal conditions in females (25). It was therefore proposed that the basal endogenous plasma levels of $PGF_{2\alpha}$ should be about 2 pg/ml and corresponding levels of 15-keto-13,14-dihydro-$PGF_{2\alpha}$ about 50 pg/ml (cf 104). The level of the latter compound agreed well with data obtained by gas-liquid chromatography-mass spectrometry (GLC-MS) (29, 83). The reason for the very high plasma or serum levels of $PGF_{2\alpha}$ and also PGE_2 published during recent years is very likely release of these compounds from platelets during collection and handling of the samples (cf 13), autoxidation of precursor acids (cf 101), and also suboptimal techniques, as reflected by wide discrepancies between levels obtained from identical samples by different methods in a collaborative study (105).

On the basis of the considerations discussed above, the levels of endogenous primary prostaglandins measured in peripheral serum or plasma are unrelated to the total endogenous production of prostaglandins in the body. Quantitation of corresponding 15-keto-13,14-dihydrometabolites as indicators for release of primary prostaglandins should be used instead (for detailed discussion, see 104 and 106).

Experiments using equilibrium dialysis indicate that primary prostaglandins can be bound to human plasma albumin (107–109), although this binding is weak and freely reversible. Addition of 1–2 μg of deuterated and tritiated $PGF_{2\alpha}$ carrier for quantitative GLC-MS results in no appreciable loss of radioactivity after extraction (110–112). Equilibration between endogenous prostaglandins and added carriers requires less than 10 min (110). It is therefore unlikely that protein binding interferes with at least those assays using deuterated carriers.

Gas Chromatography with Electron Capture Detectors

The usefulness of GLC with the electron capture detector for measuring PGE_2 after conversion to PGB_2 was demonstrated in 1970 (2) utilizing ω-nor-PGE_2 and ω-homo-PGE_1 as internal standards (lower limit of detection 1 ng). This method has been applied to measurement of PGE compounds in tissue homogenates of various species (17) and in amniotic fluid before and during labor (102, 113).

A similar technique was used to quantitate tetranor-PGE_1 and tetranor-PGB_1, metabolites of PGE_1 and PGE_2, in rat urine and to demonstrate an increased biosynthesis of prostaglandins of the E type during cold stress (22). Somewhat better sensitivity was obtained due to more specific purification and different gas chromatographic conditions. Evidence for a lower limit of detection of PGB compounds in the 200 pg range has been presented (114).

The usefulness of the electron capture detector in quantitation of PGB compounds (prepared by alkali treatment of PGA or PGE compounds) in biological samples is mainly due to the electron capturing properties inherent in the PGB structure. A number of studies on the use of halogenated derivatives of prostaglandins (mainly PGF) in connection with electron capture detection have also appeared (102, 115–117).

However, these methods have the disadvantage that many impurities are also converted into electron trapping derivatives.

Gas Chromatography—Mass Spectrometry

Combined GLC-MS has been used in developing sensitive, specific, and accurate methods for quantitation of prostaglandins. Deuterated carriers were used as internal standards during purification and gas chromatography. Two principally different techniques have been used for measurement of H/D ratios, i.e. the accelerating voltage alternator (AVA) method and repetitive magnetic scanning.

Deuterium-labeled carriers have been prepared for a large number of prostaglandins and their metabolites (23, 24, 29, 111, 112, 118–124). For a detailed discussion on carriers and techniques, see (125).

A number of techniques for quantitating prostaglandins utilizing an accelerating voltage alternator, with or without computer, have been developed (29, 111, 112, 120, 123, 126–128).

The methods developed for PGE_2, $PGF_{2\alpha}$, and their initial metabolites have been applied in many biological studies (29, 112, 123). Pharmacokinetic studies on $PGF_{2\alpha}$ and the initially formed metabolites in plasma have been performed (83, 103, 129–131). The endogenous levels of 15-keto-13,14-dihydro-PGE_2 and 15-keto-13, 14-dihydro $PGF_{2\alpha}$ in human peripheral plasma (5–92 pg/ml) have been measured (29, 83). The appearance of increasing amounts of $PGF_{2\alpha}$ in amniotic fluid after extra-amniotic hypertonic saline was shown (132). The increase of the peripheral plasma level of the latter compound parallel to progress of human labor has also been demonstrated (30). The lack of significant influence of an intrauterine device on the levels of $PGF_{2\alpha}$ and 15-keto-13,14-dihydro $PGF_{2\alpha}$ in human endometrium was shown (133). The same technique was utilized to demonstrate increased production of $PGF_{2\alpha}$ and PGE_2 in virus-transformed hamster fibroblasts relative to normal cells (16).

The major metabolites of $PGE_2 + PGE_1$ and $PGF_{2\alpha} + PGF_{1\alpha}$ in human urine have been quantitated, giving important information on the endogenous production rates of those compounds, and it was also demonstrated that aspirin and indomethacin reduced the excretion considerably (24, 25). It was shown that the production of $PGF_{1\alpha} + PGF_{2\alpha}$ in man gradually increased 4–5 times during pregnancy with an abrupt rise during labor (28). In the guinea pig anaphylaxis caused an increment in the excretion of metabolites of $PGF_{1\alpha} + PGF_{2\alpha}$ (26). It has also been possible (127, 134) to demonstrate the presence of small amounts of PGE_2 and $PGF_{2\alpha}$ in human urine.

Quantitative mass spectrometry using repetitive scanning (122, 135) has also been applied in a number of biological studies. A metabolite of $PGF_{1\alpha}$ in rat urine has thus been quantitated (118) and the occurrence of $PGF_{2\alpha}$ in sheep uterine vein plasma was demonstrated, establishing its role as luteolytic hormone (119, 136). The presence of endogenous $PGF_{2\alpha}$ in amniotic fluid (midtrimester patients) and in plasma during infusion of $PGF_{2\alpha}$ has been investigated (121, 122, 137). Measurements of PGE_2 and $PGF_{2\alpha}$ in pharmacological experiments in the monkey have also been carried out using this technique with computer control (110).

Radioimmunoassay

In recent years, a number of radioimmunoassays have been developed for prostaglandins. Most of these have been aimed at prostaglandins of the F type, mainly $PGF_{2\alpha}$ (see 138–148 and references therein), but several assays have also been developed for prostaglandins of the E, A, or B types (149–156 and references therein) and for some prostaglandin metabolites (106, 157–164). Related methods are a viro-immunoassay for $PGF_{2\alpha}$ (165) and a membrane binding assay for PGE (166).

For production of antisera, the prostaglandins have been rendered antigenic by coupling the carboxyl group to a protein molecule or a polypeptide. The specificity of the antisera obtained varies in different laboratories and with different compounds. However, some general features can be seen: a high degree of specificity is commonly exhibited toward structures in the cyclopentane ring and the C-12 side chain. In contrast, antibodies often cannot distinguish between compounds of the PG_1 and PG_2 type because of the vicinity of the 5,6 position to the site of coupling.

Comparatively few problems have been encountered in developing assays for PGF. The sera obtained generally have high specificities, with the exception that $PGF_{1\alpha}$ and $PGF_{2\alpha}$ often are assayed together for the reason mentioned above. However, attempts to develop radioimmunoassays for PGE or PGA have in most cases led to antisera which crossreact to a considerable degree with each other and also with PGB (149, 152, 153, 160). The general impression has been that it is very difficult to obtain antisera specifically directed against prostaglandins of the A, B, or E type, and in most laboratories a chromatographic separation of AB, E, and F groups has been included in the processing of the sample to allow reliable determinations. The low degree of specificity was probably due to dehydration of PGE to PGA, caused by the carbodiimide used in the conjugation procedure (167). A further conversion of A to B in the protein complex might then take place in vivo, catalyzed by the isomerases present in plasma (90–92).

The problem with the initial dehydration might be overcome by the use of N,N^1-carbonyldiimidazole as the coupling reagent (168). Some data also indicate that the nature of the protein carrier is of importance: thus the use of *Pneumococcus* cells as a second carrier (151, 155) or coupling to thyroglobulin (169, 170) resulted in antisera against A and E of high specificities. An alternative in the case of PGE_2 is reduction with $NaBH_4$ followed by the separate estimation of the formed $PGF_{2\alpha}$ and $PGF_{2\beta}$ (171).

Considerable effort has been spent on assaying levels of primary prostaglandins, mainly F, in peripheral plasma or serum in the human or other species, and relating the concentrations to various physiological conditions. The data obtained have been conflicting in many cases. So-called normal plasma or serum levels of prostaglandins have been reported from many laboratories and range from about 10 pg/ml to several thousand pg/ml (139, 144–146, 149, 150, 172–176). There are doubtless several reasons for these large discrepancies, and to eliminate differences caused by variations in sampling techniques, a comparative study was organized in Stockholm in 1972. A standard of $PGF_{2\alpha}$ and plasma and serum samples from a pool were sent to 10 laboratories. However, large variations in the $F_{2\alpha}$ levels were obtained even with these standardized samples. The results were presented at the International Conference on Prostaglandins in Vienna, 1972 (105).

To illustrate the large discrepancies between the results from different laboratories, a few examples will be given. Peripheral plasma or serum PGF levels in the human during pregnancy have been reported to be unchanged or to increase slightly (145, 146), to increase to a peak during the second trimester (173), or to reach the lowest values during this period (176, 177). However, a continuous increase in PGF production throughout pregnancy, measured by analyzing a urinary metabolite, has been unequivocally established (28, see p. 678). Most investigators report increased F levels during labor (146, 176, 178, 179, however, cf also 180). Attempts to establish a closer correlation between F levels and the stage of labor have given variable results [178, 179, 181 (bioassay)]. However, using GLC-MS it was found that neither the level of $PGF_{2\alpha}$ nor the level of 15-keto-13,14-dihydro-$PGF_{2\alpha}$ varied in relation to the uterine contractions (30). Furthermore, no correlation was found between the $PGF_{2\alpha}$ level and the stage of labor, but there was a pronounced increase in the plasma concentration of the metabolite as labor progressed.

Several reports have appeared on levels of PGA in human blood (149, 150, 153, 172, 182). In some of these, comparisons have been made with E and F levels, and generally considerably higher values have been reported for PGA (150, 153, 172). These findings were considered consistent with the concept that PGA traverses the lungs without metabolic degradation. However, it must be emphasized that a failure of the lungs to metabolize PGA has never been proved. For a discussion, see the section on metabolism in this review.

As pointed out earlier (104 and p. 677) the measured levels of primary prostaglandins in peripheral plasma do not reflect the rate of endogenous synthesis in vivo in tissues and organs. Thus the assays should preferably be aimed instead at the main metabolites, the 15-keto-13,14-dihydro compounds. A few papers reporting radioimmunoassays for 15-keto-13,14-dihydro-$PGF_{2\alpha}$ have appeared (157, 159–161) and also one for 15-keto-13,14-dihydro-PGE_2 (106). The main problem encountered in this type of assay was the preparation of the radio-labeled ligand. Thus, the antigen has been prepared from ^3H-$PGF_{2\alpha}$ by incubation with a dog lung preparation (159), however, this method must inevitably result in a considerable dilution of the labeled ligand with endogenous material and will lower the sensitivity of the assay. Others have used a different method (157, 161): incubation with a swine kidney preparation in the presence of inhibitors of prostaglandin biosynthesis (112). This method gave essentially no dilution by endogenous material. A synthetically prepared heterologous antigen, ^3H-15-keto-13,14-dihydro-$PGF_{1\alpha}$, has also been used (160). The assays were used for estimation of normal levels of the metabolite in peripheral human plasma. From one laboratory an average level of 4800 pg/ml (range 1,000–26,000 pg/ml) was reported (159). A second group found values around 1700 pg/ml (161). Values obtained by others are, however, considerably lower: 0–240 pg/ml (157), 45–100 pg/ml (160), and 10–80 pg/ml (183). These latter data agree with those obtained earlier by GLC-MS (29, 104).

An alternative solution to the problem of estimating prostaglandin production is to monitor the urinary metabolites. A radioimmunoassay for the main urinary metabolite of $PGF_{2\alpha}$ in the human was recently published (162). However, the authors used carbodiimide for coupling, and since this metabolite is a dicarboxylic acid, a mixed conjugate with resulting low specificities of the antiserum must have been the result. Unfortunately, crossreactions with prostaglandin metabolites structurally more closely related to the main metabolite were not reported. A different method has been used by two other groups (163, 164), who selectively coupled the ω-carboxyl group to the protein molecule.

RECEPTORS

E and A Type Prostaglandin Receptors

The first experimental evidence for a prostaglandin receptor was reported in 1972 (184). Specific binding of $[5,6\text{-}^3H]PGE_1$ was demonstrated in a particulate fraction obtained from a homogenate of isolated rat adipocytes. A binding assay based on filtration on columns of glass wool was used. Competition experiments with unlabeled prostaglandins showed that PGE_1 followed by PGE_2 and 13,14-dihydro-PGE_1 had the highest affinities for the receptor and that 13,14-dihydro-15-keto-PGE_1, $PGF_{2\alpha}$, and PGA_1 had considerably lower affinities. A dissociation constant (K_d) of 3.3 nM was reported for the PGE_1-receptor interaction. Evidence for both high affinity ($K_d = 4.9$ nM) and low affinity ($K_d = 40$ nM) binding sites for PGE_1 and, in addition, two types of PGA_1 binding sites in rat adipocyte ghosts was later reported (185) using a Millipore filtration assay. Depending on the ionic composition of the incubation medium the K_ds for the PGA_1-receptor interactions varied from 1.2 to 83 nM and 0.57 to 8.6 μM for the high and low affinity sites, respectively.

Reports on PGE_1 receptors have subsequently appeared for rat forestomach (186), beef thyroid membranes (187), rat thymocytes (188), hamster uterus (189), rat liver membranes (190), and membranes from bovine corpus luteum (191), and there is one report on a prostaglandin receptor A in rabbit kidney (192). Using sucrose gradient centrifugation as a binding assay, a PGE_1 receptor was demonstrated in thyroid membranes (187). Both specific and nonspecific binding were investigated. The former was optimal at pH 7 and the latter at pH 3.6. The specific binding had an absolute requirement for Ca^{2+} and, in the presence of 5 mM Ca^{2+}, had a K_d of 26 nM for PGE_1. The nonspecific binding was enhanced by, but did not require, Ca^{2+}. At concentrations that enhance PGE_1 activation of adenylate cyclase, ITP, GTP, and dGTP displaced bound ^3H-labeled PGE_1 from the receptor. A similar effect was obtained with several proteins, including thyrotropin.

Kinetic as well as thermodynamic data have been reported for the interaction between PGE_1 and the receptor in rat thymocytes (188). The association reaction was unusually rapid with a half-time for the establishment of equilibrium at 37°C of about 45 sec. The K_d values for PGE_1 were 2 nM at 37° and 0.07 nM at 0°. The subcellular distribution of PGE_1 receptors in rat liver has been investigated (190). High affinity binding sites were found exclusively in the plasma membrane fraction and not in cell fractions containing nuclei, rough microsomes, Golgi complexes, or mitochondria. Specificity studies with regard to the prostaglandin ligand were carried out for the receptor in plasma membranes by determining K_d values for unlabeled prostaglandins from competition experiments. The values obtained for $PGF_{1\alpha}$, $PGF_{2\alpha}$, PGB_1, PGB_2, PGA_1, and PGA_2 were 1100, 100, 300, 180, 16, and 16 nM, respectively. No K_d was reported for PGE_2, which makes it unclear whether the receptor is not, in fact, a PGE_2 rather than a PGE_1 receptor.

Prostaglandin $F_{2\alpha}$ Receptors

Prostaglandins have a luteolytic effect in many mammalian species, including monkeys (for reviews, see 193 and 194). A physiological role of prostaglandin $F_{2\alpha}$ as a luteolytic hormone in the sheep seems to be well established (119). At the time of luteolysis, prostaglandin $F_{2\alpha}$ is released by the uterus and transported to the ovaries via a vascular pathway. Specific receptors for $PGF_{2\alpha}$ have been demonstrated in particulate fractions from ovine (195), bovine (196), and human (197) corpora lutea. Binding assays based on Sephadex G-50 chromatography or centrifugation gave identical results. The reaction was reversible, and optimal binding was obtained at pH 6.3. Dissociation constants determined from Scatchard plots of binding data at equilibrium (195–197) and from the rate constants for association and dissociation (196) agreed well. Scatchard plot analyses were used to determine K_d values for unlabeled prostaglandins (195, 196) and prostaglandin analogs (198) from competition experiments. Of the naturally occurring prostaglandins, $PGF_{2\alpha}$ had the highest affinity for the receptors ($K_d = 50$–100 nM). Specificity studies showed that a carboxyl group in the 1 position as well as 9α-,11α-, and $15(S)$-hydroxyl groups and a 5,6-cis double bond were essential for binding to the receptors (195, 196). A 13,14-$trans$ double bond was relatively nonessential in this respect. The affinities of a number of synthetic prostaglandin analogs for the bovine receptor agreed well with their antifertility properties in vivo when the different rates of metabolism by prostaglandin 15-hydroxy dehydrogenase were taken into account (198). The results also showed that the alkyl side chain of prostaglandins could be modified extensively without diminishing the affinity for the receptor. On the other hand, changes in the carboxyl side chain caused considerable increases in the dissociation constants. A clear luteolytic effect of prostaglandins has not been demonstrated in man (194). The demonstration of a $PGF_{2\alpha}$ receptor in human corpora lutea (197), however, indicates that this prostaglandin has a physiological role also in human luteolysis.

The reaction between ^3H-labeled $PGF_{2\alpha}$ and the corpus luteum receptor was inhibited by low concentrations of several detergents (199). However, the preformed $PGF_{2\alpha}$-receptor complex could be quantitatively solubilized by sodium deoxycholate. Removal of the detergent by Sepharose chromatography caused reaggregation of the complex. In the presence of sodium deoxycholate the complex was homogeneous on columns of Sepharose-6B, having a K_{AV} value of 0.35. Subcellular distribution of the $PGF_{2\alpha}$ receptor has been investigated by means of sucrose gradient centrifugations of homogenates from bovine corpora lutea (200). The results indicated that the receptor was present in plasma membranes and not in mitochondria or endoplasmic reticulum.

In addition to the prostaglandin $F_{2\alpha}$ receptors in corpora lutea, specific binding of $PGF_{2\alpha}$ has also been reported for hamster uterine tissue (189) and segments of rabbit oviduct (201). Neither dissociation constants nor specifities for the binding reactions were given in these reports.

PROSTAGLANDINS AND CYCLIC NUCLEOTIDES

Prostaglandins interact with adenyl cyclase in a variety of tissues. This may be the reason for their wide spectrum of pharmacological effects. Since the previous review in this series (1), the number of publications on prostaglandin-cyclic AMP interactions has increased rapidly, and it is possible to describe briefly only some of the most pertinent findings in this review. There are, however, several comprehensive reviews on the subject (1, 202–206).

Endocrine Glands

The ability of thyrotropin (TSH) to stimulate release of thyroid hormones depends on cyclic AMP (cAMP) (202). Like TSH, E-type prostaglandins stimulated adenyl cyclase in various thyroid preparations (207–213). It has been suggested that a physiological role of the prostaglandins might be as messengers between the classical hormones and cAMP. Some evidence supporting this hypothesis has appeared: 1. Both TSH and long acting thyroid stimulator (LATS) increased the concentration of prostaglandins in isolated thyroid cells (214–216) and thyroid glands (217). 2. The antagonists of prostaglandin action, 7-oxa-13-prostynoic acid, its 15-hydroxy analog, and polyphloretin phosphate inhibited the stimulatory effect of TSH, LATS, and PGE_2 on thyroid cell adenyl cyclase (207, 210, 213). However, there is also evidence against the hypothesis, since indomethacin and aspirin at concentrations that inhibited TSH-induced synthesis of thyroidal prostaglandins (214) did not prevent the stimulatory effect of TSH on various thyroid preparations (212). The properties of thyroid adenyl cyclase in purified plasma membranes have been studied (208). An interesting feature of this enzyme was that stimulation by TSH or PGE_1 was potentiated by purine nucleotides. ITP was the most effective nucleotide followed by dGTP, GTP, and XTP for TSH stimulation. For PGE_1 stimulation, GTP, dGTP, and ITP were equally effective and XTP was less effective (211). Potassium ion potentiated the effect of TSH, but not of PGE_1, on adenyl cyclase (211). However, it also prevented the effects of the elevated cAMP in the thyroid (209). Lithium ion also inhibited the effects of TSH strongly (208, 218).

PGE_1 increased cAMP levels and growth hormone secretion in ox and rat pituitary glands (219, 220). Furthermore, intravenous administration of PGE_1 (221, 222) also increased growth hormone secretion. This may not have been a direct effect of PGE_1, however, since this compound is rapidly metabolized in the circulation. Growth hormone secretion was unaffected by PGE_1 but stimulated by dibutyryl cAMP in pituitary tumor cells (223). Prostaglandins E_2 and E_1 stimulated luteinizing hormone (LH) secretion when injected into the third ventricle but not when injected into the anterior pituitary (224). This indicated that the prostaglandins acted via a hypothalamic releasing factor. On the other hand, pituitary prostaglandins might be involved in the action of thyrotropin releasing hormone (TRH) on the pituitary since this was inhibited by indomethacin (225).

Whole ovaries responded with increased cAMP contents to PGE_2 and LH (226). The same was true for luteal cells although in this case the response to PGE_1 was only one sixth of that to LH (227). Rat ovaries did not respond to LH until the age of 10 days but responded to PGE_2 already in newborn animals (226). This shows that the LH and PGE_2 activations of ovarian adenyl cyclase are independent processes which might involve separate receptor mechanisms. LH and PGE_2 stimulated adenyl cyclase also in rat corpora lutea (228). In corpora lutea the sensitivity to LH decreased with increasing age of the corpus luteum whereas the sensitivity to PGE_2 remained constant. This agrees with the separate mechanisms of PGE_2 and LH action just mentioned. It was neither possible to demonstrate an increase in ovarian prostaglandin content in response to LH nor to prevent the stimulatory effect of LH on cAMP production in rat ovaries with indomethacin or aspirin (229).

PGE_2 induced meiotic division of follicle-enclosed rat oocytes, probably by mediation of cAMP (230), and stimulated corticosterone production in adrenal bisects (231). PGE_1 increased cAMP in rat testes (232) and stimulated adenyl cyclase in pancreatic islets (233).

Exocrine Glands

PGEs and cholera toxin stimulate intestinal mucosal adenyl cyclase and small intestinal fluid secretion (234–236). The effect of PGEs on the stomach was opposite to that in the intestine, with decreased gastric secretion and decreased cAMP production (237). In the toad bladder, E-type prostaglandins inhibited vasopressin-induced water flow (238–241). This effect did not seem to be mediated by a decrease in cAMP (240) although PGE_1 decreased cAMP slightly at certain concentrations of Mg^{2+} (239). The inhibition by PGE_1 on vasopressin-induced water flow was antagonized by the hypoglycemic agent chlorpropamide (241). In kidney cortex, PGE_1 inhibited parathyroid-stimulated adenyl cyclase (242, 243), and in kidney medulla it activated basal (244) but inhibited vasopressin stimulated adenyl cyclase (245, 246). PGE_1 stimulated adenyl cyclase also in mammary gland (247).

Smooth Muscle Organs

At a concentration of 10^{-11} M, PGE_1 increased cAMP levels in myocardial preparations (248). PGE_2 and PGA_1 had a similar effect at a concentration of 10^{-4} M whereas $PGF_{1\alpha}$ and $PGF_{2\alpha}$ were inactive. Concomitant with the stimulatory effect on adenyl cyclase, the prostaglandins had positive chronotropic and inotropic effects (248, 249). Myocardial adenyl cyclase has been solubilized. The soluble enzyme was stimulated by prostaglandins (250) but had lost its responsiveness to catecholamines, glucagon, and histamine. The hormone responsiveness could be restored by the addition of phospholipids to the soluble enzyme (for review, see 251).

Reports on a stimulatory effect of PGE_1 on adenyl cyclase in the uterus and in various parts of the oviduct have appeared (252–255). PGE_2 had an effect similar to cholecystokinin on the gall bladder and the sphincter of Oddi, causing contraction and relaxation respectively (256). These effects were accompanied by a decrease in gall bladder and an increase in sphincter cAMP concentration.

The F-type prostaglandins usually do not interact with adenyl cyclase. It has been suggested that cyclic GMP might be a mediator of PGF_α action. This has been supported by a report on increased cGMP levels during $PGF_{2\alpha}$-induced veno-constriction (257).

Blood and Bone Marrow Cells

E-type prostaglandins stimulated adenyl cyclase in lymphocytes (258–264), leucocytes (265), turkey erythrocytes (266), and platelets (267–269, cf also the platelet section) and potentiated the effect of erythropoietin in bone marrow cell cultures probably by activating adenyl cyclase (270). The increased cAMP levels prevented lymphocyte transformation (258), blocked the lymphocyte cytotoxic effect (260), inhibited platelet aggregation (271), and prevented immediate as well as delayed hypersensitivity in vitro (272). In thymic lymphocytes PGE_1 increased cAMP, stimulated DNA synthesis, and promoted cell division. Calcium ion at a concentration of 5 mM did not affect cAMP accumulation of DNA synthesis but prevented the mitogenic effect of PGE_1 completely (262).

Fat Cells

At low concentrations PGE_1 inhibits hormone-induced lipolysis by decreasing the cAMP levels (for review, see 273). Two recent reports confirm these results (274, 275). 7-Oxa-13-prostynoic acid and indomethacin enhanced hormone-stimulated and basal lipolysis in fat cells (276), indicating that endogenous prostaglandins had a regulatory effect on lipolysis. It has also been reported that indomethacin had no effect on lipolysis in fat cells (277). Norepinephrine, theophylline, and dibutyryl cyclic AMP increased the concentration of PGE_2 in fat cells (278). Although PGE_1 decreases cAMP levels in intact fat cells, a stimulatory rather than an inhibitory effect was obtained on adenyl cyclase in broken fat cell preparations (279). On the other hand, PGH_2 (cf section on prostaglandin endoperoxides) inhibited adenyl cyclase in fat cell ghosts (70).

Neurotumor Cells

PGE_1 increased cAMP levels in neuroblastoma cell lines (280, 281) and induced morphological differentiation of these cells (282, 283). The differentiating effect was inhibited by cytochalasin B and vinblastine (284), indicating that microtubule assembly was involved. X irradiation also induced morphological differentiation and increased catechol-O-methyltransferase activity. The latter effect did not accompany PGE_1-induced differentiation (285). PGE_1 had neglible effect on adenyl cyclase in rat glioma and human astrocytoma cells (286), and the growth of rat pituitary cells was not affected by either PGE_1 or indomethacin (223). Somatic cell hybrids between neuroblastoma, glioma, and fibroblast cell lines had lost norepinephrine-sensitive adenyl cyclase but retained PGE_1-sensitive adenyl cyclase (287). For each hybrid cell line both parental lines responded to PGE_1 but only one parental line responded to norepinephrine.

Fibroblasts

Adenyl cyclase has been studied in several fibroblast cell lines (288–291). With the exception of SV40 and polyoma virus-transformed 3T3 cells, all cell lines investigated responded to PGEs with increased adenyl cyclase activity (288). L929 cells had elevated cAMP levels 2–5 hr after the addition of PGE_1. The levels then started to decline even though more PGE_1 was added, probably because PGE_1 increased not only cAMP but also phosphodiesterase (290). Trypsin and insulin decreased PGE elevated cAMP levels (292). A mutant cell line lacking cAMP binding protein showed a much higher response to PGE_1 than did the parent cell line (293).

Miscellaneous Organs

Rat brain cortical adenyl cyclase responded to PGE_1 and PGE_2 whereas the corresponding enzymes from human and rabbit cortex did not respond (294, 295). Furthermore, cAMP levels in mouse brain increased after intravenous administration of PGE_2 or PGE_1 (296). PGE activation has also been reported for adenyl cyclase from rabbit ciliary process (297) and guinea pig lung (298, 299). Bradykinin elevated both cAMP and cGMP in lung. The stimulatory effect on cAMP production was inhibited whereas the effect on cGMP was not influenced by indomethacin (299). Moreover, the positive effect of acetylcholine on cGMP in lung was antagonized by PGE_1 (300). PGE_1 prevented induction of tyrosine aminotransferase in liver, probably by decreasing cAMP (301, 302)

PROSTAGLANDIN AND PLATELET FUNCTION

It was shown early that PGE_1 is a very active inhibitor of platelet aggregation (303, 304). PGE_2 was found to have a qualitatively different effect, i.e. this prostaglandin stimulated ADP-induced aggregation of rat platelets, whereas the PGF compounds were inactive.

Platelet adenyl cyclase is stimulated by PGE_1 (271), and both cyclic 3',5'-adenosine monophosphate (cAMP) and its dibutyryl derivative inhibit aggregation (271). Since a number of inhibitors of platelet aggregation increased the level of cAMP in the platelet while agents that produce or augment aggregation reduced the level of cAMP, it was proposed that platelet aggregation is favored by a decrease in cAMP and inhibited by an increase in cAMP (305, 306). This correlation has been put in doubt, however, since thrombin increased platelet cAMP (307) and ADP; epinephrine and collagen did not have any consistent effect on basal cAMP concentration (308). The latter workers (308) confirmed that PGE_1 increases platelet cAMP and also found that when the cAMP level of platelets treated with PGE_1 declined aggregation did not occur. These observations suggest that cAMP is not a direct inhibitor of the action of aggregating agents on platelets but appears to trigger a mechanism that results in a relatively persistent inhibitory effect on platelet function. PGE_1-mediated inhibition of ADP-induced aggregation has also been claimed to occur by another mechanism, i.e. by interference with ADP binding to specific receptors on the platelet membrane (309).

PGE_2 was found to have at least two effects on aggregation of human platelets by ADP (310): initially a weak PGE_1-like inhibitory effect on the first phase of aggregation and then a stimulatory effect on the second phase of aggregation.

Platelets produce prostaglandins on treatment with thrombin (11) and other aggregating agents (12). Aspirin and indomethacin effectively block prostaglandin synthesis by platelets (49, 311) and inhibit the second phase of platelet aggregation (312–314). The effects of PGE_1 and PGE_2 upon platelet aggregation and the release of prostaglandins during aggregation suggested the involvement of prostaglandins in platelet function. However, the effect of aspirin did not support this, since one would not expect inhibition of platelet aggregation to follow in the presence of a drug that blocks formation of both PGE_2, which stimulates aggregation only moderately, and PGE_1, which inhibits aggregation.

In recent work arachidonic acid (315) and the prostaglandin endoperoxides, PGG_2 and PGH_2 (13), were found to induce platelet aggregation. Formation of an unstable factor with aggregating properties upon incubation of arachidonic acid with platelets (316) and phenol-activated acetone powder of the sheep vesicular gland (317) has also been reported. Recent detailed studies on the transformations of labeled arachidonic acid by human platelets have led to information on the mechanism of the aggregating effect of arachidonic acid and on the involvement of this mechanism in platelet aggregation induced by various agents. Arachidonic acid added to human platelets yielded two metabolites of PGG_2, i.e. the hemiacetal derivative of 8-(1-hydroxy-3-oxopropyl)-9,12L-dihydroxy-5,10-heptadecadienoic acid (PHD) and 12L-hydroxy-5,8,10-heptadecatrienoic acid (HHT), and in addition

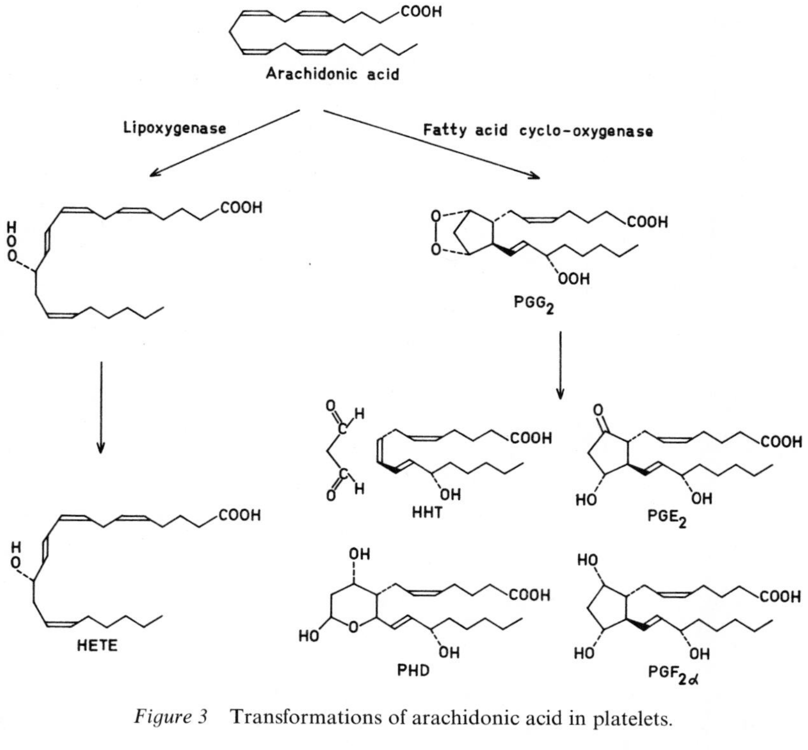

Figure 3 Transformations of arachidonic acid in platelets.

12L-hydroxy-5,8,10,14-eicosatetraenoic acid (HETE) formed by action of a novel lipoxygenase (Figure 3) (61). Apparently PGG_2 was extensively metabolized within the platelet into predominantly nonprostaglandin derivatives. Aggregation of washed platelets was accompanied by release of large amounts of PHD, HHT, and HETE (14), whereas in accordance with earlier results (13, 318) the amounts of PGE_2, $PGF_{2\alpha}$, and intact endoperoxides were much smaller (Figure 4) (14). Aspirin inhibited the conversion of labeled arachidonic acid into the endoperoxide metabolites (61) and also inhibited the release of endogenous metabolites during thrombin-induced aggregation (14). The results demonstrating that the prostaglandin endoperoxides are potent inducers of platelet aggregation; that endoperoxides and large amounts of endoperoxide metabolites are released during aggregation; and that aspirin, a drug that inhibits the second phase of platelet aggregation, blocks the synthesis of endoperoxides indicate that endogenous endoperoxides play a role in platelet aggregation. Thus, the effect of aspirin on platelets can be ascribed to blockage of the formation of endoperoxides (13, 14, 319). Recent experiments have demonstrated that the endoperoxide PGG_2 induces the so-called release reaction in human platelets (320).

A new concept concerning prostaglandin action and release has emerged from these studies. Thus, the prostaglandins can exert their biological action through the

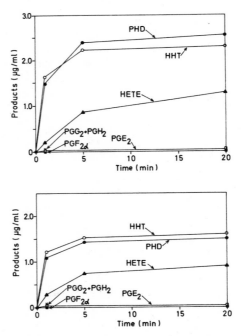

Figure 4   Release of endoperoxide metabolites, PHD and HHT; endoperoxides, PGE_2 and $PGF_{2\alpha}$; and HETE during thrombin-induced aggregation of washed human platelets.

endoperoxides, which can be almost exclusively transformed to and released as nonprostanoate metabolites. It has also been found that these endoperoxide metabolites can be released from several other tissues in response to various stimuli (321) and that the endoperoxides, PGG_2 and PGH_2, in addition to their effects on platelets have unique biological actions on, e.g., airway and vascular smooth muscle (68) and adipocyte ghosts (70). It is obvious that many biological systems should be reinvestigated to establish whether this new model of prostaglandin action and release, demonstrated for human platelets, occurs more generally for regulation of various cell functions.

Acknowledgments

Studies from the authors' laboratory were supported by the Swedish Medical Research Council and the World Health Organization.

Literature Cited

1. Hinman, J. W. 1972. *Ann. Rev. Biochem.* 41:161
2. Jouvenaz, G. H., Nugteren, D. H., Beerthuis, R. K., van Dorp, D. A. 1970. *Biochim. Biophys. Acta* 202:231
3. Takeguchi, C., Kohno, E., Sih, C. J. 1971. *Biochemistry* 10:2372
4. Takeguchi, C., Sih, C. J. 1972. *Prostaglandins* 2:169
5. Flower, R. J., Cheung, H. S., Cushman, D. W. 1973. *Prostaglandins* 4:325
6. Ziboh, V. A., Hsia, S. L. 1971. *Arch. Biochem. Biophys.* 146:100
7. Jonsson, C. E., Änggård, E. 1972. *Scand. J. Clin. Lab. Invest.* 29:289
8. Greaves, M. W., McDonald-Gibson, W. 1972. *Brit. J. Pharmacol.* 46:172
9. Mathur, G. P., Gandhi, V. M. 1972. *J. Invest. Dermatol.* 58:291
10. Tan, W. C., Privett, O. S. 1973. *Lipids* 8:166
11. Smith, J. B., Willis, A. L. 1970. *Brit. J. Pharmacol.* 40:545P
12. Smith, J. B., Ingerman, C., Kocsis, J. J., Silver, M. J. 1973. *J. Clin. Invest.* 52:965
13. Hamberg, M., Svensson, J., Wakabayashi, T., Samuelsson, B. 1974. *Proc. Nat. Acad. Sci. USA* 71:345
14. Hamberg, M., Svensson, J., Samuelsson, B. 1974. *Proc. Nat. Acad. Sci. USA* 71:3824
15. Levine, L., Hinkle, P. M., Voelkel, E. F., Tashjian, A. H. 1972. *Biochem. Biophys. Res. Commun.* 47:888
16. Hammarström, S., Samuelsson, B., Bjursell, G. 1973. *Nature New Biol.* 243:50
17. Christ, E. J., van Dorp, D. A. 1972. *Biochim. Biophys. Acta* 270:537
18. Weinheimer, A. J., Spraggins, R. L. 1969. *Tetrahedron Lett.* 5185
19. Light, R. J., Samuelsson, B. 1972. *Eur. J. Biochem.* 28:232
20. Corey, E. J., Nashburn, W. N., Chen, J. C. 1973. *J. Am. Chem. Soc.* 95:2054
21. Hamberg, M., Samuelsson, B. 1971. *J. Biol.Chem.* 246:6713
22. Gréen, K., Samuelsson, B. 1971. *Eur. J. Biochem.* 22:391
23. Hamberg, M., Samuelsson, B. 1972. *J. Biol. Chem.* 247:3495
24. Hamberg, M. 1972. *Biochem. Biophys. Res. Commun.* 49:720
25. Hamberg, M. 1973. *Anal. Biochem.* 55:368
26. Strandberg, K., Hamberg, M. 1974. *Prostaglandins* 6:159
27. Hámberg, M., Jonsson, C. E. 1973. *Acta Physiol. Scand.* 87:240
28. Hamberg, M. 1974. *Life Sci.* 14:247
29. Gréen, K., Samuelsson, B. 1974. *Biochem. Med.* 11:298
30. Gréen, K., Bygdeman, M., Toppozada, M., Wiqvist, N. 1974. *Am. J. Obstet. Gynecol.* 120:25
31. Granström, E. 1972. *Eur. J. Biochem.* 27:462
32. Cox, R. I., Thorburn, G. D., Currie, W. B., Restall, B. J., Schneider, W. 1973. *Advan. Biosci.* 9:625
33. Samuelsson, B., Granström, E., Hamberg, M. 1967. *Nobel Symp. 2 Prostaglandins,* ed. S. Bergström, B. Samuelsson, 31. Stockholm: Almqvist & Wiksell
34. Wallach, D. P., Daniels, E. G. 1971. *Biochim. Biophys. Acta* 231:445
35. Sih, C. J., Takeguchi, C., Foss, P. 1970. *J. Am. Chem. Soc.* 92:6670

36. Foss, P., Sih, C. J., Takeguchi, C., Schnoes, H. 1972. *Biochemistry* 11:2271
37. Wlodawer, P., Samuelsson, B. 1973. *J. Biol. Chem.* 248:5673
38. Pace-Asciak, C. R. 1972. *Biochim. Biophys. Acta* 280:161
39. Lee, R. E., Lands, W. E. M. 1972. *Biochim. Biophys. Acta* 260:203
40. Maddox, I. S. 1973. *Biochim. Biophys. Acta* 306:74
41. Hamberg, M., Samuelsson, B. 1967. *J. Biol. Chem.* 242:5336
42. Hamberg, M., Samuelsson, B. 1973. *Proc. Nat. Acad. Sci. USA* 70:899
43. Smith, W. L., Lands, W. E. M. 1972. *Biochemistry* 11:3276
44. Smith, W. L., Lands, W. E. M. 1972. *J. Biol. Chem.* 247:1083
45. Lands, W. E. M., LeTellier, P. R., Rome, L. H., Vanderhoek, J. Y. 1973. *Advan. Biosci.* 9:15
46. LeTellier, P. R., Smith, W. L., Lands, W. E. M. 1973. *Prostaglandins* 4:837
47. Vane, J. R. 1971. *Nature New Biol.* 231:232
48. Ferreira, S. H., Moncada, S., Vane, J. R. 1971. *Nature New Biol.* 231:237
49. Smith, J. B., Willis, A. L. 1971. *Nature New Biol.* 231:235
50. Tomlinson, R. V., Ringold, H. J., Qureshi, M. C., Forchielli, E. 1972. *Biochem. Biophys. Res. Commun.* 46:552
51. Lee, R. L. 1974. *Prostaglandins* 5:63
52. Smith, W. L., Lands, W. E. M. 1971. *J. Biol. Chem.* 246:6700
53. Flower, R. J., Vane, J. R. 1972. *Nature* 240:410
54. Ferreira, S. H., Vane, J. R. 1974. *Ann. Rev. Pharmacol.* 14:57
55. Flower, R. J., Vane, J. R. 1974. *Biochem. Pharmacol.* 23:1439
56. Ahern, D. G., Downing, D. T. 1970. *Biochim. Biophys. Acta* 210:456
57. Nugteren, D. H. 1970. *Biochim. Biophys. Acta* 210:171
58. Wlodawer, P., Samuelsson, B., Albonico, S. M., Corey, E. J. 1971. *J. Am. Chem. Soc.* 93:2815
59. McDonald-Gibson, R. G., Flack, J. D., Ramwell, P. W. 1973. *Biochem. J.* 132:117
60. Downing, D. T., Ahern, D. G., Bachta, M. 1970. *Biochem. Biophys. Res. Commun.* 40:218
61. Hamberg, M., Samuelsson, B. 1974. *Proc. Nat. Acad. Sci. USA* 71:3400
62. Samuelsson, B. 1972. *Fed. Proc.* 31:1442
63. Nugteren, D. H. 1972. Presented at the 15th Int. Conf. on the Biochemistry of Lipids, The Hague
64. Hamberg, M., Samuelsson, B. 1972.

Presented at the Int. Conf. on Prostaglandins, Vienna
65. Nugteren, D. H., Hazelhof, E. 1973. *Biochim. Biophys. Acta* 326:448
66. Sood, R., Nagasawa, M., Sih, C. J. 1974. *Tetrahedron Lett.* 423
67. Samuelsson, B., Hamberg, M. 1974. *Proc. Int. Symp. Prostaglandin Synthetase Inhibitors, New York,* ed. J. R. Vane, H. J. Robinson, 107
68. Hamberg, M., Hedqvist, P., Strandberg, K., Svensson, J., Samuelsson, B. 1975. *Life Sci.* 16:451
69. Nishizawa, E., Svensson, J., Hamberg, M. 1975. *Prostaglandins.* In press
70. Gorman, R., Hamberg, M., Samuelsson, B. 1975. *J. Biol. Chem.* In press
71. Granström, E., Samuelsson, B. 1971. *J. Biol. Chem.* 246:5254
72. Granström, E., Samuelsson, B. 1971. *J. Biol. Chem.* 246:7470
73. Granström, E., Samuelsson, B. 1972. *J. Am. Chem. Soc.* 94:4380
74. Granström, E. 1972. *Eur. J. Biochem.* 25:581
75. Gréen, K. 1971. *Biochim. Biophys. Acta* 231:419
76. Gréen, K. 1971. *Biochemistry* 10:1072
77. Granström, E., Samuelsson, B. 1969. *Eur. J. Biochem.* 10:411
78. Kindahl, H., Granström, E. 1972. *Biochim. Biophys. Acta* 280:466
79. Svanborg, K., Bygdeman, M. 1972. *Eur. J. Biochem.* 28:127
80. Dimov, V., Gréen, K. 1973. *Biochim. Biophys. Acta* 306:257
81. Hamberg, M., Wilson, M. 1973. *Advan. Biosci.* 9:39
82. Sun, F. 1974. *Biochim. Biophys. Acta* 348:250
83. Gréen, K. 1973. *Advan. Biosci.* 9:91
84. Hamberg, M., Samuelsson, B. 1971. *J. Biol. Chem.* 246:1073
85. Änggård, E., Larsson, C. 1971. *Eur. J. Pharmacol.* 14:66
86. Hamberg, M., Israelsson, U. 1970. *J. Biol. Chem.* 246:5107
87. Leslie, C. A., Levine, L. 1973. *Biochem. Biophys. Res. Commun.* 52:717
88. Hensby, C. N. 1974. *Brit. J. Pharmacol.* 50:462P
89. Schneider, W. P., Murray, H. C. 1973. *J. Org. Chem.* 38:397
90. Piper, P. J., Vane, J. R., Wyllie, J. H. 1970. *Nature* 225:600
91. Horton, E. W., Jones, R. L. 1969. *Brit. J. Pharmacol.* 37:705
92. Nakano, J., Änggård, E., Samuelsson, B. 1969. *Eur. J. Biochem.* 11:386
93. Jones, R. L. 1972. *J. Lipid Res.* 13:511
94. Jones, R. L., Cammock, S., Horton,

E. W. 1972. *Biochim. Biophys. Acta* 280 : 588
95. Polet, H., Levine, L. 1971. *Biochem. Biophys. Res. Commun.* 45 : 1169
96. Jones, R. L., Cammock, S. 1973. *Advan. Biosci.* 9 : 61
97. Smith, J. B., Kocsis, J. J., Ingerman, C., Silver, K. J. 1973. *Pharmacologist* 15 : 208
98. Jarabak, J. 1972. *Proc. Nat. Acad. Sci. USA* 69 : 533
99. Schlegel, W., Demers, L. M., Hildebrandt-Stark, H. E., Behrman, H. R., Greep, R. O. 1974. *Prostaglandins* 5 : 417
100. Alam, N. A., Clary, P., Russell, P. T. 1973. *Prostaglandins* 4 : 363
101. Nugteren, D. H., Vonkeman, H., van Dorp, D. A. 1967. *Rec. Trav. Chim. Pays Bas* 85 : 1237
102. Jouvenaz, G. H., Nugteren, D. H., van Dorp, D. A. 1973. *Prostaglandins* 3 : 175
103. Beguin, F. Bygdeman, M., Gréen, K., Samuelsson, B., Toppozada, M., Wiqvist, N. 1972. *Acta Physiol. Scand.* 86 : 430
104. Samuelsson, B. 1973. *Advan. Biosci.* 9 : 7
105. Samuelsson, B., Axen, U., Behrman, H., Granström, E., Gréen, K., Jaffe, B. M., Kirton, K., Levine, L., Skarnes, R. C., Speroff, L., Wolfe, L. S. 1973. *Advan. Biosci.* 9 : 121
106. Gréen, K., Granström, E. 1973. In *Prostaglandins in Fertility Control, 3,* ed. S. Bergström, 55. WHO Research and Training Center on Human Reproduction. Stockholm : Karolinska Inst.
107. Raz, A. 1972. *Biochem. J.* 130 : 631
108. Unger, W. G. 1972. *J. Pharm. Pharmacol.* 24 : 470
109. Attallah, A. A., Schussler, G. C. 1973. *Prostaglandins* 4 : 479
110. Axen, U., Baczynskyj, L., Duchamp, D. J., Zieserl, J. F. 1972. *J. Reprod. Med.* 9 : 372
111. Gréen, K., Granström, E., Samuelsson, B. 1972. In *Prostaglandins in Fertility Control, 2,* ed. S. Bergström, K. Gréen, 92. WHO Research and Training Center on Human Reproduction. Stockholm : Karolinska Inst.
112. Gréen, K., Granström, E., Samuelsson, B., Axen, U. 1973. *Anal. Biochem.* 54 : 434
113. Keirse, M. J. N. C., Turnbull, A. C. 1973. *J. Obstet. Gynaecol. Brit. Commonw.* 80 : 970
114. Sweetman, B. J., Frölich, J. C., Watson, J. T. 1973. *Prostaglandins* 3 : 75

115. Änggård, E., Samuelsson, B. 1971. *Ann. NY Acad. Sci.* 180 : 200
116. Levitt, M. J., Josimovich, J. B., Broskin, K. D. 1972. *Prostaglandins* 1 : 121
117. Middleditch, B. S., Desiderio, D. M. 1972. *Prostaglandins* 2 : 195
118. Gréen, K., Samuelsson, B. 1968. *Prostaglandin Symposium of the Worcester Foundation for Experimental Biology,* ed. P. W. Ramwell, J. Shaw, 389. New York : Wiley
119. McCracken, J. A. et al 1972. *Nature New Biol.* 238 : 129
120. Samuelsson, B., Hamberg, M., Sweeley, C. C. 1970. *Anal. Biochem.* 38 : 301
121. Wolfe, L. S., Pace-Asciak, C. See Ref. 111, 201
122. Gillett, P. G., Kinch, R. A. H., Wolfe, L. S., Pace-Asciak, C. 1972. *Am. J. Obstet. Gynecol.* 112 : 330
123. Axen, U., Gréen, K., Hörlin, D., Samuelsson, B. 1971. *Biochem. Biophys. Res. Commun.* 45 : 519
124. Strandberg, K., Hamberg, M. 1974. *Prostaglandins* 6 : 159
125. Gréen, K. See Ref. 106, 44
126. Holland, J. F., Sweeley, C. C., Thrush, R. E., Teets, R. E., Bieber, M. A. 1973. *Anal. Chem.* 45 : 308
127. Sweetman, B. J., Watson, J. T., Carr, K., Oates, J. A., Frölich, J. C. 1973. *Prostaglandins* 3 : 385
128. Kelly, R. W. 1973. *Anal. Chem.* 45 : 2079
129. Gréen, K., Beguin, F., Bygdeman, M., Toppozada, M., Wiqvist, N. See Ref. 111, 189
130. Granström, E., Gréen, K., Bygdeman, M., Toppozada, M., Wiqvist, N. 1973. *Life Sci.* 12 : 219
131. Bygdeman, M., Gréen, K., Toppozada, M., Wiqvist, N., Bergström, S. 1974. *Life Sci.* 14 : 521
132. Gustavii, B., Gréen, K. 1972. *Am. J. Obstet. Gynecol.* 114 : 1099
133. Gréen, K., Hagenfeldt, K. 1975. In press
134. Frölich, J. C., Sweetman, B. J., Carr, K., Splawinski, J., Watson, J. T., Änggård, E., Oates, J. A. 1973. *Advan. Biosci.* 9 : 321
135. Baczynskyj, L., Duchamp, D. J., Zieserl, J. F., Axen, U. 1973. *Anal. Chem.* 45 : 479
136. McCracken, J. A., Barcikowski, B., Carlson, J. C., Gréen, K., Samuelsson, B. 1973. *Advan. Biosci.* 9 : 599
137. Pace-Asciak, C., Wolfe, L. S., Gillett, P. G., Kinch, R. A. 1972. *Prostaglandins* 1 : 469
138. Levine, L., Gutierrez-Cernosek, R. M., van Vunakis, H. 1973. *Advan. Biosci.* 9 : 71

139. Caldwell, B. V., Speroff, L., Brock, W. A., Auletta, F. J., Gordon, J. W., Andersen, G. G., Hobbins, J. C. 1972. *J. Reprod. Med.* 9:361
140. Cornette, J. C., Kirton, K. T., Barr, K. L., Forbes, A. D. 1972. *J. Reprod. Med.* 9:355
141. Jubiz, W., Frailey, J., Bartholomew, K. 1972. *Clin. Res.* 20:178
142. Sharma, S. C. 1972. *J. Physiol. London* 226:74P
143. Orczyk, G. P., Behrman, H. 1972. *Prostaglandins* 1:3
144. Van Orden, D. E., Farley, D. B. 1973. *Prostaglandins* 4:215
145. Dray, F., Charbonnel, B. 1973. INSERM Symp. *Les Prostaglandines,* 133
146. Patrono, C. J. 1973. *J. Nucl. Biol. Med.* 17:25
147. Stylos, W., Burstein, S., Rivetz, B., Gunsalus, P., Skarnes, R. 1972. *Intrasci. Chem. Rep.* 6:67
148. Hillier, K., Dilley, S. R. 1974. *Prostaglandins* 5:137
149. Jubiz, W., Frailey, J., Child, C., Bartholomew, K. 1972. *Prostaglandins* 2:471
150. Jaffe, B., Behrman, H., Parker, C. 1973. *J. Clin. Invest.* 52:398
151. Stylos, W., Rivetz, B. 1972. *Prostaglandins* 2:103
152. Yu, S., Burke, G. 1972. *Prostaglandins* 2:11
153. Zusman, R., Caldwell, B., Speroff, L. 1972. *Prostaglandins* 2:41
154. Bauminger, S., Zor, U., Lindner, H. R. 1973. *Prostaglandins* 4:313
155. Raz, A., Stylos, W. 1973. *FEBS Lett.* 30:21
156. Attallah, A. A., Lee, J. B. 1973. *Circ. Res.* 23:696
157. Granström, E., Samuelsson, B. 1972. *FEBS Lett.* 26:211
158. Levine, L., Gutierrez-Cernosek, R. M. 1972. *Prostaglandins* 2:281
159. Levine, L., Gutierrez-Cernosek, R. M. 1973. *Prostaglandins* 3:785
160. Stylos, W. A., Burstein, S., Rosenfeld, J., Ritzi, E. M., Watson, D. J. 1974. *Prostaglandins* 4:553
161. Cornette, J. C., Harrison, K. L., Kirton, K. T. 1974. *Prostaglandins* 5:155
162. Ohki, S., Hanyu, T., Imaki, K., Nakazawa, N., Hirata, F. 1974. *Prostaglandins* 6:137
163. Granström, E., Kindahl, H., Samuelsson, B. 1975. To be published
164. Kirton, K. T. 1975. To be published
165. Andrieu, J. M., Mamas, S., Dray, F. 1974. *Prostaglandins* 6:15
166. Smigel, M., Frölich, J. 1974. *Prostaglandins* 6:537
167. Levine, L., Gutierrez-Cernosek, R. M., Van Vunakis, H. 1971. *J. Biol. Chem.* 246:6782
168. Axen, U. 1974. *Prostaglandins* 5:45
169. Raz, A., Kenig-Wakshal, R. 1973. *Isr. J. Med. Sci.* 9:552
170. Stylos, W., Howard, L., Ritzi, E., Skarnes, R. 1974. *Prostaglandins* 6:1
171. Lindgren, J. Å., Kindahl, H., Hammarström, S. *FEBS Lett.* In press
172. Zusman, R. M. et al 1974. *J. Clin. Invest.* 52:1093
173. Gutierrez-Cernosek, R. M., Levine, L. 1972. *Prostaglandins* 1:331
174. Kirton, K. T., Cornette, J. C., Barr, K. L. 1972. *Biochem. Biophys. Res. Commun.* 47:903
175. Wilks, J. W., Wentz, A. C., Jones, G. S. 1973. *J. Clin. Endocrinol. Metab.* 37:469
176. Hennam, J. F., Johnson, D. A., Newton, J. R., Collins, W. P. 1974. *Prostaglandins* 5:531
177. Brummer, H. C. 1973. *Prostaglandins* 3:3
178. Brummer, H. C. 1972. *Prostaglandins* 2:185
179. Sharma, S. C., Hibbard, B. M., Hamlett, J. D., Fitzpatrick, R. J. 1973. *Brit. Med. J.* 1:709
180. Hillier, K., Calder, A. A., Embrey, M. P. 1974. *J. Obstet. Gynaecol. Brit. Commonw.* 81:257
181. Karim, S. M. M. 1968. *Brit. Med. J.* 4:618
182. Pletka, P., Hickler, R. B. 1974. *Prostaglandins* 7:107
183. Eneroth, P., Granström, E., Kindahl, H., Samuelsson, B. 1975. To be published.
184. Kuehl, F. A., Humes, J. L. 1972. *Proc. Nat. Acad. Sci. USA* 69:480
185. Gorman, R. R., Miller, O. V. 1973. *Biochim. Biophys. Acta* 323:560
186. Miller, O. V., Magee, W. E. 1973. *Advan. Biosci.* 9:83
187. Moore, W. V., Wolff, J. 1973. *J. Biol. Chem.* 248:5705
188. Schaumburg, B. P. 1973. *Biochim. Biophys. Acta* 326:127
189. Wakeling, A. E., Kirton, K. T., Wyngarden, L. J. 1973. *Prostaglandins* 4:1
190. Smigel, M., Fleischer, S. 1974. *Biochim. Biophys. Acta* 332:358
191. Rao, Ch. V. 1973. *Prostaglandins* 4:567
192. Attallah, A. A., Lee, J. B. 1973. *Prostaglandins* 4:703
193. Kirton, K. T., Gutknecht, G. D.,

Bergström, K. K., Wyngarden, L. J., Forbes, A. D. 1972. *J. Reprod. Med.* 9:266

194. Labhsetwar, A. P. 1974. *Fed. Proc.* 33:61

195. Powell, W. S., Hammarström, S., Samuelsson, B. 1974. *Eur. J. Biochem.* 41:103

196. Powell, W. S., Hammarström, S., Samuelsson, B. 1975. *Eur. J. Biochem.* In press

197. Powell, W. S., Hammarström, S., Samuelsson, B., Sjöberg, B. 1974. *Lancet* 1120

198. Powell, W. S. et al 1975. *Eur. J. Biochem.* In press

199. Hammarström, S., Kyldén, U., Powell, W. S. Samuelsson, B. 1975. *FEBS Lett.* 50:306

200. Powell, W. S., Hammarström, S., Samuelsson, B. 1975. *Eur. J. Biochem.* In press

201. Wakeling, A. E., Spilman, C. H. 1973. *Prostaglandins* 4:405

202. Robison, G. A., Butcher, R. W., Sutherland, E. W. 1971. *Cyclic AMP.* New York: Academic

203. Ramwell, P. W., Rabinowitz, I. 1972. In *Effects of Drugs on Cellular Control Mechanisms,* ed. B. R. Rabin, R. B. Freedman, 207. Baltimore: Univ. Park Press

204. Hittelman, K. J., Butcher, R. W. 1973. In *The Prostaglandins. Pharmacological and Therapeutic Advances,* ed. M. F. Cuthbert, 151. London: Heinemann

205. Horton, E. W. 1972. *Prostaglandins.* Berlin: Springer

206. Kahn, R. H., Lands, W. E. M., Eds. 1973. *Prostaglandins and Cyclic AMP.* New York: Academic

207. Burke, G., Sato, S. 1971. *Life Sci.* 10 (II):969

208. Wolff, J., Jones, A. B. 1971. *J. Biol. Chem.* 246:3939

209. Yamashita, K., Bloom, G., Field, J. B. 1971. *Metabolism* 20:943

210. Kowalski, K., Sato, S., Burke, G. 1972. *Prostaglandins* 2:441

211. Wolff, J., Cook, G. H. 1973. *J. Biol. Chem.* 248:350

212. Wolff, J., Moore, W. V. 1973. *Biochem. Biophys. Res. Commun.* 51:34

213. Sato, S., Szabo, M., Kowalski, K., Burke, G. 1972. *Endocrinology* 90:343

214. Burke, G. 1972. *Prostaglandins* 2:413

215. Yu, S. C., Chang, L., Burke, G. 1972. *J. Clin. Invest.* 51:1038

216. Burke, G., Chang, L. L., Szabo, M. 1973. *Science* 180:872

217. Burke, G. 1973. *Prostaglandins* 3:291

218. Kendall-Taylor, P. 1972. *J. Endocrinol.* 54:137

219. Cooper, R. H., McPherson, M., Schofield, J. G. 1972. *Biochem. J.* 127:143

220. Ratner, A., Wilson, M. C., Peake, G. T. 1973. *Prostaglandins* 3:413

221. Ito, H., Momose, G., Katayama, T., Takagishi, H., Ito, L., Nakajima, H., Takei, Y. 1971. *J. Clin. Endocrinol. Metab.* 32:857

222. Hertelendy, F., Todd, H., Ehrhart, K., Blute, R. 1972. *Prostaglandins* 2:79

223. Hertelendy, F., Keay, L. 1974. *Prostaglandins* 6:217

224. Harms, P. G., Ojeda, S. R., McCann, S. M. 1973. *Science* 181:760

225. Dupont, A., Chavancy, G., Labrie, F. 1972. *Advan. Biosci.* 9:34a (Suppl.)

226. Lamprecht, S. A., Zor, U., Tsafriri, A., Lindner, H. R. 1973. *J. Endocrinol.* 57:217

227. Humes, J. L., Sigal, L. H., Kuehl, F. A. Jr. 1973. *Fed. Proc.* 32:536 (Abstr.)

228. Herlitz, H., Rosberg, S. 1973. *Acta Endocrinol.* 177:316 (Suppl. 9a)

229. Zor, U., Bauminger, S., Lamprecht, S. A., Koch, Y., Chobsieng, P., Lindner, H. R. 1973. *Prostaglandins* 4:499

230. Tsafriri, A., Lindner, H. R., Zor, U., Lamprecht, S. A. 1972. *J. Reprod. Fert.* 31:39

231. Flack, J. D., Ramwell, P. W. 1972. *Endocrinology* 90:371

232. Keichline, L. D., Hagen, A. A. 1973. *Fed. Proc.* 32:298 (Abstr.)

233. Johnson, D. G., Thompson, W. J., Williams, R. H. 1973. *Fed. Proc.* 32:801 (Abstr.)

234. Kimberg, D. V., Field, M., Johnson, J., Henderson, A., Gershon, E. 1971. *J. Clin. Invest.* 50:1218

235. Sharp, G. W. G., Hynie, S., Lipson, L. C., Parkinson, D. 1971. *Clin. Res.* 19:577

236. Chen, L. C., Rohde, J. E., Sharp, G. W. G. 1972. *J. Clin. Invest.* 51:731

237. Bieck, P. R. 1972. *Advan. Cyclic Nucleotide Res.* 1:149

238. Wong, P. Y. D., Bedwani, J. R., Cuthbert, A. W. 1972. *Nature New Biol.* 238:27

239. Zarday, Z., Gouaux, J., Hays, R. M. 1972. *J. Clin. Invest.* 51:106a

240. Besley, G. T. N., Snart, R. S. 1973. *FEBS Lett.* 31:269

241. Ozer, A., Sharp, G. W. G. 1973. *Eur. J. Pharmacol.* 22:227

242. Beck, N. P., DeRubertis, F. R., Michelis, M. F., Fusco, R. D., Field, J. B., Davis, B. B. 1972. *J. Clin. Invest.* 51:2352

243. Beck, N. P., Eichenholz, A. E., Reed, S. W., Davis, B. B. 1972. *Clin. Res.* 20: 586

244. Beck, N. P., Reed, S. W., Murdaugh, H. V., Davis, B. B. 1972. *J. Clin. Invest.* 51: 939

245. Kalisker, A., Dyer, D. C. 1972. *Eur. J. Pharmacol.* 20: 143

246. Beck, N. P., Kaneko, T., Zor, U., Field, J. B., Davis, B. B. 1971. *J. Clin. Invest.* 50: 2461

247. Bär, H. P. 1973. *Biochim. Biophys. Acta* 321: 397

248. Klein, I., Levey, G. S. 1971. *Metabolism* 20: 890

249. George, W. J., Paddock, R. J., Kadowitz, P. J. 1972. *Abstr. 5th Int. Congr. Pharmacol., 23–28 July, San Francisco,* 80

250. Levey, G. S., Klein, I. 1973. *Life Sci.* 13: 41

251. Levey, G. S. 1973. *Recent Progr. Horm. Res.* 29: 361

252. Harbon, S., Clauser, H., 1971. *Biochem. Biophys. Res. Commun.* 44: 1496

253. Beatty, C. H., Young, M. K., Bocek, R. M. 1973. *Biol. Reprod.* 9: 67

254. Lerner, L. J., Carminati, P., Rubin, B. L. 1973. *Proc. Soc. Exp. Biol. Med.* 143: 536

255. Kirton, K. T., Wyngarden, L. J. 1972. *Advan. Biosci.* 9: 108 (Suppl.)

256. Andersson, K. E., Andersson. R.. Hedner, P., Persson, C. G. A. 1973. *Acta Physiol. Scand.* 87: 41A

257. Dunham, E. W., Haddox, M. K., Goldberg, N. D. 1973. *Pharmacologist* 15: 158

258. Bourne, H. R., Epstein, L. B., Melmon, K. L. 1971. *J. Clin. Invest.* 50: 10A

259. Franks, D. J., MacManus, J. P., Whitfield, J. F. 1971. *Biochem. Biophys. Res. Commun.* 44: 1177

260. Strom, T. B., Deisseroth, A., Morganroth, J., Carpenter, C. B., Merrill, J. P. 1972. *Proc. Nat. Acad. Sci. USA* 69: 2995

261. Whitfield, J. F., MacManus, J. P. 1972. *Proc. Soc. Exp. Biol. Med.* 139: 818

262. Whitfield, J. F., MacManus, J. P., Braceland, B. M., Gillan, D. J. 1972. *Horm. Metab. Res.* 4: 304

263. Whitfield, J. F., MacManus, J. P., Braceland, B. M., Gillan, D. J. 1972. *J. Cell. Physiol.* 79: 353

264. Polgar, P., Vera, J. C., DesForges, J., Rutenburg, A. M. 1973. *Fed. Proc.* 32: 881 (Abstr.)

265. Bourne, H. R., Lehrer, R. I., Cline, M. J., Melmon, K. L. 1971. *J. Clin. Invest.* 50: 920

266. Shaw, J., Gibson, W., Jessup, S., Ramwell, P. 1971. *Ann. NY Acad. Sci.* 180: 241

267. Krishna, G., Harwood, J. P., Barber, A. J., Jamieson, G. A. 1972. *J. Biol. Chem.* 247: 2253

268. Murphy, D. L., Donnelly, C., Moskowitz, J. 1973. *Clin. Pharmacol. Ther.* 14: 810

269. Frazer, A., Wang, Y. C., Pandy, G., Mendels, J. 1973. *Clin. Res.* 21: 265

270. Dukes, P. P. 1971. *Blood* 38: 822

271. Marquis, N. R., Vigdahl, R. L., Tavormina, P. A. 1969. *Biochem. Biophys. Res. Commun.* 36: 965

272. Lichtenstein, L. M., Henney, C. S., Bourne, H. R. 1972. *J. Allergy Clin. Immunol.* 49: 87

273. Butcher, R. W. 1970. *Advan. Biochem. Psychopharmacol.* 3: 173

274. Moskowitz, J., Harwood, J. P., Forn, J., Krishna, G., Rodgers, B., Morrow, A. 1971. *Nature New Biol.* 230: 214

275. Carlson, L. A., Butcher, R. W. 1972. *Advan. Cyclic Nucleotide Res.* 1: 569

276. Illiano, G., Cuatrecasas, P. 1971. *Nature New Biol.* 234: 72

277. Fain, J. N., Psychoyos, S., Czernik, A. J., Frost, S., Cash, W. D. 1973. *Endocrinology* 93: 632

278. Dalton, C., Hope, W. C. 1974. *Prostaglandins* 6: 227

279. Frank, H., Braun, T. 1971. *Fed. Proc.* 30: 625 (Abstr.)

281. Prasad, K. N., Gilmer, K., Kumar, S. 1973. *Proc. Soc. Exp. Biol. Med.* 143: 1168

282. Prasad, K. N. 1972. *Proc. Am. Assoc. Cancer Res.* 13: 16

283. Prasad, K. N. 1972. *Proc. Soc. Exp. Biol. Med.* 140: 126

284. Prasad, K. N. 1972. *Cytobiologie* 5: 265

285. Prasad, K. N., Mandal, B. 1972. *Exp. Cell Res.* 74: 532

286. Gilman, A. G., Nirenberg, M. 1971. *Proc. Nat. Acad. Sci. USA* 68: 2165

287. Hamprecht, B., Schultz, J. 1973. *Z. Physiol. Chem.* 354: 1633

288. Peery, C. V., Johnson, G. S., Pastan, I. 1971. *J. Biol. Chem.* 246: 5785

289. Makman, M. H. 1971. *Proc. Nat. Acad. Sci. USA* 68: 2127

290. Manganiello, V., Vaughan, M. 1972. *Proc. Nat. Acad. Sci. USA* 69: 269

291. Kelly, L. A., Butcher, R. W. 1974. *J. Biol. Chem.* 249: 3098

292. Otten, J., Johnson, G. S., Pastan, I. 1972. *Fed. Proc.* 31: 410 (Abstr.)

293. Daniel, V., Bourne, H. R., Tomkins, G. M. 1973. *Nature New Biol.* 244: 167

294. Berti, F., Trabucchi, M., Bernareggi, V., Fumagalli, R. 1972. *Pharmacol. Res. Commun.* 4: 253

295. Berti, F., Trabucchi, M., Bernareggi, V., Fumagalli, R. 1973. *Advan. Biosci.* 9:475

296. Wellman, W., Schwabe, U. 1973. *Brain Res.* 59:371

297. Waitzman, M. B., Woods, W. D. 1971. *Exp. Eye Res.* 12:99

298. Weinryb, I., Michel, I. M., Hess, S. M. 1973. *Arch. Biochem. Biophys.* 154:240

299. Stoner, J., Manganiello, V. C., Vaughan, M. 1973. *Proc. Nat. Acad. Sci. USA* 70:3830

300. Kuo, W. N., Kuo, J. F. 1973. *Fed. Proc.* 32:773 (Abstr.)

301. Kajita, Y., Hayaishi, O. 1972. *Biochim. Biophys. Acta* 261:281

302. Levine, R. A. 1973. *Gastroenterology* 64:186

303. Kloeze, J. See Ref. 33, 241

304. Kloeze, J. 1969. *Biochim. Biophys. Acta* 187:285

305. Salzman, E. W., Levine, L. 1971. *J. Clin. Invest.* 50:131

306. Marquis, N. R., Becker, J. A., Vigdahl, R. L. 1970. *Biochem. Biophys. Res. Commun.* 39:783

307. Droller, M. J., Wolfe, S. M. 1972. *N. Engl. J. Med.* 286:948

308. McDonald, J. W. D., Stuart, R. K. 1973. *J. Lab. Clin. Med.* 81:838

309. Boullin, D. J., Green, A. R., Price, K. S. 1972. *J. Physiol.* 221:415

310. Shio, H., Ramwell, P. 1972. *Nature New Biol.* 236:45

311. Kocsis, J. J., Hernandovich, J., Silver, M. J., Smith, J. B., Ingerman, C. 1973. *Prostaglandins* 3:141

312. Weiss, H. J., Aledort, L. M. 1967. *Lancet* 2:495

313. O'Brien, J. R. 1968. *Lancet* 1:779

314. Zucker, M. B., Peterson, J. 1968. *Proc. Soc. Exp. Biol. Med.* 127:547

• 315. Ingerman, C., Smith, J. B., Kocsis, J. J., Silver, M. J. 1973. *Fed. Proc.* 32:219 (Abstr.)

316. Vargaftig, B. B., Zirinis, P. 1973. *Nature New Biol.* 244:114

317. Willis, A. L. 1974. *Science* 183:325

318. Smith, J. B., Ingerman, C., Kocsis, J. J., Silver, M. J. 1974. *J. Clin. Invest.* 53:1468

319. Hamberg, M., Svensson, J., Samuelsson, B. 1974. *Lancet* 2:223

320. Malmsten, C., Hamberg, M., Svensson, J., Samuelsson, B. 1975. *Proc. Nat. Acad. Sci. USA* 72: In press

321. Hamberg, M., Samuelsson, B. 1974. *Biochem. Biophys. Res. Commun.* 61:942

COMPLEMENT[1]　　　　　　　　　　*899

Hans J. Müller-Eberhard[2]

Department of Molecular Immunology, Scripps Clinic and Research Foundation,
La Jolla, California 92037

CONTENTS

INTRODUCTION

Complement is a multimolecular, self-assembling biological system which constitutes the primary humoral mediator of antigen-antibody reactions. Activation of

[1] This is publication number 891.

[2] Cecil H. and Ida M. Green Investigator in Medical Research, Scripps Clinic and
Research Foundation.

697

complement may have two distinct biological consequences: 1. irreversible structural and functional alterations of biological membranes and cell death, and 2. activation of specialized cell functions. The second mode of action is evoked by complement reaction products. Two of the activation peptides, for instance, cause release of histamine from mast cells, chemotactic attraction of polymorphonuclear leukocytes, and contraction of smooth muscle. The biomedical significance of complement has been established in recent years, both in regard to participation of the system in host resistance to infections and in certain disease mechanisms (1, 2).

Operationally the complement system may be subdivided into two pathways, each comprising several functional units. The first or classical pathway is activated by IgG- and IgM-type complexes. Its 11 proteins have been grouped into three functional units: the recognition unit, C1q, C1r, C1s; the activation unit, C2, C3, C4; and the membrane attack system, C5, C6, C7, C8, and C9 (3, 4). The alternative or properdin pathway (5) is activated by aggregates of IgA and naturally occurring polysaccharides and lipopolysaccharides (3). It is composed of at least five serum proteins, one of which, C3, is also operative in the classical pathway. The alternative pathway by-passes C1, C2, and C4 and acts on C5–9 in a manner analogous to that of the classical activation mechanism. In addition to these 15 proteins, the complement system includes three regulators, one enzyme inhibitor, and two inactivating enzymes, so that the entire system is composed of at least 18 distinct serum proteins.

In this review our current knowledge of the molecular basis of the biological activities of complement will be enunciated. One of the unusual characteristics of complement proteins is their inherent ability to undergo transition from soluble molecules to peripheral or perhaps integral membrane constituents. This ability depends on the generation of binding regions on these molecules through enzymatic removal of activation peptides. The responsible enzymes have a complex quaternary structure, are indigenous to the system, and represent examples of highly specialized proteases. Their action results in the assembly of distinct multimolecular complexes from components which in their native state exhibit reversible interactions with each other. Complement-dependent cell membrane damage is accomplished by a multi-subunit complex which impairs membrane function presumably by physicochemical attack. Noncytolytic reaction products also interact with cell membranes and produce diverse biological phenomena. At least two of the low molecular weight peptides have a high content of α-helical structure, which may be a reflection of their affinity for membrane receptors. Because of the many membrane-oriented functions, complement may be regarded as an extracellular effector and modulator of biological membranes.

NOMENCLATURE

The components of the classical system are designated numerically, C1, C2, C3, C4, C5, C6, C7, C8, C9, and the three subcomponents of C1 are referred to as C1q, C1r, and C1s (6). Individual polypeptide chains of proteins with quaternary structure are denoted with Greek letters, e.g. C3α and C3β. Physiological fragments of components resulting from cleavage by enzymes that are indigenous to the complement system are distinguished by small arabic letters, e.g. C3a and C3b. A transiently activated binding site of a nascent fragment may be indicated by an asterisk, e.g. C3b*, and its inactivated state by subscript i, e.g. C3b$_i$. Enzyme activity may be indicated by placing a bar above the designation of the component in which the activity resides. Thus the symbol for enzymatically active C1 is $\overline{C1}$. If reference is made to a composite reaction product, fragments may be indicated. For example, the systematic notation of the enzyme C3 convertase which reflects its subunit composition is $\overline{C4b,2a}$. However, for reasons of simplicity, reference to fragments is avoided where not essential.

No convention has been established as yet for the designation of the components of the alternative mechanism of complement activation. However, a trend has developed among the active workers in the field according to which a tentative terminology will utilize trivial names and symbols in parallel. Trivial names have either historical (properdin) or functional (C3 proactivator) meaning. Symbols are capital letters and will be used for the writing of formulae: P for properdin, B for C3 proactivator, and D for C3 proactivator convertase. The respective activated forms are indicated by a bar, e.g. $\overline{B}$. The Committee on Complement Nomenclature of the World Health Organization has expressed the intention to introduce a definitive, biochemical terminology as soon as the essential components of the system and their functions have been recognized.

THE PROTEINS OF THE CLASSICAL COMPLEMENT SYSTEM

Comparative Description of the Eleven Proteins

These 11 proteins differ widely in physical parameters and in their concentration in serum (Table 1). As far as known, all these proteins are glycoproteins and none is a lipoprotein (7). C1r, C1s, and C2 are proenzymes (8). C1s, C2, C3, C4, and C5 are highly susceptible to tryptic attack, which reflects the mode of their physiological

Table 1   Proteins of the classical human complement system

Protein	Serum Concentration (μg/ml)	Sedimentation Coefficient (S)	Molecular Weight	Relative Electrophoretic Mobility	Number of Chains
C1q	180	11.1	400,000	γ_2	18
C1r	—	7.5	180,000	β	2
C1s	110	4.5	86,000	α	1
C2	25	4.5	117,000	β_1	—
C3	1600	9.5	180,000	β_2	2
C4	640	10.0	206,000	β_1	3
C5	80	8.7	180,000	β_1	2
C6	75	5.5	95,000	β_2	1
C7	55	6.0	110,000	β_2	1
C8	80	8.0	163,000	γ_1	3
C9	230	4.5	79,000	α	—

activation. C7 was reported to exhibit tributyrase activity (9). This report has not yet been confirmed by others. Although proteins composed of three different poly-peptide chains are rare, C1q (10), C4 (11), and possibly C8 (12) show such a chain structure.

C1q: A Collagen-Like Glycoprotein

C1q is a truly unusual protein. Being a normal plasma constituent with important physiological functions, it is chemically collagen like (13). Per 100 amino acid residues it contains five residues of hydroxyproline, two residues of hydroxylysine, and 18 residues of glycine. Its sizable carbohydrate moiety (9.8%, wt/wt) consists primarily of equimolar amounts of glucose (3.35%) and galactose (3.10%). In addition, small amounts of glucosamine, neuraminic acid, mannose, and fucose are also present. As in collagen (14) and basement membrane (15), glucose and galactose in the C1q molecule are linked as a disaccharide to the hydroxyl group of hydroxylysine; glucosyl galactosyl hydroxylysine could be isolated from an alkaline hydrolysate of C1q(13). This carbohydrate moiety as well as hydroxylated lysine and proline have not been reported to occur in any other plasma protein.

C1q recognizes immune complexes and thereby initiates the complement reaction. The valence of C1q for IgG could be determined by analytical ultracentrifugation because C1q forms reversible complexes with native, monomeric IgG (16). A Scatchard plot of the experimental data indicates six binding sites for IgG per molecule of C1q (17).

Treatment of C1q with sodium dodecylsulfate (SDS) results in reduction of the molecular weight from 400,000 to 65,000–70,000. This observation suggests that the molecule consists of six noncovalently bonded subunits of similar or identical size. Each of these subunits is believed to carry one immunoglobulin binding site. Upon reduction in SDS, three distinct polypeptide chains are observed which have molecular weights ranging between 21,000 and 24,000. (10). Each chain consists of three distinct regions: N- and C-terminal regions with the amino acid composition of globular proteins and an intermediate region with a collagen-like sequence. The N-terminal region consists of three to nine residues and the C-terminal portion of 110 residues. The collagenous portion comprises 78 residues of the repeating sequence X-Y-glycine. All the hydroxylysine and hydroxyproline residues are found in the collagenous sequence and occupy only the Y position (18).

The unusual chemical structure of C1q is reflected in its ultrastructure as revealed by electron microscopy (19, 20). Six peripheral globular portions are connected by fibril-like strands to a common central portion. The overall diameter of this delicately structured molecule is about 350 Å. On the basis of presently available informa-tion, Reid & Porter (21) proposed a molecular model according to which C1q consists of six identical, noncovalently linked subunits. Each subunit contains three similar but not identical chains joined by disulfide bonds. The collagen-like regions of the three chains form a triple helical strand, and the C-terminal, noncollagenous regions a random coil globular arrangement. The six subunits associate through contact regions near the end of their helical portions and thus give rise to a bouquet-like structure as visualized by electron microscopy.

An alternative model (22, 23) envisions six peripheral units (69,000) to be connected by fibrillar strands to a central subunit (54,000). One peripheral subunit consists of two different disulfide-linked chains with molecular weights of 31,000 and 34,000, respectively. The central piece is composed of two identical 26,000 dalton chains. This model is based on the observation of two subunits on dissociation of ^{125}I-labeled C1q with SDS and subsequent acrylamide gel electrophoresis in presence of SDS. Since only the smaller subunit (54,000) bore the radiolabel, it was concluded that it is not a derivative of the larger and unlabeled subunit. Further, reduction in urea gave rise to three types of chains with molecular weights of 26,000, 31,000, and 34,000. Only the 26,000 dalton chain was radioactive. Obviously, both models are compatible with the ultrastructural and functional evidence. Formulation of the precise quaternary structure of C1q may have to await completion of the sequence analysis.

That the immunoglobulin binding sites of C1q may be located in the peripheral globular structures is strongly suggested by experiments with ^{125}I-C1q fragments produced with collagenase (24). One of several fragments thus obtained exhibited the ability to attach to immune precipitates. It was unlabeled and had a molecular weight of 40,000, which are properties ascribed to the C-terminal globular portions.

C2: Enhanced Activity upon Oxidation of SH Groups

C2, a trace protein in serum, is one of the two precursor molecules of the complex enzyme C3 convertase. Treatment of the protein with p-chloromercuribenzoate (p-CMB) results in loss of its hemolytic activity (25). The opposite was observed upon treatment of C2 with iodine: C2 hemolytic activity was 10- to 20-fold enhanced (26). Because two moles of radioactive p-CMB were bound per mole of C2, it was concluded that the molecule contains two free, reactive SH groups. In exploring the functional relevance of the SH groups and the nature of the chemical modification responsible for the enhancement of activity, the following observations were made: 1. iodine treatment led to iodine uptake, but the latter did not correlate with enhancement of activity; 2. iodine-treated C2 could not be inactivated by p-CMB and did not bind it; 3. ^{14}C-p-CMB-treated C2, upon exposure to iodine, not only was reactivated, but underwent a 10- to 20-fold enhancement of hemolytic activity and released all the previously bound radioactivity; 4. upon mild reduction of iodine-treated C2 the 10- to 20-fold enhanced hemolytic activity was lowered to the original level and both SH groups became again available for binding of p-CMB; and 5. iodine treatment did not influence the molecular weight of C2 (27). The data are consistent with the interpretation that C2 contains two SH groups which are positioned in close proximity to each other and form an intramolecular disulfide bond upon oxidation with iodine. Through this chemical modification, C2 acquires the extraordinary increase in hemolytic activity.

C3, C4, C5: Chains, Fragments, and Functional Sites

These three proteins have structural and functional properties in common and it is possible that they have arisen from a common ancestral molecule. C3, C4, and C5 have molecular weights of, respectively, 180,000 (28), 206,000 (11, 29), and 180,000

(30). They have similar electrophoretic mobilities. C3 and C5 contain two chains linked by disulfide bridges and noncovalent forces (30, 31). In both molecules the α chain has an approximate molecular weight of 110,000 and the β chain 70,000. C4 contains three chains, α, β, and γ, with molecular weights of 95,000, 78,000, and 33,000, respectively (11). Schematic drawings of the proposed chain structure of the three molecules are shown in Figure 1.

The three proteins are substrates of three different complement enzymes. The initial enzymatic attack is exceedingly limited, involving probably only a single peptide bond located in the α chain of the respective molecule (11, 28, 30, 31). Dislocation of the arising a fragments endows the b fragments with transiently activated binding sites through which attachment of these fragments to suitable acceptors may occur (16). Bound C3b (32) and C4b (33) are endowed with a second, stable binding site, which reacts with the immune adherence receptor on the surface of a variety of mammalian cells. This reaction facilitates phagocytosis of comple-

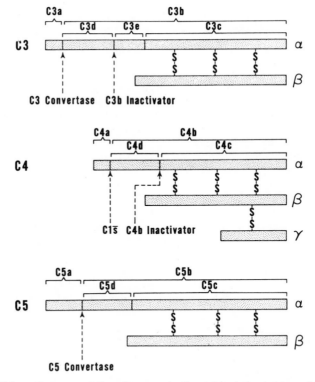

Figure 1   Schematic representation of proposed polypeptide chain structure of C3, C4, and C5. The topological relationships between chains, fragments, functional sites, and enzymatic attack regions are indicated as presently envisaged. For further explanation, see text.

ment-coated particles by polymorphonuclear leukocytes, monocytes, and macrophages (34).

C3b, C4b, and probably also C5b are subject to a secondary enzymatic attack which again involves the α chains (31, 35). The attack is performed by the serum enzyme C3b and C4b inactivators (36, 37) or by a set of similar enzymes and results in fission of the b fragments to the corresponding c and d fragments (38). Concomitantly, all b fragment-dependent functions are abrogated. The function of the C3b inactivator disclosed that cell-bound C3b is attached to the cell membrane by virtue of its d portion: whereas C3c is released by the enzyme, C3d is retained by the cell (31, 38). The composition of the enzymatically produced fragments was studied by SDS acrylamide gel electrophoresis and is indicated in Figure 1. Apparently, fission of C3b releases a third fragment of about 16,000 dalton, C3e (31). The relation of this peptide to the leukocyte mobilization factor (39) is under investigation. The factor was described as an acidic peptide of low molecular weight derived in unknown fashion from C3.

C8: An Unusual Subunit Structure

C8 (mol wt 163,000) consists of three polypeptide chains, two chains being covalently linked and the third chain noncovalently linked. Upon SDS acrylamide gel electrophoresis of nonreduced C8, two subunits become separable, C8a (93,000 daltons) and C8b (70,000 daltons). When reduced C8 is subjected to the same procedure, three subunits become detectable, C8α ($\sim$83,000 daltons), C8β (70,000 daltons), and C8γ ($\sim$10,000 daltons). These observations indicate that the small γ chain is linked by disulfide bonds to the α chain. Although the position of the noncovalently bonded β chain in the molecule is unknown, the following model is proposed: β---α-S-S-γ (12). The possibility of a fourth chain of low molecular weight has not been ruled out.

GENERAL MOLECULAR DYNAMICS

Reversible Interactions

The 11 proteins of the classical pathway arrange themselves into three functional units initially through reversible protein-protein interactions. Molecules of different proteins that function together exhibit affinity for each other, thereby conferring a certain degree of structure on the system in free solution which reflects the molecular organization of the activated system. Reversible association products are formed by 1. C1q, C1r, and C1s; 2. C2 and C4; and 3. C5, C6, C7, C8, and C9. These reversible associations have been observed on sucrose density gradient ultracentrifugation of mixtures of isolated proteins. C1q forms loose complexes with both C1r and C1s (40, 41). A stable 10S complex is formed by C1r and C1s in absence of C1q. A stable association between all three proteins requires Ca^{2+} (40, 42). These observations suggest that in the native C1 complex all three subcomponents are in physical contact with each other. The association product of C2 and C4 varies in stability with environmental conditions (43). At low pH (5.5) and high ionic strength (0.15) it is labile, and at high pH (8.0) and low ionic strength (0.05) it is stable, giving rise to a

12S complex. C5, C6, C7, C8, and C9 form a reversible association product which sediments with a velocity of approximately 15S, considerably slower than the stable 26S C5b-9 complex (see below) (44). In these studies it was possible to show that C5 has binding regions for C6, C7, and C8 and that C8 has binding regions for C5 and C9.

Transition of Soluble Proteins to Membrane Proteins

When complement is activated by antibody-coated cells, it undergoes a self-assembling process in the course of which the entire set of molecules transfers from solution to the solid phase of the target cell surface. This is an unusual and perhaps unique process in molecular biology. All 11 proteins are soluble in aqueous solutions. None has an apparent affinity for membranes in its unaltered, native form. Yet the potential of entering into direct and firm contact with membrane constituents is found in C3 (45), C4 (46), and C5,6,7 (47). There is strong reason to believe that C8 also possesses this potential, as discussed below. In contrast, C1, C2, and C9 bind to sites that are not originally part of the target membrane. Three different tactics are operative in the transfer: 1. reversible ionic interaction, which applies to C1; 2. enzymatic activation of binding sites, which is characteristic for C2, C3, C4, and C5; and 3. adsorption, which applies to C6, C7, C8, and C9.

Enzymatic Activation of Labile Binding Sites

This is the most complex of the three tactics that govern the transfer of complement from solution to the surface of a membrane. A binding region previously concealed in the interior of a molecule is exposed by cleavage of a peptide bond and subsequent dissociation or dislocation of a low molecular weight fragment. This binding region is not comparable to the substrate binding site of an enzyme. It possesses no known specificity and is not associated with a catalytic function. Rather, it has the potential of entering into strong, probably hydrophobic, interaction with acceptors of unknown chemical nature. The activated state of an unbound complement molecule is of short duration and probably corresponds to a thermodynamically unfavorable conformation. If diffusion effects collision with a membrane receptor, the molecule is firmly and indefinitely bound and enabled to fulfill its characteristic function in the cytolytic reaction. Failing collision with a receptor within less than 0.1 sec after enzymatic activation (48), the molecule undergoes secondary changes and loses its binding region. As a result it remains unbound and cytolytically inactive in the fluid phase.

Assembly of Complex Enzymes: C3 and C5 Convertase

The C1 complex does not constitute a complex enzyme. C$\overline{1r}$ and C$\overline{1s}$ fulfill their characteristic enzymatic functions even after they are isolated from the complex. In contrast, two of the enzymes of the classical pathway and, in all probability, several enzymes of the alternative pathway depend, with respect to their functionality in the complement system, on a defined, complex subunit structure.

C3 convertase (C$\overline{4b,2a}$) is assembled by the enzyme C$\overline{1s}$ from the two precursor molecules, C2 and C4 (43). As long as complement research centered on the exploration of cell-bound intermediate complement assemblies, it was difficult, if not impossible, to determine the physicochemical properties and subunit composition of active entities. It was shown, however, that cell-bound C2 and C4 generate an enzyme that turns over C3 (45, 49). The nature of the enzyme was elucidated when it could be formed in cell-free solution (43). A mixture of isolated C2 and C4 treated with isolated $C\overline{1}s$ gave rise to C3 cleaving activity. $C\overline{1}s$ is known to cleave C2 into C2a (80,000 daltons) and C2b (37,000 daltons) (50, 51) and C4 into C4a (8000 daltons) and C4b (198,000 daltons) (11, 29, 52). It was concluded that C3 convertase represents a complex of C2 and C4 in which a stable product of C4 serves as acceptor and modulator of activated C2. This conclusion was based on the finding that C3 cleaving activity 1. did not appear on treatment of either C2 or C4 alone with $C\overline{1}s$; 2. did not appear on addition of C4 to a previously incubated mixture of C2 and $C\overline{1}s$; but 3. was generated on addition of C2 to a previously incubated mixture of C4 and $C\overline{1}s$. Initial attempts to demonstrate the postulated enzymatically active complex after zone ultracentrifugation met with failure. Such demonstration was accomplished, however, using oxidized C2 (26). The C3-cleaving enzyme was shown to have a sedimentation velocity of 11 and a molecular weight of approximately 300,000. This molecular size is indicative of a bimolecular complex consisting of C4b (with C4a possibly noncovalently attached) and C2a. The quaternary structure of the enzyme comprises therefore two subunits and at least four polypeptide chains, assuming that C2a constitutes a single chain. The theoretical molecular weight of the complex is 280,000–288,000. By radiochemical and immunochemical means it was possible to demonstrate the incorporation of C2- and C4-derived protein into the enzyme (51). Whereas the natural substrate of the enzyme is C3, it also hydrolyzes the ester bond of acetyl-glycyl-lysine methyl ester. The catalytic site is thought to reside in the C2a subunit. After spontaneous dissociation from C4b, cytolytically inactive C2a retains esterase activity but not C3-cleaving activity (8).

Oxidation of C2 with iodine, which greatly enhances its cytolytic activity, augments three different C2 functions: the binding of activated C2 to C4b, the catalytic function of $C\overline{4,2}$, and the stability of the enzyme (53). The half-life at 37°C is 10 min for $C\overline{4,2}$ and 200 min for $C\overline{4,2}$, containing oxidized C2 (26). The fact that three functional parameters of the $C\overline{4,2}$ complex are affected by the presence of the iodine-induced intramolecular S–S bond in C2 strongly suggests that the location of this bond is extraneous to the active site. Instead, it appears to be located in a region of the molecule where it can influence the function of both the enzymatic and the C4b contact region.

C5 convertase is a derivative of the $C\overline{4,2}$ enzyme. Addition of C3 to cell-bound $C\overline{4,2}$ results not only in turnover of C3 and binding of C3b to the cell surface (45), but also in the generation of a cell-bound enzyme that turns over C5 (54, 55). It was assumed therefore that a product of the substrate of $C\overline{4,2}$ modulates the enzyme such that its substrate binding region becomes adapted to C5. The modulation was envisaged to result from a physical association of C3b with $C\overline{4,2}$.

A cell-free solution of $C\overline{4,2}$ and C3 did exhibit C5-cleaving activity. Upon zone ultracentrifugation of the mixture, this activity could be detected in fractions containing 15–18S material (56). C2 was utilized in oxidized form. These preliminary data suggest that a relatively firm complex is formed between C$\overline{4,2}$ and nascent C3b, giving rise to the C$\overline{4,2,3}$ enzyme. This enzyme consists of a minimum of three subunits and six polypeptide chains. The nature of the forces that unite the subunits is unknown. The catalytic site of C5 convertase is probably the same as that of C3 convertase and located in the C2 subunit of the enzyme.

Thus C$\overline{4,2}$ and C$\overline{4,2,3}$ represent highly specialized proteases: C3 is the only known protein substrate for C$\overline{4,2}$ and C5 is the only known substrate for C$\overline{4,2,3}$. Whether C5 convertase retains a limited ability to act on C3 has not been explored. Both enzymes hydrolyze in their respective natural substrates a peptide bond which involves the carboxyl group of an arginyl residue. C$\overline{4,2}$ thereby produces the fragments C3a (mol wt 8900) and C3b (mol wt 171,000), and C$\overline{4,2,3}$ the fragments C5a (mol wt 17,000) and C5b (mol wt 163,000).

Activation Peptides and Anaphylatoxins

Highly restricted proteolysis constitutes the mechanism by which the binding sites of C2, C3, C4, and C5 are activated. As a consequence the peptides C2b, C3a, C4a, and C5a are formed. These peptides do not participate in the formation of the multi-molecular functional units of the cytolytic reaction. Instead, they tend to dissociate themselves from their respective precursors C2b and C5a instantly, C3a and C4a gradually. Whereas no documented biological activity resides in either C2b or C4a, the fragments C3a and C5a have pronounced phlogogenic activity (57–59). For historical reasons they are called the anaphylatoxins.

C3a and C5a have similar or identical biological activities in that each causes release of histamine from mast cells; directed, chemotactic migration of polymorphonuclear leukocytes; and contraction of smooth muscle. When injected into the human skin, they evoke an immediate edema and erythema reaction, the minimal effective dose being 2×10^{-12} mol for C3a (60, 61) and 1×10^{-15} mol for C5a (62). Some of the physical and biological properties of the anaphylatoxins are listed in Table 2.

The mode of action of these peptides on responsive cells is not well understood.

Table 2 Properties of human anaphylatoxins

	C3a	C5a
Electrophoretic mobility		
pH 8.5 ($\times 10^{-5}$ cm^2 V^{-1} sec^{-1})	+2.1	−1.7
Diffusion coefficient ($\times 10^{-7}$ cm^2 sec^{-1})		12.1
Frictional ratio		1.05
Molecular weight	8,900	16,500
Biologic activities:		
Minimal effective concentration in vitro[a]	1.3×10^{-8} M	7.5×10^{-10} M
Minimal effective dose in vivo[b]	2.1×10^{-12} mol	1×10^{-15} mol

[a] Guinea pig ileum contraction.

[b] Wheal and erythema in human skin.

However, pertinent observations have been reported. C3a and C5a activate a proesterase of polymorphonuclear leukocytes. Activation of this enzyme was shown to be an obligatory step in the chemotaxis caused by these two peptides (63). Using an immunofluorescent technique, it was possible to show that C3a binds specifically to the plasma membrane of mast cells (64). The distribution of fluorescent stain on the cell surface suggests the presence of distinct receptors for C3a. Uptake of radiolabeled C3a by mast cells correlated well with the extent of histamine release (65). Although some functions of C3a and C5a may be mediated by histamine, e.g. contraction of guinea pig ileum, other functions are not (66). The contraction of estrous rat uterus caused by C3a is histamine independent (67). The effect of C3a on the microvasculature is also independent of histamine. Topical application of C3a to rabbit omentum caused reduction of pressure and diameter of terminal arterioles. Because the effect was inhibited by the α-adrenergic blocker, phentolamine, it was suggested that C3a in this case acted via α-receptors (68).

C3a is derived from the NH_2-terminal end of the α chain of C3 (30, 31) by cleavage of the bond between residues 77 and 78 (69). The bond is selectively attacked by two distinct complement enzymes, C3 convertase of the classical pathway and C3 convertase of the alternative pathway. The NH_2-terminal position is occupied by serine and the COOH terminus by arginine (28, 70). The primary amino acid sequence of the 77 residue peptide has recently been established by Hugli (69). The molecule contains three methionine and six half-cystine residues and it lacks tryptophan, carbohydrate, and free sulfhydryl groups. The circular dichroism spectrum exhibited pronounced minima at 208 and 222 nm (Figure 2). From the negative mean residue ellipticity it was calculated that C3a contains 40–45% α-helical structure (71). C3 and C3b contain respectively 10% and 0% α-helix (J. L. Molenaar, personal communication). Thus in terms of charge and conformation, C3a constitutes an unusual portion of the native C3 molecule.

Two essential structural requirements have been recognized to date for the expression of C3a biological activity: the COOH-terminal arginine and the ordered secondary structure. Removal of the arginine residue by pancreatic or serum carboxypeptidase B results in total loss of biological activity without affecting the characteristic circular dichroism spectrum (67). Treatment with guanidinium chloride and mercaptoethanol also leads to total inactivation and loss of negative ellipticity at 208 and 222 nm. Upon removal of the denaturing agents full activity is restored and the circular dichroism spectrum assumes the characteristics of un-treated C3a (71) (Figure 2).

C5a is derived from the α chain of C5 (30) by C5 convertase of either pathway. Arginine was found in COOH-terminal position and, as in the case of C3a, removal of this residue by carboxypeptidase B abrogates its activity (72, 73). Like C3a, the C5a peptide has a substantial content of α-helical structure, approximately 40%. Treatment of C5a with mercaptoethanol progressively diminishes the ellipticity at 208 and 222 nm and reduces its anaphylatoxin activity to limiting values. Removal of the reducing agent restores both the characteristic circular dichroism spectrum and the biological activity (74).

Porcine C5a differs from human C5a with respect to the susceptibility of the C-terminal residue to carboxypeptidase B (CPB) action. Porcine C5a generated in absence of epsilon aminocaproic acid (EACA) has leucine at the COOH terminus (75), but it contains COOH-terminal arginine when prepared in the presence of EACA (72, 76). The latter was active at a concentration of 3×10^{-10} M (72) and the former at 7×10^{-9} M (80), suggesting that C5a with COOH-terminal leucine is 90% inactivated. Porcine C5a, which contained COOH-terminal arginine, was resistant to CPB, and its COOH-terminal arginine could only be removed by CPB after boiling of the peptide. It is proposed therefore that nascent C5a in porcine serum is susceptible to inactivation by serum CPB, and that due to a subsequent conformational change, the residual, active C5a becomes resistant to the enzyme and remains active (72). A similar mechanism may apply to rat (77) and guinea pig (78) C5a, which are also resistant to CPB (79).

Both C3a and C5a are controlled by serum carboxypeptidase B, also called anaphylatoxin inactivator. The enzyme is an α-globulin with a molecular weight of 300,000 (67). It is responsible for total suppression of anaphylatoxin activity in human serum. It inactivates the peptides by efficient removal of the COOH-terminal arginine residue (67). However, inhibition of the enzyme with EACA allows demonstration of C3a and C5a activity in whole human serum following complement

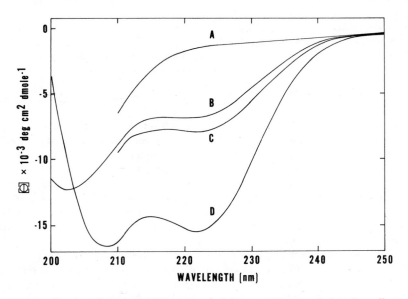

Figure 2 Circular dichroism (CD) spectra of human C3a anaphylatoxin: effect of reducing agent and denaturant. C3a dissolved in 6 M guanidinium chloride and 0.02 M mercaptoethanol (A); C3a after reduction and alkylation (B); C3a dissolved in 6 M guanidinium chloride (C); and C3a dialyzed after treatment with 6 M guanidinium chloride and 0.02 M mercaptoethanol (D). The CD spectrum of untreated C3a was identical to curve D (71).

activation (76). The use of the inhibitor has afforded large scale isolation of intact C3a from activated whole human and porcine serum for structural studies (70).

Three Topologically Distinct Target Membrane Sites

Figure 3 depicts a three site model of complement transfer from solution to the solid phase of the target cell surface. It proposes that at completion of the process the functional units have assumed stable supramolecular organizations at three topologically distinct sites, SI, SII, and SIII. C1 is bound to antibody molecules (SI), where it catalyzes the assembly of the $\overline{C4,2}$ complex which binds to SII (chemical nature unknown) and converts itself to $\overline{C4,2,3}$. From SII this enzyme initiates the assembly of the C5b-9 complex which binds to SIII (chemical nature unknown) and proceeds to convert this site to a membrane lesion. The model implies that several components must transfer from the site of enzymatic activation of their

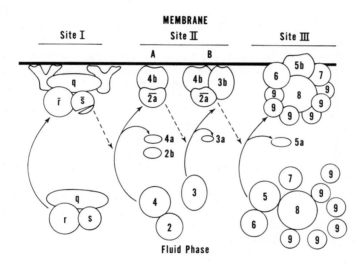

Figure 3 Pictorial representation of the three site model of complement transfer from solution to the solid phase of the target cell surface. The C1 complex is reversibly bound to antibody molecules (SI) through its C1q subunit. An internal reaction leads to activation of C1s. $\overline{C1s}$ activates labile binding sites in C2 and C4 by enzymatic removal of activation peptides. The $\overline{C4,2}$ enzyme (C3 convertase) is thereby enabled to assemble at a membrane site (SII) topologically distinct from SI. $\overline{C4,2}$ converts itself to $\overline{C4,2,3}$ (C5 convertase) by cleavage of C3 and assimilation of C3b. Many monomeric C3b fragments are bound in the process within the microenvironment of a $\overline{C4,2}$ complex. These clusters of C3b bound at a fourth category of membrane sites (not indicated in the drawing) are responsible for the immune adherence reaction. $\overline{C4,2,3}$ in cleaving C5 initiates the self assembly of the C5b-9 complex which attacks the membrane at a third, topologically distinct site (SIII). The membrane lesion at SIII is caused, in all probability, by the insertion into the membrane of a small specialized portion of one of the subunits of the complex. C3a and C5a are activation peptides with anaphylatoxin activity and a high content of α-helical structure.

binding region to the site of their binding at the cell surface. This applies in sequence of action to C4, C2, C3, and C5.

In brief, the model is based on the following experimental observations: 1. Binding of C1 to a cell is antibody dependent. 2. C4 can bind to a cell in absence of antibody and C1, provided $C\overline{1s}$ is present in the fluid phase (46). 3. It can then proceed to bind C2 and thereafter C3 in absence of $C\overline{1s}$ (45). 4. Binding of C5–9 to a cell A can be catalyzed by $C\overline{4,2,3}$ located at the surface of cell B (48, 81). Cells carrying C5–9 undergo lysis irrespective of whether earlier acting components are on their surface.

THE CLASSICAL COMPLEMENT REACTION AND MEMBRANE DAMAGE

Activation of the Recognition Unit, C1

C1, a Ca^{2+}-dependent complex of three different proteins, C1q, C1r, and C1s (42), combines reversibly with complexes of IgG or IgM. The recognition function with respect to immunoglobulins resides in the C1q subcomponent (4, 16). The C1q binding site is located in the Fc fragments of IgG and IgM (82–84). The fifth cyanogen bromide fragment of the μ chain represents a major portion of the site in IgM (85), and a 6800 dalton piece of this fragment, C_H4, was recently shown to be active in C1 binding (M. M. Hurst, personal communication). For IgG the complement binding site was found in the C_H2 domain of the γ chain (86, 87), which is located near the hinge region, and in a peptide derived from the COOH-terminal portion of the chain (88). It was concluded that the C1q binding site of IgG is composed of amino acid residues of different regions of the Fc fragment (88).

C1q has six binding sites for IgG and probably also for IgM. The relative binding affinities of C1q for IgG subclasses follow the order: G3 > G1 > G2 > G4. The single site dissociation constants ranged from 1.4×10^{-5} M for IgG3 to 1.1×10^{-4} M for IgG4 (17). Binding of C1 usually leads to activation, although both processes can be separated. IgG with modified tryptophan (89) and the subclass IgG4 (17) are capable of binding C1q but do not activate C1. Internal activation of C1 may be visualized as initiated by a critical conformational change of C1q, which in turn induces a change in C1r. As a result, the proenzyme C1r (41) acquires enzymatic activity through which C1s is converted to $C\overline{1s}$ (40, 90, 91). Conversion of C1s involves cleavage of a peptide bond and results in formation of two chains (mol wt 56,000 and 30,000) held together by a disulfide bond (40, 91). $C\overline{1s}$ is a serine protease and is inhibitable by diisopropylphosphofluoridate (DFP) (92). The active site serine is located in the 30,000 dalton chain, as was shown with radiolabeled DFP (91, 93). The amino acid sequence around the reactive serine was found to be very similar to sequences at the active site of plasmin, thrombin, and trypsin (J. E. Fothergill, personal communication).

Assembly of the Activation Unit, C4,2,3

Assembly of the activation unit proceeds in two steps and involves the sequential formation of two related enzymes: C3 convertase ($C\overline{4,2}$) and C5 convertase ($C\overline{4,2,3}$).

Enzymatic activation by $\overline{C1}$ of binding sites in C2 and C4 allows the major fragment of the C2 molecule (C2a) to fuse with the major fragment of the C4 molecule (C4b) (43). The resulting $\overline{C4,2}$ complex has the transient ability to attach to target cell receptors through a site in the C4 portion of the complex (46). The membrane binding site of C4 is activated by attack of $\overline{C1s}$ on the α chain of C4 (11). A similar study has not been performed on the mode of activation of the C4 binding site in the C2 molecule.

During the second step, C3 convertase cleaves C3 into C3a and C3b, allowing nascent C3b to associate itself with its activating enzyme. This reaction gives rise to the formation of C5 convertase which appears to be a trimolecular complex (54–56). The modulation of C3 convertase by C3b adapts the enzyme to C5 as its natural substrate.

Assembly of the Membrane Attack Complex, C5b-9

Attack of C5 by C5 convertase ($\overline{C4,2,3}$) initiates the self-assembling process which, without further enzymatic action, results in the formation of the stable C5b-9 complex (94). In this reaction the low molecular weight activation peptide, C5a (59, 95), is cleaved from the α chain of C5 (30). The resulting C5b fragment thereby acquires the transient ability to bind C6 and C7 (47, 81). The trimolecular C5b,6,7 complex constitutes the molecular arrangement for adsorptive binding of C8. The C8 molecule of the tetramolecular C5b-8 complex functions as a binding region of C9. The fully assembled membrane attack mechanism represents a stable complex of 10 molecules, one each of C5b, C6, C7, and C8, and six of C9 (94). The nascent C5b-9 complex has the ability to attach to membranes and cause membrane damage. Since the binding capacity of the unbound complex decays rapidly, cytolytically inactive complex accumulates as a byproduct in the fluid phase (96, 97).

The complex was deduced, with respect to size and composition, from experiments with cell-bound, radiolabeled C5b-9 (94). These were the early experimental results: 1. Erythrocytes uncoated with antibody or any complement proteins can be lysed by C5–C9 from the fluid phase provided cells carrying $\overline{C4,2,3}$ are in close proximity to the target cells (48). This type of experiment defined C5–C9 as the membrane attack mechanism of complement. 2. C5, C6, C7, and C8 are bound to target cells in equimolar amounts and at saturation of all C9 binding sites, six C9 molecules are bound per one C8 molecule (94). 3. A minimum of three C9 molecules are required for the production of one membrane lesion (98). 4. Antibodies to either C5, C6, or C7 inhibited binding of C8 to cells bearing C5b,6,7 sites, and antibody to C8 strongly inhibited binding of C9 to cells bearing C5b,6,7,8 sites. A model was therefore proposed which consists of a trimolecular C5b,6,7 complex resting on the target cell membrane surface. This complex forms the binding site of one C8 molecule giving rise to a tetramolecular C5b-8 complex with the geometry of a tetrahedron. Maximally six C9 molecules then bind to binding sites of the single C8 molecule in the complex, forming a decamolecular assembly with a cumulative molecular weight of one million (94) (Figure 4).

The general validity of this model could be examined when it was found that a

Figure 4 Photograph of a model of the C5b-9 decamolecular membrane attack mechanism of complement showing three stages of its assembly. The spheres were made of modeling clay, the relative weights being proportional to the molecular weights of the proteins. The numerals refer to the corresponding complement components. (*a*) Model of the membrane-bound C5b,6,7 trimolecular complex displaying triangular geometry, and constituting the proposed binding site for C8. (*b*) Model of the tetramolecular complex C5b,6,7,8 having the geometry of a tetrahedron. (*c*) Model of the fully assembled decamolecular C5b-9 complex, exhibiting two C9 trimers bound in triangular arrangement to the C8 portion of the tetrahedron (94).

soluble, stable C5–9 complex accumulates as a byproduct in the fluid phase during the cytolytic complement reaction (96). The isolated complex has a molecular weight of 1,040,000, the electrophoretic mobility of an α-globulin, and it is inactive in cytolysis. It can be dissociated into its subunits by SDS. Subjected to SDS gel electrophoresis, the complex gives rise to seven protein bands. The corresponding proteins could be identified, in order of molecular size, as C5b, C7, C6, C8a, X, C9, and C8b (99). The unidentified protein X has a molecular weight of 88,000. Studies are underway to determine whether it represents a new complement component, an inhibitor of nascent C5b-9, or the serum equivalent of a membrane acceptor of the nascent complex. Estimates of the relative concentrations of the components in the complex revealed that C5b, C6, C7, C8a, and C8b are represented in equimolar amounts, whereas X and C9 are present in multiples thereof. The sequence of the cytolytic complement reaction is summarized in Table 3.

Table 3   The cytolytic complement reaction: functional units and membrane sites[a]

First Site: Activation of Recognition Unit

1. $S_I A + C1q \begin{smallmatrix} r \\ \\ s \end{smallmatrix} \to S_I A - C1q \begin{smallmatrix} \bar{r} \\ \downarrow \\ \bar{s} \end{smallmatrix}$

Second Site: Assembly of Activation Unit

2. $C4 \xrightarrow{S_I A \,\overline{C1}} C4a + C4b^*$

3. $C2 \xrightarrow{S_I A \,\overline{C1}} C2a^* + C2b$

4. $S_{II} + C4b^* + C2a^* \to S_{II}\, \overline{C4b,2a}$

5. $C3 \xrightarrow{S_{II}\, \overline{C4b,2a}} C3a + C3b^*$

6. $S_{II}\, \overline{C4b,2a} + C3b^* \to S_{II}\, \overline{C4b,2a,3b}$

Third Site: Assembly of Membrane Attack Mechanism

7. $C5 \xrightarrow{S_{II}\, \overline{C4b,2a,3b}} C5a + C5b^*$

8. $S_{III} + C5b^* + C6 + C7 + C8 + C9 \to S_{III}\, C5b,6,7,8,9$

[a] S_I, S_{II}, S_{III}: topographically distinct sites on target cell surface. A: antibody to cell surface constituent; bar denotes active enzyme; asterisk denotes enzymatically activated, labile binding site.

Action of C5b-9 on Membranes

The accumulated evidence strongly suggests that C8 represents the molecule that executes the cytolytic function of the C5b-9 complex. The assembly of the complex at the surface of a target cell may be experimentally dissected into several reaction steps. Using isolated complement proteins, an intermediate complex may be prepared in which the cell bears C5b,6,7 sites. Whereas the membrane of this cell is completely intact, it becomes leaky upon attachment of C8. In absence of C9, highly purified C8 causes the cell to undergo protracted low grade lysis (100–102). C8-dependent lysis is greatly enhanced by 1,10-phenanthroline or 2,2'-bipyridine (103, 104). Lipid bilayers of cholesterol and sphingomyelin also can be impaired by activated complement; and as in the case of cells, impairment commences with C8 action (105). The function of C9 may be regarded as an enhancement mechanism for the expression of C8 activity.

Two questions arise: What is the manner in which C8 causes membrane impairment, and how can C8 act on membrane constituents if it is separated from them by its trimolecular binding site as shown in Figure 4? The possibility of an enzymatic attack of membrane lipids by the terminal complement components has been considered repeatedly in the past. The recent demonstration of an association between C7 and tributyrinase activity (9) tends to support the enzyme hypothesis. However, impairment of synthetic lipid bilayers by complement was not accompanied by the appearance of enzymatic degradation products of the lipids used (106). While the enzyme hypothesis has not been dismissed entirely, an alternative mode of action is increasingly considered likely. It has been proposed that the C5b-9 complex inserts itself in toto into the membrane and that it allows exchange of intra- and extracellular constituents by means of an internal hydrophilic channel (107). The so-called doughnut hypothesis utilizes the extensive experience with membrane insertion substances (108) and is patterned after the valinomycin model (109). It fails to explain how the soluble proteins of the C5b-9 complex might enter the hydrophobic interior of a membrane. Electron microscopy has revealed characteristic ultrastructural lesions of cells lysed by complement (110). Formation of the lesions is partly a function of the C5 molecule (111). Applying the freeze-etching technique, it was possible to show that these ultrastructural alterations are confined to the outer leaflet of the membrane and that the inner leaflet is not detectably affected (112). This observation strongly suggests that the postulated hydrophilic channels produced by complement across the membrane are of such a nature and size that they are not detected by present ultrastructural techniques. This would preclude the C5b-9 complex as insertion mechanism and favor the hypothesis that a small portion of the complex is extended through the membrane. While searching for unusual structural properties of the complex that might provide a clue for its function, the unusual quaternary structure of C8 came to light (see above) (113). On the basis of these findings, the following mechanism is proposed for the cytolytic function of C8. Upon attachment of C8 to C5b,6,7 the short γ chain and part of the α chain are inserted into a channel formed by the trimolecular complex on the membrane surface. This insertion allows parts of the C8 molecule, primarily the α-γ chains, to enter into direct contact with the membrane. The α-γ chains then extend into the membrane interior causing breakdown of orderly structure, or they form a helical transmembrane channel in analogy to gramicidin A (114). The C8 insertion hypothesis not only takes into account all available experimental results, but it is also amenable to rigorous testing.

Molecular Economy

The molecular economy of the assembly process is determined by the mode of binding that is operative at a given stage. Within the C1 complex, apparently one molecule of $\overline{C1r}$ is limited to activating one or perhaps two molecules of C1s (41). However, one $\overline{C1}$ complex can generate multiple $\overline{C4,2}$ complexes (115) and one $\overline{C4,2,3}$ complex can assemble multiple C5b-9 complexes (54). Further, one $\overline{C4,2}$ can give rise to only one $\overline{C4,2,3}$, however, it can catalyze binding of multiple C3b molecules to a fourth category of sites in its vicinity (45, 116). Thus at three stages the complement reaction is subject to amplification mechanisms. One of the consequences of enzymatic amplification is a high degree of unfruitful turnover of C2, C3, C4, and C5. Only 1–20% of these molecules are becoming bound to the target cell surface. This inherent lack of economy does not apply to C1, C6, C7, C8, or C9, components bound by mechanisms other than prior enzymatic activation of binding sites.

Control

The complement reaction is subject to rigid constraints in time and space. The rapid decay of activated binding sites restricts complement action to the immediate environment of the activation site. Decay of the $\overline{C4,2}$ and the $\overline{C4,2,3}$ enzymes due to dissociation of the C2 subunit limits the temporal function of the activation mechanism (117). In addition, the actions of the $\overline{C1s}$ inhibitor and the C3b inactivator destroy the activities of $\overline{C1}$ and the $\overline{C4,2,3}$ enzymes, respectively (118). $\overline{C1}$ inhibitor inhibits stoichiometrically the hemolytic (119) and the esterolytic (120) activity of $\overline{C1}$. Inhibition of $\overline{C1s}$ is associated with the formation of a 1:1 molar complex between the inhibitor and the enzyme (121). The inhibitor-enzyme complex is stable in the presence of SDS and urea. Upon reduction the heavy chain of $\overline{C1s}$ is dissociated, while the light chain remains associated with the inhibitor. This observation provides independent evidence for the contention that the active site of $\overline{C1s}$ resides in its light chain. The C3b inactivator splits cell-bound or soluble C3b into the two antigenically distinct fragments, C3c and C3d (38). As a result, all known functions of C3b are abrogated: the enzyme $\overline{C4,2,3}$ is inactivated, the immune adherence function is lost (36), and the C3b-dependent activation of C3 proactivator is abolished (122, 123).

THE ALTERNATIVE OR PROPERDIN PATHWAY

History

The second pathway of complement activation came to light largely as a result of two entirely independent approaches in serology, one studying the effect of yeast cell walls on certain serum activities, and the other, that of cobra venom.

In 1954 Pillemer and his associates (5) put forth a new concept of nonspecific resistance to infections, according to which the normal serum protein, properdin, was able to react nonspecifically with a wide variety of naturally occurring polysaccharides and lipopolysaccharides. Without participation of specific antibody, properdin was thought to activate complement and to be instrumental in the destruction of susceptible bacteria, the neutralization of susceptible viruses, and the lysis of erythrocytes from patients with paroxysmal nocturnal hemoglobinuria. The action of properdin required Mg^{2+}, a hydrazine-sensitive activity, called Factor A, and a heat-labile activity, called Factor B (124).

The validity of the properdin concept was disputed, and the activities attributed to the properdin system were explained as a function of natural antibody and complement (125). Subsequently, the problem lay dormant, except for the demonstration that highly purified and biologically active properdin is immunochemically clearly distinct from immunoglobulins and thus is not identical with natural antibody (126). In recent years several groups of investigators pointed out the ability of certain complement activators to consume C3–C9 without affecting C1, C2, and C4 (127, 128). Work with genetically C4-deficient guinea pigs corroborated these results (129). A second mechanism of complement activation was therefore postulated (128). It was at this point in time that isolation of the C3 proactivator (C3PA) had been completed and its function recognized (130). C3PA was not discovered, however, through efforts of exploring the properdin system, but in pursuit of the mechanism of action of cobra venom factor.

In 1903 Flexner & Noguchi (131) demonstrated abolition of the hemolytic activity of serum by cobra venom. The responsible principle of the venom was later utilized to generate anaphylatoxin from serum in vitro, to remove the precursor of anaphylatoxin in vivo (132), and to inactivate C3 in vitro and in vivo (133, 134). Subsequently it became clear that the factor activated the cytolytic potential of the terminal components (48, 135, 136). Highly purified cobra venom factor (CVF) was found to be a glycoprotein without detectable enzymatic activity, which had no effect on structure and function of isolated C3. For activation of C3 in whole serum, the factor was shown to recruit at least one normal serum constituent, C3 proactivator, with which it entered into an enzymatically active complex (16, 137). The complex, which was interpreted to consist of one molecule of each protein, cleaves C3 into C3a and C3b and initiates assembly of the cytolytic C5b-9 complex (59, 96). For the formation of a highly active CVF complex, the requirement of Factor D (C3 proactivator convertase) was pointed out (138). A recent comprehensive study showed that (a) isolated C3PA and CVF interact reversibly in the absence of C3PA convertase, generating some C3 activating activity; (b) the stable, highly active complex is formed only in the presence of C3PA convertase and Mg^{2+}; and (c) $\frac{1}{25}$ mole of C3PA convertase is sufficient to form one mole of complex, indicating that this enzyme is not an integral constituent (139).

In order to explore the physiological role of C3 proactivator, the fate of the protein was studied in serum treated with bacterial or plant polysaccharides. Particulate inulin, a polyfructose, was selected because, like cobra venom factor, it liberates anaphylatoxin in serum without significant consumption of early acting complement components. This work led to the description of the "C3 activator system," which included a heat-labile (C3PA) and a hydrazine-sensitive factor (C3) as well as C3PA convertase and C3b (122, 130). Because of these characteristics and because the C3 activator system was shown to be operative in the natural bactericidal activity of serum and in the lysis of erythrocytes of paroxysmal nocturnal hemoglobinuria (140), a relationship to the properdin system became apparent. Although the properdin factors originally were very vaguely defined, Factor A could be identified with C3 (122) and Factor B with C3PA (141). It is now possible for the first time to formulate a tentative molecular concept of the anatomy and the dynamics of the properdin system. Some of the properties of the proteins of the alternative pathway are listed in Table 4 and a tentative scheme of their sequence in Figure 5.

The Proteins of the Alternative Pathway

EARLY ACTING FACTORS   To date the C1 equivalent of the alternative pathway has not been identified. Properdin itself does not establish the initial contact with activators such as inulin or IgA. A candidate for this function is the nephritic factor

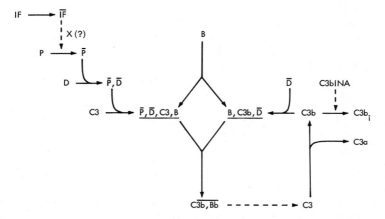

Figure 5   Schematic representation of the current tentative concept of the mode of action of the alternative pathway. The reactions outlined on the left constitute the properdin (P)-dependent initiation of the pathway. The reactions on the right are part of the C3b-dependent positive feedback. Both reaction sequences recruit C3 proactivator (B) and lead to the formation of the alternative pathway C3 convertase which has the probable subunit composition, C3b,Bb. Contact with activators (e.g. immunoglobulin A aggregates) converts the initiating factor, IF, to its functional form, IF, which together with additional serum factors converts P to P. The latter was reported to activate the precursor of C3 proactivator convertase (D) to D. At least four components, P, D, C3, and B then interact to generate C3 convertase. With the attack on C3 by this enzyme, C3b is liberated and the C3b-dependent feedback is set in motion. C3b is controlled by the C3b inactivator (C3b INA). In an as yet unknown manner, C3b,Bb is converted to C5 convertase of the alternative pathway, which catalyzes the assembly of the potentially cytolytic C5b-9 complex. C3a and C5a anaphylatoxin are byproducts of the reaction.

Table 4   Proteins of the alternative pathway of complement activation

Protein	Symbol	Serum Concentration (μg/ml)	Sedimentation Coefficient (S)	Molecular Weight	Relative Electrophoretic Mobility
Nephritic factor analog	IF	—	7	150,000	γ_1
Properdin	P	25	5.4	184,000	γ_2
C3	C3	1600	9.5	180,000	β_2
C3b	C3b	—	9.0	171,000	α_2
C3 proactivator	C3PA or B	200	5–6	93,000	β
C3 activator	C3A or $\overline{\text{B}}$	—	4	63,000	γ
C3 proactivator convertase	C3PAse or $\overline{\text{D}}$	—	3	24,000	α

analog, which may precede properdin in the reaction sequence. Nephritic factor occurs in the serum of patients with hypocomplementemic nephritis (142). Added to normal serum it consumes C3 and initiates the formation of the C5b-9 complex (143). The function of nephritic factor is independent of C1, C2, and C4, but requires C3, C3PAse, C3PA, and possibly properdin. The isolated factor was shown to be a 7S nonimmunoglobulin γ-globulin (144). Antiserum prepared to the factor was used in insolubilized form to adsorb normal human serum. The treated serum no longer reacted with inulin, but its alternative pathway was activated by isolated, active nephritic factor. These observations indicate the existence in normal serum of a protein that constitutes the inactive precursor of nephritic factor and a normal component of the properdin system. The factor may represent the initiating factor (IF). In addition to IF, there may be other early acting factors (145, 146).

PROPERDIN   The protein has a molecular weight of 184,000 and contains 9.8% carbohydrate, which consists of neutral hexose, hexosamine, and neuraminic acid. In guanidine the molecule dissociates into four noncovalently linked subunits of similar molecular weight [46,000 (147) or 53,000 (148)]. Upon removal of the denaturing agent the subunits recombine to form a dimer with a molecular weight of 88,000. The dimer possessed approximately 75% of the antigenic properties and 50% of the biological activity of the native protein. The activity of activated properdin, $\overline{\text{P}}$, could not be inhibited with 5×10^{-3} M DFP or 10^{-2} M tosyl-L-lysine-chloromethyl-ketone (149).

C3 PROACTIVATOR   This β-globulin is the precursor of a C3-cleaving enzyme (16, 130, 137). It has the unusual ability to form an enzymatically active complex with a glycoprotein from cobra venom. Following its isolation, the physiological mode of activation could be elucidated. The enzyme C3 proactivator convertase splits the molecule into two antigenically distinct fragments, one behaving as an α-globulin and the other as a γ-globulin. The γ fragment in isolated form retains some C3-cleaving activity. It was therefore designated C3 activator or C3A. This fragment also hydrolyzes N-α-acetyl-glycyl-L-lysine-methyl ester (8). C3PA consists of a single polypeptide chain with a molecular weight of 93,000. The active and inert fragments have molecular weights of 63,000 and 30,000, respectively (148). C3PA could be immunochemically identified with glycine-rich β-globulin (GBG) (150) and functionally with properdin Factor B (141). C3A is antigenically identical with glycine-rich γ-globulin (GGG) (151) and β_2-glycoprotein II (152), two serum constituents which together with GBG were previously described as protein without known physiological function.

C3 PROACTIVATOR CONVERTASE   This enzyme, which is a trace protein in serum, was first recognized as an essential component of the properdin system in 1972 (122). Its action on C3PA depends on the presence of C3b. Cobra venom factor has a C3b-like effect on C3PAse (138, 139, 153). C3 proactivator convertase (C3PAse) has recently been isolated and found to be composed of a single polypeptide chain with a molecular weight of 24,000 (148). C3PAse is identical with Factor $\overline{D}$ (138), which was reported to be a serine protease because its activity can be inhibited by 5×10^{-3} M DFP (154). The enzyme occurs in serum in active and in precursor form (D), the latter being unaffected by treatment with DFP. Factor D may be activated by trypsin or $\overline{P}$ without detectable change in molecular size (149). Activation of D by $\overline{P}$ has not yet been confirmed by others.

The C3b-Dependent Positive Feedback

The alternative pathway may be viewed as a composite of two independent reactions which both utilize C3 proactivator in the formation of an enzyme with C3 convertase activity. The first reaction is properdin dependent and C3b independent. The second reaction is C3b dependent and properdin independent.

Assembly of the alternative pathway C3 convertase involves fission of C3PA (Factor B) into two immunochemically distinct fragments, Ba and Bb. The enzymatic function resides in Bb, which in its nascent state is endowed with a labile binding site. The responsible enzyme, C3PA convertase (C3PAse, $\overline{D}$), does not by itself act on C3PA. It is felt at present that C3b modulates the enzyme-substrate interaction between C3PAse and C3PA, and secondly that C3b serves as acceptor of activated C3PA. Thus, a product of C3 (C3b) becomes instrumental in activating and assembling an enzyme system which acts on C3, and thereby establishes a positive feedback mechanism (Figure 5) (122). The reaction may be written as follows

$$C3b + B \rightleftharpoons C3b,B \xrightarrow[Mg^{2+}]{\overline{D}} \overline{C3b,Bb} + Ba$$

Alternative pathway C3 convertase $\overline{C3b,Bb}$ is thought to be very labile and to dissociate rapidly to C3b and Bb (W. Vogt, personal communication). The latter fragment, which has the electrophoretic mobility of γ-globulin, retains some C3-cleaving activity (130) and the ability to hydrolyze acetyl-glycyl-lysine methyl ester (8). It is for these reasons that the Bb fragment of C3PA was originally proposed as the entity that bears the enzymatic site of C3 convertase of the alternative pathway, a concept that has prevailed. The enzyme $C3b,\overline{Bb}$ can be assembled on the surface of particles such as yeast cell walls (155) or animal cells. Bound to the surface of a cell it is potentially cytotoxic. Lymphoblastoid cells with a high affinity C3b receptor adsorb monomeric, preformed C3b through the immune adherence mechanism and are then killed by late acting complement components in a C3PA-dependent reaction (156). More directly, C3b,Bb can be assembled on the surface of erythrocytes that bear C3b by addition of B and D̄. These cells are lysed by C3–C9 (157). The assembly of C3b,Bb by D̄ is reminiscent of the assembly of C4b,2a by C1̄s, in that C3b and C4b act as stable acceptors of the labile activation products Bb* and C2a*, respectively, and the activation enzymes are apparently both serine proteases.

Being an activator of the feedback mechanism, C3b is controlled by the serum enzyme C3b inactivator. Removal of the enzyme from serum results in activation of alternative pathway C3 convertase, due to uncontrolled, slow accumulation of C3b (158). A disease in man resulting from an inherited deficiency of C3b inactivator (159) has been described (160).

Properdin-Dependent Initiation of the Alternative Pathway

The concept of C3b-dependent activation of C3 convertase and of the ensuing positive feedback gives no clue as to how the alternative pathway is initiated. What is the link between the described feedback mechanism and activators such as inulin?

Serum depleted of properdin by immune adsorption did not sustain activation of C3PA and C3 by inulin, but activation of both proteins was observed when the depleted serum was treated with isolated properdin (161). It was shown in these experiments that properdin acts via C3, C3PAse, and C3PA and that specifically no C3PA conversion occurred in absence of properdin, C3PAse, or native C3. The following conclusions were drawn: (a) properdin plays an essential role in the early events of the alternative pathway; (b) it was isolated in activated form (P̄); and (c) P̄ requires native C3 in addition to C3PAse to effect activation of C3PA. The requirement of native C3 for P̄ action distinguishes the C3 function from that of C3b, which is independent of P. It also illuminates the earlier finding that C3 represents the hydrazine-sensitive factor required for C3PA activation by inulin (122). Native C3 may therefore be regarded as the molecular equivalent of Factor A activity of the properdin system. Properdin is not only instrumental in the generation of C3 convertase in the absence of C3b, but it also converts precursor D to D̄ (C3PAse), as was reported by Fearon et al (149).

The manner in which P is converted to P̄ is unknown. However, evidence is accumulating to show that in the reaction sequence P is preceded by one or more serum factors. One line of evidence stems from work on the so-called nephritic factor (C3NeF) (142). It occurs in the serum of patients with hypocomplementemic, chronic glomerulonephritis, and consumes C3 and later acting complement components through activation of the alternative pathway. When normal serum is treated with antiserum to C3NeF, the alternative pathway is impaired, although the other established ingredients remain functionally intact (144). These observations indicate the existence in normal serum of a protein that constitutes the inactive precursor of nephritic factor and a normal component of the properdin system. Since it acts prior to properdin, it may represent the initiating factor of the system. Figure 5 represents a schematic summary of the current tentative concept of the mechanism of action of the alternative pathway of complement activation. Emphasis is placed on the tentative nature of all interpretations that pertain to reaction mechanisms of the properdin system. Much more work is needed to elevate our understanding of this system to the level of comprehension of the classical pathway.

CONCLUSION

As the chemical basis of complement activity becomes elucidated, some unusual molecular and functional features are recognized. The occurrence of a collagen sequence in C1q, which is interpolated between two noncollagenous sequences, is without precedent for plasma proteins. The multifunctionality of a single protein, C3, is noteworthy. C3 serves as an early acting factor in the properdin pathway and constitutes the precursor of two physiological fragments. The low molecular weight fragment C3a has anaphylatoxin activity and the large C3b fragment functions as an essential subunit or modulator of four distinct complement enzymes. It also mediates opsonizing and immune adherence reactions. The membrane attack complex, C5b-9, is without analogy in other biological systems. Through a self-assembling process, five different soluble proteins become peripheral or even integral membrane constituents.

At least four different membrane-oriented complement functions may be distinguished. First, through enzymatic activation of labile binding sites, C3 and C4 are enabled to bind to nonspecific receptors on the surface of membranes without compromising membrane integrity. Second, bound C3b and C4b react with highly specific membrane receptors of a limited variety of cells, thereby evoking non-cytolytic biological phenomena. Third, the activation peptides C3a and C5a, which are endowed with a high degree of α-helical structure, address receptors on mast cells, leukocytes, and smooth muscle cells, activating diverse cellular functions. Fourth, following enzymatic cleavage of C5, the C5b-9 complex is enabled to assemble, to attach to the surface of a cell, and to kill the cell. Elucidation of the various modes of interaction between complement molecules and cell membranes, the forces involved, the nature of the relevant membrane receptors, and the corresponding chemical structures that confer membrane-oriented functions on complement molecules will constitute a new and telling chapter of biology.

Literature Cited

1. Müller-Eberhard, H. J. 1975. *Textbook of Immunopathology,* ed. P. A. Miescher, H. J. Müller-Eberhard. New York: Grune & Stratton. 2nd ed. In press
2. Austen, K. F. 1974. *Transplant. Proc.* 6:1
3. Müller-Eberhard, H. J. 1971. *Progr. Immunol.* 1:553
4. Müller-Eberhard, H. J. 1972. *Harvey Lect.* 66:75
5. Pillemer, L. et al 1954. *Science* 120:279
6. *Bull. WHO* 1968. 39:935
7. Dalmasso, A. P., Müller-Eberhard, H. J. 1966. *J. Immunol.* 97:680
8. Cooper, N. R. 1971. *Progr. Immunol.* 1:567
9. Delage, J. M., Lehner-Netsch, G., Simard, J. 1973. *Immunology* 24:671
10. Reid, K. B. M., Lowe, D. M., Porter, R. R. 1972. *Biochem. J.* 130:749
11. Schreiber, R. D., Müller-Eberhard, H. J. 1974. *J. Exp. Med.* 140:1324

12. Kolb, W. P., Müller-Eberhard, H. J. 1975. *J. Exp. Med.* In press
13. Calcott, M. A., Müller-Eberhard, H. J. 1972. *Biochemistry* 11:3443
14. Butler, W. T., Cunningham, L. W. 1966. *J. Biol. Chem.* 241:3882
15. Spiro, R. G. 1967. *J. Biol. Chem.* 242:4813
16. Müller-Eberhard, H. J., Nilsson, U. R., Dalmasso, A. P., Polley, M. J., Calcott, M. A. 1966. *Arch. Pathol.* 82:205
17. Schumaker, V. N., Calcott, M. A., Spiegelberg, H., Müller-Eberhard, H. J. 1975. *Biochemistry.* In press
18. Reid, K. B. M. 1975. *Biochem. J.* In press
19. Shelton, E., Yonemasu, K., Stroud, R. M. 1972. *Proc. Nat. Acad. Sci. USA* 69:65
20. Svehag, S.-E., Manhem, L., Bloth, B. 1972. *Nature New Biol.* 238:117
21. Reid, K. B. M., Porter, R. R. 1975. *Biochem. J.* In press
22. Yonemasu, K., Stroud, R. M. 1972. *Immunochemistry* 9:545
23. Heusser, C., Boesman, M., Nordin, J. H., Isliker, H. 1973. *J. Immunol.* 110:820
24. Isliker, H., Allan, R., Boesman, M., Heusser, C., Knobel, H. 1974. *Advan. Biosci.* 12:270
25. Leon, M. A. 1965. *Science* 147:1034
26. Polley, M. J., Müller-Eberhard, H. J. 1967. *J. Exp. Med.* 126:1013
27. Polley, M. J., Müller-Eberhard, H. J. 1975. *J. Immunol.* In press
28. Budzko, D. B., Bokisch, V. A., Müller-Eberhard, H. J. 1971. *Biochemistry* 10:1166
29. Budzko, D. B., Müller-Eberhard, H. J. 1970. *Immunochemistry* 7:227
30. Nilsson, U., Mapes, J. 1973. *J. Immunol.* 111:293
31. Bokisch, V. A., Müller-Eberhard, H. J. 1975. *Proc. Nat. Acad. Sci. USA.* In press
32. Nishioka, K., Linscott, W. D. 1963. *J. Exp. Med.* 118:767
33. Cooper, N. R. 1969. *Science* 165:396
34. Gigli, I., Nelson, R. A. Jr. 1968. *Exp. Cell Res.* 51:45
35. Cooper, N. R. 1975. *J. Exp. Med.* In press
36. Tamura, N., Nelson, R. A. Jr. 1967. *J. Immunol.* 99:582
37. Lachmann, P. J., Müller-Eberhard, H. J. 1968. *J. Immunol.* 100:691
38. Ruddy, S., Austen, K. F. 1971. *J. Immunol.* 107:742
39. Rother, K. 1972. *Eur. J. Immunol.* 2:550
40. Valet, G., Cooper, N. R. 1974. *J. Immunol.* 112:339
41. Valet, G., Cooper, N. R. 1974. *J. Immunol.* 112:1667
42. Naff, G. B., Pensky, J., Lepow, I. H. 1964. *J. Exp. Med.* 119:593
43. Müller-Eberhard, H. J., Polley, M. J., Calcott, M. A. 1967. *J. Exp. Med.* 125:359
44. Kolb, W. P., Haxby, J. A., Arroyave, C. M., Müller-Eberhard, H. J. 1973. *J. Exp. Med.* 138:428
45. Müller-Eberhard, H. J., Dalmasso, A. P., Calcott, M. A. 1966. *J. Exp. Med.* 123:33
46. Müller-Eberhard, H. J., Lepow, I. H. 1965. *J. Exp. Med.* 121:819
47. Arroyave, C. M., Müller-Eberhard, H. J. 1973. *J. Immunol.* 111:536
48. Götze, O., Müller-Eberhard, H. J. 1970. *J. Exp. Med.* 132:898
49. Shin, H. S., Mayer, M. M. 1968. *Biochemistry* 7:2997
50. Stroud, R. M., Mayer, M. M., Miller, J. A., McKenzie, A. T. 1966. *Immunochemistry* 3:163
51. Polley, M. J., Müller-Eberhard, H. J. 1968. *J. Exp. Med.* 128:533
52. Patrick, R. A., Taubman, S. B., Lepow, I. H. 1970. *Immunochemistry* 7:217
53. Cooper, N. R., Polley, M. J., Müller-Eberhard, H. J. 1970. *Immunochemistry* 7:341
54. Cooper, N. R., Müller-Eberhard, H. J. 1970. *J. Exp. Med.* 132:775
55. Goldlust, M. B., Shin, H. S., Hammer, C. H., Mayer, M. M. 1974. *J. Immunol.* 113:998
56. Müller-Eberhard, H. J. 1975. *J. Immunol.* In press
57. Jensen, J. A. 1967. *Science* 155:1122
58. Dias Da Silva, W., Eisele, J. W., Lepow, I. H. 1967. *J. Exp. Med.* 126:1027
59. Cochrane, C. G., Müller-Eberhard, H. J. 1968. *J. Exp. Med.* 127:371
60. Lepow, I. H., Willms-Kretschmer, K., Patrick, R. A., Rosen, F. S. 1970. *Am. J. Pathol.* 61:13
61. Wuepper, K. D., Bokisch, V. A., Müller-Eberhard, H. J., Stoughton, R. B. 1972. *Clin. Exp. Immunol.* 11:13
62. Müller-Eberhard, H. J., Vallota, E. H. 1971. *Biochemistry of the Acute Allergic Reactions,* ed. K. F. Austen, E. L. Becker, 217. Oxford: Blackwell
63. Becker, E. L. 1972. *J. Exp. Med.* 135:376
64. ter Laan, B., Molenaar, J. L., Feltkamp-Vroom, T. M., Pondman, K. W. 1974. *Eur. J. Immunol.* 4:393
65. Johnson, A. R., Hugli, T. E., Müller-Eberhard, H. J. 1975. *Immunology.* In press
66. Vogt, W., Zeman, N., Garbe, G. 1969. *Arch. Exp. Pathol. Pharmakol.* 262:399
67. Bokisch, V. A., Müller-Eberhard, H. J. 1970. *J. Clin. Invest.* 49:2427
68. Mahler, F., Intaglietta, M., Hugli, T. E.,

Johnson, A. R. 1975. *Microvasc. Res.* In press
69. Hugli, T. E. 1975. *J. Biol. Chem.* In press
70. Hugli, T. E., Vallota, E. H., Müller-Eberhard, H. J. 1975. *J. Biol. Chem.* 250:1472
71. Hugli, T. E., Morgan, W. T., Müller-Eberhard, H. J. 1975. *J. Biol. Chem.* 250:1479
72. Vallota, E. H., Hugli, T. E., Müller-Eberhard, H. J. 1975. *J. Immunol.* In press
73. Hugli, T. E. 1975. *Cold Spring Harbor Symp. Cell Proliferation,* Vol. II. In press
74. Morgan, W. T., Vallota, E. H., Müller-Eberhard, H. J. 1974. *Biochem. Biophys. Res. Commun.* 57:572
75. Liefländer, M., Dielenberg, D., Schmidt, G., Vogt, W. 1972. *Z. Physiol. Chem.* 353:385
76. Vallota, E. H., Müller-Eberhard, H. J. 1973. *J. Exp. Med.* 137:1109
77. Wissler, J. H. 1972. *Eur. J. Immunol.* 2:73
78. Vogt, W., Liefländer, M., Stalder, K.-H., Lufft, E., Schmidt, G. 1971. *Eur. J. Immunol.* 1:139
79. Stegemann, H., Bernhard, G., O'Neil, J. A. 1964. *Z. Physiol. Chem.* 339:9
80. Vogt, W. 1968. *Biochem. Pharmacol.* 17:727
81. Lachmann, P. J., Thompson, R. A. 1970. *J. Exp. Med.* 131:643
82. Taranta, A., Franklin, E. C. 1961. *Science* 134:1981
83. Ishizaka, T., Ishizaka, K., Borsos, T., Rapp, H. J. 1966. *J. Immunol.* 97:716
84. Augener, W., Grey, H. M., Cooper, N. R., Müller-Eberhard, H. J. 1971. *Immunochemistry* 8:1011
85. Hurst, M. M., Volanakis, J., Bennett, J. C., Stroud, R. M. 1974. *Fed. Proc.* 33:759
86. Kehoe, J. M., Fougereau, M. 1970. *Nature* 224:1212
87. Ellerson, J. R., Yasmeen, D., Painter, R. H., Dorrington, K. J. 1972. *FEBS Lett.* 24:318
88. Allan, R., Isliker, H. 1974. *Immunochemistry* 11:243
89. Allan, R., Isliker, H. 1974. *Immunochemistry* 11:175
90. Sakai, K., Stroud, R. M. 1973. *J. Immunol.* 110:1010
91. Sakai, K., Stroud, R. M. 1974. *Immunochemistry* 11:191
92. Becker, E. L. 1956. *J. Immunol.* 77:462
93. Baskas, T., Scott, G. K., Fothergill, J. E. 1973. *Biochem. Soc. Trans.* 1:31
94. Kolb, W. P., Haxby, J. A., Arroyave, C. M., Müller-Eberhard, H. J. 1972. *J. Exp. Med.* 135:549

95. Shin, H. S., Snyderman, R., Friedman, E., Mellors, A., Mayer, M. M. 1968. *Science* 162:361
96. Kolb, W. P., Müller-Eberhard, H. J. 1973. *J. Exp. Med.* 138:438
97. Koethe, S. M., Austen, K. F., Gigli, I. 1973. *J. Immunol.* 110:390
98. Kolb, W. P., Müller-Eberhard, H. J. 1974. *J. Immunol.* 113:479
99. Kolb, W. P., Müller-Eberhard, H. J. 1975. *J. Exp. Med.* In press
100. Stolfi, R. L. 1968. *J. Immunol.* 100:46
101. Kolb, W. P., Haxby, J. A., Manni, J. A., Müller-Eberhard, H. J. 1975. *J. Immunol.* In press
102. Manni, J. A., Müller-Eberhard, H. J. 1969. *J. Exp. Med.* 130:1145
103. Hadding, U., Müller-Eberhard, H. J. 1967. *Science* 157:442
104. Hadding, U., Müller-Eberhard, H. J. 1969. *Immunology* 16:719
105. Haxby, J. A., Götze, O., Müller-Eberhard, H. J., Kinsky, S. C. 1969. *Proc. Nat. Acad. Sci. USA* 64:290
106. Inoue, K., Kinsky, S. C. 1970. *Biochemistry* 9:4767
107. Mayer, M. M. 1972. *Proc. Nat. Acad. Sci. USA* 69:2954
108. Kinsky, S. C. 1970. *Ann. Rev. Pharmacol.* 4:119
109. Pressman, B. C. 1965. *Proc. Nat. Acad. Sci. USA* 53:1076
110. Humphrey, J. H., Dourmashkin, R. R. 1969. *Advan. Immunol.* 11:75
111. Polley, M. J., Müller-Eberhard, H. J., Feldman, J. D. 1971. *J. Exp. Med.* 133:53
112. Iles, G. H., Seeman, P., Naylor, D., Cinader, B. 1973. *J. Cell Biol.* 56:528
113. Kolb, W. P., Müller-Eberhard, H. J. 1975. *Proc. Nat. Acad. Sci. USA* In press
114. Urry, D. W. 1972. *Ann. NY Acad. Sci.* 195:108
115. Cooper, N. R., Müller-Eberhard, H. J. 1968. *Immunochemistry* 5:155
116. Mardiney, M. R. Jr., Müller-Eberhard, H. J., Feldman, J. D. 1968. *Am. J. Pathol.* 53:253
117. Mayer, M. M. 1970. *Immunochemistry* 7:485
118. Gigli, I. 1974. *Transplant. Proc.* 6:9
119. Gigli, I., Ruddy, S., Austen, K. F. 1968. *J. Immunol.* 100:1154
120. Pensky, J., Levy, L. R., Lepow, I. H. 1961. *J. Biol. Chem.* 236:1674
121. Harpel, P. C., Cooper, N. R. 1975. *J. Clin. Invest.* In press
122. Müller-Eberhard, H. J., Götze, O. 1972. *J. Exp. Med.* 135:1003
123. Alper, C. A., Rosen, F. S., Lachmann, P. J. 1972. *Proc. Nat. Acad. Sci. USA*

69 : 2910

124. Lepow, I. H. 1961. *Immunochemical Approaches to Problems in Microbiology,* ed. M. Heidelberger, O. J. Plescia, Chap 19, 280. New Brunswick, NJ : Rutgers Univ. Press

125. Nelson, R. A. Jr. 1958. *J. Exp. Med.* 108 : 515

126. Pensky, J. et al 1968. *J. Immunol.* 100 : 142

127. Gewurz, H., Shin, H. S., Mergenhagen, S. E. 1968. *J. Exp. Med.* 128 : 1049

128. Sandberg, A. L., Osler, A. G., Shin, H. S., Oliveira, B. 1970. *J. Immunol.* 104 : 329

129. Frank, M. M., May, J., Gaither, T., Ellman, L. 1971. *J. Exp. Med.* 134 : 176

130. Götze, O., Müller-Eberhard, H. J. 1971. *J. Exp. Med.* 134 : 90s

131. Flexner, S., Noguchi, H. 1903. *J. Exp. Med.* 6 : 277

132. Vogt, W., Schmidt, G. 1964. *Experientia* 20 : 207

133. Klein, P. G., Wellensiek, H. J. 1965. *Int. Rev. Exp. Pathol.* 4 : 245

134. Nelson, R. A. Jr. 1966. *Surv. Ophthamol.* 11 : 498

135. Pickering, R. J., Wolfson, M. R., Good, R. A., Gewurz, H. 1969. *Proc. Nat. Acad. Sci. USA* 62 : 521

136. Ballow, M., Cochrane, C. G. 1969 *J. Immunol.* 103 : 944

137. Müller-Eberhard, H. J., Fjellström, K. E. 1971. *J. Immunol.* 107 : 1666

138. Hunsicker, L. G., Ruddy, S., Austen, K. F. 1973. *J. Immunol.* 110 : 128

139. Cooper, N. R. 1973. *J. Exp. Med.* 137 : 451

140. Götze, O., Müller-Eberhard, H. J. 1972. *N. Engl. J. Med.* 286 : 180

141. Goodkofsky, I., Lepow, I. H. 1971. *J. Immunol.* 107 : 1200

142. Vallota, E. H., Forristal, J., Spitzer, R. E., Davis, N. C., West, C. D. 1970.

J. Exp. Med. 131 : 1306

143. Arroyave, C. M., Vallota, E. H., Müller-Eberhard, H. J. 1974. *J. Immunol.* 113 : 764

144. Vallota, E. H. et al 1974. *J. Exp. Med.* 139 : 1249

145. May, J. E., Frank, M. M. 1973. *Proc. Nat. Acad. Sci. USA* 70 : 649

146. Stitzel, A. E., Spitzer, R. E. 1974. *J. Immunol.* 112 : 56

147. Minta, J. O., Lepow, I. H. 1974. *Immunochemistry* 11 : 361

148. Götze, O. 1975. *Cold Spring Harbor Symp. Cell Proliferation,* Vol. II. In press

149. Fearon, D. T., Austen, K. F., Ruddy, S. 1974. *J. Exp. Med.* 140 : 426

150. Boenisch, T., Alper, C. A. 1970. *Biochim. Biophys. Acta* 221 : 529

151. Boenisch, T., Alper, C. A. 1970. *Biochim. Biophys. Acta* 214 : 135

152. Haupt, H., Heide, K. 1965. *Clin. Chim. Acta* 12 : 419

153. Vogt, W., Dieminger, L., Lynen, R., Schmidt, G. 1974. *Z. Physiol. Chem.* 355 : 171

154. Fearon, D. T., Austen, K. F., Ruddy, S. 1974. *J. Exp. Med.* 139 : 355

155. Brade, V., Lee, G. D., Nicholson, A., Shin, H. S., Mayer, M. M. 1973. *J. Immunol.* 111 : 1389

156. Theofilopoulos, A. N., Bokisch, V. A., Dixon, F. J. 1974. *J. Exp. Med.* 139 : 696

157. Fearon, D. T., Austen, K. F., Ruddy, S. 1973. *J. Exp. Med.* 138 : 1305

158. Nicol, P. A. E., Lachmann, P. J. 1973. *Immunology* 24 : 259

159. Abramson, N., Alper, C. A., Lachmann, P. J., Rosen, F. S., Jandl, J. H. 1971. *J. Immunol.* 107 : 19

160. Alper, C. A., Abramson, N., Johnston, R. B., Jandl, J. H., Rosen, F. S. 1970. *N. Engl. J. Med.* 282 : 349

161. Götze, O., Müller-Eberhard, H. J. 1974. *J. Exp. Med.* 139 : 44

CHROMOSOMAL PROTEINS AND CHROMATIN STRUCTURE

✷900

Sarah C. R. Elgin

Department of Biochemistry and Molecular Biology, Harvard University, Cambridge, Massachusetts 02138

Harold Weintraub

Department of Biochemical Sciences, Princeton University, Princeton, New Jersey 08540

CONTENTS

INTRODUCTION

One important aspect of gene regulation during cellular differentiation in eukaryotes occurs at the level of gene transcription (for examples, see 1–4; see 5 for review). Transcription of eukaryotic chromatin is tissue specific and represents a small fraction of the total number of nucleotide sequences in DNA. This has now been established by examination of the products of both repetitious and unique gene sequences; further, this specificity is preserved in chromatin isolated and transcribed in vitro (6–8). Most convincing are the in vitro results demonstrating tissue-specific control of transcription of the hemoglobin gene (9–13). The control of template activity must be encoded in some way by particular nucleotide sequences in DNA. It is possible that these sequences are read by specific activator or repressor proteins, as has been shown in prokaryotic systems (14–16). However, in the eukaryotic genome the effects of primary and higher order chromosome structures must also be considered.

There is no a priori reason that there should be structural differences in the packaging of active and inactive regions of the genome. Selective gene transcription or translation can in principle be accomplished by only "transacting" soluble factors which modify the activity of the transcribing or translating machinery. That this is unlikely for all forms of gene control in higher organisms is best illustrated by the classical observations dealing with the inactivation of the X chromosome (17). Here it is quite clear that some structural change, as visualized in the light microscope, occurs with the loss of function of an entire chromosome. Examples of structural effects with *cis* control have been observed in *Drosophila* and are termed "position effects" (18, 19). In this instance the activity of a particular gene depends on its proximity to inactive heterochromatin and presumably on the structural environment established by the heterochromatin. These position effects often vary in the different cells of the organism. Structural changes as an early step in gene activation have been observed in the giant polytene chromosomes of *Drosophila* (20, 21). Our main purpose in mentioning these phenomena (and there are other examples) is to illustrate that at least some, and perhaps all, types of transcriptional controls in higher organisms have a structural basis, which is best explained as an altered DNA packaging pattern. The packaging pattern appears to be a consequence of the population of associated protein molecules. Thus we wish to consider the problem of differential gene expression from the point of view of chromatin structure. We will consider the chemistry of chromosomal proteins, some parameters influencing their interaction with DNA, basic repeating units of chromatin structure, and possible structure/function relationships in transcriptional control.

THE HISTONES

The histones, the small basic proteins found in association with DNA, were the first chromosomal proteins identified (22). The histones are major general structural proteins of chromatin and can act as repressors of template activity (see the later

Table 1 Characterization of the histones[a]

Class	Fraction	Lys/Arg Ratio	Total Residues	Molecular Weight	N terminal	C terminal
Very lysine rich	H1 (I, f1, KAP)	22.0	~215	~21,500	Ac–Ser	Lys
Lysine rich	H2a (IIb1, f2a2, ALG)	1.17	129	14,004	Ac–Ser	Lys
	H2b (IIb2, f2b, KSA)	2.50	125	13,774	Pro	Lys
Arginine rich	H3 (III, f3, ARE)	0.72	135	15,324	Ala	Ala
	H4 (IV, f2a1, GRK)	0.79	102	11,282	Ac–Ser	Gly

[a] All data for histones of calf thymus. Compiled from (112) and (118) and the references cited therein.

section on template activity). They have now been isolated, cnaracterized, and sequenced; their salient chemical characteristics and some of the different nomenclatures in use are summarized in Table 1. (We will use the Ciba Symposium nomenclature, H1, H2a, H2b, H3, and H4.) There are a small number of different histones, as defined by amino acid sequence criteria. Studies to date have indicated that the histones are very highly conserved proteins, there being very little variation in the amino acid sequences of histones from widely differing creatures. The histones are among the most highly modified proteins; the modifications include acetylation, methylation, and phosphorylation. We do not know the functional consequences of these variations; however, since the amino acid sequences are so highly conserved, such changes are likely to have significant effects on chromatin structure (23). Because the histones are encoded in the genome by repetitious DNA sequences, an additional dimension is involved in the problem of histone variability.

Primary Structure

The sequences of the calf thymus histones are presented in Figures 1–5. Amino acid positions where substitutions have been observed are underlined; deletions are indicated by an overline, and insertions by *. Sources of partial sequence data are indicated on the figure legends by a superscript p. Only apparent genetic polymorphisms (within the species) are marked for histones 1 and 2b; genetic and evolutionary polymorphisms (all known substitutions) are marked for histones 2a, 3, and 4. Unless otherwise discussed, the amino acid substitutions are conservative ones. Post-transcriptional modifications are not shown.

The conserved nature of the histones is apparent from the data presented. The estimated mutation rate of histone 4 is 0.06 per 100 amino acid residues per 100 million years, clearly the lowest mutation rate yet observed (23). Histone 3 is

<u>Histone 4</u>

10   20
Ac-Ser-Gly-Arg-Gly-Lys-Gly-Gly-Lys-Gly-Leu-Gly-Lys-Gly-Gly-Ala-Lys-Arg-His-Arg-Lys-

30   40
Val-Leu-Arg-Asp-Asn-Ile-Gln-Gly-Ile-Thr-Lys-Pro-Ala-Ile-Arg-Arg-Leu-Ala-Arg-Arg-

50   60
Gly-Gly-Val-Lys-Arg-Ile-Ser-Gly-Leu-Ile-Tyr-Glu-Glu-Thr-Arg-Gly-Val-Leu-Lys-Val-

70   80
Phe-Leu-Glu-Asn-Val-Ile-Arg-Asp-Ala-Val-Thr-Tyr-Thr-Glu-His-Ala-Lys-Arg-Lys-Thr-

90   100
Val-Thr-Ala-Met-Asp-Val-Val-Tyr-Ala-Leu-Lys-Arg-Gln-Gly-Arg-Thr-Leu-Tyr-Gly-Phe-

Gly-Gly-COOH

Figure 1   Calf histone 4 (381–383). Comparative amino acid sequence data have been obtained for H4 of rat (384), pig (385), bovine lymphosarcoma[p] (386), Novikoff hepatoma[p] (386), trout[p] (58), sea urchin[p] (387), and pea (382).

Histone 3

```
                        10                                          20
H₂N-Ala-Arg-Thr-Lys-Gln-Thr-Ala-Arg-Lys-Ser-Thr-Gly-Gly-Lys-Ala-Pro-Arg-Lys-Gln-Leu-

                        30                                          40
Ala-Thr-Lys-Ala-Ala-Arg-Lys-Ser-Ala-Pro-Ala-Thr-Gly-Gly-Val-Lys-Lys-Pro-His-Arg-Tyr-

                        50                                          60
Arg-Pro-Gly-Thr-Val-Ala-Leu-Arg-Glu-Ile-Arg-Arg-Tyr-Gln-Lys-Ser-Thr-Glu-Leu-Leu-Ile-

                        70                                          80
Arg-Lys-Leu-Pro-Phe-Gln-Arg-Leu-Val-Arg-Glu-Ile-Ala-Gln-Asp-Phe-Lys-Thr-Asp-Leu-Arg-

                        90                                         100
Phe-Gln-Ser-Ser-Ala-Val-Met-Ala-Leu-Gln-Glu-Ala-Cys-Glu-Ala-Tyr-Leu-Val-Gly-Leu-Phe-

                       110                                         120
Glu-Asp-Thr-Asn-Leu-Cys-Ala-Ile-His-Ala-Lys-Arg-Val-Thr-Ile-Met-Pro-Lys-Asp-Ile-Gln-

                       130
Leu-Ala-Arg-Arg-Ile-Arg-Gly-Glu-Arg-Ala-COOH
```

Figure 2   Calf histone 3 (388–390). Comparative amino acid sequence data have been obtained for H3 of chicken (391–393), carp (394), trout[P] (279), shark (395), *Drosophila*[P] (S. C. R. Elgin, R. Goodfleisch, and L. Hood, unpublished observations), sea urchin[P] (396), mollusc (Patella)[P] (396), pea (397), and cycad[P] (396).

Histone 2b

```
                                        10
Calf         HN-Pro-Glu-Pro-Ala-Lys-Ser-Ala-Pro-----Ala-Pro-Lys-Lys-----Gly-Ser-Lys-Lys-Ala-----
Trout        HN-Pro-Glx-Pro-Ala-Lys-Ser-Ala-Pro-------------Lys-Lys-----Gly-Ser-Lys-Lys-Ala-----
Drosophila   HN-Pro-----Pro-----Lys-Thr-Ala-Gly-Lys-Ala-Ala-Lys-Lys-Ala-Gly---------Lys-Ala-Glx-

                   20                                        30
Calf         -------Val-Thr-Lys-----Ala-Gln-Lys-Lys-Asp-Gly-Lys-Lys-Arg-Lys-Arg-Ser-Arg-Lys-Glu-
Trout        -------Val-Thr-Lys-Thr-Ala-Gly-Lys-
Drosophila   Lys-Asx-Ilu-Thr-Lys-Thr----Asx-Lys-Lys-

                   40                                        50
Calf         Ser-Tyr-Ser-Val-Tyr-Val-Tyr-Lys-Val-Leu-Lys-Gln-Val-His-Pro-Asp-Thr-Gly-Ile-Ser-Ser

                   60                                        70
Calf         Lys-Ala-Met-Gly-Ile-Met-Asn-Ser-Phe-Val-Asn-Asp-Ile-Phe-Glu-Arg-Ile-Ala-Gly-Glu-Ala

                   80                                        90
Calf         Ser-Arg-Leu-Ala-His-Tyr-Asn-Lys-Arg-Ser-Thr-Ile-Thr-Ser-Arg-Glu-Ile-Gln-Thr-Ala-Val

                   100                                       110
Calf         Arg-Leu-Leu-Leu-Pro-Gly-Glu-Leu-Ala-Lys-His-Ala-Val-Ser-Glu-Gly-Thr-Lys-Ala-Val-Thr

             120          125
Calf         Lys-Tyr-Thr-Ser-Ser-Lys-COOH
```

Figure 3   Calf, trout[P], and *Drosophila*[P] histone 2b (279, 398; S. C. R. Elgin, R. Goodfleisch, and L. Hood, unpublished observations).

also highly conserved, whereas H2a and H2b have evolved at a more rapid rate. Comparative gel electrophoresis of the reduced and oxidized forms of H3 has shown that many mammals, including rodents and those more highly developed, have a major H3 component containing two cysteines, while all other eukaryotes have only H3 containing one cysteine (24–26). In all cases examined this alteration represents the replacement of cysteine at position 96 with serine (see references of Figure 2). Originally it was thought that this replacement was a significant change in the histone sequence. However, physical chemical evidence indicates that the cysteine at residue 96 is buried in the interior of the molecule; consequently, replacement with serine is a conservative change (27).

Histone 1 is the most divergent histone, both in terms of the number of subfractions within any given tissue and species and in terms of its evolution. From one to eight subfractions of histone 1 (different amino acid sequences) have been observed in various species. Within a species the quantitative amounts of the subfractions will vary in different tissues (28–34). Post-translational modifications also occur; see the section on metabolism of chromosomal proteins. The available data suggest that histone 1 may vary as much as 15% by amino acid substitutions among the subfractions of a given organism. The comparison between rabbit and trout H1 (Figure 5) suggests a variation of 15–28% between species. In both instances the substitutions are not conservative; they frequently involve interchanges of lysine, alanine, proline, and serine. Some of the substitutions affect functional sites. For example an alanine/serine substitution is observed at position 37; this serine is known to be specifically phosphorylated by a cAMP-dependent enzyme following hormone stimulation in the rat liver (35, 36). In addition to simple substitutions, there is also considerable heterogeneity in the size of this histone, both from a given

Histone 2a

```
                                   10                                        20
Ac-Ser-Gly-Arg-Gly-Lys-Gln-Gly-Gly-Lys-Ala-Arg-Ala-Lys-Ala-Lys-Thr-Arg-Ser-Ser-Arg

                                   30                                        40
Ala-Gly-Leu-Gln-Phe-Pro-Val-Gly-Arg-Val-His-Arg-Leu-Leu-Arg-Lys-Gly-Asn-Tyr-Ala-

                                   50                                        60
Glu-Arg-Val-Gly-Ala-Gly-Ala-Pro-Val-Tyr-Leu-Ala-Ala-Val-Leu-Glu-Tyr-Leu-Thr-Ala-

                                   70                                        80
Glu-Ile-Leu-Glu-Leu-Ala-Gly-Asn-Ala-Ala-Arg-Asp-Asn-Lys-Lys-Thr-Arg-Ile-Ile-Pro-

                                   90                                       100
Arg-His-Leu-Gln-Leu-Ala-Ile-Arg-Asn-Asp-Glu-Glu-Leu-Asn-Lys-Leu-Leu-Gly-Lys-Val-

                                  110                                       120
Thr-Ile-Ala-Gln-Gly-Gly-Val-Leu-Pro-Asn-Ile-Gln-Ala-Val-Leu-Leu-Pro-Lys-Lys-Thr-

                                 *129
Glu-Ser-His-His-Lys-Ala-Lys-Gly-Lys-COOH
```

Figure 4   Calf histone 2a (399, 400). Comparative amino acid sequence data have been obtained for H2a of rat (401) and trout (402).

Histone I

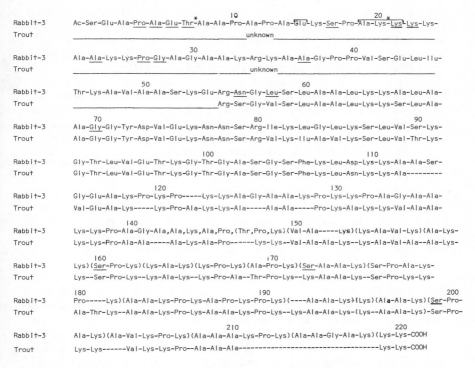

Figure 5 Rabbit histone I, subfraction 3 (34, 403; M. Hsiang and R. D. Cole, personal communication; M. Hsiang, C. Largman, and R. D. Cole, personal communication) and trout histone I (58; A. R. Macleod and G. H. Dixon, personal communication). Where overlaps are missing between tryptic peptides parentheses are shown; residues not yet sequenced are separated by commas.

organism and from different creatures (25, 37–39). The existence of subfractions in histone 1 makes it difficult to assign variability to polymorphisms per se, as opposed to varying expression of different subfractions represented in the genome.

Some general conclusions may be drawn from considering the primary sequences of the histones. A prominent and early observation was that of a skewed distribution of the basic residues in histones. In histones 2a, 2b, 3, and 4, a predominance of the basic residues occurs in the N-terminal region, with a secondary cluster of basic residues occurring again in the C-terminal region, the intermediate region being dominated by hydrophobic and acidic amino acids. The pattern is to some degree reversed in histone 1, with the dominant basic region at the C-terminal end. One may note a statistical dominance of a spacing of basic residues every fourth position rather than every third or fifth in the N-terminal region of the smaller histones. No characteristic spacing is noted for histone 1. The positioning of basic residues does not seem to be absolutely critical for the functioning of the histones; for example, a deletion of the peptide Ser-His-His has occurred in the C-terminal region of histone 2b of trout relative to that of calf. Some sequence homologies, generally of 4 to 10 residues, can be observed in direct comparisons of sequences from H2a, H3, and H4 and again for sequences in H2b and H1 (23).

There is little information available on the secondary and tertiary structure of the individual histones. Physical and conformational studies, including estimates of α-helices and β-structures, have been reviewed by Bradbury & Crane-Robinson (40). The considerable variability in post-synthetic modification may be contributing to our failure to obtain histone crystals for X-ray diffraction analysis. Perhaps the unit cell will have to be the chromatin subunit, discussed in the section on chromatin structure.

Tissue, Species, and Developmental Specificity

Many years ago, Stedman & Stedman (41) proposed that tissue-specific histones would be observed and that the interaction of these proteins with the DNA in a specific manner would lead to differential repression of gene activity. This early prediction was based on the observation of specific histones in two terminally differentiated cell types, histone 5 in chick erythrocytes and protamines in sperm. However, these are the only two salient examples of tissue-specific histones that have been observed. Histone 5, found in the nucleated erythrocytes of birds, amphibians, and fish, has been partially sequenced (42, 43). The appearance of histone 5 per se does not correlate well with repression of RNA synthesis. However, recent data indicate that newly synthesized histone 5 is phosphorylated, and the subsequent appearance of dephosphorylated H5 does correlate well with the decrease in RNA synthesis in developing erythroid cells. Thus it seems likely that this histone plays a critical role in establishing and maintaining the highly repressed state of chromatin in erythrocytes (44–47, 457). It is not clear why a specific histone is used for this purpose when simpler mechanisms for inactivation would be adequate.

Major histones specific to meiosis have been reported (48–49a). During spermato-genesis the male gametes undergo a condensation of the chromatin, frequently becoming associated with other small basic proteins instead of or in addition to the somatic histones. A number of these proteins have been isolated and characterized; examples with up to 75% arginine (from fish) and 50% lysine (from mollusc) are known (50, 51; for review see 52–54). Some of these proteins from mammalian sperm contain a number of cysteine residues; disulfide linkages appear during the condensation process (55). The process of the transition from a cell with chromatin having associated somatic histones and nonhistone chromosomal (NHC) proteins to the mature sperm, with DNA associated with basic sperm proteins only, has been studied extensively in the trout by Dixon and his co-workers. The somatic histones are gradually removed from the DNA and replaced by the protamines. Histones 2a, 2b, 3, and 4 are removed from the DNA by a process that involves acetylation followed by limited proteolysis; at the end of the process histone 1 is displaced by competition from the protamine, which is present in increasing concentration. The level of nonhistone chromosomal proteins is reduced to very low amounts (56–58). There is no information concerning removal of the protamines from the DNA following fertilization of an egg by the sperm. With these exceptions, virtually no qualitative changes in the histone fraction have been documented as events in tissue differentiation (e.g. 59).

Recently, studies have been carried out with antibody techniques to examine the question of tissue and species specificity of the histones. In all instances studied the tissue specificity has been of lower order than the species specificity, as judged by the antibody-antigen interactions (60–63). Species-specific histones are noted only rarely in the higher organisms, always as minor components. A good example, HT from the trout, has been isolated and partially characterized by Dixon and his colleagues. HT is $\sim 0.5\%$ of the trout histone fraction (64, 65).

Some interesting patterns of quantitative alteration in the histone complement have been reported, particularly regarding the subfractions of histone 1. Regular somatic histones are synthesized very early in embryogenesis from both maternal and newly transcribed messenger (66–75). In the case of sea urchin this occurs as early as the first cleavage. Paternal, as well as maternal, genes are expressed (76). A specific histone 1 subfraction is synthesized in the early sea urchin embyro, and a different subfraction is observed (by polyacrylamide gel analysis) in the late or hatching blastula. Preliminary evidence suggests that this switch may be regulated at the translational as well as the transcriptional level (72, 74, 75, 77). A marked increase in the relative amount of H1 during early embryogenesis has been reported for other systems (78, 79). Alterations in the pattern of synthesis of histone 1 subfractions have also been observed as a consequence of hormonal stimulation of the mouse mammary gland (80–82). The functional consequences of these changes remain to be established.

The Histones of Primitive Eukaryotes and Animal Cell Viruses

Histones are not found in association with the DNA of prokaryotes in vivo (83, 84), although they are capable of forming regular complexes with bacterial and viral DNAs in vitro, as shown by X-ray diffraction and nuclease studies (85; T. Maniatis, personal communication). The histones of unicellular and other primitive eukaryotes have been studied to examine a possible role of histones in chromosome replication, mitosis, and meiosis, as contrasted to a role in cellular differentiation. Slime molds and fungi appear to have an incomplete complement of histones; commonly H2a and H2b are present, H1 and H3 absent. However, negative evidence is not entirely convincing, particularly given the susceptibility of histones to proteases. Different but histone-like proteins are frequently observed in association with the DNA (86–91). Under the electron microscope yeast nucleohistone fibers appear to be organized as are those of higher creatures, despite the apparent absence of H1 and perhaps other histones (92). *Volvox* and *Euglena* likewise appear to have only a partial complement of histones (93, 94), whereas *Tetrahymena* and *Stylonychia* appear to have all the normal histones of higher eukaryotes (95–97).

Recently, information has been obtained on a protein associated with the DNA of *Thermoplasma acidophilum,* a mycoplasm that normally grows at 59°C, pH 2. The protein has the same mobility as histone 4; it contains 23% basic amino acids, 20% acidic amino acids, and no cysteine, tryptophan, tyrosine, histidine, or methionine (D. Searcy, personal communication). It is possible that this protein, which is capable of stabilizing the DNA to heat denaturation, could be an evolutionary precursor of the histones or the lysine-rich nonhistone chromosomal proteins (see Table 3).

The DNAs of the mammalian viruses SV40 and polyoma are associated with histones (98–100). It is thought that the basic proteins bound to the viral DNA in the capsid and in the nucleoprotein complex are histones 2a, 2b, 3, and 4, on the basis of analyses reflecting amino acid labeling patterns, behavior on disc gel electrophoresis, column chromatography behavior, and tryptic peptide maps (101). It will be extremely interesting to see if these viral systems can be exploited to study the role of histones in repressing transcription of specific genes (see the section on template activity).

THE HISTONE GENES

Histone Genes are Repetitious and Clustered

One of the exciting discoveries of the past few years has been the finding that the histones are coded for by repetitious genes in the eukaryotic DNA (102). This finding and the further accomplishment of isolation of the histone genes (103) have been made possible by the availability of histone mRNA in significant quantities, isolated primarily from early embryos during stages of rapid nuclear replication. Considerable evidence that the 7–12S mRNA so isolated does indeed code for the histones has now been compiled; see the section on metabolism of chromosomal proteins. Recently it has been possible to obtain three purified subfractions of message, presumably coding for histones 1, 2a–2b–3, and 4, respectively (104–106, 471). The histone mRNA subfraction designated C3 by Grunstein et al (106) is 370–400 nucleotides long. This is an appropriate length to code for histone 4 assuming some processing or regulatory sequences. Data obtained from translation experiments in cell-free systems indicate, and partial sequencing of this RNA is consistent with the hypothesis that this is the histone 4 messenger (106).

Most studies on the histone genes have been carried out in the sea urchin. Studies of the histone genes of four different species of sea urchin indicated from 400 copies in *Lytechinus pictus* to 1200 copies in *Psammechinus milaris* of each histone gene (104, 105, 107). In the most extreme case, that of *P. milaris,* from 0.5 to 0.8% of the genome is thus required to encode the histone genes and their associated spacer DNA (103). While it is clear from cross species hybridization studies that all eukaryotic organisms examined to date contain multiple copies of the histone genes in their DNA (107), the numbers are generally lower: on the order of 50 to 100 copies for both *Drosophila* and mammals (M. L. Pardue; and M. Obinata and B. J. McCarthy, personal communcations). The reiteration is characteristic of the portion of the messenger coding for the protein and not for some general processing sequence. In experiments examining the distribution of DNA sequences hybridizable with histone mRNA in cesium chloride gradients, the histone messages are all found to hybridize exclusively with a specific cryptic satellite DNA and not with the bulk of the DNA (102, 103, 106). Hybridization experiments utilizing purified messenger for individual histones indicate that about the same number of genes exists for each histone in a given haploid genome (104, 105). It seems likely that this number is also a constant for all tissues, including meiotic ones; this has been directly determined for sea urchin sperm and embryos (102).

As noted above, the histone genes band together in cesium chloride and other density gradients. Hybridization assays with the individual messenger RNAs give the same distribution curves within the density gradients (102–104). The GC composition of the individual messengers varies considerably, from 51 to 58% (105); thus the evidence suggests that the individual histone genes are interspersed among each other. Only one region of in situ hybridization, 2L 39D-E, is observed in experiments in which *Drosophila melanogaster* polytene chromosomes are hybridized with tritiated total histone mRNA (108). In experiments carried out with individual histone mRNA fractions, it appears that all the mRNAs hybridize throughout this region, which contains two large bands and possibly several small ones (M. L. Pardue, personal communication). Thus there is now some evidence that the genes for the smaller histones (2a, 2b, 3, and 4) are organized in a system of complex, tandem repeats in a fashion analogous to that of the ribosomal genes, although as yet there is no information on the regularity of the pattern. Questions concerning the locus and distribution or interspersion of the H1 genes are at present unresolved because of difficulty in obtaining pure H1 mRNA. One might anticipate that special histones, such as H5 of reticulocytes or the histone of meiosis, will not be an integral part of the "locus" because of the differences in the regulation of their expression.

Sequence Homology

Early studies on the hybrids formed by the annealing of DNA with homologous histone mRNA indicated that the hybrids were of very high fidelity with little mismatch, as the RNA-DNA hybrids melted at the temperature anticipated for that GC composition (104). In general, it could be said that within a given species there was no indication of variance among the histone genes. Recently more sensitive assays have suggested that some variance is possible. The histone genes have now been isolated in the DNA form. Melting and re-annealing of these DNA sequences suggest from 1 to 3% mismatch (103). Evidence obtained in studying the histone mRNAs per se also suggests some variation. Fractionation of the histone messengers by polyacrylamide gel electrophoresis indicates several subfractions of each. For example, the messenger C (which codes for H4) includes three size subfractions. These subfractions show small differences in the oligonucleotide maps produced from them (106). However, it is not known whether this heterogeneity is a true indicator of gene heterogeneity, a consequence of genetic polymorphism within the population, an indicator of a precursor-product relationship, i.e. a processing of the mRNA through the three forms, or the consequence of artifactual degradation. Substantial random synonymous codon substitutions (third position base changes) have not occurred in these repetitious genes within a given species.

Substitution at the theoretically possible level would lead to genes that hybridize as single copy rather than as multicopy genes (104). One may conclude that some preventive or corrective mechanism must intervene to conserve and maintain the coding sequence within a species.

Evolution and Inheritance of the Histone Genes

Comparison of the histone genes at the nucleic acid level in closely and distantly related organisms indicates that, as in the case of the ribosomal genes, the histone genes of any given species are homologous whereas divergence has occurred during the evolution of species. The homology of the total histone genes of different organisms follows patterns anticipated in terms of relationships in evolutionary time established on other bases (109). It appears that the rates of divergence are greater for the AT-rich than for the GC-rich portions of the histone genes. It is thought that the AT-rich region represents spacer between the GC-rich histone messengers; thus the situation may be analogous to that of the ribosomal genes (103, 107). Note in addition that the evolution rate of the different histones at the amino acid level varies considerably. While estimates of the evolution rate are difficult to make from hybridization studies, the data at hand suggest that the rate for the total histone genes is approximately one half that for total DNA (109, 110). The results support the notion that, while equivalent codon substitutions have certainly occurred, these changes have not accumulated at such a rate as to imply that they are entirely neutral.

The histones represent a unique opportunity to study the inheritance and evolution of multicopy genes because they are the only proteins known to be so encoded. Amino acid sequence investigations are capable of much finer analysis than can be achieved with conventional DNA/DNA or DNA/RNA hybridization experiments. Work is just now beginning on the study of histone mutants. Several have been discovered; these are presented in Table 2. In general, the genetics has not yet been examined. The histone 5 variant appears to be inherited as a single allele; however, the investigation involved only one generation (42). In a second instance, where two subfractions of H1 were studied, each variant showed independent segregation of

Table 2   Known histone polymorphisms

Histone	Organism	Amino Acid Position	Alternative Amino Acid	References
H3	Pea	96	alanine/serine	397
H3	Calf	96	cysteine/serine	438; L. Patthy and E. L. Smith, personal communication
H2a	Rat	16	serine/threonine	401
H2a	Rat, calf	99	arginine/lysine	401; P. Sautiere, personal communication
H5	Chicken	15	arginine/glutamine	42

the other (111). (Note that this observation supports the suspicion that the H1 genes are not interspersed with the rest of the histone genes.) It is to be hoped that in the near future two types of experiments will provide further evidence on this question. In the first instance, it should soon be possible to generate data on the relative position of the specific histone messenger RNAs to each other on the DNA using electron microscopy techniques. In the second instance, searches are now under way in the laboratories of M. L. Pardue and others to find histone mutants (as temperature-sensitive lethals) in *Drosophila*. One can immediately envision genetic experiments that should shed further light on the inheritance of such mutations and on the speed with which these mutations could spread throughout the gene family.

THE NONHISTONE CHROMOSOMAL PROTEINS

Considerable progress has been made in the last few years in the isolation, fractionation, and chemical and functional characterizations of the NHC proteins. These proteins play structural, enzymic, and regulatory roles in the chromatin complex. Here we will merely summarize the more salient conclusions of the last several years of work; for more detailed reviews see (23, 112–118).

Definition and Isolation

The nonhistone chromosomal proteins are defined as those proteins (excluding the histones) that isolate together with DNA in purified chromatin or chromosomes. This class of proteins probably overlaps, but is not identical with the classes of acidic nuclear proteins and nuclear phosphoproteins, which are also under intensive study (119–125). That most, if not all, of the NHC proteins isolated in chromatin are true constituents of the complex as it exists in vivo is now substantiated by considerable evidence (reviewed in 126). The major NHC proteins have proved to be difficult to isolate and study because of their tendency to aggregate with the histones and with each other. A number of protocols are in use or under development but none have been found that are satisfactory in all aspects (total yield, recovery of all NHC proteins, complete separation of NHC proteins from DNA and histones, avoidance of harsh denaturing reagents, widespread applicability, possibility of further fractionation of the NHC proteins). The NHC proteins may be obtained in solution in 1% sodium dodecylsulfate (SDS) from chromatin treated with dilute mineral acids to remove the histones (e.g. 127). Such preparations have been used extensively in comparative studies utilizing SDS disc gel electrophoresis. A frequently used, less denaturing method of preparation is to dissociate chromatin in 5 M urea–2 M NaCl, remove the DNA by hydroxylapatite chromatography or by centrifugation, and separate the histones and NHC proteins by hydroxylapatite or by ion exchange chromatography (e.g. 128). Such preparations are typically used in reconstitution experiments. See the reviews cited above for further discussion and references on isolation techniques.

The class of NHC protein is considerably more complex than the histone class. The number of NHC proteins (individual polypeptide chains) observed on analysis by SDS polyacrylamide gel electrophoresis is usually on the order of 20–115, depending on the chromatin source and the resolving power of the analytical system. [The lower limit of detection at present using disc gel electrophoresis is $\sim 5 \times 10^3$ protein molecules per mammalian haploid genome (129).] Fractionation and analysis suggest that about 15–20 major proteins comprise 50 to 70% of the NHC protein fraction on a quantitative basis (129, 130).

Chemical Characterization

The NHC proteins possess many chemical, physical, and biological properties that contrast sharply with those of the histones. The amino acid compositions of total NHC protein fractions generally indicate a ratio of acidic to basic residues of 1.2 to 1.6 (calculated without the consideration of amides) (121, 128, 131). The individual polypeptide chains range from 10,000 to several hundred thousand daltons and from less than 3.7 to 9.0 in isoelectric point (e.g. 130, 132). Several subclasses and individual major NHC proteins have now been isolated and chemically character-ized. These have been grouped accordingly and are presented in Table 3. In addition, several NHC proteins of rat liver have been purified by Douvas & Bonner (439). This listing is by no means exhaustive; as the work progresses it is anticipated that further subclasses will be characterized.

Class D, that of the lysine-rich NHC proteins, is particularly intriguing. Pre-liminary sequence data indicate that the charge distribution within these proteins is irregular, as is the case for the histones, with a basic N-terminal region and an acidic region elsewhere in the molecule (133; G. H. Dixon and J. Walker, personal communication). The D proteins bind extensively to DNA with little effect on template activity (133, 134). Further studies of the chemistry of the NHC proteins are not only of intrinsic interest, particularly in studying peculiar proteins such as the lysine-rich NHC proteins, but also aid in devising further experiments and models of chromatin structure.

Several methods that have been used extensively in the past few years to try to isolate and characterize specific nonhistone chromosomal proteins rely on their binding interactions with eukaryotic DNA. While the DNA column methodology holds promise for the future, there are many technical difficulties at present. Several recent studies indicate that some NHC proteins, particularly low molecular weight ones, and some nuclear phosphoproteins bind extensively and in species-specific fashion to DNA (125, 135–139). Membrane filter assays analogous to those developed to study the *lac* repressor also hold promise for studies of NHC protein/DNA interactions (140, 440).

Tissue Specificity

In contrast to the histones, the nonhistone chromosomal protein fraction shows a limited tissue specificity. On comparative electrophoretic gel analysis most of the major nonhistone chromosomal proteins are observed in the NHC protein fraction from all tissues of an organism. A few tissue-specific proteins are observed, as well as quantitative variations. Such analyses have been carried out using both one- and two-dimensional techniques, the latter in general utilizing charge electrophoresis or

Table 3 Major nonhistone chromosomal proteins

Class	Organism and Tissue	Number of Bands on SDS Gels	Molecular Weight	pI	(Asx+Glx)/(Arg+His+Lys) (% Lys)	Estimated Number of Subfractions	N terminal	References
A1	Rat liver	1	43,000	<3.7	2.7 (4.2)	1	Glu	130, 146
A2	Dog liver	1	7,900		2.6 (5.6)		Ser	437
B	Rat liver	4	43,000–81,000	5.4–6.6	1.7 (6.4)	6	Gly, Ala, several minor	130, 146
	Drosophila embryo	1	53,000		2.0 (5.2)	1		S. C. R. Elgin, unpublished observations
C	Rat liver	2	10,000–20,000	6.4–8.0	1.2 (8.3)	7		130
	Rat liver	3	12,000–18,000		1.1 (7.9)	3		153
	Calf thymus	1	10,000		1.2 (4.5)	1		128
D	Rat liver	1	28,000	5.6	1.4 (13.3)	1	Gly	130
(lysine rich)	Rat thymus	1	20,000		1.1 (23.9)	1		434
	Calf thymus	1	27,000	6.0–8.5	1.1 (21.8)		Gly	133, 435, 436
	Calf thymus	1	28,000	6.0–8.5	0.9 (21.8)		Gly	133, 435, 436

isoelectric focusing in conjunction with SDS gel electrophoresis. The conclusion of limited tissue specificity appears valid for a number of nuclear protein preparations, including the nonhistone chromosomal proteins, the acidic nuclear proteins, the nuclear phosphoproteins, and the DNA binding proteins (125, 127, 128, 132, 141–149). For example, considerable similarities are seen in the gel pattern of the NHC protein fraction of rat liver and rat kidney, whereas the rat brain fraction shows a number of unique, high molecular weight proteins (128, 146, 150). Such limited tissue specificity is nonetheless compatible with this fraction having a role in the control of specific gene expression. This would be accomplished either if those elements involved in the specific regulatory mechanisms were present in small amounts, or if the controlling elements act in a combinatorial system, as suggested by Gierer (151). Studies of the nonhistone chromosomal protein fractions by immunological techniques often suggest a considerable amount of tissue specificity. This has been reported in particular for DNA-protein complexes of the low molecular weight NHC proteins which show high affinity for DNA (139, 152–155). Nonetheless, analyses indicate that the bulk of the nonhistone chromosomal proteins are common to all tissues, and thus imply that these proteins are involved in common structural and enzymatic roles.

Biological Roles

ENZYME ACTIVITIES   Many enzymes of chromosomal metabolism, including nucleic acid polymerases, nucleases, and enzymes of histone metabolism, are integral components of chromatin. For example, it is estimated that from 10 to 50% of the cell's RNA polymerase is bound to the chromatin in vivo (156, 157). A representative sample of enzyme activities that have been identified in isolated chromatin is given in Table 4.

In addition, several proteins that bind to and affect the conformation of DNA have been isolated and characterized, including proteins that stabilize single stranded DNA (158–160) and that unwind superhelices (161, 162). Certain chromosomal structures, such as the meiotic synaptonemal complex, probably are made up of NHC protein elements. Since it has been observed that recombination frequencies in eukaryotic chromosomes can be both structurally dependent and genetically determined, it has been suggested that specific NHC proteins play a role in bringing about the chromosome pairing observed in meiosis, etc (163, 164); however, such proteins have not been isolated as yet.

CHROMOSOME STRUCTURE AND TEMPLATE ACTIVITY   NHC proteins no doubt play general structural and enzymatic roles in the process of gene activation, and a subset may be involved in determining the specificity of transcription. A number of studies have been published in which the presence of nonhistone chromosomal proteins is observed to have a mitigating effect on the repression of template activity by the histones (e.g. 6 and 165; see 112 and 114 for a review of this question). Recent experiments in which chromatin is fractionated into DNA, histones, and NHC protein components and subsequently reassociated by salt-urea gradient dialysis implicate the NHC protein fraction as including positive regulators of gene

Table 4 Enzyme activities associated with chromatin

Enzyme	Organism and Tissue	Reference
1. RNA polymerase	Rat liver	404, 405
	Mouse myeloma	406
	Hen oviduct	157
	Coconut	407, 408
2. Poly(A) polymerase	Wheat	409
3. DNA polymerase	Rat liver	410–412
	Rat ascites hepatoma	413, 414
	Sea urchin	415
	Calf thymus	416
4. DNA endonuclease	HeLa cells	417
5. DNA ligase	Rabbit bone marrow	418
6. DNase	Rat liver	419
7. Terminal DNA-nucleotidyltransferase	Calf thymus	420
	Tobacco	421
8. Poly(adenosine diphosphate ribose) polymerase	Rat liver	422, 423
9. Poly ADP-ribose glycohydrolase	Rat liver	424
10. Histone acetyltransferase	Rat thymus	425–427
11. Histone methylase	Calf thymus	431
12. Histone kinase (acid-labile phosphate)	Rat Walker 256 carcinosarcoma cells	433
	Rat liver	432
13. Histone protease	Rat liver	428, 441
	Calf thymus	429
14. NHC protein kinases	Rat liver	430

expression (12, 166). In all such work the test assay has been correct transcription of the hemoglobin gene as detected by hybridization of product RNA to a reverse transcript of hemoglobin mRNA. The NHC protein fractions used include small RNAs, which have also been suggested as positive regulators of gene expression (7, 167, 168).

It is thought that the NHC proteins play an important role in the interaction of steroid hormones with target cell nuclei and the subsequent specific gene activation (e.g. 4, 169). However, a successful in vitro inductive system has yet to be established. [See (170) for a review of this field.] Hopefully, the use of mutant cells deficient in the activity of the cytoplasmic receptor, the ability to bind hormone in the nucleus, or in subsequent steps will help to resolve the many interesting questions. Such mutants of a lymphoma cell line, normally killed by glucocortoids, have been isolated by Sibley & Tomkins (171).

The polytene chromosomes of *Drosophila* present a unique opportunity for combined biochemical and cytological analysis of gene activation. Specific bands puff either normally in development or in response to stimuli such as the hormone ecdyson or heat shock. In puffing a band takes on a diffuse appearance and becomes the site of new RNA synthesis. A critical early event in the transition to the active state is a 100% increase in nonhistone chromosomal protein at the puff locus; no decrease in histone is observed (172–174). Early reports suggested that new, stimulus-specific nuclear proteins could be detected in this system (175, 176). Unfortunately, this could not be substantiated in a study of the polytene chromosomal proteins (459). This is not surprising considering the percentage of the genome involved, less than 1%. Proteins as activators are implicated, however, by the finding that the induction of puffs occurring late in the hormone response can be blocked by inhibitors of protein synthesis (177). It is likely that the detailed questions will remain unresolved until they can be studied with a more specific, restricted chromatin system, wherein a greater percentage of the genomic material undergoes the transition in question. See Berendes (20, 21) for a review of the polytene chromosome system.

Structural features are important clues to the mechanisms involved in chromosomal replication as well as transcriptional activity. Certain NHC proteins may play a major role in the shift from the inactive quiescent cell state (G_0) to the active growing state (G_1). Specific synthesis of certain NHC proteins is observed and may be required for this transition and the concomitant increase in template activity (178–182, 453). Comparative studies of the chromatins of G_0 and G_1 cells and of normal and transformed cells suggest that a fraction of NHC proteins dissociable in 0.25 M NaCl is responsible for the differences in the structural and functional properties of these chromatins (183, 184). See Baserga (455) for a review of this topic.

Parallel observations on the quantity of certain NHC proteins in relation to the growth state have been made on a primitive eukaryote, *Physarum,* and cells from an advanced eukaryote, HeLa (181). LeStourgeon has recently suggested that myosin, actin, and a "tropomyosin" are major NHC proteins in *Physarum polycephalum*; an increase in the actin: "tropomyosin" ratio is observed to be characteristic of both the metaphase and the quiescent nonproliferative G_0 states (185). Actin and myosin-like filaments have previously been identified in the dividing cell and are thought to be important in movement of chromosomes (186, 466). LeStourgeon, on the basis of the quantitative observations, has suggested that these proteins play an additional role in chromosome condensation per se. NHC proteins with the molecular weights of actin and myosin have been observed by others as characteristic of mitosis (see the following section). Perhaps the use of fluorescent antibody techniques to study the distribution of chromosomal proteins in situ will resolve the question in the next few years.

NUCLEAR MEMBRANE PROTEINS   Chromatin as it is isolated contains some lipid, suggesting associated or contaminating nuclear membrane material (187–190). This is hardly surprising because much evidence indicates that the chromatin fibers are associated with the annuli of the nuclear membrane (e.g. 191, 192). Certain specific proteins may be involved in this membrane DNA interaction and as such may be legitimate chromosomal proteins as well as nuclear membrane proteins. The question remains to be resolved.

RNA TRANSPORT PARTICLE PROTEINS Since the early work of Georgiev and his associates, there has been a considerable effort to characterize the proteins of the RNA transport particles, or informosomes (see 193 for a more detailed review of early work). These particles are thought to assist in packaging, transport, and otherwise processing of mRNA. The particles possess a major protein component of ~40,000 daltons; reports vary concerning the amount and characteristics of other proteins present (194–196, 454). Certain preparations, such as the lampbrush chromosomes of amphibian oocytes, are particularly enriched for these particles and it seems likely that such proteins will be major nonhistone chromosomal protein components in this instance (197–199). Fluorescent antibodies prepared to proteins from the *Triturus* oocyte nuclear ribonucleoprotein particles will specifically stain the loops of the lampbrush chromosomes; in fact, an antibody preparation to a specific size protein selectively stains only 10 of the loop pairs (198, 200). The proportion of such nonhistone proteins in chromatin may increase as a consequence of increased template activity in response to stimuli such as hormones (201). Such particles may turn out in some way to be locus or stimulus specific, as indicated by work in the polytene chromosomes (202). Studies concerning such proteins have been recently reviewed by Williamson (203).

THE METABOLISM OF CHROMOSOMAL PROTEINS

Histone Synthesis

HISTONE mRNA TRANSCRIPTION It is now well established that the bulk of the histone synthesis is synchronized with DNA synthesis in the cell cycle. The relationship has been tested in a number of cell systems synchronized by artificial or natural means, e.g. in *Euplotes* (204), synchronized tissue culture cells (205, 206), and the regenerating rat liver (207). This coupling of synthesis operates in both directions, in that an inhibition of DNA synthesis will depress histone synthesis, while an inhibition of histone synthesis will slow DNA synthesis to about half its former level (208–210; see below). There are also several reports of small amounts of histone synthesis at other stages of the cell cycle (e.g. 206, 211), and histone synthesis is known to be uncoupled from DNA synthesis in certain situations, e.g. erythropoeisis (212) and oogenesis (213). However, it is safe to conclude that the bulk of histone synthesis occurs in S phase during the replication of the chromatin. The observation of this fact has allowed experimenters to successfully isolate histone mRNA fractions from synchronized cells in S phase and from early embryos, which are undergoing rapid nuclear divisions.

The histone mRNA has been separated from cellular RNA as a peak sedimenting between 7 and 12S on sucrose gradients (208, 214, 215). The mRNA fractions may be further separated for preparative as well as analytical purposes on polyacrylamide gels into three to five bands, as discussed above (106). That such RNA is indeed messenger for histones and is not significantly contaminated by mRNA for other proteins has been shown by its translation in cell-free systems. The presumptive messenger peak, isolated from synchronized HeLa cells in S phase, is able to direct the synthesis of all five histones in both the rabbit reticulocyte system and the mouse ascites cell-free system (214–216). Studies of tyrosine and tryptophan incorporation indicate that over 90% of the protein synthesized in the mouse ascites system is histone (214).

The histone mRNA and the histone synthesizing system possess several interesting characteristics apparently oriented towards the production of a large amount of protein within a short time. At least 20,000 histone molecules are synthesized by the cell per minute in S phase (210). The fact that the histones are coded for by repetitious DNA sequences is obviously beneficial in producing a great deal of mRNA in a very short time. In addition, Schochetman & Perry (217) have demonstrated that histone mRNA is processed and transported to the cytoplasm within several minutes of its synthesis, a very short time relative to that of other mRNAs. Histone synthesis is not inhibited by cordecypin (219); this is consistent with the observation that the mRNA appears not to contain the 3′-terminal poly(A) sequence characteristic of other mRNAs (218).

HISTONE TRANSLATION   The histone message is found in cytoplasmic polysomes that are free or loosely associated with membranes (220). Synthesis of the protein is initiated, as for other proteins, with the methionyl $tRNA_f$ (221). Current evidence suggests that this residue is cleaved from the nascent polypeptide chain, and the N-terminal serine is subsequently acetylated (222). The newly synthesized histones are transported quite rapidly (within 10 sec) into the nucleus (210). The extranuclear free histone pool is very low and is estimated to be 0.2% of the total (223).

There appears to be no gross translational control exercised on the synthesis of histones. The histone messenger can be translated equally well in cell-free translation systems isolated from S phase cells and from G_1 cells; the latter are not normally active in histone synthesis. Likewise, cells in which DNA synthesis has been inhibited—a condition that normally results in a loss of biologically active histone mRNA from polysomes—possess translational systems that can use histone mRNA with full efficiency (219, 224–226).

COUPLING OF HISTONE AND DNA SYNTHESIS   The interacting controls of histone and DNA synthesis present a very intriguing system. The lifetime of histone mRNA in the cytoplasm is much shorter after inhibition or cessation of DNA synthesis, although histone synthesis is relatively resistant to inhibition by cycloheximide or actinomycin D (210, 225, 227, 228). The lifetime of histone mRNA in mouse L cells has been estimated to be roughly equal to S phase, in contrast to other mRNAs whose lifetime is estimated to be approximately equal to the cell generation time (229). The evidence suggests that a specific degradation of histone mRNA occurs in the absence of DNA synthesis (219, 226, 229, 467). Conversely, cycloheximide inhibition of protein synthesis during S phase causes an inhibition of ∼50% in the rate of DNA synthesis (210, 230, 231). It has been suggested that the histones play a role as chain elongation proteins in the eukaryotic chromatin (210, 232). The reduced rate of DNA synthesis would then be a consequence of the availability of only the parental, and no newly synthesized, histones. The simplest hypothesis to explain this control circuitry is that free histone (that not bound to DNA) inhibits further histone synthesis in a way that results in rapid histone mRNA degradation as well as inhibiting production of new histone mRNA (228, 233).

Nonhistone Chromosomal Protein Synthesis

The synthesis of NHC proteins stands in contrast to that of histones in that it generally occurs throughout the cell cycle (234–236), although some increase in the rate has been reported in G_1 in mammalian tissue culture cells (237, 238). More detailed studies have indicated that specific NHC proteins are synthesized at specific times in the cell cycle (239, 240).

Replication of Chromatin

An intriguing problem which is just now beginning to be studied is chromatin replication. Whereas the transfer of the information in the DNA takes place by the well-known mechanisms indicated by the Watson-Crick structure of the DNA double helix, there is no obvious model for the replication of the information contained in the distribution of proteins on the DNA. The tight coupling of DNA and histone synthesis and the rapid binding of histone to newly synthesized DNA indicate that this replication of information takes place at the growing fork (233). However, the molecular mechanism remains obscure.

It has recently been shown that following the inhibition of new histone synthesis with cycloheximide, the old histone becomes associated exclusively with only one of the daughter DNA double helices (233). Under conditions where protein synthesis is not inhibited, double label experiments indicate that new histone is associated with the newly made single strands of DNA. When cells are pulse-labeled concurrently with H^3-BUdR (5-bromodeoxyuridine) and C^{14}-leucine, the chromatin isolated and then denatured, the individual DNA strands separate but the histones are only partly removed. If the separated chromatin strands are irradiated to break the BUdR strand (which is then resolved from the parental strand by its lower sedimentation rate on sucrose gradients), one finds that newly made histone migrates together with the newly made BUdR strand of the DNA (241).

Other Chromosome Forms

MITOSIS   It is of interest to look for changes in the protein population that might be indicative of particular states of chromatin. The histones are extremely stable throughout the cell cycle and are synthesized mostly in tandem with the new DNA. It is not surprising, then, that the histones of mitotic chromosomes are essentially the same as those of the interphase chromatin (242). It has been suggested that histone 3 plays a role in condensing the chromosomes by forming intermolecular Cys–Cys disulfide bonds. However, as discussed above, one of the cysteines (position 96) is interior and nonreactive in the H3 of isolated chromatin (and can be replaced by serine), suggesting that at most only one cysteine (position 110) could be involved in such intermolecular linkages. Several studies of the question have found no evidence for an increase in the amount of histone 3 in a dimer or other oxidized form in metaphase chromosomes (243, 465; R. Chalkley, personal communication).

A comparison of the NHC proteins of metaphase chromosomes with those of interphase chromatin shoŵs that this protein fraction also remains much the same (126, 243, 244). Certain significant quantitative shifts in the NHC protein pattern were found to be common to the metaphase chromosomes of HeLa and Chinese hamster ovary cells. These were a decrease in the amount of two 78,000–80,000 dalton proteins and an increase in proteins of 50,000 and 200,000 daltons in metaphase chromosomes. The latter NHC proteins thus could be involved in the metaphase chromosome condensation (see section on NHC proteins). A number of minor bands are observed in either G or M for each tissue (243).

POLYTENY   Polytene chromosomes are observed in certain tissues of *Diptera,* most notably *Drosophila.* By virtue of the differential underreplication of much of the repetitious DNA, they are greatly enriched in euchromatin (245). The chemical composition of the polytene chromosomes is similar to that of diploid chromatin, being reported as DNA:histone:NHCP:RNA of 1.0:0.97:1.03:0.03 (246). The histones of the polytene chromosomes are identical with those of *Drosophila* diploid chromatin (38); H4 is relatively quantitatively deficient (459). Certain relative deficiencies are observed in the analytical gel pattern of the nonhistone chromosomal proteins. These deficiencies may indicate proteins normally associated with the highly repetitious and heterochromatic DNA, as this DNA is severely underrepresented in the polytene chromosomes (126).

Turnover and Differential Synthesis of Chromosomal Proteins

The histones and NHC proteins stand in contrast to one another in their metabolic stability or estimated half-lives. Various labeling experiments indicate that the histones have very long half-lives, on the order of that of the DNA (247–249), although exceptions to this are noted in particular circumstances (212, 250, 251). In general, the NHC proteins possess shorter half-lives, similar to those of the cytoplasmic proteins; however, considerable variation within this fraction is observed (252–254).

Alterations in the patterns of synthesis of the histones and the NHC proteins have been observed when new genes are being expressed. Such observations include (*a*) the new synthesis of specific NHC proteins in target tissues in response to hormones (124, 255, 256), (*b*) a general stimulation of the synthesis of NHC proteins and DNA binding proteins during the transition from a quiescent to a growing cell population (see the previous section) (257, 258), and (*c*) changes in the NHC protein fraction during embryogenesis (79, 259–263). Alterations in the pattern of synthesis of histone 1 subfractions have also been observed and were discussed above.

Unfortunately, most stimuli affect cell division rates as well as differentiation, so cause and effect relationships are difficult to sort out. An increase in the cell division rate will lead to an increase in the percentage of time a given cell spends in S and M; thus one might expect to see an increase in proteins specifically associated with either of these states of the chromatin. Indeed, this is suggested in the case of alterations of the NHC protein pattern during early embryogenesis in *Drosophila.* In this instance the rate of nuclear division goes from once every 10 min to once every several hours or less during the 18 hr period covered by the study; a con-comitant decline in the amount of an NHC protein of 48,000 daltons is observed (79). Minor histones have been reported as associated with cells having higher or lower rates of cell division (264, 265). Rubio et al (266) have observed consistent changes in the DNA binding protein pattern for transformed and normal mouse cells, and such a difference for the NHC protein fraction has been confirmed using im-munological techniques (155, 267). No such differences have been noted for the histones (268–270). Much work remains to be done to sort out roles and cause and effect relationships in this field [see (53, 271, 272) for further review].

Modifications of Chromosomal Proteins

HISTONES   Considerable postsynthetic modification occurs in the chromosomal proteins, particularly in the histones. Among the unusual amino acids observed in histones are ε-N-methyllysine in the mono-, di-, and trimethyl forms, ω-N-methylarginine, 3-methylhistidine, α-N-acetylserine, ε-N-acetyllysine, O-phospho-serine, O-phosphothreonine, N-phospholysine, and N-phosphohistidine. We will review the subject only very briefly here; for more extensive discussion one may refer to other reviews (23, 58, 112, 118, 273–275). Here we wish to emphasize two aspects of histone modification. First, it is important to note the nontriviality of these modifications: they occur at specific times in the cell cycle and at specific sites on the histones as a consequence of the action of specific enzymes. In no case are all the histone molecules modified in the same way at the same time; the percentage of histone modified can range from very little to almost all. Second, such modifica-tions significantly alter the DNA-protein interaction. (Note that there is no evidence for the notion that such modifications can promote sequence specificity in the histone/DNA interaction.) It is of interest to contrast the extremely conservative amino acid sequence of the histones with the variability of the postsynthetic modification. Clearly if a conservative amino acid substitution, e.g. from leucine to isoleucine, is a forbidden one, the introduction of a methyl group or change of charge within the sequence must have a profound effect on some aspect of histone interactions. Both considerations combine to convince one that these postsynthetic modifications are important in determining the activities of the histones (23, 118). As the best studied example of histone modification, we will discuss here the case of H1 phosphorylation.

The phosphorylation of H1 has been observed as a biochemical event of S phase, mitosis, and certain hormone responses. Several investigators have reported a large increase in H1 phosphorylation during S phase, the time of most histone synthesis. A detailed analysis of histone modification during and immediately after their synthesis during trout spermatogenesis by Dixon and his colleagues indicates that considerable modification, phosphorylation, and/or acetylation occurs for all the histones and for the protamines at this time. They suggest that these modifications may aid in the transport and correct association of the histones with the DNA. In particular, Louie & Dixon (277) have proposed that during chromosome assembly histone 4 is "fitted" into its proper configuration with the DNA by a mechanism involving primarily histone acetylation. They suggest that acetylation will decrease the charge interactions between DNA and protein, and as a consequence a particular type of interaction will take place. After this association is achieved, deacetylation (which is observed) would allow electrostatic interactions which could "lock in" that conformation. Similarly, the phosphorylation of protamines may inhibit tight binding to DNA at a time when the chromatin is still being transcribed. The protamine is subsequently dephosphorylated, and presumably it is this form that is responsible for turning off the sperm nuclei (57, 58, 276–279). A similar mechanism appears to apply to the inactivation of avian erythrocyte nuclei in terms of histone 5 phosphorylation and dephosphorylation (see the section on the histones).

It has been established in other diverse systems that phosphorylation of H1 also occurs during metaphase (280, 281). In most eukaryotic cells, two to four sites of H1 phosphorylation are present in S and additional sites are phosphorylated in M. The degree of overlap between the S and M phase sites is not completely known at this time, but it appears most likely that the sites of phosphorylation are independent (282). These phosphorylation sites are threonine as well as serine residues; they are clearly different from those sites involved in H1 phosphorylation in hormone response (283) (see below). A significant amount, half or more of H1, is phosphorylated in rapidly dividing cells (270). The increased phosphorylation at M is most likely the consequence of a six- to tenfold elevation of a specific ATP : histone phosphotransferase in the mitotic mammalian cells (284, 285). A rapid dephosphorylation of H1 occurs as cells move into early G_1, suggesting that the phosphorylation of H1 is necessary for the condensed metaphase state (270, 286, 287). The slime mold *P. polycephalum* is particularly advantageous for studying this role of phosphorylation of H1, in that in the slime mold S phase and M phase are clearly separated in time. Using *Physarum,* Bradbury and his colleagues (288–290) have obtained intriguing results suggesting that the phosphorylation of histone 1, in response to a steady increase in H1 phosphokinase levels, constitutes the mitotic trigger and leads to the condensation of chromosomes in this system.

That the specific phosphorylation of H1 is an early response to glucagon stimulation in the rat liver is now well established. The response is mediated by the glucagon-induced increase in cellular cAMP, which in turn stimulates the rate of histone phosphorylation. The cAMP effect can be demonstrated in vitro with the isolated enzyme as well as in the rat and the isolated perfused rat liver (291–295). The H1 phosphorylation is a very early event in response to glucagon, occurring in about 15 min, whereas enzyme induction occurs between 30 and 240 min (35, 292, 296). The enzyme responsible for this phosphorylation is clearly different from those discussed above, which are not affected by the cAMP level. The cAMP-dependent enzyme phosphorylates H1 at position 37. Thus two enzymes are clearly different, and it is likely that there are other enzymes showing specificities for histone substrates (35, 291–293, 297). Rat liver contains a phosphatase specific for phosphorylated histones and protamine. This enzyme has been found in all eukaryotic tissues examined, but not in any prokaryotic cells (298, 299). The existence of this complete and specific H1 phosphorylation/dephosphorylation system has led to speculation concerning a possible role in gene activation.

That such modifications do affect the interactions of histone with DNA has been shown by physical chemical studies. Phosphorylated H1 is eluted from a DNA-cellulose column at lower salt concentrations than nonphosphorylated H1 (S. Fisher and U. Laemmli, personal communication). Circular dichroism techniques show that the modified histones are less able to produce the conformational distortions of DNA that histones normally cause. This is true both of histone phosphorylation at the specific sites discussed and for histone acetylation. A random change in the charge of H1 is not effective in this regard (300, 301). Recently Watson & Langan (302) have presented evidence suggesting that H1 and phosphorylated H1 differ in their ability to repress template activities of chromatin following selective reassociation of these histone fractions to H1-depleted chromatin. It should be noted that in in vivo hormone stimulation, only 1% of the total H1 is phosphorylated. This is in accordance with expectations in that only a very limited portion of the genome should be activated in response to hormones. However, this makes it difficult or impossible to detect the physical chemical change, if any, in the total chromatin.

NONHISTONE CHROMOSOMAL PROTEINS   The only postsynthetic modification of nuclear nonhistone proteins that has been studied in detail is phosphorylation. It has been shown that within the mammalian nucleus there exist a large number of enzymes specifically responsible for phosphorylation and dephosphorylation of nonhistone proteins (including RNA polymerase and associated factors) at serine and threonine residues. There is evidence that, in parallel with the observations on histones, their activity and response may be highly specific (116, 254, 275, 303–312). Unfortunately, it has not been possible to correlate the nuclear phosphoproteins with the known nonhistone chromosomal proteins. Interesting questions remaining to be studied concern the relatedness of these two fractions, the binding coefficient to DNA of the proteins in their phosphorylated and nonphosphorylated forms, as well as the interaction of these proteins and enzymes with the histone component. See Kleinsmith (312) for a review of recent progress in this complex field.

CHROMATIN STRUCTURE

The essential features describing the gross structure of interphase eukaryotic chromatin will probably be elucidated within the next few years. The field is developing rapidly and we will therefore try to anticipate many of the probable conclusions that are emerging as well as enumerate some of the basic questions that will probably remain for quite some time. Several extensive reviews and discussions of the literature regarding chromosome structure and protein-DNA interactions have been presented recently (40, 313–315, 442, 450). Consequently we will concentrate on the more important observations of the last few years. These observations have changed our concept of chromatin from that of a smooth, linear fiber dominated by a regular (316) or irregular (317) supercoiling of the DNA double helix, to one where the DNA folds around histone complexes spaced fairly regularly along the chromosome fiber. This "beads on a string" model is visualized in the electron micrograph shown in Figure 6. We do not yet know how the DNA of each particle is packed. Some elements of the supercoil model may be applicable to the arrangement of the DNA in these particles or to the arrangement of the particles themselves.

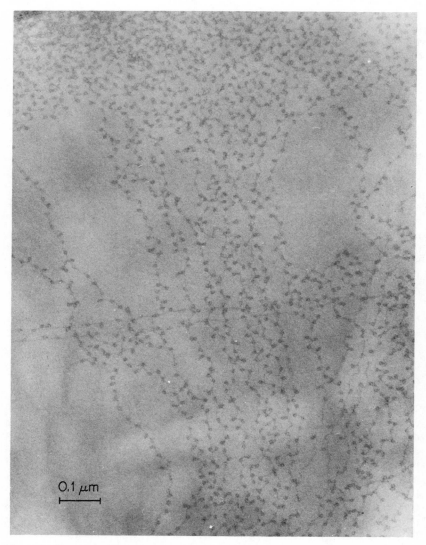

Figure 6 Chromatin fibers streaming out of a chicken erythrocyte nucleus. The spheroid chromatin units (nu bodies) are about 70 Å in diameter. Negative stain : 0.5 mM uranylacetate in methanol. Preparation as described in (333). 167,000 ×. (Courtesy of A. L. Olins and D. E. Olins, 1974.)

Periodicity of Histone Along the DNA Fiber

The most important recent advance has been the use of acrylamide gels to analyze in detail the products of nuclease digestion of chromatin. Hewisch & Burgoyne (318) showed that an endogenous nuclease from rat liver cuts nuclear DNA into a series of discrete bands that migrated aₒ if they were multiples of a basic repeat. The size of this repeat was later quoted as about 200 base pairs (315). One difficulty in interpreting these data comes from the fact that exhaustive treatment fails to digest all the nuclear DNA into pieces of 200 base pairs, much of the DNA being present in higher multiples of 200. The reason for this is still unclear; however, subsequent experiments using staphylococcal nuclease have made this distinction somewhat less important. Low concentrations of staphylococcal nuclease (or DNase I) digest nuclear DNA into pieces consistent with a repeating unit(s) of between 140 and 200 base pairs (319–321; B. Shaw and K. Van Holde, personal communication). [A similar treatment of yeast nuclei yields an analogous pattern of resistant DNA fragments (B. Shaw and K. Van Holde, personal communication). This is particularly striking because yeasts are purported to be missing some histones.] After the initial nuclease attack, less than 15% of the DNA is acid soluble. Upon additional digestion with the nuclease, there is a decrease in DNA size from 200 to 170 base pairs (319). Extensive digestion with staphylococcal nuclease yields a limit-digest of at least eight DNA bands that range in size between 45 and 145 base pairs (320, 321). Under these conditions, approximately half of the DNA is rendered acid soluble, while the remainder is present in these discrete DNA fragments. The acid-soluble DNA reflects the so-called open regions in chromatin, while the eight limit-digest DNA fragments reflect nuclease-resistant closed regions (322).

At very low concentrations of nuclease, discrete chromatin fragments can be separated on sucrose gradients (319, 321; R. Axel and G. Felsenfeld, and B. Shaw and K. Van Holde, personal communications). These fragments sediment in a modal distribution consisting of monomer, dimer, trimer, etc. The DNA in each peak migrates on gels as if it were a multiple of a monomer repeat of 150–200 base pairs. When the monomer chromatin fragment is treated with nuclease, approximately 43% of the DNA is rendered acid soluble (320) and the resistant DNA consists of approximately eight DNA fragments between 45 and 145 base pairs which are characteristic of the staphylococcal nuclease limit-digest of whole chromatin. It has been proposed that extensive digestion with nuclease gives rise to artifacts due to the redistribution of histones during the digestion (315, 323, 324). While this is undoubtedly true for exhaustive digestion with pancreatic DNase (325), it seems extremely unlikely for staphylococcal nuclease for two major reasons: 1. the staphylococcal nuclease limit-digest yields most of the DNA in extremely discrete fragments and 2. redigestion of the monomer chromatin fragment with nuclease results in the same DNA limit-digest as found with whole chromatin. The latter finding indicates that if transfer of protein occurs it must happen *between* particles; yet, such a hypothetical transfer has not been detected by a number of sensitive assays. It is likely, therefore, that different degrees of nuclease digestion reveal different characteristics of chromatin structure (see 452 and references cited above).

The clear advantage of staphylococcal nuclease is that it digests essentially all the chromosome into discrete units, whereas the liver enzyme, although giving a rather clear indication that chromatin is digestible into discrete units, senses additional structural information since it is unable to cleave *all* the DNA into a single unit. An explanation offered by Hewisch & Burgoyne (318) is an enticing and testable one, namely that nonhistone protein can protect certain open regions from the liver endonuclease; however, as yet there is no evidence for this. (Note that there is no formal proof that staphylococcal nuclease and liver nuclease generate the *same* repeat.) Indeed, although it is becoming increasingly clear that the histones are organized into defined "repeats" along the chromosome and that the DNA itself is organized into repeats of reiterated and nonreiterated sequences (326), the organization of nonhistone proteins in this general scheme is unknown.

The repeat size obtained from nuclease digestion may be in very good agreement with the 110 Å repeat observed from X-ray diffraction studies (315). Unfortunately, it is not clear how the two repeats are related. While the nuclease digestion data indicate that a nuclease-sensitive site, one that is presumably deficient in protein, exists at regular intervals along the chromosome, the X-ray data (at least in the most superficial analysis) indicate only a structural repeat about every 110 Å. It is not clear whether it is the protein or DNA component that gives rise to this reflection, what portion of the chromosome structure gives rise to this repeat, or whether the lower angle reflections at 55, 37, 27, and 22 Å are higher orders of the 110 Å repeat or attributable to independent, repeating structures. Reconstitution with selected histones (327–329) demonstrates that H2a, H2b, H3, and H4 are *all* required for the generation of the normal X-ray pattern and that H1 is not (328–330). In addition, selective histone removal by chymotrypsin indicates that only histone 4 is needed to maintain some of the X-ray structure (331). Clearly maintenance and generation of the structure are two separate phenomena.

Much of the X-ray diffraction data can be better understood in terms of recent data obtained using neutron scattering (332). The basic method involves matching the scattering of the solvent with that of either DNA or histone. Because DNA and histone have different scattering properties and either can be theoretically eliminated by increasing the concentration of D_2O in the solvent (scattering from DNA is matched by a solvent concentration of 63% D_2O while that from histone is matched at 37.5% D_2O), it is possible to determine which diffraction spacings arise from which macromolecule. By varying the D_2O-to-H_2O ratio of the medium and analyzing the changes in the reflections at 110, 55, 37, and 27 Å, Bradbury and his colleagues (332) have found that all the spacings do not change coordinately, and consequently all chromatin reflections *do not* come from higher orders of the same basic repeat. The data indicate that the 110 and 37 Å reflections come from protein and that the 55 and 27 Å reflections come from DNA. Their interpretation of these findings is in remarkable agreement with recent electron microscopic observations (333, 334) and with results from nuclease and trypsin digestions of chromatin, i.e. a "particles on a string" model. It is proposed that the repeat of 110 Å corresponds to a histone core and that the folding of the DNA within the particle corresponds to the DNA rings at 55 and 27 Å. The 37 Å repeat may come from the way the histones are packaged within each particle.

Electron Microscopy

Recent electron microscopic data are strikingly consistent with the emerging physical and chemical picture of chromatin. Olins & Olins (333, 473) and Woodcock (334) have observed that chromatin, when prepared according to their methods, appears as a chain of beads on a string (see Figure 6). The beads ("nu" bodies) have a diameter of about 70 Å (333) (approximate molecular weight 160,000) and their spacing is roughly consistent with a DNA repeat of about 200 base pairs (335), although this is only very approximate. Although it is difficult to prove that the "particles on a string" view of chromatin as visualized by electron microscopy is not an artifact, it is nevertheless very consistent with the suggestions from nuclease experiments. Indeed, it is most likely that this picture is a meaningful "vestige" (333) of the original structure within the nucleus.

There are four major questions raised by the recent electron microscopic findings. The first is the composition of each nu body. Are they homogeneous? The second is the composition of the threadlike substances (about 15 Å in diameter) between them. Is it sensitive to nuclease? If so, is it sensitive at a particular point or throughout its length? Does it contain histones? nonhistones? The third and most difficult question is the nature of the higher order structure that was probably destroyed in obtaining these pictures; or equivalently, how do the nu bodies interact to produce the next order of packing? Do they interdigitate between fibers? Do they fold back on themselves? And finally, what is the internal structure of the DNA and protein within the nu body? Clearly the electron microscopic findings are quite compatible with the original nuclease digestion experiments of Clark & Felsenfeld (322), which indicated that chromatin exists as a series of alternating segments of protein-covered and protein-depleted DNA, each about 75,000 daltons. Although not yet proven, we shall assume that staphylococcal nuclease is attacking sites between individual nu bodies.

Senior et al (335) have shown that extensive sonication of formaldehyde-fixed, chick erythrocyte chromatin yields a subpopulation consisting almost entirely of single nu bodies. An analysis of these particles indicates a DNA strand of about 200 base pairs and about an equal weight of protein. Interestingly, the authors report preliminary evidence indicating that the isolated nu bodies give the same X-ray reflections as intact chromatin. Since nu bodies are about 80 Å in diameter, it seems unlikely that they themselves give rise to the 110 A repeat. An explanation offered by Senior et al (335) is that the repeat is actually a consequence of the packing of the nu bodies, in much the same way as the packing of spheres might give rise to a higher order reflection.

In contrast to the features of chromosome structure detected by nucleases over rather short distances of DNA (50–200 base pairs), some structural properties present over longer distances appear to be quite variable. Georgiev and his colleagues (336, 337) have investigated the quantity of free DNA generated by shearing chromatin and banding according to density in CsCl after formaldehyde fixation. Their data clearly show that as the ionic strength of the buffer increases, there is a corresponding increase in free DNA, from essentially zero at low ionic strength to approximately 25% in 2 mM $MgCl_2$ or 225 mM NaCl. Interestingly, the digestion pattern from staphylococcal nuclease is hardly changed in similar ionic conditions (H. Weintraub, unpublished observations). As intimated by the authors, these findings may be relevant to the problems of differential gene activity; thus, they are presently doing the appropriate hybridization experiments to determine whether the free DNA generated at high ionic strength is in any way related to transcriptional activity. In any event, the behavior of histones in chromatin is highly dependent on ionic conditions and it would not be surprising if subtle, ionic effects (from small or large molecules) occurring in particular regions of the nucleus were used to impose specific *local* chromosome conformations. One of the models that the authors propose to explain their data is that of a long DNA loop. One side of the loop is extended and not tightly bound to histone, whereas the other side is highly compacted by histones. It should be emphasized that this model represents a higher order of structure than that indicated by nuclease experiments and it is in no way inconsistent with that data.

Nu Body Structure

Accepting the general model that seems to be emerging, that of a globular, protein-rich structure associated with every 150–200 base pairs of the chromosome fiber and bordered by short protein-poor, nuclease-sensitive regions, it is possible to ask certain questions about the internal structure of these nu bodies. One insight into their substructure comes from the fact that the limit-digest with staphylococcal nuclease yields very small but discrete DNA bands on electrophoresis (320). The size of these bands is significantly less than the 150–200 base pair repeat for nu bodies. Consequently, exhaustive digestion with nuclease must be hydrolyzing less accessible sites within each nu body. If this is so, then since there are a minimum of eight limit-digest DNA fragments of 45–145 base pairs, there are likely to be some differences between individual nu bodies. These differences would determine the variable points at which nuclease is hydrolyzing in the limit-digestion conditions.

Sahasrabuddhe & Van Holde (338) have shown that with mild nuclease digestion chromatin falls apart into fairly homogeneous particles that sedimented at 11–12S, and (by equilibrium sedimentation) have a molecular weight of about 176,000 (remarkably consistent with the 160,000 dalton estimate of the size of a nu body obtained from electron microscopy). These particles were therefore almost spherical and it is now likely that they correspond to nu bodies (456). When these particles were digested with trypsin, there was a rapid shift in sedimentation to about 5S. The shift was stable over long periods of trypsin digestion and the resulting particles had a molecular weight of about 158,000, indicating that trypsin did not remove very much protein. The rather meager loss of material with trypsin digestion allowed the interpretation that the altered sedimentation was due to an *unfolding* of the nu body.

These observations have been extended by Weintraub & Van Lente (339), who have further analyzed the digestion of chromosomal histones by trypsin. They have shown that exhaustive digestion of chick erythrocyte chromatin results in the complete digestion of histones 1 and 5 and the cleavage of only about 20–30 amino acid residues from the positively charged N-terminal parts of histones 3, 4, 2b, and (possibly) 2a. The resistance of C-terminal histone fragments (from 80–100 amino acids in length) seems to be due to their mutual interaction, since removal of the histones from the DNA by 2 M NaCl fails to alter the trypsin digestion pattern. The sensitivity of the basic N-terminal regions was interpreted in terms of a model in which these basic "arms" extend from a trypsin-resistant histone complex and define the binding sites for the folding of a DNA fiber about this complex. This type of structure is consistent with the nu bodies observed in the electron microscope. When trypsin-digested chromatin was treated with nuclease (339), the resistant DNA contained some, but not all of the eight DNA fragments characteristic of the staphylococcal nuclease limit-digest. The authors interpreted this as indicating that histone N terminals protected the DNA corresponding to some DNA fragments, whereas histone C terminals protected the DNA corresponding to other DNA fragments. In addition, they postulated that histones could fold the chromosome by crosslinking the DNA regions that corresponded to the different DNA fragments.

Subsequent experiments, repeating and extending the work of Sahasrabuddhe & Van Holde (338), showed that the transition from the 11S nuclease particles to the 5S trypsin-treated nuclease particle occurred when the N termini of histones 3 and 4 were being digested (340). If the 11S particles are generated by slightly higher nuclease concentrations than used by Van Holde (so that the chromatin is digested to the limit with nuclease) and then trypsinized and run on 6% acrylamide gels, pH 7.8, the resulting 5S particles are resolved by the gel into about eight extremely discrete subparticles which are visualized as bands when stained with ethidium bromide. In addition, five of these bands stained with Coomasie blue, while three or four did not. Clearly, trypsinization of the 11S particles generated by extensive nuclease treatment results in the production of protein-free and protein-associated DNA fragments of defined size.

These observations have specific implications for the internal structure of the 11S nuclease particles, or nu bodies. Since trypsinization results in the cleavage of 20–30 amino acids from the N-terminal ends of the arginine and slightly lysine-rich histones, it is likely that the protein-free DNA fragments released by trypsin treatment of the nuclease particles were originally bound to amino-terminal parts of the histones. The protein associated with the DNA fragments released after these procedures were shown by peptide analysis to be the trypsin-resistant histone C-terminal segments. The model derived from this data is that the general structure of a nu body consists of a group of histones which fold the DNA in the nu body by binding to a stretch of DNA of about 80 base pairs with their N termini and another stretch of about 80 base pairs with their C termini, the intervening DNA defining a nuclease-sensitive site.

To summarize, by cleaving histones at their N termini, trypsin in effect splits the 11S nuclear particles into two subparticles, one of which contains protein and the other does not. The evidence that histones crosslink the DNA supports previous work and speculations from a number of laboratories.

Are the individual nu bodies homogeneous? The finding from peptide analysis that most subchromosomal particles obtained by nuclease and trypsin treatment of chromatin contain trypsin-resistant fragments from all four major histones would argue that these particles are qualitatively homogeneous (340). Why then are so many DNA fragments generated by nuclease treatment, and why so many sub-chromosomal particles? Several explanations are possible. One is that there are quantitative differences in the ratios of the different histones in each of the nu bodies. There is some evidence for this, and the finding of one subchromosomal particle (generated by nuclease and trypsin treatment of chromatin and isolated by gel electrophoresis) with only histones 3 and 4 further supports this idea (340). A second possibility is that the ratio of the histones is the same, but their conformations are different. This could be a consequence of 1. their association with particular DNA sequences, 2. their entrapment in specific metastable energy states, or 3. their selective modification. The first explanation is unlikely because reconstitution experiments generate the same limit-digest DNA bands with staphylococcal nuclease whether the DNA is from calf thymus or phage λ (320; T. Maniatis, personal communication). Also, three of the eight DNA bands of the nuclease limit-digest (all that were examined) hybridize with tracer amounts of labeled cellular DNA with kinetics identical to those obtained when the reaction is driven by total cellular DNA, indicating that they contain a random collection of DNA sequences (H. Weintraub, unpublished observation). In addition, three of the remaining bands contain a pattern of pyrimidine tracts which are indistinguishable from total cellular DNA (H. Weintraub, unpublished observation). There is no evidence for or against the second two possibilities suggested to explain the number of subchromosomal particles generated by nuclease; however, since reconstitution experiments appear to generate a structure very similar to the original one, both of these possibilities can be tested.

What is the structure of the DNA in the nu bodies? Assuming that nu bodies are equivalent to nuclease particles, the work of Sahasrabuddhe & Van Holde (338) indicates that they have a circular dichroism (CD) spectra very similar to C-type DNA. In this context, it is worth discussing the proposals made by Crick (341). He suggested that histones crosslinked the DNA and that this produced a strain on the double helix conformation, resulting in the unpairing of double helices. The unpaired bases could serve as specific recognition signals. These predictions are consistent with much of the data emerging about chromosome structure. Thus, there is growing evidence that the histones fold the chromosome by crosslinking DNA within a given nu body, and the recent CD data (338) indicate that as a consequence of this crosslinking the conformation of the DNA is altered. This is not surprising from what is known about the physical chemistry of DNA because the type of bending needed to package about 150 base pairs of DNA into a particle the size of a nu body would probably require distortion of the normal base pairing schemes. Since we do not know the path taken by the DNA within or around the nu body, it is impossible to say just how much of the double helix should be distorted. As far as we are aware, no one has shown that nu bodies are sensitive to single stranded nucleases; however, this is a difficult experiment because the single stranded regions may not be accessible and it is possible that the denatured DNA relaxes as soon as it is nicked.

One prediction made by Crick (341) will clearly be very difficult to prove, i.e.

that the single stranded regions are specific recognition signals. The evidence, previously mentioned, indicating that histones are bound to chromosomal DNA independent of base sequence, argues against the possibility that histones fold the DNA to expose specific signals. On the other hand, most of this work was done with bulk chromosomal DNA and it is clear that one must look at particular genes. In this respect, it is extremely interesting that Axel, Cedar & Felsenfeld (10, 446) have shown that a reverse transcriptase probe to globin mRNA hybridized only partially to DNA obtained from nuclease-sensitive parts of the genome. Addition of nuclease-insensitive sequences led to a protection of the entire probe. These experiments are extremely important and clearly indicate that along the globin gene, some portions of the globin sequence in every reticulocyte are covered and other portions are exposed. Whether the same results would be obtained with other cell types, whether the exposed sequences in the globin gene are single stranded, whether they are involved in control (they may not be because the probe is detecting largely those sequences common to the globin structural gene), and whether they are generated by a base sequence specificity peculiar to histones, certain combinations of histones, or combinations of histones and nonhistone chromosomal proteins remain to be determined.

Histone: Histone Interactions

In general, the experiments described thus far indicate that specific histone clusters are periodically located along the chromosome fiber and that these clusters are intimately involved in folding chromosomal DNA. The nature of the histone interactions that may be responsible for generating these clusters has been extensively studied in solution by Isenberg and his colleagues over the past few years (342, 343, 447, 448). Using fluorescence anisotropy (to detect the rotational mobility of the transition moments of tyrosine residues) and circular dichroism, they have shown that specific interactions occur between the histones. Very strong associations exist between histones 3 and 4, 2b and 2a, and 2b and 4. For many of these interactions, there is an increased amount of α-helix produced upon complex formation. Weaker associations also exist between 2a and 4, and it would not be surprising if many other types of interactions could occur in different types of environments. This work extends earlier studies (344, 449) showing that histone 2b undergoes a progressive association with increasing salt concentration. The nuclear magnetic resonance (NMR) studies of Bradbury and colleagues (40, 345), which demonstrated salt-induced changes in resonance due to the altered mobility of amino acids in particular regions of histones 4, 1, and 2b, have recently been reviewed and we will mention only that they support the growing consensus that histone:histone interactions are important for establishing the folding parameters of chromosomal DNA.

Additional evidence comes from the work of Kelley (346), who showed that histones 2b and 2a can form a 1:1 complex. In contrast, using the crosslinking reagent dimethylsuberimidate, Kornberg & Thomas (329) have shown that H2b and H2a can form polymers of the form $(2b-2a)_n$. In addition, they have demonstrated by both sedimentation analysis and crosslinking that histones 3 and 4 can form a specific association in solution, probably as a tetramer. The same general conclusions from similar types of studies come from the work of Rouark et al (347), who have additionally presented data indicating that at low concentrations the 3–4 complex may be a dimer. Both groups have questioned the validity of information obtained by using histones that have been extracted and purified with acid; however, other types of experiments indicate that the acid denaturation of histone is reversible (e.g. 458).

To summarize, studies with isolated histones in solution show that a great number of interactions are possible; given the appropriate ionic condition and methods of histone preparation, a great many more will no doubt be observed. The primary question of course is which of these interactions occur in the chromosome?

With respect to the types of histone:histone interactions occurring in chromatin, several observations are of importance. It has been shown that most particles obtained after nuclease and trypsin treatment of chromatin contain the C-terminal fragments from all four major histones (340). The stoichiometry was such that depending on the particle, anywhere from 6–10 histones were bound to DNA fragments between 60 and 140 base pairs. One particle, however, appeared to contain only histones 3 and 4. A second type of observation involved the use of crosslinking reagents to produce high molecular weight protein products from chromatin. Kornberg & Thomas (329) have reported the generation of oligomers up to pentamers using dimethylsuberimidate, but have not yet identified the specific crosslinked products.

In addition, Olins & Wright (348) have used glutaraldehyde to crosslink chick erythrocyte chromatin. They found that at low concentrations of glutaraldehyde several polymeric bands appeared on their SDS gels. By their amino acid composition, the authors were able to infer that these crosslinked products were probably derived from the red cell-specific histone 5. From their apparent molecular weight, a minimum of up to eight molecules could be crosslinked and consequently, it was suggested that histone 5 appeared in clusters either because of its contiguous alignment along a particular stretch of DNA or because of the tertiary folding of the chromosome fiber. What is disturbing about the glutaraldehyde data and what appears to be emerging as a general phenomenon with reagents that crosslink chromosomal histones is the fact that only a small percentage (5–10%) of a particular histone ever gets crosslinked into a resolvable product. When attempts are made to increase this percentage by increasing the concentration of reagent, then essentially all the material fails to enter the gel. Thus, it may prove difficult to show that a crosslinked product representing about 10% of the protein is indeed representative of a general structure and not a specific type of variant.

Recently, Martinson & McCarthy (349), using the tyrosine-specific crosslinking agent, tetranitromethane, have presented evidence consistent with the notion that all four major histones are intimately associated in chromatin. They demonstrated that the protein product of tetranitromethane treatment of chromatin was a dimer. By reconstituting DNA with specific histones, they showed that only histones 2b, 4, and 2a were required to produce a crosslinked product with the same electrophoretic properties as that obtained from chromatin. After reconstituting DNA with specifically labeled histone species, they were able to identify the crosslinked product as a dimer of 2b and 4 and argued that the crosslinked product from native chromatin had the same composition. Since 2a was required for the proper reconstitution of 2b and 4, it was proposed that this histone was also in the complex, but because of the rigid specificity of the crosslinking reagent (tetra-nitromethane requires intimate contact between the activated tyrosine and a neighboring molecule) 2a was not crosslinked into the product. Relying on the strong evidence from the solution chemistry which indicated that 3 and 4 were tightly complexed, they interpret their studies in terms of a model where all four histones interact to produce a basic chromosomal unit. As with the other crosslinking experiments, only about 10% of the histones appear in the product. J. F. Jackson, F. Van Lente, and H. Weintraub (unpublished observations) have observed dimer products, H4 + H2b and H2a + H2b, on crosslinking with formaldehyde or glutaraldehyde. To summarize, the X-ray evidence, the crosslinking data, and the analysis of the particles released after digestion with nuclease and trypsin support the notion that all four major histones are involved as a group in generating the final chromosomal structure.

Histone Accessibility

Specific labels have been used to map the accessibility of specific histone residues. This approach is extremely valuable with histones because their sequences have been so well characterized. One of the first attempts in this direction was by Simpson (350), who showed that at very high concentrations of acetic anhydride, only about 25% of the potential lysine residues in chromatin become acetylated. By analyzing the electrophoretic mobility of the histones after treatment, he was able to show that all histone species were accessible to the probe. No attempt was made, however, to localize the sites where acetylation was occurring. In vivo, of course, the acetylating enzymes in chromatin appear to be very specific for histone N-terminal regions. This work was extended by Malchy & Kaplan (351), who used an internal standard to compare the acetylation of particular histones. In essence, they were able to show that at rather low concentrations of acetic anhydride their marker was much more reactive and they concluded that all the potential lysine residues in chromatin were essentially inaccessible to acetylation. This is in contrast to the pH titration studies (451) which reveal that about 25% of the lysines in chromatin are freely titratable.

A second type of label available to measure accessibility is iodination. Using ^{125}I to label histone tyrosines, F. Van Lente (unpublished observations) has shown that in chromatin no tyrosine in H2a is accessible, even at very high concentrations of ^{125}I. When the chromatin is treated with increasing concentrations of urea, H2a readily becomes iodinated. Van Lente has also shown that histone 3, although iodinated in chromatin, is much more available when free in solution. Histone 3 is the only histone with tyrosine in its N-terminal region and hence these tyrosines are intimately associated with basic residues. By mapping all the iodinatable peptides from all the histones, Van Lente has demonstrated that the histone 3 peptides become relatively more accessible to ^{125}I with increasing ionic strength. The data are used to support the notion that histone 3 N-terminal segments are involved in DNA binding.

Potentially the most useful histone probes will be antibodies of different specificity. Bustin (352) has used antibodies made against specific histones and has observed that antigenic determinants of H1 and H2b are more available in chromatin to interact with homologous antibody than those in H3 and H4, and that determinants of H2a are the least available. Most striking was the additional observation that antibodies bound to any one histone class failed to interfere with subsequent antibody binding to a second histone class. The implication of these findings is that antibody determinants from one histone class are spatially separated from those of another. Hopefully, it will be possible to show where the antibody determinants reside in the individual histone molecules.

Conclusions

It is increasingly clear that histone:histone interactions are determining the folding parameters of chromosomal DNA. The basic unit involved in these interactions is probably the nu body. It is likely that specific histone complexes form the core of this structure and that positively charged N-terminal arms extend from this core, defining the path taken by the folding DNA fiber, although it is also clear that other regions toward the C termini can bind DNA very tightly. We do not really know how homogeneous nu bodies are, we do not know the conformation and path of the DNA within (or around) these structures, and we know nothing about the structures between these nu bodies or what determines their spacing. Most important, we do not even know whether it is possible to expect solutions of histones to behave in the same manner as histones in chromatin. What role does DNA have in forcing histones to form the "proper" associations? What function might histone modification play in generating these conformations? Finally, what function, if any, might be served by nonequilibrium associations between the various histones?

Clearly, there are many questions to be answered, but perhaps the ones that are most outstanding involve the possible differences in structure between active and inactive chromatin. There is a great deal of evidence indicating that differences do occur; however, further characterization of these differences and, most importantly, an explanation for how these differences are generated are required. Finally, we should emphasize that our attempts in this section have been focused on events of the past few years. Consequently, we have not discussed the less recent work involving X-ray diffraction, cation binding to DNA, flow dichroism, circular dichroism, viscosity, and sedimentation. All these data are documented in the reviews cited earlier in this section.

TEMPLATE ACTIVITY

Histones as General Repressors

If RNA polymerase is specifically restricted by the conformation of DNA in chromatin, this implies that tissue-specific genes are accessible in the chromatin of one tissue and inaccessible in that of another. Transcription from isolated duck chromatin from reticulocytes and from liver has been examined. The chromatin was transcribed with RNA polymerase from *Escherichia coli* and the RNA isolated and hybridized to a labeled reverse transcript probe for globin sequences. The results showed that the addition of RNA polymerase and nucleoside triphosphates to reticulocyte chromatin led to the production of RNA that significantly increased protection from a single stranded nuclease for the labeled probe. In contrast, addition of RNA polymerase and nucleoside triphosphates to the liver chromatin did not lead to an increase in RNA sequences providing probe protection (9, 10). The interpretation of these findings is that the structure of the globin DNA sequences allowed transcription in reticulocyte chromatin but not in liver chromatin. The same results have been obtained in studies of erythropoietic and brain chromatin of mouse (11, 12). How these differences are generated during development and what other factors may amplify this differential signal in the reticulocyte are far from being worked out. Nevertheless, a basic difference between cells that make hemoglobin and cells that do not is that the "accessibility" of the globin sequences is different. This basic approach has recently been extended to the transcription of chromatin from viral-infected myeloblasts (353) and SV40-transformed cells (468, 469).

The limited transcription potential of eukaryotic chromatin with RNA polymerase appears to be largely a consequence of the association of chromosomal DNA with histones. The removal of these proteins from chromatin with any of several reagents leads to an increase in template activity (354–356). There is some indication that histone 1 plays a special role in this generalized repression because its removal has been reported in some instances to result in a large increase in transcription (357, 358). It is not at all clear how the histones are blocking transcription. Recent work by Cedar & Felsenfeld (359) shows that *E. coli* RNA polymerase binds to about 10% the number of sites in chromatin as in DNA. The *E. coli* enzyme elongates RNA chains using a chromatin template at about 33% the rate observed with purified DNA. Assuming the endogenous enzyme works at least approximately in the same way as the *E. coli* enzyme, these experiments demonstrate that the major restriction imposed on the DNA by the histones involves the blocking of sites for RNA chain initiation. There is no indication from any published work that the polymerase, once properly initiated, cannot transcribe histone-associated DNA. Whether the histones actually come off or slide during this process remains to be determined. There appears to be no a priori reason to assume that the histones must sterically block the polymerase, although they certainly could do so, for example, by actually crosslinking the individual strands of the double helix. Again, there is no evidence for or against this possibility.

Chromatin Fractionation

Biochemical attempts to isolate so-called active chromatin presuppose that such a fraction is structurally different from the bulk of the chromatin. Such an "active" subfraction can readily be evaluated: it should contain a unique subset of DNA sequences; it should contain sequences for genes that are known to be active, for example, rRNA, tRNA, globin (if the chromatin is from erythroblasts), and histone genes (if the chromatin is from S phase cells). Somewhat less definitive assays include copurification of endogenous RNA polymerase or nascent RNA chains, assays of template activity with exogenous RNA polymerase, etc.

The earliest fractionation attempts used differential centrifugation of sheared chromatin on sucrose gradients (360, 361). A similar approach has been adopted by McCarthy et al (362, 443, 444) and others (363) who have separated active fractions on the basis of their hydrodynamic properties, either as being preferentially excluded by agarose columns or as slower sedimenting in sucrose gradients. Both fractionations succeed in isolating a subfraction of sheared chromatin that is more extended than the bulk chromatin.

A second type of fractionation takes advantage of a possible difference in charge between active and inactive chromatin. The fractionation involves chromatography on epichlorohydrin-coupled tris(hydroxymethyl)amino methane (ECTHAM) cellulose (364) where the putative active fraction is the last to be eluted, indicating that it is probably more negatively charged than the bulk chromatin. Active chromatin isolated in this manner sediments more slowly than bulk chromatin and is enriched in low melting (relatively histone-free) sequences. Moreover, it is highly enriched in *E. coli* RNA polymerase binding sites, containing approximately half as many as protein-free DNA (365).

A third type of fractionation involves very limited cutting of chromatin with DNase II followed by a differential precipitation in 2 mM $MgCl_2$ (366–368). The resulting soluble material, the active fraction, contains about 11% of the total chromatin. It fractionates together with nascent RNA chains and has a high relative template activity with exogenous polymerase. The DNA of the active fraction hybridizes with a kinetic complexity that is about one tenth that of total DNA, which indicates that this fraction contains a unique subset of single copy DNA sequences (and probably a unique subset of middle repetitive DNA as well). The observation that bulk cellular RNA preferentially hybridizes with DNA from this fraction further supports the notion that this fraction is indeed active in in vivo transcription.

The bulk of the work on chromatin fractionation has indicated that active chromatin is relatively enriched in nonhistone chromosomal proteins and deficient in histones, particularly H1. Specific major NHC proteins have been suggested to be present exclusively in the euchromatin (362, 368, 369). Active chromatin is consistently found to melt at lower temperatures, implying less stabilization of the DNA double helix by proteins, presumably because of the depletion of histone. McCarthy and his colleagues have fractionated chromatin by thermal chromatography on hydroxyapatite. Here the lowest melting chromatin fraction has been shown to be enriched for DNA sequences that hybridized to homologous cellular RNA (370). This reinforces the idea of a structural basis of gene activity that can be exploited for chromatin fractionation. It will be important to determine whether or not the chromatin of the active fraction is in the nu body form, and if so, whether any differences in frequency, spacing, or substructure exist.

Much of the present data are consistent with the idea that the histones can block the accessibility of the polymerase to initiation sites, but only marginally affect the accessibility of the polymerase to the DNA bases once the chain is initiated. Thus it may be not the accessibility of the polymerase to the bases that is being hampered, but some other type of block preventing initiation. What immediately comes to mind is the conformation of the DNA, especially since it is increasingly clear that proper initiation with both the *E. coli* RNA polymerase and mammalian RNA polymerase requires a particular promoter configuration (371–374, 445). It is possible that by folding the DNA, the histones alter the structure of potential promoter sites and hence initiation fails to occur. Assuming that all histones participate in the folding of the DNA chain, then the removal of any one would presumably destroy the structure and allow the polymerase to initiate properly. Indeed, experiments involving selective histone removal have led Smart & Bonner (355, 356) to precisely this conclusion. It remains to be discovered how desired promoters are identified and protected. The nonhistone chromosomal proteins may play such a role. In some instances histones or similar basic proteins have been reported to have a stimulatory effect on transcriptional activity, perhaps implying the removal of nonpromoter binding sites in competition for the RNA polymerase (375, 376).

An indication of broad levels of structural organization related to activity has been obtained from studies of metaphase chromosomes. Unique and reproducible chromosome banding patterns can be obtained by treating metaphase chromosomes with a number of reagents and dyes, e.g. quinacrine (377, 378). The banding patterns appear to reflect the state and accessibility of the DNA, and thus to be a consequence of differential protein-DNA interactions along the chromatids. Experiments by Gottesfeld et al (379) show that the fluorescence of quinacrine associated with DNA in active chromatin (euchromatin) is quenched very effectively relative to that in inactive chromatin (heterochromatin), and suggest that this difference is due to the difference in DNA conformation caused by the associated proteins. Thus the banding pattern may correlate roughly with those structural features that help to define heterochromatin and euchromatin. It has further been suggested that certain types of heterochromatin that stain differently, such as the N bands, may do so as a consequence of a specific association of NHC proteins with DNA (380).

Model Systems

It is clear that the animal viruses are going to be helpful in future analyses of chromatin structure and activity. As mentioned previously, SV40 viruses contain all histones with the possible exception of histone 1. The SV40 complex can exist in two major forms, as the packaged virion released from the cell and as the replicating and transcribing nucleoprotein complex found in the nucleus. The former may be a prototype for inactive chromatin and the latter a prototype for active chromatin. Preliminary studies by B. Polisky and B. J. McCarthy (personal communication), using the packaged particle, have shown that all six DNA fragments produced by the Hin 3 restriction endonuclease appear when the particle is treated with the enzyme; however, only about 30% of the total DNA is in these fragments, the remaining DNA being only partially digested. The data imply that, as in chromatin, there is a random arrangement of "open" and "closed" or protected regions of approximately equal lengths along the SV40 DNA. A given Hin digestion site has a particular probability of being either open or closed. This probability depends on a number of factors; the position of the Hin sites relative to each other, the average distance between closed regions, and the average length of a closed and an open region, to mention only a few. The actual analysis of the data from the Hin cutting will no doubt yield a great deal of information about a number of packaging parameters. Similar conclusions have been reached from studies with the R_1 restriction enzyme. Finally, it has been observed that approximately 50% of the nucleoprotein complex (the replicating form of SV40) is sensitive to staphylococcal nuclease, a nonspecific enzyme (J. Beltz, A. J. Levine, and H. Weintraub, unpublished observations). Moreover, the same limit-digest sized DNA bands are obtained as were found from eukaryotic chromatin. These experiments indicate that the nucleoprotein complex and virion are good structural models of chromatin. Many aspects of SV40 replication are also quite similar to those of eukaryotic chromatin. In addition, the in vivo transcription behavior of the virus is very well worked out. Taken as a whole, it is likely that SV40 will be a very useful prototype for active genes in higher cells.

CONCLUSIONS

It is hoped that just as the solving of the DNA structure yielded insights into the way the genetic information was replicated and transcribed, so the solving of the structure of higher chromosomes will be equally informative about the way chromosomes are replicated and information transfer is controlled during eukaryotic growth and differentiation. Our present knowledge of chromatin structure is derived from the work on primary structure and chemical physical characterization of the histones. This information is being used to determine the way in which the histones interact with each other and fold the DNA; in the future, additional work on the histones will undoubtedly provide valuable insights into the evolutionary changes in chromosome structure and control processes. Biological evidence implies that the nonhistone chromosomal proteins play an equally important role. It will be necessary to achieve an equivalent chemical analysis of these proteins if fruitful experiments are to be designed. The ultimate goal, that of understanding gene expression in terms of chromatin structure, appears attainable, but much work remains to be done.

NOTE ADDED IN PROOF   D'Anna & Isenberg (460) have extended and summarized their results showing that specific interactions between all pairs of histones (excluding histone 1) can occur in solution. The energies of some of these interactions are enormous (for example, the binding coefficient of H3 to H4 is on the order of 10^{21} M^{-3}), while some of these interactions are comparatively weak (for example, the binding coefficient of H3 to H2a is about 10^{-6} M). Several very important papers concerning nuclease digestion of chromatin have also appeared recently. Burgoyne, Hewish & Mobbs (461) have shown that the endogenous nuclease of rat liver cuts chromatin into single stranded multiples of a regular repeat of about 55 to 60 bases. In addition, Noll (462) has shown that when nuclei are digested with pancreatic DNase and the resistent DNA fragments are analyzed under denaturing conditions, a series of bands is observed between about 40 and 200 bases. Each successive band occurs as an integer multiple of 10 base pairs. These very striking results were interpreted in terms of a model where pancreatic DNase nicks exposed DNA regions, which occur every 10 base pairs (or one turn of a double helix) on the outside of the nu body. Finally, the experiments of Honda, Baillie & Candido (463) have shown that during the development of trout sperm there is an inverse relationship between the percentage of chromatin digested into 11S particles by staphylococcal nuclease and the amount of DNA associated with protamine. The nuclease particles obtained contain only histones and negligible quantities of protamine.

In a very novel approach, Mirzabekov & Melnikova (464) have used H^3-dimethylsulfate as a probe of DNA accessibility in chromatin. This reagent methylates the N-7 atom of guanine (located in the major DNA groove), the N-3 atom of adenine (in the minor groove), and the N-1 atom of adenine (available only in single stranded regions). The results suggest that 1. the minor groove of DNA is freely available to methylation in chromatin, 2. the major groove is only marginally protected by histones, and 3. less than 0.5% of the DNA is in a single stranded conformation. The surprising level of accessibility of both DNA grooves in chromatin is consistent with the idea that the DNA may be on the outside of the nu body (332, 462). All these recent data support the numerous and similar "beads on a string" models of chromatin structure that have recently been proposed.

Wilson et al (470) have recently reported that the human genome contains 10–20 copies of each histone gene. This is the lowest reiteration frequency reported to date.

The characteristic electron microscope pictures of chromatin fibers showing nu body structure have recently been obtained from preparations of *Drosophila* polytene chromosomes (C. L. F. Woodcock, J. P. Safer, and J. E. Stanchfield, personal communication). This is of particular interest in that the polytene chromosomes are almost totally euchromatin, the heterochromatin having been severely under-replicated during polytenization. The nucleoplasmic form of SV40 has similarly been visualized as a "minichromosome" with one nu body per 200 base pairs of DNA (472).

Using a filter binding assay, Renz has shown that histone 1 prefers to bind to purified native mammalian DNA relative to *E. coli* DNA. The data indicate that there are preferential binding sites on the mammalian DNA which appear to be distributed at intervals of approximately 0.5 to 2×10^6 daltons of DNA (474).

J. Thomas and R. D. Kornberg (personal communication) have recently shown that dimethylsuberimidate can produce crosslinked products up to octamers from histones in solution. In chromatin at pH 9, a similar pattern of crosslinked histones is produced, while at lower pH, higher order crosslinked products appear. In addition, using a reversible crosslinking reagent, they have been able to identify the histones present in several different dimer products. These data support the proposal of a repeating chromosomal unit of eight of the smaller histones, H2a, H2b, H3, and H4.

ACKNOWLEDGMENTS

We would like to thank the many scientists who have provided us with copies of work in press and allowed us to quote their unpublished observations, as well as those who have helped by criticizing the manuscript. Work in our laboratories is supported by the Jane Coffin Childs Memorial Fund for Medical Research and USPHS Grant R01 GM 20779-01 from the National Institutes of Health (S.C.R.E.) and by the American Cancer Society and Grant GB 40148 from the National Science Foundation (H.W.). We thank L. Silver and N. Niles for technical assistance.

Literature Cited

1. Grouse, L., Chilton, M. D., McCarthy, B. J. 1972. *Biochemistry* 11:798–805
2. Brown, I. R., Church, R. B. 1972. *Dev. Biol.* 29:73–84
3. Palmiter, R. D. 1973. *J. Biol. Chem.* 248:8260–70
4. Chan, L., Means, A. R., O'Malley, B. W. 1973. *Proc. Nat. Acad. Sci. USA* 70: 1870–74
5. Davidson, E. H., Britten, R. J. 1973. *Quart. Rev. Biol.* 48:565–613
6. Paul, J., Gilmour, R. S. 1968. *J. Mol. Biol.* 34:305–16
7. Bekhor, I., Kung, G. M., Bonner, J. 1969. *J. Mol. Biol.* 39:351–64
8. Smith, K. D., Church, R. B., McCarthy, B. J. 1969. *Biochemistry* 8:4271–77
9. Axel, R., Cedar, H., Felsenfeld, G. 1973. *Proc. Nat. Acad. Sci. USA* 70:2029–32
10. Axel, R., Cedar, H., Felsenfeld, G. 1973. *Cold Spring Harbor Symp. Quant. Biol.* 38:773–83
11. Gilmour, R. S., Paul, J. 1973. *Proc. Nat. Acad. Sci. USA* 70:3440–42
12. Paul, J., Gilmour, R. S., Affara, N., Birnie, G., Harrison, P., Hell, A., Humphries, S., Windass, J., Young, B. 1973. *Cold Spring Harbor Symp. Quant. Biol.* 38:885–90
13. Steggles, A. W., Wilson, G. N., Kantor, J. A. 1974. *Proc. Nat. Acad. Sci. USA* 71:1219–23
14. Gilbert, W., Maizels, N., Maxam, A. 1973. *Cold Spring Harbor Symp. Quant. Biol.* 38:845–56
15. Maniatis, T., Ptashne, M., Maurer, R. 1973. *Cold Spring Harbor Symp. Quant. Biol.* 38:857–68
16. Maniatis, T., Ptashne, M., Barrell, B. G., Donelson, J. 1974. *Nature* 250:394–97
17. Lyon, M. F. 1961. *Nature* 190:372–73
18. Lewis, E. B. 1950. *Advan Genet.* 3:73–115
19. Baker, W. K. 1968. *Advan. Genet.* 14:133–69
20. Berendes, H. D. 1971. *Symp. Soc. Exp. Biol.* 25:145–61
21. Berendes, H. D. 1973. *Int. Rev. Cyto.* 35:61–116
22. Kossel, A. 1884. *Z. Physiol. Chem.* 8:511–15
23. DeLange, R. J., Smith, E. L. 1975. *Ciba Found. Symp.* 28:59–70
24. Panyim, S., Chalkley, R., Spiker, S., Oliver, D. 1970. *Biochim. Biophys. Acta* 214:216–21
25. Panyim, S., Bilek, D., Chalkley, R. 1971. *J. Biol. Chem.* 246:4206–15
26. Panyim, S., Sommer, K. R., Chalkley, R. 1971. *Biochemistry* 10:3911–17
27. Palau, J., Padros, E. 1972. *FEBS Lett.* 27:157–60
28. Kinkade, J. M. Jr., Cole, R. D. 1966. *J. Biol. Chem.* 241:5790–97
29. Kinkade, J. M. Jr. 1968. *Fed. Proc.* 27:773
30. Bustin, M., Cole, R. D. 1968. *J. Biol. Chem.* 243:4500–5
31. Bustin, M., Cole, R. D. 1969. *J. Biol. Chem.* 244:5286–90
32. Evans, K., Hohmann, P., Cole, R. D. 1970. *Biochim. Biophys. Acta* 221:128–31
33. Stellwagen, R. H., Cole, R. D. 1969. *Ann. Rev. Biochem.* 38:951–90
34. Rall, S. C., Cole, R. D. 1971. *J. Biol. Chem.* 246:7175–90
35. Langan, T. A. 1971. *Ann. NY Acad. Sci.* 185:166–80
36. Langan, T. A., Rall, S. C., Cole, R. D. 1971. *J. Biol. Chem.* 246:1942–44
37. Subirana, J. A., Palau, J., Cozcolluela, C. Ruiz-Carrillo, A. 1970. *Nature* 228:992–93
38. Cohen, L. H., Gotchel, B. V. 1971. *J. Biol. Chem.* 246:1841–48
39. Sherod, D., Johnson, G., Chalkley, R. 1974. *J. Biol. Chem.* 249:3923–31
40. Bradbury, E. M., Crane-Robinson, C. 1971. *Histones and Nucleohistones,* 85–134. New York: Plenum
41. Stedman, E., Stedman, E. 1950. *Nature* 166:780–81
42. Greenaway, P. J., Murray, K. 1971. *Nature New Biol.* 229:233–38
43. Garel, A. et al 1975. *FEBS Lett.* 50:195–99
44. Bradbury, E. M., Crane-Robinson C., Johns, E. W. 1972. *Nature New Biol.* 238:262–64
45. Billett, M. A., Hindley, J. 1972. *Eur. J. Biochem.* 28:451–62
46. Moss, B. A., Joyce, W. G., Ingram, V. M. 1973. *J. Biol. Chem.* 248:1025–31
47. Brasch, K., Adams, G. H. M., Neelin, J. M. 1974. *J. Cell Sci.* 15:659–77
48. Sheridan, W. F., Stern, H. 1967. *Exp. Cell Res.* 45:323–35
49. Strokov, A., Bogdanov, Yu. F., Reznikova, S.A. 1973. *Chromosoma* 43:247–60
49a. Liapunovo, N. A., Babadjanian, D. P. 1973. *Chromosoma* 40:387–99
50. Ando, T., Watanabe, S. 1969. *Int. J. Protein Res.* 1:221–24
51. Subirana, J. A., Cozcolluela, C., Palau, J., Unzeta, M. 1973. *Biochim. Biophys. Acta* 317:364–79
52. Bloch, D. P. 1969. *Genet. Suppl.* 61:93–111
53. Hnilica, L. S., McClure, M. E., Spelsberg,

T. C. See Ref. 40, 187–240
54. Phillips, D. M. P. See Ref. 40, 47–83
55. Marushige, Y., Marushige, K. 1974. *Biochim. Biophys. Acta* 340:498–508
56. Marushige, K., Dixon, G. H. 1971. *J. Biol. Chem.* 246:5799–5805
57. Dixon, G. H. 1972. *Karolinska Symp.* 5:130–54
58. Dixon, G. H. et al 1975. *Ciba Found. Symp.* 28:229–50
59. Kischer, C. W., Hnilica, L. S. 1969. *Exp. Cell Res.* 48:424–30
60. Stollar, B. D., Ward, M. 1970. *J. Biol. Chem.* 245:1261–66
61. Bustin, M., Stollar, B. D. 1973. *J. Biol. Chem.* 248:3506–10
62. Hekman, A., Sluyser, M. 1973. *Biochim. Biophys. Acta* 295:613–20
63. Sluyser, M., Bustin, M. 1974. *J. Biol. Chem.* 249:2507–11
64. Wigle, D. T., Dixon, G. H. 1971. *J. Biol. Chem.* 246:5636–44
65. Huntley, G. H., Dixon, G. H. 1972. *J. Biol. Chem.* 247:4916–19
66. Thaler, M. M., Cox, M. C. L., Villee, C. A. 1970. *J. Biol. Chem.* 245:1479–83
67. Byrd, E. W. Jr., Kasinsky, H. E. 1973. *Biochim. Biophys. Acta* 331:430–41
68. Farquhar, M. N., McCarthy, B. J. 1973. *Biochem. Biophys. Res. Commun.* 53:515–22
69. Seale, R. L., Aronson, A. I. 1973. *J. Mol. Biol.* 75:633–45
70. Skoultchi, A., Gross, P. R. 1973. *Proc. Nat. Acad. Sci. USA* 70:2840–44
71. Gross, K. W., Jacobs-Lorena, M., Baglioni, C., Gross, P. R. 1973. *Proc. Nat. Acad. Sci. USA* 70:2614–18
72. Gross, P. R., Gross, K. W. 1973. *The Role of RNA in Reproduction and Development,* 4–10. New York: Elsevier
73. Destree, O. H. J., d'Adelhart Toorop, H. A., Charles, R. 1973. *Cell Differ.* 2:229–42
74. Ruderman, J. V., Baglioni, C., Gross, P. R. 1974. *Nature* 247:36–38
75. Ruderman, J. V., Gross, P. R. 1974. *Dev. Biol.* 36:286–98
76. Easton, D. P., Chamberlain, J. P., Whiteley, A. H., Whiteley, H. R. 1974. *Biochem. Biophys. Res. Commun.* 57:513–19
77. Ruderman, J. V., Gross, P. R. 1973. *J. Cell Biol.* 59:296a
78. Fambrough, D. M., Fujimura, F., Bonner, J. 1968. *Biochemistry* 7:575–84
79. Elgin, S. C. R., Hood, L. E. 1973. *Biochemistry* 12:4984–91
80. Stellwagen, R. H., Cole, R. D. 1968. *J. Biol. Chem.* 243:4456–62
81. Stellwagen, R. H., Cole, R. D. 1969. *J. Biol. Chem.* 244:4878–87
82. Hohmann, P., Cole, R. D. 1971. *J. Mol. Biol.* 58:533–40
83. Raaf, J., Bonner, J. 1968. *Arch. Biochem. Biophys.* 125:567–79
84. Makino, F., Tsuzuki, J. 1971. *Nature* 231:446–47
85. Baldwin, J. P., Bradbury, E. M., Butler-Browne, G. S., Stephens, R. M. 1973. *FEBS Lett.* 34:133–36
86. Mohberg, J., Rusch, H. P. 1969. *Arch. Biochem. Biophys.* 134:577–89
87. Mohberg, J., Rusch, H. P. 1970. *Arch. Biochem. Biophys.* 138:418–32
88. Coukell, M. B., Walker, I. O. 1973. *Cell Differ.* 2:87–95
89. Hsiang, M. W., Cole, R. D. 1973. *J. Biol. Chem.* 248:2007–13
90. Wintersberger, U., Smith, P., Letnansky, K. 1973. *Eur. J. Biochem.* 33:123–30
91. Franco, L., Johns, E. W., Navlet, J. M. 1974. *Eur. J. Biochem.* 45:83–89
92. Gray, R. H., Peterson, J. B., Ris, H. 1973. *J. Cell Biol.* 58:244–47
93. Bradley, D. M., Goldin, H. H., Claybrook, J. R. 1974. *FEBS Lett.* 41:219–22
94. Netrawali, M. S. 1970. *Exp. Cell Res.* 63:422–26
95. Iwai, K., Hamana, K., Yabuki, H. 1970. *J. Biochem.* 68:597–601
96. Gorovsky, M. A., Pleger, G. L., Keevert, J. B., Johmann, C. A. 1973. *J. Cell Biol.* 57:773–81
97. Lipps, H. J., Sapra, G. R., Ammermann, D. 1974. *Chromosoma* 45:273–80
98. Roblin, R., Härle, E., Dulbecco, R. 1971. *Virology* 45:555–66
99. Frearson, P. M., Crawford, L. V. 1972. *J. Gen. Virol.* 14:141–55
100. Lake, R. S., Barban, S., Salzman, N. P. 1973. *Biochem. Biophys. Res. Commun.* 54:640–47
101. *Cold Spring Harbor Symp. Quant. Biol.* 1974. Vol. 39
102. Kedes, L. H., Birnstiel, M. L. 1971. *Nature New Biol.* 230:165–69
103. Birnstiel, M. L., Telford, J., Weinberg, E., Stafford, D. 1974. *Proc. Nat. Acad. Sci. USA* 71:2900–4
104. Weinberg, E. S., Birnstiel, M. L., Purdom, I. F., Williamson, R. 1972. *Nature New Biol.* 240:225–28
105. Grunstein, M., Schedl, P., Kedes, L. 1973. *Symposium on Molecular Cytogenetics,* 115–23. New York: Plenum
106. Grunstein, M., Levy, S., Schedl, P., Kedes, L. 1973. *Cold Spring Harbor Symp. Quant. Biol.* 38:717–24
107. Farquhar, M. N., McCarthy, B. J. 1973. *Biochemistry* 12:4113–22
108. Pardue, M., Weinberg, E., Kedes, L., Birnstiel, M. L., 1972. *J. Cell Biol.* 55:199a

109. McCarthy, B. J., Farquhar, M. N. 1972. *Brookhaven Symp. Biol.* 23:1–43
110. Birnstiel, M. L., Weinberg, E., Pardue, M. L. See Ref. 105, 75–93
111. Stout, J. T., Phillips, R. L. 1973. *Proc. Nat. Acad. Sci. USA* 70:3043–47
112. Elgin, S. C. R., Froehner, S. C., Smart, J. E., Bonner, J. 1971. *Advan. Cell Mol. Biol.* 1:1–57
113. Spelsberg, T. C., Wilhelm, J. A., Hnilica, L. S. 1972. *Subcell. Biochem.* 1:107–45
114. MacGillivray, A. J., Paul, J., Threlfall, G. 1972. *Advan. Cancer Res.* 15:93–162
115. Elgin, S. C. R., Bonner, J. 1973. *The Biochemistry of Gene Expression in Higher Organisms,* 142–63. Sydney: Aust. New Zealand Book Co.
116. Johnson, J. D., Douvas, A., Bonner, J. 1974. *Int. Rev. Cytol.* Suppl. 4
117. Elgin, S. C. R., Stumpf, W. 1975. *Ciba Found. Symp.* 28:113–23
118. DeLange, R. J., Smith, E. L. 1974. *Proteins* 4: Chap 2, 3rd ed.
119. Langan, T. A. 1967. *Biochim. Biophys. Acta Libr. Ser.* 10:233–50
120. Holoubek, V., Crocker, T. T. 1968. *Biochim. Biophys. Acta* 157:352–61
121. Benjamin, W., Gellhorn, A. 1968. *Proc. Nat. Acad. Sci. USA* 59:262–68
122. Kleinsmith, L. J., Allfrey, V. G. 1969. *Biochim. Biophys. Acta* 175:123–35
123. Gershey, E. L., Kleinsmith, L. J. 1969. *Biochim. Biophys. Acta* 194:331–34
124. Shelton, K. R., Allfrey, V. G. 1970. *Nature* 228:132–34
125. Teng, C. S., Teng, C. T., Allfrey, V. G. 1971. *J. Biol. Chem.* 246:3597–3609
126. Elgin, S. C. R., Boyd, J. B., Hood, L. E., Wray, W., Wu, F. C. 1973. *Cold Spring Harbor Symp. Quant. Biol.* 38:821–33
127. Elgin, S. C. R., Bonner, J. 1970. *Biochemistry* 9:4440–47
128. MacGillivray, A. J., Cameron, A., Krauze, R. J., Rickwood, D., Paul, J. 1972. *Biochim. Biophys. Acta* 227:384–402
129. Garrard, W. T., Pearson, W. R., Wake, S. K., Bonner, J. 1974. *Biochem. Biophys. Res. Commun.* 58:50–57
130. Elgin, S. C. R., Bonner, J. 1972. *Biochemistry* 11:772–81
131. Marushige, K., Brutlag, D., Bonner, J. 1968. *Biochemistry* 7:3149–55
132. MacGillivray, A. J., Rickwood, D. 1974. *Eur. J. Biochem.* 41:181–90
133. Johns, E. W., Goodwin, G. H., Walker, J. M., Sanders, C. 1975. *Ciba Found. Symp.* 28:95–107
134. Shooter, K. V., Goodwin, G. H.,
135. Kleinsmith, L. J., Heidema, J., Carroll, A. 1970. *Nature* 226:1025–26
136. van den Broek, H. W. J., Nooden, L. D., Sevall, J. S., Bonner, J. 1973. *Biochemistry* 12:229–36
137. Kleinsmith, L. J. 1973. *J. Biol. Chem.* 248:5648–53
138. Patel, G. H., Thomas, T. L. 1973. *Proc. Nat. Acad. Sci. USA* 70:2524–28
139. Wakabayashi, K., Wang, S., Hord, G., Hnilica, L. S. 1973. *FEBS Lett.* 32:46–48
140. Sheehan, D. M., Olins, D. E. 1974. *Biochim. Biophys. Acta* 353:438–46
141. Platz, R. D., Kish, V. M., Kleinsmith, L. J. 1970. *FEBS Lett.* 12:38–40
142. MacGillivray, A. J., Carroll, D., Paul, J. 1971. *FEBS Lett.* 13:204–8
143. Wang, J. C. 1971. *J. Mol. Biol.* 55:523–33
144. Wu, F. C., Elgin, S. C. R., Hood, L. 1975. *J. Mol. Evol.* In press
145. Gronow, M., Thackrah, T. 1973. *Arch. Biochem. Biophys.* 158:377–86
146. Wu, F. C., Elgin, S. C. R., Hood, L. E. 1973. *Biochemistry* 12:2792–97
147. Yeoman, L. C., Taylor, C. W., Jordan, J. J., Busch, H. 1973. *Biochem. Biophys. Res. Commun.* 53:1067–76
148. Barrett, T., Gould, H. J. 1973. *Biochim. Biophys. Acta* 294:165–70
149. Fujitani, H., Holoubek, V. 1974. *Experientia* 30:474–76
150. Tashiro, T., Mizobe, F., Kurokawa, M. 1974. *FEBS Lett.* 38:121–24
151. Gierer, A. 1973. *Cold Spring Harbor Symp. Quant. Biol.* 38:951–61
152. Chytil, F., Spelsberg, T. C. 1971. *Nature New Biol.* 233:215–18
153. Wakabayashi, K., Wang, S., Hnilica, L. S. 1974. *Biochemistry* 13:1027–32
154. Wakabayashi, K., Hnilica, L. S. 1973. *Nature New Biol.* 242:153–55
155. Chiu, J. F., Craddock, C., Morris, H. P., Hnilica, L. S. 1974. *FEBS Lett.* 42:94–97
156. Liao, S., Sagher, D., Fang, S. M. 1968. *Nature* 220:1336–37
157. Cox, R. F. 1973. *Eur. J. Biochem.* 39:49–61
158. Hotta, Y., Stern, H. 1971. *Dev. Biol.* 26:87–99
159. Hotta, Y., Stern, H. 1971. *Nature New Biol.* 234:83–86
160. Herrick, G., Alberts, B. 1973. *Fed. Proc.* 32:497
161. Champoux, J. J., Dulbecco, R. 1972. *Proc. Nat. Acad. Sci. USA* 69:143–46
162. Basse, W. A., Wang, J. C. 1974. *Bio-*

chemistry 13:4299–4303
163. Putrament, A. 1971. *Genet. Res.* 18: 85–95
164. Comings, D. E., Riggs, A. D. 1971. *Nature* 233:48–50
165. Shea, M., Kleinsmith, L. J. 1973. *Biochem. Biophys. Res. Commun.* 50: 473–77
166. Barrett, T., Maryanka, D., Hamlyn, P. H., Gould, H. J. 1974. *Proc. Nat. Acad. Sci. USA* 71:5057–61
167. Britten, R. J., Davidson, E. H. 1969. *Science* 165:349–57
168. Steggles, A. W., Spelsberg, T. C., O'Malley, B. W. 1971. *Biochem. Biophys. Res. Commun.* 43:20–27
169. Holmes, D. S., Mayfield, J. E., Sander, G., Bonner, J. 1972. *Science* 177:72–74
170. O'Malley, B. W., Means, A. R. 1974. *The Cell Nucleus,* ed. H. Busch, 3: 379–416. New York: Academic
171. Sibley, C. H., Tomkins, G. M. 1974. *Cell* 2:213–27
172. Gorovsky, M., Woodard, J. 1967. *J. Cell. Biol.* 33:723–28
173. Berendes, H. D. 1968. *Chromosoma* 24: 418–37
174. Holt, Th. K. H. 1971. *Chromosoma* 32:428–35
175. Helmsing, P. J., Berendes, H. D. 1971. *J. Cell Biol.* 50:893–96
176. Helmsing, P. J. 1972. *Cell Differ.* 1: 19–24
177. Ashburner, M. 1974. *Dev. Biol.* 39:141–57
178. Rovera, G., Farber, J., Baserga, R. 1971. *Proc. Nat. Acad. Sci. USA* 68:1725–29
179. Rovera, G., Baserga, R. 1973. *Exp. Cell Res.* 78:118–26
180. LeStourgeon, W. M., Rusch, H. P. 1971. *Science* 174:1233–35
181. LeStourgeon, W. M., Wray, W., Rusch, H. P. 1973. *Exp. Cell Res.* 79:487–92
182. LeStourgeon, W. M., Goodman, E. M., Rusch, H. P. 1973. *Biochim. Biophys. Acta* 317:524–28
183. Baserga, R., Bombik, B., Nicolini, C. 1975. *Ciba Found. Symp.* 28:269–79
184. Lin, J. C., Nicolini, C., Baserga, R. 1974. *Biochemistry* 13:4127–33
185. LeStourgeon, W. M., Forer, A., Yang, Y. Z., Bertram, J. S., Rusch, H. P. 1975. *Biochim. Biophys. Acta* 379:539–52
186. Hinkley, R., Telser, A. 1974. *Exp. Cell Res.* 86:161–64
187. Jackson, V., Earnhardt, J., Chalkley, R. 1968. *Biochem. Biophys. Res. Commun.* 33:253–59
188. Harlow, R., Tolstoshev, P., Wells, J. R. E. 1972. *Cell Differ.* 2:341–49
189. Tata, J. R., Hamilton, M. J., Cole, R. D. 1972. *J. Mol. Biol.* 67:231–46
190. Shelton, K. R. 1973. *Can. J. Biochem.* 51:1442–47
191. Comings, D. E., Okada, T. A. 1970. *Exp. Cell Res.* 62:293–302
192. Aaronson, R. P., Blobel, G. 1974. *J. Cell Biol.* 62:746–54
193. Georgiev, G. P., Samarina, O. P. 1971. *Advan. Cell Biol.* 2:47–110
194. Samarina, O. P., Lukanidi, E. M., Molnar, J., Georgiev, G. P. 1968. *J. Mol. Biol.* 33:251–63
195. Krichevskaya, A. A., Georgiev, G. P. 1969. *Biochim. Biophys. Acta* 194:619–21
196. Martin, T. E., Swift, H. 1973. *Cold Spring Harbor Symp. Quant. Biol.* 38: 921–32
197. Hill, R. J., Maundrell, K. G., Callan, H. G. 1973. *Molecular Cytogenetics,* 147–53. New York: Plenum
198. Sommerville, J. 1973. *J. Mol. Biol.* 78: 487–503
199. Sommerville, J., Hill, R. J. 1973. *Nature New Biol.* 245:104–6
200. Scott, S. E. M., Sommerville, J. 1974. *Nature* 250:680–82
201. Pederson, T. 1974. *Proc. Nat. Acad. Sci. USA* 71:617–21
202. Derksen, J., Berendes, H. D., Willart, E. 1973. *J. Cell Biol.* 59:661–68
203. Williamson, R. 1973. *FEBS Lett.* 37: 1–6
204. Prescott, D. M. 1966. *J. Cell Biol.* 31: 1–9
205. Robbins, E., Borun, T. W. 1967. *Proc. Nat. Acad. Sci. USA* 57:409–16
206. Sadgopal, A., Bonner, J. 1969. *Biochim. Biophys. Acta* 186:349–57
207. Takai, S., Borun, T. W., Muchmore, J., Lieberman, I. 1968. *Nature* 219:860–61
208. Kedes, L. H., Gross, P. R. 1969. *Nature* 223:1335–39
209. Yarbro, J. W. 1967. *Biochim. Biophys. Acta* 145:531–34
210. Weintraub, H. 1972. *Nature* 240:449–53
211. Gurley, L. R., Walters, R. A., Tobey, R. A. 1972. *Arch. Biochem. Biophys.* 148:633–41
212. Appels, R., Wells, J. R. E. 1972. *J. Mol. Biol.* 70:425–34
213. Adamson, E. D., Woodland, H. R. 1974. *J. Mol. Biol.* 88:263–85
214. Jacobs-Lorena, M., Baglioni, C., Borun, T. W. 1972. *Proc. Nat. Acad. Sci. USA* 69:2095–99
215. Breindl, M., Gallwitz, D. 1973. *Eur. J. Biochem.* 32:381–91
216. Gallwitz, D., Breindl, M. 1972. *Biochem.*

Biophys. Res. Commun. 47:1106–11
217. Schochetman, G., Perry, R. P. 1972. *J. Mol. Biol.* 63:577–90
218. Adesnik, M., Darnell, J. E. 1972. *J. Mol. Biol.* 67:397–406
219. Breindl, M., Gallwitz, D. 1974. *Mol. Biol. Rep.* 1:263–68
220. Zauderer, M., Liberti, P., Baglioni, C. 1973. *J. Mol. Biol.* 79:557–86
221. Jacobs-Lorena, M., Baglioni, C. 1973. *Mol. Biol. Rep.* 1:113–17
222. Pestana, A., Pitot, H. C. 1974. *Nature* 247:200–2
223. Oliver, D., Granner, D., Chalkley, R. 1974. *Biochemistry* 13:746–49
224. Gallwitz, D., Mueller, G. C. 1969. *J. Biol. Chem.* 244:5947–52
225. Jacobs-Lorena, M., Gabrielli, F., Borun, T. W., Baglioni, C. 1973. *Biochim. Biophys. Acta* 324:275–81
226. Breindl, M., Gallwitz, D. 1974. *Eur. J. Biochem.* 45:91–97
227. Borun, T. W., Scharff, M. D., Robbins, E. 1967. *Proc. Nat. Acad. Sci. USA* 58:1977–83
228. Butler, W. B., Mueller, G. C. 1973. *Biochim. Biophys. Acta* 294:481–96
229. Perry, R. P., Kelley, D. E. 1973. *J. Mol. Biol.* 71:681–96
230. Hershey, H., Stieber, J., Mueller, G. C. 1973. *Biochim. Biophys. Acta* 312:509–17
231. Balhorn, R., Tanphaichitr, N., Chalkley, R., Granner, D. K. 1973. *Biochemistry* 12:5146–50
232. Woese, C. R. 1973. *J. Mol. Evol.* 2:205–8
233. Weintraub, H. 1973. *Cold Spring Harbor Symp. Quant. Biol.* 38:247–56
234. Cave, M. D. 1968. *Chromosoma* 25:392–401
235. Stein, G., Baserga, R. 1970. *Biochem. Biophys. Res. Commun.* 41:715–22
236. Cross, M. E. 1972. *Biochem. J.* 128:1213–19
237. Stein, G. S., Borun, T. W. 1972. *J. Cell Res.* 52:292–307
238. Gerner, E. W., Humphrey, R. M. 1973. *Biochim. Biophys. Acta* 331:117–27
239. Bhorjee, J. S., Pederson, T. 1973. *Biochemistry* 12:2766–73
240. Borun, T. W., Stein, G. S. 1972. *J. Cell Biol.* 52:308–15
241. Tsanev, R., Russev, G. 1974. *Eur. J. Biochem.* 43:257–63
242. Sadgopal, A., Bonner, J. 1970. *Biochim. Biophys. Acta* 207:227–39
243. Wray, W., Elgin, S. C. R. 1975. *J. Cell Biol.* In press
244. Comings, D. E., Tack, L. O. 1973. *Exp. Cell Res.* 82:175–91
245. Gall, J. G. 1973. *Symposium on Molecular Cytology,* 59–74. New York: Plenum
246. Helmsing, P. J., van Eupen, O. 1973. *Biochim. Biophys. Acta* 308:154–60
247. Bloch, D. P., Macquigg, R. A., Brack, S. D., Wu, J. R. 1967. *J. Cell Biol.* 33:451–67
248. Byvoet, P. 1966. *J. Mol. Biol.* 17:311–18
249. Hancock, R. 1969. *J. Mol. Biol.* 40:457–66
250. Appels, R., Ringertz, N. R. 1974. *Cell Differ.* 3:1–8
251. Garrard, W. T., Bonner, J. 1974. *J. Biol. Chem.* 249:5570–79
252. Bondy, S. C. 1971. *Biochem. J.* 123:465–69
253. Dice, J. F., Schimke, R. T. 1973. *Arch. Biochem. Biophys.* 158:97–105
254. Karn, J., Johnson, E. M. Vidali, G., Allfrey, V. G. 1974. *J. Biol. Chem.* 249:667–77
255. Teng, C. S., Hamilton, T. H. 1970. *Biochem. Biophys. Res. Commun.* 40:1231–38
256. Enea, V., Allfrey, V. G. 1973. *Nature* 242:265–67
257. Stein, G., Baserga, R. 1970. *J. Biol. Chem.* 245:6097–6105
258. Choe, B. K., Rose, N. R. 1974. *Exp. Cell Res.* 83:261–80
259. Cognetti, G., Settineri, D., Spinelli, G. 1972. *Exp. Cell Res.* 71:465–68
260. Hill, R. J., Poccia, D. L., Doty, P. 1971. *J. Mol. Biol.* 61:445–62
261. Sevaljevic, L. 1973. *Dev. Biol.* 34:267–73
262. Pipkin, J. L. Jr., Larson, D. A. 1973. *Exp. Cell Res.* 79:28–42
263. Chytil, F., Glasser, S. R., Spelsberg, T. C. 1974. *Dev. Biol.* 37:295–305
264. Panyim, S., Chalkley, R. 1969. *Biochem. Biophys. Res. Commun.* 37:1042–49
265. Lea, M. A., Youngworth, L. A., Morris, H. P. 1974. *Biochem. Biophys. Res. Commun.* 58:862–67
266. Rubio, V., Tsai, W. P., Rand, K., Long, C. 1973. *Int. J. Cancer* 12:545–50
267. Zardi, L., Lin, J. C., Baserga, R. 1973. *Nature New Biol.* 245:211–14
268. Johns, E. W., Davies, A. J. S., Barton, M.E. 1970. *Biochim. Biophys. Acta* 213:537–38
269. Hohmann, P., Cole, R. D., Bern, H. A. 1971. *J. Nat. Cancer Inst.* 47:337–41
270. Balhorn, R., Chalkley, R., Granner, D. 1972. *Biochemistry* 11:1094–98
271. McClure, M. E., Hnilica, L. S. 1972. *Subcell Biochem.* 1:311–32
272. Stein, G., Baserga, R. 1972. *Advan.*

Cancer Res. 15:287–330

273. DeLange, R. J., Smith, E. L. 1971. *Ann. Rev. Biochem.* 40:279–314

274. Shepherd, G. R., Noland, B. J., Hardin, J. M., Byvoet, P. See Ref. 115, 164–76

275. Johnson, E. M., Karn, J., Allfrey, V. G. 1974. *J. Biol. Chem.* 249:4990–99

276. Ingles, C. J., Dixon, G. H. 1967. *Proc. Nat. Acad. Sci. USA* 58:1011–18

277. Louie, A. J., Dixon, G. H. 1972. *Proc. Nat. Acad. Sci. USA* 69:1975–79

278. Louie, A. J., Dixon, G. H. 1973. *Nature New Biol.* 243:164–68

279. Candido, E. P. M., Dixon, G. H. 1972. *Proc. Nat. Acad. Sci. USA* 69:2015–19

280. Lake, R. S., Goidl, J. A., Salzman, N. P. 1972. *Exp. Cell Res.* 73:113–21

281. Gurley, L. R., Walters, R. A., Tobey, R. A. 1973. *Biochem. Biophys. Res. Commun.* 50:744–50

282. Balhorn, R., Jackson, V., Granner, D., Chalkley, R. 1975. *Biochemistry.* In press

283. Langan, T. A., Hohmann, P. 1974. *Fed. Proc.* 33:1597

284. Lake, R. S., Salzman, N. P., 1972. *Biochemistry* 11:4817–26

285. Lake, R. S. 1973. *Nature New Biol.* 242:145–46

286. Marks, D. B., Park, W. K., Borun, T. W. 1973. *J. Biol. Chem.* 248:5660–67

287. Gurley, L. A., Walters, R. A., Tobey, R. A. 1974. *J. Cell Biol.* 60:356–64

288. Bradbury, E. M., Inglis, R. J., Matthews, H. R., Sarner, N. 1973. *Eur. J. Biochem.* 33:131–39

289. Bradbury, E. M., Inglis, R. J., Matthews, H. R. 1974. *Nature* 247:257–61

290. Bradbury, E. M., Inglis, R. J., Matthews, H. R., Langan, T. A. 1974. *Nature* 249:553–56

291. Langan, T. A. 1968. *Science* 162:579–80

292. Langan, T. A. 1961. *Fed. Proc.* 28:600

293. Langan, T. A. 1969. *J. Biol. Chem.* 244:5763–65

294. Langan, T. A. 1969. *Proc. Nat. Acad. Sci. USA* 64:1276–83

295. Mallette, L. E., Neblett, M., Exton, J. H., Langan, T. A. 1973. *J. Biol. Chem.* 248:6289–91

296. Takeda, M., Ohga, Y. 1973. *J. Biochem.* 73:621–29

297. Langan, T. A., Smith, L. 1967. *Fed. Proc.* 26:603

298. Meisler, M. H., Langan, T. A. 1967. *J. Cell Biol.* 35:91a

299. Meisler, M. H., Langan, T. A. 1969. *J. Biol. Chem.* 244:4961–68

300. Adler, A. J., Langan, T. A., Fasman, G. D. 1972. *Arch. Biochem. Biophys.* 153:769–77

301. Adler, A. J., Fasman, G. D., Wangh, W., Allfrey, V. G. 1974. *J. Biol. Chem.* 249:2911–14

302. Watson, G., Langan, T. A. 1973. *Fed. Proc.* 32:588

303. Kleinsmith, L. J., Allfrey, V. G., Mirsky, A. E. 1966. *Science* 154:780–81

304. Gershey, E. L., Kleinsmith, L. J. 1969. *Biochim. Biophys. Acta* 194:519–25

305. Johnson, E. M., Allfrey, V. G. 1972. *Arch. Biochem. Biophys.* 152:786–94

306. Platz, R. D., Stein, G. S., Kleinsmith, L. J. 1973. *Biochem. Biophys. Res. Commun.* 51:735–40

307. Platz, R. D., Hnilica, L. S. 1973. *Biochem. Biophys. Res. Commun.* 54:222–27

308. Rickwood, D., Riches, P. G., MacGillivray, A. J. 1973. *Biochim. Biophys. Acta* 299:162–71

309. Kish, V. M., Kleinsmith, L. J. 1974. *J. Biol. Chem.* 249:750–60

310. Trevithick, J. R. 1974. *Can. J. Biochem.* 52:406–13

311. Fleischer-Lambropoulos, H., Sarkander, H. I., Brade, W. P. 1974. *FEBS Lett.* 45:329–32

312. Kleinsmith, L. J. 1975. *J. Cell Physiol.* In press

313. Huberman, J. A. 1973. *Ann. Rev. Biochem.* 42:355–78

314. Simpson, R. T. 1973. *Advan. Enzymol.* 38:41–108

315. Kornberg, R. D. 1974. *Science* 184:868–71

316. Pardon, J. F., Wilkins, M. H. F. 1972. *J. Mol. Biol.* 68:115–24

317. Bram, S., Ris, H. 1971. *J. Mol. Biol.* 55:325–36

318. Hewisch, D., Burgoyne, L. 1973. *Biochem. Biophys. Res. Commun.* 52:504–10

319. Noll, M. 1974. *Nature* 251:249–52

320. Axel, R., Melchior, W., Sollner-Webb, B., Felsenfeld, G. 1974. *Proc. Nat. Acad. Sci. USA* 71:4101–5

321. Weintraub, H., Van Lente, F. 1975. *Ciba Found. Symp.* 28:291–307

322. Clark, R. J., Felsenfeld, G. 1971. *Nature New Biol.* 229:101–6

323. Itzhaki, R. F. 1971. *Biochem J.* 125:221–24

324. Itzhaki, R. F. 1974. *Eur. J. Biochem.* 47:27–33

325. Mirsky, A. E. 1971. *Proc. Nat. Acad. Sci. USA* 68:2945–48

326. Davidson, E. H., Graham, D. E., Neufield, B. R., Chamberlin, M. E., Amenson, C. S., Hough, B. R., Britten,

R. J. 1973. *Cold Spring Harbor Symp. Quant. Biol.* 38:295–302
327. Garrett, R. A. 1960. *J. Mol. Biol.* 38: 249–50 plus plate
328. Bradbury, E. M., Molgaard, H. V., Stephens, R. M. 1972. *Eur. J. Biochem.* 31:474–82
329. Kornberg, R. D., Thomas, J. O. 1974. *Science* 184:865–68
330. Richards, B. M., Pardon, J. F. 1970. *Exp. Cell Res.* 62:184–96
331. Skidmore, C., Walker, I. O., Pardon, J. F., Richards, B. M. 1973. *FEBS Lett.* 32:175–78
332. Baldwin, J. P., Boseley, P. G., Bradbury, E. M., Ibel, K. 1975. *Nature* 253:245–49
333. Olins, A. L., Olins, D. E. 1974. *Science* 183:330–32
334. Woodcock, C. L. F. 1973. *J. Cell Biol.* 59:368a
335. Senior, M. B., Olins, A. L., Olins, D. E. 1975. *Science* 187:173–75
336. Georgiev, G. P., Varshavsky, A. J., Church, R. B., Ryskov, A. P. 1973. *Cold Spring Harbor Symp. Quant. Biol.* 38:869–84
337. Varshavsky, A. J., Ilyin, Y. V., Georgiev, G. P. 1974. *Nature* 250:602–6
338. Sahasrabuddhe, C. G., Van Holde, K. E. 1974. *J. Biol. Chem.* 249:152–56
339. Weintraub, H., Van Lente, F. 1974. *Proc. Nat. Acad. Sci. USA* 71:4249–53
340. Weintraub, H. 1975. *Proc. Nat. Acad. Sci. USA.* In press
341. Crick, F. H. C. 1971. *Nature* 234: 25–27
342. D'Anna, J. A. Jr., Isenberg, I. 1972. *Biochemistry* 11:4017–25
343. Smerdon, M. J., Isenberg, I. 1974. *Biochemistry* 13:4046–49
344. Edwards, P. A., Shooter, K. V. 1969. *Biochem. J.* 114:53
345. Bradbury, E. M., Rattle, H. W. E. 1972. *Eur. J. Biochem.* 27:270–81
346. Kelley, R. I. 1973. *Biochem. Biophys. Res. Commun.* 54:1588–94
347. Roark, D. E., Geoghegan, T. E., Keller, G. H. 1974. *Biochem. Biophys. Res. Commun.* 59:542–47
348. Olins, D. E., Wright, E. B. 1973. *J. Cell Biol.* 59:304–17
349. Martinson, H., McCarthy, B. J. 1975. *Biochemistry* 14:1073–78
350. Simpson, R. T. 1971. *Biochemistry* 10: 4466–70
351. Malchy, B., Kaplan, H. 1974. *J. Mol. Biol.* 82:537–45
352. Bustin, M. 1973. *Nature New Biol.* 245:207–9
353. Jacquet, M., Groner, Y., Monroy, G.,

Hurwitz, J. 1974. *Proc. Nat. Acad. Sci. USA* 71:3045–49
354. Shih, T. Y., Bonner, J. 1970. *J. Mol. Biol.* 48:469–87
355. Smart, J. E., Bonner, J. 1971. *J. Mol. Biol.* 58:661–74
356. Smart, J. E., Bonner, J. 1971. *J. Mol. Biol.* 58:675–84
357. Koslov, Y. V., Georgiev, G. P. 1970. *Nature* 228:245–47
358. Limborska, S. A., Georgiev, G. P. 1972. *Cell Differ.* 1:245–51
359. Cedar, H., Felsenfeld, G. 1973. *J. Mol. Biol.* 77:237–54
360. Frenster, J. H., Allfrey, V. G., Mirsky, A. E. 1963. *Proc. Nat. Acad. Sci. USA* 50:1026–32
361. Frenster, J. H. 1965. *Nature* 206:680–83
362. McCarthy, B. J., Nishiura, J. T., Doenecke, D., Nasser, D. S., Johnson, C. B. 1973. *Cold Spring Harbor Symp. Quant. Biol.* 38:763–71
363. Murphy, E. C. Jr., Hall, S. H., Shepherd, J. H., Weiser, R. S. 1973. *Biochemistry* 12:3843–53
364. Reeck, G. R., Simpson, R. T., Sober, H. A. 1972. *Proc. Nat. Acad. Sci. USA* 69:2317–21
365. Simpson, R. T. 1974. *Proc. Nat. Acad. Sci. USA* 71:2740–43
366. Marushige, K., Bonner, J. 1971. *Proc. Nat. Acad. Sci. USA* 68:2941–44
367. Billing, R. J., Bonner, J. 1972. *Biochim. Biophys. Acta* 281:453–62
368. Gottesfeld, J. M., Garrard, W. T., Bagi, G., Wilson, R. F., Bonner, J. 1974. *Proc. Nat. Acad. Sci. USA* 71:2193–97
369. Simpson, R. T., Reeck, G. R. 1973. *Biochemistry* 12:3853–58
370. McConaughy, B. L., McCarthy, B. J. 1972. *Biochemistry* 11:998–1003
371. Saucier, J. M., Wang, J. C. 1972. *Nature New Biol.* 239:167–70
372. Travers, A., Baillie, D. L., Pederson, S. 1972. *Nature New Biol.* 243:161–63
373. Travers, A. 1974. *Cell.* 3:97–104
374. Hossenlopp, P., Oudet, P., Chambon, P. 1974. *Eur. J. Biochem.* 41:367–78
375. Ghosh, S., Echols, H. 1972. *Proc. Nat. Acad. Sci. USA* 69:3660–64
376. Hall, B. D., Brzezinska, M., Hollenberg, C. P., Schultz, L. D. See Ref. 105, 217–26
377. Hsu, T. C. 1972. *Methods Cell Physiol.* 5:1–36
378. Hsu, T. C. 1973. *Ann. Rev. Genet.* 7: 153–76
379. Gottesfeld, J. M., Bonner, J., Radda, G. K., Walker, I. O. 1974. *Biochemistry*

13:2937–45
380. Faust, J., Vogel, W. 1974. *Nature* 249: 352–53
381. DeLange, R. J., Fambrough, D. M., Smith, E. L., Bonner, J. 1969. *J. Biol. Chem.* 244:319–34
382. DeLange, R. J., Fambrough, D. M., Smith, E., Bonner, J. 1969. *J. Biol. Chem.* 244:5669–79
383. Ogawa, Y. et al 1969. *J. Biol. Chem.* 244:4387–92
384. Sautiere, P., Tyrou, D., Moschetto, Y., Biserte, G. 1971. *Biochimie* 53:479–83
385. Sautiere, P., Lambelin-Breynaert, M. D., Moschetto, Y., Biserte, G. 1971. *Biochimie* 53:711–15
386. Desai, L., Ogawa, Y., Mauritzen, C. M., Taylor, C. W., Starbuck, W. C. 1969. *Biochim. Biophys. Acta* 181:146–53
387. Strickland, M., Strickland, W. N., Brandt, W. F. 1974. *FEBS Lett.* 40:346–48
388. DeLange, R. J., Hooper, J. A., Smith, E. L. 1972. *Proc. Nat. Acad. Sci. USA* 69:882–84
389. DeLange, R. J., Hooper, J. A., Smith, E. L. 1973. *J. Biol. Chem.* 248:3261–74
390. Olson, M. O. J., Jordan, J., Busch, H. 1972. *Biochem. Biophys. Res. Commun.* 46:50–55
391. Brandt, W. F., van Holt, C. 1972. *FEBS Lett.* 23:357–60
392. Brandt, W. F., van Holt, C. 1974. *Eur. J. Biochem.* 46:407–17
393. Brandt, W. F., van Holt, C. 1974. *Eur. J. Biochem.* 46:419–29
394. Hooper, J. A., Smith, E. L., Sommer, K. R., Chalkley, R. 1973. *J. Biol. Chem.* 248:3275–79
395. Brandt, W. F., Strickland, W. N., von Holt, C. 1974. *FEBS Lett.* 40:349–52
396. Brandt, W. F., Strickland, W. N., Morgan, M., von Holt, C. 1974. *FEBS Lett.* 40:167–72
397. Patthy, L., Smith, E. L., Johnson, J. 1973. *J. Biol. Chem.* 248:6834–40
398. Iwai, K., Hayashi, H., Ishikawa, K. 1972. *J. Biochem.* 72:357–67
399. Yeoman, L. C. et al 1972. *J. Biol. Chem.* 247:6018–23
400. Sautiere, P., Tyrou, D., Laine, B., Mizon, J., Ruffin, P., Biserte, G. 1974. *Eur. J. Biochem.* 41:563–76
401. Sautiere, P. 1975. *Ciba Found. Symp.* 28:77–88
402. Bailey, G. S., Dixon, G. H. 1973. *J. Biol. Chem.* 248:5463–72
403. Jones, G. M., Rall, S. C., Cole, R. D. 1974. *J. Biol. Chem.* 249:2548–53
404. Weiss, S. B. 1960. *Proc. Nat. Acad. Sci.*

USA 46:1020–30
405. Tata, J. R., Baker, B. 1974. *Exp. Cell Res.* 83:111–24, 125–38
406. Lentfer, D., Lezius, A. G. 1972. *Eur. J. Biochem.* 30:278–84
407. Mondal, H., Ganguly, A., Das, A., Mandal, R. K., Biswas, B. B. 1972. *Eur. J. Biochem.* 28:143–50
408. Ganguly, A., Das, A., Mondal, H., Mandal, R. K., Biswas, B. B. 1973. *FEBS Lett.* 34:27–30
409. Sasaki, K., Tazawa, T. 1973. *Biochem. Biophys. Res. Commun.* 52:1440–49
410. Patel, G., Howk, R., Wang, T. Y. 1967. *Nature* 215:1488–89
411. Howk, R., Wang, T. Y. 1969. *Arch. Biochem. Biophys.* 133:238–46
412. Howk, R., Wang, T. Y. 1970. *Eur. J. Biochem.* 13:455–60
413. Tsuruo, T., Tomita, Y., Satoh, H., Ukita, T. 1972. *Biochem. Biophys. Res. Commun.* 48:776–82
414. Tsuruo, T., Ukita, T. 1974. *Biochim. Biophys. Acta* 353:146–59
415. Loeb, L. A. 1970. *Nature* 226:448–49
416. Chang, L. M. S. 1973. *J. Biol. Chem.* 248:3789–95
417. Urbanczyk, J., Studzinski, G. P. 1974. *Biochem. Biophys. Res. Commun.* 59:616–22
418. Gaziev, A. I., Kuzin, A. M. 1973. *Eur. J. Biochem.* 37:7–11
419. O'Connor, P. J. 1969. *Biochem. Biophys. Res. Commun.* 35:805–10
420. Wang, T. Y. 1968. *Arch. Biochem. Biophys.* 127:235–40
421. Srivastava, B. I. S. 1972. *Biochem. Biophys. Res. Commun.* 48:270–73
422. Yamada, M., Sugimura, T. 1973. *Biochemistry* 12:3303–8
423. Sugimura, T. 1973. *Progr. Nucl. Acid Res. Mol. Biol.* 13:127–51
424. Miyakawa, N., Veda, K., Hayaishi, O. 1972. *Biochem. Biophys. Res. Commun.* 49:239–45
425. Racey, L. A., Byvoet, P. 1971. *Exp. Cell Res.* 64:366–70
426. Racey, L. A., Byvoet, P. 1972. *Exp. Cell Res.* 73:329–34
427. Gallwitz, D., Sures, I. 1972. *Biochim. Biophys. Acta* 263:315–328
428. Garrels, J. I., Elgin, S. C. R., Bonner, J. 1972. *Biochem. Biophys. Res. Commun.* 46:545–51
429. Bartley, J., Chalkley, R. 1970. *J. Biol. Chem.* 245:4286–92
430. Takeda, M., Yamamura, H., Ohga, Y. 1971. *Biochem. Biophys. Res. Commun.* 42:103–10
431. Comb, D. G., Sarkar, N., Pinzino, C. J. 1966. *J. Biol. Chem.* 241:1857–62

432. Chen, C., Smith, D. L., Bruegger, B., Halpern, R. M., Smith, R. A. 1974. *Biochemistry* 13:3785–89
433. Smith, D. L., Chen, C. C., Bruegger, B. B. 1974. *Biochemistry* 13:3780–85
434. Smith, J. A., Stocken, L. A. 1973. *Biochem. J.* 131:859–61
435. Goodwin, G. H., Johns, E. W. 1973. *Eur. J. Biochem.* 40:215–19
436. Goodwin, G. H., Sanders, C., Johns, E. W. 1973. *Eur. J. Biochem.* 38:14–19
437. Patel, N. T., Holoubek, V. 1974. *FEBS Lett.* 46:154–57
438. Marzluff, W. F. Jr., Sanders, L. A., Miller, D. M., McCarty, K. S. 1972. *J. Biol. Chem.* 247:2026–33
439. Douvas, A., Bonner, J. 1974. *J. Cell Biol.* 63:89a
440. Sevall, J. S., Cockburn, A., Savage, M., Bonner, J. 1975. *Biochemistry* 14:782–89
441. Chong, M. T., Garrard, W. T., Bonner, J. 1974. *Biochemistry* 13:5128–34
442. Simpson, R. T. 1974. *Curr. Top. Biochem.* 1973:135–85
443. Duerksen, J. D., McCarthy, B. J. 1971. *Biochemistry* 10:1471–78
444. Janowski, M., Nasser, D. S., McCarthy, B. J. 1972. *Karolinska Symp.* 5:112–29
445. Botchan, P., Wang, J. C., Echols, H. 1973. *Proc. Nat. Acad. Sci. USA* 70:3077–81
446. Felsenfeld, G., Axel, R., Cedar, H., Sollner-Webb, B. 1975. *Ciba Found. Symp.* 28:29–41
447. D'Anna, J. A., Isenberg, I. 1973. *Biochemistry* 12:1035–43
448. D'Anna, J. A., Isenberg, I. 1974. *Biochemistry* 13:2098–2104
449. Edwards, P. A., Shooter, K. V. 1970. *Biochem. J.* 120:61–66
450. von Hippel, P. H., McGhee, J. D. 1972. *Ann. Rev. Biochem.* 41:231–300
451. Walker, I. O. 1965. *J. Mol. Biol.* 14:381–98
452. Clark, R. J., Felsenfeld, G. 1974. *Biochemistry* 13:3622–28
453. Hill, B. T., Baserga, R. 1974. *Biochem. J.* 141:27–34
454. Pederson, T. 1974. *J. Mol. Biol.* 83:163–83
455. Baserga, R. 1974. *Life Sci.* 15:1057–72
456. Van Holde, K. E., Sahasrabuddhe, C. G., Shaw, B. R., van Bruggen, E. F. J., Arnberg, A. C. 1974. *Biochem. Biophys. Res. Commun.* 60:1365–70
457. Sung, M., Harford, J., Bundman, M., Vidalakis, G. 1974. *Fed. Proc.* 33:1597
458. D'Anna, J. A. Jr., Isenberg, I. 1974. *Biochem. Biophys. Res. Commun.* 61:343–47
459. Elgin, S. C. R., Boyd, J. B. 1975. *Chromosoma.* In press
460. D'Anna, J. A. Jr., Isenberg, I. 1974. *Biochemistry* 13:4992–97
461. Burgoyne, L. A., Hewish, D. R., Mobbs, J. 1974. *Biochem. J.* 143:67–72
462. Noll, M. 1974. *Nucl. Acid Res.* 11:1573–78
463. Honda, B. M., Baillie, D. L., Candido, E. P. M. 1974. *FEBS Lett.* 48:156–59
464. Mirzabekov, A., Melnikova, K. 1974. *Mol. Biol. Rep.* 1:35–42
465. Garrard, W. T., Nobis, P., Hancock, R. 1975. *Fed. Proc.* 34:611
466. Jockusch, B. M., Becker, M., Hindennach, I., Jockusch, H. 1974. *Exp. Cell Res.* 89:241–46
467. Borun, T. W., Gabrielli, F., Ajiro, K., Zweidler, A., Baglioni, C. 1975. *Cell* 59–67
468. Astrin, S. M. 1973. *Proc. Nat. Acad. Sci. USA* 70:2304–8
469. Swetly, P., Watanabe, Y. 1974. *Biochemistry* 13:4122–26
470. Wilson, M. C., Melli, M., Birnstiel, M. L. 1974. *Biochem. Biophys. Res. Commun.* 61:354–59
471. Levy, S., Wood, P., Grunstein, M., Kedes, L. 1975. *Cell* 4:239–48
472. Griffith, J. D. 1975. *Science* 187:1202–3
473. Olins, A. L., Carlson, R. D., Olins, D. E. 1975. *J. Cell Biol.* 64:528–37
474. Renz, M. 1975. *Proc. Nat. Acad. Sci. USA* 72:733–36

POST-TRANSLATIONAL CLEAVAGE OF POLYPEPTIDE CHAINS: ROLE IN ASSEMBLY

Avram Hershko and Michael Fry

Department of Clinical Biochemistry, Technion—Israel Institute of Technology,
The Aba Khoushy School of Medicine, Haifa, Israel

CONTENTS

INTRODUCTION

"You don't know how to manage looking-glass cakes," the Unicorn remarked, "Hand it round first, and cut it afterwards."

Lewis Carrol, *Through the Looking Glass*

An increasing body of recent evidence indicates that postsynthetic processing of polypeptide and polynucleotide chains has widespread occurrence and considerable biological significance. Post-transcriptional cleavage mechanisms play an important role in the formation of many RNA species in a variety of biological systems (1, 2).

775

Some well-known examples of specific scission of polypeptides are the activation of digestive enzyme precursors (3) and the cascade mechanisms involved in blood coagulation (4) and complement action (5). More recently, post-translational cleavage mechanisms have been shown to be involved in the formation of proteins as different as insulin (6), collagen (7), and possibly albumin (8) and immunoglobulin light chains (9, 10). In addition, proteolytic cleaving enzymes of high specificity play a role in the inactivation (11, 12) and activation (13) of specific intracellular enzymes. Thus, post-translational cleavage mechanisms participate in a wide variety of biological processes, ranging from the formation of polypeptides from large biosynthetic precursors, to the modification of activity of both intracellular and extracellular enzyme systems.

In this review, we discuss still another role of polypeptide cleavage, which has been discovered recently in several viral systems. The numerous cleavage phenomena of virus-specific proteins can be classified, in general, into two distinct categories, which we shall call "formative" and "morphogenetic" cleavages. In the first class, discovered in small RNA-containing animal viruses, the formation of all functional proteins proceeds through the cleavage of high molecular weight biosynthetic precursors. This process is probably a mechanism substituting for internal initiation in the synthesis of proteins on a polycistronic message in animal cells (14). The second class of viral protein cleavage, which is widespread in many animal viruses as well as in certain complex bacteriophages, is the specific scission of some structural proteins in the course of viral assembly. These morphogenetic cleavages seem therefore to play an integral part in the assembly of relatively complex viral structures in which simple self-assembly mechanisms may not be sufficient.

We shall try to illustrate current progress in this field by describing in some detail two selected, relatively well-studied viral systems (i.e. picornaviruses and T4 bacteriophage). Post-translational cleavages occurring in other viruses will be discussed more briefly, with emphasis on biological principles which may be better illuminated in these systems.

Part of the earlier literature concerning picornaviruses has been reviewed (15–20). Various other aspects of virus assembly are discussed in another review in this volume (21).

PROCESSING OF PICORNAVIRUS PROTEINS

Since the discovery of the participation of post-translational cleavages in the formation of picornavirus proteins (14, 22, 23), this subject has been reviewed several times (15–20). We summarize here some of the more recent advances and outline the various stages in the processing of picornavirus proteins according to the current status of our knowledge. Although most studies were performed with poliovirus, remarkably similar features of polypeptide processing were observed in other picornaviruses, such as encephalomyocarditis (EMC) virus (24–28) or rhinovirus (27, 28).

Three different classes of cleavages seem to occur during the processing of picornavirus proteins, of which the first two may be regarded as different types of formative cleavages. The first is involved in the formation of the primary products arising directly from the translational process, and most probably represents cleavage of nascent polypeptide chains; this has been termed "nascent" cleavage (18). Another class, which we call "secondary" cleavages, is a series of transformations that convert the primary precursors to the final functional proteins. The third type is evidently a "morphogenetic" cleavage, since it is intimately associated with the final steps of picornavirus assembly. We shall discuss these various types of polypeptide cleavages separately.

Formation of Primary Translation Product(s)

The translation of poliovirus-specific proteins takes place on large polysomes, in which the whole viral RNA appears to function as mRNA (29, 30). Baltimore and his associates have suggested that the total genome of poliovirus is translated as a single unit to yield a large precursor polypeptide ("polyprotein"), from which all viral proteins are produced by subsequent cleavages (14, 16, 19). The size of poliovirus RNA is about 2.6×10^6 daltons (31, 32) and would thus code for a protein of $\sim 270,000$ daltons. Giant polypeptides of almost this size have been observed in poliovirus-infected cells in the presence of amino acid analogs (14, 33, 34), protease inhibitors (35, 36), and in some temperature-sensitive mutants at the restrictive temperature (37, 38). Under normal conditions, however, such a giant precursor cannot be seen, and the earliest detectable products are polypeptides NCVP1a, NCVP1b, and NCVPX, with molecular weights of 110,000, 93,000, and 34,000, respectively[1] (see Figure 1). It could be imagined that a polyprotein is formed but is cleaved thereafter at an extremely rapid rate. However, available evidence indicates that such a large precursor is not released from the polysome as such. Jacobson et al (33) have measured the size distribution of polysome-bound nascent polypeptides of poliovirus-infected cells, and could not detect poly-peptides of higher molecular weight than 130,000. Similar findings were obtained in EMC virus (26). Furthermore, kinetic experiments have shown that the precursor of capsid proteins NCVP1a can be labeled and released even in a short pulse of 1–2 min (26, 42). Since it takes 10–12 min for the ribosome to traverse the entire length of the viral RNA molecule (25, 42), and since capsid proteins map near the 5' terminus (see below), the capsid protein precursor must be released independently soon after its synthesis.

---

[1] The nomenclature of poliovirus proteins is somewhat discouraging. Summers et al (39), in their initial work describing poliovirus-specific proteins, designated the four capsid virion proteins VP1–VP4, and the 10 noncapsid virus-specific proteins were named NCVP1–NCVP10, in order of their migration in sodium dodecylsulfate acrylamide gel electrophoresis. With the refining of the resolution techniques, more poliovirus-specific proteins were found, and these were marked by additional letters. For example, band NCVP1 has been resolved to the two proteins NCVP1a and NCVP1b (17), NCVP3 has been split to NCVP3a and NCVP3b (40), etc. To make matters worse, different groups of investigators are using different terminologies for some proteins. Thus, NCVP1b is called NCVP1$\frac{1}{2}$ by Baltimore and his associates (33), NCVP6 is also VP0 (41), and so on.

Although the hypothetical polyprotein is not released normally in its entirety, several lines of experimental evidence indicate that initiation of translation occurs only at one site on picornavirus RNA. Analysis of the products of cell-free protein synthesis directed by EMC virus or poliovirus RNA indicates the existence of a single initiation site (43–45). As in some other eucaryotic systems (9, 10), translation starts with a short "lead-in" peptide which is subsequently cleaved off (43, 46, 47). The single initiation site hypothesis also requires that an equimolar production of the three primary proteins (together with their respective cleavage products) will be found in vivo. Such equimolar proportions were found in EMC virus (25, 28). Other in vivo experiments utilizing pactamycin, a specific inhibitor of initiation, are also consistent with a single site initiation mechanism (see below).

The sum of these experiments is consistent with the notion that picornavirus translation is initiated at a single site, and that the three primary products are formed by the cleavage of nascent polypeptides.

Secondary Cleavages

Figure 1 summarizes the presently known cleavage steps in the formation of poliovirus proteins, including the secondary cleavages. The methodology used for elucidating this scheme of proteolytic conversions requires some comment.

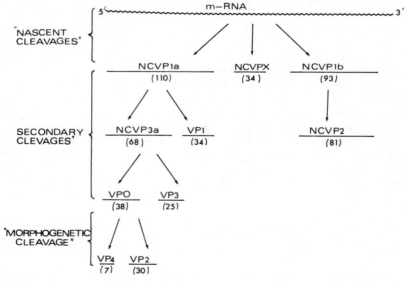

Figure 1   Post-translational cleavages in the formation of poliovirus proteins. Molecular weights ($\times 10^{-3}$) are indicated in the brackets and are the average of the values reported in (28, 33, and 186). The locations NCVP1b and NCVP3a can be regarded as tentative; they are based mainly on pactamycin mapping (28, 40). Polypeptides NCVP3b and NCVP4 (not shown) map near the 3' end (28) and therefore are probably derived from NCVP2. The positions and relationships of the other virus-specific proteins, not included in this scheme, are unknown. Further details are described in the text.

The gene order of the primary, intermediate, and stable viral proteins has been determined mainly by the use of pactamycin, a specific inhibitor of protein synthesis initiation (25, 28, 40, 42, 48). When proteins are labeled several minutes after the addition of pactamycin, previously initiated nascent chains are completed, and the label is selectively incorporated into the part farthest from the NH_2 terminus. Assuming a single initiation site in the translation of picornavirus proteins (see above) and that protein synthesis proceeds from the 5' to 3' end of mRNA, the labeling of polypeptides encoded near the 5' end of the genome will be most extensively inhibited. When initiation was selectively affected by treatment with hypertonic medium (49), essentially similar results were obtained. The validity of the assumptions involved in these methods (mainly the unique initiation site) is confirmed by the agreement of the results with genetic mapping data, where all capsid proteins are placed at one end of the genome (50).

In determining the relationships between the various precursors and their cleavage products, several methods have been used. Kinetic pulse-chase experiments in conjunction with molecular weight determinations may give a strong clue for possible precursor-product relationships in many instances. A mathematical model has been developed for such multistep cleavage systems (26). The method of pactamycin mapping mentioned above is also valuable in defining precursor-product relationships when used together with the other data. In some instances, the use of specific viral mutants may also be helpful. For example, in the defective interfering particles isolated by Cole et al (51), part of the genome coding for capsid proteins is deleted and the normal precursor of capsid proteins, NCVP1a, is missing as well (52). The final and most conclusive proof of precursor-product relationships is the demonstration of similarity in amino acid sequences, as studied by the comparison of tryptic peptides or cyanogen bromide fragments of postulated precursors and products.

The precursor-product relationships of poliovirus-specified proteins, as elucidated by the above methods, is depicted in Figure 1. Polypeptide NCVP1a is clearly the precursor of the four capsid proteins VP1-VP4, as determined by kinetic studies (14, 17, 22), pactamycin technique (28), genetic methods (52), and tryptic peptide analysis (17, 33). The role of VP0 as the immediate precursor of proteins VP2 and VP4 in the last steps of the assembly (see below) is also well established (17, 33). The gene order of the capsid proteins is 5' → 3' VP4-VP2-VP3-VP1 (28, 42). Much less is known about the formation, or even the identity, of nonstructural functional picornavirus-specific polypeptides. Such proteins, probably derived from NCVP2 and NCVPX, are those involved in viral RNA replication. Genetic (53) as well as recent biochemical (54) evidence indicates the existence of two distinct RNA replication complexes, one concerned with the formation of template ("minus") RNA strands and another that synthesizes progeny viral RNA. However, the polypeptide composition of the replication complexes is very heterogeneous, although they are enriched in nonstructural virus-specific proteins (50, 55). Some purification of EMC virus polymerase has been achieved (56), but its virus-specific components have not yet been identified. The suppression of host macro-molecular synthesis is another essential viral function (57). However, this function may not require a separate gene product. Cooper and his associates have recently proposed (58–60) that the inhibition of host protein synthesis is carried out by a precursor of viral structural unit.

It should be pointed out that the presently known proteolytic conversions, as described in Figure 1, account for less than one half the multitude of virus-specific polypeptides that can be detected in picornavirus-infected cells. Most of these are probably intermediates or waste fragments of the various cleavage processes. In addition, it is not known whether or not some proteolytic conversions may include the disposal of whole fragments by total proteolysis to acid-soluble products. For example, the cumulative mass of the three primary products is lower than the total coding capacity of picornavirus genome (25).

Proteolytic Cleavage in Picornavirus Assembly

The cleavage of VP0 to the capsid proteins VP2 and VP4 is associated with the final stages of poliovirus assembly (33, 41, 61). The icosahedral shell of picornaviruses is constructed of 60 identical structural units, or protomers; each protomer is assumed to be composed of the four nonidentical capsid proteins (62–64). According to the postulates of Rueckert, Baltimore, Phillips, and their associates (16, 62, 65), the morphogenesis of picornaviruses starts with a stepwise assembly of the "immature protomer" (64), which is a 5S particle containing one molecule each of VP0, VP1, and VP3, into increasingly larger structures. Thus, the assembly of five 5S particles is assumed to form a 14S structure, and twelve 14S particles associate to produce 73S empty capsids ("procapsids") (16, 62). Viral RNA then combines with the empty shell to make a 125S "provirion" structure (61), and morphogenesis is culminated by the cleavage of VP0 with the production of mature virions (41, 61).

Evidence for this postulated sequence of events in picornavirus morphogenesis is based mainly on kinetic and pulse-chase experiments. During the course of poliovirus infection, 5S material can be detected about $\frac{1}{2}$ hr before the appearance of empty capsids and virions (66). Upon chase with unlabeled amino acids, radioactivity in 5S particles decreases rapidly (66, 67), whereas it continues to increase in 14S particles (67) and in the higher aggregates (66). In these types of experiments radioactivity in 14S particles begins to decline only after about 30 min of chase (68), while empty capsids are quite resistant to chase (66); this may be due to the presence of large pools of labeled precursor particles. The postulate that 73S empty capsids are the precursors of virions is based mainly on experiments with guanidine, an inhibitor of viral RNA formation. In the presence of guanidine, empty capsids containing uncleaved VP0 accumulated (41). When guanidine was removed, radioactivity was lost from empty capsids, with a parallel increase in virions and a concomitant cleavage of VP0 to VP2 and VP4 (41). The role of the recently discovered 125S particle (the provirion) (61, 69) in virus morphogenesis is also based on pulse-chase experiments in the presence of guanidine and following its removal (61).

Experimental evidence for some features of this proposed assembly mechanism is not conclusive. This model requires that the various proteins will be present in the intermediary structures in equimolar amounts, but the reported relative amounts of VP0, VP1, and VP3 in 5S, 14S, and 73S structures of poliovirus are not equimolar (70). Most importantly, precursor-product relationships between the various particles cannot be unambiguously defined by pulse-chase experiments in this system, due to the high number and apparently large pool sizes of the intermediary structures. In addition, it has to be taken into account that some particles may be the product of abortive assembly, rather than intermediary steps in virus morphogenesis. Thus, Ghendon et al (67) have shown that in poliovirus-infected MiO cells (as opposed to HeLa cells), 73S empty capsids are not formed, and even in the presence of guanidine, 14S particles rather than empty capsids accumulate. This casts some doubt on the obligatory role of empty capsids in poliovirus morphogenesis. Obviously, more definite conclusions can be obtained only by in vitro experiments. To date, attempts to reproduce in vitro poliovirus assembly have met only limited success. Phillips et al (68, 71) have discovered that 14S particles can be assembled in vitro into 73S structures in the presence of extracts from virus-infected, but not from uninfected cells. Subsequently, it was demonstrated that this activity is localized in the rough membranes (65), and that 14S can be self-assembled to 73S material even in the absence of cellular extracts, provided that the concentration of the 14S particles is high enough (72). It was suggested that virus-modified membranes may facilitate this assembly reaction by their ability to adsorb and concentrate 14S particles (72). It should be noted, however, that extracts from MiO cells can also carry out the 14S → 73S in vitro conversion, whereas the 73S particle is not formed in vivo in these cells (67).

Although the complete details of picornavirus assembly have not yet been elucidated, there seems to be no doubt about the close association of the cleavage of VP0 to VP2 and VP4 with viral morphogenesis. This cleavage reaction occurs only in highly assembled structures and, furthermore, there seems to be a definite requirement for the association of RNA with the assembled structure to allow this cleavage. Therefore, this type of morphogenetic cleavage reaction seems to fulfill a role biologically different from the formative cleavages described earlier.

Enzymic Mechanisms

Very little is known about the nature of proteolytic enzyme(s) participating in the cleavages of picornavirus proteins. Several protease inhibitors have been utilized to block these cleavages in vivo. Diisopropylfluorophosphate (DFP) at rather high concentrations partially inhibits the processing of viral proteins in cells infected with poliovirus (33) and EMC virus (73). When poliovirus proteins were labeled in the presence of DFP, radioactivity had accumulated in proteins larger than NCVP1a (33). Similarly, in poliovirus-infected cells treated with the inhibitor of chymotrypsin tolylsulfonyl-phenylalanyl-chloromethyl ketone (TPCK), the accumulation of high molecular weight polypeptides ($> 200,000$) has been found (36). Korant (35) has reported that the action of the inhibitor varied with the cell line used: in monkey kidney cells, cleavage of poliovirus proteins was inhibited by TPCK, whereas in HeLa cells, only the trypsin inhibitor tolylsulfonyl-lysyl-chloromethyl ketone (TLCK) was effective. The latter finding is at variance with the results of Summers et al in HeLa cells (36); the reason for this discrepancy is not clear. A different inhibitor, iodoacetamide, prevents the formation of capsid proteins and causes the accumulation of NCVP1a, NCVP2, and NCVPX (74). It has been concluded that iodoacetamide blocks a later cleavage step than do serine protease inhibitors (74). In addition, zinc ions also inhibit the processing of picornavirus proteins by an as yet unknown mechanism (75). Although these experiments provide further evidence for the existence of cleavage mechanisms in the formation of poliovirus proteins, the action of these protease inhibitors on in vivo systems can hardly be regarded as specific. The chloromethyl ketones, TPCK and TLCK, may alkylate a variety of cellular proteins (36), and the actions of DFP and iodoacetamide are also rather nonspecific. Thus, the effect of serine protease inhibitors on intracellular protein breakdown (76) has been shown to be due to a nonspecific inhibition of cellular energy metabolism (77). It is well known that the breakdown of intracellular proteins is energy dependent (78–80); it remains to be seen whether similar energy-requiring proteolytic mechanisms participate also in the processing of virus-specific proteins.

Available information concerning the characterization of these cleavage reactions by in vitro systems is very scanty. Holland et al have reported that purified virions contain some protease activities (81), but since this activity degrades mature capsid proteins, its relationship to the cleavage of precursor proteins is not clear. Korant (35) has shown that extracts from poliovirus-infected cells degraded high molecular weight precursor proteins to smaller polypeptides with molecular sizes similar to NCVP2, VP0, VP1, and VP3. In contrast, the cleaving activity contained in extracts of uninfected cells produced only proteins resembling NCVP1a and NCVP2. Cleavage of NCVP1a, accumulated by prior treatment with iodoacetamide, is carried out by infected but not uninfected extracts (74). It was suggested that the initial nascent cleavages are catalyzed by a host enzyme, while the subsequent splicing of NCVP1a is carried out by a second protease which is either virus specified or activated by infection (74). Opposing these findings, high molecular weight precursors from EMC virus-infected Krebs carcinoma cells were degraded in vitro by extracts from both infected and uninfected cells, but the products of degradation had no resemblance to the proteins formed in vivo (73). In addition, while the processing of precursor proteins in vivo was inhibited by DFP, the in vitro degradation was not affected by this drug. The authors concluded that the cleavages observed in the in vitro system were probably due to nonspecific proteases liberated during homogenization of cells (73).

It thus seems that much remains to be elucidated about the nature and mode of action of the enzyme systems participating in post-translational cleavages of viral proteins.

POLYPEPTIDE CLEAVAGE IN VARIOUS ANIMAL VIRUSES

Apart from picornaviruses, cleavage of proteins has been described in other widely different types of animal viruses. In the arboviruses, Sindbis virus and Semliki forest virus, both formative and morphogenetic polypeptide cleavages apparently occur. Viruses of this group consist of an RNA-containing nucleocapsid surrounded by a lipoprotein envelope (82). There are only three proteins in the virion, one nucleocapsid and two envelope proteins (83–85). On the other hand, there are as many as 9–13 additional virus-specific proteins in Sindbis virus-infected cells, some of them considerably larger than virion proteins (83). In pulse-chase experiments, the label is shifted from high molecular weight proteins to the smaller structural polypeptides (86–89). As in the case of poliovirus, extremely large Sindbis virus-specific proteins accumulate in the presence of protease inhibitors (90, 91) or in temperature-sensitive mutants at the nonpermissive temperature (92–94). It is believed, therefore, that Sindbis virus proteins originate from a large common precursor (91, 93). In fact, tryptic peptide similarity between the large protein accumulated in a temperature-sensitive mutant of Sindbis virus and all three virion proteins has been demonstrated (93). In addition, a large molecular weight peptide which contains tryptic peptides of the capsid protein is synthesized in a cell-free system with Sindbis virus mRNA (95). There is some evidence for a single initiation site in the translation of Sindbis virus proteins, since pactamycin treatment inhibits the labeling of capsid proteins much more than that of envelope proteins (87). Another type of cleavage appears to be the conversion of a precursor of intermediary size to one of the envelope proteins in both Sindbis virus (87) and Semliki forest virus (85). This conversion occurs at the plasma membranes and may therefore be associated with virus morphogenesis (87, 96). Although evidence is still incomplete, these findings indicate that the formation of group A arbovirus proteins is quite analogous to the processing of picornavirus polypeptides.

In the large DNA-containing poxviruses, protein cleavages are closely linked to morphogenetic events. At least two major proteins of the vaccinia virus core are derived from larger precursors by post-translational processing (97–100). These cleavages are rather slow and occur at a time coincident with the maturation of virus particles (98). In the morphogenesis of this complex virus, assembly begins with the formation of envelope units (101). Rifampicin, which inhibits vaccinia virus morphogenesis at the stage of envelope formation (102, 103), also prevents cleavage of these precursor proteins (97). Upon removal of the drug, the formation of normal envelope units precedes restoration of the cleavage of core protein precursors (97, 103). Thus, the influence of the drug on polypeptide cleavage is probably secondary to the block in envelope assembly (103). In a rifampicin-resistant strain of vaccinia virus, the drug had no influence on either envelope formation or protein cleavage (104). The action of rifampicin is rather specific to this stage of morphogenesis, since the processing of a third protein, which is located outside the virus core, is not influenced by the drug (100).

In the morphogenesis of another group of relatively large DNA viruses, the adenoviruses, proteolytic cleavage mechanisms play a somewhat similar role. Coincidentally with the conversion of empty capsids to mature adenovirus-2 virions, five different polypeptides are converted to smaller virion proteins (105–107). The precursors are only slightly larger than the products, and there is no evidence for the existence of high molecular weight precursors (107). It appears that the processing mechanisms of adenovirus, as those of vaccinia virus and T4 bacteriophage, are restricted to morphogenetic events. Post-translational proteolytic conversions have been reported in viral systems as different as influenza virus (108–110), paramyxovirus (111), avian RNA tumor virus (112), and possibly polyoma virus (113). However, these conversions do not play a significant role in viral development in all cases. For example, in influenza virus, two glycoproteins located on the surface of the viral envelope are derived by the cleavage of a large precursor (108–110, 114). It has been shown more recently, however, that this cleavage is due to the action of serum plasmin (115) and is not required for virus assembly or for any other biological properties of the virion (116).

ROLE OF PROTEIN CLEAVAGE IN THE MORPHOGENESIS OF COMPLEX BACTERIOPHAGES

Cleavage of proteins during the assembly of large DNA-containing bacteriophages is a widespread phenomenon, although it may not be a general mechanism common to the assembly of all phages of this type. Since cleavage of coliphage T4 head components was first described in 1970 (117–120), proteolytic cleavages were found to occur during the morphogenesis of the head of bacteriophage P2 (121) and in the maturation of the head and tail of coliphage λ (122–125) and T5 (126). On the other hand, no proteolytic cleavage phenomena could be detected in the morphopoiesis of other DNA containing bacterial viruses, such as Salmonella phage P22 (127, 128) and coliphage T3 (129).

Cleavage of Head Proteins During the Assembly of Bacteriophage T4

MATURATION OF THE HEAD   The assembly of the head of T4 has been recently reviewed (130) and is described in full elsewhere in this volume (21). We limit our discussion to the outline of the assembly process as related to the proteolytic cleavage events associated with it.

The head of the phage appears to be based on an icosahedral design with a triangulation number of 21 (131, 132). The mature T4 DNA, which is a linear uninterrupted duplex molecule with a cyclically permuted, terminally repetitious nucleotide sequence (133, 134), occupies most of the internal volume of the head (135, 136). About 11 different protein species compose the capsid of this phage (119). Three additional internal proteins designated IP I, IP II, and IP III* (137) and three acid-soluble internal peptides (138) are located inside the head and are found in association with the DNA. The assembly of T4 head is controlled by some 18 genes (139, 140). Genes 20, 21, 22, 23, 24, and 40 are required for determination of the size and shape of the head (139, 141). Genes 2, 4, 13, 14, 16, 17, 49, 50, 64, and 65 control the subsequent steps of the head maturation (140). Gene 31 is thought to control the solubility of the major head protein (119, 142).

The events leading to the packaging of the DNA molecule within its protein shell are still largely unknown. Three possible models for the assembly of T4 head have been offered. One model proposes that an empty capsid is formed first and is then converted into mature phage head with the gradual penetration and packaging of the DNA (143, 144). An alternative model suggests the formation of a DNA condensate around which the protein shell is later assembled (145). A third model hypothesizes that capsid formation and DNA packaging are concomitant and that a small piece of DNA serves as a core for head formation (130). Available experimental evidence is most consistent with a pathway of head maturation that involves formation of a precursor protein shell largely devoid of DNA, into which nucleic acid is packaged at a later stage. Identification of intermediates in the pathway of T4 head maturation was attempted through experiments of three types: (*a*) ultrastructural studies of head-related structures formed at various stages

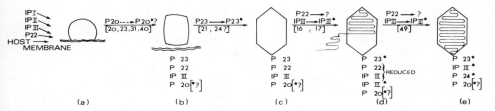

Figure 2   Protein cleavages associated with a tentative pathway of the assembly of the head of bacteriophage T4. The numbers in brackets underneath the arrows designate genes that are probably involved in particular steps of the assembly of the head. The cleavage events are indicated above the main arrows. The prefix P designates a gene product. Asterisk marks a cleaved form of the gene product. The various "prohead" designations are according to Laemmli & Favre (143).

(*a*) *Assembly core*   Biochemical (146) and ultrastructural (144, 157) studies suggest the in vivo existence of such early structures. It was suggested (146) that the internal proteins, shortly followed by P22, become localized in a complex with a bacterial membrane and serve as a nucleation site around which the capsid proteins are later assembled. Since the internal proteins seem to be dispensable (146, 187), the assembly core probably serves as an initiation complex rather than a shape-determining structure (187).

(*b*) *Prohead I* (400S particles)   These structures were identified and partially isolated by Laemmli & Favre (143) and observed electron microscopically by Simon (144). It is a short lived particle composed mainly of uncleaved P23, P20, P22, IP III, and perhaps other minor proteins (143). It is possible that P20 is modified prior to the prohead I stage (164). Prohead I is membrane bound and contains less than 1% of the phage DNA complement (143). This structure is analogous to gene 24 defective τ particle (147–149, 188).

(*c*) *Prohead II* (350S particles)   The major event leading to the conversion of prohead I into prohead II is the cleavage of P23 to form P23* (143). It is possible that P24 joins the capsid at this stage (143, 147). This particle does not seem to contain DNA (143).

(*d*) *Prohead III*   These intermediates, containing 10–30% (152) to 40–50% (143) of the phage DNA complement and sedimenting at 320–550S, are bound to the replicative DNA (143, 152). During the conversion of prohead II into prohead III, about 50% of P22 and IP III are cleaved into an unknown product and IP III*, respectively (143). A structure somewhat analogous to prohead III is the gene 49 defective head-related particle, which is also partially filled with DNA (143, 152). It should be emphasized, however, that genes 16, 17, and 49 defective particles contain the capsid polypeptides in their cleaved form (143).

(*e*) *Mature Head*   The mature T4 head, sedimenting at 1100S, contains a full DNA complement and all the capsid proteins in their final cleaved forms.

of the phage development (144); (b) biochemical (119, 146–156) and ultrastructural (141, 147, 157, 158) studies of head-related particles accumulated in cells infected by phages defective in genes controlling various steps of head assembly; and (c) pulse-chase experiments in cells infected with either wild-type or mutant phages establish the percursor-product relationships between successive inter-mediates in the pathway of head formation (143, 147, 148). However, all these types of experiments are not conclusive; it was argued that empty capsids could arise from previously filled intermediates which, by virtue of their fragility, tend to lose their DNA (121, 145, 147, 157, 159; see also 146 for a possible association of the "assembly" core with T4 DNA). In addition, a recent paper by Chao et al (160) indicated the involvement of host DNA repair enzymes in the assembly of T4 heads. Such connection between DNA metabolism and the morphogenesis of the head suggests that DNA does play a morphopoietic role in the assembly of the phage T4 head. A postulated pathway of T4 head maturation based on the model of the formation of empty heads and their sequential filling with DNA is shown diagrammatically in Figure 2. This scheme is based mainly on the tentative path-way of T4 head maturation presented by Laemmli & Favre (143). More details on the various steps of the assembly process are given elsewhere in this volume (21). The role of the host membrane in T4 morphogenesis was reviewed (161). Additional data are supplied in several recent papers (143, 146, 147). The functions assumed for the various phage genes that control head development were discussed by several authors (141, 143, 146, 147, 149, 150, 152).

POST-TRANSLATIONAL CLEAVAGE EVENTS　Four head-associated phage-specified proteins are cleaved during the morphogenesis of coliphage T4. Three of the head proteins cleaved are structural components of the capsid coded for by genes 22, 23, and 24 and synthesized late in infection (119). The fourth is internal protein IP III, which is an "early" product of an unidentified T4 gene (119, 137, 162).

Cleavage of head-related precursor proteins was first described in 1970 by four separate groups of investigators whose experimental approaches were similar (117–120). Essentially, proteins of the purified capsid and total lysates of wild-type and head defective T4-infected cells were compared by means of sodium dodecyl-sulfate (SDS) acrylamide gel electrophoresis. Comparative studies of this type allowed identification of the products of genes 18, 19, 20, 22, 23, and 24, which were all present in wild-type lysates and missing from the respective defective gene lysates (119). In the process of assigning polypeptides to their genes of origin, several authors have noted significant differences between the molecular weights of certain gene products present in wild-type and head defective lysates and capsids (117–120). Specifically, it was reported that during wild-type infection, P23 is converted from a protein of 55,000–56,000 daltons to the major head protein P23* with a molecular weight of 46,000–48,000 (117–120). Similarly, Laemmli (119) and Hosoda & Cone (117) have shown that P22 of 31,000 mol wt is cleaved during wild-type infection to form an unidentified product. Laemmli (119) has also demonstrated that P24 (molecular weight 45,000) is converted during normal T4 maturation into a minor head component designated P24* with a molecular weight of 43,000. Likewise, IP III (23,500 daltons) is cleaved to form the internal protein IP III* (21,000 daltons) (119, 137). All the precursor proteins described are accumulated in their uncleaved form during infection with T4 mutants defective in any of the head genes 20, 21, 22, 23, 24, and 31 (117–120). In addition to the above well-defined proteolytic cleavage events, several additional cleavages of T4 polypeptides were proposed: Coppo et al (163) and Laemmli & Favre (143) have reported that a T4 protein designated B1 (163) appears to be a cleavage product of an unknown precursor. Recently, Bolin & Cummings (164) have shown that the arginine analog L-canavanine inhibits the cleavage reactions of P22, P23, P24, and IP III. This drug also prevents the appearance of P20 (head protein) and P10, P12, and P18 (tail proteins), indicating that these poly-peptides may also arise by proteolytic splicing during normal T4 assembly. However, cleavage of P20 and any tail protein was not demonstrated in infected cells unexposed to amino acid analogs (119, 143, 165–167). Also, the identities of the precursors to these proteins are unknown.

The kinetics of the proteolytic reactions was established by pulse-chase experiments. This technique, in conjunction with the use of specific mutants, allowed the establishment of precursor-product relationships between P23 and P23*, P24 and P24*, and IP III and IP III*. In the case of P22, although it was found to disappear rapidly during the chase, no band that could be derived from this protein has been traced in the gel pattern (117, 119). The cleavage reactions for P22 and P23 were found to be very rapid; about 50% of the precursor proteins disappeared within 2–3 min following chase of the label. In contrast, the cleavage of IP III is much slower (119).

Cleavages bring about the loss of small fragments from P23, IP III, and P24 of about 10,000, 2,500, and 1,500 daltons, respectively (117–120). Attempts to find these fragments have failed, possibly because peptides of such small size were not sieved on the acrylamide gels at the concentrations used (119). It is also possible that the fragments are further broken down to undetectable sizes. A possibility was raised that the internal peptides II and VII having molecular weights of 3900 and 2500, respectively, could be the products of the extensive cleavage and degradation of P22 (146, 168). This contention is supported by the finding that P22 is an internal component associated with the assembly core (146) and also by a study inferring the sharing of common antigenic determinants by P22 and the internal peptides (169).

Several lines of evidence suggest that proteolytic cleavages are intimately linked to the assembly process. As mentioned above, the cleavage of P22, P23, and IP III is blocked in mutants defective in any of the head genes 20, 21, 22, 23, 24, and 31. As a result, mature heads are not formed and aberrant head-related structures accumulate instead (120, 141, 147–149). Also, IP III, which is an early protein synthesized 4–5 min after infection, starts to be cleaved only at late times (about 14–17 min), indicating the coupling of its proteolytic cleavage to the assembly process (119, 170). Recently, Laemmli & Favre (143) were able to differentiate between the cleavages of P23 and those of other three head proteins and to correlate these events with defined stages of the head assembly process.

These investigators reported that prohead I, which contains the uncleaved P22, P23, and IP III, is converted into prohead II with the concomitant cleavage of P23 to P23*, whereas the cleavage of other proteins occurs at later stages of the morphogenesis (see Figure 2). Bijlenga et al (155) have also shown that under permissive conditions, P23 of the gene 24 defective head-related τ particle is transformed in situ into P23* of the capsid. This transformation is presumably analogous to the cleavage of P23 during the conversion of prohead I to prohead II, and thus genes 21 and 24 defective τ particles are analogs of prohead I (147–149). However, whereas the gene 24 defective particles are maturable, gene 21 defective structures are abortive nonmaturable particles (147). Since both proheads I and II are thought to be devoid of DNA (143, 147–149), it was concluded that the cleavage of P23 precedes the DNA packaging step. The cleavage of P22 and IP III, on the other hand, occurs concomitantly with the packaging of phage DNA: Prohead III, into which only part of the DNA complement is packed, contains P22 at a reduced amount and both IP III and IP III* (Figure 2) (143). Cleavage of P22 and IP III is completed with the conversion of prohead III into mature T4 head, which contains a full complement of DNA, P23*, P24*, and IP III* and no uncleaved precursor polypeptides (119, 143). These results strongly suggest a close link between the cleavage of P22, IP III, and possibly P24 and the process of packaging the DNA molecule within its protein shell. In addition, the above set of experiments confirms that all the cleavage reactions occur in large multimeric structures as originally proposed by Laemmli (119).

MECHANISMS OF THE CLEAVAGE REACTION   Very little is known about the mechanism of the proteolytic cleavage reactions associated with T4 head assembly. The origins, number, and identity of the cleavage enzymes, their mechanisms of action, and specific requirements for activity are completely unknown. Some insight into the mode of the proteolytic digestion that converts P23 into P23* was gained in the work of Celis et al (171), which demonstrated that fragments synthesized by several *amber* mutants in gene 23 contained tryptic peptides present in the uncleaved P23 but not in the modified P23*. It was concluded therefore that it is the N-terminal portion of P23 that is cleaved during T4 maturation. Some further aspects of the cleavage reactions were recently described in cell-free proteolytic systems (170, 172, 173). It was reported (172) that T4 precursor proteins could be cleaved in vitro with the formation of a peptide that was indistinguishable from internal peptide II—a presumed in vivo cleavage product with a molecular weight of 3900 (172, 174). The in vitro formation of this short peptide was dependent on the presence of a factor present in extracts of T4-infected cells but absent from extracts of uninfected cells. It was also shown that both native and acid-denatured precursors could serve as substrates for cleavage in vitro. This result is somewhat surprising, since it would be expected that it is the native conformation of precursor proteins that determines their recognition by the cleaving system. Bachrach & Benchetrit (170) have recently reported that lysates of T4-infected cells possess proteolytic activity which converts an early protein—presumably IP III—into a polypeptide with an electrophoretic mobility similar to that of IP III*. This putative cleavage enzyme appears in T4-infected lysates at 14–16 min postinfection, in accordance with earlier observations on the cleavage of IP III at a late stage of the infection (119). Proteins labeled late in infection were not cleaved under the same conditions, indicating that the activity described is not involved in the cleavage of P22, P23, and P24. These authors have also reported that whereas the in vitro cleaving activity is present in mutants defective in genes 20, 22, 23, and 24, it is missing in *amber* mutants of gene 21. It was proposed, therefore, that the product of gene 21 is involved in the cleavage of IP III. It should be noted, however, that this differs from the in vivo findings, in which mutation in any of these genes prevents the cleavage of IP III (117–120). In a more recent communication (173), the above authors have demonstrated that only the cleaved form of the internal proteins, spliced either in vivo or in vitro, binds to T4 DNA and not with heterologous DNA. Poglazov & Levshenko (175) have found that a T4-induced trypsin-like proteolytic activity reaches a maximum 18 min postinfection. The relevance of this activity to phage assembly is not clear as yet.

Amino acid analogs which have been widely used for the characterization of cleavage reactions occurring in animal viruses (14, 33, 34) were used to a limited extent in the study of T4 protein cleavage. Couse et al (176) have shown that the arginine analog L-canavanine prevents the intracellular development of T4 and induces the formation of aberrant tubular heads and polyheads. This analog did not markedly inhibit total protein and RNA synthesis but DNA synthesis and cleavage of P23 were blocked (177, 178). It is not clear as yet whether the primary effect of the analog is to block synthesis of DNA or to inhibit the scission of capsid proteins. Interestingly, exposure of cells infected by T4 to L-canavanine, followed by addition of arginine, causes the production of normal phages together with enormous infectious particles named "lollipops," which contain P23*, P24*, and IP III* and no precursor proteins (179).

A recent study on the effects of L-canavanine on the maturation and utilization of specific T4 gene products (164) revealed that this analog prevents the cleavage reactions of P22, P23, P24, and IP III but, surprisingly, it also inhibits the appearance of P20 (head protein) and P10, P12, and P18 (tail proteins). As a consequence of these effects on specific phage proteins, tail assembly as well as head assembly are inhibited. It was suggested that in addition to the already known cleavage events, P20, P10, P12, and P18 are also processed. However, it remains to be seen whether the proposed protein modification events can be demonstrated in T4-infected cells not exposed to L-canavanine.

ROLE OF THE HOST CELL   Proteolytic cleavage of phage-specified proteins could be either a phage function, a host function, or both. As mentioned, the in vitro cleavage of some T4 precursor proteins requires the presence of infected cell extracts (170, 172). However, the isolation of mutant strains of *Escherichia coli,* in which cleavage and assembly are specifically blocked, may allude to a possible host participation in the cleavage of phage-specified proteins. Such host mutants, isolated in several laboratories, were designated *tab B* (163), *mop* (180), and *gro E* (181) and seem to be closely related (163, 181). While phage adsorption, DNA injection, and tail production are unaffected by the bacterial mutation (180, 181), the cleavage of the phage precursor proteins and capsid formation are blocked (163, 180, 181). The development of other phages, such as all T-even phages (163, 180), ϕ80 (180, 181), and 434 (181), is also often inhibited in these mutant strains. The uncleaved precursor polypeptides P23 (163, 181), P22, P24, and IP III (163) accumulate in membrane-associated lump-like aggregates similar to those found in wild-type bacteria infected with gene 31 defective phage (163, 180, 181). Indeed, phage mutations Com^B (163), "R" (180), and T4ε (181), mapped in the region of gene 31, enable the phage to overcome the block exerted by the host mutation and allow the precursor proteins to be cleaved and head assembly to proceed normally (163, 180). It was suggested (163, 180, 181) that the bacterial mutation impairs the function of T4 gene 31 and that a modified P31 produced by the mutant phage overcomes the bacterial lesion. Thus, solubilization of precursor proteins of the head, their modification, and ordered assembly require interaction between a wild-type bacterial gene product and P31. It is not known, however, whether the bacterial mutation primarily affects the cleavage reaction or some other critical event of the head morphogenesis. There are two possible mechanisms by which phage and host products may interact: 1. P31 together with the bacterial product, or under its influence, could be part of the proteolytic system responsible for the cleavage of T4 precursor proteins (180); or 2. P31 together with the host factor are involved in some early step of T4 morphogenesis, such as the interaction between the head components and the bacterial membrane on which they are assembled (163, 180). There is some evidence that the membranes of the *mop* mutant are indeed biochemically and functionally modified (180). Thus, the possibility of a lesion in the bacterial membrane as a primary cause for inhibition of the assembly of T4 head was favored in that case.

Cleavage of Head and Tail Proteins in Some Other Large DNA-Containing Bacteriophages

Several authors have recently described proteolytic cleavage events occurring during the assembly of bacteriophages T5 (126), λ (122–125), and P2 (121). Although these reactions are not as well characterized as those associated with the morphogenesis of coliphage T4, their study adds new insight into the role and scope of protein cleavage in the assembly of bacteriophages. In two cases, T5 and λ, precursor proteins of the tail undergo cleavage during maturation of the phage (124, 126). Since assembly of the tail does not involve any interaction between DNA and proteins, these cases indicate that post-translational cleavage may also play a role in the protein-protein interactions responsible for ordered assembly of the tail.

BACTERIOPHAGE T5   Coliphage T5 contains at least 15 different structural polypeptides, among them a major head component with a molecular weight of 32,000 and a major tail component that has a molecular weight of 55,000 (182). Zweig & Cummings (126) have recently shown that at least three of the phage-specified proteins undergo proteolytic cleavage during the assembly of T5. Two of the polypeptides cleaved are related to the morphogenesis of the head, whereas the third is a precursor to a minor component of the tail. Specifically, the following cleavage events have been reported: A tail-related precursor protein with a molecular weight of 135,000 is cleaved to form a minor tail component having a molecular weight of 128,000. A second precursor protein with a molecular weight of 50,000 undergoes cleavage to form the major protein component of the head, which has a molecular weight of 32,000. A third T5-specified minor component of the head (molecular weight 43,000) also seems to be a cleavage product of a slightly larger precursor protein.

The close linkage between the scission of the tail-specific precursor protein band *a* and the morphogenesis of the tail is indicated by the finding that all studied tail-defective phages fail to induce cleavage of the band *a* protein under restrictive conditions. A block in the synthesis of the major head protein, on the other hand, does not interfere with the proteolytic modification of the tail-related precursor protein or with the assembly of the tail. Interestingly, cleavage of the tail-related polypeptide is blocked in *E. coli* mutants *groEA639* and *groEA36* infected with wild-type T5. This block is specific to tail proteins since cleavage of the head-related proteins proceeds normally and normal heads are produced in these mutant hosts. A T5 mutant designated *T5ε6* is able to overcome the bacterial lesion and to propagate normally in the *groEA* hosts. The above finding raises the possibility of a requirement for interaction between host and phage products for cleavage and assembly of the tail-related proteins. Cleavage of the head-related proteins and assembly of the head are specifically inhibited by the arginine analog L-canavanine. Although the drug interferes to some extent with the assembly of the tail, a significant amount of competent tails is produced in its presence. It seems, therefore, that the head- and tail-specific cleavage events are independent and under different controls.

BACTERIOPHAGE P2 In a recent paper, Lengyel et al (121) have described the cleavage of two head-related precursor proteins during the morphogenesis of coliphage P2. The product of gene *N*, having a molecular weight of 44,000, is cleaved to form mainly the major capsid protein with a molecular weight of 36,000. In addition, four other minor head components with molecular weights ranging between 37,400 and 42,200 seem also to be cleavage products of the same precursor protein. The product of gene *O* with a molecular weight of 30,000 also undergoes cleavage with approximately the same kinetics as that of the *N* protein. The product of the cleavage of the *O* protein is unidentified but it is certainly smaller than 17,000 daltons. The cleavage reactions of the *N* and *O* proteins are coupled: All *amber* mutants in the *O* gene produce measurable amounts of the *N* gene product but this precursor protein does not undergo cleavage. Conversely, an *amber* mutation in the *N* gene blocks cleavage of the *O* protein. In contrast to the cleavage of the T4 precursor proteins, which is closely linked to DNA synthesis, both DNA and protein synthesis are not required for the cleavage of the two P2-specified precursor head proteins.

BACTERIOPHAGE λ Several instances of post-translational scission of head and tail proteins of bacteriophage λ have been described. Two proteins of molecular weights 35,000 and 12,000, specified by λ genes E and D respectively, make up about 94% of the protein mass of the λ capsid (183). Of the several remaining minor protein components of the head, a polypeptide designated h3 of molecular weight 56,000 accounts for 2–3% of the shell protein mass (122). The kinetics of the formation of h3 suggests that it is a conversion derivative rather than a primary product of any λ gene (122). Murialdo & Siminovitch (123) and Georgopoulos et al (122) have proposed that the precursor to h3 is either the product of head gene B (pB) or the product of head gene C (pC). More recently, Hendrix & Casjens (184) have shown by tryptic peptide analysis that pB is the precursor of h3. It was also demonstrated that in addition to pB and pC, the product of gene E (pE) is needed for this conversion to occur (123). Georgopoulos et al (122) have found that a host function is also required for either the production or the functioning of h3. During normal λ head morphogenesis, h3 is prominent in fast sedimenting head-related structures. In contrast, this protein is missing from the fast sedimenting particle in extracts of $groE$ mutant strains that specifically block the assembly of λ heads. It was proposed therefore, that the $groE$ mutation either inhibits the cleavage reaction forming h3 or blocks the association of h3 with the fast sedimenting structure. Since some $groE$ mutants called $groEB$ can be compensated by mutations in the λ head genes E or B, it was suggested that the proteins specified by λ genes E, B, and possibly C interact with a bacterial component defined by the $groE$ mutation. Thus, the bacterial gro function could be directly or indirectly involved in the formation of h3.

A novel cleavage and possibly a fusion reaction involved with λ head assembly has been described recently by Hendrix & Casjens (125). Bacteriophage λ heads contain two additional minor proteins, X_1 and X_2, having molecular weights of 31,000 and 29,000, respectively. On the basis of the tryptic fingerprints of ^{35}S labeled X_1 and X_2, it was concluded that X_2 is a proteolytic cleavage product of X_1. In addition, both X_1 and X_2 contain tryptic peptides also present in either pE or pC. Only part of the different sequences of pE and pC appears in X_1 and X_2 and these sequences are found in the products in equimolar amounts. This evidence indicates, therefore, that X_1 and hence X_2 may be the products of a fusion reaction between pE and pC. The covalent bond joining pE and pC does not seem to be a disulfide bridge. The recent study of Hendrix & Casjens (184) indicates that the processing of pC, as well as the cleavage of the head protein pB, occur on nascent multimeric structures rather than prior to the assembly of the protein subunits.

Hendrix & Casjens (124) have reported recently that a λ tail protein is also cleaved prior to the attachment of tails to heads. The tryptic digest of a minor tail protein designated t2 with a molecular weight of 78,000 was found to share many peptides in common with a digest of another protein pH, which is found in λ-infected lysates but not in phage particles and has a molecular weight of 90,000. It was proposed, therefore, that t2 is derived from pH by proteolytic cleavage. It was also found that tails that accumulate in head-defective λ mutants contain polypeptide t2 rather than pH. The authors concluded, therefore, that the cleavage of pH to t2 occurs independently of the attachment of tails to heads (124).

CONCLUDING REMARKS

In considering the possible functions of post-translational cleavage of polypeptides in various viral systems, a clear distinction should be made between formative and morphogenetic proteolytic scission. Formative cleavage of proteins is an integral part of the translation mechanism of certain animal viruses. The most plausible assumption concerning its role seems to be that originally proposed by Jacobson & Baltimore (14), namely that the protein synthesizing machinery of animal cells, in contrast to that of bacteria, is incapable of carrying out internal initiation in the translation of polycistronic viral messages. Since post-translational scissions are characterized by a high degree of specificity and precision, they can effectively replace translational punctuation signals. There may be some biological advantages in this mechanism, such as the equimolar production of polypeptides required in identical amounts (for example, picornavirus capsid proteins). On the other hand, it is evident that this mechanism does not enable the regulation of synthesis of individual polypeptides, in contrast to other viral systems in which coat proteins are synthesized in much larger amounts than polypeptides involved in RNA replication. Since the yield of viral RNA production in picornaviruses appears to be well balanced with the rate of formation of capsid proteins, additional and as yet unidentified regulatory mechanisms have to exist which influence either the activity or the stability of picornaviral replication complexes (20, 59, 185).

In contrast to formative cleavages, the much more widespread morphogenetic proteolytic scissions are characterized by their occurrence in multimeric structures that are intermediary stages in viral assembly. Furthermore, these proteolytic cleavages occur at rather specific steps of viral morphogenesis, and they are invariably blocked when the normal course of morphogenesis is prevented by several specific agents and viral or host cell mutations. However, the exact roles of the proteolytic scissions in viral assembly are still largely unknown. It appears that protein cleavages associated with viral morphogenesis are not limited to a single type of molecular interaction and that the cleaved polypeptides may interact with either nucleic acids or with other proteins. Thus, in the maturation of the head of bacteriophage T4, cleavage of the core proteins P22 and IP III seems to be intimately linked to the packaging of DNA, and it has been suggested that in this case protein cleavage could either provide DNA binding sites or bring about condensation of the DNA through the production of acidic internal peptides (143). In contrast to the cleavage of the core proteins, the scission of P23 is completed before packaging of the DNA begins (see Figure 2). That protein cleavage is not limited to DNA-protein interactions is also illustrated by the finding that such reactions occur in the assembly of the tails of bacteriophages T5 and λ. In the latter case, it has been proposed that protein cleavage may produce a higher energy configuration of tail proteins, providing the energy required for DNA injection (124). On the other hand, in poliovirus morphogenesis, the cleavage of VP0, which converts the provirion to mature virion, may have a role in the stabilization of the RNA-containing particle, since the former but not the latter structure can be disrupted by denaturing agents (61).

A common feature of these various types of molecular interactions might be that all proteolytic modifications convert the assembly reactions in which they participate to irreversible processes (130). This characteristic and the dependence of protein cleavage on correct particle conformation may provide a mechanism for the sequential assembly of the components of relatively complex viral structures. Possibly, structural proteins remain in their uncleaved form until a specific intermediary structure is assembled and only then does cleavage occur, making that specific assembly step irreversible and enabling the association of the particle with additional components. Obviously, further advance in elucidating the pathways of morphogenesis of complex viruses is needed in order to gain a better understanding of the role of proteolytic cleavages in viral assembly.

ACKNOWLEDGMENTS

The authors wish to thank Drs. B. E. Butterworth, S. R. Casjens, P. D. Cooper, R. W. Hendrix, E. Kellenberger, G. Koch, and B. A. Phillips for making their results available to us prior to publication, and Drs. A. Cohen, H. Engelberg, B. D. Korant, and B. Moss for helpful comments. Special thanks are due to Mrs. Rachel Neiger for devoted secretarial assistance.

Literature Cited

1. Burdon, R. H. 1971. *Progr. Nucl. Acid Res. Mol. Biol.* 11:33–79
2. Bautz, E. K. F. 1972. *Progr. Nucl. Acid Res. Mol. Biol.* 12:129–60
3. Neurath, H., Walsh, K. A., Winter, W. P. 1967. *Science* 158:1638–44
4. Davie, E. W., Kirby, E. P. 1973. *Curr. Top. Cell. Regul.* 7:51–86
5. Müller-Eberhard, H. J. 1971. *Harvey Lect.* 66:75–104
6. Steiner, D. F. et al 1969. *Recent Progr. Horm. Res.* 25:207–82
7. Bellamy, G., Bornstein, P. 1971. *Proc. Nat. Acad. Sci. USA* 68:1138–42
8. Judah, J. D., Gamble, M., Steadman, J. H. 1973. *Biochem. J.* 134:1083–91
9. Milstein, C., Brownlee, G. G., Harrison, T. M., Mathews, M. B. 1972. *Nature New Biol.* 239:117–20
10. Mach, B., Faust, C., Vassali, P. 1973. *Proc. Nat. Acad. Sci. USA* 70:451–55
11. Katunuma, N. 1973. *Curr. Top. Cell. Regul.* 7:175–203
12. Holzer, H. et al 1973. *Advan. Enzyme Regul.* 11:53–60
13. Cabib, E., Ulane, R. 1973. *Biochem. Biophys. Res. Commun.* 50:186–91
14. Jacobson, M. F., Baltimore, D. 1968. *Proc. Nat. Acad. Sci. USA* 61:77–84
15. Baltimore, D. 1969. *The Biochemistry of Viruses,* ed. H. B. Levy, 101–76. New York: Dekker
16. Baltimore, D. 1971. *From Molecules to Man, Perspectives in Biology,* ed. M. Pollard, 1–14. New York: Academic
17. Summers, D. F., Roumiantzeff, M., Maizel, J. V. 1971. *Strategy Viral Genome, Ciba Found. Symp., 1971,* 111–24
18. Baltimore, D. 1971. *Bacteriol. Rev.* 35:235–41
19. Baltimore, D. 1971. *Trans. NY Acad. Sci.* 33:327–32
20. Sugiyama, T., Korant, B. D., Lonberg-Holm, K. K. 1972. *Ann. Rev. Microbiol.* 26:467–502
21. Casjens, S., King, J. 1975. *Ann. Rev. Biochem.* 44:555–611
22. Summers, D. F., Maizel, J. V. 1968. *Proc. Nat. Acad. Sci. USA* 59:966–71
23. Holland, J. J., Kiehn, E. D. 1968. *Proc. Nat. Acad. Sci. USA* 60:1015–22
24. Butterworth, B. E., Hall, L., Stoltzfus, C. M., Rueckert, R. R. 1971. *Proc. Nat. Acad. Sci. USA* 68:3083–87
25. Butterworth, B. E., Rueckert, R. R. 1972. *J. Virol.* 9:823–28
26. Butterworth, B. E., Rueckert, R. R. 1972. *Virology* 50:535–49
27. McLean, C., Rueckert, R. R. 1973. *J. Virol.* 11:341–44
28. Butterworth, B. E. 1973. *Virology* 56:439–53
29. Penman, S., Becker, Y., Darnell, J. E. 1964. *J. Mol. Biol.* 8:541–55
30. Summers, D. F., Levintow, L. 1965. *Virology* 27:44–53

31. Granboulan, N., Girard, M. 1969. *J. Virol.* 4:475–79
32. Tannock, G. A., Gibbs, A. J., Cooper, P. D. 1970. *Biochem. Biophys. Res. Commun.* 38:298–304
33. Jacobson, M. F., Asso, J., Baltimore, D. 1970. *J. Mol. Biol.* 49:657–69
34. Kiehn, E. D., Holland, J. J. 1970. *J. Virol.* 5:358–67
35. Korant, B. D. 1972. *J. Virol.* 10:751–59
36. Summers, D. F., Shaw, E. N., Stewart, M. L., Maizel, J. V. 1972. *J. Virol.* 10:880–84
37. Garfinkle, D. B., Tershak, D. R. 1971. *J. Mol. Biol.* 59:537–41
38. Cooper, P. D., Summers, D. F., Maizel, J. V. 1970. *Virology* 41:408–18
39. Summers, D. F., Maizel, J. V., Darnell, J. E. 1965. *Proc. Nat. Acad. Sci. USA* 54:505–13
40. Taber, R., Rekosh, D., Baltimore, D. 1971. *J. Virol.* 8:395–401
41. Jacobson, M. F., Baltimore, D. 1968. *J. Mol. Biol.* 33:369–78
42. Rekosh, D. 1972. *J. Virol.* 9:479–87
43. Öberg, B. F., Shatkin, A. J. 1972. *Proc. Nat. Acad. Sci. USA* 69:3589–93
44. Boime, I., Leder, P. 1972. *Arch. Biochem. Biophys.* 153:706–13
45. Villa-Komaroff, L., Baltimore, D., Lodish, H. F. 1973. *Fed. Proc.* 32:531 (Abstr.)
46. Smith, A. E. 1973. *Eur. J. Biochem.* 33:301–13
47. Öberg, B. F., Shatkin, A. J. 1974. *Biochem. Biophys. Res. Commun.* 57:1186–91
48. Summers, D. F., Maizel, J. V. 1971. *Proc. Nat. Acad. Sci. USA* 68:2852–56
49. Saborio, J. L., Pong, S. S., Koch, G. 1974. *J. Mol. Biol.* 85:195–211
50. Cooper, P. D., Geissler, E., Scotti, P. D., Tannock, G. A. 1971. *Strategy Viral Genome, Ciba Found. Symp., 1971,* 75–95
51. Cole, C. N., Smoler, D., Wimmer, E., Baltimore, D. 1971. *J. Virol.* 7:478–85
52. Cole, C. N., Baltimore, D. 1973. *J. Mol. Biol.* 76:325–43
53. Cooper, P. D., Stanček, K., Summers, D. F. 1970. *Virology* 40:971–77
54. Caliguiri, L. A. 1974. *Virology* 58:526–35
55. Caliguiri, L. A., Mosser, A. G. 1971. *Virology* 46:375–86
56. Rosenberg, H., Diskin, B., Oron, L., Traub, A. 1972. *Proc. Nat. Acad. Sci. USA* 69:3815–19
57. Willems, M., Penman, S. 1966. *Virology* 30:355–67
58. Steiner-Pryor, A., Cooper, P. D. 1973.

59. Cooper, P. D., Steiner-Pryor, A., Wright, P. J. 1973. *Intervirology* 1:1–10
60. Wright, P. J., Cooper, P. D. 1974. *Virology* 59:1–20
61. Fernandez-Tomas, C. B., Baltimore, D. 1973. *J. Virol.* 12:1122–30
62. Rueckert, R. R., Dunker, A. K., Stoltzfus, C. M. 1969. *Proc. Nat. Acad. Sci. USA* 62:912–19
63. Dunker, A. K., Rueckert, R. R. 1971. *J. Mol. Biol.* 58:217–35
64. Stoltzfus, C. M., Rueckert, R. R. 1972. *J. Virol.* 10:347–55
65. Perlin, M., Phillips, B. A. 1973. *Virology* 53:107–14
66. Scharff, M. D., Maizel, J. V., Levintow, L. 1964. *Proc. Nat. Acad. Sci. USA* 51:329–37
67. Ghendon, Y., Yakobsen, E., Mikhejeva, A. 1972. *J. Virol.* 10:261–66
68. Phillips, B. A., Summers, D. F., Maizel, J. V. 1968. *Virology* 35:216–26
69. Fernandez-Tomas, C. B., Guttman, N., Baltimore, D. 1973. *J. Virol.* 12:1181–83
70. Phillips, B. A., Fennel, R. 1973. *J. Virol.* 12:291–99
71. Phillips, B. A. 1969. *Virology* 39:811–21
72. Phillips, B. A. 1971. *Virology* 44:307–16
73. Ginevskaya, V. A., Scarlat, I. V., Kalinina, N. O., Agol, V. I. 1972. *Arch. Gesamte Virusforsch.* 39:98–107
74. Korant, B. D. 1973. *J. Virol.* 12:556–63
75. Korant, B. D., Kauer, J. C., Butterworth, B. E. 1974. *Nature* 248:588–89
76. Prouty, W. F., Goldberg, A. L. 1972. *J. Biol. Chem.* 247:3341–52
77. Shechter, Y., Rafaeli-Eshkol, D., Hershko, A. 1973. *Biochem. Biophys. Res. Commun.* 54:1518–24
78. Steinberg, D., Vaughan, M. 1956. *Arch. Biochem. Biophys.* 65:93–105
79. Mandelstam, J. 1958. *Biochem. J.* 69:110–19
80. Hershko, A., Tomkins, G. M. 1971. *J. Biol. Chem.* 246:710–14
81. Holland, J. J., Doyle, M., Perrault, J., Kingsbury, D. T., Etchison, J. 1972. *Biochem. Biophys. Res. Commun.* 46:634–39
82. Harrison, S. C., David, A., Jumblatt, J., Darnell, J. E. 1971. *J. Mol. Biol.* 60:523–28
83. Strauss, J. H., Burge, B. W., Darnell, J. E. 1969. *Virology* 37:367–76
84. Schlesinger, M. J., Schlesinger, S., Burge, B. W. 1972. *Virology* 47:539–41
85. Simons, K., Keränen, S., Kääriänen, L.

1973. *FEBS Lett.* 29:87–91
86. Burrell, C. J., Martin, E. M., Cooper, P. D. 1970. *J. Gen. Virol.* 6:319–23
87. Schlesinger, S., Schlesinger, M. J. 1972. *J. Virol.* 10:925–32
88. Snyder, H. W., Sreevalsan, T. 1973. *Biochem. Biophys. Res. Commun.* 53:24–31
89. Snyder, H. W., Sreevalsan, T. 1973. *J. Virol.* 13:541–44
90. Pfefferkorn, E. R., Boyle, M. K. 1972. *J. Virol.* 9:187–88
91. Morser, M. J., Burke, D. C. 1974. *J. Gen. Virol.* 22:395–409
92. Scheele, C. M., Pfefferkorn, E. R. 1970. *J. Virol.* 5:329–37
93. Schlesinger, M. J., Schlesinger, S. 1973. *J. Virol.* 11:1013–16
94. Waite, M. R. F. 1973. *J. Virol.* 11:198–206
95. Cancedda, R., Schlesinger, M. J. 1974. *Proc. Nat. Acad. Sci. USA* 71:1843–47
96. Jones, K. J., Waite, M. R. F., Rose, H. R. 1974. *J. Virol.* 13:809–17
97. Katz, E., Moss, B. 1970. *Proc. Nat. Acad. Sci. USA* 66:677–84
98. Katz, E., Moss, B. 1970. *J. Virol.* 6:717–26
99. Pennington, T. H., 1973. *J. Gen. Virol.* 19:65–79
100. Moss, B., Rosenblum, E. N. 1973. *J. Mol. Biol.* 81:267–69
101. Dales, S., Mosbach, E. H. 1968. *Virology* 35:564–83
102. Moss, B., Rosenblum, E. N., Katz, E., Grimley, P. M. 1969. *Nature* 224:1280–84
103. Grimley, P. M., Rosenblum, E. N., Mims, S. J., Moss, B. 1970. *J. Virol.* 6:519–33
104. Moss, B., Rosenblum, E. N., Grimley, P. M. 1971. *Virology* 45:135–48
105. Ishibashi, M., Maizel, J. V. 1974. *Virology* 57:409–24
106. Walter, G., Maizel, J. V. 1974. *Virology* 57:402–8
107. Anderson, C. W., Baum, P. R., Gesteland, R. F. 1973. *J. Virol.* 12:241–52
108. Lazarowitz, S. G., Compans, R. W., Choppin, P. W. 1971. *Virology* 46:830–43
109. Skehel, J. J. 1972. *Virology* 49:23–36
110. Klenk, H. D., Rott, R. 1973. *J. Virol.* 11:823–31
111. Samson, A. C. R., Fox, C. F. 1973. *J. Virol.* 12:579–87
112. Vogt, V. M., Eisenman, R. 1973. *Proc. Nat. Acad. Sci. USA* 70:1734–38
113. Friedmann, T. 1974. *Proc. Nat. Acad. Sci. USA* 71:257–59

114. Klenk, H. D., Wöllert, W., Rott, R., Scholtissek, C. 1974. *Virology* 57:28–41
115. Lazarowitz, S. G., Goldberg, A. R., Choppin, P. W. 1973. *Virology* 56:172–80
116. Lazarowitz, S. G., Compans, R. W., Choppin, P. W. 1973. *Virology* 52:199–212
117. Hosoda, J., Cone, R. 1970. *Proc. Nat. Acad. Sci. USA* 66:1275–81
118. Dickson, R. C., Barnes, S. L., Eiserling, F. A. 1970. *J. Mol. Biol.* 53:461–74
119. Laemmli, U. K. 1970. *Nature* 227:680–85
120. Kellenberger, E., Kellenberger-Van der Kamp, C. 1970. *FEBS Lett.* 8:140–44
121. Lengyel, J. A., Goldstein, R. N., Marsh, M., Sunshine, M. G., Calendar, R. 1973. *Virology* 53:1–23
122. Georgopoulos, C. P., Hendrix, R. W., Casjens, S. R., Kaiser, A. D. 1973. *J. Mol. Biol.* 76:45–60
123. Murialdo, H., Siminovitch, L. 1972. *Virology* 48:785–823
124. Hendrix, R. W., Casjens, S. R. 1974. *Virology* 61:156–59
125. Hendrix, R. W., Casjens, S. R. 1974. *Proc. Nat. Acad. Sci. USA* 71:1451–55
126. Zweig, M., Cummings, D. J. 1973. *J. Mol. Biol.* 80:505–18
127. Botstein, D., Waddel, C. H., King, J. 1973. *J. Mol. Biol.* 80:669–95
128. King, J., Lenk, E. V., Botstein, D. 1973. *J. Mol. Biol.* 80:697–731
129. Matsuo, H., Fujisawa, H. 1973. *Virology* 54:313–17
130. Eiserling, F. A., Dickson, R. C. 1972. *Ann. Rev. Biochem.* 41:467–502
131. Walker, D. H., Mosig, G., Bayer, M. E. 1972. *J. Virol.* 9:872–75
132. Boy de la Tour, E., Kellenberger, E. 1965. *Virology* 27:222–25
133. Streisinger, G., Emrich, J., Stahl, M. M. 1967. *Proc. Nat. Acad. Sci. USA* 57:292–95
134. Matthews, C. K. 1971. *Bacteriophage Biochemistry.*, Am. Chem. Soc. Monogr. New York: Reinhold. 373 pp.
135. Tikchonenko, T. I. 1970. *Advan. Virus Res.* 15:201–90
136. Mosig, G., Renshaw, J. 1972. *J. Virol.* 9:857–71
137. Black, L. W., Ahmad Zadeh, C. 1971. *J. Mol. Biol.* 57:71–92
138. Eddleman, H. L., Champe, S. P. 1966. *Virology* 30:471–81
139. Epstein, R. H. et al 1963. *Cold Spring Harbor Symp. Quant. Biol.* 28:375–93
140. Edgar, R. S., Wood, W. B. 1966. *Proc. Nat. Acad. Sci. USA* 55:498–505
141. Laemmli, U. K., Mölbert, E., Showe,

M., Kellenberger, E. 1970. *J. Mol. Biol.* 49:99–113
142. Laemmli, U. K., Beguin, F., Gujer-Kellenberger, G. 1970. *J. Mol. Biol.* 47:69–85
143. Laemmli, U. K., Favre, M. 1973. *J. Mol. Biol.* 80:575–99
144. Simon, L. D. 1972. *Proc. Nat. Acad. Sci. USA* 69:907–11
145. Kellenberger, E., Sechaud, J., Ryter, A. 1959. *Virology* 8:478–98
146. Showe, M. K., Black, L. W. 1973. *Nature New Biol.* 242:70–75
147. Bijlenga, R. K. L., Scraba, D., Kellenberger, E. 1973. *Virology* 56:250–67
148. Laemmli, U. K., Johnson, R. A. 1973. *J. Mol. Biol.* 80:601–11
149. Luftig, R. B., Lundh, N. P. 1973. *Proc. Nat. Acad. Sci. USA* 70:1636–40
150. Frankel, F. R., Batcheler, M. L., Clark, C. K. 1971. *J. Mol. Biol.* 62:439–63
151. Luftig, R. B., Ganz, C. 1972. *J. Virol.* 10:545–54
152. Luftig, R. B., Ganz, C. 1972. *J. Virol.* 9:377–89
153. Luftig, R. B., Lundh, N. P. 1973. *Virology* 51:432–42
154. Luftig, R. B., Wood, W. B., Okinaka, R. 1971. *J. Mol. Biol.* 57:555–73
155. Bijlenga, R. K. L., Broek, R. V. D., Kellenberger, E. 1974. *J. Supramol. Struct.* 2:45–59
156. Hohn, B., Hohn, T. 1974. *Proc. Nat. Acad. Sci. USA* 71:2372–76
157. Kellenberger, E., Eiserling, F. A., Boy de la Tour, E. 1968. *J. Ultrastruct. Res.* 21:335–60
158. Aebi, U. et al *J. Supramol. Struct.* In press
159. Granboulan, P., Sechaud, J., Kellenberger, E. 1971. *Virology* 46:407–25
160. Chao, J., Chao, L., Speyer, J. F. 1974. *J. Mol. Biol.* 85:41–50
161. Siegel, P. J., Schaechter, M. 1973. *Ann. Rev. Microbiol.* 27:261–82
162. Black, L. W., Gold, L. M. 1971. *J. Mol. Biol.* 60:365–88
163. Coppo, A., Manzi, A., Pulitzer, J. F., Takahashi, H. 1973. *J. Mol. Biol.* 76:61–87
164. Bolin, R. W., Cummings, D. J. 1974. *J. Virol.* 13:1378–91
165. King, J., Laemmli, U. K. 1973. *J. Mol. Biol.* 75:315–37
166. King, J., Mykolajewycz, N. 1973. *J. Mol. Biol.* 75:339–58
167. Dickson, R. C. 1974. *Virology* 59:123–38
168. Hosoda, J., Levinthal, C. 1968. *Virology* 34:709–27
169. Showe, M., Onotaro, L. 1972. *Proc. Am. Soc. Microbiol.* (Abstr.) (mentioned in 146)
170. Bachrach, U., Berichetrit, L. 1974. *Virology* 59:51–58
171. Celis, J. E., Smith, J. D., Brenner, S. 1973. *Nature New Biol.* 241:130–32
172. Goldstein, J., Champe, S. P. 1973. *J. Virol.* 13:419–27
173. Bachrach, U., Benchetrit, L. 1974. *Virology* 59:443–54
174. Champe, S. P., Eddleman, H. 1967. *The Molecular Biology of Viruses,* ed. J. S. Colter, W. Paranchych, 55–70. New York: Academic. 184 pp.
175. Poglazov, B. F., Levshenko, M. T. 1974. *J. Mol. Biol.* 84:463–67
176. Couse, N. L., Cummings, D. J., Chapman, V. A., De Long, S. S. 1970. *Virology* 42:590–602
177. Couse, N. L., Haworth, P., Moody, W., Cummings, D. J. 1972. *Virology* 50:765–71
178. Bolin, R. W., Cummings, D. J. 1974. *J. Virol.* 13:1368–77
179. Cummings, D. J., Chapman, V. A., De Long, S. S., Couse, N. L. 1973. *Virology* 54:245–61
180. Takano, T., Kakefuda, T. 1972. *Nature New Biol.* 239:34–37
181. Georgopoulos, C. P., Hendrix, R. W., Kaiser, A. D., Wood, W. B. 1972. *Nature New Biol.* 239:38–41
182. Zweig, M., Cummings, D. J. 1973. *Virology* 51:443–53
183. Murialdo, H., Siminovitch, L. 1971. *The Bacteriophage Lambda,* ed. A. D. Hershey, 711–23. Cold Spring Harbor, NY: Cold Spring Harbor Lab
184. Hendrix, R., Casjens, S. R. In preparation
185. Baltimore, D., Girard, M., Darnell, J. E. 1966. *Virology* 29:179–89
186. Abraham, G., Cooper, P. D. *J. Gen. Virol.* In press
187. Black, L. W. 1974. *Virology* 60:166–79
188. Bijlenga, R. K. L., Broek, R. V. D., Kellenberger, E. 1974. *Nature* 249:825–27

BASIC MECHANISMS IN BLOOD COAGULATION [1]

Earl W. Davie and Kazuo Fujikawa

Department of Biochemistry, University of Washington, Seattle, Washington 98195

CONTENTS

INTRODUCTION

A major focus on hereditary disorders of coagulation occurred when hemophilia appeared in the royal families of Europe. In 1853, Queen Victoria, a carrier of hemophilia, gave birth to her fifth son, Leopold, who suffered from this disease. Leopold died at the age of 31 from a brain hemorrhage after a minor blow to the head. Two of Queen Victoria's daughters were carriers of hemophilia and they passed the disease to other royal families in Europe. Thus, the coagulation disorders had an important role in the history of the Western world. This was particularly true in the case of Queen Victoria's granddaughter, Alexandra, a carrier of hemophilia, who married Nicholas II, Czarevitch of Russia (1).

The defect in the hemostatic mechanism found in hemophilia is only part of the complex series of reactions that participate in arresting bleeding from ruptured blood

[1] References available to the authors prior to August 1, 1974 have been included in this review. A number of excellent reviews have been written in recent years on topics discussed in this article. These are identified in the appropriate section. They provide many outstanding references of earlier work which could not be included in this review.

vessels. Indeed, vascular damage leads to the participation of many physiological processes that are important in controlling blood loss. First, there is a platelet adhesion reaction where the platelets become sticky and bound to the endothelial connective tissue structures, including the basement membrane and collagen fibers. This leads to platelet plug formation and the platelet release reaction. If the lesion is minor and the vessel wall is small, the formation of the platelet plug may be sufficient to halt the loss of blood from the vessel (2). Second, the platelet release reaction gives rise to vasoactive amines, such as serotonin which causes vaso-restriction of the injured vessel. The third major effect is the triggering of the coagulation process, involving both intrinsic and extrinsic systems. These reactions involve collagen, elastin, and other cellular components such as tissue factor. The fourth important effect is the activation of the fibrinolytic system which leads to the degradation of the fibrin clot as healing and regeneration of the vessel wall occur.

The major emphasis in this review will be placed on the coagulation proteins themselves and how they interact with each other in reactions that eventually lead to the formation of the fibrin clot.

COAGULATION FACTORS AND THEIR NOMENCLATURE

In the last twenty years, a large number of plasma proteins have been discovered that participate in the coagulation process. Most of these proteins have been isolated and fairly well characterized. They are inactive or absent in individuals with various coagulation diseases (3). The coagulation proteins participate in two closely related clotting mechanisms that lead to the formation of the fibrin clot. These mechanisms have been referred to as the intrinsic and the extrinsic coagulation pathways (Figure 1). The intrinsic pathway refers to those reactions that lead to thrombin formation by utilization of factors present only in plasma. The extrinsic pathway involves plasma factors as well as components present in tissue extracts. As dis-cussed below, both pathways play an important physiological role in mammalian hemostasis. Furthermore, both of these processes are closely interrelated and apparently function simultaneously in vivo.

Table 1 lists the various plasma proteins and several other components that influence coagulation. Most of these components have been assigned a roman numeral (4). In recent years, the majority of investigators working in the field of coagulation use the roman numeral nomenclature for factors V through XIII. Fibrinogen, prothrombin, calcium ions, and tissue factor are still generally referred to by their common names.

MOLECULAR EVENTS IN BLOOD COAGULATION

In the test tube, the intrinsic and extrinsic pathways of blood coagulation can be separated, at least functionally, depending on the conditions of the experiment. In the intrinsic system, coagulation is initiated by the activation of factor XII (Hageman factor) (Figure 1). On the other hand, the initiation of blood coagulation in the extrinsic system involves the interaction of tissue factor and factor VII. The tissue factor pathway bypasses a large number of the reactions found in the intrinsic system and enters in the coagulation cascade at the factor X level (Figure 1). It is important to remember, however, that both of these systems are triggered by their initial interaction with components of the cell wall.

Under normal physiological conditions, little or no intravascular coagulation occurs. This is due to three or more effects. The first is that in flowing blood the concentration of any given coagulation factor that becomes activated is greatly reduced by dilution. This is evident, for instance, by the fact that massive fibrin thrombus does not readily occur in arteries having a relatively fast blood flow. Fibrin thrombus, however, may be substantial in disorders with venous stasis where

INTRINSIC PATHWAY

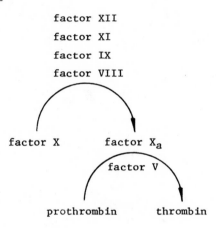

EXTRINSIC PATHWAY

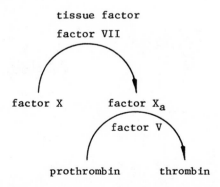

Figure 1   Abbreviated intrinsic and extrinsic pathways for blood coagulation.

Table 1 Properties of some of the well-defined coagulation factors

Coagulation Factor		Coagulation Pathway		Molecular weight[b]	Molecular weight references
Roman numeral designation[a]	Common name	Intrinsic system	Extrinsic system		
Factor I	Fibrinogen	+		340,000 (human, bovine)	294, 295
Factor II	Prothrombin	+	+	68,700 (human)	188
				72,000 (bovine)	296, 297
Factor III	Tissue factor		+	220,000; 330,000 (bovine)	139
Factor IV	Calcium ions	+	+		
Factor V	Proaccelerin	+	+	290,000–400,000 (bovine)	159, 160
Factor VII	Proconvertin		+	63,000 (human)	290
Factor VIII	Antihemophilic factor	+		1.1 million (human, bovine)	78, 79
Factor IX	Christmas factor	+		55,400 (bovine)	58
Factor X	Stuart factor	+	+	55,000 (bovine)	120, 121
Factor XI	Plasma thromboplastin antecedent	+		160,000 (human, bovine)	24, 44
Factor XII	Hageman factor	+		90,000 (human)	24
				82,000 (bovine)	29
Factor XIII	Fibrin stabilizing factor	+		300,000 (bovine plasma)	298
				320,000 (human plasma)	252
				146,000 (human platelets)	254

[a] Activated factor V was originally thought to be a new clotting factor and was assigned Roman numeral VI.

[b] Some disagreement exists regarding the exact molecular weight of many of the coagulation factors, as noted in the text. Thus, some modification of the values listed may be necessary.

the blood flow is slower. A second controlling factor is the presence in plasma of numerous natural inhibitors of blood coagulation (5–9). These inhibitors inactivate coagulation enzymes, such as thrombin or factor X_a. Although these reactions are relatively slow under physiological conditions, they lead to the inactivation of activated coagulation factors and thus terminate their prolonged participation in the coagulation process. Thirdly, activated clotting factors are rapidly removed by the liver (10). Proteins, such as factor X_a, are removed by the hepatocytes, in contrast to the precursor form. Thus, the activity of activated coagulation factors in flowing blood is rapidly decreased, not only by dilution or reaction with inhibitors but also by their clearance in the liver.

At the present time, the interaction of the various coagulation factors in the intrinsic and extrinsic systems is fairly well understood. A scheme for these reactions in mammalian plasma in the intrinsic system is shown in Figure 2. This scheme is a modification of two similar mechanisms proposed ten years ago (11, 12). Other mechanisms have also been suggested to explain fibrin formation, along with supporting data (13). These mechanisms, however, do not take into account experiments published by other investigators. As shown in Figure 2, fibrin formation results from the stepwise interaction of a large number of proteins present in plasma in precursor or zymogen forms. These proenzymes are shown on the left side of Figure 2 and include factor XII, factor XI, factor IX, factor X, prothrombin, and factor XIII. Contact of factor XII with a surface such as collagen apparently initiates the coagulation process in vivo by converting factor XII to factor XII_a (14). Factor XII_a then activates factor XI, converting it to an enzyme, factor XI_a. Factor XI_a will in turn activate factor IX in the presence of calcium ions (15–17). Thus, a series of stepwise reactions is triggered, leading to the formation of thrombin and a fibrin clot. This series of reactions has considerable potential for amplification in that a few molecules of factor XII_a can activate hundreds of molecules of factor XI, which in turn will activate thousands of factor IX molecules (12). The requirements and restrictions for this type of amplification have been reviewed (18, 19).

In vitro, the first known reaction occurs by the interaction of factor XII with a surface such as kaolin or glass (20–22). This reaction is relatively slow and inefficient. Cochrane et al (23), Wuepper (24), and Weiss et al (25), however, have shown the participation of another protein in this early phase of coagulation, which is often called the contact phase. These investigators have demonstrated the activation of prekallikrein (Fletcher factor) by factor XII_a. Furthermore, kallikrein is capable of activating factor XII (23). These experiments suggest that collagen, glass, or some similar surface may trigger the conversion of factor XII to factor XII_a, which will in turn activate prekallikrein to kallikrein. The kallikrein will then activate more factor XII, converting it to factor XII_a. This series of reactions provides an unusual amplification of the contact phase of coagulation. The role of kallikrein in these reactions has been supported by the use of antibodies to prekallikrein, which inhibit the contact phase of coagulation (26). The identification of prekallikrein with Fletcher factor and its participation in the intrinsic coagulation scheme clarify the earlier experiments of Hathaway et al, who first described Fletcher factor deficiency (27). In vitro, the contact reactions occur rather slowly.

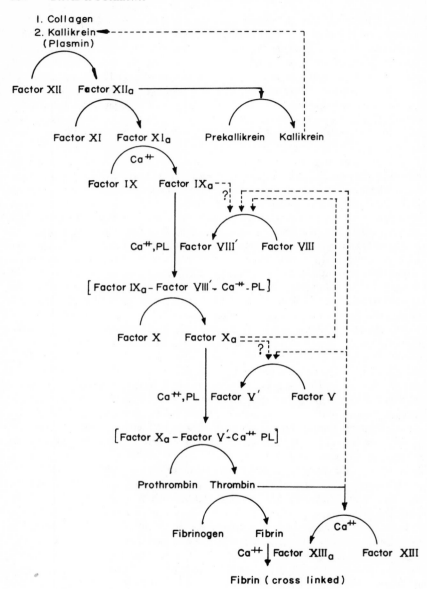

Figure 2 Tentative mechanism for the initiation of blood clotting in mammalian plasma in the intrinsic system [modified from Davie & Ratnoff (11) and Macfarlane (12)]. PL refers to phospholipid.

Indeed, platelet-rich plasma placed in a glass test tube may require incubation of 5 to 15 min before a visible fibrin clot can be observed. Trace amounts of tissue extract from organs such as the lung or brain, however, can cause extremely rapid fibrin formation. In this pathway, called the extrinsic system, clot formation occurs in vitro within 10 to 12 sec. Both the intrinsic and extrinsic pathways are important, however, in the activation of the coagulation process in vivo. This is evident from the fact that patients who lack a coagulation factor that participates in one or the other pathway usually have a bleeding disorder.

Biochemical Properties of Factor XII and Its Mechanism of Activation

Factor XII has been extensively purified in a number of different laboratories from bovine, human, and rabbit sources (23, 28–32). Extensive physical-chemical characterization, however, has not been possible thus far since only trace amounts of highly purified protein have been available. The human preparation has a molecular weight of about 110,000–120,000 as estimated by gel filtration (32). The molecular weight as determined by sodium dodecylsulfate (SDS) polyacrylamide gel electrophoresis, however, is 90,000 (24). Factor XII isolated from rabbit sources is composed of three similar polypeptide chains, each having a molecular weight of about 30,000. These chains are held together by disulfide bonds. The human and rabbit preparations sediment in sucrose density gradient with an s value of 4.5–4.6. This value is similar to that of Donaldson & Ratnoff (33), who reported a value of 4.5–5.5S for their human preparation.

Schoenmakers et al (29) reported the isolation of a highly purified preparation of bovine factor XII. The first step in their preparation involved the adsorption and elution of the protein from glass powder. They achieved an overall purification of about 20,000-fold. This preparation was homogeneous in the ultracentrifuge and by electrophoretic analysis. A molecular weight of 82,000 was calculated by using an s value of 7.08. This preparation contained about 15% carbohydrate, including hexose, hexosamine, and neuraminic acid. It presumably contains some or substantial amounts of factor XII$_a$, since it demonstrated considerable esterase activity toward substituted arginine esters and was inhibited by lima bean trypsin inhibitor and diisopropylfluorophosphate (DFP).

More recently, Komiya et al (30) extensively purified bovine factor XII using a defibrinated euglobulin fraction as starting material. They were unable, however, to obtain a preparation of factor XII that was entirely in the precursor form since it changed progressively to an active form during the isolation procedure. Their final preparation showed one band on polyacrylamide disc gel electrophoresis and had a molecular weight of 95,000. It was also inhibited by DFP and lima bean trypsin inhibitor. These investigators concluded that clotting activity, esterase activity, and prekallikrein activating activity reside in the same molecule.

The mechanism of activation of factor XII by the various pathways shown in Figure 2 is not known. Collagen and vascular basement membranes readily bind factor XII and lead to its activation (23). Other insoluble substances such as kaolin, diatomaceous earth, charcoal, and barium.carbonate will also activate factor XII. Ellagic acid at concentrations as low as 1×10^{-8} M will convert factor XII to an activated form (34). The activation of factor XII by surfaces apparently does not result in any major change in the molecular weight. Cochrane et al (35) eluted factor XII_a labeled with I^{125} from kaolin with 0.3% SDS and found no change in its SDS polyacrylamide gel electrophoresis pattern when compared to the precursor form. Both human and rabbit preparations were used in these experiments. Factor XII appeared to be fully activated since exposure to trypsin or plasmin failed to augment the clotting activity.

Factor XII activated in the fluid phase by kallikrein or plasmin also shows no change in sedimentation rate or migration in SDS polyacrylamide gels. Furthermore, no change in size of the 30,000 mol wt subunits of factor XII_a is noted after reduction of the protein with 2-mercaptoethanol (35).

Treatment of factor XII with trypsin also leads to a rapid formation of factor XII_a. In this case, however, the sedimentation coefficient decreases from 4.5 to 3. This gives rise to 30,000 mol wt fragments formed in the absence of reducing agent. These fragments possess both prekallikrein activating activity as well as clot promoting activity (35). Thus, factor XII appears to consist of three polypeptide chains of about equal size, and these chains are held together by disulfide linkages. The disulfide bonds apparently are located at the end of each of the 30,000 mol wt chains since the activation of the molecule with trypsin yields three polypeptide chains of comparable size. However, the true physiological role of factor XII still requires clarification since patients lacking this plasma protein appear to have no serious coagulation disorders.

Biochemical Properties of Factor XI and Its Mechanism of Activation

Once factor XII_a is formed, it converts factor XI to an enzyme (15–17) (Figure 2). This enzyme, referred to as factor XI_a, has endopeptidase activity toward factor IX (36) and esterase activity toward benzoyl arginyl ethyl ester (37). Factor XI can also be activated by other proteases such as trypsin (24, 38, 39). Some investigators have proposed that factor XII_a forms a complex with factor XI and it is this complex that is active in the initiation of coagulation (40). This proposal is inconsistent, however, with studies using antibodies to factor XI or factor XII (41). Furthermore, it is inconsistent with experiments demonstrating the direct activation of factor IX by factor XI_a using homogeneous preparations of each (36).

Factor XI has been extensively purified from both human and bovine sources (24, 37, 38, 42, 43). It is a glycoprotein with a molecular weight of about 160,000 as determined by sedimentation equilibrium and SDS polyacrylamide gel electrophoresis (24, 44). It is composed of two similar or identical polypeptide chains with a molecular weight of 80,000, and these two chains are held together by disulfide bonds (24, 44). The bovine preparation contains 12% carbohydrate, including 4.2% hexose, 4.0% hexosamine, and 3.6% neuraminic acid (44).

The mechanism of activation of factor XI by factor XII_a or trypsin has been examined with human and bovine preparations. Factor XI and factor XI_a both sediment in sucrose gradients at 6.9S (24). Furthermore, no difference has been noted in factor XI and factor XI_a by immunoelectrophoresis or SDS gel electrophoresis in the absence of 2-mercaptoethanol (24, 44).

The cleavage of internal peptide bonds in factor XI during the activation reaction, however, can be shown following reduction of the subunits of factor XI_a. In this case, two polypeptide chains are observed with molecular weights of about 50,000 and 30,000 (24, 44). Furthermore, labeling of factor XI_a with ^{32}P DFP has shown that the active site or reactive serine residue is present in the light chain or 30,000 mol wt polypeptide (44).

The nature of the bonds that are split during this reaction has not been established. Furthermore, the number of active sites per mole of enzyme has not been clarified. If the two subunits are identical, factor XI_a may contain two active sites. Factor XI_a may, however, fall into the "half-of-the-site reactivity" classification with only one functional site per mole of enzyme (45).

Recently, Schiffman & Lee purified human factor XI about 2800-fold (43). This preparation was readily activated by trypsin but not by kaolin-activated factor XII. When factor XII-deficient plasma was added to kaolin-activated factor XII, however, a marked activation of factor XI occurred. These authors suggest that an additional factor is required in the activation of factor XI by factor XII_a, and this factor is not prekallikrein.

Biochemical Properties of Factor IX and Its Mechanism of Activation

Once factor XI is activated, it in turn activates factor IX (15–17, 36, 42, 46–49) (Figure 2). This reaction has an absolute requirement for a divalent cation such as calcium (42, 50). Factor IX has been extensively purified in recent years from human and bovine sources (51–58). The methods of isolation for the human and bovine preparations have been recently reviewed (59, 60). A particularly useful step in the purification of factor IX has been the utilization of heparin-agarose (58, 61, 62).

Bovine factor IX is a glycoprotein of molecular weight 55,400 (58). It contains 26% carbohydrate, including 10.6% hexose, 6.5% hexosamine, and 8.7% neuraminic acid. Bovine factor IX is composed of a single polypeptide chain with an amino-terminal tyrosine residue. The amino-terminal region of bovine factor IX is homologous with prothrombin and the light chain of factor X (Table 2) (63). Six of the first 14 amino acids are identical in all three proteins. Also, two amino acids (residues 7 and 8) are similar or perhaps identical in all three proteins. These two residues, as discussed in the prothrombin section, may well be γ-carboxyglutamic acid residues, which play an important role in calcium binding. The amino-terminal regions of the heavy chain of factor IX_a, factor X_a, and thrombin as well as their active site regions are also homologous (Table 2) (36, 64–66). Thus, it is apparent that these three coagulation proteins have evolved from a common ancestral gene (63, 64, 66).

During coagulation, factor IX is converted to factor IX_a, a glycoprotein with a molecular weight of 46,500 (Figure 2) (36). Recently, it has been shown that the activation of bovine factor IX occurs in a two step reaction (Figure 3) (36). In the first step, an internal peptide bond in factor IX is cleaved, leading to the formation of an intermediate containing two polypeptide chains held together by a disulfide bond(s). This intermediate has no clotting activity. It is composed of a light chain

Table 2 Sequence homologies for bovine prothrombin, factor IX, and factor X (63–66)

Amino-terminal sequence of the proenzymes[a]

Prothrombin	Ala Asn	Lys Gly Phe Leu Gla Gla	—	Val Arg	Lys	Gly Asn Leu
Factor IX	Tyr Asn Ser Gly	Lys	Leu Glu Glu	Phe	Val Arg	—	Gly Asn Leu
Factor X	Ala Asn Ser	—	Phe Leu Glu Glu	—	Val	Lys	Gln	Gly Asn Leu

Amino-terminal sequence of the heavy chains of the active enzymes

Thrombin	Ile	Val	Glu	Gly	Gln Asp Ala Glu Val	Gly	Leu Ser	Pro Trp Gln
Factor IXa	Val	Val	Gly	Gly Glu Asp Ala Glu	Arg	Gly Glu Phe	Pro Trp Gln
Factor Xa	Ile	Val	Gly Gly	Arg	Asp	Cys Ala Glu Glu Glu	Cys	Pro Trp Gln

Sequence of the active site[b]

Thrombin	Phe Cys Ala Gly Tyr	Lys Pro Glu Glu	Gly Lys	Arg Gly	Asp Ala	Cys	Glu	Gly Asp SER Gly Gly Pro	Phe	Val	Met Lys
Factor IXa	Phe Cys Ala Gly Tyr His	—	Glu Gly Lys	—	—	Asp	Ser	Cys Gln Gly Asp SER Gly Gly Pro His Val Thr Glx
Factor Xa	Phe Cys Ala Gly Tyr Asp	—	Thr Gln Pro Glu	—	—	Asp Ala Cys Gln Gly Asp SER Gly Gly Pro His Val Thr Arg

[a] Gla refers to γ-carboxyglutamic acid. Dashes (—) refer to spaces that have been inserted to bring the three proteins into alignment for greater homology.

[b] The reactive serine is shown in all capital letters.

of molecular weight 16,000 and a heavy chain with molecular weight 38,800. The amino-terminal sequence of the light chain is Tyr-Asn-Ser-Gly- and the amino-terminal sequence of the heavy chain is Ala-Glu-Thr-Ile-. Thus, the light chain arises from the amino-terminal portion of the precursor protein. In the second step of the reaction, an activation peptide of molecular weight 9000 is split from the amino-terminal end of the heavy chain. This activation peptide contains a large amount of carbohydrate, reducing the total carbohydrate of the precursor protein by approximately 50%. This gives rise to factor IX_a, a protein that is also composed of two chains held together by disulfide bonds. The light chain of factor IX_a has a molecular weight of 16,000 and contains the amino-terminal sequence Tyr-Asn-Ser-Gly-; the heavy chain has a molecular weight of 27,300 and contains the amino-terminal sequence Val-Val-Gly-Gly-. These results demonstrate directly that factor XI_a contains or possesses endopeptidase activity. Furthermore, it is consistent with earlier results that factor IX and factor IX_a have different electrophoretic mobilities and different elution properties when examined by gel filtration (67, 68).

Factor IX may also be activated by a factor(s) present in platelets. Schiffman et al have shown that platelets contain a factor XI-like activity which bypasses factors XII and XI (69). This platelet activity is distinct from the phospholipid provided by the platelets for factor X activation and prothrombin activation. The importance of this pathway is not known, but it could represent a major mechanism for initiating the intrinsic system.

Factor IX_a is a serine protease with esterase activity towards benzoyl arginyl ethyl ester (70). It also has endopeptidase activity toward factor X where it cleaves an arginyl–isoleucine peptide bond as described below (71, 72).

Biochemical Properties of Factor VIII and Its Interaction with Factor IX_a

In the next reaction of the intrinsic coagulation pathway, factor IX_a interacts with factor VIII, apparently by forming a complex which then acts as a catalyst in the activation of factor X (Figure 2) (73–75). This reaction requires the presence of calcium ions and phospholipid (76). Under physiological conditions, the phospholipid is supplied by the platelets. As mentioned earlier, factor VIII is a plasma protein which is inactive in individuals with classic hemophilia. It is a sex-linked recessive disorder resulting from a mutation of a gene in the X chromosome. It is

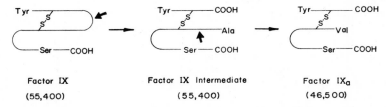

Figure 3 The mechanism of activation of bovine factor IX [modified from Fujikawa et al (36)].

by far the most common of all the coagulation disorders, occurring with a frequency of about 1 in every 10,000 births. Accordingly, it has been of major interest to investigators for many years. The physiology of factor VIII has been reviewed recently by Barrow & Graham (77).

Factor VIII has been purified about 10,000-fold from human and bovine plasma (78–84). It is a glycoprotein with a molecular weight of about 1.1 million (78, 79). Human and bovine factors VIII differ substantially in their amino acid composition and their carbohydrate content. The human preparation contains 6% carbohydrate, including 2.2% hexose, 2.7% hexosamine, and 0.9% neuraminic acid (79). The bovine preparation contains about 9% carbohydrate, including 3.8% hexose, 4.2% hexosamine, and 0.6% neuraminic acid (85). This carbohydrate content is lower than that originally published for the bovine preparation (78). The discrepancy may be due, in part, to contamination of factor VIII by methyl-α-D-glucopyranoside which was used in the original preparations to dissociate factor VIII-concanavalin A complexes (78). In the presence of reducing agents such as 2-mercaptoethanol or dithiothreitol, the native protein is reduced to similar or identical subunits with a weight average molecular weight of about 200,000 (78, 79, 84), as determined by sedimentation equilibrium. The molecular weight of the subunits also appears to be about 200,000 as determined by SDS gel electrophoresis (78, 79, 84). These data suggest that factor VIII is composed of similar or identical subunits held together by disulfide bonds. These preparations have both coagulant activity and platelet aggregating activity (85, 86).

Factor VIII activity is inhibited by antibodies prepared in rabbits against these proteins. Also, antibodies to bovine factor VIII inhibit the human preparation and antibodies to human factor VIII inhibit the bovine preparation (78, 79). The potency of the respective antibodies against the homologous protein, however, is much greater than that toward the heterologous protein.

With antibodies prepared against human factor VIII, it has been possible to demonstrate the presence of an inactive crossreactive factor VIII molecule in the plasma of individuals with classic hemophilia (87–93). Furthermore, the subunits of abnormal factor VIII have been isolated and partially characterized (84, 94). These data suggest that an abnormal factor VIII protein is synthesized in individuals with classic hemophilia and that the protein is probably inactive due to an amino acid replacement in some essential portion of the molecule.

Low levels of factor VIII activity are also found in another coagulation disorder called von Willebrand's disease (95). This disease, in contrast to classic hemophilia, is inherited as an autosomal dominant bleeding disorder in which there is often a decrease in the plasma factor VIII coagulant activity as well as abnormal platelet aggregating activity. The abnormal platelet aggregating activity may be due to the lack of another plasma protein. which is closely related to but distinct from factor VIII coagulant protein. This protein has been called platelet aggregating factor or von Willebrand factor. Patients with von Willebrand's disease and low factor VIII activity, however, lack the 1.1 million mol wt glycoprotein as measured by immunological techniques (89, 92). Thus, classic hemophilia is closely related to but distinct from von Willebrand's disease.

Recently, a number of investigators have reported the dissociation of factor VIII coagulant activity from the high molecular weight protein fraction in the presence of 1.0 M NaCl or 0.25 M CaCl$_2$ (96–103). In these experiments, normal plasma or partially purified factor VIII was used. This dissociation, which occurs in the absence of reducing agents, gives rise to a factor VIII coagulant protein with an apparent molecular weight of 25,000 (99). More recently, this value has been revised upward to 100,000 (100). The separation of the high molecular weight protein from the factor VIII coagulant activity has been shown by sucrose gradient centrifugation and by gel filtration, and under these conditions the platelet aggregating activity remains associated with the high molecular weight protein. At the present time, it is difficult to reconcile the data concerning the low molecular weight factor VIII coagulant protein and the highly purified factor VIII described by others (78, 79, 84). The latter protein has a molecular weight of 1.1 million, is composed of subunits held together by disulfide bonds, and contains both coagulant activity and platelet aggregating activity. Furthermore, this protein is not reduced in size by high concentrations of salt (85). Minor proteolysis of the 1.1 million mol wt protein in the presence of high concentrations of salt or CaCl$_2$ may be responsible for a low molecular weight factor VIII coagulant protein. If proteolysis is occurring, however, it does not appear to involve plasmin, since the addition of 0.1 M ε-amino-n-caproic acid did not influence the dissociation step (100). Another explanation is that the highly purified preparations of bovine and human factor VIII contain only a small amount of factor VIII coagulant protein of low molecular weight and contain primarily von Willebrand factor of high molecular weight (104). Factor VIII activity, however, is directly proportional to the high molecular weight protein present in various normal and abnormal plasmas (87, 105). Thus, a clear explanation for low molecular weight factor VIII coagulant activity is not evident.

Recently, Brown and co-workers (106) have obtained factor VIII coagulant activity and platelet aggregating activity in separate fractions by using QAE Sephadex A25 column chromatography at high ionic strength. With bovine factor VIII, essentially all the human platelet aggregating activity was found in the first peak, while factor VIII coagulant activity eluted in a second peak. The coagulant activity under these conditions was very unstable. We have confirmed these experiments in our laboratory using highly purified preparations of bovine factor VIII (107). The recovery of protein and coagulant activity, however, was extremely low in our experiments. It should be pointed out that the platelet aggregating activity in the bovine plasma is assumed to be identical with the human von Willebrand factor, although this has not been fully established. These data suggest that factor VIII coagulant protein and platelet aggregating factor are extremely similar in their physical-chemical properties since the two activities are readily purified together. A clear understanding of the relationship between factor VIII deficiency and von Willebrand's disease, however, will require a detailed characterization of the low molecular weight coagulant protein and a comparison of its biological and chemical properties with the high molecular weight protein.

The original cascade or waterfall proposal suggested that factor VIII was converted to an enzyme, factor VIII$_a$, by factor IX$_a$ during the coagulation process

(11, 12). Factor VIII_a was then thought to convert factor X to factor X_a. Present evidence, however, indicates that factor VIII probably participates as a regulatory factor and not as an enzyme. The study of the role of factor VIII in coagulation is complicated by the fact that the coagulant activity of factor VIII is markedly increased by traces of thrombin and other proteolytic enzymes such as plasmin or trypsin (85, 108–111). The increase in coagulant activity by thrombin, plasmin, or trypsin is short lived, however, since proteolysis apparently continues, leading to a loss of the coagulant activity. The effect of thrombin appears to be associated with only the coagulant activity since it has no influence on the platelet aggregating activity (112). Thrombin, however, does not stimulate the low molecular weight factor VIII described by Owen & Wagner (99).

Recent evidence indicates that factor X_a will also dramatically increase the coagulant activity of factor VIII (113). In this case, the coagulant activity is far more stable than that formed by thrombin, trypsin, or plasmin.

The molecular basis for the proteolytic activation of factor VIII coagulant activity is unclear. Presumably it involves the hydrolysis of an internal peptide bond(s) in factor VIII, and this bond(s) probably contains an arginine residue. With thrombin activation, no change has been noted in the SDS polyacrylamide gel electrophoresis pattern of the factor VIII subunits following reduction with 2-mercaptoethanol (79, 84, 85). The increase in coagulant activity is dramatic, however, and Rapaport and co-workers concluded that factor VIII as such is essentially inactive in coagulation and thrombin modification is required for its participation in the intrinsic coagulation scheme (108, 110). Accordingly, Nemerson & Pitlick have made the suggestion that in vivo it is necessary to produce thrombin by the extrinsic pathway before factor VIII and the intrinsic pathway become operative (114). The finding that factor X_a and perhaps other serine plasma proteases will also activate factor VIII indicates, however, that thrombin may not necessarily be a medium of communication between the intrinsic and the extrinsic pathways.

It has been suggested by a number of different investigators that a complex between factor IX_a and factor VIII is formed during coagulation in the presence of phospholipid and calcium ions, and it is this complex that is the factor X activator (73–75, 115) (Figure 2). Several lines of evidence support the concept of an enzymatic role for factor IX_a and a regulator role for factor VIII. First, antibodies to either factor VIII or factor IX inhibit the activation of factor X (74). This indicates that both factor IX_a and factor VIII are required for the activation of factor X. These results are inconsistent with the original proposal that factor IX_a activates factor VIII, which in turn activates factor X. These experiments, however, do not rule out the possibility that the activation of factor X might occur in two steps, one catalyzed by factor IX_a and the other by factor VIII_a. This possibility appears unlikely, however, since factor X_a can be formed from factor X by the cleavage of a single peptide bond, as discussed below (72). Second, factor IX_a alone will activate factor X in the presence of calcium ions and phospholipid (113). This reaction is accelerated approximately 1000-fold by the addition of factor VIII. Factor VIII in the presence of calcium ions and phospholipid, however, will not activate factor X. Third, factor IX_a has esterase activity toward benzoyl arginyl ethyl ester, whereas factor VIII has no known esterase activity or other enzymatic activity (113).

Biochemical Properties of Factor X and Its Activation by Factor IX_a and Factor VIII

Factor X is a glycoprotein which has been isolated from bovine and human plasma in a number of different laboratories. Esnouf & Williams extensively purified bovine factor X in 1962 using $BaSO_4$ adsorption and elution and DEAE-cellulose column chromatography (116). This procedure has been modified in recent years in a number of different laboratories (117–123). Jackson & Hanahan separated factor X into two forms called factors X_1 and X_2 (124). Both proteins have a molecular weight of about 55,000 and contain 10% carbohydrate, including 3.8% neuraminic acid, 2.9% hexose, and 3.6% hexosamine (120, 121). Following reduction of factor X with 2-mercaptoethanol, a heavy chain with molecular weight 38,000 and a light chain with molecular weight 17,000 are formed. The heavy chain from either factor X_1 or factor X_2 has an amino-terminal sequence of Trp-Ala-Ile-His- (120). This chain contains essentially all the carbohydrate present in the protein (120), and the carbohydrate chains are probably bound to two amino acid residues (125). The light chain from factor X_1 or factor X_2 has an amino-terminal sequence of Ala-Asn-Ser-Phe- and contains little or no carbohydrate. These two chains are held together by a disulfide bond(s). Fujikawa et al found that the amino acid composition and carbohydrate content for factors X_1 and X_2 are the same (120). Jackson, however, reported a small difference in the carbohydrate content of these two glycoproteins (121).

Recently, the total amino acid sequence of the light chain of bovine factor X has been determined (126). A portion of this is shown in Table 2 (64). Mattock & Esnouf reported the presence of a single chain factor X in bovine plasma (127). This single chain preparation, however, was not activated by the protease from Russell's viper venom or by the extrinsic clotting system (128). Thus, additional studies are necessary to clarify the relationship of this protein to factor X, which is composed of two chains.

Factor X is readily converted to factor X_a by factor IX_a and factor VIII in the presence of calcium ions and phospholipid (Figure 2) (73–76, 129–132). Factor X is also activated by tissue factor and factor VII in the extrinsic system (133–141) (Figure 1). Also, other proteolytic enzymes, such as trypsin and a protease from Russell's viper venom, will activate factor X (117, 123, 142–145). Fujikawa et al have shown that the activation of factor X by the protease from Russell's viper venom involves the cleavage of a single arginyl-isoleucine peptide bond in the amino-terminal region of the heavy chain (71, 72). This gives rise to an activation peptide (a glycopeptide of molecular weight 11,000) and factor $X_{a z}$ (a glycoprotein of molecular weight 44,000). This reaction is shown in Figure 4. Jesty & Esnouf have also studied the activation of factor X by the protease from Russell's viper venom and found a molecular weight of 40,000–42,000 for their factor X_a preparation (146). Bajaj & Mann reported that factor X_a formed in the presence of trypsin has a molecular weight of 52,000 and concluded that this is due to the cleavage of a small peptide of molecular weight 4000 from the heavy chain (123). Jesty & Nemerson investigated the activation of factor X by tissue factor and factor VII and concluded that more than one peptide bond is cleaved when it is converted to factor X_a (141). Radcliffe & Barton, however, reported that the various activation

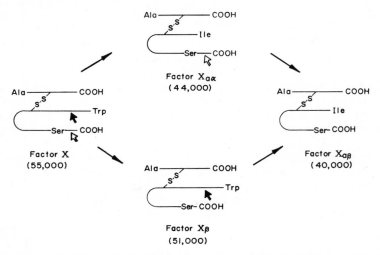

Figure 4   The mechanism of activation of bovine factor X [modified from Fujikawa et al (72)].

mechanisms for factor X yield essentially the same molecular form of factor X_a (147, 148). Furthermore, the factor X_a that they characterized contained no carbohydrate.

Recent experiments by Fujikawa et al (72) have shown that the conversion of factor X to factor $X_{a\alpha}$ occurs in the presence of factor IX_a and factor VIII, tissue factor and factor VII, trypsin, or the protease from Russell's viper venom, and each of these reactions involves the cleavage of the same activation peptide from the amino-terminal end of the heavy chain (72) (Figure 4). Factor $X_{a\alpha}$ can also be converted to factor $X_{a\beta}$, a protein with equivalent coagulant activity. This is due to the cleavage of a glycopeptide(s) from the carboxyl-terminal end of the heavy chain. This reduces the molecular weight from 44,000 to 40,000 and removes the residual carbohydrate present in factor $X_{a\alpha}$. The second reaction, catalyzed by factor IX_a and factor VIII or tissue factor and factor VII, is very slow in the presence of trace amounts of the protease from Russell's viper venom.

Factor IX_a and factor VIII, tissue factor and factor VII, or trypsin will also convert factor X to factor X_β (molecular weight 51,000) by the initial cleavage of the glycopeptide fragment(s) from the carboxyl-terminal region of the heavy chain (Figure 4). The peptide bond cleaved in this reaction appears to be the same as that involved in the conversion of factor $X_{a\alpha}$ to factor $X_{a\beta}$. Factor X_β is a partially degraded protein which can also be converted to an enzyme by cleavage of the same arginyl-isoleucine bond in the amino-terminal region of the heavy chain. This reaction leads to the formation of an Ile-Val-Gly-Gly- sequence in the heavy chain of the activated enzyme. This enzyme, factor $X_{a\beta}$, appears to be identical to the factor $X_{a\beta}$ formed by degradation of factor $X_{a\alpha}$. It appears probable that factor X_β is identical to the protein that Bajaj & Mann identified in the presence of trypsin (123).

Factor X_a is a serine protease and is inhibited by DFP (64, 71, 123, 149). It also has esterase activity toward synthetic substrates such as tosyl arginyl ethyl ester (116, 124, 150, 151). The amino acid sequence surrounding the active serine is shown in Table 2 (64). Factor X_a is also inhibited by p-nitrophenyl-p'-guanidinobenzoate (152), phenylmethylsulfonyl fluoride (71), antithrombin-heparin cofactor (5), and soybean trypsin inhibitor (124, 129, 153). It is not affected by N-tosyl-L-phenylalanylchloromethane and N-tosyl-L-lysylchloromethane (71).

Factor X_a appears to be identical with thrombokinase, which was extensively purified and characterized many years ago by Milstone as an extremely potent activator of prothrombin (154).

Biochemical Properties of Factor V and Its Interaction with Factor X_a

In the next reaction in the coagulation scheme, factor X_a interacts with factor V to form a complex that acts as a catalyst in the activation of prothrobin (Figure 2) (146, 154–158). This reaction also requires the presence of calcium ions and phospholipid.

Factor V is a plasma glycoprotein with a molecular weight of 290,000 (159). Other estimates have ranged as high as 400,000 (160). It is stabilized by calcium ions and rapidly loses activity in the presence of chelating agents (159, 161–165). Philip et al and Colman et al have suggested that it may exist in plasma in several different forms (166, 167). Day & Barton have presented evidence, however, to indicate that the presence of more than one form of plasma factor V is due to an artifact of the preparative procedure (162).

The bovine preparation has been purified about 6,000–10,000-fold from plasma by $BaSO_4$ adsorption and column chromatography on TEAE-cellulose and phosphocellulose (159). This preparation is relatively stable. The human preparation, however, appears to be far more labile and has been difficult to isolate and characterize.

Incubation of factor V with trace amounts of thrombin increases its specific activity as much as 10- to 15-fold (160, 165, 168, 169). The extent of this increase, however, varies with the preparation of factor V, and some investigators found no increase in factor V activity in the presence of thrombin (157, 170). From the data of Newcomb & Hoshida, it appears that the extent of increase of factor V activity by thrombin is inversely related to the specific activity of the original factor V (168).

Thrombin activation of factor V decreases the molecular weight from 400,000 or more to around 200,000, as estimated by gel filtration on Sephadex G-200 (160). Colman et al, however, reported no change in the molecular size of their factor V preparation in the presence of thrombin (167).

Other proteases will also activate factor V (134). Schiffman et al have fractionated Russell's viper venom on Sephadex G-200 (171). One fraction that activates factor X has a molecular weight of approximately 145,000. A second fraction that activates factor V was isolated, and this protein has a molecular weight of 10,000–20,000. The latter enzyme will increase factor V activity 20–26-fold (172). It also markedly changes the behavior of factor V on gel filtration, decreasing its apparent molecular weight from a value greater than 400,000 to one near 205,000 (172). In contrast, factor V isolated from serum shows only a two- to fourfold increase in activity upon incubation with the purified venom protease. Furthermore, the molecular weight of approximately 230,000 for serum factor V does not change on treatment with the venom activating enzyme.

Recently, it has been shown that the factor V activator enzyme in Russell's viper venom is an arginine esterase and is inhibited by DFP (173, 174). These findings are consistent with the concept that both thrombin and the factor V activator from Russell's viper venom activate factor V by minor proteolysis.

Other enzymes, including factor X_a, have been implicated in the activation of factor V (175, 176). The role of factor X_a in factor V activation is difficult to study, however, since factor X_a will also activate traces of prothrombin which may be present as a contaminant in the factor V preparation (177). The thrombin thus formed makes it difficult to interpret experiments of this type.

During the coagulation process, prothrombin is converted to thrombin in the presence of a lipoprotein complex containing factor X_a, factor V, calcium ions, and phospholipid (Figure 2). In the absence of factor V, the rate of this reaction is markedly decreased. Present evidence indicates that factor X_a is the enzyme in this reaction, since it alone will activate prothrombin (154), while factor V alone has no effect on prothrombin (156). The formation of the lipoprotein complex is reversible in that removal of calcium ions dissociates factor X_a from factor V and phospholipid. Furthermore, factor V can be recovered unchanged from this mixture (156).

Phospholipid plays an important role in the formation of this complex. Both factor X_a and factor V bind to the phospholipid micelle in the presence of calcium ions (155). The substrate for this lipoprotein complex is prothrombin which also readily binds to phospholipid. Thrombin, which is the product of this reaction, does not bind to the phospholipid and is released from the complex during the activation reaction (178).

The exact role of factor V in this reaction is not known. Esmon et al have reported that thrombin-activated factor V, in contrast to native factor V, will bind to prothrombin (179). This suggests that thrombin-activated factor V participates as a substrate binding protein in the lipoprotein complex favoring a more optimal orientation of factor X_a and prothrombin, and thus enhancing the activation reaction.

Biochemical Properties of Prothrombin and Its Activation by Factor X_a and Factor V

Prothrombin is a plasma glycoprotein which has been extensively purified from a number of different sources. The bovine and human proteins have a molecular weight of 68,000–72,000 and are composed of a single polypeptide chain. The amino-terminal sequence of bovine prothrombin is Ala-Asn-Lys-Gly- (63, 180–182). Recently, the total amino acid sequence has been determined by Magnusson and co-workers (182a). Human prothrombin has an amino-terminal sequence of Ala-Asn-Pro-Phe- (183). Bovine prothrombin contains 10–14% carbohydrate, including hexose, hexosamine, and neuraminic acid (63, 184, 185). The carbohydrate has been reported to be present in four branched chains per mole of protein, and these are probably linked to asparagine (185). Based upon this information, it was concluded that three of the carbohydrate chains are located in the amino-terminal region of the molecule and the fourth near the carboxyl-terminal end of the protein (186). From the structural studies of Magnusson et al, however, it is clear that only three carbohydrate chains are present and these are linked to asparagine residues 77, 101, and 376 (182a). Human prothrombin contains 8.2% carbohydrate, including 2.8% hexose, 3.1% hexosamine, and 2.3% neuraminic acid (187). These values differ slightly from those published by Lanchantin et al (188). Removal of the carbohydrate has no effect on the activity of prothrombin, although it does increase the rate of hepatic uptake of the molecule in vivo (189).

Prothrombin is one of four clotting proteins that require vitamin K for biosynthesis. Recent evidence indicates that vitamin K plays an important role in the formation of specific calcium binding sites in the molecule (190–192). At 1 mM calcium, prothrombin binds four moles of calcium per mole of protein (191). Animals treated with dicumarol, a vitamin K antagonist, synthesize an abnormal prothrombin molecule which is biologically inactive (193, 194). This protein has a different electrophoretic mobility than the normal molecule. In addition, it does not adsorb quantitatively to barium citrate. The amino acid and carbohydrate contents and the amino- and carboxyl-terminal residues of the normal and abnormal molecules are the same (195–201). Abnormal prothrombin, however, binds less than one calcium ion per mole of protein at 1 mM calcium concentration (191, 192). Stenflo and co-workers have recently reported that the strong calcium binding sites in normal prothrombin are due to the presence of adjacent γ-carboxyglutamic acid residues in positions 7 and 8 from the amino-terminal end of the protein (202). The abnormal molecule contains glutamic acid in these two positions. These investigators have also noted another peptide constituting residues 12–34 which also contains modified glutamic acid residues. Magnusson had noted earlier the presence of two extra negative groups, probably carboxyl groups, in tryptic peptides from the amino-terminal portion of prothrombin (181, 202a). Furthermore, these investigators have identified 10 γ-carboxyglutamic acid residues in positions 7, 8, 15, 17, 20, 21, 26, 27, 30, and 33 (202b). Thus, three pairs of adjacent γ-carboxyglutamic acid residues are present in normal prothrombin and each pair very likely binds one calcium ion in a manner analogous to the binding of calcium to EDTA. Present evidence indicates that all the calcium binding sites are located in the amino-terminal region of prothrombin in fragment 1, a glycopeptide released during the activation reaction (198, 200, 201, 203).

During the activation of prothrombin, the molecule is bound to phospholipid and this binding is mediated by calcium ions (178, 204). Thus, abnormal prothrombin apparently lacks biological activity since it lacks this calcium-mediated binding to phospholipid (202). Abnormal prothrombin can be converted to thrombin when activated by nonphysiological systems, such as trypsin (195). This indicates that the adjacent γ-carboxyglutamic acids are not required for the enzymatic activity of thrombin. These experiments provide an exciting and important clue to the biochemical role of vitamin K, suggesting that it plays a direct or indirect role in a carboxylation reaction of glutamic acid.

Factors VII, IX, and X also require vitamin K for their biosynthesis. Furthermore, abnormal factors IX and X also fail to bind calcium (205). These two proteins also contain adjacent glutamic acid residues in the amino-terminal portion of their molecule (Table 1). Thus, it appears probable that these coagulation factors contain adjacent γ-carboxyglutamic acid residues required for calcium binding during the coagulation process.

The activation of bovine and human prothrombin in vitro occurs in several steps. Present evidence suggests the scheme shown in Figure 5 (146, 182, 186, 203, 206–214). The first step in the activation reaction appears to be the cleavage of a large glycopolypeptide (fragment 1·2) from the amino-terminal end of prothrombin by factor X_a. Fragment 1·2 has a molecular weight of about 33,500 (30,730 g protein plus two carbohydrate chains). This reaction also gives rise to a single chain protein called intermediate II. Bovine intermediate II has an amino-terminal threonine and a molecular weight of about 38,000 (35,386 g protein plus one carbohydrate chain). In the second step of the activation reaction, intermediate II is cleaved by factor X_a to form thrombin. Thrombin contains a light chain and a heavy chain, and these chains are held together by a disulfide bond. The light chain has a molecular weight of 5721 and contains an amino-terminal threonine residue; the heavy chain has a molecular weight of about 32,000 (29,683 g protein plus one carbohydrate chain) and contains an amino-terminal isoleucine residue (Table 2). The heavy chain also contains the active serine residue (Table 2). This pathway was first demonstrated by Stenn & Blout, who used low concentrations of DFP to preferentially inactivate thrombin and not factor X_a (206). In the absence of DFP, the first traces of thrombin formed react with prothrombin to form fragment 1 and intermediate I (step 3). Fragment 1 has a molecular weight of about 23,000 (17,973 g protein plus two carbohydrate chains) and contains an amino-terminal alanine residue. Intermediate I has a molecular weight of about 50,500 (48,143 g protein plus one carbohydrate chain) and contains an amino-terminal serine residue. Inter-

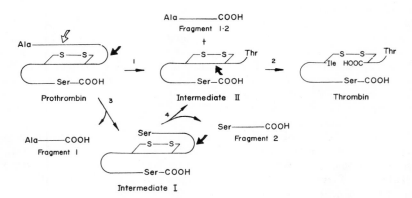

Figure 5   The mechanism of activation of bovine prothrombin [modified from Stenn & Blout (206) and Heldebrant et al (207)].

mediate I is readily converted to intermediate II in the presence of factor X_a (step 4). This releases fragment 2, a polypeptide with an amino-terminal serine residue and a molecular weight of 12,775. Intermediate II is then converted to thrombin by factor X_a, as shown in step 2. Because the first traces of thrombin lead to the formation of intermediate I, very little fragment $1 \cdot 2$ is formed. The molecular weights for the various fragments and intermediates have been kindly provided by Magnusson and co-workers prior to publication (182a).

The fragments released during the activation of bovine prothrombin are not inert. Recently, Jesty & Esnouf (146) have reported the inhibition of bovine prothrombin activation by fragment 1, and this has also been observed with the human preparation (209). These data are consistent with the experiments of Kandall et al, who suggested the formation of an inhibitor during prothrombin activation which inhibits the activation reaction after about 50% conversion to thrombin (215).

Prothrombin can also be activated by other proteolytic enzymes such as trypsin (216). The activation mechanism appears to be essentially the same as that obtained with factor X_a (209). The total yield of thrombin, however, is smaller since trypsin continues to degrade and inactivate thrombin.

The physical-chemical properties of thrombin have been reviewed by Magnusson, who has included a preliminary report of the complete amino acid sequence for the bovine molecule (65). Thrombin is a serine protease which is readily inhibited by DFP (65, 217–219). It also has strong esterase activity toward arginine esters such as tosyl arginine methyl ester (220, 221).

Thrombin is inhibited by a number of natural protein inhibitors, including hirudin (222) and antithrombin III (heparin cofactor) (5, 7). The latter inhibitor forms a 1:1 stoichiometric complex with thrombin, and this complex is not dissociated with denaturing or reducing agents. Heparin specifically accelerates the rate of complex formation without altering the stoichiometry.

In recent years, a number of excellent methods have been described for purifying thrombin (223). Several of these techniques have used affinity chromatography (224, 225).

Biochemical Properties of Fibrinogen and Its Interaction with Thrombin

When thrombin is formed, it in turn converts fibrinogen, a soluble plasma protein, to an insoluble gel called fibrin (Figure 2). Fibrin formation is due to the release of four fibrinopeptides from the amino-terminal ends of the fibrinogen molecule by limited proteolysis. The reaction is stimulated in the presence of calcium ions. The fibrinogen-to-fibrin conversion has recently been reviewed in an excellent article by Doolittle (226). Accordingly, only a few major points will be cited in this manuscript.

Human and bovine fibrinogen are large glycoproteins with molecular weights of about 340,000. The dimension of fibrinogen appears to be about 90×450 Å (227). Fibrinogen is composed of three pairs of nonidentical polypeptide chains (228, 229). These chains are held together by disulfide bonds. Estimates of the total number of disulfide bonds have ranged from 21–22 to as high as 32–34 (230–232). The α, β, and γ chains of bovine fibrinogen contain amino-terminal glutamic acid, pyrrolidone carboxylic acid, and tyrosine, respectively. The α, β, and γ chains of human fibrinogen contain amino-terminal alanine, pyrrolidone carboxylic acid, and tyrosine, respectively (233). The molecular weights for the α, β, and γ chains of human fibrinogen are 63,500, 56,000, and 47,000, respectively, as determined by sedimentation equilibrium (234). Thus, the formula for fibrinogen can be written as $\alpha_2\beta_2\gamma_2$. Approximately 20–30% of the primary sequence of the α, β, and γ chains has been determined primarily by Blomback and co-workers (235, 236). Carbohydrate is present in all three chains (237) and asparagine residue 52 has been shown to be an attachment site in the γ chain (238). The total carbohydrate in various mammalian fibrinogen is 4–5%, including hexose, hexosamine, and neuraminic acid (239).

Most fibrinogen preparations that have been isolated are heterogeneous. This heterogeneity may be due in part to variable carbohydrate composition, minor proteolysis in vivo or during isolations, genetic polymorphism, or all three.

When fibrinogen is attacked by thrombin, two fibrinopeptides A and two fibrinopeptides B are liberated (240, 241). These peptides make up about 3% of the original fibrinogen molecule (242–244). The thrombin effect is due to the hydrolysis of four specific arginyl-glycine bonds in the α and β chains of fibrinogen. The fibrinopeptides A are released from the amino-terminal end of the α chains and the fibrinopeptides B are released from the amino-terminal end of the β chains. The γ chains are not hydrolyzed during the formation of fibrin. The fibrinopeptides A are released at a faster rate than the fibrinopeptides B (245), and this may be one of the critical steps leading to fibrin formation (246). In recent years, the amino acid sequence has been determined for the fibrinopeptides from a large number of different species (233). These peptides show a great deal of variability in their sequence and have been of considerable interest from the standpoint of evolution. The fibrinopeptides also contain several modified amino acids. Tyrosine-O-sulfate is present in a number of different fibrinopeptides B (247), while phosphoserine occurs in the fibrinopeptides A of human and dog fibrinogen (248–250).

When the fibrin monomer is formed, it undergoes polymerization, apparently involving both end-to-end and lateral aggregation. Polymers formed in the absence of calcium ions are soluble in dilute acid and base and 6 M urea solutions. This polymer is also very susceptible to degradative enzymes such as plasmin. It is stabilized, however, in the presence of another plasma enzyme, factor XIII_a. These reactions are discussed in the next section.

Biochemical Properties of Factor XIII, its Activation by Thrombin, and Its Interaction with Fibrin

Factor XIII is a plasma glycoprotein which has been isolated and characterized from a number of different sources. It is the subject of a recent review on transglutaminase by Folk & Chung (251). Factor XIII is composed of two pairs of nonidentical polypeptide chains, and these chains (a and b) are held together by noncovalent bonds (252). The a chains have a molecular weight of 75,000 and the b chains have a molecular weight of 88,000, as determined by sedimentation equilibrium (252). The intact protein is a tetramer with molecular weight 320,000 (253, 254). Thus, the formula for the plasma protein can be written as a_2b_2.

Factor XIII has also been isolated and characterized from platelets (252, 255) and placenta (256). These proteins, which have been crystallized, have a molecular weight of 146,000–165,000 (256) and are composed of two a chains. These preparations have no b chains. The a subunits from plasma and platelet factor XIII appear to be identical in size, amino acid composition, and SDS gel electrophoresis patterns (252, 254, 256).

The amino-terminal group in the a subunit of plasma and platelet factor XIII is blocked (257) and this is due to the presence of a N-acetyl serine (258). The b subunit contains an amino-terminal glutamic acid (or glutamine) (257). Each of the a chains contains six –SH groups, while the b chain has no –SH groups (259). Plasma factor XIII contains 5% carbohydrate located in the b chain (252, 256).

In the presence of thrombin and calcium ions, factor XIII is converted to an enzyme with transglutaminase activity (260, 261). The activation reaction with plasma factor XIII is as follows

$$\text{factor XIII } (a_2 b_2) \xrightarrow{\text{thrombin}} \text{intermediate } (a'_2 b_2) \xrightarrow{\text{Ca}^{2+}} \text{factor XIII}_a \ (a'_2)$$

$$(320{,}000) \qquad\qquad + \qquad\qquad\qquad (140{,}000)$$

$$\text{two activation peptides} \qquad +$$

$$b_2$$

In the first step, thrombin cleaves an arginyl-glycine peptide bond in each of the a chains, liberating two activation peptides from the amino-terminal portion of the precursor molecule (252, 254, 257). This reaction occurs in the absence of calcium ions (259) and decreases the molecular weight of the factor XIII tetramer by about 8000. The activation peptide from human factor XIII has a molecular weight of about 4000. It contains 37 amino acids with a N-acetyl serine as its amino-terminal residue. Its complete amino acid sequence has recently been reported by Takagi & Doolittle (258). The activation peptide from bovine factor XIII contains 36 residues and shows a marked similarity in its amino acid sequence with the activation peptide isolated from human factor XIII (258). Furthermore, the activation peptide from the human plasma factor XIII is indistinguishable from that obtained from platelet factor XIII (258).

The thrombin reaction gives rise to an inactive tetrameric intermediate composed of two modified a chains with amino-terminal glycines and two b chains with amino-terminal glutamic acids (or glutamine). This structure can be written as $a'_2 b_2$.

In the next step, the tetrameric intermediate dissociates into a catalytic dimer (a'_2) and a noncatalytic dimer (b_2). This reaction requires the presence of calcium ions (259). The active enzyme is readily inhibited by selective alkylation of a single cysteine –SH group in only one of the catalytic subunits, and this reaction does not occur in the absence of calcium. These results suggest that factor XIII$_a$ is one of the "half-of-the-site reactivity" enzymes (45).

Thrombin activation of factor XIII from platelets occurs at a significantly faster rate than that of the plasma enzyme, suggesting that the b subunits control the rate of enzyme formation by thrombin (259).

Factor XIII is also activated by other proteolytic enzymes, including trypsin, reptilase, and papain (262–268). These enzymes are also endopeptidases and appear to activate factor XIII by splitting the same specific arginyl-glycine bond that is hydrolyzed by thrombin (254).

When factor XIII$_a$ is formed, it forms crosslinkages in fibrin to create a tough insoluble clot. The crosslinkage is due to the formation of $\varepsilon(\gamma$-glutamyl) lysine linkages between a glutamine side chain in one fibrin monomer with a lysine side chain in another fibrin monomer (262, 269, 270). The crosslinkage results in the formation of intermolecular γ-γ linkages (271, 272). The presence of these bonds makes the crosslinked fibrin resistant to dispersing solvents. The γ crosslinking reaction is followed by a slower α-α chain crosslinking reaction (273). Eventually, about six crosslinkages are formed per mole of fibrin (274), four of which involve the α chains and two involve the γ chains (275). After the crosslinking reaction, the fibrin clot is very insoluble and resistant to lysis (276).

Factor XIII will also crosslink fibrinogen, but this reaction is extremely slow (277). Indeed, factor XIII will even crosslink itself, forming a_2 or a'_2 dimers which are now covalently linked (278).

Other transglutaminase preparations will also crosslink fibrinogen as well as fibrin. For instance, guinea pig liver transglutaminase readily crosslinks the α chains, the β chains, and the γ chains of human fibrinogen to form a very high molecular weight polymer (254, 273, 279). The pattern of crosslinking with fibrin is much different, however, than that with factor XIII$_a$. The most striking difference is the absence of crosslinking between two γ chains.

Biochemical Properties of Factor VII, Tissue Factor, and the Extrinsic System of Coagulation

Trace amounts of tissue extract cause a dramatic increase in the rate of plasma coagulation. This reaction is due to the interaction of factor VII, a plasma protein with a tissue lipoprotein called tissue factor. These two proteins activate factor X in the extrinsic coagulation system (Figure 1) and bypass several plasma proteins such as factors VIII, IX, and XI which participate in the intrinsic system. These reactions have been reviewed recently by Nemerson & Pitlick (114).

Tissue factor has been found in a large number of different tissues. Brain, lung, and placenta microsomes are particularly rich in tissue factor (280, 281). It is also found in other organs such as liver, spleen, and kidney (282). Tissue factor is present in large vessels such as aorta and vena cava, where it is located most abundantly in the intima. It has also been reported in small vessels, leukocytes, and the plasma membranes of endothelial cells (283). Thus, trauma to the vascular endothelium presumably results in the release of tissue factor, initiating the extrinsic coagulation process at the site of injury.

Crude tissue factor preparations contain 30–45% phospholipid by weight (284, 285). Separation of tissue factor into lipid and protein fractions was reported by Deutsch et al in 1964 (137) and later by Hvatum & Prydz (284, 285). Nemerson & Pitlick purified the apoprotein 800–1500-fold from bovine lung (139). The final preparation contained two fractions with molecular weights of 330,000 and 220,000.

In the presence of phospholipid, tissue factor activity increased about 950-fold. Direct binding of the phospholipid to the protein was demonstrated by sucrose gradient centrifugation (286).

Tissue factor prepared from different organs exhibits significant peptidase and endopeptidase activity whether particle bound or solubilized (287). Furthermore, coagulant and peptidase activities of the soluble apoprotein are coincident following gel filtration, electrofocusing, and disc gel electrophoresis. Tissue factor, however, is not sensitive to DFP (288).

Extensive purification has been achieved for factor VII. As mentioned earlier, this coagulant factor also requires vitamin K for its biosynthesis. Nemerson & Esnouf reported a purification of about 2000-fold for factor VII from bovine plasma with a yield of 7–12% (288). This preparation, which showed at least three bands on SDS polyacrylamide gel electrophoresis, was inactivated by DFP. These authors concluded that factor VII circulates in blood as a DFP-sensitive protein. The preparation of factor VII also formed a complex with highly purified tissue factor, and the appearance of coagulant activity was accompanied by the cleavage of a protein thought to be factor VII. More recently, however, Jesty & Nemerson reported a 150,000-fold purification of factor VII using affinity chromatography on benzamidine-Sepharose (141). A similar purification has been achieved independently by Fujikawa et al (72). The highly purified factor VII preparation of Jesty & Nemerson was also DFP sensitive, but showed no change in molecular weight when incubated with tissue factor (141). These investigators also reported that the tissue factor-factor VII complex, which was DFP sensitive, activated factor X directly. As mentioned in the section on factor X, this reaction occurs by the cleavage of a single peptide bond in the amino-terminal portion of the heavy chain of factor X (72) (Figure 4).

Gladhaug & Prydz purified factor VII from human serum about 10,000-fold with an overall recovery of about 1% (289). This preparation had a molecular weight of 44,700, had no esterolytic or caseinolytic activity, and was somewhat smaller than human plasma factor VII, which has a molecular weight of 63,000 (290). Østerud et al also reported that incubation of human factor VII with tissue factor was necessary prior to its participation in blood coagulation (140). These investigators, however, found that a factor VII_a activity could be obtained free of tissue factor activity. This was accomplished by treating the complex with phospholipase C to destroy the tissue factor activity or by passing the complex through $0.1\ \mu$ filters to remove tissue factor activity. The factor VII_a thus obtained activated factor X. These experiments suggest that tissue factor may indeed convert factor VII to factor VII_a, which in turn activates factor X. Thus, tissue factor may function mainly as an accelerator for factor X activation by factor VII_a.

Gjonnaess has published evidence for the activation of factor VII by another pathway (291, 292). His experiments indicate that factor XII_a activated with kaolin at low temperature will activate prekallikrein, and kallikrein in turn will activate factor VII. This may explain the observations of Alexander & Landwehr that the one stage prothrombin time is shortened when plasma is stored at low temperature (293). These experiments also provide another link between the intrinsic and extrinsic coagulation systems. These reactions are extremely slow, however, and therefore their physiological importance remains to be established.

These data leave open the question of factor VII activation and the mechanism of this reaction. It should be emphasized, however, that three of the vitamin K-dependent clotting factors, i.e. prothrombin, factor IX, and factor X, show striking similarities in their primary structure, calcium binding properties, activation by proteolysis, and participation in coagulation as serine esterases. Thus, it is tempting to suggest that factor VII may have similar properties. It is possible that high concentrations of factor VII_a will activate factor X and that this reaction is accelerated in the presence of tissue factor. This would be analogous to both the activation of factor X by factor IX_a in a reaction accelerated by factor VIII and the activation of prothrombin by factor X_a in a reaction accelerated by factor V.

It is clear that a complete characterization of factor VII will be necessary before these questions can be answered.

Literature Cited

1. McKusick, V. A. 1965. *Sci. Am.* 213: 88–95
2. Murer, E. H., Day, H. G. 1972. In *Platelets and Thrombosis,* ed. S. Sherry, A. Scriabine, 1–22. Baltimore: Univ. Park Press
3. Ratnoff, O. D. 1972. *Progr. Hemostasis Thromb.* 1:39–74
4. Wright, I. 1959. *J. Am. Med. Assoc.* 170:325
5. Yin, E. T., Wessler, S., Stoll, P. J. 1971. *J. Biol. Chem.* 246:3694–3702
6. Abildgaard, U. 1967. *Scand. J. Clin. Lab. Invest.* 19:190–95
7. Rosenberg, R. D., Damus, P. S. 1973. *J. Biol. Chem.* 248:6490–6505
8. Rimon, A., Shamash, Y., Shapiro, B. 1966. *J. Biol. Chem.* 241:5102–7
9. Heimburger, N., Haupt, H., Schwick, H. G. 1971. *Proc. Int. Res. Conf. Proteinase Inhibitors,* ed. H. Fritz, H. Tschesche, 1–21. Berlin: W. de Gruyter.
10. Deykin, D., Cochios, F., DeCamp, G., Lopez, A. 1968. *Am. J. Physiol.* 214:414–19
11. Davie, E. W., Ratnoff, O. D. 1964. *Science* 145:1310–12
12. Macfarlane, R. G. 1964. *Nature* 202: 498–99
13. Seegers, W. H. 1967. In *Blood Clotting Enzymology,* ed. W. H. Seegers, 1–21. New York: Academic
14. Niewiarowski, B., Bankowski, E., Rogowicka, I. 1965. *Thromb. Diath. Haemorrh.* 14:387–400
15. Hardisty, R. M., Margolis, J. 1959. *Brit. J. Haematol.* 5:203
16. Ratnoff, O. D., Davie, E. W., Mallett, D. L. 1961. *J. Clin. Invest.* 10:803
17. Soulier, J.-P., Prou-Wartelle, O., Menache, D. 1958. *Rev. Fr. Etud. Clin. Biol.* 3:263
18. Davie, E. W., Kirby, E. P. 1973. *Curr. Top. Cell. Regul.* 51–86
19. Davie, E. W., Hougie, C., Lundblad, R. L. 1969. In *Recent Advances in Blood Coagulation,* ed. L. Poller, 13–28. London: Churchill
20. Ratnoff, O. D., Rosenblum, J. M. 1958. *Am. J. Med.* 25:160–68
21. Biggs, R. et al 1958. *Brit. J. Haematol.* 4:177
22. Nossel, H. L. 1964. *The Contact Phase of Blood Coagulation.* Philadelphia: Davis
23. Cochrane, C. G., Revak, S. D., Aikin, B. S., Wuepper, K. D. 1972. In *Inflammation: Mechanisms and Control,* ed. I. H. Lepow, P. A. Ward, 119–29. New York: Academic
24. Wuepper, K. D. Ibid, 93–117
25. Weiss, A. S., Gallin, J. I., Kaplan, A. P. 1974. *J. Clin. Invest.* 53:622–33
26. Saito, H., Ratnoff, O. D. 1974. *Nature* 248:597–98
27. Hathaway, W. E., Belhasen, L. P., Hathaway, H. S. 1965. *Blood* 26:521
28. Schiffman, S., Rapaport, S. I., Ware, A. G., Mehl, J. W. 1960. *Proc. Soc. Exp. Biol. Med.* 105:453
29. Schoenmakers, J. G. G., Matze, R., Haanen, C., Zilliken, F. 1965. *Biochim. Biophys. Acta* 101:166–76
30. Komiya, M., Nagasawa, S., Suzuki, T. 1972. *J. Biochem.* 72:1205–18
31. Ratnoff, O. D., Davie, E. W. 1962. *Biochemistry* 1:967–75
32. Cochrane, C. G., Wuepper, K. D. 1971.

J. Exp. Med. 134:986–1004

33. Donaldson, V. H., Ratnoff, O. D. 1965. *Science* 150:754–56

34. Ratnoff, O. D., Crum, J. D. 1964. *J. Lab. Clin. Med.* 63:359–77

35. Cochrane, C. G., Revak, S. D., Wuepper, K. D. 1973. *J. Exp. Med.* 138:1564–83

36. Fujikawa, K., Legaz, M. E., Kato, H., Davie, E. W. 1974. *Biochemistry* 13:4508–16

37. Kingdon, H. S., Davie, E. W., Ratnoff, O. D. 1964. *Biochemistry* 3:166–73

38. Kato, H., Fujikawa, K., Legaz, M. E. 1974. *Fed. Proc.* 33:1505

39. Saito, H., Ratnoff, O. D., Marshall, J. S., Pensky, J. 1973. *J. Clin. Invest.* 52:850–61

40. Haanen, C., Morselt, J., Schoenmakers, J. G. G. 1967. *Thromb. Diath. Haemorrh.* 17:307–20

41. Ratnoff, O. D. 1972. *J. Lab. Clin. Med.* 80:704–10

42. Ratnoff, O. D., Davie, E. W. 1962. *Biochemistry* 1:677–85

43. Schiffman, S., Lee, P. 1974. *Brit. J. Haematol.* 27:101–14

44. Kato, H., Legaz, M. E., Davie, E. W. Unpublished results

45. Levitzki, A., Stallcup, W. B., Koshland, D. E. Jr. 1971. *Biochemistry* 10:3371–78

46. Waaler, B. A. 1959. *Scand. J. Clin. Lab. Invest.* 11(Suppl. 37):1

47. Schiffman, S., Rapaport, S. I., Patch, M. J. 1963. *Blood* 22:733

48. Cattan, A. D., Denson, K. W. E. 1964. *Thromb. Diath. Haemorrh.* 11:155

49. Yin, E. T., Duckert, F. 1961. *Thromb. Diath. Haemorrh.* 6:224

50. Kingdon, H. S., Davie, E. W. 1965. *Thromb. Diath. Haemorrh.* Suppl. 17:15–22

51. White, S. G., Aggeler, P. M., Glendening, M. B. 1953. *Blood* 8:101

52. Biggs, R. et al 1961. *Brit. J. Haematol.* 7:349

53. Denson, K. W. E. 1967. *The Use of Antibodies in the Study of Blood Coagulation.* Philadelphia: Davis

54. Somer, J. B., Castaldi, P. A. 1970. *Brit. J. Haematol.* 18:147

55. Swart, A. C. W., Hemker, H. C. 1970. *Biochim. Biophys. Acta* 222:692–95

56. Chandra, S., Pechet, L. 1973. *Biochim. Biophys. Acta* 328:456

57. Pechet, L., Smith, J. A. 1970. *Biochim. Biophys. Acta* 200:475–85

58. Fujikawa, K., Thompson, A. R., Legaz, M. E., Meyer, R. G., Davie, E. W. 1973. *Biochemistry* 12:4938–45

59. Davie, E. W., Fujikawa, K., Kato, H., Legaz, M. E. 1975. *Ann. NY Acad. Sci.* 240:34–42

60. Kingdon, H. S., Lundblad, R. L. 1975. In *Handbook of Hemophilia,* ed. K. M. Brinkhous, A. F. H. Britten. Amsterdam: Excerpta Med. In press

61. Gentry, P. W., Alexander, B. 1973. *Biochem. Biophys. Res. Commun.* 50:500–9

62. Andersson, L.-O., Borg, H., Miller-Andersson, M. 1973. *Abstracts IVth Int. Congr. Thrombosis Haemostasis,* 109

63. Fujikawa, K. et al 1974. *Proc. Nat. Acad. Sci. USA* 71:427–30

64. Titani, K. et al 1972. *Biochemistry* 11:4899–4903

65. Magnusson, S. 1971. *Enzymes* 3:277

66. Enfield, D. L., Ericsson, L. H., Fujikawa, K., Titani, K., Walsh, K. A., Neurath, H. 1974. *FEBS Lett.* 47:132–35

67. Schiffman, S., Rapaport, S. I., Patch, M. J. 1964. *J. Clin. Res.* 12:110

68. Kingdon, H. S. 1969. *J. Biomed. Mater. Res.* 3:25

69. Schiffman, S., Rapaport, S. I., Chong, M. M. Y. 1973. *Brit. J. Haematol.* 24:633–42

70. Fujikawa, K., Davie, E. W. Unpublished results

71. Fujikawa, K., Legaz, M. E., Davie, E. W. 1972. *Biochemistry* 11:4892–99

72. Fujikawa, K., Coan, M. H., Legaz, M. E., Davie, E. W. 1974. *Biochemistry* 13:5290–99

73. Hougie, C., Denson, K. W. E., Biggs, R. 1967. *Thromb. Diath. Haemorrh.* 18:211

74. Østerud, B., Rapaport, S. I. 1970. *Biochemistry* 9:1854–61

75. Hemker, H. C., Kahn, M. J. P. 1967. *Nature* 215:1201–2

76. Lundblad, R. L., Davie, E. W. 1964. *Biochemistry* 3:1720–25

77. Barrow, E. M., Graham, J. B. 1974. *Physiol Rev.* 54:23–74

78. Schmer, G., Kirby, E. P., Teller, D. C., Davie, E. W. 1972. *J. Biol. Chem.* 247:2512–21

79. Legaz, M. E., Schmer, G., Counts, R. B., Davie, E. W. 1973. *J. Biol. Chem.* 248:3946–55

80. Pool, J. G., Hershgold, E. J., Pappenhagen, A. R. 1964. *Nature* 203:312

81. Hershgold, E. J., Davison, A. M., Janszen, M. E. 1971. *J. Lab. Clin. Med.* 77:185

82. Marchesi, S. L., Shulman, N. R., Gralnick, H. R. 1972. *J. Clin. Invest.* 51:2151

83. Kass, L., Ratnoff, O. D., Leon, M. A. 1969. *J. Clin. Invest.* 48:351

84. Shapiro, G. A., Anderson, J. C., Pizzo, S. V., McKee, P. A. 1973. *J. Clin. Invest.* 52:2198

85. Legaz, M. E., Weinstein, M. J., Helde-

brant, C. M., Davie, E. W. 1975. *Ann. NY Acad. Sci.* 240:43–61

86. Forbes, C. D., Prentice, C. R. M. 1973. *Nature New Biol.* 241:149–50

87. Zimmerman, T. S., Ratnoff, O. D., Powell, A. E. 1971. *J. Clin. Invest.* 50: 244–54

88. Bennett, E., Huehns, E. R. 1970. *Lancet* 2:956–58

89. Shapiro, G. A., McKee, P. A. 1970. *Clin. Res.* 18:615

90. Stites, D. P., Hershgold, E. J., Perlman, J. D., Fudenberg, H. H. 1971. *Science* 171:196–97

91. Bouma, B. N., Wiegerinck, Y., Sixma, J. J., van Mourik, J. A., Mochtar, I. A. 1972. *Nature New Biol.* 236:104–6

92. Davie, E. W., Legaz, M. E., Schmer, G. 1974. *Thromb. Diath. Haemorrh.* 59 (Suppl.):3–9

93. Meyer, D., Lavergne, J.-M., Larrieu, M.-J., Josso, F. 1972. *Thromb. Res.* 1: 183–96

94. Bennett, B., Forman, W. D., Ratnoff, O. D. 1973. *J. Clin. Invest.* 52:2191–97

95. Weiss, H. J. 1968. *Blood* 32:667–79

96. Thelin, G. M., Wagner, R. H. 1961. *Arch. Biochem. Biophys.* 95:70–76

97. Weiss, H. J., Kochwa, S. 1970. *Brit. J. Haematol.* 18:89

98. Weiss, H. J., Phillips, L. L., Rosner, W. 1972. *Thromb. Diath. Haemorrh.* 27:212

99. Owen, W. G., Wagner, R. H. 1972. *Thromb. Diath. Haemorrh.* 27:502

100. Cooper, H. A., Griggs, T. R., Wagner, R. H. 1973. *Proc. Nat. Acad. Sci. USA* 70:2326–29

101. Griggs, T. R., Cooper, H. A., Webster, W. P., Wagner, R. H., Brinkhous, K. M. 1973. *Proc. Nat. Acad. Sci. USA* 70:2814–18

102. Donati, M. B., de Gaetano, G., Vermylen, J. 1973. *Thromb. Res.* 2:97

103. Rick, M. E., Hoyer, L. W. 1973. *Blood* 42:737

104. Zimmerman, T. S., Edgington, T. S. 1973. *J. Exp. Med.* 138:1015

105. Weinstein, M. J., Davie, E. W. Unpublished results

106. Brown, J. E., Braugh, R. F., Sargeant, R. B., Hougie, C. 1975. *J. Exp. Med.* In press

107. Legaz, M. E., Heldebrant, C. M., Davie, E. W. Unpublished results

108. Rapaport, S. I., Schiffman, S., Patch, M. J., Ames, S. B. 1963. *Blood* 21:221

109. Macfarlane, R. G., Biggs, R., Ash, B. J., Denson, K. W. E. 1964. *Brit. J. Haematol.* 10:530

110. Rapaport, S. I., Hjort, P. F., Patch, M. J. 1965. *Scand. J. Clin. Lab. Invest.* 17(Suppl. 84):88

111. Ozge-Anwar, A. H., Connell, G. E., Mustard, J. F. 1965. *Blood* 26:500

112. Weiss, H. J., Hoyer, L. W. 1973. *Science* 182:1149

113. Legaz, M. E., Fujikawa, K., Davie, E. W. Unpublished results

114. Nemerson, Y., Pitlick, F. A. 1972. *Progr. Hemostasis Thromb.* 1:1–37

115. Chuang, T. F., Sargeant, R. B., Hougie, C. 1972. *Biochim. Biophys. Acta* 273: 287–91

116. Esnouf, M. P., Williams, W. J. 1962. *Biochem. J.* 84:62–71

117. Papahadjopoulos, D., Yin, E. T., Hanahan, D. J. 1964. *Biochemistry* 3:1931–39

118. Jackson, C. M., Johnson, T. F., Hanahan, D. J. 1968. *Biochemistry* 7:4492–4505

119. Aronson, D. L., Mustafa, A. J., Mushinski, J. F. 1969. *Biochim. Biophys. Acta* 188:25–30

120. Fujikawa, K., Legaz, M. E., Davie, E. W. 1972. *Biochemistry* 11:4882–91

121. Jackson, C. M. 1972. *Biochemistry* 11: 4873–81

122. Esnouf, M. P., Lloyd, P. H., Jesty, J. 1973. *Biochem. J.* 131:781–89

123. Bajaj, S. P., Mann, K. G. 1973. *J. Biol. Chem.* 248:7729–41

124. Jackson, C. M., Hanahan, D. J. 1968. *Biochemistry* 7:4506–17

125. Enfield, D. L., Ericsson, L. H., Fujikawa, K., Titani, K. 1974. *Fed. Proc.* 33:1473

126. Enfield, D. L., Ericsson, L. H., Walsh, K. A., Neurath, H., Titani, K. 1975. *Proc. Nat. Acad. Sci. USA* 72:16–19

127. Mattock, P., Esnouf, M. P. 1973. *Nature New Biol.* 242:90

128. Mattock, P., Esnouf, M. P. 1973. *Abstracts, IVth Int. Congr. Thrombosis Haemostasis,* 143

129. Lundblad, R. L., Davie, E. W. 1965. *Biochemistry* 4:113–20

130. Biggs, R., Macfarlane, R. G. 1965. *Thromb. Diath. Haemorrh.* Suppl. 17: 23

131. Schiffman, S., Rapaport, S. I., Chong, M. M. Y. 1966. *Proc. Soc. Exp. Biol. Med.* 123:736

132. Barton, P. G. 1967. *Nature* 215:1508–9

133. Flynn, J. E., Coon, R. W. 1953. *Am. J. Physiol.* 175:289

134. Hjort, P. F. 1957. *Scand. J. Clin. Lab. Invest.* 9(Suppl. 27):7–183

135. Hougie, C. 1959. *Proc. Soc. Exp. Biol. Med.* 101:132

136. Straub, W., Duckert, F. 1961. *Thromb. Diath. Haemorrh.* 5:402

137. Deutsch, E., Irsigler, K., Lomoschitz, H. 1964. *Thromb. Diath. Haemorrh.* 12:12

138. Williams, W. J., Norris, D. G. 1966. *J. Biol. Chem.* 241:1847–56

139. Nemerson, Y., Pitlick, F. A. 1970. *Biochemistry* 9:5100–5
140. Østerud, B., Berre, A., Otnaess, A.-B., Bjorklid, E., Prydz, H. 1972. *Biochemistry* 11:2853–57
141. Jesty, J., Nemerson, Y. 1974. *J. Biol. Chem.* 249:509–15
142. Ferguson, J. H., Wilson, E. G., Iatridis, S. G., Rierson, H. A., Johnston, B. R. 1960. *J. Clin. Invest.* 39:1942
143. Pechet, L., Alexander, B. 1960. *Fed. Proc.* 19:64
144. Yin, E. T. 1964. *Thromb. Diath. Haemorrh.* 12:307
145. Rimon, A., Alexander, B., Katchalski, E. 1966. *Biochemistry* 5:792–98
146. Jesty, J., Esnouf, M. P. 1973. *Biochem. J.* 131:791–99
147. Radcliffe, R. D., Barton, P. G. 1972. *J. Biol. Chem.* 247:7735–42
148. Radcliffe, R. D., Barton, P. G. 1973. *J. Biol. Chem.* 248:6788–95
149. Leveson, J. E., Esnouf, M. P. 1968. *Brit. J. Haematol.* 17:173
150. Aronson, D. L., Menache, D. 1968. *Biochim. Biophys. Acta* 167:378–87
151. Adams, R. W., Elmore, D. T. 1971. *Biochem. J.* 124:66P–67P
152. Smith, R. L. 1973. *J. Biol. Chem.* 248:2418–23
153. Breckenridge, R. T., Ratnoff, O. D. 1964. *Clin. Res.* 12:221
154. Milstone, J. H. 1964. *Fed. Proc.* 23:742–48
155. Papahadjopoulos, D., Hanahan, D. J. 1964. *Biochim. Biophys. Acta* 90:436–39
156. Barton, P. G., Jackson, C. M., Hanahan, D. J. 1967. *Nature* 214:923–24
157. Jobin, F., Esnouf, M. P. 1967. *Biochem. J.* 102:666–74
158. Hemker, H. C., Esnouf, M. P., Hemker, P. W., Swart, A. C. W., Macfarlane, R. G. 1967. *Nature* 215:248–51
159. Esnouf, M. P., Jobin, F. 1967. *Biochem. J.* 102:660–65
160. Papahadjopoulos, D., Hougie, C., Hanahan, D. J. 1964. *Biochemistry* 3:264–70
161. Leikin, S., Bessman, S. P. 1956. *Blood* 11:916
162. Day, W. C., Barton, P. G. 1972. *Biochim. Biophys. Acta* 261:457–68
163. Blomback, B., Blomback, M. 1963. *Nature* 198:886–87
164. Weiss, H. J. 1965. *Thromb. Diath. Haemorrh.* 14:32
165. Barton, P. G., Hanahan, D. J. 1967. *Biochim. Biophys. Acta* 133:506–18
166. Philip, G., Moran, J., Colman, R. W. 1970. *Biochemistry* 9:2212–18
167. Colman, R. W., Moran, J., Philip, G. 1970. *J. Biol. Chem.* 245:5941–46
168. Newcomb, T. F., Hoshida, M. 1965. *Scand, J. Clin. Lab. Invest.* Suppl. 84:61
169. Colman, R. W. 1969. *Biochemistry* 8:1438–45
170. White, N. B., Ferguson, J. H., Hitsumoto, S. 1968. *Am. J. Med. Sci.* 255:143
171. Schiffman, S., Theodor, I., Rapaport, S. I. 1969. *Biochemistry* 8:1397–1405
172. Hanahan, D. J., Rolfs, M. R., Day, W. C. 1972. *Biochim. Biophys. Acta* 286:205–11
173. Esmon, C. T., Jackson, C. M. 1973. *Thromb. Res.* 2:509–24
174. Jackson, C. M., Gordon, J., Hanahan, D. J. 1971. *Biochim. Biophys. Acta* 252:255–61
175. Breckenridge, R. T., Ratnoff, O. D. 1965. *J. Clin. Invest.* 44:302–14
176. Breckenridge, R. T., Ratnoff, O. D. 1966. *Blood* 27:527
177. Bergsagel, D. E., Nockolds, E. R. 1965. *Brit. J. Haematol.* 11:395–410
178. Barton, P. G., Hanahan, D. J. 1969. *Biochim. Biophys. Acta* 187:319–27
179. Esmon, C. T., Owen, W. G., Duiguid, D. L., Jackson, C. M. 1973. *Biochim. Biophys. Acta* 310:289–94
180. Stenflo, J. 1973. MD thesis. Univ. of Malmo, Malmo, Sweden
181. Magnusson, S. 1972. *Folia Haematol. Leipzig* 98:4, S.385
182. Heldebrant, C. M., Noyes, C., Kingdon, H. S., Mann, K. G. 1973. *Biochem. Biophys. Res. Commun.* 54:155–60
182a. Magnusson, S., Sottrup-Jensen, L., Petersen, T. E., Claeys, H. Personal communication
183. Pirkle, H., McIntosh, M., Theodor, I., Vernon, S. 1973. *Thromb. Res.* 2:461–66
184. Magnusson, S. 1965. *Ark. Kemi* 23:285
185. Nelsestuen, G. L., Suttie, J. W. 1972. *J. Biol. Chem.* 247:6096–6102
186. Heldebrant, C. M., Mann, K. G. 1973. *J. Biol. Chem.* 248:3642–52
187. Kisiel, W., Hanahan, D. J. 1973. *Biochim. Biophys. Acta* 304:103–13
188. Lanchantin, G. F., Hart, D. W., Friedman, J. A., Saavedra, N. V., Mehl, J. W. 1968. *J. Biol. Chem.* 243:5479–85
189. Nelsestuen, G. L., Suttie, J. W. 1971. *Biochem. Biophys. Res. Commun.* 45:198–203
190. Ganrot, P. O., Nilehn, J. E. 1968. *Scand. J. Clin. Lab. Invest.* 22:23–28
191. Nelsestuen, G. L., Suttie, J. W. 1972. *Biochemistry* 11:4961–64
192. Stenflo, J., Ganrot, P. O. 1973. *Biochem. Biophys. Res. Commun.* 50:98–104
193. Stenflo, J. 1970. *Acta Chem. Scand.* 24:3762–63

194. Josso, F., Lavergne, J. M., Goualt, M., Prou-Wartelle, O., Soulier, J. P. 1968. *Thromb. Diath. Haemorrh.* 20:88
195. Nelsestuen, G. L., Suttie, J. W. 1972. *J. Biol. Chem.* 247:8176–82
196. Stenflo, J., Ganrot, P. O. 1972. *J. Biol. Chem.* 247:8160–66
197. Stenflo, J. 1972. *J. Biol. Chem.* 247: 8167–75
198. Stenflo, J. 1973. *J. Biol. Chem.* 248: 6325–32
199. Stenflo, J. 1974. *J. Biol. Chem.* 249: 5527–35
200. Nelsestuen, G. L., Suttie, J. W. 1973. *Proc. Nat. Acad. Sci. USA* 70:3366–70
201. Howard, J. B., Nelsestuen, G. L. 1974. *Biochem. Biophys. Res. Commun.* 59: 757–63
202. Stenflo, J., Fernlund, P., Egan, W., Roepstorff, P. 1974. *Proc. Nat. Acad. Sci. USA* 71:2730–33
202a. Magnusson, S., Sottrup-Jensen, L., Petersen, T. E., Klemmensen, P., Kouba, E. 1975. *Thromb. Diath. Haemorrh. Suppl.* In press
202b. Magnusson, S., Sottrup-Jensen, L., Petersen, T. E., Morris, H. R., Dell, A. 1974. *FEBS Lett.* 44:189
203. Benson, B. J., Kisiel, W., Hanahan, D. J. 1973. *Biochim. Biophys. Acta* 329: 81–87
204. Bull, R. K., Jevons, S., Barton, P. G. 1972. *J. Biol. Chem.* 247:2747–54
205. Reekers, P. P. M., Lindhout, M. J., Kop-Klaassen, B. H. M., Hemker, H. C. 1973. *Biochim. Biophys. Acta* 317:559–62
206. Stenn, K. S., Blout, E. R. 1972. *Biochemistry* 11:4502–15
207. Heldebrant, C. M., Butkowski, R. J., Bajaj, S. P., Mann, K. G. 1973. *J. Biol. Chem.* 248:7149–63
208. Fass, D. N., Mann, K. G. 1973. *J. Biol. Chem.* 248:3280–87
209. Kisiel, W., Hanahan, D. J. 1973. *Biochim. Biophys. Acta* 329:221–32
210. Owen, W. G., Esmon, C. T., Jackson, C. M. 1974. *J. Biol. Chem.* 249:594–605
211. Esmon, C. T., Owen, W. G., Jackson, C. M. 1974. *J. Biol. Chem.* 249:606–11
212. Engel, A. M., Alexander, B. 1973. *Biochim. Biophys. Acta* 320:687–700
213. Lanchantin, G. F., Friedman, J. A., Hart, D. W. 1968. *J. Biol. Chem.* 243: 476–86
214. Aronson, D. L., Menache, D. 1966. *Biochemistry* 5:2635–40
215. Kandall, C., Akinbami, T. K., Colman, R. W. 1972. *Brit. J. Haematol.* 23: 655–68
216. Alexander, B., Pechet, L., Kliman, A.

1962. *Circulation* 26:596–611
217. Miller, K. D., van Vunakis, H. 1956. *J. Biol. Chem.* 223:227–37
218. Gladner, J. A., Laki, K. 1956. *Arch. Biochem. Biophys.* 62:501–3
219. Laki, K., Gladner, J., Folk, J. E., Kominz, D. R. 1958. *Thromb. Diath. Haemorrh.* 2:205
220. Sherry, S., Troll, W. 1954. *J. Biol. Chem.* 208:95–105
221. Weinstein, M. J., Doolittle, R. F. 1972. *Biochim. Biophys. Acta* 258:577–90
222. Markwardt, F. 1963. *Blutgerinnungs-hemmende Wirkstoffe aus Blutsaugenden Tieren,* 1–122. Jena: VEB Gustav Fischer
223. Lundblad, R. L. 1971. *Biochemistry* 10:2501–6
224. Schmer, G. 1972. *Z. Physiol. Chem.* 353:810–14
225. Thompson, A. R., Davie, E. W. 1971. *Biochim. Biophys. Acta* 250:210–15
226. Doolittle, R. F. 1973. *Advan. Protein Chem.* 27:1–109
227. Tooney, N. M., Cohen, C. 1972. *Nature* 237:23–25
228. Henschen, A. 1962. *Acta Chem. Scand.* 16:1037
229. Clegg, J. B., Bailey, K. 1962. *Biochim. Biophys. Acta* 63:525–27
230. Henschen, A. 1964. *Ark. Kemi* 22:355
231. Cartwright, T., Kekwick, R. G. O. 1971. *Biochim. Biophys. Acta* 236:550–62
232. Latallo, Z. S., Dudek, G. A., Kloczewiak, M. 1971. *Scand. J. Haematol.* Suppl. 13:37
233. Dayhoff, M. O. 1969. *Atlas of Protein Sequence and Structure.* Silver Spring, Md.: Nat. Biomed. Res. Found.
234. McKee, P. A., Rogers, L. A., Marler, E., Hill, R. L. 1966. *Arch. Biochem. Biophys.* 116:271–79
235. Blomback, B., Hessel, B., Iwanaga, S., Reuterby, J., Blomback, M. 1972. *J. Biol. Chem.* 247:1496–1512
236. Blomback, B., Grondahl, N. J., Hessel, B., Iwanaga, S., Wallen, P. 1973. *J. Biol. Chem.* 248:5806–20
237. Henschen, A. 1964. *Ark. Kemi* 22:375
238. Iwanaga, S., Blomback, B., Grondahl, N. J., Hessel, B., Wallen, P. 1968. *Biochim. Biophys. Acta* 160:280–83
239. Blomback, B. 1958. *Ark. Kemi* 12:99
240. Laki, K., Gladner, J. A. 1964. *Physiol. Rev.* 44:127
241. Laki, K. 1968. In *Fibrinogen,* ed. K. Laki, 1–24. New York: Dekker
242. Lorand, L. 1951. *Nature* 167:992
243. Lorand, L. 1952. *Biochem. J.* 52:200–3
244. Bettelheim, F. R., Bailey, K. 1952. *Biochim. Biophys. Acta* 9:578–79
245. Bettelheim, F. R. 1956. *Biochim.*

Biophys. Acta 19:121–30

246. Herzig, R. H., Ratnoff, O. D., Shainoff, J. R. 1970. *J. Lab. Clin. Med.* 76:451

247. Bettelheim, F. R. 1954. *J. Am. Chem. Soc.* 76:2838–39

248. Blomback, B., Blomback, M., Edman, P., Hessel, B. 1962. *Nature* 193:883–84

249. Doolittle, R. F., Wooding, G. L., Lin, Y., Riley, M. 1971. *J. Mol. Evol.* 1:74

250. Osbahr, A. J. Jr., Colman, R. W., Laki, K., Gladner, J. A. 1964. *Biochem. Biophys. Res. Commun.* 14:555–58

251. Folk, J. E., Chung, S. I. 1973. *Advan. Enzymol.* 38:109–91

252. Schwartz, M. L., Pizzo, S. V., Hill, R. L., McKee, P. A. 1971. *J. Biol. Chem.* 246:5851–54

253. Loewy, A. G., Dahlberg, A., Dunathan, K., Kriel, R., Wolfinger, H. L. Jr. 1961. *J. Biol. Chem.* 236:2634–43

254. Schwartz, M. L., Pizzo, S. V., Hill, R. L., McKee, P. A. 1973. *J. Biol. Chem.* 248:1395–1407

255. Bohn, H. 1970. *Thromb. Diath. Haemorrh.* 23:455

256. Bohn, H. 1972. *Ann. NY Acad. Sci.* 202:256–72

257. Mikuni, Y., Iwanaga, S., Konishi, K. 1973. *Biochem. Biophys. Res. Commun.* 54:1393–1402

258. Takagi, T., Doolittle, R. F. 1974. *Biochemistry* 13:750–56

259. Chung, S. I., Lewis, M. C., Folk, J. E. 1974. *J. Biol. Chem.* 249:940–50

260. Buluk, K., Januszko, T., Olbromski, J. 1961. *Nature* 191:1093–94

261. Lorand, L., Konishi, K. 1964. *Arch. Biochem. Biophys.* 105:58–67

262. Lorand, L., Downey, J., Gotoh, T., Jacobsen, A., Tokura, S. 1968. *Biochem. Biophys. Res. Commun.* 31:222–30

263. Konishi, K., Takagi, T. 1969. *J. Biochem.* 65:281–84

264. Kopec, M., Latallo, Z. S., Stahl, M., Wegrzynowicz, A. 1969. *Biochim. Biophys. Acta* 181:437–45

265. Buluk, K., Zuch, A. 1967. *Biochim. Biophys. Acta* 147:593–94

266. Josso, Z. 1963. In *Fibrinogen and Fibrin Turnover of Clotting Factors,* ed. A. Koller, 138. Stuttgart: Schattauer

267. Dvilansky, A., Britten, A. F. H., Loewy, A. G. 1970. *Brit. J. Haematol.* 18:399–410

268. Tyler, H. M. 1970. *Biochim. Biophys. Acta* 222:396–404

269. Matacic, S., Loewy, A. G., 1968. *Biochem. Biophys. Res. Commun.* 30:356–62

270. Pisano, J. J., Finlayson, J. S., Peyton, M. P. 1968. *Science* 160:892–93

271. Chen, R., Doolittle, R. F. 1969. *Proc.*

Nat. Acad. Sci. USA 63:420–27

272. Takagi, T., Iwanaga, S. 1970. *Biochem. Biophys. Res. Commun.* 38:129–36

273. McKee, P. A., Mattock, P., Hill, R. L. 1970. *Proc. Nat. Acad. Sci. USA* 66:738–44

274. Pisano, J. J., Finlayson, J. S., Peyton, M. P., Nagai, Y. 1971. *Proc. Nat. Acad. Sci. USA* 68:770–72

275. Ball, A. P., Hill, R. L., McKee, P. A. 1972. *Abstracts, 3rd Congr. Int. Soc. Thromb. Haemostasis,* 62

276. Gormsen, J., Fletcher, A. P., Alkjaersig, N., Sherry, S. 1967. *Arch. Biochem. Biophys.* 120:654–65

277. Lorand, L., Ong, H. H. 1966. *Biochemistry* 5:1747–53

278. Takagi, T., Doolittle, R. F. 1973. *Biochem. Biophys. Res. Commun.* 51:186–91

279. Chung, S. I. 1972. *Ann. NY Acad. Sci.* 202:240–55

280. Williams, W. J. 1964. *J. Biol. Chem.* 239:933–42

281. Williams, W. J. 1966. *J. Biol. Chem.* 241:1840–46

282. Astrup, T. 1965. *Thromb. Diath. Haemorrh.* 14:401

283. Zeldis, S. M., Nemerson, Y., Pitlick, F. A., Lentz, T. L. 1972. *Science* 175:766–68

284. Hvatum, M., Prydz, H. 1966. *Biochim. Biophys. Acta* 130:92–101

285. Hvatum, M., Prydz, H. 1969. *Thromb. Diath. Haemorrh.* 21:217

286. Pitlick, F. A., Nemerson, Y. 1970. *Biochemistry* 9:5105–13

287. Pitlick, F. A., Nemerson, Y., Gottlieb, A. J., Gordon, R. G., Williams, W. J. 1971. *Biochemistry* 10:2650–57

288. Nemerson, Y., Esnouf, M. P. 1973. *Proc. Nat. Acad. Sci. USA* 70:310–14

289. Gladhaug, A., Prydz, H. 1970. *Biochim. Biophys. Acta* 215:105–11

290. Prydz, H. 1965. *Scand. J. Clin. Lab. Invest.* 17(Suppl. 84):78

291. Gjonnaess, H. 1972. *Thromb. Diath. Haemorrh.* 28:182–93

292. Gjonnaess, H. 1972. *Thromb. Diath. Haemorrh.* 28:194–205

293. Alexander, B., Landwehr, G. 1949. *Am. J. Physiol.* 159:322–31

294. Caspary, E. A., Kekwick, R. A. 1957. *Biochem. J.* 67:41–48

295. Shulman, S. 1953. *J. Am. Chem. Soc.* 75:5846–52

296. Mann, K. G., Heldebrant, C. M., Fass, D. N. 1971. *J. Biol. Chem.* 246:6106–14

297. Cox, A. C., Hanahan, D. J. 1970. *Biochim. Biophys. Acta* 207:49–64

298. Takagi, T., Konishi, K. 1972. *Biochim. Biophys. Acta* 271:363–70

PROTEIN PHOSPHORYLATION ✴903

Charles S. Rubin and Ora M. Rosen

Departments of Neuroscience, Molecular Biology, and Medicine,
Albert Einstein College of Medicine, Bronx, New York 10461

CONTENTS

INTRODUCTION

In this article we review recent developments regarding the regulation and function of protein kinases and protein phosphorylation. In certain cases, such as glycogen metabolism, detailed information pertaining to the regulatory role of protein

831

phosphorylation at the molecular level has been ascertained. Similarly, the classical phosphoproteins casein and phosvitin have been intensively studied for years. Our aim is to first present some well-defined examples of protein phosphorylation and then organize and abstract a large mass of correlative data in an attempt to discern the direction of future research on the roles of protein phosphorylation in the functional regulation of membranes, ribosomes, and nuclei. Although these subjects have not been thoroughly studied, we think that the potential importance of protein modification by phosphorylation at all levels of cellular organization justifies their presentation even at this early stage of development.

Because of space limitations we have covered only some of the articles published in the last few years and have not attempted a comprehensive review of the literature. We have also not included adenylylation of proteins, phosphoryl enzyme intermediates, and phosphoproteins involved in bacterial transport systems. Readers are referred to a number of excellent recent reviews on protein kinases and their substrates (1–4), glycogen metabolism (5, 6), enzymatic interconversion of active and inactive forms of enzymes (7, 8), protein adenylylation reactions (9), ATPases (10), and transport in bacteria (11).

CYCLIC NUCLEOTIDE-DEPENDENT PROTEIN KINASES[1]

Cyclic AMP-Dependent Protein Kinases: Activation by Cyclic AMP

Protein kinases catalyze the transfer of the γ-phosphate of ATP to serine or threonine hydroxyl groups in proteins. Their activity is regulated by cyclic nucleotides and they are referred to as "cyclic nucleotide-dependent" kinases. The term "cyclic nucleotide-dependent," which is in widespread use, should be taken to mean dependent for maximal activity because most protein kinases exhibit some activity in the absence of added cyclic nucleotide. Phosphoprotein kinases also catalyze phosphotransferase reactions. Their activity, however, is unaffected by cyclic nucleotides, and they may further be distinguished from protein kinases by their ability to utilize GTP as a phosphate donor and their failure to interact with either the cyclic nucleotide binding subunit of protein kinases or the protein inhibitor of protein kinase. (For discussion, see the section on cyclic AMP-independent protein kinases.)

Most protein kinases can be activated by more than one cyclic 3',5'-mono-nucleotide. Investigations have focused on the interaction with cyclic AMP (cAMP) and cyclic GMP (cGMP), however, because these are the only (or predominant) cyclic nucleotides found in vivo.

As originally proposed by Brostrom et al (12) and Gill & Garren (13), cyclic nucleotide-dependent protein kinases are composed of two dissimilar subunits. In

---

[1] We, as others, have operationally divided the kinases that catalyze the formation of phosphoester bonds with serine and threonine residues in proteins into those whose activity is regulated by cyclic nucleotides (cyclic nucleotide-dependent) and those whose activity is unaffected by cyclic nucleotides (phosphoprotein kinases). We do not intend this simple categorization to exclude the existence of other classes of protein kinases either within these two groups or in addition to them.

the absence of cyclic nucleotides, protein kinases exhibit a low basal activity. Cyclic nucleotides activate protein kinases in vitro by binding to their regulatory or cyclic nucleotide binding proteins (R), effecting the release of active, catalytic subunits (C), which exhibit maximal activity and are unaffected by cyclic nucleotides. Upon removal of cAMP, the R and C components reassociate to form a cAMP-dependent holoenzyme (see 4).

$$CR \xrightarrow{\text{cAMP}} C + R\text{-}cAMP \qquad\qquad 1.$$
$$\text{(inactive)} \underset{\text{cAMP}}{\overset{}{\longleftarrow}} \text{(active)}$$

Dissociation has been monitored by resolving the cyclic nucleotide binding protein from cyclic nucleotide-independent catalytic subunits by gel filtration, ion-exchange chromatography, affinity chromatography, sucrose density gradient centrifugation, electrophoresis, or isoelectric focusing. Activation concomitant with dissociation also occurs in vivo. Soderling, Corbin & Park (14, 15) studied protein kinase activity in rat epididymal fat pads following the administration of lipolytic hormones in vivo. The generation of C was measured by determining the ratio of protein kinase activity in the absence of cAMP to that measured in its presence and this was corroborated by isolating kinase subunits by gel filtration at 0°C in 0.5 M NaCl. Dissociation of protein kinase correlated with the increase in intracellular cAMP induced by lipolytic hormones. Insulin, which inhibited epinephrine-stimulated cAMP formation, decreased the formation of free C while having no effect on the basal protein kinase activity ratio. The equilibrium of equation 1 was stabilized by the presence of NaCl, and cyclic nucleotide phosphodiesterase activity was inhibited by methylxanthines. Protein kinase activity ratio has also been shown to rise after addition of follicle stimulating hormone (FSH) to seminiferous tubules (16). Appearance of free C was correlated temporally with binding of FSH and a rise in intracellular cAMP. Similar evidence for the regulation of protein kinase by dissociation in vivo has been obtained with hepatic protein kinase (17). Following a glucagon-induced rise in intracellular cAMP and the subsequent dissociation of protein kinase, total protein kinase activity measured in the presence of cAMP fell. The decrease in activity might be attributable to translocation of the enzyme from the cytosol to a particulate fraction of the cell, inactivation of C by an inhibitory protein, or lability of the free C subunit. Takeda & Ohga (18) administered a single intraperitoneal injection of glucagon to rats and monitored cAMP levels, tyrosine aminotransferase (TAT) activity, protein kinase activity (measured in the absence of added cAMP), and histone phosphorylation. A rapid increase in intracellular cAMP and protein kinase activity was followed by phosphorylation of a specific site in lysine-rich (f1) histone. These results agree with the important earlier observations of Langan (19) and their subsequent confirmation in studies on isolated perfused liver (20). In the cited study of Takeda & Ohga (18), phosphorylation was correlated with an increased intracellular cAMP level which preceded the increase (and presumed induction) of TAT. Since the chronology is compatible with current dogma and the same site on histone is phosphorylated in vivo and in vitro, the authors conclude that histone f1 is a substrate in vivo for hepatic cAMP-dependent protein kinase. The possibility that hormones may alter the protein kinase activity in their target cells by mechanisms other than cAMP-mediated dissociation has been c scussed in connection with the interactions of insulin and liver (21) and adrenocorticotropin and adrenal (22). The exact nature of these putative alterations in protein kinase activity has not been determined. There is no evidence for an active ternary complex of cAMP, R, and C.

Regulation by Other Mechanisms

Enzymes involved in a host of bioregulatory phenomena may be influenced by multiple effectors. A protein inhibitor I, which occurs in many mammalian tissues (23), binds to the catalytic subunit of cAMP-dependent protein kinase and inhibits enzymic activity. It has been purified from rabbit skeletal muscle (24). The inhibitor does not compete with the substrates of C, ATP, and casein, and inhibits C to approximately the same extent as R. It also prevents recombination of C with R-cAMP and the concomitant release of cAMP from R. Although clearly different from R by virtue of its molecular weight (26,000), heat stability, and inability to bind cyclic nucleotides, it remains possible that I is a modified R that has retained a site for interaction with C (24). A recent review (3) discusses the mechanism of action of this protein inhibitor in detail. A similar protein found in lobster tail muscle (25, 26) affects the protein substrate specificity of the protein kinases from this and other tissues and has been termed protein kinase modulator, M. The M protein of lobster tail muscle is a heat-stable acidic protein with a minimum molecular weight of 34,000 (25). It can be purified by conventional procedures as well as by the rather drastic heat and acid precipitation steps used to purify the inhibitor protein from skeletal muscle (24). The modulator inhibits or stimulates protein kinase activity depending upon substrate protein, metal ion, cyclic nucleotide, and source of the modulator and protein kinase used in the assay (26). The modulator stimulated the activity of the lobster muscle cGMP-dependent protein kinase with arginine-rich histone or protamine as substrates but inhibited the enzyme when lysine-rich histone or histone mixture were used. Modulator also inhibited the activity of a number of cAMP-dependent protein kinases with histone but stimulated their activity with protamine (25). The authors suggest that M may enhance the activity of cGMP-dependent protein kinases and inhibit the activity of cAMP-dependent protein kinases in tissues where both cGMP content and cGMP-dependent protein kinase activity are lower than the content of cAMP and cAMP-dependent protein kinase. Although the biological function of these protein inhibitors is unknown, they have been used to distinguish cyclic nucleotide-independent protein kinases that are subunits of cyclic nucleotide-dependent protein kinases from other kinds of kinases (phosphoprotein kinases) that are completely independent of cyclic nucleotide control (see 1) by their ability to specifically combine with and inhibit C.

Two heat-stable modulators in lactating rat mammary gland (27) alter the substrate specificity of homologous protein kinase (28). A heat-labile, nondialyzable lysosomal fraction of polymorphonuclear leukocytes interferes with the binding of cyclic nucleotides and inhibits protein kinase activity measured in leukocyte extracts (29). The factor was neither purified nor characterized, but the observation reveals another pitfall in the measurement of protein kinase and cyclic nucleotide binding activities in crude cytosol. cAMP- and cGMP-dependent protein kinases can be dissociated and activated by interaction with their protein substrates (30–32). If one can extrapolate from the conditions used in these experiments to those used in standard enzyme assays, it is possible that this dissociation may contribute to the basal kinase activity in the absence of added cyclic nucleotides. Nonsubstrate polypeptides such as polyarginine (33) also enhance protein kinase activity with certain protein substrates such as bovine serum albumin, although it is not clear whether the stimulation is due to interaction of the polypeptide with the enzyme or the substrate. The ability of substrate and nonsubstrate proteins to influence protein kinase activity and the localization of protein kinase activity in particulate subcellular fractions (see below) suggest that the topographic relationship of protein kinases and their substrates may confer specificity that cannot be reproduced in vitro. It has been difficult to assess protein substrate specificity of protein kinases because most studies use both inhomogeneous enzyme and protein substrates, and the physiological protein substrates are in many cases unknown. The extent of endogenous phosphorylation and denaturation of the protein substrate as well as the precise conditions of the assay (1) may significantly influence the results of such analyses.

Brostrom et al (34) showed that the addition of ATP-Mg^{2+} enhanced the dissociation of cAMP from the R-cAMP complex in the presence of C. The addition of micromolar concentrations of ATP-Mg^{2+} to a partially purified protein kinase from skeletal muscle inhibited cAMP binding and kinase activation by cAMP (35). ATP appeared to be bound to either protein kinase itself or a protein that behaved like it during isoelectric focusing and sedimentation in sucrose gradients. Since the enzyme used in these studies was impure, the binding of ATP to the protein kinase subunits could not be established. In vivo, the concentration of intracellular ATP is approximately millimolar; protein kinase may therefore contain bound ATP in its native form. If so, it would have a lower affinity for cAMP in vivo than that predicted from binding studies performed with purified preparations in vitro.

Beavo et al (36, 37) have calculated that the concentration of protein kinase in skeletal muscle (0.23 μM) is about the same as the basal concentration of cAMP in this tissue. The equivalence of enzyme and ligand would make the apparent activation constant for cAMP higher than that derived from studies using a low ratio of enzyme to ligand. The high enzyme concentration coupled with the presence in vivo of Mg^{2+}-ATP and sufficient I to inhibit 20% of the total protein kinase activity (38) would enable protein kinase to respond with great sensitivity to the physiological range of cAMP concentrations in muscle.

Another possible mechanism for regulating protein kinase activity is suggested by the finding that bovine heart protein kinase may exist as a phosphoprotein (39). Addition of [γ-^{32}P] ATP to homogeneous enzyme led to the rapid incorporation of 2 mol ^{32}P as phosphorylserine per dimer of R. The phospho form of the enzyme could be dissociated by cAMP, whereas the dephospho form bound the cyclic nucleotide but was not easily dissociated by it. Nearly identical observations have now been made for protein kinase from swine heart muscle (40). Although nothing is known about the ratio of phospho to dephospho forms of protein kinase in vivo, it appears that under basal conditions the enzyme is principally in the phospho form because enzyme in unpurified homogenates is easily dissociated by cAMP and can be rendered more resistant to such dissociation by treatment with phosphoprotein phosphatase (39). During purification the enzyme may become dephosphorylated (due to the action of contaminating phosphoprotein phosphatases) and progressively less sensitive to dissociation by cAMP unless it is rephosphorylated by the addition of ATP and Mg^{2+}. The cyclic nucleotide binding protein of the human erythrocyte membrane may also be a substrate for autophosphorylation (41, 42), as is the R subunit of a protein kinase purified from bovine brain (43). In the latter case approximately 0.5 mol phosphate was incorporated per regulatory chain and, as reported for the enzyme from cardiac muscle (39), phosphorylation was rapid and did not affect the ability of R to bind cyclic nucleotides. Brain cytosol R migrated on SDS acrylamide gel electrophoresis together with a phosphoprotein found in synaptic membrane fractions of rat and bovine cerebral cortex (44), but the identity of the membrane protein has not been established.

A unique type of activation has been described for one of the protein kinases purified from bovine corpus luteum by DEAE-cellulose chromatography (45). The kinase is directly stimulated in vitro by the addition of leuteinizing hormone (LH), a hormone known to stimulate the adenylate cyclase of the corpus luteum. LH does not influence cAMP binding and is not a substrate for protein kinase. Although the amount of LH needed to demonstrate direct activation is greater than that found under physiological conditions, the requirement for biologically active LH raises the possibility that some peptide hormones (or similar molecules) may penetrate target cells and directly alter the activity of cytoplasmic or membrane-bound protein kinases. Another kind of protein kinase regulation was proposed by Majumder & Turkington (46, 47), who concluded that the enzyme in mouse mammary gland is induced by the combined actions of insulin and prolactin. A still undefined modification of net protein kinase activity may be involved in the mechanism of action of insulin (21). Activation of glycogen synthase by insulin in perfused livers of normal rats is accompanied by a fall in protein kinase activity without a decrease in the apparent hepatic concentration of cAMP.

There is some evidence that protein kinases may be translocated between cytosol and subcellular organelles. Korenman et al (48) incubated uteri with isoproterenol and found a rise in both uterine cAMP and cAMP-independent protein kinase activity. Total soluble protein kinase activity in the $20,000 \times g$ supernatant fluid (measured in the presence of cAMP) decreased. Activity lost from the cytosol could be recovered in the $20,000-145,000 \times g$ microsomal fraction after extraction with 0.7% Triton X-100, suggesting translocation of kinase from cytosol to microsomes.

Some of the pharmacological agents that influence cAMP-mediated biological processes may turn out to have direct effects on protein kinase. There is, however, little current information on this subject. Tolbutamide, a sulfonylurea that is a clinically useful oral hypoglycemic agent, inhibits both basal and cAMP-stimulated protein kinase activity in rat epididymal fat pads (49). The authors suggest that this may explain its antilipolytic effects in vitro.

Properties

The biochemical properties of different protein kinases are very similar (see 1) and there is little new information. The ability of $[\gamma\text{-}^{32}P]$ GTP to serve as substrate for crude preparations of protein kinase in a number of rat tissues is probably attributable to the formation of $[\gamma^{32}P]$ and $[\beta,\gamma^{32}P]$-ATP in these preparations via exchange and transfer reactions (50), although cyclic nucleotide-independent phosphoprotein kinases transfer phosphate from GTP to a variety of protein substrates (see below). CTP, GTP, UTP, and ITP were found to be poor substitutes for ATP as the phosphoryl donor in skeletal muscle protein kinase-catalyzed reactions (1). Lemaire et al (51) used purified catalytic subunit derived from pituitary protein kinase to determine some of the structural requirements for the ATP site. Specificity for the adenine nucleus probably accounts for the seemingly paradoxical inhibition exerted by high ($> 10^{-5}$ M) concentrations of cAMP. Similar studies showing competitive inhibition by adenine, adenosine, AMP, and ADP for the ATP site of the catalytic subunits and holoenzymes of rat liver protein kinase (52) and bovine thyroid gland (53) have also been reported.

Analogs of cyclic nucleotides have been synthesized and used to activate and dissociate protein kinases (54–56). In an extensive review of these compounds (57), Simon et al made the following generalizations about the effects of 18 analogs on activation of bovine brain protein kinase: 1. Substitution at C-2′ or change of configuration of the 2′-OH leads to loss of activity; 2. Formation of anhydro-nucleoside abolishes activity as does change in configuration at C-3′; 3. Substitution of a methylene group at C-3′ or C-5′ nearly abolishes activity; 4. A sulfur in place of an oxygen in the exocyclic phosphate ring abolishes activity but such substitution at C-5′ of the ribose does not influence activity; and 5. Replacement of the OH on the phosphorus of the 3′,5′-cyclic ring with a di- or methylamino group abolishes activity, whereas replacement of the N^7 of the purine ring with a C-H or C-CN group has little effect on protein kinase activity.

Cyclic GMP-Dependent Protein Kinases

Cyclic GMP-specific kinases, first described in arthropods (58), have recently been found in mammalian pancreas (59) and cerebellum (60). The protein kinase activity in the high speed supernatant fluid of a rat cerebellar extract is activated by cGMP and cAMP with activation constants of 3×10^{-8} and 10^{-7} M, respectively (60). Activation by cGMP was observed only in the presence of phosphate buffer. Attempts to purify a cGMP-specific enzyme were unsuccessful, so it could not be determined whether there is one kinase stimulated by both cyclic nucleotides or two protein kinases with differing cyclic nucleotide specificities. Since the activity of protein kinases with cAMP and cGMP can be significantly influenced by other proteins (25, 26) and assay conditions such as pH (61), it is difficult to draw conclusions about cyclic nucleotide specificity from studies performed in relatively crude extracts. Fractionation of an extract of rat pancreas on DEAE-cellulose yielded three protein kinase activities, one of which was stimulated only by cGMP and bound [³H]cGMP but not [³H]cAMP (59).

Subunit Structure and Multiplicity of Protein Kinases

Studies of homogeneous protein kinases purified from bovine cardiac muscle (62) and rabbit skeletal muscle (36) suggest a very similar model: a holoenzyme consisting of two catalytic subunits and one R dimer. Utilizing Stokes radii calculated from gel filtration data and sedimentation constants derived from sucrose gradient and sedimentation velocity ultracentrifugation, it was concluded that the holoenzyme, cyclic nucleotide binding protein, and catalytic subunit of the bovine cardiac enzyme had molecular weights of 174,000, 98,000, and 38,000, respectively (62). Molecular weights determined by sedimentation equilibrium were in excellent agreement with these data. The molecular weight of the cyclic nucleotide binding monomer, determined by sodium dodecylsulfate (SDS) acrylamide gel electrophoresis, was 55,000 (63). The axial ratios of both the holoenzyme and cAMP binding dimer were 12, suggesting that these molecules are more asymmetric than the catalytic subunit, which has an axial ratio of three. The proposed asymmetry may explain the difficulties encountered in estimating molecular weights solely by gel filtration (62). Recent reports[2] indicate that protein kinase from bovine cardiac muscle binds 2 mol [³H]cAMP per mole holoenzyme (i.e. per cyclic nucleotide binding dimer), as does the protein kinase purified from rabbit skeletal muscle (36).

One of the difficulties encountered in defining the molecular structure of protein kinases is their propensity to undergo structural modification in vitro with retention of enzymic activity (1, 63). Tao & Hackett (31), for example, reported three protein kinases associated with cyclic nucleotide binding activity in rabbit erythrocytes. The three species, estimated by gel filtration to have molecular weights of 170,000, 120,000, and 240,000, had s values of 7.4, 5.2, and 7.2, respectively. The form with the s value of 7.2, however, gradually converted to a 5.2S protein on storage. A similar change in physical properties with generation of new, active protein kinase species has been seen upon storage of the bovine cardiac enzyme (63). The three cyclic nucleotide-dependent protein kinases found in bovine epididymal sperm (64) have molecular weights determined by gel filtration of 120,000, 78,000, and 56,000. Their cyclic nucleotide binding components have molecular weights of 78,000, 35,000–40,000, and 17,000–18,000, respectively. The authors suggest that the large R is a dimer and that the smallest R, which is absent in freshly prepared extracts, may derive from the R monomer by proteolysis. A cyclic nucleotide-dependent protein kinase from bakers' yeast (65) has a substantially lower molecular weight than that usually found for the enzyme in higher eukaryotes. The holoenzyme had a molecular weight of 58,000 estimated by gel filtration, and the C and R components have molecular weights of 30,000 and 28,000, respectively. The stoichiometry of cyclic nucleotide binding was not ascertained. Free catalytic and cyclic nucleotide

---

[2] Erlichman, J., Rosen, O. M. Manuscript in preparation.

binding components have been detected in extracts of a variety of tissues, as would be expected from the in vivo activation studies previously cited (14, 15). Other forms of cyclic nucleotide-dependent protein kinases which may represent partial dissociation or conformational alterations have been detected by sedimentation in sucrose gradients and gel filtration (39, 66).

The C and R components of protein kinases from a wide variety of organisms and tissues interact to form hybrid enzymes. The catalytic component of a cGMP-dependent protein kinase from lobster muscle can interact with the cAMP binding protein of either lobster muscle cAMP-dependent protein kinase or bovine brain kinase yielding holoenzyme with the cyclic nucleotide specificity of the donor R (30). The protein kinase subunits from cells as different as yeast and rat liver also combine to form enzymes that are activated by cyclic nucleotides (65).

The kinetics of dissociation and association of cAMP has been studied using purified endometrial cAMP-dependent protein kinase and its subunits (67, 68). The binding of cAMP to R was different kinetically from the binding of cAMP to the holoenzyme. Association of R and cAMP exhibited second order kinetics, was temperature dependent, and influenced by the initial concentration of cAMP. The association of cAMP with the intact protein kinase could not be described with linear second order plots and was not temperature dependent. Dissociation of the R-cAMP complex obeys first order kinetics in the presence or absence of C. The reactions between cAMP and R in the presence or absence of C were qualitatively the same at pH 5 (67) and 7.4 (68). The physiological corollaries of these complex interactions have not been unraveled.

Many tissues appear to have at least two and sometimes multiple cyclic nucleotide-dependent protein kinases which are separable by ion-exchange chromatography and/or gel filtration (69–75 and reviewed in 1). A calf thymus protein kinase may have two kinds of R, although one could be a dimer of the other (76). Dastugue et al (77) reported that the protein kinase activity in rat liver cytosol can be resolved into three species by DEAE-cellulose chromatography. On the basis of electrophoretic mobility of the subunits derived from these forms, they conclude that there are two distinct Cs and two Rs. Chen & Walsh (75) previously demonstrated three protein kinase activities in rat liver cytosol which were resolved by chromatography on DEAE-Sephadex. The 4S cAMP-independent form and the 4S catalytic subunits derived from the principal 6.8S holoenzyme were distinguishable by isoelectric focusing. Tao & Hackett (31), using gel filtration and sucrose gradient centrifugation, showed that the three protein kinases of rabbit erythrocytes have similar or identical Cs and dissimilar Rs. Catalytic subunits from the multiple protein kinases of rabbit skeletal muscle and rat liver phosphorylate the same seryl and threonyl residues in histone, protamine, muscle glycogen phosphorylase kinase, and glycogen synthase (78–80). Although the Cs of these enzymes exhibit microheterogeneity not apparent on gel filtration, the principal differences between them were thought to reside in their R components. The nature and meaning of the apparent heterogeneity of cyclic nucleotide-dependent protein kinases are difficult to understand in view of the similarity of their biochemical properties, substrate specificities, and crossreactivity of subunits. Somatic cell genetics is a potentially important tool for probing the molecular regulation of protein kinases and the role of phosphorylation reactions in the mediation of cyclic nucleotide effects. Taking advantage of the lympholytic effect of $N^6,O^{2'}$-dibutyryl cAMP (DBC), a population of lymphoma cells was developed which was DBC resistant (81). Cells of this line had four- to eightfold less cytoplasmic cAMP binding activity than the parental line and decreased cAMP-dependent protein kinase activity. It was proposed that, analogous to glucocorticoid-resistant mutants (82), resistance to DBC is related to modification of the cyclic nucleotide binding component (R) of cyclic nucleotide-dependent protein kinases. The cyclic nucleotide binding protein and protein kinase of these cells must be purified and characterized before the precise lesion in the kinase molecule can be identified.

In synchronized cultures of Chang liver cells, the cAMP-dependent and cAMP-independent protein kinase activities increased in parallel with each other and with overall cell protein throughout the cell cycle, i.e. both types of subunits appeared to be coordinately expressed (83). Protein kinase activity in rat liver, heart, and skeletal muscle is high prenatally and falls postnatally (84). This pattern was different from that exhibited by three other glycogenolytic enzymes as well as the contents of cAMP and glycogen.

Although there is substantial evidence that the synthesis and metabolism of cyclic nucleotides are altered in some transformed cells (see 85), little is known about changes in protein kinase activity or specific protein phosphorylation reactions. A preliminary report (86) suggests some alterations in protein kinases found in mouse hepatoma compared to normal rat liver. Mackenzie & Stellwagen (87) found that extracts of hepatoma (HTC) cells lacked one of the two major cAMP-stimulated protein kinases present in liver. Comparing analogous protein kinase fractions, those from the hepatoma were less responsive to cAMP and had lower cAMP binding affinity than those present in liver. Granner (88) reported that, compared to normal rat liver, HTC cells contain more (threefold) protein kinase activity which is, however, less responsive to activation by cAMP. The HTC cells contain approximately one tenth the cyclic $[^3H]$ AMP binding activity exhibited by extracts of normal liver, even though they have little adenylate cyclase activity and presumably low levels of endogenous cAMP. Addition of hepatic cAMP binding protein to extracts of HTC cells decreased kinase activity and rendered it more sensitive to activation by cAMP. Thus, these hepatoma cells seem to have a defect in cAMP binding protein which may be responsible for both their increased cAMP-independent protein kinase and decreased cAMP binding activities. This kind of comparison is exceedingly important particularly when there is a defined genetic relationship between the normal and transformed cells. No difference was found in the levels of cAMP-dependent or cAMP-independent protein kinases using histone as substrate in normal and feline sarcoma virus-transformed cells (89). Sensitivity to cAMP was similarly unaltered. Uninfected African green monkey kidney cells have been reported to contain a protein kinase that is activated by cGMP and not by cAMP (90). Infection with SV40 (90) did not alter the activity or substrate specificity of this enzyme in vitro.

PROTEIN KINASES AND SUBSTRATES IN VIRUSES, PROKARYOTES, AND SIMPLE EUKARYOTES

Although a preliminary report of a cAMP-dependent protein kinase activity in *Escherichia coli* was made several years ago (91), there has been a paucity of subsequent reports on protein kinases in prokaryotes. Protein kinase and cAMP binding activities have been found in oral streptococci (92). Histones and protamine served as protein acceptors but the phosphorylated product of the reaction was not chemically characterized. This is a critical point because basic proteins can activate other kinds of kinase reactions including polyphosphate kinase, which transfers phosphate from ATP to polyphosphate (93), and the phosphorylation of nuclear acidic proteins by nuclear protein kinases (94).

In contrast, some of the simple eukaryotes appear to have unequivocal protein kinase activity. The slime mold *Physarum polycephalum* contains cAMP-activated and cAMP-inhibited kinases (95, 96). Inhibition, like activation, occurred at low concentrations of cAMP (less than 10^{-5} M). Sensitivity to inhibition by cAMP was maximal during G2 and sharply diminished at the onset of mitosis (95). In the absence of cAMP, protein kinase activity did not change during the cell cycle. *Euglena* also has cAMP-dependent protein kinase associated with a $30,000 \times g$ particulate fraction (97). The yeast *Saccharomyces cerevisiae* has at least three kinds of protein kinases. A cyclic nucleotide-dependent protein kinase purified 100-fold has kinetic properties similar to those of mammalian protein kinases (65). The product of the reaction has been identified as a phosphoserine residue in histone but the pattern of phosphorylated tryptic peptides of histone differs from that produced by a protein kinase from rat liver. This yeast kinase appears to be distinct from both cAMP-independent casein phosphoprotein kinase (98, 99) and another cAMP-independent kinase which has a preference for protamine rather than histone (65) and is not a catalytic subunit of cAMP-dependent protein kinase. A cAMP binding protein unassociated with protein kinase activity has been isolated from *Saccharomyces fragilis* (100). This protein might be an R subunit of a cyclic nucleotide-dependent protein kinase or an analog of the cyclic nucleotide binding proteins found in prokaryotes (101, 102).

A cAMP-dependent protein kinase has also been purified (116-fold) from cell-free extracts of the mycelia of wild-type *Neurospora crassa* (103). It has a molecular weight of 60,000 on gel filtration, an *s* value of 3.8, and catalyzes the phosphorylation of serine residues in casein and phosvitin. The authors of this study (103) also describe the preparation of casein peptides that competitively inhibit the phosphorylation of casein and function as dead-end inhibitors rather than alternate substrates for the enzyme. Although the biological function of protein phosphorylation reactions has not been established in any of these microorganisms, *Neurospora* does possess a cAMP-sensitive glycogen phosphorylase system (104).

A cyclic nucleotide-independent protein kinase appears in *E. coli* following infection with bacteriophage T$_7$ (105). The product of the reaction has the chemical characteristics of phosphoserine, distinguishing the enzyme from acyl phosphate

(106) and polyphosphate (93) kinases and documenting endogenous phosphoester formation by a prokaryotic kinase for the first time. The kinase maps in the early region of the phage genome. Protein synthesis is required for its appearance, and ultraviolet irradiation of the phage but not the host bacterium prevents its appearance. The enzyme is clearly coded for by the phage but the substrates are principally host ribosomal proteins which are phosphorylated in vivo and in vitro by the induced enzyme. The authors conclude that there is no protein kinase activity in uninfected *E. coli* and that the phosphoproteins found in uninfected cells are primarily phosphoenzyme intermediates. This conclusion is in agreement with the earlier observation that there is much less phosphoserine and phospho-threonine in *E. coli* than in eukaryotes (107). The biological role of the kinase during infection is unknown. Mutant phages, unable to induce protein kinase activity, seem to develop normally. Despite this, the system may ultimately provide new insights into the function of protein phosphorylation in microorganisms.

Protein kinase activity, phosphoproteins, and substrates for phosphorylation have been found in many RNA- and DNA-containing animal cell viruses. In general, the enzymes are cyclic nucleotide independent and in most cases neither their function nor the source of the nucleic acid which codes for them (host cell or virus) has been ascertained. Protein kinase activity was first found in Rauscher leukemia and vesicular stomatitis viruses by Strand & August (108). Purified C-type viruses from mouse, cat, hamster, and viper were subsequently shown to incorporate ^{32}P from $[\gamma\text{-}^{32}P]$ATP into virion proteins (109). Protein kinase activity of equine herpes virus is stimulated by protamine and histone and catalyzes the phosphorylation of all 17 viral proteins seen on polyacrylamide gel electrophoresis in SDS (110). Kinase activity was demonstrated only after removal of a loosely bound phosphohydrolase from the virus. The reaction product has the chemical characteristics of a serine or threonine phosphate ester. Vaccinia virus cores also exhibit cyclic nucleotide-independent protein kinase activity that phosphorylates viral proteins (111, 112) and added histones and protamine (112). A viral phosphoprotein with a molecular weight of 11,000–12,000 (113) is labeled in vivo with $^{32}P_i$. Intermediates in vaccinia virus morphogenesis also contain phospho-proteins (114). Protein kinase activity associated with vaccinia virus was solubilized with 0.1% sodium deoxycholate and 0.25 M NaCl or KCl and was stimulated by basic proteins but did not phosphorylate them (115).

A heat-stable protein(s) in NP40 (a non-ionic detergent) extracts obtained during preparation of poxvirus cores activates partially purified protein kinase. This activation could not be mimicked by cAMP. Vesicular stomatitis virus has a protein, NS, which is phosphorylated in vivo (116–119) and in vitro (116, 118). Virions purified from cells pretreated with cycloheximide and actinomycin had diminished protein kinase activity and diminished specific infectivity although, as the authors (116) point out, it cannot be concluded that these two observations are causally related. A case is made (116) that the origin of the vesicular stomatitis-associated protein kinase is cellular rather than viral, since (*a*) the kinetic data obtained with protein kinases of virions derived from different cells are slightly different, (*b*) actinomycin pretreatment of the host cells diminishes the protein kinase activity of the virions derived from these cells, (c) the bulk of the viral proteins are separated from the virion kinase by phosphocellulose chromatography, and (d) there is no evidence for viral induction of protein kinase. Additionally, there is no evidence that the viral-associated protein kinase is the enzyme that catalyzes the phosphorylation of NS protein in vivo. A similar conclusion was reached after finding that both cytoplasm and membranes of the host cell exhibit protein kinase activity similar to that found in purified virions (118). Intact viruses are also phosphorylated by unrelated, exogenous protein kinases (120).

Purified virions of frog virus 3, an enveloped DNA virus that replicates in the cytoplasm, contain a protein kinase of high specific activity that catalyzes the phosphorylation of more than 10 endogenous polypeptides as well as exogenously added substrate proteins (121). The solubilized enzyme has a molecular weight of 65,000 and is cAMP independent. Viral polypeptides that are good substrates for this enzyme are poor substrates for cAMP-dependent protein kinase from beef heart. It has been suggested that protein kinase associated with frog polyhedral cytoplasmic deoxyribovirus may be involved in the phenomenon of nongenetic reactivation, i.e. reactivation of denatured virus by cultivation in susceptible cells together with UV-inactivated virus (122). Phosphoproteins have been described in SV40 (123), Semliki Forest virus (124), Sindbis virus (124, 125), and adenovirus (126) but only Semliki Forest and Sindbis virus (125) displayed protein kinase activity. Conversely, two oncornaviruses (127) exhibit protein kinase activity but none of the virion proteins appeared to be phosphorylated in vivo. cAMP-dependent protein kinase derived from bovine cardiac muscle phosphorylated all the structural proteins of SV40 in vitro. Interestingly, the rate of phosphorylation using either partially phosphorylated or alkali-treated viral proteins was unaffected by cAMP (128).

In summary, although many purified viruses contain phosphoproteins and cAMP-independent protein kinase activity, the relationship between these two sets of observations is unclear. Similarly unsettled is the origin of the protein kinase and the biological function of the phosphorylations. It is possible that the complete viral phosphorylation system may require both cellular and viral components.

OTHER CYCLIC AMP-INDEPENDENT PROTEIN KINASES

Cyclic AMP-dependent protein kinases can be cAMP independent under certain conditions. It is important, therefore, to distinguish true cAMP-independent kinases from either modified forms of the protein kinase holoenzyme (RC) or its C subunit. Only the C subunit is inhibited by the addition of R or I (1). Inhibition by R can be overcome by the addition of cAMP.

A cyclic nucleotide-independent protein kinase that catalyzes the phosphorylation of phosvitin and casein as well as calf brain and rat liver nuclear phosphoprotein has been purified from calf brain (129, 130). Because it is not activated by cAMP and does not catalyze the phosphorylation of histone, the enzyme appears to be different from the cyclic nucleotide-dependent histone kinase activity also present in brain. A similar phosvitin kinase has been extensively purified (greater than 8000-fold) from the cytosol of liver of estrogen-treated roosters (131). It catalyzes the phosphorylation of phosvitin at 10 times the rate observed with casein and does not use histone or protamine as substrates. It can utilize GTP or ATP as phosphoryl donors and its activity is unaffected by the estrogen treatment which induces phosvitin synthesis. Density gradient sedimentation studies indicate a molecular weight of about 160,000, although the enzyme appears much larger on gel filtration. The principal protein band, seen after SDS-acrylamide gel electrophoresis, has a molecular weight between 40,000 and 50,000 and is thought to be a subunit of kinase. Phosvitin kinases (or phosphoprotein kinases) from avian and nonavian sources differ in a number of properties from the cyclic nucleotide-activated protein kinases. Their substrates are restricted to acidic proteins, often containing multiple pre-existing phosphoamino acid residues, and their activity is unaffected by cyclic nucleotides; they are stimulated by Mg^{2+} but inhibited by Co^{2+} and Mn^+ (many of the cyclic nucleotide-dependent protein kinases are stimulated by all three cations) and GTP can serve as a phosphoryl donor (131; see also 132–134). A kinase with preference for unphosphorylated casein has been found in the Golgi apparatus of lactating mammary gland (135). The enzyme works less well on the milk proteins β-lactoglobulin, α-lactalbumin, and native casein. Histones, phosvitin, and lysozyme are not appreciably phosphorylated. ATP is the phosphoryl donor, Ca^{2+} and Mg^{2+} are able to support the reaction, and there is no effect of either cyclic AMP or I. The properties and location of the enzyme are compatible with the general supposition that casein phosphorylation is a post-translational event. (See also 136.)

A cyclic nucleotide-independent kinase has been purified from dogfish skeletal muscle (137). It has a minimal molecular weight of 46,000 determined by electrophoresis in SDS and phosphorylates proteins with the following relative rates: protamine, 100; an endogenous, parvalbumin-like phospho-acceptor protein, 75; phosvitin, 20; histone f2a, 10; and casein, 5. Curiously, this protein kinase does not activate dogfish phosphorylase kinase.

A variety of protein kinase activities have been found in nuclei (138–141). Most of these kinases phosphorylate nuclear proteins in a cyclic nucleotide-independent fashion and their relationship to either cyclic nucleotide-dependent or -independent protein kinases of the cytosol is unknown. A chromatin-associated f1 histone phosphokinase has been studied in Chinese hamster (CHO) cells (141, 142). It has a molecular weight of more than 90,000 and is increased six- to eightfold in metaphase arrested cells. The enzyme phosphorylates f1 histone in vitro, yielding the same major $[^{32}P]$tryptic peptides that can be derived from in vivo labeling with $[^{32}P]$ during the G_2-M part of the cell cycle.

A different kind of cAMP-independent kinase which catalyzes the formation of an acid-labile, $[^{32}P]$-3-phosphohistidine bond on histone has been found in nuclei derived from rat tissue and Walker 256 carcinosarcoma (139). A microsomal fraction of rat brain phosphorylates microsomal protein, histone, and purified brain ribosomal proteins (143). With histone as substrate, the cAMP dependency of this activity decreased somewhat from birth to adulthood. The protein phosphorylation reactions that occur in cabbage leaf preparations are unaffected by cyclic nucleotides but modulated by cytokinins which are N^6-substituted purines (144).

Purified pig brain tubulin is associated with a protein kinase activity capable of phosphorylating casein and lysine-rich histone (145). The activity in partially purified rat brain tubulin preferentially phosphorylates the B chain of tubulin and is unaffected by cAMP, colchicine, or vinblastine. Eipper (146) demonstrated that protein kinase associated with aggregated tubulin had a low specific activity (compared to rat brain cytosol), was not enriched during tubulin purification, could be partially solubilized, and was not present in soluble tubulin dimer. The evidence suggests that protein kinase activity associated with rat brain tubulin results from contamination with the cAMP-independent protein kinase or free C of rat brain cytosol (147) from which tubulin was purified. Although soluble tubulin dimer is not phosphorylated in vitro, it can be labeled with ^{32}P in vivo (148). It remains unclear whether the phosphorylation of tubulin is biologically significant and whether the association of protein kinase with tubulin is more than fortuitous.

THE ROLE OF PHOSPHORYLATION IN THE REGULATION OF GLYCOGEN METABOLISM

Phosphorylase Kinase

Phosphorylase kinase, a component of the enzyme system that controls the metabolism of glycogen in skeletal muscle, plays an essential role in coupling physiological phenomena in muscle (including nervous and hormonal stimuli) to the regulation of glycogenolysis (5). It mediates the activation of phosphorylase b by catalyzing the phosphorylation of a single, specific serine hydroxyl group (149) in a highly basic region (5, 150) in each of the identical subunits of phosphorylase b (151, 152). Phosphorylase kinase activity is dependent on both millimolar concentrations of Mg^{2+} and micromolar levels of Ca^{2+} (153, 153a, 153b).

In the case of rabbit muscle phosphorylase b, phosphorylation is accompanied by the conversion of an inactive dimer to a physiologically active dimer or tetramer (phosphorylase a) (154, 154a) which catalyzes the phosphorolytic cleavage of glucose 1-phosphate from glycogen. Muscle contraction (155) and the administration of epinephrine (156) elicit conversion of phosphorylase b to phosphorylase a and concomitant glycogenolysis in vivo.

Phosphorylase kinase comprises 0.5–1.0% of the soluble proteins of rabbit muscle and its intracellular concentration has been estimated to be 4 μM (5).

Recently, phosphorylase kinase was purified to homogeneity (157, 158) using a slight modification of the procedure of DeLange et al (159). In the final step a column of Sepharose 4B was used to separate high molecular weight, polydisperse material of low activity, and the buffers contained 10% sucrose to stabilize the enzyme (157, 158).

Table 1 summarizes the physical parameters determined for phosphorylase kinase in two independent investigations (157, 158). The native enzyme is an exceptionally large, acidic protein and the frictional ratio indicates a spherical shape (157). During storage the homogeneous enzyme slowly polymerizes to active 37 and 48S aggregates, which presumably represent dimers and trimers of the native kinase (158).

Three dissimilar polypeptide subunits were detected when phosphorylase kinase was subjected to SDS-acrylamide electrophoresis (157, 158). The inclusion of

Table 1   Physical properties of phosphorylase kinase

	Reference	
	Hayakawa et al (157)	Cohen (158)
Molecular weight	1.33×10^6	1.28×10^6
$S_{20,w}$	26.1	23
f/fo	1.17	—
Isoelectric point	5.77	—
$A_{280}^{1\%}$	11.8	12.4
$A_{280} : A_{260}$	1.75	1.90
Partial specific volume	0.730 ml/g	0.735 ml/g

protease inhibitors throughout the purification and dissociation procedures and the carboxymethylation or performic acid oxidation of sulfhydryls prior to denaturation and reduction did not alter the pattern observed on the SDS-acrylamide gels (158), suggesting that neither the number nor the size of the subunits was the result of artifacts. The molecular weights of the subunits (Table 2) were estimated by electrophoresis in SDS-acrylamide gels (157, 158) and confirmed in sedimentation equilibrium experiments (158, 160).

Cohen found the molar ratios among the subunits to be unity (1A : 1B : 1C) by (a) quantitative densitometry, (b) direct determination of protein concentration in subunits fractionated by chromatography in SDS, and (c) comparing the radioactivity observed in the subunits following exhaustive carboxymethylation with iodo-[^{14}C]-acetate and SDS-acrylamide electrophoresis with that expected from amino acid analyses (158). The excellent agreement among these three determinations supports the suggestion that the minimum molecular weight is 318,000 and the native enzyme is a dodecamer with the composition $A_4B_4C_4$ (158). Hayakawa et al (157), using only densitometry, proposed a native enzyme with 16 subunits $A_4B_4C_8$.

Phosphorylase kinase occurs in two interconvertible molecular forms in skeletal muscle (161). One form, designated activated phosphorylase kinase, exhibits a high level of enzymic activity at pH 6.8, whereas the second form, nonactivated phosphorylase kinase, has little or no activity at this pH (162).

Activation of phosphorylase kinase is accomplished by phosphorylation (159) of serine hydroxyl groups in the enzyme (163). In the activation process, phosphorylase kinase serves as a cosubstrate in a reaction catalyzed by cAMP-dependent protein kinase (164, 165).

The rate of protein kinase-mediated phosphorylation and activation of phosphorylase kinase was enhanced tenfold by an optimal concentration of cAMP, and the concentration of cyclic nucleotide required for half-maximal stimulation was 7×10^{-8} M (164). An increased intracellular level of cAMP presumably causes the sequential activation of protein kinase, phosphorylase kinase, and phosphorylase (165), which account for approximately 0.03, 1, and 5–10% of the total soluble

Table 2 Subunit structure of phosphorylase kinase

	Reference	
	Hayakawa et al (157, 160)	Cohen (158)
Molecular weight		
Subunit A	118,000	145,000
Subunit B	108,000	128,000
Subunit C	41,000	45,000
Minimum molecular weight	308,000(ABC$_2$)	318,000(ABC)
Composition of native enzyme	A$_4$B$_4$C$_8$	A$_4$B$_4$C$_4$

protein in skeletal muscle, respectively (4, 5, 157). Based on enzyme activities measured in vitro, Cohen & Antoniw (166) estimated that the time required for the conversion of 50% of muscle phosphorylase *b* to phosphorylase *a* via this cascade of phosphorylation is on the order of 2 sec. This calculation is consistent with the rate of activation of phosphorylase observed when isoproterenol is administered to skeletal muscle in vivo (167).

The mechanism of phosphorylase kinase activation was studied in a homogeneous preparation of the enzyme by monitoring the increase in enzyme activity and the pattern of phosphorylation as functions of time in the presence of $[\gamma\text{-}^{32}P]$ ATP and either cAMP-dependent protein kinase (158) or its catalytic subunit (160). The B subunit (see Table 2) of phosphorylase kinase was rapidly phosphorylated, reaching its half-maximal value in 1–2 min (158, 160). The kinetics of conversion of phosphorylase kinase from the nonactivated to the activated species coincided with phosphorylation of the B subunit. No significant phosphorylation of the A subunit was observed until B was 50% phosphorylated. After this lag, the A subunit was phosphorylated at a rate equal to 25% of the initial rate of B subunit phosphorylation but this phosphorylation was not associated with an increase in the activity of phosphorylase kinase (158, 160). The C subunit was not phosphorylated by cAMP-dependent protein kinase. Hayakawa et al (160) observed that 0.45 mol ^{32}P-phosphate could be incorporated into the B subunit while more than 1 mol was transferred to A. In similar experiments, Cohen (158) found a maximum of 1 mol ^{32}P-phosphate transferred to both the A and B subunits. The discrepancy between these two studies may be due to the content of serine phosphate present in different preparations of nonactivated phosphorylase kinase (168).

The A and B subunits are rapidly phosphorylated during the activation of phosphorylase kinase in protein-glycogen complexes that are isolated from rabbit muscle (169) and thought to be the physiological matrices for enzymes that regulate glycogen metabolism (170, 171). After maximal phosphorylation and depletion of ATP, the process is completely reversed: the enzyme is inactivated and inorganic phosphate is released from the A and B subunits.

The mechanism and regulation of dephosphorylation were scrutinized utilizing a preparation of essentially homogeneous phosphorylase kinase that was fortuitously contaminated with phosphorylase kinase phosphatase(s) (166). This preparation was phosphorylated by protein kinase for various periods of time to produce phosphorylase kinases containing variable amounts of phosphate in the A and B subunits.

Three distinct phases of dephosphorylation were discerned: (a) when the enzyme contained 1 mol phosphate per mole ABC, mol wt = 318,000 (0.75 mol in subunit B and 0.23 mol in A), little dephosphorylation occurred; (b) when the enzyme possessed 2 mol phosphate (1.0 mol in A and B) the B subunit was rapidly and completely dephosphorylated, whereas no dephosphorylation of the A subunit was apparent; and (c) when an intermediate level of phosphate was present (e.g. 0.8 mol in B and 0.6 mol in A) the B subunit was slowly dephosphorylated. A perfect correlation was observed between the loss of phosphate from the B subunit and the conversion of phosphorylase kinase from the active to the inactive form. The addition of Mg^{2+} or Mn^{2+} activated the release of phosphate from the A subunit following dephosphorylation of B or the release of phosphate from A and B when the enzyme contained an amount of phosphate insufficient to activate the metal-independent phosphatase (see case a above). The K_m values for Mg^{2+} and Mn^{2+} were 5–10 mM. No change in enzyme activity was associated with dephosphorylation of the A subunit.

These results demonstrate that the activation of phosphorylase kinase by cAMP-dependent protein kinase may be completely reversed by a phosphorylase kinase phosphatase (see below). The phosphorylation of subunit A appears to be a signal for the initiation of dephosphorylation of the B subunit and inactivation of phosphorylase kinase. The mechanism of activation of the phosphatase may involve cooperative effects because the phosphorylation of at least 50% (two per mole enzyme, $A_4B_4C_4$) of the A subunits is required before the initiation of dephosphorylation occurs (166). Conversely, the phosphorylation of the A subunit in the reverse reaction does not begin until 50% of the B subunits are phosphorylated (158, 160). Thus, the dephosphorylation of subunit B and inactivation of phosphorylase kinase are controlled by the secondary phosphorylation of a separate site on a different subunit. This newly described form of regulation via phosphorylation has been labeled "regulation by second site phosphorylation" (166). There is sufficient phosphorylase kinase phosphatase in skeletal muscle to catalyze the inactivation of 50% of the total muscle phosphorylase kinase in 2 sec. The possibility that dephosphorylation may also be regulated by the availability of Mg^{2+} or Mn^{2+} ions requires further study.

Investigations on the regulation of dephosphorylation were preliminary and somewhat ambiguous because a trace contaminant was the source of "specific phosphatase." Studies will ultimately have to be performed using a purified preparation of phosphorylase kinase phosphatase and a phosphatase-free phosphorylase kinase. Finally, detailed studies on the specificity, multiplicity, and metal ion requirements of the putative phosphorylase kinase phosphatase will be required to substantiate the physiological relevance of the observations described above.

Phosphorylase kinase is activated by incubation with trypsin (162, 172). In the course of proteolysis (158, 160, 173) subunit A is rapidly broken down and after prolonged incubation, subunit B vanishes but enzyme activity remains at its peak value. Subunit C is resistant to proteolysis. These data make the possibility that subunit C is the catalytic subunit of phosphorylase kinase an attractive but as yet unsubstantiated proposal (158, 160, 173).

In summary, a tentative working hypothesis concerning the roles of the three subunits of phosphorylase kinase has emerged. Subunit C possesses the catalytic center of phosphorylase kinase whereas subunits B and A function in the regulation of enzyme activity. Subunits A and B apparently lock the catalytic subunit into an inactive conformation in the dephosphorylated enzyme. The phosphorylation of subunit B at a specific site by cAMP-dependent protein kinase produces a conformational modification that results in the expression of catalytic activity. The subsequent phosphorylation of subunit A activates a specific phosphatase or makes the B subunit accessible to a previously activated phosphatase. This serves as a signal for the dephosphorylation of subunit B and inactivation of the enzyme.

Role of Phosphorylase Kinase in the Phosphorylation of Troponin

Troponin, a protein composed of three dissimilar subunits (designated TNC, TNI, and TNT), is present in myofibrils and confers Ca^{2+} sensitivity on actomyosin ATPase (174, 175). Its TNC subunit (mol wt = 18,000) binds Ca^{2+}; its TNI subunit (mol wt = 24,000) inhibits actomyosin ATPase in the absence of Ca^{2+}, and its TNT subunit (mol wt = 37,000) is required for binding troponin to tropomyosin and Ca^{2+} sensitization of the ATPase (176–180). Phosphorylase kinase catalyzes the phosphorylation of the TNI and TNT subunits of troponin in reactions analogous to the conversion of phosphorylase b to a (181–183). Ca^{2+} is required for catalysis and a maximum of 2 mol phosphate can be incorporated per mole troponin (183) after endogenous phosphate is removed by pretreatment of troponin with phosphorylase phosphatase (see below). Incorporated phosphate is equally distributed between the TNI and TNT subunits (183). The Ca^{2+} binding subunit (TNC) is not a substrate for phosphorylase kinase (181–183). Activated and nonactivated phosphorylase kinase phosphorylate isolated TNI subunits at approximately the same rate, indicating that regulation of troponin phosphorylation by Ca^{2+} flux is feasible (181, 183). In an additional parallel to the phosphorylase b-phosphorylase a interconversions, phosphotroponin was completely dephosphorylated by phosphorylase phosphatase (183). K_m and V_{max} values obtained in kinetic studies indicate that TNI and phospho-TNI may be physiologically appropriate substrates for phosphorylase kinase and phosphorylase phosphatase, respectively. Phospho-TNI also functions as a competitive inhibitor of the dephosphorylation of phosphorylase a by phosphorylase phosphatase and vice versa (183, 184).

The phosphorylation of TNI and TNT could be physiologically significant because troponin is intimately associated with the process of muscle contraction at the level of actomyosin (175, 185) and is also an excellent substrate for two highly specific muscle enzymes that coordinate contractile activity and glycogen metabolism (181–184). Consistent with this proposal are preliminary studies that have shown that freshly isolated troponin contains 0.5–1.0 mol endogenous P_i/mole protein (182, 183), and that actomyosin ATPase is markedly stimulated by Ca^{2+} in the presence of phosphorylated troponin, but enzyme activity remains unchanged when dephosphotroponin is used (183). Stull et al (181) speculated that the accumulation of phosphotroponin may be controlled by Ca^{2+} translocation and correlated with a positive inotropic effect in cardiac muscle. TNI and TNT are also phosphorylated by cAMP-dependent protein kinase (183, 186, 187). Protein kinase may make only a minor contribution to the phosphorylation of troponin in muscle, however, because (a) its intracellular concentration is lower than that of phosphorylase kinase by an order of magnitude (4, 5), (b) its turnover number (using TNI as a substrate) is only 5% of that obtained with phosphorylase kinase (183), and (c) its K_m for TNI is higher by a factor of 4 (183).

The functional significance of troponin phosphorylation has been studied in a strain of mice I/LnJ which carries a sex-linked deficiency in phosphorylase kinase activity (188). Skeletal muscle extracts from these animals exhibit less than 1% normal activity in converting phosphorylase b to a (188, 189), but contain a normal amount of protein that crossreacts with antiserum prepared against rabbit muscle phosphorylase kinase (190). The mice have no abnormalities in the contractile process and I/LnJ muscle extracts are capable of phosphorylating troponin B at 30–60% of the rate observed in extracts from normal animals (191). The conservation of muscle contraction and troponin phosphorylation in the absence of phosphorylase activation is inconsistent with the suggestion (181, 183) that the phosphorylase kinase-mediated phosphorylation of troponin serves an important physiological function in muscle contraction. The observations that the skeletal muscle of I/LnJ mice is capable of catabolizing glycogen in response to electrical stimulation or epinephrine administration in vitro and in vivo but is unable to convert phosphorylase b to a (188, 189, 192) conflict with the concept that action of phosphorylase kinase is absolutely required for the coupling of glycogenolysis to nervous and hormonal stimuli (158). It is possible that phosphorylase b is activated by the production of 5'-AMP in I/LnJ muscle (193), but it is not known if such a mechanism operates in the presence of native (wild-type) phosphorylase kinase.

Although troponin is an excellent substrate for highly purified phosphorylase kinase in vitro (181, 183), the physiologic mediator of troponin phosphorylation has not been identified. The near normal rate of troponin B phosphorylation in I/LnJ mouse muscle raises the possibilities that this activity might be ascribed to (a) a second catalytic site in phosphorylase kinase, (b) cAMP-dependent protein kinase, or (c) a specific "troponin kinase" distinct from the other kinases.

A causal relationship has yet to be established between troponin phosphorylation and muscle contraction. If troponin phosphorylation proves to be an integral part of muscle contraction, then mutations resulting in the impairment of the TNI- or TNT-phosphorylating capacity of phosphorylase kinase could account for some dysfunctions of muscle contraction.

Regulation of Glycogen Synthase by Phosphorylation

Glycogen synthase (or synthetase) catalyzes the repetitive transfer of glucosyl moieties from UDPG to glycogen. It is of primary importance in tissues that normally maintain significant stores of glycogen, such as liver and muscle. (See 6 and 194 for reviews on glycogen synthase.)

Glycogen synthase has been purified to homogeneity (165, 195, 196) and accounts for 0.2% of the soluble muscle proteins. Its intracellular concentration, approximately 4 μM, is considerably lower than the concentration of glycogen phosphorylase (100 μM) (5).

Native glycogen synthase is a multichain protein (165, 196) which can be dissociated into identical subunits (mol wt = 90,000–100,000) by exposure to SDS and mercaptoethanol. The number of subunits in the intact enzyme has not been unequivocally established. Soderling et al (165) have determined a molecular weight of 400,000, indicating a tetrameric structure, but the molecular weight of 250,000 obtained by Smith, Brown & Larner (195, 196) is in accord with a trimeric enzyme.

Two interconvertible forms of glycogen synthase have been identified (197): synthase D generally exhibits a minimal level of catalytic activity and synthase I is the active enzyme species. Glycogen synthase D exhibits the same specific activity as synthase I when it is activated in vitro by a high concentration (1–10 mM) of glucose 6-phosphate. This property of the enzyme provides a quantitative, plus-minus assay for determining the proportions of the total synthase in the I and D forms (198). Friedman & Larner (197) demonstrated that the transformation of synthase I to the D form involved the formation of a phosphoester bond at a serine residue. The phosphorylation and concomitant inactivation of muscle synthase I is mediated by the same cAMP-dependent protein kinase that phosphorylates and activates phosphorylase kinase (165, 199). Prolonged incubation of synthase I with endogenous protein kinase and cAMP resulted in the incorporation of 6 mol phosphate per mole synthase subunit (196). This is at variance with a previous report (200) describing the sequence of a unique hexapeptide containing a single phosphoserine residue, which accounted for all the covalently bound phosphate, unless this sequence appears with a sixfold redundancy in each subunit. It is interesting that glycogen synthase and phosphorylase have identical hexapeptide sequences at their phosphorylated sites (150, 200); however, each of these enzymes is phosphorylated by a different and specific kinase, suggesting that the primary sequence around the phosphoserine residue does not determine the recognition of substrates by protein kinase and phosphorylase kinase. The action of cAMP-dependent protein kinase is more direct in inhibiting glycogen synthesis than in the activation of glycogen breakdown, which requires the additional amplification provided by activated phosphorylase kinase. This difference in mechanism is in accord with the facts that glycogenolysis is a rapid process requiring significant amplification between the binding of hormones at a small number of receptor sites and the activation of phosphorylase which is present in high concentration (~ 0.1 mM), whereas the inhibition of glycogen synthesis is much slower and the intra-cellular concentration of glycogen synthase is relatively low (4 μM). Protein kinase is, therefore, a key determinant in the regulation of glycogen metabolism because it simultaneously activates the breakdown of glycogen and inhibits glycogen synthesis.

Synthase D (Phosphorylase Kinase) Phosphatase

The intracellular concentrations of glycogen synthase D and activated phosphorylase kinase are determined by the balance between the activities of cAMP-dependent protein kinase reactions and specific phosphoprotein phosphatases.

Glycogen synthase phosphatase catalyzes the dephosphorylation and activation of synthase D, a transformation that is a prerequisite for the glycogenic effect of insulin (6, 194). Although the regulatory properties of synthase phosphatase are of considerable interest, the enzyme has received only cursory attention since its discovery in 1963 (197, 201). Two interconvertible forms have been tentatively identified in crude extracts of liver (202, 203). Synthase phosphatase b is dependent upon high concentrations (1–5 mM) of Mg^{2+} and 5'-AMP for the expression of its activity in vitro and is probably inactive under physiological conditions. Synthase phosphatase a functions independently of the concentrations of divalent cations and 5'-AMP in vitro and is apparently the agent responsible for the conversion of synthase D to synthase I in vivo (203).

Preliminary studies have suggested that insulin participates in the regulation of the activity of synthase phosphatase in vivo. The infusion of hormone into insulin-deficient animals stimulated the rapid (5 min) formation of synthase phosphatase a and the dissipation of synthase phosphatase b while the total, Mg^{2+}-dependent phosphatase activity remained unchanged (202). This activation was readily reversed by the administration of glucagon. These results indicated that synthase phosphatase may be controlled by a reversible modification in the structure of the enzyme and a "synthase phosphatase-activating enzyme" has been postulated.

The contention that insulin regulates glycogen synthesis by modulating the interconversion of synthase phosphatases a and b is still highly speculative. Aside from kinetic studies carried out on crude extracts of liver (202, 203), no physical or chemical evidence has been obtained to support the assertion that synthase phosphatase exists in interconvertible active and inactive forms. In addition, a more recent report (204) suggests that the apparent interconversion of two forms of synthase phosphatase in liver extracts may actually be attributed to fluctuations in the concentration of phosphorylase a, which is a potent inhibitor of synthase phosphatase (204). There is, however, a striking parallel to this hypothesis in the previously cited studies of Cohen & Antoniw (166) on a near homogeneous preparation of phosphorylase kinase. This comparison is particularly pertinent to synthase phosphatase, since Zieve & Glinsmann (205) have recently suggested that the dephosphorylations of synthase D and phosphorylase kinase are catalyzed by a single phosphoprotein phosphatase.

Little is understood about the biochemical mechanism by which insulin promotes glycogen synthesis. The status of our knowledge concerning the roles of epinephrine and glucagon in regulating glycogenolysis has recently advanced to the point where an understanding of the mechanism of insulin action is essential for achieving a comprehension of the integrated, hormonal regulation of glycogen metabolism. It is clear that insulin elicits a decrease in the intracellular concentration of cAMP (and the activity of protein kinase) in some circumstances (206–209) and the activation of synthase phosphatase under other conditions, but intensive investigations designed to probe the sequence of biochemical events responsible for these phenomena have not been carried out. Such studies may also help to define the regulatory function of protein dephosphorylation in cellular metabolism.

Kato & Bishop (210) partially purified synthase phosphatase from rabbit skeletal muscle. The enzyme required 5 mM Mn^{2+} for maximum activity and catalyzed the simultaneous activation and dephosphorylation of purified ^{32}P-labeled glycogen synthase D. Using the same kind of preparation, Zieve & Glinsmann (205) demonstrated that the enzyme also catalyzed the inactivation of phosphorylase kinase. They concluded that a single phosphoprotein phosphatase catalyzed the dephosphorylation of glycogen synthase D and phosphorylase kinase because (a) both reactions were reversed by incubation with cAMP-dependent protein kinase, (b) the ratio of the two phosphatase activities was constant throughout the five step purification procedure, and (c) phosphorylated phosphorylase kinase was a competitive inhibitor of the synthase D to synthase I conversion, whereas de-phosphorylated phosphorylase kinase was ineffective. This conclusion was expanded to the general working hypothesis that a unique phosphoprotein phosphatase dephosphorylates proteins that are phosphorylated by cAMP-dependent protein kinase in vivo (205).

The conclusions of Zieve & Glinsmann (205) must be viewed with some reservation because the phosphatase was only partially purified (205, 210). Thus, the possibility that numerous phosphoprotein phosphatases were present has not been excluded. Ultimate confirmation of their proposal will require the demonstration that a homogeneous preparation of synthase phosphatase can catalyze the de-phosphorylation of phosphorylase kinase, synthase D, and other substrates of cAMP-dependent protein kinase.

Phosphorylase Phosphatase

The serine phosphoester bonds in phosphorylase a can be cleaved by phosphorylase phosphatase, an enzyme that exhibits a high degree of specificity for phosphorylase a (211). It will not dephosphorylate low molecular weight phosphate esters or other naturally occurring phosphoproteins, with the single exception of the TNI and TNT subunits of troponin (183).

Phosphorylase phosphatase is present in low concentration (1 μM) in muscle (212), but its activity is sufficient to catalyze the rapid conversion of phosphorylase a to b in vitro in protein-glycogen complexes which contain all the glycogenolytic and glycolytic enzymes (213) and in vivo during muscle relaxation (214).

Recently, Gratecos et al (212) purified phosphorylase phosphatase 6000-fold by taking advantage of the enzyme's high affinity for polylysine and its resistance to urea denaturation. The enzyme was quantitatively adsorbed to polylysine-Sepharose, purified by a high salt wash, and eluted with 6 M urea. Subsequent dialysis, concentration with ammonium sulfate, and a second dialysis resulted in the irreversible precipitation of contaminating proteins (212). Physicochemical characterization of the phosphatase has been hampered by its tendency to form multiple, active aggregates (mol wt = 30,000–120,000). Gel filtration studies indicate a minimal molecular weight of 32,000 for the native enzyme (212). No information is available on the subunit composition of phosphorylase phosphatase under denaturing conditions.

Other properties of the enzyme include a remarkable resistance to proteolytic enzymes, susceptibility to product inhibition by phosphorylase b and P_i, and potent inhibition by low concentrations (10–250 μM) of 5'-AMP, which probably results from the interaction between the nucleotide and the substrate, phosphorylase a (150, 212).

In a general sense phosphorylase phosphatase may be regulated synchronously and inversely with respect to phosphorylase kinase (5); i.e. conditions that favor activation of phosphorylase kinase simultaneously lead to the inactivation of phosphorylase phosphatase and vice versa. Synchronous regulation of these two enzymes has been observed in protein-glycogen complexes believed to be functional units in intact muscle cells (170, 171, 213).

OTHER SUBSTRATES FOR PHOSPHORYLATION

Enzymes

The regulation of enzyme action through the mechanism of interconversion of active and inactive forms by enzyme-catalyzed covalent modification (for recent reviews, see 7, 8) provides the potential for rapid, versatile adaptation of cells to their environment and to the needs of the whole organism. Enzymes that are clearly regulated by phosphorylation and dephosphorylation (see 7) include liver and muscle phosphorylase, phosphorylase b kinase, muscle and liver glycogen synthases (see below), hormone-sensitive lipase (see 215), and pyruvate dehydrogenase (see 216, 217).

It has been claimed that the pyruvate dehydrogenase from $N.$ $crassa$ is regulated, like mammalian pyruvate dehydrogenases, by an associated kinase and phospho-protein phosphatase (218). A protein kinase from rat liver nuclei has been reported to activate homologous nuclear RNA polymerases I and II. The ability of the kinase to catalyze phosphorylation of the polymerase could not be established, however, because of the impurity of both the polymerase and protein kinase (219). Partially purified protein kinase from calf ovary activated ovarian nuclear RNA polymerase activity in vitro (220). Since both kinase and polymerases were impure, the mechanism of this activation was not defined. The authors suggest that gonadotropins stimulate adenylate cyclase in the sexually immature ovary, ultimately leading to cAMP-dependent translocation of cytoplasmic protein kinase to the nucleus where phosphorylation and consequent activation of nuclear RNA polymerase occurs.

Rat hepatic tyrosine aminotransferase (TAT) is phosphorylated in vivo and in cultured hepatoma cells (221). Subunits of TAT, isolated by SDS-acrylamide gel electrophoresis following immunoprecipitation, contained [^{32}P] phosphoserine. The effect of phosphorylation on enzyme activity is unknown; treatment of the phosphorylated enzyme with alkaline phosphatase removed most of the radioactivity but had no effect on enzymic activity.

A homogeneous preparation of a bovine erythrocyte carbonic anhydrase isoenzyme was a substrate for cAMP-dependent protein kinases from bovine brain and hog muscle (222). Phosphorylation was accompanied by enzyme activation. The other impure isoenzyme was also phosphorylated by protein kinase but enzymic activation did not occur. Acetyl CoA carboxylase, a rate-limiting enzyme in the synthesis of long chain fatty acids, has been partially purified from liver and shown to be activated by Mg^{2+} and inactivated by phosphorylation (223).

Regulation by adenylylation, a mechanism so elegantly worked out by Stadtman and his colleagues for the glutamine synthetase of E. coli (see 9), also occurs in the lysine-sensitive aspartyl kinase from E. coli (224) in vitro. The possible regulatory significance of this latter modification is not yet known.

Soluble Proteins

Several proteins from E. coli are substrates for mammalian protein kinase, including certain histone-like basic proteins (225), the sigma factor of RNA polymerase (226), and initiation factor IF-2 (227). These proteins are probably not phosphorylated in vivo (see above), but phosphorylation in vitro may be useful in studying transcription and translation. In a similar vein, Kinzel & Mueller (228) have demonstrated that cytoplasmic cAMP-dependent protein kinase can be used to label 15 external proteins of the HeLa cell plasma membrane. The [^{32}P]-labeled phosphoproteins may serve as useful markers in plasma membrane isolation and studies on membrane protein turnover. An acidic, calcium binding protein has been purified to homogeneity from beef adrenal medulla (229). Each mole of protein (mol wt = 11,000) contains one mole of phosphate and binds one mole of calcium. Interestingly, a similar or identical protein is also found in beef brain but not in a number of other tissues, suggesting that it may have something to do with adrenergic neurons. Another calcium binding phosphoprotein has been isolated from dogfish skeletal muscle (137). It contains one acid-stable, alkali-labile phosphate per subunit of molecular weight 11,000–13,000 but only the aggregated forms (mol wt = 25,000, 75,000, and 350,000) serve as substrates for the dogfish cAMP-independent protein kinase. This substrate protein is similar in physical, chemical, and immunological properties to the parvalbumin of fish and amphibian muscle. Its function and the relationship of calcium binding to phosphorylation are unknown.

A component of the light chain of skeletal muscle myosin (230, 231) and human platelet myosin (232) is phosphorylated by protein kinase. Human platelets contain a kinase that catalyzes the transfer of ^{32}P from [γ-^{32}P] ATP to a specific myosin component of molecular weight 20,000. The incorporated phosphate is alkali labile and acid stable. When rabbit skeletal muscle myosin is isolated, the light chain component ML_2 contains close to 1 mol phosphate per chain (mol wt = 18,000) (231). Treatment with alkaline phosphatase converts ML_2 into a protein with the properties of another light chain component, ML_3. ML_3 was reconverted to ML_2

by the incorporation of ^{32}P from [γ-^{32}P] ATP catalyzed by either impure phosphorylase b kinase or protein kinase from skeletal muscle. Preliminary information suggested that the kinase responsible for the phosphorylation may be a specific, calcium-dependent myosin light chain kinase (231). Only one serine-phosphate site was found by analysis of chymotryptic peptides, and the sequence around the phosphorylation site bore no obvious similarity to the site on phosphorylase b that is phosphorylated by phosphorylase kinase (150). The physiological significance of myosin light chain phosphorylation is not yet known.

The addition of specific information-bearing molecules to target cells can lead to the phosphorylation of selected proteins presumably via generation of cAMP and activation of protein kinases. In a number of cases, the proteins are components of membranes or subcellular organelles and their function is not known (see below). An octopamine and serotonin-stimulated phosphorylation in vivo of a specific protein in the abdominal ganglion of *Aplysia* has been reported (233) but phosphorylation was not observed in larger animals (greater than 1300 g) or in experiments carried out for less than 22 hr. Insulin in physiological concentrations led to the phosphorylation of a protein of molecular weight 140,000 in rat epididymal fat pad and in isolated fat cells (234). The relationship of this phosphorylation to alterations in cyclic nucleotide concentrations in fat cells or to the ultimate function of insulin remains to be clarified.

The total phosphoprotein content of tissues has been monitored during various physiological states. It is clear, however, that this approach may not be sufficiently sensitive to detect changes in one or a few proteins, and phosphoprotein intermediates may not be distinguished from protein kinase-catalyzed phosphoesters. The protein phosphate content of rat mammary tissue has been used to measure alterations in casein during lactogenesis (235). Protein phosphorylation has been studied in the developing ova of *Rana pipiens* (236). Release of prophase block at ovulation was associated with a marked increase in protein phosphorylation, much of which probably represented phosphorylation of phosvitin. Release of metaphase block at fertilization coincided with protein dephosphorylation.

Nuclear Basic Proteins

The possibility that nuclear functions might be regulated by phosphorylation and dephosphorylation of the histones in chromatin has stimulated considerable investigation over the last decade (see 237–240). A number of years ago (241, 242) it was shown that ^{32}P$_i$ could be incorporated into rat liver histones in vivo and in isolated thymus nuclei. The principal substrate for phosphorylation was the very lysine-rich histone f1.[3]

Langan (245) first demonstrated that glucagon but not ACTH or hydrocortisone elicited incorporation of ^{32}P into a specific serine residue in the f1 histone of liver. Insulin also elicited the phosphorylation of f1, but this effect has been recently attributed to insulin-induced hypoglycemia which causes a secondary increase in

---

[3] A chart comparing the currently used designations for the major histone fractions is presented in (243). We use the nomenclature of Johns & Butler (244).

glucagon secretion (20). Histone phosphorylation could be catalyzed in vitro by a hepatic cAMP-dependent histone kinase or by a less specific cAMP-independent kinase (246, 247). The same $[^{32}P]$ phosphoserine peptide was isolated from histones phosphorylated in vitro by histone kinase or in vivo following the administration of hormones that elevate intracellular cAMP (19, 245, 247). To complete the picture, Meisler & Langan (248) purified a hepatic phosphoprotein phosphatase which exhibited specificity for histones and protamine. Maeno & Greengard (249) have resolved three histone/protamine phosphatases from rat cerebral cortex cytosol by ion-exchange chromatography. Each enzyme was stimulated twofold by 2.5 mM $MnCl_2$, and individual phosphatases displayed unique ratios of protamine dephosphorylating activity to arginine-rich histone dephosphorylating activity. More recently (250), Langan reported that the cAMP-independent kinase found in liver cytosol catalyzes the phosphorylation of a site in f1 histone that is different from the principal site phosphorylated by the cAMP-dependent protein kinase.

In an excellent review of this subject, Hnilica (251) points out some of the difficulties encountered in documenting the phosphorylation of histones in vivo. Contamination of nuclear histone fractions by nonhistone phosphorus-containing molecules necessitates rigorous chemical identification of the phosphorylated product including, where feasible, sequence analysis. In evaluating studies of histone phosphorylation during the cell cycle, the methods used to synchronize cells, the degree of synchrony achieved, and the cell type may influence the results obtained. Resolution of histone fractions also presents difficulties. Utilizing preparative polyacrylamide gel electrophoresis in 8 M urea, 8 mM Triton X-100, and 0.9 N acetic acid, Gurley & Walters (252) clearly separated f2a2 and f2b histones and showed that the most highly phosphorylated histones in interphase CHO cells were fractions f1 and f2a2.

Phosphorylation of lysine-rich histones is greater in growing than in resting cells and is particularly active during the S phase of the cell cycle (253–256) although some phosphorylation continues after this phase is completed (142, 257–259). For example, when histones of several mouse tissues and Ehrlich ascites tumor cells were compared (260), the more rapidly dividing cells exhibited more lysine-rich histone phosphorylation. In phytohemagglutinin or dibutyryl cAMP-activated lymphocytes (261, 262), rates of f1 phosphorylation and total f1 phosphate content increase in the period immediately prior to early DNA synthesis and during the phase of accelerated DNA synthesis.

Shepherd et al (263) reported that incorporation of phosphate into f1 histones of CHO cells continuously increased during the entire cell cycle, whereas histone f2b was phosphorylated in only G1 and S. In studies of histone phosphorylation during the cell cycle of CHO cells synchronized by isoleucine deprivation, Gurley et al (255, 257) found that phosphorylation of f1 histone depended upon the cell cycle position. Histone f1 was not phosphorylated in cells arrested in G1 or G1-traversing cells but actively incorporated phosphate in S. Phosphorylation of f2a2, however, occurred throughout the cycle and was independent of its own synthesis, DNA synthesis, and the phosphorylation of f1. In cells X-irradiated with 800 rad (264), the turnover and phosphorylation of f1 were inhibited but the phosphorylation of f2a2 and DNA replication continued. Thus, phosphorylation of f1 does not seem to be directly involved in DNA replication and the results suggest that phosphorylation of f1 and f2a2 serve different biological functions. When cells were resynchronized at the G1/S boundary with hydroxyurea, phosphorylation in both f1 and f2a2 occurred as the cells entered S and the relative rates of phosphorylation of both fractions increased in G2-rich and metaphase-rich cultures. The authors concluded that once f1 phosphorylation has been initiated, it continues through G2 into M and is not dependent upon continued DNA replication. When CHO cells were arrested in metaphase by vinblastine sulfate treatment (258), the phosphorylated form of f1 was retained despite a decline in the total f1 phosphokinase activity. Lake et al (265) compared the histone phosphorylation pattern of metaphase (M) and interphase (I) CHO cells, HeLa S_3 cells, and rat nephroma cells by assessing the mobility of histones in polyacrylamide gel electrophoresis before and after removal of protein-bound phosphate by alkaline phosphatase. The extent of f1 phosphorylation was independent of the time elapsed during the vinblastine-induced metaphase arrest. They also found phosphorylation in mitotic cells collected without the use of mitotic poisons. Histone f1 was dephosphorylated when M cells were allowed to enter G1.

When macromolecular synthesis in synchronized hepatoma (HTC) cells was inhibited by hydroxyurea or cycloheximide, DNA synthesis, histone synthesis, and histone phosphorylation were shown to be interrelated but not tightly coupled events (253, 266). In the presence of sufficient hydroxyurea to completely inhibit DNA synthesis, the capacity to phosphorylate f1 did not change although the rate of phosphorylation declined slowly. Cycloheximide treatment caused a marked decrease in f1 phosphorylation which resulted primarily from a diminished supply of fresh histone. The rate constant for phosphorylation of pre-existing histone appeared to be unchanged. When cycloheximide was removed, newly synthesized histone was rapidly phosphorylated. Phosphorylation of histones f1 and f2a2 coincides with DNA synthesis in synchronized cultures of HTC cells pulsed with $[^{32}P_i]$ (267). When rat liver and HTC cells were studied under conditions of rapid proliferation (regeneration and exponential growth, respectively) or nonproliferation, major differences were found in the phosphorylation of lysine-rich histones (269). Both normal (liver) and abnormal (HTC) dividing cells showed extensive phosphorylation of the lysine-rich histones. Nondividing cells showed little phosphorylation of this fraction. The phosphorylation of f1 in HTC cells during exponential growth was largely abolished when the cells entered stationary phase. A lag of 30–60 min occurred between the time of histone f1 synthesis and its subsequent phosphorylation (270). These studies (267, 269, 270) were interpreted as indicating that the phosphorylation of f1 histone is important in the control of chromosome replication rather than modification of transcription.

A correlation between histone phosphorylation and chromosome condensation has been made in a variety of cells (256, 265, 271). Bradbury et al (268) proposed that cell division in the slime mold *P. polycephalum* is triggered by f1 phosphorylation. The ability of sonicated nuclei to incorporate ^{32}P from $[\gamma\text{-}^{32}P]ATP$ into histone f1 increased from a minimum near metaphase to a maximum in late G2. The increase in phosphorylating ability preceded the rise in f1 phosphate content. A high f1 phosphate content in metaphase chromosomes has also been found in mammalian cells (141). Although it is evident from this discussion that the role of phosphorylation of f1 in chromosome function has not been established, phosphorylated f1 has been shown to have a smaller effect than dephosphorylated histone on the conformation of DNA as measured by circular dichroism (272).

Siebert et al (273) reported that the nuclear content of a cAMP-independent kinase increased sixfold 22 hr after partial hepatectomy in the rat. Sung et al (274) also studied rat liver regeneration after partial hepatectomy and demonstrated enhanced phosphorylation of f2a2 15 hr after surgery. The $^{32}P_i$ incorporated into f1 and f2b histones increased 6 hr after partial hepatectomy, coinciding with the appearance of new transcriptional activity. Gutierrez-Cernosek & Hnilica showed (275) that histone synthesis increases before the onset of DNA replication in regenerating liver. Maximal incorporation of $^{32}P_i$ into histone f3 occurs at the time of DNA synthesis (24 hr post-hepatectomy), whereas maximal incorporation of fractions f1 and f2a occurs in 12 hr (275). When the relative rates of $^{32}P_i$ and [^{14}C]-lysine incorporation into histones were compared, intensive phosphorylation of histones f2b, f1, and f2a occurred during 6 hr following partial hepatectomy while phosphorylation of f3 remained unaltered. During the phase of rapid DNA replication, all the histones were phosphorylated in proportion to their rates of biosynthesis (275).

An analysis of the phosphorylation of acid-soluble nucleolar proteins of Novikoff hepatoma cells 2 hr after the administration of $^{32}P_i$ in vivo showed that 40 of the 97 proteins resolved by two-dimensional gel electrophoresis and autoradiography were labeled (276). Thirty of these were analyzed and shown to contain [^{32}P] phosphorylserine.

Dixon and his colleagues have performed a series of elegant experiments on the modification of protamines and histones in the trout testis (see 238, 243 for reviews of some of their work). Trout testis is a particularly good biological system for this kind of analysis because testes can be obtained at different stages of development throughout the year; there is a transition from histones to protamines at midstage of spermatid development and the various testicular cell types can be obtained in relatively homogeneous form. When testicular cells were labeled with [^3H] arginine or [^3H] lysine and $^{32}P_i$, most of the histones were found to be synthesized and phosphorylated in diploid cells (277). The ratio of synthesis to phosphorylation was about the same for each cell type. During spermatogenesis (243) newly synthesized molecules of protamines are sequentially phosphorylated [at four serine residues (278)] and dephosphorylated over 5–10 days. During this period histones are lost and condensation of the spermatid nucleus occurs. The chronology of events is compatible with the proposition that phosphorylation of protamine may be necessary for the correct binding of newly synthesized protamine to DNA and that dephosphorylation of protamine may play a role in the condensation of spermatid chromatin. All five major histone fractions in testis are phosphorylated, principally in cells actively engaged in DNA synthesis. These modifications occur in a distinct pattern and may be involved in the proper binding of histones to DNA. Shortly after synthesis, histone f2a1 is sequentially acetylated and deacetylated, whereas f2a2 is phosphorylated and then dephosphorylated. Newly synthesized histones f1 and f2a1 are phosphorylated after a lag period. Histone f1 has four sites which are also sequentially phosphorylated and dephosphorylated. Both newly synthesized and "old" f1 are phosphorylated and dephosphorylated during the cell cycle. The authors propose that modifications occurring shortly after synthesis may be involved in the correct binding of histones to DNA, whereas those occurring after a lag may regulate some gross physical structure or activity of chromosomes. The first histone replaced by protamine in the spermatocyte chromatin is f2a1 (279). During this displacement, four lysyl residues are acetylated and one seryl residue (at the N terminus) is acetylated and phosphorylated. Histone f2a2 is similarly modified at its N-terminal serine, which resides in a 6 amino acid sequence identical with that found in f2a1. Such modifications might be involved in dissociating the histone from DNA to make way for the binding of protamine. They could also serve to make the molecule more accessible to proteolytic degradation. Studies correlating the physical state and biological function of chromatin with the state of phosphorylation of individual nuclear proteins in this system will undoubtedly continue to provide important clues about the role phosphorylation plays in modulating nucleic acid-protein interactions.

Nuclear Acidic Proteins

The nonhistone proteins of the nucleus comprise a heterogeneous array of acidic polypeptides with subunit molecular weights ranging from 5000 to approximately 200,000 (280–283b). These proteins are integral components of chromatin and have been implicated in the positive control of gene expression at the level of RNA transcription. The evidence supporting this hypothesis has been recently reviewed (284, 285) and only those findings germane to a discussion of protein phosphorylation will be mentioned here. In brief, nuclear acidic proteins (a) exhibit tissue-specific patterns on SDS-acrylamide gel electrophoresis; (b) are synthesized throughout the cell cycle and rapidly turned over; (c) undergo qualitative and quantitative changes during differentiation; (d) are present at elevated concentrations in metabolically active cells; and (e) are required for tissue-specific RNA synthesis in reconstituted chromatin in vitro. (For references, see 284, 285.) In contrast to histones, therefore, nuclear acidic proteins exhibit sufficient variability, specificity, and heterogeneity to participate in the regulation of transcription.

Rat liver nuclear acidic proteins have been resolved into 115 polypeptide components by SDS-acrylamide gel electrophoresis, and it is likely that many minor proteins escaped detection by staining (280a). Heterogeneity is further compounded because nearly all the nuclear acidic proteins are phosphorylated, containing an average of 30 mol serine (or threonine) phosphate/100,000 g protein (133, 242, 286).

Acidic proteins in both rat liver and kidney nuclei are rapidly labeled subsequent to an intraperitoneal injection of $^{32}P_i$, but the profiles of molecular weights and specific radioactivities of isolated phosphoproteins from the two tissues were remarkably dissimilar (280). Purified acidic nuclear phosphoproteins specifically bind to homologous DNA (280, 287, 288) and enhance RNA synthesis in an in vitro system (280, 286, 289, 290). Partial dephosphorylation by alkaline phosphatase treatment abolished the increase in RNA synthesis (290). Neither specific binding nor stimulation of transcription was observed when heterologous DNA was used (280, 287, 290). Addition of a purified, homologous nuclear acidic phosphoprotein fraction to condensed rat liver chromatin significantly augmented transcription (291). DNA-RNA hybridization analyses showed that RNA sequences synthesized in the presence of added homologous acidic phosphoproteins clearly differed from RNA produced in control experiments without additions or from chromatin treated with heterologous nuclear phosphoproteins. Thus, phosphorylation of nuclear acidic proteins may control recognition of and association with certain polynucleotide sequences in DNA and, ultimately, specific gene transcription.

Among the acidic proteins of chromatin are protein kinases and their endogenous substrates (133, 286, 289, 292, 293). Nuclear protein kinase and bulk acidic proteins were purified in parallel during preparation and extraction of liver nuclei and subsequent chromatography on DEAE-cellulose (294) but were cleanly separated by chromatography on phosphocellulose (289, 292, 293). Most endogenous substrates remained in the bulk acidic protein fraction. When the acidic protein fraction was extensively phosphorylated by pre-incubation with purified nuclear kinase and then utilized in the reconstitution of chromatin, RNA synthesis was enhanced by 40 to 65% above the level obtained with chromatin containing untreated acidic proteins (295). Purified nuclear kinase alone had no effect on in vitro RNA transcription from reconstituted chromatin. These results suggest a correlation between gene activation and nuclear kinase-mediated phosphorylation of acidic proteins.

Rickans & Ruddon (296) resolved four peaks of protein kinase and two peaks of cAMP binding activity from rat liver acidic nuclear proteins by chromatography on phosphocellulose. Addition of cyclic nucleotide binding protein from either skeletal muscle or the acidic nuclear protein preparation inhibited the major peak of nuclear protein kinase. Inhibition was slight, however, and only partially overcome by the addition of cAMP.

Twelve distinct protein kinases have been isolated from the acidic protein fraction of beef liver nuclei by chromatography on phosphocellulose (140). Individual kinases were further resolved into multiple components by acrylamide gel electrophoresis, demonstrating an unusually high degree of enzyme heterogeneity. The possibilities that unique protein kinases undergo dissociation, aggregation, or degradation during purification and electrophoresis have not been excluded. Each enzyme fraction contains endogenous phosphate acceptors and some kinases are unable to phosphorylate exogenously added histones or casein. The latter observation may explain why the majority of these enzymes escaped detection in earlier investigations (138, 289, 292, 293). cAMP stimulates phosphorylation of endogenous substrates by five nuclear kinases, has no effect on two enzyme fractions, and inhibits four others, thus revealing a further degree of complexity among the acidic phosphoproteins. A different profile of protein kinase activities and endogenous substrates was observed when the acidic proteins of kidney nuclei were fractionated (140). The multiplicity and heterogeneity of nuclear protein kinases are consonant with the complex distribution of nuclear acidic proteins in chromatin and suggests that these enzymes could play a role in the tissue-specific regulation of RNA transcription.

One of the nuclear kinases was compared to a cytoplasmic histone kinase and major differences between the two enzymes were apparent (140). cAMP stimulated the phosphorylation of histone f1 by the cytosol enzyme, but inhibited the nuclear kinase. Both enzymes catalyzed the cAMP-regulated phosphorylation of acidic nuclear proteins, but analyses on SDS-acrylamide gels demonstrated divergent enzyme specificities: histone kinase phosphorylated the entire spectrum of acidic proteins, whereas the nuclear enzyme phosphorylated a single polypeptide (140).

cAMP selectively stimulated the phosphorylation of specific nuclear acidic proteins in vivo. Intraperitoneal injection of dibutyryl cAMP stimulated the phosphorylation of total acidic nuclear proteins in liver by 30% within 15 min (281). However, the specific activities of two groups of proteins with average molecular weights of approximately 30,000 and 70,000 were elevated by 200%. Similar results were observed when cAMP was added to isolated nuclei (281). The possibilities that some cytoplasmic protein kinases originate in the nucleus and leak out during cell fractionation or that cytoplasmic protein kinases may be translocated to the nucleus in response to hormones or other stimuli have not been ruled out.

In fact, a cAMP binding protein has been found in ovarian cytosol which binds to isolated nuclei or chromatin when complexed with cAMP and concomitantly enhances chromatin-associated protein kinase activity (297, 298). The mechanism of kinase activation apparently involves the transfer of soluble protein kinase present in the "binding protein" preparation to specific sites in chromatin (298), but direct activation of nuclear kinase has not been eliminated as an alternative possibility. Treatment of intact cells with cAMP elicited a reduction in cytosol protein kinase and a corresponding increase in cAMP-independent activity in the nucleus. Jungmann et al (298) have speculated that chorionic gonadotropin (which is known to stimulate ovarian adenylate cyclase and enhance acidic protein phosphorylation) may induce the translocation of the cAMP binding protein to the nucleus accompanied by either a catalytic subunit of a cAMP-dependent protein kinase or a cAMP-independent protein kinase. Preliminary experiments have shown that cytosol protein kinase taken up by nuclei (in the presence of cAMP binding protein complex) specifically phosphorylates two nonhistone chromosomal proteins, suggesting that a cytoplasmic cAMP-dependent protein kinase could activate specific gene transcription in a manner analogous to the action of steroid hormones (299).

The influence of histones on acidic protein phosphorylation has been examined in isolated nuclei supplemented with ATP (283). Highly purified histone fractions are quantitatively taken up by nuclei and bound to the chromatin (283). Addition of a large excess of histones has no effect on the content of free DNA phosphate groups or template activity (300), indicating that histones combine with nuclear acidic phosphoproteins (301). Histone f2a1 enhanced the phosphorylation of a low molecular weight, acidic protein fraction (average mol wt = 22,000) by 70%, whereas f1 had a similar effect on polypeptides with an average molecular weight of 40,000; both histones completely inhibited the phosphorylation of a third component which exhibited a molecular weight of only 7000 (283). Histones f2a2, f2b, and f3 had no significant effects on nuclear phosphorylation. One nuclear acidic phosphoprotein that binds the acetylated histones, f2a1 and f1, is a histone deacetylase (301). This observation prompted the speculation (283) that there may be an interlocking control between acidic protein phosphorylation and the recognition of acetyl groups in histones.

Histones also modulate the endogenous phosphorylation of chromatin-free acidic proteins. At a histone-to-acidic protein ratio of 15, f2a (i.e. f2a1 + f2a2) and f1 were potent stimulators of nuclear protein kinases, while f3 and f2b were less effective (94). Under optimal conditions f1 histone increased the rate and extent of phosphorylation by a factor of 10. Phosphorylation was uniformly increased across the entire spectrum of acidic nuclear proteins in the presence of histones. Since the histone fractions were neither phosphorylated by endogenous kinase nor contaminated with other kinases or phosphoprotein phosphatases, it seems likely that the binding of histones by nuclear acidic proteins promotes an increase in the number of accessible phosphate-acceptor sites (94). Direct activation of nuclear protein kinase by histones is also feasible. Thus, the histones exhibit a similar order of potency in stimulating the phosphorylation of nuclear acidic proteins in vitro (94) and in intact nuclei (283); however, the data suggest that the rates and specificity of phosphorylation may be controlled by concerted interactions among histones, acidic proteins, and DNA in their appropriate proportions and conformations in native chromatin.

The phosphorylation of nuclear acidic proteins during the cell cycle has been studied in synchronized HeLa S_3 cells (288, 302, 303). Karn et al (302) observed that the rate of phosphate incorporation into nuclear acidic proteins was maximal in early S and early G_1 and depressed in late S, G_2, and M. The periods of active phosphorylation correspond to those portions of the cycle where the rate of RNA synthesis (304–306), the intracellular cAMP concentration (307), and cAMP-dependent protein kinase activity (302) also reach their peak values. The rates of phosphorylation of all nuclear acidic proteins were coordinately increased during S and G_1 and no proteins that exhibited exceptionally high or low specific activities were found (302). Platz et al (303) also observed enhanced nuclear acidic protein phosphorylation in S and G_1 but reported maximal phosphorylation in G_2 and specific phosphorylation of a unique protein in G_1. Both laboratories (302, 303) used synchronized HeLa S_3 cells and the reasons for the discrepancies in their results are not apparent. The techniques used for nuclear acidic protein extraction were different (280, 308), which could account for the disagreement on phosphorylation of a unique protein, but not for the divergent observations on the labeling of all nuclear acidic proteins in G_2.

Phosphoryl group turnover was measured after labeling the cells with $^{32}P_i$ and ^{14}C-amino acids for 23 hr (302). The half-life for disappearance of ^{32}P was 6.7 hr vs 25 hr for ^{14}C, demonstrating that phosphate turnover was not dependent on protein degradation. This is consistent with previous observations indicating that phosphorylation was independent of protein synthesis (133, 242). Differential turnover of phosphate was noted among the nuclear acidic proteins with $t_{1/2}$ values ranging from a low of 5.5 hr for a major component with a molecular weight of 55,000 to 12 hr for a smaller protein (mol wt = 28,000) (302). Total phosphate in the acidic protein fraction remained constant throughout the cell cycle.

The cell cycle studies suggest that differential turnover of protein phosphoryl groups may be a mechanism for effecting the modification of the structural and functional properties of acidic proteins in chromosomes during cell growth and division.

MEMBRANE PHOSPHORYLATION[4,5]

Central Nervous System

Mammalian brain is a rich source of protein-bound phosphoserine (309–311). Cytoplasmic and particulate subcellular fractions from cerebral cortex contain approximately 10 nmol phosphoserine (and 0.5–1 nmol phosphothreonine) per milligram protein, an amount sufficient to provide a mole of phosphoamino acid per mole of brain protein, assuming an average protein molecular weight of 100,000 (312, 313).

A significant fraction (10–25%) of membrane-bound phosphoserine turns over rapidly in respiring slices of cerebral cortex (132, 314) and the turnover rate is elevated when the tissue is electrically depolarized (311, 315). The remaining phosphoproteins appear to be metabolically stable in vivo and in brain slices and membrane fractions incubated under a variety of conditions (132, 314, 316).

Weller & Rodnight (314) have speculated that some proteins may be phosphory-lated by a cytoplasmic cAMP-independent phosphoprotein kinase (98, 317) prior to being incorporated into nerve cell membranes. Brain cytosol phosphoprotein kinase utilized soluble proteins containing multiple phosphoserine moieties as preferred substrates and also catalyzed the phosphorylation of proteins solubilized from cerebral membrane fragments (312). Cytoplasmic phosphoprotein kinase is readily distinguished from cAMP-dependent protein kinases by its utilization of both ATP and GTP as interchangeable phosphate donors (317) and by the absence of a stimulatory effect of cAMP (318). A phosphoprotein phosphatase, which hydrolyzed phosphoester bonds in casein and phosvitin and reversed the action of phospho-protein kinase, has been partially purified from soluble extracts of whole brain (319). Although this phosphatase and phosphoprotein kinase received considerable attention for a decade (309, 310, 312, 317, 318), their regulatory properties, physiological functions, and membrane-associated substrates have not been defined.

The discovery of a species of seryl-tRNA that can be phosphorylated on the serine hydroxyl subsequent to aminoacylation (320–322) suggests a possible trans-

---

[4] In the following discussion the terms "protein" and "polypeptide" will often be used operationally to denote the positions of peaks of protein and/or radioactivity on poly-acrylamide gels. These apparently unique components can frequently be fractionated into multiple species by electrophoresis in two dimensions, in gradient gels, or by other techniques (e.g. 283b, 420). No conclusions on the homogeneity of proteins can be reached solely on the basis of one-dimensional gels and no such interpretation is intended.

[5] All phosphorylation reactions discussed in the following sections refer to the formation of serine and threonine phosphoesters unless otherwise noted.

lational mechanism for incorporating phosphoesters into proteins. Phosphorylation of cytoplasmic (136, 323, 324) and nuclear (242) proteins is primarily a post-translational event, but a possible role for phosphoseryl tRNA in the synthesis of membrane phosphoproteins has not been evaluated.

Proteins in subcellular organelles obtained from homogenates of whole cerebrum or cerebral cortex are also phosphorylated by cytoplasmic or intrinsic, membrane-associated cAMP-dependent protein kinases (318, 325). In contrast to the pre-dominance of soluble protein kinases in most other tissues, approximately 60% of brain cAMP-dependent protein kinase activity is associated with particulate subcellular fractions (326).

When the crude mitochondrial fraction is subjected to hypo-osmotic lysis and subfractionation by differential and gradient sedimentation, a significant portion (15–20%) of the protein kinase activity purifies together with synaptic plasma membranes (326). Particulate protein kinases are not exclusively localized in synaptic plasma membranes. In fact, the enzyme is enriched only twofold in synaptic membranes and comparable specific activities of protein kinase have been observed in nuclei, microsomes, synaptic vesicles, and purified myelin (326).

The distribution of phosphate-acceptor proteins among subcellular membranes is similar to that of cAMP-dependent protein kinase (325). Purified synaptic plasma membranes are enriched in kinase substrates, accounting for approximately 15% of the total phosphate-acceptor capacity of cerebral tissue (325). In the presence of 0.1% Triton X-100, polypeptides in synaptic membranes were phosphorylated by both homologous, cytoplasmic protein kinase and the intrinsic, membrane-bound enzyme. The accessibility of endogenous substrates to the cytoplasmic kinase was not studied in intact synaptic membranes, leaving open the possibility that soluble and particulate protein kinases could modify specific membrane proteins simul-taneously or sequentially. Extraction of synaptic plasma membranes with 0.1% Triton X-100 solubilized most of the cAMP-dependent protein kinase, but 75% of the phosphate-accepting polypeptides remained associated with the insoluble pellet (325, 326). Detergent treatment increased total protein kinase activity (with respect to added histones) twofold, presumably by releasing the partially occluded enzyme from the membrane matrix (326).

Neither the specificity nor the cAMP stimulation of phosphorylation can be evaluated in intact, multicomponent membranous structures, because a significant amount of cAMP-independent phosphorylation of proteins and lipids may occur at this level of organization. Generally, these parameters have been analyzed by dissolving isotopically labeled membranes in SDS and resolving the component polypeptides by electrophoresis in SDS-acrylamide gels (327–329). Electrophoretic analyses of synaptic plasma membranes, which were labeled (via endogenous protein kinase) by incubation with $[\gamma\text{-}^{32}P]$ ATP of high specific activity, revealed several phosphoproteins (44, 330). The rates of phosphorylation of only two polypeptides, designated proteins I (mol wt = 86,000) and II (mol wt = 48,000), were regulated by cAMP (44). An optimal concentration of cAMP (5 μM) enhanced the phosphorylation of proteins I and II by factors of 6 and 3, respectively. Protein II, however, incorporated twice as much phosphate as I. The phosphotransferase reactions proceeded rapidly, reaching maximal levels within 5 sec at 20°C. Ehrlich & Routtenberg (331) confirmed these observations and showed that cAMP increased the phosphorylation of a third protein (mol wt = 47,000) by 50%.

Synaptic membrane protein kinase was characterized using proteins I and II as substrates (44). cIMP maximally stimulated the phosphorylation of proteins I and II at concentrations 10–100-fold higher than cAMP; cGMP was ineffective. Values for optimal and half-maximal concentrations of ATP, cAMP, and Mg^{2+} were similar to those determined for cytoplasmic, cAMP-dependent protein kinases (see above). Substitution of Mn^{2+} or Co^{2+} for Mg^{2+} increased both basal and cAMP-stimulated kinase activity by 1.5–3-fold. When Zn^{2+} was used, protein II was maximally phosphorylated in the absence of cAMP, but the phosphotransferase reaction was strongly inhibited by the cyclic nucleotide. This effect of Zn^{2+} was not observed with protein I. It is not known if the conversion of the membrane-associated protein kinase to a cAMP-inhibited enzyme by Zn^{2+} is a physiologically relevant process, but these findings underscore the need to consider modes of protein kinase regulation other than cyclic nucleotide stimulation. Adenosine, 5'-AMP, and adenine were potent inhibitors of the synaptic kinase, but two other effective inhibitors of soluble protein kinases, FMN and protein kinase modulator protein (25, 332), failed to affect the activity of the particulate enzyme.

The tissue distribution of proteins I and II is both restricted and specific—all four possible permutations have been observed (44): intrinsic phosphorylation of proteins similar to I and II (determined solely by their apparent molecular weights upon electrophoresis in SDS-acrylamide gels) was observed in synaptic membranes from cerebral cortex and caudate nucleus (44) and the plasma membranes of human erythrocytes (41, 42, 333); only protein I appeared in cerebellar synaptic membranes, while protein II, but not I, was observed in mammalian heart, kidney, vas deferens, uterus, intestine (44, 334), and toad bladder membranes (335, 336); neither substrate was present in membranes obtained from synaptic vesicles, brain microsomes, lingual nerve, liver, lung, or spleen (44). Additional variability was introduced by the observation that cAMP decreased the phosphorylation of protein II in membranes prepared from caudate nucleus (44) and toad bladders (335, 336). The survey of tissues described above did not report the presence of membrane phosphoproteins other than I and II, but other investigators have described the cAMP-regulated phosphorylation of membrane proteins larger and smaller than I and II (see below).

Approximately 75% of the ^{32}P incorporated in synaptic membrane proteins I and II was present as phosphoserine, 10% was found in phosphothreonine, and the remainder was associated with phosphopeptides and inorganic phosphate (44). Phosphorylated proteins I (331) and II (44) were rapidly dephosphorylated by an intrinsic, synaptic phosphatase. Phosphoprotein phosphatases were present in all the subcellular fractions obtained from cerebral homogenates, and their pattern of distribution closely paralleled that of cAMP-dependent protein kinase and its endogenous substrates (337). Synaptic plasma membranes were enriched in phosphatase when compared to other particulate subfractions. Synaptic phosphoprotein phosphatase exhibits maximal activity at neutral pH; has no requirement for divalent metals; is completely inhibited by 2.5 mM Zn^{2+}, Ca^{2+}, and Fe^{2+}; and may also be inhibited 50% by 10 mM phosphate or fluoride (249). In synaptic membranes and most other membranes, the observed phosphorylation of endogenous acceptors represents a net differential between the actions of protein kinase and phosphoprotein phosphatase, and not the actual rate of the phosphotransferase reaction. The available data indicate an apparent initial rate of phosphorylation of 2 nmol/min/mg membrane protein in purified synaptic plasma membranes (44). Intrinsic ATPase, adenylate cyclase and cyclic nucleotide phosphodiesterase activities, the variable degree of pre-existing occupancy of the phosphoserine acceptor sites, and steric factors in the membrane matrix may all participate in the regulation of phosphorylation-dephosphorylation reactions in membranes, further complicating precise quantitation.

A body of electrophysiological, biochemical, and pharmacological evidence has been amassed (for reviews, see 337–339) in support of the proposal that cAMP modulates synaptic transmission at postsynaptic sites in the central and peripheral nervous systems. In view of (*a*) the hypothesis (340) and supporting data (337–339) that all biological effects of cAMP are specified and mediated by protein kinases (see above), (*b*) the high specific activities and parallel enrichment of protein kinase, endogenous substrates, phosphoprotein phosphatase, adenylate cyclase, and cyclic nucleotide phosphodiesterase in synaptic plasma membranes (325, 341), and (*c*) the identification of two membrane polypeptides that undergo rapid and reversible, cAMP-dependent phosphorylation (44, 330), Greengard and colleagues (337, 338) have proposed that the cAMP-regulated phosphorylation of membrane proteins alters the permeability of the postsynaptic membrane to inorganic cations, thereby effecting a transient hyperpolarization which modulates the neuronal response to subsequent impulses.

Proof of this provocative hypothesis will depend on the demonstration of a direct causal relationship between synaptic membrane phosphorylation and alterations in postsynaptic potential and an unequivocal involvement of proteins I and II in controlling ionic fluxes across the postsynaptic membrane. Aside from their minimal molecular weights (44, 331), no information is currently available regarding the topographic location, the native structure, or the physiological function of proteins I and II.

Glial cells are intimately associated with neurons in the central nervous system and comprise more than 80% of the brain cell population. Cloned glial tumor cells generate high intracellular concentrations of cAMP in response to catecholamine neurotransmitters (342, 343). Approximately 50% of the cAMP-dependent protein kinase in a human astrocytoma line (344) and nearly 80% of the enzyme in rat glioma cells (345) were found in particulate subcellular fractions. Endogenous substrates for these enzymes were not identified.

Another phosphoprotein of neural origin is a constituent of the myelin sheath membrane. Myelin basic protein (EAE, or experimental allergic encephalomyelitis protein) served as an excellent substrate for cAMP-dependent protein kinase from brain cytosol, having a total phosphate-accepting capacity of 4 mol phosphate per mole of protein (346). Highly purified myelin membrane possesses an endogenous, cAMP-independent protein kinase which catalyzes a limited phosphorylation (0.02 mol phosphate per mole protein) of EAE protein (346, 347). An endogenous phosphoprotein phosphatase is also apparent in myelin preparations (346). As isolated by standard procedures (348) a mole of basic protein contains 0.2 mol P_i equally distributed between phosphoserine and phosphothreonine (346). In vivo phosphorylation of myelin basic protein was observed following injection of $^{32}P_i$ into rat brain ventricle (346, 347). Myelin basic protein rapidly achieved a level of high specific radioactivity, after which radioactive phosphate was rapidly lost, indicating considerable turnover at serine and threonine acceptor sites. These results permit the speculation that following its biosynthesis, myelin basic protein could be phosphorylated by a cytosol protein kinase. Transient alterations in its structural or functional properties could then be effected by changes in its degree of phosphorylation mediated by myelin-associated kinase and phosphatase.

Rhodopsin is localized in the disk membranes of the outer segments of retinal rods (349). Extensively purified rod outer segments catalyze the transfer of the γ-phosphate of ATP to specific serine and threonine residues in rhodopsin (350–353), as indicated by the migration of a ^{32}P-labeled polypeptide and rhodopsin covalently labeled with ^3H-retinaldehyde by borohydride reduction (353). Identification of ^{32}P-labeled rhodopsin is also facilitated because it is the major protein component of the disk membrane (mol wt = 35,000 on SDS gels) and is readily extracted in 1% aqueous digitonin. Intrinsic kinase activity is not affected by cAMP, but the rate and degree of phosphorylation are proportional to the percentage of visual pigment bleached by exposure to light (352, 353). Protein kinase activity was clearly dissociated from rhodopsin by extraction with dilute Tris-EDTA (pH 8.4) and was equally active in the light or dark (352). Other preliminary results suggest that light absorption induces a conformational change in rhodopsin, resulting in an increase in the accessibility of its phosphate-acceptor sites to protein kinase (352, 353). Rhodopsin incorporates 0.5–1 mol phosphate per mole protein in vitro (352, 353), but it is not known if rhodopsin is phosphorylated in vivo.

Phosphorylation of Erythrocyte Membranes

The plasma membranes contain more than 70% of the cAMP-dependent protein kinase activity of human erythrocytes (354). When assayed in the presence of exogenous protamine or histones, this enzyme closely resembles cytoplasmic cAMP-dependent protein kinases in its kinetic properties. Most of the catalytic activity of erythrocyte membrane protein kinase was solubilized by extraction of the membranes with 1 M NH_4Cl and separated from cAMP binding activity which remained associated with the particulate residue, suggesting that the binding protein is more firmly integrated into the cell membrane (354). Treatment of the membranes with cAMP (10^{-8}–10^{-4} M) did not induce the solubilization of either subunit.

Three polypeptide components of the erythrocyte membrane, proteins II (mol wt = 210,000), III (mol wt = 88,000), and IVc (mol wt = 50,000), served as substrates for intrinsic protein kinase (41, 42, 333). The rates of phosphorylation of proteins III and IVc were significantly increased (threefold or more) by an optimal concentration of cAMP (1–5 μM). cAMP also slightly stimulated phosphorylation of protein II (41, 333). Approximately 80% of the [^{32}P]-phosphate transferred from [γ-^{32}P] ATP to membrane proteins by endogenous protein kinase was recovered as phosphoserine (70%) and phosphothreonine (10%) (42, 333).

Phosphoprotein II accounts for 15% of the membrane protein (329) and is a subunit of spectrin (335–357). Spectrin subunits polymerize to form fibrils at the inner membrane surface, suggesting a cytoskeletal function (355–358). There is a preliminary report (359) that the amount of phosphate incorporated into spectrin in membranes prepared from normal erythrocytes was twice the amount incorporated by an equivalent quantity of spectrin in erythrocyte membranes obtained from patients with hereditary spherocytosis, suggesting that the phosphorylation of spectrin plays a role in maintaining cell shape.

Protein III is the major polypeptide component of the membrane, comprising 30% of the total protein (329). This protein spans the membrane (360, 361) and contains a specific, high affinity binding site for another membrane protein, glyceraldehyde 3-phosphate dehydrogenase (G3PD), at the inner surface of the membrane (362). It would be of obvious interest to determine if phosphorylation of protein III influences the binding of G3PD to the membrane. Considering the stoichiometry of ^{32}P incorporation [0.02 mol ^{32}P$_i$/mole protein (42)], it seems possible that protein III contains a high pre-existing level of phosphorylated sites or that the actual substrate is a minor protein component that migrates with III. The SDS-acrylamide gel system (329) utilized for resolving the membrane proteins in previous studies (42) does not clearly differentiate between protein III and the major erythrocyte glycoprotein [glycophorin (363)]. Recent studies have shown that no ^{32}P-labeled material was isolated in the aqueous phase of a chloroform-methanol buffer (364) extract of phosphorylated membranes which contained nearly 100% of the major erythrocyte glycoproteins (365). In addition, a peak of ^{32}P-labeled protein exhibited the same R_f as protein III (identified by staining) in SDS gels at a variety of acrylamide concentrations, but no such correlation was seen with the glycoproteins (365). Thus, protein III appears to be a substrate for membrane-associated protein kinase.

Protein IVc is a minor component of the membrane, comprising less than 2% of the protein (329). In preliminary experiments Guthrow et al (366) labeled the erythrocyte membrane cAMP binding protein by using a specific photoaffinity analog of cAMP, [^{3}H]-N^6-ethyl-2-diazomalonyl cAMP (367), and proposed that phosphoprotein IVc and the binding protein were identical, based solely on the migration of peaks of ^{32}P and ^{3}H in separate experiments (41, 366). These results have been confirmed in part, utilizing double labeling techniques, but selective solubilization of some membrane proteins with Triton X-100 in borate buffer (368) clearly separated all the cAMP binding activity from ^{32}P-labeled IVc (365).

Membrane-associated, cAMP-dependent protein kinases are often viewed as displaced cytoplasmic enzymes and studied accordingly (kinetics, metal ion requirements, etc). Another rewarding approach is to consider protein kinase as an integral component of the membrane and examine possible regulatory interactions between the kinase and other (nonsubstrate) membrane components. For instance, the topographic location of protein kinase in the membrane may be an important factor in determining the biological role and specificity of the enzyme. The localization of the subunits of protein kinase was determined by comparing their accessibilities to the accessibility of enzymes having previously determined orientations in sealed, inside-out and right-side-out membrane preparations (369, 370). Both the catalytic and cAMP binding subunits were located on the inner membrane surface (371, 372), which is consistent with the internal disposition of protein II (373), the transmembrane orientation of protein III (360, 361), and the absence of cAMP binding activity in intact erythrocytes (371). The internal location of the catalytic and cAMP binding subunits and the ability of the enzyme to phosphorylate endogenous substrates of specific orientation suggest that protein kinase may be used as an intrinsic probe for studying structure-function relationships in the erythrocyte membrane.

Roses & Appel (374) have found that all intrinsic substrates in erythrocyte membranes obtained from subjects with myotonic muscular dystrophy undergo phosphorylation at 50% of the rate observed in control membranes after both membrane preparations were stored at $-20°C$. This type of assay may provide a marker for the disease, although it is not clear whether the defect is in protein kinase or related to a decrease in accessibility of substrates resulting from a dysfunction in the structural assembly of the membrane.

A few studies have focused on the phosphorylation of membranes in intact erythrocytes. After incubating whole blood with $^{32}P_i$, Palmer & Verpoorte (375) observed the phosphorylation of a membrane component corresponding to protein II. When rat or human erythrocytes were prelabeled with $^{32}P_i$ and then exposed to 0.1 mM norepinephrine or 1 μM PGE$_2$, a 20–40% increase in the total phosphate incorporated into membrane proteins was observed (376). Dibutyryl cAMP stimulated phosphorylation by 50%. Unfortunately, the distribution of ^{32}P-phosphate among the membrane polypeptides was not analyzed, but the data are consistent with the presence of an adenylate cyclase-cAMP-protein kinase system in mature erythrocytes. Rat erythrocyte membranes are known to possess a hormone-sensitive adenylate cyclase (377) and human erythrocytes exhibit low, but significant, fluoride-stimulated adenylate cyclase activity (42). The effects of prostaglandin and catecholamines on membrane phosphorylation may be related to the abilities of these compounds to alter the shape of erythrocytes (378).

In turkey erythrocytes, isoproterenol elicits an elevated concentration of intracellular cAMP and a corresponding increase in the passive permeability of the membrane to Na$^+$ (379). Isoproterenol simultaneously stimulated the incorporation of $^{32}P_i$ into a single membrane polypeptide (mol wt = 240,000), corresponding to protein II of the human erythrocyte membrane, thus providing a correlation between phosphorylation and a functional alteration in the membrane (380).

Membrane Phosphorylation and Transport

The possible interdependence of transport and phosphorylation in cell membranes has been explored in adipocytes and heart microsomes. The phosphorylation of two small membrane proteins (mol wt = 16,000 and 22,000) by an intrinsic protein kinase in fat cells has been associated with the inhibition of insulin-stimulated glucose transport (381). Phosphorylation occurred at the external membrane surface in intact adipocytes and was stimulated by cAMP in purified plasma membrane preparations. Phloretin blocked the phosphorylation of the larger protein, intimating a possible carrier function for this polypeptide.

Cardiac microsomes manifest intrinsic, cAMP-stimulated phosphorylation (382, 383) of a unique membrane protein (383), having a molecular weight of 20,000. cAMP-dependent phosphorylation of this protein preceded a transient increase in ATP-dependent Ca^{2+} uptake (383), indicating a potential contractile regulatory mechanism. An active phosphoprotein phosphatase removed phosphate from the microsomes, adding the element of rapid reversibility to the system. Soluble cAMP-dependent protein kinases also phosphorylated cardiac microsomes and enhanced Ca^{2+} uptake (384).

In contrast to heart muscle, the particulate protein kinase of skeletal muscle is localized in the plasma membrane and catalyzes the phosphorylation of three polypeptides with molecular weights less than 30,000 (385). Kinase activity was not enhanced by cAMP and the physiological significance of this intrinsic phosphorylation is unknown. Muscle sarcolemma also served as a substrate for cytoplasmic, cAMP-dependent protein kinase, and the phosphorylated membranes exhibited a slight increase in the rate of Ca^{2+} uptake (386). Because the preparation of skeletal muscle membranes involves prolonged treatment with solutions of high ionic strength (e.g. 0.4 M LiBr), it is improbable that the isolated membranes are truly representative of the native sarcolemma (385, 387).

Membrane Phosphorylation and Secretory Processes

Phosphorylation of adenohypophyseal plasma membranes and secretory granule membranes is particularly intriguing because these membranes undergo fusion during the secretion of hormones from the anterior pituitary. Labrie and colleagues (388) have demonstrated the cAMP-stimulated (30–100%) phosphorylation of nine polypeptide subunits in adenohypophyseal plasma membranes as well as the cyclic nucleotide-independent incorporation of phosphate into secretory granules (389). Localization of specific phosphoproteins in secretory granule membranes was not carried out, and the possible phosphorylation of soluble granule proteins was not excluded. There is no direct evidence that phosphorylation of these two membranes regulates membrane fusion and hormone release.

Studies on pancreatic fragments and mammary gland explants have also implicated membrane phosphorylation in secretory processes. Pancreozymin and the related decapeptide, caerulein, stimulate the secretion of hydrolases from pancreatic acinar cells. cAMP has been implicated in enzyme release but the concentrations of hormone required for adenylate cyclase activation are 200-fold higher than those that induce enzyme secretion (390, 391), and enhanced exocytosis has not always been associated with an increased intracellular concentration of cAMP (392, 393). Nevertheless, both pancreozymin and caerulein significantly stimulate (2.7-fold) the phosphorylation of secretory granule membranes in intact acinar cells (394). Isolated

zymogen granules exhibit endogenous protein kinase activity (59) but phosphorylation by the more active cytoplasmic kinases has not been excluded. Identification
of the granule membrane phosphoproteins has not been reported.

Plasma membranes derived from differentiating mammary epithelial cells can
also be phosphorylated (47). In the presence of hydrocortisone, insulin stimulated
the phosphorylation of 63% of the polypeptide chains that comprise the protein
portion of the plasma membrane of developing mammary cells (47). Treatment with
prolactin generated an additional increase in the rates of phosphorylation of all the
membrane phosphoproteins. A membrane-associated, cAMP-independent protein
kinase, as well as a soluble, cAMP-activated kinase, catalyzed the phosphorylation
of the same endogenous substrates in purified plasma membranes. Although the
presence of multiple phosphate acceptors is similar to the occurrence of numerous
substrates in anterior pituitary plasma membrane (388), the phosphorylation of
mammary cell plasma membranes is unique in that it is coupled to RNA and
protein synthesis (47). Among the earliest proteins induced by the synergistic action
of insulin and prolactin are the catalytic and cAMP binding subunits of cytoplasmic
protein kinase (46). In contrast, cycloheximide inhibition of protein synthesis in
pancreatic acinar cells had no effect on protein phosphorylation or the secretion
of hydrolases (394).

Membrane Phosphorylation and Permeability

Antidiuretic hormone (ADH) regulates the flow of water and Na^+ across a number
of epithelial membranes. The hormone binds to a specific receptor in the basal-
lateral (serosal) membrane of the mammalian renal collecting duct, concomitantly
activating adenylate cyclase at the same surface (395, 396). Subsequently, the
permeability of the apical (mucosal) membrane is increased in response to an
elevated intracellular concentration of cAMP (395–398). Recently, these two
functionally distinct domains (i.e. the apical and basal-lateral membranes) of the
highly polar renal epithelial cells have been separated by homogenization, differential centrifugation, and free flow electrophoresis (399). Adenylate cyclase was
purified together with Ca^{2+}-ATPase, a marker for the basal-lateral membrane,
whereas cAMP-stimulated protein kinase and its endogenous substrates were found
in membrane fractions enriched for HCO_3^--ATPase, an indicator of the apical
portion of the cell surface. Similar results were obtained with rat renal cortical
epithelial cells (400). These observations suggest that cAMP generated at the basal-
lateral surface in response to ADH mediates hormone action by specifying the
phosphorylation of apical membrane proteins, which in turn permit an increased
flow of Na^+ and H_2O from the mucosal to the serosal surface of the cell. In the
experiments described above, specific apical membrane phosphoproteins were not
identified, an analysis for phosphoserine and phosphothreonine was not performed,
and the possible contribution of acyl phosphate intermediates of ATPases to the
total phosphate incorporated by the membranes was not evaluated. The absence
of a membrane-associated phosphoprotein phosphatase (399) leaves the reversibility
and, therefore, the physiological significance of apical membrane phosphorylation
open to conjecture. Investigations aimed at correlating ADH-induced permeability

changes and the phosphorylation of apical membrane polypeptides in $^{32}P_i$-labeled, intact nephrons would provide a good test of the validity of the foregoing hypothesis.

When intact toad bladders were prelabeled with $^{32}P_i$ and then treated with ADH or N^6 monobutyryl cAMP, the net phosphorylation of a single, particulate protein (protein D, mol wt = 50,000) was reduced by 50% (335). This decrease in phosphorylation preceded (by 1–2 min) a change in electrical potential across the bladder which was indicative of increased Na^+ flux. Decreased phosphorylation of protein D was also noted in membrane fractions incubated with cAMP and $[\gamma^{-32}P]$ ATP as compared to treatment with $[\gamma^{-32}P]$ ATP alone (335, 336). By using EDTA and adenosine to inhibit protein kinase and Zn^{2+} to inhibit phosphoprotein phosphatase, DeLorenzo & Greengard (336) concluded that cAMP effected a decrement in the phosphorylation of protein D by stimulating the activity of a membrane-associated phosphoprotein phosphatase, rather than inhibiting a particulate kinase. Thus, cAMP may mediate the action of ADH in toad bladder in a unique fashion: by stimulating dephosphorylation of a specific membrane protein which facilitates Na^+ permeability as an inverse function of its degree of phosphorylation. cAMP could act by (a) stimulating a phosphatase kinase, thereby providing a mechanism that does not contradict the postulated involvement of protein kinases in all the actions of the cyclic nucleotide (340), (b) direct activation of the phosphatase, or (c) increasing the accessibility of phosphoproteins to the active site of the phosphatase.

The differences observed in the amphibian and mammalian systems may be due to the evolution of different mechanisms of response to ADH which achieve the same end result with similar degrees of complexity and control. At the present time it is difficult to compare the data obtained in the two systems because studies on the mammalian medullary collecting duct used highly purified apical and basal-lateral membrane fragments whereas the toad bladder studies were carried out with whole homogenates and crude microsomal "membranes" (335, 336) which are known to contain a variety of contaminating organelles in addition to plasma membranes (401, 402). Thus, the assignment of an essential role for protein D dephosphorylation in the apical membrane of toad bladder epithelial cells cannot presently be made.

Another observation on the amphibian system which requires further exploration is the finding that cytosol contains a protein that is apparently identical with protein D and accounts for approximately 50% of the ^{32}P incorporated into toad bladder proteins (335). The physiological function of this protein and its relationship to its membrane-associated counterpart are unknown.

Membrane Phosphorylation and Adenylate Cyclase

Membranes obtained from polymorphonuclear granulocytes and platelets underwent self-phosphorylation in the presence of ATP (403). Phosphorylation was correlated with a decrease in adenylate cyclase activity, but normal cyclase activity was restored by treatment with F^- or PGE_1. These two agents also stimulated the release of phosphate from endogenous substrates, presumably by activating specific phosphoprotein phosphatases (404). These preliminary results suggest that the level of membrane phosphorylation may partially regulate adenylate cyclase activity (405). This conclusion must be regarded as speculative because neither the membrane preparations nor the endogenous substrates were carefully characterized, and the effects of hormones on the adenylate cyclase activity of phosphorylated or de-phosphorylated membranes were not considered.

Cyclic GMP-Dependent Protein Kinase in Membranes

Physiological and pharmacological studies have implicated cGMP in the mediation of muscarinic cholinergic effects in smooth muscle (406, 407). Microsomal fractions obtained from ductus deferens, uterus, and small intestine displayed intrinsic protein kinase and endogenous phosphate-acceptor activities which were greatly stimulated by low concentrations of cGMP (K_a for cGMP = 3×10^{-8} M) (334). The principal phosphate acceptor was a polypeptide having a molecular weight of 130,000. cGMP also regulated the phosphorylation of a slightly smaller protein (mol wt = 100,000) (334). It seems possible, therefore, that membrane-associated, cGMP-dependent protein kinases may mediate the effects of cholinergic agents in a manner quite analogous to the actions of particulate and soluble cAMP-dependent protein kinases in mediating cellular responses to polypeptide hormones, adrenergic agents, and neurotransmitters.

PHOSPHORYLATION OF RIBOSOMES

Numerous eukaryotic ribosomal proteins are phosphorylated on serine and threonine residues by cytoplasmic and ribosome-associated protein kinases (80, 134, 408–416). Cytosolic cAMP-dependent and -independent protein kinases which phosphorylate ribosomal proteins also utilize histones, protamine, phosphorylase kinase, and hormone-sensitive lipase as substrates and they cannot be distinguished from other cytoplasmic kinases by their cyclic nucleotide specificity, metal ion requirements, or kinetic properties (80, 134, 142, 417, 418). Since most soluble protein kinases used for ribosome phosphorylation have been only partially purified from adrenal cortex (13, 149), liver (409, 417), or reticulocyte cytosol (134, 418), it is likely that multiple species of cAMP-dependent and -independent protein kinases were actually present. Traugh & Traut (418) have demonstrated the occurrence of multiple ribosomal kinases in rabbit reticulocyte cytosol, including two cAMP-dependent protein kinases; a free catalytic subunit, inhibitable by cAMP binding protein or inhibitor protein; and a casein phosphokinase which is not affected by binding protein or inhibitor. A similar group of kinases is present in rat liver cytosol (409, 420).

Some protein kinases are tightly bound to intact ribosomes (409–416, 421) and cannot be extracted with buffers of moderate ionic strength (<0.5 M KCl). Ribosome-bound kinases exhibit properties very similar to those of the cytoplasmic enzymes: they are stimulated by cAMP (but their sensitivity to the nucleotide is reduced) (409, 412, 413); they also phosphorylate the same ribosomal proteins to the same final ^{32}P-specific activities (412), suggesting that neither soluble nor endogenous kinases operate on unique sites. Tightly bound kinase activity is extracted from ribosomes by washing with 0.5–1.0 M KCl (409–416, 421) and by other procedures that promote dissociation of 80S ribosomes into 60 and 40S subunits (134, 410, 412, 417). All these observations lead to the conclusion that protein kinases are probably not integral ribosomal proteins and allow for the speculation that ribosomal proteins may be phosphorylated by a number of cAMP-dependent and -independent cytoplasmic kinases. Some portion of these kinases may be adventitiously bound to ribosomes during cell fractionation procedures, thereby accounting for the ribosome-associated kinase activity.

Physiological responsiveness requires phosphorylation to be reversible. Phospho-protein phosphatases capable of removing 65% of incorporated ^{32}P from ribosomes and 40 or 60S subunits have been noted in liver cytosol (422) but have not been the subject of serious investigation. E. coli alkaline phosphatase has proven to be a useful analytical tool for splitting phosphate esters in ribosomal and other proteins (410, 423).

Phosphate-acceptor proteins have been found in both 60 and 40S subunits cf eukaryotic ribosomes (134, 409, 412, 417) incubated in vitro with cytoplasmic or endogenous protein kinases, cAMP, and $[\gamma\text{-}^{32}P]$ ATP. In general, the pattern of phosphoproteins in 60S particles is quite complex. In rabbit reticulocyte 60S subunits there are nine phosphoproteins with molecular weights ranging from 14,000 to 53,000 (134). Comparable 60S subunits from liver and adrenal cortex possess ten and seven phosphorylated proteins, respectively (412, 417). In contrast, the 40S subunit from reticulocytes contains only two phosphate acceptors with molecular weights of 35,000 and 50,000 (134), and the smaller subunit of adrenal cortical ribosomes has a single phosphorylated protein constituent (412).

The total number of phosphate acceptor sites was determined by treating 60 and 40S subunits with excess protein kinase and ATP until saturation was achieved. Data on ribosomal subunits obtained from adrenal cortex (412) and liver (419) were in accord: 40S subunits displayed a capacity for 2–3 mol phosphate/mole subunit, while 60S particles had the potential for accepting 10–12 mol phosphate/mole subunit. When the number of phosphoproteins, their relative specific activities, and their total phosphate capacity are taken into account, it becomes apparent that, on the average, acceptor proteins of ribosomal subunits are phosphorylated in a stoichiometric fashion (412, 417). Prokaryotic ribosomes have no integral phospho-proteins (424). In the foregoing studies, no correction was made for pre-existing phosphoamino acids in purified ribosomes; therefore the results represent minimal estimates of the number of sites. Walton & Gill (412) determined that there were 18 mol protein phosphate/mole 80S ribosome in freshly isolated adrenal cortical ribosomes. An accurate quantitation of the apparent number of acceptor sites will necessitate the utilization of untreated and dephosphorylated ribosomes to discriminate between accessible proteins and proteins phosphorylated prior to ribosome assembly and not subject to turnover.

Two distinct types of ribosomal protein kinases which have decidedly different substrate specificities have been separated from rabbit reticulocyte cytosol by chromatography on DEAE- and phosphocellulose (134). The first category includes intact, cAMP-dependent protein kinases and their free catalytic subunits. These enzymes phosphorylate six proteins in 60S subunits and a single polypeptide in the smaller subunit. The second type resembles classical phosvitin or casein kinases in its properties and catalyzes the phosphorylation of three 60S proteins and one 40S protein, utilizing either ATP or GTP as a phosphate donor. A GTP (or ATP) phosphotransferase present in rat liver cytosol phosphorylates four 40S proteins (420). Only one ribosomal protein appears to be phosphorylated in common by both types of enzyme in reticulocytes (134). The physiological significance of these findings cannot be evaluated until it can be determined whether ribosomal substrates are available to both or either of these enzymes in vivo. The possibility that these enzymes phosphorylate ribosomal entities in the same nonspecific manner as other basic proteins, such as histones and protamine, has not been excluded.

Ribosome phosphorylation has also been studied in intact cells and in vivo. Treatment of rats with glucagon produced a two- to threefold increase in phosphorylation of a single protein in liver ribosomes (425). Thyroidectomy caused a 35% decrease in the phosphate content of ribosomes and a concomitant fall in protein kinase and cAMP binding activity in liver cytosol (426). Injection of thyroid hormone reversed these effects. These studies indicate a possible physiological role of protein phosphorylation at the level of translation because glucagon induces hepatic enzymes (427, 428) and thyroid hormone elicits a general increase in liver protein synthesis (429), but the observations may be criticized on the grounds that the ribosomes were not subjected to extraction with 0.5 M KCl which solubilizes nonribosomal (but tightly bound) proteins that can account for an amount of phosphate equal to that associated with integral ribosomal proteins (412).

Insulin stimulated the phosphorylation of eight ribosomal proteins in cultured mouse mammary cells (47). Addition of prolactin to insulin-treated cells synergistically increased the incorporation of phosphate in four of these phosphoproteins. Enhancement of phosphorylation by prolactin required RNA and protein synthesis and was preceded by the induction of cytoplasmic cAMP-dependent protein kinase (46). Ribosomal subunits were not separated prior to analysis.

Kabat and colleagues (423, 430, 431) examined the phosphorylation of ribosomal proteins in intact rabbit reticulocytes. The autoradiographic profile of phosphoproteins in 40 and 60S subunits was similar, but not identical with the pattern observed in cell-free preparations (134, 431). Four phosphate acceptors on the larger subunit and two on the smaller were readily discerned (431). Conservation of ribosomal structure was disclosed by the demonstration that mouse sarcoma 180S ribosomal phosphoproteins were identical with those seen in rabbit reticulocytes (432). Detailed analyses of phosphoryl group turnover (430) showed that all phosphoryl moieties turned over slowly (3% per minute) relative to the rate of protein synthesis, and that there were 11 phosphoryl sites turning over on single ribosomes, but only seven on polysomal ribosomes. The latter observation suggested that there was little or no interchange between these two pools and is consistent with the proposal that single ribosomes do not enter the polysome subunit cycle (433–435).

When reticulocytes were incubated with $^{32}P_i$, cAMP or its dibutyryl derivative elicited a generalized twofold increase in the specific activities of all ribosomal phosphoproteins, presumably by facilitating phosphate uptake (431). In addition, the cyclic nucleotides specifically enhanced the phosphorylation of a single protein (mol wt = 27,500) in the 40S subunit by an additional factor of 2.5 (431).

Eil & Wool (422) examined the functional properties of ribosomes that were phosphorylated in vitro by cytoplasmic protein kinases. They found that phosphorylated and control ribosomes were identical in their abilities to interact with elongation factors EF-1 and EF-2, initiation factors EIF-1 and EIF-2, and to translate encephalomyocarditis virus RNA (422), suggesting that ribosomal protein phosphorylation does not influence aminoacyl tRNA binding, peptide bond formation, translocation of peptidyl tRNA or the initiation of translation of a naturally occurring template. No correlation has yet been established between phosphorylation and physiological function in eukaryotic ribosomes. Of course, a variety of other ribosomal functions have not been scrutinized in regard to ribosome phosphorylation, including the translation of cell-specific mRNA, the termination of protein synthesis, and the binding of ribosomes to endoplasmic reticulum. It is also possible that the studies carried out in vitro (422) did not utilize optimal conditions for discriminating between phosphorylated and nonphosphorylated ribosomes. Furthermore, the substrate and biological specificity of the protein kinase which was used has not been defined, and the possibility remains that important regulatory proteins were phosphorylated prior to the isolation of ribosomes. Despite the negative data, it seems probable that the active turnover of ribosomal phosphoryl groups in intact cells (430) and the sensitivity of the level of phosphorylation in certain acceptors to hormones (47, 425, 426) and cAMP (431) are reflections of physiologically significant alterations in ribosome function or structure.

SUMMARY AND CONCLUSIONS

Protein phosphorylation reactions catalyzed by protein kinases and regulated by cyclic nucleotides are probably universal and essential links in the chain of biochemical events that converts the binding of a variety of polypeptide hormones, catecholamines, and other activators of adenylate cyclase at target cell surfaces into specific metabolic responses. Cytoplasmic cAMP-dependent protein kinases have been studied intensively since their discovery seven years ago, and a significant body of evidence has been amassed in support of the proposition that these enzymes are primary receptors for cyclic nucleotides in eukaryotic cells. Protein kinases from cardiac and skeletal muscle have been purified to homogeneity and physically characterized; the cAMP-mediated dissociation of kinases into free C and cAMP-R components has been demonstrated in vitro and in vivo; and a model for the subunit structure of "protein kinase" has been proposed. In contrast, considerably less is known about the naturally occurring substrates for the ubiquitous protein kinases and the phosphoprotein phosphatases responsible for reversing the actions of cAMP-dependent protein kinases. Investigations concerned with the identification, purification, and characterization of phosphate-acceptor proteins promise to yield new insights regarding the multiple physiological functions of protein kinases, since the nature of the acceptors determines the specific cellular response to hormones and other stimuli.

Phosphoprotein phosphatases participate in the regulation of phosphate turnover and insure the rapid reversibility of bioregulatory phosphorylation reactions, yet these enzymes have not been subjected to intensive investigation and much remains to be learned about their structural, kinetic, and regulatory properties.

The emerging importance of protein phosphorylation in regulation should shift the study of phosphoprotein phosphatases from an ancillary position to a focal point of attention.

Elegant studies on the hormonal control of glycogen metabolism have provided well-defined examples of the physiological roles of protein phosphorylation and dephosphorylation in coordinately regulating a complex group of interdependent enzymes. The protein kinases, phosphate acceptors, and phosphoprotein phosphatases that control glycogen metabolism have all been purified and studied in depth in isolated, nonphysiological systems. Parallel investigations, however, were carried out on the same enzymes in vivo and in protein glycogen particles which closely resemble the in vivo functional units of glycogen metabolism. Only through the synthesis of these two approaches have we come to appreciate the exquisite specificity and sensitivity (provided in part by protein phosphorylation reactions) at the molecular level in the hormonal regulation of glycogen synthesis and degradation. Analogous approaches will be valuable in delineating the functional significance of protein phosphorylations in response to other hormones (and other stimuli) in a variety of tissues.

Among the phosphotransferases, phosphorylase kinase is a complex and unique enzyme in that it (a) is activated by the cAMP-dependent phosphorylation of its B subunits, (b) is inactivated by a specific phosphatase subsequent to the phosphorylation of its A subunits, and (c) catalyzes the transfer of phosphate from ATP to phosphorylase b and troponin.

The metabolic roles of the cyclic nucleotide-independent phosphoprotein kinases have not been established, aside from catalyzing the phosphorylation of "nutritive" phosphoproteins such as casein and phosvitin. These enzymes can and should be distinguished from the cyclic nucleotide-independent catalytic subunit of protein kinases by their failure to be inhibited by cyclic nucleotide binding protein R or the protein kinase inhibitor I. Ultimately, with purification and elucidation of their molecular structure, the relationship, if any, of these kinases to the cyclic nucleotide-dependent protein kinases will be ascertained.

cAMP-dependent and -independent protein kinases and phosphate-acceptor proteins are integral components of a number of subcellular organelles including plasma membranes, sarcoplasmic reticulum, secretory vesicles, nuclei, and perhaps ribosomes. Numerous correlations have been made which suggest that phosphorylation of particulate proteins may be linked to specific functional modifications in subcellular structures. At present, however, no direct causal relationship has been established between the phosphorylation of polypeptides in cellular organelles and a defined functional alteration. Proof of such a relationship entails the exceedingly difficult tasks of isolating and purifying particulate phosphoproteins, establishing their functional identities in vitro, and finally, reconstituting physiologically relevant systems from their individual components. Current technology permits the partial reconstitution of ribosomes and chromatin, whereas the dissociation and reassembly of functional plasma membranes are not yet attainable. Nevertheless, the widespread occurrence of protein kinases and associated phosphate-acceptor proteins in sub-cellular organelles, the in vitro and in vivo turnover of phosphate in these acceptors, and the correlations between phosphorylation of organelles and functional alterations described in this review suggest possible roles for protein phosphorylation in the regulation of transport, permeability, protein synthesis, RNA transcription, and secretory processes.

Finally, particulate protein kinases should be considered from the perspective of their associated organelles in order to evaluate the influence of structural organization on the interactions among the kinases, phosphate acceptors, and other components of the organelle.

ACKNOWLEDGMENTS

We are indebted to Drs. E. G. Krebs, T. A. Langan, and H. Silberstein for their critical evaluation of parts of this review.

Literature Cited

1. Walsh, D. A., Krebs, E. G. 1973. *Enzymes* 8:555
2. Langan, T. A. 1973. *Advan. Cyclic Nucleotide Res.* 3:99
3. Walsh, D. A., Ashby, C. D. 1971. *Recent Progr. Horm. Res.* 29:329
4. Krebs, E. G. 1972. *Curr. Top. Cell Regul.* 5:99
5. Fischer, E. H., Heilmeyer, L. M. G. Jr., Haschke, R. H. 1971. *Curr. Top. Cell Regul.* 4:211
6. Larner, J., Villar-Palasi, C. 1971. *Curr. Top. Cell. Regul.* 3:196
7. Segal, H. L. 1973. *Science* 180:25
8. Holzer, H., Duntze, W. 1971. *Ann. Rev. Biochem.* 40:345
9. Stadtman, E. R. 1973. *Enzymes* 8:2
10. Dahl, J. L., Hokin, L. E. 1974. *Ann. Rev. Biochem.* 43:327
11. Roseman, S. 1972. *Metab. Pathways* 6:42
12. Brostrom, M. A., Reimann, E. M., Walsh, D. A., Krebs, E. G. 1970. *Advan. Enzyme Regul.* 8:191
13. Gill, G. N., Garren, L. D. 1970. *Biochem. Biophys. Res. Commun.* 39:335
14. Soderling, T. R., Corbin, J. D., Park, C. R. 1973. *J. Biol. Chem.* 248:1822
15. Corbin, J. D., Soderling, T. R., Park, C. R. 1973. *J. Biol. Chem.* 248:1813
16. Means, A. R., MacDougall, E., Soderling, T. R., Corbin, J. D. 1974. *J. Biol. Chem.* 249:1231
17. Sudilovsky, O. 1974. *Biochem. Biophys. Res. Commun.* 58:85

18. Takeda, M., Ohga, Y. 1973. *J. Biochem.* 73:621
19. Langan, T. A. 1969. *J. Biol. Chem.* 244:5763
20. Mallette, L. E., Neblett, M., Exton, J. H., Langan, T. A. 1973. *J. Biol. Chem.* 248:6289
21. Miller, T. B. Jr., Larner, J. 1973. *J. Biol. Chem.* 248:3483
22. Shima, S., Mitsunaga, M., Kawashima, Y., Taguchi, S., Nakao, T. 1974. *Endocrinology* 94:650
23. Walsh, D. A., Ashby, C. D., Gonzales, C., Calkins, D., Fischer, E. H., Krebs, E. G. 1971. *J. Biol. Chem.* 246:1977
24. Ashby, C. D., Walsh, D. A. 1972. *J. Biol. Chem.* 247:6637
25. Donnelly, T. E. Jr., Kuo, J. F., Reyes, P. L., Liu, Y.-P., Greengard, P. 1973. *J. Biol. Chem.* 248:190
26. Donnelly, T. E. Jr., Kuo, J. F., Miyamoto, E., Greengard, P. 1973. *J. Biol. Chem.* 248:199
27. Majumder, G. C. 1974. *Biochem. Biophys. Res. Commun.* 58:756
28. Turkington, R. W., Majumder, G. C., Kadohama, N., MacIndoe, J. H., Frantz, W. L. 1973. *Recent Progr. Horm. Res.* 29:417
29. Tsung, P. K., Weissmann, G. 1973. *Biochem. Biophys. Res. Commun.* 51:836
30. Miyamoto, E., Petzold, G. L., Kuo, J. F., Greengard, P. 1973. *J. Biol. Chem.* 248:179
31. Tao, M., Hackett, P. 1973. *J. Biol. Chem.*

248 : 5324
32. Murray, A. W., Froscio, M., Kemp, B. E. 1972. *Biochem. J.* 129 : 995
33. Rosen, O. M., Rubin, C. S., Erlichman, J. 1973. In "Protein Phosphorylation in Control Mechanisms," *Miami Winter Symp.* 5 : 67
34. Brostrom, C. O., Corbin, J. D., King, C. A., Krebs, E. G. 1971. *Proc. Nat. Acad. Sci. USA* 68 : 2444
35. Haddox, M. K., Newton, N. E., Hartle, D. K., Goldberg, N. D. 1972. *Biochem. Biophys. Res. Commun.* 47 : 653
36. Beavo, J. A., Bechtel, P. J., Krebs, E. G. 1975. *Advan. Cyclic Nucleotide Res.* 5 : In press
37. Beavo, J. A., Bechtel, P. J., Krebs, E. G. 1974. *Proc. Nat. Acad. Sci. USA* 71 : 3580
38. Ashby, C. D., Walsh, D. A. 1973. *J. Biol. Chem.* 248 : 1255
39. Erlichman, J., Rosenfeld, R., Rosen, O. M. 1974. *J. Biol. Chem.* 249 : 5000
40. Rosen, O. M., Erlichman, J., Rubin, C. S. 1975. *Advan. Enzyme Regul.* 13 : In press
41. Guthrow, C. E. Jr., Allen, J. E., Rasmussen, H. 1972. *J. Biol. Chem.* 247 : 8145
42. Rubin, C. S., Rosen, O. M. 1973. *Biochem. Biophys. Res. Commun.* 50 : 421
43. Maeno, H., Reyes, P. L., Ueda, T., Rudolph, S. A., Greengard, P. 1974. *Arch. Biochem. Biophys.* 164 : 551
44. Ueda, T., Maeno, H., Greengard, P. 1973. *J. Biol. Chem.* 248 : 8295
45. Menon, K. M. J. 1973. *J. Biol. Chem.* 248 : 494
46. Majumder, G. C., Turkington, R. W. 1971. *J. Biol. Chem.* 246 : 5545
47. Majumder, G. C., Turkington, R. W. 1972. *J. Biol. Chem.* 247 : 7207
48. Korenman, S. G., Bhalla, R. C., Sanborn, B. M., Stevens, R. H. 1974. *Science* 183 : 430
49. Wray, H. L., Harris, A. W. 1973. *Biochem. Biophys. Res. Commun.* 53 : 291
50. Kuo, J. F. 1974. *J. Biol. Chem.* 249 : 1755
51. Lemaire, S., Labrie, F., Gauthier, H. 1974. *Can. J. Biochem.* 52 : 137
52. Iwai, H., Inamasu, M., Takeyama, S. 1972. *Biochem. Biophys. Res. Commun.* 46 : 824
53. Yamashita, K., Field, J. B. 1972. *Metabolism* 21 : 150
54. Kuo, J. F., Miyamoto, E., Reyes, P. L. 1974. *Biochem. Pharmacol.* 23 : 2011
55. Neelon, F. A., Birch, B. M. 1973. *J. Biol. Chem.* 248 : 8361
56. Meyer, R. B. Jr., Miller, J. P. 1974. *Life Sci.* 14 : 1019
57. Simon, L. N., Shuman, D. A., Robbins, R. K. 1973. *Advan. Cyclic Nucleotide Res.* 3 : 225
58. Kuo, J. F., Greengard, P. 1970. *J. Biol. Chem.* 245 : 2493
59 Van Leemput-Coutrez, M., Camus, J., Christophe, J. 1973. *Biochem. Biophys. Res. Commun.* 54 : 182
60. Hoffman, F., Sold, G. 1972. *Biochem. Biophys. Res. Commun.* 49 : 1100
61. Haddox, M. K., Nicol, S. E., Goldberg, N. D. 1973. *Biochem. Biophys. Res. Commun.* 54 : 1444
62. Erlichman, J., Rubin, C. S., Rosen, O. M. 1973. *J. Biol. Chem.* 248 : 7607
63. Rubin, C. S., Erlichman, J., Rosen, O. M. 1972. *J. Biol. Chem.* 247 : 36
64. Garbers, D. L., First, N. L., Lardy, H. A. 1973. *J. Biol. Chem.* 248 : 875
65. Takai, Y., Yamamura, H., Nishizuka, Y. 1974. *J. Biol. Chem.* 249 : 530
66. Sands, H., Meyer, T. A., Rickenberg, H. V. 1973. *Biochim. Biophys. Acta* 302 : 267
67. Sanborn, B. M., Bhalla, R. C., Korenman, S. G. 1973. *J. Biol. Chem.* 248 : 3593
68. Sanborn, B. M., Korenman, S. G. 1973. *J. Biol. Chem.* 248 : 4713
69. Tsang, B. K., Singhal, R. L. 1973. *Can. J. Physiol. Pharmacol.* 51 : 942
70. Zapf, J., Froesch, E. R. 1972. *FEBS Lett.* 20 : 141
71. Wombacher, H., Reuter-Smerdka, M., Körber, F. 1973. *FEBS Lett.* 30 : 313
72. Hoskins, D. D., Casillas, E. R., Stephens, D. T. 1972. *Biochem. Biophys. Res. Commun.* 48 : 1331
73. Takats, A., Farago, A., Antoni, F. 1972. *Biochim. Biophys. Acta* 268 : 77
74. Kumar, R., Tao, M., Piotrowski, R., Solomon, L. 1973. *Biochim. Biophys. Acta* 315 : 66
75. Chen, L.-J., Walsh, D. A. 1971. *Biochemistry* 10 : 3614
76. Pierre, M., Loeb, J. E. 1972. *Biochim. Biophys. Acta* 284 : 421
77. Dastugue, B., Tichonicky, L., Kruh, J. 1973. *Biochimie* 55 : 1021
78. Yamamura, H., Nishiyama, K., Shimomura, R., Nishizuka, Y. 1973. *Biochemistry* 12 : 856
79. Kumon, A., Nishiyama, K., Yamamura, H., Nishizuka, Y. 1972. *J. Biol. Chem.* 247 : 3726
80. Yamamura, H., Inoue, Y., Shimomura, R., Nishizuka, Y. 1972. *Biochem. Biophys. Res. Commun.* 46 : 589
81. Daniel, V., Litwack, G., Tomkins, G. M. 1973. *Proc. Nat. Acad. Sci. USA* 70 : 76
82. Rosenau, W., Baxter, J. D., Rousseau, G. G., Tomkins, G. M. 1972. *Nature*

New Biol. 237:20
83. Makman, M. H., Klein, M. I. 1972. *Proc. Nat. Acad. Sci. USA* 69:456
84. Novak, E., Drummond, G. I., Skala, J., Hahn, P. 1972. *Arch. Biochem. Biophys.* 150:511
85. Pastan, I., Johnson, G. S. 1974. *Advan. Cancer Res.* 19:303
86. Criss, W. E., Morris, H. P. 1973. *Biochem. Biophys. Res. Commun.* 54:380
87. Mackenzie, C. W. III, Stellwagen, R. H. 1974. *J. Biol. Chem.* 249:5755
88. Granner, D. K. 1972. *Biochem. Biophys. Res. Commun.* 46:1516
89. Troy, F. A., Vijay, I. K., Kawakami, T. G. 1973. *Biochem. Biophys. Res. Commun.* 52:150
90. Tan, K. B., Sokol, F. 1974. *J. Virol.* 13:234
91. Kuo, J. F., Greengard, P. 1969. *J. Biol. Chem.* 244:3417
92. Khandelwal, R. L., Spearman, T. N., Hamilton, I. R. 1973. *FEBS Lett.* 31:246
93. Li, H.-C., Brown, G. G. 1973. *Biochem. Biophys. Res. Commun.* 53:875
94. Kaplowitz, P. B., Platz, R. D., Kleinsmith, L. J. 1971. *Biochim. Biophys. Acta* 229:739
95. Kuehn, G. D. 1972. *Biochem. Biophys. Res. Commun.* 49:414
96. Kuehn, G. D. 1971. *J. Biol. Chem.* 246:6366
97. Keirns, J. J., Carritt, B., Freeman, J., Eisenstadt, J. M., Bitensky, M. W. 1973. *Life Sci.* 13:287
98. Rabinowitz, M., Lipmann, F. 1960. *J. Biol. Chem.* 235:1043
99. Kemp, B. E., Froscio, M., Murray, A. W. 1973. *Biochem. J.* 131:271
100. Sy, J., Richter, D. 1972. *Biochemistry* 11:2784
101. Pastan, I., Perlman, R. 1970. *Science* 169:339
102. Zubay, G., Schwartz, D., Beckwith, J. 1970. *Proc. Nat. Acad. Sci. USA* 66:104
103. Gold, M. H., Segel, I. H. 1974. *J. Biol. Chem.* 249:2417
104. Tellez-Inon, M. T., Torres, H. N. 1970. *Proc. Nat. Acad. Sci. USA* 66:459
105. Rahmsdorf, H. J., Pai, S. H., Ponta, H., Herrlich, P., Roskoski, R. Jr., Schweiger, M., Studier, F. W. 1974. *Proc. Nat. Acad. Sci. USA* 71:586
106. Agabian, N., Rosen, O. M., Shapiro, L. 1972. *Biochem. Biophys. Res. Commun.* 49:1690
107. Rask, L., Wålinder, O., Zetterquist, O., Engström, L. 1970. *Biochim. Biophys. Acta* 221:107
108. Strand, M., August, J. T. 1971. *Nature New Biol.* 233:137
109. Hatanaka, M., Twiddy, E., Gilden, R. V. 1972. *Virology* 47:536
110. Randall, C. C., Rogers, H. W., Downer, D. N., Gentry, G. A. 1972. *J. Virol.* 9:216
111. Downer, D. N., Rogers, H. W., Randall, C. C. 1973. *Virology* 52:13
112. Paoletti, E., Moss, B. 1972. *J. Virol.* 10:417
113. Rosemond, H., Moss, B. 1973. *J. Virol.* 11:961
114. Sarov, I., Joklik, W. K. 1973. *Virology* 52:223
115. Kleiman, J., Moss, B. 1973. *J. Virol.* 12:684
116. Imblum, R. L., Wagner, R. R. 1974. *J. Virol.* 13:113
117. Sokol, F., Clark, H. F. 1973. *Virology* 52:246
118. Moyer, S. A., Summers, D. F. 1974. *J. Virol.* 13:455
119. Sokol, F., Clark, H. F. 1973. *Virology* 52:246
120. Tao, M., Doerfler, W. 1972. *Eur. J. Biochem.* 27:448
121. Silberstein, H., August, J. T. 1973. *J. Virol.* 12:511
122. Gravell, M., Cromeans, T. L. 1972. *Virology* 48:847
123. Tan, K. B., Sokol, F. 1972. *J. Virol.* 10:985
124. Tan, K. B., Sokol, F. 1974. *J. Virol.* 13:1245
125. Waite, M. R. F., Lubin, M., Jones, K. J., Bose, H. R. 1974. *J. Virol.* 13:244
126. Russell, W. C., Skehel, J. J., Machado, R., Pereira, H. G. 1972. *Virology* 50:931
127. Ziegel, R. F., Clark, H. F. 1969. *J. Nat. Cancer Inst.* 43:1097
128. Tan, K. B., Sokol, F. 1973. *J. Virol.* 12:696
129. Wålinder, O. 1972. *Biochim. Biophys. Acta* 258:411
130. Wålinder, O. 1973. *Biochim. Biophys. Acta* 293:140
131. Goldstein, J. L., Hasty, M. A. 1973. *J. Biol. Chem.* 248:6300
132. Rodnight, R., Lavin, B. E. 1966. *Biochem. J.* 93:84
133. Kleinsmith, L. J., Allfrey, V. G. 1969. *Biochim. Biophys. Acta* 175:123
134. Traugh, J. A., Mumby, M., Traut, R. R. 1973. *Proc. Nat. Acad. Sci. USA* 70:373
135. Bingham, E. W., Farrell, H. M. Jr. 1974. *J. Biol. Chem.* 249:3647
136. Schirm, J., Gruber, M., Ab, G. 1973. *FEBS Lett.* 30:167

137. Blum, H. E., Pocinwong, S., Fischer, E. H. 1974. *Proc. Nat. Acad. Sci. USA* 71:2198
138. Ruddon, R. W., Anderson, S. L. 1972. *Biochem. Biophys. Res. Commun.* 46: 1499
139. Smith, D. L., Bruegger, B. B., Halpern, R. M., Smith, R. A. 1973. *Nature* 246: 103
140. Kish, V. M., Kleinsmith, L. J. 1974. *J. Biol. Chem.* 249:750
141. Lake, R. S., Salzman, N. P. 1972. *Biochemistry* 11:4817
142. Lake, R. S. 1973. *J. Cell Biol.* 58:317
143. Schmidt, M. J., Sokoloff, L. 1973. *J. Neurochem.* 21:1193
144. Ralph, R. K., McCombs, P. J. A., Tener, G., Wojcik, S. J. 1972. *Biochem. J.* 130: 901
145. Soifer, D., Laszlo, A. H., Scotto, J. M. 1972. *Biochim. Biophys. Acta* 271:182
146. Eipper, B. A. 1974. *J. Biol. Chem.* 249: 1398
147. Inoue, Y., Yamamura, H., Nishizuka, Y. 1973. *Biochem. Biophys. Res. Commun.* 50:228
148. Eipper, B. A. 1974. *J. Biol. Chem.* 249: 1407
149. Krebs, E. G., Fischer, E. H. 1956. *Biochim. Biophys. Acta* 20:150
150. Nolan, E., Novoa, W. B., Krebs, E. G., Fischer, E. H. 1966. *Biochemistry* 3:542
151. Seery, V. L., Fischer, E. H., Teller, D. C. 1967. *Biochemistry* 6:3315
152. Seery, V. L., Fischer, E. H., Teller, D. C. 1970. *Biochemistry* 9:3591
153. Brostrom, C. O., Hunkeler, F. L., Krebs, E. G. 1971. *J. Biol. Chem.* 246: 1961
153a. Meyer, W. L., Fischer, E. H., Krebs, E. G. 1964. *Biochemistry* 3:1033
153b. Ozawa, E., Hosoi, K., Ebashi, S. 1967. *J. Biochem.* 61:531
154. Krebs, E. G., Fischer, E. H. 1962. *Advan. Enzymol.* 24:263
154a. Graves, D. J., Wang, J. H. 1972. *Enzymes* 7:435
155. Cori, C. F. 1956. In *Enzymes: Units of Biological Structure and Function,* ed. O. H. Gaebler, 573. New York: Academic
156. Sutherland, E. W. 1951. *Symp. Phosphorus Metab.* 1:53
157. Hayakawa, T., Perkins, J. P., Walsh, D. A., Krebs, E. G. 1973. *Biochemistry* 12:567
158. Cohen, P. 1973. *Eur. J. Biochem.* 34:1
159. DeLange, R. J., Kemp, R. G., Riley, W. D., Cooper, R. A., Krebs, E. G. 1968. *J. Biol. Chem.* 243:2200
160. Hayakawa, T., Perkins, J. P., Krebs, E. G. 1973. *Biochemistry* 12:574
161. Krebs, E. G., Graves, D. J., Fischer, E. H. 1959. *J. Biol. Chem.* 234:2867
162. Krebs, E. G., Love, D. S., Bratvold, G. E., Trayser, K. A., Meyer, W. L., Fischer, E. H. 1964. *Biochemistry* 3: 1022
163. Riley, W. D., DeLange, R. J., Bratvold, G. E., Krebs, E. G. 1968. *J. Biol. Chem.* 243:2209
164. Walsh, D. A., Perkins, J. P., Krebs, E. G. 1968. *J. Biol. Chem.* 243:3763
165. Soderling, T. R., Hickenbottom, J. P., Reimann, E. M., Hunkeler, F. L., Walsh, D. A., Krebs, E. G. 1970. *J. Biol. Chem.* 245:6317
166. Cohen, P., Antoniw, J. F. 1973. *FEBS Lett.* 34:43
167. Stull, J. T., Mayer, S. E. 1971. *J. Biol. Chem.* 246:5716
168. Mayer, S. E., Krebs, E. G. 1970. *J. Biol. Chem.* 245:3153
169. Cohen, P. 1973. *Biochem. Soc. Symp.* 39:51
170. Meyer, F., Heilmeyer, L. M. G., Haschke, R. H., Fischer, E. H. 1970. *J. Biol. Chem.* 245:6642
171. Heilmeyer, L. M. G., Meyer, F., Haschke, R. H., Fischer, E. H. 1970. *J. Biol. Chem.* 245:6649
172. Huston, R. B., Krebs, E. G. 1968. *Biochemistry* 7:2116
173. Graves, D. J., Hayakawa, T., Horvitz, R. A., Beckman, E., Krebs, E. G. 1973. *Biochemistry* 12:580
174. Ebashi, S., Kodama, A. 1966. *J. Biochem.* 60:733
175. Katz, A. M. 1970. *Physiol. Rev.* 50:63
176. Greaser, M. L., Gergely, J. 1971. *J. Biol. Chem.* 246:4226
177. Hartshorne, D. J., Drizen, P. 1973. *Cold Spring Harbor Symp. Quant. Biol.* 37: 225
178. Greaser, M. L., Yamaguchi, M., Brekke, C., Potter, J., Gergely, J. 1973. *Cold Spring Harbor Symp. Quant. Biol.* 37: 235
179. Perry, S. V., Cole, M. A., Head, J. F., Wilson, F. J. 1973. *Cold Spring Harbor Symp. Quant. Biol.* 37:251
180. Ebashi, S., Ohtsuki, I., Mikoshi, K. 1973. *Cold Spring Harbor Symp. Quant. Biol.* 37:215
181. Stull, J. T., Brostrom, C. O., Krebs, E. G. 1972. *J. Biol. Chem.* 247:5272
182. Perry, S. V., Cole, M. A. 1973. *Biochem. J.* 131:425
183. England, P. J., Stull, J. T., Huang, T. S., Krebs, E. G. 1974. In *Metabolic Interconversion of Enzymes,* ed. E. H. Fischer, E. G. Krebs, H. Neurath, E. R. Stadtman, 175. Berlin: Springer
184. England, P. J., Stull, J. T., Krebs, E. G.

1972. *J. Biol. Chem.* 247:5275
185. Ebashi, S., Endo, M., Ohtsuki, I. 1969. *Quart. Rev. Biophys.* 2:351
186. Bailey, C., Villar-Palasi, C. 1971. *Fed. Proc.* 30:1147
187. Pratje, E., Heilmeyer, L. M. G. 1972. *FEBS Lett.* 27:89
188. Lyon, J. B. 1970. *Biochem. Genet.* 4:169
189. Lyon, J. B., Porter, J. 1963. *J. Biol. Chem.* 238:1
190. Cohen, P. T. W., Cohen, P. 1973. *FEBS Lett.* 29:113
191. Gross, S. R., Mayer, S. E. 1973. *Biochem. Biophys. Res. Commun.* 54:823
192. Danforth, W. H., Lyon, J. B. 1964. *J. Biol. Chem.* 239:4047
193. Ullmann, A., Vagelos, P. R., Monod, J. 1964. *Biochem. Biophys. Res. Commun.* 17:86
194. Stalmans, W., Hers, G. 1973. *Enzymes* 9:309
195. Brown, N. E., Larner, J. 1971. *Biochim. Biophys. Acta* 242:69
196. Smith, C. H., Brown, N. E., Larner, J. 1971. *Biochim. Biophys. Acta* 242:81
197. Friedman, D. L., Larner, J. 1963. *Biochemistry* 2:669
198. Rosell-Perez, M., Larner, J. 1964. *Biochemistry* 3:75
199. Schlender, K. K., Wei, S. H., Villar-Palasi, C. 1969. *Biochim. Biophys. Acta* 191:272
200. Larner, J., Sanger, F. 1965. *J. Mol. Biol.* 11:491
201. Hizukuri, S., Larner, J. 1964. *Biochemistry* 3:1783
202. Bishop, J. S. 1970. *Biochim. Biophys. Acta* 208:208
203. DeWulf, H., Stalmans, W., Hers, H. G. 1970. *Eur. J. Biochem.* 15:1
204. Stalmans, W., DeWulf, H., Hers, H. G. 1971. *Eur. J. Biochem.* 18:582
205. Zieve, F. J., Glinsmann, W. H. 1973. *Biochem. Biophys. Res. Commun.* 50:872
206. Butcher, R. W., Sneyd, J. G. T., Park, C. R., Sutherland, E. W. 1966. *J. Biol. Chem.* 241:1651
207. Manganiello, V. C., Murad, F., Vaughan, M. 1971. *J. Biol. Chem.* 246:2195
208. Jefferson, L. S., Exton, J. H., Butcher, R. W., Sutherland, E. W., Park, C. R. 1968. *J. Biol. Chem.* 243:1031
209. Exton, J. H., Lewis, S. B., Ho, R. J., Robison, G. A., Park, C. R. 1971. *Ann. NY Acad. Sci.* 185:85
210. Kato, K., Bishop, J. S. 1972. *J. Biol. Chem.* 247:7420
211. Graves, D. J., Fischer, E. H., Krebs, E. G. 1960. *J. Biol. Chem.* 235:805

212. Gratecos, D., Detwiler, T., Fischer, E. H. See Ref. 183, 43
213. Haschke, R. H., Heilmeyer, L. G., Meyer, F., Fischer, E. H. 1970. *J. Biol. Chem.* 245:6657
214. Danforth, W. H., Helmreich, E., Cori, C. F. 1962. *Proc. Nat. Acad. Sci. USA* 48:1191
215. Steinberg, D. 1973. *Protein Phosphorylation in Control Mechanisms,* ed F. Huijing, E. Y. C. Lee, 47. New York: Academic
216. Reed, L. J., Pettit, F. H., Roche, T. E., Butterworth, P. J., Barrera, C. R., Tsai, C. S. See Ref. 183, 99
217. Linn, T. C., Pettit, F. H., Reed, L. J. 1969. *Proc. Nat. Acad. Sci. USA* 62:234
218. Wieland, O. H., Hartmann, U., Siess, E. A. 1972. *FEBS Lett.* 27:240
219. Martelo, O. J., Hirsch, J. 1974. *Biochem. Biophys. Res. Commun.* 58:1008
220. Jungmann, R. A., Hiestand, P. C., Schweppe, J. S. 1974. *J. Biol. Chem.* 249:5444
221. Lee, K.-L., Nickol, J. M. 1974. *J. Biol. Chem.* 249:6024
222. Narumi, S., Miyamoto, E. 1974. *Biochim. Biophys. Acta* 350:215
223. Carlson, C. A., Kim, K. H. 1973. *J. Biol. Chem.* 248:378
224. Niles, E. G., Westhead, E. W. 1973. *Biochemistry* 12:1723
225. Kuo, C.-H., August, J. T. 1972. *Nature New Biol.* 237:105
226. Martelo, O. J. 1973. In "Protein Phosphorylation in Control Mechanisms," *Miami Winter Symp.* 5:199
227. Fakunding, J. L., Traugh, J. A., Traut, R. R., Hershey, J. W. B. 1972. *J. Biol. Chem.* 247:6365
228. Kinzel, V., Mueller, G. C. 1973. *Biochim. Biophys. Acta* 322:337
229. Brooks, J. C., Siegel, F. L. 1973. *J. Biol. Chem.* 248:4189
230. Perrie, W. T., Smillie, L. B., Perry, S. V. 1972. *Cold Spring Harbor Symp. Quant. Biol.* 37:17
231. Perrie, W. T., Smillie, L. B., Perry, S. V. 1973. *Biochem. J.* 135:151
232. Adelstein, R. S., Conti, M. A., Anderson, W. Jr. 1973. *Proc. Nat. Acad. Sci. USA* 70:3115
233. Levitan, I. B., Barondes, S. H. 1974. *Proc. Nat. Acad. Sci. USA* 71:1145
234. Benjamin, W. B., Singer, I. 1974. *Biochim. Biophys. Acta* 351:28
235. Kuhn, N. J. 1972. *J. Endocrinol.* 55:219
236. Morrill, G. A., Murphy, J. B. 1972. *Nature* 238:282
237. Allfrey, V. G., Johnson, E. M., Karn,

J., Vidali, G. 1973. In "Protein Phosphorylation in Control Mechanisms," *Miami Winter Symp.* 5:217

238. Dixon, G. H., Candido, E. P. M., Louie, A. J. Ibid, 5:279

239. Chalkley, R., Balhorn, R., Oliver, D., Granner, D. Ibid, 5:251

240. Langan, T. A. Ibid, 5:287

241. Ord, M. G., Stocken, L. A. 1966. *Biochemistry* 98:888

242. Kleinsmith, L. J., Allfrey, V. G., Mirsky, A. E. 1966. *Proc. Nat. Acad. Sci. USA* 55:1182

243. Louie, A. J., Candido, E. P. M., Dixon, G. H. 1973. *Cold Spring Harbor Symp. Quant. Biol.* 38:803

244. Johns, E. W., Butler, J. A. V. 1962. *Biochem. J.* 82:15

245. Langan, T. A. 1969. *Proc. Nat. Acad. Sci. USA* 64:1276

246. Langan, T. A. 1968. In *Some Regulatory Mechanisms for Protein Synthesis in Mammalian Cells,* ed. A. San Pietro, M. Lamberg, F. T. Kenney, 101. New York: Academic

247. Langan, T. A. 1968. *Science* 162:579

248. Meisler, M. H., Langan, T. A. 1969. *J. Biol. Chem.* 244:4961

249. Maeno, H., Greengard, P. 1972. *J. Biol. Chem.* 247:3269

250. Langan, T. A. 1971. *Fed. Proc.* 30:1089A

251. Hnilica, L. S. 1972. *The Structure and Biological Functions of Histones,* 84. Cleveland: CRC

252. Gurley, L. R., Walters, R. A. 1973. *Biochem. Biophys. Res. Commun.* 55:697–703

253. Tanphaichitr, N., Balhorn, R., Granner, D., Chalkley, R. 1974. *Biochemistry* 13:4245–54

254. Balhorn, R., Bordwell, J., Sellers, L., Granner, D., Chalkley, R. 1972. *Biochem. Biophys. Res. Commun.* 46:1326

255. Gurley, L. R., Walters, R. A., Tobey, R. A. 1973. *Arch. Biochem. Biophys.* 154:212–18

256. Marks, D. B., Paik, W. K., Borun, T. W. 1973. *J. Biol. Chem.* 248:5660

257. Gurley, L. R., Walters, R. A., Tobey, R. A. 1973. *Biochem. Biophys. Res. Commun.* 50:744

258. Lake, R. S. 1973. *Nature New Biol.* 242:145

259. Oliver, D., Balhorn, R., Granner, D., Chalkley, R. 1972. *Biochemistry* 11:3921

260. Sherod, D., Johnson, G., Chalkley, R. 1970. *Biochemistry* 9:4611

261. Cross, M. E., Ord, M. G. 1970. *Biochem. J.* 118:191

262. Cross, M. E., Ord, M. G. 1971. *Biochem. J.* 124:241

263. Shepherd, G. R., Noland, B. J., Hardin, J. M. 1971. *Arch. Biochem. Biophys.* 142:299

264. Gurley, L. R., Walters, R. A. 1971. *Biochemistry* 10:1588

265. Lake, R. S., Goidl, J. A., Salzman, N. P. 1972. *Exp. Cell Res.* 73:113

266. Balhorn, R., Tanphaichitr, N., Chalkley, R., Granner, D. 1973. *Biochemistry* 12:5146–50

267. Balhorn, R., Bordwell, J., Sellers, L., Granner, D., Chalkley, R. 1972. *Biochem. Biophys. Res. Commun.* 46:1326–33

268. Bradbury, E. M., Inglis, R. J., Matthews, H. R. 1974. *Nature* 247:257–61

269. Balhorn, R., Chalkley, R., Granner, D. 1972. *Biochemistry* 11:1094

270. Oliver, D., Balhorn, R., Granner, D., Chalkley, R. 1972. *Biochemistry* 11:3921–25

271. Bradbury, E. M., Inglis, R. J., Matthews, H. R., Sarner, N. 1973. *Eur. J. Biochem.* 33:131

272. Adler, A. J., Schaffhausen, B., Langan, T. A., Fasman, G. D. 1971. *Biochemistry* 10:909

273. Siebert, G., Ord, M. G., Stocken, L. A. 1971. *Biochem. J.* 122:721

274. Sung, M. T., Dixon, G. H., Smithies, O. 1971. *J. Biol. Chem.* 246:1358

275. Gutierrez-Cernosek, R. M., Hnilica, L. S. 1971. *Biochim. Biophys. Acta* 247:348

276. Olson, M. O. J., Orrick, L. R., Jones, C., Busch, H. 1974. *J. Biol. Chem.* 249:2823

277. Louie, A. J., Dixon, G. H. 1972. *J. Biol. Chem.* 247:5498

278. Marushige, K., Ling, V., Dixon, G. H. 1969. *J. Biol. Chem.* 244:5935

279. Sung, M. T., Dixon, G. H. 1970. *Proc. Nat. Acad. Sci. USA* 67:1616

280. Teng, C. S., Teng, C. T., Allfrey, V. G. 1971. *J. Biol. Chem.* 246:3597

281. Johnson, E. M., Allfrey, V. G. 1972. *Arch. Biochem. Biophys.* 152:786

282. Vidali, G., Boffa, L. C., Littau, V. C., Allfrey, K. M., Allfrey, V. G. 1973. *J. Biol. Chem.* 248:4065

283. Johnson, E. M., Vidali, G., Littau, V. C., Allfrey, V. G. 1973. *J. Biol. Chem.* 248:7595

283a. Garrard, W. T., Pearson, W. R., Wake, S. K., Bonner, J. 1974. *Biochem. Biophys. Res. Commun.* 58:50

283b. MacGillivray, A. J., Rickwood, D. 1974. *Eur. J. Biochem.* 41:181

284. Stein, G., Baserga, R. 1972. *Advan. Cancer Res.* 15:287

285. Stein, G., Spelsberg, T. C., Kleinsmith,

L. 1974. *Science* 183:817
286. Langan, T. A. 1967. In *Regulation of Nucleic Acid and Protein Biosynthesis*, ed. V. V. Konigsberger, L. Bosch, 233. Amsterdam: Elsevier
287. Kleinsmith, L. J. 1973. *J. Biol. Chem.* 248:5648
288. Allfrey, V. G., Inoue, A., Karn, J., Johnson, E. M., Vidali, G. 1973. *Cold Spring Harbor Symp. Quant. Biol.* 38:785
289. Kamiyama, M., Dastugue, B., Kruh, J. 1971. *Biochem. Biophys. Res. Commun.* 44:1345
290. Shea, M., Kleinsmith, L. J. 1973. *Biochem. Biophys. Res. Commun.* 50:473
291. Kostraba, N. C., Wang, T. Y. 1972. *Biochim. Biophys. Acta* 262:169
292. Takeda, M., Yamamura, H., Ohga, Y. 1971. *Biochem. Biophys. Res. Commun.* 42:103
293. Desjardins, P. R., Lue, P. F., Liew, C. C., Gornall, A. G. 1972. *Can. J. Biochem.* 50:1249
294. Kamiyama, M., Dastugue, B. 1971. *Biochem. Biophys. Res. Commun.* 44:29
295. Kamiyama, M., Dastugue, B., Defer, N., Kruh, J. 1972. *Biochim. Biophys. Acta* 277:576
296. Rickans, L. E., Ruddon, R. W. 1973. *Biochem. Biophys. Res. Commun.* 54:387
297. Hiestand, P. C., Eppenberger, U., Jungmann, R. A. 1973. *Endocrinology* 93:217
298. Jungmann, R. A., Hiestand, P. C., Schwepp, J. S. 1974. *Endocrinology* 94:168
299. O'Malley, B. W., Means, A. R. 1974. *Science* 183:610
300. Paul, J., More, I. R. 1972. *Nature New Biol.* 239:134
301. Vidali, G., Boffa, L., Allfrey, V. G. 1972. *J. Biol. Chem.* 247:7365
302. Karn, J., Johnson, E. M., Vidali, G., Allfrey, V. G. 1974. *J. Biol. Chem.* 249:667
303. Platz, R. D., Stein, G. S., Kleinsmith, L. J. 1973. *Biochem. Biophys. Res. Commun.* 51:735
304. Pfeiffer, S. E., Tolmach, L. J. 1968. *J. Cell Physiol.* 71:71
305. Farber, J., Stein, G., Baserga, R. 1972. *Biochem. Biophys. Res. Commun.* 47:790
306. Johnson, T. C., Holland, J. J. 1965. *J. Cell Biol.* 27:565
307. Zeilig, C. E., Johnson, R. A., Friedman, D. L., Sutherland, E. W. 1972. *J. Cell Biol.* 55:296a
308. Gershey, E. L., Kleinsmith, L. J. 1969. *Biochim. Biophys. Acta* 194:331

309. Heald, P. J. 1960. *Phosphorous Metabolism of Brain.* Oxford: Pergamon
310. Heald, P. J. 1958. *Biochem. J.* 68:580
311. Jones, D. A., Rodnight, R. 1971. *Biochem. J.* 121:597
312. Heald, P. J. 1961. *Biochem. J.* 78:340
313. Trevor, A. J., Rodnight, R., Schwartz, A. 1965. *Biochem. J.* 95:883
314. Weller, M., Rodnight, R. 1971. *Biochem. J.* 124:393
315. Trevor, A. J., Rodnight, R. 1965. *Biochem. J.* 95:889
316. Rodnight, R. 1971. *Handb. Neurochem.* 5:141
317. Rodnight, R., Lavin, B. E. 1964. *Biochem. J.* 93:84
318. Weller, M., Rodnight, R. 1970. *Nature* 225:187
319. Rose, S. P. R., Heald, P. J. 1961. *Biochem. J.* 81:339
320. Mäenpää, P. H., Bernfield, M. R. 1970. *Proc. Nat. Acad. Sci. USA* 67:688
321. Bernfield, M. R., Mäenpää, P. H. 1971. *Cancer Res.* 31:684
322. Hatfield, D., Portugal, F. H., Caicuts, M. 1971. *Cancer Res.* 31:697
323. Sanger, F., Hocquard, E. 1962. *Biochim. Biophys. Acta* 62:606
324. Turkington, R. W., Topper, Y. 1966. *Biochim. Biophys. Acta* 127:366
325. Johnson, E. M., Maeno, H., Greengard, P. 1971. *J. Biol. Chem.* 246:7731
326. Maeno, H., Johnson, E. M., Greengard, P. 1971. *J. Biol. Chem.* 246:134
327. Shapiro, A. L., Vinuela, E., Maizel, J. V. 1967. *Biochem. Biophys. Res. Commun.* 28:815
328. Weber, K., Osborn, M. 1969. *J. Biol. Chem.* 244:4406
329. Fairbanks, G., Steck, T. L., Wallach, D. F. H. 1971. *Biochemistry* 10:2606
330. Johnson, E. M., Ueda, T., Maeno, H., Greengard, P. 1972. *J. Biol. Chem.* 247:5650
331. Ehrlich, Y., Routtenberg, A. 1974. *FEBS Lett.* 45:237
332. Kuo, J. F., Krueger, B. K., Sanes, J. R., Greengard, P. 1970. *Biochim. Biophys. Acta* 212:79
333. Roses, A. D., Appel, S. H. 1973. *J. Biol. Chem.* 248:1408
334. Casnellie, J. E., Greengard, P. 1974. *Proc. Nat. Acad. Sci. USA* 71:1891
335. DeLorenzo, R. J., Walton, K. G., Curran, P. F., Greengard, P. 1973. *Proc. Nat. Acad. Sci. USA* 70:880
336. DeLorenzo, R. J., Greengard, P. 1973. *Proc. Nat. Acad. Sci. USA* 70:1831
337. Greengard, P., McAfee, D. A., Kebabian, J. W. 1972. *Advan. Cyclic Nucleotide Res.* 1:337
338. Greengard, P., Kebabian, J. W. 1974.

Fed. Proc. 33 : 1059
339. Kalix, P., McAfee, D. A., Shorderet, M., Greengard, P. 1974. J. Pharmacol. Exp. Ther. 188 : 676
340. Kuo, J. F., Greengard, P. 1969. Proc. Nat. Acad. Sci. USA 64 : 1349
341. De Robertis, E., Rodriguez de Lores Arnaiz, Alberici, M., Butcher, R. W., Sutherland, E. W. 1967. J. Biol. Chem. 242 : 3487
342. Gilman, A. G., Nirenberg, M. 1971. Proc. Nat. Acad. Sci. USA 68 : 2165
343. Schimmer, B. P. 1971. Biochim. Biophys. Acta 252 : 567
344. Perkins, J. P., Macintyre, E. H., Riley, W. D., Clark, R. B. 1971. Life Sci. 10 : 1069
345. Jard, S., Premont, J., Benda, P. 1972. FEBS Lett. 26 : 344
346. Miyamoto, E., Kakiuchi, S. 1974. J. Biol. Chem. 249 : 2769
347. Steck, A. J., Appel, S. H. 1974. J. Biol. Chem. 249 : 5416
348. Oshiro, Y., Eylar, E. H. 1970. Arch. Biochem. Biophys. 138 : 392
349. Arden, G. B. 1969. Progr. Biophys. Mol. Biol. 19 : 371
350. Kühn, H., Dreyer, W. J. 1972. FEBS Lett. 20 : 1
351. Bownds, D., Dawes, J., Miller, I., Stahlman, M. 1972. Nature New Biol. 237 : 125
352. Kühn, H., Cook, J. H., Dreyer, W. J. 1973. Biochemistry 12 : 2495
353. Frank, R. N., Cavanagh, H. D., Kenyon, K. R. 1973. J. Biol. Chem. 248 : 596
354. Rubin, C. S., Erlichman, J., Rosen, O. M. 1972. J. Biol. Chem. 247 : 6135
355. Marchesi, V. T., Steers, E. 1968. Science 159 : 203
356. Marchesi, V. T., Steers, E., Tillack, T. W., Marchesi, S. L. 1969. In Red Cell Membrane Structure and Function, ed. G. A. Jamieson, T. J. Greenwalt, 117. Philadelphia : Lippincott
357. Marchesi, S. L., Steers, E., Marchesi, V. T., Tillack, T. W. 1970. Biochemistry 9 : 50
358. Rosenthal, A. S., Kregenow, F. M., Moses, H. L. 1970. Biochim. Biophys. Acta 196 : 254
359. Greenquist, A. C., Shohet, S. B. 1973. Blood 42 : 997
360. Bretscher, M. S. 1971. J. Mol. Biol. 59 : 351
361. Bretscher, M. S. 1971. Nature New Biol. 231 : 229
362. Kant, J. A., Steck, T. L. 1973. J. Biol. Chem. 248 : 8457
363. Segrest, J. P., Kahane, I., Jackson, R. L., Marchesi, V. T. 1973. Arch. Biochem.

Biophys. 155 : 167
364. Hamaguchi, H., Cleve, H. 1972. Biochem. Biophys. Res. Commun. 47 : 459
365. Rubin, C. S. In preparation
366. Guthrow, C. E., Rasmussen, H., Brunswick, D. J., Cooperman, B. S. 1973. Proc. Nat. Acad. Sci. USA 70 : 3344
367. Brunswick, D. J., Cooperman, B. S. 1971. Proc. Nat. Acad. Sci. USA 68 : 1801
368. Yu, J., Fischman, D. A., Steck, T. L. 1973. J. Supramolecular Struct. 1 : 233
369. Kant, J. A., Steck, T. L. 1972. Nature New Biol. 240 : 26
370. Steck, T. L., Kant, J. A. 1974. Methods Enzymol. 31 : 172
371. Rubin, C. S., Rosenfeld, R. D., Rosen, O. M. 1973. Proc. Nat. Acad. Sci. USA 70 : 3735
372. Kant, J. A., Steck, T. L. 1973. Biochem. Biophys. Res. Commun. 54 : 116
373. Nicholson, G. L., Marchesi, V. T., Singer, S. J. 1971. J. Cell Biol. 51 : 265
374. Roses, A. D., Appel, S. H. 1973. Proc. Nat. Acad. Sci. USA 70 : 1855
375. Palmer, F. B. St. C., Verpoorte, J. A. 1971. Can. J. Biochem. 49 : 337
376. Duffy, M. J., Schwarz, V. 1974. Biochem. Pharmacol. 23 : 2464
377. Sheppard, H., Burghardt, C. 1970. Mol. Pharmacol. 6 : 425
378. Allen, J. E., Rasmussen, J. 1971. Science 174 : 512
379. Gardner, J. D., Klaeveman, H. L., Bilezikian, J. P., Aurbach, G. D. 1973. J. Biol. Chem. 248 : 5590
380. Rudolph, S. A., Greengard, P. 1974. J. Biol. Chem. 249 : 5684
381. Chang, K.-J., Cuatrecasas, P. 1974. J. Biol. Chem. 249 : 3170
382. Wray, H. L., Gray, R. R., Olsson, R. 1973. J. Biol. Chem. 248 : 1496
383. La Raia, P. J., Morkin, E. 1974. Circ. Res. 35 : 298
384. Kirchberger, M. A., Tada, M., Repke, D. I., Katz, A. 1972. J. Mol. Cell Cardiol. 4 : 673
385. Andrew, C. G., Roses, A. D., Almon, R. R., Appel, S. H. 1973. Science 182 : 927
386. Sulakhe, P. V., Drummond, G. I. 1974. Arch. Biochem. Biophys. 161 : 448
387. Sulakhe, P. V., Drummond, G. I., Ng, D. C. 1973. J. Biol. Chem. 248 : 4150
388. Lemay, A., Deschenes, M., Lemaire, S., Poirier, G., Poulin, L., Labrie, F. 1974. J. Biol. Chem. 249 : 323
389. Labrie, F., Lemaire, S., Poirier, G., Pelletier, G., Boucher, R. 1971. J. Biol. Chem. 246 : 7311
390. Bauduin, H., Rochus, L., Vincent, D.,

Dumont, J. E. 1971. *Biochim. Biophys. Acta* 252:171

391. Rutten, W. J., De Pont, J. J., Bonting, S. L. 1972. *Biochim. Biophys. Acta* 274:201

392. Case, R. M., Scratcherd, T. 1972. *J. Physiol.* 223:649

393. Benz, L.., Eckstein, B., Matthews, E. K., Williams, J. A. 1972. *Brit. J. Pharmacol.* 46:66

394. Lambert, M., Camus, J., Christophe, J. 1973. *Biochem. Biophys. Res. Commun.* 52:935

395. Orloff, J., Handler, J. S. 1967. *Am. J. Med.* 42:757

396. Grantham, J. J., Burg, M. B. 1966. *Am. J. Physiol.* 211:255

397. Leaf, A. 1967. *Am. J. Med.* 42:745

398. Ganote, C. E., Grantham, J. J., Moses, H. L., Burg, M. B., Orloff, J. 1968. *J. Cell Biol.* 36:355

399. Schwartz, I. L., Schlatz, L. J., Kinne-Saffron, E., Kinne, R. 1974. *Proc. Nat. Acad. Sci. USA* 71:2595

400. Schlatz, L. J., Kinne, R., Kinne-Saffron, E., Schwartz, I. L. 1973. *Physiologist* 16:451

401. Wallach, D. F. H., Lin, P. S. 1973. *Biochim. Biophys. Acta* 300:211

402. Steck, T. L. 1972. In *Membrane Molecular Biology*, ed. C. F. Fox, A. Keith, 76. Stamford, Conn.: Sinauer Assoc.

403. Constantopoulos, A., Najjar, V. A. 1973. *Biochem. Biophys. Res. Commun.* 53:794

404. Layne, P., Constantopoulos, A., Judge, J. F. X., Rauner, R., Najjar, V. A. 1973. *Biochem. Biophys. Res. Commun.* 53:800

405. Najjar, V. A., Constantopoulos, A. 1973. *Mol. Cell. Biochem.* 2:87

406. Lee, T. P., Kuo, J.-F., Greengard, P. 1972. *Proc. Nat. Acad. Sci. USA* 69:3287

407. Schultz, G., Hardman, J. G., Schultz, K., Davis, J. W., Sutherland, E. W. 1973. *Proc. Nat. Acad. Sci. USA* 70:1721

408. Loeb, J. E., Blat, C. 1970. *FEBS Lett.* 10:105

409. Eil, C., Wool, I. G. 1971. *Biochem. Biophys. Res. Commun.* 43:1001

410. Kabat, D. 1971. *Biochemistry* 10:197

411. Li, C., Amos, H. 1971. *Biochem. Bio-phys. Res. Commun.* 45:1398

412. Walton, G. M., Gill, G. N. 1973. *Biochemistry* 12:2604

413. Fontana, J. A., Picciano, D., Lovenberg, W. 1972. *Biochem. Biophys. Res. Commun.* 49:1225

414. Pavlovic-Hournac, M., Delbouffe, D., Virion, A., Nunez, J. 1973. *FEBS Lett.* 33:65

415. Jergil, B. 1972. *Eur. J. Biochem.* 28:546

416. Jergil, B., Ohlsson, R. 1974. *Eur. J. Biochem.* 46:13

417. Eil, C., Wool, I. 1973. *J. Biol. Chem.* 248:5122

418. Traugh, J. A., Traut, R. R. 1974. *J. Biol. Chem.* 249:1207

419. Gill, G. N., Garren, L. D. 1971. *Proc. Nat. Acad. Sci. USA* 68:786

420. Ventimiglia, F. A., Wool, I. G. 1974. *Proc. Nat. Acad. Sci. USA* 71:350

421. Walton, G. M., Gill, G. N., Abrass, I. B., Garren, L. D. 1971. *Proc. Nat. Acad. Sci. USA* 68:880

422. Eil, C., Wool, I. G. 1973. *J. Biol. Chem.* 248:5130

423. Kabat, D. 1970. *Biochemistry* 9:4160

424. Gordon, J. 1971. *Biochem. Biophys. Res. Commun.* 44:579

425. Blat, C., Loeb, J. 1971. *FEBS Lett.* 18:124

426. Correze, G., Pinell, P., Nunez, J. 1972. *FEBS Lett.* 23:87

427. Wicks, W. D., Kenney, F. T., Lee, K. L. 1970. *J. Biol. Chem.* 244:6008

428. Jost, J. P., Hsil, A. W., Rickenberg, H. V. 1969. *Biochem. Biophys. Res. Commun.* 34:748

429. Tata, J. R. 1964. In *Action of Hormones on Molecular Processes*, ed. G. Litwack, D. Kritchevsky, 58. New York: Wiley

430. Kabat, D. 1972. *J. Biol. Chem.* 247:5338

431. Cawthon, M. L., Bitte, L., Krystosek, A., Kabat, D. 1974. *J. Biol. Chem.* 249:275

432. Bitte, L., Kabat, D. 1972. *J. Biol. Chem.* 247:5345

433. Joklik, W. K., Becker, Y. 1965. *J. Mol. Biol.* 13:496

434. Girard, M., Latham, H., Penman, S., Darnell, J. E. 1965. *J. Mol. Biol.* 11:187

435. Kabat, D., Rich, A. 1969. *Biochemistry* 8:3742

CHEMICAL STUDIES OF ENZYME ACTIVE SITES

David S. Sigman
Department of Biological Chemistry and Molecular Biology Institute,
University of California at Los Angeles School of Medicine,
Los Angeles, California 90024

Gregory Mooser
Department of Biochemistry, University of Southern California
School of Dentistry, Los Angeles, California 90033

CONTENTS

INTRODUCTION

The primary focus of the present review will be the use of chemical modification of proteins as a technique for investigating enzyme mechanism. Although chemical modification has historically played a central role in understanding biological catalysis, the recent contributions of X-ray crystallography, magnetic resonance techniques, stereochemistry, and kinetics have tended to dilute the reliance on chemical modification in the elucidation of enzymic mechanisms. However, in many cases, chemical modification still provides the best available evidence for defining the structure of essential intermediates or identifying catalytically important residues. In all cases, chemical modification can confirm suggestions originated by other techniques of mechanistic investigation.

There are several different types of chemical modification experiments that have revealed the presence and/or function of reactive groups at the active sites of enzymes. They include the use of group-specific reagents, affinity labels, pseudosubstrates, suicide substrates, the trapping of reactive intermediates, deletion of amino acid residues from the carboxyl or amino termini, and the chemical synthesis of either peptide fragments or entire proteins in which an amino acid of interest is replaced by another at a given position in the sequence.

The intrinsic information content of these approaches varies. For example, the trapping of reactive intermediates probably provides the greatest insight into a catalytic mechanism in the absence of a large body of information regarding the mechanism of action of the target enzyme. Reactions of pseudosubstrates and suicide substrates also possess high inherent information content. Pseudosubstrates react with their target enzyme via mechanisms with strong homologies to the enzyme-catalyzed reaction. Suicide substrates, which like pseudosubstrates do not react readily with free amino acids, are transformed by the enzyme's catalytic site into highly reactive species which then modify active site amino acid residues. On the other hand, site-specific modifications with group-specific reagents or affinity labels often do not provide unambiguous information by themselves as to the mechanism of catalysis. Even though numerous examples of stoichiometric modifications of active sites by group-specific reagents and affinity labels are available, in many cases it is still difficult to determine if the residue modified is involved in catalysis. The loss of biological activity is often not due to alteration of a vital catalytic group. It can be due to either the disruption of the conformation of the active site or steric hindrance resulting from the introduction of a bulky substituent. But in the context of an X-ray structure, for example, results from these approaches to the chemical modification of proteins can provide invaluable insights.

Recent contributions of various chemical modifications to the elucidation or confirmation of the mechanism of action of different enzymes will be discussed below. By necessity, coverage cannot be exhaustive. The most glaring area of omission is the many important and exciting contributions in the study of NAD/NADP-dependent dehydrogenases. An attempt has been made to avoid treatment of subjects included in two recent excellent reviews of enzyme mechanisms in this series (1, 2). Many general and laboratory-oriented monographs and reviews on the chemical modification of proteins are available (3–20).

OXIDOREDUCTASES

Lactate Dehydrogenase and Lactate Oxidase

Several irreversible inhibitors of flavoenzymes have been recently studied in which the enzyme transforms a relatively unreactive molecule into a potent inhibitor via a mechanism closely analogous to that occurring in the normal catalytic process. A full understanding of the mechanism of action of these suicide substrates on their target enzymes promises to reveal important features of these enzymes' catalytic mechanism and at the same time permit the design of extraordinarily specific inhibitors which might be of pharmacological value (21).

Three different enzymes that catalyze the oxidation of lactate, bakers' yeast L-(+)-lactate dehydrogenase (cytochrome b_2) (22), D-lactate dehydrogenase from *Escherichia coli* (23), and L-lactate oxidase from *Mycobacterium smegmatis* (24) are inhibited by a racemic mixture of 2-hydroxy-3-butynoate (I). Only the enantiomeric

$$H - C \equiv C - \overset{\overset{\displaystyle HO}{|}}{\underset{\underset{\displaystyle H}{|}}{C}} - \overset{\overset{\displaystyle O}{\|}}{C} - O^{\ominus}$$

I

form of I that is consistent with the stereochemistry of the target enzyme serves as a specific inhibitor. Lactate oxidase from *M. smegmatis,* which catalyzes the reaction summarized in equation 1, is inactivated by I under either aerobic or

$$R - \overset{\overset{\displaystyle OH}{|}}{\underset{\underset{\displaystyle H}{|}}{C}} - CO_2^{\ominus} + O_2 \xrightarrow{\;\;E-FMN\;\;} RCOO^{\ominus} + CO_2 + H_2O \qquad\qquad 1.$$

anaerobic conditions. The flavin spectrum of the modified holoenzyme is characteristic of the reduced form of the coenzyme. Confirmation of the flavin as the site of modification by I was provided by the demonstration that all the radioactivity incorporated into the holoenzyme by $(4-^3H)$ I was associated with the coenzyme after the flavin was resolved from the apoprotein. Although the precise structure of the flavin adduct was not determined, the lack of any radioactivity incorporation into the coenzyme when enzyme was inactivated with $(2-^3H)$ I suggests that loss of the α-proton precedes the formation of inactive enzyme. The absence of an exchangeable acetylenic proton in the flavin adduct suggests that the acetylenic linkage has been altered in the irreversibly inhibited product (24).

An important feature of the reaction of I with lactate oxidase is that under anaerobic conditions a stoichiometric amount of I, based on the assumption that only the L form is reactive, is sufficient to completely inactivate the enzyme. However, under aerobic conditions, some oxygen consumption is observed prior to complete inactivation of the enzyme. Structure I is therefore both a substrate and an inhibitor of lactate oxidase. Since the amount of I consumed prior to the complete inactivation of the enzyme depends on the concentration of oxygen, the partitioning of a common intermediate (see 2-IIb) between oxidation and irreversible inactivation is suggested. One possible kinetic scheme consistent with both reactions is indicated in equation 2. It includes an intermediate (2-III) implicit in a previous

2.

description (24). The isolatable flavin derivative obtained from lactate oxidase could arise from: 1. nucleophilic attack on the incipient allene (2-IIb) (or its fully protonated form in which another proton is added to C-4) derived from an enzyme-bound carbanion (2-IIa) (process A, equation 2) in analogy to the reaction of acetylenic inhibitors with β-OH-decanoyl thioester dehydrase (25); or 2. rapid nucleophilic attack by reduced flavin at the acetylenic moiety of 2-keto-3-butynoic acid (process B, equation 2). This latter possibility would imply an as yet unreported spectro-scopically identifiable transient intermediate in the inactivation reaction.

A key question in the mechanism of action of flavoenzymes is whether or not electron transfer takes place via covalent intermediates (26). It is possible that some type of adduct exists in the overall process designated 2 in equation 2. If this is true, another pathway for the formation of inactive enzyme would be rearrangement of such a covalent adduct. No evidence is available to either support or exclude this possibility. The ability of lactate oxidase to dehydrohalogenate β-chlorolactate to form pyruvate under anaerobic conditions and to form pyruvate and chloro-pyruvate under aerobic conditions suggests the rate-limiting formation of a common carbanion intermediate (similar to 2-IIa in equation 2) in both processes, but does not exclude the existence of obligatory covalent intermediates for reduction of FMN to $FMNH_2$ (27).

Monoamine Oxidase and D-Amino Acid Oxidase

Both these flavoenzymes catalyze the oxidative deamination of amines (equation 3). The three readily identifiable partial reactions leading to the net reaction are

$$R—CH_2NH_2 + O_2 + H_2O \rightarrow RCHO + NH_3 + H_2O_2$$ 3.

summarized in equations 4–6, where EnFl and $EnFlH_2$ are the oxidized and reduced forms of the flavin enzyme, respectively (28). The demonstration that the hydrolysis of the imine (equation 6) generated by oxidation (equation 4) takes place in solution

$$R—CH_2—NH_2 + EnFl \rightarrow R—CH=NH + EnFlH_2 \qquad 4.$$

$$EnFlH_2 + O_2 \rightarrow EnFl + H_2O_2 \qquad 5.$$

$$R—CH=NH + H_2O \rightarrow RCHO + NH_3 \qquad 6.$$

is based on the production of equimolar amounts of α-(^{3}H)-D- and L-methionine upon addition of (^{3}H)-NaBH$_4$ to a reaction mixture composed of D-amino acid oxidase and D-methionine. Further evidence that the imino acid freely dissociates from the enzyme is that ε-N-(1-carboxyethyl)-L-lysine is generated by D-amino acid oxidase from D-alanine and L-lysine in the presence of sodium borohydride at pH 8.3 (29). Apparently, α-imino acids readily undergo transaldimination reactions with amines and the resulting Schiff bases are rapidly reduced by sodium borohydride. Previous workers have demonstrated that (^{14}C) D-alanine is incorporated into D-amino acid oxidase as ε-N(1-carboxyethyl)-L-lysine in the presence of sodium borohydride (30), but in light of the facile transaldimination reaction, these results do not indicate the presence of an essential enzyme-bound Schiff base intermediate.

In addition to the production of a rapidly dissociable α-imino acid intermediate, formation of a carbanion, at least in the case of D-amino acid oxidase, seems essential prior to the reduction of the flavin (equation 4). In analogy to studies with lactate oxidase and β-chlorolactate, the most convincing evidence for this conclusion is the ability of the enzyme to catalyze elimination reactions using β-chloro-L-alanine and α-amino-β-chlorobutyrate as substrates to yield pyruvate and α-ketobutyrate, respectively (31, 32). The spectra of transients in the enzyme-catalyzed dehydrogenation of α-amino-β-chlorobutyrate resemble those of oxidized flavins. Since kinetic isotope studies reveal that their rate of formation depends on the cleavage of the α-C–H bond, formation of the intermediate carbanion probably does not involve a flavin adduct. However, since α-amino-β-chlorobutyrate is not oxidatively deaminated to β-chloro-α-ketobutyrate, this does not exclude the formation of flavin adducts after carbanion formation. The D-amino acid oxidase-catalyzed oxidation of the carbanion of nitroethane has provided strong but not compelling evidence for adduct formation at N-5 (33).

Several specific irreversible inhibitors of monoamine oxidase have been described whose mode of action depends on the prior generation of a reactive intermediate in an enzyme-catalyzed process. They can profitably be discussed in the context of the presumed mechanism of action of this enzyme, assuming its similarity to D-amino acid oxidase. 3-Bromoallylamine (II) is an effective inhibitor whose

II

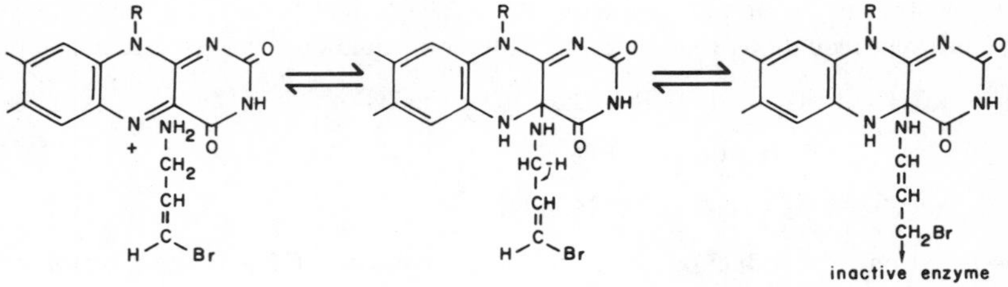

inactive enzyme

7.

suggested mode of action (equation 7) (34) parallels the postulated mechanism for the reduction of flavin by alcohols (26). In view of the unproven formation of covalent intermediates prior to production of the carbanion, the adduct suggested may not form. However, the central feature of the scheme, namely the enzyme-catalyzed isomerization of the double bond to yield a potent alkylating agent capable of modifying a vicinal amino acid residue, is undoubtedly correct. One argument presented in support of the formation of a covalent adduct is that rapid dissociation of the inhibitor from the enzyme surface could lessen the likelihood of reaction. However, since the dissociation of products is a limiting step in catalysis by D-amino acid oxidase, covalent anchoring of the inhibitor may not be necessary for modification of the enzyme.

One case where the anchoring of an enzyme-generated inhibitor is clearly unnecessary for modification is the inhibition of monoamine oxidase by phenylhydrazine (28). The enzyme reacts with phenylhydrazine under anaerobic conditions to yield a completely reduced flavin which, upon introduction of oxygen, is completely reoxidized with total restoration of activity. However, if oxygen is initially present, an irreversibly inhibited enzyme with reduced-type flavin spectrum, stable to autoxidation, is generated. Incubation with $(1-^{14}C)$-phenylhydrazine yielded a protein where 1.4 mol of inhibitor were incorporated per mole of flavin. Three lines of evidence suggest that phenyldiazene (III), the product formed upon

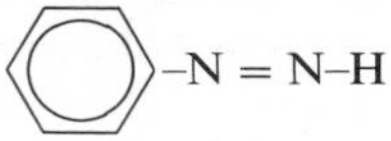

III

initial oxidation of phenylhydrazine, is the true inhibitor of monoamine oxidase. The most compelling is that III generated by in situ decarboxylation of phenylazoformate inhibits the enzyme to yield an inactive derivative that has the same flavin spectrum obtained from inhibition by phenylhydrazine. Secondly, the rates of oxidase inhibition by a series of p-substituted phenylhydrazines, yield a ρ value in a Hammett plot of -1.9. Phenylhydrazine reduction of the flavin is therefore the rate-determining step for inhibition. Finally, 2-phenylethylhydrazine, although a substrate, is a poor inhibitor because the diazene generated from it rapidly rearranges to yield phenylacetaldehyde hydrazone, which has independently been found to be an ineffective inhibitor.

Monoamine oxidase is also irreversibly inhibited by pargyline (N-benzyl-N-methyl-2-propynylamine), an acetylenic compound presently used in antihypertensive therapy. Pargyline inhibits monoamine oxidase in a 1:1 stoichiometry and forms a stable flavin adduct (35, 36). Although the precise structure of the flavin adduct has not yet been determined, it is tempting to assume that monoamine oxidase catalyzes the formation of the carbanion of pargyline (IVa), which is reactive in its incipient allene form in analogy to the reaction of 2-OH-3-butynoate with lactate oxidase and lactate dehydrogenase from bakers' yeast and *E. coli*. The

inhibition of monoamine oxidase by pargyline provides indirect evidence that carbanion formation is as important for this flavoenzyme as it is in the others discussed.

TRANSFERASES

Thymidylate Synthetase

The one carbon transfer from 5,10-methylene tetrahydrofolate to deoxyuridine 5'-phosphate to form thymidylic acid is catalyzed by thymidylate synthetase and is one of the slow steps in DNA synthesis (37). Since effective cancer chemotherapeutic agents such as 5-fluorouracil and 5-trifluoromethyl-2'-deoxyuridine inhibit thymidylate synthetase (38–43), the mechanism of inhibition of this enzyme by uracil analogs has been carefully studied.

5-Fluoro-2'-deoxyuridine 5'-phosphate (V) inhibition of thymidylate synthetase has been particularly intensively investigated. The enzymes most widely used in

V

these studies were isolated from an amethopterin-resistant strain of *Lactobacillus casei* (44–47) or from T-2 bacteriophage-infected *E. coli* (48). All available evidence indicates that V is a suicide substrate which reacts with the enzyme via a mechanism closely analogous to the normal enzyme-catalyzed process. In the presence of 5,10-methylene tetrahydrofolate, V forms an irreversible ternary complex composed of V, coenzyme, and enzyme. The absorption spectrum of the complex indicates that the 5,6 double bond in the pyrimidine is no longer present (44, 45, 47). The ternary complex is stable in guanidine hydrochloride (45, 48), sodium dodecylsulfate, and urea (46), and does not dissociate upon digestion with trypsin (47) or pronase (45). Therefore, a covalent bond between enzyme and V, which most likely arises from the nucleophilic addition of an amino acid residue to either the 5 or 6 position of V, must exist. The requirement of coenzyme for tight binding of V to the enzyme, coupled with the isolation of a single peptide containing both coenzyme and V (49), indicate that the coenzyme as well as the enzyme are covalently attached to V (47, 49).

At present the precise structure of the adduct is unknown. However, since the 6 position of various uridine derivatives is very susceptible to nucleophilic attack (50–52), addition at this position is most likely responsible for the abolition of the double bond between C-5 and C-6. Nucleophilic addition to C-6 should activate C-5 for electrophilic substitution. For example, model system studies have shown that 2′,3′-O-isopropylidine uridine in alkaline solution exhibits a greatly enhanced rate of deuterium exchange at C-5 in D_2O relative to the corresponding 5′-deoxy compound because of addition of the 5′-hydroxy group to the 6 position of the pyrimidine ring (53) (equation 8). The coenzyme, rather than a proton, must be the

8.

electrophile capable of reacting with the nucleophilic center generated at C-5 as a consequence of the addition of active site amino acid residue to C-6 of V. At present, the form of the coenzyme that reacts is unclear, but the fluorescence and absorption spectra of the isolated peptide containing V and the coenzyme are characteristic of

5-alkyltetrahydrofolates (49). However, before this spectral evidence is used as proof of the reactive form of the coenzyme, it should be noted that 10-methyltetrahydrofolate, a potent competive inhibitor, facilitates the formation of a tight but non-covalent complex with V, whereas 5-methyltetrahydrofolate does not (47). In addition, the enzyme–10-methyltetrahydrofolate complex causes exchange of C-5 protons of 2'-deoxyuridylate 5'-phosphate (47).

At present, the active site nucleophilic amino acid that initially adds to C-6 is unknown. Even though the enzyme does possess an active site cysteine residue whose modification by iodoacetamide is protected by V (47), the absence of a cysteine in the active site peptide containing covalently bound V excludes this amino acid as the enzyme-bound nucleophile that adds to C-6 (49). Since the only nucleophilic amino acid residues present in the isolated peptide are threonine and histidine (49), either a hydroxyl group or an imidazole residue adds to the C-6 position to initiate the formation of the irreversible complex.

Details of the inhibition of thymidylate synthetase by V are directly relevant to the mechanism of action of this enzyme. Apparently, the stability of the carbon-fluorine bond effectively traps and causes accumulation of a complex of closely analogous structure to the steady-state intermediate. Although the general structure of the adduct or complex formed with V is likely represented by VI, the complete solution of its structure requires identification of the essential nucleophilic amino acid residue as well as the isomeric form of coenzyme bound to C-5 of the pyrimidine.

VI

Choline-O-Acetyltransferase and Acetyl-CoA : Arylamine N-Acetyltransferase

Choline-O-acetyltransferase and acetyl-CoA : arylamine N-acetyltransferase, which catalyze the reactions summarized in equations 9 and 10, respectively, exhibit steady-state kinetics and exchange patterns indicative of a covalent intermediate (54–57).

choline + acetyl-CoA → acetylcholine + CoA   9.

arylamine + acetyl-CoA → N-acetylarylamine + CoA   10.

For both enzymes, the expected acetyl-enzyme intermediate has been isolated (56, 57). Because both enzymes are inhibited by sulfhydryl reagents (58, 59), the resistance of the isolated acetyl-enzyme intermediates to N-ethylmaleimide inactivation provides excellent evidence for a thioester structure of the essential intermediate (56, 60).

The complete protection of a reactive group by the formation of a covalent intermediate permits the use of group-specific chemical modification reagents, such as iodoacetamide, to determine if the intermediate's formation or decomposition is rate determining under a given set of experimental conditions where all intermediates have achieved their steady-state concentrations. This technique, which is applicable to impure enzyme preparations, was originally devised in the study of papain (61) and has recently been applied to choline-O-acetyltransferase (60) and acetyl-CoA : arylamine-N-acetyltransferase (62). Studies on the latter enzyme were designed to confirm that, at saturating substrate concentrations, acylation is rate limiting with strongly basic aniline acceptors (e.g. p-toluidine), whereas deacylation is rate limiting with weakly basic anilines (e.g. p-cyanoaniline) (55).

p-Nitrophenylacetate, which can be used as an acetyl donor, retards the rate of inactivation by iodoacetamide by 58% at concentrations as low as 1.7×10^{-8} M even though the observed K_s for this substrate in steady-state kinetic measurements is 8.6×10^{-3} M. Therefore, protection is not afforded by the reversible binding of substrate but must involve formation of the acetyl-enzyme intermediates as indicated in equation 11. Addition of aniline acceptors, which by themselves do not protect

$$E-SH + CH_3-\overset{\overset{\textstyle O}{\|}}{C}-O-\!\!\bigcirc\!\!-NO_2 \rightleftharpoons E-S-\overset{\overset{\textstyle O}{\|}}{C}-CH_3 + HO-\!\!\bigcirc\!\!-NO_2$$

11.

against iodoacetamide inactivation, decreases the extent of protection of the enzyme provided by p-nitrophenylacetate. The explanation for this loss of protection is that deacylation of the enzyme by aniline increases the concentration of the free enzyme, the form of the enzyme susceptible to inhibition. This observation suggests that the protected acetyl-enzyme is an intermediate in the enzymic reaction (62).

As support for this qualitative conclusion, the ability of the various anilines to cause deprotection of the enzyme should show the same concentration dependence as the normal enzyme-catalyzed process. Therefore, when the second order rate constant for inactivation of the enzyme by iodoacetamide is measured as a function of aniline at a constant concentration of p-nitrophenylacetate, the concentration of aniline that provides 50% of the maximal deprotection should correspond to the apparent K_m measured by steady-state kinetic techniques. Further, the extent of deprotection afforded by the various anilines at high concentrations should be a measure of the fraction of enzyme in the iodoacetamide-resistant acetyl-enzyme form under steady-state conditions. As a result, a low degree of deprotection indicates the accumulation of acetyl-enzyme and rate-limiting deacylation, whereas a large extent of deprotection indicates little acetyl-enzyme accumulation and rate-limiting acylation. Since strongly basic amines effectively eliminate the p-nitrophenylacetate protection towards iodoacetamide and weakly basic amines do not, the rate-limiting step for the enzyme reaction depends on amine structure in the manner predicted by steady-state kinetic measurements (55). The use of protection towards inactivation in the manner described above provides an additional way to investigate details of enzyme mechanisms using group-specific reagents.

β-Oxoacyl-CoA Thiolases

β-Oxoacyl thiolases catalyze the thiolytic cleavage of β-ketoacyl-CoA esters (equation 12). Enzymes differ with respect to their specificity to β-ketoacyl-CoA

$$R-CH_2-\overset{\overset{\displaystyle O}{\|}}{C}-CH_2-\overset{\overset{\displaystyle O}{\|}}{C}-S-CoA + CoA$$

$$\rightarrow CH_3-\overset{\overset{\displaystyle O}{\|}}{C}-S-CoA + R-CH_2-\overset{\overset{\displaystyle O}{\|}}{C}-S-CoA \qquad 12.$$

substrates (63). For example, enzymes isolated from yeast (64) and rat liver (65) are specific for acetoacetyl-CoA, but an enzyme isolated from pig heart is able to cleave longer chain β-ketoacyl-CoA esters (66). Steady-state kinetics and exchange reactions indicate that the reaction proceeds by a two step mechanism as indicated in equations 13 and 14 (65, 67–69), where the acyl-enzyme intermediate is assumed to be a thioester in view of the sensitivity of the enzyme to thiol reagents and the

$$RCH_2-\overset{\overset{\displaystyle O}{\|}}{C}-CH_2-\overset{\overset{\displaystyle O}{\|}}{C}-S-CoA + HSE \rightleftharpoons RCH_2-\overset{\overset{\displaystyle O}{\|}}{C}-SE + CH_3-\overset{\overset{\displaystyle O}{\|}}{C}-S-CoA$$
$$13.$$

$$R-CH_2-\overset{\overset{\displaystyle O}{\|}}{C}-SE + HS-CoA \rightleftharpoons R-CH_2-\overset{\overset{\displaystyle O}{\|}}{C}-S-CoA + HSE \qquad 14.$$

isolation of an active site peptide following inactivation of the enzyme with iodoacetamide (67, 70). In analogy to the studies of acetyl-CoA-arylamine transferase, the effective protection of rat liver cytoplasmic acetoacetyl-CoA thiolase by acetoacetyl-CoA with respect to inhibition by iodoacetamide strongly supports the formation of a thioester intermediate (65). Acetyl-CoA provides some protection, but desulpho-CoA by itself is ineffective in preventing inhibition by iodoacetamide.

The pig heart enzyme is effectively inhibited by 3-pentynoyl-, 3-butynoyl-, 4-bromocrotonyl-, and 2-bromoacetyl-CoA but is not readily modified by 2-butynoyl- or 4-pentynoyl-CoA (66). As might be expected from the protection of the enzyme by acetoacetyl-CoA towards iodoacetamide inhibition, this substrate prevents inactivation by all the above acyl-CoA esters, whereas acetyl-CoA only prevents modification by 4-bromocrotonyl- and 2-bromoacetyl-CoA. Since the products of modification by these various reagents have not been identified, it is quite possible that the active site cysteine is not the sole amino acid modified. The protection by exogenous thiols, such as dithiothreitol, against modification by these reagents indicates that all affinity labels are capable of reacting with thiol groups but does not indicate that a cysteinyl residue is modified (66).

The inhibition by the three acetylenic derivatives is perhaps of greatest interest because, under the experimental conditions used, they exhibit the least reactivity to free glutathione but are still very effective inhibitors. In view of the ineffective inhibition of the enzyme by 2-butynoyl- or 4-pentynoyl-CoA, unassisted addition of an active site nucleophile to the acetylenic inhibitor appears unlikely. If a basic group is vicinal to the α-carbon and capable of removing an α-proton, in analogy to the proposed mechanism of inhibition of other enzymes by acetylenic derivatives, the modification of pig heart thiolase could proceed through an incipient allene anion (equation 15), where X, the attacking nucleophile, could be the active site cysteine. However, since the appropriate protein chemistry has not been performed, this conclusion must be regarded as tentative (66).

15.

Yeast and pig heart thiolases are inhibited by incubation of acetoacetyl-CoA in the presence of sodium borohydride (64, 66). These results have been interpreted as evidence for an amino group at the active site of the enzyme, whose function may be to form a Schiff base with the β-keto group of the substrate, which is then transformed to product in several steps. The product of the sodium borohydride inactivation with acetoacetyl-CoA has not been determined. In view of the ability of acetyl-CoA to substitute for acetoacetyl-CoA and the reduction by sodium borohydride of an enzyme-γ-glutamyl-CoA intermediate of succinyl-CoA:acetoacetate coenzyme A transferase (71), we feel a reasonable explanation for the sodium borohydride inhibition observed for thiolase is the reduction of the thioester acetyl-enzyme intermediate of thiolase to form the corresponding inactive thiohemiacetal form of the enzyme. In the absence of further work, therefore, the presence of a lysyl residue at the active site should not be assumed (63). In this context, it is pertinent to note the recent use of sodium borohydride to trap acylphosphate intermediates (72).

Creatine Kinase

Creatine kinase has an active site cysteine that is readily modified by many sulfhydryl reagents (73). Substrate protection is generally possible with appropriate substrate combinations (74–77). By itself, these data are not sufficient to imply an essential catalytic role for the sulfhydryl, since the introduction of a new group can alter the conformation and accessibility of substrates at the active site. Indeed, isozymic forms of the enzyme from a number of mammalian and nonmammalian sources have shown half-site reactivity of the native dimer relative to alkylation, but not to inactivation (78–81).

A recent approach to the solution of this persistent problem in the interpretation of chemical modification results, at least for the modification of essential sulfhydryl groups, has been the introduction of two new reagents, VII and VIII, which permit the introduction of small, uncharged, and nonhydrogen bonding alkane thio (RS-) groups (82, 83) according to equations 16 and 17. The disulfides formed by

$$E\text{—}SH + R'\text{—}SO_2\text{—}S\text{—}R \rightarrow E\text{—}S\text{—}S\text{—}R + R'SO_2H \qquad\qquad 16.$$

VII

$$E\text{—}SH + R'\text{—}O\overset{\overset{\textstyle O}{\parallel}}{\text{—}C}\text{—}S\text{—}SR \rightarrow E\text{—}S\text{—}S\text{—}R + OCS + R'OH \qquad 17.$$

VIII

modification with VII and VIII can be fully reactivated by treatment with mercaptoethanol or dithiothreitol. The advantage of temporary masking of cysteinyl residues is that it allows the modification of other amino acid residues by group-specific reagents which may have some partial reactivity with cysteinyl residues (3).

Treatment of creatine kinases with VII ($R = R' = CH_3$) results in modification of one cysteine residue per active site. Significantly, the modified enzyme retains 18% activity which cannot be decreased by treatment with iodoacetamide at concentrations sufficient to cause complete inactivation of the native enzyme (83). Complete activity can be restored upon incubation with mercaptoethanol. Therefore, despite the fact that 5,5'-dithiobis-(2-nitrobenzoic acid) (DTNB) can cause complete inactivation of the enzyme, the reactive cysteine of creatine kinase has no apparent essential catalytic role.

A single ε-amino group of lysine at the active site of creatine kinase is modified upon treatment with dansyl chloride (84). The magnetic resonance techniques of internuclear double resonance spectroscopy (INDOR) and the intermolecular nuclear Overhauser effect (NOE) were used to monitor the interaction of formate in both the native and dansylated enzymes (85), since previous magnetic resonance studies had shown that formate as well as nitrate bind tightly to an enzyme creatine-ADP complex in a position consistent with that of the transferring phosphoryl group (86, 87). A possible role for the protonated lysine residue would be to bind formate in the inactive complex presumed to be analogous to the transition state. A large NOE for formate was observed only in the presence of both creatine and ADP and indicated that the structure comparable to the transition state is achieved only in the presence of the full complement of substrate or substrate analogs. However, the disappearance of the NOE for formate in the modified enzyme, coupled with a chemical shift appropriate for the ε-CH_2 group of lysine in the native enzyme as measured with the INDOR technique, suggests that formate and, by inference, the phosphoryl group interact with the ε-amino group of lysine (85). Site-specific chemical modification, in conjunction with magnetic resonance techniques, therefore provides insights not accessible with either technique alone.

Aspartate Transcarbamylase

Aspartate transcarbamylase from *E. coli,* which catalyzes the synthesis of N-carbamylaspartate from carbamylphosphate and aspartate, contains 12 peptide chains which may be dissociated into two trimeric catalytic (C) subunits and three dimeric regulatory (R) subunits. The C subunits can be isolated free of the R subunits and have full catalytic activity but do not respond to allosteric effectors (88). Each of the three peptides in the C subunit has a single cysteine residue that is sensitive to a limited number of sulfhydryl modification reagents. N-Ethylmaleimide, iodoacetic acid, and some organomercurials react slowly or not at all (89, 90). However, certain negatively charged reagents, including DTNB, p-hydroxymercuri-benzoate (89, 91, 92), 2-chloromercuri-4-nitrophenolate (93, 94), and potassium permanganate (90), react with the free sulfhydryl with concomitant inactivation. As might be expected, negatively charged reversible inhibitors of the enzyme, phosphate, and the aspartate analog, succinate, protect the enzyme against inactivation by these sulfhydryl reagents.

Despite the inactivation of the enzyme accompanying its modification, in analogy with the studies on creatine kinase discussed above, this cysteine residue is not directly involved in catalysis. This conclusion is based on an important and general method involving DTNB first used on aspartate transcarbamylase (89). The mixed disulfide composed of 5-thio-2-nitrobenzoate and cysteine, formed from DTNB and an enzymic sulfhydryl group, is susceptible to nucleophilic attack by cyanide, sulfite, and 2-mercaptoethanol or any organic thiol. Because 5-thio-2-nitrobenzoate is an effective leaving group, the nucleophiles will displace 5-thio-2-nitrobenzoate from the cysteine residue of the enzyme to yield either S-cyano, S-sulfo, or mixed disulfides. For aspartate transcarbamylase, the S-sulfo or S-2-mercaptoethanol derivatives are inactive, but the S-cyano derivative is fully active. The results with the small and uncharged S-cyano group clearly indicate no essential catalytic function for the sulfhydryl group modified (89).

Pyridoxal 5-phosphate, but not pyridoxal, is an effective reversible inhibitor of aspartate transcarbamylase, which competes with carbamylphosphate (95). It forms a Schiff base with an active site lysine residue that has a characteristic absorption maximum at 430 nm. Three moles of pyridoxal 5-phosphate bind per trimer of catalytic subunit. Consistent with its observed behavior as a competitive inhibitor, the absorption maximum is lost upon addition of carbamylphosphate or the effective reversible inhibitor (N-phosphonacetyl)-L-aspartate. The efficacy of pyridoxal 5-phosphate as a modification reagent, as opposed to pyridoxal, indicates that the phosphate moiety must interact specifically with the enzyme and that the Schiff base contributes only a part of the favorable free energy of binding. Many of the enzymes that are potently inhibited by pyridoxal 5-phosphate catalyze reactions involving phosphorylated substrates (3).

Of particular interest for aspartate transcarbamylase is the ability of the Schiff base of pyridoxal phosphate to serve as a specific photosensitizing agent which, upon irradiation, yields an inactive enzyme derivative in which two histidine residues are destroyed per active site. These results demonstrate that a lysine residue and two histidine residues are located at the active center. Since photo-inactivated enzyme does not regain activity upon dissociation of pyridoxal 5-phosphate, at least one of the histidine residues oxidized probably has a direct role in catalysis (95).

Aspartate Aminotransferase

The basic kinetic scheme of aspartate aminotransferase, a pyridoxal phosphate-dependent enzyme, is summarized in equation 18 (96, 97). Aspartate aminotransferase

18.

in mammalian tissue exists in at least two isozymic forms which differ in cellular location, kinetics, physiochemical, and immunochemical properties (98). One isozyme is of cytoplasmic origin and the other of mitochondrial origin. Both the mitochondrial and cytoplasmic enzymes are relatively insensitive to iodoacetate, iodoacetamide, α-bromopropionate, α-bromobutyrate, and bromosuccinate (99). However, β-bromopropionate selectively inhibits the mitochondrial enzyme and not the cytoplasmic enzyme. The site of modification is the ε-amino group of the lysine involved in coenzyme binding (100).

An active site sulfhydryl in the pig heart cytoplasmic aspartate aminotransferase can be modified with concomitant inactivation by tetranitromethane, N-ethylmaleimide, and DTNB (101–105). However, this residue does not appear to be critical for activity. The S-cyano derivative, prepared as described above for aspartate transcarbamylase (89), had 60% activity (106). The bulky or charged derivatives prepared with glutathione, sulfite, mercaptoethanol, or methanethiol had less than 20% activity (106).

Experiments with aspartate aminotransferase have revealed that conformational changes during the course of catalysis can be reflected by changes in the reactivity of an amino acid residue to a group-specific reagent. The reactivity of the active site cysteine residue discussed above toward DTNB and other reagents is lowest in the free holoenzyme but increases slightly in the presence of α-methylaspartate and erythro-β-hydroxyaspartate. The aldimine (18-II) accumulates with the former amino acid (107), whereas the quinoid intermediate (18-III) predominates with the latter (108, 109). Reactivity is enhanced two orders of magnitude in the presence of the substrate pair, glutamate and α-ketoglutarate, after the equilibrium concentrations of all covalent intermediates are established. Therefore, the active site sulfhydryl, which plays no direct catalytic role, is most reactive in the ketimine form (18-IV) to modification reagents such as DTNB (106, 110, 111).

Bifunctional compounds such as bromopyruvate (112, 113), L-serine-O-sulfate (114), β-chloroalanine (115, 116), and β-bromoalanine (112) are potent irreversible inhibitors of the cytoplasmic and mitochondrial forms of aspartate aminotransferase. The amino acid derivatives react with the pyridoxal form of the enzyme, whereas bromopyruvate is reactive with only the pyridoxamine form. These modification reagents also serve as substrates and yield pyruvate as product. In addition, transamination reactions have also been reported for β-chloroalanine (115) and L-serine-O-sulfate (114).

The amino acids at the active site modified by β-chloroalanine and bromopyruvate are different. β-Chloroalanine modifies a lysine residue which can be isolated from the inactivated enzyme following sodium borohydride reduction as the lysine derivative, IX. A cysteine residue is modified by bromopyruvate (113) although

$$
\begin{array}{c}
\text{COO}^- \\
| \\
\text{H}_3{}_+\text{N}\!\!-\!\!\text{CH} \\
| \\
(\text{CH}_2)_4 \qquad \text{H} \\
| \qquad\qquad | \\
\text{N}\!\!-\!\!\text{CH}_2\!\!-\!\!\text{C}\!\!-\!\!\text{COO}^- \\
| \qquad\qquad | \\
\text{H} \qquad\quad \text{NH} \\
| \\
\text{CH}_2
\end{array}
$$

IX

because the appropriate protein chemistry has not been performed, it is not known if this is the same residue that is rapidly modified by DTNB in the presence of α-ketoglutarate and glutamate. The sequence of reactions presented in equation 19 summarizes the reaction of bromopyruvate with the enzyme. Although two pathways (A and B) could account for the modification of the cysteine residue, the identical maximal rates for pyruvate formation and the inactivation reaction indicate a common intermediate for both processes and suggest that pathway B is responsible for the alkylation of the sulfhydryl group.

The sequence of steps most likely responsible for the inactivation reaction by β-chloroalanine is summarized in equation 20, where pathway B probably leads to the alkylation of the lysine residue. The lysine residue modified may be involved initially in the Schiff base linkage to the coenzyme in the holoenzyme and serve as a general base catalyst for the labilization of the α-hydrogen of the substrate during the course of the reaction (97).

19.

20.

HYDROLASES

Endopeptidases—Serine Proteases

Chemical modification has played a central role in establishing the mechanism of action of chymotrypsin, trypsin, and elastase from the points of view of identifying reactive groups and establishing the existence of essential intermediates. The use of diisopropylfluorophosphate and phenylmethane sulfonyl fluoride to modify an essential serine residue, the use of haloketone affinity labels to alkylate active site histidine residues, and the application of pseudosubstrates such as p-nitrophenyl-acetate and cinnamoylimidazole to establish the acyl-enzyme mechanism in equation 21 have been extensively reviewed (117–121). However, it is pertinent to note that

$$
\text{E—OH} + \underset{\displaystyle}{\text{R—}\overset{\displaystyle \overset{O}{\|}}{\text{C}}\text{—X}} \underset{k_{-1}}{\overset{k_1}{\rightleftharpoons}} \text{EOH} \cdot \text{R—}\overset{\displaystyle \overset{O}{\|}}{\text{C}}\text{—X} \overset{k_2}{\rightarrow} \text{E—O—}\overset{\displaystyle \overset{O}{\|}}{\text{C}}\text{—R} \underset{H_2O}{\overset{k_3}{\rightarrow}} \text{E—OH} \qquad 21.
$$

$$
\begin{array}{cc}
+ & + \\
X^- & O \\
 & \| \\
 & R\text{—}C\text{—}O^-
\end{array}
$$

the experimental approaches largely pioneered in the studies of serine proteases have been useful in such diverse areas as classifying glycosidases (122) and lipases (123) as to basic kinetic scheme and in devising sensitive fluorimetric techniques to titrate the active site normality of acetylcholinesterase (124, 125).

Several new reactive groups have been introduced which exhibit high reactivity toward serine proteases. Alkylisocyanates react rapidly and irreversibly in a 1:1 stoichiometry with the serine residue of chymotrypsin and elastase (126). The carbamyl derivatives formed are sufficiently stable to permit their isolation using standard techniques of protein chemistry (127). In view of the reactivity of alkyl-isocyanates with other nucleophilic amino acid residues (e.g. cysteine) (128, 129), the site of inactivation of alkylisocyanates cannot be assumed in the absence of appropriate confirmatory experiments.

Certain aldehydes, arsonates, and boronates have recently been demonstrated to be potent specific inhibitors of serine proteases (130–134). The inactive complexes formed can be reactivated by dialysis even though it is likely that the active site serine residue adds to the electrophilic centers of the various inhibitors. The probable reason for the potent inhibition is that the adducts generated resemble the tetrahedral intermediate through which both the formation and hydrolysis of the acyl-enzyme proceed. Because X-ray crystallographic analysis of subtilisin and chymotrypsin have revealed a hydrogen bonding network of appropriate geometry to bind a tetrahedral intermediate (or a transition state closely resembling the tetrahedral intermediate), the inhibitors are probably interacting with the enzyme in a manner closely analogous to a true substrate (135). The importance of tetrahedral structures in inhibiting serine proteases is emphasized by the discovery that several naturally occurring peptides terminating in arginylaldehyde are potent inhibitors of trypsin and plasmin (136). Further, in the bovine trypsin-pancreatic trypsin inhibitor complex, a tetrahedral adduct is formed by the addition of the reactive serine of the enzyme to a carbonyl group of a specific lysine residue of the inhibitor (137).

Another use of pseudosubstrates and specific inhibitors of serine proteases has been to study the zymogens of pancreatic proteolytic enzymes. X-ray crystallographic results have indicated that a surprisingly minor degree of reorganization in tertiary structure accompanies the transformation of chymotrypsinogen to active chymotrypsin (138). The spatial arrangement of the essential charge relay system in chymotrypsin composed of a hydrogen bonding network between the side chains of a serine, histidine, and aspartic acid residues (139, 140) is effectively intact in the zymogen (138). Nuclear magnetic resonance studies have confirmed the presence of charge relay system in the zymogen (141). One of the few pronounced and readily interpretable structural changes in the generation of enzyme activity from chymotrypsinogen is the movement of the side chain of a methionyl residue (Met-192) from a deeply buried position in the zymogen to the surface of the enzyme, where it serves as a flexible lid of the substrate binding cavity partially created by this movement.

Studies on the reactivity of the zymogen with active site titrants and substrates are consistent with the absence of effective binding sites on the zymogen. A measurable reaction of chymotrypsinogen and trypsinogen with diisopropylfluorophosphate, even though the rate constants are 10^3 to 10^4 less than their respective enzymes, shows that a reactive serine residue pre-exists in the zymogen (142). Further, chymotrypsinogen catalyzes the hydrolysis of p-nitrophenyl guanidinobenzoate 10^6 to 10^7 times less rapidly than the enzyme, despite the fact that the deacylation of the acyl zymogen is $\frac{1}{70}$ that of the enzyme (143). The significantly lower acylation rate but roughly comparable deacylation rate indicate that the ineffective catalytic properties of the zymogen are due primarily to an inefficient or poorly developed binding site rather than to a distorted catalytic site.

There are at least three further lines of evidence supporting the pre-existence of a competent active site in trypsinogen and chymotrypsinogen (144). First, acetyltrypsinogen, which cannot be activated by either trypsin or acetyltrypsin, causes measurable activation of chymotrypsinogen, even though the rate is 10^5 less than that of trypsin (145). Second, specific oxidation of the Met-192 of chymotrypsinogen to methionine sulfoxide (146) by hydrogen peroxide yields a zymogen derivative that is phosphorylated by diisopropylfluorophosphate two times more rapidly and acylated by p-nitrophenyl-p-guanidinobenzoate eight times more rapidly than the native zymogen (147). The deacylation rate for the acyl zymogen formed from the latter titrant is the same for the native and modified zymogen. These results suggest that the oxidation of Met-192 has induced an alteration in the position of the methionine similar to that which occurs in the zymogen-to-enzyme conversion. The binding of the pseudosubstrates, and hence their reactivity, are facilitated (147). Finally, methane sulfonyl fluoride, which because of its relatively small size does not exhibit high binding specificity, reacts with trypsinogen only 40 times less rapidly than it does with the enzyme (148).

Endopeptidases—Acid Proteases

The unusually low pH optimum of pepsin has long indicated that this enzyme has a structure fundamentally different from the serine proteases and that carboxylate groups are probably directly involved in catalysis (149). Determination of the amino acid sequence of a peptide from the carboxyl terminus of both porcine and human pepsin has supported the former view (150, 151). Of particular interest is the apparent common evolutionary origin of calf renin and hog pepsin as suggested by the high degree of homology of their carboxyl-terminal sequences (152).

The chemical modification of pepsin with a series of reagents has supported the presence of carboxylate residues, in particular aspartyl residues, at the active site. Reagents that either have been proven to or are likely to esterify active site carboxylate residues include p-bromophenacyl bromide (153, 154), 1,2-epoxy-3(p-nitrophenoxy)-propane (155, 156), L-1-diazo-4-phenyl-3-tosylamido butanone-2 (157), diazoacetyl-D,L-norleucine methyl ester (158), diazoacetyl-L-phenylalanine methyl ester (159), α-diazo-p-bromoacetophenone (160), and 1-diazo-4-phenyl-2-butanone (161). All the modification reagents bearing diazo linkages require cupric ion for rapid stoichiometric modification of pepsin.

Few unambiguous conclusions can be drawn from these chemical modification experiments other than the certain existence of aspartyl residues at the active center of the enzyme. These studies indicate the inherent limitations of chemical modification experiments when the derivatization reaction does not bear a basic similarity to the enzyme-catalyzed process. Two moles of 1,2-epoxy-3(p-nitrophenoxy)-propane are incorporated per mole of pepsin (155). This stoichiometry is not altered even if the enzyme is pretreated with p-bromophenacyl bromide or diazoacetyl-D, L-norleucine methyl ester (155). One mole of 1,2-epoxy-3(p-nitrophenoxy)-propane reacts with an aspartyl residue, while the second mole reacts with a methionine residue. The amino acid sequences around the reactive methionine and aspartate are indicated in sequences I′ and II′, respectively, where the residue modified is

Phe-Glu-Gly-Met-Asp-Val-Pro-Thr-Ser-Ser-Gly

I′

Ile-Phe-Asp-Thr-Gly-Ser-Ser-Asn

II′

underlined (156, 162, 163). The aspartyl residue modified by 1-diazo-4-phenyl-2-butanone (161), diazoacetyl-L-phenylalanine methyl ester (159), and apparently α-diazo-p-bromoacetophenone (163) is located in sequence III′. The carboxylate group modified by diazoacetyl-D,L-norleucine methyl ester has not yet been reported nor

Ile-Val-Asp-Thr-Gly-Thr-Ser-Leu

III′

has that modified by L-1-diazo-4-phenyl-3-tosylamido butanone-2 (157). The aspartyl residue modified by α-bromo-p-bromoacetophenone is not in sequences II′ and III′ because it does not inhibit the incorporation of either 1,2-epoxy-3(p-nitrophenoxy)-propane or α-diazo-p-bromoacetophenone into the enzyme (155, 160). Most likely this residue is not essential for activity in any case, since enzyme modified by α-bromo-p-bromoacetophenone still retains substantial activity toward denatured hemoglobin (160). In addition, gastricsin, which is homologous in structure to pepsin, is not inhibited by α-bromo-p-bromoacetophenone, although it is inhibited by α-diazo-p-bromoacetophenone and diazoacetyl-D,L-norleucine methyl ester (164).

The assignment of a catalytic function to the aspartyl residue in either sequence II′ or III′ has been complicated by the report that pepsin modified by either α-diazo-p-bromoacetophenone or 1,2-epoxy-3(p-nitrophenoxy)-propane retains its sulfite esterase activity but not its peptidase activity using hemoglobin as its substrate (163). However, there is evidence that a single active site is responsible for both activities. The initial report on the sulfite esterase activity of pepsin demonstrated that both the peptidase and sulfite esterase activity were lost upon modification with diazoacetyl-D,L-norleucine methyl ester (165). Furthermore, the K_i calculated for the inhibition of sulfite ester hydrolysis by N-carbobenzoxy-L-phenylalanyl-L-tyrosine is the same as the K_m calculated using the dipeptide as a substrate (165).

If, as seems likely, diazoacetyl-D,L-norleucine methyl ester reacts with the aspartyl residue in either sequence II′ or III′ and the mechanisms of sulfite ester and peptide hydrolysis are the same, there is no unequivocal evidence that an aspartyl residue participates directly in bond making and breaking at the active site. Inhibition could be due to steric effects resulting from the introduction of a bulky substituent in the active site. On the other hand, sulfite ester and peptide hydrolysis may proceed by different mechanisms. Identification of the acidic residue esterified by diazoacetyl-D,L-norleucine methyl ester would help resolve this question, as would a detailed analysis of the peptide products and catalytic activities of pepsin inhibited by the bifunctional reagents, 1,1-bis-diazoacetyl-2-phenylethane (X) or D,L-1-diazoacetyl-1-bromo-2-phenylethane (XI) (166). Esterification of active site carboxy-

lates, using reagents that introduce small substituents, such as trimethyloxonium borate, might be an alternative approach (167).

Just as diisopropylfluorophosphate has been a useful reagent to classify enzymes as serine proteases, it appears that diazoacetyl-D,L-norleucine methyl ester might be valuable to determine if proteases are homologous to pepsin. In addition to pepsin from a variety of sources, enzymes inhibited in a cupric ion-dependent reaction by diazoacetyl-D,L-norleucine include renin from mouse submaxillary glands (168), penicillopepsin from *Penicillium janthinillum* (169), and human gastricsin (164).

Exopeptidases—Carboxypeptidases

Bovine carboxypeptidase A, a zinc metalloenzyme, has been extensively studied by chemical modification and X-ray crystallography (e.g. 170, 171). The X-ray structure of the carboxypeptidase A-glycyl-tyrosyl complex has revealed that the carbonyl group of the peptide coordinates to the active site zinc ion, and two amino acid residues, Glu-270 and Tyr-248, are close enough to the scissile peptide bond to be catalytically important. The positive charge of the guanidinium group of Arg-145 interacts with the terminal carboxylate ion and is responsible for the specificity of carboxypeptidase as an exopeptidase. The role of the zinc ion is to polarize the carbonyl group and enhance its susceptibility to nucleophilic attack. The glutamate residue either attacks the coordinated peptide carbonyl group to form an anhydride intermediate which rapidly hydrolyzes or serves as a general base catalyst for the nucleophilic attack of water on the peptide carbonyl group. The presumed role of the tyrosine is to serve as a general acid and donate a proton to the nitrogen of the susceptible peptide bond (170, 172).

Although affinity labeling experiments have confirmed that Glu-270 is at the active site, they cannot, by their intrinsic nature, resolve the ambiguity with respect to its precise catalytic function. Both the esterase and peptidase activities of carboxypeptidase A are inhibited by incubation with N-bromoacetyl-N-methyl-L-phenylalanine (XII) (173). Two moles of XII are incorporated per mole of enzyme,

$$CH_2Br$$
$$|$$
$$C=O$$
$$|$$

$$CH_3—N \qquad O$$
$$| \qquad \|$$

⟨phenyl⟩—CH_2—$\overset{\displaystyle |}{\underset{\displaystyle H}{C}}$—$C$—$O^-$

XII

but in the presence of D-phenylalanine, a competitive inhibitor of the enzyme, only one mole of XII is incorporated and there is no loss of activity. Therefore, there are two loci of XII attachment on the enzyme, one at the active site and one at a site that does not affect enzymic activity (173). An analysis of the products reveals that the alkylation of Glu-270 to form an ester is responsible for the loss of activity (174). The second mole of XII reacts with the α-amino group of the N-terminal asparagine (0.5 residue/mol) and the imidazole moiety of a histidine residue (0.2 residue/mol) in the N-terminal portion of the molecule (174). The rate of enzyme inactivation by XII depends on an ionizable group of pK_a 7.0 which reacts in the deprotonated form. Because the inhibition of carboxypeptidase by Woodward's Reagent K shows a comparable pH dependence (175, 176) and an ionizable group of approximately this value is important in peptidase activity (177, 178), Glu-270 is most likely the residue responsible for this ionization despite the divergence of its pK_a from that of a normal carboxylic acid.

Both XII and α-N-bromoacetyl-D-arginine inhibit carboxypeptidase B (179, 180). As might be expected from the similarity of these two enzymes, the glutamate residue of carboxypeptidase B esterified is located in a peptide that is homologous to the active site peptide of carboxypeptidase A (181).

The chemical modification of tyrosine residues has provided interesting details with respect to carboxypeptidase A-catalyzed hydrolyses. Acetylation of the enzyme with acetic anhydride (182) or acetylimidazole (183) abolishes peptidase activity using carbobenzoxyglycyl phenylalanine as substrate but enhances esterase activity using hippuryl-DL-phenyllactate as substrate. The interpretation of these findings is somewhat tentative since the enzyme is susceptible to substrate and product inhibition (184–186), and the products of the modification have not been determined, although the acetylation of Tyr-248 is assumed to be primarily responsible for the observed kinetic effects. However, the results indicate that a free hydroxyl group on Tyr-248 may not be absolutely required for ester hydrolysis, even though in all cases examined it is essential for peptidase hydrolysis (172).

Experiments involving the modification of Tyr-248 in the crystalline state with the diazonium salt of p-arsanilic acid have questioned whether substrate-induced conformational changes deduced from X-ray crystallography occur in solution (187, 188). These changes involve the movement of the phenolic hydroxyl group of Tyr-248 by roughly 12 Å when the substrate (glycyltyrosine) binds to the free enzyme. These structural changes appear to be triggered by the formation of a salt link between the terminal carboxylate ion of the substrate and the guanidinium group of Arg-145. Initially, the hydroxyl group of the tyrosine is 17 Å from the zinc ion, but after binding of the substrate, it is located near the nitrogen of the scissile peptide bond (189).

The tyrosine derivative formed by modification with the diazonium salt, XIII,

XIII

is yellow in its protonated form and red in its phenolate form. The free phenolate has an absorption maximum at 485 nm but, coordinated to zinc ion, it has an absorption maximum at 540 nm (190). The α, β, and γ forms of monoarsanilazo-tyrosine-248 carboxypeptidase A in crystalline forms distinct from that used in the X-ray studies are yellow at pH 8.2 but red ($\lambda_{max} = 510$ nm) in solution at the same pH (190). Circular dichroism and absorption spectra of modified carboxypeptidase with and without the active site zinc ion indicate that the red color in solution at pH 8.2 is due to the coordination of the active site zinc ion by monoarsanilazotyrosine-248. These studies indicate that the position of Tyr-248 in this crystalline form is different from that in solution (190). This conclusion is supported by the pK_a difference of the modified tyrosine of nitrotyrosyl-248 carboxypeptidase in the crystalline state and in solution (191).

However, the monoarsanilazotyrosine-248 derivative of carboxypeptidase A_z (the form used in the X-ray studies) is red in solution at pH 8.2 and red in the crystal structure used in the X-ray analysis at pH 8.2 (192). Another important difference between the crystals used in the X-ray studies and the other crystalline form is that the former hydrolyzes carbobenzoxyglycyl-L-tyrosine 100 times more rapidly than the latter (192). At pH 7.5 and 4°C, the conditions used in the X-ray investigation, the absorption spectra of the modified enzyme retain a residual absorption maximum at 510 nm (192) both in solution and in the crystal form used in X-ray studies. These results suggest that some fraction of the carboxypeptidase A_z molecules in both the crystalline and solution states exists in a conformation in which Tyr-248 coordinates to the zinc ion. Re-examination of the electron density map of unmodified carboxypeptidase A_z reveals that roughly 15–25% of the molecules do exist in a conformation consistent with the tyrosine hydroxyl group coordinated to the active site zinc ion, although in the majority of enzyme molecules, the hydroxyl group is 17 Å from the zinc ion (189). The spectra of arsanilazotyrosine-248 carboxypeptidase therefore accurately reflect conformational states of the unmodified enzyme, and Tyr-248 of carboxypeptidase A has greater conformational flexibility than had been previously noted.

Since substrates and inhibitors such as glycyltyrosine and L-phenylalanine destroy the red color characteristic of chelation of zinc ion by arsanilazotyrosine, the direct coordination of the metal ion by the carbonyl group of the peptide deduced from the X-ray studies is confirmed. The crucial problem to resolve is whether the pattern of induced conformational changes deduced from the X-ray studies is an obligatory pathway in solution to achieve the interactions essential for catalysis or whether alternative pathways, beginning with the conformation of the enzyme in which the tyrosine is coordinated to the zinc ion, also exist. This question is complicated by the finding that the α, β, and γ forms of the enzyme, which are very similar in their catalytic properties, possess different ratios of the two conformational states (190). The kinetics of displacement of the modified tyrosine from the zinc ion by substrates and inhibitors should help determine the kinetic competence of the conformational forms of the enzyme.

Carboxypeptidases from nonmammalian sources, such as yeast and cotyledons of germinating cotton seedlings, do not appear to be homologous to the pancreatic enzyme (193–196). Both are inhibited by DFP and are therefore serine proteases. However, the sequence of the active site peptide containing the reactive serine of the yeast enzyme does not correspond to that of the serine proteases (194).

Phospholipases and Lipases

Enzymes that act on lipid-water interfaces, such as phospholipase A_2 and triglyceride lipase, are difficult to study because of the complex physiochemical structure of their substrates. Although magnetic resonance techniques have helped define the micellar structures composed of Triton X-100 and dipalmitoyl phosphatidylcholine for phospholipase A_2 from the venom of *Naja naja* (197). comparable studies have not been performed for pseudosubstrates and chemical modification reagents. Nevertheless, the efficacy of certain reagents does appear to depend on their physical state.

For example, *p*-bromophenacyl bromide alkylates a single histidine residue in both porcine pancreatic phospholipase A_2 and its zymogen, prophospholipase, at the same rate. The enzyme is inactivated and the potential activity of the zymogen is destroyed (198).

The modification of the enzyme by *p*-bromophenacyl bromide is retarded by D-dihexanoyllecithin at a constant concentration of alkylating agent until the critical micellar concentration is reached. At this point, *p*-bromophenacyl bromide can be incorporated into micelles of D-dihexanoyllecithin, and the rate of enzyme inhibition is enchanced. For the zymogen, increasing concentration of D-dihexanoyllecithin inhibits alkylation with the phenacyl bromide until the critical micellar concentration is reached. After this point, the rate of modification of prophospholipase is constant and independent of the phospholipid concentration (198).

The reactivity of the enzyme and zymogen toward *p*-bromophenacyl bromide parallels their reactivity toward substrates. Substrates in micelles are more effectively hydrolyzed by the enzyme than the substrate in its monomeric form. On the other hand, the zymogen is active toward the monomeric substrate but is inactive toward the substrate in micellar form (199). In analogy to the studies on the pancreatic proteolytic enzymes discussed above, the zymogen apparently possesses a competent catalytic site but an ineffective binding site for its native substrate. As a result, it is not modified by *p*-bromophenacyl bromide in micelles. The enhanced reactivity of the *p*-bromophenacyl bromide in micelles towards the enzyme is probably due partially to the increased local concentration of the alkylating agent at the active site when it is incorporated into micelles. In addition, micelles may induce conformational changes in the enzyme which enhance the histidine's nucleophilicity.

The catalytic function of the modified histidine residue in the porcine pancreatic enzyme is at present unknown. However since calcium ion, which is essential for the catalytic activity (200), protects against *p*-bromophenacyl bromide inactivation, the metal ion probably binds at or near the reactive histidine. Recent studies on the phospholipase A_2 from the venom of *Crotalus adamanteus,* which also requires calcium ion (201), have suggested a possible role for the histidine residue in the porcine enzyme. The venom enzyme is inactivated upon modification of a lysine residue by ethoxyformic anhydride (202). A possible catalytic role for this lysine residue and, by analogy, the histidine residue in the pancreatic enzyme is suggested by the primary amine-catalyzed methanolysis of phosphatidylcholine to lysophosphatidylcholine and the methyl esters of the fatty acids originally present in the lecithin (203). Since the reaction exhibits a kinetic isotope effect in both CH_3OD and CD_3OD, the primary amine probably serves as a general base catalyst. The general base-catalyzed methanolysis is further accelerated by calcium ion. In analogy to other metal ion-catalyzed transesterification reactions (204, 205), the calcium in the model reaction for phospholipase probably forms a reactive complex where the metal ion, besides serving as a template for reactants, facilitates the deprotonation of methanol by lowering its pK_a. The deprotonation of the methanol is further assisted by the amine serving as a general base catalyst. In the pancreatic enzyme, the general base catalyst could be the histidine alkylated by p-bromophenacyl bromide; in the snake venom enzyme, it could be the lysine acylated by ethoxyformic anhydride.

Although phospholipases do not possess reactive serine residues, lipases from a variety of sources appear to react with serine esterase pseudosubstrates. For example, an extracellular lipase from *Corynebacterium acnes* is inhibited by p-nitrophenylacetate and diisopropylfluorophosphate (206). In addition, pancreatic lipase is inhibited by diethyl-p-nitrophenylphosphate (207) and hydrolyzes p-nitrophenylacetate (123). In analogy with inhibition studies on phospholipase, the reactions of these pseudosubstrates are dependent on their physiochemical state and proceed more rapidly when incorporated into micelles.

Nucleases—Ribonuclease A

As methods for the chemical synthesis of peptides improve, a direct way to investigate the role of an amino acid in catalysis will be the chemical synthesis of modified enzymes in which the residue of interest is replaced by an amino acid of different structure and properties. Presently, the most practical way to achieve this goal is to first develop procedures to cleave native enzymes into well-defined peptide products which, though inactive individually, can be added together to restore enzymic activity by noncovalent association of the separated fragments. Since the synthesis of the smaller peptides is now practical (208), systematic substitution and permutation of residues in these peptides through chemical synthesis are feasible. The pioneering studies in this area, which were based on the cleavage of pancreatic ribonuclease A by subtilisin into a 20 unit and 104 unit peptide, have been recently reviewed (209), as has work on staphylococcal nuclease cleaved by trypsin (210–213).

Recent studies on bovine pancreatic ribonuclease A have involved complementation of inactive enzyme in which residues from the carboxyl terminus have been removed by limited proteolytic digestion and enzymic activity regenerated by addition of appropriate carboxyl-terminal peptides (214). One question investigated was whether Phe-120 plays an essential role in binding substrates. X-ray studies had revealed the close association of the aromatic rings of Phe-120 and the pyrimidine substrate. To evaluate the significance of this interaction, the regeneration of enzyme activity in ribonuclease deficient in amino acid residues 119–124 by synthetic tetradecapeptides corresponding to residues 111–124 was examined (215). In particular, tetradecapeptides were synthesized in which the phenylalanine at position 120 was substituted with isoleucine, leucine, and tryptophan. Although the peptide containing tryptophan yielded only a trace amount of activity upon addition to the large fragment, the peptides containing isoleucine and leucine yielded roughly 15% activity. The peptide containing phenylalanine regenerated 98% activity. The lower activity in the peptide containing amino acids other than phenylalanine at position 120 is largely due to reduced binding of the two fragments. However, since the K_m values for the active product formed from the phenylalanine, isoleucine, and leucine containing peptides were all equal within experimental error, these results demonstrate that the interaction of the phenyl ring of phenylalanine with the pyrimidine ring of the substrate is not essential for binding (215).

The enzymically active products generated by complementation of fragments apparently have multiple conformations. When native ribonuclease A is alkylated by iodoacetate at pH 5.5, two inactive protein products are formed. The major product (87%) is carboxymethylated at His-119, whereas the minor product is carboxymethylated at His-12 (216). When ribonuclease A lacking residues 121–125 is complemented with the tetradecapeptide corresponding to residues 111–124, the essential catalytic function of His-119 can be contributed by the histidine residue either from the abbreviated protein or from the peptide. To determine which histidine residue is at the catalytic site, this enzymically active complex was alkylated by iodoacetic acid. Since the histidine residue at 119 is modified in both components (4:1, ratio favoring the protein), the complemented enzyme does not possess a unique conformation and the histidine residue can be provided by either the peptide or the protein. When ribonuclease lacking residues 120–124 was alkylated in the presence of the same tetradecapeptide, His-119 in both components was modified to the same extent (217).

Nucleases—Deoxyribonucleases

The mechanisms of action of pancreatic ribonuclease and deoxyribonuclease must be significantly different because deoxyribonuclease cannot form an intermediate comparable to the cyclic 2′,3′ cyclic phosphate intermediate of ribonuclease (209). However, a related covalent enzyme phosphodiester intermediate may be involved in the deoxyribonuclease-catalyzed reaction if the hydroxyl group of a threonine, serine, or tyrosine replaces the 2′-OH group of a ribonucleotide as a nucleophile. Although no covalent intermediates have been detected in the staphylococcal nuclease-catalyzed hydrolysis of deoxythymidine 3-phosphate-5-p-nitrophenyl phosphate (218), the rapid inactivation of pancreatic DNase by methane sulfonyl chloride at pH 7.0 and 5.0 has revealed a highly reactive serine residue at the active site of this enzyme (219). Evidence for a serine residue as the site of sulfonylation is 1. the alkaline lability of the ^{35}S-methane sulfonyl label in the modified enzyme, and 2. the generation of S-aminoethylcysteine upon incubation of the modified enzyme with β-mercaptoethylamine (equation 22). Both these properties are shared by chymotrypsin, which has been specifically sulfonylated at its reactive serine (220, 221).

$$\cdots \underset{\underset{\vdots}{NH}}{\overset{\overset{O}{\parallel}}{C}} - \underset{\overset{\mid}{NH}}{\overset{H}{\underset{\mid}{C}}} - CH_2 - O - \underset{\underset{O}{\parallel}}{\overset{\overset{O}{\parallel}}{S}} - CH_3 + HS - CH_2CH_2 - NH_2$$

$$\rightarrow CH_3 - \underset{\underset{O}{\parallel}}{\overset{\overset{O}{\parallel}}{S}} - O^- + \cdots \overset{\overset{O}{\parallel}}{C} - \underset{\underset{\vdots}{NH}}{\overset{H}{\underset{\mid}{C}}} - CH_2 - SCH_2CH_2NH_2 \qquad 22.$$

Although the active serine of DNase reacts roughly 3500 times more rapidly than would be expected for a primary alcohol under comparable conditions, these data merely suggest but certainly do not prove an essential catalytic role for the serine. The generality of reactive serine residues in deoxyribonucleases is uncertain because porcine spleen deoxyribonuclease is not inhibited by diisopropylfluorophosphate (222). However, these two enzymes have fundamentally different properties. Pancreatic DNase produces 5'-phosphorylated ends, requires a divalent metal ion, and has optimal activity at neutral pH. However, the spleen deoxyribonuclease yields 3'-phosphorylated products, has no apparent metal ion requirement, and is optimally active at acidic pHs (223).

Both enzymes are inhibited by iodoacetate (222–224). In each case, the inactivation is due to the alkylation of a histidine residue to yield 3-carboxymethyl histidine. Because the complete sequence of the 257 residue pancreatic enzyme is known (225, 226), it is possible to designate His-131 as the reactive residue. Although no definitive proof of the catalytic function of the histidine residue is available, in analogy to ribonuclease, it may serve as a general base catalyst. In line with the differing metal ion requirements of the two enzymes, a divalent metal ion is essential for the inhibition of pancreatic DNase by iodoacetate but not for inactivation of the spleen enzyme (223). Inhibition of the pancreatic enzyme in the presence of cupric ion with a series of haloacetates revealed that chloroacetate was the best alkylating agent followed by the bromo and iodo derivatives (227). Since this reactivity sequence corresponds to the stability of the various cupric halide complexes and not the normal leaving group tendencies of the halide (227), the cupric ion probably serves as an electrophilic catalyst by coordinating the halide leaving group in the transition state of the alkylation reaction. If correct, this conclusion implies that the binding site of the metal ion is at or near the reactive histidine. Since metal ions protect the enzyme against inactivation of the enzyme by methylsulfonyl chloride (219), the reactive serine and histidine residues may be adjacent to each other at the active site.

LYASES

Pyridoxal 5-Phosphate-Dependent Decarboxylases

Although important mechanistic features of pyridoxal 5-phosphate-dependent decarboxylases remain to be determined, the general structure of reactive intermediates and their sequence of formation are well understood (96). The sodium borohydride reduction of Schiff bases formed between pyridoxal 5-phosphate and lysine residues pioneered in the study of phosphorylase (228), has played a key role in elucidating the active site chemistry of all pyridoxal 5-phosphate-dependent enzymes. For example, the demonstration of a Schiff base or aldimine linkage between a lysine residue and the coenzyme, prior to addition of substrate, has indicated that the first step in the catalytic reaction involves a transaldimination reaction in which the amino group of the substrate displaces the ε-amino group of the lysine from the pyridoxal phosphate (see equation 18).

Sodium borohydride reduction of the holoenzyme has permitted the isolation of pyridoxyl peptides from the active sites of these pyridoxal 5-phosphate-dependent enzymes (229, 230). Striking sequence homologies in the peptides from E. *coli* lysine and arginine decarboxylases suggest a common evolutionary origin of these enzymes despite the fact that the two proteins are sufficiently distinct to show no immunological crossreactivity (230). A basic residue, either histidine or lysine, is frequently adjacent to the pyridoxyllysine residue in many pyridoxal phosphate-dependent enzymes. No conclusive statement can be made as to the role, if any, of these residues, but molecular models indicate that interaction with the phosphate moiety of pyridoxal 5-phosphate is possible. Alternatively, if the neighboring basic group is charged, it may lower the pK_a and therefore increase the reactivity of the active site lysine (229).

β-Chloroalanine is both a substrate and an irreversible inhibitor of L-aspartate-β-decarboxylase from *Alcaligenes faecalis* (231). In analogy to aspartate transaminase (115, 116), the decarboxylase also catalyzes the transformation of β-chloroalanine into pyruvate. Since β-chloro-L-alanine does not alkylate the apoenzyme or the 4'-deoxypyridoxine 5-phosphate-enzyme complex (231), the catalytic and inactivation reaction most likely proceeds through intermediates comparable to those summarized in equation 20 for the reaction of β-chloroalanine with aspartate transaminase. All the data are consistent with a glutamate residue as the sole site of modification. The catalytic function of this glutamate is unclear, although its location within the active site suggests that it could serve as the donor of the proton that replaces the β-carboxyl group (232).

Tryptophan Synthetase

The use of pyridoxal 5-phosphate to effect the photosensitized oxidation of a histidine residue proximal to it in the three-dimensional structure of the β-subunit of E. *coli* tryptophan synthetase has recently been described (233–235). Analysis of the peptide products indicates that the modified histidine residue either immediately precedes the pyridoxyllysine in the primary sequence or is five residues away in the direction of the N terminus. The suggested catalytic function of the histidine is to serve as a general base catalyst to assist the labilization of the α-carbon proton, although it is possible that the lysine displaced from the coenzyme by substrate in some cases might also serve in this catalytic role (229). Pyridoxal 5-phosphate-induced photoxidation of amino acids may be a general and useful technique to probe the environment in close proximity to lysine residues capable of forming Schiff bases with the coenzyme. As noted elsewhere in this review, this technique has been successfully applied to aspartate transcarbamylase and aldolase. Pyridoxal phosphate is not the only cofactor capable of serving as a photosensitizing agent. Protoporphyrin has been used in the photoxidation of apoperoxidase (236).

Non-Pyridoxal Phosphate-Dependent Decarboxylases

Sodium borohydride has played an important role in elucidating the active site chemistry of decarboxylases lacking pyridoxal phosphate. For example, it has been used to trap a covalent Schiff base intermediate formed in decarboxylation reactions catalyzed by acetoacetate decarboxylase from *Clostridium acetobutylicum* (237). Since the inactive enzyme formed upon reduction contains an isopropyllysine residue, the ketimine of acetone and not of acetoacetate is reduced (238). These experiments suggest that electrostatic effects may play an important role in the ability of sodium borohydride to trap imine intermediates. Reduction of the acetoacetate imine may be inhibited by repulsive electrostatic interactions between the negatively charged carboxylate and borohydride. In agreement with this conclusion, acetonylsulfonate, an effective inhibitor, is reduced onto the enzyme only one fifth as rapidly as acetone (239), and acetonylphosphonate cannot be reduced onto the enzyme by borohydride (240). Removal of one negative charge from the phosphonate by forming the monomethyl ester yields a derivative that is readily reduced onto the enzyme even though it is a poorer reversible inhibitor (240). For acetoacetate decarboxylase, the addition of cyanide ion to the enzymic imine shows a greater sensitivity to electrostatic repulsion than borohydride because it fails to facilitate the binding of acetonylphosphonate and acetonylsulfonate to the enzyme (240).

Phosphatidylserine decarboxylase from *E. coli* is inhibited by sodium borohydride but apparently is not a pyridoxal phosphate-dependent enzyme (241). This enzyme may be similar to histidine decarboxylase from *Lactobacillus,* in which a pyruvylphenylalanyl residue at the N terminus provides the carbonyl cofactor necessary for Schiff base formation (242, 243). The use of borohydride to trap intermediates has therefore helped to identify a new type of cofactor.

Fructose 1,6-Diphosphate Aldolase—Class I

There are two types of fructose 1,6-diphosphate aldolases (244). Class I aldolases form Schiff base intermediates and are therefore inhibited by sodium borohydride. Class II aldolases require metal ions, are not inhibited by borohydride, and proceed via ene-diolate intermediates similar to those observed in triosephosphate isomerase (244). The inhibition of class II but not class I aldolases by phospho-glycolohydroxamate (XV), a stable analog of the proposed ene-diolate intermediate (XIV), provides further evidence for this distinction (245). Class I aldolases are

XIV   XV

generally found in animals and higher plants, whereas class II aldolases are found in bacteria and molds (244). Recent studies have indicated that this distinction is not absolute (246).

The sequence of intermediates formed in class I aldolase-catalyzed reactions is summarized in equation 23 (244). Unambiguous evidence for the formation of Schiff

$$
\begin{array}{c}
CH_2-OPO_3^= \; + \; NH_2{-}E \quad \xrightarrow[\;\textcircled{1}\;]{H_2O} \quad CH_2-OPO_3^= \quad \xrightarrow[\;\textcircled{2}\;]{H^+} \quad CH_2-OPO_3^= \\
C=O \qquad\qquad\qquad\qquad C=\overset{+}{N}{-}E \qquad\qquad\qquad C=\overset{+}{N}{-}E \\
CH_2 \qquad\qquad\qquad\qquad CH_2 \qquad\qquad\qquad\qquad HO-\overset{\ominus}{C}H \\
OH \qquad\qquad\qquad\qquad OH \\
\qquad\qquad\qquad\qquad 23\text{-}I \qquad\qquad\qquad\qquad 23\text{-}II \quad \textcircled{3}
\end{array}
$$

23.

$$
\begin{array}{c}
CH_2-OPO_3^= \qquad\qquad\quad \xleftarrow[\;\textcircled{5}\;]{H_2O} \quad CH_2-OPO_3^= \quad \xleftarrow[\;\textcircled{4}\;]{RC(=O)-H} \quad CH_2-OPO_3^= \\
C=O \qquad\qquad\qquad\qquad C=\overset{+}{N}{-}E \qquad\qquad\qquad C{-}N{-}E \\
HOC-H \qquad\qquad\qquad\qquad HOC-H \qquad\qquad\qquad CH \\
HC-OH \;+\; NH_2{-}E \qquad\qquad HC-OH \qquad\qquad\qquad OH \\
R \qquad\qquad\qquad\qquad R \\
\qquad 23\text{-}IV \qquad\qquad\qquad\qquad 23\text{-}III
\end{array}
$$

base intermediates includes 1. the trapping by borohydride of 23-I (247–249) and 23-IV (249); 2. rates of fructose 1,6-diphosphate carbonyl ^{18}O exchange which exceed the rate of the cleavage reaction (250); and 3. formation of inactive aminonitrile enzyme derivatives by addition of cyanide ion to 23-I (251). Examination of the reactive properties of the essential active site lysine residue of rabbit muscle aldolase, using a competitive labeling technique involving (^3H)-acetic anhydride (252), has demonstrated that this residue is less readily acetylated than a normal surface lysine in the pH range 7–12 (253). Its apparent pK_a is greater than 11.5 even though the pH optimum of class I aldolase is 6.5–8.0. Since the deprotonated form of lysine is essential for Schiff base formation, a basic group of the enzyme must facilitate the formation of 23-I or 23-IV after initial formation of the non-covalent enzyme-substrate complex (253).

The pK_a of the lysine residue of aldolase contrasts with that of the essential lysine residue at the active site of acetoacetate decarboxylase. 2,4-Dinitrophenyl propionate acylates this group, causing inactivation. The pH rate profile for this reaction is governed by an ionizing group of pK_a 5.9, which is reactive in its basic form. Although alternative explanations are possible, the observed pK_a is most likely that of the active site lysine residue which is acylated (254). These studies indicate that imine formation at the active sites of aldolase and acetoacetate decarboxylase takes place by substantially different mechanisms.

Pyridoxal 5-phosphate inhibits rabbit muscle aldolase reversibly (255). Reduction of the inactive complex with sodium borohydride and isolation of the peptide products reveal that the lysine modified is distinct from that involved in forming the essential Schiff base intermediates. In view of the ability of the reduced pyridoxal phosphate derivative of aldolase to form a Schiff base with dihydroxyacetone phosphate, the lysine residue of aldolase which reacts with pyridoxal phosphate may serve as a binding site for the 6-phosphate of fructose 1,6-diphosphate (255). Pyridoxal phosphate has proven to be a remarkably specific affinity label for the binding site of phosphorylated substrates (3, 256).

The Schiff base formed between pyridoxal phosphate and either rabbit muscle or spinach aldolase can serve as a photosensitizer for the oxidation of histidine residues below pH 8.5 (257) and for the oxidation of a tyrosine residue above this pH (258). The enzyme derivative in which histidine is oxidized retains its ability to form a Schiff base with dihydroxyacetone phosphate. In addition, the modified enzyme retains transaldolase activity. Aldolase alkylated at a histidine two residues from the C-terminal tyrosine by N-bromoacetylethanolamine (259, 260) and altered by carboxypeptidase A treatment with concomitant release of the C-terminal tyrosine (261, 262) exhibits similar partial reactions as the photo-oxidized enzymes. These and other results (263) indicate that the C-terminal region of the molecules plays some role in the chemistry of the carbanion intermediate (23-II). The histidine and tyrosine in the C-terminal region of the rabbit muscle enzyme may serve conformational rather than direct catalytic roles, since liver aldolase is not affected by carboxypeptidase treatment nor is the C-terminal tyrosine residue modified by photo-oxidation of the enzyme-pyridoxal phosphate complex (258).

β-Hydroxydecanoyl Thioester Dehydrase

β-Hydroxydecanoyl thioester dehydrase permits the synthesis of long chain unsaturated fatty acids in bacteria by an anaerobic pathway (25). The enzyme from *E. coli* catalyzes the dehydration of β-hydroxydecanoyl thioesters to *trans*-α-β-decenoyl and *cis*-β-γ-decenoyl thioesters as well as the direct interconversion between the two unsaturated fatty acids. At equilibrium, the relative concentrations of the β-hydroxydecanoyl, α-β-decenoyl, and β-γ-decenoyl thioester derivatives are 70: 27:3. Kinetic and isotopic studies have indicated that all dehydrase reactions proceed through an enzyme-α,β-decenoyl intermediate (equation 24) (25).

$$\beta\text{-OH (free)} \rightleftharpoons (\alpha\text{-}\beta)\text{-Enzyme} \underset{\alpha,\beta \text{ (free)}}{\overset{\beta,\gamma \text{ (free)}}{\rightleftharpoons}} \qquad\qquad 24.$$

The enzyme is inhibited by alkylation of an active site histidine and the nitration of a tyrosine by tetranitromethane (264). More importantly, all the activities of the enzyme are irreversibly and stoichiometrically inhibited by the N-acetylcysteamine (NAC) thioester of 3-decynoyl acid (XVI) at concentrations less than 10^{-7} M (265). The site of inactivation is a histidine residue because an acid hydrolysate of the

$$CH_3—(CH_2)_5—C\equiv C—CH_2—\overset{\overset{\displaystyle O}{\|}}{C}—S—CH_2—CH_2—\overset{\overset{\displaystyle H}{|}}{N}—\overset{\overset{\displaystyle O}{\|}}{C}—CH_3$$

XVI

modified enzyme contains one less histidine than that of the native enzyme (264). Three structural properties of the 3-decynoyl-NAC contribute to its inhibitory ability. First, a thioester linkage is essential. 3-Decynoic acid and its methyl ester are ineffective inhibitors (265). Second, the carbon chain of ten, which exactly corresponds to the substrate specificity of the enzyme, is important. Although comparable derivatives of nine and eleven carbons are potent inhibitors, derivatives with eight and twelve carbons are substantially weaker inhibitors. Finally, and most significantly, the triple bond must be positioned between the β and γ carbons. The N-acetylcysteamine thioesters of 2-decynoic and 4-undecynoic acids are ineffective inhibitors.

The mechanism of enzyme inhibition by 3-decynoyl-NAC proceeds via an enzyme-generated allenic derivative according to equations 25–27 (266). The process summarized in equations 26 and 27 is analogous to the enzyme-catalyzed isomerization of cis-3-decenoyl-NAC to trans-2-decenoyl-NAC and is consistent with

$$E + CH_3(CH_2)_5 \overset{O}{\underset{\|}{-C \equiv C - CH_2 - C - NAC}}$$

$$\rightleftharpoons E \overset{O}{\underset{\|}{-CH_3(CH_2)_5 - C \equiv C - CH_2 - C - NAC}} \qquad 25.$$

$$E \overset{O}{\underset{\|}{-CH_3(CH_2)_5 - C \equiv C - CH_2 - C - NAC}}$$

$$\rightleftharpoons E \overset{H \quad H \quad O}{\underset{| \quad | \quad \|}{-CH_3(CH_2)_5 - C = C = C - C - NAC}} \qquad 26.$$

$$E \overset{H \quad H \quad O}{\underset{| \quad | \quad \|}{-CH_3(CH_2)_5 - C = C = C - C - NAC}} \rightarrow E' \text{ (inactive)} \qquad 27.$$

the requirement for a β-γ triple bond in the inhibitor. Direct proof for the inactivation scheme involved the demonstration that 2,3-decadienyl-NAC (XVII) (shown in its

$$CH_3(CH_2)_5 \overset{H \quad H \quad O}{\underset{| \quad | \quad \|}{-C = C = C - C - SNAC}}$$

XVII

complexed form in equation 27) inhibits the enzyme more rapidly than 3-decynoyl-NAC at equal concentrations (266). The residue modified is a histidine at the active site. Further, whereas α-dideutero-3-decynoyl-NAC reacted slower than its hydrogen analog, the inactivation of the enzyme by α-deutero-2,3-decadienoyl-NAC exhibited no isotope effect in its inactivation reaction with the enzyme. Therefore, the slow step in the inactivation by acetylenic inhibitors is the removal of an α-proton to form the enzyme-bound allene (equation 26). A comparable process is not required for inhibition by XVII. The absolute requirement of thioester derivatives for acetylenic but not dienic acids is explicable in these terms, since thioesters are known to facilitate labilization of α-protons. In addition, the thioester may be essential in binding the acetylenic inhibitor in the proper orientation for proton abstraction from the α-carbon (266).

Although the dienic acids have high intrinsic reactivity, the enzyme is most potently inhibited by dienic acids of ten carbons. This correspondence to the enzyme specificity, also observed for inhibition by acetylenic derivatives, indicates that inhibition by the dienic acids proceeds via reversible binding followed by covalent bond formation (267). Comparison of the various C-10 dienic acid derivatives shows that the effectiveness of inhibition decreases in the series thioester, oxygen ester, free acid, and amide. This pattern of reactivity would be expected if a Michael-type addition reaction (equation 28) was responsible for covalent bond

28.

formation, since electrophilicity would be greatest for those derivatives where the resonance form, $^-O\text{–}C = X$, is least important. Although the precise structure of the histidine adduct has not been determined, model studies have demonstrated that the derivative with the β-γ double bond (28-I) is more likely than the α-β isomer (267). It is not yet known which nitrogen in the imidazole nucleus of histidine serves as the nucleophile. However, these chemical modification experiments unambiguously demonstrate the essential role of a histidine residue as a general base catalyst at the active site of β-hydroxydecanoyl thioester dehydrase.

ISOMERASES

Triose Phosphate Isomerase

The mechanism of triose phosphate isomerase involves proton abstraction from dihydroxyacetone phosphate by a basic group at the active site to form an enediolate intermediate (29-I) which, in turn, is reprotonated to yield 3-phosphoglyceraldehyde (equation 29) (268, 269). The proton abstracted readily exchanges

29.

29-I

with solvent and very little, if any, is added back to the ene-diolate (270). The effective inhibition of the enzyme by phosphoglycollate (271) and its hydroxamic acid (245) supports the presence of an ene-diolate intermediate.

Active site-directed inhibitors have implicated the γ-carboxyl group of a glutamate residue as the participating base in the reaction. Halohydroxyacetone phosphates (XVIII) (272–277) and glycidol phosphate (XIX) (278–281) inhibit the enzyme by alkylating a single glutamate residue. The isolation of homologous active site

X = Cl, Br, I

XVIII   XIX

peptides containing the reactive glutamate residue from rabbit muscle (275, 279), chicken muscle (273, 282), human erythrocyte (272), and yeast (274) triose phosphate isomerase underscores the essential role of glutamate in the catalytic mechanism.

Analysis of the reaction products formed by the inactivation of chicken muscle triose phosphate isomerase by bromohydroxyacetone phosphate revealed that this affinity label can serve as a crosslinking reagent at the active site of this enzyme (273). After triose phosphate isomerase is inactivated by bromohydroxyacetone ^{32}P-phosphate, the inactive enzyme product loses its label upon overnight dialysis without regain of enzymic activity. Inactive enzyme prepared from ^{14}C-labeled inhibitor does not lose its radioactivity under comparable conditions. Reduction by sodium borohydride prevents the loss of the ^{32}P label.

Isolation of a labeled hexapeptide from inactive enzyme that had been reduced by borohydride showed that glutamate is the site of alkylation by the affinity label. The sequence of the active site peptide is Ala-Tyr-Glu-Pro-Val-Trp. When the same

30.

peptide was isolated from inactive enzyme that had not been reduced with boro-hydride, the adjacent tyrosine was modified. A migration of affinity label from the initial site of alkylation, the glutamate residue, to the neighboring tyrosine residue as indicated in equation 30 is therefore consistent with all experimental results. Sodium borohydride reduction limits the intramolecular migration since the phosphate in the glycerol derivative is less susceptible to nucleophilic attack than the phosphate on the α-carbon of an acetone derivative. These results indicate that the phenolic side chain of the tyrosine residue is adjacent to the γ-carboxylate of the glutamate in the tertiary structure of the enzyme. They also illustrate the complexity of product analysis using multifunctional affinity labels.

Mandelate Racemase

Mandelate racemase from *Pseudomonas putida* is irreversibly and stoichiometrically inhibited by D,L-α-phenylglycidate (XX), a structural analog of mandelate (XXI) (283). The release of β-phenylglyceric acid from the modified enzyme provides tentative support for the modification of an aspartate or glutamate residue (283). Acidic residues seem to be the preferred, but not the exclusive, site of attack by affinity labels containing epoxides (3).

XX   XXI

Both the inactivation of the enzyme by XX and the enzyme-catalyzed racemization of mandelate require a tightly bound magnesium ion (284). If one assumes that magnesium ion is coordinated by the oxygen of XX and XXI, internally consistent mechanisms for both processes can be offered (283, 284). In the inactivation reaction with XX, the function of the metal ion would be to polarize the epoxide in order to make it more susceptible for nucleophilic attack by the active site aspartate or glutamate (283). In the enzyme-catalyzed reaction, the coordination by the oxygen would facilitate the formation of the probable carbanion intermediate by making the α-proton more acidic (284).

LIGASES

Aminoacyl-tRNA Synthetase

Enzymes that utilize macromolecular substrates, such as the aminoacyl-tRNA synthetases, present unique problems in the study of their active site chemistry by chemical modification. Clearly, the synthesis of affinity labels using tRNA as the carrier ligand will be limited by the available techniques of nucleic acid chemistry.

Two approaches that have been successful in the selective modification of aminoacyl-tRNA synthetase include photo-induced covalent bond formation of the enzyme-tRNA complex and acylation of the α-amino group of aminoacyl-tRNA with reactive groups.

The photo-induced formation of a covalent nucleic acid-enzyme complex was first reported with DNA and RNA polymerase from *E. coli* using alternating dA-dT copolymers (285). Irradiation of tRNA^{tyr} in the presence of *E. coli* tyrosyl-tRNA, followed by selective nuclease digestion, demonstrated that covalent bonds with the enzyme were formed with the dihydrouridine arm, the anticodon, and the extra loop of tRNA^{tyr} (286).

p-Nitrophenyl-carbamylmethionyl-tRNA has been used to modify methionyl-tRNA synthetases from *E. coli* (287). A single lysyl residue at the active site was modified and the sequence of the active site peptide containing it was determined. Similarly, N-bromoacetyl-isoleucyl-tRNA inhibits isoleucyl-tRNA synthetase from *E. coli* (288, 289). N-Bromoacetyl-isoleucine is not an effective inhibitor. Irreversible inhibition of the enzyme could be blocked with tRNA^{ile} but not by tRNA^{phe}. The amino acid residue modified has not yet been identified.

CONCLUDING REMARKS

The studies cited in this review illustrate different ways that chemical modification can be used to study enzyme mechanisms. The increasing use of suicide substrates merits particular attention. The various suicide substrates discussed above for lactate oxidase, monoamine oxidase, thymidylate synthetase, aspartate transaminase, aspartate β-decarboxylase, and β-hydroxydecanoyl thioester dehydrase have revealed and/or confirmed details of the active site chemistry of these enzymes in a satisfying manner. Because of their inherent specificity, the design of suicide substrates may represent the best strategy in developing pharmacologically useful enzyme inhibitors.

Sodium borohydride and pyridoxal 5-phosphate have emerged as two reagents of impressive generality. In addition to the extensive use of sodium borohydride in reducing imines, recent work has shown that this reducing agent can also be used to trap thioester and acylphosphate intermediates. Pyridoxal 5-phosphate has exhibited remarkable specificity for lysine residues at the binding sites for phosphorylated substrates of many enzymes. The ability of pyridoxal 5-phosphate in its Schiff base linkage with several of these enzymes to cause the photosensitized oxidation of neighboring amino acid residues has greatly extended its utility.

The accurate interpretation of the results of chemical modification experiments by group-specific and affinity labeling reagents remains a central problem. However, examination of modified enzymes by magnetic resonance and X-ray crystallographic techniques has permitted in some cases precise structural explanations for the loss of biological activity caused by these reagents. The introduction of new reagents and procedures to assess the importance of steric effects associated with modification of a given amino residue has also facilitated the interpretation of chemical modification experiments. Parallel chemical modification studies on the same enzyme from different biological sources also help to evaluate whether a reactive amino acid is of fundamental importance to the catalytic process.

Perhaps, in the future, the chemical synthesis of modified enzymes will replace chemical modification experiments as a method to identify which amino acid residues play functional roles in catalysis. However, certain types of chemical modification experiments will still be essential to define the precise nature of these catalytic functions.

ACKNOWLEDGMENTS

We wish to thank Ms. Carole Feingold and Ms. Katherine Kanamori for their assistance in the preparation of this manuscript.

Literature Cited

1. Mildvan, A. S. 1974. *Ann. Rev. Biochem.* 43:357
2. Kirsch, J. F. 1973. *Ann. Rev. Biochem.* 42:205
3. Glazer, A. N., DeLange, R. J., Sigman, D. S. 1975. *Selected Procedures for the Chemical Modification of Proteins.* Amsterdam: Elsevier. In press
4. Hirs, C. H. W. 1967. *Methods Enzymol.* Vol. 11
5. Hirs, C. H. W., Timasheff, S. N. 1972. *Methods Enzymol.* Vol. 25
6. Means, G. E., Feeney, R. E. 1971. *Chemical Modification of Proteins.* San Francisco: Holden-Day
7. Singer, S. J. 1967. *Advan. Protein Chem.* 22:1
8. Baker, B. R. 1967. *Design of Active Site Directed Irreversible Enzyme Inhibitors.* New York: Wiley
9. Knowles, J. R. 1972. *Accounts Chem. Res.* 5:155
10. Cohen, L. A. 1968. *Ann. Rev. Biochem.* 37:695
11. Cohen, L. A. 1970. *Enzymes* 1:147
12. Freedman, R. B. 1971. *Quart. Rev.* 25:431
13. Glazer, A. N. 1970. *Ann. Rev. Biochem.* 39:101
14. Glazer, A. N. 1975. *Proteins 2.* In press
15. Riordan, J. F., Sokolovsky, M. 1971. *Accounts Chem. Res.* 4:353
16. Stark, G. R. 1970. *Advan. Protein Chem.* 24:261
17. Spande, T. F., Witkop, B., Degani, Y., Patchornik, A. 1970. *Advan. Protein Chem.* 24:97
18. Shaw, E. 1970. *Enzymes* 1:91
19. Shaw, E. 1970. *J. Physiol. Rev.* 506:244
20. Vallee, B. L., Riordan, J. F. 1969. *Ann. Rev. Biochem.* 38:733
21. Rando, R. 1974. *Science* 185:320
22. Lederer, F. 1974. *Eur. J. Biol.* 46:393
23. Walsh, C. T., Abeles, R. H., Kaback, H. R. 1971. *J. Biol. Chem.* 247:7858
24. Walsh, C. T., Schonbrunn, A., Lockridge, O., Massey, V., Abeles, R. H. 1972. *J. Biol. Chem.* 247:6004
25. Bloch, K. 1969. *Accounts Chem. Res.* 2:193
26. Hamilton, G. 1971. *Progr. Bioorg. Chem.* 1:83
27. Walsh, C., Lockridge, O., Massey, V., Abeles, R. 1973. *J. Biol. Chem.* 248:7049
28. Patek, D. R., Hellerman, L. 1974. *J. Biol. Chem.* 249:2373
29. Hafner, E. W., Wellner, D. 1971. *Proc. Nat. Acad. Sci. USA* 68:987
30. Hellerman, L., Coffey, D. S. 1967. *J. Biol. Chem.* 242:582
31. Walsh, C. T., Schonbrunn, A., Abeles, R. H. 1971. *J. Biol. Chem.* 246:6855
32. Walsh, C. T., Krodel, E., Massey, V., Abeles, R. H. 1973. *J. Biol. Chem.* 248:1946
33. Porter, D. J. T., Voet, J. G., Bright, H. J. 1973. *J. Biol. Chem.* 248:4400
34. Rando, R. R. 1973. *J. Am. Chem. Soc.* 95:4438
35. Hellerman, L., Erwin, V. L. 1968. *J. Biol. Chem.* 243:5234
36. Chuang, H. Y. K., Patek, D. R., Hellerman, L. 1974. *J. Biol. Chem.* 249:2381
37. Blakley, R. L. 1969. *Biochemistry of Folic Acid and Related Pteridines,* 219. New York: Wiley
38. Heidelberger, C. 1970. *Cancer Res.* 30:1549
39. Heidelberger, C. et al 1957. *Nature* 179:663
40. Heidelberger, C., Parsons, D. G., Remy, D. G. 1964. *J. Med. Chem.* 7:1
41. Heidelberger, C. et al 1960. *Cancer Res.*

20:903

42. Hartmann, K. U., Heidelberger, C. 1961. *J. Biol. Chem.* 236:3006

43. Cohen, S. S. et al 1958. *Proc. Nat. Acad. Sci. USA* 44:1004

44. Santi, D. V., McHenry, C. S. 1972. *Proc. Nat. Acad. Sci. USA* 69:1855

45. Santi, D. V., McHenry, C. S., Sommer, H. 1974. *Biochemistry* 13:471

46. Langenbach, R. J., Danenberg, P. V., Heidelberger, C. 1972. *Biochem. Biophys. Res. Commun.* 48:1565

47. Danenberg, P. V., Langenbach, R. J., Heidelberger, C. 1974. *Biochemistry* 13:926

48. Galivan, J., Maley, G. F., Maley, F. 1974. *Biochemistry* 13:2282

49. Sommer, H., Santi, D. V. 1974. *Biochem. Biophys. Res. Commun.* 57:689

50. Fox, J. J., Miller, N. C., Cushley, R. J. 1966. *Tetrahedron Lett.* 40:4927

51. Otter, B. A., Falco, E. A., Fox. J. J. 1969. *J. Org. Chem.* 34:1390

52. Reist, E. J., Benitez, A., Goodman, L. 1964. *J. Org. Chem.* 29:554

53. Santi, D. V., Brewer, C. F. 1968. *J. Am. Chem. Soc.* 90:6236

54. Jenne, J. W., Boyer, P. D. 1962. *Biochim. Biophys. Acta* 65:121

55. Riddle, B., Jencks, W. P. 1971. *J. Biol. Chem.* 246:3250

56. Steinberg, M. S., Cohen, S. N., Weber, W. W. 1971. *Biochim. Biophys. Acta* 235:89

57. Roskoski, R. Jr. 1973. *Biochemistry* 12:3709

58. Reisberg, R. B. 1954. *Biochim. Biophys. Acta* 14:442

59. Tabor, H., Mehler, A. H., Stadtman, E. R. 1953. *J. Biol. Chem.* 204:127

60. Roskoski, R. Jr. 1974. *J. Biol. Chem.* 249:2156

61. Sluyterman, L. A. 1968. *Biochim. Biophys. Acta* 151:178

62. Jencks, W. P. et al 1972. *J. Biol. Chem.* 247:3756

63. Gehring, U., Lynen, F. 1972. *Enzymes* 7:391

64. Kornblatt, J. A., Rudney, H. 1971. *J. Biol. Chem.* 246:4417

65. Middleton, B. 1974. *Biochem. J.* 139:109

66. Holland, P. C., Clark, M. G., Bloxham, D. P. 1972. *Biochemistry* 12:3309

67. Lynen, F. 1953. *Fed. Proc.* 12:683

68. Goldman, D. S. 1954. *J. Biol. Chem.* 208:345

69. Gehring, U., Riepertinger, C., Lynen, F. 1968. *Eur. J. Biochem.* 6:264

70. Gehring, U., Harris, J. I. 1970. *Eur. J Biochem.* 16:4192

71. Solomon, F., Jencks, W. P. 1969. *J. Biol. Chem.* 244:1079

72. Degani, C., Boyer, P. D. 1973. *J. Biol. Chem.* 248:8222

73. Watts, D. C. 1973. *Enzymes* 8:383

74. Lui, N. S. T., Cunningham, L. 1966. *Biochemistry* 5:144

75. Milner-White, E. J., Watts, D. C. 1971. *Biochem. J.* 122:727

76. O'Sullivan, W. J., Diefenbach, H., Cohn, M. 1966. *Biochemistry* 5:2666

77. Watts, D. C., Rabin, B. R. 1962. *Biochem. J.* 85:507

78. Kumudavalli, I., Moreland, B. H., Watts, D. C. 1970. *Biochem. J.* 117:513

79. Hooton, B. T. 1968. *Biochemistry* 7:2063

80. Dawson, D. M., Eppenberger, H. M., Kaplan, N. O. 1967. *J. Biol. Chem.* 242:210

81. Atherton, R. S., Laus, J. F., Thomson, A. R. 1970. *Biochem. J.* 118:903

82. Smith, D. J., Kenyon, G. L. 1974. *J. Biol. Chem.* 249:3317

83. Smith, D. J., Maggio, E. T., Kenyon, G. L. In press

84. Kassab, R., Rouston, C., Pradel, L. A. 1968. *Biochim. Biophys. Acta* 167:316

85. James, T. L., Cohn, M. 1974. *J. Biol. Chem.* 249:2599

86. Reed, G. H., Cohn, M. 1972. *J. Biol. Chem.* 247:3073

87. Reed, G. H., McLaughlin, A. C. 1974. *Ann. NY Acad. Sci.* 222:118

88. Jacobson, G. R., Stark, G. R. 1973. *Enzymes* 9:225

89. Vanaman, T. C., Stark, G. R. 1970. *J. Biol. Chem.* 245:3565

90. Benisek, W. F. 1971. *J. Biol. Chem.* 246:3151

91. Gerhart, J. C., Schachman, H. K. 1968. *Biochemistry* 7:538

92. Jacobson, G. R., Stark, G. R. 1973. *J. Biol. Chem.* 248:8003

93. Evans, D. R., McMurray, C. H., Lipscomb, W. N. 1972. *Proc. Nat. Acad. Sci. USA* 69:3638

94. McMurray, C. H., Evans, D. R., Sykes, B. D. 1972. *Biochem. Biophys. Res. Commun.* 48:572

95. Greenwell, P., Jewett, S. L., Stark, G. R. 1973. *J. Biol. Chem.* 248:5994

96. Snell, E. E., Di Mari, S. J. 1970. *Enzymes* 2:335

97. Snell, E. E. 1962. *Brookhaven Symp. Biol.* 15:32

98. Braunstein, A. E. 1973. *Enzymes* 9:379

99. Okamoto, M., Morino, Y. 1972. *Biochemistry* 11:3188

100. Morino, Y., Okamoto, M. 1972. *Biochemistry* 11:3196

101. Wilson, K. J., Birchmeier, W., Christen,

P. 1974. *Eur. J. Biol.* 41:471
102. Bocharov, A. L., Demidkina, T. V., Karpeisky, M. Ya., Polyanovsky, O. L. 1973. *Biochem. Biophys. Res. Commun.* 50:377
103. Polyanovsky, O. L. et al 1973. *FEBS Lett.* 35:322
104. Torchinsky, Yu. M., Zufarova, R. A., Agalarova, M. B., Severin, E. S. 1972. *FEBS Lett.* 28:302
105. Cournil, I., Arrio-Dupont, M. 1971. *Biochem. Biophys. Res. Commun.* 43:40
106. Birchmeier, W., Wilson, K. J., Christen, P. 1973. *J. Biol. Chem.* 248:1751
107. Hammes, G. G., Haslam, J. L. 1968. *Biochemistry* 7:1519
108. Jenkins, W. T. 1964. *J. Biol. Chem.* 239:1742
109. Hammes, G. G., Haslam, J. L. 1969. *Biochemistry* 8:1591
110. Birchmeier, W., Christen, P. 1971. *FEBS Lett.* 18:209
111. Birchmeier, W., Wilson, K. J., Christen, P. 1972. *FEBS Lett.* 26:113
112. Morino, Y., Okamoto, M. 1970. *Biochem. Biophys. Res. Commun.* 40:600
113. Okamoto, M., Morino, Y. 1973. *J. Biol. Chem.* 248:82
114. John, R. A., Fasella, P. 1969. *Biochemistry* 8:4477
115. Morino, Y., Okamoto, M. 1972. *Biochem. Biophys. Res. Commun.* 47:498
116. Morino, Y., Okamoto, M. 1973. *Biochem. Biophys. Res. Commun.* 50:1061
117. Hess, G. P. 1971. *Enzymes* 3:213
118. Keil, B. 1971. *Enzymes* 3:249
119. Shaw, E. 1971. *Enzymes* 1:91
120. Bender, M. L., Kezdy, F. J. 1965. *Ann. Rev. Biochem.* 34:49
121. Hartley, B. S. 1960. *Ann. Rev. Biochem.* 29:45
122. Fink, A. L., Good, N. E. 1974. *Biochem. Biophys. Res. Commun.* 58:126
123. Semeriva, M., Chapus, C., Bovier-Lapierre, C., Desnuelle, P. 1974. *Biochem. Biophys. Res. Commun.* 58:808
124. Rosenberry, T. L., Bernhard, S. 1971. *Biochemistry* 10:4114
125. Mooser, G., Schulman, H., Sigman, D. S. 1972. *Biochemistry* 11:1595
126. Brown, W. E., Wold, F. 1973. *Biochemistry* 12:828
127. Brown, W. E., Wold, F. 1973. *Biochemistry* 12:835
128. Twu, J.-S., Wold, F. 1973. *Biochemistry* 12:381
129. Twu, J.-S., Chin, C. C. Q., Wold, F.

1973. *Biochemistry* 12:2856
130. Thompson, R. C. 1973. *Biochemistry* 12:47
131. Glazer, A. N. 1968. *Proc. Nat. Acad. Sci. USA* 59:996
132. Glazer, A. N. 1968. *J. Biol. Chem.* 243:3693
133. Philipp, M., Bender, M. 1971. *Proc. Nat. Acad. Sci. USA* 68:478
134. Lienhard, G. E., Secemski, I. I., Koehler, K. A., Lindquist, R. N. 1971. *Cold Spring Harbor Symp. Quant. Biol.* 36:45
135. Robertus, J. D., Kraut, J., Alden, R. A., Birktoft, J. 1972. *Biochemistry* 11:4293
136. Aoyagi, T. et al 1969. *J. Antibiot.* 22:558
137. Ruhlmann, A. et al 1973. *J. Mol. Biol.* 77:417
138. Kraut, J. 1971. *Enzymes* 3:165
139. Blow, D. M., Birktoft, J., Hartley, B. S. 1969. *Nature* 221:337
140. Steitz, T. A., Henderson, R., Blow, D. M. 1969. *J. Mol. Biol.* 46:337
141. Robillard, G., Shulman, R. G. 1972. *J. Mol. Biol.* 71:507
142. Morgan, P. H., Robinson, N. C., Walsh, K. A., Neurath, H. 1972. *Proc. Nat. Acad. Sci. USA* 69:3312
143. Gertler, A., Walsh, K. A., Neurath, H. 1974. *Biochemistry* 13:1302
144. Kassell, B., Kay, J. 1973. *Science* 180:1022
145. Kay, J., Kassell, B. 1971. *J. Biol. Chem.* 246:6661
146. Wasi, S., Hofmann, T. 1973. *Can. J. Biochem.* 51:797
147. Gertler, A., Walsh, K. A., Neurath, H. 1974. *FEBS Lett.* 38:157
148. Morgan, P. H., Walsh, K. A., Neurath, H. 1974. *FEBS Lett.* 41:108
149. Fruton, J. S. 1971. *Enzymes* 3:119
150. Dopheide, T. A., Moore, S., Stein, W. H. 1967. *J. Biol. Chem.* 242:1833
151. Huang, W. Y., Tang, J. 1969. *Fed. Proc.* 28:2753
152. Foltmann, B., Hartley, B. S. 1967. *Biochem. J.* 104:1064
153. Erlanger, B. F., Vratsanos, S. M., Wasserman, N., Cooper, A. G. 1965. *J. Biol. Chem.* 240:3447
154. Gross, E., Morell, J. 1966. *J. Biol. Chem.* 241:3638
155. Tang, J. 1971. *J. Biol. Chem.* 246:4510
156. Chen, K. C. S., Tang, J. 1972. *J. Biol. Chem.* 247:2566
157. Delpierre, G. R., Fruton, J. S. 1966. *Proc. Nat. Acad. Sci. USA* 56:1821
158. Rajagopalan, T. G., Stein, W. H., Moore, S. 1966. *J. Biol. Chem.* 241:4795
159. Bayliss, R. S., Knowles, J. R.,

Wybrandt, G. B. 1969. *Biochem. J.* 113: 377

160. Erlanger, B. F., Vratsanos, S. M., Wasserman, N., Cooper, A. G. 1967. *Biochem. Biophys. Res. Commun.* 28: 203

161. Fry, K. T., Kim, O.-K., Spona, J., Hamilton, G. A. 1970. *Biochemistry* 9: 4624

162. Hartsuck, J. A., Tang, J. 1972. *J. Biol. Chem.* 247: 2575

163. Chen, H. J., Kaiser, E. T. 1974. *J. Am. Chem. Soc.* 96: 625

164. Hunkapiller, M., Heinze, J. E., Mills, J. N. 1970. *Biochemistry* 9: 2897

165. Reid, T. W., Fahrney, D. 1967. *J. Am. Chem. Soc.* 89: 3941

166. Husain, S. S., Ferguson, J. B., Fruton, J. S. 1971. *Proc. Nat. Acad. Sci. USA*: 68: 2765

167. Rawn, J. D., Lienhard, G. E. 1974. *Biochem. Biophys. Res. Commun.* 56: 654

168. Inagami, T., Misuno, K., Michelakis, A. M. 1974. *Biochem. Biophys. Res. Commun.* 56: 503

169. Sodek, J., Hofmann, T. 1968. *J. Biol. Chem.* 243: 450

170. Hartsuck, J. A., Lipscomb, W. N. 1971. *Enzymes* 3: 1

171. Vallee, B. L., Riordan, J. F. 1968. *Brookhaven Symp. Biol.* 21: 91

172. Kaiser, E. T., Kaiser, B. L. 1972. *Accounts. Chem. Res.* 5: 219

173. Hass, G. M., Neurath, H. 1971. *Biochemistry* 10: 3535

174. Hass, G. M., Neurath, H. 1971. *Biochemistry* 10: 3541

175. Petra, P. H. 1971. *Biochemistry* 10: 3163

176. Petra, P. H., Neurath, H. 1971. *Biochemistry* 10: 3171

177. Carson, F. W., Kaiser, E. T. 1966. *J. Am. Chem. Soc.* 88: 1212

178. Auld, D. S., Vallee, B. L. 1970. *Biochemistry* 9: 4352

179. Hass, G. M., Govier, M. A., Grahn, D. T., Neurath, H. 1972. *Biochemistry* 11: 3787

180. Plummer, T. H. Jr. 1971. *J. Biol. Chem.* 246: 2930

181. Kimmel, M. T., Plummer, T. H. Jr. 1972. *J. Biol. Chem.* 247: 7864

182. Vallee, B. L., Riordan, J. F., Coleman, J. E. 1963. *Proc. Nat. Acad. Sci. USA* 49: 109

183. Simpson, R. T., Riordan, J. F., Vallee, B. L. 1963. *Biochemistry* 2: 616

184. Vallee, B. L. et al 1968. *Biochemistry* 7: 3547

185. Barber, A. K., Fisher, J. R. 1972. *Proc. Nat. Acad. Sci. USA* 69: 2970

186. French, T. C., Yu, N.-T., Auld, D. S. 1974. *Biochemistry* 13: 2877

187. Johansen, J. T., Vallee, B. L. 1971. *Proc. Nat. Acad. Sci. USA* 68: 2532

188. Johansen, J. T., Livingston, D. M., Vallee, B. L. 1972. *Biochemistry* 11: 2584

189. Lipscomb, W. N. 1973. *Proc. Nat. Acad. Sci. USA* 70: 3797

190. Johansen, J. T., Vallee, B. L. 1973. *Proc. Nat. Acad. Sci. USA* 70: 2006

191. Riordan, J. F., Muszynska, G. 1974. *Biochem. Biophys. Res. Commun.* 57: 447

192. Quiocho, F. A., McMurray, C. H., Lipscomb, W. N. 1972. *Proc. Nat. Acad. Sci. USA* 69: 2850

193. Hayashi, R., Moore, S., Stein, W. H. 1973. *J. Biol. Chem.* 248: 2296

194. Hayashi, R., Moore, S., Stein, W. H. 1973. *J. Biol. Chem.* 248: 8366

195. Ihle, J. N., Dure, L. S. 1972. *J. Biol. Chem.* 247: 5034

196. Ihle, J. N., Dure, L. S. 1972. *J. Biol. Chem.* 247: 5041

197. Dennis, E. A. 1973. *Arch. Biochem. Biophys.* 158: 485

198. Volwerk, J. J., Pieterson, W. A., de Haas, G. H. 1974. *Biochemistry* 13: 1446

199. Pieterson, W. A., Vidal, J. C., Volwerk, J. J., de Haas, G. H. 1974. *Biochemistry* 13: 1455

200. Pieterson, W. A., Volwerk, J. J., de Haas, G. H. 1974. *Biochemistry* 13: 1439

201. Wells, M. A. 1974. *Biochemistry* 13: 2265

202. Wells, M. A. 1973. *Biochemistry* 12: 1086

203. Wells, M. A. 1974. *Biochemistry* 13: 2258

204. Sigman, D. S., Jorgensen, C. T. 1972. *J. Am. Chem. Soc.* 94: 1724

205. Werber, M. M., Shalitin, Y. 1973. *Bioorg. Chem.* 2: 221

206. Fulton, J. E. Jr., Noble, N. L., Bradley, S., Awad, W. M. Jr. 1974. *Biochemistry* 13: 2320

207. Maylie, M. F., Charles, M., Desnuelle, D. 1972. *Biochim. Biophys. Acta* 276: 162

208. Marglin, A., Merrifield, R. B. 1970. *Ann. Rev. Biochem.* 39: 841

209. Richards, F. M., Wyckoff, H. W. 1971. *Enzymes* 4: 647

210. Anfinsen, C. B., Cuatrecasas, P., Taniuchi, H. 1971. *Enzymes* 4: 177

211. Chaiken, I. M., Sanchez, G. R. 1972. *J. Biol. Chem.* 247: 6743

212. Chaiken, I. M. 1972. *J. Biol. Chem.* 247: 1999

213. Sanchez, G. R., Chaiken, I. M., Anfinsen, C. B. 1973. *J. Biol. Chem.* 248:3653
214. Lin, M. C., Gutte, B., Moore, S., Merrifield, R. B. 1970. *J. Biol. Chem.* 245:5169
215. Lin, M. C., Gutte, B., Caldi, D. G., Moore, S., Merrifield, R. B. 1972. *J. Biol. Chem.* 247:4768
216. Crestfield, A. M., Stein, W. H., Moore, S. 1963. *J. Biol. Chem.* 238:2413
217. Gutte, B., Lin, M. C., Caldi, D. G., Merrifield, R. B. 1972. *J. Biol. Chem.* 247:4763
218. Dunn, B. M., DiBello, C., Anfinsen, C. B. 1973. *J. Biol. Chem.* 248:4769
219. Poulos, T. L., Price, P. A. 1974. *J. Biol. Chem.* 249:1453
220. Weiner, H., White, W. N., Hoare, D. G., Koshland, D. E. Jr. 1966. *J. Am. Chem. Soc.* 88:16
221. Gold, A. M. 1965. *Biochemistry* 4:897
222. Melzer, M. S. 1969. *Can. J. Biochem.* 47:987
223. Oshima, R. G., Price, P. A. 1973. *J. Biol. Chem.* 248:7522
224. Price, P. A., Moore, S., Stein, W. H. 1969. *J. Biol. Chem.* 244:924
225. Salnikow, J., Liao, T.-H., Moore, S., Stein, W. H. 1973. *J. Biol. Chem.* 248:1480
226. Liao, T.-H., Salnikow, J., Moore, S., Stein, W. H. 1973. *J. Biol. Chem.* 248:1489
227. Plapp, B. V. 1973. *J. Biol. Chem.* 248:4896
228. Fischer, E. H., Kent, A. B., Snyder, E. R., Krebs, E. G. 1958. *J. Am. Chem. Soc.* 80:2906
229. Huang, Y. Z., Snell, E. E. 1972. *J. Biol. Chem.* 247:7358
230. Sabo, D. L., Fischer, E. H. 1974. *Biochemistry* 13:670
231. Tate, S. S., Relyea, N. M., Meister, A. 1969. *Biochemistry* 8:5016
232. Relyea, N. M., Tate, S. S., Meister, A. 1974. *J. Biol. Chem.* 249:1519
233. Miles, E. W., Kumagi, H. 1974. *J. Biol. Chem.* 249:2843
234. Miles, E. W., McPhie, P. 1974. *J. Biol. Chem.* 249:2852
235. Miles, E. W. 1974. *Biochem. Biophys. Res. Commun.* 57:849
236. Mauk, M. R., Girotti, A. W. 1974. *Biochemistry* 13:1757
237. Fridovich, I., Westheimer, F. H. 1962. *J. Am. Chem. Soc.* 84:3208
238. Warren, S., Zerner, B., Westheimer, F. H. 1966. *Biochemistry* 5:817
239. Autor, A. P., Fridovich, I. 1970. *J. Biol. Chem.* 245:5214
240. Kluger, R., Nakaoka, K. 1974. *Biochemistry* 13:910
241. Dowhan, W., Wickner, W. T., Kennedy, E. P. 1974. *J. Biol. Chem.* 249:3079
242. Riley, W. D., Snell, E. E. 1968. *Biochemistry* 7:3520
243. Recsei, P., Snell, E. E. 1970. *Biochemistry* 9:1492
244. Horecker, B. L., Tsolas, O., Lai, C. Y. 1972. *Enzymes* 7:213
245. Collins, K. D. 1974. *J. Biol. Chem.* 249:136
246. Lebherz, H. G., Bradshaw, R. A., Rutter, W. J. 1973. *J. Biol. Chem.* 248:1660
247. Grazi, E., Cheng, T., Horecker, B. L. 1962. *Biochem. Biophys. Res. Commun.* 7:250
248. Horecker, B. L., Rowley, P. T., Grazi, E., Cheng, T., Tsolas, O. 1963. *Biochem. Z.* 338:36
249. Avigard, G., Englord, S. 1972. *Arch. Biochem. Biophys.* 153:337
250. Model, P., Ponticorvo, L., Rittenberg, D. 1968. *Biochemistry* 7:1339
251. Cash, D. J., Wilson, I. B. 1966. *J. Biol. Chem.* 241:4290
252. Kaplan, H., Stevenson, A. K. J., Hartley, B. S. 1974. *Biochem. J.* 124:289
253. Anderson, P. J., Kaplan, H. 1974. *Biochem. J.* 137:181
254. Schmidt, D. E. Jr., Westheimer, F. H. 1971. *Biochemistry* 10:1249
255. Shapiro, S., Enser, M., Pugh, E., Horecker, B. L. 1968. *Arch. Biochem. Biophys.* 128:554
256. Rippa, M., Spanio, L., Pontremoli, S. 1967. *Arch. Biochem. Biophys.* 118:48
257. Davis, L. C., Brox, L. W., Gracy, R. W., Ribereau-Gayon, G., Horecker, B. L. 1970. *Arch. Biochem. Biophys.* 140:215
258. Davis, L. C., Ribereau-Gayon, G., Horecker, B. L. 1971. *Proc. Nat. Acad. Sci. USA* 68:416
259. Hartman, F. C., Welch, M. H. 1974. *Biochem. Biophys. Res. Commun.* 57:85
260. Hartman, F. C., Suh, B., Welch, M. H., Barker, R. 1973. *J. Biol. Chem.* 248:8233
261. Drechsler, E. R., Boyer, P. D., Kowalsky, A. G. 1959. *J. Biol. Chem.* 234:2627
262. Rose, I. A., O'Connell, E. L., Mehler, A. H. 1965. *J. Biol. Chem.* 240:1758
263. Grazi, E. 1974. *Biochem. Biophys. Res. Commun.* 56:106
264. Helmkamp, G. M. Jr., Bloch, K. 1969. *J. Biol. Chem.* 244:6014
265. Helmkamp, G. M. Jr., Rando, R. R., Brock, D. J. H., Bloch, K. 1968. *J. Biol. Chem.* 243:3229
266. Endo, K., Helmkamp, G. M. Jr., Bloch, K. 1970. *J. Biol. Chem.* 245:4293

267. Morisaki, M., Bloch, K. 1972. *Biochemistry* 11:309
268. Rose, I. A. 1962. *Brookhaven Symp. Biol.* 15:293
269. Rose, I. A. 1970. *Enzymes* 2:290
270. Reider, S. V., Rose, I. A. 1959. *J. Biol. Chem.* 234:1007
271. Wolfenden, R. 1972. *Accounts Chem. Res.* 5:10
272. Hartman, F. C., Gracy, R. W. 1973. *Biochem. Biophys. Res. Commun.* 52:388
273. De La Mare, S., Coulson, A. F. W., Knowles, J. R., Priddle, J. D., Offord, R. E. 1972. *Biochem. J.* 129:321
274. Norton, I. L., Hartman, F. C. 1972. *Biochemistry* 11:4435
275. Hartman, F. C. 1970. *J. Am. Chem. Soc.* 92:2170
276. Coulson, A. F. W., Knowles, J. R., Offord, R. E. 1970. *Chem. Commun.* 7
277. Coulson, A. F. W., Knowles, J. R., Priddle, J. D., Offord, R. E. 1970. *Nature* 227:180
278. Schray, K. J., O'Connell, E. L., Rose, I. A. 1973. *J. Biol. Chem.* 248:2214
279. Miller, J. C., Waley, S. G. 1971. *Biochem. J.* 123:163
280. Waley, S. G., Miller, J. C., Rose, I. A., O'Connell, E. L. 1970. *Nature* 227:181
281. Rose, I. A., O'Connell, E. L. 1969. *J. Biol. Chem.* 244:6548
282. Priddle, J., Offord, R. E. 1974. *FEBS Lett.* 39:349
283. Fee, J. A., Hegeman, G. D., Kenyon, G. L. 1974. *Biochemistry* 13:2533
284. Fee, J. A., Hegeman, G. D., Kenyon, G. L. 1974. *Biochemistry* 13:2529
285. Markovitz, A. 1972. *Biochim. Biophys. Acta* 281:522
286. Schoemaker, H. J. P., Schimmel, P. R. 1974. *J. Mol. Biol.* 84:503
287. Bruton, C. J., Hartley, B. S. 1970. *J. Mol. Biol.* 52:165
288. Santi, D. V., Marchant, W., Yarus, M. 1973. *Biochem. Biophys. Res. Commun.* 51:370
289. Santi, D. V., Cunnion, S. O. 1974. *Biochemistry* 13:481

AMINO ACID METABOLISM IN MAN   �ష905

Philip Felig

Department of Internal Medicine, Yale University School of Medicine,
New Haven, Connecticut 06510

CONTENTS

Much of the interest in amino acid metabolism has centered on the intermediary steps involved in catabolism, the regulation of biosynthesis and transport, and genetic disorders resulting in abnormal urinary and/or plasma concentrations of amino acids. This review will focus on recent studies in which amino acid exchange between various organs has been examined in a variety of physiologic and pathologic conditions in intact humans. Particular emphasis, reflecting my own interest, will be focused on the relationship of amino acid metabolism to gluconeogenesis, its interaction with glucoregulatory hormones, and its role in nitrogen exchange. The technique used in determining tissue exchange of amino acids in many of the studies to be reviewed involves simultaneous sampling of arterial and venous blood across various organs and measuring blood flow. Where applicable, reference will be made to studies in experimental animals, although the data in humans form the overall basis of this review.

AMINO ACID EXCHANGE IN THE POSTABSORPTIVE STATE

The concentrations of free amino acids in plasma have been well defined in normal subjects and show relatively little inter- or intra-individual variation (1, 2). Maintenance of these steady-state concentrations is dependent upon the net balance between release from endogenous protein stores and utilization by various tissues. Since muscle accounts for well over 50% of the total body pool of free amino acids (3) and since the liver is the repository of the urea cycle enzymes necessary for nitrogen disposal, these two organs would be expected to play a major role in determining the circulating levels and turnover of amino acids.

Muscle

Examination of amino acid exchange across the deep tissues of the human forearm demonstrates that in normal man in the postabsorptive state (i.e. following a 12–14 hr overnight fast), there is a net release of amino acids from muscle tissue, as reflected in consistently negative arteriovenous differences (4–6). The pattern of this release is quite distinctive, the output of alanine and glutamine exceeding that of all other amino acids and accounting for over 50% of total alpha amino nitrogen release (5, 7). The same pattern is observed if one examines amino acid exchange across the leg by determining arteriofemoral venous differences (8, 9). This net negative balance is also demonstrable in cardiac muscle. In the heart, the output of alanine shows even greater predominance than in skeletal muscle tissue, accounting for over 80% of total amino acid release (10). In contrast to the net outputs from muscle tissue observed for most amino acids, small but consistent uptakes by peripheral tissues are demonstrable for serine (5, 8), cystine (5, 8), and glutamate (7).

Splanchnic Tissues

Complementing the negative balance of amino acids in muscle tissue is the consistent uptake of amino acids across the splanchnic bed. Since the portal vein is not readily accessible for catheterization in normal man, studies directed at evaluating the role of the liver in human amino acid exchange have primarily involved measurements of arterial-hepatic venous differences (11–13). Such studies provide data on overall splanchnic balances, including contributions from the liver as well as extrahepatic tissues (the upper gastrointestinal tract, spleen, and pancreas). As in the case of peripheral output, alanine and glutamine predominate in the uptake of amino acids by splanchnic tissues. In fact, there is fairly close correspondence between the relative outputs of most amino acids from the periphery and their uptake by splanchnic tissues (Figure 1). A notable exception is serine, which is extracted by peripheral tissues as well as the splanchnic bed. The source of the serine consumed by these tissues is the kidney (see below). In contrast to all other amino acids, a consistent release from the splanchnic bed is observed only for citrulline.

Data on the relative contributions of hepatic and extrahepatic tissues to net splanchnic balances are available to a limited extent in humans (8, 9). The importance of distinguishing hepatic from gut uptakes derives from the fact that the liver is the key gluconeogenic organ in man, whereas extrahepatic splanchnic tissues do not contribute to glucose production. Measurements of arterial-portal venous differences in subjects studied at the time of elective surgery reveal that the gut in fasting man releases alanine and, to a lesser extent, other amino acids as well (8, 9). Thus, the splanchnic uptake of alanine, as determined by arterial-hepatic venous differences, represents an underestimate of true hepatic uptake, perhaps by as much as 50%. In marked contrast, there is a net uptake of glutamine by gut tissues (9, 14). In fact, comparison of arterial-portal and arterial-hepatic venous differences for glutamine (9) suggests that the gut, rather than the liver, is entirely responsible for the net uptake of this amino acid by the splanchnic bed.

Studies in experimental animals have confirmed and extended these observations on the relative contributions of the gut and hepatic tissues to net splanchnic amino acid balances. In the fasted rat (15, 16) as well as the dog (17), alanine is released from the gut and contributes approximately 50% to the total hepatic uptake of this amino acid (the remainder representing output from peripheral tissues). With respect to glutamine, a consistent uptake is observed by nonhepatic splanchnic tissues (15, 16). The overall importance of the gut as a site of glutamine catabolism is indicated by the

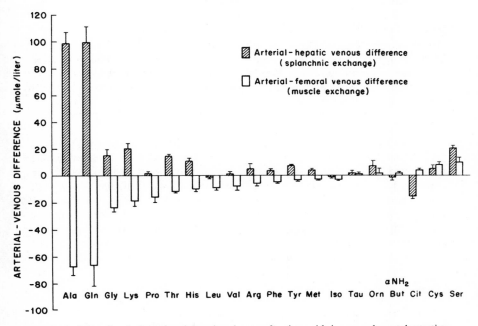

Figure 1 Net splanchnic and peripheral exchange of amino acids in normal, postabsorptive man. Splanchnic exchange is shown in the hatched bars as arterial-hepatic venous differences, and peripheral exchange is shown in open bars as arterial-femoral venous differences. The data are based on the studies of Felig & Wahren (8) and (for glutamine) Felig et al (9).

fact that evisceration causes a far greater accumulation of plasma glutamine than does nephrectomy (18). Studies in the rat also suggest that the glutamine extracted by the gut is the source of nitrogen utilized for the synthesis of the large amounts of alanine released by these tissues. Following the intravenous administration of glutamine, a specific increase in the output of alanine from the gut is observed (18).

From the foregoing data on extrahepatic splanchnic amino acid balances, the following conclusions regarding hepatic amino acid exchange may be drawn. Net hepatic uptake of alanine is approximately twice the measured rate of splanchnic consumption of this amino acid inasmuch as the gut contributes alanine in amounts almost equal to that released from the periphery. Since glutamine uptake by the splanchnic bed occurs entirely in nonhepatic tissues, alanine accounts for over 50% of total hepatic amino acid uptake. In addition to alanine, uptakes by the liver are demonstrable for serine, threonine, and glycine and, to a lesser extent, virtually all other glycogenic amino acids (Figure 1). Notable exceptions, however, are the branched chain amino acids, valine, leucine, and isoleucine, for which no consistent uptake (8, 12) and occasionally even an output from the splanchnic bed is observed (13, 19). The lack of hepatic uptake of the branched chain amino acids is consistent with studies in experimental animals, which indicate that extrahepatic tissues, rather than the liver, are the major sites of catabolism of valine, leucine, and isoleucine (20).

Data on the net balance of glutamine across the human liver are not available. Although there is little glutamine release from the perfused rat liver (21), in vivo studies demonstrate a brisk hepatic output in the intact fasted rat (16, 22). Indirect evidence suggesting an hepatic (or other extramuscular) site of glutamine production in man derives from studies of patients with a metabolic or respiratory acidosis. In these conditions, uptake of glutamine by the kidney is markedly increased, yet muscle output of this amino acid is not augmented (23).

Kidney

The net balance of amino acids across the kidney consists of an uptake of glutamine, proline, and glycine and an output of serine and alanine (24). As noted above, serine is extracted by liver as well as peripheral tissues, indicating that the kidney is the major source of release of this amino acid into the systemic circulation. In contrast, the contribution by the kidney to total alanine release (10–20 μmol/min) (24) is far smaller than that released from muscle (100 μmol/min) (5, 6). Precise data on the relative importance of the kidney and gut in the uptake and catabolism of glutamine are not available for man. However, studies with eviscerated, nephrectomized rats suggest that the gut may be quantitatively more important than the kidney in the clearance of glutamine from plasma (18).

Brain

An uptake of most amino acids by human brain tissue has been demonstrated by measuring arterial-jugular venous differences in normal subjects (25). The uptake of the branched chain amino acids, particularly valine, exceeds that of all other amino acids. In this regard, in vivo studies in the mouse indicate that valine uptake is more rapid than that of other amino acids (26). Furthermore, the capacity of rat brain to

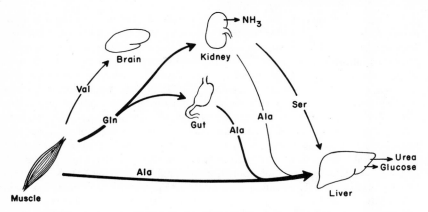

Figure 2   Interorgan amino acid exchange in normal postabsorptive man. The key role of alanine in amino acid output from muscle and gut and uptake by the liver is shown.

oxidize the branched chain amino acids is fourfold greater than that of muscle and liver (27). Inasmuch as significant amounts of valine, leucine, and isoleucine are released from muscle (5, 6) but not extracted by liver (8, 12, 13, 19), it is likely that the brain constitutes an important site of utilization of these amino acids.

Summary

The observations on the net balances of amino acids across muscle, liver, gastrointestinal tissues, kidney, and brain in normal man in the postabsorptive state clearly demonstrate the key role of alanine and glutamine in the overall flux of amino acids between tissues (Figure 2). Free amino acids are released from muscle and gut and are extracted by the liver. Alanine and glutamine account for more than 50% of the total alpha amino nitrogen released by muscle. The gut also releases substantial amounts of alpha amino nitrogen, primarily as alanine. The major site of alanine uptake is the liver, where its extraction exceeds that of all other amino acids. The kidney and gut are the major sites of glutamine uptake, which provides the nitrogen source for alanine synthesis in the gut (18) and ammoniagenesis in the kidney (28). The branched chain amino acids, particularly valine, escape hepatic uptake and are utilized by the brain. The key role of alanine in this overall formulation is underscored by observations in the rat (15, 16, 18), dog (17), sheep (29), and reptiles (30), demonstrating the universality of this amino acid as a vehicle of nitrogen transport and as an endproduct of nitrogen catabolism in a variety of species.

ALANINE AND GLUCONEOGENESIS

Since mammalian liver lacks the capacity to convert even-chain fatty acids to glucose, gluconeogenesis primarily involves the conversion of amino acids and the recycling of glucose-derived carbon skeletons (lactate, pyruvate, glycerol) to glucose. The primacy of alanine in the overall availability and uptake of amino acids by the liver immediately suggested the importance of this amino acid as the key protein-derived glucose precursor (12, 31). Support for this hypothesis was obtained from in vitro and in vivo studies in experimental animals as well as from observations in intact man. In the perfused liver, the rate of glucose synthesis from alanine and serine is far higher than that observed from all other amino acids (32). Gluconeogenesis from alanine has been demonstrated to increase in proportion to the availability of this amino acid (33). Furthermore, whereas glucose production from a mixture of amino acids is saturated at three times normal concentrations, gluconeogenesis from alanine does not reach saturation until a concentration of 9 mM, which is 20–30 times the normal physiological level (34). In vivo incorporation of ^{14}C-labeled alanine into blood glucose in the rat occurs six times more rapidly than labeled serine (35) and increases linearly as alanine levels are raised to tenfold above physiologic concentrations (36).

In human subjects, prompt incorporation of ^{14}C-labeled alanine into blood glucose has been demonstrated in the overnight and prolonged fasted state (37, 38). The proportion of the injected dose of labeled alanine recoverable in blood glucose (39) is comparable to that observed with lactate (40). Bolus injections of unlabeled alanine in glycogen-depleted, fasted subjects result in a rise in blood glucose 30–60 min after infusion of the alanine (41). In contrast to alanine, studies with ^{14}C-arginine reveal a much slower rate of incorporation into blood glucose and a failure to raise blood glucose levels when a bolus injection of arginine is administered to glycogen-depleted, fasted subjects (42).

It is clear from the above data that with respect to availability from muscle tissue,

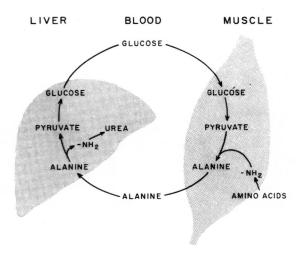

Figure 3   The glucose-alanine cycle. Alanine is synthesized in muscle by transamination of glucose-derived pyruvate, released into the bloodstream, and taken up by the liver. In the liver the carbon skeleton of alanine is reconverted to glucose and released into the bloodstream where it is available for uptake by muscle and resynthesis of alanine.

rate of uptake by liver, and rapidity of conversion to glucose, alanine is pre-eminent among amino acids as a glucose precursor. Thus, while virtually all amino acids other than leucine may be categorized as glycogenic with respect to their *potential* metabolic fate (43), quantitatively alanine accounts for more than half the total amino acid utilization for hepatic gluconeogenesis. With respect to the overall contribution of amino acids to glucose production, in the postabsorptive, overnight fasted state approximately 75% of glucose output is due to glycogenolysis rather than gluconeogenesis (44). Of the residual hepatic glucose output that is not glycogen-derived, 25–40% can be accounted for by hepatic uptake of alanine (44). Thus, it is primarily in circumstances in which glycogen stores are depleted and glucose output is dependent on gluconeogenesis that alanine availability assumes clinical importance in the regulation of blood glucose. Such conditions include fasting that extends beyond 24 hr (12), pregnancy (45), ketotic hypoglycemia (46), and prolonged exercise (47).

Since splanchnic uptake of glutamine is comparable to alanine, the possibility that glutamine, in addition to alanine, functions as a key glucose precursor has been suggested (7). However, as noted above, the site of splanchnic glutamine uptake is the gut rather than the liver (9, 14). Furthermore, in vitro studies with labeled glutamine reveal that incorporation into glucose is one tenth that observed with alanine (36). Thus, glutamine does not appear to be an important glucose precursor in man.

THE GLUCOSE-ALANINE CYCLE

The predominance of alanine in the outflow of amino acids from muscle cannot be explained on the basis of its availability in constituent cellular proteins. Alanine comprises no more than 7–10% of muscle proteins (48, 49) yet accounts for 30% or more of the net flow of alpha amino nitrogen from muscle to the splanchnic bed. This discrepancy led to the suggestion that alanine is synthesized de novo in muscle tissue by transamination of pyruvate (5, 6). On the basis of the evidence implicating alanine as a glucose precursor, Felig et al (5, 8) and Mallette et al (34) proposed a "glucose-alanine cycle" in which alanine is synthesized in muscle by transamination of glucose-derived pyruvate and is transported to the liver where its carbon skeleton is reconverted to glucose (Figure 3). The branched chain amino acids were suggested as the origin of the amino groups for muscle alanine synthesis (8) inasmuch as extra-hepatic tissues, particularly muscle, had been demonstrated as the site of their oxidation (20, 50). A variety of studies in man and experimental animals have subsequently appeared supporting the existence of the glucose-alanine cycle (44).

In normal man, a direct linear relationship between plasma alanine and pyruvate levels has been observed; such a relationship does not exist between pyruvate and other amino acids (8). In association with the increased glucose utilization of muscular exercise, a specific increase in muscle production and circulating levels of alanine is observed (8, 51). In acquired or congenital disorders of lactate and pyruvate metabolism characterized by hyperpyruvicemia, plasma alanine levels are elevated (52–55). In contrast, in patients with McArdle's syndrome in whom glucose consumption by exercising muscle is limited and pyruvate levels fall during exercise, a simultaneous fall in alanine concentration has been reported (56). The addition of glucose or insulin to incubated rat diaphragm results in a two- to three-fold increase in the release of alanine (49, 57) but fails to increase the release of other amino acids (58). Furthermore, studies using ^{14}C-glucose indicate that 60% of the alanine released by diaphragm is derived from exogenous glucose (49). These data thus provide strong support for the thesis that alanine synthesis in muscle is intimately dependent upon glucose breakdown to pyruvate.

With regard to the suggested role of the branched chain amino acids as the nitrogen source for alanine formation in muscle (8), it is noteworthy that the output of valine, leucine, and isoleucine from forearm muscle (5, 6) and isolated rat diaphragm (49) accounts for only 10–12% of total alpha amino nitrogen released, yet these amino acids represent 20% of the residues in muscle proteins (48, 49), suggesting substantial in situ catabolism. Studies with the isolated rat diaphragm have, in fact, repeatedly demonstrated that muscle tissue is capable of oxidizing the branched chain amino acids (27, 50, 59, 60) and probably accounts for 50% of total leucine oxidation (27). The relationship of the breakdown of these amino acids to alanine formation is indicated by the fact that production of alanine (but not that of glutamine or other amino acids) and the incorporation of ^{14}C from glucose into alanine by rat diaphragm are increased by physiologic increments of the branched chain amino acids in the incubation medium (49). Furthermore, amino groups released on oxidation of the branched chain amino acids can account for all the nitrogen recovered in alanine. In contrast, the addition of other amino acids at twice their plasma concentrations fails to influence alanine output (49). Although other amino acids have been reported to stimulate alanine output from muscle (61, 62), these data have been criticized (49) with respect to their physiologic import and the specificity of the enzymatic alanine assay used, inasmuch as amino acids were supplied at concentrations 25–200 times the normal plasma levels. The intimate relationship between the oxidation of the branched chain amino acids and alanine synthesis in muscle has led to the suggestion that a branched chain amino acid cycle involving transfer of these amino acids from liver to muscle may complement the glucose-alanine cycle (49, 60).

Precise, direct quantification of the relative importance of the glucose-alanine cycle in overall alanine turnover in man is not available. However, by comparing alanine output with that of other amino acids, gross estimates may be made (Table 1). Since lysine does not undergo synthesis or catabolism in muscle, its release from muscle serves as an index of proteolysis. In addition, the concentration of alanine in muscle protein is similar to that of lysine (48, 49). Thus, by comparing alanine output with that of lysine, the proportion of total alanine release that may be ascribed to in situ, de novo synthesis (by transamination of pyruvate) can be calculated. As shown in Table 1, the estimated rate of glucose-derived alanine release accounts for 67% of the total alanine release from forearm muscle. This value is in excellent agreement with studies using ^{14}C-glucose, which indicate that 60–80% of alanine released by rat diaphragm is derived from glucose (49).

It is also interesting to determine the relative contribution of the alanine cycle to glucose turnover and to compare its quantitative importance with the Cori (lactate)

Table 1 The contribution of the glucose-alanine cycle to total alanine release and to total glucose consumption by forearm muscle in postabsorptive man[a]

A-DV Alanine[b] = -111 mol/l
A-DV Lysine = -37 mol/l
A-DV Glucose-derived alanine = -74 mol/l
Proportion of total alanine release that is glucose derived = $74/118 = 67\%$
A-DV Glucose = 211 mol/l
Proportion of glucose uptake accounted for by alanine production[c] = $37/211 = 18\%$

[a] Based on the data of Felig et al (5).
[b] A-DV: Arterial-deep venous difference.
[c] Calculated as: 100 ($\frac{1}{2}$ glucose-derived alanine V-A difference/glucose A-V difference).

cycle. As shown in Table 1, glucose-derived alanine accounts for 18% of basal glucose uptake by muscle. As noted above, alanine extraction can account for a minimum of 6–12% of hepatic glucose output in the postabsorptive state (44). With regard to lactate turnover, production of this glycolytic endproduct accounts for 20–40% of muscle glucose uptake (63, 64), while lactate uptake by liver is responsible for 15–20% of glucose output (40, 65). Thus, with respect to its role as an endproduct of peripheral glucose utilization as well as its contribution to hepatic glucose production, recycling of carbon skeletons along the glucose-alanine cycle occurs at a rate approximately 50% of that observed for the Cori (lactate) cycle.

Although the glucose-alanine cycle does not yield new carbon skeletons for de novo glucose synthesis, its importance may relate not only to glucose homeostasis but to nitrogen and energy metabolism as well. Inasmuch as the branched chain amino acids are preferentially catabolized in muscle, alanine provides a nontoxic alternative to ammonia in the transfer of amino groups from the periphery to the liver. This carrier role of alanine in nitrogen metabolism may be of particular importance in circumstances characterized by augmented formation or inadequate disposal of ammonia. Ammonia production by contracting muscle has long been recognized (66). During exercise, increased transfer of amino groups to pyruvate may serve to limit ammonia formation in muscle. Hyperalaninemia is also observed in a variety of disorders of urea cycle enzymes (67–69), where it may serve to mitigate the degree of hyperammonemia.

The glucose-alanine cycle may also be useful with respect to ATP production (49). Conversion of glucose to alanine provides 8 mol ATP compared to 2 mol provided by conversion to lactate. Furthermore, to the extent that alanine formation facilitates the oxidation of the branched chain amino acids, an additional 30–40 mol ATP will be generated per mole of amino acid oxidized (43).

PROTEIN INGESTION

The discussion of amino acid exchange between muscle, gut, liver, and kidney and its influence on gluconeogenesis and nitrogen transport has to this point dealt with the postabsorptive state (i.e. the condition that exists after a 10–12 hr fast). Clearly, one would anticipate marked changes in amino acid metabolism following the ingestion of a protein meal. Although it has been recognized since the studies over 60 years ago of VanSlyke & Meyer in dogs that protein ingestion results in increased levels of amino acids in the circulation (70), data are still not available on the effect of protein meals on interorgan amino acid exchange in normal man. The published observations in humans relate primarily to changes in systemic amino acid concentrations. Frame reported that ingestion of 55–60 g protein as eggnog resulted in a 20% elevation in total alpha amino nitrogen, which reached its peak at 4 hr (71). The most marked increments (100–150%) were observed for the branched chain amino acids which remained elevated for 8 hr. A similar increase in the branched amino acids has been observed in subjects following milk ingestion (72). In marked contrast, alanine levels failed to rise after a protein meal and decreased below basal levels at 5–8 hr (71).

With respect to the effects of protein ingestion on the exchange of amino acids across the gut and liver, data are not available in man. However, studies in dogs (17), rats (73), and sheep (29) have shown a uniform response, suggesting that a similar pattern may occur in all mammals. In each of these species, the outflow of amino acids from the gut into the portal vein following protein ingestion is characterized by a predominance of alanine and the absence of glutamate and aspartate (despite the fact that the latter amino acids account for 20–30% of the constituent amino acids in the ingested protein). Uptake of alanine by the liver exceeds gut release, indicating continued output of alanine from peripheral tissues during the absorptive period. Hepatic uptake of the branched chain amino acids is substantially below gut output so that a net release of these amino acids from the splanchnic bed occurs. For the remainder of the amino acids, hepatic uptake increases in concert with gut output so that systemic concentrations and net splanchnic exchange show little change between the absorptive and postabsorptive periods. These data in experimental animals thus indicate an active role for the gut in converting ingested nitrogen to alanine to increase the uptake of this amino acid by the liver. The responses of the liver may be characterized as active with respect to buffering the effect of protein ingestion on the systemic levels of most amino acids, and passive in permitting the escape from the splanchnic bed of the branched chain amino acids absorbed in the gut. The disproportionate rise in systemic levels of valine, leucine, and isoleucine and the fall in alanine observed in man (71) would suggest that similar splanchnic balances occur in humans.

With respect to the effect of protein ingestion on muscle amino acid exchange, studies in the rat have demonstrated a net uptake by peripheral tissues of the branched chain amino acids in the absorptive period (73). Recent studies undertaken in normal humans also reveal selective splanchnic release of branched chain amino acids and uptake by muscle tissue which persists for at least 3 hr after protein ingestion (73a). In contrast, alanine output from muscle continues unabated or is reduced for only 1 hr after protein intake (73, 73a, 74).

When amino acids are infused intravenously rather than ingested, thus bypassing the gut and liver, the importance of alanine in amino acid catabolism and nitrogen disposal is also readily apparent. In the rat, infusion of each of 20 amino acids results in increases in the alanine content of a variety of tissues (liver, kidney, muscle) (75). Similar increments are not observed for glutamine (75).

From these observations it is clear that alanine is the major vehicle of alpha amino nitrogen output from gut and muscle in the fed as well as the fasted state. Repletion of nitrogen stores in muscle, which releases alanine fairly continuously in the absorptive as well as the postabsorptive states, occurs primarily by transfer from the gut to the periphery of branched chain amino acids contained in ingested protein. The latter observation provides further evidence for the importance of the branched chain amino acids as the nitrogen source for alanine synthesis in muscle (see above). It should be noted, however, that despite the release of these amino acids from the splanchnic bed after protein intake, the total exchange of alpha amino nitrogen across the liver shows a net gain over the entire course of the day (29, 76). To account for this imbalance, which would result in depletion of extrahepatic nitrogen stores, it has been suggested that plasma proteins synthesized in the liver undergo hydrolysis in peripheral tissues and thereby contribute to the interorgan transport of amino acids and repletion of muscle protein (76).

THE ROLE OF ERYTHROCYTES IN INTERORGAN AMINO ACID TRANSPORT

Because of the slow equilibration time of amino acid transport across the erythrocyte membrane indicated by in vitro studies (77), it has generally been believed that plasma rather than the erythrocyte is the vehicle of amino acid exchange between tissues (3). Consequently, measurements of circulating levels of amino acids have traditionally involved determining plasma rather than whole blood concentrations. The studies of Elwyn et al in dogs fed beef meals first called attention to the possible role of erythrocytes in interorgan amino acid flux (17, 78). In humans, blood cells have recently been shown to play a pivotal role in insulin-mediated uptake of glutamate by muscle (79). More extensive studies in postabsorptive subjects have demonstrated that transport by way of plasma could not account for total amino acid release from muscle and gut or uptake by the splanchnic bed (9). For alanine, serine, threonine, methionine, leucine, isoleucine, and tyrosine, significant tissue exchange occurs via the blood cellular elements (presumably erythrocytes), whose direction parallels the net shifts observed to occur via plasma. The overall contribution of blood cells accounts for 30% of the alanine output from muscle and gut and 20% of the uptake by the splanchnic bed. Only in the case of glutamine is transport via plasma and blood cells in opposite directions: an uptake by portal-drained viscera from plasma accompanied by a simultaneous output from red cells into plasma (4, 80). In a single human subject studied after 6 wk starvation, one third to one half of the muscle uptake of the branched chain amino acids and lysine that followed ingestion of a beef meal occurred by way of transfer from the blood cellular compartment (74).

To account for the apparent discrepancy between these in vivo studies and the slow transfer of amino acids between plasma and erythrocytes observed in vitro, direct exchange of amino acids from erythrocytes to tissue cells has been suggested (78). Regardless of the mechanism involved, these observations suggest that measurements with plasma underestimate overall interorgan amino acid exchange. Nevertheless, the direction of transfer and the relative contributions of individual amino acids are not obscured by measurements restricted to the plasma compartment.

GLUCOREGULATORY HORMONES

Insulin

The effect of insulin on amino acid metabolism has received considerable attention with respect to stimulation of protein synthesis (81), transport of amino acids into cells (82), and more recently, inhibition of protein catabolism (57). An effect of insulin in lowering the systemic concentration of circulating amino acids has long been recognized (83, 84), but the influence of insulin on interorgan exchange of amino acids and its relationship to gluconeogenesis has only recently been examined.

The demonstration by Mirsky (85) and subsequently by Russell (86) that insulin effectively lowered amino acid concentration in eviscerated animals suggested that muscle tissue was the site of insulin-mediated amino acid uptake. Studying the response to insulin administration in dogs, Lotspeich noted that the pattern of decrease of individual amino acids corresponded to their relative concentrations in muscle protein (87). Subsequently, numerous reports have confirmed the effectiveness of endogenous or exogenous insulin in lowering the amino acid content of plasma in humans (88–91). The pattern of response in these studies is such that the magnitude and consistency of the decline in amino acid concentration is most marked for valine, leucine, isoleucine, methionine, tyrosine, and phenylalanine. Studies in the human forearm in which physiologic increments in insulin were infused into the brachial artery have demonstrated that the specificity of this response is a consequence of the pattern of net inhibition of amino acid output from muscle (6). A consistent inhibition in net output from forearm muscle was observed for leucine, isoleucine, methionine, tyrosine, phenylalanine, and threonine (6). Using whole blood, an uptake of glutamate has also been demonstrated (74).

Alanine is notably absent from the group of amino acids affected by insulin with respect to lowering systemic levels or inhibiting muscle output. In fact, following stimulation of endogenous insulin secretion by oral glucose administration, a rise in arterial alanine levels has been observed in normal man (92). Furthermore, in the isolated diaphragm, insulin has been demonstrated to increase alanine output from muscle while decreasing the release of all other amino acids (57). In addition, only in the case of alanine does the presence of glucose or pyruvate reduce the stimulatory effect of insulin on the incorporation of ^{14}C into protein from labeled amino acids (93, 94). Since, under appropriate circumstances, insulin stimulates intracellular transport and incorporation of alanine into muscle protein (93, 95, 96), these unique aspects of alanine behavior cannot be ascribed to a primary resistance on the part of this amino acid to the action of insulin. Rather, the seemingly anomalous behavior of this amino acid can be explained on the basis of insulin-stimulated synthesis of this amino acid from glucose-derived pyruvate, as proposed in the glucose-alanine cycle (5, 44).

The action of insulin on amino acid metabolism is also interesting with respect to the regulation of gluconeogenesis. It is well recognized that insulin not only increases glucose utilization by fat and muscle tissue but also inhibits the release of glucose from the liver (97). In insulin-deficient diabetics, a correlation between urinary glucose output and nitrogen loss in the urine has long been recognized (82, 98). An inhibition of gluconeogenesis by insulin was suggested by the early clinical observations of increased muscle mass accompanying blood glucose regulation in juvenile diabetics treated with insulin preparations. As to the mechanism of insulin-mediated inhibition of gluconeogenesis, this could be a consequence of a peripheral effect on precursor supply or a result of an hepatic action on precursor uptake and disposal. Studies on splanchnic amino acid balance in humans in whom endogenous insulin secretion is stimulated by intravenous (13) or oral glucose (92) suggest an hepatic effect. Following glucose administration, splanchnic uptake of alanine declines by 30–75% (13, 92). This diminution in uptake is entirely a consequence of decreased fractional extraction of alanine (rather than reduced peripheral delivery) inasmuch as arterial alanine levels are unchanged (13) or increased (92). Similarly, in the case of other glycogenic substrates such as lactate and pyruvate, splanchnic uptake is inhibited despite an increase in arterial concentration (92). Direct evidence of an hepatic effect of insulin in regulating gluconeogenesis is provided by the demonstration that insulin inhibits incorporation of ^{14}C-alanine into glucose in the perfused liver (99). To the extent that some degree of alanine uptake by the splanchnic bed continues in the face of hyperinsulinemia (albeit at a reduced rate), the regulatory effect of insulin on gluconeogenesis may involve alterations in hepatic disposal as well as a reduction in hepatic uptake of glycogenic substrates. Both effects, however, point to the liver as the locus of insulin action in decreasing gluconeogenesis.

Glucagon

Although glucagon was discovered shortly after the isolation of insulin, its role in the regulation of amino acid metabolism remains to be established. Much of the difficulty in interpreting the available data relates to the use of pharmacologic rather than physiologic doses of glucagon in experimental studies. Until the development of a specific radioimmunoassay for pancreatic glucagon, the precise levels of glucagon in systemic blood and the secretory rate of this hormone were unknown. The studies of Unger, using a specific antibody for pancreatic glucagon, indicate basal levels of approximately 100 pg/ml, which increase to 150–200 pg/ml after a protein meal or during starvation (100). The concentration of glucagon in portal venous blood is only 30% higher than in the peripheral circulation (101). Physiologic increments in glucagon are achieved with infusion rates no greater than 3–4 ng/kg per min (M. Fisher and P. Felig, unpublished data), suggesting a 24 hr secretory rate of glucagon well below 1 mg in a 70 kg man.

With respect to its action on amino acids, pharmacologic doses of glucagon have a hypoaminoacidemic effect (102) involving virtually all amino acids (103, 104). Interestingly, the magnitude of this decline is greatest for alanine (104). With respect to the factors responsible for this decrease, increased splanchnic uptake of total amino acids (105, 106) and specifically of alanine has been observed (107). Furthermore, in the perfused liver, glucagon increases the production of glucose from alanine (34, 108) and stimulates the intracellular transport and utilization of glycogenic amino acids (109). The latter effect is also most pronounced with respect to alanine (109). Supporting the gluconeogenic action of glucagon is the negative nitrogen balance induced in fed (110) and fasted man (111) and in human (112) and experimental diabetes (113). In muscle, glucagon increases the oxidation of branched chain amino acids (114), decreases amino acid incorporation into protein (115), and increases the output of amino acids from isolated muscle tissue (116). The overall thrust of these data points to a catabolic effect characterized by augmented transfer of amino acids from muscle to liver for conversion to glucose. The decline in systemic amino acid levels suggests that hepatic uptake and utilization of amino acids are stimulated to a greater extent than muscle output.

The significance of these studies for human physiology and disease may be seriously questioned, however, inasmuch as the amounts of glucagon injected and the concentrations used in incubation media were generally far in excess of physiologic levels. Furthermore, since pharmacologic doses of glucagon result in insulin secretion (117), it is difficult to distinguish direct effects of glucagon from those resulting from accompanying hyperinsulinemia. For example, with respect to the generalized hypoaminoacidemia induced by glucagon in intact man, the decline in the branched chain amino acids is insulin dependent and extrasplanchnic in origin (107). In contrast, in the isolated perfused liver, glucagon causes an increase in hepatic release of branched chain amino acids (109).

Physiologic increments (80–100 pg/ml) in glucagon were induced by Marliss et al in prolonged fasting subjects, resulting in small decrements (15%) in a variety of amino acids (111). However, inasmuch as urea nitrogen excretion failed to increase, these data do not suggest a catabolic or gluconeogenic effect. In the human forearm, infusion of glucagon in high physiologic amounts has no effect on muscle balance of amino acids in either postabsorptive or briefly fasted subjects (118). In a recent report, increased incorporation of alanine into glucose has been demonstrated in the intact dog following physiologic increments in plasma glucagon (119).

Glucocorticoids

The glucocorticoids are generally recognized as having catabolic and gluconeogenic effects. Studies in experimental animals treated with these hormones have documented decreased protein synthesis in muscle (120), augmented hepatic uptake and accumulation of amino acids (3), and a rise in circulating amino acid levels following evisceration (121). These data thus suggest catabolism of protein in muscle and transfer of amino acids to liver for gluconeogenesis. Direct studies of the effects of glucocorticoids on muscle or splanchnic amino acid exchange are not available in man. However, both short term and long term hypercortisicism are associated with increased levels of plasma alanine, while other amino acids are unchanged (122). These findings agree with the data in experimental animals indicating that the rise in intracellular alanine in muscle and liver tissue induced by hydrocortisone exceeds that of other amino acids (123, 124). Corticosteroids also increase glucagon secretion (122, 125), suggesting that in vivo stimulation of gluconeogenesis by adrenal steroids may involve a variety of mechanisms (80). On the other hand, following prolonged starvation, glucocorticoids fail to increase protein catabolism (126), suggesting that the status of nutrition and pre-existing protein stores influences the response to these hormones.

Amino Acids as Secretagogues

The interaction of glucoregulatory hormones and amino acids includes the stimulatory action of these substrates on hormone secretion. Intravenous administration of a variety of amino acids or ingestion of a protein meal results in increased secretion of insulin (127, 128), glucagon (100), and growth hormone (129). In fact, it has been suggested that protein ingestion may be the most important physiologic stimulus for secretion of glucagon in preventing the hypoglycemia that would otherwise accompany the hyperinsulinemia induced by a protein meal (130, 131). In addition to the effects of exogenous amino acids, altered levels of endogenous amino acids have been implicated as having a feedback role in the regulation of hormonal secretion. An increase in the plasma concentrations of the branched chain amino acids has been suggested as the signal responsible for the hyperinsulinemia of obesity (90). Similarly, the pancreatic alpha cell is exquisitely sensitive to small increments in plasma alanine (132). Accordingly, it has been suggested that hyperalaninemia may be a contributory factor in the hyperglucagonemia observed in exercise (47, 133) and following glucocorticoid administration (122).

STARVATION

The metabolic response to starvation has been described as biphasic, the changes in body fuel metabolism differing in the early and late stages of fasting (31). The initial response is directed at maintaining hepatic glucose output by increasing gluconeogenesis, whereas the late response is directed at maintaining body protein reserves by minimizing protein catabolism. Since liver glycogen stores are rapidly depleted in fasting (134), initially there is an augmented hepatic uptake of glucose precursors, notably alanine, to maintain hepatic glucose output in the absence of significant glycogenolysis (12). This increase in gluconeogenesis observed during the first 3 days of starvation is a consequence of augmented splanchnic fractional extraction of alanine (12) as well as increased release of this amino acid from muscle (118). These changes in hepatic and muscle alanine metabolism are probably related to the fall in insulin (135) and the rise in glucagon observed early in starvation (100, 111). A striking increase in the plasma concentration of the branched chain amino acids is also noted at this time (12, 136) and has been ascribed to the hypoinsulinemia of starvation inasmuch as a similar rise in the branched chain amino acids is observed in diabetes (137, 138). Since the oxidation of valine, leucine, and isoleucine by muscle tissue is augmented in fasting (59, 114), while the output of these amino acids from muscle (118) and liver (12) is unchanged, the tissue source of this hyperaminoacidemia has not been established.

As starvation progresses from days to weeks, survival is dependent upon minimizing protein catabolism inasmuch as the loss of one third to one half of body protein stores results in death (31). A progressive decline in protein catabolism as evidenced by a fall in urinary nitrogen excretion (139), a decrease in hepatic gluconeogenesis (12), and replacement of glucose by ketone acids as the major oxidative fuel utilized by the brain (140) characterize the late response to starvation. A generalized decline in plasma amino acids is observed, in which the fall in alanine is most prominent (12). Hepatic uptake of alanine is markedly reduced (12) as a consequence of decreased output of this amino acid from muscle tissue (5). In contrast, splanchnic fractional extraction of alanine is unchanged from the postabsorptive state. Furthermore, intravenous (38, 41) or oral administration of alanine (141) results in a prompt increase in blood glucose. These findings thus suggest that substrate availability rather than inhibition of hepatic enzymes is the rate-limiting step in the decrease in hepatic gluconeogenesis observed in prolonged fasting.

The precise mechanism of the decreased release of alanine from muscle in prolonged starvation has not been identified, but it cannot be accounted for by changes in plasma insulin or glucagon. Recent studies have shown that infusion of ketone acids results in a prompt and specific decline in plasma alanine levels (142). Furthermore, oxidation of the branched amino acids, the source of the amino groups for alanine synthesis in muscle tissue (49, 60) (see above), is inhibited by physiologic increments in ketone acids (60). These observations thus suggest that in prolonged starvation, ketones not only replace glucose as the major oxidative fuel for the brain (140), but also contribute to protein conservation by limiting amino acid (alanine) availability for gluconeogenesis (142).

In contrast to the effects of total starvation, restriction or elimination of protein intake in subjects who continue to ingest some carbohydrate results in increased levels of plasma alanine. In patients with protein-calorie malnutrition and kwashiorkor (143) or in normal adults fed a low protein diet (136, 144), plasma alanine concentrations are markedly higher than those observed in starvation. Nevertheless, as in prolonged starvation, release of alanine from muscle is greatly reduced (143, 145). In addition, absolute uptake and fractional extraction of alanine by the splanchnic bed are decreased by 90%, indicating an overall decline in alanine turnover (145). It has been suggested that intermittent ingestion of carbohydrate serves to reduce the requirement for gluconeogenesis, resulting in a decrease in alanine uptake by the liver and consequent accumulation of this amino acid in blood despite a reduction in output from muscle (145).

DIABETES MELLITUS

The metabolic millieu in diabetes mellitus is, in a number of respects, comparable to that observed in short term starvation, inasmuch as insulin levels are reduced and plasma glucagon concentration exhibits a relative or absolute increase (100). Similarities in the circulating levels and splanchnic exchange of amino acids are also observed. An increase in the plasma concentration of the branched chain amino acids and a fall in plasma alanine and other glycogenic amino acids have been repeatedly reported in nonketotic (138, 146) as well as ketotic diabetic patients (137). Splanchnic uptakes of alanine, lactate, and pyruvate are increased by 50–100% so that the contribution by gluconeogenic precursors to total hepatic glucose output approaches

40%, as compared to 15–20% in normal postabsorptive man (138). Inasmuch as arterial alanine levels are reduced, augmented gluconeogenesis is a consequence of a two- to threefold increase in splanchnic fractional extraction of alanine and other aminogenic glucose precursors (138). These findings thus suggest that increased gluconeogenesis in diabetes is a consequence of altered intrahepatic events as opposed to a primary increase in availability of glucose precursors. It should be recalled that hyperinsulinemia in normal man (13, 92) results in the exact reversal of events (decreased gluconeogenesis due to inhibition of hepatic alanine uptake despite normal to increased circulating alanine levels). With respect to the elevation in plasma concentration of the branched chain amino acids, as in short term starvation, there is no evidence in human diabetes of an increase in peripheral or hepatic release of these amino acids (138). On the other hand, studies with the intact diabetic rat (147) and the perfused rat liver (109, 148) have suggested an hepatic origin for the systemic accumulation of these amino acids.

OBESITY

In obesity, the plasma pattern and splanchnic uptake of amino acids are similar to those observed in brief starvation and diabetes. In obese subjects, plasma valine, leucine, isoleucine, phenylalanine, and tyrosine are elevated (19, 90, 136). Furthermore, compared to nonobese subjects, insulin infused into the forearm is less effective in inhibiting the outflow of these amino acids from muscle tissue (149). With respect to splanchnic balances, an increase in alanine uptake comparable to that observed in diabetes and brief starvation has been demonstrated and is primarily a consequence of augmented fractional extraction of gluconeogenic substrates (19). Using labeled precursors, increased incorporation of alanine into glucose has also been demonstrated in obese animals (150). Inasmuch as insulin levels in obesity considerably exceed those observed in nonobese subjects (19, 90), these changes in amino acid metabolism have been interpreted as reflecting resistance on the part of muscle (90, 144) and liver tissue (19) to the action of insulin.

Amino acids are potent stimuli of insulin secretion. The increase in plasma insulin in obesity is directly proportional to the rise in specific plasma amino acids. Accordingly, it has been suggested that hyperaminoacidemia may provide the feedback signal responsible for the hyperinsulinemia of obesity (90).

EXERCISE

The effects of muscular exercise on amino acid metabolism vary with the duration and intensity of activity. Studies have consequently been undertaken to examine the influence of brief periods of mild to severe exercise (10–40 min), prolonged exercise (4 hr), and the recovery phase after cessation of exercise (8, 47, 151).

During brief exercise, alanine is the only amino acid for which an elevation in arterial concentration as well as a consistent output from the exercising limb is observed (8, 51). The increase in alanine output varies from 50–500%, depending upon the intensity of the exercise. The rise in arterial levels varies from 25–100% and is directly proportional to the elevation in arterial pyruvate (8). Contrasting with the response in normal subjects are the changes observed in McArdle's syndrome, a disorder characterized by a lack of muscle phosphorylase in which exercise-induced glycolysis and pyruvate formation are markedly limited (56). In these patients, arterial levels and muscle concentrations of pyruvate fall during exercise. In a like manner, arterial alanine decreases and, rather than an increase in peripheral output, alanine is taken up by muscle where it serves as an ancillary fuel during exercise (56). These observations in exercise thus have provided important support for the hypothesis that alanine is synthesized in muscle and that its formation is dependent in part on the availability of glucose-derived pyruvate (8). With respect to the increased source of amino groups for alanine synthesis in exercise, stimulation of branched chain amino acid oxidation (152) and aspartate conversion to oxaloacetate (66, 153) has been demonstrated in exercise. Exercise also causes increased ammonia production by muscle (66). It is thus likely that in providing an alternative means of disposing of amino groups, alanine formation is of particular importance in limiting the hyperammonemic response to exercise.

With respect to splanchnic alanine exchange, the absolute rate of uptake of this amino acid during brief exercise is unchanged from the basal state (8). However, inasmuch as hepatic glucose output increases three- to fourfold, the relative contribution of gluconeogenesis to total hepatic glucose release is markedly reduced.

During prolonged exercise (4 hr), changes are observed in splanchnic as well as peripheral alanine metabolism (47). As exercise continues beyond 40 min, peripheral output of alanine remains elevated but splanchnic alanine uptake rises at an even more rapid rate, resulting in a decline in arterial alanine concentration. The increase in splanchnic uptake of alanine is primarily a consequence of a rise in splanchnic fractional extraction of this amino acid. Overall uptake of glucose precursors increases threefold so that gluconeogenesis can account for 45% of total hepatic glucose output (47), a situation analogous to that observed after a 3 day fast (12), in diabetes (138), and in obesity (19). The fall in insulin and rise in glucagon observed with prolonged exercise have been suggested as possible mediators of this gluconeogenic response (47).

In addition to influencing alanine metabolism, prolonged exercise also alters the interorgan exchange and disposal of the branched chain amino acids so that they assume importance as a fuel for exercising muscle. A rise in arterial concentration, augmented splanchnic production, and a reversal from the net output observed in the resting state to a consistent uptake by the exercising leg are demonstrable for valine, leucine, and isoleucine (47). In fact, there is close correspondence between the absolute rates of splanchnic production and the peripheral uptake of these amino acids (47). With respect to the mechanism of these changes, glucagon has been shown to stimulate branched chain amino acid release from liver (109), while glucagon, epinephrine, and fatty acids have been demonstrated to stimulate the oxidation of these amino acids by muscle tissue (114).

During the recovery phase following cessation of relatively brief exercise (40 min), splanchnic uptake of alanine and other glycogenic precursors increases by 50–100%, primarily as a consequence of augmented splanchnic fractional extraction (151). Total precursor consumption can account for 45% of hepatic glucose output, suggesting that during the recovery phase, gluconeogenesis serves as a means of repleting and sparing hepatic glycogen stores, which are mobilized during exercise. Uptake of glucose and output of alanine by muscle also remain elevated (above basal levels) during the recovery phase, indicating an increased turnover of carbon skeletons in the glucose-alanine cycle (151).

ALANINE DEFICIENCY AND HYPOGLYCEMIA

The key role of alanine as a glucose precursor and the limitation to gluconeogenesis resulting from hypoalaninemia were initially suggested on the basis of observations in prolonged starvation (5, 12). Subsequent investigations have documented a variety of conditions in which a deficiency in circulating alanine levels is responsible for, or contributes to, fasting hypoglycemia. The conditions most thoroughly evaluated are pregnancy (45), ketotic-hypoglycemia of childhood (46), and alcohol hypoglycemia (39).

It is recognized in both humans (154) and experimental animals (155, 156) that normal pregnancy results in an exaggeration and acceleration of the hypoglycemic, hyperketonemic, and hypoinsulinemic response to starvation. This pattern is due to the failure of maternal glucose production to keep pace with the increased glucose requirements of the conceptus as a consequence of maternal substrate deficiency. In pregnancy, an augmented rather than diminished capacity for maternal hepatic gluconeogenesis from exogenous precursors has been demonstrated (157). On the other hand, in both humans and experimental animals, endogenous alanine levels are significantly reduced and fall more rapidly during starvation in the pregnant state (45, 158). Furthermore, when plasma alanine levels are increased by infusing this amino acid, a prompt rise in blood glucose is observed (45). These findings thus indicate that the fasting hypoglycemia of pregnancy is due to substrate limitation.

Ketotic hypoglycemia is a disorder of childhood characterized by recurrent episodes of hypoglycemia and ketosis, associated with hypoinsulinemia, normal adrenal and pituitary function, and intact hepatic gluconeogenic mechanisms (46). A deficiency in circulating alanine levels and prompt restoration of blood glucose to normal following alanine administration or corticosteroid-induced hyperalaninemia have been demonstrated (46, 159). Nevertheless, the possibility has not been excluded that such children have as their primary disorder an increased rate of glucose utilization resulting secondarily in depletion of alanine.

Ethanol-induced hypoglycemia is a well-established clinical entity in which altered hepatic redox potential interferes with hepatic gluconeogenesis from a variety of glucose precursors, particularly lactate (160). Substrate lack has generally not been considered an important aspect of this disorder inasmuch as blood lactate levels are increased. Recent studies, however, have shown that alanine deficiency may also be a contributory factor. In both normal (39) and diabetic subjects (161), ethanol administration results in a prompt decline in plasma alanine concentration. The mechanism of this hypoalaninemic effect remains to be established.

Finally, alanine availability may be a key factor in the age- and sex-determined variability in tolerance to starvation observed among healthy subjects. It has been shown that during fasting, normal women develop a greater degree of hypoglycemia than men (162) and that healthy children have an exaggerated glucose fall (163) compared to adults. A recent report indicates that alanine availability during fasting is highest in adult men, intermediate in women, and lowest in children (164).

ACKNOWLEDGMENTS

Dr. Felig is recipient of a Research Career Development Award (AM 70219) from the National Institutes of Health. Original investigations in the author's laboratory cited in this paper were supported by USPHS Grants AM 13526 and RR 125.

Literature Cited

1. Scriver, C. R., Clow, C. L., Lamm, P. 1971. *Am. J. Clin. Nutr.* 24:876–90
2. Holden, J. T., Ed. 1961. *Amino Acid Pools.* Amsterdam: Elsevier. 850 pp.
3. Munro, H. N. 1970. *Mammalian Protein Metab.* 4:299–388
4. London, D. R., Foley, T. H., Webb, C. G. 1965. *Nature* 208:588–89
5. Felig, P., Pozefsky, T., Marliss, E., Cahill, G. F. Jr. 1970. *Science* 167:1003–4
6. Pozefsky, T., Felig, P., Tobin, J., Soeldner, J. S., Cahill, G. F. Jr. 1969. *J. Clin. Invest.* 48:2273–82
7. Marliss, E. B., Aoki, T. T., Pozefsky, T., Most, A. S., Cahill, G. F. Jr. 1971. *J. Clin. Invest.* 50:814–17
8. Felig, P., Wahren, J. 1971. *J. Clin. Invest.* 50:2703–14
9. Felig, P., Wahren, J., Raf, L. 1973. *Proc. Nat. Acad. Sci. USA* 70:1775–79
10. Carlsten, A., Hallgren, B., Jagenburg, R., Svanborg, A., Werko, L. 1961. *Scand. J. Clin. Lab. Invest.* 13:418–28
11. Carlsten, A., Hallgren, B., Jagenburg, R., Svanborg, A., Werko, L. 1967. *Acta Med. Scand.* 181:199–207
12. Felig, P., Owen, O. E., Wahren, J., Cahill, G. F. Jr. 1969. *J. Clin. Invest.* 48:584–93
13. Felig, P., Wahren, J. 1971. *J. Clin. Invest.* 50:1702–11
14. Felig, P., Wahren, J., Karl, I., Cerasi, E., Luft, R., Kipnis, D. M. 1973. *Diabetes* 22:573–76
15. Ishikawa, E., Aikawa, T., Matsutaka, H. 1972. *J. Biochem Tokyo* 1097–99
16. Aikawa, T., Matsutaka, H., Yamamoto, H., Okuda, T., Ishikawa, E., Kawano, T., Matsumura, E. 1973. *J. Biochem. Tokyo* 74:1003–17
17. Elwyn, D. H., Parikh, H. C., Shoemaker, W. C. 1968. *Am. J. Physiol.* 215:1260–75
18. Matsutaka, H., Aikawa, T., Yamamoto, H., Ishikawa, E. 1973. *J. Biochem. Tokyo* 74:1019–29
19. Felig, P., Wahren, J., Hendler, R., Brundin, T. 1974. *J. Clin. Invest.* 53:582–90
20. Miller, L. L. See Ref. 2, 708–21
21. Lund, P. 1971. *Biochem. J.* 124:653–60
22. Addae, S. K., Lotspeich, W. D. 1968. *Am. J. Physiol.* 215:269–77
23. Weisswange, A., Cuendet, G. S., Bopp, P., Suter, P., Stauffacher, W., Marliss, E. B. 1973. *Clin. Res.* 21:642 (Abstr.)
24. Owen, E. E., Robinson, R. R. 1963. *J. Clin. Invest.* 42:263–76
25. Felig, P., Wahren, J., Ahlborg, G. 1973. *Proc. Soc. Exp. Biol. Med.* 142:230–31
26. Battistin, L., Grynbaum, A., Lajtha, A. 1971. *Brain Res.* 29:85–99
27. Odessey, R., Goldberg, A. L. 1972. *Am. J. Physiol.* 223:1376–83
28. Cahill, G. F. Jr., Owen, O. E. 1970. *Mammalian Protein Metab.* 4:559–81
29. Wolf, J. E., Bergman, N., Williams, H. H. 1972. *Am. J. Physiol.* 223:438–46
30. Coulson, R. A., Hernandez, T. 1967. *Am. J. Physiol.* 213:411–17
31. Felig, P., Marliss, E., Owen, O. E., Cahill, G. F. Jr. 1969. *Arch. Intern. Med.* 123:293–98
32. Ross, B. D., Hems, R., Krebs, H. A. 1967. *Biochem. J.* 102:942–51
33. Herrera, M. G., Kamm, D., Ruderman, N., Cahill, G. F. Jr. 1966. *Advan. Enzyme Regul.* 4:225–35
34. Mallette, L. E., Exton, J. H., Park, C. R. 1969. *J. Biol. Chem.* 244:5713–23
35. Ishikawa, E., Aikawa, T., Matsutaka, H. 1972. *J. Biochem. Tokyo* 71:1093–95
36. Aikawa, T., Matsutaka, H., Takezawa, K., Ishikawa, E. 1972. *Biochim. Biophys. Acta* 279:234–44
37. Felig, P., Marliss, E., Pozefsky, T., Cahill, G. F. Jr. 1970. *Am. J. Clin. Nutr.* 23:986–92

38. Felig, P. 1972. *Isr. J. Med. Sci.* 8 : 262–70
39. Kreisberg, R. A., Siegal, A. M., Owen, W. C. 1972. *J. Clin. Endocrinol.* 34 : 876–83
40. Kreisberg, R. A., Pennington, L. F., Boshell, B. R. 1970. *Diabetes* 19 : 53–63
41. Felig, P., Marliss, E., Owen, O. E., Cahill, G. F. Jr. 1969. *Advan. Enzyme Regul.* 7 : 41–46
42. Felig, P., Marliss, E. 1972. *Diabetes* 21 : 308–10
43. Krebs, H. A. 1964. *Mammalion Protein Metab.* 1 : 125–76
44. Felig, P. 1973. *Metabolism* 22; 179–207
45. Felig, P., Kim, Y. J., Lynch, V., Hendler, R. 1972. *J. Clin. Invest.* 51 : 1195–1202
46. Pagliara, A. S., Karl, I. E., DeVivo, D. C., Feigin, R. D., Kipnis, D. M. 1972. *J. Clin. Invest.* 51 : 1440–49
47. Ahlborg, G., Felig, P., Hagenfeldt, L., Hendler, R., Wahren, J. 1974. *J. Clin. Invest.* 53 : 1080–90
48. Kominz, D. R., Hough, A., Symonds, P., Laki, K. 1954. *Arch. Biochem. Biophys.* 50 : 148–59
49. Odessey, R., Khairallah, E., Goldberg, A. L. 1975. *J. Biol. Chem.* 249 : 7623–29
50. Manchester, K. L. 1965. *Biochim. Biophys. Acta* 100 : 295–98
51. Carlsten, A., Hallgren, B., Jagenburg, R., Svanborg, A., Werko, L. 1962. *Scand. J. Clin. Lab. Invest.* 14 : 185–91
52. Sussman, K. E., Alfrey, A., Kirsch, W. M., Zweig, P., Felig, P., Messner, F. 1970. *Am. J. Med.* 48 : 104–12
53. Marliss, E. B., Aoki, T. T., Toews, C. J., Felig, P., Connon, J. J., Kyner, J., Huckabee, N. E., Cahill, G. F. Jr. 1972. *Am. J. Med.* 52 : 474–81
54. Lonsdale, D., Faulkner, W. R., Price, J. W., Smeby, R. R. 1969. *Pediatrics* 43 : 1025–33
55. Brunette, M. G., Delvin, E., Hazel, B., Scriver, C. R. 1972. *Pediatrics* 50 : 702–11
56. Wahren, J., Felig, P., Havel, R. J., Jorfeldt, L., Pernow, B., Saltin, B. 1973. *N. Engl. J. Med.* 288 : 774–77
57. Fulks, R. M., Li, J. B., Goldberg, A. L. 1975. *J. Biol. Chem.* In press.
58. Odessey, R. 1973. PhD thesis. Harvard Univ., Cambridge, Mass. 210 pp.
59. Goldberg, A. L., Odessey, R. 1972. *Am. J. Physiol.* 223 : 1384–91
60. Buse, M. G., Biggers, S. J., Friderici, K. H., Buse, J. F. 1972. *J. Biol. Chem.* 247 : 8085–96
61. Ruderman, N., Lund, P. 1972. *Isr. J. Med. Sci.* 8 : 295–302
62. Garber, A. J., Karl, I. E., Kipnis, D. M. 1973. *J. Clin. Invest.* 52 : 31a (Abstr.)
63. Rabinowitz, D., Zierler, K. L. 1962. *J. Clin. Invest.* 41 : 2173–81
64. Jorfeldt, L., Wahren, J. 1970. *Scand. J. Clin. Lab. Invest.* 26 : 73–81
65. Reichard, G. A. Jr., Moury, N. F. Jr., Hochella, N. J., Patterson, A. L., Weinhouse, S. 1963. *J. Biol. Chem.* 238 : 495–501
66. Lowenstein, J. 1972. *Physiol. Rev.* 52 : 382–414
67. Malmquist, J., Jagenburg, R., Lindstedt, G. 1971. *N. Engl. J. Med.* 284 : 997–1002
68. Mohyuddin, F., Rathbun, J. C., McMurray, W. C. 1967. *Am. J. Dis. Child.* 113 : 152–56
69. Levin, B., Oberholzer, V. G., Sinclair, L. 1969. *Lancet* 1 : 170–74
70. VanSlyke, D. D., Meyer, G. M. 1913. *J. Biol. Chem.* 16 : 197–212
71. Frame, E. G. 1958. *J. Clin. Invest.* 37 : 1710–23
72. Armstrong, M. D., Stave, U. 1973. *Metabolism* 22 : 549–60
73. Yamamoto, H., Aikawa, T., Matsutaka, H., Okuda, T., Ishikawa, E. 1974. *Am. J. Physiol.* 226 : 1428–33
73a. Felig, P., Wahren, J. 1975. *Clin. Res.* In press
74. Aoki, T. T., Muller, W. A., Brennan, M. F., Cahill, G. F., Jr. 1973. *Diabetes* 22 : 768–75
75. Coulson, R., Hernandez, T. 1968. *Am. J. Physiol.* 215 : 741–46
76. Elwyn, D. H. 1970. *Mammalian Protein Metab.* 4 : 523–58
77. Winter, C. G., Christensen, H. N. 1964. *J. Biol. Chem.* 239 : 872–78
78. Elwyn, D. H., Launder, W. J., Parikh, H. C., Wise, E. M. Jr. 1972. *Am. J. Physiol.* 222 : 1333–42
79. Aoki, T. T., Brennan, M. F., Muller, W. A., Moore, F. D., Cahill, G. F. Jr. 1972. *J. Clin. Invest.* 51 : 2889–94
80. Felig, P. 1974. Diabetes, Proc. 8th Congr. Int. Diabetes Fed., ed. W. J. Malaise, J. Priart, 217–30. Amsterdam: Excerpta Med. 860 pp.
81. Wool, I. G., Castles, J. J., Leader, D. P., Fox, A. 1972. *Handbook of Physiology, Endocrine Pancreas,* ed. D. F. Steiner, N. Freinkel, 385–94. Washington DC: Am. Physiol. Soc. 721 pp.
82. Cahill, G. F. Jr., Aoki, T. T., Marliss, E. B. Ibid, 563–78
83. Wiechmann, E. 1924. *Z. Gesamte Exp. Med.* 44 : 158–67
84. Luck, J. M., Morrison, G., Wilbur, L. F. 1928. *J. Biol. Chem.* 77 : 151–56
85. Mirsky, I. A. 1938. *Am. J. Physiol.* 124 : 569–75
86. Russell, J. A. 1955. *Fed. Proc.* 14 : 696–706

87. Lotspeich, W. D. 1949. *J. Biol. Chem.* 179:175–80
88. Crofford, O. B., Felt, P. W., Lacy, W. W. 1964. *Proc. Soc. Exp. Biol. Med.* 117:11–14
89. Zinneman, H. H., Nuttall, F. Q., Goetz, F. C. 1966. *Diabetes* 15:5–8
90. Felig, P., Marliss, E., Cahill, G. F. Jr. 1969. *N. Engl. J. Med.* 281:811–16
91. DeBarnola, F. V. 1965. *Acta Physiol. Lat. Am.* 15:260–65
92. Felig, P., Wahren, J., Hendler, R. 1975. *Diabetes.* In press
93. Sinex, F. M., MacMullen, J., Hasting, A. B. 1952. *J. Biol. Chem.* 198:615–19
94. Manchester, K. L., Young, F. G. 1958. *Biochem. J.* 70:353–58
95. Manchester, K. L., Young, F. G. 1970. *Biochem. J.* 117:457–65
96. Wool, I. G. 1965. *Fed. Proc.* 24:1060–70
97. Madison, L. L. 1969. *Arch. Int. Med.* 123:284–92
98. Benedict, F. G., Joslin, E. P. 1912. Carnegie Inst. Washington Publ. No. 176
99. Rudorff, K. H., Albrecht, G., Staib, W. 1970. *Horm. Metab. Res.* 2:49–51
100. Unger, R. H. 1974. *Metabolism* 23:581–93
101. Felig, P., Gusberg, R., Hendler, R., Gump, F. E., Kinney, J. M. 1974. *Proc. Soc. Exp. Biol. Med.* 147:88–90
102. Bondy, P. K., Cardillo, L. R. 1956. *J. Clin. Invest.* 35:494–501
103. Landau, R. L., Lugibihl, K. 1969. *Metabolism* 18:265–76
104. Felig, P., Brown, W. V., Levine, R. A., Klatskin, G. 1970. *N. Engl. J. Med.* 283:1436–40
105. Kibler, R. F., Taylor, W. J., Myers, J. D. 1964. *J. Clin. Invest.* 43:904–15
106. Shoemaker, W. C., VanItallie, T. B. 1960. *Endocrinology* 66:260–68
107. Lacy, W. W., Lewis, S. B., Liljenquist, J. E., Bomboy, J. D., Felts, P. W., Sinclair-Smith, B. C., Crofford, O. B. 1972. *Diabetes* 21:340 (Abstr.)
108. Garcia, A., Williamson, J. R., Cahill, G. F. Jr. 1966. *Diabetes* 15:188–93
109. Mallette, L. E., Exton, J. H., Park, G. R. 1969. *J. Biol. Chem.* 244:5724–28
110. Salter, J. M., Ezrin, C., Laidlaw, J. C., Gornall, A. G. 1960. *Metabolism* 9:753–68
111. Marliss, E. B., Aoki, T. T., Unger, R. H., Soeldner, J. S., Cahill, G. F. Jr. 1970. *J. Clin. Invest.* 49:2256–70
112. Izzo, J. L., Roncone, A., Paliani, R. A. 1957. *Fed. Proc.* 16:200 (Abstr.)
113. Rocha, D. M., Santeusanio, F., Faloona, G. R., Unger, R. H. 1972. *N. Engl. J. Med.* 288:700–3
114. Buse, M. G., Biggers, J. F., Crier, C., Buse, J. F. 1973. *J. Biol. Chem.* 248:697–706
115. Hait, G., Kypson, J., Massih, R. 1972. *Am. J. Physiol.* 222:404–8
116. Peterson, R. D., Beatty, C. H., Bocek, R. M. 1963. *Endocrinology* 72:71–77
117. Samols, E., Marri, G., Marks, V. 1965. *Lancet* 2:415–17
118. Pozefsky, T., Tancredi, R. G., Moxley, R. T., Dupre, J., Tobin, J. 1974. *J. Clin. Invest.* 53:61a (Abstr.)
119. Chiasson, J. L., Cook, J., Liljenquist, J. E., Lacy, W. W. 1974. *Am. J. Physiol.* 227:19–23
120. Manchester, K. L. 1970. *Mammalian Protein Metab.* 4:229–98
121. Smith, O. K., Long, C. N. H. 1967. *Endocrinology* 80:561–66
122. Wise, J. K., Hendler, R., Felig, P. 1973. *J. Clin. Invest.* 52:2774–82
123. Ryan, W. L., Carver, M. J. 1963. *Proc. Soc. Exp. Biol. Med.* 114:816–19
124. Betheil, J. J., Feigelson, M., Feigelson, P. 1965. *Biochim. Biophys. Acta* 104:92–97
125. Marco, J., Calle, C., Roman, D., Diaz-Fierros, M., Villanueva, M., Valverde, I. 1973. *N. Engl. J. Med.* 128–31
126. Owen, O. E., Cahill, G. F. Jr. 1973. *J. Clin. Invest.* 52:2596–2605
127. Floyd, J. C. Jr., Fajans, S. S., Conn, J. W., Knopf, R. F., Rull, J. 1966. *J. Clin. Invest.* 45:1487–1502
128. Floyd, J. C., Fajans, S. S., Conn, J. W., Knopf, R. F., Rull, J. 1966. *J. Clin. Invest.* 45:1479–86
129. Merimee, T. J., Rabin, D. 1973. *Metabolism* 22:1235–51
130. Unger, R. H., Ohneda, A., Aguilar-Parada, E., Eisentraut, A. M. 1969. *J. Clin. Invest.* 48:810–22
131. Cahill, G. F. Jr. 1973. *N. Engl. J. Med.* 288:157–58
132. Muller, W. A., Faloona, G. R., Unger, R. H. 1971. *J. Clin. Invest.* 50:2215–18
133. Felig, P., Wahren, J., Hendler, R., Ahlborg, G. 1972. *N. Engl. J. Med.* 287:184–85
134. Hultman, E., Nielson, L. H. 1971. *Advan. Exp. Med. Biol.* 11:143–51
135. Cahill, G. F. Jr., Herrera, M. G., Morgan, A. P., Soeldner, J. S., Steinke, J., Levy, P. L., Reichard, G. A. Jr., Kipnis, D. M. 1966. *J. Clin. Invest.* 45:1751–69
136. Adibi, S. A. 1968. *J. Appl. Physiol.* 25:52–57
137. Felig, P., Marliss, E., Ohman, J. L.,

Cahill, G. F. Jr. 1970. *Diabetes* 19:727–29

138. Wahren, J., Felig, P., Cerasi, E., Luft, R. 1972. *J. Clin. Invest.* 51:1870–78

139. Owen, O. E., Felig, P., Morgan, A. P., Wahren, J., Cahill, G. F. Jr. 1969. *J. Clin. Invest.* 48:574–83

140. Owen, O. E., Morgan, A. P., Kemp, H. G., Sullivan, J. M., Herrera, M. G., Cahill, G. F. Jr., 1967. *J. Clin. Invest.* 46:1589–95

141. Genuth, S. M. 1973. *Metabolism* 22:927–37

142. Sherwin, R., Hendler, R., Felig, P. 1975. *J. Clin. Invest.* In press

143. Smith, S. R., Pozefsky, T., Chetri, M. K. 1974. *Metabolism* 23:603–18

144. Young, V. R., Scrimshaw, N. S. 1968. *Brit. J. Nutr.* 22:9–20

145. Carlberger, G., Einarsson, K., Felig, P., Hellstrom, K., Wahren, J., Wengle, B., Zetterstrom, J. 1971. *Nutr. Metab.* 13:100–13

146. Carlsten, A., Hallgren, B., Jagenburg, R., Svanborg, A., Werko, L. 1966. *Acta Med. Scand.* 179:361–70

147. Bloxam, D. L. 1972. *Brit. J. Nutr.* 27:249–59

148. Schimassek, H., Gerok, W. 1965. *Biochem. Z.* 343:407–15

149. Felig, P., Horton, E. S., Runge, C. F., Sims, E. A. H. 1971. *Experimental Obesity in Man: Hyperaminoacidemia and Diminished Effectiveness of Insulin in Regulating Peripheral Amino Acid Release.* Presented at the 53rd Ann. Meet. Endocrine Soc., San Francisco

150. Shigeta, Y., Oji, N., Hoshi, M., Kang, M. 1965. *Metabolism* 15:761–63

151. Wahren, J., Felig, P., Hendler, R., Ahlborg, G. 1973. *J. Appl. Physiol.* 34:838–45

152. Turner, L. V., Manchester, K. L. 1973. *Biochim. Biophys. Acta* 320:352–56

153. Randle, P. J., England, P. J., Denton, R. M. 1971. *Biochem. J.* 117:677–95

154. Felig, P., Lynch, V. 1970. *Science* 170:990–92

155. Scow, R. O., Chernick, S. S., Brinley, M. S. 1964. *Am. J. Physiol.* 206:796–804

156. Freinkel, N. 1965. *On the Nature and Treatment of Diabetes,* ed. B. S. Leibel, G. A. Wrenshall, 679–91. Amsterdam: Excerpta Med. 804 pp.

157. Herrera, E., Knopp, R. H., Freinkel, N. 1969. *J. Clin. Invest.* 48:2260–72

158. Metzger, B. E., Hare, J. W., Freinkel, N. 1971. *J. Clin. Endocrinol. Metab.* 33:869–72

159. Haymond, M. W., Karl, I. E., Pagliara, A. S. 1974. *J. Clin. Endocrinol. Metab.* 38:521–30

160. Kreisberg, R. A. 1972. *N. Engl. J. Med.* 287:132–37

161. Kalkhoff, R. K., Kim, H. J. 1973. *Diabetes* 22:372–80

162. Merimee, T. J., Fineberg, S. E. 1973. *J. Clin. Endocrinol. Metab.* 37:698–702

163. Chaussain, J. L. 1973. *J. Pedia.* 82:438–43

164. Santiago, J., Haymond, M., Karl, I., Clarke, W., Pagliara, A., Kipnis, D. 1974. *Comparative Substrate and Hormone Responses to Fasting in Normal Adults and Children.* Presented at 56th Ann. Meet. Endocrine Soc., Atlanta

AUTHOR INDEX

SUBJECT INDEX

A

Acetoacetate decarboxylase
  intermediate in, 917, 918
2-Acetylaminofluorene
  activation of, 94, 95
  DNA binding of, 97, 98
  RNA binding of, 96, 97
Acetyl CoA-ACP transacylase,
  316
Acetyl-CoA: arylamine N-
  acetyltransferase
  covalent intermediate in,
  897
  rate-limiting step in, 898
Acetyl-CoA carboxylase
  guanosine tetraphosphate
  effect, 335
  mutants in, 320
  reaction of, 316
N-Acetylgalactosamine sulfatase
  deficiency of, 358, 361,
  362
  and Maroteaux-Lamy syn-
  drome, 358, 361, 362
N-Acetyl-β-galactosaminidase
  deficiency of, 361
N-Acetyl-α-glucosaminidase
  deficiency of, 358, 361
  and Sanfilippo syndrome,
  358, 362
Acholeplasma laidlawii
  membranes of
  and ATPase, 333
  function of, 331, 332
  phase transition in, 331
  sterols in, 334
Acid lipase
  and Wolman's disease, 359
Acid phosphatase
  deficiency of, 359, 370
ACTH
  and adenylate cyclase, 511
  analog of, 480
  conformation of, 479, 480
Actin
  in contraction, 176, 177
  electron microscopy of, 176,
  177
S-Adenosylmethionine
  as methyl donor
  and methyltetrahydrofolate,
  436
  to polysaccharides, 445-
  47
  to proteins, 439-42
  sole, 436-38, 440
  to sugars, 444, 445
Adenosylmethionine trans-

ferase, 447, 448
Adenoviruses
  assembly of, 581-83, 783
  and chemical carcinogens,
  109
  classes of, 581
  DNA of
  cleavage map of, 280
  encapsulation of, 581
  terminal repetitions in,
  280
  transcriptional map of,
  281
  morphology of, 581
Adenyl cyclase
  and ACTH, 511
  and antidiuretic hormone,
  872
  and cell density, 506
  cholera toxin effect, 493,
  500, 501, 509, 684
  in hepatoma cells, 840
  insulin effect, 507, 515
  in leukemia, 511
  levels of
  and cell cycle, 506
  and cell density, 506
  and luteinizing hormone,
  684
  in malignant transformation,
  509-11
  measurement of, 513
  and membrane phosphorylation, 873, 874
  in platelets, 686
  phytohemagglutinin effect,
  508
  and prostaglandins
  and cell growth, 493
  and endoperoxides, 673
  in fibroblasts, 506, 507,
  685, 686
  as messengers, 683
  reviews of, 683
  stimulation of, 683-86
  regulation of
  activation in, 508, 512
  and cell proliferation, 508,
  509
  desensitization in, 507,
  508
  by hormones, 506-10,
  514
  by prostaglandins, 493,
  506, 507
  in transformed cells, 509-
  13
  solubilized, 684
  stimulators of, 500, 501
  and virus transformation,

511-13
Adenyl kinase
  and membrane structure,
  322
  secondary structure of,
  454
Adenylosuccinase
  complementation of, 296
Adenylylation
  of aspartyl kinase, 855
  and glutamine synthetase,
  855
  reviews of, 832
Adrenodoxin, 384, 385
Aequorin, 261, 262
Aerobacter aerogenes
  tryptophan synthetase of,
  304
Affinity labeling, 642, 643,
  663
Aflatoxins
  activation of, 103, 104
  nature of, 103
Alanine
  carrier role of, 941, 943
  in diabetes mellitus, 948
  exchange of
  and absorption, 942
  and exercise, 949, 950
  from gut, 935, 942
  insulin effect, 944
  with kidney, 936
  with liver, 935, 936,
  942
  with muscle, 934, 942,
  944
  role of, 937, 943
  with splanchnic tissues,
  934, 935
  and gluconeogenesis
  glucose-alanine cycle in,
  938-41
  insulin effect, 944, 945
  in liver, 937-41
  role of, 938-41
  in starvation, 947, 948
  and hypoglycemia, 951
  synthesis of
  and gluconeogenesis, 938-
  41
  in muscle, 939
  nitrogen source, 939, 940
Albumin
  post-translational cleavage
  of, 776
Alcohol dehydrogenase
  in bile acid formation,
  239
Aldehyde oxidase
  and superoxide formation,

extrahepatic, 241
formation, 234-39
hormone effects, 244
by intestinal microorgan-
isms, 240
methods in, 248, 249
regulation of, 235, 238,
242-45
reviews of, 233, 240
vitamin effects, 244
radioimmunoassay of, 248
Bioluminescence
in bacteria
flavin mononucleotide in,
263-66
NADH in, 263
stoichiometry of, 264
in coelenterates
blue light emission, 257-
59
control of, 260, 261
green light emission, 259,
260
mechanism of, 258, 259
occurrence of, 256
photoprotein in, 261-63
control of
calcium in, 261
and luciferase, 260, 262
and luciferin binding protein, 260-62
emitting chromophore in,
266, 268
in firefly
reactions of, 267
reviews of, 267
luciferin in
in firefly, 267, 268
mechanism of, 258, 259
oxidation of, 257-59
and photoproteins, 261,
262
in Renilla, 257, 258
synthetic, 257
in lumisomes, 261, 263
mechanism of
flavin mononucleotide in,
263-66
and green fluorescent protein, 259, 260
in oxidation, 258, 259
in Pholad, 268, 269
photoproteins in
action of, 260-62
definition of, 256
quantum yield in
bacterial, 263
definition of, 256
and luciferin, 258, 259
reviews of, 255, 267
terminology of, 255, 256
Biopterin, 389
Blood coagulation
activation of
factor V, 815
factor VII, 823, 824
factor IX, 807, 809

factor X, 812-14
factor XI, 806, 807
factor XII, 804-6
factor XIII, 820-22
fibrinogen, 802, 804, 819,
820
proteolysis in, 806, 807,
809, 811-16, 818-22
prothrombin in, 804-19
extrinsic pathway of
and factor VII, 822
factors in, 800, 801
and thrombin production,
812
factors in
amino acid sequence of,
808, 813
factor III, 802
factor IV, 802, 804
factor V, 801, 802, 805,
815, 816
factor VI, 802
factor VII, 801, 802, 818,
822-24
factor VIII, 801, 802, 804,
809-12
factor IX, 801, 802, 804,
807-9, 818
factor X, 801, 802, 804,
808, 812-16, 818
factor XI, 801, 802, 804,
806, 807
factor XII, 801, 802, 804-
6
factor XIII, 802, 804, 820-
22.
fibrin in, 804, 819, 820,
822
fibrinogen, 802, 804, 819,
820
and hemophilia, 809, 810
interaction of, 803, 804
list of, 802
nomenclature of, 800
prothrombin, 801, 802,
804, 808, 815-19
removal of, 803
thrombin, 804, 812, 818-
22
tissue factor, 822, 823
and hemophilia
factor VIII in, 809, 810
in royal families, 799
inhibitors of, 803
intrinsic pathway of
factors in, 800, 801
mechanism of, 804
pathways for, 800-4
phospholipid in, 804, 809,
816, 817, 822
proteases in
and factor V, 815, 816
and factor VIII, 811, 812
and factor IX, 807, 809
and factor X, 813-16
and factor XI, 806, 807
and factor XII, 805, 806

and factor XIII, 821, 822
and fibrinogen, 819, 820
and prothrombin, 815-19
and vitamin K, 817, 818,
823
and von Willebrand's dis-
ease, 810, 811
Bluetongue virus, 590
Bradykinin, 483
Burkitt's lymphoma, 585
γ-Butyrobetaine hydroxylase,
443

C

Camphor methylene hydroxylases
components of
cytochrome, 380-83
flavoprotein dehydrogenase,
380, 381
putidaredoxin, 380-83
regulation of, 382, 383
Carbon fixation
and aspartate aminotrans-
ferase, 128
in chloroplasts
lag in, 125
magnesium effect, 125,
126
rates of, 126, 127
ferredoxin in, 125
and NADP-malate dehydro-
genase, 128
phosphoenolpyruvate carboxylase in
localization of, 130
mechanism of, 127-29
ribulose diphosphate carboxylase in
intermediates in, 124
localization of, 130
low K_m form, 126
regulation of, 125
and transport, 127, 128
Carbonic anhydrase
localization of, 130
phosphorylation of, 855
unfolding of, 457
Carboxypeptidase A
mechanism of, 910-12
solution structure of, 911,
912
X-ray studies of, 909-12
zinc in, 910-12
Carboxypeptidase B, 910
Carcinogenesis
by aflatoxins, 103, 104
by aromatic amines
activation of, 94, 95
and DNA binding, 96-98
and RNA binding, 96, 97
in cell cultures
and cell cycle, 108
and epithelial cell transformation, 106, 107
of hamster embryo, 104,

in nu bodies, 756
phosphorylation of
   acid labile, 844
   and DNA binding, 859,
      860
   and DNA synthesis, 858,
      859
   and histone synthesis, 858
   hormone effects, 833, 834,
      856, 857
   and protein kinases, 833,
      834, 844, 857
   primary structure of, 728-
      32, 736
   of primitive eukaryotes,
      733
   properties of, 726-28
   and protamines, 732
   as repressors, 760, 761
   synthesis of
      and DNA synthesis, 743,
         744, 858
      transcription, 743-45
      translation, 734, 743,
         744
   tissue-specific, 732, 733
   turnover of, 746, 747
   of viruses, 734
   see also Chromatin; Chro-
      mosomal proteins
HMG CoA reductase
   diet effect on, 245
   regulation of, 242
Hormones
   see specific hormones
Hunter syndrome
   clinical phenotype of, 362,
      363
   deficiency in, 358, 361,
      362
   detection of, 370
   mucopolysaccharide accum-
      ulation in, 361
Hurler syndrome
   clinical phenotype of, 362
   deficiency in, 358, 361,
      362
   detection of, 370
Hyaluronidase, 361
Hydrogen exchange
   NMR measurements in,
      482
   in peptides, 481, 482
   in proteins
      and dynamic accessibility,
         469
      and folding, 454, 469
   review of, 481
β-Hydroxyacyl dehydrase,
   316
m-Hydroxybenzoate-6-hydrox-
   ylase, 402
p-Hydroxybenzoate hydroxylase
   flavin intermediates in,
      399
   mechanism of, 398, 399
   uncoupling of, 400

β-Hydroxydecanoyl thioester
   dehydrase
   inhibition of, 920-22
   intermediate in, 920
Hydroxylamine, 20
11β-Hydroxylase
   components of, 384, 385
   review of, 384
17α-Hydroxylase, 385
19-Hydroxylase, 385, 386
20α, 22R-Hydroxylase
   mechanism of, 384
   reaction of, 383
21-Hydroxylase, 386
ω-Hydroxylase
   components of, 394
   iron of, 378, 393, 394
Hydroxylases
   see Oxygenases
Hydroxylysine, 700
p-Hydroxyphenylpyruvate
   hydroxylase, 392, 393
Hydroxyproline, 700
15-Hydroxyprostaglandin
   dehydrogenase, 676
Hyperlipoproteinemia
   type II
      bile acid formation in,
         247, 248
      cholestyramine effect, 247,
         248
   type III, 188, 189
   type IV, 247, 248
Hypoglycemia, 951
Hypothyroidism
   lipoproteins in, 188, 189

I

Iduronate sulfatase
   deficiency of, 358, 369
   and Hunter syndrome, 358,
      361, 362
   inheritance of, 369
Iduronidase
   deficiency of, 358, 361
   and Hurler syndrome, 358,
      361, 362
   and Scheie syndrome, 358,
      361, 362
Immunoglobulins
   and affinity labeling, 633,
      642, 643
   classes of, 640
   and complement binding,
      700, 701, 710
   electron microscopy of,
      643, 644
   Fab fragments of
      binding site of, 650
      X-ray studies of, 650,
         651
   Fc fragments of, 643, 659,
      660
   heavy chains of
      and binding site, 663
      domains of, 641, 642, 653,

654, 657, 658
      groups of, 642
      hypervariable regions in,
         642, 651, 662
      structure of, 641, 642
   heterogeneity of, 640
   J chains of, 642
   light chains of
      and Bence-Jones proteins,
         641
      domains in, 645, 646, 650,
         653, 657, 658
      groups of, 640, 641
      hypervariable regions,
         642, 647, 649, 651, 662
      structure of, 640, 641,
         645-47
      X-ray studies, 645-49
   and myeloma proteins
      antigen binding by, 640,
         652-55, 662
      domains of, 653, 654
      quaternary structure of,
         654, 655
      X-ray studies on, 652-55
   proteolysis of, 640, 776
   structure of
      Bence-Jones proteins,
         644, 645, 648-50, 660
      binding site of, 648, 650,
         651, 653-55, 662-64
      chemical, 640-43
      comparisons of, 655-64
      hypervariable regions in,
         642, 643, 647, 649, 651,
         662, 663
      J chain of, 642
      quaternary, 654, 655,
         659-62
      schematic, 641
      solvent channel in, 647,
         648
      β-structure in, 646, 649,
         654
      tertiary, 656-58
      three-dimensional, 643-
         64
   X-ray crystallography of
      Bence-Jones proteins,
         644, 645, 648-50, 660
      binding sites in, 648, 650,
         651, 653-55, 662-64
      contacts in, 647-49
      domains in, 645, 646, 650,
         653, 654, 656-59
      Fab fragments, 650, 651
      Fc fragments, 643, 659,
         660
      high resolution, 644, 655
      light chain dimer, 645-49
      low resolution, 643, 644
      myeloma proteins, 652-
         55
   and tertiary structure,
      656-58
   see also Antibodies
IMP dehydrogenase, 308, 309

CUMULATIVE INDEXES

CONTRIBUTING AUTHORS VOLUMES 40-44

CHAPTER TITLES VOLUMES 40-44